To Paul,
With best skeptical wishes,

The *Skeptic* Encyclopedia of Pseudoscience

Santa Barbara, California   Denver, Colorado   Oxford, England

THE *SKEPTIC* ENCYCLOPEDIA of PSEUDOSCIENCE

Michael Shermer, Editor

Pat Linse, Contributing Editor

VOLUME TWO

Michael Shermer, Skeptics Society, *Skeptic* magazine, P.O. Box 338, Altadena, CA 91001
URL: http://www.skeptic.com, email: skepticmag@aol.com
(626) 794-3119 (phone), (626) 794-1301 (fax)

Library of Congress Cataloging-in-Publication Data
The *Skeptic* encyclopedia of pseudoscience / edited by Michael Shermer.
p. cm.
Includes index.
ISBN 1-57607-653-9 (set : hardcover : alk. paper) – ebook ISBN 1-57607-654-7
1. Pseudoscience–Encyclopedias. I. Shermer, Michael. II. Skeptic.
Q172.5.P77 S44 2002
503–dc21
2002009653

02 03 04 05 06 10 9 8 7 6 5 4 3 2 1

This book is also available on the World Wide Web as an e-book.
Visit abc-clio.com for details.

ABC-CLIO, Inc.
130 Cremona Drive, P.O. Box 1911
Santa Barbara, California 93116-1911

This book is printed on acid-free paper.
Manufactured in the United States of America

To James "the Amazing" Randi,
our hero, colleague, friend, and inspiration

Contents

Volume One

Volume Two

SECTION 3
CASE STUDIES IN PSEUDOSCIENCE FROM SKEPTIC MAGAZINE

The *Skeptic* Encyclopedia of Pseudoscience

3

CASE STUDIES IN PSEUDOSCIENCE FROM SKEPTIC MAGAZINE

The Alien Archetype

The Origin of the "Grays"

JOHN ADAMS

In the aftermath of the Heaven's Gate mass suicide, the task of debunking erroneous and unwholesome notions of the "alien" acquires a previously unanticipated level of seriousness. By "alien" I mean the now-classic and ever-popular "Gray," the alien archetype that was depicted on the Heaven's Gate web site, and featured in nearly every film, television program, book and magazine article, from *Communion* to *The X-Files,* and even recently satirized on *The Simpsons.* While artistic and theatrical productions often use this image, the Gray has been popularized even more by those who wish to create the belief that aliens actually exist (e.g., reported abductees and peddlers of "autopsy" footage). Given the absence of any clarity or consistency in UFO folklore concerning the exact origins or motives of the Grays, the striking consistency between purported eyewitness reports of their physical appearance is perhaps the greatest comfort to those preoccupied with proving their authenticity.

I would like to argue that the historical context of those who envision aliens is not only crucial to establishing how they should look but is also quite plausibly the sole basis of popular extraterrestrial stereotypes. Two centuries ago, aliens were often visualized as Native Americans or Blacks in both eyewitness recollection and in straightforward fiction accounts. In the past century, with the advent of Darwinism, a new stereotype of the alien began to take shape. Space invaders were recurrently both "witnessed" and artistically represented as hairless, with enormous, potent eyes, enlarged craniums, and light complexions. The Gray may reveal little about any actual world beyond Earth but a great deal about the imaginative content of the human mind in the western world at this point in our history.

In the summer of 1996, NASA revealed compelling evidence that Mars may once have sustained life. In the spring of 1997, a mass suicide occurred at the Heaven's Gate Temple in California, inspired, at least to some degree, by a longing for communion with alien beings. The juxtaposition of these two events shows that the spectre of extraterrestrial life looms forth both as an object of serious inquiry and as a catalyst to morbid delusion.

In such an atmosphere, the task of dealing with the subject of extraterrestrial life demands a level of seriousness previously unneeded. As the Heaven's Gate example attests, some characterizations of aliens formed in the absence of any actual contact are far less wholesome than others. To address the issue of life beyond Earth sensibly, we should demystify some common conceptions of the anatomy of extraterrestrials by exploring their historical and cultural roots.

One particular image of the alien currently dominates popular culture in both abduction testimonials and science fiction accounts. Dr. John E. Mack, professor of psychiatry at

Harvard and well-known advocate of alleged abduction victims, uses the widely known term *Grays* when referring to these alien visitors, describing them as "by far the most common entity observed" (Mack, 1995, 22). They are typically pale, hairless, and genderless, with prominent eyes and enormous heads balanced on diminutive bodily frames. While there is no hint of agreement concerning precise alien origins or motives, the consensus on their physical appearance in many separate reports is so broad that UFO enthusiasts often cite it as corroborative proof of authenticity.

Common sense calls for us to challenge these testimonies from the get go: Why come so far across the vast distances of interstellar space merely to gather, probe, and traumatize a few humans (although some believers put the number into the millions), and then why bother—ever so discreetly—to return them? Is overexposure to *Star Trek* and *X-Files* a likely culprit here, and don't the abductions seem to be straight out of the late night horror show? Believers may counter such swipes with their own, inquiring: But what of the degree of concurrence among so many eyewitnesses? Since this cannot be denied, why should it be ignored? They may also propose that Hollywood is not inspiring a collective fantasy, but rather quietly employing knowledgeable researchers to assure the authenticity of sci-fi productions like the historical consultants who inform the directors of the great epics.

In demystifying the Gray the first issue to address is the matter of color. "Little green men" was the color of choice for some B-movie aliens, but serious believers all agree that actual ETs lack skin pigmentation. Much closer to white than to charcoal, a Gray is by definition a traveler whose scientific competence vastly surpasses humanity's. In considering this, we should not ignore the European genealogy of modern science fiction, which was once steeped in racial prejudice.

Few consistencies in alien appearance have endured since the cultures that accepted the Copernican model of the universe first began to speculate about how extraterrestrials should appear. One of the only assumptions lasting over the centuries is that the voyagers capable of making the trip must have lighter skin than the awe-stricken spectators who greet their arrival. From early modern Europe, representative interplanetary fantasies include Francis Godwin's *The Man in the Moone* (1638) and Ralph Morris' *A Narrative of the Life and Astonishing Adventures of John Daniel* (1751). In such narratives, first contact is achieved through the genius and initiative of European aeronauts inspired by the exploits of Christopher Columbus and Sir Francis Drake (Adams, 1995, 71–73). Until the 1890s it was taken for granted that the aliens would not come to Earth, but instead be "discovered." True to the Columbian legacy, they are envisioned in early modern European fantasy narratives as American Indians or Africans (Adams, 1995, 70, 73–81). Testimony about actual contact, it should be emphasized, closely conformed to fictional stereotypes. In 1758 Emanuel Swedenborg, a foremost scientist and theologian of the Enlightenment (and founder of the Swedenborgian Church), chronicled aliens inhabiting his own solar system which he encountered in a trance state reportedly bestowed upon him by the grace of God (Adams, 1995, 78; Swedenborg, 1787, 1). Martians resembled the dark-skinned races of his own world and clothed themselves in tree bark (Adams, 79; Swedenborg, 107). On Jupiter, the aliens lived in conical tents (Adams, 80; Swedenborg, 63). His eyewitness account bears the title "De telluribus in mundo nostro solari, quae vocanter planetae," or "The Earths in our universe which are called planets."

Nineteenth century German racists not surprisingly imagined that Martian invaders capable of subduing the Nordic states must be more white than Europeans. In 1897, the year that H. G. Wells' *War of the Worlds* first ap-

The geneaology of the Gray may be found in the earth's historical or cultural past. (left to right) A lega sculpture from the Congo (Thompson, 1974, 120). The familiar gray (Rosenblatt, 1996, 44). Yoruba Earth cult brass figure from Nigeria (Thompson, 1974, 70).

Occultist Aleister Crowley claimed he had contacted Lam, an alien, and drew this picture of him. Note the uncharacteristically small eyes (Grant, 1980, 160–161).

peared as a magazine serial, a book was published in Germany titled *Auf Zwei Planeten* or *Two Planets*. Its author Kurd Lasswitz was an acclaimed science fiction writer in his own country. While Wells' invaders perish shortly after their arrival (from exposure to common germs), Lasswitz's aliens survive Earth's microorganisms and conquer the planet. Regarded as heroic imperialist problem-solvers by the author, they are similar to humans, especially (not surprisingly) to Germans (Lasswitz, 1971, 27; 55). They privilege Europeans in their global colony with coveted positions as collaborators. The book contains revealing images. The Martians have "large heads" and "large shining eyes." Their hair is noteworthy solely because "nearly all" of them had "very light, nearly white hair" (Lasswitz, 1971, 16, 25). (Blondness taken to the next level, perhaps?) Strolling down a German lane, a lightly veiled Martian who knew the ranges of Earth's climactic conditions notes with apparent relief (Lasswitz, 1971, 323): "And how pleasantly one can walk here in the sunshine without being burnt!"

A vast proportion of the aliens portrayed in every form of mass media since the genre of science fiction became popular have had light skin. Those who currently perpetuate the image in artistic endeavors or through other forms of self-expression are by now almost certainly oblivious to the possible roots of this conception in the social and political context of the times when images of aliens first permeated popular culture.

Another prominent feature of the Gray is the bulbous, hairless head. Not surprisingly, this too is traceable to European history. One of the first famous Europeans of the 20th century to lay claim to privileged communication

Science fiction writer Hugo Gernsback's imaginative conception of how an alien might look, illustrating an article from December 1937 bearing the title "Can We Signal Mars by Short Wave?" The "Hugo" Awards for science fiction are named after Gernsback (Dick, 1993, 101).

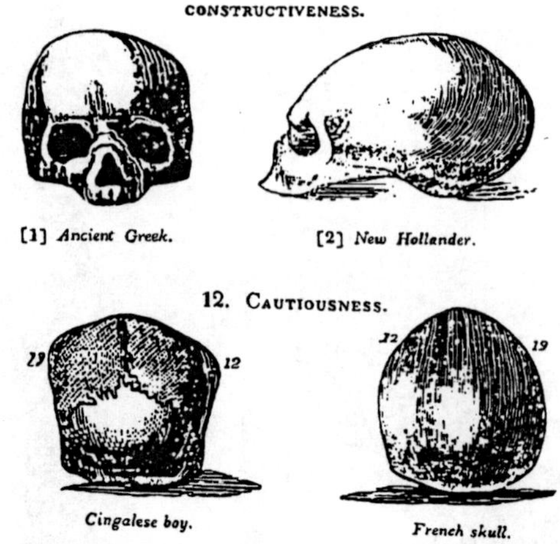

The concept that the Western skull's supposed rounder forehead shape and larger forehead size is directly correlated with preferred traits is illustrated in this comparison between Western skulls and skulls from groups considered more primitive. From George Combe's *Lectures on Phrenology* (Combe, 1854, 173; 185).

with aliens was occultist Aleister Crowley. Crowley claimed contact with Lam, an extraterrestrial being that he believed functioned as a "link" between Sirius and the Andromeda constellations (Grant, 1980, 281). He himself drew Lam between 1918 and 1919 as a humanoid closely resembling a white man with a grossly expanded bald head (Grant, 1980, 95, 160, 281). The piercing eyes which gaze upon the viewer so knowingly, however, are far too small to match the current Gray model. Perhaps this disparity accounts for the absence of the portrait in the standard evangelistic UFOlogy tracts.

Just as it is no surprise that Swedenborg saw Indians, it is to be expected that an Englishman of the early 20th century would visualize Lam. Over the past few generations those immersed in the culture of the West, whether they reside in its bosom or view its images in movie houses, are likely to have some familiarity with Darwinism. The idea that "bigger is better" insofar as the brain is concerned, though prone to some ridicule in scientific circles in recent times, gained wide exposure and appeal even before Darwinism and still maintains some credibility in the general populace today. We have all seen the famous Time-Life book "march of progress" diagram, reproduced countless times in numerous ways: as the incarnations of pre-human typologies "advance" over time, the skull becomes more bulbous and prominent while the jaws and nasal passages recede. The body diminishes as the hair thins to reveal a typical Caucasian man.

Sculptures of Newton and Voltaire that embellished the Victorian drawing rooms of Crowley's world often seem to grossly accentuate the prominence of the forehead. In marked contrast to these white marble busts were living displays of large muscular Blacks afflicted with microcephalism presented as "missing links" in the 1860s by P. T. Barnum. As purported evolutionary throwbacks, the latter would signify a virtual inversion of the Gray ideal.

The celebration of the expanded cranium has a long history in western history. Jean-

Professor Cavor, an English adventurer from H. G. Wells's 1901 novel, *The First Men in the Moon*, is captured by moon-dwelling Selenites. This story, as well as Wells's better-known *War of the Worlds*, was first published in popular magazines before the turn of the 20th century. Illustrated by lavish and imaginative paintings, they provided prototypes in the public mind for aliens more than a hundred years ago. The illustrator of these particular aliens seems to have mixed traits of both high and low cast moon-dwellers from Wells's text.

Jacques Rousseau (Rousseau, 1915, 150), Charles Darwin (Darwin, 1974, 452–3), Lewis Henry Morgan (Morgan, 1967, 25), Friedrich Engels (Engels, 1990, 138), and Franz Boas (Boas, 1974, 232–3) would only commence a list of influential thinkers who either unequivocally affirmed a correlation between the measure of brain size and the quality of the intellect or at some point seriously toyed with the notion as a significant issue worthy of serious scientific consideration. *The First Men in the Moon,* written by H. G. Wells in 1901, conformed to the expectations of its age by making a direct correlation between large cranial

Ancient representations of mythic spirit beings: Vinca figures from the Balkans. Such spiritual art dates back to the fifth millennium B.C.E. (Gimbutas, 1992, 63, 124).

size and natural superiority. The moon was populated by Selenites—humanoids of varied anatomies organized according to a caste system in which cranial magnitude determined moral and political authority. After violent encounters with brutish lower-caste cattle herders, an earthling observes: "But presently I came upon a body of Selenites led by two who were curiously different, even in form, from any of those we had seen hitherto, with larger heads and smaller bodies and much more elaborately wrapped about." The moon is ruled by a creature called the Grand Lunar, surrounded by a retinue of worthy advisors with "swollen heads." The Grand Lunar is in certain respects the ultimate Gray. Having "no face," his head is several yards long, reposed upon a tiny "shriveled" body. While Wells has little to say concerning the complexion of the majority of the aristocratic Selenites and places their eyes to the sides, comparing them to hens, the Grand Lunar is described as "white," with "little eyes" that "stared down at" the narrator (Wells, 1901, 133, 140, 145, 150–151).

The connection between cranial size and

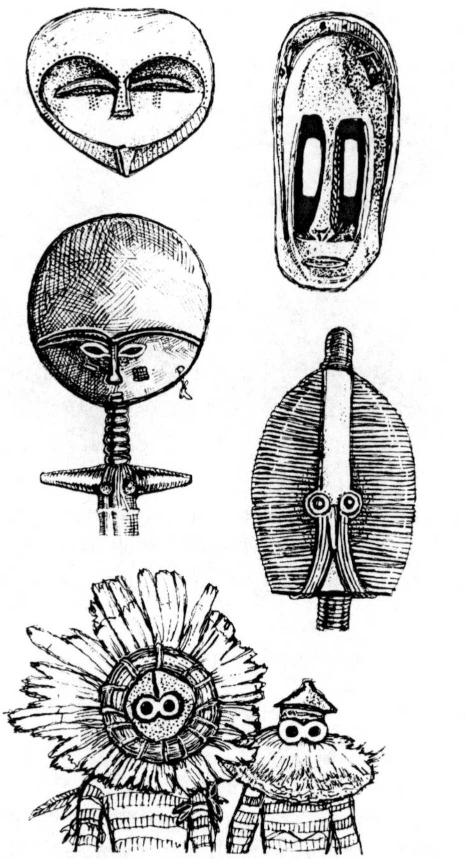

The eyes have it. African folk art often exaggerates eyes over other facial features when depicting the spiritual. (Top left) Bakwele dance mask, Congo; (Top right) Dogan dance mask, Mali; (Middle left) Ashanti fertility doll, Ghana; (Middle right) Bakota reliquary, Gabon; (Bottom) Bapended masked dancers, Congo. (Redrawn from Trowell, N.D., 23, 46, 50, 92, 152).

genius lost favor in the latter half of this century. Quack anthropology and racist biology were silenced by disgust and embarrassment over the Nazi legacy. Among the topics of concern for scientific professionals in Nazi Germany was "the persistence of the 'Cromagnon' racial type in certain populations" (Proctor, 1988, 41). It is amusing to imagine that an abductee some day may claim to be spirited away by furry pin-headed aliens with compact brains that surpass bulkier models. But we should not hold our breath.

A sixth-century Christ. Byzantine artists exaggerated the eyes in portrayals of human subjects, particularly in representations of heroes and saints.

In Asia sacred buddhist monuments known as stupas dot the landscape. They are often painted with two large human eyes which are unaccompanied by any other sensory organs.

An examination of the belief systems of the past two centuries may suggest the origins of the light skin color and large head, but it is not as useful in explaining the size of the eyes. While a Darwinian imagination would not likely posit minuscule eyes on any creature evolved beyond human capacities, to banish ears, noses, and mouths from the visage would make even less sense. While some insights into the riddle may be found in the modern history of the West, others are equally or best addressed by turning to representations of spiritually superior beings from the ancient world. (Spiritual associations may also explain the depictions of baldness, which may have been inspired by Darwinism but also seems rooted in more universal and global conceptions. "Wild men" are universally by definition hirsute beasts. Tonsuring is quite common in monastic settings as a mark of spiritual advancement; note the practice at Heaven's Gate.)

The idea that extraordinary beings should see rather than hear, smell, touch, or taste is one that is universal and widespread. All-seeing presumes all-knowing—a capacity widely hoped for in god-like beings. In Western culture, too,

God is described as "all-seeing" rather than "all-hearing." A quest for terms that pertain to the remaining four senses as powerful as the English words "visionary," "illumination," and "enlightenment" would entail quite a scavenger hunt. These associations of the visual with all-knowing god-like wisdom could provide an explanation for the tendency to depict spiritually superior beings with enlarged eyes.

A wide variety of cultures depict gods or goddesses as hairless beings with huge eyes but no other distinct facial sensory organs. Exaggerated eyes are found in such diverse sources as primitive African art, Byzantine mosaics, Buddhist sacred sites and prehistoric icons from the Balkans. Although they confirm nothing more than the certainty that many societies have accentuated the eyes of their divinities, one look at some of them would almost necessarily prompt an association with the Gray (Thompson, 1974, 70, 120; Gimbutas, 1992, 63, 124, 126, 182).

The alien as Gray, like the other parts of the alien abduction and UFO belief system, clearly has a terrestrial rather than extraterrestrial origin. It would be interesting to come back in another century or two to read what historians have to say about the UFO/alien craze that swept America in the latter half of the 20th century.

References:

Adams, J. 1995. "Outer Space and the New World in the Imagination of Eighteenth-Century Europeans." *Eighteenth-Century Life.* 19 (1): 70–83.

Boas, F. 1974 (1894). "Human Faculty as Determined by Race." In *The Shaping of American Anthropology: 1883–1911. A Franz Boas Reader.* George Stocking (ed.). New York: Basic Books.

Booth, W., and W. Clailborne. 1997. "Group's Leader among 39 Cult Suicides: Families Get News of Members Lost Many Years Ago." *Washington Post,* March 29. A: 1, 9.

Combe, G. 1854. *Lectures on Phrenology.* New York: Fowlers and Wells.

Dart, J., J. Rainey and L. Stammer. 1997. "Tract Offers Clues about Group's Theology, Motives." *Los Angeles Times,* March 28. A: 1, 15.

Darwin, C. 1974 (1838). "Extracts from the B-C-D-E Transmutation Notebooks." In *Darwin on Man: A Psychological Study of Scientific Creativity.* Howard E. Gruber (ed.). Paul Barrett (trans. and annotated). New York: E.P. Dutton & Co.

Dick, S. 1993. "The Search for Extraterrestrial Intelligence and the NASA High Resolution Microwave Survey (HRMS): Historical Perspectives." *Space Science Reviews.* 64: 93–139.

Engels, F. 1990 (1884). "The Origin of the Family, Private Property and the State." In *Karl Marx & Frederick Engels: Collected Works.* (Vol. 26). New York: International Publishers.

Gimbutas, M. 1992. *The Goddesses and Gods of Old Europe.* Berkeley: University of California Press.

Grant, K. 1980. *Outside the Circles of Time.* London: Frederick Muller, Ltd.

Lasswitz, K. 1971 (1897). *Two Planets.* Carbondale: Southern Illinois University Press.

Mack, John. 1995. *Abduction: Human Encounters with Aliens.* New York: Ballantine Books.

Morgan, L. H. 1967 (1877). *Ancient Society.* New York: World Publishing Company.

Proctor, R. 1988. *Racial Hygiene: Medicine under the Nazis.* Cambridge: Harvard University Press.

Rosenblatt, R. 1996. "Classifying the Unknown: Types of Alien Beings." *Fate.* June: 44–7.

Rousseau, J. J. 1915 (1755). "Discours sur l'origine et les fondements de l'inégalité parmi les hommes." In *The Political Writings of Jean-Jacques Rousseau.* (Vol. 1). Cambridge: Cambridge University Press.

Swedenborg, E. 1787 (1758). "De telluribus in mundo nostro solari, quae vocanter planetae." London: R. Hindmarsh.

Thompson, R. 1974. *African Art in Motion.* Berkeley: University of California Press.

Trowell, M., and Neverman, H. (N.D.). *African and Oceanic Art.* New York: Harry N. Abrams, Inc.

Wells, H. G. 1965 (1901 Publishing Co.). *The First Men in the Moon.* New York: Airmont.

Anastasia

A Case Study in the Myth of the Miraculous Survival

TIM CALLAHAN

"From a word to a word was I led."
—Havamal (The Sayings of the High One), from the Poetic Edda of Iceland. (The speaker is the god Odin.)

In the process of researching an upcoming book, I came across a passage in the Gospel of John which indicated that the author saw the end of the world as taking place in his own generation. Jesus says (John 5:25–29):

> Truly, truly, I say to you, the hour is coming, and now is, when the dead will hear the voice of the Son of God, and those who hear will live. For as the Father has life in himself, so he has granted the Son also to have life in himself, and has given him authority to execute judgment, because he is the Son of man. Do not marvel at this; for the hour is coming when all who are in the tombs will hear his voice and come forth, those who have done good, to resurrection of life, and those who have done evil to the resurrection of judgment.

As is often the case, an examination of the key words in the original Greek clarifies the meaning of the passage. The Greek word translated as "hour" is *hora,* which if it doesn't mean exactly an hour does mean a fleeting period of time. This is further emphasized in that the form of the verb "to come" (erchomai) is present imperfect, meaning that "is coming" is an exact translation. The word for "now," *nun,* also expresses immediacy. It is not so much "now" as "right now." The word for the dead here is *nekros,* deriving from *nekus,* meaning a corpse. So the dead in question are dead bodies. These bodies will experience resurrection or anastasis, meaning literally to stand again (*ana* again + *stasis* stand), either to life or krisis (judgment or damnation). So John has Jesus saying that the hour or fleeting period of time is coming and in fact is here right now, when dead bodies will come forth from their tombs and stand again at the sound of Jesus' voice to face judgment.

While each of these Greek words is important in conveying the immediacy and physicality of the imminent last judgment—both of which are somewhat less forceful in the English translation—the word *anastasis* in particular stood out for me since it is obviously the source of the name of the famous Princess Anastasia (1901–1918), who alone of all the family of Czar Nicholas II is so often fabled to have survived the Bolshevik firing squad. A name which means "resurrection" or literally to "stand again" is naturally going to be popular among Christians, particularly those of the Eastern Orthodox persuasion, where Greek, rather than Latin, was at least the initial liturgical language. Thus it is not surprising that there are four popes and two Byzantine emperors named Anastasius. Since Russia was Christianized by Greek Orthodox missionaries, several Russian names have Greek origins. Among these are Feodor (Theodore, Gr. *theodoros* "gift of God"), Vassily (Basil, Gr. *basilios* "king") and, of course, Anastasia. It is my contention that the reason the princess bearing this name was chosen by those hoping for a restoration of the Romanovs as the survivor of the firing squad is based, consciously or unconsciously, on the symbolism of her name. There is no other reason for choosing her as the sole survivor.

One might argue that, if what we are dealing with here is myth rather than a tale that is at least possibly historical, the logical survivor would be the Czarevitch Alexis. However, his malady, hemophilia, made it so unlikely that he would survive multiple gunshot wounds, that one of his sisters had to be chosen. It is not surprising that the one chosen to have survived being shot at close range by a firing squad, partially burned, and subsequently buried happened to be named "Resurrection." That there must be a survivor, in spite of the horrendous odds against such a possibility, is demanded by myth as surely as it is militated against by reality. In fact this story has been told over and over from ancient times and is a powerful legend. Anastasia's mythic survival was assured far more by the emotional needs it fulfilled than by the dubious protection the family jewels, supposedly sewn into the bodices of the princesses to conceal them, provided against a hail of bullets. (The deflection of most of the bullets by the bejeweled bodices is the usual explanation of Anastasia's survival.)

Not only is the legend of the miraculous survival of a supposedly murdered royal scion an ancient story, it has a precedent in Russian history, literature and music in the character of False Dmitry from the story of Boris Godounov, which, ironically enough, provides a prelude to the accession of the Romanov dynasty. The story of Boris Godounov was immortalized in an epic poem by Alexander Pushkin, which subsequently became the basis of the opera by Modest Moussorgsky. Boris rose to power on the death of Ivan IV ("the Terrible") in 1584. He was advisor to Feodor, Ivan's eldest surviving son, who was mentally retarded. Even before Feodor died without leaving an heir, Ivan's seven-year old son, Dmitry, died in 1591. Though this seems to have occurred as a result of an epileptic seizure, it was rumored that he was murdered and it was widely held that Boris, who was elected Czar by the Zemsky Sobor (privy council) on Feodor's death in 1598 and who had everything to gain by extinguishing Ivan's line, had ordered the child's assassination. At first popular, Boris had increasing problems because of opposition from the boyars, compounded by crop failures in 1601–2. Many of the common people saw this as a sign of divine wrath for the supposed murder of Dmitry. In 1603 a young monk by the name of Grigory Otrepyev declared that he was Prince Dmitry and that he had in fact survived the "attempted" assassination. Naturally, such a claim was not welcomed by Boris, and Grigory was forced to flee to Poland. There his tenuous claim was taken seriously, since it was to the benefit of Poland to recognize him as pretender to the Russian throne. He married Marina Mniszech, the 15-year old daughter of one of the chief Polish nobles supporting his claim, and returned to Russia with a Polish army in 1604. This force was swelled by various Russian malcontents, and though it was defeated, Boris died in 1605, and "Dmitry" was made czar.

Of course, he had to give considerable territorial concessions to the Poles, and a Polish garrison was established in Moscow. Under the protection of Polish forces, the Roman Catholic church, intent on bringing Orthodox Russia under the sway of the Pope, sent Jesuit missionaries into Russia to convert the people. This and other high-handed policies of the new czar provoked a violent reaction. The Jesuits were murdered, and a period of anarchy called the "Time of Troubles" ensued. Moscow was retaken, and Grigory was put to death in 1606 in a coup led by Vassily Shuisky, who was then elected czar. However, there were many in Russia opposed to Shuisky's rule, and in 1607 another False Dmitry appeared, claiming this time to have survived the coup. Though he did not physically resemble Grigory Otrepyev, the malcontents rallied around him, and Marina even acknowledged him to be her husband. He was eventually killed by his own followers in 1610. But the very next year yet another False Dmitry appeared and held sway until he was captured and executed in Moscow in 1612. The general chaos, including the Polish military presence in Russia, wasn't ended until after the accession of Michael Romanov as Czar in 1613.

The story of False Dmitry echoes the history of another interregnum, this one resulting from the assassination of Nero Caesar in C.E. 68. This is the Nero redivivus legend, the belief that Nero either survived his assassination or rose from the dead afterward and escaped to the east where he was marshaling an army of Parthians with which to invade the Roman Empire. This superstition was strong enough that between C.E. 69 and 88 three different pretenders posing as Nero attempted to seize control of the Roman Empire. It was also a probable source of much of the imagery of the Beast (the Antichrist), and particularly of the Number of the Beast, in Revelation. Since the letters of the Greek alphabet, like those of the Hebrew alphabet, have numerical values, names can be converted into numbers and numbers into names. The number 666, the number of the beast, converts into "Neron Caesar." In some early manuscripts the number of the beast is 616, which gives us "Nero Caesar" in the Greek alphabet.

Yet another assassinated prince whose death spawned pretenders was Bardiya, son of Cyrus the Great and brother of Cambyses. Cambyses succeeded Cyrus as ruler of the Persian Empire in 529 B.C.E. and had his brother Bardiya murdered before setting out to conquer Egypt in 525. As brutal as such an act was, it was probably also quite prudent in that Bardiya might well have taken the throne in Cambyses' absence. As it was, when Cambyses died on his way back from Egypt in 522, a Mede noble by the name of Gaumata claimed to be Bardiya and seized the throne. Darius, Cambyses' son-in-law, managed to murder Gaumata and lay claim to the throne in 519. Since he was only distantly related to the royal family, the legitimacy of this claim was tenuous at best, and Darius had to put down a series of revolts before he could become Darius I, ruler of the Persian Empire. Among those he had to dispose of was another false Bardiya, this time a Persian noble named Vahyazdata.

Thus it can be seen that Anastasia is among the latest of murdered royal scions whose death gave rise to legends of a miraculous survival. That she is not the last is evidenced in our own legends of the survival of John F. Kennedy, Elvis Presley, and now it appears, even Princess Diana. As trivial as Elvis sightings may make the legend seem, the ancient lineage of the tale shows that it is a myth to be understood rather than merely dismissed out of hand. The irrational hope that the king, kingly line or beloved leader (or even cultural icon) didn't really die a premature sordid death but is waiting for the propitious moment in which to return, avenge his would-be destroyers and dispense justice is related to the myth of the sleeping king of old. This may be

Arthur, who supposedly sleeps with his knights at Avalon, or it might be Charlemagne or even Frederick Barbarosa. In all these cases the great king of old did not actually die but instead lies in a death-like sleep in a cave surrounded by his loyal retainers, waiting by divine mandate for the time of his nation's greatest peril, when he and his knights will awake, ride forth, deliver his land, and possibly even reestablish Camelot. The emotional hold of such legends can be seen from the fact that despite the failure of Charlemagne to drive the Nazis out of France or that Britain owed more to the R.A.F. than to Arthur in repelling Hitler, the myths do not die.

Another variant of this myth is that the heir to the rightful kingly line is in hiding and will one day drive out the invader. Historically this has a bit more validity than the legend of miraculous survival. Alfred the Great (849–899) managed to go from hiding in the swamps to overthrowing the Danes in 871, and Robert the Bruce was likewise able to deliver Scotland from the English invaders at the Battle of Bannockburn in 1314. Both of these situations also recall the plight of David and his outlaw band being pursued by King Saul. For all that, the Stuart cause and that of the Scottish Highlanders died on the field of Culloden in 1746 with the final defeat of Bonnie Prince Charlie (1720–1788). The myth of the king in hiding who rides out of obscurity to claim his rightful throne, drive out the vile invaders, and revive the ancient realm was given new life by that greatest of all modern myth makers, J.R.R. Tolkien, in *The Lord of the Rings* trilogy, the final volume of which is aptly named *The Return of the King*. In Tolkien's modern epic, Aragorn, rightful heir of the kings of the west, wanders the land as the ranger Strider. The other rangers are his knights, and as their title implies, they provide protection from evildoers in the wild lands that have grown up since the fall of the kingdom of the west. When we first meet Strider, he bears a broken sword. The "Sword that was Broken" has been carried by generations of Aragorn's forbearers and cannot be reforged until the one destined to regain the lost throne is given the proper sign. With the reforging of the sword the king will reclaim his realm.

The sword that is either lost, broken or virtually unattainable is a motif which connects the lost but rightful king to the myth of the hero in general. Arthur proves that he is the rightful heir to the throne of Uther Pendragon by pulling the sword Excalibur from the stone, which none but the rightful king can do. Likewise, in the Icelandic Volsunga Saga only Sigmund can pull the sword Nothung from the sacred oak, Branstock. And just as the stone represents the earth, Branstock represents Yggdrasil, the world tree or axis mundi of Norse myth. At its roots lies the underworld; in its branches are the heavens and the gods; and in its trunk is the world of human beings. At the hero's death Nothung is broken and must be reforged by Sigmund's son Sigurd (the German Siegfried). In Greek myth Theseus, raised in Troezen in the Peloponnesian peninsula, but son of King Aegeus of Athens, must roll a great boulder away to gain the sword and sandals that will establish his identity as the true son of Aegeus. Once again the sword is symbolically held by the earth, which will only yield it up to the rightful heir. In all three of these myths we find the elements of the kingly family that has lost power, the kingdom that must be reclaimed and the rightful heir emerging from obscurity. Since the death of Uther Pendragon, Britain had been divided into warring petty kingdoms. Arthur reunites them and establishes Camelot. The power of the Volsungs has been broken by a treacherous attack. Sigurd reforges the broken sword his father had pulled from the world tree and regains the glory of the Volsungs. While Aegeus hasn't lost his throne, he has been unable to procure an heir from two marriages. Theseus is conceived while Aegeus is visiting Troezen and, being drunk, ends up in the bed of a princess named Aethra, who had originally been promised to another hero. Thus the origin of Theseus, like that of Arthur (the product of Uther deceiving Igraine, wife of the duke of Cornwall), is through an illicit affair. As a result both Arthur and Theseus are raised in obscurity—the same obscurity in which the miraculous survivor exists until he or she remembers that they are in reality Bardiya, Nero, Dmitry or Anastasia. In the final case most women who claimed to be the Romanov princess, including Anna Anderson, the most famous pretender, said that amnesia induced by the trauma of their near assassination prevented them from pressing their claims earlier.

That the hero is commonly ignorant of his true parentage until it is revealed to him at puberty and that he is often raised by humble step-parents or as an illegitimate child was the case with not only Arthur and Theseus, but with Moses, Sargon the Great, Perseus, Romulus and Remus, and a host of others as well. This has been seen in psychoanalytic terms by Otto Rank and Sigmund Freud as a neurotic rejection on the part of the hero of the low station of his true parents or his position of illegitimacy. He claims instead the fantasy of a father who is a king or even a god. In the latter case, though Herakles (Hercules) is raised by his supposed parents, he is in reality the son of Zeus. In the psychoanalytic view, the hero's vain fantasy is that he will prove himself worthy of his divine or kingly father, who will then recognize him as his true son. However, we might just as well turn the psychoanalytic explanation around and see it as a tale which had to be invented by others for the lineages of great men, since their greatness obviously betrayed either kingly or even divine origins. For example, eventually, even great philosophers were given this status. Thus, after his death, Plato was made the son of Apollo, who assumed the guise of Plato's mortal father to impregnate his mother. This was commonly given as the reason the pharaohs of Egypt could seemingly be descendants of mortal fathers, while in reality being the sons of a god. The deity, usually Amon-Re, assumed the likeness of the reigning pharaoh to engender the new crown prince, who as ruler of Egypt was required to have divine parentage. In the case of Moses the situation is seemingly reversed in that he is in reality the child of a people held in bondage and is raised in the Egyptian court. However, as a prince of Egypt, he is fatherless, and the Israelites are God's chosen people, while the Egyptians are destined to be overthrown as a demonstration of God's power. Moses' real royalty in this myth is his position as a Levite, one who does indeed have a real father, and who is as well the chosen deliverer of his people. Another possible source of the hero's recognition at puberty that he is a person of different parentage from what he previously thought may involve a source that is more anthropological than psychological, namely the rite of passage from childhood and ignorance into adulthood and revelation of the tribal secrets, at which time the newly initiated adult often received a new name and as such a new identity.

While there is often a very pragmatic reason for the miraculous survivor to put forth his or her claim—in the case of the false Bardiyas, Nero redivivus and the False Dmitries the rule of a kingdom, in the case of Anastasia a considerable fortune held for the Romanovs in a Swiss bank—there are also powerful emotional, psychological, anthropological and mythic/religious forces which support these recurring claims. The elements common to the myths of the miraculous survival, the sleeping king and the hero in general are as follows:

1. The central character is lost, disinherited, estranged from his kingly or divine father, raised in obscurity and deceived as to his true position or placed under a

spell, such as a deathlike sleep or its modern variant, amnesia.

2. Though they have been wronged and denied their heritage, they are in fact destined for greatness. This will be shown by a supernatural sign, omen or the passage of an ordeal: the recovery of the lost or broken sword, the sign of the burning bush, or even the recovery of a lost memory.
3. Just as the central character begins the story in a state of ignorance and estrangement, so the kingdom or even the world itself is in a state of disorder for lack of the true king. Either a state of anarchy exists or an impostor, possibly even an invader, sits upon the the throne: Sauron menaces Middle Earth; Uther's kingdom is in shambles; the Volsungs are in hiding and denied their rightful lands; the Danes have overrun Wessex; a fratricidal murderer rules Persia; Rome is in chaos; Boris Godounov has usurped the throne of the line of Rurik; the Bolsheviks rule with an iron fist.

That the world or at least the kingdom is in chaos for want of the true king involves not only myth, but the ritual of sacred marriage, in which the king is symbolically married to the goddess who is the earth or at least the land through a ritual sexual union with a priestess representing the goddess. That the true king was reigning was evidenced by the fertility of the land. Thus, the crop failures in the reign of Boris Godounov which convinced the people that he was unfit to rule tapped into a surviving stratum of pre-Christian belief. The chaos of the kingdom in the hero myth is a divine sign that the land awaits the true king. The sacred kings of this ancient belief system were often ceremonial, their power being more symbolic than real. However, attempts to replace them often provoked violent reprisals. Thus when the Germanic war leader Arminius (Herman or "war man"), who had kept the Romans from conquering Germany, attempted to make himself king of the Cherusci, he was summarily assassinated. One of the likely reasons his claim to kingship was seen as invalid might well have been that he was a warrior. The sacred king was often forbidden to profane himself by shedding human blood. The tension between the character of the sacred king who must not go to war and the deliverer of the kingdom who must can be seen in the two titles claimed for Jesus by his followers. In apocalyptic terms the "Son of God," a title previously applied to the Davidic kings (see Psalms 2:7), was a war leader who would drive out the invader and restore the line of David to the throne of Israel. Yet in such apocalyptic works as the Book of Daniel or the non-canonical but extremely influential 1 Enoch, the "Son of man" is a holy king without weapons who, because of his holiness, inherits the millennial kingdom after the apocalyptic destruction of evil and rules it in peace.

4. In a grand climactic confrontation, possibly even an apocalyptic battle between ultimate good and ultimate evil, the king (or even the lost princess) will return and set all things right. By now many readers will have observed that the elements of divine intervention to right a cosmic wrong by the restoration of the rightful king, with justice held in abeyance until that ruler reclaims the throne by way of a final battle, are central to the Christian mythos. Jesus, the divine son, is born and raised in obscurity and humility while his enemy Herod seeks his death. The entire world is in a state of travail, awaiting the

> righting of old wrongs when Christ, the true king—variously characterized in terms of the warlike Son of God and the holy Son of man—returns. And so we are brought back to John 5:25–29 with Jesus saying that the hour is coming when the dead will rise and stand again (anastasis) to face judgment.

That the miraculous survival, whether it be of Anastasia, JFK or even Elvis ("the King"), is part and parcel of the same mythos that is the source of the heroic quest in general and Christianity in particular not only establishes its roots deep in the psyche and often beyond the reach of reason, it also has profound political implications. For example, consider Islam. Like Christianity it was initially spread with messianic zeal, but unlike Christianity it did not make even the pretense that conversion should be by peaceful persuasion. Coercion to compel right thinking is not seen by many of the pious as being anything but proper. It also, like Christianity, has a linear view of time, that is that human history has a logical beginning, middle and end, and that the end is divinely ordained. Now let us consider that many of the Islamic states have a national view which mirrors that of the hero or miraculous survivor. That is, they can reasonably see themselves as wrongly deprived of their central position. Geographically located on lands where civilization had its origins and being descendants of the peoples who gave us not only our alphabet but our numerical system as well, they see the world—which is clearly an unjust place—in the grip of largely secular powers, Johnny-come-lately foreigners and infidels who often support corrupt regimes. In other words, the same deep psychological motivations that gave us the hero myth may well operate as the motivating force of a holy war. It is therefore important for us to engage the Islamic world, to work to raise its standard of living and to encourage within it democracy and the free flow of ideas.

That the psychological elements of the hero myth might interact with the sense of injustice in Islamic nations should alert us to the potential danger in general posed by an appeal to its deep and powerful emotional roots. Consider some of the elements common to many totalitarian regimes:

1. The Charismatic Leader, or "the man on the white horse." This is the hero himself or even the sacred king. Whether it be Hitler, Stalin or Mao, he embodies the ruler who will set things right, often by his mere presence. By bringing fertility to the land, the sacred king of ancient times was a cosmic functionary.
2. The Organic State, in which all social functions are subsumed into state organizations. This is usually seen as a requirement to unify the social structure in order to mobilize the nation in service to the great cosmic goal (see # 4).
3. The enemy within and the enemy without. Historically the enemy within has been anything and everything from homosexuals, "liberals" and opposition parties, to any ethnic group which either hasn't been willing to assimilate (such as the Gypsies) or hasn't been allowed to (such as the Jews). The enemy without is whatever power opposes the totalitarian state.
4. The Cosmic Goal: Whether it be the Thousand Year Reich, the eventual "withering away of the state" or even the millennial kingdom, this is the justification and focus for the subordination of the individual and often the basis for external aggression.

Let us consider all of these points together. The mythos of the totalitarian state is essentially as follows: We seek to fulfill a great goal, a world of ultimate good and justice. In order to reach it we must submerge our individual differences and all pull together under the guidance of our great leader. The only reason we cannot immediately achieve our great goal is that the world is an imperfect place, filled with small minded people, some undermining our society within and others attacking it from without. Until these enemies are overcome, we cannot achieve our cosmic goal. This mythos contains within it the noble quest which is nothing short of the restoration of the cosmic ideal, the profound sense of being a victim of injustice and, embodied in both the state and the leader, a semi-divine father figure whose recognition is eagerly sought. This last aspect was noted by the late Wilhelm Reich, who, before he became entangled in his pseudoscientific orgone therapy, had many profound psychological insights. In *The Mass Psychology of Fascism* he noted that the fascist state reproduced in macrocosm the authoritarian patriarchal family. He argued that the lower middle class, from whom the Nazis gained their core support, would have been better served by the political agenda of the Socialists, but that they voted for the Nazi Party based on an emotional appeal which bypassed reason. While Reich concentrated on the patriarchal family, the emotional appeal implicit in the totalitarian mythos is far broader in that it embodies those elements common to the myths of the miraculous survivor, the hero and the sacred king. (Indeed, as representative of the sacred king, the leader must triumph if the cosmos is to function properly.) That is to say that the same grand myth which provided the emotional basis for the high ideals of the Arthurian cycle and the works of J.R.R. Tolkien also helped fuel the rise of the Third Reich.

It is commonly noted by the apologists of religion that, as bad as theocratic states such as Calvin's Geneva or Cromwell's England were, secular states such as Nazi Germany and the Soviet Union have far surpassed them in dehumanizing brutality, the implication being that modern totalitarianism is the logical end of rational humanism. Yet, as I noted above, the emotional appeal of the totalitarian ideal parallels that of myth and religion. It is thus hardly the logical end point of rational secularism. In any case the fact of the matter is that in every country in Europe in which the fascists took power in the 1930s they did so in concert and coalition with religious conservatives. Thus the Catholic church supported, either implicitly or explicitly, Mussolini in Italy, Franco in Spain, Salazar in Portugal, and Ante Pavalic in Croatia. Likewise, the Orthodox church supported Metaxas in Greece, and both Protestants and Catholics supported the rise of Adolf Hitler. Likewise, in Japan the role of the emperor as the focus of national identity was part and parcel of the national religion. Of course the dictatorships of Stalin and Mao were largely anti-religious. Thus it would seem that, while religious authorities can be used to back dictators—particularly those who crusade against "Godless communism"—totalitarianism is less a function of relative belief or disbelief in a god than of the modern technology of mass media married to ancient mythic themes during periods of national crisis.

At the present time, the collapse of the Soviet Union has deprived the Anastasia myth of much of its power by removing the brutal, unjust usurper required by the hero myth. Anna Anderson, the most famous of the Anastasia pretenders, died in 1984, and DNA tests on a portion of intestinal tissue removed before her death and subsequent cremation proved that she could not have been Anastasia. In any case, even had the Russian princess survived, she would be 97 by now. The Romanov romance could conceivably be revived by a son of Anastasia, but such a pretender would not be a miraculous survivor, and even in its present disordered state, Russia is hardly as susceptible now as it was in the Time of Troubles to the appeal of that myth. Nevertheless, the hero myth retains its power even in our largely secularized society, as evidenced by the popularity of the *Star Wars* movies. It is important to know and understand the dynamics of the hero myth as a way of not only understanding the recurring variations of the miraculous survivor but as well to know how to debunk incipient dictators and others of their ilk. Yet, as we have seen over and over again, rational exposure of the falsehood of cherished beliefs often provokes a fierce defense rather than an acceptance of the proof. Witness the response to the appearance of Emily Rosa on Larry Mantle's "Air Talk" show (KPCC 5/7/98). Every one of those who called in supported therapeutic touch and had rationalizations for why Emily's simple and elegant experiment didn't really prove the pseudoscientific practice invalid. At least one caller was incensed.

It would appear then that an emotional basis for opposing the use of the hero myth in the service of a totalitarian ideal must be sought. To some degree this can be achieved by encouraging people to be their own heroes, rather than by subordinating themselves to another hero. However, I would attack the emotional desire for closure as the prime defense against dictators, and I would do this by way of making people comfortable with the chaos and imperfection of the cosmos and the human condition. I would argue that a certain level of chaos is necessary for the dynamic functioning of the universe. In a state of crystalline perfection any change, any level of dynamism must be viewed as evil. One argument against a perfect god is that he, she or it would have no reason to create a universe, since the deity's own perfection would be sullied by the creation. Viewing the deity as an artist, one driven to create, implies imperfection. Fortunately, the idea of the cosmos as necessarily flawed is actually implicit in the hero myth itself in the form of the usurper whom the hero must overcome. The oppression of the kingdom by the evil usurper can be seen in psychological terms as a neurotic failure to deal with a less than ideal world and the focused projection of the hero's lowly state on the deliberate frustration of his rightful position by a malevolent personage. Yet, once again there is as well an equally valid anthropological reason for the mythic character of the usurper. This is particularly evident in the myth of Osiris and Set, where Set, representing death, kills Osiris the rightful king. Horus, whose magical conception and imperiled infancy are not only classic hero motifs but precursors of the Christian nativity myths, not only avenges his father's death but restores Osiris to life. But this is not the end of the story, for the myth is cyclical. In fact the year is divided between the rule of Osiris and the rule of Set. In short the deity or hero representing the sun or the grain must be destroyed by his rival and be reborn seasonally for the world to function. Thus the usurper is a functionary of a dynamic, cyclic cosmic order. In some variants of the Osiris myth Horus and Set are eventually reconciled. Thus, if understood properly, the hero myth need not be a dangerous source of neurosis and totalitarianism. If, along with the appreciation of the power of myth, we can teach people to strive for improvement of the human condition while still accepting imperfection and chaos, in short lack of final closure, as valid aspects of both the human condition and the cosmos itself, we can greatly diffuse the source of the totalitarian mythos to which the myth of the miraculous survival belongs.

References:

Bury, J. B., S. A. Cook and F. E. Adcock, eds. 1970. *The Cambridge Ancient History*, Vol III. Cambridge and London: Cambridge University Press.

Campbell, Joseph. 1964. *The Masks of God: Occidental Mythology*. New York: Penguin Books U.S.A. Inc.

Ellis Davidson, H. R. 1964. *Gods and Myths of Northern Europe*. Harmondsworth, Middlesex, U.K.: Penguin Books Inc.

Graves, Robert. 1955. *The Greek Myths*. Harmondsworth, Middlesex, U.K.: Penguin Books Inc.

Hastings, James, ed. 1934. *Encyclopaedia of Religion and Ethics*. 4th edition. Edinburgh: T. & T. Clark (New York: Charles Scribner's Sons).

Helms, Randel. 1988. *Gospel Fictions*. Buffalo: Prometheus Books.

Pushkin, Alexander (trans. Alfred Hayes). 1982. *Boris Godounov*. New York: The Viking Press.

Reich, Wilhelm. 1942. *The Mass Psychology of Fascism*. New York: Farrar, Straus and Giroux.

Strong, James. 1890. *Strong's Exhaustive Concordance of the Bible*. Gordonville, TN: Dugan Publishers Inc.

Tolkien, J. R. R. 1965. *The Lord of the Rings*. New York: Ballantine Books Inc.

Ancient Astronauts

Zecharia Sitchin as a Case Study

ERIC WOJCIEHOWSKI

In 1968, Erik von Däniken released his book *Chariots of the Gods?*, which touched off a fire storm of debate between his admirers and those who had found his claims wanting. To refresh your memory, its thesis was that Extraterrestrials had come to planet Earth in ancient times and have been remembered in myths and legends as well as from the architecture they allegedly left behind.

Despite the fact that von Däniken has lost much of his audience (at least in America), he still seems to have left an impression on the public that has not faded. Although academics have spent many hours showing the errors in von Däniken's reasoning, the Ancient Astronaut notion remains alive. Why? Well, for starters, critics have spent most of their time on von Däniken's theory of ancient astronauts and not on the general concept of extraterrestrial visitations. By this, I mean that although Erik von Däniken popularized the idea, the theory itself has largely been defined by a number of other people.

Sadly the usual approach taken against the Ancient Astronaut theory is akin to dismissing the "UFO's are space aliens" theory by only criticizing one of the many authors who have promoted this viewpoint. The skeptical community is very familiar with old claims resurfacing with different packaging. This is why each case must be addressed separately. It is related to what UFO researcher Jacques Vallee has called "The Ratchet Effect" ("most amateurs of the paranormal never went back to a baseline of normal belief once they had become convinced of a certain weird fact, even if it was later proven to be false," 1991, 85).

This essay is not intended to be an exhaustive critique of everyone who has ever proposed the existence of ancient astronauts. It evaluates instead one author who seems to inspire a continued following even though von Däniken has faded.

Out of all the people who have ever claimed that aliens have been to planet Earth in the distant past, Zecharia Sitchin is the one man who deserves the most attention. To date, he has suffered little or no criticism (except Oberg, 1978). In following the Ancient Astronaut theory for many years (as well as being a member of the Ancient Astronaut Society), I have observed that Sitchin is cited time and again by believers as the definitive "expert" in this field (there are now even "study groups" forming in the hope of the construction of "Sitchin Centers" to continue his legacy).

In 1976, Zecharia Sitchin released the first of his books, *The 12th Planet.* Subsequently, more books followed along the same theme, including *The Stairway to Heaven* (1980), *The Wars of Gods and Men* (1985), *The Lost*

Realms (1990), *Genesis Revisited* (1990), *When Time Began* (1993), and *Divine Encounters* (1995). Taken together, Sitchin has chosen to call his work "The Earth Chronicles." Although all his books should be considered in evaluating his work, the first three are the most important.

Briefly, Sitchin believes that approximately 450,000 years ago an alien race came to Earth from an as-yet-undiscovered 10th planet in our solar system. (The sun and moon were counted as 10 and 11, making this undiscovered planet the "12th" planet. Sitchin displays pictures from the ancient Near East where 11 or 12 orbs or "stars" appear in a circle. He then makes the claim that this, along with speculative references found in ancient scriptures, represents the fact that ancient people knew of all the planets we do now; see Oberg, 1978, for a critique of Sitchin's claims in regards to astronomy.) These beings came here to mine gold and other materials. Approximately 250,000 years ago the aliens interbred with *Homo erectus* to create modern *Homo sapiens,* to be used as slave labor in mines, on farms, and in the homes of the aliens. As time went on, the aliens began to give privileges to humans as well as allowing them to run their own lives and affairs.

During these events, an immense flood of biblical proportions occurred, Egyptian and Near Eastern civilizations were established, wars involving aliens and humans commenced (where flying machines and a nuclear missile were involved), and the pyramids of Giza and other monumental structures (including some of those in the Americas) were built.

According to the blurb on the dust jacket of the hardcover edition of *The 12th Planet,* "Zecharia Sitchin was raised in Palestine where he acquired a profound knowledge of modern and ancient Hebrew, other Semitic and European languages, the Old Testament, and the history and archaeology of the Near East. He attended the London School of Economics and Political Science and graduated from the University of London, majoring in economic history. A leading journalist and editor in Israel for many years, he now lives and writes in New York." According to the program of the Ancient Astronaut Society's 16th anniversary world conference, "Mr. Sitchin spent nearly 40 years in gathering and synthesizing the data (for his books). Mr. Sitchin is a member of the Israel Exploration Society, the American Oriental Society, and the Middle East Studies Association of North America." It should also be noted that Sitchin is one of a handful who can read the Sumerian language and cuneiform script. This alone suggests long hours of study in ancient Near Eastern history and culture. It is no wonder that his work is trotted out by believers. He appears to be an educated man who has sided with them. His opinions have weight and therefore deserve attention. So with such credentials, what brought Sitchin to the conclusion that aliens had shaped much of human history?

Sitchin answers this by stating, "It was at school in Tel-Aviv; we reached in our bible studies Chapter VI of Genesis—the story of the Great Flood or Deluge. It begins with several enigmatic verses, undoubtedly the remnant of a longer text, that describe the circumstances on Earth prior to the Deluge. They tell us—in the familiar King James translation":

> when men began to multiply on the face of the earth, and daughters were born unto them, that the sons of God saw the daughters of men that they were fair; and they took them wives of all which they chose. . . . There were giants in the earth in those days, and also after that, when the sons of God came in unto the daughters of men and they bear children to them, the same became mighty men who were of old, men of renown (Freer, 1987, iii).

But Sitchin was studying the Bible in its original Hebrew and he noticed that the word

"Nephilim" literally meant, "Those who had come down" and not (as translated in the King James version of the Bible) "giants" (Freer, 1987, iii). From this Sitchin began his quest to find out who the Nephilim were, taking their presence in the Bible as a literal truth of existence.

Sitchin traced this word back to the names of ancient Sumerian and Babylonian gods. By concentrating on these and Near Eastern texts Sitchin concluded that the Nephilim were really an alien race that literally "came down" to Earth thousands of years ago.

So let us begin where Sitchin did and see if his claims bear out. Our starting point is in the ancient Near East, with the Sumerian words and pictographs they left behind to describe their gods.

According to Sitchin, if we trace back the word *Nephilim* we come to the Sumerian equivalent of DIN-GIR. The first syllable, DIN, according to Sitchin, means "righteous," "pure," or "bright" (Sitchin, 1976, 169); the second syllable, GIR, "was a term used to describe a sharp-edged object" (Sitchin, 1976, 168). Expanding on this, Sitchin states that by putting these syllables together, "DIN-GIR as 'gods' or 'divine beings' conveyed the meaning of 'the righteous ones of the bright, pointed objects' or more explicitly, 'the pure ones of the blazing rockets'" (Sitchin, 1976, 169). What is important here is how Sitchin came to his final translation. A literal translation of the word DIN-GIR should read (based on Sitchin's translation) "pure sharp-edged object" or "bright sharp-edged object." This should lead one to the conclusion that the DIN-GIR was one object. But Sitchin claims that DIN-GIR should be read as "pure ones of the blazing rockets," which insinuates two things: the "pure ones" and the "blazing rockets." This part of Sitchin's rendition does not justify his final translation. So why did Sitchin go from the DIN-GIR being only one object to two? And why did he choose to translate the syllable GIR from the more accurate description of "sharp-edged object" to the rather dramatic "blazing rockets"? It seems he was influenced by the pictorial signs for each syllable.

According to Sitchin (Sitchin, 1976, 170), the pictorial signs for DIN and GIR, which resemble a multistage rocket ship, combined with the textual references to gods roaming and flying from heaven to Earth leads to the conclusion that they were indeed rocketships: "Sumerian and Akkadian texts leave no doubt that the peoples of the ancient Near East were certain that the Gods of Heaven and Earth were able to rise from Earth and ascend into the heavens, as well as roam Earth's skies at will" (Sitchin, 1976, 128). Attentive readers will notice that Sitchin interprets the DIN to be two things at once: the "pure ones" (aliens) and part of the multistage rocket. This despite his pointing out that the word and the pictograph are supposed to represent one and the same thing! This shows that Sitchin is finding and using many unfounded meanings as well as creating double ones for each of these words and pictographs.

All we can really say is that we are dealing with a "pure" or "bright" "sharp-edged object." All the other elements Sitchin applies to the DIN-GIR, like making each word stand for two things at once, are not justified. The pictograph for DIN-GIR does not necessarily represent a rocket ship from antiquity. This only occurs through speculation.

It has been noted that DIN-GIR is somewhat like the Egyptian word for god which is *Neter* (Morenz, 1960, 19). It is clear that although many different interpretations have been given for Neter, none are absolute. However, the most likely explanation seems to be that the Neter (complete with its own pictograph) could be nothing more than a sort of flag (Morenz, 1973, 9). This would then suggest a clear sign of totemism, a sort of "banner" which stood for each group of people who rallied around it. If DIN-GIR can be said to be similar in meaning to Neter, then it is possible that the word and its pictograph are of the same sort. Only in this case, the "banner" would be a "bright" or "pure" sharp-edged object somewhat akin to an obelisk or spear-like construction. Sumerologist Samuel Noah Kramer states that the peoples of the ancient Near East thought of their gods as the assumed powers which operate behind the natural order of the world (Kramer, 1981, 77–78). Thus it is possible that the DIN-GIR was a totem used to symbolically represent these assumed forces and nothing more.

Sitchin repeatedly argues throughout his work that the aliens were anthropomorphic and human-like in design. And it is true that some cylinder seals reveal the gods in this type of form. But aside from the difficulties in explaining how two species, separated on different planets, came to be so similar (so similar that they were able to interbreed), the engravings which display the anthropomorphism of the gods are not the oldest forms. The majority of the oldest existing drawings show us the ancient Near Eastern gods are more animalistic in design. It was only in later times that the gods began to be drawn as erect-standing, two-armed, two-legged beings with a body and a head (Jacobsen, 1976, 9). If the aliens were really the gods, and if they were human-like in characteristics, then the oldest drawings should bear this out. They do not. (Jacobsen does state that in the early periods the human forms may have been a competing characterization of the gods. Regardless, this was not the dominant form. Some may say that this is purely a metaphorical way of expressing the attributes of real historical beings. But the error in this thinking is easily discerned. By claiming such, one would have already concluded that these beings did exist. But since we are dealing with a multitude of forms, we must treat them as a whole when attempting to evaluate the religious aspects of the ancients.)

To fully understand Sitchin's claims about aerial objects, we must also examine his interpretation of the Sumerian word *MU*. According to Sitchin, this word (equal to the Hebrew word *shem*) should be properly translated as a "skyborne vehicle" (Sitchin, 1976, 139–167). He spills much ink telling the reader that the MU was described as "lights up as a fire" and of an enclosure specifically created to "protect" the MU "which in a fire comes forth." He also quotes from a text which describes the goddess Inanna flying in her MU (Sitchin, 1976, 42). He then shows how the word MU evolved in later times to describe obelisk-type structures and believes that the obelisks were erected in memory of multistage rocket ships that humans once saw when the aliens were here. However, as with the DIN GIR, Sitchin has not made a very strong case. We do not have to assume that just because the gods were said to have been flying in their MU's or because the MU's looked like rocket ships (notice that an obelisk gives such an appearance) that they actually were. We must ask, how were the MU's perceived to have been used by the ancients? Were they used in connection with rituals performed to create symbolic flights as shamans worldwide have been known to conduct? Or were they really technological wonders? The fact remains that until the spade of an archaeologist uncovers the corroded remains of a buried rocket, a more mundane, orthodox interpretation should be sought in connection with the DIN, GIR, and MU.

We can now move on to some of Sitchin's other reworkings of the historical texts to reveal his other blunders. The biggest problem with Sitchin's work is that although he lists an extensive bibliography at the end of his books, he rarely gives specific references to individual works when he quotes a particular text. I have attempted to track down many of his references to see if his retelling of the tales matches what was actually written. For those that I have found, some of the texts that Sitchin uses seem to have been taken out of context, or abbreviated, leading to a loss of the intended meaning.

As with other Ancient Astronaut theorists Sitchin carefully selects evidence that matches his preconceived notions. He documents the texts that seem to support his claims but fails to mention those that contradict them. For instance, in *The 12th Planet* Sitchin argues that the extraterrestrials genetically created modern humans for slave labor from the already existing *Homo erectus*. He quotes from various texts to show that the ancients believed that the gods created man and then launches into more word play as he substitutes modern technological terms for the descriptions of the actions of the gods who performed this alleged feat of genetic engineering. What Sitchin fails to mention is that many different versions of how mankind came to be exist among the ancient sources. There is not one coherent belief system working here. Religious scholar Mircea Eliade has noted, "There are at least four Sumerian narratives that explain the origin of man. They are so different that we must assume a plurality of traditions" (1978, 59). Not only do we have stories that humans were created by gods, we also have stories that suggest that humans sprouted from the Earth like plants! Also, within the Babylonian creation story known as the Epic of Creation, also known as the Enuma Elish, it is said that the god Ea (Enki) created humans from the blood of the god Kingu. In Sitchin's discussion of the Enuma Elish, he considers Kingu to be our present day moon.

Additional texts reveal Sitchin's short-comings for lack of attention to detail. There is a tale from the ancient Near East entitled the Etana myth. It begins where the gods are looking for some human worthy of sitting on the throne of the city of Kish. A man by the name Etana is chosen for such an honor. The tale then takes a strange turn when an eagle and a snake make an oath not to operate outside the rules and laws handed down by the sun god Shamash. Shortly after, the eagle breaks this part of the bargain and snatches the snake's young to feed his own babies. Because of this crime, the eagle is punished by being imprisoned in a deep pit until its natural death. The story then returns to the life of Etana. He pleads to Shamash, asking that the god help him procreate, for what concerns Etana most is his inability to have children. This is where the two, seemingly independent, stories converge. Shamash tells Etana where to find the entrapped eagle. With the eagle's help, Etana may journey to heaven to obtain the "plant of birth." As they travel higher and higher, the eagle repeatedly asks Etana to look back and see how the land and the sea look smaller. As can probably be guessed, Sitchin argues that the eagle was actually a spacecraft that took Etana to the god's planet. His proof seems to come from Etana's observation, preserved in the texts, of how the land and sea seem to grow smaller with distance (Sitchin, 1976, 161–163). What is most interesting is that Sitchin completely ignores the fact that the eagle is never described as anything other than an actual eagle. No references appear in the Etana myth to indicate anything technological. For instance, when Etana "boards" the eagle and prepares for flight, the eagle instructs Etana and Etana follows. The eagle states: "put your chest over my breast, put your hands over the quills of my wings, put your arms over my sides." And Etana follows, "He puts his chest over its breast, put his hands over its feathers. . . ."

It is not difficult to see that Etana supposedly travels into space on the outside of the eagle! What about Etana's observation of the appearance of the receding land and sea as he flew higher? Well, so what? The peoples of the ancient Near East were surely aware that as something moved further away, it appeared to be smaller. The writer (or writers) of the Etana myth probably assumed (correctly) that the same illusion would occur if one were to travel skyward. This observation of distance is no proof of an actual journey to the sky.

Finally, Sitchin's blatant "pick and choose" method is illustrated by his silence about the serpent's role in the story, as well as the conflict between it and the eagle. The story must be dealt with by examing all elements of it. And this story presents us with a very poetic account of Etana and the snake-eagle oath which is purely a mythological-poetic account, not a historical one.

In Sitchin's book *The Stairway to Heaven*, we see another example of his selective methods. Sitchin charts what he thinks was the journey and final destination Egyptian Pharaohs believed they would undertake after death. He uses the Egyptian writings and texts and concludes that Pharaohs believed they would exit their tombs, travel east and then proceed through an underground base made up of 12 levels and end up aboard a rocket ship bound for the "Imperishable Star" (which Sitchin identifies as the aliens' home planet). In one instance Sitchin quotes Utterance 422 from the Egyptian Pyramid Texts. When I checked this quote against the one found in R. O. Faulkner's book, *The Ancient Egyptian Pyramid Texts* (which Sitchin lists in his bibliography and which is known to be the best English translation of these writings), the actual Utterance 422 is almost five times as long as Sitchin's quote! He never reveals that he is abbreviating. From reading both versions, it can be shown that Sitchin's lack of attention to detail damages his position. Within the original Utterance 422, it is proclaimed to the king, "may he do what he was wont to do among the spirits, the Imperishable Stars." Note that the "Imperishable Stars" appears in plural form. This would suggest that the king is to be among the many Imperishable Stars (or the stars we see in the sky today) and not on any particular star. In fact, many of the Utterances speak of a plurality of Imperishable Stars. This detail, which Sitchin fails to document, completely changes where the Pharaoh was believed to go after death. Sitchin's claim that Pharaohs envisioned themselves boarding a rocket ship to fly to space is also flawed. In Utterance 508, it states, "O Re, I have laid down for myself this sunshine of yours as a stairway under my feet on which I will ascend to that mother of mine, the living Uraeus which should be upon me. . . ." Here Re is clearly identified as the sun and the sun's rays are what the Pharaoh plans to use to get to him. Where was the king believed to go among the stars? Part of Utterance 471 states, "and I (Pharaoh) ascend the sky, I will go aboard this bark of Re, it is I who will command on my own account those gods who row him. Every god will rejoice at meeting me just as they rejoice at meeting Re when he ascends from the eastern side of the sky in peace. . . ." In other words, Re (the sun) is said to go from east to west, carried on a boat. It is this mythological boat that Pharaohs are said to have gone to, not another planet. These concepts are purely mythological and nothing more.

We should also note that despite Sitchin's interpretation, the 12 levels of underground passages through which Pharaoh supposedly traveled after death were most likely not a real place. The main reason for this is that we clearly have not uncovered such an immense underground base nor have we found any rocket launch pads or anything of the sort.

This points out the most damaging flaw in Sitchin's theory—the lack of physical evidence. Not one trace of evidence exists anywhere in the world for such a high technology in the not too distant past. Some have claimed that since all this supposedly happened a long time ago it is no wonder that this technology has probably eroded away or been destroyed by the natural process of time and the deeds of men. However, Sitchin has argued in *The 12th Planet* that our alien visitors arrived here approximately 450,000 years ago and in his book *The Wars of Gods and Men*, he argues that the aliens were still on Earth at the time of Alexander the Great, circa 333 B.C.E. Even if our visitors left around 300 B.C.E., that means that they were here for about 448,000 years! Yet there is none of the "waste" one would expect to find from such a highly advanced civilization residing on this planet for so long. By comparison, we have only had a space program since the 1960s and hundreds of pieces of junk material remain in orbit around the Earth. We have only been a technological society for about 100 years and we can see the scars upon the planet from the extensive mining, farming and building. If a technological society existed on this planet for about 448,000 years and left a mere 2,000 years ago, we would know it from more than rock pyramids and legendary tales.

Some other fine points to note are Sitchin's interpretations of various pictures and art from the ancient sites. On page 93 of his book *The Stairway to Heaven* he remarks that the picture labeled Figure 49 displayed on page 94 is that of the sun, sky, and the aliens' home planet. However, it probably better represents the sun, sky and the moon instead. On page 35 of *The 12th Planet*, in referring to Figure 15 on page 36, Sitchin claims that this is a picture of "a man lying on a special bed; his face is protected by a mask, and he is being subjected to some kind of radiation." There really is no proof that this is what is occurring. Not only that, but the picture does not suggest a "mask" but suggests instead that the man lying on the table has two heads!

Although Zecharia Sitchin is an educated man in a different category from most authors promoting the Ancient Astronaut theory, he still employs the same faulty logic as the rest. Sitchin's work delves into astronomy, archaeology, anthropology, ancient history, geology, genetics, biology, mythology, linguistics and more. I have chosen to deal particularly with Sitchin's use of legend and myth and other texts because he quotes them extensively as "proofs" of his thesis. The elements I have examined, specifically the lack of physical evidence to support Sitchin's claims, demonstrate the pseudoscientific nature of his work.

References:

Anthes, Rudolf. "Mythology in Ancient Egypt." 1961. In Kramer, Samuel Noah (ed.) *Mythologies of the Ancient World.* New York: Doubleday.

Dalley, Stephanie (trans.). 1989. *Myths from Mesopotamia.* New York: Oxford University Press.

Eliade, Mircea. (Willard R. Trask, trans.). 1978. *A History of Religious Ideas,* Vol. I. Chicago: University of Chicago Press.

Faulkner, R. O. (n.d.). *The Ancient Egyptian Pyramid Texts.* Oak Park: Bolchazy Carducci Publishers.

Freer, Neil. 1987. *Breaking the Godspell.* Phoenix: New Falcon Publications.

Harrold, Francis B., et al. 1995. "Cult Archaeology and Creationism in the 1990's and Beyond." In Harrold, Francis B. and Eve, Raymond A. (eds.). *Cult Archaeology and Creationism: Understanding Pseudoscientific Beliefs about the Past.* Iowa City: University of Iowa Press.

Jacobsen, Thorkild. 1976. *The Treasures of Darkness.* Westford: Murray Printing Company.

Kramer, Samuel Noah. 1981. *History Begins at Sumer.* Philadelphia: University of Pennsylvania.

Morenz, Siegfried. 1960. (Ann E. Keep, trans.). *Egyptian Religion.* New York: Cornell University Press.

Oberg, James E. 1978. "*The 12th Planet* book review." *Skeptical Inquirer.* Spring/Summer pp. 116–118.

Sitchin, Zecharia. 1976a. *The 12th Planet.* Santa Fe: Bear & Company, Inc.

—. 1976b. *The 12th Planet.* New York: Avon Books.

—. 1980. *The Stairway to Heaven.* New York: Avon Books.

—. 1985. *The Wars of Gods and Men.* New York: Avon Books.

—. 1990. *The Lost Realms.* New York: Avon Books.

Vallee, Jacques. 1991. *Revelations.* New York: Ballantine Books.

Holistic Medicine

The Case of Caroline Myss

PHIL MOLÉ

Philosopher Paul Kurtz, in his masterful book *The Transcendental Temptation,* identified two traits common to all types of supernormal thinking: (1) All varieties of magical thinking thrive when there is ignorance of the natural causes responsible for a phenomenon; (2) This ignorance leads the magical thinker to hypothesize the existence of unknown, miraculous causes (1986, 455).

This type of thinking is especially prevalent in alternative medicine, where practitioners with poor understanding of physiology or standards of medical evidence offer mystical explanations for diseases and therapies. The defining traits of alternative, or holistic, models of health are their untestability, their abundant use of metaphor in lieu of scientific evidence, and their belief in intuition as a means of obtaining medical knowledge.

Caroline Myss is currently one of the most prominent authors in this lucrative field. Thanks to the success of her 1996 book *Anatomy of the Spirit,* Myss has been virtually impossible to ignore. She commands top dollar for her workshops and appearances, and has promoted her teachings on *Oprah.* Her most recent book, *Why People Don't Heal and How They Can,* has been an overwhelming best-seller in both its hardcover and paperback editions. Obviously, a large number of people like what Caroline Myss has to say.

Myss draws upon several strains of spirituality, including Roman Catholicism and Judaism, but her teachings have three distinct characteristics. First, she uses a hypothetical system of energy centers in the human body to explain the development of disease, and to link each illness with a specific emotional issue. Second, she maintains that her unaided intuition can provide detailed and accurate diagnoses of a subject's illnesses, even if the subject is not physically present in the room with her. Third, she stresses a vague type of holism that champions the very duality it claims to transcend, and ultimately wishes for the subservience of all other viewpoints to her own. This article will closely examine each of these characteristics, and assess the merits of Myss's teachings through the perspective of conventional medicine.

Chalking It up to Chakras

Caroline Myss bases most of her teachings on her belief in seven energy centers, or chakras, located within the human body. In traditional Tantric practice, the chakras are represented as lotus flowers positioned along the spinal cord. Specifically, the chakras are thought to correspond with the following anatomical positions:

1. First chakra—between the anus and genitals
2. Second chakra—lower abdomen
3. Third chakra—solar plexus
4. Fourth chakra—chest cavity
5. Fifth chakra—throat
6. Sixth chakra—center of the forehead
7. Seventh chakra—top of the head

The vertical arrangement of the seven chakras represents a hierarchy of increasing spiritual awareness, and each chakra is thought to be activated by a particular set of emotions. For example, the energy of the first chakra is activated by base feelings of tribal affiliation, while the seventh chakra resonates to our quest for spiritual wisdom. However, negativity can cause our chakras to lose energy. This, she says, is how we become sick. When we fail to be as positive and spiritually aware as we should be, we withdraw energy "directly from the basic energy level we need to run our physical bodies." This depletion "is the mechanism through which the physical body becomes weakened" (Shealy and Myss, 1988, 93).

Not surprisingly, Myss also maintains that imbalances in the chakras always manifest themselves as afflictions in very specific parts of the body. Since the chakras themselves are thought to respond to particular emotions, this essentially assigns each illness a unique emotional cause. Myss dubs this model of the body the "human energy system." In this system, heart attacks are caused by excessive guilt and fear, AIDS is attributed to having a "victim consciousness," and syphilis is chalked up to feelings of hostility toward oneself (Shealy and Myss, 1988). Thus, illnesses are the cumulative history of our emotional lives or, as Myss repeats with mantra-like purpose, "our biography becomes our biology."

All of the above claims are completely unfounded scientifically. First, the chakra system, while undoubtedly a valuable component of a cultural and religious practice, has no proven relationship with the anatomy or physiology of the human body. Nothing resembling the energy of the chakras has ever been detected, despite the exquisite sensitivity of modern instruments. Myss tries to brush past this detail by alluding to the mysterious nature of this energy, assuring us that "conventional medical tests have no way of measuring energy loss" (1996, 10). This waffling won't do: Energy is energy, and if it has discernible physical effects, it should also be measurable.

To make matters worse, there is no agreement about how many chakras there are. The traditional kundalini yogic system recognizes seven major chakras (Campbell, 1974, 331), and this remains the most widely accepted number. However, some mystics recognize other minor chakras, and some maintain there are other major ones as well. Myss takes the latter position. In *Why People Don't Heal and How They Can,* she claims to have learned of an eighth chakra external to the physical body while doing intuitive readings during the 1980s (Myss, 1997, 89). This begs the question of why this bonus chakra wasn't mentioned in *Anatomy of the Spirit,* published in 1996. Why did Myss take so long to discover this chakra, and even longer to incorporate it into her teachings? One may also ask why she doesn't mention the additional four external chakras recognized in some yogic systems (Dale, 1996, 47). Is there an objective criterion Myss used to exclude these chakras? If so, what is it?

If this inconsistency doesn't trouble you, imagine what would happen if conventional medicine showed a similar absence of objective guidelines. Imagine, for instance, that your physician couldn't decide if you had one kidney or two. Flabbergasted, you'd go to another doctor. What if he informed you that there is a third kidney floating outside your body? If you're anything like me, you'd be out of there faster than you can say "muladhara chakra."

Even if we could agree how many chakras there are, there's no guarantee that we'd understand the origins of our illnesses any better. In much of her writings, Myss stresses that unresolved emotional issues often lead directly to illness. "Refusing to let go of past events, whether positive or negative, means throwing away some part of your daily energy budget. If you start losing energy and don't do anything about it, you'll inevitably develop a weakness in your physical body" (1997, 19). From this perspective, disease develops because of our inability to accept life's lessons—it is at least partly our own creation. Yet, this emotional model of illness fails to account for how a person who's perfectly content with life can suddenly collapse from heart failure, or how a happy, innocent child can be diagnosed with cancer. In situations like these, Myss assures us it is simply the will of God, who wants us to learn "certain lessons that our soul needs to discover" (1997, 28). But how do we know if an illness is our fault or God's? Myss offers no clear guidelines. Furthermore, unresolved emotional issues can hardly explain how disease develops in animals and plants. Only the conventional medical model can account for these anomalous data.

Myss also fails to explain—in terms other than reckless metaphor—how her hypothetical human energy system is understandable in terms of our knowledge of modern medicine. This deficiency is particularly apparent in *The Creation of Health*, her collaboration with Dr. Norman Shealy. In one chapter, Myss supplies "energy analyses" of illnesses to complement Shealy's traditional medical descriptions. But the two authors rarely find a tangible intersection; Myss cannot reconcile her system's untestable mystical claptrap with the established canon of medical thought. Here is her explanation for blocked arteries (Shealy and Myss, 1988, 161):

> Blocked arteries, as a rule, are created through the warehousing of guilt feelings and fears related to disappointing the expectations of others. Guilt weighs heavily on a person's consciousness, and like cement being poured slowly into someone's body, eventually hardens.

This comes just two pages after Shealy describes coffee and cigarette use, high fat diets, high density lipoproteins (HDLs) and other agents with at least partially understood causal mechanisms leading to blocked arteries. We are not told what, if anything, these mechanisms have to do with Myss's explanation.

When discussing AIDS in another section of the same book, Myss speculates that "the AIDS virus has spontaneously emerged into our global atmosphere in response to the massive victimization of all forms of life, including the planet itself" (1988, 200). She goes on to explain, in a particularly ridiculous passage, that the Earth itself is showing symptoms of AIDS. Her reasoning goes like this: (1) AIDS usually results in diseases such as pneumocystis and Kaposi's sarcoma which affect the lungs; (2) Forests around the world are being destroyed; (3) Forests can be considered "the lungs of the earth;" (4) The "lungs of the earth" are thus being destroyed; Therefore (5) the Earth has AIDS. Conspicuous by their absence are any discussions of HIV, intravenous drug use, or any other known mode of contracting AIDS.

As these examples show, Myss makes almost no attempt to incorporate medical knowledge into her chakra system, and seems unaware that this kind of synthesis is the only way she could possibly validate her theory. The burden of proof is on Myss to demonstrate how her ideas improve on our established understanding of illnesses, and how the chakras can add to the study of human physiology. Theories which provide no new information are useless. Unless Myss can establish the objective existence of chakras, prove they have measurable energy, and establish unambiguous mechanisms for how this energy affects the body, her ideas will remain empty of content.

Energy Analysis: The Power of Intuitive Diagnosis

When you imagine you've developed a whole new model for explaining illnesses and the human body, it's only natural to offer a new methodology for working with your model. For Myss, this new methodology takes the form of intuitive diagnosis, an ability she claims to have honed nearly to perfection over years of practice.

Myss alleges that, using nothing but pure intuition, she can "see" the energy shortages in a person's chakras and accurately diagnose their illnesses. She claims to have discovered this ability in the autumn of 1982, while working at Stillpoint Books—a publishing company she co-founded to publish books about alternative medicine. Myss, mind you, didn't actually believe in all of that metaphysical hocus pocus; she was a skeptical materialist who "smoked while drinking coffee by the gallon" and "developed an absolute aversion to wind chimes, New Age music, and conversations on the benefits of organic gardening" (1996, 1).

Nonetheless, Myss could not hide her gifts from the world for much longer. Soon, she discovered she had an uncanny ability to gain insights about the causes of her friends' illnesses. She describes them as being "like impersonal daydreams that start to flow as soon as I receive a person's permission, name and age. Their impersonality, the nonfeeling sensation of the impressions, is extremely significant because it is my indicator that I am not manufacturing or projecting these impressions" (1996, 2). Why was Myss chosen to receive this magical ability? Here are her thoughts on this question:

> While I can teach you up to a certain point about how to become intuitive, I'm actually not sure how I learned it myself. I suspect that I became extremely intuitive as a consequence of my curiosity about spiritual matters, combined with a deep frustration I felt when my life didn't unfold according to plan. On the other hand, it's equally possible that my medical intuition was simply the result of something I ate. Knowing how the gods work, I would not find it surprising in the least (1997, 5–6).

This passage, with its profound mixture of silliness (in considering her ability the possible result of something she ate) and arrogance (in claiming to be privy to divine knowledge), is a fine example of the intellectual value to be found in Myss's books.

Myss claims to be most accurate in her diagnoses when she has no information about the medical subject except for her name and birthdate. She explains that this method allows her to "receive information that a more personal connection would otherwise tend to block," such as information about the spread of cancer in a subject's body (Shealy and Myss, 1988, 85). She claims to accomplish this by projecting "emotional energy" toward a patient by intensely concentrating on her name and age.

How accurate is Myss? It depends on what standard of evidence you demand. If you need almost no evidence you can embrace the statement by her collaborator, Dr. Norman Shealy, that she is 93% accurate (Shealy and Myss, 72). Those of us who need good reasons for believing medical statements, however, may find Shealy's statement—and the evidence it's based on—a tad unconvincing.

Let's consider how Shealy "tested" his colleague's accuracy. According to Shealy, he would speak to Myss on the telephone while he had a patient in his office. He allegedly would inform Myss only of the patient's name and age, record her impressions, and compare them against his own "traditional" medical diagnosis.

There are valid reasons not to accept results obtained using Shealy's methodology. First, there did not seem to be any control for experimenter bias. Shealy spoke to Myss during her process of intuitive diagnosis; we have no way of knowing that he did not subconsciously provide clues to lead Myss to the right answer. This possibility seems especially important to control for because of indications that Shealy believed in Myss's alleged ability immediately upon meeting her at a conference in 1985, and he never seemed to question the validity of her claims. "Norm never tested me from the position of wanting me to prove to him that this skill existed. He already knew it was possible to develop exceptional perceptual abilities, and thus his interest in my work was from the position of whether or not I could be accurate enough for his purposes" (Shealy and Myss, 1988, 86). What "accurate enough for his purposes" means is open to interpretation, especially when the accuracy is determined by an experimenter who already assumes that the phenomenon under investigation exists, and fails to control for his bias. How accurate would Myss have been if Shealy had given her a written list of the patients and their ages, sequestered her in a room until she decided on her diagnoses, and then compared her diagnoses with those made by impartial physicians? Alas, we do not know, for no experiment like this appears to have been performed.

Second, we do not know what criterion of accuracy Shealy used to assess Myss's abilities, or the total number of tests he performed. For example, did Myss always give the specific diagnosis of a patient (e.g., "blocked arteries") or did she sometimes merely state a vague impression couched in symbolism (e.g., "I see an energy blockage in the fourth chakra")—with Shealy translating into medical terms? Would an unbiased physician have read Myss's diagnoses the same way Shealy did? And how many readings did Shealy base his accuracy estimate upon? Could he have subconsciously kept a record of "hits" and ignored the "misses"? These are not trivial questions; any reputable medical journal would require this information before even considering Shealy and Myss's claims for publication.

In the absence of evidence for her abilities, Myss tries to establish proof indirectly by claiming a historical precedent for intuitive diagnosis. She and Shealy discuss the emergence of intuitive medicine through such pioneers as Franz Mesmer and Edgar Cayce. We are told that Mesmer—a German physician who founded an odd pseudoreligion based on the mysterious properties of "animal magnetism" in the late 18th century—"laid the foundations for psychiatry and psychology" and helped later practitioners to learn the value of intuition (1988, 62). Cayce, a would-be prophet who thought himself the reincarnation of an angel who graced the Earth before Adam and Eve (Randi, 1995, 42), is said to have "laid the groundwork for all intuitive diagnosticians to follow" by giving thousands of mystically divined diagnoses during his lifetime (66).

The average person, reading about Mesmer and Cayce for the first time in *The Creation of Health*, would have no way of knowing that both men have been almost universally recognized as quacks. Mesmer was investigated in 1784 in France by a Royal Commission containing, among others, Benjamin Franklin and Antoine Lavoisier. The commission ingeniously and conclusively demonstrated that "animal magnetism" did not elicit the physiological effects Mesmer had claimed—the power of suggestion had done it all (see Mesmerism entry in section 5). Cayce, similarly, has been shown to have erred very significantly in many of his diagnoses, and his reputation was built entirely through anecdotal evidence and the faith of true believers (Gardner, 1952). The fact that Myss has foisted off this pair of known charlatans as genuinely important contributors to medical science does not speak highly of her scholarship—or her intentions.

Mindless Mergers: The Holistic Hodgepodge

Insignificant evidence and unrestrained speculation cannot stand on their own; the careful self-help guru must also provide a philosophy comprehensive and uncritical enough to support even her wildest metaphysical musings. And what philosophy could possibly be as intimidatingly encompassing or as thoroughly accepting of any and every vaguely pleasing notion as New Age holism?

"Holism," as usually described in popular metaphysical books, is more than just the well-known wisdom that "the whole is greater than the sum of its parts." Rather, the most common and virulent strains of holism state that any distinction between the whole and its parts is unfathomable because we simply cannot obtain any real understanding about something by taking it apart into smaller units and analyzing it. In other words, the traditional scientific method of reductionism is a big no-no, and we are fools to think otherwise. We should simply accept the universe in all its glorious infinity, and realize that such an immense whole cannot ever be truly subordinated to rational investigation. If we could overcome the confinement of reason, we would see that categories are meaningless, opposites are illusory, and—as Myss is fond of saying—"all is one."

Myss ushers in the new era of holism with great enthusiasm, explaining that we've been living in the Age of Pisces for the last 2000 years, an age characterized by categorical thinking. "The Piscean age was a time of dualism," she writes, "when human consciousness divided in a powerful way into polarities, such as those between Western and Eastern culture, church and state, body and spirit (in a split epitomized by Manicheanism), the science of magnetics, even political polarities of left and right" (Myss, 1997, xiv). However, the coming millennium will mark the beginning of the Age of Aquarius, and we'll put an end to all of this typological silliness. "The energy of this emergent age pulls us to create a culture in which spirit and energy have a higher priority than matter and the body, and to understand that the energy within our minds, bodies, and spirits is the same as that of God or the greater divinity" (Myss, 1997, xv). In this new system, we'll finally understand the body as an energy system, and "healing will then be a much gentler process of delicately manipulating the etheric body through the use of crystals, sound and color" (Shealy and Myss, 1988, 370).

To the uncritical, all of this may seem as harmless and pleasing as a Yanni album. However, there are glaring deficiencies in Myss's holistic philosophy. First, her historical justification for considering the last 2000 years an age of dualism is a gross oversimplification. And as far as "the science of magnetics" goes: It's true that we've learned that all magnets maintain a type of "duality," since every magnet has a "north" and "south" pole. However, the study of magnets also led the great physicist James Clerk Maxwell to show that light, magnetism, and electricity were related electromagnetic phenomena. This was a substantial step toward unity in the laws of nature, not dichotomy.

Second, Myss's holism is far too unselective to ever be of any scientific use. It seems as if Myss will incorporate just about anything into her philosophy if it sounds pleasing enough and has sufficient superficial similarity to other components of her philosophy. For example, in *Anatomy of the Spirit,* she links the Christian sacraments to her chakras—apparently because (could you guess?) there are seven of each. In both *The Creation of Health* and *Why People Don't Heal and How They Can,* she assigns great value to astrology, a pseudoscience boasting a long and spectacular history of failure. What's her reason for putting stock in such a worthless practice?

> For me, astrological influences are authentic, but not as commonly thought of by people who assume that astrology is a form of fortune telling. It is not. It is the study of the influences of the energies of the planets on the entire system of life, including human life. That we are part of a whole is a given. That individual parts of that whole radiate certain qualities is natural. Astrological influences do not control one's life; they merely indicate potentials and possibilities (Shealy and Myss, 365–366).

So the stars influence us because we're all inseparable parts of a whole. Why, then, can't the magnet here on my desk fetch the can of Foster's I left upstairs on the counter? Because it's too far away, you say, and my magnet doesn't exert enough force? But the planets and stars are also much too far away to exert a measurable gravitational force on us. Why does Myss predict that planets and stars will influence us, but she doesn't seem concerned that my magnet cannot retrieve my cold beverage? This holism stuff is really confusing.

Third, the more Myss talks about holism, the more she reveals the hopeless categorization in her own thinking. Myss has a rather novel conception of holism; she seems to think that her system is different from dualism because she ultimately forces all phenomena to be arbitrarily subordinated to a single philosophical outlook. Yet, there is no such difference, because Myss is really a strict dualist who ruthlessly imposes her self-made categories on the world. Much of her philosophy is phrased in terms of opposing pairs: "energy" versus "matter," "mind" versus "body," and "holistic" medicine versus "allopathic" medicine. Myss generally holds the first term in each of these pairs in higher esteem, and predicts it will triumph over its assigned adversary. Her distinction between matter and energy is almost trivially easy to discredit, since Einstein did all the work for us at the beginning of this century. However, her other two dichotomies indicate important failures in logic.

Myss clearly considers mind to be composed of a different substance than the physical body, but simultaneously seems to think that our mind permeates every cell of our body, as new age physiologist Candace Pert has often stated (Myss, 1996, 35). This "mind" is considered closely allied with "an energy field that extends as far out as your outstretched arms" which acts as "an information center and a highly sensitive perceptual system" (33). Thus, the mind-body dichotomy is linked with the "energy-matter" dichotomy Einstein so elegantly debunked long ago.

This system is essentially identical to the dualistic theories of mind promulgated centuries ago by philosophers such as Descartes. However, Descartes did not have access to the knowledge of neuroscience we have today. In the light of work by thinkers like Paul and Patricia Churchland and Nobel Laureate Francis Crick, we are beginning to understand the mysteries of consciousness in terms of the relationships between neurons. In this perspective, consciousness is an emergent property resulting from the functions of these neurons, and there is no need to postulate a transcendental material as the underlying "stuff" of thought. Consciousness is a process, as Ian Stewart and Jack Cohen remind us, and can best be understood through a contextual analysis of its functions (1998, 211). Emotions are an important part of this context, but only when included in testable hypotheses about the psychology and neurology of thought. Myss's vague mystical notions clearly do not belong in this developing model.

The presented dichotomy between alternative and conventional medicine is even more spurious, and potentially damaging. Myss claims she has no dislike for conventional medicine, and repeatedly stresses her desire to see holistic medicine incorporated into mainstream practice, but she does not hide her preference to see the latter system subordinated to the former. "Holistic and conventional medicine take two different attitudes toward power: active and passive," she tells us. On the same page she observes that "the language of conventional medicine sounds more military than that of energy medicine: 'The patient was attacked by a virus' or 'A substance contaminated the cell tissue, resulting in a malignancy'" (1996, 48). The implication is clear: mainstream medicine is cold-spirited and trivializes the patient's power to heal, while holistic medicine is gentle and celebrates autonomy in the face of sickness.

Myss also assures us that holistic medicine has its own methodology and standards, and should not be subject to the rigid appraisal afforded mainstream medicine.

> The traditional medical community, which includes physicians, nurses, psychiatrists and psychologists, has specific professional standards and requirements. Within the holistic field, there are numerous forms of therapy that do not require the same intense periods of education. The training needed to become a massage, color or polarity therapist, for instance, is not as formal a process as is medical or nursing school training. That is not to say the work therapists do in these alternative fields is not of immense value or that the training they receive is inadequate. The training involved for several categories of holistic therapies, however, is not as formally organized in traditional academia, and that difference is cause for much of the lack of appreciation for the work of holistic practitioners (Shealy and Myss, 1988, 24).

Unfortunately, the reason so many of these holistic therapies aren't "formally organized in traditional academia" is that they are utterly without merit. Without some method of separating effective therapies from useless therapies and strict training standards from lax standards, there is simply no way to perform medicine competently.

The best antidote for Myss's brand of holistic doublespeak has been provided by *Journal of the American Medical Association (JAMA)* editors Phil B. Fontanarosa, M.D. and George D. Lundberg, M.D. In a special issue of *JAMA* dedicated to alternative medicine, the editors issued the following statement: "There is no alternative medicine. There is only scientifically proven, evidence-based medicine supported by solid data or unproven medicine, for which scientific evidence is lacking" (1998). It doesn't matter what the origin of a therapy is, or whether we choose to think of it as "holistic" or "allopathic." We are concerned only with its effectiveness and safety.

This attitude directly contradicts the viewpoint of holists like Myss, who continually stress that Western medicine is averse to "natural" remedies. However, a close look at the evidence clearly shows that the holists are wrong. Treatments derived from nature or passed down through folklore are subjected to the same experimental evaluations as treatments engineered in the laboratory, and history records the careful integration of worthy therapies. For instance, one of the most effective cancer treatments used today is derived from a plant found in Madagascar (Morell, 1999, 17). The heart medicine digoxin was extracted from the foxglove plant, aspirin was obtained from a compound in the bark and leaves of willow trees, and antimalarial drugs have been taken from the bark of the cinchona tree (Mestel, 1999, 74). And as bacteria develop increasing resistance to antibiotics, scientists are searching every habitat from the ocean floors to the sediments of coastal mangrove forests for the next generation of antibacterial agents.

Many alternative medicine practitioners, including Myss, seem unaware of the importance naturally obtained remedies have held in the history of medicine. They imagine a kind of warfare between themselves and the medical establishment. They prefer to dig their trenches, crouch out of sight, and make the occasional sneer at their adversary. As a result, many people are misled into unfair judgments about the attitudes of modern medicine, and are compelled to uncritically accept the "alternatives" offered by holistic practitioners. This situation should not and cannot continue. Medicine is far too important to be turned into a game of pernicious accusations in which the truth is obscured. The smokescreen of holism must be cleared, and alternative treatments must be exposed to the light of critical scrutiny. As the editors of *JAMA* concluded, "for patients, for physicians and other health care professionals, and for alternative medicine practitioners—indeed, for all who share the goal of improving the health of individuals and of the public—there can be no alternative" (Fontanarosa and Lundberg, 1998, 1619).

The Healing Power of Truth

Caroline Myss offers no tangible evidence to support any of her claims. Her hypothetical energy system cannot be detected, her intuitive diagnostic abilities are unproven, and her holistic philosophy is riddled with inconsistencies and unsubstantiated judgments. I predict medical progress will continue quite nicely without even the slightest help from Myss or her ideas.

Myss, undaunted, will undoubtedly press on in search of spiritual insights into the nature of healing. On May 4, 2000, she and some of her faithful will travel on a "healing journey" to Peru to visit the Incan ruins, where they will "fully experience the healing nature" of these sacred sites. May I offer some advice for those going? First, ask yourself this: If these sites had so much sacred, healing power, why aren't there still Incas there? Next, read a few books about the history of the Incas, and consider the very likely possibility that their civilization was destroyed by poor understanding of their environment's capacity to support their population density (Shermer, 1997, 76–77). Despite their amazing cultural accomplishments, the Incas' religious beliefs and practices couldn't provide the knowledge needed to keep their society intact.

Is Caroline Myss really offering factually based claims, or is she simply appealing to mysticism? What would the consequences be if mainstream medicine adopted her standards of evidence? The results, you may well conclude, would be disastrous; but that is a fate we are empowered to avoid. We can counter mystical claims with objective evidence, and use modern scientific inquiry to guide us through troubled times. In fact, it is our responsibility to do so. To follow Myss and to forsake scientific knowledge for a haphazard system of unwarranted speculations is to choose the road to ruin.

References:

Appleton, Elaine. 1997. "Solo Flight: Medical Intuitive Caroline Myss Teaches Spiritual Healing." *Natural Health*. 27:1, 84–89.

Campbell, Joseph. 1974. *The Mythic Image*. New York: MJF Books.

Chambers, Veronica. 1997. "Heal Thyself: New Age Author Caroline Myss Hits the Big Time." *Newsweek*, December 22.

Churchland, Patricia S., and Churchland, Paul M. 1998. *On the Contrary: Critical Essays, 1987–1997.* London: MIT Press.

Cohen, Jack, and Stewart, Ian. 1998. *Figments of Reality: The Evolution of the Curious Mind.* New York: Cambridge University Press.

Crick, Francis. 1994. *The Astonishing Hypothesis.* New York: Touchstone.

Dale, Cyndi. 1996. *New Chakra Healing: The Revolutionary 32-Center System.* St.Paul: Llewellyn.

Fontanarosa, Phil B., and Lundberg, George D. 1998. "Alternative Medicine Meets Science." *Journal of the American Medical Association.* 280:18, 1618–1619.

Gardner, Martin. 1952. *Fads and Fallacies in the Name of Science.* New York: Dover.

Kurtz, Paul. 1986. *The Transcendental Temptation.* Buffalo: Prometheus Books.

Mestel, Rosie. 1999. "Drugs from the Deep." *Discover.* 20:3, 70–75.

Morell, Virginia. 1999. "The Variety of Life." *National Geographic.* 195:4, 6–32.

Myss, Caroline. 1996. *Anatomy of the Spirit: The Seven Stages of Power and Healing.* New York: Harmony Books.

——. 1997. *Why People Don't Heal and How They Can.* New York: Harmony Books.

Randi, James. 1995. *An Encyclopedia of Claims, Frauds and Hoaxes of the Occult and Supernatural.* New York: St. Martin's Press.

Shealy, C. Norman, and Myss, Caroline. 1988. *The Creation of Health: The Emotional, Psychological and Spiritual Responses that Promote Health and Healing.* Walpole: Stillpoint Publishing.

Shermer, Michael. 1997. "The Beautiful People Myth: Why the Grass Is Always Greener in the Other Country." *Skeptic,* Vol. 5, No. 1.

Police Psychics

Noreen Renier as a Case Study

GARY P. POSNER

On April 3, 1996, the skeletal remains of 76 year-old Norman Lewis, missing for two years, were recovered from the murky waters of a limestone quarry in the tiny Florida town of Williston, located just southwest of Gainesville. The April 5 Associated Press story, as headlined in the *St. Petersburg (Florida) Times*, revealed: "Psychic tip leads to missing man's body." Although she was not present during the search or recovery, the "tipster" was Florida "psychic detective" Noreen Renier, who boasts of a successful history of assisting in hundreds of police investigations into unsolved homicides and missing-person cases.

Before specializing as a "psychic detective," Renier, age 60, was credited with having predicted the 1981 assassination attempt on President Ronald Reagan, and the assassination of Egyptian President Anwar Sadat later that year. Through the years, she has appeared on numerous national television programs including the *Joan Rivers Show*, *Geraldo*, *Sightings*, and even the CBS newsmagazine *48 Hours*. In the classic textbook *Practical Homicide Investigation*, used by the FBI and many other police academies, the author identifies Renier as "a psychic and recognized authority on the phenomena of extrasensory perception."

According to press accounts, on March 24, 1994, after telling his girlfriend that he would be right back, the elderly Mr. Lewis drove off from home, leaving behind his wallet and respiratory inhaler, and (along with his truck) was never seen or heard from again. In its April 11, 1994, edition, the *Ocala (Florida) Star-Banner* quoted Williston Police Chief Olin Slaughter as observing, "It's like he fell off the edge of the earth."

After spending more than a year following up on "hundreds" of leads and conducting numerous land and aerial searches, all to no avail, the Williston police, and the Lewises, decided to enlist the aid of a psychic. Investigator Brian Hewitt suggested Noreen Renier, having previously been impressed by a performance of hers. The Lewis family reportedly provided the $650 fee for her services (the police department was said not to have had sufficient funds).

On July 17, 1995, three weeks after Hewitt's initial phone call to her, Renier performed her "psychic" reading, at her home. Clutching one of Mr. Lewis' possessions, she tuned into his "vibrations" and provided a number of specific clues intended to help lead the police to his body. The *Williston Pioneer* (on April 4 and June 27, 1996) quoted Chief Slaughter as saying that Renier indicated Lewis had traveled "east from his home to an area where there is . . . water in something like a pit." The *Chiefland (Florida) Citizen* (April 11, 1996) quotes Slaughter: "She could see he was surrounded by metal. . . . She could see a cliff wall, and loose bricks, a railroad track, and a bridge." The numbers "45" and "21" were also said to have been offered as helpful clues.

A subsequent look by the police into several bodies of water proved as fruitless as the earlier searches. But because of Renier's reading, the police called in a team of Navy divers from Jacksonville to search one particular limestone quarry among many scattered throughout the area. Although about eight months elapsed before the team could arrive, on April 3, 1996, with the assistance of a $70,000 detection device, the divers did indeed locate the missing truck containing Lewis' remains, submerged in 20 feet of murky water.

When the Williston police announced that the case had been solved largely as a result of Renier's psychic clues, the story quite naturally captured the attention of the media. In addition to regional newspapers and television, the Associated Press and national radio icon Paul Harvey reported upon it, and the TV show *Sightings* devoted a segment to it in November 1996. The "Williston Case" quickly became the pinnacle of Renier's storied career.

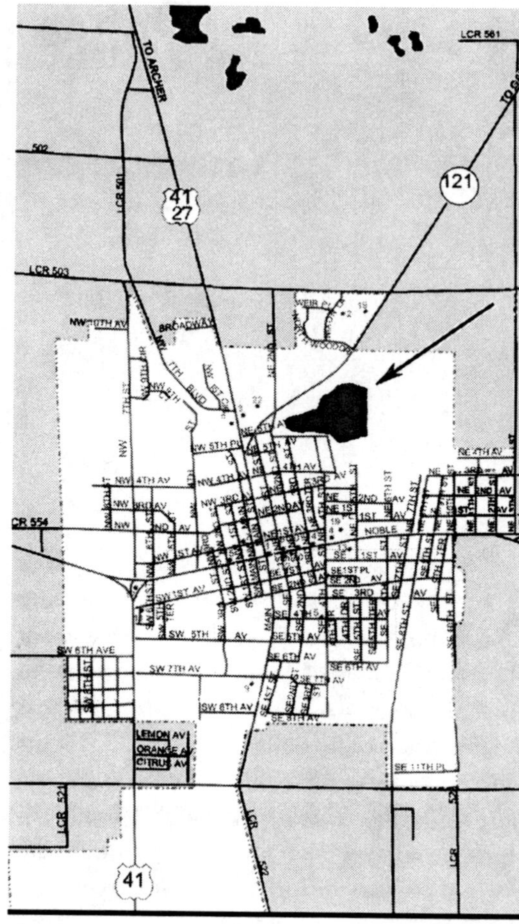

Figure 1

Enter the Skeptic

My involvement in the Williston case began in May 1996, when I received a telephone call from a researcher for Towers Productions, which was producing a series called *The Unexplained* for the A & E Network. Their program on "Psychic Detectives" (which first aired in January 1997) would feature several individuals, including Renier, and would specifically cover Williston. My participation was requested to insure a balanced presentation.

Two months later, the police and I were interviewed for the show (in Williston), as was Renier the following day (at her Orlando home). By then, I had accumulated a number of newspaper articles and maps and had come to an unexpected and provocative conclusion: Norman Lewis' remains appeared to have been found not because the police had the Navy divers search the body of water best fitting Renier's psychic clues, but because they had the Navy search the wrong watery pit!

Scanning my roadmap of Williston, I immediately noticed its most striking feature—a blue body of water nearly in the heart of town, less than one mile east of Mr. Lewis' home. (Figure 1) This limestone quarry, when approached from the west, is located adjacent to the inter-

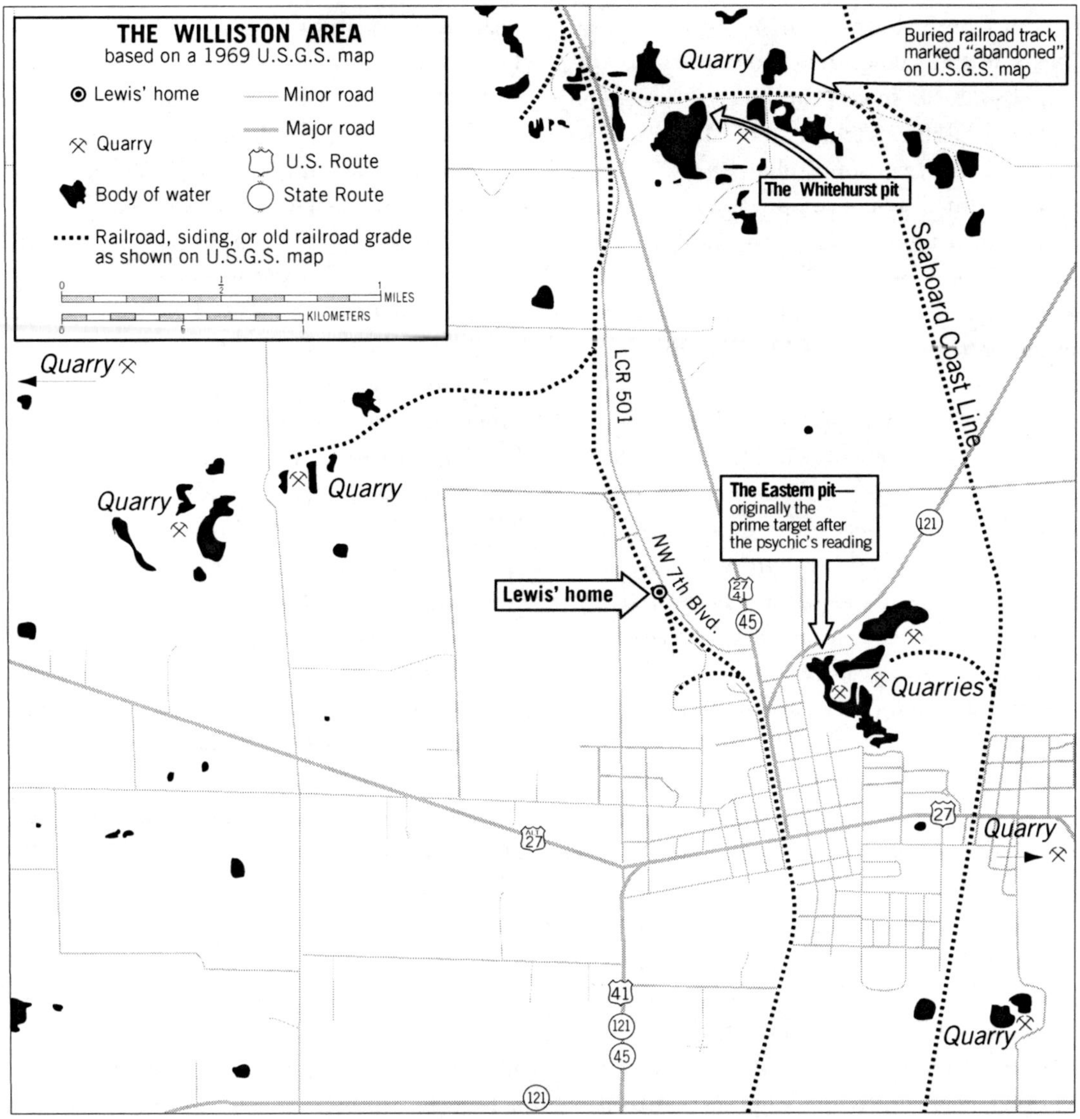

Figure 2

section of U.S. 41 and State Route 121. Flipping that map over I saw that the map on the opposite side reveals that U.S. 41 is also known in Williston as State Route 45. If Lewis had indeed traveled east from his home to a watery pit, as Chief Slaughter indicated Renier had seen in her psychic vision, he would have encountered such a quarry just beyond the junction of State Routes 45 and 121. Renier's two numerical clues were reportedly "45" and "21"—had she offered "45" and "121," someone might have cynically accused her of having researched the case and consulted a map!

The U.S. Geological Survey's "Williston Quadrangle" map, which I purchased at a Tampa map store, shows this clearly marked "Quarry" area in more detail. (Figure 2) Of note is the Seaboard Coast Line's north/south railroad track 3/4 of a mile east of the quarry's eastern circumference, with a branch directed westward into the heart of the quarry area. One of Renier's clues was "railroad track."

As I told the Towers producer, I cannot be certain if Renier's clues were the result of "psychic" power or some other, purely natural, process. But, I added, forget about "psychic" power for a moment and just employ "ordinary" detective-style reasoning and common sense. Consider that the intensive ground and aerial searches had turned up nothing. If Mr. Lewis and his truck were somewhere within the potential reach of the Williston police, where could they possibly be? In the middle of a densely wooded area? In an abandoned building? (Either, perhaps, if only a body was missing. But a truck?) Only one possibility even comes to mind—submerged in a body of water.

Chief Slaughter, it seems, had had the right idea all along, even if he was not consciously aware of it. It did indeed appear "like [Lewis] fell off the edge of the earth"—and into a bottomless, or at least very deep, watery pit. A quick glance at the Williston roadmap revealed an obvious potential site, as confirmed by the U.S.G.S. map.

One minor problem: The logical site, the one that Renier's clues seem tailored to—the limestone quarry less than a mile east of Lewis' home, at the junction of State Routes 45 and 121, serviced by a railroad track—was not where Lewis' truck and remains were ultimately found! Rather, they were located in a different limestone pit, one nearly due north of Lewis' home and more than twice as far away! The recovery site, known as the Whitehurst pit, is also located adjacent to State Route 45, but not Route 121.

Renier's "21" clue, in fact, played no beneficial role whatsoever in assisting in the location of Lewis' body. Yet, this clue has been hailed by the police as perhaps her most eerily precise of all. Why? Because, after Lewis' body had been recovered, it was announced that he had been found "2.1" miles from his home—even though, as the crow flies, the distance appears to measure only about 1.6 miles on the U.S.G.S. map.

Nor was her "railroad track" clue of any value in deciding which of these two quarries to have the Navy divers search. Although the U.S.G.S. map clearly shows an "abandoned" track traversing the Whitehurst quarry east/west, the police did not become aware of the buried track until a portion of it was unearthed after the divers had already been called in.

Nor did her "bridge" clue offer any assistance in targeting this particular pit, or in helping narrow down the search area within the 30-acre Whitehurst quarry. But, as WTVT-TV 13 (Tampa) reported on April 19, 1996, "Another clue that amazed [Chief] Slaughter was that the psychic saw a bridge nearby. Turned out [after the fact] that he'd passed it countless times and never saw it—on the access road to the quarry—an old, wooden truck scale that smacks for all the world of a bridge, if you take the time to stare at it."

And as for her apparently precise State Route "45" clue, read on.

"Hits" and Misses

In July 1996, a skeptical Tampa attorney made a Florida Public Records Act request of the Williston police department to provide him with a copy of its entire file on this case, which he then forwarded to me. Investigator Hewitt responded by sending copies of all the paperwork, which included two items of immediate interest: a May 12, 1995, report (supplemented on June 15) filed by Hewitt, and the "clues" jotted down by Hewitt from Renier's July 17, 1995, "psychic" reading. In his two-page May 12/June 15 report (I have corrected a few spelling errors), Hewitt notes that a

> handyman . . . had recently told [a client] that [Lewis] had told him that if [Lewis] were not

able to take care of himself because of illness, he would find a river or pit rather than the [retired] sailors home. . . . Four days before his disappearance, [Lewis] told [the handyman] that if his health were failing, he would never be cared for by relatives or submit to the sailors home, that there were too many pits and canals. . . . [The handyman later] arrived at the police station . . . and he related [to Hewitt] the last conversation he had with Norman Lewis . . . indicating it [actually] took place approx. three weeks before his disappearance. He stated Norman seemed agitated and dissatisfied with . . . his life [including having] problems at the house with his girlfriend, relating she did not make him feel needed. . . . Told [handyman] not to get old, and made some reference to knowing every rock pit in the county. . . . (Figure 3.)

This "smoking gun" document had been previously unknown to me and to the A & E producer. But it was now apparent that as a result of his failing health and other personal problems (an early newspaper article had also described him as "despondent" over financial matters), Lewis had threatened to commit suicide in a "river" or a "rock pit." Further, word of this had begun to spread through his tiny community and had become known to the police two months prior to their session with Renier. Might Renier have actually learned of this, in advance, from the police?

The Tampa attorney had also specifically requested any video/audiotapes of Renier's reading. After inquiring as to why only written records were released, Hewitt advised him that an audiotape did in fact exist and would be provided. As for a videotape, Hewitt wrote: "As I have advised you in several telephone conversations, the only [video]tape contained in the requested file . . . is of the recovery, which you indicated you did not want."

When even the promised audiotape failed to arrive, the attorney threatened a lawsuit "if a copy of the [audio]tape is not presented to us within seven (7) days." Hewitt finally responded by delivering what he termed "a copy of the field audiotape [which] contains portions of the session with Noreen Renier." To my dismay, upon playing the tape, it was evident that there was a cut/edit after nearly every sentence spoken by Renier (and often in mid-sentence or mid-word). Further, the entire tape runs for a mere five minutes and forty-three seconds. Yet, it does contain some "clues" worth discussing:

- "A lot of rocks. . . . Swallowed up [down there in the water] but there's hardness higher up. . . . We have a lot of things that go straight down. No one really knows what's down there because it's so hazardous and dangerous and people don't go down there. . . . There's a railroad track that goes through there." [Did she know about the suicide threat? Or consult maps, as I did?]
- "Let me have a starting place. . . . We want to get you in the quadrant from 9 to 12 . . . into that pie-shaped area." And from Hewitt's handwritten notes from Renier's reading: "Where do you want me to start? At his house. . . ." [Starting from Lewis' house, his body was found in the 12:00 to 3:00 quadrant, not "9 to 12."]
- "Speedometer is zero in front of the house. . . . Maybe 4, maybe 5. If it's 45 miles, if it's 4.5 miles. I want to go to my left. I want to go to 9. . . . I feel 45, 45 degrees. You know how they have that little baby circle up there? [i.e., 45°]. . . . Looking for H and 45." [This is the "45" clue being credited as a "hit" because Lewis was found near State Route 45!]
- "Must be still somehow in the vehicle. I feel the metal very, very strongly." [Renier had been told in advance that Lewis' truck was also missing.]
- "We're not too far from an old bridge. Either it's been decayed or it's broken or it's not

WILLISTON POLICE DEPARTMENT — Continuation / X Supplement — X REPORT A / REPORT B

1. CASE REPORT #: 94 | | | 01581 (Yr Mo Day No)

2. CRIME TYPE	3. LOCATION OF OCCURRENCE	4. DATE OF ORIG. REPORT	5. DATE OF THIS REPORT
MISSING PERSONS	752 NW 7TH BLVD	03-25-94	05-12-95

6. X V / C — 7. NAME (First, Middle, Last): NORMAN LEWIS

8. YES NO UK: X JUVENILE INVOLVED; X SUSPECT NAMED; YES NO: X SUSPECT VEHICLE DESCRIBED; X ADDITIONAL PROPERTY LISTED

9. YES NO: X CRIME TYPE RECLASSIFIED

10. DISPOSITION: A CBA/SWORN; B EX. CL; C UNF.; D INA; E C/C; X INV. CONT

11. BLOCK NO — INDICATE BLOCK NUMBER IN LEFT MARGIN

ON 05-12-95, AT APPROXIMATELY 1143 HRS, RECEIVED MESSAGE VIA DISPATCH TO PHONE VIRGINIA LEWIS. ATTEMPTED CONTACT

A CONVERSATION SHE HAD WITH SHIRLEY YAGER (~~[illegible]~~) IN WHICH SHE WAS TOLD BY A BLACK MAN WHO WAS INTRODUCED TO SHIRLEY BY NORMAN TO DO SOME HANDYMAN WORK FOR HER, HAD RECENTLY TOLD HER NORMAN HAD TOLD HIM THAT IF HE WERE NOT ABLE TO TAKE CARE OF HIMSELF BECAUSE OF ILLNESS HE WOULD FIND A RIVER OR PITT RATHER THAN THE SAILORS HOME. THANKED JOE FOR

AND THAT THE AFOREMENTIONED BLACK MAN (ED ROBINSON) HAD ALSO BEEN CLOSE WITH NORMAN. SHE RELATED THAT ROBINSON HAD RECENTLY TOLD HER THAT FOUR DAYS BEFORE HIS DISAPPEARANCE HE TOLD ROBINSON THAT IF HIS HEALTH WERE FAILING HE WOULD NEVER BE CARED FOR BY RELATIVES OR SUBMITT TO THE SAILORS HOME THAT THERE WERE TO MANY PITTS AND CANALS TO MET WITH. SHE ALSO

06-15-95 AT APPROXIMATELY 1240 HRS, MR ROBINSON ARRIVED AT THE POLICE STATION. I BROUGHT HIM INTO MY OFFICE AND HE RELATED THE LAST CONVERSATION HE HAD WITH NORMAN LEWIS, INDICATING IT TOOK PLACE APPROX 3 WEEKS BEFORE HIS DISAPPEARANCE. HE STATED NORMAN SEEMED AGITATED AND DISSATIFIED WITH THE THINGS GOING ON IN HIS LIFE. THAT WAS THE GIST OF HIS CONVERSATION WHICH WAS VERY ERRATIC AS HE JUMPED FROM TOPIC TO TOPIC. HE RELATED TO PROBLEMS AT THE HOUSE WITH HIS GIRLFRIEND, RELATING SHE DID NOT MAKE HIM FEEL NEEDED. HOW HE HAD MARRIED THE SAME WOMAN THREE TIMES, AND WOULD NEVER MARRY AGAIN. MR ROBINSON STATED THAT NORMAN HAD LIVED PART OF THE YEAR NORTH AND PART IN FLORIDA AND HE WAS RELATING HE WAS DISSATISFIED WITH BOTH LOCATIONS, AND WOULD NEVER COME BACK TO FLORIDA AGAIN. STATED HE WAS NOT GOING TO DEPEND ON ANYBODY AGAIN. TOLD MR ROBINSON NOT TO GET OLD AND MADE SOME REFERENCE TO KNOWING EVERY ROCK PIT IN THE COUNTY. (THE AFOREMENTIONED RECOLLECTIONS ARE IN NO PARTICULAR ORDER OR SEQUENCE ACCORDING TO ROBINSON)

MR ROBINSON STATED HE HAD BEEN FRIENDS WITH NORMAN FOR APPROX THE LAST 5 OR 6 YEARS AND RELATED HE HAD NOT SEEN HIM THAT TROUBLED BEFORE

PAGE 2 OF 2

12. REPORTING OFFICER	ID #	SHIFT	13. SUPERVISOR	I.D. #	YES NO	PRIORITY 1 2 3 4 5
E. Brian [signature]	6044	1			REPORT APPROVED / FOLLOW-UP NEEDED	

Figure 3

used. . . . It's called the old bridge or is an old bridge." [The old truck scale was nearby, although it was certainly not known as "the old bridge."]

- "One point, or one-one point two. I see two-two-I [the letter "I"]. I believe a very strong H, 'Ha'-sounding or an H in it." And from Hewitt's notes: "221 . . . 22 . . . 21 . . . 2I . . . H . . . EML . . . E . . . 11.2" [Renier was credited with an eerily accurate "hit" because Lewis was supposedly found "2.1" miles from his home! But what about "45 miles" and "4.5 miles"?]

Among the pages in the police file is a map of Williston with a 90° (L-shaped) area from 11:00 to 2:00 (not "9 to 12") drawn on it and labeled "Noreen's quadrant." The point of convergence of the two lines is correctly marked "Norman's House," and the quadrant includes the northern Whitehurst pit where the body was found (at about 1:00) but not the eastern pit that her clues appear to more closely fit (at about 3:30–4:00).

In stark contrast to A & E's balanced coverage of the case on *The Unexplained,* the Sci-Fi Channel's *Sightings* coverage included no skeptical input. The *Sightings* narrator asks, but is not able to answer, the question: "Why did Norman disappear?" No mention is made of Investigator Hewitt's report, filed two months prior to Renier's "psychic" reading, regarding Lewis' "rock pit" suicide plan. But in fairness to *Sightings,* the police had also withheld this crucial information from the A & E producer.

In Renier's re-created reading for the *Sightings* cameras, with her eyes closed, feigning a trance-like state, she strays from her original reading so as to now specifically associate the number "21" with miles: "Numbers—21. I feel miles." On the edited audiotape, there is no mention of "21" in any context (although, as shown earlier, "21" does appear in Hewitt's notes in the midst of a stream of numbers/letters with no particular regard to mileage), and the only numbers associated with "miles" are "45" and "4.5." Nor is there a "21" clue on the edited videotape (yes, it did finally materialize—see below).

The "45 miles" clue is especially puzzling, as Renier has been credited with correctly determining that Lewis would be found a short distance from his home. From her *Sightings* re-creation: "I'm driving for a short distance, and then something happens, and I see him in the air, going downward." And from the edited videotape of her original reading:

> "Norman's house is here [gesturing to the right with her right arm]. Here's the road [gesturing straight ahead with her left arm]. We go this way [pointing straight ahead with her left hand]. . . . But we don't go very far that way, we're going to veer off here [pointing left with her left hand] . . . towards the river. And for some reason the river is down below [as if describing Lewis' arrival at the pit/quarry's sheer cliff]."

This passage on the videotape appears to be a second "smoking gun," this time with regard to the particular body of water to which Renier's directions actually lead. As I earlier indicated, Renier's clues (as I understood them even before receiving this video) seemed to lead not to the Whitehurst pit (located north of Lewis' home) where the body was ultimately found, but rather to another rock pit much closer to, and nearly due east of, his home.

During my two visits to Williston, I viewed the former Lewis residence, located on N.W. 7th Blvd. With the home on the right side of the street, proceeding straight ahead (as per Renier's "psychic" vision) leads southeasterly for approximately one-third mile, at which time the road curves left to a due east bearing, until N.W. 7th Blvd. ends at its intersection with U.S. 41, approximately one-half mile from the Lewis home. Another quarter-mile or so due east, dead ahead (no pun intended), is the massive "eastern" rock quarry, the most prominent feature on the Williston roadmap.

Summarizing Renier's role in this case, the *Sightings* narrator says, "Investigator Hewitt put all of Renier's clues together, used some gut instinct of his own, and came up with one word—'Quarry.'" But we now know that Hewitt had actually learned two months earlier of Lewis' plans for ending his life in a quarry. And in the edited video of Renier's actual reading, she refers to the body of water not as a "pit" or "quarry," but as a "river" (although she appears puzzled as to why it goes "down" such a sheer cliff). The word "quarry" is heard once on the videotape, not after Hewitt has a chance to digest all of Renier's clues and apply his "gut instinct," but in the midst of the session, by an unidentified male questioner present with Hewitt in Renier's living room: "Now look at that quarry. As you're looking at it and looking at it from the entrance there. . . ."

Following Renier's reading, did the police zero in on one quarry to which Noreen's directions pointed? Hewitt says on *Sightings* that he "walked around probably 30 quarries" before deciding that the Whitehurst pit most closely matched the totality of Renier's clues. Perhaps that was his reason for having the Navy divers scour that one pit, which did result in Lewis' body and truck being recovered. But his initial rationale for concentrating on the Whitehurst pit was described this way in his report filed six days after Renier's reading: ". . . the Whitehurst pits are an obvious first impression . . . being the closest and the most accessible from the Lewis residence." (Although the "eastern" pit was fenced off by this time, it had been easily accessible when Lewis disappeared, and it is half as far from Lewis' home as is Whitehurst.)

As for this "eastern" pit, a person with some inside knowledge of the police investigation (who allowed me to tape our conversation but requests anonymity) told me that this had been the "prime target for the investigation" immediately following Renier's reading. "They didn't think there was a [railroad] track [at Whitehurst]."

At the conclusion of the *Sightings* report, the narrator explained how Renier's "22" clue (remember that stream of numbers?) had also been remarkably accurate: When Lewis' body was recovered, "the calendar date on Norman's diving watch was stopped on the number 22." For the record, he had disappeared and presumably committed suicide on the 24th of March, 1994. As a clue to the location of Mr. Lewis' missing body, "22" was utterly useless.

The Videotape's Curious Arrival

The ultimate arrival of the edited videotape came as a complete surprise. When the audiotape turned out to have been heavily and crudely edited, the attorney wrote back to Hewitt requesting "a complete copy of the audiotape." Hewitt's reply explained that the tape "is the only audio tape I have regarding Noreen Renier's session [and] was expressly made [from a more lengthy original] for field use with regard to the location of Mr. Lewis." Most curiously, the letter continued: "You are requesting additional material. . . . We are under no obligation to provide you with any material without prepayment. Therefore, with your payment of [an additional $14.00] . . . I will forward to you the only remaining tape I have regarding this case."

The attorney assumed that the "only remaining tape" was a video of the recovery of the truck and body, as Hewitt had previously indicated. Nonetheless, he decided to fork over the $14.00. Incredibly, a month later, he received from Hewitt the videotape which, despite having been edited down to about 14 minutes, still contained the "smoking gun" segment.

In a letter accompanying the videotape, Hewitt informed the attorney that he had "filed for mediation with the State Attorney General's office . . . to assure you [that] we are in full compliance under the Florida Public Records Act." Through the mediator, the attorney then posed several questions, including these: "Why did the police department initially deny having a videotape and thereafter send us one?" "One map . . . depicts an area labeled 'Noreen's quadrant.' Who drew this quadrant on the map?" [see earlier discussion]. "What is the personal relationship, if any, between Detective Hewitt and Noreen Renier?"

This third question was prompted by two peculiar circumstances—the apparent initial withholding of information by Hewitt, and a stunning move by Renier: After living in Orlando for more than 20 years, she has now packed her bags and relocated to Williston!

Another question relates to an undated police report, filed by Hewitt, which does not appear to comport with Renier's reading, at least as excerpted on the tapes. Writes Hewitt, "She picked out [L]CR 501 on local map which I provided, indicating it was the road Lewis had traveled after leaving his residence, in a northerly direction." LCR 501 is the northern extention of Lewis' street, but according to the video's "smoking gun" segment, Renier actually indicated that he headed south.

But the answers to these questions have not been forthcoming. The mediator has written back informing the attorney that the Public Records Act does not compel Hewitt to respond. And the City Attorney for the town of Williston has sternly weighed in: "[Y]ou have [already] received all public records in possession of the City relating to [this] investigation."

After-the-Fact Reasoning

Two final questions, fundamental to the very nature of "psychic" phenomena, require consideration. In my chapter on Renier for the book *Psychic Sleuths* (edited by Joe Nickell, Prometheus Books, 1994), I showed at the time how Renier (like the rest of the psychics profiled in the book) had yet to convincingly demonstrate genuine "psychic" power under proper observing conditions. Has Renier now become the first psychic to successfully do so? Or might her "success" in the Williston case be explainable in more mundane terms, perhaps as the result of a combination of factors such as advance research, common sense/intuition, feeding back information gleaned from the police themselves, and "retrofitting"—interpreting ambiguous clues, after the fact, as having been remarkably accurate and valuable "hits"?

To those who believe in "psychic" power and other supernatural phenomena, the answers to these two questions no doubt remain "crystal ball" clear. And they remain equally clear, though through quite another prism, to those skeptics of the paranormal who demand extraordinary proof of such extraordinary claims.

Pseudoarchaeology

Native American Myths as a Test Case

KENNETH FEDER

A little less than 15 years ago, I was invited to participate in a radio talk show at a local station in Hartford, Connecticut. I was a last-minute addition to a panel that included a local museum curator and three Native Americans. The curator's museum housed a collection of ancient Indian artifacts including material related to at least one human burial that was on display. The curator originally had agreed to be the lone spokesperson on the broadcast arguing for the importance and legitimacy of the excavation, analysis, display, and curation of archaeological objects. However, sensing an "ambush," the curator had requested that an archaeologist be included on the panel. I was available and agreed, perhaps naively, to participate.

Like most of us conducting field archaeology of prehistoric sites in the United States, I had been attracted to the discipline because of an abiding interest in the human past. Also, like most North American prehistorians, though I am not an Indian, I became an archaeologist equally because of a fascination with and intense admiration for the cultures of Native America.

Knowing this, it was a terrible irony to me, that, even 15 years ago, the relationship between Native Americans and archaeologists could be characterized as an uneasy and eroding truce. Many Native Americans viewed archaeologists as interlopers from the dominant culture, outsiders who exploited native peoples for their own purposes. The common, and often reasonable perspective of many Indians was that archaeologists were scientists who studied Indian ancestors, but who had little interest in and no accountability to the descendants of the people who had produced the cultures and sites upon which these scientists focused. Many Native Americans believed that archaeologists had merely updated, to a degree, the old racist saw: "The only good Indian is a dead (i.e., prehistoric) Indian." Many Native Americans believed, often justifiably, that archaeologists were concerned about only the ancient ancestors of Indians, and cared little or nothing about living native peoples or those peoples' perspectives of their own history. For many, as archaeologist Randall McGuire (1997) points out, archaeology represented yet another instance in which outsiders had appropriated something that belonged to Native Americans—their history: ". . . the archaeologist's authority over Indian pasts is simply one other aspect of their lives that has been taken from their control" (McGuire 1997, 65). Archaeologist Larry Zimmerman goes even further, indicating that to some Native Americans, the pursuit of archaeology is a kind of "scientific colonialism" (1997, 108).

I had agreed to participate in the radio panel for two fundamental reasons. I felt some level of general responsibility for the bad behavior of some members (by no means all or even a majority) of my discipline and I believed, innocently I suppose, that I could expiate my personal feeling of guilt and exonerate my field of study if only I could explain my work and the work of most of my fellow archeologists.

Unfortunately, my museum colleague had foreseen the scenario of the radio panel correctly. It was a set up; the goal all along had been to exploit the growing controversy within Native American communities about archaeology. No real dialogue took place. None had been intended. The museum curator and I had been invited to serve as effigies of our disciplines. We were the representatives of evil western culture, ghouls of science who desecrated and then displayed the graves of Native Americans for fun and profit.

As depressing as this was 15 years ago, the relationship between at least some Native Americans and some archaeologists has deteriorated, if anything, since that radio broadcast. It is a shame and is based more on political issues and less on any genuine conflict between what archaeologists actually do and what some natives find objectionable.

Archaeologists as Desecrators of the Sacred

Archaeologists are sometimes depicted as exploiters and despoilers of native culture. There is a popular perception that archaeologists spend much of their time looking for and then desecrating tombs, looting them of their fabulous and sacred treasures placed there to accompany the deceased to the afterlife, all in the name of museums willing to spend huge sums of money for such objects. But is this really what archaeologists do? Certainly it conforms to a commonly held stereotype, but does this reflect the kind of archaeology conducted by anthropologically trained archaeologists in the late 20th century?

In fact, it does not. Certainly I could understand the Indians on the radio panel objecting strenuously to the excavation of the bones of their immediate ancestors, but I have never excavated a human burial and know of very few archaeologists who have. The passage at the federal level in 1990 of the Native American Graves Protection and Repatriation Act (NAGPRA) has resulted in the removal of large collections of human remains and their associated grave goods from museums and laboratories and has made excavation and curation of the human remains of Native Americans all but impossible. State regulations are also in place to control quite rigorously the excavation of human remains. Archaeologists may debate the wisdom of this policy and many may decry the inestimable loss to science that accompanied NAGPRA (Haederle 1997; Meighan 1994), while others feel that the obligation is to the sensitivities of the people most directly concerned and not some idealized notion of "science" (Zimmerman 1994), but the argument is moot. These days, burials most often come to light only as the result of natural erosion or construction, and most municipalities have rules that tightly regulate the disposition of human remains so exposed. In many places these rules were drawn up with substantial input from native peoples.

This is not to say that conflicts do not arise, but, again, it seems that this occurs because of misunderstandings on both sides of the issue, and such conflicts are exacerbated by the degree of animosity that has developed as a result. For example, in an interview in the *New York Times* (Johnson 1996), noted archaeologist Rob Bonnichsen recounted the following horror story. Bonnichsen was excavating at the 10,000 year old Mammoth Meadow site in Montana when, much to his surprise and delight, human hair turned up in the most ancient levels. I am aware of, at most, one other example of human hair from a site of this age in North America, and the potential for DNA analysis must have been terribly exciting to the researchers.

One might have reasonably assumed that only an archaeologist or paleoanthropologist could get all that worked up over a handful of ancient hair. However, when word got out about the hair, two local Indian tribes demanded that the research stop and that the hair be returned for reburial under the provisions of NAGPRA! As of October 1996, the hair was still in limbo, research on an important site had been held up for two years, but at least the final regulations of NAGPRA now exclude "portions of remains that may reasonably be determined to have been freely given or naturally shed by the individual from whose body they were obtained" (NAGPRA regulations, section 10.1 (d) (1)). In the Lewis Carroll world (or is it Franz Kafka?) of federal regulations regarding archaeology, this new wording can be viewed as a major step forward. As attorney Alan Schneider (1996) points out, now archaeologists can legally hold on to and analyze human hair, toenail clippings, and coprolites (ancient, preserved feces) without the wrath of NAGPRA being visited upon them.

Of course, it isn't only hair, toenails, and the like that divides Indians and archaeologists. Not just the intentional excavation but even the analysis of human remains exposed by natural processes has become a point of contention. The most recent and unfortunate example of this is the so-called Kennewick skeleton found in Washington state. Before word got out about the remains, radiocarbon dating was performed and the bones turn out to be more than 9,000 years old. This date surprised researchers because the skeleton exhibited gross morphological characteristics more in line with a European rather than a Native American population. Subsequent to the dating, however, the local Umatilla tribe demanded it be returned to them for reburial and they further demanded that no additional analysis be conducted on this well-preserved skeleton. The disposition of the skeleton is still up in the air (it has spawned a court case), but in another instance, in Idaho, the Shoshone-Bannoks allowed the radiocarbon dating of a skeleton found in their historical territory—it was 10,600 years old—but the tribe then vetoed DNA analysis (Johnson 1996).

Neither the Umatilla nor the Shoshone-Bannoks can prove any direct or intimate biological connection with these very ancient skeletons. The irony here is that with the analysis of mitochondrial DNA (if any is preserved in the skeletons) it might be possible to prove that, indeed, these modern Indians are the lineal descendants of the individuals represented, strengthening their demand for stewardship of the remains. Of course, this is a two-edged sword—it might also turn out that the modern Indians claiming stewardship are not closely related to the ancient person, thereby reducing the strength of their claim. In the case of the Shoshone-Bannoks, for example, the ancestors of these modern Indians probably migrated into their current territory less than a thousand years ago, so their connection to the person represented by the skeletal remains found in their modern territory is weak.

Many Indians, however, seem unconcerned with such historical particulars, asserting kinship with and demanding control over any Indian remains found in their modern territory. From a scientific perspective, this makes no sense. We end up with remarkable instances in which modern natives assert stewardship of ancient bones of their ancestral enemies simply because those bones are now located within the recently demarcated boundaries of their reservation. Concern for the bones of immediate ancestors might be understandable, but desiring control over the very ancient bones of individuals who were not immediately ancestral is perplexing. I count among my ancestors Germans, Russians, and Poles, but I feel no great kinship with or reverence for the bones of Upper Paleolithic people unearthed in those modern nations.

Nevertheless, it is understandable from an anthropological perspective how members of different segments of a beleaguered minority, often treated as a monolithic group by the majority, might feel a broad solidarity with members of their larger group, transcending economic, political, tribal, or even temporal boundaries. For example, we do not hear of African Americans expressing solidarity only with other descendants of the particular African tribes from which they can trace their ancestors taken into slavery. Ordinarily, they draw their boundaries more broadly, to include all people in a similar circumstance—the descendants of people taken into slavery, originating anywhere on the African continent. It is not surprising, therefore, that Native Americans do the same, even claiming kinship with and demanding stewardship of enormously ancient human remains that can be connected only in the most tenuous way to any particular modern tribe. When good science meets legitimate emotionalism there seems little room for compromise, with archaeologists and Indians possessed of fundamentally different and equally defensible perspectives. The law now stands on the side of Native Americans and, like it or not, archaeology in North America has changed as a result.

Though archaeologists are adjusting to the restrictions of NAGPRA's rules concerning human remains, there is a broader and potentially more devastating issue. A low point in the radio dialogue mentioned above had to have been when one panel member informed me that everything buried in the ground had been placed there for a spiritual reason by his ancestors, and I had no right to disturb these "sacred objects." If this were true, archaeology faces extinction, but what "sacred objects" could he have meant? Gnawed on deer bones? Sherds of a shattered cooking pot? A spear point snapped in two when it struck an animal? Minuscule flakes shattered off a stone core or a simple, sharp-bladed utilitarian tool? These are the materials most commonly recovered during archaeological excavations in North America; these are the "treasures" we most commonly unearth, not anything that can possibly be construed as "sacred."

Beyond the mundane nature of the vast majority of the material archaeologists regularly excavate, it should be added that most of this material has not been intentionally hidden away by ancient people but consists, instead, of objects that have simply been abandoned and that have, through any combination of entirely natural processes—alluviation, soil formation through organic decay, etc.—simply been covered up. The vast majority of what we excavate is "garbage" in the literal sense; food remains, waste products from manufacturing processes (for example, unusable flakes of stone produced when stone tools were made), or pieces of tools that had broken, been used up, worn out, and then simply discarded.

Those who assert that everything we excavate was sacred to ancient people have bought into the romantic, popular media caricature of archaeologists mentioned above where we dig up mostly treasures intentionally hidden away under the ground for ceremonial reasons. In reality, most of what we dig up is stuff ancient people cared so little about they simply tossed it on the ground, in a trash pit, or on a pile of other garbage. Native Americans might have a reasonable argument when they complain that archaeologists care more for what trash can tell them about Indian history than what their own oral history tells them. Most archaeologists are convinced that garbage represents objective truth and that self-conscious histories—oral and written—often are far more subjective and biased. Nevertheless, the claim that we regularly and intentionally extract objects from the ground that the ancestors of modern Indians placed there with the intention that these things remain buried is a gross exaggeration and a distortion. This belief is untenable from either a scientific or emotionalist perspective. What can sometimes result is the paradox that material not sacred to a people in antiquity becomes so in the present simply because archaeologists dug it up! How else can we explain the recent case in Florida where, not pursuant to NAGPRA but following state regulations, the excavated paleontological remains of an extinct elephant (a mastodon) were "returned" to a local Native American group for reburial (as cited in Lepper 1996)?

Indian Origins

Just when I thought the radio panel discussion was proving to be a waste of everyone's time, I spotted a book brought along by one of the Native Americans. The book was titled *American Genesis,* written by Jeffrey Goodman (1981), a writer who advertised himself as an academically trained anthropologist, fully armed with a Ph.D.

Trying to deflect the conversation from archaeology and museums, I asked the others on the panel what they thought about Goodman's book—which, coincidentally, at the time I was in the process of reviewing (Feder 1983b) and also for which I was writing a detailed and scathing deconstruction (Feder 1983a). I was, again rather naively, shocked at the response: "It's a great book. Dr. Goodman recognizes that we Indians didn't come from somewhere else. We've always been here. Not like you archaeologists. You think we are foreigners. You claim we were latecomers."

Until that moment I had no idea that Goodman had garnered some interest among Indians as the result of the major theme of *American Genesis.* Archaeologists believe that the ancestors of modern Native Americans originated in northeast Asia and migrated across the Bering Land Bridge sometime toward the end of the Pleistocene epoch. They accomplished this during a period when sea level was depressed as a result of the binding up of an enormous quantity of the earth's seawater in ice fields called glaciers that covered much of the higher latitudes and altitudes of North and South America, Europe and Asia. *American Genesis* represented a categorical rejection of this scientific orthodoxy.

It seemed to me that Indian support for Goodman's thesis was yet another irony in an already spectacularly ironic situation since it was based on an ignorance of what Goodman had stated explicitly about the origins of Native Americans in his previous book. Though Goodman made a major issue of disputing the accepted Bering Land Bridge migration scenario in *American Genesis,* and while the title of that book itself seemed to indicate it, he did not explicitly support the claim that Indians had originated in the New World, as the Native American on the radio panel seemed to believe. In fact, in a previous book (*Psychic Archaeology*), Goodman (1977) had been quite explicit. Based on information provided to him by a self-proclaimed psychic, Goodman claimed that New World native peoples had not originated in the New World but, instead, had migrated from, of all places, the Lost Continent of Atlantis, thus creating a rather remarkable nexus of pseudoscientific claims about the human past.

After the radio broadcast we all went our separate ways. Goodman's work lost much of its sheen—or, at least, its currency—and I heard little or nothing of him. Also, controversy about the Bering Land Bridge migration scenario seemed to disappear. Specific versions and especially the timing of the migration or migrations certainly have been argued: was an interior route across the land bridge more significant than a coastal route; did the initial influx of people occur around 12,000 years ago, 15,000 years ago, or before even 20,000 years ago? However, the general notion of a movement of human beings from northeast Asia across the land bridge into North America has not been disputed in the popular media or professional journals in the last two decades. I thought, or at least, hoped that this point of contention between Indians and archaeologists had been disposed of and that more important issues could be discussed. Unfortunately, this assumption and hope were in vain. The issue of the origin of Native Americans has again become a topic of popular debate. And, interestingly, not just where they came from but, even more fundamentally, how we should approach the question and, essentially, how we can know anything about their past (including their origins) are now subject to debate.

A recent book, *Red Earth, White Lies: Native Americans and the Myth of Scientific Fact*, by Indian activist, scholar, writer, and university professor Vine Deloria Jr., attacks archaeology rather viciously and in particular assails those who support the Bering Land Bridge scenario. It compounds the irony to report that (if my small personal sample is representative) many of us who went into archaeology in the 1960s and 1970s read and applauded one of Deloria's (1969) previous books, *Custer Died for Your Sins*. We likely are more sensitive to the issues being discussed here at least in part because of having read it. In a recent compendium of papers (Biolsi and Zimmerman 1997; see especially Grobsmith 1997), a number of anthropologists agree that the anthropological study of Native Americans as it is practiced today is partially a result of Deloria's criticisms of the discipline in *Custer*.

With the publication of *Red Earth, White Lies*, however, not just a few of us have taken lately to scraping the remnants of our "Custer Died for Your Sins" bumper stickers off of our aging automobiles (see Whittaker [1997] for a review of *Red Earth, White Lies*). Deloria rejects any claim that the ancestors of modern American Indians came from somewhere else and proposes, instead, that, based on Native American creation stories, American Indians have always been in the New World since the time of their creation.

One must understand Deloria's rejection of the almost certainly historically accurate land bridge scenario within a broader historical context. The belief that Native Americans must have come from somewhere in the Old World can be traced back to almost immediately after it was recognized that Columbus had not made landfall on Cathay (China) or Cipangu (Japan). It must be admitted that this belief was based on biblical exegesis and not on any particular scientific evidence or reasoning. In 1537 Pope Paul III had decreed that "the Indians are truly men and that they are not only capable of understanding the catholic faith but, according to our information, desire exceedingly to receive it" (as cited in Hanke 1937, 72). Therefore, as Spanish clerics Gregoria Garcia and Joseph de Acosta (see Huddleston 1967) pointed out in their works written barely one hundred years after Columbus's voyages, the Indians must be traceable to one of Noah's three sons because all other people had been killed in the flood. Because the ark landed on "the mountains of Ararat" in southwest Asia, the descendants of Noah who were to become the ancestors of Native Americans must have traveled to the New World, either by ocean-going vessels (Garcia) or by traversing on foot a land connection between the Old and New Worlds (Acosta).

Beyond simply accounting for Native Americans in a way that conformed to the Bible, some 16th-century writers cited biological evidence for an Old World source for the native peoples of the Americas. For example, Giovanni de Verrazzano, an Italian navigator sailing for France in 1524, made landfall at what is today the border of North and South Carolina and then traveled north, looking for a sea route to the west and, it was hoped, a way past the New World and to Asia. He entered Delaware Bay and the mouth of the Hudson River, sailed along Connecticut's coast, entered and explored Narragansett Bay, followed the shore of Cape Cod and then went home, unsuccessful in his attempt to find a passage to the west. Verrazzano spent several weeks exploring the interior of Rhode Island and had an opportunity to examine local natives closely. He concluded: "They tend to be rather broad in the face. . . . They have big black eyes. . . . From what we could tell in the last two respects they resemble the Orientals."

Today, this kind of gross, morphological comparison is no longer the only biological datum on which we base the assertion of a connection between Asians and Native Americans. For example, based on his analysis of 200,000 teeth, physical anthropologist Christy Turner (1987) has shown the clear affiliation of northeast Asians and Native Americans; their teeth share far more in common than either group's teeth share with the dentition of Africans, Europeans, or native Australians. More recently, analysis of mitochondrial DNA (mtDNA) has reaffirmed what Turner's analysis of teeth indicated (Gibbons 1993; Stone and Stoneking 1993). These researchers have shown that four mtDNA variants are found among Native Americans. All four of these variants are found in Asia, and they are not found in Europe, Africa, or Australia.

So, how can Native Americans question these seemingly indisputable data and why would they want to? After all, what does it matter that science can show that neither the ancestors of Native Americans nor anyone else's ancestors are truly "native" to the New World—or, for that matter, to Europe, Asia, or Australia? The hominid family and the species of anatomically modern *Homo sapiens* are native to Africa. We are all, ultimately, natives of Africa; everywhere else in the world, people are immigrants. So what?

At the same time that Europeans were attempting to trace the source of New World native peoples, there also was a great and transparent desire to somehow diminish the legitimacy of the claim of these natives to the lands of the Western Hemisphere. One way in which this was done was to deny the depth of the antiquity of their presence here. As writer Robert Silverberg (1989, 48) puts it, it was "comforting to the conquerors" to believe that, though the Indians may have had some temporal priority, they hadn't really made it to the New World all that long before Columbus.

One major challenge to this belief was the seemingly ancient ruins found in Central America and, especially, the remnants of a geographically extensive, technologically sophisticated culture of "mound builders" who had been responsible for the construction of thousands of burial tumuli and enormous, truncated pyramids of earth that were nearly ubiquitous throughout the Ohio, Illinois, Missouri, and Mississippi River valleys. European thinkers responded to this challenge by denying any cultural or biological connection between Indians and the mound builders, asserting, instead, that the "Moundbuilder" civilization had been the product of a greatly ancient, pre-Indian migration of perhaps even Europeans to the New World. In this historical fantasy, the peaceful and complex Moundbuilder culture had been wiped out before the arrival of Europeans in the 16th century, almost certainly by an only slightly pre-European influx of marauding, aggressive, and warlike savages. These latecoming savages were the ancestors of, of course, American Indians.

One can understand and empathize with a negative reaction on the part of some modern Indians to the more recent scientific assertion that Native Americans arrived here from somewhere else in the measurable past. The claim that the native peoples of the New World came from someplace else was viewed by the Native American on the radio panel and, I believe, is viewed by Vine Deloria Jr. and many other modern Indians, as just another attempt in a history of attempts to contradict or somehow reduce the rightful native claim to the New World. As Randall McGuire (1997, 77) puts it, the archaeological view of Indian origins represents, to many natives, the self-serving "viewpoint of the conquerors of the continent."

My response to this today is the same as it was 15 years ago. Modern archaeology shows that, by the most conservative of estimates, the ancestors of American Indians arrived in the New World 13,000 years ago and, in all likelihood, made the trek across the Bering Land Bridge 15,000 or, perhaps, 20,000 years ago. That would be a minimum of 650 and as many as 1000 generations (at 20 years per generation) of a human presence in the New World. By any definition, that would make quite firm any hypothetical claim of ownership of the New World by American Indians. No archaeologist disputes this; American Indians were here first, and their roots run very deep, orders of magnitude deeper than Europeans.

Deloria's perspective on Native American origins is unabashedly creationist, but not the fundamentalist Christian variety that most of us are familiar with. This should not be surprising, and scientists have long seen this coming. In debating creationists, scientists have often pointed out the fact that the so-called two-model approach of evolution on the one hand and creationism on the other is predicated on a false dichotomy. Of course, there is no such thing as the "creation model," because this presupposes that there is a single—i.e., Judeo-Christian—creationist view. As scientists have constantly pointed out to creationists, there are as many creationist perspectives as there are cultures that have pondered the origins of the universe, the world, life, and people—and very few cultures have not so pondered. We have used this fact to argue against a "two-model" approach in education, because this, in reality, establishes the Judeo-Christian origin myth as the single representative of creationism when, by the very argument of the creationists for fairness, we ought to be devoting equal time in our biology classrooms to Hindu, Navajo, Azande, Egyptian, Iroquois, etc., creation views as well.

There is, of course, a significant contradiction within Deloria's variety of Native American creationism. Deloria is a Standing Rock Sioux and, I presume, the creation story he personally accepts comes from his culture. Fair enough. However, in the Outline of World Cultures of the Human Relations Area Files (HRAF), a broad but by no means exhaustive database of ethnographic studies covering the world, about 250 separate and distinct native culture groups in North and Middle America are inventoried and close to an additional 250 separate and distinct culture groups in South America are listed. The federal government officially recognizes more than 550 Indian tribes and native Hawaiian groups. Some linguists argue that there were close to 1500 different languages and dialects in the Americas aboriginally, so one could argue that there may have been about that many cultural groups.

Examining the HRAF database for New World origin stories or myths, we find literally hundreds of very different stories concerning the creation of people. To compound the problem, as a member of a tribal group, in *Red Earth, White Lies* Deloria expresses solidarity with other tribal peoples elsewhere in the world. These tribal groups also have their own creation stories, adding further to the variety. Michael Shermer (1997, 129–130) presents a taxonomy of some of these myriad creation stories: slain monster version, primordial parents version, cosmic egg version, spoken edict version, sea or water version, and even the no creation/the world and people have always been here version. There is as much disagreement among these many stories as there is between any one of them and scientists adhering to evolution in general and to the Bering Land

Bridge migration scenario specifically in the case of Native Americans.

Under most prosaic patterns of thought and reasoning, one would assume that these stories can't all be right. Either the scientific conclusion based on evidence and logic is correct, or one of the origin myths based on faith and oral history must be correct. Nevertheless, Deloria appears to take the opposite approach. In his view, only one of the explanations is wrong—that is, of course, the explanation given by science—and all of the others, regardless of the fact that they are contradictory—are correct.

(Deloria rejects the assertion that the ancestors of the American Indians migrated to the New World from Asia partially because none of their origin stories say that they did so. Considering that this migration likely involved a small number of people at least 13,000 years ago, I am perplexed why this should be significant. For example, I doubt that many modern Parisians have had stories of painting the fabulous images on the walls of Lascaux cave passed down in their families. Nevertheless, it is almost certainly the case that some of the direct ancestors of some modern Parisians were the actual Lascaux artists.)

Deloria recognizes this apparent flaw in his reasoning. In response, he is explicit about his rejection of the notion of objective historical or scientific fact: "Tribal elders did not worry if their version of creation was entirely different from the scenario held by a neighboring tribe. People believed that each tribe had its own special relationship to the superior spiritual forces which governed the universe. . . . Tribal knowledge was not fragmented and was valid within the historical and geographical scope of the people's experience" (Deloria 1995, 51–2).

Remember, Deloria sub-titles his latest book *Native Americans and the Myth of Scientific Fact*. Deloria is not merely accusing scientists of making up myths about Native Americans (though, certainly, he does this). More significantly, he asserts that the very concept of scientific fact itself is a myth. So, we are left with Deloria's apparent belief that each tribal or traditional culture's reality is different, yet each is "valid" or correct, and that this is a useful and legitimate way to view the world.

Much of this confusion can be traced to the fact that Deloria ignores the reality that myth and science are two different things and approach explanation in entirely different ways. As scientists, rationalists, and even Pope John-Paul II have pointed out, the creation stories of religion instruct people in what their relation is to the "creator" and how, flowing from that, they should live good and moral lives. The Lakota story of the ancestral Buffalo People emerging from the Earth's interior, no less so than Genesis, tells people "how one goes to heaven," but not literally "how heaven was made" (Pope John-Paul II, referring to the Bible, as cited in Lieberman and Kirk 1996). Maintaining that the Lakota creation story is historical truth is no different from claiming that Genesis is literally true and makes inevitable an otherwise avoidable clash between religion and science.

Is There a Future for the Science of the Past?

It is easy to be pessimistic about the future of American archaeology. The rift between myth and science, between emotionalism and rationalism, seems so great, so fundamental, so defining, that it would appear that there is very little common ground possible on which both Indians and archaeologists can stand together.

However, there is at least some hope, of not only a rapprochement, but cooperation. Some natives have written in support of archaeological research, recognizing its contribution to the history of their tribes (see the volume edited by Swindler et al. 1997). Furthermore, there is a small, but dedicated cadre of anthropologists and archaeologists who are, in fact, Native Americans. Those who find themselves straddling both worlds may be the discipline's best hope to communicate to Indian people the significance, potential, and rationale of what we do and for natives to communicate their concerns to archaeologists.

For example, Dorothy Lippert (a Choctaw working on her Ph.D. in anthropology) has written in a wonderfully eloquent piece: "For many of our ancestors, skeletal analysis is one of the only ways that they are able to tell us their stories . . . these individuals have found one last way to speak to us about their lives" (1997, 126). Though many Indians might disagree, Lippert feels "appropriate reverence" for her ancestors can be maintained while scientists study their physical remains to enable her ancestors to use their "voice made of bone."

Even for the many natives who would disagree with Lippert, the excavation, analysis, and curation of demonstrably non-sacred objects are possible in many circumstances. For example, Rose Kluth and Kathy Munnell (both Chippewa) make an absolute distinction between burial and non-burial sites and agree that: "Archaeological sites contain the history of our people, in different stages of their lives, according to the seasons of the year. I believe that useful information can be recovered from these types of sites that will be helpful and interesting to Native Americans" (1997, 117).

Beyond this, some tribes have sponsored their own programs of archaeological research on their reservations. The Navajo, Zuni, and Hopi are good examples. A particularly positive example of Indian recognition of the benefits of archaeology comes from Connecticut where the Pequot tribal nation initiated its own archaeology program (McBride 1990). This tribe obtained federal recognition only recently and with the enormous revenues generated by their wildly successful casino, sought to reconstruct their history and recognized the value of archaeology in that pursuit. Archaeological excavations are nearly continuous on Pequot reservation land and the tribe is currently building a state of the art museum in which the archaeology they have sponsored will be a major element.

It might be suggested that at least part of the success of the relationship between the Pequot and archaeology rests in this simple fact: the archaeology of the Pequot is something that the Pequot wanted, initiated, paid for, and control. In terms of access to sites as well as who signs the checks, the archaeologists necessarily are accountable to them. This is a situation unlikely to be repeated terribly often elsewhere in North America, but it is a clear reflection of the significance of Indian control of their own past in the dispute between Indians and archaeologists.

Conclusions

Many Native Americans may find the pursuit of archaeology unnecessary, redundant, trivial, and, at best, a "necessary evil" for complying with federal regulations (Johnson 1996). They may view the results of our research as antagonistic to their personally held religious beliefs. They may find insulting the very notion that ancient trash may be more accurate than their oral histories. Nevertheless, archaeology may survive anyway, only because, though they may feel they have no use for it, many Indians do not find at least some of our activities to be fundamentally objectionable. This may be the best we can currently hope for. The suspicion some Native Americans feel about archaeology is thoroughly understandable, but this does not diminish the irony that the people whose cultures archaeologists hope to illuminate and, in fact, celebrate may find the entire thing at worst a desecration and at best a peculiar waste of time.

References:

Biolsi, T., and L. Zimmerman, eds. 1997. *Indians and Anthropologists: Vine Deloria Jr. and the*

Critique of Anthropology. Tucson: University of Arizona Press.

Deloria Jr., Vine. 1969. *Custer Died for Your Sins: An Indian Manifesto*. New York: Macmillan.

——. 1995. *Red Earth, White Lies: Native Americans and the Myth of Scientific Fact*. New York: Scribners.

Feder, Kenneth L. 1983a. "Absurdist Archaeology: A Review of Jeffrey Goodman's *American Genesis*." *Bulletin of the Archaeological Society of Connecticut* 45:89–92.

——. 1983b. "American Disingenuous: Goodman's *American Genesis*–A New Chapter in Cult Archaeology." *The Skeptical Inquirer* 7(4):36–48.

Gibbons, A. 1993. "Geneticists Trace the DNA Trail of the First Americans." *Science* 259:312–313.

Goodman, Jeffrey. 1977. *Psychic Archaeology: Time Machine to the Past*. New York: Berkley.

——. 1981. *American Genesis*. New York: Berkley.

Grobsmith, Elizabeth. 1997. "Growing up on Deloria: The Impact of His Work on a New Generation of Anthropologists." In *Indians and Anthropologists: Vine Deloria Jr. and the Critique of Anthropology*. T. Biolsi and L. J. Zimmerman, eds. Pp. 35–49. Tucson: University of Arizona Press.

Haederle, Michael. 1997. "Burying the Past." *American Archaeology* 1(3):14–18.

Hanke, L. 1937. "Pope Paul III and the American Indians." *Harvard Theological Review* 30:65–102.

Johnson, George. 1996. "Indian Tribes' Creationists Thwart Archaeologists." *New York Times*. Pp. C1, C13. October 22.

Kluth, Rose, and Kathy Munnell. 1997. "The Integration of Tradition and Scientific Knowledge on the Leech Lake Reservation." In *Native Americans and Archaeologists: Stepping Stones to Common Ground*. N. Swindler, K. E. Dongoske, R. Anyon, and A. S. Downer, eds. Pp. 112–119. Walnut Creek, California: Altamira.

Lepper, B. T. 1996. "Hidden History, Hidden Agenda. A Review of Hidden History of the Human Race." *Skeptic*, Vol. 4, No. 1: 98–100.

Lieberman, Leonard, and Rodney C. Kirk. 1996. "The Trial Is Over: Religious Voices for Evolution and the 'Fairness' Doctrine." *Creation/Evolution* 16(2):1–9.

Lippert, Dorothy. 1997. "In Front of the Mirror: Native Americans and Academic Archaeology." In *Native Americans and Archaeologists: Stepping Stones to Common Ground*. N. Swindler, K. E. Dongoske, R. Anyon, and A. S. Downer, eds. Pp. 120–127. Walnut Creek, California: Altamira.

McBride, Kevin. 1990. "The Historical Archaeology of the Mashantucket Pequots, 1637–1900." In *The Pequots in Southern New England: The Fall and Rise of an American Indian Nation*. L. M. Hauptman and J. D. Wherry, eds. Pp. 96–116. Norman: University of Oklahoma Press.

McGuire, Randall. 1997. "Why Have Archaeologists Thought the Real Indians Were Dead and What Can We Do About It?" In *Indians and Anthropologists: Vine Deloria Jr. and the Critique of Anthropology*. T. Biolsi and L. J. Zimmerman, eds. Pp. 63–91. Tucson: University of Arizona Press.

Meighan, Clement W. 1994. "Burying American Archaeology." *Archaeology* 47(6):64, 66, 68.

Schneider, Alan L. 1996. "Recent NAGPRA Developments." *Current Research in the Pleistocene* 13:9.

Shermer, Michael. 1997. *Why People Believe Weird Things*. New York: W. H. Freeman.

Silverberg, Robert. 1989. *The Moundbuilders*. Athens: Ohio University Press.

Stone, Anne C., and Mark Stoneking. 1993. "Ancient DNA From a Pre-Columbian Amerindian Population." *American Journal of Physical Anthropology* 92:463–471.

Swindler, Nina, Kurt E. Dongoske, Roger Anyon, and Alan S. Downer, eds. 1997. *Native Americans and Archaeologists: Stepping Stones to Common Ground*. Walnut Creek, California: Altamira Press.

Turner, Christy G. 1987. "The Tell-Tale Teeth." *Natural History*. January, 6–10.

Whittaker, John C. 1997. "Red Power Finds Creationism." *Skeptical Inquirer* 21(1):47–50.

Zimmerman, Larry J. 1994. "Sharing Control of the Past." *Archaeology* 47(6):65, 67, 68.

——. 1997. "Anthropology and Responses to the Reburial Issue." In *Indians and Anthropologists: Vine Deloria Jr. and the Critique of Anthropology*. T. Biolsi and L. J. Zimmerman, eds. Pp. 92–112. Tucson: University of Arizona Press.

Pseudoarchaeology

Precolumbian Discoverers of America as a Test Case

RONALD FRITZE

Who discovered America? It seems like an innocuous question. We all know that Columbus "discovered" America, in the sense that Europeans first heard about a New World through him. And we all also know that the Indians were here first, and thus they "discovered" America before anyone, if one considers migrating peoples discoverers.

A 1992 CNN poll, however, revealed that only 20% of Americans thought that Columbus was the first to discover America. An overwhelming majority of the respondents (70%) thought that other people had preceded Columbus in reaching the Americas, while 10% did not know. The problem is that a question which simply asks—who first discovered America?—is badly posed. Among that 70% of people who deny Columbus's priority in the discovery of the Americas are undoubtedly many people who possess a sophisticated understanding of the pre-Columbian history of the Americas. They are right. Columbus was not first. The prehistoric hunters who were the ancestors of the Native Americans and crossed the Bering Land Bridge some 15,000 years ago were the true discoverers of the Americas. Furthermore, Leif Ericsson and other Norse seafarers reached the Americas in the decades after 1000 CE (Common Era). The testimony of the Norse sagas has been confirmed by the discovery of a genuine Norse archaeological site at L'Anse aux Meadows on Newfoundland which may have even been Leif Ericsson's own camp.

Unfortunately many other people who deny the priority of Columbus are not thinking of either the prehistoric wanders who crossed the Bering Land Bridge or Leif Ericsson. Instead, they credit the first discovery of the Americas to various peoples from the ancient and medieval eras, including: Egyptians, Phoenicians, Africans, Trojans, Carthaginians, Romans, Arabs, Irish, Welsh, Germans, Poles, and various groups of Jews such as the wandering Hebrews, one or more of the Ten Lost Tribes, and refugees from the Bar Kokhba revolt. All of these people have been proposed as having crossed the Atlantic Ocean well before 1492. On the other side of the world various Chinese, Japanese, Hindu, Polynesian, and Mongol explorers, and travellers along with a lost fleet of Alexander the Great, all supposedly crossed the Pacific and found the Americas prior to Columbus.

Numerous books and articles have been published which advocate one or more of these dubious theories of pre-Columbian contacts between the Old World and the Americas. In 1990 the Foundation for Ancient Research and Mormon Studies published *Pre-Columbian Contact with the Americas across the Oceans: An Annotated Bibliography*, edited by John L. Sorenson and Martin

Raish. This massive two volume work lists 5,613 items and is not exhaustive. New works are being published all the time. Sadly, the vast majority of these works are poor pieces of scholarship in which the same errors of method and fact keep appearing again and again, year after year. It is a situation that professional anthropologists, archaeologists, and historians all find to be quite discouraging. Furthermore, these professional scholars often find their own writings and opinions rejected and disdained by these advocates of various pseudohistories of the pre-Columbian Americas. The distinguished anthropologist Robert Wauchope described the situation as follows:

> Lay writers on these subjects [pre-Columbian contacts] have one great bias in common: they all scorn, ridicule, and complain bitterly about the professional anthropologists of American museums and universities, whom they regard variously as stupid, stubborn, hopelessly conservative, and very frequently plain dishonest.

It is a claim all too familiar to skeptics, who are frequently told by pseudoscientists that those who oppose them are ignorant or fraudulent. At the same time, these very same people profess to be following the strictest scholarly standards in their own work. That claim is not true. The following is a survey of the types of errors committed by the adherents of various pre-Columbian contact theories. While it covers most of the main ones, it is by no means comprehensive, let alone exhaustive. (See my book *Legend and Lore of the Americas before 1492.*)

Diffusionism Made Simple

1. *Diffusionism and Hyper-Diffusionism.* Diffusionism is an anthropological concept that seeks to explain cultural change on the basis of unilateral or reciprocal borrowing between different cultural groups that occurs as a result of trade, migration, or conquest. All theories that explain the rise of higher civilizations and their various cultural traits primarily on the basis of supposed contacts with the Old World are inherently diffusionist.

Anthropologists universally accept the phenomenon of diffusion as a partial explanation for cultural change. Some advocates of diffusionism, however, have been extreme in their claims about the extent of cultural exchanges between different societies. As a result they have been labeled hyper-diffusionists. Hyper-diffusionists deny that parallel evolution or independent invention of tools or ideas took place to any great extent at all throughout prehistory. They claim that humans were remarkably uninventive and that history never repeats itself. During the early 20th century, the British anthropologist W. J. Perry and the anatomist Grafton Elliot Smith took hyper-diffusionist theory to its ultimate extreme by tracing the origins of all higher civilizations throughout the world back to one source—ancient Egypt. Both men wrote numerous books and articles postulating the influence of ancient Egyptian culture on various societies throughout the world. Though hyper-diffusionist theories never dominated anthropological and archaeological thinking, moderate diffusionism did in the early 20th century. Therefore, it is not surprising that various fringe theories postulating visits to the New World by one or another group from the Old World (e.g., the Ten Lost Tribes of Israel, Mongols) found support in the rise of diffusionist concepts. After all, Grafton Elliot Smith's theory that Egypt was the source of all other ancient civilizations was simply a somewhat more restrained version of Augustus Le Plongeon's earlier theories about the ancient Maya being the mother culture of world civilization including the Egyptians.

The development of radiocarbon dating after 1946, and its calibration using correlations with dendrochronology (tree-ring dating) during the 1960s, completely undermined the hyper-diffusionist reconstructions of prehistory. These techniques revealed that cultures once thought to be the beneficiaries of cultural diffusion from ancient Egypt were actually as old or older than the oldest Egyptians. Archaeological thinking was revolutionized. The independent invention of various cultural traits had obviously taken place far more frequently than diffusionists had supposed. But hyperdiffusionists have refused to give up and continue to revive the same flawed evidence, demonstrating that they are really doing pseudohistory, not scientific history.

2. *Pyramids and Statues.* Egypt is famous for its pyramids but so is Central America with its great pyramids at Teotihuacán, Chichén Itzá, and other places. A casual observer might easily conclude that the ancient Egyptians and Americans were in contact because these great structures look so much alike. Indeed, the general similarity of the pyramids, and the "negroid-like" features of the Olmec statues of Mexico, have led extreme Afrocentrists to conclude that black Africans (they also claim the Egyptians were all black) were the first to discover America. Unfortunately two basic problems make any Egyptian-American contacts impossible. The first objection is chronology. Many centuries separated the Pyramid Age of Egypt from the time when the pyramids of Teotihuacán and Chichén Itzá were constructed. Second, while the form of the pyramids may be similar, the functions are totally different. Egyptian pyramids primarily served as tombs while the American pyramids were temples. Furthermore, archaeological research has reconstructed the independent evolution of the pyramids in both regions, leaving no room for diffusionist explanations. Finally, while features on Olmec statues do indeed resemble those of African peoples, they also look similar to those of native Americans. What one "sees" in a statue, however, is hardly historical evidence of origin, since one can easily see what one wants or expects to see, especially in such generalized forms as artwork.

3. *Supposed Pre-Columbian Diffusion of Plants.* If a cultivated plant of Old World origin could be traced to the Americas before 1492 or vice versa, it would be strong evidence for human contact between the two hemispheres. Many such claims are associated with cotton, maize, and the sweet potato, but they have proven in most cases to be fallacious.

There are over 20 species of cotton of which four are cultivated for their fibers. Two of the cultivated species are *Gossypium arboreum* and *Gossypium herbaceum*, which have 13 chromosomes in their cells and are known as the Old World cottons. The other two species, *Gossypium hirsutum* and *Gossypium bardadense*, possess 26 chromosomes and are known as the New World cottons. Genetically, the two cultivated New World cottons are hybrids that contain the 13 chromosomes of another wild species of New World cotton and the 13 chromosomes of the cultivated Old World cottons. The wild New World cottons are not capable of producing useful fibers. But when these two sets of chromosomes are combined, a cotton plant is created that produces lush clumps of useable fibers. Obviously somehow and sometime in the past the cultivated Old World cottons came into contact with the wild New World cottons and the result was the hybrid, cultivated cottons of the New World. The mystery is whether this process occurred naturally or was assisted by humans.

The creation of the cultivated New World cottons definitely took place a long time ago. Archaeologists have found remains of cotton at Mohenjo-Daro in the Indus River valley dating from 3,000 BCE. In the Americas, cotton fabrics dating from 2,000–3,000 BCE have been recovered from archaeological sites at the Tehuacán Valley in Mexico. Obviously the creation of

Gossypium hirsutum and *Gossypium bardadense* took place in the distant past. So distant, in fact, that human assistance by means of transoceanic contact between the Old World and the Americas seems very unlikely. Instead, natural means seem to have produced the cultivated New World cottons. Scholars have developed two possibilities for how this process occurred. First, they suggest that the cultivated Old World cottons *Gossypium arboreum* and *Gossypium herbaceum* had grown in the Americas at one time but became extinct sometime before 1492. No archaeological evidence has yet been found to support this theory. Second, they suggest that the unopened cotton bolls of the Old World cottons are capable of floating across the oceans. The prolonged exposure to salt water will not always destroy the seeds' ability to germinate successfully. Either of these scenarios brings Old World cottons into contact with wild New World cottons so that hybridization can take place. Neither depends on human travellers to carry the seeds.

Maize, or corn as it is more commonly known in North America, is almost universally accepted by the scholarly world to have originated in ancient America and later spread throughout the world after Christopher Columbus's voyage of 1492. At the same time, many diffusionist writers have suggested that maize actually originated in Asia or that it was of American origin but travelled to Africa, Asia, and Europe before 1492, thus indicating the existence of pre-Columbian contacts between the Americas and the Old World. George F. Carter, the distinguished geographer, has made such claims for pre-Columbian maize in China. Extensive research into the voluminous and detailed botanical literature of pre-Columbian China, however, has failed to reveal any evidence of the cultivation of maize before the early 16th century. The archaeological and historical record for South Asia also has provided no indication of the existence of maize in that region prior to 1492. The same observation applies to Europe where maize first received notice in 1532 in a herbal written by Jerome Buck. From that point onward, maize appeared regularly on the pages of European herbals and botanical works from the 16th century. No such mentions occurred in European botanical works written during the 14th and 15th centuries. This omission would be highly suspicious if maize had already reached Europe before 1492, which it apparently did not do. The literature concerning pre-Columbian maize in Africa is extensive, although the chief exponent of that theory is the South African anthropologist M. D. W. Jeffreys. Jeffreys believes that Arabic–Black African contacts with the Americas took place about 900 CE and after. But as Paul Mangelsdorf, the leading authority on the evolution and history of maize/corn, has suggested, the ambiguities in the terminologies used by Jeffreys' historical sources appear to have caused a confusion between maize and the similar sorghums that did grow in pre-Columbian Africa.

Mangelsdorf has also pointed out that the most telling evidence for the post-Columbian introduction of maize into the Old World is the total absence of pre-Columbian corn cobs outside of the western hemisphere. Pre-Columbian corn cobs are very commonly found in archaeological sites throughout the Americas. They survive readily under many climates and conditions, but so far none that can be convincingly dated to before 1492 have been found in the Old World. Maize cannot be cited as evidence that pre-Columbian contacts took place between the Old World and the Americas because no pre-Columbian maize appears to have existed in the Old World.

There are two divergent claims regarding the sweet potato as evidence for pre-Columbian contact between the Americas and the Old World. One theory is that the sweet potato originated in Africa and was carried to the Americas. The other places the origin of the sweet potato in the Americas but claims that it was carried into Polynesia during the era before European contact.

The sweet potato (*Ipomoea batatas*) is considered by the vast majority of scholars to be a native of the Americas. It is a member of the morning glory family of plants, and research indicates that it evolved from a wild plant in tropical Central America with the scientific name of *Ipomoea trifida*. In 1954, the botanist Elmer Drew Merrill speculated about a possible African origin for the sweet potato, although other botanists have either rejected his idea as unfounded or ignored it. That has not stopped some diffusionist writers from occasionally using Merrill's theory of an African origin for the sweet potato to bolster their own ideas about African voyages to pre-Columbian America. It should be remembered, however, that the botanical and archaeological evidence overwhelmingly puts the original home of the sweet potato in the Americas.

Except for a few Spanish landings in the 16th century, sustained European contact with Polynesia began in the 18th century with Jacob Roggeveen's discovery of Easter Island in 1722, and Captain James Cook's visits to the Hawaiian Islands in 1778 and New Zealand in 1769. When the Europeans arrived, the natives of these islands were all cultivating the sweet potato. Obviously the plant came from the Americas, but how and when did it get to Polynesia? Some people have suggested that a natural transfer occurred in which a sweet potato seed or tuber floated from the Americas to the various Polynesian islands by accident. Most experts, however, feel that the sweet potato's seeds or tubers were not capable of floating such vast distances across the Pacific Ocean. Furthermore, prolonged exposure to salt water would also destroy the fertility of the seeds and tubers. As a result, the presence of sweet potatoes in Polynesia would seem to indicate that Polynesian-American contacts similar to Thor Heyerdahl's *Kon Tiki* voyage occurred during the pre-Columbian era.

Besides the physical presence of the sweet potato in Polynesia, supporters of Polynesian-American contacts also cite linguistic evidence. They claim that in the Lima region of Peru, the native Quechua word for sweet potato is *kumar* or *kumal*. The Polynesians know the sweet potato by variations of these Quechua words so that it is called *uwala* in Hawaii, *kumara* in New Zealand and Easter Island, *umara* in Tahiti, and *unala* in Samoa. Unfortunately, this impressive linguistic evidence is inaccurate. *Kumar* or *kumal* was not the Quechua word for sweet potato. In reality, the Quechua word for sweet potato is *apichu*. Kumar does not refer to sweet potato anywhere along the coastal region of Peru. So, the best linguistic evidence does not support the occurrence of Polynesian-American contacts.

Donald D. Brand, a geographer from the University of Texas, has advanced a subtle theory that claims that the spread of the sweet potato occurred entirely during the post-Columbian times. According to his scenario, Spanish settlers carried the sweet potato home to Spain. From there it reached Portugal in 1500. The Portuguese then carried it to their trading stations in India before 1505. From there Asian traders—Persians, Arabs, and Hindus—took the sweet potato into the Moluccas, or the Spice Islands. At that point, the sweet potato entered a trading network connected to Melanesia. After spreading quickly across these islands, the sweet potato then reached Polynesia before any Europeans set foot on those islands.

Flawed Methodologies

1. *The Wordlist Game.* In 1846 the future historian Francis Parkman made the following observation while travelling on the great plains:

> The Indians raised in concert their cries of lamentation over the corpse, and among [which was] . . . clearly distinguished those strange sounds resembling the word "Hallelujah," which together with some other accidental coincidences, has given rise to the absurd theory that the Indians are descended from the ten lost tribes of Israel.

What seemed an "accidental coincidence" or "absurd" to Parkman, however, has seemed to be sound evidence to many, more credulous theorists of pre-Columbian contacts between the Americas and the Old World. Algonquins and Irish, Maya and Egyptians, or Peruvians and Polynesians are among the groups for which fallacious wordlists have been compiled. Countless other lists of similar sounding words with similar meanings between one Native American language and another Old World language have appeared to prove various theories of pre-Columbian contacts.

Unfortunately the compiling of such wordlists is an overly simplistic form of comparative linguistics. Linguistic scholars consider the study of grammar to be the more reliable way to make comparisons between different but possibly related languages. Grammatical structures tend to change slowly. In comparison, the words used in any language change over time, often quite rapidly. If two languages contained significant numbers of words borrowed from each other, it would indicate a fairly recent contact. Otherwise, wordlists are fairly useless as evidence of contact in the more distant past.

It is possible to compile lists of similar sounding words with similar meanings for every language in the world with every other language in the world. These lists can be larger or smaller depending on how generously one allows the words to sound alike or have similar meanings. In the end, however, all these lists really prove is that there are a limited number of sounds (phonemes) or combinations of sounds that human beings can make to form words. That number may be a large one but when compared to the vastly larger number of words in all the languages of the world that exist or have existed, there will be many cases where the same sounds have roughly the same meaning in two different languages even though there are no historical connections between those two languages. The similarity is not merely a coincidence but one that has a high probability of happening one way or another. Truly significant connections between words in different languages can only be determined by studying the etymology (the changing history of a word's usage) of the individual words. When proper linguistic methods are applied to the problem of pre-Columbian contacts between the Americas and the Old World they invariably show that nothing significant took place.

2. *Inscription Mania and Illegitimate Epigraphy.* Epigraphy is the study of inscriptions left by ancient peoples. It is one of the major sources of information that historians use to study the ancient Mediterranean world. Various ancient cultures in Central America, notably the Maya, also produced large numbers of inscriptions of use to epigraphers. Outside of Central America, the various cultures of Native Americans did not possess systems of writing and so would have left no inscriptions for epigraphers. Recently, however, the Harvard marine biologist and amateur archaeologist Barry Fell has theorized that various ancient Celtic peoples and other groups of Mediterranean people colonized North America in the pre-Christian era. He claims that numerous inscriptions in the ancient Celtic script called Ogam are scattered throughout New England and other regions. He and his supporters are constantly on the lookout for such inscriptions and they claim to have been quite successful. The problem is that Ogam script basically consists of combinations of straight lines. So what Fell and his supporters claim is an ancient

Celtic inscription looks like natural scratching and wear on rocks to mainstream archaeologists. Fell's case is further compromised by his regarding such proven archaeological frauds as the Davenport Tablets as a genuine artifact left by his "ancient colonists." Basically Fell and other amateur epigraphers are guilty of seeing what they want to see among the weathered rocks of New England. Their unquestioning belief in the existence of these pseudo-inscriptions has been labeled "inscription mania" by professional archaeologists.

3. *The Game: Patolli-Pachesi Parallels.* A frequently discussed and superficially compelling evidence for pre-Columbian contacts between Asia and America is the similarities between the Aztec game of patolli and the Hindu game of pachesi. In both games the players move pieces around boards with cross-shaped tracks divided into segments. The number of moves a piece can make is determined by throwing lots; the Hindus used cowrie shells while the Aztecs used beans. Similarities between these and other games have been noted as early as 1724. Later in 1879 the great English anthropologist E. B. Tylor (1832–1917) wrote a paper suggesting that patolli has actually been derived from pachesi as a result of ancient contacts between Asia and the Americas. Tylor added the authority of probability to his argument in 1896 and stated that it was highly improbable that two such similar games could have been invented independently.

Tylor's contemporaries, the American scholars Stewart Culin and Daniel Brinton, rejected his conclusion that patolli came to ancient America as a result of cultural diffusion from Asia. They stressed that it was independently invented in the Americas without any Asian influences and went on to cite evidence of geographical distribution and variations to bolster their contention. On the other hand, in the next generation of scholars A. L. Kroeber (1876–1960), the doyen of American anthropology, supported Tylor's conclusions for many years although not because of any particularly sound reasons.

Some anthropologists have developed a theory of limited possibilities to explain similarities between different cultures as an alternative to diffusion. Basically, this theory states that the number of cultural choices may not be large in some cases. Seemingly complex and similar institutions and artifacts could develop independently because the probabilities against it happening are not all that great. In the case of patolli and pachesi, the dice or lots must have at least two flat sides to be functional, while the cross shape of the board is really quite a common and universal shape. Furthermore, with cultures all over the world engaged in gaming, it is not surprising for similar games to appear independently. The anthropologist John Charles Erasmus has cautioned against the facile calculating of possibilities or probabilities for the development of similar cultural traits. Large numbers of people at all times and all over the world are engaged in the process of cultural evolution. That variable, however, is seldom taken into consideration when the probabilities of independent invention are discussed. Furthermore, patolli is the only aspect of Aztec culture that shows any indication of possible Hindu contact. The absence of other Mexican cultural traits of probable Hindu origin is another strong evidence against any pre-Columbian contacts between Mexico and India.

4. *Coin Finds.* Over the years, 41 documented reports have appeared of Old World coins with pre-Columbian dates being found in the Americas, particularly North America. There may be others. These finds have been used to argue for pre-Columbian visits by Canaanites, Phoenicians, Hebrews, Greeks, Romans, and Norse sailors although so far only the Norse find has managed to stand up to scholarly scrutiny.

Lucio Marineo Siculo (1460–1533), a somewhat credulous Italian humanist, in 1533 reported the finding of a Roman coin from the time of Caesar Augustus in a gold mine in Panama. He concluded that the presence of this coin proved that the Romans had reached the Americas before the Spanish. Gonzalo Fernandez d Oviedo y Valdes touched on Siculo's story in his *Historia general y natural de las Indias* of 1535 and showed that it was ridiculous.

Significantly, no one found any more pre-Columbian coins in the Americas until several Roman coins from the imperial era were found in the Fayetteville area of Tennessee between 1818 and 1823. The early archaeologist Caleb Atwater was immediately skeptical and suspected that the coins were deliberate plants. The Tennessee antiquarian John Haywood, however, considered the find to be authentic. It is interesting that even Haywood reported that after a Mr. Colter, a man known to possess Roman coins, left Tennessee for Alabama in 1823 that no more coins have ever been found in Tennessee. Modern archaeologists generally agree with Atwater's original assessment and think that the Tennessee coins were part of a hoax.

Only one other documented coin find took place in the 19th century. It occurred in 1880 on an Illinois farm and involved the finding of a Seleucid Greek coin from c. 173–64 BCE. Otherwise all of the remaining 32 coin finds took place in the 20th century, and of that number, 24 were found after 1945. With the exception of the Norse penny found in Maine, all of these coins appear to have been brought to the Americas after 1492. Some of the coins have actually turned out to be forgeries such as the three Bar Kokhba coins found at various places in Kentucky in 1932, 1952, and 1967.

Other genuine ancient coins have been located in archaeological situations that indicate they may be losses from modern collections rather than remains from the distant past. Many coins have been found on the surface of the soil rather than having been dug up. The most common natural tendency for a coin on the ground would be slowly to sink down into the soil rather than to work its way up to the surface after it was buried. It is estimated that some one million Roman coins are in the coin collections of the late 20th century United States. Most of those were brought back from Europe after World War II. Many of these Roman coins are only worth $10.00 or less and so are not looked after all that carefully. The possibility of accidental losses is quite real, and that appears to be what has happened in most of these 20th century finds of pre-Columbian coins.

Jeremiah F. Epstein's 1980 study of coin finds basically concluded that none of them provide legitimate evidence for pre-Columbian contacts. One exception to his conclusion, however, is the Norse penny from the reign of Olav Kyrre (1066–1093) of Norway found in Maine in 1957. Tests have established that it is genuine. But since no one now denies that the medieval Norse reached Newfoundland, it is not implausible that they visited Maine as well.

Fake Artifacts

1. *Kensington Rune Stone.* This famous but fraudulent Norse artifact was first discovered in Minnesota in 1898 and still has supporters of its authenticity in spite of considerable debunking scholarship to the contrary.

In 1898, Olof Ohman "discovered" the Kensington Rune Stone while clearing trees from his farm in Douglas County, near the town of Kensington, Minnesota. It contained an inscription in runic characters, the ancient alphabet of Scandinavia. Unfortunately, the physical appearance of the inscription belied its supposed antiquity; e.g., its cuts showed none of the weathering associated with a stone carving over 300 years old. It has even been suggested that the inscription was added after Ohman first unearthed the stone.

The Rune Stone's inscription told of a party of Norse making its way through the wilderness during 1362 and suffering the loss of 10 of its members from attacks by hostile Indians. Such a find would have been of immense interest to the Scandinavian immigrant community of the Upper Midwest. In 1898 they were anxious to find proof of Norse precedence over Christopher Columbus in the European discovery of America. The World Columbian Exposition at Chicago during the 1890s had aroused their ethnic ire. Scandinavian Americans wanted to believe that the Kensington Rune Stone was authentic so local support was strong. The scholarly reception of the Kensington Rune Stone, however, was negative from the start on the basis of anachronistic usages of both runic characters and Norse words. Eventually enthusiasm for the Rune Stone stalled and Ohman took it back to his farm where he used it as a steppingstone. True believers in the Scandinavian community, however, continued to claim the Kensington Rune Stone was genuine.

In 1907 Hjalmar R. Holand, a young researcher, came to Douglas County to gather material on the Norwegian immigration to the United States. During his researches, the locals told him about Ohman's rune stone and a curious Holand went to see it. Rejecting earlier scholarly opinion, Holand decided it was a true Norse artifact and Ohman even gave it to him. Starting in 1908, for the rest of his life, Holand attempted to prove that the Kensington Rune Stone was really a medieval Norse inscription. He even got the Minnesota Historical Society so interested that when they issued a report on the stone's authenticity, they ignored additional scholarly opinions to the contrary and pronounced it genuine. Efforts by Holand in 1911 to secure favorable judgments from European scholars met with failure as they all considered the stone to be a hoax. But Holand remained undaunted. In 1932 he published his first book, *The Kensington Stone,* which defended the stone's authenticity. The stone travelled to the Smithsonian Institute in 1948 for further scholarly investigation which again produced negative results. Holand, however, continued to believe that the Kensington Rune Stone was a true medieval Norse artifact.

The 1950s saw the beginning of a wave of scholarly publications denying the authenticity of the Kensington Rune Stone. The two most devastating attacks came from books by experts on Norse studies—Erik Wahlgren in 1958 and Theodore C. Blegen in 1968. Their studies convincingly showed that the Kensington Rune Stone was a fake. Blegen even suggests how Olof Ohman may have collaborated with his neighbor Sven Fogelbad to produce the inscription. None of this scholarly activity has managed to stop some true believers from continuing to have faith in the Kensington Rune Stone. For the vast majority of historians and archaeologists, the Kensington Rune Stone is no more than one of the most persistent hoaxes in the history of American archaeology.

2. *Paraiba Stone.* In 1872, the most enigmatic of the supposed Phoenician artifacts in the Americas came to light—the Paraiba Stone. A man named Joaquim Alves da Costa claimed to have found, "near the Paraiba" river, a broken stone which had an inscription in a strange alphabet carved on it. After transcribing the inscription, Costa sent the copy to Rio de Janeiro for study. But Brazil had no experts in ancient semitic languages. Instead the conscientious naturalist Ladislau Netto took up the assignment, learned Hebrew, and ultimately determined that the writing on the stone was Phoenician and then translated it. His translation described how 10 Phoenician ships were blown by storms to the coast of Brazil in 534 BCE. Immediately the French scholar Ernest Renan attacked the Paraiba inscription as a fake and others soon joined him. By 1885 the hapless Netto felt compelled to

publish a retraction of his original conclusions and even suggested five possible suspects who might have engineered the hoax. Meanwhile Costa disappeared with the stone and no accredited scholar ever saw it first hand. Even the original location of the find was in great doubt since Brazil had two different Paraiba regions. During the 1960s, Cyrus Gordon, a professor of semitic languages and an ardent diffusionist, revived the Paraiba Stone's claims to authenticity. Basically, Gordon had asserted that the Paraiba inscription contains Phoenician grammatical constructions that were unknown in 1872. Other equally qualified specialists in semitic languages disagree with his conclusions and continue to declare the Paraiba Stone to be a hoax. That judgment is the opinion of archaeologists and prehistorians in general.

Historical Fallacies

1. *Portuguese Policy of Secrecy or Silence.* This controversial historical thesis, formulated in the first quarter of the 20th century by various historians, primarily Portuguese, states that Portugal made many voyages and discoveries in the Atlantic Ocean, including the discovery of the Americas sometime before 1492, but chose to keep those discoveries secret.

Before the 19th century, the historical record contains many gaps and breaks. This condition certainly applies to the surviving records from the Great Age of Discovery. None of the original logs for Christopher Columbus's four voyages survived, although a partial transcript exists for the first voyage. John Cabot's voyages to North America in 1497 are practically without any contemporary documentation and the same situation applies to Bartolomeu Dias's discovery in 1487 of the Cape of Good Hope. Such losses of primary sources are tragic but all too common and they usually occur quite innocently as the result of accidents or neglect. Some historians, however, have questioned whether the gaps in the Portuguese records are all that random. They suggest that some design or policy may lie behind the disappearance of some documents.

The thesis of a deliberate and systematic Portuguese government policy of secrecy concerning overseas exploration is a product of 20th-century historians. Jaime Cortesao, a Portuguese historian, first formulated the thesis in 1924. He contended that the surviving Portuguese chronicles about overseas explorations show definite signs of truncation and censorship.

If one is inclined to believe Cortesao, quite a lot of information was suppressed, including a Portuguese discovery of America prior to 1448. Jaime Cortesao was not alone in his support for the existence of a policy of secrecy. In Portugal the thesis has become a historical orthodoxy and a pillar of national pride. School textbooks at all levels teach it as fact. Lisbon's city government has even decorated its Avenida de Liberdade with a mosaic inscription which reads "Descoberta da America 1472 Joao Vaz Corte-Real Descobridor da America."

Outside of Portugal, historians, including Samuel Eliot Morison, generally reject Cortesao's thesis of a policy of secrecy and its various claims of monumental but previously uncredited Portuguese achievements during the 15th century. Dissent exists even in the Portuguese historical community where the respected historian Duarte Liete attacked Cortesao's theory as early as 1936. But in spite of all the controversy, the thesis of a Portuguese policy of secrecy still possesses enthusiastic supporters, and so continues to attract equally determined opponents.

The basic complaint of skeptical historians concerning the policy of secrecy is the almost complete absence of solid evidence for its existence. Historians admit that monarchs and countries throughout history have attempted to protect their overseas commerce by maintaining secrecy about the how and the where of their sources. But ultimately these efforts have failed. Supporters of the policy of secrecy reply that the lack of evidence is in itself evidence of the existence of a policy of secrecy that was extremely effective. Of course, their opponents, particularly Samuel Eliot Morison, find such an argument both circular and ridiculous. Ultimately Morison feels that Cortesao's thesis requires the Portuguese to maintain their secrets apparently for the sake of secrecy alone and often against their own best interests. He rightly argues that the Portuguese government's pursuit of a policy of secrecy needs to make sense and be of benefit to the national interests. If Portugal already knew about the Americas before 1492, why did Joao II abdicate virtually all of that new land to Spain in the Treaty of Tordesillas?

Another argument repeatedly brought to bear against the existence of such a policy of secrecy is the well-documented and sustained participation of a substantial number of foreigners in Portugal's overseas explorations. Martin Behaim of Germany and Christopher Columbus of Genoa are simply the best known of a host of foreigners who served in Portugal's overseas ventures. With so many foreigners involved in Portugal's overseas enterprises, it would have been impossible to keep important discoveries a secret. Details of Portugal's jealously guarded African trade leaked out with amazing rapidity. Furthermore, little attempt was made to keep secret Bartolomeu Dias's discovery of the Cape of Good Hope in 1487 or Vasco da Gama's voyage to India in 1497. Why did the Portuguese let these important discoveries become public knowledge if they had such an effective policy of secrecy? Not surprisingly, outside of Portugal, the thesis of the policy of secrecy and its accompanying suppression of information about various discoveries, most notably a pre-Columbian discovery of America, has found little support among historians.

2. *White God Legends.* This group of Native American myths purportedly describes vague memories of pre-Columbian visitors from the Old World. Most of these legends supposedly relate to peoples from the ancient Mediterranean or Western European cultures. Some adherents of pre-Columbian contacts between the Old World and the Americas claim that these same legends actually refer to visitors from Africa or China, which would more accurately make them yellow or black god legends.

The Native American gods commonly identified as white gods are Quetzalcoatl, Kukulcan, Itzamna, Votan, Viracocha, and Sume. According to various popular writers, all of these deities were bearded, white-skinned, departed from the Americas with a promise to return, and established civilization and higher humanitarian values during the time they ruled over the various indigenous tribes and kingdoms. It is claimed that these legends of white gods are almost universal among the aboriginal peoples of both North and South America. These legends supposedly aided the Spanish conquest of the Aztecs, the Incas, the Mayas, the Chibchas, and various other peoples since they mistakenly took the Spanish conquistadors to be their returning white gods.

Although there were many supposed white gods among the various groups of Native Americans, there are even more candidates to serve as the inspiration for the white god legends among the supposed pre-Columbian visitors to the Americas. The list includes St. Thomas, St. Brendan, Prince Madoc of Wales, and even Jesus Christ.

The problem with all of these theories is that they are not based on original and authentic Native American legends. Most of the so-called white gods are actually humans who filled the role of being culture-heroes. Like the Greek culture-hero Prometheus who brought civilizing fire to humanity, the Native American culture heroes brought the benefits of agriculture, writing, the calendar, and true religion to their peoples. Generally these gods are described as bearded but that is no proof of their being white. Native Americans can sometimes grow beards, and these beards, such as the Aztec emperor Moctezuma II's, were observed by the Spanish. The problem is that many versions of these legends have been contaminated with post-Columbian additions by the Spanish. The whiteness of these white gods is not mentioned in the most authentic versions of the culture-hero legends. Quetzalcoatl is actually described as having a black or a black and yellow striped face. It also appears that the white god's departure from and promise to return to the Americas are usually post-Columbian additions. In the case of Quetzalcoatl, some historians, such as Nigel Davies, think that the belief that Hernán Cortés was the returned Quetzalcoatl was a delusion concocted by the nervous Aztec emperor Moctezuma II. There was no general belief among the Aztecs that Quetzalcoatl would return. David Carrasco, a historian of religion, disagrees and instead claims that during its final years the Aztec empire lived in dread anticipation of Quetzalcoatl's return. But in the case of the other Native American gods—Votan, Viracocha, and Sume—the legend of the white gods was a Spanish fraud.

Other problems with linking white god legends to historic persons or peoples are chronological. Quetzalcoatl lived sometime during the years 900–1100 CE, which eliminates most of the supposed ancient pre-Columbian visitors, including Jesus Christ, as candidates for inspiring his legend. Furthermore, the white god legends, like most tales of pre-Columbian visitors to the Americas, lack a convincing foundation in the archaeological and documentary evidence. Close study of the Native American myths simply makes the white god legends seem less and less credible.

Why Pseudohistory?

Why do people continue to believe in dubious theories about pre-Columbian contacts between the Old World and the Americas? One reason is that it is a common characteristic of human nature to have a fascination with the strange and fantastic and these theories are, for the most part, very strange and utterly fantastic. They also claim to be based on lost or even suppressed knowledge which provides yet a further source of fascination. There are hints and even outright claims of some sort of conspiracy to suppress such knowledge. Ultimately pre-Columbian contact theorists and their adherents can believe that they are embattled intellectual heroes. Since it is difficult, if not impossible, to disprove a secret conspiracy (it is, in essence, a nonfalsifiable claim), adherents are fairly safe in their belief.

Sadly, there is also an element of racism inherent in many of the theories of pre-Columbian contact. The 19th-century supporters of the theory of a lost white race of moundbuilders were basically denying that the Native Americans possessed the ability to create a higher civilization. But any modern theory that attributes the fundamental development of higher civilization in the Americas to visiting Egyptians, Hebrews, Phoenicians, Romans, Africans, Chinese, Japanese, or some other ancient Old World peoples is also unfairly downplaying the manifest creativity and intelligence of the Native Americans. Such theories ignore a substantial archaeological record which fully documents the achievements of the Native Americans. Too many theorists of pre-Columbian contacts have their own racial or ethnic agenda which ignores the legitimate achievements of the pre-Columbian

Native Americans and is insensitive to the feelings of their descendants.

In spite of their logical and scholarly problems, theories about pre-Columbian contacts between the Americas and the Old World continue to thrive, while books supporting those theories are steadily proliferating. Sloppy and inappropriate methodologies and inadequate or non-existent evidence have never stood in the way of the concoction or the survival of the most preposterous theories about pre-Columbian contacts. Just in the past few years several new books concerning this realm of pseudohistory have appeared or are scheduled to appear. In 1992 two books appeared which surveyed the whole gamut of theories about pre-Columbian contacts: Patrick Huyghe, *Columbus was Last: From 200,000 B.C. to 1492, A Heretical History of Who Was First* (Hyperion), and Gunnar Thompson, *American Discovery: The Real Story* (Misty Isles Press). Apparently the various theories of pre-Columbian contacts can mutually coexist in relative peace with each other, at least in the pages of these two tomes. Meanwhile in the same year R. J. Jairazbhoy published *Rameses III: Father of Ancient America* (Karnak House) which continues his earlier efforts to establish the role of travellers from ancient Egypt in the rise of higher civilization in the Americas. Publication of two additional books is expected at any time. Ivan Van Sertima, the author of *They Came before Columbus: The African Presence in Ancient America* (Random House, 1977), is supposed to be close to publishing *African Voyages before Columbus.* Even more imminent is Jim Bailey's *Sailing to Paradise: The Discovery of America in 5,000 B.C.* (Simon & Schuster, forthcoming) which appears to extend the theories he first put forward in *The God-Kings and the Titans: The New World Ascendancy in Ancient Times* (St. Martin's, 1973). But in spite of all the hype, these books are all plowing or will be plowing the same old, tired, and infertile fields of evidence. It is truly a never ending story.

References:

Nigel Davies, *Voyagers to the New World* (1979).

Kenneth L. Feder, *Frauds, Myths, and Mysteries: Science and Pseudoscience in Archaeology* (1990).

Ronald Fritze, *Legend and Lore of the Americas before 1492.* (1993).

Robert Wauchope, *Lost Tribes and Sunken Continents: Myth and Method in the Study of American Indians* (1962).

Stephen Williams, *Fantastic Archaeology: The Wild Side of North American Prehistory* (1991).

Pseudoscience and the Paranormal

JAMES RANDI

I am in a very peculiar business. I appear on stages around the world as a conjurer. The American term for it is magician. It's not a good expression because if you look in the dictionary the strict definition of a magician is one who uses magic. And magic, at least by the definition I prefer from a leading dictionary, is the attempt to control nature by means of spells and incantations. Now, ladies and gentlemen, in my time, as you might have guessed, I have tried spells and incantations. No good. You can spell and incant all you want; the lady will still be on the couch, waiting patiently to float into the air, or will be imprisoned in the box with the saw blade descending upon her unprotected midriff, and in some danger of being severely scratched, if not worse! Spells and incantations don't work. You have to use skulduggery. And let me make it very clear what the magical trade—the conjuring trade—is with a precise definition. It is the approximation of the effect of a true magician using means of subterfuge and trickery.

The magician, in the American usage, is an actor playing the part of a wizard. We are entertainers. I don't think that there are many folks—but there are some out there by David Copperfield's own admission to me—who still believe that they really can do the things they purport to do. After a magical performance we've all undergone the same experience, all of us in the trade; you get people coming to you afterwards and saying: "I really enjoyed what you did; thank you so much for coming." And you say, "Well, it's great to be here. I'm happy that you were pleased with it." Then they say, "You know, the business with the bottles that multiplied. Obviously, that's a trick. And the one where you did the thing with the rings and the ropes. That's a trick too. But the one where you told the lady what word she'd chosen out of the newspaper—that, of course, can't be a trick." I'd say, "Yes, that's a trick, too, but it's disguised as a miracle of a semi-religious nature." And they wink at you and they say, "Sure." Then they walk away and tell their friends afterwards, "Well, he won't admit it, but we all know."

There is a hunger, a very strong hunger, within us all to believe there is something more than what the laws of nature permit. I'm not just saying audiences that watch the magician. I mean within us all. We'd like to have a certain amount of fantasy in our lives, but it's a very dangerous sort of temptation to immediately assume that it must be supernatural or occult or paranormal if we don't have an explanation for it. I can tell you that in my life I've spent a great deal of time investigating and observing and carefully noting and making use of psychology. I am not a psychologist; I have no academic credentials whatsoever, so I come to you today absolutely unencumbered by any responsibilities of that nature. There is no dean who will call me on the carpet tomorrow morning and say, "You shouldn't have said that." You see, I'm in the business of giving opinions from an uninformed point of view, except from the point of view of a skeptical person who knows how people's minds work and often don't work.

Historians of science have calculated that at the current rate of scientific growth, in a certain number of years scientists will consist of every human being on earth, as well as all the animals—the donkeys, the burros, the whole thing. Well, my friend David Alexander remarked to me, in a cruel aside, that even today certain parts of certain horses have become scientists. And that is quite true; I have met many of them and though they have Ph.D.s, you'd hardly know it. I've just come back from a project that's ongoing at the moment and I've seen that principle at work. I must share with you another thing in passing. I have a theory; this is only a theory, and it is at present unproven. But observations so far tend to support its possible validity, with my advance apologies to Ph.D.s in the room. I have a theory about Ph.D.s and the granting of the degree itself. I am outside the field, not an academic, so as a curious observer I have many times seen films of, and in a couple of cases actually attended, ceremonies where Ph.D.s are created. They are created, you know. The Ph.D. itself is earned, of course, but then the person who has passed all the tests and done all the right things in the right way and has been approved doesn't become a Ph.D. until one significant moment where a roll of paper, usually with a red or a blue ribbon around it, is pressed into his or her hand. At that moment that person becomes a very special class of being known as Ph.D.

Now, I have noted at those ceremonies, and perhaps you have observed it as well, that the man who gives out those rolls of paper wears gloves. Why? Why would he want to wear gloves? Is the paper dirty? I don't think so. Is there something about that roll of paper, or perhaps the ribbon, that he doesn't want to contaminate him, and he doesn't want to touch his skin? I'm going to postulate—just an idea—that perhaps there is a secret chemical that has been genetically engineered which is on the surface of that paper so that when the Ph.D. candidate receives that roll of paper this chemical is absorbed by the skin, goes into the bloodstream and is conducted directly to the brain. This is a very carefully engineered chemical which goes directly—please don't laugh; this is science—goes directly to the speech center of the brain and paralyzes the brain in such a way that two sentences from then on, in any given language, are no longer possible to be pronounced by that person. Those two sentences are "I don't know" and "I was wrong."

I honestly don't know about that; however, my observations of the situation are that I have never heard any Ph.D. utter either one of those sentences. I have never heard them say, "I'd like to marry a lobster" either, but that doesn't mean they can't say it. But those two sentences never seem to pass their lips.

I am being exceedingly facetious, of course. I have every respect not only for science, but for those who pursue the various disciplines of science. It takes a great deal of courage, application, study, sacrifice, and in many cases, some outrageous attacks on your integrity and your ability in order to maintain a point of view in science which may or may not be popular. I have been with many prominent scientists who have, from time to time, had to stick their professional necks out, and sometimes their necks get pretty badly beaten up in the process. It's not an easy thing to speak against what is generally accepted.

What then is generally accepted? I'm afraid, due to the media impact on our civilization, that a great number of things are easily swallowed because they are repeated so often. They are endlessly presented to the public, and eventually make their way to the academic community as well. Any number of times I have spoken to scientists who, when I ask them a critical question about some belief in some sort of parapsychological, supernatural or occult claim, have said, "You know, I hear a good deal about it and Professor so-and-so did make a statement about it. Perhaps, Professor so-and-so, based upon the small amount of data which he has presently gathered, compared to what should be gathered, in order to establish a satisfactory statistical picture, an amount of data on which conclusions could be drawn by one of the various statistical pictures available to him, has come upon conclusions that are prematurely expressed. Therefore, furthermore, and moreover, on further examination. . . ." That's the academic's reply. When they ask me, I simply say, "In my layman's non-academic opinion, I think that Professor so-and-so is not rowing with both oars in the water." It's simple, direct, and an honest expression of my opinion.

I am presently faced with a situation, again unnamed, where I am going to have to show a number of dedicated, honest, hard-working people that they have made a colossal error of judgment. I have to do this in a resounding manner, simply because to not do so could result in a great deal of personal damage, grief, and considerable heartbreak and discomfort to a great number of people who are already laboring under certain disadvantages and burdens that they did not bring upon themselves. I hate to be so mysterious about it, but it is an ongoing work of investigation. I am not often involved in that serious a situation. Usually my circumstances are more open—I am looking into an astrologer's claims or into some sort of pseudoscientific thing. But I always have to remember an experience that occurred to me.

It is easy, when faced with an apparently supernatural phenomenon, to say "I guess it's a ghost" or "It must be paranormal," or "It could be poltergeists," and we walk away from it because we can't or won't look a little further into it. Some years ago, when I lived in New Jersey my house was a sort of a wayside stop for itinerant magicians, conjurers, mountebanks—various characters of ill repute who would come by to visit for a while. One time I came home after a couple of days away, very tired, and came in on my foster son, Alexis, who was in the kitchen helping a couple of magicians drink up the beer. I walked in and said, "Guys, I'm very, very tired; I'm going to bed. I'll see you in the morning."

I guess they carried on until late that night. I fell asleep, woke up the next morning, came staggering into the kitchen in time to see them eating up more of my groceries in the form of breakfast at this time. I sat down, got a half cup of coffee into me and straightened up the table. Alexis looked at me and said, "What's with you?" I said, "I think last night I might have actually had a classic example of the O.B.E." That's the out-of-body experience. It means that somehow you find yourself out of your body and looking down on it or from a distance. Alexis looked at me and said, "Sure. You?" I said, "Yes, I have to be honest. It appears to me as if I did undergo such an experience."

"OK, give us a description," they replied. The two magicians at the table leaned closer over their bacon and eggs and wanted to hear what I had to say. "Well, I remember waking up in the middle of the night—I couldn't get to sleep at first because I was so tired, so I turned on the television. The program went on and on and I eventually fell asleep. I remember waking up in the middle of the night, and I felt that I was spread-eagled against the ceiling of my bedroom, looking down at the bed. Alice, my black cat, was curled up in a ball in the exact center of the bed so that I had to be way over to one side. And I was, of course, trying not to disturb the cat! As I was up against the ceiling I noticed that the room was lit in sort of a grayish light. I looked down toward the television set and saw nothing but static on the screen and heard nothing but white noise. What I saw was startling. I saw myself, in bed, scrunched over to one side, a chartreuse bedspread on it, with Alice the cat in the middle. I noted that as she opened her eyes they were green. It almost looked like two holes punched through her head. She looked at me and went, 'hmmph' and went right back to sleep."

Now that was a very strong experience for me. I really believed, from the evidence presented to me, that I had an out-of-body experience that matches the description that we've all heard about so many times. But, fortunately for me, I'm not really dead-set against having my belief structure disturbed or having new facts come in that would disturb my previous convictions. And, fortunately, I am able to tell you what actually happened. Alexis looked at me and said, "I've got two things to show you." He went to the foot of the stairs and came up with a big, transparent laundry bag. He had taken it half way down to the laundry room. He brought it all the way up the stairs, and inside noted sheets, pillowcases, and the chartreuse bedspread. He said, "That's been there since yesterday." The bedspread hadn't been on the bed last night! I dashed to the bedroom door, looked in, and the spread I used when the other one was in the laundry lay on the bed. They looked nothing alike. Alexis then called my attention to the patio, noting that he had put Alice outside yesterday afternoon because one of the magician guests was highly allergic to cats. She had remained outside, very unhappily, through the night and into this morning. She could not have been curled up in the middle of the bed last night.

It was a dream—a hallucination, if you will. I had two very good pieces of evidence that it could not have happened. That's important in that if I did not have one or both of those pieces of evidence, I would now have to say to you that, to the best of my knowledge, I had an out-of-body experience. But, all the other out-of-body experiences we hear of, we have to wonder. Those folks are not quite as skeptical about the subject as I am, in most cases. If they don't have some convincing evidence to the contrary, what's to stop them from saying, "I'm absolutely certain I've had an out-of-body experience?" There is no other explanation for it except the possible and rather parsimonious conclusion that they were either dreaming or had a hallucination. It might have been a bad pork chop, for all we know. Please consider that carefully, and don't forget it, because it's a good example of how even the arch-skeptic could possibly have been taken in.

I have had a number of small experiences like that, including the déjà-vu type experiences that so many people have had. (I love the line from the fellow who says, "I keep having the same déjà-vu, over and over again.") But I have resisted the temptation to merely say, "Well, at last I've got proof of it." I'm highly skeptical, but what is that skepticism based on? If you're skeptical as well, have you asked yourself, "Upon what do I base my skepticism?" Are you just plain ornery? Do you just not want to go along with the status quo? Do you know some people who believe in it who are really pretty dense and you don't want to join their group?

You must have a reason, I think, for yourself and for others as to why you are skeptical. These things are not likely to be true; therefore, you need proof of them. We're not required to prove a negative; we can't do that. I can't prove telepathy doesn't exist. I remember getting a question years ago. A lady stood up in the audience and said, "Can you prove to me that ESP doesn't exist?" I said, "No, I can't." She sat down with her arms folded and replied "Ah ha." That was a victory for her. I went on to explain that I can't prove a negative. My question is, "Do you believe in it?" She said, "Absolutely." I asked if she could prove that it is so. She said, "Well, I'm quite convinced of it."

"That's not my question," I responded. "Can you prove that it is so? You're the one making the claim." We skeptics, as Michael Shermer clearly pointed out, are not in the business of debunking. If I were in the business of debunking, and I've often had that label pinned on me and I've always resented it and denied it—it means I would go into an investigation convinced that "this ain't so and I'm going to show you that it isn't." I'm not a lawyer; I don't have an advocacy position to take. I go into a situation as an investigator. To be perfectly fair, I can't prove a negative, but I go into this thing prepared to be shown. Am I prejudiced against it? Oh, yes! I have to admit that. But if you've been sitting by a chimney for 63 years on the evening of December 24 and a fat man in a red suit has never bounced down that chimney, you can say, "One hundred percent of my evidence shows me that this claim is not necessarily so. I cannot prove that it isn't, but it's not very likely to be true, based on what we know."

The Santa Claus example may seem trivial and a little inappropriate, but it is actually a good metaphor for so many paranormal and pseudoscientific claims. Another is flying reindeer. This one we can actually test. (Please don't tell the SPCA about this.) I don't really want to do the experiment, but let's walk through it as if I were doing it. It's a thought experiment. Let's select, by some randomizing process, a thousand reindeer. We'll number them and get them all together in a reindeer truck (I don't know what you put reindeer in) and take them to the top of the Empire State Building in New York. We are going to test whether or not reindeer can fly. You have your reindeer all lined up, a video-camera operator standing by, lots of pads of paper and pens at work. The time is now ten past ten in the morning. OK, first experiment. Number one reindeer, please, up to the edge. Camera going? Good. Push. Uhh, write down "no." Really NO! Number two. Push. I don't know what the result of the experiment will be; I suspect strongly what it will be, based upon my meagre knowledge of the aerodynamics of the average reindeer, though I'm not an expert on it. But based upon previous accounts of what reindeer can and cannot do, I think we are going to end up with a pile of very unhappy and broken reindeer at the foot of the Empire State Building. And probably a couple of policemen will be standing by a squad car saying, "I don't know, but here comes another one."

What have we proven with this experiment? Have we proven that reindeer cannot fly? No, of course not. We have only shown that on this occasion, under these conditions of atmospheric pressure, temperature, radiation, at this position geographically, at this season, that these 1000 reindeer either could not or chose not to fly. (If the second is the case, then we certainly know something of the intelligence of the average reindeer.) However, we have not, and can not, prove the negative that reindeer cannot fly, technically, rationally, and philosophically speaking. People will often look at this example and say, "Well, how many reindeer would you have to test?" I'm not going to get into the statistics of the argument; I will only tell you that you cannot prove a negative. The other folks who claim that something is so are required to prove it. It is what we call the burden of proof. In this case, if it's so it's very easy to prove. Just show me one flying reindeer. Then they rationalize, saying, "Oh, no. It's only the eight tiny reindeer that live at the North Pole who can, and will, on the evening of December 24, fly to do that specific job." In that case you have to throw up your hands and say, "Well, I don't think your hypothesis is very testable." Don't spin your wheels!

Thought experiments like this one only go so far. As an example of a real experiment testing unusual claims, I just came back from Hungary where I was invited to Budapest by the Academy of Sciences. They are very concerned about the fact that now that many of these countries are freed from the burdensome and onerous yoke of Communism and have the freedom to receive all kinds of scientific information in the form of journals and lectures, nonsense comes in as well. The astrologers, the faith healers, the ESP artists, the people with the pendulums, the water dowsers—they're lined up and pouring across the border because they see a new market. The scientists of Hungary were concerned about this. A well-known member of parliament who is also a well-respected brain scientist of international repute said to me, "Mr. Randi, have you seen any of the publicity on the magnetic ladies?" I had.

In case you're not familiar with the magnetic ladies of Hungary, I will relieve you of that ignorance immediately. You may have seen a picture that made all the wire services in this country last year, of the magnetic man from (then) Leningrad. There was a picture of a middle-aged man standing like this, naked from the belt up, with a flatiron stuck here, a hammer there, nails, razor blades—all kinds of metal clinging to his body. The caption said that he attracted these things. They just jumped, willy nilly, onto his body. He was somehow magnetic. I'll bet his wristwatch was a mess! Don't bring him near your computers! I can just imagine him going through a steel door. Bam! Right into it!

Well, I took that with the proverbial grain of salt about the size of a basketball, and just put it in the scrapbook and forgot about it. But the professor asked me about the magnetic ladies of Hungary, and he said, "Their reputation is such that objects, not necessarily metallic ones, cling to their bodies with such tenacity that a strong man cannot tear them loose." Now, wait a minute! Suppose you have some instant glue, and we take a tennis ball and stick it on the lady's neck, on the side. If a strong man can't tear it off, he's going to tear her skin off—or her head! Something has to give! My scientist friend and sponsor of this trip looked at me and said, "How can they make claims like this?" I said, "Well, show me the magnetic ladies." He said that the following day, after the press conference, they were scheduled to arrive. I could hardly wait.

One of the parapsychologists had suggested that he could bring me some instrumentation for detection of their magnetism. He promised that we'd take the two ladies down to the laboratory (hand in hand, clinging to one another, no doubt). I declined to go to the laboratory because the laymen reading the report in the newspaper wouldn't understand "laboratory." What are you going to do? Put a cyclotron on her ear? No. I equipped myself with a scientific device and I went along. The device was called a compass. It's a scientific instrument and an easy way to perform the test. If a woman is magnetic, the compass is going to point right at her. The two ladies showed up. I told my friend in advance that he must understand that the claim is one thing; the event itself will often be something totally different. It won't be half as entertaining or amusing, or true, as the actual demonstration.

One lady literally did this: she took her wristwatch off her wrist and did this. [Randi put his watch on his forehead and it stays there without falling.] "How do you explain that?" she said. I looked at both ladies, who were wearing very greasy, high gloss makeup. It was obviously sticky, mixed with a little perspiration. She said, "We have no explanation for it." I said I didn't find it terribly difficult to explain.

The second lady had an even better demonstration. She took a small ceramic saucer from her purse, stuck it on her forehead, where it remained. "And how do you explain that?" she parroted the other woman. I pulled it off her forehead, and stuck it on the foreheads of the first four people standing on my right. It stuck very effectively to the foreheads of all of them! Then we tested under controlled conditions. (By the way, the compass test failed miserably; it pointed to North, obstinately refusing to point at them.) I asked for soap and water and, through the interpreter, asked the first lady if I could wash her forehead to remove the makeup and any perspiration that might be there. She informed me that if she washed her forehead it wouldn't work because water is absorbed into the skin and water and electricity, or magnetism, don't mix. She denied that that would be a satisfactory test, as did the second lady, and they left. "You've learned your first lesson in the scientific investigation of unusual claims," I told the Professor. "Don't start to give theories on how it might work until you've seen whether it meets the claims of the newspaper account, or is really something much less impressive."

To be fair to these women, I can see how that account might have ended up in a newspaper. I'm sure those ladies didn't say to the newspaper reporters, "A strong man can't pull it away from me." But a reporter is a human being and maybe his story doesn't look all that great when he writes that things cling to their bodies. Then perhaps he thinks: "Um, how about 'with such tenacity that a strong man can't pull them loose'"? Now, he has a story! What I'm saying is that the media are as much to blame for the spread of nonsense and pseudoscience as the claimants themselves. For example, a few years ago the *New York Daily News*, on page three where they put the "heavy news" and sensational stuff, announced that a student at Duke University had successfully in great detail described not only an aircraft accident 24 hours in advance of the event, he even gave the number of people who would be killed. He was short by only two. He even described the location of the crash in the Canary Islands. That was picked up by news services and was featured on television programs; it was on every newscast for quite some time. It was received by the press as a genuine example of prophecy, and the director of the program in which this was involved at Duke University actually made a statement that he had a sealed envelope 24 hours before in his safe that was not touched by this gentleman until after the episode had taken place. It was allegedly torn open at that time and it contained the prediction.

To explain this phenomenon I will take you into a different world, for just a moment, so you will understand something. Magicians know how this young man could very easily have done this trick. I won't go into all the details; you can imagine some of them yourself. But the effect is exactly as described—a sealed signed envelope, put into a safe, later carefully opened. Inside you either find a tape cassette or a sealed letter with all kinds of security on it, signed, maybe genuinely notarized, as of the day before. It contains the prediction. A miracle of a semi-religious nature? No, it's a trick. It can be done by any good magician.

Now, let us return to article in the *Daily News*, first edition. It came out in the afternoon. It had the story on page 3, and a box in the middle of it describing the mechanics of how it had been locked up in a safe and it had a final paragraph which quoted the student at Duke University who made the prediction with this disclaimer: "It's all part of the publicity for my magic show, which is happening tomorrow night. Don't take it seriously." The second and third editions of the *Daily News* had everything except that one sentence.

Another example of how the media distorts claims comes from a young fellow who lived a few doors down from me when I was living in Rumson, New Jersey. He was one of the local characters who did adventurous things like going out on rafts and sailboats. I thought he was a nice kid. One day on the front page of the *New York Times* there was a little box showing a map of the Bermuda Triangle with a Maltese Cross on it. The headline read: "Rumson boy lost at sea in Bermuda Triangle." I read the short article, continued on another page, which said he had taken off in his one-man sailboat, sailed into the Triangle carrying a radio transmitter, and hadn't been heard from since. The Coast Guard was searching for him. No sooner had I finished reading this and had called some friends in New York to tell them about it, I went out to pick up the mail and to my shock there was this kid waving hello to me. He was perfectly all right! I said to him, "You're in the *Times* this morning." He said, "Yeah, they picked me up late last night and they brought me in. They want me to be observed in the hospital, but I feel perfectly all right. We had a bit of a storm; I lost the radio overboard and they finally picked me up very, very early, around 2:30 this morning and they flew me in."

Though the first story made a big splash, the followup never appeared in the *New York Times* or any other paper of which I know. It's still part of the mythology about the Bermuda Triangle. So far as we know, from reading accounts that were published in newspapers, that kid is still someplace out in a sailboat in the Bermuda Triangle or perhaps taken off to Mars. You've got to learn that newspaper editors and reporters are subject to the same kinds of pressures that we all are. We all want something successful. Often the choice is between a story and a non-story. We have to realize that we cannot depend on the media to always represent the facts as they actually are. That's not a great deal of news to you, but you must bear it in mind at all times. Be careful about accepting what appears in print; don't let them say to you, "They put it in the paper; it must be so," or "Someone wrote a book on it. It must be so."

Furthermore, books are often published which, before they actually reach the stands and are on sale, have been completely refuted because they are based on false information. Do they immediately withdraw them? No. I'll give you a good example. There's a book called *Learning to Use Extrasensory Perception*. It's published by Charles Tart, a respected psychologist at University of California, Davis. I heard Dr. Tart give a talk in Casper, Wyoming. I'm going to tell you exactly what he said and see if your reaction is the same as mine. I recorded it on tape so I know exactly the words he said; this is not a case of interpretation or faulty recollection. He said, in speaking to the audience:

> There was a time, years ago, when I was highly skeptical of any paranormal claims of any kind. One of the things that convinced me that there must be something to this is a strange experience that I personally went through. It was wartime. I was at Berkeley, California, and everybody was working overtime. We worked until very late hours of the night and the young lady who was my assistant at the time worked with me until very late this one night. She finally went home; I went home. Then the very next day she came in, all excited. She reported this event. It was wartime; they did work overtime. They often were very, very tired when they went home. It was understandable they would fall into a deep sleep and get as much sleep as they possibly could during the night. She reported that during this night she had suddenly sat bolt upright in her bed, convinced that something terrible had happened. "I had a terrible sense of foreboding," she said, but she did not know what had happened. "I immediately swung out of bed and went over to the window and looked outside to see if I could see anything that might have happened like an accident. I was just turning away from the window and suddenly the window shook violently. I couldn't understand that. I went back to bed, woke up the next morning and listened to the radio." A munitions ship at Port Chicago had exploded. It literally took Port Chicago off the map. It levelled the entire town and over 300 people were killed. Whether it was an accident or sabotage, no one ever found out. She said she had sensed the moment when all these people were snuffed out in this mighty explosion. How would she have suddenly become

> terrified, jumped out of bed, gone to the window, and then—from 35 miles away, the shock wave had reached Berkeley and shook the window?

Indeed, she remembered looking at the clock to see what time it was—right to the minute. Well, when I heard this, I said to myself, "There's something wrong here." I see a couple of smiles around the audience; maybe you've spotted the same thing I did. I had a geologist friend sitting three or four seats away; I handed him a note. He winked, smiled, got up and left the room. He came back in, handed it to me, and it just said on it, "8 seconds." What question did I ask him? [Answer from audience: What is the difference in time of propagation over a distance of 35 miles of a shock wave through the air, compared to a shock wave through the ground? The difference is 8 seconds.] So 8 seconds before that window shook, she had been startled by the room itself shaking, not by the airwave, but by the groundwave. My theory is this: the groundwave which shook the bed startled her, she swung out of bed, went over to the window, looked outside, didn't see anything, went to turn away from the window and suddenly the pane shook in front of her.

The next morning I went to Professor Tart where he was having breakfast by himself. I had known him through correspondence and phone conversations but had never met him personally. I went over, introduced myself, sat down for a moment and gave him this bit of theory. I said there would be 8 seconds difference in the time. He didn't look up from his scrambled eggs for the longest time. Finally, when he did, he smiled and said, "Mr. Randi, that may be the explanation that you prefer." I think he had just decided that he wasn't going to entertain that idea very solidly. But I don't know that he ever made that statement subsequent to that, so maybe he did come to the conclusion that what I offered as an explanation was more likely to be true. But it is so typical of the field! Again, I'm involved in some stuff that I can't tell you about, and I apologize for that, where I have a number of prominent scientists who are absolutely ignoring, refusing to look at very good evidence in this case that I'm investigating. They can come up with rationalizations for it that you wouldn't believe, unless you've been through this process before. It is incredible how they can ignore good evidence to show that there is a prosaic, rational, and very probable explanation for what they are observing.

I want to close this presentation with some parallel examples of scientific claims that turned out to be so much nonsense. Let's go back to 1903 in France. You may have heard of this; if not it really is something you should look up. A prominent scientist—a physicist named Rene Blondlot—startled the world of science with his announcement of the discovery of N-rays. A very well respected man who had won many prizes in science and justifiably so, he was doing experiments by today's standards that were very simple—such as finding the speed of electricity in a conductor. It sounds easy today, but in those days it was a very sophisticated experiment and not all that easily done. Blondlot was in his 70s at the time when he discovered N-rays, named after the town of Nancy, where he was head of the Department of Physics at the University of Nancy.

What were N-rays? N-rays were allegedly radiation exhibiting impossible properties emitted by all substances with the exception of green wood (wood not dried out) and anesthetized metal. (Metal that had been dipped in ether or chlorophorm did not give out N-rays!) Within a matter of six to eight months of the announced discovery of N-rays, 30 papers had come in from all over Europe confirming the existence of N-rays. Reports were published in journals despite the fact that there were many laboratories reporting failure after failure in replicating the results. Such acceptance was understandable considering that X-rays, which also exhibited unsuspected properties, were by then firmly established.

What Blondlot had was a basic spectroscope with a prism (not glass, but aluminum) on the inside, and a thread. The narrow stream of N-rays was refracted through the prism and coming out produced a spectrum on a field. The N-rays were reported to be invisible, except when viewed when they hit a treated thread (for example, treated with calcium sulfide). They moved the thread across the gap where the N-rays came through and when it was illuminated that was reported as the detection of the N-rays.

Before long N-rays were established as factual. *Nature* magazine was skeptical of the N-rays since laboratories in England and Germany were unable to find them. (Germany had just discovered X-rays the decade before and the French were annoyed that they didn't have a ray.) *Nature* sent an American physicist named Robert W. Wood from Johns Hopkins University to investigate. Now, I've been accused of skulduggery in my time, but what Wood did was brilliant. When no one was looking he removed the prism from the N-ray detection device and put it in his pocket. Without the prism the machine could not possibly work because it was dependent on the refraction of N-rays by the aluminum-treated prism. Yet, when the assistant conducted the next experiment he found N-rays! He swore they were there.

When the experiment was over Wood knew it was really over. He was prepared to make his report, and when he went to replace the prism back in the machine, one of the other assistants saw him do this and thought he was actually removing it, and he decided to show Wood up. Thinking Wood had removed the prism (when he had actually put it back), he set up the experiment, could find no lines, and opened the box to show that the prism was not there and to his dismay, there it was! The whole incident blew up. Papers were withdrawn, those that were in the mail were retracted, and N-rays disappeared from the scene.

How did this happen? How did over 30 papers get published? Not because the scientists who wrote the papers were stupid. Not because they were lying. But because they were deceiving themselves. Irving Klotz made this observation in *Scientific American:*

> According to Blondlot and his disciples, then, it was the sensitivity of the observer rather than the validity of the phenomena that was called into question by criticisms such as Wood's, a point of view that will not be unfamiliar to those who have followed more recent controversies concerning extrasensory perception.

By 1905, when only French scientists remained in the N-ray camp, the argument began to acquire a somewhat chauvinistic aspect. Some proponents of N-rays maintained that only the Latin races possessed the sensitivities (intellectual as well as sensory) necessary to detect manifestations of the rays. It was alleged that Anglo-Saxon powers of perception were dulled by continual exposure to fog and Teutonic ones blunted by constant ingestion of beer.

Yet science does not always learn from these mistakes. Visiting Nancy recently and speaking on the subject of pseudoscience, I discussed this example and though I was in the city that gave the name to N-rays, no one in the audience had ever heard of them, or of Blondlot, not even the professors from the University of Nancy!

Now let's go to modern Germany, after the fall of Communism, and compare N-rays to the newly discovered "E-rays." They are actually called Erdestrallen, or "Earth-rays," but I've gotten the media all over the world to call them E-rays, a sort of parallel to N-rays. E-rays are even sillier than the N-rays. What are they? First of all they cannot be detected by any known means, except by water dowsers. They cause cancer. They supposedly come from the center of the Earth. The West German government spent over 400,000 marks, or about $200,000, to pay dowsers to go around to hospitals that were federally funded and federal office buildings to move beds and desks that were in the way of these deadly E-rays. I offered to go over for nothing and conduct a very simple two-part test: 1. Can one dowser find the same spot twice? and 2. Can two dowsers find the same spot once? I told them about this and their response was, "We don't need to do the experiment because we know dowsing works. It's been around since the Middle Ages and the historical tradition validates its truthfulness."

I challenge all the dowsers in a similar way. Since 94 percent of the Earth's surface has water within drillable distance my challenge is to find a dry spot! They don't want to do it. Why? Because they only have a six percent chance of success. Dowsing is an idiomotor reaction that is very deceptive. It is an unconscious motion that you cannot detect and it looks for all the world like some mysterious force.

In a similar fashion, a few years ago I was in France investigating the results of experiments done by Jacque Benveniste on water with memory. He managed to get his article published in *Nature*, who put a disclaimer in the middle of the paper that perhaps "vigilant members of the scientific community with a flair for picking holes in other people's work may be able to suggest further tests of the validity of the conclusions." *Nature* sent a team of investigators over to his laboratory, of which I was a part. (The other two were John Maddox and Walter W. Stewart.) We showed that there were serious problems with the protocol, as well as the fudging of data. When controls were tightened, the experimenter could not replicate the results.

Then there is the theory of homeopathy born in the nineteenth century, the brainchild of one Samuel Hannaman. Medicine was in its infancy. Poor people could not afford doctors and recovered more often than the aristocracy who received all sorts of substances, which often killed them. Samuel Hannaman gave sick poor people water, which was suppose to contain a curative agent. Since these people did not go to doctors, they tended to survive, and this supported his belief in the curative power of his special water.

The first principle of homeopathy is that an extract of some substance in water will help cure you. The second principle is that an attenuated or diluted solution will work even better. How diluted were these? If you take a solution and dilute it with 10 parts of water for every one part of itself, you've got what is called a "one solution." If you take one part of that and put it in 10 parts of water, now one part in 100, it's called a "two solution." If you have a "five solution," you have one part in 100,000. When you get to "Avogadro's limit" there is a chance of there being one molecule in the solution. One more dilution and you have one chance in 10 of there being one molecule in the solution. Well, the homeopathy people start off with a solution of 10 to the power of 50 (a one followed by 50 zeros)! Since there are 10 to the power of 23 stars in the known universe, that's what I call dilute. But that's nothing. They go all the way to 10 to the power of 1500!!!

That is so diluted that I could not conceive of what 10 to the 1500 really means, so I called Martin Gardner and asked for an example with which to illustrate it. He called me back and said that an equivalent is to take one grain of rice, crush it up in a teaspoon and dissolve that powder in a sphere of water the size of the solar system, then repeat that process two billion times!! (The technical problems of mixing such a solution are obvious!)

The critical point of homeopathy—the point of all this diluting—is that every molecule of water that comes into contact with the homeopathy water retains the memory of that special water! Thus a little substance can go a long way. I have a simple question from a layman's perspective. Since water has been around for "billions and billions of years," in this process it must have come into contact with every organic and inorganic molecule on Earth. That being the case, why not just give the patient ordinary tap water?

In fact, these homeopathic waters are so diluted that the homeopathy doctors and scientists can't even tell the difference between the water with iron and the water with gold. Come on folks, let's get real. There is no evidence that this stuff works, yet people go right on believing anyway. Here's a typical response, this from a letter written by Boaz Robinzon of the Faculty of Agriculture: "I want you to know that no matter what the *Nature* investigating committee has written, I am still confident that the phenomenon observed is a real and reproducible one and it is only a matter of time until we shall be proven right."

That's a classic example of someone who does not wish to face reality. I've been going around the world telling people to get real for years. That's the peculiar business that I'm in. I shouldn't have to be in that business but someone has to do it. Will it ever end? Probably not, but perhaps with the efforts of the skeptics and scientists we can "dilute" it a little!

Psi and Psi-Missing

TODD C. RINIOLO AND
LOUIS A. SCHMIDT

When investigating group results of extrasensory performance, most researchers find chance results (Kurtz, 1985, 508–9). However, a discrepancy exists, as some parapsychologists consistently find results that statistically vary from chance. Consistent variations above chance (i.e., extrasensory perception) and below chance (i.e., psi-missing) are both interpreted as evidence of a psi mechanism (Rhine, 1952, 91; Schmeidler, 1966, 387). Most skeptics believe this discrepancy exists not because of extrasensory influence, but because of poor experimental controls and/or improper randomization (Marks, 1986, 121; Hyman, 1994, 19). Currently, there exists no scientifically credible evidence that demonstrates a psi mechanism (Krauss, 1998, 51).

In contrast, many parapsychologists attribute the discrepancy in extrasensory performance not to methodological issues, but to differences in "test conditions." Test conditions are hypothesized to influence the outcome of parapsychological studies (Bem and Honorton, 1994, 14–15; Schmeidler, 1966, 387). Bem and Honorton (1994) suggest that "psi performances should covary with experimental and subject variables in psychologically sensible ways" (15). For example, "good rapport" and a confident attitude in the testing environment should facilitate extrasensory perception (ESP), while "negativism" and "hostility" in the testing environment should result in psi-missing (Schmeidler, 1966, 396).

In support of the test conditions hypothesis, parapsychologists have identified variables believed to influence psi performance:

1. Participants with a belief in ESP (sheep) score above chance while skeptical participants (goats) score below chance (Broughton, 1991, 109; Schmeidler, 1943, 212; 1966, 389).
2. Subjects who interact with a "positive" (e.g., friendly and supportive) experimenter show ESP while subjects who interact with a "negative" (e.g., abrupt and unfriendly) experimenter show psi-missing (Honorton, Ramsey, and Cabibbo, 1975, 137–8).
3. Experimenter attitude towards psi is believed to alter performance (Schmeidler, 1997, 83). For example, a psi-positive researcher is more likely to index ESP while a skeptical researcher increases the chances of psi-missing (Rhine, 1952, 108).
4. A reduction in sensory input facilitates ESP (Bem and Honorton, 1994, 5–6) while distractions in the testing environment have been associated with psi-missing (Sharp and Clark, 1937, 136).

5. Implementing scientific controls or subjecting psi phenomena to observation is hypothesized to sometimes cause "stage fright" that results in psi-missing (Rhine, 1952, 108).
6. Marks (1986, 120) reports that some parapsychologists have theorized that the readership of the journal (if skeptical) can alter psi performance through backward causality. Therefore, a psi-positive readership should increase the chances of ESP while a skeptical readership should increase the chances of psi-missing.

While this literature is consistent with the theory that test conditions alter psi performance in predictable ways, support for the test conditions hypothesis is often speculative, can be questioned on methodological grounds, and has not met the scientific standard of independent verification.

Note that the mentioned literature implies that skeptics are not the appropriate experimenters to test for above chance claims of psi (e.g., ESP). One reviewer of Bem and Honorton (1994, 14) worried that this emphasis upon test conditions provides "an escape clause." In other words, if a skeptic were to replicate a parapsychological study and find no variation from chance, results could be dismissed due to inappropriate test conditions. While this literature implies that skeptics are not the ideal researchers to test for ESP, it also suggests that skeptics are appropriate to elicit psi-missing. Thus, one way for skeptics to empirically evaluate the test conditions hypothesis is for skeptics to test for psi-missing.

The purpose of this paper is to evaluate if test conditions can elicit a reliable reduction from chance (i.e., psi-missing) on a card-guessing task. The test conditions implemented in this study incorporate the above parapsychological literature. If test conditions do alter psi-results in predictable ways (and a psi mechanism exists), a psi-hostile testing environment should elicit psi-missing. If psi-missing can be reliably demonstrated with proper research controls (i.e., double blind) and randomization by skeptics, this would strengthen the test conditions hypothesis and help to explain the consistent failure of skeptics to replicate above chance psi findings. In contrast, results consistent with chance would question the validity of the research used to support the test conditions hypothesis.

Methods

Experiment 1 Participants

Our subject pool of 100 females (Mean 21.4 years; Standard Deviation 6.87) and 52 males (M 21.5 years; SD 4.21) met the a priori criteria to participate (i.e., goats) in this study from introductory psychology courses at Adams State College (Alamosa, Colorado). An additional 45 students participated, but were excluded because they believed they possessed ESP.

Procedures. Prior to the experiment, a brand new set of regular playing cards was purchased. The cards were divided into five piles of 10 cards each with 5 red and 5 black cards in each pile. The five piles each were shuffled thoroughly by an independent "goat" blind to the study, and subsequently placed into five opaque envelopes.

Also prior to the experiment, participants were asked (on a handout) if they believed they personally possessed any psychic ability. Participants circled "yes" or "no." Only those who indicated no (i.e., the goats) were included in any subsequent data analysis. Participants were instructed they would be required to carry out two simultaneous tasks to evaluate ESP ability. First, participants would perform a mental counting task (i.e., an environmental distraction). Second, participants would guess the color of the card (red or black) selected by the experimenter from a pile of 10 cards (5 red and 5 black cards). The card was selected as the experimenter simultaneously read numbers at a slow and steady pace and shuffled cards. As the experimenter read the last number, the top card was placed face down and the experimenter stated "Please guess the correct card and total the numbers." Participants were told this process would be repeated 5 times.

The numbers used for the mental arithmetic tasks were randomly generated in length from four to ten total numbers. The numbers ranged from 1 to 10 (randomly determined). Random generation was implemented so participants could not accurately anticipate when the mental arithmetic task would end. A practice example (8 + 1 + 6 + 3 + 9 + 1 "Please guess the correct card and total the numbers") was given. No card was drawn for the practice trial.

No attempt was made to create a "warm" experimental environment and the experimenter behaved in a cold, formal, and impersonal manner. The experimenter also did not believe in psi. The following statements were read by the experimenter immediately before testing:

1. After tens of thousands of experiments, no one has been able to convincingly demonstrate Extra Sensory Perception (ESP). Thus, the majority of psychologists agree that ESP does not exist.
2. The Central Intelligence Agency (CIA) has recently abandoned their psychic spy program after spending millions of dollars and classifying the project as "useless."
3. Currently, a 1.1 million dollar reward exists on the internet (www.randi.org) for anyone who can demonstrate genuine ESP or any other psychic ability. While many people claiming Extra Sensory abilities have been tested, nobody has yet claimed the reward.

The experiment began immediately after these statements were read. The process of reading numbers, shuffling cards, and selecting a card for each trial was repeated five times during the experiment. Only five seconds were given between each trial. During the experiment, at no time did either the experimenter or the participants know the color of the card selected (i.e., a double-blind format). To ensure that no inadvertent cues were transmitted to the participants, the experimenter shuffled and selected the cards behind a podium. The use of new cards and the practice of simultaneously reading the numbers for the counting tasks ensured that the experimenter was also "blind." After the five trials were completed, answer sheets were collected and the experimenter recorded the correct answers for the five trials.

Data Analysis. In order to ensure that participants were attending to the counting task, it was determined in advance that only those trials in which participants had accurately totaled the counting tasks would be included for statistical analysis. The data sheets were "blindly" tallied and loaded into a spreadsheet. The statistical analysis was limited to one a priori test in order to control for statistical errors (Riniolo and Schmidt, in press).

Results. In response to the counting task, participants scored correctly an average of 4.08 (out of five). The criteria indicating psi-missing defined a statistically significant reduction from 2.04 correct identifications (50% of 4.08). A one sample t-test (one-tailed) was performed using a 95% confidence interval. Group results indicated that participants (N=152) identified the card color correctly an average of 2.16 times. Results did not statistically differ from what was expected by chance

(95% confidence a chance result would fall between 2.0128 and 2.3162).

Experiment 2 Participants

Our subject pool consisted of 124 females (M 21.2 years; SD 1.59) and 32 males (M 21.5 years; SD 1.98) from a large introductory psychology course at McMaster University, Ontario, Canada. Of these, 76 were excluded from data analysis because of a belief they personally possess psychic abilities.

Procedures. For the replication study, two minor modifications were made. First, Dr. Schmidt implemented the replication, whereas Dr. Riniolo implemented the initial study. Second, the mental counting task was eased by reducing the length to randomly vary from 4 to 7 numbers.

Results. Participants scored correctly an average of 4.70 (out of five) on the counting task. The criterion to indicate psi-missing for the group average is a statistically significant reduction from 2.35 correct identifications (50% of 4.70). A one sample t-test (one-tailed) was performed using a 95 confidence interval. Group results indicated that participants (N=155) identified the card color correctly an average of 2.32. Results did not statistically differ from what was expected by chance (CI95, 2.1792, 2.4531).

Discussion

The purpose of this paper was to empirically evaluate if psi-hostile test conditions could elicit psi-missing. Results were consistent with chance expectation despite implementing multiple variables previously identified by parapsychologists as increasing the chances of psi-missing. Several interpretations of the results deserve attention. First, our results are consistent with the non-existence of a psi-mechanism. Simply put, test conditions cannot alter a phenomenon that does not exist when proper experimental controls and randomization procedures are implemented. However, because it is impossible to disprove a negative, other interpretations are possible.

Specifically, perhaps the test conditions were not sufficiently hostile to elicit psi-missing, or our experiments lacked adequate statistical power. This is unlikely as parapsychologists have reported psi-missing using much less "psi-hostile" conditions (e.g., Honorton, Ramsey and Cabibbo, 1975, 136–7) and with much less statistical power (e.g., Sharp and Clark, 1937, 136) than implemented here. For example, we analyzed the data provided by Sharp and Clark (1937, 136, Table VI) using the same statistical approach above. Results indicated statistical evidence of psi-missing and ESP (depending on test conditions) with only four and 11 participants, respectively. This inconsistency raises the possibility that inadequate methodology or random error was responsible for previous findings of psi-missing.

In addition, others may argue that the presence of skeptics (i.e., the authors) would not facilitate psi-missing, but rather would inhibit any demonstration of psi performance (both above and below chance). This belief that some researchers are psi-conducive (can find reliable variations from chance) and others are psi-inhibitory (repeatedly find chance results) is an endless cycle that makes psi "untestable" (Blackmore, 1985, 429). After-the-fact explanations to find a psi-inhibitory link responsible for chance results can be invoked endlessly. More important, the scientific standard of independent verification of results is impossible. As psi-research has a long history of fraud (Hansen, 1990, 25) and methodological error (Marks, 1986, 120–1), reliance upon a few "psi-conducive" individuals to establish an extraordinary claim (i.e., a psi-mechanism exists) is unacceptable. Perhaps the only consistent finding in parapsychological research the last 100 plus years is that irrespective of test conditions, when proper methods and randomization procedures are used (by both believers and nonbelievers), participants score at chance expectation over repeated evaluations. Being "psi-inhibitory" may simply reflect the experimenter's ability to prevent bias from influencing testing results.

Our results are inconsistent with the hypothesis that test conditions can alter psi-performance in predictable ways. We are unconvinced of the validity of the test conditions hypothesis that is often used post-hoc to dismiss results inconsistent with a psi-mechanism. Unfortunately, there is a long and continuing history of after-the-fact rationalizations for failures to scientifically demonstrate paranormal phenomena. For example, Kurtz (1985, 180–1) reports that when the Fox sisters could not produce "rappings" during an empirical evaluation in 1851, they claimed the presence of skeptics caused the spirits to retire. Recently, therapeutic touch practitioners attempted to rationalize their failure by questioning the test conditions despite agreeing in advance that the experimental paradigm was fair (Rosa, Rosa, Sarner, and Barrett, 1998, 1008). To our knowledge, there currently exist as many scientifically credible studies supporting the test conditions hypothesis as there are for other psi phenomena—zero.

References:

Bem, D. J., and C. Honorton. 1994. "Does Psi Exist? Replicable Evidence for an Anomalous Process of Information Transfer." *Psychological Bulletin*, 115, 4–18.

Blackmore, S. 1985. "The Adventures of a Psi-inhibitory Experimenter." In P. Kurtz (Ed.), *A Skeptic's Handbook of Parapsychology* (425–448). Buffalo, NY: Prometheus Books.

Broughton, R. S. 1991. *Parapsychology: The Controversial Science*. New York: Ballantine Books.

Hansen, G. P. 1990. "Deception by Subjects in Psi Research." *The Journal of the American Society for Psychical Research*, 84, 25–80.

Honorton, C., M. Ramsey, and C. Cabibbo. 1975. "Experimenter Effects in Extrasensory Perception." *The Journal of the American Society for Psychical Research*, 69, 135–149.

Hyman, R. 1994. "Anomaly or Artifact? Comments on Bem and Honorton." *Psychological Bulletin*, 115, 19–24.

Krauss, L. M. 1998. "May the Force Be with You." *Skeptical Inquirer*, 22, 49–53.

Kurtz, P. 1985a. "Spiritualists, Mediums, and Psychics: Some Evidence of Fraud." In P. Kurtz (Ed.), *A Skeptic's Handbook of Parapsychology* (177–223). Buffalo, NY: Prometheus Books.

——. 1985b. "Is Parapsychology a Science?" In P. Kurtz (Ed.), *A Skeptic's Handbook of Parapsychology* (503–518). Buffalo, NY: Prometheus Books.

Marks, D. F. 1986. "Investigating the Paranormal." *Nature*, 320, 119–124.

Rhine, J. B. 1952. "The Problem of Psi-missing." *The Journal of Parapsychology*, 16, 90–129.

Riniolo, T. C., and L. A. Schmidt. (In press). "Searching for Reliable Relationships with Statistics Packages: An Empirical Example of the Potential Problems." *The Journal of Psychology*.

Rosa, L., E. Rosa, L. Sarner, and S. Barrett. 1998. "A Close Look at Therapeutic Touch." *Journal of the American Medical Association*, 279, 1005–1010.

Schmeidler, G. R. 1943. "Predicting Good and Bad Scores in a Clairvoyance Experiment: A Final Report." *The Journal of the American Society for Psychical Research*, 37, 210–227.

——. 1966. "The Influence of Attitude on ESP Scores." *International Journal of Neuropsychiatry*, 2, 387–397.

——. 1997. "Psi-conducive Experimenters and Psi-permissive Ones." *European Journal of Parapsychology*, 13, 83–94.

Sharp, V., and C. C. Clark. 1937. "Group Tests for Extrasensory Perception." *The Journal of Parapsychology*, 1, 123–142.

Recovered Memory Therapy and False Memory Syndrome

A Father's Perspective as a Test Case

MARK PENDERGRAST

"I can't believe that!," said Alice.

"Can't you?" the Queen said in a pitying tone. "Try again: draw a long breath, and shut your eyes."

Alice laughed. "there's no use trying," she said. "One can't believe impossible things."

"I daresay you haven't had much practice," said the Queen. "When I was your age, I always did it for half-an-hour a day. Why, sometimes I've believed as many as six impossible things before breakfast."

–Lewis Carroll,
Through the Looking-Glass

A few decades ago some bright agronomist imported a nifty Japanese vine called "kudzu" to my native Georgia, hoping to halt erosion and provide cheap cow fodder. The insidious kudzu, with its broad, shiny green leaves, now covers entire forests, swallowing trees whole. While cows may indeed eat the stuff, I suspect a few of them have been enveloped, too, along the way. I have come to regard the initial incest suspicion that fuels the repressed memory movement as being a kind of mental kudzu seed—perhaps a perverse analogue to Jesus' parable of the sower and the seed.

Repressed memories seem to grow in the same way. It does not take much—just a small seed, planted in your fertile brain by a television program, a book, a friend, or a therapist. Maybe, just maybe, all of your problems stem from childhood incest. Maybe you have forgotten it. Maybe that is why you are uncomfortable at family reunions. Maybe. No, no, that's insane! Forget it, not Dad, not Mom! You try to dismiss the idea. But it won't go away. It takes root, sends out creepers, and grows. Soon the mental kudzu is twining out of your ears, sending roots down to your gut, taking over your life. It's true! Your worst fears were justified!

Given that our memories can fool us sometimes, it is still hard to understand why or how people would want to believe that their parents committed such awful acts upon them. Numerous types of "evidence" are used to provoke and "prove" the reality of repressed memories. These include hypnotic regression, sodium Amytal, dreams, visualizations, bodily pangs or marks, panic attacks, or just general unhappiness. Once the seed is planted, once the idea takes hold, it does not matter what method is employed. The results are almost foreordained. I should know. I am a victim of the "recovered memory" movement. I am an accused parent.

Lost Daughters

"Stacey" and "Christina" (my daughters have changed their surnames, and I have changed their first names to protect their identities) are exceptionally attractive, intelligent, creative, caring young women. Both have graduated with high marks from fine Ivy League schools. And both, through therapy, have recently retrieved "memories" of sexual abuse which they think I inflicted on them. I do not know exactly what I am supposed to have done, because they will not tell me. In fact, they do not communicate with me at all, and I am forbidden to call or write.

It all started five years ago, when Christina, my youngest daughter, was in college and went to a counselor. In therapy, she uncovered a repressed memory of being molested by my housemate when she was nine years old. Within the next year, without accusing me, she nonetheless cut off all contact. Then, in the fall of 1992, she apparently "remembered" something terrible I did to her, though she has never directly confronted me. She told Stacey, who in turn entered therapy and wrote me a letter, which began: "I'm sorry if you aren't ready for this letter, but it must be written. I have recently recalled some memories I have of you . . ."

The letter was filled with what I now recognize as recovered memory jargon. I had violated her "boundaries" and made her and Christina my "surrogate wives." I was "abusive" and "manipulative," and had probably been sexually abused myself as a child. "I know what you did to my sister," she wrote. "You have to recall what happened and deal with this on your own." She ended the letter by forbidding me to contact her.

I have not heard from my children in over two years now. It breaks my heart, and I am deeply concerned for them, especially after conducting the research for my book *Victims of Memory*. Though accused parents certainly suffer terribly, I have become convinced that the real victims of memory are the children, who have been sucked into a destructive belief system that strips them of their identities, pasts, and families.

In the rest of this article I will detail a few of the methods used to create such a belief system.

Hypnosis: Memory Prod or Production?

After Stacey wrote me that awful letter, I thought that maybe I really had done something horrible to my children and had repressed the memory myself. So I went to a hypnotist. Like most people, I thought that when you sank into a deep hypnotic trance, you could magically tap into your dormant subconscious, unlocking long-forgotten memories. Fortunately, I went to an ethical hypnotist who did not lead me into believing I had committed incest on my children. She failed, however, to tell me how questionable memories are when "uncovered" in hypnosis. I discovered that fact during my research.

From its inception, hypnosis has caused considerable controversy and spawned innumerable myths. One thing that experts agree on, however, is that memories retrieved under hypnosis are often contaminated mixtures of fantasy and truth. In many cases, outright "confabulations"—the psychologists' term for illusory memories—result.

The reason that memories retrieved under hypnosis are suspect goes to the very definition of the process, which invariably includes the concept of suggestion. Clark Hull and A. M. Weilzenhoffer defined hypnosis simply as "a state of enhanced suggestibility." When a subject agrees to be hypnotized, he or she tacitly agrees to abide by the suggestions of the hypnotist. This state of heightened suggestibility can work quite well if the goal is to stop smoking, lose weight, enhance self-esteem, reduce perceived pain, or improve one's sex life. But it is not an appropriate method for retrieving supposedly repressed memories, as psychiatrist Martin Orne and psychologist Elizabeth Loftus have repeatedly stressed in courtroom settings.

The hypnotized subject is not the only one who is deluded. The hypnotist who believes that he or she is delving for hidden memories takes an active part in the shared belief system. Both hypnotist and subject are engaged in a tacitly accepted mini-drama in which they act out prescribed roles.

I am not trying to imply that "hypnosis," whether a real state or not, does not have a profound effect. The human imagination is capable of incredible feats, and herein lies the potential problem. Similarly, the "guided imagery" exercises that trauma therapists employ to gain access to buried memories can be enormously convincing, whether we choose to call the process hypnosis or not. When someone is relaxed, willing to suspend critical judgment, engage in fantasy, and place ultimate faith in an authority figure using ritualistic methods, deceptive scenes from the past can easily be induced.

Hypnotism entails a powerful social mythology. Just as those "possessed" by demons believed in the process of exorcism, most modern Americans believe that in a hypnotic state, they are granted magical access to the subconscious, where repressed memories lie ready to spring forward at the proper command. Hollywood movies have reinforced this mythology, beginning with a spate of amnesia-retrieval dramas, such as Hitchcock's *Spellbound*, in the 1940s. A good hypnotic subject therefore responds to what psychologists call "social demand characteristics." As psychologist Robert Baker puts it, there is a "strong desire of the subject to supply the information demanded of him by the hypnotist." Psychiatrist Herbert Spiegel says it more graphically: "A good hypnotic subject will vomit up just what the therapist wants to hear."

The hypnotist is often completely unaware that he is influencing the inductee, but what psychologists term "inadvertent cuing" can easily occur, often through tone of voice. "It is incredible," wrote French psychologist Hippolyte Bernheim in 1888, "with what acumen certain hypnotized subjects detect, as it were, the idea which they ought to carry into execution. One word, one gesture, one intonation puts them on the track." Simply urging the subject to "go on" at a crucial point, or asking, "How does that feel to you?" can cue the desired response. A person who agrees to play the role of the hypnotized subject is obviously motivated to believe in that role and act it properly. This goes double for clients in psychotherapy who are desperately seeking to locate the source of their unhappiness. If the therapist has let them know, either subtly or directly, that they can expect to find scenes of sexual abuse while under hypnosis or through guided imagery, they are likely to do so.

One of the characteristics of well-rehearsed hypnotic confabulations is the over confidence with which they are eventually reported. Such memories tend to become extraordinarily detailed and believable with repetition. "The more frequently the subject reports the event," Martin Orne has written, "the more firmly established the pseudomemory will tend to become." As a final caution, he warns that "psychologists and psychiatrists are not particularly adept at recognizing deception," adding that, as a rule, the average hotel credit manager is a far better detective.

Dream Work

Ever since Joseph saved Egypt by properly interpreting the Pharaoh's dreams—and probably long before that—humans have sought deep meaning from the strange stories they picture in their sleep. In our dreams, anything is possible. We can fly, jump through time, read other people's thoughts. Animals can talk, objects appear and disappear quickly, one thing metamorphoses quickly into something else. Sometimes our dreams are exciting, sexy, or soothing. Often, they are bizarre and frightening. What are we to make of them?

No one really knows, not even the most renowned dream researchers who shake people awake to ask what they're experiencing when their REM (rapid eye movements) indicate that they are in an active dreaming state. Some interpreters, including Freud, have asserted with great authority that dream ingredients symbolize certain objects, emotions, or events. For example, a skyscraper represents a penis. In the second century, Artemidorus used the same kind of logic. For him, a foot meant a slave, while a head indicated a father. The kinky ancient Egyptians apparently dreamed frequently of sexual congress with various animals. One papyrus explained, "If an ass couples with her, she will be punished for a great fault. If a he-goat couples with her, she will die promptly."

Modern trauma therapists also use sexual dreams as a form of interpretation. They tell their clients to be particularly aware of any night visions that could be interpreted as sexual abuse. This is called "dream work." Not too surprisingly, such dreams are often forthcoming. "Oh, my God!" the woman reports in therapy. "It's all true! In my dream last night, my Dad and uncle were taking turns having sex with me. And I was just a little kid!" Such dreams are taken as recovered memories and are presumed to represent literal truth, even though some events seem unlikely—in one well-publicized case, for instance, a daughter recalled being raped by her mother, who was equipped with a penis.

But if these dreams don't necessarily stem from repressed memories of actual events, where do they come from? From the same place that spawns hypnotically guided fantasies—the fertile and overwhelmed imagination. Here is someone feverishly working on her memory recovery, reading books describing horrible abuse, her life consumed with the possibility that her father did something to her. As Calvin Hall noted in *The Meaning of Dreams*, "It has been fairly well established that some aspects of the dream are usually connected with events of the previous day or immediate past." It is not surprising that someone with an obsession about incest would dream about it. Hall also warned that "dreams should never be read for the purpose of constructing a picture of objective reality," but therapists and patients eager for repressed memories ignore such advice.

The role of expectation in all aspects of memory recovery is crucial. What we expect to see, we see, as Joseph Jastrow observed in his 1935 classic, *Wish and Wisdom:* "Everywhere, once committed by whatever route, the prepossessed mind finds what it looks for." Elizabeth Loftus tells the true story of two bear hunters at dusk, walking along a trail in the woods. Tired and frustrated, they had seen no bear. As they rounded a bend in the trail, they spotted a large object about 25 yards away, shaking and grunting. Simultaneously, they raised their rifles and fired. But the "bear" turned out to be a yellow tent with a man and woman making love inside. The woman was killed. As psychologist Irving Kirsch notes, "response expectancy theory" explains how "when we expect to feel anxious, relaxed, joyful, or depressed, our expectations tend to produce those feelings." At its extreme, such a mindset can even lead to self-induced death, as has been well-documented among tribes in which those under a powerful curse fulfill it by wasting away and dying, unless some way to reverse the curse can be found.

Similarly, when we expect to have a particular type of dream, we tend to perform accordingly. As Jerome Frank notes in *Persuasion and Healing*, patients routinely give their therapists the dreams they want. "The dream the therapist hears is, of course, not necessarily the one the patient dreamed," Frank explains, "since considerable time has usually elapsed between the dream and its report. One study compared dreams reported immediately upon awakening with the versions unfolded before a psychiatrist in a subsequent interview. Any material the patient anticipated would not be approved was not recalled." In his classic 1957 text, *Battle for the Mind*, psychiatrist William Sargant described an acquaintance who had entered first Freudian, then Jungian therapy. "His contemporary notes show that dreams he had under Freudian treatment varied greatly from those he had under Jungian treatment; and he denies having experienced the same dreams before or since." Sargant concluded: "The increased suggestibility of the patient may help the therapist not only to change his conscious thinking, but even to direct his dream life."

Therapist Renee Fredrickson certainly believes in such directives. "You can also prime your dream pump, so to speak," she writes in *Repressed Memories*. "Before you go to sleep at night, visualize yourself as a little child. Then suggest that your inner child show you in a dream what you need to know about the abuse." Nor does the dream abuse have to be obvious. Fredrickson describes how Diane reported a dream in which "she was on her hands and knees in a kitchen, washing the floor. Floating in the air were green U-shaped neon objects. Her father was standing next to a large mirror over the sink, watching her." Eventually, Diane interpreted her dream as follows:

> My father raped me in the evenings when I was cleaning the kitchen. He would make me crawl around naked while he watched in the mirror. I also believe the green neon things are about a time he put a cucumber in me.

Sleep Paralysis

Another fascinating form of semi-dream, which typically occurs in the twilight state between waking and sleeping, accounts for many "repressed memories." The psychological term is either a "hypnogogic" or "hypnopompic" state, respectively referring to the time just before sleep or prior to waking, but more commonly it is just called "sleep paralysis." During this curious in-between semiconscious state, people often report chilling visions.

Robert Baker describes the phenomenon: "First, the hallucinations always occur just before or after falling asleep. Second, the hallucinator is paralyzed or has difficulty moving. . . . Third, the hallucination is usually bizarre. . . . Finally, the hallucinator is unalterably convinced of the reality of the entire event." The vision's content is often related to the dreamer's current concerns. In one study, as many as 67% of a normal sample population reported at least one experience of sleep paralysis, with its attendant hallucinations. Many people experience sleep paralysis during the day, particularly if they take afternoon naps. Those with narcolepsy—a relatively common disorder characterized by brief involuntary periods of sleep during the day, with difficulties resting at night—are particularly prone to these frightening hallucinations. The word "nightmare" actually stems from sleep paralysis. A "mare," or demon, was supposed to terrorize people—mostly women—by sitting on their breasts, making it difficult to breathe. Often, the mare was a Satanic incubus or succubus who also forced the frightened sleeper into sexual intercourse. The following is a 1763 description of the phenomenon:

> The nightmare generally seizes people sleeping on their backs, and often begins with frightful dreams, which are soon succeeded by a difficult respiration, a violent oppression on the breast, and a total privation of voluntary motion. In this agony they sigh, groan, utter indistinct sounds [until] they escape out of that dreadful torpid state. As soon as they shake off that vast oppression, and are able to move the body, they are affected by strong palpitation, great anxiety, languor, and uneasiness.

David Hufford has written an entire book about sleep paralysis, *The Terror That Comes in the Night*. His 1973 interview with Caroline, a young graduate student, sounds quite similar to the reports of many "incest survivors." When Caroline woke up one day, she reports, "I felt like there was a man next to me with his arm underneath my back, and holding my left arm." His smell was quite distinct, "all sweaty and kind of dusty." When she tried to move, he gripped her arm tighter. "Now if I move again, he's going to rape me," she thought. She tried to scream, but she could make no sound. "Then he was on top of me, and I tried to look up to see who it was or something—I could just see this—it looked like a white mask, like a big white mask." After several minutes of this horrible experience, Caroline "felt sort of released, you know. And I—I could sit up, and I got the feeling there was nobody there." In the 1990s, such experiences are frequently interpreted as "flashbacks" or "body memories," and women are encouraged to visualize a face to fill in the blank mask. Other "evidence" of repressed memories also relates to sleep—or its lack. In *The Courage to Heal*, Ellen Bass and Laura Davis quote one typical woman's experience as she obsessed over possible repressed memories: "I just lost it completely. I wasn't eating. I wasn't sleeping." Sleep deprivation is a well-established technique used in brain-washing. As sleep expert Alexander Borbely writes, chronic lack of sleep blurs the borderline between sleeping and waking, "so that the kind of hallucinations that often occur at the moment of falling asleep now begin to invade the waking state as well . . . the floor appears to be covered with spider webs, faces appear and disappear. Auditory illusions also occur." In addition, "when sleep deprivation experiments last more than four days, delusions can manifest themselves, in addition to the disturbances of perception. The participants grow increasingly suspicious and begin to believe that things are going on behind their backs."

Body Memories and Panic Attacks

People who are trying to recover repressed memories are often told that "the body remembers what the mind forgets," particularly in cases of abuse suffered as a pre-verbal infant. These "body memories" can take the form of virtually any form of physical ailment, from stomach aches to stiff joints. Psychosomatic complaints such as these have always been common in Western culture and almost invariably accompany general unhappiness and anxiety. Add to this the "expectancy effect," and it isn't surprising that during the "abreaction" or reliving of an event, a woman might feel terrible pelvic pain, or a man might experience a burning anus.

Those in search of memories often submit to massages by experienced "body workers," who can trigger feelings either by light touch or deeper muscle manipulation. "An area of your body may get hot or feel numb," Renee Fredrickson assures readers in *Repressed Memories*. "Powerful emotions may sweep over you, causing you to weep or even cry out." It is certainly true that people can experience profound, inexplicable emotions while they are being massaged, particularly if they are tense and unhappy in general. When they let down their guards and relax, allowing intimate touch by a stranger, they often do weep. Given the admonition to be on the look-out for any stray sensation, many subjects have no difficulty locating and interpreting various body memories. Fredrickson gives two examples: "She [Sarah] was undergoing a passive form of body work involving laying on of hands when she had a slowly burgeoning sense of rage at her father for abusing her." Later on, Sarah discovered that the "exquisite sensitivity" of her toes was caused by her grandfather having shoved a wood chip under her toenail.

Some "body memories" take the form of rashes or welts that fit particular memory scenarios. The mind can apparently produce remarkable and sometimes quite specific effects on the body. It has been demonstrated that hypnotic suggestion can actually remove warts, while some people can consciously control their pulse rates, respiration, or blood flow. In *Michelle Remembers*, Michelle Smith evidently possessed similar powers, producing a red rash on her neck that her psychiatrist interpreted as a welt left by the devil's tail.

Nothing so dramatic need account for most "body memories," however. One of the most common was recounted by A. G. Britton in her article, "The Terrible Truth." She experienced a choking sensation and interpreted that as evidence that her father had forced his penis into her mouth when she was a baby. It turns out, though, that a constricted throat is one nearly universal human reaction to fear and anxiety. In fact, the word "anxious" derives from the Latin word meaning "to strangle." This classic symptom—an inability to swallow and the feeling of being choked—is now one of the diagnostic symptoms for panic disorders. For hundreds of years it was called, among other things, *globus hystericus*, because it felt as though a ball were rising from the abdomen and lodging in the throat.

Many people who fear that they may have been abused suffer repeated panic attacks at unexpected moments and, with their therapists' encouragement, interpret them as repressed memories surging forth from the subconscious. Yet these little-understood episodes are extremely common. As psychologist David Barlow points out in his comprehensive text, *Anxiety and Its Disorders*, "Anxiety disorders represent the single largest mental health problem in the country, far outstripping depression." In Western cultures, reports of this affliction are much more common among women than among men, although that is not so in Eastern countries. Recent surveys indicate that 35% of Americans report having experienced panic attacks. Unfortunately, those seeking help for severe anxiety disorders are frequently misdiagnosed, seeing an average of 10 doctors or therapists before receiving appropriate help. As listed in the third revised edition of the *Diagnostic and Statistical Manual of Mental Disorders*, familiarly known as DSM-III-R, the symptoms experienced during panic attacks (four or more being sufficient by the official definition) sound like a check-list for what trauma therapists interpret as body memories:

(1) shortness of breath (dyspnea) or smothering sensations; (2) dizziness, unsteady feelings, or faintness; (3) palpitations or accelerated heart rate (tachycardia); (4) trembling and shaking; (5) sweating; (6) choking; (7) nausea or abdominal distress; (8) depersonalization or derealization (the feeling that you don't really exist or that nothing is real); (9) numbness or tingling sensations (paresthesias); (10) flushes (hot flashes) or chills; (11) chest pain or dis-

comfort; (12) fear of dying; (13) fear of going crazy or of doing something uncontrolled.

Surprisingly, Barlow reports that "the overwhelming evidence is that many phobias and the majority of fears are not learned through a traumatic experience." Instead, panic attacks appear to stem from contemporarily stressful life situations and a fearful mindset—though biological factors and early childhood trauma may contribute to a predisposition to anxiety disorders. Psychologists Aaron Beck and Gary Emery give an example of a typical episode involving a 40-year-old man who, while on the ski slopes, began to feel shortness of breath, profuse perspiration, and faintness. He thought he was having a heart attack. In the midst of this, he had a vivid image of himself lying in a hospital bed with an oxygen mask. It transpired that this man's brother had just died of a heart attack, and he feared the same might happen to him.

Similarly, people who think they may have repressed memories fear that they may be like others they know (or have read about or seen on television). They, too, may be unknowing incest victims who will have flashbacks. For such people, panic attacks are often triggered when they become over-tired or over-stressed and spontaneously envision images of their worst fears, which, in turn, provoke even more anxiety. "Once the fear reaction has started," Beck and Emery write, "it tends to build on itself." These "autonomous" images then persist "without the patient's being able to stop them," and they seem utterly real, "as though the traumatic episode were actually occurring in the present."

After the first attack of this inexplicable fear, a vicious cycle can commence in which the very fear of another episode provokes it. This would be particularly likely for a woman who is extremely stressed by the idea that she might have been sexually abused and is minutely aware of every bodily and emotional twinge. As David Barlow notes, "self-focused attention greatly increases sensitivity to bodily sensations and other aspects of internal experience. Furthermore, this sensitivity . . . quickly spreads to other aspects of the self, such as self-evaluative concerns." Barlow calls this process a "negative feedback cycle" which leads to a chronic feeling of helplessness, dependence, and self-absorption. As Ann Seagrave and Faison Covington—two women who have overcome their panic attacks—write in *Free from Fears*, "We can become frightened to such a degree that we learn to monitor every twitch, every ache, and it is in that way that we often scare ourselves needlessly."

One final point related to panic attacks seems quite puzzling. Attacks are often triggered by deep relaxation exercises such as those which induce hypnosis or guided imagery sessions. In one study, 67% of a group of panic-disorder patients experienced three or more symptoms while listening to a relaxation tape. As David Barlow notes, "relaxation is surely the strangest of panic provocation procedures." He hypothesizes that it may be caused by a fear of losing control. Whatever the reason, this finding certainly relates to therapy clients who are led to a "safe place" during deep relaxation exercises. It contributes to our understanding of why they might experience panic attacks during the process.

The Contexts of Insanity

In conclusion, a vicious cycle of social influence, combined with a widespread belief in massive repression of sexual abuse memories, has produced an epidemic of Survivors. In the current situation, it is sometimes difficult to ascertain who is fulfilling whose expectations. A woman enters therapy, already afraid that her problems may stem from repressed memories. Her therapist plays into those fears, and between the two of them, they find "evidence" in the form of dreams, flashbacks, or body memories. They see dysfunction everywhere, and when the client sinks into a hypnotic trance, she pictures horrifying events from her childhood.

In 1993, a CNN reporter took a hidden camera into a counseling session with a therapist known to have convinced at least six other women that they were Survivors. The reporter said that she had been "kind of depressed" for a few months, and that her marital sex life had worsened. At the end of the first session, the therapist suggested that she might have been sexually abused as a child. When the reporter said she had no such memories, the therapist stated that many women completely forget incest. "They have no idea, in fact. I mean, what you've presented to me, Lee-Anne, is so classic that I'm just sitting here blown away, actually." Once a therapist labels someone an Incest Survivor, everything the client says is perceived as evidence to validate the diagnosis. And the client, having accepted the possibility that the label might be accurate, quickly falls into the trap of seeing the same life problems as symptoms of a childhood full of sexual abuse. Once that belief system is in place, "memories" are not far behind.

Recovered Memory Therapy and False Memory Syndrome

A Patient's Perspective as a Test Case

LAURA PASLEY

With the exception of my former counselor, the names in my story are real. My attorney's name and firm have been used with his permission.

It was Monday, November 18, 1991. My appointment was for 4:00 P.M. I arrived early as I always do. Simpson & Dowd is a law firm in Dallas, Texas, specializing in mental health issues. I was to meet with Skip Simpson, Attorney at Law, along with a couple of other families who had been polluted by a perverse group of therapists. Here I was, meeting a family that I had heard for years were Satanists. Imagine my shock when I read their story in a popular magazine—false accusations, devastation, hurt, pain, humiliation, the separation from their only daughter, a daughter they professed much love for, a daughter I knew well. She was a woman in the same sort of circumstances I was in, needing a reason why she felt so "abnormal." She was a daughter that I watched accuse these people before the rest of the group, to her therapists, to anyone who would listen, just as I had done. Now here I was with her parents in the office of an attorney, attempting to sort out the mess and to help put an end to this senseless destruction of the family system.

As I look back, I wonder how it got this far. How could a relationship with a therapist become the sole focus of my life for four long years? How could I have sold my soul to a mere human being?—a man who, it turns out, has untouched problems in his own life; a man so sick he needed me and other women to stay "sick" in order for him to excel. I trusted this man with my innermost soul. I shared my dreams with him, confessed my sins to him. "Steve" was my mother, my father, my brother, my sister, my best friend, my husband, boyfriend, decision maker, choice maker, teacher and pastor. He had become everything to me. If Steve said it, it was so. My life became so enmeshed and intertwined with his life, my ability to think for myself disappeared. I thought what he wanted me to think. I believed what he wanted me to believe. I became what he wanted me to become. Skeptics might call this a "therapy cult." By any other name it was destructive. How in the world did I allow therapy to become the most important function in my life?

My ordeal began on Friday, December 20, 1985. Steve was supposed to be a specialist in treating eating disorders and I had one in a big way. Since I was ten years old I would eat and then force myself to throw up. By the time I got to Steve I was nearly 32 years of age. For 22 years I had been forcing myself to vomit. When I began therapy, I was binging and purging sometimes 15 to 20 times a day! I would gain weight, lose weight, then gain weight again. I abused laxatives, diuretics and diet pills. I could not deal with feelings of any kind. *Any* emotion would trigger a binge, then a purge. Food was my best friend and my worst enemy. My parents did not know I had bulimia. I did not even know it had a name until 1981. I read an article in the paper and it said this disorder was coming out of the closet and was a widespread problem. At first I was relieved because I had felt so alone and different from other people. Then I became frustrated because there seemed to be no one out there who knew how to treat it.

Then I heard about Steve. He was supposed to be the expert. I was told, "Steve will save your life." "Steve is your answer." "Go to him, trust him, do whatever he says and you will get well." God knows how badly I wanted to be well, how badly I wanted to feel "normal."

I began my journey with Steve by sitting on the couch in his office and spending the next hour with him staring at me. He was overweight and balding but seemed very confident and sure of himself. He seemed to be looking right into my soul. I was very uncomfortable. What few things I was able to tell him did not even seem to faze him. He seemed cold and uncaring and unfeeling. I told him I did not like him staring at me and he asked, "Why is that?"

I snapped back at him, "Hell, I don't know, I just don't like it." After that he only seemed to stare harder. I left my session feeling confused but I was so desperate and determined to end this terrible disorder that had plagued my life since childhood that I was ready to do anything to get my life in order. "Trust him, believe him, he is your answer." So, I put all my energy, all my money, everything into this therapy. Although much of the time Steve was staring, he also did something else. He was listening.

I was so hungry for someone to listen to me, just listen. To hear what I had to say, no matter what it was. Nobody had ever done that. If I felt something when expressing my feelings I was used to hearing such answers as "You don't really feel that way." "That's not the 'right' way to feel." "You don't really think that." "If you think about those kinds of things, you're gonna make God mad." "He's ashamed of you, I'm ashamed of you, you should be ashamed of you."

Now I had met a man, a parental figure, an authoritative figure who would listen to anything I had to say and not once did he say, "You should be ashamed." With this strategy he won my trust. I began seeing him every week, then twice a week. Steve would have me close my eyes. He would make me keep them closed throughout most of the session. Before long I was saying anything and everything that came into my mind. There were thoughts, ideas, images, and feelings that I had never shared with anyone until now. I never believed I was worth listening to. My heart was so empty and lonely, and for so many years the only comfort I had found was in binging and purging and then binging and purging some more. But now it appeared that someone who could help me cared.

In the beginning of my therapy, I brought with me some very real hurts and disappointments. I had spent five years of my life with a man, loved him deeply, had his child and then he was gone. Not only gone, but he discounted what we had shared for five years. The loss of this relationship alone had put me into a deep depression for several years. Add to that recurring female problems, financial difficulties, raising a child as a single parent and many other things that had my life out of control. Steve was not concerned with those things. In four years of therapy, we never dealt with issues that had occurred in my adult life. Steve was not concerned with those and discounted their importance when I brought them up. He said the pain was "deeper," and that it had been buried, or "repressed." According to Steve, my bulimia was "slow suicide." To have such a "death wish" to the magnitude I had, Steve explained, I had to have repressed something so horrible and so traumatic that only a lengthy therapy, hypnosis, and hard work were going to make me better.

By this time, Steve controlled me. He had bought my loyalty and dependence by giving me the one thing that I was starving for—attention. It was attention with absolutely no boundaries, but plenty of control. I called him anytime I wanted to, day or night, and we talked as long as I needed it, unless he got mad at something I said and hung up in my face. If I showed any concern for my family, he got mad. He said I was hurting myself to protect them. If I was at home when I called him and I was upset or crying, he would have me take a broom and beat the hell out of my bed while he listened on the phone. At times, I would voice concern that my six-year-old daughter was in the house and it might frighten her. He told me I was "showing her how to exhibit anger in a healthy way." I found out years later that this behavior terrified her.

Week after week, session after session, through hypnosis and going deep within myself, strange images began to appear. At first they were images of this tiny blond with the biggest and saddest eyes I have ever seen. Steve said it was my "little girl"—the child within. It was as if I were sitting on a chair as high as the ceiling watching her. Steve wanted me to reveal to him each and every image or movement the "little girl" made.

My first "flashback " came while I was home vacuuming the floor. I had been to therapy earlier in the day. All of a sudden, I broke out into a sweat and I could not breathe. I was in a total panic. It was like a nightmare, only I was awake. I had images of a young boy holding a pillow over the face of an infant. It was a terrifying experience. I called Steve and he "walked" me through the "flashback." After I was calmed down, he literally put me to sleep on the phone. I went to see Steve the next day and my session was very uncomfortable. Steve kept drilling me, "When are you going to accept the fact that your brother tried to kill you?" I argued with him that this was not my brother, it could not have happened in my family. Over and over he said, "You'll have to accept the fact that your brother tried to kill you."

This flashback got Steve's attention, as did all the others. The images in my head got more and more bizarre. I began going to therapy more. I was going to the group room to write, a place where Steve said I would be "safe." Every flashback I had was judged to be actual, factual data from my past. Every dream, no matter how bizarre, was what had actually happened to me. The images grew. The scenes became more and more horrific. Had all of this junk really been hidden in my mind? Were these horrible scenes things that really took place in my family? Was this reality? What was reality? I got caught up in a full circle of flashbacks. They would reach out and snatch me up and engulf me in them at almost any moment. I cannot say where my logical mind was at this point. The flashbacks took control.

Steve told me to ask my doctor for a drug called Xanax, a sedative. I did. I began taking them, as Steve put it, "to take the edge off." I was swallowing them left and right. Soon I needed two, then three, then more. I was playing Russian roulette with my life. I would take a few too many pills and end up in the emergency room and guess who I called? Steve.

What was Steve giving me? The worse the flashback, the more self-destructive I was, the more attention I was getting from the main source for all things in my life. Steve kept telling me, "You have to get worse before you get better." Well, I was definitely getting worse. I was overmedicating myself, vomiting more and more, my weight was climbing. I got no exercise, and my life seemed more out of control than ever before.

In addition, no matter how many times I overmedicated myself or ended up in an emergency room, my doctor kept prescribing Xanax to me. Not only Xanax, but numerous other pills. There were pills to help me sleep, pills to relieve depression, pills to "mellow out my rage." If I had it there was a pill for it, and I took them all. My therapist would goad me, make me angry and push me over the edge and then the doctor would step in and medicate me so I would not be in such a rage. The therapy group in the hospital (I was hospitalized twice in a psychiatric hospital for 30 days each time) would get on a subject and harass me until I was livid and then the nurse would come get me and put me in a little room because I was angry. The nurses at the hospital said they had to take the "control" away from me; yet when I did what they said, I was tagged with being "over compliant." My mind was apparently gone, although at the time I was convinced this was the only way I would ever get well.

I lost control on so many occasions and Steve was the only one who could calm me down, make me "think right" again. I wanted more than anything in the world to be well, to be "normal." In spite of the still small skeptical voice inside of me, doubting, questioning, and wondering, I trusted this man to know the truth. That voice would soon fade over time. I believed in him so deeply I began telling other people, "Trust him, believe in him, he will make you whole." I trusted him so completely, in fact, that in 1986 I spent five months coordinating a retreat for women suffering from bulimia. In that period, I spoke or corresponded with over 350 people suffering from this disorder. I wanted them all to know about Steve. The retreat was held in a beautiful wooded retreat campground in East Texas. There were 77 women and one man in attendance. They came to hear the "truth." I wanted the world to come and hear Steve speak. If he said it I believed it.

It was not long until the "repressed memories" of child abuse began to come up. The visions in my head were of severe physical and sexual abuse. The images were so incredibly bizarre, yet they seemed so real. My picture of my family became distorted. Was it the drugs the doctors had me on, was it television shows or traumatic events I had witnessed over the years, or was it actual memories? I did not know, but Steve said they were fact and to deny them meant that I did not want to get well. He said I was in denial, I was running, I was "protecting" my family, I was staying sick to "cover up" for my family. He always had an answer. He was always right.

I was put into a group therapy situation. This is where my therapy team grew to include Steve's partner, Dave. I did not want to go but Steve said I was just transferring the fear of my family onto the group. He said I must go. At first we all just talked and I found a common ground with the other women. Then, slowly, right before my very eyes, the group emerged into a room full of "victims." We began as Eating Disorders (EDs), then on to Sexual Abuse Victims (SA), then on to Incest Victims (where family members became the perpetrators), then there was Satanic Ritual Abuse Victims (SRAs), and then on to Multiple Personality Disorders (MPD). It was a veritable "disease-of-the-month" laundry list. All of the women systematically had similar flashbacks, uncovered repressed memories and severe abuse. It was eerie at times. Each week we sat in a group and the stories were enough to make a strong stomach sick. One woman might have a flashback one week about her parents or someone else in the family and then the next week another one would have a similar memory come up. My mind became so confused and tormented. It was not long before my own flashbacks got even more bizarre. There was "group sexual abuse," a dead man hanging from a rope, killed by my grandfather, being sexually abused by animals, and much more.

Most of the time, members of the group were advised to stay away from their families and/or anyone who challenged their therapy. There was much anger aimed at all of the parents. If someone had some doubt that a flashback or memory was reality, Steve and Dave would goad them, then the whole group would join in, "You're in denial," "You want to stay sick for your family," "You don't want to get well." This type of input from people we trusted so very much and were so very dependent on kept us enmeshed in their treatment program.

There were many times when a group member was instructed to write her parents (the perpetrators) very hostile and mean letters, divorcing them, accusing them of terrible acts they believed they had done to them. These letters were coached by the group and group leaders. They were always read out loud to the group to get support. In many cases, such as mine, Steve said it would be too dangerous to send my mom a letter with accusations. Some were encouraged to send them and cut off all ties with their families. In my case, because I lived so close to my parents and refused to move, Steve and Dave felt I was in more danger than some of the others.

Once Steve instructed me to write my mother and list every mean thing she had done to me (that is, what I believed she had done at the time). Then, he stood beside me reading every horrible word in the most hateful, hostile tone imaginable. I was standing there with balls of clay, throwing them against the wall. The louder and meaner he read, the harder I threw those balls. It was a very intense session. This was supposed to release my repressed anger. After each session such as this one I was exhausted. I believe if you constantly fill your head with vile images, it will spit out vile images. Being placed in that situation had my mind being filled with a constant flow of it. Drinking blood, killing babies, sexual abuse of everything imaginable—incest, torture, murder, you name it.

Out of the women in my particular Monday night group, nearly all of them have since realized their "flashbacks" were not reality. Most will not speak out. I am not sure if it is loyalty to Steve and Dave or maybe lack of courage, or an inability to stand up for something that is right. Whatever the reason, it makes me angry because if they would come forward and be outspoken, more people would come out of this delusional state much more quickly.

One woman who was one of my very favorites accused her family of being Satanist. She "divorced" her parents, and her in-laws helped her through the toughest parts of her therapy. She had some horrible flashbacks, including of a baby, supposedly her twin, being hung in a tree and one of herself severely abused by most of her family members. She did question Steve and Dave about the fact that her birth certificate had "single birth" on it. Steve said that the coven had people who took care of all of those things to cover up reality. Later on in her therapy when she seemed to be doing well, she said she wanted to drive to the nearby state her parents lived in and talk to them about all of her "memories." Steve was livid in group and kept trying to talk her out of going. "What about the coven?" he said. He was furious and yelled at her that her life was in danger. This beautiful, petite woman said, "I don't care, I've got to find out." She went home to her parents, talked everything out and made peace with them. Shortly afterward, her mom died of a heart attack. I talked with her just recently and she told me when she went home that time there was absolutely nothing to substantiate her claims of Satanic Ritual Abuse. She said to me, "You know, I live with guilt each and every day of my life about what I did to mom."

My relationship with my family became extremely troubled. My sister would not allow my nephew to spend the night in my home. I looked at my parents with suspicion. Steve had me believing my mother had been trying to kill me for years. Not in an obvious attempt, but in the things she would do for me. I was bulimic. If Mama bought us groceries and any of them were easily ingested "binge foods," Steve said it was to kill me. At one point, I took some badly needed groceries back to her, threw the bag and asked if she was trying to kill me because there were some cookies and chips in the bag. I looked at her with disgust. I suspected her every move, her every motive. I questioned every remark. I missed many family functions and at the ones that I did attend, I was cold and suspicious of everyone there.

For years, I was consumed with suspicion, anger, fear, confusion. Could anyone in the world be trusted? Even my pastor, who was also my dear friend, became suspect when he began "doubting" my therapy. I called him when I was admitted into the hospital in 1988 and he was really upset. He said, "Pasley, you don't belong in a nut house and I will not support this therapy any longer." After that, Steve began telling me that he was using me and wanted to keep me sick. I was losing everyone and everything who meant anything to me.

Police officers who were friends of mine that I worked with (I am an employee of a large police department in Texas) would tell me I was turning into a "pill head." One officer took my purse one night and dumped all of the pills out into the trash. I became so enraged, I jerked the phone out of the wall in the jail and threw it at him. I screamed, "I have to get worse before I get better. This therapy is going to save my life." He told me they were quacks and after my money. Other officers told me I was not acting normally, I was not myself any longer, and that they missed the person I used to be. Steve and Dave would tell me, "The group is your 'new family.' Move away from your family of origin, divorce them, they are dangerous, you will never get well living near them." They even wanted me to quit my job with the police department because they said I was trying to shut my "little girl" up with the violence.

Desperate to be normal, feeling so abnormal, I was in a constant rage for years. I was furious with every single thing that had ever happened to me, or that did not happen to me. My family members had become my enemies—people placed on this earth to destroy me. I could not distinguish memory from reality. Nothing seemed real anymore.

To be sure, my parents made mistakes—plenty of them. But, let's be real. Is there any human being, parent or child, who has not made mistakes? I make them every day with my daughter. I believe the key is to acknowledge them, ask for forgiveness, and move forward. I also believe it is important for our children to see us as human, not to continually profess perfection. The question here in my case is, were my parents intentionally trying to destroy me? Of course not. But this is precisely what my therapy team, my group family made me believe.

My family's response to accusations I made would have mattered. If they said nothing, it was because they were guilty. If they cried innocence, they were trying to hide something. If they did not remember something the way I remembered it, they were in denial. There was always an answer. This was ingrained into every conversation and thought I had. I was told to read books about evil, sexual abuse, dysfunctional families, co-dependency, etc. Some of the required reading was *People of the Lie*, *Courage to Heal*, *Healing the Shame That Binds You*, *On Becoming a Person*, and *The Child Within*. I "lived" therapy seven days a week, 24 hours a day, 365 days a year. When I was not at therapy, I was calling my therapist.

When I was not talking to my therapist, I was thinking about my therapy. The entire ordeal consumed every ounce of energy I had and every penny I could get my hands on. All of this was "necessary" for me to "get well." Steve repeatedly told me, "You have to get worse before you get better." I continued to get worse believing this was progress.

One lesson from this experience is that we can never underestimate what a desperate person will do. Any person, no matter how bright or intelligent, if they are desperate enough, can fall into the same pit I fell into. I had worked in a jail for a large city in Texas since I was 19 years old. I knew the correct name for every charge in the Texas Penal Code, the Penal Code number and the penalty class. I could tell you what kind of time you could get for nearly every crime listed in our penal code. I could catch an error on an arrest report with a simple glance, book a drunk in 30 seconds, and usually determine the elements of arrest if I chose to read the report. I mastered county and city computers. I could research a criminal history and "find" just about anyone. I know literally hundreds of police officers, most of their badge numbers, and most of them would do nearly anything for me. Before entering this therapy situation, I had many commendations and was nominated by my sergeant for Non-Sworn Employee of the Year. After getting into therapy, I was still good at what I did, but my work, the officers, my daughter, everything took a back seat. By the time I left therapy, I had expended all of my sick time, my vacation time and came close to being fired over one of my stays in the hospital. I was also on the verge of losing my home. Was this progress?

I believe the worst part of this type of therapy is living through the flashbacks. It was frightening and left me empty and drained. I would literally "feel" pain of the things I was seeing in my head. My mother became my sexual abuser, then my brother and grandfather and a neighbor. The sexual abuse was vivid and seemed so real. Ordinary objects terrified me because they were sexual abuse tools in my flashbacks. It started out with simple fondling or molestation; it ended with torture, torment and indescribable pain.

I would emerge from one of these flashbacks and feel such rage. At times, I believe I was homicidal. My nostrils would flare and I would throw things, rant and rave, chain smoke, sometimes two cigarettes at one time, lock myself in the bedroom and pace back and forth. I used to scream and pray to God, "Why did you let these things happen to me?" "What did I ever do to anyone to merit this kind of pain?" Confusion at this point was a way of life for me.

My anger was constant. My therapy also included "rage reduction." It consisted of throwing things like clay, bean bags, etc. I was ripping phone books, beating with bataaka bats and screaming into pillows. I personally got more relief from breaking glass. I would drive down the street and throw coke bottles into the ground. When they would shatter, it was like a sedative, temporarily. These things were supposed to decrease my "repressed anger." In essence, the more anger I expressed, the madder I got. I was in a constant state of rage. After a flashback, Steve would have me direct that rage at Mama. He literally hated my mother. He would insult her, distorting everything that she said or did. Once, she wrote a check for my therapy because I simply did not have the money and he tore it up in my face. "I don't want her money," he said. (He then added it to my bill.) My mother knew better than to speak against Steve. I would not tolerate that. He was going to save me.

I spent four years with this therapy team. After four years, I wanted to do more. I wanted to be more. I was at the point of feeling like I would never get well. There was no hope for me, I was too far gone. I wanted to make the best of my life. I called Steve on December 20,

1989, exactly four years after I had walked into his office. I said I wanted to write a book about my experiences in the jail. I had contacted an author of a book about police and felt sure he would help me get started. Steve was quiet. I asked him if he thought I could do this. I waited, listening like a child waiting for approval from a parent. The words that followed tore into me, stinging me to the core of my being. "You are not through with flashbacks."

Disenchanted. Angry. Frustrated. I terminated my therapy. I grieved so much for them I had to enter therapy with another counselor to get through it. I went to her, telling her the same stories I had come to believe in therapy about my family. I spent the next 22 months still convinced these things had happened to me.

In October, 1991, I picked up an article on a family who had been accused of horrible abuse by their daughter in therapy. I was at Kroger and never left the parking lot until I had read every single word. The daughter was in therapy with me. I had listened to her pain and suffering. Now, I was getting another side to this picture. Steve and Dave insisted these people were Satanists—the cruelest, meanest people in the world. They had committed indescribable acts on their children. What really interested me was that some of the "stories" I had heard from Steve and Dave were presented differently than those in the article. Could Steve and Dave maybe have lied to me? Lied to us all? I was glued to the article. Then, after I read it, I drove home and read it again. I wanted to know these people. I wanted to meet them and see for myself that they were not really what I had heard. In meeting them and seeing the severe contrast to what I had heard, I was able to begin to discern my reality. They had lied to me—the con job of all con jobs.

This therapy has snatched something from me that I can never get back. I lost years of my life where I was emotionally distant from my family and my daughter. There was pain, despair, humiliation, fear, and frustration. It caused me to be paranoid. I have had trouble trusting anyone. Professionals had me scared to death, even ones there to help me. My daughter and I had no financial security and nearly lost our home. I did not have a car that ran. All my energy, all my money, everything I had went to them. When I woke up, my daughter was 12 years old and I missed it. I missed some of her most precious years while searching endlessly for the next "memory."

With the help of Skip Simpson, his law firm, my faith in God and the support of family and friends, I have held these two men accountable for what they did to me and my daughter. They were responsible for unethical, unprofessional treatment of me and my child. They injured us and it will take a long time to undo the damage. On December 19, 1991, Skip filed a lawsuit on behalf of myself and my daughter, Jennifer. We sued them, in part, for creating false memories, for giving me substandard care, for therapeutic negligence, and for fraud. It was extremely hard to trust anyone, especially an attorney. It was quite a while before I felt I could trust Skip but through his being trustworthy, I am learning to trust again. Now, however, I do it with my thinking cap on. I have learned through all of this that no one, not one single person in this world, has all of the answers. One of the quickest ways to turn me off is for someone to tell me "This is the only answer, the only way." I am now into critical thinking and proper skepticism. I look back now and see so many things that were just not logical. I will never again allow another person to control my mind or my life.

On June 25, 1993, Skip Simpson called me at work. He told me they were having a meeting to possibly settle my case and for me to stay by the phone. When the call came, I went to his office. We talked and he gave me two options for a settlement. We decided which would be the best one for my particular situation and that of my daughter. He went downstairs and moments later came back up. He said, "It's over." Tears were streaming down my face. We hugged. I looked at this man who had taken my case before he knew it was a national problem, believed in me before I could believe in myself, and I said, "You helped me get my power back from those who took it from me. I have my mind back and for this, I cannot thank you enough." (As a condition of my settlement I cannot disclose the settlement amount, the location where I was treated, or the names of my counselors.)

My life has changed so drastically this past year. Since my case broke on the news I have been talking with people all over the country who have lost children to this therapy, and adults who have absolutely had their lives destroyed and lost everything by being in it. I was speaking in Illinois and we got picketed. The signs read, "We believe the children." I would like to ask at what point do they believe the children? Is it when they are insisting nothing happened, or, after they place them with a social worker, or therapist with an agenda, who spends hours, days, weeks and months trying to get them to say what they want them to say?

What happened to me is not about sexual abuse or child abuse by a parent. It is about therapeutic negligence and fraud. We must begin to think critically about this situation. If we do not do something to stop this, the family structure as we know it will be gone. Families have been shattered and homes destroyed because troubled, hurting, vulnerable people sought out help and those who have taken an oath to do no harm abused the trust placed in them and did, in fact, harm. They not only hurt the patient, they destroy the patient's parents, siblings, their own children, and virtually anyone else who has been in their lives.

My life now is only getting better because I am not into the blame game any longer. I am no longer searching out "memories." After only one year with a good competent counselor, and two years working with an attorney who refuses to treat me like a "mental patient," I have begun to rebuild my life. Skip Simpson had faith in me and recognized my strength before I could see it. I responded to him because he treated me like I had a brain. He expected me to use it. The pain of what I went through is still there; however, I now take responsibility for my own life, for changing it the way I want it to be. I could sit forever and worry about the past and what this one or that one did or did not do, but the ultimate choice for my life is mine to make. I now take that challenge.

If you have been affected by this type of therapy, or are interested in further information on the subject, please contact the False Memory Syndrome Foundation, 3401 Market, Suite 130, Philadelphia, PA 19104. 800/568-8882.

To obtain a copy of *True Stories of False Memories* by Eleanor Goldstein and Kevin Farmer, in which Laura Pasley's story appears, contact the SIRS Publishing Company: 800/232-7477.

Recovered Memory Therapy and False Memory Syndrome

A Psychiatrist's Perspective as a Test Case

JOHN HOCHMAN

Thousands of patients (mostly women) in the United States have undergone or are undergoing attempted treatment by psychotherapists for a non-existent memory disorder. As a result, these same therapists have unwittingly promoted the development of a real memory disorder: False Memory Syndrome. To make sense of this unfortunate situation, I need to offer a few definitions.

Some psychotherapists believe that childhood sexual abuse is the specific cause of numerous physical and mental ills later in life. Some term this Incest Survivor Syndrome (ISS). There is no firm evidence that this is the case, since even where there has been documented sexual abuse during childhood, there are numerous other factors that can explain physical or emotional complaints that appear years later in an adult.

These therapists believe that the children immediately repress all memory of sexual abuse shortly after it occurs, causing it to vanish from recollection without a trace. The price for having repressed memories is said to be the eventual development of ISS.

Therapists attempt to "cure" ISS by engaging patients in recovered memory therapy (RMT), a hodge-podge of techniques varying with each therapist. The purpose of RMT is to enable the patient to recover into consciousness not only wholly accurate recollections of ancient sexual traumas, but also repressed body memories (such as physical pains) that occurred at the time of the traumas.

In actuality, RMT produces disturbing fantasies which are misperceived by the patient and misinterpreted by the therapist as memories. Mislabeled by the therapist and patient as recovered memories, they are actually false memories.

The vast majority of false memory cases developing from RMT are in women, which is why this article assumes patients to be female.

Initiation of Patients into RMT

A woman consults a psychotherapist for relief of various emotional complaints. The therapist informs her that she may have been molested as a child and does not know it, and this could explain her symptoms. Some patients think this idea is absurd and go to another therapist; others accept the therapist's suggestions and stay on. More than a few women have heard about repressed memories from talk shows or tabloids even prior to coming to the therapist's office, and may even make the appointment believing they too could be "victims."

Though the patient has no memories of abuse, she becomes motivated for "memory recovery" since she is told this will cure her symptoms. The therapist will offer encouragement that "memories" will return. Suggestive dreams or new pains are interpreted by the therapist as proof that repressed memories are lurking.

The therapist may refer the patient to a "survivor recovery group." There she will meet women who further encourage her to keep trying to remember. Attendance at these support groups, as well as assigned reading in self-help books, surrounds the patient with validation for the therapist's theories.

The vast majority of women with FMS are white, middle class, and above average in education. This corresponds to the profile of a typical woman who enters long term psychotherapy, and who perceives such activity as an important way to solve life's problems.

Generating False Memories

Unlike courts of law which obtain objective evidence where allegations of evil-doing are made, RMT solely directs the patient to attend toward her inner world for "proof" she was sexually abused. Such RMT techniques may include:

- Meditation on fantasy production, such as pictures drawn in "art therapy," dreams, or stream of consciousness journal writing.
- Hearing or reading about the "recovered memories" of other women which can serve as inspirations.
- Amytal interviews ("truth serum") and/or hypnosis (including "age regression" where the patient is told she is temporarily being transformed into the way she was when she was five years old).
- Telling the patient to review family albums; if she looks sad in some of her childhood photos, she is told this is further confirmation that abuse occurred.

The Dark Side of "Recovery"

Patients start out RMT with the hope that things will be better once they recover their repressed memories. But usually life becomes far more complicated.

The FMS patient will often become estranged from the "perpetrator" (most often her father). If the patient has small children, they will be off limits to "perpetrators" as well. Relationships with other family members become contingent on their not challenging the patient's beliefs.

Therapists may urge parents to come for a "family conference" in order to allow the patient to surprise the "perpetrator" with a rehearsed confrontation. Family members are usually too shocked and disorganized to coherently respond to accusations. The rationale for this scenario is that since "survivors" feel powerless, they need "empowerment."

FMS patients may file belated crime reports with local law enforcement agencies and may go on to sue "perpetrators." Such lawsuits demand compensation for bills from psychotherapists and possibly other doctors who treated adult medical problems that therapists somehow link to childhood traumas. Of course, there may be demands for "punitive damages." Spouses of "perpetrators" (usually the patient's mother) may be sued as well for being negligent, thus making householder's insurance into a courtroom piggy bank. Since FMS patients sincerely believe they have been victimized, more than a few juries have given verdicts sympathetic to them.

Preoccupied with the continuing chores of "memory recovery," the FMS patient may come to ignore more pressing problems with her marriage, family, schooling, or career. Often the time demands and expense of the therapy itself become a major life disruption.

Some patients during the course of RMT develop "multiple personality disorder" (MPD). RMT therapists have claimed that they need to not only recover repressed memories, but also to uncover repressed personality fragments; some women come to believe they are repositories of dozens of hidden personalities ("alters"). "Alters" have their own names and characteristics, and may identify themselves as men or even animals. An increasing number of psychiatrists and psychologists are coming to view MPD as a product of environmental suggestion and reinforcement, since the diagnosis was hardly made prior to ten years ago. One area where there is no controversy: once MPD is diagnosed, therapy bills become astronomical.

Some FMS patients become convinced that their abuse was actually "satanic ritual abuse" (SRA), due to participation by relatives in a secret satanic cult. Some therapists believe SRA is the work of a vast underground cult network in these United States. No evidence beyond "recovered memories" has ever been offered as proof that satanic cults exist at this claimed level of frequency. Therapists who lecture on the topic have explained away the lack of evidence that such cults exist by claiming that no defectors speak out due to iron-clad secrecy via brainwashing and terror.

The Care and Maintenance of False Memories

FMS involves a combination of mistaken perceptions and false beliefs. The fledgling FMS patient is encouraged to "connect" with an environment that will reinforce the FMS state, and is encouraged to "disconnect" from people or information that might lead her to question the results of RMT.

The FMS subculture is victim-oriented. Even though they have not undergone anticancer chemotherapy or walked away from airplane crashes, FMS patients are told they too are "survivors." This becomes a kind of new identity, giving FMS patients the feeling of a strong bond with other "survivors" of abuse. Patients will often start attending "survivor" support groups, subscribe to "survivor" newsletters, or even attend "survivor" conventions (sometimes with their therapists).

They will read books found in "recovery" sections of bookstores. The best known book, *The Courage to Heal*, is weighty, literate, and thus appears authoritative. Authors Laura Davis and Ellen Bass have no formal training in psychology, psychiatry, or memory. This paperback, modestly priced at $20, has sold over 700,000 copies.

Patients are told to shy away from dialogue with skeptical friends or relatives, since this will hinder their "recovery." "Perpetrators" who proclaim their innocence cannot be taken seriously since they are "in denial" and incapable of telling the truth.

Aside from these social influences, people by nature often resist seeing themselves as being in error. It can be terribly painful to acknowledge having made a big mistake, particularly when harmful consequences have resulted.

RMT exploits the tendency within each of us to blame others for our problems, and to latch onto simple answers for life's complicated problems. RMT therapists suggest that aside from entirely ruining childhoods, childhood sexual abuse can explain anything and everything that goes wrong during adulthood. RMT becomes the ultimate crybaby therapy.

How Memory Really Works

In Freud's theory of "repression" the mind automatically banishes traumatic events from memory to prevent overwhelming anxiety. Freud further theorized that repressed memories cause "neurosis," which could be cured if the memories were made conscious. While all this is taught in introductory psychology courses and has been taken by novelists and screenwriters to be a truism, Freud's repression theory has never been verified by rigorous scientific proof.

Freud, were he alive today, would be traumatized to see how RMT has redefined his pet concept. While Freud talked of the repression of single traumatic episodes, today's therapists maintain that dozens of similar traumatic episodes occurring over years are repressed with 100% efficiency.

The well known syndrome of Post Traumatic Stress Disorder shows us that verifiable traumatic events, rather than disappearing from memory, leave trauma victims haunted by intrusive memories in which the victim relives the trauma. For those who were in Nazi concentration camps or underwent torture as POWs in Vietnam, this can become a serious lifelong problem.

People forget most of what occurs to them, including some events that were pleasant or significant to them at the time. If an event is lost from memory, there is no scientific way to prove whether it was "repressed" or simply forgotten. And there is no reason that memories of sexual abuse should be handled any differently than childhood memories of physical abuse or of emergency surgery.

Events that have slipped away from memory cannot be recalled with the accuracy of a videotape. Individuals forget not only insignificant events in their entirety, but also significant events. Some events (traumatic or not) are recalled, but with significant details altered.

A study of children whose school was attacked by a sniper showed that some who were not on the school grounds later insisted they had personal recollections of being in school during the attack. These false memories apparently were inspired by exposure to the stories of those who truly experienced the trauma.

Memories can be deliberately distorted in adults by presenting a display of visual information, and later exposing subjects to verbal disinformation about what they saw. This disinformation often becomes incorporated into memory, contaminating the ultimate memories that are recalled.

To be sure, some who enter therapy were abused as children, but they have always remembered this abuse. They do not need special help in "memory recovery" to tell the therapist what happened to them.

Why Recovered Memory Therapy Is Bad Therapy

RMT purportedly is undertaken to help patients recover from the effects of sexual abuse from childhood; however, at the onset of RMT there is no evidence that such abuse ever occurred. Thus, instead of a therapist having some evidence for a diagnosis and then adopting a proper treatment plan, RMT therapists use the "treatment" to produce their diagnosis.

Some RMT therapists over-attribute common psychological complaints as signs of forgotten childhood sexual abuse. In their zeal to find memories, these therapists overlook any and all alternative explanations for the patient's complaints.

RMT therapists ignore basic psychological principles that all individuals are suggestible, and that patients in distress seeking psychotherapy are particularly likely to adopt beliefs and biases of their therapist.

Many RMT therapists have studied neither basic sciences related to memory nor the diagnosis of actual diseases of memory. Their knowledge is often based on a single weekend seminar, as opposed to years of formal training in any graduate program they attended to get their licenses.

Hypnosis and sodium amytal administration ("truth serum") are unacceptable procedures for memory recovery. Courts reject hypnosis as a memory aid. Subjects receiving hypnosis or amytal as general memory aids (even in instances where there is no question of sexual abuse) will often generate false memories. Upon returning to their normal state of consciousness, subjects assume all their refreshed "memories" are equally true.

RMT therapists generally make no attempt to verify "recovered memories" by interviewing third parties, or obtaining pediatric or school records. Some have explained that they do not verify the serious allegations that arise from RMT because their job is simply to help the patient feel "safe" and "recover."

Many patients who have known all their lives that they were mistreated or neglected by their parents decide as adults to be friends with the offending parents. By contrast, RMT therapists encourage their patients, on the basis of "recovered memories," to break off relationships with the alleged "perpetrators" as well as other relatives who disagree with the patient's views. This is completely at odds with the traditional goals of therapists: to allow competent patients to make their own important decisions, and to improve their patient's relationships with others.

Patients undergoing RMT often undergo an increase of symptoms as their treatment progresses, with corresponding disruption in their personal lives. Few therapists will seek consultation in order to clarify the problem, assuming instead that it is due to sexual abuse having been worse than anyone might have imagined.

Other Kinds of FMS

Some individuals come to believe that they lived "past lives" as a result of having undergone "past life therapy." This phenomenon generally develops in participants who are grounded in the New Age zeitgeist and already open to "discovering" their past lives. They enroll in seminars which can run up to an entire weekend and will involve some measure of group hypnotic induction and guided meditations. This sort of FMS also involves continuing group reinforcement. In contrast to horrific images of sexual abuse, recollections of "past lives" are generally pleasant and interesting. Few participants will recall spending prior lives in lunatic asylums or dungeons. The whole experience is assumed to be therapeutic by helping participants better understand the situation of their present lives.

A small number of individuals develop "recovered memories" of being abducted by aliens from outer space. Almost always these individuals had some curiosity about this area and were hardly skeptics before they fell into an alien abduction FMS.

In contrast to women who are plagued with concerns that they were sexually abused, these varieties of FMS are of a much more benign nature and do not disrupt personal functioning or family life. While some of these individuals suffer the ignominy of being perceived as "kooks," they may receive compensating group support from those who share their beliefs.

A Word about the Future

Increasing numbers of women who claimed to have recovered memories of sexual abuse have retracted their claims and now see themselves as having had FMS. This may spontaneously occur when women relocate to another locale and lose contact with their prior therapists and support group. Without the "positive reinforcement" from others to encourage false memory development and maintenance, some women begin to doubt the veracity of what they had believed was true. While some remain suspended in a twilight of doubt, others have fully recanted.

These retractors may have a profound influence on getting women with an active FMS to re-evaluate their situation. While FMS patients learn from the FMS culture to dismiss critics as either "perpetrators" or their apologists, the voice of a woman who says she is recovering from FMS is more easily heard.

Although most influential among family counselors and social workers, RMT affected the practices of some licensed psychologists and psychiatrists, some of whom were practicing in special "dissociative disorders units" in psychiatric hospitals. These activities have gone on with little challenge, until recently.

The number of women with FMS who have become retractors is increasing. Some have sued their former therapists for malpractice (see Laura Pasley's story in the previous entry), and others are weighing the possibilities of doing so. One malpractice insurance carrier for clinical psychologists in California recently tripled its rates without explanation; this has led to speculation that the carrier is anticipating increasing numbers of lawsuits alleging that psychologists caused FMS.

The False Memory Syndrome Foundation, formed in 1991, has been contacted by over 7,000 families in the U.S. and Canada who believe their grown children have FMS, and these families let their views be known to state licensing boards and professional organizations. Managed care administrators are starting to question megabills submitted by RMT therapists, some of whom see their patients through lengthy psychiatric hospitalizations. Understandably, all of this has gained the attention of the American Psychiatric Association and American Psychological Association, who are setting up task forces to try to examine the whole phenomenon.

Meanwhile, there is a large FMS subculture consisting of women convinced that their "recovered memories" are accurate, therapists keeping busy doing RMT, and of authors on the "recovery" lecture and talk show circuits. In addition, there are some vocal fringes of the feminist movement that cherish RMT since it is "proof" that men are dangerous and rotten, unless proven otherwise. Skeptical challenges to RMT are met by emotional rejoinders that critics are front groups for perpetrators, and make the ridiculous analogy that "some people even say the Holocaust did not happen."

RMT will eventually disappear, but it will take time.

4

SCIENCE AND PSEUDOSCIENCE—FOR AND AGAINST

Evolutionary Psychology as Good Science

FRANK MIELE

Is "the fault, dear Brutus, not in our stars but in ourselves?" In our genes? Or in our jeans? Why do some "bestride the narrow world like a Colossus" while other "petty men [most of us] peep about to find ourselves dishonorable graves"? Are not men, as Shakespeare suggested in *Julius Caesar*, at least sometimes "masters of their fates"? Or, as Jack Nicholson's "average horny little devil" asks about the differences between men and women, in the film version of Updike's *The Witches of Eastwick:*

> Do you think God knew what he was doing . . . or do you think it was just another of his minor mistakes—like tidal waves, earthquakes, floods. . . . When we make mistakes, they call it evil; God makes mistakes, they call it nature.

A mistake? Or did he do it on purpose? Because if it's a mistake, maybe we can do something about it—find a cure; invent a vaccine; build up our immune system.

Throughout most of human history, the answers to these questions have come from myth or literature. Starting with the Enlightenment, however, the answers have usually been couched in the allegedly "objective findings" of either history or science. Since the end of World War II, the "standard model of social science," as summarized by Robert Wright in his very readable introduction to evolutionary psychology, skeptically (if not cynically) titled *The Moral Animal*, has held that "the uniquely malleable human mind, together with the unique force of culture, has severed our behavior from its evolutionary roots; . . . [and] there is no inherent human nature driving events . . . our essential nature is to be driven" (1994, 5).

For example, Emile Durkheim, the patriarch of modern sociology, referred to human nature as "merely the indeterminate material that the social factor molds and transforms." He argued that even such deeply felt emotions as sexual jealousy, a father's love of his child, or the child's love of the father are "far from being inherent in human nature." Robert Lowie, a founding father of American cultural anthropology, argued that "the principles of psychology are as incapable of accounting for the phenomena of culture as is gravitation to account for architectural styles." Ruth Benedict, one of the founding mothers of American anthropology, and a crusader against the theory of racial differences (which was the norm in pre–World War II days), wrote that "we must accept all the implications of our human inheritance, one of the most important of which is the small scope of biologically transmitted behavior, and the enormous role of the cultural process of transmission of tradition." (All quotes from Wright, 1994.) B. F. Skinner founded the school of behavioral psychology, dominant in American psychology in the 1950s and 1960s, on the bedrock assumption that human and animal behavior could be accounted for in terms of rewards and punishments.

To all of this, evolutionary psychologists reply with the gusto of a Wayne and Garth "NOT!" Human nature is real, it is important, and it isn't going to go away. Here is a sampling of the sorts of questions evolutionary psychologists ask and attempt to answer:

- Are we all naturally the same or naturally different?
- Is our mind all of one piece or is it composed of modules?
- Are we naturally moral and good and only become evil through circumstance, or are we naturally evil and only made good through enforced circumstance?
- Why are men and women so different?
- Do men naturally want young and beautiful women—and as many as they can get?
- Do women naturally want rich and powerful men—and a bonded, monogamous, caring relationship?
- Are men naturally turned on (maybe too turned on) by the sight of a woman—or even a silhouette or cartoon of one?
- Do men like sex more than women do? If so, why?
- Just how much does a man's or woman's looks tell a member of the opposite sex about them and their value as a potential mate?
- Why do men get turned on by "lips like rubies, eyes like limpid pools, and skin like silk"? And why do women spend so much time and money trying to achieve and reinforce that appearance?
- Why do human males have such large penises relative to our nearest primate relatives the great apes?
- Do some human groups, on average, have larger (and therefore less ape-like) penises than other groups? If so, why?
- Why do human females have such large breasts relative to our nearest primate relatives the great apes?
- Do dominant Alpha Males have all the fun and leave the most descendants, or do "Sneaky Fuckers" beat them at their own game?
- Why do women have orgasms?
- Why do cute, lovable children so quickly transmogrify into wild, ungrateful teenagers?
- Why, as we grow old, do we feel, in the words of retiring Supreme Court Justice Thurgood Marshall, that we're "just fallin' apart"?
- Do men naturally form power pyramids and hierarchies while women naturally form cliques?
- Do we naturally partition the world into US v. THEM?
- Does maternal instinct explain why moms usually act like moms, while dads all too often act like cads?
- Do we naturally prefer those who physically resemble us and find them to be more like us in other ways as well?

Are such questions even scientifically meaningful or do they more properly fall in the realm of religion, literature, or politics? They are certainly great openers to liven up even the dullest party. But the new and emerging field of evolutionary psychology, building on work from Charles Darwin's *Descent of Man* and *The Expression of Emotion in Man and Animals*, tells us that the answers to these age-old questions, dear Brutus, are in our evolutionary history and our genes. And they claim they've got the "bloody daggers" to prove it!

This introduction cannot examine the evolutionary argument on each of these points. Instead, it merely outlines the case and describes the type of evidence and the nature of the arguments to be placed before you, the skeptical jury. The references in the bibliography provide a more complete "transcript." The article that follows presents a case against evolutionary explanations of human behavior.

From Survival of the Fittest to Inclusive Fitness

The fundamental theorem upon which evolutionary psychology is based is that behavior (just like anatomy and physiology) is in large part inherited and that every organism acts (consciously or not) to enhance its inclusive fitness—to increase the frequency and distribution of its selfish genes in future generations. And those genes exist not only in the individual but in his or her identical twin (100%), siblings (on average, 50%), cousins (on average, 25%) and so on down the kinship line. Thus, aid to and feelings for relatives make evolutionary sense.

This revision and extension of Darwinian evolution, from "survival of the fittest" to inclusive fitness, was worked out primarily by George Williams (in the US) and by William Hamilton and John Maynard Smith (in the UK) in the 1960s, with some clever twists added by Robert Trivers (in the US) in the 1970s. How efficiently can the Darwinian mill grind? they asked. It largely depends on the type of grain fed in. Darwinian selection operates most effectively if the units on which it is working:

1. are more, rather than less, variable;
2. have shorter, rather than longer, lifetimes;
3. are more heritable, rather than environmental.

Richard Alexander (1979) has argued convincingly that "genes are the most persistent of all living units, hence on all counts the most likely units of selection. One may say that genes evolved to survive by reproducing, and they have evolved to reproduce by creating and guiding the conduct and fate of all the units above them" (38).

Implicit in this reasoning is the conclusion that species and populations (races) are very unlikely units of selection. Hence, all talk of individuals doing things, especially dying, for the good of the species or the race appears improbable if not downright impossible. But if that is the case, then how could any sort of cooperative behavior, of which there are as many examples all around us as there are of competitive behavior, have ever evolved?

Well, humans, like most complex species, don't pass on their genes by simply dividing and producing exact replicas of themselves the way amoebas do. It takes at least two, not only to tango, but to reproduce. While you need not share any genes with your mate, you must share some, but not necessarily all of them with your relatives (except in the interesting case of an identical twin, who shares all your genes). Work out the arithmetic and it produces some interesting consequences in terms of whom you should help and when, as summarized in Figure 1 (adapted from Alexander, 1979). Rather than anything so simple as either "every man for himself" or "all for one and one for all," Figure 1 shows that, like it or not, you're stuck in a complex, time-directed matrix of cooperation, competition, trust, and deception with all your blood relatives and even those you might think are blood relatives.

Appropriately enough, you watch out for Number 1 first; your parents, children, and full siblings next; and so on in order of decreasing genetic similarity. But given that time's arrow flies in one direction only, you have a better chance of passing on your genes by helping your children than by helping your aging parents.

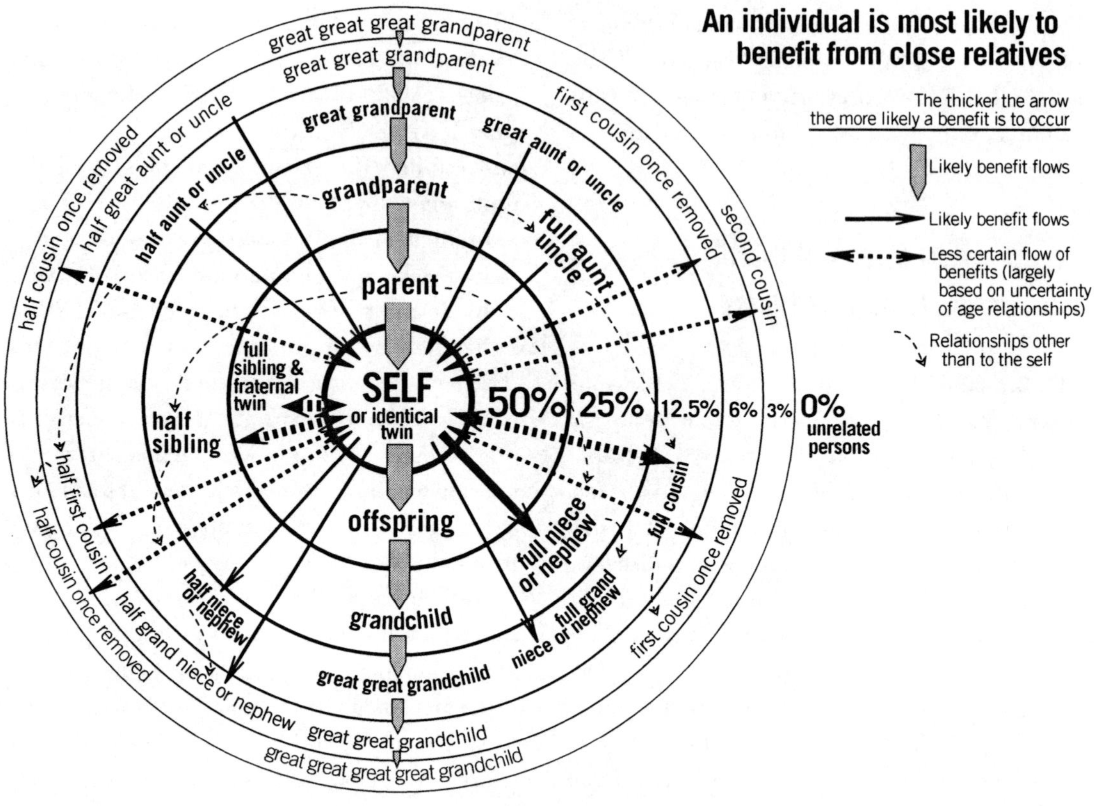

Figure 1

Symons Says

What does evolutionary theory predict you should expect from your mates? The answer is even more disconcerting. A corollary to the fundamental theorem is that the differences between males and females in humans, just as in most mammalian species, are readily explainable in terms of differential parental investment. That is, the male contributions to the reproductive process—lots of sperm and a few minutes of light work—are plentiful and cheap, short and pleasurable; while the female contributions—eggs and months of pregnancy—are rare and expensive, long, dangerous, and often painful. Given that, the best way for a male to maximize his inclusive fitness is to . . . well, diversify his genetic portfolio; while the best way for a female to insure the survival of the baby she has invested so much time and effort in is to try and get that guy to meet his monthly payments.

In *The Evolution of Human Sexuality* (27, 1979), anthropologist Donald Symons provides evolutionary psychology's point-by-point reply to "the horny little devil's" soliloquy on men and women:

1. Intrasexual competition generally is much more intense among males than among females, and in preliterate societies competition over women probably is the single most important cause of violence.
2. Men incline to polygyny, whereas women are more malleable in this respect and,

depending on the circumstances, may be equally satisfied in polygynous [one male–multiple females], monogamous, or polyandrous [one female–multiple males] marriages.

3. Almost universally, men experience sexual jealousy of their mates. Women are more malleable in this respect, but in certain circumstances, women's experience of sexual jealousy may be characteristically as intense as men's.
4. Men are much more likely to be sexually aroused by the sight of women and the female genitals than women are by the sight of men and the male genitals. Such arousal must be distinguished from arousal produced by the sight of, or the description of, an actual sexual encounter, since male-female differences in the latter may be minimal.
5. Physical characteristics, especially those that correlate with youth, are by far the most important determinants of women's sexual attractiveness. Physical characteristics are somewhat less important determinants of men's sexual attractiveness; political and economic prowess are more important; and youth is relatively unimportant.
6. Much more than women, men are predisposed to desire a variety of sex partners for the sake of variety.
7. Among all peoples, copulation is considered to be essentially a service or favor that women render to men, and not vice versa, regardless of which sex derives or is thought to derive greater pleasure from sexual intercourse.

To many, this sets a new standard in arguing for the inherent and therefore inescapable nature of the double standard. What evidence is there to support the argument that male-female differences are so deeply rooted in our nature? Anthropologists Lionel Tiger and Robin Fox argued in 1971 in *The Imperial Animal* that if "we look at enough primates to see what we all have in common, we'll get some idea of what it was we evolved from. If we see what we had to change from to get to be what we are now, it might help to explain what we in fact are."

Of Belles and Balls

Figures 2 and 3 are adapted from Jared Diamond's *The Third Chimpanzee* (73–74). They compare the relevant male and female anatomy for humans and our nearest living relatives, the great apes.

First look at the amount of sexual dimorphism in the four species. As Diamond notes, "chimps of both sexes weigh about the same; men are slightly larger than women, but male orangutans and gorillas are much bigger than females" (73). These are interesting facts from comparative anatomy, but what do they have to do with behavior? Throughout the animal kingdom, polygynous species (i.e., those in which each dominant male breeds with multiple females) are sexually dimorphic. This makes sense from an evolutionary point of view. The only way a male can pass on his genes is to breed with a female, and to better the odds, the more the merrier. But since there are only so many females to go around, from day one males are in competition with other males for those females. An arms race begins in which males are selected for their ability to win out against other males for access to the females. And since nothing escalates like an arms race, you end up with male gorillas and orangs that are not only twice the size of the females, but armed with huge canines, and loaded with secondary sexual characteristics like crested heads and silver backs that are easily recognizable at a distance and help to attract mates.

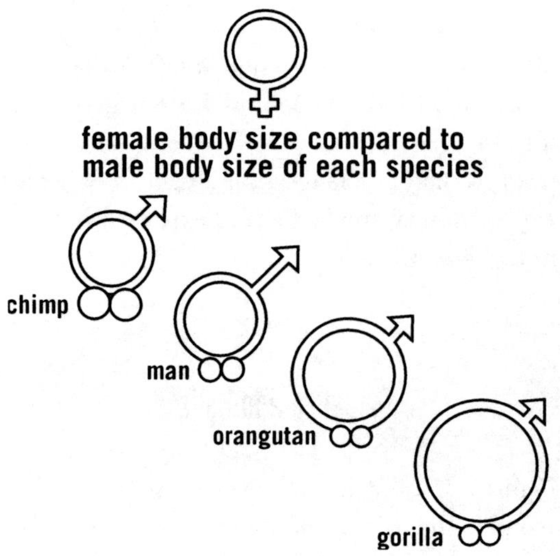

Figure 2

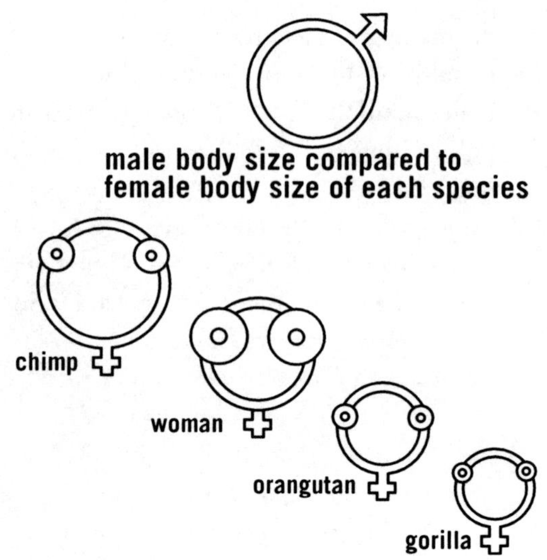

Figure 3

Chimps, on the other hand, show little sexual dimorphism, less even than humans. The gibbon (an ape, but not a great one) shows the least sexual dimorphism. Males and females look identical at a distance and the gibbons' strict adherence to monogamy should win an award from the Moral Majority (though that would mean acknowledging man's common primate ancestry and therefore ditching creationism). Going simply by the dope sheet of sexual dimorphism, an evolutionary handicapper would bet the rent that *Homo sapiens* would, by nature, be mildly polygynous. And he'd walk away from the pay window a big winner. A cross-cultural analysis of 853 societies revealed that 83% of them are polygynous. Polygyny occurs frequently, even when legally prohibited. There are an estimated 25,000 to 35,000 polygynous marriages in the US; a study of 437 financially successful American men found that some maintained two separate families, each unknown to the other (Buss, 177–178). Polyandry (one female with multiple males), on the other hand, is "virtually absent" among hunter/gatherers and confined to "agriculturalists and pastoralists living under very difficult economic conditions" and disappears quickly "when more usual conditions are present" (Symons, 225).

To move on from gross anatomy to gross discourse, if the male gorilla is so big and tough, how come he has such small balls? How does evolutionary theory account for those differences in testicle, penis, and breast size? It may be a tough climb to the top of the male gorilla dominance pyramid, but once there, things become quieter. Until dethroned, you have virtually uncontested access to all the females, so sex is no big thing. In fact, the dominant male with a harem of females "experiences sex as a rare treat: if he is lucky, a few times a year" (Diamond, 73). So just a little bit of sperm goes a long way to insuring the male gorilla's inclusive fitness.

For the minimally sexually dimorphic chimp, things get a little dicier. Chimps do have power pyramids. Compared to the gorilla and the orang, their hierarchies are so complex that Frans de Waal entitled his study of them *Chimpanzee Politics*. Getting to the top and staying there calls more for the skills of a Machiavelli than of a Mike Tyson. Dominant males have frequent though not exclusive access to the females. Rather than simply their bodies, it is their sperm that must compete against those of their fellow dominants, as well as those of the occasional "sneaky fucker." And all of this follows directly from one of the triumphs of evolutionary biology—the Theory of Testicle Size. To wit, "species that copulate more often need bigger testes; and promiscuous species in which several males routinely copulate in quick sequence with one female need especially big testes (because the male that injects the most semen has the best chance of being the one to fertilize the egg). When fertilization is a competitive lottery, large testes enable a male to enter more sperm in the lottery" (Diamond, 72).

Humans, according to evolutionary theory, should therefore be intermediate between chimps and gorillas both in polygyny and in promiscuity—and the data fit the prediction. I leave it to the reader to speculate as to what the evolutionary result would be if groups of religious cultists (in which the leader tries to monopolize the females) and outlaw biker gangs (who after all gave us the term "gang bang") were to each pursue their own evolutionary path, separate from the rest of human society.

Diamond provides more hard anatomical data (75):

> The length of the erect penis averages 1 1/4 inches in a gorilla, 1 1/2 inches in an orangutan, 3 inches in a chimp, and 5 inches in a man. Visual conspicuousness varies in the same sequence: a gorilla's penis is inconspicuous even when erect because of its black color, while the chimp's pink erect penis stands out against the bare white skin behind it. The flaccid penis is not even visible in apes.

To date, however, there is no adequate evolutionary explanation of the between-species differences in penis size. J. P. Rushton has offered a very controversial explanation of the mean differences in penis size between various racial groups within the human species. His letter to *Skeptic* (Vol. 3, No. 4, 22–25), with an accompanying table, summarizes his argument that there is a "tradeoff" between cognitive assets (brain size and IQ score) and reproductive assets (penis size and gamete production). Both neurons and gametes are expensive and Rushton's data are replicable, but most evolutionary biologists and psychologists do not accept his interpretation.

Rushton's work highlights two important differences among evolutionary explanations of behavior. Evolutionary explanations of genetic differences between individuals, and especially between groups of individuals, have an air of an earlier Social Darwinism which many today find downright offensive. Which is not to say that they are, for that reason, factually wrong. But most of today's evolutionary psychologists are concerned with the universals of human nature, not the differences. They argue that "genetic differences among individuals surely play a role, but perhaps a larger role is played by genetic commonalities: by a generic, species-wide developmental program that absorbs information from the social environment and adjusts the maturing mind accordingly." They therefore believe that "future progress in grasping the importance of environment will probably come from thinking about genes" (Wright, 9).

And whereas Rushton and others, located on the pro side of *The Bell Curve* controversy, argue for a unitary view of the mind (usually manifested in a single trait variously referred to as intelligence, IQ, cognitive ability, or psychometric *g*) on which all individuals (and even groups) can be measured and ranked from top to bottom ("alphabetically by height" as legendary New York Yankee manager Casey Stengel once put it), most of today's evolutionary psychologists argue that evolution would rather select for distinct mental modules. In their view, evolution can give males a "love of offspring" module, and make that module sensitive to the likelihood that the offspring in question is indeed the man's. But the adaptation cannot be foolproof. Natural selection can give women an "attracted to muscles" module, or an "attracted to status" module, and . . . it can make the strength of those attractions depend on all kinds of germane factors. . . . As Tooby and Cosmides say, human beings aren't general purpose "fitness maximizers." They are "adaptation executors." The adaptations may or may not bring good results in any given case, and success is especially spotty in environments other than a small hunter-gatherer village (Wright, 106–107).

In the view of most evolutionary psychologists, the modules may differ in effectiveness from one individual to another, but given the number of different modules, their effect is to "average out" individual differences to the point where any attempt to "line everyone up" on a single dimension is as nebulous as Casey's syntax.

Now let's look at the females. "Human females are unique in their breasts, which are considerably larger than those of apes even before the first pregnancy" (Diamond, 74). Since the female gorilla and her baby are comparable in size to their human counterparts, the bulk of the huge (by primate standards) human female breast consists of fat, not milk glands, and breast size varies greatly among human females without affecting their ability to nurse young. Thus, the explanation cannot be based on the need to nurse infants. Rather, human female breasts are secondary sexual characteristics that evolved to attract mates. According to Desmond Morris (1967), this took place along with the switch from front-to-rear to front-to-front mating, the pendulous shape and cleavage of the breasts mimicking the pre-existing attractiveness of the female buttocks. This also, according to the theory, explains why men find other pendulous shapes (like ear lobes) and other cleavages (like toes in low-vamped shoes) such a turn-on.

And while we're on the subject, what other female attributes turn men on? Gentlemen prefer young, nubile women, with lips like rubies, eyes like limpid pools, skin like silk, breasts like a milch cow, and legs like a race horse. According to evolutionary theory, this is not the result of either Hollywood or Madison Avenue, but because all of these features have served as cues to a female's health, reproductive potential and sexual availability over the course of human evolutionary history. Evolution has built into every red-blooded male a desire to find "Pornotopia"—the fantasy land where "sex is sheer lust and physical gratification, devoid of more tender feelings and encumbering relationships, in which women are always aroused, or at least easily arousable, and ultimately are always willing" (Symons, 171). The entire cosmetics, fashion, and pornography industries are attempts to create Pornotopia here on Earth.

Figure 4, adapted from Daly and Wilson (1988), depicts human female reproductive value, calculated in terms of expected live births among hunter/gatherers, as a function of female age. This curve parallels the curve for men's preferences in females as determined in cross-cultural studies (Buss, 49–60; Symons, 187–200).

Men naturally prefer young women because

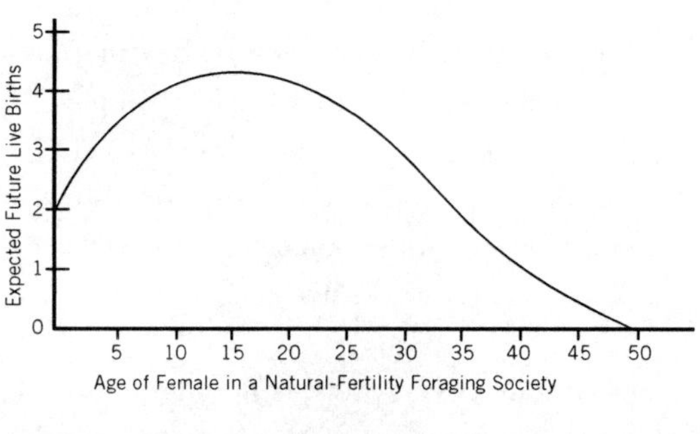

Figure 4

they provide the most reproductive potential for passing on the male's genes. If anything, males are biased toward selecting females before reproductive age in order to insure that no other male has beaten them to the finish line. From an evolutionary perspective, the least wise thing a male can do is to divert his hard-earned resources to rearing another man's child. Indeed, evolutionary psychologists would argue that this is why cuckolds are universally held in such low regard.

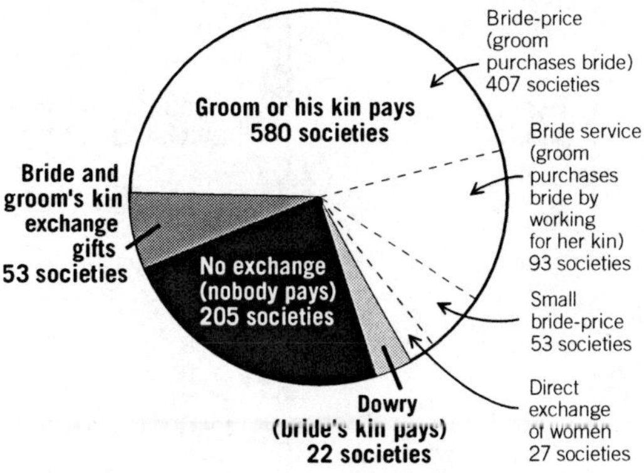

Figure 5

Murder 1, Incest 0

According to evolutionary theory, sex is a service women provide to men in return for resources. Evolutionary psychologists Martin Daly and Margo Wilson note that (188, emphasis theirs):

> marriage is a contract not between husband and wife, but between *men*, a formalized transfer of a woman as a commodity. And indeed when one examines the material and labor exchanges that surround marriage, it does begin to look like a trafficking in women. In our society, as in many, a father *gives* his daughter in marriage. Men purchase wives in the majority of human societies, and they often demand a refund if the bargain proves disappointing. Although the relatively rare practice of dowry might be construed to mean that who pays whom is arbitrary and reversible, dowry and bride-price are not in fact opposites: A bride-price is given as compensation to the bride's kin, whereas a dowry typically remains with the newlyweds.

Figure 5 (adapted from Daly and Wilson, 189) summarizes the exchange considerations at marriage in a cross-cultural comparison of 860 societies and emphasizes the universality of compensation for rights to female reproductive capacity.

Even worse from the point of view of the male and his family than failure by the female to live up to her part of the contract is the thought that the male's investment in resources may be going into a competitor's product. Figures 6 and 7 (adapted from *Homicide* by Daly and Wilson) show that child abuse and even murder are much more common for adoptive parents than for natural parents.

While evolutionary theory predicts a certain level of parent-child and sibling rivalry, its predictions are contrary to another mainstay of social science—the Freudian Oedipus Complex. Under evolutionary theory, fathers have a strong vested interest in their son's well-being; provided, of course, it is their son. As sons mature, they may in fact compete with their fathers for status and for females (as daughters may compete with their mothers for males), but not for their *own* mother (or father). Many evolutionists argue that, given the decreased viability of children born out of incest, selection has created an incest taboo, especially against mother-son incest. The comparative ethnographic data support the existence of the incest taboo, not the Oedipus complex (Alexander, 165; Wright, 315–316).

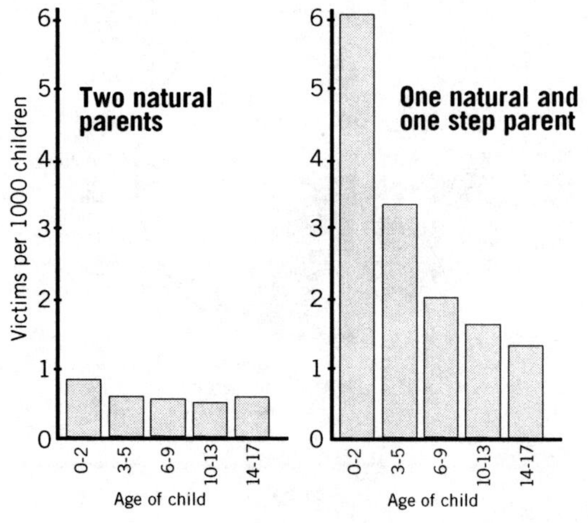

Figure 6

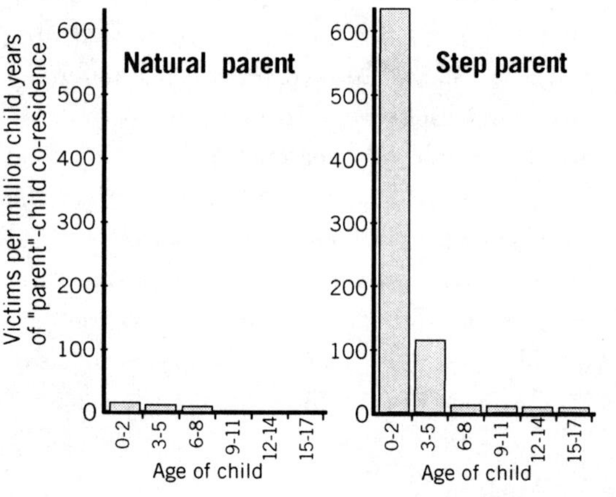

Figure 7

They Say That Breaking up Is Hard to Do: Fisher's Divorce Law Says It Isn't

Evolutionary psychology provides explanations not only of why we pair up, but why we split up. Conservative social critics have decried the alarming increase in divorce in the US since the 1960s, and variously attribute it to removing Bible reading from the public schools, rock 'n' roll, TV and movies, liberal social welfare programs, decriminalization of abortion, women's lib, and even the teaching of evolution. The evolutionary perspective, on the other hand, leads one to see lifetime monogamy as the exceptional result of an increased level of social pressure rather than as the rule for humans.

Anthropologist Helen Fisher has gathered divorce data from 62 societies around the world (Figures 8 and 9). She finds that "human beings in a variety of societies tend to divorce between the second and fourth years of marriage, with a divorce peak during the fourth year" (360). She also finds that the divorce statistics for the US in 1986, well past the sexual revolution of the 1960s, fit the same pattern, with most divorces taking place between the second and third year of marriage (362).

Fisher's evolutionary explanation attributes the universality of the divorce statistics to the "remarkable correlation between the length of human infancy in traditional societies, about four years, and the length of many marriages, about four years. Among the traditional !Kung, mothers hold their infants near their skin, breast-feed regularly through the day and night, nurse on demand, and offer their breasts as pacifiers. As a result of this constant body contact and nipple stimulation, as well as high levels of exercise and a low-fat diet, ovulation is suppressed and the ability to become pregnant is postponed for about three years" (153). She therefore concludes (154):

> The modern divorce peak—about four years—conforms to the traditional period between human successive births—four year . . . . Like pair-bonding in foxes, robins, and many other species that mate only through a breeding season, human pair-bonds originally evolved to last only long enough to raise a single dependent child through infancy, the first four years, unless a second child was conceived.

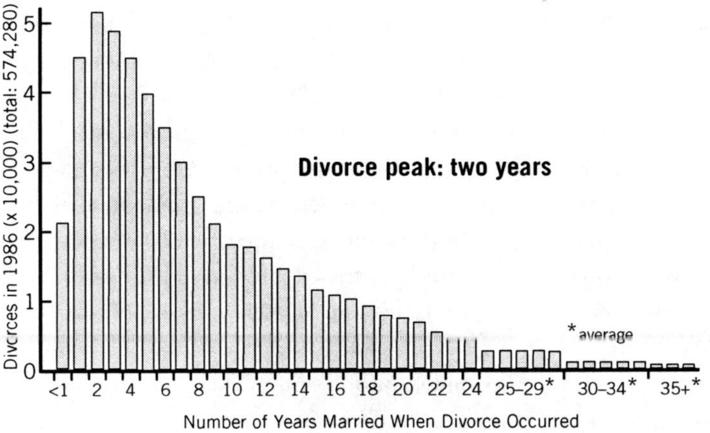

Figure 8

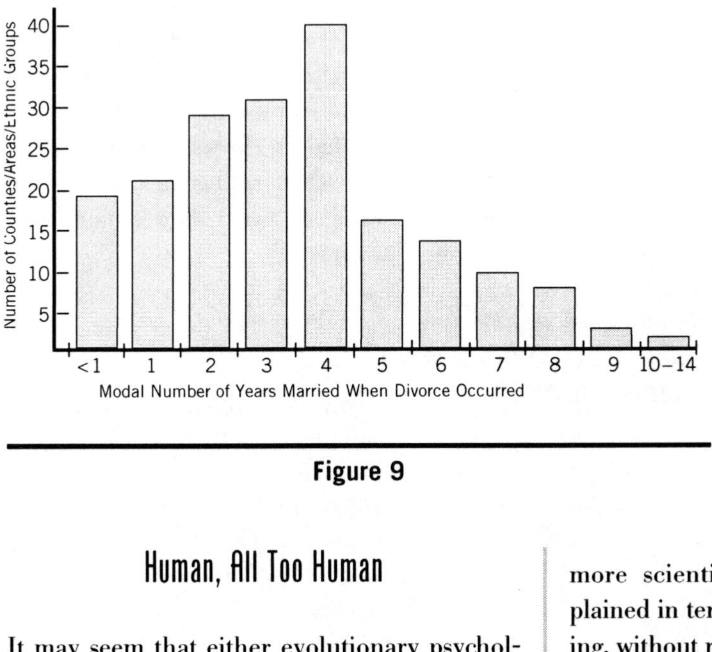

Figure 9

Human, All Too Human

It may seem that either evolutionary psychology or the examples selected for this quick and dirty summary are more suited to tabloid TV than to *Skeptic* magazine. Are we trying to increase circulation by slumming to the lowest common denominator of human behavior? Well, evolutionary psychology has an answer for that one too. It is precisely because of our evolutionary history and the importance of maximizing inclusive fitness that humans in all cultures, throughout history, have found such lurid tales so irresistible. Some may prefer them told with British accents on *Masterpiece Theatre,* rather than in the dialect of Rap or the twang of Country & Western, but the archetypal themes are the same and evolutionary psychology tells us that they will never go away.

But just how scientific are these attempts to explain human behavior in evolutionary terms? To what extent do the questions we ask automatically set up the answers we get? After all, as Cassius taunted Brutus, we are sometimes masters of our own fate! To what extent are human nature and individual and group differences scientifically meaningful concepts, rather than the social constructions of learning and experience, political and economic conditions? Is there any scientific there there?

In *Skeptic,* vol. 4, no. 1, Harry Schlinger, a psychologist at Western New England College, critically analyzes evolutionary theories and argues that human behavior can be more scientifically and parsimoniously explained in terms of the verifiable laws of learning, without recourse to evolutionary or genetic arguments. Harmon Holcomb, a philosopher of science at the University of Kentucky, skeptically examines the theories of evolutionary psychology and finds that for the most part, at this point, they are neither pseudoscience nor hard science, but protoscience, that is, science in the making. To graduate to the status of true science evolutionary psychology must put forth hypotheses that are capable of being critically disproven, rather than just reinforced or reconfirmed. He is a fair skeptic. Edward O. Wilson wrote on the cover of Holcomb's book *Sociobiology, Sex, and Science,* "Holcomb is now clearly the leading authority on sociobiology among philosophers of science" and (the book) "can and should be the standard reference on the subject." Reviewing the papers presented at the 1996 meeting of the Human Behavior and Evolution Society, he shows which research has reached the level of real science. Frank Salter of the Max Planck Institute supplies a biological counterattack. He critically examines sociology by taking us on a skeptical browse through *The Oxford Dictionary of Sociology,* and finds that its studied avoidance of basic human nature amounts to little more than modern alchemy.

Also in that issue of *Skeptic,* we matched pairs of interviews and book reviews. Lionel Tiger and Robin Fox, two of the grand old men of evolutionary theories of behavior, look back on what's taken place in the field in the 25 years since they published their groundbreaking and controversial book *The Imperial Animal. Skeptic* advisory board member Stephen Jay Gould, a longtime critic of excessive appeals to evolution and genetics in the explanation of human behavior, offers his thoughts on evolution, his own revision of Darwinism, the problems with ultra-Darwinism, and the politics of science. Philosopher of science Michael Ruse, an expert on the nexus between philosophy and biology, reviews one of the most controversial new books in this field—Daniel Dennett's *Darwin's Dangerous Idea*—which is very critical of those who would revise basic Darwinian explanations, such as Gould with his theory of punctuated equilibrium. *Skeptic* publisher Michael Shermer also reviews Dennett's book, though from a different perspective than Ruse, in his analysis of "Gould's Dangerous Idea"—contingency, necessity, and the nature of history. And lest we be accused of presenting only the evolutionary side of the argument, we conclude with some comic relief as anthropologist and long-time creationist observer, Tom McIver, takes us on "A Walk through Earth History: All Eight Thousand Years" in his skeptical tour of the Institute for Creation Research's museum.

So here then, ladies and gentlemen of the jury, is the issue at hand: Should we accept as a default hypothesis that human behavior, and the similarities and differences in behavior between individuals and groups, are the result of a complex interaction of the genes that reflect our evolutionary history as well as the environment in which we find ourselves? Or should we opt for the statistically null hypothesis that any invocation of genes and evolution to explain human behavior must be proved beyond a reasonable doubt? If nothing else, I think you will be forced to conclude, in the words of Nobel Prize Winner and co-discoverer of DNA James Watson, that "Charles Darwin will eventually be seen as a far more influential figure in the history of human thought than either Jesus Christ or Mohammed."

References:

Alexander, Richard. 1979. *Darwinism and Human Affairs.* Seattle, WA: University of Washington Press.

Buss, David. 1994. *The Evolution of Desire.* New York: Basic Books.

Daly, Martin, and Wilson, Margo. 1988. *Homicide.* New York: Aldine.

de Waal, Frans. 1982. *Chimpanzee Politics: Power and Sex among the Apes.* Baltimore: Johns Hopkins.

Diamond, Jared. 1992. *The Third Chimpanzee.* New York: Harper.

Fisher, Helen. 1992. *The Anatomy of Love: The Natural History of Monogamy, Adultery, and Divorce.* New York: Norton.

Morris, Desmond. 1967. *The Naked Ape.* New York: McGraw Hill.

Rushton, J. P. 1995. *Race, Evolution, and Behavior.* New Brunswick, NJ: Transaction.

Symons, Donald. 1979. *The Evolution of Human Sexuality*. New York: Oxford University Press.

Tooby, John, and Cosmides, Leda. "The Psychological Foundations of Culture" in Barkow, Jerome; Cosmides, Leda; and Tooby, John. 1992. *The Adapted Mind: Evolutionary Psychology and the Generation of Culture*. New York: Oxford.

Wright, Robert. 1994. *The Moral Animal: The New Science of Evolutionary Psychology*. New York.: Vintage.

Evolutionary Psychology as Pseudoscience

HENRY SCHLINGER JR.

In 1902 Rudyard Kipling published a children's book of stories and poems with the curious title *Just So Stories.* They included such natural curiosities as "How the Elephant Got Its Trunk," "How the Rhinoceros Got Its Skin," and "How the Leopard Got Its Spots." The stories, of course, are pure fantasy, and "just so stories" has become a critical cliche for similarly fanciful tales that attempt to explain nature. The new field of evolutionary psychology, while different in many respects from its predecessor sociobiology, is still subject to the accusation of telling just so stories.

As a sampling from this new science, the following are headlines from recent articles or reviews of various books appealing to evolutionary explanations of human behavior:

> Cheating Husband: Blame It on His Genes?
> Is There a Gene for Compassion?
> Is Prejudice Hereditary?
> A Scientist Weighs Evidence That the X Chromosome May Carry a Gene for Gayness.
> IQ: Is It Destiny?

Headlines such as these are meant to capture the attention and imagination of readers, and they usually do. They suggest that the books to which they refer are going to offer serious scientific evidence for their claims of an evolutionary explanation of much human social and intellectual behavior. Do these claims reflect the results of serious science or just more "pop sociobiology," as Kitcher (1985) calls it?

Most books on sociobiology appeared in the decade between about 1975 and 1985. Barash's 1977 *Sociobiology and Human Behavior*, Lumsden and Wilson's 1983 *Promethean Fire,* and especially E. O. Wilson's two great works, *Sociobiology: The New Synthesis* (1975) and *On Human Nature* (1978), created a new field of study of human behavior that forcefully challenged the hegemony of behavioral psychology that had reigned so long. Despite the existence of serious critical analyses of sociobiology (e.g., Bock, 1980; Futuyma, 1979; Gould, 1981; Kitcher, 1985; Sahlins, 1976), in the past few years, there has been an explosion of books offering evolutionary explanations for a variety of human behaviors, including intelligence, morality, mating, sexual preference, aggression, xenophobia, prejudice, and even our tendency to seek out various forms of nature, such as trips to zoos and visits to national parks. These books may be classified according to two distinct but related arguments about the evolution of human behavior: (a) individuals and groups that differ behaviorally in some way (e.g., IQ) do so because of underlying genetic differences; and (b) invariant, universal human traits (e.g., morality, aggression) represent fixed expressions of the human genome (Futuyma, 1979).

Recent books that argue for genetic differences between groups of humans with respect to such characteristics as intelligence include *The Bell Curve: Intelligence and Class Structure in American Life* (1994) by Herrnstein and Murray, *Race, Evolution, and Behavior* (1995) by Rushton, and *The Decline of Intelligence in America: A Strategy for National Renewal* (1994) by Itzkoff. Books that make the case that there are distinctly human behaviors—collectively called human nature—that reflect a uniquely human evolutionary history, include *Homicide* (1988) by Daly and Wilson *The Biophilia Hypothesis* (1993) edited by Kellert and Wilson, *The Moral Animal* (1994) by Wright, *The Evolution of Desire* (1994) by Buss, *Why We Get Sick: The New Science of Darwinian Medicine* (1994) by Nesse and Williams, *Eve's Rib: The Biological Roots of Sex Differences* (1994) by Poole, *The Science of Desire: The Search for the Gay Gene and the Biology of Behavior* (1994) by Hamer and Copeland, and *The Adapted Mind* (1992) by Barkow, Cosmides, and Tooby.

Both arguments on the evolution of human behavior rely to varying degrees on a combination of three types of supporting evidence:

1. Evolutionary logic supported by casual observations or statistical data.
2. Behavioral analogies and comparisons with animals.
3. Statistical analyses of data generated by non-experimental research methods.

Each of these types of evidence, while sometimes compelling and frequently interesting, is often flawed scientifically. This does not mean that the explanations themselves are wrong, only that the supporting evidence is insufficient. In many instances, an alternative and much more plausible approach to understanding human behavior is that rather than selecting for specialized behavioral traits, human evolutionary history has selected for behavioral plasticity, or learning capacity (Futuyma, 1979). Experimental evidence from the literature on learning shows overwhelmingly the powerful influence of the environment in shaping human behavioral similarities and differences.

In the present essay I describe the three types of evidence with supporting examples from both evolutionary positions on human behavior and then critique them according to certain methodological criteria. I argue that, in most cases, a much more cautious and scientifically defensible position on the origin of many human behaviors is that they are a function of individual environmental, and not evolutionary, history.

Evolutionary Logic

One of the hallmarks of the scientific method is the interpretation of phenomena that have not been subjected to experimental analysis. Scientific interpretation is the use of already established principles of science to explain novel instances of the subject matter. Hence, the logical or mathematical use of Darwinian principles of selection to interpret human behavior could have a sound basis in science. The main questions are (a) whether the data presented for interpretation are both valid and reliable, and (b) whether the interpretations of human behavior as presented in recent books and articles represent an appropriate extension of Darwinian theory.

Theorists from both positions on the evolution of human behavior cite examples of evolutionary logic and supporting data that are problematic. Theorists who emphasize genetic differences between groups of humans (races) have employed evolutionary logic to explain differences in intelligence (Herrnstein and Murray, 1994; Itzkoff, 1994; Rushton, 1995a), brain and head size and aggressiveness (Rushton, 1995a), among other traits. Evolutionary psychologists have used evolutionary logic to explain, among other things, why people kill one another (Daly and Wilson, 1988), why mothers who have just given birth seem to mention their neonate's resemblance to the father more than to themselves (Daly and Wilson, 1982), why social rejection may produce feelings of insecurity (Wright, 1995), and why people seek out zoos and parks and easily develop phobias to natural objects, like spiders (Wilson, 1993). The data cited by these theorists consist of casual observation, personal reflection, and anecdote, as well as statistics derived from non-experimental studies. To illustrate, consider an example of the use of evolutionary logic from each of the two positions on the evolution of human behavior.

Rushton (1995a) uses evolutionary logic to support his claim that human racial groups evolved under conditions where different environmental pressures selected for differences in a wide range of physical and intellectual characteristics. Rushton suggests that an r-K reproductive strategy analysis combined with information on human evolution can be used to understand important behavioral differences between Mongoloids, Caucasoids, and Negroids, as he calls them. The r-strategies are those with high reproductive rates, and the K-strategies are those with high levels of parental investment in offspring. According to Rushton (1995a), "Mongoloid people are more K-selected than Caucasoids, who, in turn, are more K-selected than Negroids" (xiii). In other words, Mongoloids invest relatively more in the care of their offspring than Caucasoids who invest relatively more in the care of their offspring than Negroids. Rushton appeals to evolutionary logic to explain the presence of these different r-K strategies in different human racial groups. Specifically, Rushton claims that the selection pressures in the hot African savanna, where Negroids evolved, were far different in terms of the required relationship between parental investment and high reproductive rates than selection pressures in the cold Arctic environment where Mongoloids evolved. Presumably, higher reproductive rates and lower rates of parental investment are more favorable in hotter climates, whereas the opposite is true in colder climates. According to Rushton, this is the evolutionary basis for the differences in r-K reproductive strategies supposedly observed in humans.

The first problem with Rushton's analysis concerns the reliability of the data offered to support his evolutionary logic. For example, he provides a table of the relative ranking of races on diverse variables such as physical maturation rate, including age of first sexual intercourse and pregnancy; reproductive effort, including relative frequency of two-egg twinning and of intercourse; personality, including aggressiveness and impulsivity; brain size; and intelligence (Rushton, 1995a, 1995b). The data for these rankings were generated by non-experimental research methods where average differences between groups were often very small. Moreover, there is no scientific evidence, other than correlations, to support many of Rushton's assumptions, including his assumption that brain size is functionally related to cognitive ability.

Rushton often relies on statistical analyses of aggregated data to bolster his claim that small differences between groups are significant. Even if we assume that the data cited by Rushton were derived from well-designed and well-controlled studies—a questionable assumption—his evolutionary interpretation of the data has several attendant problems. First, there is no way to test and thereby falsify his claim that these characteristics represent evolutionary adaptations. Rushton's evolutionary logic is not too dissimilar from that used by his sociobiological predecessors, as summarized by Futuyma (1979). He has simply imagined that higher reproductive rates and lower rates of parental investment must have conferred differential fitness in different climates, compared the predicted outcome with observations from correlational studies, and then concluded that these characteristics represent adaptive genetic traits. A second problem with Rushton's hypothesis is that his extension of the r-K reproductive strategy analysis (usually used to compare large differences between different species) to the small variations between groups within the human species represents a "fatal scientific error" by assuming that behavioral differences between groups within one species can be accounted for by genetic differences (Tavris, 1995). It is not even clear that behavioral differences between individuals reflect genetic differences or, if they do, to what extent (Futuyma, 1979). A third problem is that Rushton's concept of race, which reflects that of Western culture—based on a few physical features such as skin color, hair form, and the epicanthal fold—is subjective (Futuyma, 1979). And finally, any reliable differences in Rushton's data are just as likely to be due to environmental variables as genetic ones. Still, Rushton (1995a) boldly contends that his book will offer "new truths about racial group differences."

Consider, now, an example of how evolutionary logic might be used to interpret some human characteristic from the perspective of evolutionary psychology. Robert Wright, a science journalist, writing in *The New Yorker* (March, 1995), illustrates how evolutionary psychologists would approach the explanation of some presumably universal human behavioral trait. Suppose, Wright asks, that social rejection early in a person's life results in an enduring insecurity. According to Wright, we should ask whether this pattern "might have had a genetic payoff during evolution" (71). Presumably, our ancestors who faced such rejection were less likely to reproduce unless they became more socially vigilant about nourishing their social ties as a result of the insecurity. Insecurity as a response to social rejection, then, may have been reproductively advantageous for humans. The assumptions inherent in Wright's argument can be stated as follows: (a) human evolutionary history has selected a genetic "program" that is somehow sensitive to environmental input called "social rejection," (b) this genetic program is especially sensitive to input early in an individual's life, and (c) the behavioral response called insecurity is essentially the same for all people to this input.

There are several obvious problems with this example that are relevant to many such examples cited by evolutionary psychologists. The first problem is with the validity of the behavioral data. Wright simply assumes that insecurity, which is not objectively defined, is a general human response to early social rejection, which is also not objectively defined. Wright offers no evidence that his evolutionary model is based on precise behavioral observations. Rather, his analysis is based on common sense assumptions about human nature which have no scientific basis. A second problem deals with Wright's evolutionary interpretation of the data. Even if such a reaction could be precisely measured and were observed in most humans as a result of a precisely defined set of environmental inputs, an evolutionary interpretation that it was adaptive is untestable because there is no crucial test that can falsify the hypothesis (Futuyma, 1979). Finally, an evolutionary explanation of the pattern of behavior in Wright's example may not be the most parsimonious one. For example, it might be that the reaction to rejection that we refer to as "feelings of insecurity" might be a more general physiological response to the withholding or withdrawal of reinforcement following some behavior. The effect of such environmental operations is to simultaneously produce physiological responses and to alter the stimuli that define the situation such that they suppress the behavior under similar circumstances. These are the scientific principles of operant extinction and punishment. The "feeling of insecurity" may be a by-product of the withholding or withdrawal of reinforcement with no special selective advantage of its own.

Cross-Species Comparisons

A second type of evidence frequently used to support evolutionary explanations of human similarities and differences consists of analogies or comparisons between nonhuman and human behavior. It is common linguistic practice among humans, including scientists, to give names to things. When two or more forms of behavior are given the same name, it may see reasonable to assume that they are alike functionally as well. Kitcher (1985) points out that because we have such a rich vocabulary for describing human behavior, it is easy to use this vocabulary to describe nonhuman behavior that resembles it. Once described in similar ways, it becomes easier to then move freely from the nonhuman instance back to the superficially similar human instance and to assume that both result from similar processes. According to Kitcher (1985), "vulgar anthropomorphism" is the original sin of pop sociobiologists, in that they neglect "to investigate the kinship of forms of behavior that are superficially similar" (185). Even if scientists discovered a genetic basis for a behavior in an animal, which is rare, this does not mean that the human behavior that appears to be similar also has a genetic basis. As evolutionary biologists know, phenotypic similarity does not necessarily imply genotypic similarity.

Social theorists, like Rushton, who emphasize genetic differences between groups of humans typically point to between-species differences that are more than likely a function of differences in genes to make the case that within-species differences in humans are also a function of differences in genes. Rushton (1995a) employs an interesting kind of cross-species analogy to make a case for the genetic basis of human racial differences. First, he points out that significant differences in learning ability between species are due to genetic differences. Thus, mammals with larger brains, such as chimpanzees, rhesus monkeys, and spider monkeys, learn faster than mammals with smaller brains, such as marmosets, cats, gerbils, rats, and squirrels. Rushton then uses these comparisons to argue that within-species differences in human brain or head size are related to differences in intelligence, at least as measured by standardized IQ tests, and are likewise related to genetic differences. Rushton's ultimate point is that blacks have statistically smaller heads (and brains) than whites and that this correlates positively with differences in intelligence between the two groups, at least as measured by standardized tests. It is interesting to note that of the 32 studies summarized by Rushton on head size and intelligence in humans, most found low correlations.

Rushton takes a reasonable between-species example and extends it to an insupportable within-species difference. Even if the measurements of brain size and intelligence can be defended as reliable, Rushton's explanation of the behavioral differences is not the most parsimonious one, especially when one considers the myriad differences in environments on average between black and white children. Before genetic explanations of differences in learning ability between individuals or groups are proffered, environmental factors, such as nutrition, prenatal care, learning, and educational opportunities, should be investigated if for no other reason than the variables are easier to test.

Another example of questionable cross-species analogizing by Rushton (1995a) concerns the r-K reproductive strategies described previously. According to Rushton, the great apes exemplify the extreme end of the K-strategy because they produce one infant every five or six years and provide much parental care. At the other extreme are oysters who exemplify the r-strategy, producing 500 million eggs a year but providing no parental care. Although this scale is generally used to compare the life histories of widely disparate species, Rushton (1995a) applies it to the much smaller variations within the human species. Although Rushton believes that all humans are K-selected relative to other species, he also believes that some humans may be more so than others. He cites data showing that, compared to white women, black women average a shorter period of ovulation and produce more eggs per ovulation which is evidenced by their comparatively higher rate of two-egg twinning. His data also show that black women have comparatively lower intelligence than white women as measured by standardized tests. Rushton claims that the correlation between IQ and biological variables related to reproduction supports his view that the within-species variations in humans can be accounted for in the same way that between-species variations can. Even if the correlation can be proven to be valid, there are serious problems with Rushton's cross-species comparison. First, there is no biological justification for extending an analysis of between-species differences to within-species differences. Second, Rushton provides no evidence other than correlations that differences in IQ and certain biological variables between women represent adaptations resulting from natural selection. Third, simply demonstrating a correlation between two or more variables in no way clarifies causal relations.

Evolutionary psychologists, like their sociobiological predecessors, frequently employ cross-species analogies and comparisons to argue their case for the existence of universal human characteristics. For example, Daly and Wilson (1988) use an analogy with female ground squirrels to show how the concept of inclusive fitness may be used to understand sibling rivalry in humans. They argue that genetic relationship should be important to solidarity and social conflict. In other words, the closer the family relationship between two individuals, the more solidarity and the less conflict should exist between them. Daly and Wilson point out that such a theory has been tested in female ground squirrels who discriminate between their full sisters and half sisters when occupying adjacent territories as adults. Full sisters will apparently help each other whereas half sisters will exhibit more territorial aggression. They then suggest that the same prediction can be made with regard to human siblings; namely, that the intensity of sibling rivalry should reflect the likelihood of common paternity. In other words, full siblings should show less competition than half siblings. In their own words, "we might have evolved specialized psychological mechanisms whose function is to assess the likelihood of common paternity and to adjust the intensity of sibling competition accordingly," and some "psychologist should check it out" (1988, 11).

Cross-species analogies, such as the one offered by Daly and Wilson (1988), are intriguing, suggesting as they do that certain human characteristics that we seem to have in common with other species may be understood as part of our deeper human nature. There are serious problems with such analogies, however. The first problem is that the similarity between human and nonhuman behaviors is subjective and is only suggested after it is believed that there may be a common genetic basis for both. In other words, behavioral similarity is often in the eyes of the beholder. Who is to say that territorial aggression among ground squirrels is anything but superficially similar to disagreements or fights among human half-siblings? The causes of these similar behaviors could be completely different. A second problem is that even if the behavior of human siblings could be compared to female ground squirrels, there is no independent evidence for the existence of an evolved "psychological mechanism" or any suggestion as to how it would work to "assess the likelihood of common paternity and to adjust the intensity of sibling competition accordingly." In the absence of such a suggestion, based on some kind of objective scientific evidence rather than inferences, Daly and Wilson's explanation is simply hypothetical.

Futuyma (1979) has pointed out several other problems with cross-species analogies. For example, even if behavioral generalizations could be supported by reliable observations, we are still left with the nagging question of whether behaviors between species that are superficially similar are functionally similar; that is, whether the same processes are responsible for both. If we discover the genetic bases of territorial aggression in female ground squirrels, does this mean that behaviors we refer to as "human sibling rivalry" also have a genetic basis? A simpler approach would be to consider first whether other factors, such as environmental ones, could produce the human behaviors of interest. Such an approach might lead us to ask, for example, whether there is as much sibling rivalry between half-siblings who are raised together from birth or infancy and who are not aware of their genetic relationship to each other as there is between siblings who know they are half-siblings. Other than the interesting evolutionary theorizing that superficially similar behaviors in different species may be functionally similar, evolutionary psychologists offer no direct scientific evidence that they are.

Correlative Analysis

It should be noted that social evolutionary theorists typically do not conduct experiments, nor do they, in most instances, cite experimental data. Rather, they rely almost exclusively on a combination of anecdotal and statistical evidence to make their case that there are species-specific behaviors in humans. Moreover, in almost no case is direct genetic evidence used to support evolutionary theories of human behavior (see below). Since genes are identified as playing a causal role in important similarities and differences between humans, a true experimental test of the hypothesis would necessarily involve direct manipulation of genes as independent variables. Such manipulations are only carried out by geneticists and, for obvious reasons, they have been constrained in such endeavors to working with relatively simple organisms, such as fruit flies with extremely short gestation periods, where the focus is more on structural than behavioral characteristics. Those who write about the genetic bases of human behaviors are typically not geneticists, however. And because they cannot make their genetic case experimentally, these evolutionary theorists must rely on data generated by non-experimental, usually correlational, research methods. There are several problems with the ways in which some evolutionary theorists use correlative analysis.

Validity and Reliability of the Data. The first problem is, the validity and reliability of the methods used to generate the actual data are often questionable. E. O. Wilson (1993) states that one mode of testing an evolutionary hypothesis "is the correlative analysis of knowledge and attitudes of peoples in diverse cultures" (34). Knowledge and attitudes, poorly defined as they are, must be obtained from surveys and questionnaires. Methodological problems with such devices are well known among researchers. For instance, there are numerous ways in which researcher bias may affect the outcome, such as the sampling procedure used and the way in which questions on surveys and questionnaires are worded. Even when safeguards are included, inferences to larger populations (the ultimate goal of surveys or questionnaires) are questionable. Also, as most good researchers know, the reliability of verbal self reports is notoriously poor.

In addition to surveys and questionnaires, evolutionary theorists may use psychological tests to assess more general and presumably universal characteristics of populations. Rushton (1995a) provides an example of the use of such a test. His thesis of racial differences is based on the assumption that there is "a core of human nature" or character traits "around which individuals and groups consistently" differ. To wit, he cites a study conducted in the 1920s by Hartshome and May called the "Character Education Enquiry" in which 11,000 elementary and high school students were given a battery of 33 different tests of altruism, self-control, and honesty in various contexts (home, school, church, etc.). Children's reputations with teachers and classmates were also obtained and then correlated with the scores on the battery of tests. Notwithstanding the problems with questionnaires, the only behavior measured by such tests is that of answering questions on the test. The actual behaviors called "altruistic" or "honest" are not measured in the context wherein one would normally call them altruistic or honest. This is not to say that we cannot discern something of value with such tests, but only that the test may correlate poorly with the behaviors of interest, and only a direct experimental approach can potentially yield a scientific understanding of the behaviors.

Of course, the most notorious type of test cited in the literature on evolutionary theories of human behavior is the IQ test. Volumes have been written on problems with intelligence tests, and I will not repeat them here. Suffice it to say that one problem with such tests is what they purport to measure. Rather than measuring some qualitatively distinct structure or process as defenders of such tests would have us believe, intelligence tests literally measure only the correctness of a variety of learned behaviors—answers to questions on the test—in a contrived context—the test taking situation (Schlinger, 1992). Alfred Binet knew this when he developed the first modern intelligence test (although he eschewed the use of the term "intelligence" in favor of the more descriptive and neutral "intellectual level"). The challenge for serious scientists is to ask about the variables that affect the broad range of behaviors we describe as intelligent; and only an experimental analysis can answer such questions.

The Use of Statistics. A second problem with the use of correlative analyses by evolutionary theorists concerns the complex statistical tests employed to "make sense" of the data generated by surveys, questionnaires, psychological tests, and the like. The importance of correlative analyses in making the argument for genetic explanations of human behavior is underscored in the following quotation by Sir Francis Galton, which Rushton twice cited (1995a, b):

> General impressions are never to be trusted. Unfortunately when they are of long standing they become fixed rules of life, and assume a prescriptive right not to be questioned. Consequently, those who are not accustomed to original inquiry entertain a hatred and a horror of statistics. They cannot endure the idea of submitting their sacred impressions to cold-blooded verification. But it is the triumph of scientific men to rise superior to such superstitions, to devise tests by which the value of beliefs may be ascertained, and to feel sufficiently masters of themselves to discard contemptuously whatever may be found untrue.

The most obvious problem with this quote and the approach to the study of individual differences that it fostered is the equation of statistics, in the absence of experimentation, with scientific practice. Although we may debate the role of inferential statistics in the natural sciences, it is true that Galton's quote predated the application of the experimental method to the behavior of organisms by psychologists (e.g., Skinner, 1938). Rushton (1995a) and Herrnstein and Murray (1994), however, consider Galton to be the intellectual and scientific father of their genetic theories of racial differences. Rushton calls Galton "the originator of scientific research on individual differences" (1995a, 10). Herrnstein and Murray, who refer to the Galtonian tradition of intelligence testing as "the classic tradition," claim: "By accepted standards of what constitutes scientific evidence and scientific proof, that classic tradition has in our view given the world a treasure of information . . ." (1994, 19). This is especially interesting coming from a scientist such as Herrnstein whose own scientific output consists almost exclusively of the use of within-subject experimental designs.

Authors such as Herrnstein, Murray, and Rushton point out that while individual scores on behavioral or psychological tests, for instance IQ tests, correlate poorly, the correlations become much higher when scores are aggregated. The principle of aggregation, according to Herrnstein and Murray (1994), is where the classic (Galtonian) tradition has the most to offer. The rationale for aggregating data is that "randomness in any one measure (error and specificity variance) is averaged out . . . leaving a clearer view of what a person's true behavior is like" (Rushton, 1995a, 19). Also, relationships between individual tests or between scores on tests are more likely to emerge. Thus, aggregating data is supposed to correct for any errors in the actual measurement of the variable(s) in question. The contradiction in this line of reasoning is that the further away one gets from the behavior of the individual, the less can be said about the individual. Herrnstein and Murray acknowledge that the practice of aggregating data does not necessarily permit the prediction, much less the understanding, of individual behavior. More importantly, aggregating data from different tests, or, worse, from different studies, is fraught with so many methodological problems as to render the results meaningless. For example, aggregating data masks differences in methodology (e.g., time, place, populations, sampling procedures, control procedures, measurement tools, etc.). Aggregating data, especially from different studies, can only mean that the results of any individual study were so equivocal that no conclusions could be drawn. Pooling data from different studies is only valid if the studies are methodologically interchangeable which, as I have implied, is a questionable assumption in the present case. Nevertheless, Rushton (1995a) describes instances where low correlations between individual tests were raised by aggregating data from many different tests as if this were sound scientific practice.

In criticizing formalized methods of research and statistics, B. F. Skinner (1972) advocated the use of the experimental method in the study of human behavior. Each approach leads to a different strategy for dealing with measurement error. In contrast to the strategy of aggregating scores from many individuals to increase the statistical reliability of the measurement device (e.g., IQ test) or the sensitivity of the statistical method (e.g., t-Test), Skinner (1972) argued for refining direct experimental control over the behavior of individual subjects. In this way, the reliability of the independent variables is enhanced and sources of variability are eliminated *before* measurements are made rather than after, as is the case when researchers aggregate data. As Skinner (1972) wrote tongue-in-cheek, "No one goes to the circus to see the average dog jump through a hoop significantly oftener than untrained dogs raised under the same circumstances . . ." (114).

Interpreting the Data. A third problem with the use of correlative analyses involves the interpretation of the data. Demonstrating that a correlation exists between two or more variables does not in any way clarify causal relations, although it may hint at possible ones. There is an oft-cited dictum among researchers: "Correlation does not imply causation" (Neale and Liebert, 1973). A correlation between two or more variables is often due to an unspecified process, or "third variable." Those who argue for an evolutionary explanation of human behavior appeal to a third variable—the human genome. Although it is theoretically possible that some human social and intellectual behaviors represent fixed expressions of the human genome, a better explanation for the behaviors in question is one in which a different third variable is implicated—the environmental histories of individuals. In many of the examples cited by social evolutionary theorists, any one or more of the multitudinous environmental variables found in the individual histories of the subjects studied may produce the reported correlations. Just as behavioral similarities between individuals may reflect genotypical similarity, they may just as easily reflect environmental similarity. The correlational evidence offered by evolutionary theorists is simply insufficient to distinguish the biological from the environmental position. The challenge for scientists is to tease apart these possible determinants of behavior, and this cannot be accomplished using correlational methods. Only an experimental analysis can potentially reveal the variables of which human behavior is a function. Galton got it wrong. The "triumph of scientific men" occurs not when human behavior can be subjected to statistical correlation, but rather when it can be subjected to direct experimentation.

Whether one conducts experimental or correlational research in the first place reflects fundamental differences in the types of questions asked. And the types of questions asked reveal differences in the motivations of the researchers. Many authors who either conduct and/or cite correlational research on the relation between behavioral and genetic differences and similarities between groups of humans do so to show what they already believe—that genetics plays a significant role in such characteristics as intelligence, aggression, and reproductive behavior. Hubbard and Wald (1994) have noted that "scientists only look for genetic components in behaviors which their society considers important and probably hereditary" (93). For instance, they point out that even though European peoples read from left to right, whereas Semitic peoples read from right to left, no one has suggested that these are inherent racial differences. As Futuyma said (473):

> The history of scientists' pronouncements on human genetics and behavior is, to a distressing extent, a history of the conventional societal attitudes on these subjects; science has served more as a defense of the status quo than as a force for change.

Genes

I have referred to the social theorizing discussed in this paper as evolutionary; and such a conception implicitly recognizes that what has evolved due to natural selection is a particular genotype that is different from other possible genotypes. In short, evolutionary theories are genetic theories and, as such, we should expect some supporting genetic evidence. According to Kitcher (1985), physical characteristics most susceptible to rigorous genetic analysis are not those that social evolutionary theorists find most interesting. For example, it was recently reported that scientists at the University of Basel in Switzerland have discovered the master control gene responsible for eye development in fruit flies. The scientists have been able to manipulate the gene directly so as to produce eyes in unusual places, like on the legs and thorax. Human geneticists, by comparison, are relegated to studying genetic variation that produces deleterious effects, such as metabolic disorders and defects in color vision. In other words, human geneticists are unable to manipulate the actual genes and must wait for natural genetic variation to produce outcomes that they can then investigate. The genetic evidence most often cited by social evolutionary theorists comes from the field of behavior genetics. Contrary to their name, behavior geneticists do not directly study genes. Rather, they are constrained to examining correlations between poorly defined variables such as scores on intelligence or other psychological tests and family relationships. The reliability of the observations and measurements reported by behavior geneticists is questionable because of the many methodological problems inherent in such research. For example, several authors have pointed out problems with subject selection in research on separated identical twins (e.g., Horgan, 1993; Hubbard and Wald, 1993; Kamin, 1974; Lewontin, Rose, and Kamin, 1984). Moreover, the fact that conclusions about the differences in genes must be based on family resemblance introduces a well-known confound: Family members resemble each other not only because they share genes but also because they share environments. Despite the perception that behavior geneticists have made impressive gains in demonstrating the genetic bases for a wide range of human conditions, such as aggression, homosexuality, intelligence, schizophrenia, and alcoholism, there have been an equal number of serious methodological critiques which, at the very least, temper the claims by behavior geneticists (e.g., Byne, 1994; Horgan, 1993; Kamin, 1974; Hubbard and Wald, 1993; Lewontin, Rose, and Kamin, 1984).

Some social evolutionary theorists argue their case based on a flawed interpretation of evolutionary and genetic logic. For example, Itzkoff (1985), who is neither an evolutionary biologist nor a geneticist, presents a case for the evolution of human intelligence as a function of the natural selection of the human brain. Itzkoff reasons that because so many biochemical combinations are involved in the growth and patterns of brain structure, slight variations can exist between close relatives and large variations between relatively isolated groups of humans. He concludes: "The brain evolved along a wide diversity of lines" producing differences in both "the quantity and quality of intelligence" (23). He presents this rationale to support his claim that different groups of humans (blacks and whites) come into the world with different genetic potentials for intelligence. There are serious flaws in Itzkoff's reasoning, the most fatal of which is that there is simply insufficient evidence to support his conclusions that normal variation in intelligence has a genetic basis. Moreover, his argument is based on the assumption that there exists genetic variation within populations of humans, and that selection has operated differently in different human groups even though "there is insufficient evidence to conclude that normal variation in human behavioral traits has a genetic basis" (Futuyma, 1979). Finally, there is a broader principle of genetics that is often not fully appreciated by many social evolutionary theorists, as Futuyma notes (476):

> One cannot say that a universal trait . . . is either genetic or environmental, for it is the expression of genes in a series of environments. Genetics provides no means of investigating the inheritance of an invariant trait. Thus to postulate that it is genetic is to pose an untestable and meaningless hypothesis. The only question one can legitimately ask is, Is the trait highly canalized, or does it vary greatly under different environmental conditions, compared to other traits?

If certain behavioral traits, such as aggression, sibling rivalry, sex-role behavior, or intelligence, were highly canalized, then, according to Futuyma, we would not expect them to be modifiable by environmental factors.

Environment

Contrary to most traditional conceptions of the environment, scientists who study the functional relationship between the behavior of organisms and environmental variables—behavior analysts—define environment functionally as all of the stimuli that enter into functional relationships with an organism's behavior at any one time (Schlinger, 1995). Behavior analysts view the environment as consisting of energy changes (stimuli) of various sorts that not only affect the sensory receptors of organisms but, more importantly, affect their behavior. Thus, the environment is not defined necessarily by its structure prior to the study of behavior, but rather after functional relations have been established by experimentation. In other words, behavior analysts define environment by how it functions to control behavior. The environmental history of an individual represents one category of ultimate behavioral causation; the other being the evolutionary history of the species to which the individual belongs.

Over the last 50 years, scientists who study learning have amassed volumes of testable, repeatable, experimental data demonstrating the powerful influence of environmental manipulations on a wide range of behaviors. Several scientific journals are devoted almost exclusively to direct experimentation on the effects of the environment. The *Journal of the Experimental Analysis of Behavior,* for example, has produced almost 40 years of data, including direct and systematic replication experiments. In none of these instances are data aggregated in order to achieve criteria of significance. In fact, in many experiments, little, if any, statistical analysis is needed to verify the reliability of the results. Internal validity is demonstrated time and time again by direct within-experiment refinement and control of objective independent variables. External validity of these findings has been consistently shown over the same 40-year period by successfully applying the scientific principles discovered in the experimental laboratory to problem human behavior. For example, the *Journal of Applied Behavior Analysis* has produced almost 30 years of experimental research on human behavioral problems, including compliance, crying, social interaction, cooperation, aggression, walking, reading and writing. Perhaps more convincing, numerous experiments have shown that behaviors previously thought to be impervious to environmental manipulation could be dramatically altered via operant conditioning, including psychotic behavior (Ayllon, 1963) mutism (Isaacs, Thomas, and Goldiamond, 1960), coma (Boyle and Greer, 1983; Fuller, 1949), and a wide range of physiological functions, such as diastolic and systolic blood pressure, Galvanic skin response, cardiac function, and asthma (Shapiro and Surwit, 1976), to mention a few. Moreover, the neurophysiological bases of basic learning processes have recently been uncovered, thus strengthening their status as scientific laws. For example, experimental evidence now shows that individual neurons can be operantly conditioned (Stein and Belluzzi, 1988; Stein, Xue, and Belluzzi, 1994). Such experiments demonstrate that the laws of operant conditioning discovered at the level of behavior-environment have their basis in neurophysiology.

Although volumes could be written summarizing the findings of the experimental science of behavior, suffice it to say that this is the only "cold-blooded verification" of theory that one should accept. Although not every human behavior that we find interesting can be subjected to experimental verification, a large corpus of experimental findings on basic learning processes is valuable in part because scientists can extrapolate from that foundation to novel behaviors. This is the essence of scientific interpretation (Palmer, 1991; Schlinger, 1995).

Some psychologists who espouse evolutionary theories of human behavior, however, cite non-experimental, and even non-quantitative, approaches to the understanding of certain human behaviors as evidence against a behavior analytic interpretation. For example, Cosmides and Tooby (1987 and 1992) cite Chomsky extensively to make their argument that behaviorist approaches to language have been falsified and, therefore, cannot account for the acquisition of human language. Their conclusion is that evolutionarily adapted cognitive learning mechanisms constitute the only adequate explanation of human language acquisition. It is interesting that these citations consist solely of rationalist argument and not scientific experimentation and yet they are presented as if they are scientifically conclusive. Behavior analysts, in contrast, have not only provided substantive rebukes of Chomsky's critique of behaviorist interpretations of language (MacCorquodale, 1970), but they have also argued persuasively that Chomsky's own evolutionary account of language is untenable when held to Darwinian standards (Palmer, 1986; Dennett, 1995).

The susceptibility of human language to operant conditioning is no longer a debatable issue. During the past 50 years the operant control of verbal behavior has been demonstrated numerous times, including experiments on the operant conditioning of infant vocalizations (Poulson, 1983; Whitehurst, 1972), the content of conversation (Azrin, Holz, Ulrich, and Goldiamond, 1961), fluent requests (Rosenfeld and Baer, 1970), and grammatical forms, such as prepositional phrases (Lee, 1981) and plural morphemes (Guess, 1969). Experiments have also verified Skinner's (1957) hypothesized functional verbal operants (see Oah and Dickinson, 1989, for a review). Moreover, behavior analytic principles have been used fruitfully to interpret a diverse group of studies on language development in infancy (Schlinger, 1995). The critical question regarding human language, or any complex human behavior for that matter, is whether plausible mechanisms or processes have been postulated. Operant learning principles constitute a plausible process both for verbal and nonverbal behavior, if for no other reason than they have already been shown experimentally to affect a wide range of human behaviors. Cognitive learning mechanisms, however, are not plausible in part because they are almost wholly inferred from the very behavior they are invoked to explain. Cognitive theorists cannot tell us what cognitive mechanisms look like or how they actually affect behavior.

Nature-Nurture

Perhaps it would be appropriate to conclude with a word about nature-nurture, the phrase first coined by Galton. The issue of the nature or nurture of behavior is not as meaningless as some might suppose, as Dobzhansky asked (1964, 55): "To what extent are differences observed among people conditioned by the differences of their genotypes and by the differences between the environments in which people were born, grew and were brought up?"

The question about the genesis of a given behavior is an empirical question. The only truly scientific approach is to conduct experiments in an attempt to uncover functional relations between behavior and its determinants. The amount of data demonstrating the overwhelming effects of environment on behavior establishes the plausibility of environmental interpretations not only of behavioral similarities but also of behavioral differences between humans. Evolution has obviously played an important part in human behavior. But rather than selecting for behavioral rigidity, it has selected for behavioral plasticity (Dobzhansky, Ayala, Stebbins, and Valentine, 1977). As Futuyma concluded (491):

> On balance, the evidence for the modifiability of human behavior is so great that genetic constraints on our behavior hardly seem to exist. The dominant factor in recent human evolution has been the evolution of behavioral flexibility, the ability to learn and transmit culture.

Conclusion

The problem with evolutionary explanations of behavior is that the evidence proffered to support them is so fraught with methodological problems that it is simply insufficient to warrant any conclusions about the role of genes and, thus, evolution. In contrast, there is already a wealth of experimental evidence establishing the plausibility of an environmental/learning account of much human behavior. This is not to say that genes play no role in human behavioral differences or similarities, only that the jury is still out on the verdict regarding the extent and nature of that role. The only way to truly make a case for genetic influence on behavior is to control for environmental variables and manipulate genetic variables, which, at present, are simply not possible with humans. Finally, from a practical point of view, environmental explanations are more valuable than evolutionary ones because they suggest immediate ways in which behavior can be changed.

Evolutionary theorists certainly succeed in making an interesting and often compelling case that perhaps there is some deeper core of human nature that ties us all together and around which we as individuals, and maybe even as groups, differ. It is a case that appeals to many people, including the media, all of whom are hungry for some evidence that sheds light on our nature. Unfortunately, the case is replete with evidential problems, and will have to be retried if and when more substantial evidence can be obtained. Until then, we should rely on what we know scientifically about human behavior.

References:

Ayllon, T. 1963. "Intensive Treatment of Psychotic Behaviour by Stimulus Satiation and Food Reinforcement." *Behaviour Research and Therapy.* 1:53–61.

Azrin, N. H., W. Holz, R. Ulrich, & I. Goldiamond. 1961. "The Control of the Content of Conversation through Reinforcement." *Journal of the Experimental Analysis of Behavior.* 4:25–30.

Barash, D. 1977. *Sociobiology and Human Behavior.* New York: Elsevier.

Barkow, J. H., L. Cosmides, and J. Tooby. 1992. *The Adapted Mind: Evolutionary Psychology and the Generation of Culture.* New York: Oxford University Press.

Bock, K. 1980. *Human Nature and History.* New York: Columbia University Press.

Boyle, M. E., & R. D. Greer. 1983. "Operant Procedures and the Comatose Patient." *J. of Applied Behavior Analysis.* 16:3–12.

Buss, D. M. 1994. *The Evolution of Desire: Strategies of Human Mating.* New York: Basic Books.

Byne, W. 1994. "The Biological Evidence Challenged." *Scientific American.* 270:50–55.

Cosmides, L., & J. Tooby. 1987. "From Evolution to Behavior: Evolutionary Psychology as the Missing Link." In J. Dupre (Ed.), *The Latest on the Best Essays on Evolution and Optimality.* Cambridge, MA: MIT Press.

Daly, M., & M. Wilson. 1982. "Whom are Newborn Babies Said to Resemble?" *Ethology and Sociobiology.* 3:69–78.

——. 1988. *Homicide.* New York: Aldine de Gruyter.

Dennett, D. C. 1995. *Darwin's Dangerous Idea: Evolution and the Meanings of Life.* New York: Simon & Schuster.

Dobzhansky, T. 1964. *Heredity and the Nature of Man.* New York: Harcourt, Brace & World.

Dobzhansky, T., F. J. Ayala, G. L. Stebbins, & J. W. Valentine. 1977. *Evolution*. San Francisco: Freeman.

Freedman, D. G. 1979. *Human Sociobiology*. New York: The Free Press.

Fuller, P. F. 1949. "Operant Conditioning of a Vegetative Human Organism." *American Journal of Psychology*. 69:587–590.

Futuyma, D. J. 1979. *Evolutionary Biology*. Sunderland, MA: Sinauer.

Gould, S. J. 1981. *The Mismeasure of Man*. New York: Norton.

Guess, D. 1969. "A Functional Analysis of Receptive Language and Productive Speech: Acquisition of the Plural Morpheme." *J. of Applied Behavior Analysis*. 2:55–64.

Hamer, D., & P. Copeland. 1994. *The Science of Desire: The Search for the Gay Gene and the Biology of Behavior*. New York: Simon & Schuster.

Herrnstein, R. J., & C. Murray. 1994. *The Bell Curve: Intelligence and Class Structure in American Life*. New York: The Free Press.

Horgan, J. 1993. "Eugenics Revisited." *Scientific American*. 268:123–131.

Hubbard, R., & E. Wald. 1993. *Exploding the Gene Myth*. Boston: Beacon Press.

Isaacs, W., J. Thomas, & I. Goldiamond. 1960. "Application of Operant Conditioning to Reinstate Verbal Behavior in Psychotics." *J. of Speech and Hearing Disorders*. 25:8–12.

Itzkoff, S. W. 1985. *Triumph of the Intelligent: The Creation of* Homo Sapiens Sapiens. Ashfield, MA: Paideia.

——. 1994. "The Decline of Intelligence in America: A Strategy for National Renewal." Westport, CT: Praeger.

Kamin, L. 1974. *The Science and Politics of IQ*. Potomac, MD: Erlbaum.

Kellert, S. R., & E. O. Wilson. 1993. *The Biophilia Hypothesis*. Washington, DC: Island Press.

Kitcher, P. 1985. *Vaulting Ambition: Sociobiology and the Quest for Human Nature*. Cambridge, MA: The MIT Press.

Lee, V. L. 1981. "Prepositional Phrases Spoken and Heard." *J. of the Experimental Analysis of Behavior*. 35:227–242.

Lewontin, R. C., S. Rose, L. J. Kamin. 1984. *Not in Our Genes*. New York: Pantheon Books.

Lumsden, C., & E. O. Wilson. 1983. *Promethean Fire*. Cambridge, MA: Harvard University Press.

MacCorquodale, K. 1970. "On Chomsky's Review of Skinner's Verbal Behavior." *J. of the Experimental Analysis of Behavior*. 13:83–89.

Neale, J. M., & R. M. Liebert. 1973. *Science and Behavior: An Introduction to Methods of Research*. Englewood Cliffs, NJ: Prentice-Hall.

Nesse, R. M., & G. C. Williams. 1994. *Why We Get Sick: The New Science of Darwinian Medicine*. New York: Times Books.

Oah, S., & A. M. Dickinson. 1989. "A Review of Empirical Studies of Verbal Behavior." *The Analysis of Verbal Behavior*. 7:53–68.

Palmer, D. C. 1986. "Chomsky's *Nativism*: A Critical Review." In L. J. Hayes & P. N. Chase (Eds.), *Dialogues on Verbal Behavior*. Springfield, IL: Thomas.

——. 1991. "A Behavioral Interpretation of Memory." In L. J. Hayes and P. N. Chase (Eds.), *Dialogues on Verbal Behavior*. Reno, NV: Context Press.

Poole, R. 1994. *Eve's Rib: The Biological Roots of Sex Differences*. New York: Crown.

Poulson, C. L. 1983. "Differential Reinforcement of Other-Than-Vocalization as a Control Procedure in the Conditioning of Infant Vocalization Rate." *J. of Experimental Child Psychology*. 36: 471–489.

Rosenfeld, H. M., & D. M. Baer. 1970. "Unbiased and Unnoticed Verbal Conditioning: The Double Agent Robot Procedure." *J. of the Experimental Analysis of Behavior*. 14:99–107.

Rushton, J. P. 1995a. *Race, Evolution, and Behavior: A Life History Perspective*. New Brunswick: Transaction Publishers.

——. 1995b. "J. Philippe Rushton Responds." *Skeptic*. 3:22–25.

Sahlins, M. 1976. *The Use and Abuse of Biology*. Ann Arbor: University of Michigan Press.

Schlinger, H. D. 1992. "Intelligence: Real or Artificial?" *The Analysis of Verbal Behavior*. 10:125–133.

——. 1995. *A Behavior Analytic View of Child Development*. New York: Plenum.

Shapiro, D., & R. S. Surwit. 1976. "Learned Control of Physiological Function and Disease." In H. Leitenberg (Ed.), *Handbook of Behavior Modifi-*

cation and Behavior Therapy. Englewood Cliffs, NJ: Prentice-Hall.

Skinner, B. F. 1938. *The Behavior of Organisms*. Englewood Cliffs, NJ: Prentice-Hall.

——. 1972. "A Case History in Scientific Method." In *Cumulative Record* (3rd ed.). New York: Meredith.

Stein, L., & J. D. Belluzzi 1988. "Operant Conditioning of Individual Neurons." In M. L. Commons, R. M. Church, J. R. Stellar, & A. R. Wagner (Eds.), *Quantitative Analysis of Behavior: Vol. 7.* Hillsdale, NJ: Erlbaum.

Stein, L., B. G. Xue, & J. D. Belluzzi. 1994. "In Vitro Reinforcement of Hippocampal Bursting: A Search for Skinner's Atoms of Behavior." *J. of the Experimental Analysis of Behavior.* 61:155–168.

Symons, D. 1992. "On the Use and Misuse of Darwinism in the Study of Human Behavior." In J. Barkow, L. Cosmides, & J. Tooby (Eds.), *The Adapted Mind.* New York: Oxford University Press.

Tavris, C. 1995. "A Place in the Sun." *Skeptic.* 3:62–63.

Tooby, J., & L. Cosmides. 1989. "Evolutionary Psychology and the Generation of Culture, Part II." *Ethology and Sociobiology.* 10:51–97.

——. 1992. "The Psychological Foundations of Culture." In J. Barkow, L. Cosmides, & J. Tooby (Eds.), *The Adapted Mind.* New York: Oxford University Press.

Whitehurst, G. J. 1972. "Production of Novel and Grammatical Utterances by Young Children." *J. of Experimental Child Psychology.* 13:502–515.

Wilson, E. O. 1975. *Sociobiology: The New Synthesis.* Cambridge, MA: Harvard University Press.

——. 1978. *On Human Nature.* Cambridge, MA: Harvard University Press.

——. 1993. "Biophilia and the Conservation Ethic." In Kellert & Wilson (Eds.), *The Biophilia Hypothesis.* Washington, DC: Island Press.

Wright, R. 1994. *The Moral Animal: The New Science of Evolutionary Psychology.* New York: Pantheon.

——. 1995. "The Biology of Violence." *The New Yorker,* March, 69–77.

Memes as Good Science

SUSAN J. BLACKMORE

Without the theory of evolution by natural selection nothing in the world of biology makes much sense. Without Darwin and neo-Darwinism, you cannot answer questions like Why do bats have wings? Why do cats have five claws? or Why do our optic fibres cross in front of our retinas? You can only fall back on appeals to an imaginary creator.

I am going to make a bold claim. Without the theory of evolution by memetic selection nothing in the world of the mind makes much sense. Without memetics you cannot answer questions like Why can't I get that thought out of my mind? Why did I decide to write this article and not another one? Who am I? Without memetics you can only fall back on appeals to an imaginary conscious agent.

In this article I want to lay the groundwork for a theory of memetics and see how far we can get. I shall outline the history and origins of the idea, explore how it has been used, abused, and ignored, and how it has provided new insight into the power of religions and cults. I shall then take on a meme's-eye view of the world and use this to answer five previously unanswered questions about human nature. Why can't we stop thinking? Why do we talk so much? Why are we so nice to each other? Why are our brains so big? And, finally, what is a self?

A History of the Meme Meme

In 1976 Dawkins published his best-selling *The Selfish Gene*. This book popularised the growing view in biology that natural selection proceeds not in the interest of the species or of the group, nor even of the individual, but in the interest of the genes. Although selection takes place largely at the individual level, the genes are the true replicators and it is their competition that drives the evolution of biological design. Dawkins, clear and daring as always, suggested that all life everywhere in the universe must evolve by the differential survival of slightly inaccurate self-replicating entities, which he called replicators.

Furthermore, these replicators automatically band together in groups to create systems, or machines, that carry them around and work to favour their continued replication. These survival machines, or vehicles, are our familiar bodies—and those of cats, E-coli, and cabbages—created to carry around and protect the genes inside them. At the end of the book Dawkins suggested that Darwinism is too big a theory to be confined to the narrow context of the gene. So he asked an obvious, if provocative, question. Are there any other replicators on our planet? Yes, he concluded. Staring us in the face, though still drifting clumsily about in its primeval soup of culture, is another replicator—a unit of imitation. He gave it the name *meme*, to rhyme with dream or seem. As examples he suggested—tunes, ideas, catch-phrases, clothes fashions, ways of making pots or of building arches—memes are stored in human brains and passed on by imitation.

In just a few pages Dawkins laid the foundation for understanding the evolution of memes. He discussed their propagation by jumping from brain to brain, likened them to parasites infecting a host, treated them as physically realised living structures, and showed how mutually assisting memes will group together just as genes do. He argued that once a new replicator arises it will tend to take over and begin a new kind of evolution. Above all he treated memes as replicators in their own right, chastising those of his colleagues who tended always to go back to biological advantage to answer questions about human behaviour. Yes, he agreed, we got our brains for biological (evolutionary and genetic) reasons but now we have a new replicator that has been unleashed and it need not be subservient to the old. In other words, memetic evolution can now proceed without regard to its effects on the genes.

A few years later Douglas Hofstadter wrote about viral sentences and self-replicating structures in his *Scientific American* column "Metamagical Themas." Readers replied, with examples of text using bait and hooks to ensure their own replication. They suggested viral sentences from the simplest instruction, such as "copy me!", through those with added threats ("I put a curse on you") or promises ("grant you three wishes"), to examples of virulent chain letters (Hofstadter, 1985). One reader suggested the term *memetics* for the discipline of studying memes.

Yet memetics did not really take off. Why not? The basic idea is very simple. If Dawkins is right then everything you have learned by imitation from someone else is a meme. This includes all the words in your vocabulary, the stories you know, the skills and habits you have picked up from others and the games you like to play. It includes the songs you sing and the rules you obey. So, for example, whenever you drive on the right (or on the left in my case here in England), eat a hamburger or a pizza, whistle Happy Birthday to You or Mama I Love You, or even shake hands, you are dealing in memes. Memetics is the study of why some memes spread and others do not.

The greatest proponent of memetics since Dawkins has been the philosopher Dan Dennett. In his books *Consciousness Explained* (1991) and *Darwin's Dangerous Idea* (1995) Dennett expands on the idea of the meme as replicator. In *On the Origin of Species*, Darwin (1859) explained how natural selection must happen if certain conditions are met. If there is heredity from parent to offspring, variation among the offspring, and not all the offspring can survive—then selection must take place. Individuals who have some useful relative advantage "have the best chance of being preserved in the struggle for life" (Darwin, 1859, 127) and will then pass on this advantage to their offspring. Darwin clearly saw how obvious the process of natural selection is once you have grasped it. Dennett describes evolution as a simple algorithm—a mindless procedure that when carried out must produce a result. You need three things—heredity, variation and selection—to make evolution inevitable. Evolution need not produce us, of course, or anything remotely like us; for evolution has no plans and no foresight. Nevertheless, you must get something more complex than what you started with. The evolutionary algorithm is "a scheme for creating Design out of Chaos without the aid of Mind" (Dennett, 1995, 50). This, says Dennett, is *Darwin's Dangerous Idea.* No wonder people have been terrified of it, and fought so hard against it. It is outrageously simple and terrifyingly powerful.

If evolution is an algorithm then it should

be able to run on different substrates. We tend to think of evolution as depending on genes because that is the way biology works on this planet, but the algorithm is neutral about this and will run wherever there is heredity, variation and selection; or as Dawkins puts it, a replicator. It doesn't matter which replicator. If memes are replicators then evolution will occur. So are memes replicators? There is enormous variety in the behaviors human beings emit; these behaviors are copied, more or less accurately, by other human beings, and not all the copies survive. The meme therefore perfectly satisfies the conditions of heredity, variation and selection. Just think of jokes.

Millions of variants are told by millions of people. Only a few get passed on and repeated and even fewer make it into the big time or the collections of classics. Scientific papers proliferate but only a few get long listings in the citation indexes. Only a few of the disgusting concoctions made in woks actually make it onto the TV shows that tell you how to wok things and only a few of my brilliant ideas have ever been appreciated by anyone! In other words, competition to get copied is fierce.

Of course memes are not like genes in many ways and we must be very careful in applying terms from genetics to memetics. The copying of memes is done by a kind of "reverse engineering" by one person copying another's behaviour, rather than by chemical transcription. Also we do not know just how memes are stored in human brains and whether they will turn out to be digitally stored, like genes, or not. However, the important point is that if memes are true replicators, memetic evolution must occur. Dennett is convinced they are and he explores how memes compete to get into as many minds as possible. This competition is the selective force of the memosphere and the successful memes create human minds as they go, restructuring our brains to make them ever better havens for more memes. Human consciousness, claims Dennett, is itself a huge meme-complex, and a person is best understood as a certain sort of ape infested with memes. If he is right then we cannot hope to understand the origins of the human mind without memetics.

This makes it all the more fascinating that most people interested in the human mind have ignored memetics or simply failed to understand it. Mary Midgley (1994) calls memes "mythical entities" that cannot have interests of their own; "an empty and misleading metaphor." In a 1996 radio debate, Stephen Jay Gould called the idea of memes a "meaningless metaphor" (though I am not sure one can actually have a meaningless metaphor!). He wishes "that the term cultural evolution would drop from use" (1996, 219–220).

The word *meme* does not even appear in the index of many important books about human origins and language (e.g., Donald, 1991; Dunbar, 1996; Mithen, 1996; Pinker, 1994; Tudge, 1995; Wills, 1993); nor is it in an excellent collection on evolutionary psychology (Barkow, Cosmides and Tooby, 1992); nor in books about evolutionary ethics (Ridley, 1996; Wright, 1994). Although there are many theories of the evolution of culture, almost all make culture entirely subservient to genetic fitness, as in E. O. Wilson's (1978) metaphor of genes holding culture on a leash, or Lumsden and Wilson's claim that "the link between genes and culture cannot be severed" (1981, 344). Cavalli-Sforza and Feldman (1981) treat "cultural activity as an extension of Darwinian fitness" (362), and even Durham (1991)—the only one to use the word *meme*—sticks to examples of cultural features with obvious relevance to genetic fitness such as color naming, dietary habits and marriage customs.

Perhaps Boyd and Richerson (1990) come closest to treating the cultural unit as a true replicator. However, they still view "genetic and cultural evolution as a tightly coupled coevolutionary process in humans" (Richerson and Boyd, 1992, 80). As far as I can understand them, no one except Cloak (1975) and Dawkins treats their unit of cultural exchange as a true replicator. If there is a continuum from Gould's outright rejection at one end, to Dawkins and Cloak at the other, then most scholars lie in between. They accept cultural evolution but not the idea of a second replicator. When they say adaptive or maladaptive they mean for the genes. When it comes to the crunch they always fall back on appeals to biological advantage, just as Dawkins complained his colleagues did 20 years ago.

Dawkins is clear on this issue when he says "there is no reason why success in a meme should have any connection whatever with genetic success." I agree. I am going to propose a theory of memetics that lies at the far end of this continuum. I suggest that once genetic evolution had created creatures that were capable of imitating each other, a second replicator was born. Since then our brains and minds have been the product of two replicators, not one. Today many of the selection pressures on memes are still of genetic origin (such as whom we find sexy and what food tastes good), but as memetic evolution proceeds faster and faster, our minds are increasingly the product of memes, not genes. If memetics is true then the memes have created human minds and culture just as surely as the genes have created human bodies.

Religions as Meme-Complexes

Dawkins (1976) introduced the term *co-adapted meme-complex*. By this he meant a group of memes that thrive in each other's company. Just as genes group together for mutual protection, leading ultimately to the creation of organisms, so we might expect memes to group together. As Dawkins (1993, 20) puts it, "there will be a ganging up of ideas that flourish in one another's presence." Meme-complexes include all those groups of memes that tend to be passed on together, such as political ideologies, religious beliefs, scientific theories and paradigms, artistic movements, and languages. The most successful of these are not just loose agglomerations of compatible ideas, but well-structured groups with different memes specializing as hooks, bait, threats, and immune systems. (Memetic jargon is still evolving and these terms may change but see Grant's "memetic lexicon," 1990.)

When I was about 10 years old I received a postcard and a letter that contained a list of six names that instructed me to send a postcard to the first name on the list. I was to put my own name and address at the bottom and send the new list to six more people. It promised me I would receive lots of postcards. This was a fairly innocuous chain letter as these things go, consisting just of a bait (the promised postcards) and a hook (send it to six more people). Threats are also common (send this on or the evil eye will get you) and many have far worse consequences than a waste of stamps. What they have in common is the instruction to "duplicate me" (the hook) along with co-memes for coercion. These simple little groups can spread quite well.

With the advent of computers, viral meme-groups have much more space to play in and can leap from disk to disk among "unhygienic" computer users. Dawkins (1993) discusses how computer viruses and worms use tricks to get themselves spread. Some bury themselves in memory only to pop up as a time bomb; some infect only a small proportion of those they reach, and some are triggered probabilistically. Like biological viruses they must not kill their host too soon or they will die out. Their final effect may be quite funny, such as one that makes the Macintosh computer's loudspeaker say "Don't Panic!," but some have clogged up entire networks and destroyed whole doctoral theses. My students have recently encountered a virus in WORD 6 that lives in a formatting section called "Thesis"—tempting you to get infected just when your year's work is almost finished. No wonder we now have a proliferation of anti-virus software—the equivalent of medicine for the infosphere.

Internet viruses are a relatively new arrival. Last week I received a very kind warning from someone I've never met. "Do not download any message entitled Penpal Greetings" it said—and went on to warn me that if I read this terrible message I would have let in a Trojan Horse virus that would destroy everything on my hard drive and then send itself on to every e-mail address in my mail box. To protect all my friends, and the world-wide computer network, I had to act fast and send the warning on to them.

Have you spotted the trick? The virus described does not make sense; in fact, it does not exist. The real virus is the warning. This is a very clever little meme-complex that uses both threats and appeals to altruism to get you—the silly, caring victim—to pass it on. It is not the first—Good Times and Deeyenda Maddick used a similar trick—and it probably won't be the last. However, as more people learn to ignore the warnings, these viruses will start to fail and perhaps that will let in worse viruses, as people start to ignore warnings they ought to heed. So, Watch Out!

What does this have to do with religions? According to Dawkins, a great deal. The most controversial application of memetics is undoubtedly his treatment of religions as co-adapted meme-complexes (Dawkins, 1976, 1993; Miele, 1995). Dawkins unashamedly describes religions as "viruses of the mind" and sets about analysing how they work. They work, he says, because human brains are just what info-viruses need; brains can soak up information, replicate it reasonably accurately, and obey the instructions it embodies. Dawkins uses the example of Roman Catholicism, a gang of mutually compatible memes that is stable enough to deserve a name. The heart of Catholicism is its major beliefs: a powerful and forgiving God, Jesus his son who was born of a virgin and rose again from the dead, the holy spirit, and so on. If these aren't implausible enough you can add belief in miracles or the literal transubstantiation of wine into blood.

Why should anyone believe these things? Threats of hell fire and damnation are an effective and nasty technique of persuasion. From an early age children are brought up by their Catholic parents to believe that if they break certain rules they will burn in hell forever after death. The children cannot easily test this since neither hell nor God can be seen, although He can see everything they do. So they must simply live in life-long fear until death, when they will find out for sure, or not. The idea of hell is thus a self-perpetuating meme.

Did I say test the idea? Some religious beliefs could be tested, such as whether wine really turns into blood, or whether prayer actually helps; hence the need for the anti-testing meme of faith. In Catholicism, doubt must be resisted, while faith is nurtured and respected. If your knowledge of biology leads you to doubt the virgin birth—or if war, cruelty and starvation seem to challenge the goodness of God—then you must have faith. The biblical story of Doubting Thomas is a cautionary tale against seeking evidence. As Dawkins puts it, "Nothing is more lethal for certain kinds of meme than a tendency to look for evidence" and religions, unlike science, make sure they discourage it (Dawkins, 1976, 198). Also unlike science, religions often include memes that make their carriers violently intolerant of new and unfamiliar ideas, thus protecting themselves against being ousted in favour of a different religion—or none at all.

Finally the meme-complex needs mechanisms to ensure its own spread. A kill the infidel meme will dispose of the opposition. Go forth and multiply will produce more children to pass itself on to. So will forbidding masturbation, birth control, or interfaith marriages. If fear of going blind doesn't work, there are prizes in heaven for missionaries and those who convert unbelievers (Dawkins, 1993; Lynch, 1996). Catholicism generally spreads from parent to child but celibate priests play a role too. This is particularly interesting since celibacy means a dead end for the genes, but not for the memes. A priest who has no wife and children to care for has more time to spread his memes, including that for celibacy. Celibacy is another partner in this vast complex of mutually assisting religious memes.

Dawkins (1993) gives other examples from Judaism, such as the pointlessness of rabbis testing for the kosher-purity of food, or the horrors of Jim Jones leading his flock to mass suicide in the Guyana jungle. Today he might add Heaven's Gate to the catalogue. "Obviously a meme that causes individuals bearing it to kill themselves has a grave disadvantage, but not necessarily a fatal one . . . a suicidal meme can spread, as when a dramatic and well-publicised martyrdom inspires others to die for a deeply loved cause, and this in turn inspires others to die, and so on" (Dawkins, 1982, 111).

Dawkins might equally have chosen Islam, a faith that includes the concept of the jihad (holy war), and has particularly nasty punishments for people who desert the faith. Even today the author and heretic Salman Rushdie lives in fear of his life because many Muslims consider it their holy duty to kill him. Once you have been infected with powerful memes like these you must pay a high price to get rid of them.

Lynch (1996) explores in depth some techniques used by religions and cults. "Honour thy father and mother" is an excellent commandment, increasing the chance that children will take on beliefs from their parents, including the commandment itself. As a secular meme it might not succeed very well, since kids would surely reject it if they thought it came straight from the parents. However, presented as an idea from God (who is all powerful, all-seeing and punishes disobedience) it has a much better chance—a good example of memes ganging up. Dietary laws may thrive because they protect against disease, but may also keep people in the faith by making it harder for them to adapt to other diets outside. Moral codes may enhance effective cooperation and survival but may also be ways of punishing lapses of faith. Observing "holy days" ensures lots of time for spreading the memes, and public prayers and grace at meals ensure that lots of people are exposed to them. Learning sacred texts by heart and setting them to inspiring or memorable music ensure their longevity.

In the long history of religions, most of them have spread vertically—that is, from parent to child. Even today the best predictor of your religion is your parent's religion—even if you think you rationally chose the best or truest one! Today, however, more and more new religions and cults spread horizontally—from any person to any other person. The two types use different meme tricks for their replication. As an example of the first type Lynch (1996) cites the Hutterites. They average more than 10 children per couple, a fantastic rate that is possibly helped by the way they distribute parental responsibility, making each extra child only a slightly greater burden for its natural parents. Other religions put more effort into conversion, like the evangelical faiths which thrive on instant rewards and spiritual joy on conversion.

In case I seem to be implying that people have deliberately manufactured religions this way, that is not the case. Imagine in the long, long history of human religious endeavour, all the millions and millions of different statements, ideas, and commandments that must have been uttered at some time or another. Which would you expect to have made it to the present? The answer is, of course, the ones that just happened to have included clever tricks or come together with other ideas they could gang up with. The countless millions of other ideas have simply been lost. This is memetic evolution, and extinction.

Taking the Meme's Eye View

We are now ready to take the meme's eye view. Imagine a world full of hosts for memes (e.g., brains) and far more memes than can possibly find homes. Which memes are more likely to find a safe home and get passed on again? It's that simple.

In doing this I try to follow some simple rules. First, remember that memes (like genes) do not have foresight! Second, consider only the interests of the memes, not of the genes or the organism. Memes do not care about genes or people—all they do is reproduce themselves. Shorthand statements like memes want x or memes try to do y must always be translatable back into the longer version, such as memes that have the effect of producing x are more likely to survive than those that do not. Third, memes, by definition, are passed on by imitation. So learning by trial and error or by feedback is not memetic, nor are all forms of communication. Only when an idea, behaviour, or skill is passed on by imitation does it count as a meme. Now, remembering these rules, we can ask the question and see where it leads.

Imagine a world full of brains, and far more memes than can possibly find homes. Which memes are more likely to find a safe home and get passed on again? Some of the consequences are startlingly obvious—once you see them. And some are frighteningly powerful. I shall start with two simple ones, partly as exercises in thinking memetically.

1. *Why Can't We Stop Thinking?* Can you stop thinking? If you have ever meditated you will know just how hard this is—the mind just seems to keep blithering on. If we were thinking useful thoughts, practising mental skills, or solving relevant problems there might be some point, but mostly we don't seem to be. So why can't we just sit down and not think? From a genetic point of view all this extra thinking seems extremely wasteful—and animals that waste energy don't survive. Memetics provides a simple answer. *Imagine a world full of brains, and far more memes than can possibly find homes. Which memes are more likely to find a safe home and get passed on again?* Imagine a meme that encourages its host to keep on mentally rehearsing it, or a tune that is so easy to hum that it goes round and round in your head, or a thought that just compels you to keep thinking it. Imagine in contrast a meme that buries itself quietly in your memory and is never rehearsed, or a tune that is too unmemorable to go round in your head, or a thought that is too boring to think again. Which will do better? Other things being equal, the first lot will. Rehearsal aids memory, and you are likely to express (or even sing) the ideas and tunes that fill your waking hours. What is the consequence? The memosphere fills up with catchy tunes, and thinkable thoughts. We all come across them and so we think an awful lot. The principle here is familiar from biology. In a forest, any tree that grows tall gets more light. So genes for growing tall become more common in the gene pool and the forest ends up being as high as the trees can make it.

2. *Why Do We Talk So Much? Imagine a world full of brains, and far more memes than can possibly find homes. Which memes are more likely to find a safe home and get passed on again?* Imagine any meme that encourages talking. It might be an idea like talking makes people like you or it's friendly to chat. It might be an urgent thought that you feel compelled to share, a funny joke, good news that every-

body wants to hear, or any meme that thrives inside a talkative person. Imagine in contrast any meme that discourages talking, such as the thought talking is a waste of time. It might be something you dare not voice aloud, something very difficult to say, or any meme that thrives inside a shy and retiring person. Which will do better? Put this way the answer is obvious. The first lot will be heard by more people and, other things being equal, simply must stand a better chance of being propagated. What is the consequence of this? The memosphere will fill up with memes that encourage talking and we will all talk an awful lot. And we do!

A simpler way of putting it is this: people who talk more will, on average, spread more memes. So any memes which thrive in chatterboxes are likely to spread. This makes me see conversation in a new light. Is all that talking really founded on biological advantage? Talking takes a lot of energy and we talk about some daft and pointless things. Do these trivial and stupid thoughts and conversations have some hidden biological advantage? I think not. In fact, in this case memes seem to be working against genes. This sets the stage for a more audacious suggestion.

3. *Why Are We So Nice to Each Other?* Of course we aren't always nice to each other, but human cooperation and altruism are something of a mystery—despite tremendous advances made in understanding kin selection and inclusive fitness, reciprocal altruism and evolutionarily stable strategies (Wright, 1994; Ridley, 1996; *Skeptic,* Vol. 4, Nos. 1 and 2). Human societies exhibit much more cooperation than is typical of vertebrate societies, and we cooperate with non-relatives on a massive scale (Richerson and Boyd, 1992). As Cronin puts it, human morality "presents an obvious challenge to Darwinian theory" (1991, 325). Everyone can probably think up their own favorite example. Dawkins (1989, 230) calls blood doning "a genuine case of pure, disinterested altruism." I am more impressed by charitable giving to people in faraway countries who probably share as few of our genes as anyone on Earth and whom we are unlikely ever to meet. And why do we turn in wallets found in the street, rescue injured wildlife, support eco-friendly companies, or recycle our bottles? Why do so many people want to be poorly paid nurses and counselors, social workers and psych techs, when they could live in bigger houses, attract richer mates, and afford more children if they were bankers, stock brokers, or lawyers?

Many people believe all this must ultimately be explained in terms of biological advantage. Perhaps it will, but I offer an alternative for consideration: a memetic theory of altruism. We can use our, by now, familiar tactic. *Imagine a world full of brains, and far more memes than can possibly find homes. Which memes are more likely to find a safe home and get passed on again?* Imagine the sort of meme that encourages its host to be friendly and kind. It might be a meme for throwing good parties, for being generous with the homemade marmalade, or just being prepared to spend time listening to a friend's woes. Now compare this with memes for being unfriendly and mean—never cooking people dinners or buying drinks, and refusing to give your time to others. Which will spread more quickly? The first type, of course. People like to be with nice people. So those who harbor lots of friendliness memes will spend more time with others and have more chances to spread their memes. In consequence many of us will end up harboring lots of memes for being nice to others. A simpler way of putting it is this: people who are altruistic will, on average, spread more memes. So any memes which thrive in altruistic people are likely to spread—including the memes for being altruistic.

Is this hypothesis testable? Well, research in social psychology reveals that people are more likely to adopt ideas from people they like

(Eagly and Chaiken, 1984). Whether this is a cause or a consequence of the above argument is debatable. This memetic explanation predicts that people should act in ways that benefit the spread of their memes even at some cost to themselves. We are familiar with buying useful information, and with advertisers buying their way into people's minds for the purposes of selling products, but this theory predicts that people will pay (or work) simply to spread the memes they hold—because the memes force them to. Missionaries, Mormons, and Jehovah's Witnesses come to mind.

Many aspects of persuasion and conversion to causes may turn out to involve meme-driven altruism. Altruism is yet another of the meme tricks that religions have purloined. Almost all of them thrive on making their members work for them and believe they are doing good. Of course, being generous is expensive. There will always be pressure against it, and if memes can find alternative strategies for spreading they will. For example, powerful people may be able to spread memes without being altruistic at all! However, that does not change the basic argument that altruism spreads memes.

You may have noticed that the underlying theme in all these arguments is that the memes may act in opposition to the interest of the genes. Thinking all the time may not use much energy but it must cost something. Talking is certainly expensive, as anyone who has been utterly exhausted or seriously ill will attest. And, of course, any altruistic act is, by definition, costly to the actor. I would say that this is just what we should expect if memes are true replicators. They do not care about the genes or the creatures the genes created. Their only interest is self-propagation. So if they can propagate by stealing resources from the genes, they will do so.

4. *Why Are Our Brains So Big?* Yes, I know there are lots of good answers to this old chestnut, but are they good enough? Let us not forget how mysterious this issue really is. Brains are notoriously expensive both to build and to run. They take up about 2% of the body's weight but use about 20% of its energy. Our brains are three times the size of the brains of apes of equivalent body size. Compared to other mammals our encephalisation quotient (the ratio of actual brain size to brain expected for the average animal of that body size) is even higher, up to about 25 (Jerison, 1973; Leakey, 1994; Wills, 1993). On many measures of brain capacity and behavioral complexity humans stand out alone. The fact that such intelligence has arisen in an animal that stands upright may or may not be a coincidence but it certainly adds to the problem. Our pelvises are not ideally suited for giving birth to huge brains and so childbirth is a risky process for human beings—yet we do it. Why? The mystery was deepened for me by thinking about the size of the biological advantage required for survival. In a study concerned with the fate of the Neanderthals, Zubrow (Leakey, 1994) used computer simulations to determine the effect of a slight competitive edge. He concluded that a 2% advantage could eliminate a competing population in less than a millennium. If we needed only such a tiny advantage why do we have such a huge one? Several answers have recently been proposed. For example, Dunbar (1996) argues that we need large brains in order to gossip, and gossip acts as a kind of verbal grooming to keep very large bands of people together. Christopher Wills (1993) argues that the runaway evolution of the human brain resulted from an increasingly swift gene-environment feedback loop. Miller (1993) proposes that our vast brains have been created by sexual selection; and Richerson and Boyd (1992) claim they are needed for individual and social learning, favored under increasing rates of environmental variation.

What these authors all have in common is

that their ultimate appeal is to the genes. Like Dawkins' bewailed colleagues, they always wish to go back to biological advantage. I propose an alternative based on memetic advantage. Imagine early hominids who, for good biological reasons, gained the ability to imitate each other and to develop simple language. Once this step occurred memes could begin to spread, and the second replicator was born.

Remember—once this happened the genes would no longer be able to stop the spread! Presumably the earliest memes would be useful ones, such as ways of making pots or knives, or ways of catching or dismembering prey. Let us assume that some people would have slightly larger brains and that larger brains are better copiers. As more and more people began to pick up these early memes, the environment would change so that it became more and more necessary to have the new skills in order to survive. A person who could quickly learn to make a good pot or tell a popular story would more easily find a mate, and so sexual selection would add to the pressure for big brains. In the new environment larger-brained people would have an advantage and the importance of the advantage would increase as the memes spread. It seems to me that this fundamental change in selection pressures, spreading at the rate of meme propagation, provides for the first time a plausible reason why our brains are totally out of line with all other brains on the planet. They have been meme-driven. One replicator has forced the moves of another.

5. *Who Am I?* We can now see the human mind as the creation of two replicators, one using for its replication the machinery created by the other. As Dennett pointed out, people are animals infested with memes. Our personalities, abilities, and unique qualities derive from the complex interplay of these replicators. What then of our innermost selves—the real me, the person who experiences my life?

I would like to suggest that selves are co-adapted meme complexes—though only one of many supported by any given brain (Blackmore, 1996). Like religions, political belief systems, and cults, they are sets of memes that thrive in each other's company. Like religions, political belief systems, and cults, they are safe havens for all sorts of travelling memes and they are protected from destruction by various meme tricks. They do not have to be true. In fact we know that selves are a myth. Look inside the brain and you find only neurons. You do not find the little person pulling the strings or the homunculus watching the show on an inner screen (Dennett, 1991). You do not find the place where my conscious decisions are made. You do not find the thing that lovingly holds all those beliefs and opinions. Most of us still persist in thinking about ourselves that way. But there is no one in there!

We now have a radically new answer to the question Who am I?, and a rather terrifying one at that. I am one of the many co-adapted meme-complexes living within this brain. This scary idea may explain why memetics is not more popular. Memetics deals a terrible blow to the supremacy of self.

The Future of Memes

The memes are out! For most of human history memes have evolved alongside genes. They were passed on largely vertically—from parent to child—and therefore evolved at much the same rate as genes. This is no longer true. Memes can leap from brain to brain in seconds—even when the brains are half a planet apart. While some memes hang around in brains for weeks, months, or years before being passed on, many now spread in multiple copies at the speed of light. The invention of the telephone, fax machine, and e-mail all increase the speed of meme propagation. As high speed, accurate, horizontal copying of memes increases we can expect some dramatic developments in the memosphere.

First, the faster memes spread, the weaker is the hold of natural (genetic) selection. This relative decoupling of genes and memes may mean that more than ever before memes will spread that are detrimental to their carriers. We may be seeing this already with some of the dangerous cults, fads, political systems, copy-cat crimes, and false beliefs that can now spread so quickly. Second, we may expect memes to build themselves ever better vehicles for their own propagation. Genes have built organisms to carry themselves. What is the memic equivalent? Artifacts such as books, paintings, tools, and aeroplanes might count (Dennett, 1995), but they are feeble compared with computers or the internet. Even these recent inventions are still largely dependent on humans for their functioning, and on the genes those humans are carrying—after all, sex is the most popular topic on the internet.

Can the second replicator ever really break free of the first? It might if ever we construct robots that directly imitate each other. Fortunately this is such a difficult task that it will not be achieved very soon and perhaps by then we will have a better understanding of memetics and be in a better position to cope with our new neighbors.

Conclusion

I have shown how a theory of memetics provides new answers to some important questions about human nature. If I am right, then we humans are the product of two replicators, not one. In the past century we have successfully thrown off the illusion that a God is needed to understand the design of our bodies. Perhaps in the next century we can throw off the illusion that conscious agents are needed to understand the design of our minds.

References:

Barkow, J. H., Cosmides, L., and Tooby, J. (Eds.). 1992. *The Adapted Mind: Evolutionary Psychology and the Generation of Culture.* New York: Oxford University Press.

Blackmore, S. J. 1996. "Waking from the Meme Dream." The Psychology of Awakening: An International Conference on Buddhism, Science and Psychotherapy, Dartington, Nov. 9.

Boyd, R., and Richerson, P. J. 1990. "Group Selection among Alternative Evolutionarily Stable Strategies." *Journal of Theoretical Biology,* 145: 331–342.

Cavalli-Sforza, L. L., and Feldman, M. W. 1981. *Cultural Transmission and Evolution: A Quantitative Approach.* Princeton: Princeton University Press.

Cloak, F. R. 1975. "Is a Cultural Ethology Possible?" *Human Ecology,* 3: 161–182.

Cronin, H. 1991. *The Ant and the Peacock.* New York: Cambridge University Press.

Darwin, C. 1859. *On the Origin of Species by Means of Natural Selection.* London: Murray.

Dawkins, R. 1989 (1976). *The Selfish Gene.* Oxford University Press.

——. 1982. *The Extended Phenotype.* Oxford University Press.

——. 1993. "Viruses of the Mind." In B. Dahlbohm (Ed.), *Dennett and His Critics: Demystifying Mind.* Oxford: Blackwell, 13–27.

Dennett, D. 1991. *Consciousness Explained.* Boston: Little, Brown.

——. 1995. *Darwin's Dangerous Idea.* London: Penguin.

Donald, M. 1991. *Origins of the Modern Mind: Three Stages in the Evolution of Culture and Cognition.* Cambridge: Harvard University Press.

Dunbar, R. 1996. *Grooming, Gossip and the Evolution of Language.* London: Faber & Faber.

Durham, W. H. 1991. *Coevolution: Genes, Culture and Human Diversity.* Stanford: Stanford University Press.

Eagly, A. H., and Chaiken, S. 1984. "Cognitive Theories of Persuasion." In L. Berkowitz (Ed.), *Ad-*

vances in Experimental Social Psychology, 17: 267–359. New York: Academic Press.

Gould, S. J. 1996. *Full House*. New York: Harmony Books.

Grant, G. 1990. Memetic lexicon: http://pespmc1.vub.ac.be/*memes.html

Hofstadter, D. 1985. *Metamagical Themas: Questing for the Essence of Mind and Pattern*. New York: Basic Books.

Jerison, H. J. 1973. *Evolution of the Brain and Intelligence*. New York: Academic Press.

Leakey, R. 1994. *The Origin of Humankind*. London: Weidenfeld and Nicolson.

Lumsden, C. J., and Wilson, E. O. 1981. *Genes, Mind and Culture*. Cambridge: Harvard University Press.

Lynch, A. 1996. *Thought Contagion: How Belief Spreads through Society*. New York: Basic Books.

Midgley, M. 1994. "Letter to the Editor." *New Scientist*, Feb. 12: 50.

Miele, F. 1995. "Darwin's Dangerous Disciple: An Interview with Richard Dawkins." *Skeptic*, 3: 4, 80–85.

Miller, G. 1993. *Evolution of the Human Brain through Runaway Sexual Selection*. Ph.D. Thesis, Stanford University Psychology Department.

Mithen, S. 1996. *The Prehistory of the Mind*. London: Thames and Hudson.

Pinker, S. 1994. *The Language Instinct*. New York: Morrow.

Richerson, P. J., and Boyd, R. 1992. "Cultural Inheritance and Evolutionary Ecology." In E. A. Smith and B. Winterhalder (Eds.), *Evolutionary Ecology and Human Behaviour*, 61–92. New York: Aldine de Gruyter.

Ridley, M. 1996. *The Origins of Virtue*. London: Viking.

Skeptic. 1996. Special Theme Issue: "Are We Slaves to Our Evolutionary Past?" Volume 4, #1.

———. 1996. Special Theme Issue: "Where Does Morality Come From? Did Ethics Evolve?" Volume 4, #2.

Tudge, C. 1995. *The Day before Yesterday: Five Million Years of Human History*. London: Jonathan Cape.

Wills, C. 1993. *The Runaway Brain: The Evolution of Human Uniqueness*. New York: Basic Books.

Wilson, E. O. 1978. *On Human Nature*. Cambridge: Harvard University Press.

Wright, R. 1994. *The Moral Animal*. New York: Pantheon Books.

Memes as Pseudoscience

JAMES W. POLICHAK

In his 1976 book *The Selfish Gene*, Richard Dawkins introduced the term "meme" to refer to a hypothetical unit of imitation or information that is transmitted from person to person. In Dawkins' and later memetic analyses of information processing, cultural information is treated as being analogous to genetic information—it exists in discrete self-replicating units that are subject to environmental selective forces. These forces result in differential survival of memes, much as environmental forces result in differential survival of genes. In short, memetic theorists argue that cultural evolution is analogous to, but partially independent from, biological evolution.

Proponents of memetics have made a number of extremely bold claims about the power of memetic analysis and the insights to be gained by applying such an analysis to culture and the transfer of information. The memetic approach has been called a new and revolutionary way of looking at culture and information, and even a "paradigm shift" (Brodie, 1996; Lynch, 1996). Memetics is viewed as the "missing link" that will allow researchers (specifically memeticists) to unify the social sciences (Lynch, 1996). Finally, it has been argued that "without the theory of evolution by memetic selection, nothing in the world of the mind makes much sense" (Blackmore, 1997, 43). Such a powerful approach should surely be appreciated, yet even such a strong proponent as Susan Blackmore noted that "the very idea of the meme seems to strike fear into even the most hardened evolutionist" (1996, 1). She shows (1997, and in the previous entry), in fact, that few books on evolution and culture even mention the word "meme," much less delve into analyses based on memes.

In this article I will endeavor to explain why most scientists dismiss the meme and theories based on it. In short, it is because memetic analyses are very shallow and imprecise compared to more traditional approaches, and because proponents of such analyses are all too willing to offer untested, unsupported, or incorrect assertions as proof of the value of their approach. I suggest that hardened evolutionists and social scientists are not fearful of memes. Rather, they are far more likely to be dismayed at the overzealous promotion of memes, the lack of supporting data or strong logical arguments, and the circularity of the "answers" memeticists offer to challenging questions concerning the origins and nature of culture, the human mind, and information. It is likely that these are the reasons that the memetic approach has been largely dismissed or ignored for the past 20 years. The problem can be enumerated as follows:

1. Memeticists have not done an adequate job of defining the meme, nor have they offered any examples of what a meme might be that withstand scrutiny.
2. Memeticists have failed to show that memes are necessary to understanding

culture. As a consequence they are unable to show that models based on biological selection are inadequate.

3. By largely ignoring the principles and data concerning information processing from the social sciences, especially psychology, memeticists have argued for a highly inaccurate model of information transfer, and a highly limited model of the activity of the human brain.
4. Memeticists have offered inaccurate and circular claims about what kind of explanatory power is obtained by assuming the existence of memes.

What Is a Meme?

The first major problem with meme-based approaches to understanding information processing and culture is that no one seems to be quite sure what a meme is. There is no direct evidence for the existence of any meme (i.e., no single meme has been isolated in the way that single genes have), nor does anyone know what memes might be made of (i.e., there have been no discoveries of meme-units analogous to the four base pairs of DNA). This lack of direct evidence does not doom memetic analyses to failure, however. Darwin (1859) produced his theory of evolution by natural selection and Mendel performed a number of seminal experiments on genetic transmission long before anyone knew what a gene was or what it might be made of. Indeed, it was the rediscovery of Mendel's experiments that changed the then current scientific belief that hereditary information was carried in the bloodstream, and demonstrated that there were somewhat discrete units of hereditary information.

Given the lack of direct evidence for memes, we are left with a wide assortment of analogies. Memes were originally conceived of by Dawkins (1976/1989) as a self-replicating entity consisting of information in some form. According to Dawkins, the meme is "a unit of cultural transmission, or a unit of imitation" (192), much like a gene is the unit of biological or hereditary transmission. This definition is, of course, quite imprecise, and Dawkins recognized this. The extent to which memes are analogous to genes is not clear. For this analogy to be effective, memes must be self-replicating and they must be so with the chance for error, and certain memes must be more successful than others. While memes are interested only in ensuring their transmission, there also must be some memes that offer an advantage to their hosts that is not reducible to a biological reproductive advantage, possibly an advantage that is actually reproductively detrimental but memetically advantageous. In other words, possessing a particular meme may lead one to engage in activities that lessen the chance of transmitting one's genes to future generations while increasing the chances that one will spread one's memes.

The analogy between genes and memes has been filled out or modified in various ways, so much so that proponents of meme theory have directly contradicted each other. For example, while Blackmore writes that "memes are not like genes" (1996, 3), the Meme FAQ (frequently asked questions) at Meme Central states that "memes are the basic building blocks of our minds and culture, in the same way that genes are the basic building blocks of biological life" (Brodie, 1997, 2).

Other memeticists have seemingly tried to avoid deciding how exactly memes are or are not like genes by instead trying to develop the analogy to viral transmission (which seems to be simply making the analogy less direct, given that viruses are composed mostly of genetic material). The *Memetic Lexicon* defines a meme as "a contagious information pattern that replicates by symbiotically infecting human minds and altering their behavior, causing them to propagate the pattern" (Grant,

Sandberg, and McFadzean, 1995, 2). Lynch (1996) does not offer much discussion of the meme-gene analogy, instead focusing on the various ways to spread "thought contagions," while Dawkins (1993) and Brodie (1996) look at memes as "viruses of the mind." The more precise treatments of the analogy (e.g., Dawkins, 1993) focus on computer viruses, but again suffer from the same problem as the meme-gene analogy. How exactly are memes like or not like viruses, computer or biological? One way that memes and viruses are apparently not alike is that not all memes are detrimental to their hosts. Regardless of whether or not this is an accurate conception of viruses, developers of this analogy (Brodie, 1996; Dawkins, 1993; Grant et al., 1995; Lynch, 1996) have been forced to take special pains to make the point that memes are often beneficial to their hosts. This, however, leaves us in the unsatisfactory state of saying that some memes are like viruses, including being bad for you, and others are like viruses except for not being bad for you.

At least one memeticist has tried to avoid using analogies by defining a meme as "a single unit of thought" (Nehring Bliss, 1997, 1). This definition, however, loses the benefits of the analogies to genes and viruses. It does not include the concept of self-replication with error that is necessary for meme theory to have any force. Additionally, it is far from clear what "thought" is and what one unit of it might be. Does any brain activity count as "thought," or must "thought" be conscious, whatever that is? Memeticists have largely failed to offer a precise, useful definition of the meme. Nor have they managed to fully develop either the analogy with genes or with viruses in a manner that states which parts of the analogy hold and which do not. This is an admittedly difficult task. A systematic series of controlled experiments will be necessary to fully establish whether memes are genuine self-replicating entities and how they are similar to the other known self-replicator (the gene).

Unfortunately, the attempt to get at what a meme is by looking at the examples offered also fails. Dawkins gives a number of examples of memes: "tunes, ideas, catch-phrases, clothes fashions, ways of making pots or of building arches" (1976/1989, 192). Most have followed Dawkins' lead to the extent of quoting or paraphrasing his examples (e.g., Blackmore, 1997; Grant et al., 1995; Speel, 1997). Here, to better describe a poorly defined concept, we are given a series of poorly defined terms. I doubt that Dawkins or any of those who have followed his lead have an adequate definition of "idea" or even of a "clothes fashion." Dawkins recognized this problem to some extent in his original chapter, writing, "I have said that a tune is one meme, but what about a symphony? Is each movement one meme . . . ?" (1976/1989, 195). It is still not clear what exactly a "tune" is, though. Dawkins further suggests that "if a single phrase of Beethoven's Ninth Symphony is sufficiently distinct and memorable to be abstracted from the context of the whole symphony . . . then to that extent it deserves to be called one meme" (195). While this example seems more precise, it introduces another difficulty. A chunk of information is a meme to the extent that it is distinct and memorable. This would seem to suggest that, for the time being at least, we can only offer probabilistic examples of what a meme is, and that probabilistic example rests on quite shaky ground, especially when one wants to provide evidence for a discrete entity. A chunk of information must be sufficiently distinct and memorable, but to whom? Dawkins suggests that use as a call-sign of "European broadcasting station" might be sufficient to determine if a chunk of information is a single meme (195). But if memes are defined in this manner, then what may be a meme to me or to a radio station manager may not be a meme to you. If I notice an idea in a book that you do not, is that detail then a meme or not a meme or both? I don't think Dawkins or anyone else wants to define memes in a manner that seems to be largely based on individual differences in attention or memory, but his analysis suggests just that. How many people, or which people, would have to notice and remember a chunk of information for it to be sufficiently distinct and memorable to be a meme? Perhaps for now memes can be like pornography—we may not know exactly what memes are, but we know them when we see them.

Another important issue that remains unclear from the examples in the memetics literature is whether a chunk of information has to be in a human brain or not to be considered a meme. While using terms like "idea" and "thought" seems to imply a residence in the human brain, this is not necessarily the case. The *Memetic Lexicon* defines a meme as "dormant" when it is currently without human hosts, seemingly indicating that printed or televised (and so on) information counts as memetic, but that information in one of these forms is not as good in an important way (Grant et al., 1995). Brodie (1996) and Lynch (1996) also, at the very least, imply that the human brain is the preferred meme habitat. It is not clear why, if memes are concerned only with replicating themselves as much as possible, being transmitted from computer to computer is not as good as being transmitted from brain to brain. If memes are not concerned with whether their human hosts live or die, as long as they are replicated (as the analyses suggest), why would the human brain necessarily be the best place to be? Others recognize this issue and suggest that memes do not necessarily need human hosts. For example, Blackmore (in quite a flight of science fiction) suggests that one day robots may directly imitate each other, and thus transmit memes, and furthermore that our knowledge of memetics might help us better understand our new neighbors (1997, 49). Dawkins seems to suggest that computer viruses are not quite memes yet (since they do not "strictly evolve") but may become memes in the future (1993, 18). Vajk, however, has gone so far as to state that "hula hoops, pet rocks, and Frisbees" are memes (1989, 7). Perhaps Vajk meant to refer metonymically to the idea of the Frisbee but this is not very clear from his writing. Similar questions arise about some of Dawkins's famous examples. Does "ways of making pots" refer to the ideas about making pots or the actions that people perform in order to make pots (given the unlikely assumption that one can distinguish among these in such a case)? Imprecision appears to be one of the hallmarks of memetic theory.

Memetics and the Social Sciences

While considering the inadequate definition of terms and poorly thought out analogies and examples found in the memetic literature, one must keep in mind that this is a new field, and that precision and clarity may come in the future. We may wish to examine alternate ways of justifying and supporting the memetic enterprise. As noted above, memeticists have made strong claims about the utility of memetic analyses. Recall Blackmore's assertion that "nothing in the world of the mind makes much sense" without memetic theory. This assertion implies two claims, as does meme theory in general. First, that more widely accepted approaches to the study of culture and information processing have in some important ways shown themselves incapable of reasonably accounting for what is actually happening in the world. Second, that memetics can supply the theoretical framework to account for what is actually happening in the world. Note that the two claims are logically distinct: If it were demonstrated that other approaches to the study of culture and information processing were indeed lacking, it would in no way necessarily imply that memes exist and that meme theory is adequate to explain what these other models cannot.

Scientific investigation of culture and information processing by humans is still in its infancy. Numerous attempts to examine and model how genes interact with the environment and influence cultural development have been made (e.g., Barkow, Cosmides, and Tooby, 1992; Cavalli-Sforza and Feldman, 1981; Richerson and Boyd, 1992). These works, as their authors or editors acknowledge, are only beginnings and are necessarily incomplete. We clearly do not yet understand the full extent to which genes and environment can account for human culture and human brain activity. As such is the case, it might seem premature to many to postulate an entirely new class of replicating entities to account for the as-yet-unknown inadequacies of the more widely accepted approaches to the development of the human brain and culture. Yet this has been the method of memeticists from the very start. Dawkins writes, "we do not have to look for conventional biological survival values of traits like religion, music, and ritual dancing, though these may also be present. Once the genes have provided their survival machines with brains that are capable of rapid imitation, the memes will automatically take over" (1976/1989, 200). Dawkins postulates the existence of a new class of entity, then assumes its existence and decides that we can therefore ignore the effects of genes and biological evolution, whatever they may be. It seems that we should look for conventional survival values for religion, for example, before we decide that it makes any sense to look for non-conventional survival values. Dawkins and his later followers have failed to present any strong evidence that conventional approaches are inadequate. They have instead asserted this as if it were a fact and used this assertion to then assume the existence of memes.

Memeticists have also largely focused their attention, when describing conventional approaches to studying culture and information transfer, on evolutionary biology. Examining the bibliography in a long work on memetics, for example Lynch's (1996) *Thought Contagion*, will turn up book after book on the applications of evolutionary theory to culture. Blackmore's (1997) list of references shows a similar strong bias toward books on biological evolution and culture. There are, however, a number of other fields concerned with human culture and human information processing. They are generally known collectively as the social sciences, and any research that has been done in these areas over the past hundred years to elucidate and describe information processing and culture has been largely ignored by memeticists. This has, among other things, led memeticists to argue for a highly inaccurate model of information processing.

Lynch (1996) contains the chapter "A Missing Link: Memetics and the Social Sciences" on how a memetic approach might fit in with the established social sciences. However, Lynch's review of the social sciences is far from complete and even somewhat disturbing. Lynch offers a few pages of superficial analysis about, for example, economics and memetics, or sociobiology and memetics, and so on. These fields and nearly all that Lynch discusses are interesting areas of inquiry, with well-developed methodologies and well-accepted findings. The exception is psychohistory. Some readers may be unfamiliar with psychohistory, and with good reason. Psychohistory is not a social science. Psychohistory is an idea from Isaac Asimov's (1974) highly acclaimed *Foundation* science fiction series. The basic premise is that in the far distant future humans will know enough about social change and history to predict, on a rather coarse scale, future events, similar to the way we can currently predict the weather. While Asimov's books and the idea of psychohistory are interesting and appealing, such an idea certainly does not belong on an equal footing with economics or sociobiology. More disturbingly, Lynch writes that psychohistory and memetics have "surprising similarities" (38) in their concerns and scope, though Lynch sees psychohistory as a more wide-ranging theory. I will leave it to readers to consider further the implications of this failure to distinguish fact from fantasy.

What is notably absent from Lynch's review and from the analyses of most memeticists is any mention of the research that has been done in two fields that are directly concerned with human information processing and the behaviors that result from the intake of information—cognitive and social psychology. Researchers in these fields have been systematically investigating how humans receive, process, and transfer information (Hunt, 1993). A cursory examination of some of the basic findings in these fields will show that, rather than unifying the study of the human brain and culture, memetic theory is based on an inaccurate model of information processing, is incapable of accounting for much of the activity of the human brain, and can only consider human thought in an extremely limited way.

Meme theory is concerned with the way information is transferred. To examine these issues, memeticists have chosen to focus on the information itself, treating humans as hosts who may be active to a greater or lesser extent in transmitting the information. It is the lack of emphasis on the actual activity of the human being with the information that dooms memetics to failure. Memeticists have adopted the view that information is independent of either its source or of its receiver, and can be effectively examined with little regard for either. The idea that one can examine the transfer of information without regard for the systems sending and receiving it has been challenged on a number of levels. Shuy (1993) argues, based on his linguistic training and experience as an expert linguistic witness at a number of trials, that such a position is a common misunderstanding jurors have about the way language works. Using examples from real criminal trials, Shuy demonstrates that people have the mistaken belief that they can examine verbal testimony in the absence of context because all of the necessary information is contained in the words spoken. This belief, Shuy argues, has led to wrongful convictions a number of times. Reddy (1979) argues that this inaccurate belief is based on the way the English language has developed, and refers to the mistaken idea that information is sent and received unaltered by the acts of sending and receiving as the conduit metaphor.

This model of information transfer has been shown false most powerfully by experimental psychologists studying human memory. Cognitive psychologists developed and rejected as inadequate models of memory that focused on the properties of information and ignored the activities of the receiver and the context in which the information was received. They have also rejected as inadequate to explain the experimental data models that focus solely on the properties of the information and the processing it is given at the time of reception (Craik and Lockhart, 1972; Morris, Bransford, and Franks, 1977). Kolers and Roediger (1984), after examining numerous controlled studies on human memory, conclude that it makes little sense to consider information to be remembered without considering the conditions and processes involved in receiving it and the conditions and processes involved in its retrieval (which must be considered if information is to be transmitted—information that can't be remembered can't be passed on to others). Memory researchers have shown in hundreds of studies that the match between the conditions in which a person receives information and the conditions in which that person attempts to retrieve the information has powerful effects on the amounts and kinds of information remembered. This is known as the principle of transfer appropriate processing (Morris et al., 1977). The factors that affect memory include such seemingly non-memetic influences like whether the receiving and remembering occurred under the same drug influence or not, whether they occurred in the same room or with the same experimenter, and so on (Tulving, 1983). Memetic approaches ignore the extent to which environmental factors influence human memory and determine what information will be remembered. They also ignore the important consequences of the processing that the human brain performs on information, which demonstrates the inadequacy of claiming that we can separate information from its processing.

Examining the research on false memories will effectively demonstrate the difficulties of separating information from information processing. Roediger and McDermott (1995) presented participants with study lists of words that were associates of one nonpresented word. For example, one list contained the words "bed," "rest," "awake," and nine other sleep-associated words, but the word "sleep" was never presented. During later free recall tests, participants recalled the nonpresented words (e.g., "sleep") 40% and 55% of the time, in Experiments 1 and 2, respectively. Similar results were found using word recognition tests, and participants were highly confident that the words they had recalled were on the study lists. This finding of false memories using word lists has been replicated and extended by a number of researchers. (In fact, Roediger and McDermott's study was a replication of Deese, 1959; for review, see Payne, Neuschatz, Lampinen, and Lynn, 1997.) Similar false memory data have been obtained using memory for sentences (Bransford and Franks, 1971), eyewitness testimony (Lindsay, 1990), and childhood events (for review, see Loftus, 1997). Experimental research on human memory has shown that people "remember" information that they never saw and events that never happened under a wide number of conditions and with a variety of testing methods. Payne et al. (1997) summarize their theoretical position on human memory: "the act of remembering involves the reperception of internal representations that are created from experiences with the world . . . these internal representations frequently are not separate and distinct from the sensory and perceptual processes that give rise to them" (59).

This description of human memory, while echoing that of Kolers and Roediger (1984), is clearly inconsistent with memetic ideas about information processing. People do not receive information and transmit it to others without processing and altering it in a way that is both highly sensitive to the environmental conditions at both the time the information is received and the time it is remembered, and highly dependent on the perceptual, attentional, and cognitive capabilities of those involved at both times. Given the memory research it is far from clear to what extent we can meaningfully discuss information independently of the activities of the people involved in the process of transmitting it. Memeticists must demonstrate that they can account for the sensitivity of memory to the factors identified by experimental psychologists. They must also adequately deal with the numerous false memory phenomena, which are a powerful challenge to meme theory. Presumably the word "sleep" fits the vague criterion for meme-hood, given that, in this experimental paradigm, words are presented to participants one at a time, and participants are expected to recall and rate their confidence in each individual word. Yet this word, recalled by about half of all people, was never seen. It does not seem that we can reasonably view this information as having been transmitted—who could have done so? In these and the other cases, it is better to view the memory as having been created. It is up to memeticists to challenge the dominant theory in experimental psychology—that all memories are created in a similar manner to the false memories through active reconstruction of past experiences that are heavily dependent on environmental, perceptual, and cognitive factors whose impact varies at different times. Cognitive psychologists have developed powerful models of human memory that challenge memetic theory; it is up to memeticists to show that the experimental data have been misinterpreted.

It is clear that accurate transmission of information is difficult and highly sensitive to a number of environmental and mental factors that have not been considered by memeticists. Unfortunately, in addition to ignoring decades of memory research, memeticists have also ignored the best source concerning how people learn and act with information they are exposed to on a more coarse-scaled behavioral level. Lynch (1996) uses epidemiology as a model for the way information is transferred from person to person on a relatively coarse scale (i.e., he is not concerned with perceptual, attentional, or the cognitive factors discussed above), extending the virus-meme analogy to methodology. Brodie (1996) and Dawkins (1993) pursue similar courses. It is not clear why they do this. For the past 50 years, social psychologists have studied specifically how people form and change attitudes and beliefs. Hundreds of carefully controlled experiments have been performed examining the factors that affect whether a person will be persuaded by information (or "infected" to use memetic terminology), how lasting that persuasion might be, and whether the person will actually act in response to the information to which they have been exposed (Eagly and Chaiken, 1993). One would think that this large body of research would form a much stronger starting point for memetic analyses than would an analogy to epidemiology. Yet, aside from a brief mention by Blackmore (1997), this work has been ignored by memeticists. Memeticists have neglected to consider virtually all of the experimental data, from both social and cognitive psychology, concerning information processing, and the behaviors based on this information processing, in favor of an inaccurate model of information transmission (the conduit metaphor) and an untested and underdeveloped analogy with the distantly related field of epidemiology. The emphasis on these flawed analogies has also led memeticists to adopt an extremely limited and incomplete view of human mental activity, as examination of the research will show.

Cognitive psychologists regularly hypothesize and find evidence for thought processes that are largely or entirely unavailable to conscious introspection. For example, Allbritton and Gerrig (1991) hypothesized that when people read stories with unfavorable outcomes (e.g., a bomb exploding) they are mentally generating alternate outcomes that affect their ability to recognize the actual outcome. These alternate outcomes are not generated in any way of which readers are necessarily aware. These counterfactual alternatives express themselves as a difference in reaction time to recognize the actual outcome between items that had favorable or unfavorable outcomes. This methodology is far from unusual in cognitive psychology. With regard to memetics, one can then ask: Does subconscious mental activity (which comprises most of the activity of the brain; Baars, 1988) count as memetic in any way? It does not seem to. The *Memetic Lexicon* states that "an idea or information pattern is not a meme until it causes someone else to replicate it, to repeat it to someone else. All transmitted information is memetic" (Grant et al., 1995, 2). Ignoring the inconsistency of this quotation (certainly information can be transmitted without causing someone to repeat it; most information falls into this class), it implies that the mental alternatives generated are not memes, and similarly that most of the mental activity that occurs in the human brain is not memetic. However, difficulties with this position arise when we consider that the consequences of these counterfactual thoughts were demonstrated by Allbritton and Gerrig, suggesting that they were then transmitted. Can these thoughts be called into existence as memes by the processes of measurement used by Allbritton and Gerrig, even though no individual thought has actually been transmitted? Cognitive psychology is based on methodologies of this kind, and has demonstrated the existence of many kinds of thought processes through the measurement of behaviors in a manner that seems to be inconsistent with memetics.

A brief excursion into introspection will make it clear just how limited the memetic approach to information processing is. (I am aware of the general inadequacy of this method, but I believe it will do here. See Hunt, 1993, for discussion of the role of introspection in psychology.) While you are reading this article, an incredible number of conscious and unconscious thoughts are occurring, and most of these thoughts are entirely dependent on your individual circumstances. While you read, your attention wanders, and you consider getting up for something to eat. While you read, you are inspired to think about the text in a unique way, supplementing or modifying your knowledge and experience of the text based on your prior experience and knowledge (e.g., Bartlett, 1932/1977). For example, the earlier mention of Asimov's (1974) work may have spurred you to remember your childhood love of science fiction, or the word "foundation" may have caused you to remember that your house is sinking and needs its foundation repaired. Few to none of these thoughts will be transmitted in any way, yet they comprise much of our brain activity (excluding the vast amount of brain function devoted to various autonomic and regulatory activities). They are ephemeral, existing for a moment and disappearing, only to be replaced by others. They are not memetic; nor is there any obvious way that some biological survival value can be applied to them. (Though there may be. Unlike some theorists, I am unwilling to say that the absence of intuitive biological survival value implies that there is none.) However, these thoughts do have important consequences on behavior and cognition as has been demonstrated by psychologists.

Memetic theory, even when fully developed, will not be able to account for these thoughts, and this is a problem. Given the strong evidence that the reading of a text is supplemented and modified by prior experiences in accordance with the reconstructive nature of memory, it seems that memeticists will not be able to describe how a reader obtains information from reading a book (similar concerns exist for films, conversation, observational learning, and so on). Memetic theory will not prove able to unify the social sciences when many of the concerns of social scientists about information processing and transfer cannot be addressed by memetic analyses.

With regard to how information is transmitted with potential mutation and is subject to selective forces leading to differential survival, the writings of memeticists are about as vague as their attempts to define the meme. It is also not clear to what extent we can meaningfully discuss transmission of information (as opposed to reconstruction of information). Memeticists have also not done enough to differentiate memetic transmission of information from non-memetic transmission. It is known that humans can transmit information to each other that could not reasonably be considered memetic. For example Russell, Switz, and Thompson (1980) showed that human menstrual cycles become synchronized through olfactory cues. Presumably there is some variance in the degree to which people's menstrual cycles become synchronized, but we would probably not want to say that this variability is evidence for mutation and differential survival of any particular menstrual cycle. It is up to memeticists to demonstrate that the information that they deal with is different, and this will prove difficult. Cognitive psychologists have demonstrated that learning and remembering are sensitive to environmental and perceptual factors, which are not considered in memetic analyses, and that most human thought is not likely to be memetic. They have also shown evidence for the recall of information never transmitted. Memeticists must show that, after accounting for these pieces of evidence and the psychological theories based on them, there is some form of discrete information left over that is subject to mutation (not merely variability) and differential selection (not based on perception, attention, or mental reconstruction of experience). In other words, they must demonstrate that, contrary to current psychological models, all forms of information in the human brain are not like the information discussed above before they can develop meaningful predictions and models of memetic transmission.

What Is Memetics Used For?

Memeticists have mostly focused their efforts at explaining how large-scale behavioral patterns and complex beliefs are transmitted in an attempt to demonstrate the power of memetic approaches to information transmission. These examples show that memetic analyses of information transmission are as simplistic and flawed as their attempts to define memes and their beliefs about the nature of information processing.

Lynch (1996), for example, suggests that a meme can be transmitted by giving its hosts reproductive advantages. In particular, Lynch writes that: "the 'baby doll for girls' meme replicates partly by training females to play the domestic role that leads to more children. Parents who give baby dolls to their daughters thus have their memes imparted to more grandchildren" (56). This is quite an interesting and controversial claim and Lynch unfortunately offers no evidence of the supposed greater reproductive of women who were given dolls as children and those who were not.

One wonders why, if dolls make women have more children, they would not do the same for men? Wouldn't memes get spread more effectively (in particular a more general "give baby dolls to your kids" meme) if boys were encouraged to have dolls and thus be more interested in domestic affairs? Wouldn't this shared set of interests in childrearing encourage husbands and wives to interact more and thus exchange more memes with each other? (Lynch also gives a similarly superficial analysis of "hero dolls for boys." What I hope is clear from my discussion is that these analyses amount to no more than memetic just-so stories, and are, if anything, less believable than their oft-maligned gene-based equivalents.) A second concern with this analysis is the fact that the people who give their children the most and nicest dolls—that is, the wealthiest segments of the population—have the lowest birthrates. Birthrate is generally negatively correlated with wealth across cultures, as the poorer peoples of the world tend to have the most children and also (presumably) the fewest baby dolls (cf. Dasgupta, 1995; Wattenberg, 1997). The final concern is that Lynch is content to reduce a complex pattern of behavior to a simple phrase and to act as if he has explained things. Buying dolls for one's children involves interacting with the child to determine if the child wants a doll and which one, traveling to a store in some manner, selecting and purchasing the doll, and so on. The difficulty of performing these activities is highly variable, as are the particular circumstances of each doll-buyer. It seems likely that the environment and a person's past experiences will determine whether, when, and how a person buys a doll. Memeticists may challenge this by arguing that regardless of the complexity of the overt behaviors, a person is really buying the doll because they have the meme. Given the absence of evidence for memes this objection cannot be taken very seriously. Furthermore, even if memes exist, it seems unlikely that the knowledge required for a person to buy a doll is reducible to a single meme.

Blackmore (1997) makes a series of even more controversial and sweeping claims than the above claim by Lynch (1996), designed to show how powerful "thinking memetically" is. Blackmore asserts that the memetic approach can be used "to answer five previously unanswered questions about human nature" (1997, 43). She first seeks to learn "why can't we stop thinking?" It seems to her that much of the time spent thinking is wasteful. Little of it seems to have any benefit for survival and is thus a waste of energy. Such thinking is maladaptive, Blackmore believes, writing that "animals that waste energy do not survive" (47). Blackmore's answer to this quandary is to suggest that it is memes that lead to this excessive thinking, that when we are doing (genetically) needless thinking we are actually rehearsing our memes, and that this rehearsal will result in an increased likelihood of transmission. Our excess thought is for the benefit of our memes.

There are a number of problems with this analysis. The first is that we simply do not yet know which thought processes are adaptive and which are not. We cannot yet label most mental activities as biologically adaptive or maladaptive, nor can we decide that any mental activity is beneficial to the survival of the as-yet-undocumented meme. Second, meme theory, with its emphasis on the transmission of information, cannot account for the vast amount of human thought that is unconscious and untransmittable, and these thoughts are whirring through our brains all of the time. If memetics cannot account for these thoughts, or at least explain the relationship between them and memes, it cannot offer an answer to Blackmore's question. Finally, even if we accept Blackmore's assertion that animals which waste energy do not survive, we still cannot accept Blackmore's answer. Calling the energy waste "meme rehearsal" does nothing to solve this problem. These meme-infested animals should just be rehearsing their memes and themselves to death. The only way an organism can avoid this dire fate is if the memes it possesses at least compensate for the loss of energy due to their rehearsal—in other words the activities used in rehearsing memes (i.e., thinking) must contribute to biological survival. If this is the case—that all our thinking actually does have survival value whether it is memetic or not—it seems unnecessary to refer to memes in the first place to explain the value of thinking.

Blackmore then turns to examine the issue: "why do we talk so much?" Again her answer is memes. We talk so much so that we can spread our memes. Again, Blackmore is assuming such behavior is biologically maladaptive, and, again, this assumption is not justified nor will labeling the wasteful activity "meme transmission" provide an answer. There have been numerous attempts to relate the emergence of language to biological factors. Blackmore herself cites one such attempt. Dunbar (1993, 1996) has argued that language evolved as a way to reinforce social relationships when the band size of our ancestors became too large for the earlier grooming techniques to function effectively. While Dunbar's analysis is far from achieving general scientific consensus (see the commentary following

Dunbar, 1993), it seems at least more plausible than Blackmore's assertion that it is memes that make us talk. Blackmore offers nothing to challenge Dunbar's analysis. This is unfortunate because Dunbar challenges the major point Blackmore bases her argument on. Blackmore writes that "we talk about some pretty daft and pointless things" (47) that to her mind cannot possibly have any survival value. However, in Dunbar's analysis, such talk, the gossip and so on, is used to reinforce our social bonds, discouraging aggression, and promoting food sharing and mating. Perhaps these are the "hidden biological advantages" (47) Blackmore is missing?

Blackmore's (1997) analysis of why we talk so much also conflicts, like so much in memetics, with psychological theory and research. Blackmore presents an extremely competitive model of the development of human language and its current use—we don't seem to care much what anyone else has to say because we're just waiting for our turn so that we can transmit our memes. There is a growing body of experimental evidence in psychology for a collaborative theory of language use (e.g., Clark, 1992). Numerous experiments have examined the ways that speakers work together to decide what to call ambiguous objects. Parallels to figuring out where to go to dinner, how to put together a bicycle, and so on, should be obvious. According to Clark's collaborative theory, language is used by people so that they can attain a reasonable degree of mutual understanding of their environments and intentions in order to interact effectively. Like Dunbar's (1993, 1996) analysis of language development, Clark's theory is based on the idea that language is an important way to coordinate activity among people and to effectively describe and manipulate each other and the environment. Blackmore's ideas about language use and development seem far more limited and far less likely.

Blackmore (1997) offers three similar analyses for why we are so nice to each other, why our brains are so big, and why we think we have a self. In each case, of course, the answer is memes. Throughout her analyses, Blackmore asks the reader to continually "imagine a world full of brains, and far more memes than can possibly find homes." This pattern of thought, imagining all those memes struggling to survive in the limited human brain, she suggests, will allow us to answer the difficult questions. Blackmore does not, however, offer any evidence for why and how memes might actually be the cause of our thoughts, big brains, niceness, and so on. We are asked to take our excessively big brains as evidence for the existence of memes and are expected to accept memes as a reason for our big brains existing. Blackmore, and other memeticists, are essentially asserting that memes are out there, without evidence or even an adequate example, and without regard for the conflict with psychological models. They then expect us to assume the existence of memes and insert that term as an answer to life's mysteries.

I hope that the above critique has shown that memeticists have grossly overstated the power of a memetic approach to understanding information processing and culture. They have much work to do to convince the skeptical scientist of the value of the meme, much less its existence. Memeticists should start by looking at the data from the social sciences and the models developed from them. They need to show that they can account for the objections put forth in this paper based on those psychological models and on logical grounds. Memeticists need to more clearly define the kinds of information they are going to deal with, and show that existing models are flawed when it comes to understanding this kind of information. Then they must demonstrate that the memetic approach can succeed where biological or psychological approaches have failed. Nothing presented in the memetics literature thus far suggests that memeticists will be able to accomplish this. Ill-considered examples, ignorance of relevant experimental research, and exaggerated claims of explanatory power do not make for a convincing scientific theory.

References:

Allbritton, D. W., & Gerrig, R. J. 1991. "Participatory Responses in Text Understanding." *Journal of Memory and Language*, 30, 603–626.

Asimov, I. 1974. *The Foundation Trilogy: Three Classics of Science Fiction*. New York: Avon.

Baars, B. J. 1988. *A Cognitive Theory of Consciousness*. New York: Cambridge University Press.

Barkow, J. H., Cosmides, L., & Tooby, J. 1992. *The Adapted Mind: Evolutionary Psychology and the Generation of Culture*. New York: Oxford University Press.

Bartlett, F. 1932/1977. *Remembering: A Study in Experimental and Social Psychology*. Cambridge: Cambridge University Press.

Blackmore, S. 1996. *Memes, Minds and Selves:* http://www.memes.org.uk/lectures/mms.html

Blackmore, S. 1997. "The Power of the Meme Meme." *Skeptic*, 5, 43–49.

Boyd, R., & Richerson, P. J. 1985. *Culture and the Evolutionary Process*. Chicago: University of Chicago Press.

Bransford, J. D., & Franks, J. J. 1971. "The Abstraction of Linguistic Ideas." *Cognitive Psychology*, 2, 331–350.

Brodie, R. 1996. *Viruses of the Mind: The New Science of the Meme*. New York: Integral Press.

Brodie, R. 1997. *Meme Central:* http://www.brodietech.com/rbrodie/meme.htm

Cavalli-Sforza, L. L., & Feldman, M. W. 1981. *Cultural Transmission and Evolution*. Princeton: Princeton University Press.

Clark, H. H. 1992. *Arenas of Language Use*. New York: Cambridge University Press.

Craik, F. I. M., & Lockhart, R. S. 1972. "Levels of Processing: A Framework for Memory Research." *Journal of Verbal Learning and Verbal Behavior*, 11, 671–684.

Darwin, C. 1859/1947. *On the Origin of Species by Means of Natural Selection*. New York: Oxford University Press.

Dasgupta, P. S. 1995. "Population, Poverty, and the Local Environment." *Scientific American*, 272, 41–46.

Dawkins, R. 1976/1989. *The Selfish Gene* (new edition). New York: Oxford University Press.

Dawkins, R. 1993. "Viruses of the Mind." In B. Dahlbom (Ed.), *Dennett and His Critics: Demystifying Mind*. Cambridge, MA: Blackwell, 13–27.

Dunbar, R. I. M. 1993. "Coevolution of Neocortical Size, Group Size and Language in Humans." *Behavioral & Brain Sciences*, 16, 681–735; commentaries and response follow.

Dunbar, R. I. M. 1996. *Grooming, Gossip, and the Evolution of Language*. London: Faber & Faber.

Eagly, A. H., & Chaiken, S. 1993. *The Psychology of Attitudes*. Fort Worth, TX: Harcourt Brace Jovanovich College Publishers.

Grant, G., Sandberg, A., & McFadzean, D. 1995. *Memetic Lexicon:* http://maxwell.lucifer.com/virus/memlex.html

Hunt, M. 1993. *The Story of Psychology*. New York: Doubleday.

Kolers, P. A., & Roediger, H. L. 1984. "Procedures of Mind." *Journal of Verbal Learning and Verbal Behavior*, 23, 425–449.

Lindsay, D. S. 1990. "Misleading Suggestions Can Impair Eyewitness' Ability to Recall Event Details." *Journal of Experimental Psychology: Learning, Memory, and Cognition*, 16, 1077–1083.

Loftus, E. F. 1997. "Memory for a Past That Never Was." *Current Directions in Psychological Science*, 6, 60–65.

Lynch, A. 1996. *Thought Contagion: How Belief Spreads through Society*. New York: Basic Books.

Morris, C. D., Bransford, J. D., & Franks, J. J. 1977. "Levels of Processing versus Transfer Appropriate Processing." *Journal of Verbal Learning and Verbal Behavior*, 16, 519–533.

Nehring Bliss, C. 1997. Dawn of the info-age: http://www.ice.net/%7Enehring/info1.htm

Payne, D. G., Neuschatz, J. S., Lampinen, J. M., & Lynn, S. J. 1997. "Compelling Memory Illusions: The Qualitative Characteristics of False Memories." *Current Directions in Psychological Science*, 6, 56–60.

Reddy, M. J. 1979. "The Conduit Metaphor: A Case of Frame Conflict in Our Language about Language." In A. Ortony (Ed.), *Metaphor and Thought.* New York: Cambridge University Press, 164–201.

Roediger, H. L., & McDermott, K. B. 1995. "Creating False Memories: Remembering Words Not Presented in Lists." *Journal of Experimental Psychology.* Learning, Memory, and Cognition, 21, 803–814.

Russell, Switz, & Thompson 1980. "Olfactory Influences on the Human Menstrual Cycle." *Pharmacology, Biochemistry, and Behavior,* 13, 737–738.

Shuy, R. W. 1993. *Language Crimes: The Use and Abuse of Language Evidence in the Courtroom.* Cambridge, MA: Blackwell.

Speel, H. C. 1997. *Memetics: On a Conceptual Frame Work for Cultural Evolution:* http://www.sepa. tudelft. nl/webstaf/hansseinst. htm

Tulving, E. 1983. *Elements of Episodic Memory.* New York: Oxford University Press.

Vajk, J. P. 1989. *Memetics: The Nascent Science of Ideas and Their Transmission:* http://www.geocities.com/Athens/2424/memetics.html

Wattenberg, B. J. 1997. "The Population Explosion Is Over." *The New Times Magazine,* November 23.

Race and I.Q. as Good Science

VINCE SARICH

The Bell Curve and the many commentaries on it have brought several issues into an often uncomfortably sharp focus. Though race is by no means the most important of these, the historical baggage the term carries and the reality it symbolizes require us to get past it before we are able to deal with more substantive matters. Yet that same baggage and those same realities often raise emotional barriers so powerful that they defy facts, reason, and logic.

Many commentators would have us believe that *The Bell Curve* is obsessed with race, and thereby provide a prime exemplar of pots, kettles, and blackness, evidenced in the following quote from the sociologist Alan Wolfe: "Murray and Herrnstein may not be racists, but they are obsessed by race. They see the world in group terms and must have data on group membership." This is an interesting charge, says Charles Krauthammer (1994), "given the fact that for the last two decades it is the very liberals who so vehemently denounce Murray who have been obsessed by race, insisting that every institution—universities, fire departments, Alaskan canneries—must have data on group membership."

It is the liberals who have oppressively insisted that we measure ethnic "over-" and "underrepresentation" in every possible field of human endeavor. Here is a liberal establishment forcing racial testing for every conceivable activity, and when a study comes along which does exactly that for SATs and IQ, the authors are pilloried for being obsessed by race.

No one who has actually read *The Bell Curve* could honestly document any such obsession. But, by the same token, no one even moderately conversant with the American society of the last 20 to 30 years could deny the accuracy of Krauthammer's assertion that "it is the very liberals who so vehemently denounce Murray who have been obsessed by race."

Further, it is these "very liberals" who deny that there is any significant genetic, biological, and evolutionary substance to race, and argue that it is, in effect, nothing more than a social and cultural construct. This view is epitomized in a recent story in *Time* (January 16, 1995) that carries the subtitle: "A landmark global study flattens *The Bell Curve*, proving that racial differences are only skin deep." The reference is to *The History and Geography of Human Genes*, a recent, massive compilation and analysis of human gene frequency data (Cavalli-Sforza et al., 1994). The story is an honest summation of that work—given that the genetic distances among human races are minimal, and that sections 1.5 and 1.6 of the book are entitled "Classical Attempts to Distinguish Human 'Races'" and "Scientific Failure of the Concept of Human Races." One looks in vain, however, in both the *Time* piece and the book on which it is based for any definition of the term "race." This omission is typical of race-debunking efforts. They never bother to define what it is that they are debunking. So let's start there.

The Reality of Human Races

We can begin this trip out of political correctness by noting that there is a substantial amount of agreement on both a working definition of the term "race" and on the existence of races in species other than our own. Races are populations, or groups of populations, within a species, that are separated geographically from other such populations or groups of populations, and distinguishable from them on the basis of heritable features.

We can agree that we are all members of a single species—*Homo sapiens*—and that each of us is also a unique individual. The most basic evidence that races exist is the fact that we can look at individuals and place them, with some appreciable degree of accuracy, into the areas from which they or their recent ancestors derive. The process involved is illustrated by a thought experiment where one imagines a random assortment of 50 modern humans and 50 chimpanzees. No one, chimp or human, would have any difficulty in reconstituting the original 50 member sets by simple inspection. But the same would be true within our species with, say, 50 humans from Japan, 50 from Malawi, and 50 from Norway. Again, by simple inspection, we would achieve the same 100% sorting accuracy. Granted, in the second experiment fewer sorting characteristics were available, but not nearly so few as to produce any doubt as to the placement of any individual. Extending this look-see experiment to the whole of the human species would obviously give us a substantial number of such geographical groupings. The addition of direct genetic evidence—from blood groups to DNA sequences—would provide further resolving power. But there is a real problem here that goes well beyond ideology and political correctness.

The Nature of Categories

One might clarify the problem of defining groups by reference to the issue of color categorization. We know that speakers of various languages that have a term for "red" (and who also have a comparable number of basic color terms) will also show a remarkable degree of agreement as to the range of the spectrum to which the term applies, and as to which hues are better reds than others (Berlin and Kay, 1969). We look at a rainbow and we tend to see not continuity, but rather a small number of specific colors that we have no trouble naming. This example tells us that whatever may be going on with respect to cognitive processing of the visible light spectrum, we have no operational difficulties in at least this realm with the notion that categories do not have to be discrete. Red does shade imperceptibly into orange, and orange into yellow, but we have no difficulty in agreeing as to where red becomes orange, and orange, yellow. Thus, human cognition can handle categories that are not discrete. The flip side of that is that categories can be real without necessarily being enumerable—and that is the critical matter for this article.

In other words, we can easily forget that categories do not have to be discrete. If this were not so, then why should the notion of "fuzzy sets" have been seen as so revolutionarily productive? Races are fuzzy sets.

How Many Races Are There?

One of the most commonly asked questions about race is: "How many races are there?" I contend that this is the wrong question. "How many" requires a precise integer as an answer—3, 7, 15, whatever. But the nature of the category "race" is such as to make such an answer impossible, depending as it necessarily does on the degree of sorting accuracy required in a context where the categories involved are not discrete. Races, after all, are not species, since all humans are fully interfertile. Therefore, races must necessarily grade into one another. But they do not do so evenly. Even today, for example, to drive along the road north from Aswan to Luxor (a hundred miles or so) is to cross a portion of ancient boundary between, to use old anthropological terms, Caucasians and Negroes. These two large groupings have been separated for millennia by the Sahara Desert. The Sahara has caused the populations north and south of it to evolve in substantial genetic independence from one another. And that is all one needs for race formation—geographical separation plus time.

The race quantity answer depends on the degree of sorting accuracy with respect to individuals. If it is something close to 100%, then the areas involved could become smaller and more distant from one another, with at least 20 races easily recognized, or larger and less separated, in which case one would see the few "major" races that everyone has tended to see. If, however, the criterion were something more like the 75% which has often sufficed for the recognition of races in other species, then obviously the number would be very large. In either case, if we use a straightforward definition of race, such as a population within a species that can be readily distinguished from other such populations using only heritable features, then there can be no doubt of the existence of a substantial number of human races. But, I hear you ask, don't the races all blend into one another? Yes, they are supposed to blend into one another. That's what races do. Nature's categories need not be discrete. It is not for us to impose our cognitive limitations upon Nature.

The Cause of Racial Separation

If all that is needed for racial differentiation is geographic separation and time, then why have humans remained a single species? The answer almost certainly lies in the fact of glacial cycles throughout the existence of our genus. These have necessitated major movements of human populations at fairly frequent, if irregular, intervals throughout the million years or so that *Homo* has existed outside of sub-Saharan Africa and therefore been susceptible to differentiation into races. Thus, there would have been periods of relative glacial stability (such as the last 10,000 years or so) during which racial differentiation would have become more marked, and periods of glacial movement, such as the retreat which began about 18,000 years ago, during which gene flow would have pretty much obliterated the previously developed racial boundaries. This logic also leads to the conclusion that most existing racial variation must have developed since that last period of large-scale, world-wide gene flow; that is, over the last 15,000 or so years. There is extensive evidence at a number of disciplines—anatomy, linguistics, biochemistry, archeology—which is consistent with such a scenario. The most straightforward is the fact that *Homo sapiens* fossil skulls found in areas currently populated by "Caucasians" and ranging in age from about 15,000 to 30,000 years are not more similar to those of modern "Caucasians" than they are to those of other major racial groupings.

The question of the antiquity of human racial lineages remains one of the most controversial areas of human evolution. Basically two quite opposed views predominate, neither of which takes the fact of glacial cycles into account. (1) Regional Continuity or Multiregional Evolution. *Homo erectus* populations in different areas of the world are seen as having appreciable direct genetic continuity with modern populations in those same areas. This theory sees significant aspects of modern racial variability as having separate histories for the high hundreds of thousands of years. (2) Out of Africa or African Eve. *Homo sapiens* have a single, relatively recent (something around 100,000 years ago) origin in some limited area and are characterized by some novel adaptation which enabled them to expand out of that homeland, replacing the more primitive humans they found along the way. Racial differentiation then followed. Most people in the field have tended to see #1 as implying much more significant racial differences because they would have had longer to develop. This has also been a major factor contributing to its relative lack of support.

But, as the late Glynn Isaac (perhaps the most influential archeologist involved in studies of early *Homo*) pointed out to me in a Berkeley seminar many years ago, it is the Out of Africa model, not that of regional continuity, which makes racial differences more functionally significant. It does so because the amount of time involved in the raciation process is much smaller, while, obviously, the degree of racial differentiation is the same—large. The shorter the period of time required to produce a given amount of morphological difference, the more selectively important the differences become. The Out of Africa model in its earlier formulations envisioned perhaps 40,000 years for raciation of anatomically modern *Homo sapiens*. The current formulations would nearly triple that figure, and, thus reduce the implied significance of racial differences. Obviously the model I outlined above would do the opposite, increasing that significance well beyond anything contemplated in recent years. But that might not be all. During the last 10,000 years human cultures have differentiated to a much greater extent with respect to achievement than was the case previously. Thus, not only might the time involved for raciation have been brief, but the selective demands on human cognitive capacities might have differed regionally to a substantially greater extent than could have been the case previously (see Sarich, 1995, for an extended discussion of these matters).

How Large Are Actual Racial Differences?

Current textbooks on human biology and human evolution go out of their way to deny either the reality, the significance, or both of race in our species. Their efforts would appear to be based in the hope that if we can make races disappear, racism will follow. For example:

> *Race:* In terms of biological variation, a group of populations sharing certain traits that make them different from other groups of populations. In practice, the concept of race is very difficult to apply to patterns of human variation.

The first sentence is fine. But the second implies that most human variation is not racially patterned. Which is certainly true. Most of the variation in our species, and in all other species, is found within and among individuals. But truth here has nothing to do with relevance. No one argues that race is the only dimension along which humans vary genetically. But, by the same token, there is more than enough heritable variation to produce human groupings which conform to any generally accepted definition of the term "race." This fact tells us that a substantial amount of human variation is clearly racially distributed, and leads to the question of how different from one another human races are.

The answer is, it depends on what you are looking at. At the level of morphology human races are more strongly differentiated from one another than are any other mammalian species. I first became aware of this fact when considering the arguments in the anthropological literature as to the place of the Neandertals. There one would often see statements to the effect that "Neandertals are too different from us to be part of our evolutionary history," but "too different" was never quantified. Quantifying it by using a standard set of measurements, correcting for size and calculating an average percent difference per measurement, gave some substance to the claim. Neandertals are, in fact, about twice as distant, on the average, from various extant human populations as the latter are from one another. But that exercise also demonstrated that (1) the anatomical distances among some modern races, for example, East Africans and Central Siberians, were much larger than those between Neandertals and the modern human populations most similar to them, and (2) racial morphological distances within our species are, on the average, about equal to the distances among species within other genera of mammals, as, for example, between pygmy and common chimpanzees. I am not aware of another mammalian species where the constituent races are as strongly marked as they are in ours.

The genetic distances are, in contrast, very small, and the no-races-in-our-species protagonists (such as Cavalli-Sforza) have seized on this fact to buttress their position. However, one needs to put the data into an evolutionary context to see what they really mean. The problem here lies in the fact that morphological evolution in our species has been extremely rapid, and this is not some sort of anthropocentric judgment. It can be demonstrated through two simple observations. We and our two closest living relatives—gorillas and chimpanzees—are about equidistant from one another at the DNA level with about 1.7% sequence difference seen in each of the three comparisons. Yet, morphologically chimps and gorillas are far more similar to one another than either is to us. This must mean that there has been much more morphological change along our lineage than along those leading to the African apes since the three genera last shared a common ancestor some 4.5 million years ago (the amounts of sequence change at the DNA level are the same). The current racial situation in our species is then entirely consistent with the history of our lineage: much morphological variation and change, little genetic variation and change.

Racial Differences in Athletic Ability

Another tack has been to acknowledge racial differences, but then argue that they are generally small with respect to differences among individuals within races, and, in any case, likely to be functionally irrelevant for any features of particular importance for the species. Consider the following example from sports. Every year perhaps 75 young men newly make NBA (National Basketball Association) teams. Of these, about 60 will be Black, and 15 White. (I am here using four years as the average length of an NBA career, and the current racial composition of the league as a source for these figures. "Black" means, in this country, that the individual has a substantial amount of obvious recent sub-Saharan African ancestry. "White" means no obvious ancestry other than European.) These numbers mean that the chance for a Black to play in the NBA is about one in 4,500; the corresponding figure for a White is about one in 90,000. We can then ask from how far out on their respective bell curves these 75 are drawn. Recourse to a z-score table tells us that 1 in 4,500 takes us about 3.4 SD (standard deviations) from the mean; 1 in 90,000 is about 4.3 SD from the mean. I submit that this almost one SD difference between populations in this suite of abilities based on a fundamental human trait is pretty substantial. In other words, it is simply not true that "bipedalism is such a critical aspect of the human adaptation that one would not expect to see great differences from either the individual to individual level, or between populations." Bipedalism is certainly a "critical aspect of the human adaptation," but it does not follow that therefore individual and group variation in what might be termed the quality of the bipedal adaptation would have been reduced. Indeed, it seems to me that, if anything, we might expect quite the opposite result. It took me a long time to figure this out, and thus it might prove useful to others to recount some of that process. The context is the relationship, if any, between brain size and cognitive performance.

Racial Differences in Brain Size

Discussing racial differences in athletic ability can get you into trouble, as some sportscasters have discovered. Discussing racial differences in brain size can be literally life threatening, as some psychologists have discovered. This issue ultimately divided Charles Darwin from Alfred Russel Wallace. Darwin was entirely comfortable with the notion that the human mind had evolved through natural selection, just as did the human body. Wallace, on the other hand, to the end of his much longer life, insisted that while our body had evolved, our mind must have been created. (See Michael Shermer's book, *In Darwin's Shadow: The Life and Science of Alfred Russel Wallace*.) A century later the very influential book *The Mismeasure of Man* by Stephen Jay Gould also, in effect, denied that our brains had evolved. Gould spends the first two chapters telling us that brain size and intellectual performance have nothing to do with one another, without once bothering to remind us that our brains have not always been the size they are today. Nor is that awkward fact mentioned anywhere else in the book. You could never learn from it that in our evolutionary lineage brain size had increased from around 400cc to 1300–1400cc over the last four million years. Why this omission?

I think the answer is quite straightforward. That part of Gould's psyche concerned with basic evolutionary biology knew that those large brains of ours could not have evolved unless having large brains increased fitness through minds that could do more. In other words, individuals with larger brains must have been, on the average and in the long run, slightly better off than those with smaller brains. How advantaged? Dare one say it? By being smarter. What else? If variation in brain size mattered in the past, as it must have, then it almost certainly still matters. And if you are going to argue that it does not, then you are going to have to explain why it does not. I do not think you can do this while maintaining your intellectual integrity. Thus Gould just ignored the demands of the evolutionary perspective by denying, implicitly, that our brains had evolved. I find it of some interest that no one has really challenged him on this point.

The evolutionary perspective demands that there be a relationship—in the form of a positive correlation—between brain size and intelligence. That proposition, I would argue, is not something that need derive from contemporary data (although, as we will see, those data do give it strong support). It is what we would expect given our particular evolutionary history; that is, it is the evolutionary null hypothesis, and, thus, something to be disproven. It seems to me that a demonstration of no correlation between brain size and cognitive performance would be about the best possible refutation of the fact of human evolution. It took me a long time to figure out what really ought to have been obvious: descent with modification by means of natural selection has been, and continues to be, the reality. It should be incumbent on those who would deny our evolutionary history to show that our biology is not involved. Otherwise there is an implicit creationism present in those who persist in ignoring the evolutionary perspective when they try to explain some aspect of our behavior (all too common in the social sciences). Brain size is an effective proxy for behavior, and it reminds us that evolutionary processes and evolutionary lineages are rather good data.

In other words, natural selection requires genetically based phenotypic variation to work on; thus throughout the period of change in brain size, there must have been present a substantial amount of genetic variation for brain size, and, likely, the greater the advantage of larger brains, the greater the underlying genetic variation for brain size. I had long been frustrated by the canalization argument (the more important the characteristic, the less variation) with respect to human intelligence, my teaching experiences telling me that cognitive performance was one of our most variable features. Yet at the same time I was unable to refute the logic of the argument. This lasted until 1983 when I remembered Fisher's Fundamental Theorem of Natural Selection: "The rate of increase in the fitness of any organism at any time is equal to its genetic variance in fitness at that time."

This says it all. An earlier statement of the general argument was made by the late Bernard Davis in 1976:

> Let me further emphasize that, even if no one had ever devised a test for measuring IQ, we could still be confident, on grounds of evolutionary theory, that our species contains wide genetic variance in intelligence. The reason is that natural selection cannot proceed unless it has genetic diversity, within a species, to act on; and when our species is compared with its nearest primate relatives, it is obvious that our main selection pressure has been for an increase in intelligence. Indeed, this change proceeded at an unprecedented rate (on an evolutionary time scale): in the past three million years the brain size of the hominid line increased threefold. Such rapid selection for increased intelligence could not have occurred unless the selection pressure had a large substrate of genetic variation to act on.

Brain Size and Cognitive Performance: Data Validate Theory

Any suggestion on one's part that people with bigger brains are, on the average, smarter by virtue of those bigger brains leads the listener to doubt one's intelligence, if not one's sanity. The general belief is that this inherently sexist and racist notion died an ignoble death sometime in the last century. Its recent resurrection began with a 1974 article by Leigh Van Valen. In it he reviewed the literature and concluded that the published correlations between brain size and intelligence (as measured by standardized tests) were unrealistically low because they did not allow for the fact that external measurements of head size were an imperfect indicator of brain size. Correcting for this attenuation indicated that the actual value was probably about 0.3. (*The Mismeasure of Man* does not even mention Van Valen's work.) A subsequent large-scale study of Belgian army recruits, which also used a much wider variety of tests of cognitive function, gave figures consistent with Van Valen's analyses (Susanne, 1979). Since 1987, there have been several studies on this subject in which the brain size of living individuals was measured directly and accurately using magnetic resonance imaging (e.g., Willerman et al., 1991; Andreasen, et al., 1993; Wickett et al., 1994). These suggest that Van Valen's estimate was, if anything, conservative—the consensus being in the area of 0.4 or a bit more. Although, as argued above, a positive relationship was to be expected on the basis of simple evolutionary considerations, the actual correlations found are higher than just about anyone would have predicted prior to Van Valen's pioneering effort.

A correlation of 0.4 means that of the average of 17 IQ points separating two randomly chosen individuals (within sex and population), about 7 IQ points would derive from the differences in the sizes of their brains. The same would hold for populations, and existing human populations can differ in their means by as much as 2 SD in brain size. Thus, this variable alone could lead to close to a 1 SD difference in mean intellectual performance among them. With respect to the difference between American Whites and Blacks, the one good brain size study we have (Ho et al., 1980) indicates a difference between them of about 0.8 SD; this could correspond to a difference of about 5 IQ points; that is, about one-third of the observed differential.

It should also be noted that these data strongly suggest that IQ tests are, in fact, measuring something that has been significant in human evolution, given that performance on them correlates so nicely with brain size. And what of the common accusation of circularity that intelligence is what the tests test? As Daniel Seligman notes, in *A Question of Intelligence* (1992, 15):

> [Herrnstein] said it was not at all intended as a put-down of IQ tests, certainly not as a complaint about circularity. It represented, rather, the perspective of a psychologist who believed (a) that "intelligence" needed to be anchored to some unambiguous operational definition and (b) that the cluster of abilities measured by IQ tests constituted a reasonable anchor. Fast analogy: You could define length . . . as "a distance or dimension expressed in units of linear measure." You could also define it as the thing that tape measures measure.

Individuals and Groups

So far I have tended to go from group to individual and back again without addressing the fact that any number of commentators on *The Bell Curve* have argued that: (1) individual variation within groups is generally greater than variation between groups, and (2) the existence of functionally significant genetic differences among individuals (with which most of them apparently feel comfortable) does not necessarily imply such among populations (with which they, along with most people, definitely do not). But the obvious truth of these two assertions in no sense justifies the object lesson we are supposed to draw from them—that therefore group variation is not something that need particularly concern us. First, the fact is group differences can be much greater than individual differences within them; for example, hair form in Kenya and Japan, or body shape for the Nuer and Inuit. And even when the first assertion is correct, as it is for most human characteristics, the differences between groups can, as already noted, be quite consequential. There is a much weaker case to be made for the relevance of the second assertion. While a qualification such as "does not necessarily" makes it technically correct, the statement as a whole implies that we should expect a connection between individual and group variation to be the exception, rather than the rule.

The evolutionary perspective begs to disagree. Consider again the example of brain size. Within sex and population, the coefficient of variation (standard deviation/mean x 100) is about 10%, a value typical for mass or volume characters. Two randomly chosen same-sex individuals within a population would then differ by about 12%, or about 150cc. But so can two populations. And this should not surprise us. Remember that our brain has increased in size some 1000cc in the last 3 million years. This is often termed "an explosive rate of growth," yet it works out to only 1/4 drop per generation. It could have gone faster, given what we know of individual variation and heritability for the character. That it did not implies that the huge advantages conferred by having more brain to work with must have been offset by (almost) equally large disadvantages. In other words, the adaptation here is best seen as a very slow moving compromise involving small relative differences between large forces. We should then have no expectation that those advantages and disadvantages would have balanced out in the same way in different populations at differing times and in differing ecological and cultural circumstances. But this same argument will apply to most aspects of individual variation. Given the number of characteristics in which functional variation is present, the ways in which they will balance out in two populations evolving more or less independently of one another are almost guaranteed to be different in the two. The balancing will take place at the level of individual phenotypes, and thus there is, in general, going to be a direct, inescapable connection between individual and group variation whenever evolutionary change is taking place.

Harmful Truths or Useful Lies?

Of all the thousands of words in print about *The Bell Curve*, about its data and arguments, perhaps none cut so close to the bone as those of Nathan Glazer in the October 31, 1994, issue of the *New Republic* (15–16):

> The authors project a possible utopia in which individuals accept their places in an intellectual pecking order that affects their income, their quality of life, their happiness. It may be true that we do not commonly envy the intellectual capacities of others—we allow Albert Einstein and Bobby Fischer their eminence—though I think even at this level the authors underplay the role of envy and rancor in human affairs. But how can a group accept an inferior place in society, even if good reasons for it are put forth? It cannot.
>
> Richard Wollheim and Isaiah Berlin have written: "If I have a cake, and there are ten persons among whom I wish to divide it, then if I give exactly one-tenth to each, this will not . . . call for justification; whereas if I depart from this principle of equal division I am expected to produce a special reason." Herrnstein and Murray have a very good special reason: smarter people get more and properly deserve more, and if there are more of them in one group than another, so be it. Our society, our polity, our elites, according to Herrnstein and Murray, live with an untruth: that there is no good reason for this inequality, and therefore our society is at fault and we must try harder. I ask myself whether the untruth is not better for American society than the truth.

And Bill Clinton, in a press conference of similar vintage, said:

> I haven't read it. But as I understand the argument of it, I have to say I disagree with the proposition that there are inherent, racially based differences in the capacity of the American people to reach their full potential. I just don't agree with that. It goes against our entire history and our whole tradition.

Are All Men Created Equal?

The issue here is not so much about "inherent, racially based differences in the capacity of the American people to reach their full potential." It is about inherent, racially based differences in the potentials themselves. The "entire history and our whole tradition" is, of course, encapsulated in our Declaration of Independence, where Thomas Jefferson wrote:

> We hold these truths to be self-evident, that all men are created equal, that they are endowed by their creator with certain unalienable rights, that among these are life, liberty, and the pursuit of happiness. That to secure these rights, governments are instituted among men, deriving their just powers from the consent of the governed. . . .

Which takes us back to Glazer, and the real need to ask "whether the untruth is not better for American society than the truth." The untruth in Jefferson is his first truth: "that all men are created equal." We know Jefferson did not believe that to be literally true, or perhaps more fairly, that he could not have believed it true unless one word were added to his sacred text: ". . . all men are created equal, in that they are endowed by their creator with certain unalienable rights. . . ." This addition in no way detracts from the power of the text (and only slightly from its rhythm), but does provide the advantage of literal truth—understanding, of course, that, ever since 1859, "creator" has had to be read as "the evolutionary process." Reading it that way also has the virtue of ultimately leading us to an understanding of why "the evolutionary process has made all men equal" is no better than the original text. Also note that last right—it is not "happiness" but "the pursuit of happiness"—an opportunity, not a result.

There have, in fact, been attempts to provide a justification based in evolutionary biology for a literal reading of "all men are created equal." Gould, for example, entitled one of his essays "Human Equality Is a Contingent Fact of History," and summarized his argument as such (1985, 198):

> *Homo sapiens* is a young species, its division into races even more recent. This historical context has not provided enough time for the evolution of substantial differences. But many species are millions of years old, and their geographic divisions are often marked and deep. *H. sapiens* might have evolved along such a scale of time and produced races of great age and large accumulated differences—but we didn't. Human equality is a contingent fact of history.

The problems with this line of argument are many. First, it is strange to have one of the inventors of the theory of "punctuated equilibrium" argue that human races cannot be very different from one another because they are too young. Second, nowhere in the article does Gould give us an example of a species in which races are as strongly marked as ours. The reason very likely is, as I have already noted, that there isn't any such species. Third, there are substantial racial differences present today—however they may have come about. I have already discussed two of these: athletic performance and brain size. Thus, Gould has it backwards. It is from the present that we obtain most of our knowledge of the past, and not, as most paleontologists would have it, the other way around. Finally, at least for our purposes, there is a strong tendency just about everywhere to extend the "there are no significant racial differences" argument to one which says "there are no significant gene-based differences between individuals." And as more and more groups are seen as needing some sort of official recognition, this extension becomes more and more inevitable as society becomes more sensitive to various groups.

Nature, Nurture, and the Individual

The above extension is, and for a long time has been, the prevailing point of view in the social sciences and humanities. If one takes a course at U.C. Berkeley in, say, Anthropology 3 (Introduction to Social and Cultural Anthropology), or Sociology 1 (Introduction to Sociology), one will hear an enormous amount about individuals as constituents of groups, and precious little about individuals as individuals. There will be little discussion of genes, evolution, or biology. It then goes almost without saying that you are not going to hear anything about free will or personal responsibility. The willful development of this situation in this country is very nicely documented in Carl Degler (1991). His Preface begins (and I quote directly and at length because the statement is so representative):

> Like most white Americans of my sex and class (the son of a fireman) and my generation (born in 1921) I came into a world that soon made me a racist and a sexist. And then, like most well-educated people of that generation, as I grew up I repudiated both race and sex as explanations for differences in the behavior of human beings. Indeed, I spent a good deal of my youth and adulthood arguing by voice and in print against biology as a source of human behavior, not only in regard to race and sex, but in other respects as well. How and why that sea change occurred in my thinking concerned me only peripherally. I knew there had been a time when biology was thought to be an important way of explaining why social groups differed, why some people were considered better than others. But that was another time. In my new outlook it was a given that the repudiation of biology had resulted from a penetrating, perhaps even lengthy scientific investigation of biology's inadequacy in accounting for the ways in which human groups differed. In ruling out biology as a cause for human differences, I thought of myself as defending a truth as solidly established as the heliocentric universe. Human nature, I believed, was constructed over time, not inherited from time. I had no trouble accepting Karl Marx's famous remark that man made his own history, not entirely as he pleased, meaning that history may limit us at times, but biology has little to say about our social behavior.
>
> Today, in the thinking of citizens and social scientists alike the deeply held assumption is that culture has severed for good the link between human behavior and biology. The conviction is that human beings in their social behavior, alone among animals, have succeeded in escaping biology. The irony is heavy here. For that belief is accompanied by another deeply held conviction: that human beings, like all other living things, are the products of the evolution that Charles Darwin explained with his theory of natural selection. The irony is almost palpable as Darwin entertained no doubt that behavior was as integral a part of human evolution as bodily shape. And that is where Book III enters. It seeks to tell the story of how biological explanations have begun to return to social science. . . . It is important to recognize that this "return of biology" is not simply a revival of repudiated ideas, like racism, sexism, or eugenics.

The problem here is that a "return of biology" means a return to the idea that sex and race will have consequences, and if you recognize this publicly, then you become, for many, a racist or a sexist. But the fact is, the evolutionary process cannot and does not produce equality either among individuals or groups.

Much of the furor surrounding *The Bell Curve* thus derives from a very real problem. Herrnstein and Murray are officially agnostic on the degree of genetic involvement in racial differences in intellectual performance, give gender differences one small paragraph on page 275, and mention the implications of our evolutionary history not at all—but all that does not really help. The fact is that deep down all too many of us are aware of the reality of group differences, and of the virtual certainty that genes are somehow involved in producing some of those differences. But, as Ernst Mayr pointed out in 1963: "Equality in the face of evident nonidentity is a somewhat sophisticated concept and requires a moral stature of which many individuals seem incapable."

Consider the treatment of E. O. Wilson after the publication of his masterful *Sociobiology* in 1975. Or, a more recent example, the June, 1993 issue of *Scientific American* features a lengthy essay by John Horgan, one of their staff writers, entitled "Eugenics Revisited," and teased on the cover as "The dubious link between genes and behavior." This one was so egregious—especially given the venue—that I was moved to send a long letter to the editor, publisher, and other officials of the magazine. I had no illusions that it would be publicly acknowledged by them (and it wasn't, though I did get a letter from the editor Jonathan Piel). *Scientific American* continued in this vein in its January and February, 1995 issues. In the first, Tim Beardsley, one of their staff writers, authored a piece entitled "For Whom the Bell Curve Really Tolls," and subtitled: "A tendentious tome abuses science to promote far-right policies." My thought is that you have to be pretty far left to see any of Herrnstein and Murray's "messages" as "far right." And Beardsley apparently has no compunction about penning flat-out lies, such as: ". . . numerous studies have demonstrated that early childhood surroundings have a large role in molding IQ scores—while very few studies have indicated a significant role for heredity." Anyone who could write those last 10 words presumably would also describe our national debt as composed of very few dollars. The February issue then contains a review of *The Bell Curve* by Leon Kamin, one of the authors of the 1984 book *Not in Our Genes.* His position can be inferred from the title, and from the fact that he and his coauthors were willing to state: "For all we know, the heritability (of IQ) may be zero. . . ." And, in its final paragraph:

> We should recall that the title of the article by A. R. Jensen . . . was "How Much Can We Boost IQ and Scholastic Achievement?" The answer, from cross-racial and cross-class adoption studies, seems unambiguous. As much as social organization will allow. It is not biology that stands in our way.

I submit that someone who could seriously entertain the notions that the heritability for any human performance measure could really be zero, and that our biology places no limits on a human performance, has thereby removed himself from serious consideration as a scholar of anything.

The Decline of Racism in Society

From an evolutionary perspective freedom can only mean freedom of opportunity, which, in the context of this article, necessarily leads to the question of how we are to recognize it among races and groups when we are living in a world where functionally significant, gene-based, racial and other group differences may well be the rule rather than the exception. It is here I think *The Bell Curve* makes its most meaningful single contribution (323–4). There we find the income data for young (average age = 29) year-round workers of three racial/ethnic groups: White, Black, Latino—with Latinos earning 86% and Blacks 77% as much as Whites. But when IQ is held constant (average = 100 for all three groups), both the Latino and Black figures climb to 98% of that for Whites. This result (which could be seen as remarkable only if one accepts the "this is a racist society" mantra) tells us about the degree of equality of opportunity in recent American society, and yet only one commentator of the more than 100 I have read or heard (including Murray and Herrnstein themselves) seems to have found it worthy of comment. This was Daniel Seligman, himself the author of the highly readable and most informative 1992 volume, *A Question of Intelligence*, who titled his brief column in the December 12, 1994, issue of *Fortune*, "News Nobody Noticed":

> Your servant has now read scores of reviews of *The Bell Curve*. Most have fiercely criticized the book's thesis, which emphasizes the centrality of IQ in lives and careers, and most have dwelt insistently on race and the 15-point black-white IQ gap. But, oddly, we have yet to read a review noticing the racial news built into a table on page 324. In a rational world, the news would be on the front pages. . . . The news is about racial discrimination in America. As we all keep reading, blacks earn a lot less than whites, even when you compare workers of similar ages and educational backgrounds. This table confirms this finding. But it points to something else one has never before read: that when you control for age and IQ, the black-white earnings gap just about disappears. . . .
>
> Obvious implication: At least so far as younger workers are concerned, employers no longer engage in irrational discrimination based on race. They discriminate based on IQ—which is rational, given the avalanche of data linking IQ to performance in many different job markets. Fascinating question: How can it be, in a world where racial discrimination is (properly) an object of enormous concern, that we are ignoring powerful evidence of its decline?

I would add that Seligman's comment that employers "discriminate based on IQ" has to be taken metaphorically. What they are doing is rewarding performance (as any rational employer would). The connection with IQ, as Herrnstein and Murray point out (80–81), is that it is the best single predictor of performance—better than biographical data, reference checks, education, interview, or college grades. And as to his final question? The cynic in me cannot help commenting "So what else is new?" One does not really expect our media to report anything positive about this society, does one?

The Rise of Racism on College Campuses

No society has an unstained history. The treatment of individuals of sub-Saharan African ancestry is without doubt our largest and deepest stain, and that history, as are all histories, is beyond change. Given those truths, the worst thing we could do is to repeat that past in the name of producing an equality of results, by again allowing the treatment of an individual to be influenced by that individual's race (or sex, or ethnicity, or any other grouping). Yet, increasingly over the past 30 years we have been doing just that.

My own direct experiences with such race-norming, quota-driven treatment of individuals has been at U.C. Berkeley, where, for the last 10 years or so, a substantial percentage of freshman admissions (up to about 40%) has been reserved for "underrepresented minorities," and where race, ethnicity, and sex have become major factors in the hiring of new faculty. For students, what this has done is to produce two populations separated by race/ethnicity and performance who wind up, in the main, in different courses and pursue different majors. That is only to be expected when the SAT difference between the White and Asian students on the one hand, and Black and Latino students on the other, is about 270 points (1270 v. 1000; about 1 SD difference).

This is equivalent to about three to four years of academic achievement, and U.C. Berkeley is no place to play catch-up. And, as far as anyone knows (there are no published studies on the matter), no catching-up in fact takes place. I wrote of this situation in 1990:

> The Berkeley administration has, in its admissions policies, especially over the past five years or so, ignored certain unpalatable realities, and given us an even more unpalatable set of results. They have given us a situation where the association between race/ethnicity and performance is real, obvious, and of ever-increasing strength. What we are getting at Berkeley is two communities, separable on racial/ethnic grounds, and increasingly divergent from one another academically, socially, and in ethos—a result desired, presumably, by no rational soul. It is, frankly, difficult to imagine policies more deliberately crafted or better calculated to exacerbate racial and ethnic tensions, discourage individual performance among all groups, and contribute to the decay of a magnificent educational institution.

The fact is that any group-based policies are bound to have effects of this sort. As I have already noted, the evolutionary necessity of individual variation is almost always going to lead to group variation, and statistical realities require that group differences get exaggerated as one goes toward the ends of the bell curves involved. Thus, when you look at group representations with respect to the high-visibility pluses (e.g., high-paying jobs) and minuses (e.g., criminality) in any society, one can virtually guarantee that they are not going to be equal—and that the differences will not be trivial. The problem is in recognizing and adapting to those realities, and not, as has so often been the case with responses to *The Bell Curve*, denying them. I noted this in a letter to a Berkeley Faculty Senate committee on "diversity":

> This current focus on "diversity," if continued and "successful," can only have the effect of rewarding individuals for making their primary allegiances to certain defined groups, and, thus, of tribalizing our society. It would require a mind completely closed to current realities, never mind historical ones, to remain ignorant of the disastrous effects of tribalization. One therefore has to suspect that anyone supporting policies that tribalize is either ignorant, or simply playing the very effective political game of "divide and conquer." The number of different roles to be played in a society increases as the complexity of a society increases. Ours is a very complex society that will only become more complex in the future. The number of different roles to be played will thus increase, requiring a larger number of allegiances for individuals within the society, and selecting against those whose primary allegiance is to a particular group—be it one based on biology (race, sex, age) or culture (ethnicity, religion). If one of your roles is chemist, then one set of your allegiances is to the community of chemists and chemistry. You are then a chemist, period—and not a female, or White, or Catholic, or old. To the extent that you do not look at it that way; that is, to the extent that you see yourself as some sort of hyphenated chemist, you will necessarily reduce the effectiveness of your chemistry. And this is going to be true for each of the other roles you will come to play. To the extent that you see yourself as a hyphenated anything, your achievement in that "anything" will tend to be reduced. And to the extent that a society encourages and rewards individuals for looking at themselves in such a fashion, it necessarily reduces its total level of accomplishment.

There are certain harsh realities in life. One of these is that groups, whether age, sex, race, ethnic, or whatever, are groups, and groups of anything are very likely going to differ from one another. If they didn't, then they wouldn't be groups, would they? I can then confidently guarantee that when we measure performance by groups, we are going to find group differences in performance. Some part of those differences will be nature-based, some part will be nurture-based, some will be will-based. No society has, or can have, the power to even things up. Societies are not omnipotent. They can provide opportunity; they cannot mandate individuals or groups making equal use of those opportunities, and, therefore, they cannot make either individuals or groups come out even. Individual and group variation are realities that they cannot will out of existence. They can try, and what happens then is, unfortunately, no secret: a temporary leveling-down bought at enormous cost. They can in no sense make groups equal. They cannot level up—only down—and thus any such leveling is necessarily at the expense of individual freedom and, ultimately, that society's total level of accomplishment.

Ending Racism without Ending Race

There would appear to be a substantial consensus among some of the more "conservative" commentators on *The Bell Curve* as to its policy implications, and, for better or worse, I find myself in total agreement with them. Seligman, for example, closes his *A Question of Intelligence* with: "One major message of the IQ data is that groups are different. A major policy implication of the data, I would argue, is that people should not be treated as members of groups but as individuals." Herrnstein and Murray give us the same message, but at much greater length, in their Conclusion (549–552).

I opened with a quote from Charles Krauthammer. His conclusion says it better than I can:

> I distrust all multiculturalism, liberal or conservative. The Balkans amply demonstrate the perils of balkanization. My answer is simpler: Stop counting by race. Stop allocating by race. Stop measuring by race. Let's return to measuring individuals.
>
> It seems hopelessly naive to propose this today. But it was not naive when first proposed by Martin Luther King and accepted by a white society that was finally converted to his vision of color blindness. Instead, through guilt and intimidation, a liberal establishment has since mandated that every study of achievement be broken down by race. "The Bell Curve" takes that mandate to its logical conclusion.
>
> Enough. As both Murray and Thomas Sowell explicitly state, knowing the group score tells you nothing about the individual. Well, we have seen the group score. Let's go back to counting individuals. How many of Murray's critics will agree to that?

Amen. Let's go back to counting individuals. And how do we encourage such behavior? Simple. Just remove all reference to group identity from both statutory and administrative law. Period.

References:

Andreasen, N. C., M. Flaum, V. Swayze, D. S. O'Leary, R. Alliger, G. Cohen, J. Erhardt, and W. T. C. Yuh. 1993. "Intelligence and Brain Structure in Normal Individuals." *American Journal of Psychiatry* 150:130–134.

Beardsley, T. 1995. "For Whom the Bell Curve Really Tolls." *Scientific American* 272:1:14–17 (January).

Berlin, B., and P. Kay. 1969. *Basic Color Terms: Their Universality and Evolution.* Berkeley: University of California Press.

Cavalli-Sforza, L. L., P. Menozzi, and A. Piazza. 1994. *The History and Geography of Human Genes.* Princeton, N.J.: Princeton University Press.

Davis, B. 1976. "Evolution, Human Diversity, and Society." In *Zygon* 11:2:80–95.

Degler, C. N. 1991. *In Search of Human Nature.* New York: Oxford University Press.

Gould, S. J. 1981. *The Mismeasure of Man.* New York: Norton.

——. 1985. "Human Equality Is a Contingent Fact of Evolution." In *The Flamingo's Smile.* New York: Norton.

Hardin, G. 1959. *Nature and Man's Fate.* New York: Holt, Rinehart, and Winston.

Ho, K.-C., U. Roessmann, J. V. Straumfjord, and G. Monroe. 1980. "Analysis of Brain Weight." *Archives of Pathology and Laboratory Medicine* 104:635–645.

Horgan, J. 1993. "Eugenics Revisited." *Scientific American* 270:6:122–131 (June).

Kamin, L. 1995. "Behind the Curve." *Scientific American* 272:2:99–103 (February).

Krauthammer, C. 1994. "Liberals, Obsessed by Race, Can Hardly Complain." *The News & Observer* (Raleigh, N.C.) 23 October 1994.

Mayr, E. 1963. *Animal Species and Evolution.* Cambridge: Harvard University Press.

Sarich, V. M. 1995. *Race and Language in Prehistory.* In press. For copies write: 555 Pierce, Unit 730, Albany, CA 94706.

Seligman, D. 1992. *A Question of Intelligence: The IQ Debate in America.* New York: Birch Lane Press.

——. 1994. "News Nobody Noticed." *Fortune* 12 December:255.

Smith, C. L., and K. L. Beals. 1990. "Cultural Correlates with Cranial Capacity." *American Anthropologist* 92:193–200.

Subramanian, S. 1995. "The Story in Our Genes." *Time* 145:2:54–55 (16 Jan 1995).

Susanne, C. 1979. "On the Relationship between Psychometric and Anthropometric Traits." *American Journal of Physical Anthropology* 51:421–424.

Van Valen, L. 1974. "Brain Size and Intelligence in Man." In *American Journal of Physical Anthropology* 40:417–424.

Wickett, J. C., P. A. Vernon, and D. H. Lee. 1994. "In Vivo Brain Size, Head Perimeter, and Intelligence in a Sample of Healthy Adult Females." *Personality and Individual Differences* 16: 831–838.

Willerman, L., R. Schultz, J. N. Rutledge, and E. D. Bigler. 1991. "In Vivo Brain Size and Intelligence." *Intelligence* 15:223–228.

Race and I.Q. as Pseudoscience

DIANE HALPERN

As I read *The Bell Curve* by Richard J. Herrnstein and Charles Murray I was reminded of a cartoon from the popular children's television show *Sesame Street*. As regular viewers of *Sesame Street* already know, every episode is brought to you courtesy of a number and letter. On those days when the star of the show is the letter "I," we are shown a group of hard-working cartoon characters whose job it is to polish a giant letter "I" until it glistens like an expensive jewel in the sunlight. In fact, this small army of letter polishers spend their entire day polishing the letter "I" because it is such an Important and Interesting letter. In a similar manner, Herrnstein and Murray also polish their "I"—Intelligence—and its related measure, IQ, which assume the spotlight as the best predictors of socioeconomic class and a diverse range of variables that cover the rest of the alphabet from Abusive relationships to Xenophobia and Zealotry.

Commenting on *The Bell Curve* is a lot like trying to catch a ball of jello. The arguments are slick and, like most skilled rhetoricians who are attempting to change how people think, the authors provide a veneer of fairness to cover the flaws and biases in their message. In this case, the veneer is thin—so thin that it allows their hypocrisy and social agenda to peek through. In making their points, the authors present, discredit, and then dismiss all opposing points of view. Contradictory evidence is criticized as statistically or methodologically flawed. Unfortunately, the stringent criteria that they apply to counterarguments are abandoned when they present the evidence in support of their favored conclusions. The authors shape their arguments like skilled word smiths. A factual statement like "some educational programs have not worked" is gradually morphed into a misleading statement like "educational programs have not worked," and then, "educational programs cannot work," a subtle change in wording that occurs as the authors stray from their data.

Can anyone seriously believe that Murray was shocked and dismayed when he found that he had upset many people with his pronouncements of racial inequality or the way he used IQ data to support an ultra-conservative political agenda? The authors have created the perfect medium for a growing media frenzy with a very long book in which much of the supporting evidence is relegated to a statistical appendix and extreme claims are succinctly summarized. The voracious appetite of the media is whetted by controversies, sound bites, and simple explanations of complex subjects. Even lengthy and thoughtful articles are condensed into a few words for newspaper headlines that are supposed to pique the reader's interest. This is the stuff that sells newspapers, keeps people tuned to the chatty banter that passes as television news, and sustains conversations in countless barber shops, bus stops, and kitchen tables. Despite Murray's protestations to the contrary, this is a book about race, and race is one topic in which we are all self-proclaimed experts. Each of us has an opinion about racial similarities and differences and a story to tell that shows how right our own opinions are. Cognitive psychologists who study stereotypes and prejudice have known for a long time that strongly held beliefs are difficult to change, and that people cling to their beliefs even when confronted with evidence that shows that these beliefs are wrong. We are more likely to change our interpretations of experience and our memory for events so that they fit our existing belief system than we are to abandon our beliefs. Perhaps books like this one should be sold with warning labels in which readers are urged to be alert for misleading statements, missing evidence, and biased interpretations—sort of a surgeon general's warning. The messages in *The Bell Curve* are at least as dangerous as cigarettes and alcohol.

My response to Herrnstein and Murray's thesis is organized around a brief summary of seven main points that they make in their controversial and massive tome, so that my comments and criticisms can be understood in their appropriate context even by readers of this article who have not read their book.

Intelligence Is Important

According to Herrnstein and Murray:

This is a basic underlying assumption of the authors' argument. It is difficult to disagree with the statement that intelligence is important, although I would have to add, "Important for what purpose?" Most of us would agree that it is also important to be a kind and loving person and that empathy and other socially desirable traits are at least equally important for the betterment of society or individual happiness. Although this is not the place to engage in philosophical musings about whether it is more important to be a good person or a smart one, the authors do provide a definitive answer to a similar burning question. They ask if it is better to be born intelligent or rich, which, for most of us, would seem to be a rhetorical question or one in which the answer depends on individual values. According to the authors, however, the correct answer is intelligent, and lots of intelligence is even better than lots of money. But, what is intelligence, and how can we tell who has more or less of it?

Intelligence is one of the most controversial topics in psychology even though the concept has a long history and the term is commonly used in everyday language. If I asked you to list the characteristics of an intelligent person, you would probably include terms like "reasons logically and well," "keeps an open mind," "reads with high comprehension," and "can understand complexities." In addition, most people believe that they are about average or above average in intelligence. It seems that Garrison Keillor's mythical Lake Wobegon is not the only place where the laws of mathematics are suspended so that everyone can be in the top half of the distribution.

Today's most commonly used intelligence tests, the Stanford-Binet and Wechsler Intelligence Tests, are normed so that the average score is 100 and measures of how the scores are spread out (standard deviations) are derived by transforming scores so that they conform to a mathematical formula. IQ scores greater than 100 indicate greater than average intelligence, and scores less than 100 indicate less than average intelligence. Intelligence tests are based on the idea that the more questions you answer correctly, the more intelligent you are. Tests of intelligence are like other sorts of tests, and the scores depend on all of the factors that affect performance on any other test—variables like the nature of the test questions and the test takers' motivation, knowledge of the material, health, and willingness to guess when unsure of an answer. The scores that are obtained on intelligence tests are known as "intelligence quotients" (because they used to involve forming a fraction or quotient) or, more informally, IQ scores. IQ is a number that is obtained on a test that supposedly measures intelligence—it is not a direct measure of intelligence.

Here are some examples that are similar to questions on common intelligence tests:

Verbal Test Items:

1. At what temperature does water freeze?
2. Who wrote *The Republic?*
3. How many inches are in 3 1/2 feet?
4. Explain the meaning of "strange."
5. Explain the meaning of "adumbrate."
6. Repeat a series of digits after the test administrator recites them. For example, repeat the following digits: "8175621."

Performance Test Items:

1. Use wooden cubes painted red and white to duplicate a design shown on cards.
2. Arrange a series of cartoons into a logical sequence.
3. Assemble a jig-saw puzzle.

Most psychologists believe that intelligence is a multidimensional construct, although there is much disagreement over how many different kinds of intelligence there are. One way of dividing intelligence is to consider it as made up of fluid intelligence, the kind of intelligence that you would use when you are dealing with a novel task, like writing your first computer program, and crystallized intelligence, the kind of intelligence you would use when dealing with information that you have already learned, like finding the area of a pyramid when you know the formula. There are many other ways to divide intelligence including verbal intelligence, which involves the use of words and language, and spatial intelligence, which involves the use of spatial displays like maps.

A major controversy among psychologists concerns the existence of a general intelligence factor called "g." The question is whether it makes sense to think about people as being generally "smart" or "dumb," or is it more accurate to think that people can be smart in some ways and not in others? If people can be smart in some areas and not others, then a single score on an intelligence test will not be able to measure how intelligent they are, but if people can be thought of as generally smart or generally dumb, then a single number could assess the extent to which they are intelligent. In order to answer this question, the data from intelligence tests are analyzed with mathematical procedures to determine whether a single factor, "g," emerges or whether the data are described more accurately with multiple factors. Some of the disagreements over the existence of a general factor of intelligence concern the mathematical procedures, and other disagreements concern the way that intelligence is conceptualized. The measurement of intelligence is not separable from the way it is conceptualized because the mathematics that we use influences the way psychologists think about intelligence, and the way we think about intelligence influences the mathematical procedures that we use. Many of the controversies surrounding the measurement of intelligence involve the mathematical analyses that are used in understanding the data. This is one of the reasons why it is difficult to explain to the general public why the experts cannot agree about intelligence.

When Administered Properly, Intelligence Tests Are Fair and Valid Measures of Intelligence

Although the authors have felled many trees to make this point, I do not agree that their conclusion is fair or valid. IQ is a number on a test. The test questions reflect the sort of information that most people know and the intellectual activities that most people can perform. IQ scores seem to predict academic success equally well for all racial and ethnic groups, a point that the authors make in several different places in their book, but this does not mean that they measure intelligence equally well. In addition, IQ scores can only account for a relatively small proportion of the variance in academic or job success. Success depends on many other variables like motivation, persistence, expectations, and education. The influences of variables other than intelligence are quickly dismissed by the authors, a practice that suggests that they are not important when, in fact, they are.

All intelligence tests are culturally dependent, but all people are not equally exposed to the "majority" culture. Suppose we called "intelligence tests" by some other name, such as tests of acculturation to middle-class American life. This could be a descriptive name for these tests because the questions on the tests reflect what most people in the standardization sample knew and did not know at some point in time. For example, we might expect an average American adult in 1995 to know what a disk drive is, but we would not have expected this sort of knowledge from average Americans in 1985.

It is a fact that approximately 50% of African-Americans and other groups of ethnic minority children grow up in poverty. On the average, people who grow up in poverty do not have the same experiences as people who do not grow up in poverty. It is likely that fewer individuals from low income families will know what a disk drive is than individuals from families with higher incomes. Even if the same test predicts academic success equally well for all test-takers, it does not measure intelligence equally well, unless we decide to define intelligence as synonymous with academic success. This sort of definition leads to a type of circular reasoning (intelligence = academic success and academic success = intelligence) that would not be indicative of intelligent thought.

Intelligence Is Mostly Inherited

Of course, the authors prudently claim "we are not so rash as to assert that the environment or culture is wholly irrelevant" (301); however, they definitively conclude that "IQ is substantially heritable" (105). This is an example of the sort of weasel language that I referred to earlier. I do not believe that the data support this sort of blanket conclusion. Intelligence is far too complex to decide that it is mostly any one variable. It is clear to me that intelligence is partly inherited, but it is not meaningful or possible to quantify the size of that part. Also, the role of the environment is not a linear one as we climb the IQ scale. Consider, for example, a profoundly retarded individual—someone who scores below the cut-point designated as "educably retarded." Many such individuals cannot learn to feed themselves, to talk, or to use the bathroom; they need constant custodial care, often with direct feeding through their stomachs. In these rare instances, intelligence is unaffected by environmental variables. By definition, they will not benefit from education. But, as we ascend the intelligence curve, environmental variables become increasingly important. The most brilliant rocket scientist would not be functioning at a high intellectual level if she never attended school or had an opportunity to learn to read or study science. Many of the items on intelligence tests are the sorts of items that are learned in school. How can anyone conclude that formal and informal education doesn't have a massive effect on intelligence (for those who are at least near average and above in intelligence), when we measure intelligence with information that is learned in school?

There are many other problems with the dichotomization of nature and nurture and the attempt to assign a proportional value to each side of the nature-nurture equation. Nature and nurture are not separable components because biological propensities influence the environment that we seek, and through our interactions with the environment our biology changes. We now know that changes in the environment cause changes in brain structures, and altered brain structures change how we interact with the environment. Heredity and environment are like conjoined twins who share a common heart—they cannot be separated. It is impossible to declare a winner in the age-old tug-of-war between nature and nurture.

Low Intelligence Causes a Wide Range of Social Problems Such as Poverty, Injury, Crime, "Illegitimate" Births, and Idleness

My response to this list of social ills is a less-than-intelligent "Huh?" Let's consider the evidence and reasoning that the authors marshal for this conclusion. Take some time to examine the bell curve that is shown in Figure 1.

It is apparent that its name is descriptive of its bell-like shape. The large "hump" in the middle shows that most people are around average in intelligence. The bell curve, which is more formally known as the normal curve, is ubiquitous in the sciences with variables like height, weight, IQ, petal-size in flowers, crop yields, length of pickles, and more—all showing this distribution.

There is a cluster of variables that tend to occur together at the low (left) end of the intelligence curve. They include such "socially undesirable" behaviors and characteristics as child abuse and neglect, poverty, low levels of education, unemployment, "idleness," increased injuries, "illegitimate" births, welfare, higher birth rates, and crime. The opposites of these variables cluster with high intelligence and are shown at the upper (right) portion of the curve. The variables that cluster at the low-intelligence end of the distribution are the usual indicators of low socioeconomic status. The authors then conclude that low intelligence is the cause of the other variables in this cluster. They pronounce that: "Socioeconomic status is . . . a result of cognitive ability" (286). How can they know that being unintelligent caused poverty and not the reverse, or, at least, a more reciprocal relationship in which poverty and low intelligence operate jointly and influence each other? Poor people differ from rich people in many ways—they have poorer health, poorer nutrition, and poorer living conditions. Would it not make more sense to reverse the causal arrow and hypothesize that poverty and all of its associates (lack of prenatal care, inadequate heat, ingestion of lead paint, poor diet, etc.) cause low intelligence? The statistical procedures that the authors used to establish which of these related variables was causal cannot be used to establish that low intelligence is the cause of the other variables. The variables are at least interactive or possibly even unidirectional—in the other direction.

Current Social Programs Like Welfare, Affirmative Action, and Head Start Cannot Work

Finally, I understand the reason for this book. Although the data that were used to support their conclusions are from a fairly recent data set, the arguments themselves have been made countless times before. There is nothing new in the Herrnstein and Murray treatise. *The Bell Curve* is a book about money and values

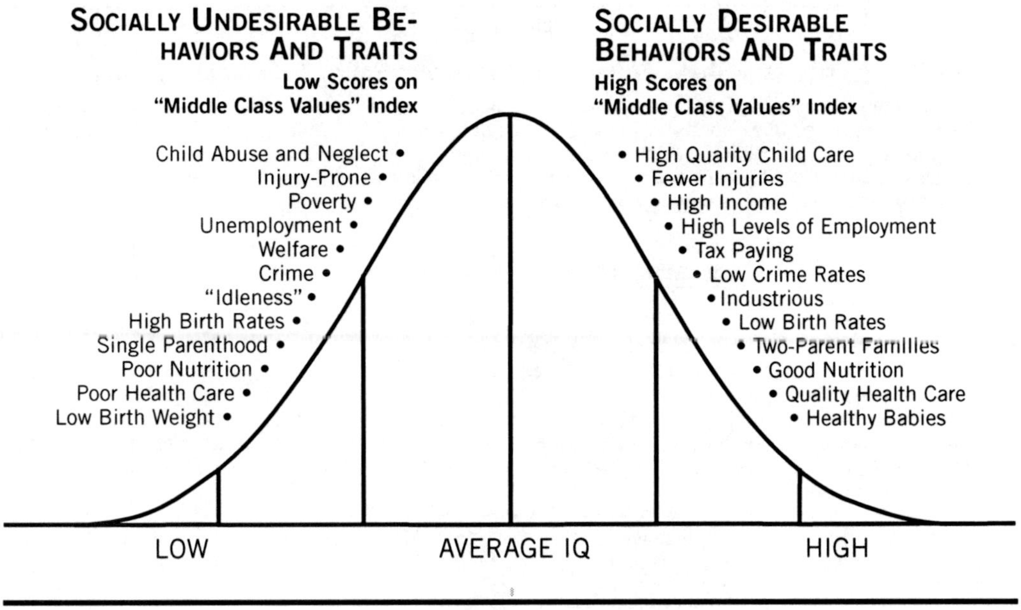

Figure 1

and how we should be spending tax dollars so that they reflect politically conservative values. Social programs like welfare are very expensive, and many, maybe even most, have not worked well. Why? Are the disappointing results because we have made many mistakes in how we set up these programs? Were our expectations too high? Did we set up the wrong contingencies or perhaps use insensitive measures of success? If so, then we should be able to find better ways to provide aid to the poor—ways that help more of them obtain jobs and move out of poverty. But, if social welfare programs cannot work because the recipients are too dumb or too idle or too criminal to benefit, then why spend money on programs that are either doomed to failure or actually increase the number on welfare by paying for out-of-wedlock babies? (Herrnstein and Murray prefer the term "illegitimate," an old-fashioned term that blames the baby for its mother's marital status. Their deliberate use of emotionally laden terms like "illegitimate" makes my skin crawl.)

Although the authors reach an opposite conclusion, it is clear that one kind of social program that has reaped considerable social benefits is education. Many studies have shown that education does improve thinking abilities, and it is these very abilities that are at the heart of any definition of intelligence. In their usual style, the authors present some of the data that show the beneficial effects of education and then dismiss these data as unreplicated, suspicious, lacking control groups, statistically flawed, etc. It is especially surprising that they arrive at this conclusion because the senior author, Herrnstein, was a contributor to a major program to improve intelligence in Venezuela. The Venezuela program has undergone careful scrutiny by international scientists, including random assignment of subjects to experimental conditions and "blind" scoring so that experimenter expectations cannot influence the outcomes, and it clearly has yielded improvements in thinking skills for those who were involved in the program.

In understanding what is at the heart of the authors' argument, it is important to distinguish between data and the interpretation of data. This relationship is shown in Figure 2.

Yes, poverty, crime, low intelligence, and

Figure 2

Misusing Data as a Shaky Foundation for Public Policy Recommendations

The data presented by Herrnstein and Murray are generally accepted as correct, although these authors slant the way the data are presented (e.g., downplay the importance of environment). Their interpretation of the data is substantially influenced by their beliefs, which reflect their prejudices and stereotypes. Herrnstein and Murray's recommendations for public policies are more dependent on their beliefs and the way they interpret the data than on the data themselves. This figure shows how this is done.

DATA	INTERPRETATION	PRE-EXISTING BELIEFS	PUBLIC POLICY RECOMMENDATIONS
IQ predicts academic success (partially).	Intelligence is a fixed quantity that remains fairly stable throughout life.	Prejudice concerning groups. Stereotypes about groups.	Eliminate Affirmative Action. End welfare for dependent children.
African Americans score, on the average, lower than white Americans on standardized intelligence tests.	Intelligence tests are fair and accurate measures of intelligence.	Poor women have babies as a way of making money from welfare.	Stop social programs that are designed to help poor people achieve.
Poverty, crime, high birth rates, etc. co-occur with low intelligence.	Low intelligence *causes* socially undesirable behaviors.	Social class is determined by intelligence.	Make it easier to convict accused criminals.
Intelligence is, in part, inherited.	Intelligence cannot be raised with education or other experiences.	Need for "breeding" programs that improve general level of intelligence.	End mandatory child support from unmarried fathers.
Some social programs have not raised intelligence or produced desirable outcomes.	Intelligence is a unitary concept ("g") that can be expressed with a single number.	Immigrants reduce the overall level of intelligence in the U.S.	Enact a competency test for immigrants as a criterion for immigration.

high birth rates occur together. These are the data, and they are not in question, although the authors often present the data in misleading ways. What is in question is the way these authors interpreted the data and the "cure" or public policy recommendations that arise from their interpretation. Their interpretation or explanation of the data is influenced by their belief system, and their explanations and beliefs intervene between the data and the public policy recommendations that are built on the data. There is good reason to believe that their interpretation of the data is "tainted" or not as pure or data-based as their academic affiliations, thick statistical appendix, and scientific-sounding language make it seem. Consider this quote from *The Bell Curve:* "The median earning of . . . workers in 1992 [was] $41,005 for white male graduates with a bachelor's degree and only $31,001 for black males with a bachelor's degree" (324). Most readers would interpret these data as evidence of persistent discrimination in the labor market. After all, how else could you explain the finding that even when African-Americans and Whites have the same education, and other variables like sex are held constant, African-Americans are paid much less? The authors conclude that this disparity in income shows how important the differences in intelligence really are. The bias in their interpretation of these data is too obvious to deserve additional comment.

Similarly, Herrnstein and Murray cite high drop-out rates for students who are admitted to college as a result of affirmative action programs as evidence that these students lack the intelligence to succeed in college, and therefore affirmative action programs cannot work.

Affirmative action admissions are almost always first-generation students from low income households. Why don't they consider other explanations for the high drop-out rate of students admitted under affirmative action programs, like the fact that these students are more likely to work while they are in college and when they work, they work more hours than their wealthier counterparts? Why don't they even consider the possibility that affirmative action students start college with deficits that are attributable to an inferior secondary education and social pressures that are not compatible with attending college? Wouldn't these facts be expected to increase drop-out rates? Like other interpretations of data in *The Bell Curve*, these conclusions do not ring true.

Follow the Money

This Watergate maxim is a good one to follow here. In deciding whom to believe, it is important to determine if the speaker or writer has an ulterior motive in convincing you that a certain conclusion is valid. For example, if the patent holder on a miracle cream that claims to "melt unsightly fat" told you that it was a wonder product, you would be less likely to believe this claim than if you had heard it from an unbiased scientific source with no potential for financial gain. The authors show a particular bias to cite studies that were funded by the infamous Pioneer Fund, which dispenses about $1 million annually to academics who support the idea that intelligence is genetically determined and that humans should be bred selectively for intelligence. I had a brief run-in with some of the academics whose work they have sponsored. In my book entitled *Sex Differences in Cognitive Abilities* (2nd ed.), I summarized a large body of research on brain size and concluded that although males have, on the average, larger and heavier brains, when these values are adjusted for body size, there is no sex difference. Following the publication of this book, I received an article from Richard Lynn, an Irish researcher, in which he says that his work shows that I am wrong. At first, I gave this rebuke very little thought because it is not unusual for researchers to come up with different findings and different conclusions, although his results were at odds with those reported by virtually all of the other researchers in this field.

I then received a copy of the Lynn article with a letter from a psychologist whom I know, Philippe Rushton, who is notable for his theory that intelligence is inversely related to penis size. He posits that those males with the largest penises have the lowest intelligence, and furthermore, there are racial differences in both penis size and intelligence. According to Rushton, the racial line-up in descending order of intelligence is Asians, Caucasians, and Africans, with the reverse order for penis size. (No, I don't know how he collected his data, nor do I know how other ethnic groups fare in this linear array.) This sort of theory is reminiscent of the penis-centered theories of Freud which posited a universal stage of development for boys and girls that he named the phallic stage. The word "phallic" means "penis," and Freud saw no reason why this stage should have a different name when it referred to female development. Rushton's penis-centric theory of intelligence suggests that some things never change since he proposes that we can learn about the intelligence of both females and males in an ethnic group by reference to the male anatomy. Much of the contemporary research funded by the Pioneer Fund is both racist and sexist. In fact, the founding fathers of this fund were also anti-Semitic with strong ties to the Nazi movement and its goal to rid the world of Jews. There are 23 separate references to Lynn in the bibliography of *The Bell Curve* and 11 to Rushton. Both of these critics of my work received high praise by Herrnstein and Murray, and, like other frequently cited researchers in *The Bell Curve*, received large amounts of money from the Pioneer Fund.

The parallels between sexist and racist theories became more apparent to me when I received a copy of Rushton's latest research, which was published after *The Bell Curve* went to press. Based on a study of helmet sizes used by the military, he concluded that African-Americans have smaller heads and therefore smaller brains than Caucasians—a result that mirrors the one by Lynn that compared male and female brains. There are many problems with these studies. Most importantly, brain size, weight, and neural structures depend upon life experiences. That is, our brains respond to our environment, so that we cannot know whether larger and heavier brains caused different life experiences or the experiences caused differences in brain size and weight. Many of the correlates of poverty, such as inadequate nutrition, alcohol and other drug use, lack of prenatal and pediatric health care, ingestion of lead-based paint and other toxins, all have negative effects on brain development during the critical prenatal and infancy periods when the brain is most vulnerable. I do not know if the brain weight data are valid, but even if they are, lower brain weight is more likely a consequence of poverty than the reverse. In addition, there is absolutely no evidence that heavy brains are found in smarter people or that skull size is a good measure of brain size. The leaps from the actual data to the conclusions are irresponsible.

Soon after *The Bell Curve* was published, I received a FAX and phone call from Linda Gottfredson, a professor at the University of Delaware, who summarized what she believed was the dominant professional view on intelligence. She asked me to sign her summary statement to indicate my support. She explained that this was important so that the media and the public had a single summary statement on intelligence to guide their understanding of the points raised by Herrnstein and Murray. I found her summary troubling as it essentially agreed with Herrnstein and Murray's conclusions. In fact, I agree with many of the statements made in *The Bell Curve*, but there are many others that I believe are wrong. I did not sign the statement that appeared in *The Wall Street Journal*, although 52 other psychologists did. I later learned that she is also supported by the Pioneer Fund. Although there is nothing morally wrong with being financed by people who share an author's ideological point of view, it is troubling when all of the research that is funded in this manner happens to support the ideology of the funding agency. If you understand the social and political agenda that has financed this work, the next conclusion made by Herrnstein and Murray should not surprise you.

Recent Immigrants Are Less Intelligent Than Immigrants Who Came to the United States Earlier This Century

The reasons in support of this conclusion are so flimsy that I cannot present them in a meaningful way. The authors argue that recent immigrants obtain special entry status because they are related to citizens; whereas immigrants at the turn of the 20th century fled persecution and were more motivated to succeed. Frankly, I cannot understand the logic in this argument. Why should we expect that recent immigrants from war torn and poverty stricken areas of the world would differ in motivation or intelligence from those who fled persecution earlier in the century? The political philosophy that the authors espouse is blatantly anti-immigration, which is as legitimate as any other political philosophy—except that this one is "dressed up" to look like a data-based conclusion, which it is not.

Herrnstein and Murray go on to argue that the recent flood of immigrants, coupled with high birth rates among the low intelligence portions of the population, have lowered the average intelligence of Americans. When the average intelligence of a country is lowered, it is less able to compete in world markets, it is less able to produce and use advanced technologies, and other dire consequences result. While this may seem to be a reasonable argument, they also present data that show that the average IQ scores have risen every decade, an effect known as the Flynn Effect, named for the individual who first hypothesized this rise. I do not know how to interpret these inconsistencies, except to say that they seem to be able to argue that average IQ is both rising and falling, depending on what is more convenient at the time.

How to Make Reasoned Judgments About Controversial Research

1. Read the original research, if possible. Second-hand accounts often distort the facts and make faulty inferences from the original research.
2. Identify the conclusions—What do the authors want you to believe and do?
3. Examine the data and other evidence that are provided to support the conclusions. Were tests of statistical significance used? Was the size of the effect considered (e.g., was the difference between groups large enough to be meaningful)? Was the sample sufficiently large? Was it representative of the population?
4. Is the conclusion a matter of opinion (e.g., euthanasia is wrong) or a matter of fact (e.g., men are taller, on average, than women)?
5. Do the authors have the expertise needed to conduct and interpret the study?
6. Are the authors unbiased or do they have a vested interest in the outcomes of the study?
7. Are appeals to emotion being used to convince readers that the authors' conclusions are correct (e.g., arguments against the person or name calling)?
8. What is missing? Would other explanations fit the data equally well or better?
9. Do the conclusions follow from the data?
10. Are the stated and unstated assumptions acceptable?
11. Are correlational data being used to make causal claims? (Random assignment of subjects to groups is needed to make strong causal claims.)
12. Can you identify fallacies in the reasoning (e.g., false dichotomy)?
13. Are valid and reliable measures being used?
14. Are the results unusual? If so, why? Why is the study controversial?
15. Overall, what is the strength of the support for the conclusions?

There Is Some Good News

Readers may be thinking that *The Bell Curve* forecasts a bleak future unless we stop welfare programs and curtail immigration so that the intelligent portion (or the "over the hill" portion on the right hand side) of the curve will have higher birth rates and the less intelligent portion stops reproducing and entering the country. Well, there is also good news. You and I are not at fault! We are all in the "over the hill" gang, a group repeatedly referred to as the "cognitive elite" because we are intelligent enough to read their massive tome and rich enough to spend $30 to buy it. We can look down on the poor unfortunates who live on the other side of the intelligence hill from us, and like responsible parents we can decide to do the right thing and eliminate social programs. The solutions that the authors offer have a very contemporary sound because they are now heard on Capitol Hill. It is not the politically conservative point of view that I am objecting to in this review—it is the misuse of data and the blatant biases in the way the data are interpreted in support of this point of view that I find objectionable. Yes, we have difficult contemporary problems with welfare and immigration, among others. Responsible social science data are needed to guide public policy on these immensely complex issues, but the authors provide blatantly biased interpretations that are closer to propaganda than responsible research. Social programs may very well be doomed to failure for economic, social, or political reasons, but they are not doomed for the reasons Herrnstein and Murray present.

They also offer other solutions. We can return to simpler times (541) when all people had a "valued place" in society (535). The authors define a "valued place" as "other people would miss you when you were gone" (535). What does this sentimental dribble really mean? Slave masters missed their slaves when they were gone; does this mean that slaves had a "valued place" in society? The call for simpler laws seems like an excellent idea. In fact, I found myself nodding frequently with many of their recommendations until I realized that "simpler" laws really meant fewer rights and safeguards for citizens. The nostalgia for the good old days when the neighborhood cop was your friend were not so good for everyone. African-American children never assumed that the local police officer was their friend, especially if they grew up in the segregated South. Have the authors really thought through their suggestion that fathers who are not married should not be required to pay child support—so-called "deadbeat fathers"? This solution is misogynist, anti-child, and fiscally foolish. How can this proposed policy discourage out-of-wedlock births or save taxpayer dollars? It certainly will not provide males with incentives to use contraception, if they have no financial responsibility for the children they father. How is this policy consistent with the creation of a "valued place" for everyone? What will we gain as a society by getting those deadbeat toddlers off welfare—a move that virtually ensures that many of America's children will be denied access to even the most basic of human needs like adequate nutrition, health care, and heat? I don't know whether to cry for a society that sacrifices its young or rage in anger against the intelligent people who forgot to care about the rest of society. Herrnstein and Murray's proposed solutions drip with hypocrisy and offer simplistic cures for society's most difficult ills. And for these solutions I don't think that even Forrest Gump, the lovable role model for those in the low-intelligence portion of the curve, would offer Herrnstein and Murray a piece of his coveted chocolates.

Race and Sports as Good Science

JON ENTINE

"If you can believe that individuals of recent African ancestry are not genetically advantaged over those of European and Asian ancestry in certain athletic endeavors, then you could probably be led to believe just about anything." Or so says biological anthropologist Vincent Sarich. To which professor of sociology Harry Edwards, also of University of California/Berkeley, provides the antithesis: "What really is being said in a kind of underhanded way is that blacks are closer to beasts and animals in terms of their genetic and physical and anatomical make up than they are to the rest of humanity. And that's where the indignity comes in."

For the synthesis, turn to Gideon Ariel, Biomechanist, former U.S. Olympic Committee scientist, former Israeli Olympic athlete: "I know that the American system is very sensitive to statements of black and white. But you cannot defy science. You cannot just say that day is night and night is day. These are facts."

In fact, in running, basketball, football, and soccer—sports in which the social and economic barriers to participation are very low, creating the most level of playing fields—the yawning performance gap between blacks and everyone else is nothing short of astonishing. Yet allegations of racism often quash the overwhelming scientific evidence which convincingly suggests that this growing on-field disparity cannot be explained by culture and environment alone.

Even a casual mention that there exist any meaningful genetic differences between races can ignite a firestorm. In a speech before the British Association for the Advancement of Science in 1995, Roger Bannister, the distinguished neurologist, retired Oxford dean, and the first man to break the four minute barrier in the mile, in 1954, was showered with ridicule for venturing his opinion "as a scientist rather than a sociologist" that all athletes are not created equal. "I am prepared to risk political incorrectness," he said, "by drawing attention to the seemingly obvious but under stressed fact that black sprinters and black athletes in general all seem to have certain natural anatomical advantages."

That's the explosive "N" word—natural. "Nurture" alone cannot explain the remarkable trends. Over the past 30 years, as sports has opened wide to athletes from almost every country, the results have become increasingly segregated. There are only 800 million blacks, or one in eight of the world population, but athletes of African origin hold every major world running record from the 100 meters to the marathon. In the United States, where African Americans make up about 13% of the population, almost 90% of professional basketball players, 70% of the National Football League, and more than a third of professional baseball is black. In Britain, with a black population of less than 2%, one in 5 professional soccer players is black. Blacks have also come to dominate world boxing.

Why do blacks of West African ancestry dominate sports in which the social and economic barriers are lowest?

Fifty years of anthropological and more recent physiological studies have documented clear, if overlapping, biologically based differences between athletes of different populations. Scientists are just beginning to isolate the genetic links to those biologically based differences (though the fact that the biology is grounded in genetics is unequivocal). That's the science. The politics is more precarious. Any suggestion of human differences is publicly and politically seen as divisive or worse in a country which sometimes gives lip service to equal opportunity and where race remains a festering sore.

African Americans understandably are suspicious about where this discussion can lead. "People feel if you say blacks are better athletically, you're saying they're dumber," Frank Deford, the respected author and sports reporter once noted. "But when Jack Nicklaus sinks a 30-foot putt, nobody thinks his IQ goes down."

Athletic achievement has long been a Catch-22 for blacks. When an athlete lost a contest, it encouraged racist notions that blacks were an inferior race, intellectually and physically. But winning reinforced the equally pernicious stereotype that blacks were closer to animals and therefore less evolved than whites or Asians. That is the fate that befell Jesse Owens after he shocked the 1936 Olympics, held in the capital of Hitler's Germany. His four gold medals were subtly devalued as a product of his "natural" athleticism.

The racist stereotype of the "animalistic black" stretches back centuries. Fascination about black physicality and black anger about being caricatured as a lesser human being, closer to a jungle beast, have been part of the dark side of the American dialogue on race, with deep historical roots in hundreds of years of European colonialism. In the 19th century, white Europeans were enraptured by pseudosciences such as phrenology. Racial and ethnic groups were ranked by skull size that supposedly proved that white males were intellectually superior. Jews, blacks, and other minorities were targets of the most egregious generalizations, usually associated with physical characteristics and intellectual prowess.

Since World War II, in an understandable reaction to extremist race theories that provided intellectual fuel for Nazism, anthropological orthodoxy has held that the very concept of race is a meaningless social construct. Discussing "race science," as it came to be called, became a taboo subject, publicly and academically. The issue took on incendiary proportions in the early 1970s when it was publicly married to findings of race differences in I.Q.

Growing up in the Sixties, it never occurred to me to judge blacks as less intelligent. And I celebrated with most liberal-thinking Americans when Muhammad Ali redefined boxing and when the raised black fist of the 1968 Mexico City Olympians became a potent symbol of freedom. I entered the shark infested waters of this debate in 1987, when Los Angeles Dodger general manager Al Campanis had been fired after commenting on national television that he believed that blacks didn't have the mental "necessities" to be a manager or general manager. The following January, Jimmy "the Greek" Snyder, a prognosticator with CBS Sports, was fired and publicly ridiculed after making an off-hand comment that slave owners had bred blacks to produce the best physical specimens and that this contributed to black success in sports. At the time,

I was producing for Tom Brokaw at *NBC Nightly News.* After much internal handwringing, we decided that maybe we should address the myths and stereotypes of blacks in sports—including the racial taboos. Perhaps dialogue could dissipate some of the noxious poison.

The end product was our 1989 documentary, *Black Athletes: Fact and Fiction.* Before it aired, it provoked intense reaction, dividing journalists, frequently along racial lines. A white columnist at *Newsday* called it "a step forward in the dialogue on race and sports" while a black writer at the same daily wrote that "NBC had scientists answer questions that none but a bigot would conjure up." Yet the public, particularly African Americans, seemed far more receptive to the balanced treatment of a heretofore untouchable subject. Even Harry Edwards, a long-time critic of the suggestion that there are any meaningful racial differences, would comment that "the NBC documentary opened the door to enlightenment on a controversial subject." *Black Athletes* went on to win numerous awards including Best International Sports Film.

Over the next few years, the science of human performance and our knowledge of human genetics barreled forward at breakneck speed. I became even more intrigued by the genetics of human performance. At the urging of my literary agent, I circulated a book proposal that offered to explore the issue in far more depth. The timing, I believed, was opportune. This was a chance to write a cutting edge, popular but scholarly book that discussed genetics and the problematic social history of race. Sports would merely be an access point for a wide-ranging conversation.

As a measure of my commitment, I assembled a "board of advisors"—top biologists, anthropologists, exercise physiologists, and sociologists, black and white, from all over the world, who offered to act as informal scholarly reviewers as the book took shape. They embraced the proposal as provocative and responsible. Perhaps that's why I was so stunned by the consistently negative response it engendered from publishers, many of whom refused to even read it—on "principle." Again and again, I heard: "This is a racist subject. By even suggesting that blacks may have a genetic edge in sports, you are opening up the Pandora's box of intellectual inferiority."

Finally, after more than a dozen rejections, an independent-minded editor at Macmillan, Rick Wolff, offered a contract for what was to become *Taboo.* The turn of good fortune proved fleeting, however. Soon after, Mr. Wolff moved to Warner Books. Though he wanted to take the book with him, Warner balked. "It was considered too dicey a subject, too controversial," Wolff recalls. "Once the other editors heard it was about racial differences, they wouldn't even let me present it at an editorial meeting."

Unfortunately, Mr. Wolff's eventual replacement as editor, Natalie Chapman, knew nothing about sports and was only vaguely sensitive to the science and politics of race. Nonetheless, I proceeded with an early draft, always staying in close contact with my advisory board and an expanding list of experts, who were sent the evolving manuscript for feedback.

By this time, I had grown quite confident of my findings. Using DNA evidence, scientists were in the process of compiling maps of the waves of human migrations that have led to today's "races." Although the move out of Africa by modern humans to Europe and Asia occurred rather recently in evolutionary time, scientists were nearly unanimous in their belief that even small, chance mutations can trigger a chain reaction with cascading consequences, possibly even the creation of new species, in relatively few generations. Economic ravages, natural disasters, genocidal pogroms, and geographic isolation caused by mountains, oceans, and deserts have deepened these differences.

As a result of evolution, every population group has some unique physical and physiological characteristics, many of which have a genetic basis (Cartmill, 1988; Chakraborty et al., 1993). Most of today's genetic research focuses on finding cures for diseases, more than 3,000 of which are genetically based (Overfield, 1995). For instance, blacks are predisposed to carry genes for sickle cell anemia and susceptibility to colorectal cancer (Weber, 1999). Beta-thalassemia is most prevalent in Mediterranean populations. A form of diabetes has been linked to a gene most commonly found among North American Indians.

So why do we so readily accept that evolution has turned out Ashkenazi Jews with a genetic predisposition to Tay-Sachs, or blonde haired and blue-eyed Scandinavians, yet find it racist to suggest that blacks of West African ancestry have evolved into the world's best sprinters and jumpers?

In fact, highly heritable characteristics such as skeletal structure, the distribution of muscle fiber types, reflex capabilities, lung capacity, and the ability to use energy more efficiently are not evenly distributed across racial groups and cannot be explained by known environment factors (Entine, 2000; Samson and Yerlès, 1988). Consider diving, gymnastics, and ice-skating, sports in which East Asians excel. Asians tend to be small with relatively short extremities, long torsos, and a thicker layer of fat. "Chinese splits," a rare maneuver demanding extraordinary flexibility, has roots in this anthropometric reality (Carter, 1982; Eveleth and Tanner, 1990; Martin and Saller, 1959; Himes, 1988; Behnke, 1974; Hirata and Kaku, 1968; Hirata, 1979).

Eurasian whites are the premier wrestlers and weight lifters in the world. Evolutionary forces have shaped a population with large, muscular upper bodies with relatively short arms and legs and thick torsos. These proportions tend to be an advantage in sports in which strength rather than speed is at a premium. This region also turns out an extraordinary number of top field athletes—javelin throwers, shot-putters, and hammer throwers.

Athletes who trace their ancestry to western African coastal states, including British, Caribbean and American blacks, are the quickest and best leapers in the world. Consequently, they almost completely monopolize the sprints up to 400 meters. No white, Asian, or East African runners have broken 10 seconds in the 100m. The top two hundred times in the 100m—all under 10 seconds—are held by athletes of West African descent. All 32 finalists in the last four Olympic men's 100-meter races were West African. The likelihood of that happening based on population numbers alone is 0.0000000000000000000000000000001. Yet there are no—not one—premier middle or long distance runners from this region in Africa.

Studies have shown that athletes of West African origin hit a biomechanical wall after about 45 seconds of intense, anaerobic activity, when aerobic skills come into play. East Africans, who have small and slender ectomorphic body types and are therefore hapless in the sprints, dominate distance running (Ama et al., 1990; Saltin, 1973; Levesque, 1995; Simoneau, 1991; Levesque, 1994).

Whereas the West African population evolved in the lowlands and remained relatively isolated, East African runners trace their ancestry to the highlands. This region in Africa is also a genetic stew, with studies indicating a mixture of genes from invading Arabs and Middle Easterners.

Kenya, with 28 million people, is the athletic powerhouse. At the Seoul Olympics in 1988, Kenyan men won the 800, 1,500, and 5,000 meters, along with the 3,000-meter steeplechase. Based on population percentages alone, the likelihood of such a performance is one in 1.6 billion. The Kalenjin people of the Great Rift Valley adjacent to Lake Victoria—who represent 1/2000th of the world population—win

40% of top international distance running honors and three times as many distance medals as athletes from any other nation in the world. One tiny district, the Nandi, with only 500,000 people, swept an unfathomable 20% of major international distance events. By almost any measure, the Nandi region is the greatest concentration of raw athletic talent in the history of sports. It's a potent example of the interacting bio-cultural forces that shape great athletes.

By this time, the draft of *Taboo* was taking shape. I sent it off to Macmillan and waited. And waited. Eight months passed without a word before I received the brush-off in a brusque letter. "Much of the manuscript is smoothly and elegantly written, and most of it is quite enjoyable to read," wrote Chapman. "[But] while I admire the goals of the book, I must regretfully inform you that [it] lacks sufficient persuasiveness . . . to avoid being torn apart by critics, reviewers, and readers."

Years of work were suddenly in mortal danger. My agent embarked on a full court press to find a new publisher, but to no avail. As before, most everyone treated the proposal (and now an early manuscript) as a skunk on the loose. Basic Books, a first-rate independent publisher affiliated with HarperCollins, appeared ready to publish *Taboo* until an African American consultant nixed the book, without reading it, as "potentially racist." One female editor lectured my agent about how insensitive he was even to propose such an idea. Would she please read the book? he responded. "I don't have time for such trash," she retorted.

Such intense personal reaction was all the more dispiriting given the lengths to which I had gone to include, in a non-polemical way, many diverse historical and ideological perspectives. To a man and woman, the board and reviewers were on record that they respected *Taboo* as fair and constructive, with racial healing as one of its messages.

"You will be accused of spouting old fashioned racism for even raising the issue of African American superiority in athletics," wrote Earl Smith, chairman of the department of sociology and ethnic studies at Wake Forest University, a leading black scholar and author of several books on race and sports, and one of my board members. "All this beating around the bush has to stop. This is a good book. I am quite excited with the arguments that are raised."

But Dr. Smith's endorsement, along with reviews and letters of support from the president of the Human Biology Association, the current editor of the *Journal of Human Biology*, a US Olympic Committee scientist, prominent African American anthropologists, and top athletes couldn't crack the political status quo. As I was learning, when it comes to race, "the cortex shuts down." No one would even read the manuscript and give *Taboo* a chance.

Public Affairs, another independent publisher with authors such as international financier George Soros, former Secretary of Defense during the Vietnam war Robert McNamara, and *60-Minutes* commentator Andy Rooney, broke the log jam when an editor read it, loved it, and assumed the rights.

Yet even with a respected publisher behind *Taboo*, the hysteria continues in some quarters. In early January, just before the book was released, *The New York Times Magazine* informed me that it was killing plans to publish an adaptation, calling the book's thesis potentially "dangerous." "Our reluctant decision to drop the project is no reflection of my regard for your work, which remains high," wrote Kyle Crichton, an editor who had championed the article. "In brief, the whole subject worries my editor. . . ."

Taboo is now finally in the hands of the public. Will it be as skittish about the contents as the publishing industry? Apparently not. As of the day I write this, *Taboo* has so far received almost unanimous if sometimes guarded praise in more than three dozen reviews. Most have been raves. The only negative comments have come from those journalists who consider themselves "liberals." For instance, writing in the *Chicago Sun-Times*, columnist Rick Telander, apparently attempting to inject some "balance" into a review that generally praised the book, wrote: "Reviews of *Taboo* have been as uptight as anything, with reviewers figuratively holding the book the way an exterminator might hold a spraying skunk."

To buttress this incendiary conclusion, Telander writes: "'Some Things Are Better Left Unsaid,' is how *USA Today* titled its review." Minor problem: The title of the article was 180 degrees the opposite: "Some Things Not Better Left Unsaid." In fact, *USA Today* columnist Christine Brennan praised the book, writing "the dialogue that [Entine] almost certainly will provoke is not the problem. It's the solution."

Telander's second citation is from a *New York Times* column by Robert Lipsyte. "Entine's research is 'simultaneously silly and dangerous,'" quoted Telander. Oops. Lipsyte wasn't referring to me or my book, but to the issue: "Sports race science can be viewed as silly and dangerous" is the real quote. The *Times* actually praised *Taboo* as "consistently interesting, readable, provocative"—hardly a skunk like renunciation.

Telander's third example—he quoted a *Washington Post* reviewer that *Taboo* "underplays the political and cultural land mines underlying the discussion"—is equally misleading. Paul Ruffins, a former editor of the NAACP's *Crisis* magazine, actually admired the book. "Because it bravely tackles the exhaustive list of ideas that must be considered in any open-minded discussion of this topic, *Taboo* could well be the most intellectually demanding sports book ever written," Ruffins wrote. "*Taboo* is an informed exploration of a fascinating phenomenon. Entine marshals such an impressive array of evidence that we should no longer be content to explain why blacks excel at certain sports by simply resorting to the old cultural argument that athletics have been the only avenues of upward mobility that were truly open to them. He's raised the argument to new heights."

A number of reviewers (every one white and supposedly liberal) apparently felt uncomfortable about being seen as praising a book that suggested that humans are indeed as diverse—culturally and biologically—as multi-culturalists claim. These white writers assumed, incorrectly it turns out, that *Taboo* would provoke widespread anger among blacks.

Claiming that *Taboo* has provoked "racial ire," Stan Hochman, an otherwise thoughtful columnist with the *Philadelphia Daily News*, wrote that "People of many hues say his science is flimsy, his conclusions are racist." He cited Harry Edwards who, he wrote, claimed that "Entine's scientific data is an underhanded way of saying that blacks were 'closer to beast . . . than they are to the rest of humanity.'" These were incendiary claims and grossly inaccurate in regard to both scientists and African Americans.

"*Taboo* is carefully researched and intellectually honest," wrote Jay T. Kearney of the U.S. Olympic Sport Science Committee. Michael Crawford, professor of genetics, former editor of the *Journal of Human Biology* and current president of the Human Biology Association, wrote that "*Taboo* provides a wonderful opportunity to share a message of the importance of human biological and cultural diversity in its myriad forms. Any dialogue between different racial groups should start with the facts."

What about that slashing quote from Edwards? The quote is actually lifted from *Taboo* itself. It is directed not against the book, which had not yet been written, but the misuse of science in the service of racism—the nefarious history that *Taboo* exposes. In fact, Edwards has publicly called my research "enlightening" and had offered to blurb the book.

What has been the reaction from others in

the black community? Intriguingly, the most effusive comments have come from African Americans. Earl Smith ended up writing the introduction. The *Journal of the African American Male* is carrying two chapters in future issues; its editor, Gary Sailes, wrote a blurb for the book in which he calls *Taboo* "Compelling, bold, comprehensive, informative, and enlightening." The black magazine *Emerge*, in its March issue, called the book "thoughtful, thorough, and sensitive. . . . *Taboo* is a good read for anyone interested in the history of black athletes in the United States and world-wide."

"*Taboo* is both provocative and informed," wrote John C. Walter, professor of history in the American Ethnic Studies Department at the University of Washington, in a review in the *Seattle Times*. "Entine has provided a well-intentioned effort for all to come clean on the possibility that black people might just be superior physically, and that there is no negative connection between that physical superiority and their IQs."

What are we to make of this phenomenon in which some whites, so quick to crow about their own racial sensitivity, recklessly inject racial divisiveness into a debate in which most African Americans see thoughtfulness? It's apparent that many blacks have become irritated to the point of anger by the patronizing censorship and condescension of many journalists and academics. To date no one has yet criticized *Taboo* for racial insensitivity or shoddy science. "I am an editorial columnist," wrote Bill Maxwell of the *St. Petersburg Times* in a personal note to me after his glowing column on *Taboo*. "I reviewed your book because I enjoyed reading it. It cut through all of the bullshit. I am black."

Although the African biological edge is not great, at the level of an elite athlete, even a small advantage can be the difference between a gold medal and finishing out of the money. On-the-field trends create a cultural advantage that forms a biosocial feedback loop, with nature and nurture fueling each other. Nevertheless, it is critical to remember that no individual athlete can succeed without the 'X-factor,' the lucky spin of the roulette wheel of genetics matched with considerable dedication and sport smarts. "It's the brain, not the heart or lungs, that is the critical organ," Sir Roger Bannister told me. "But one would have to be blind not to see a pattern here. I hope we are not at a time and place where we are afraid to talk about remarkable events. I hope not."

Popular thought is now beginning to catch up with scientific knowledge. The genetics revolution has decisively overturned the dated belief that all humans are created with equal potential, a tabula rasa, or blank slate, for experience and culture to write upon. Acknowledging human biodiversity may approach a danger zone, but pretending that there are no slippery questions does not prevent them from being asked, if only under one's breath.

Taboo is not so much a sports book as it is a thought-provoking look at what defines us as human. It debunks facile theories of race that have been used for hundreds of years to justify racism and even genocide. Most important, it shatters stereotypes that blacks or whites or any racial group are innately "superior" or "inferior." This is a book about the rich diversity of life, free of the myths of "ranking" that have plagued Western thought for centuries. That's the message of *Taboo;* for the most part, it is being heard.

"Entine understands the reasons Blacks lash out against the determination theory, knows that whatever White America gives to Black athletes in terms of athletic superiority, it takes from their mental abilities," wrote Carolyn White of *Emerge* magazine. "Great athletes, dumb jocks. And the stereotype, suggests Entine, is probably the single most important reason people have problems debating the issue."

Although it should never be far from anyone's mind that white fascination with black physicality has long framed this issue, it's more than clear that the stereotype that blacks make better athletes than whites is neither wrong nor racist. Censorship and the invocation of a taboo on issues of human diversity, biological and cultural, are not viable options.

"In human biology and clinical studies, as well as in epidemiological research, it is important to understand if age, gender, race, and other population characteristics contribute to the phenotype variation," wrote Claude Bouchard, Laval University geneticist, obesity expert and exercise physiologist, in a recent article in the *American Journal of Human Biology*. "Only by confronting these enormous public health issues head-on, and not by circumventing them in the guise of political correctness, do we stand a chance to evaluate the discriminating agendas and devise appropriate interventions. To disregard monumental public health issues is to be morally bankrupt" (Bouchard, 1988).

"Since the word *race* causes such discomfort, *ethnic groups* is often substituted, but it is inappropriate," adds Theresa Overfield, University of Utah professor of anthropology and expert on the biology of health and illness. "Race is a characteristic used most effectively to describe, rather than explain, health difference. . . . Ignoring the differences between humans is at least shortsighted and can be medically harmful" (Overfield, 1995).

Human beings are different. Limiting the rhetorical use of folk categories such as race, an admirable goal, is not going to make the patterned biological variation on which they are based disappear. The question is no longer whether these inquiries will continue but in what manner and to what end. Science is a skeptical endeavor. It is a method of interrogating reality, a cumulative process of testing new and more refined explanations, not an assertion of dry, unalterable facts. It is a way of asking questions, not of imposing answers. The challenge is in whether we can conduct the debate so that human diversity might be cause for celebration of our individuality rather than serving as fodder for demagogues.

References:

Ama, Pierre F. M. et al. "Anaerobic Performances in Black and White Subjects," *Medicine and Science in Sports and Exercise* 22, 4 (1990), 508–511.

Behnke, Albert R. *Evaluation and Regulation of Body Build and Composition* (Englewood Cliffs, N.J.: Prentice Hall: 1974), 359–386.

Bouchard, C. A., S. Leon, D. C. Rao, J. S. Skinner, and J. H. Wilmore. "Response," *American Journal of Human Biology* 10, 3 (1998), 279–280.

Carter, E. L. (ed.). *Physical Structure of Olympic Athletes, Part I: The Montreal Olympic Games Anthropological Project* (Basel, Switzerland: S. Karger, 1982).

Cartmill, Matt. "The Status of the Race Concept in Physical Anthropology," *American Anthropologist* 100 (September 1988), 651–660.

Chakraborty, Ranajit, Ranjan Deka, and Robert Ferrell, "Letter to the Editor: Reply to Baer," *American Journal of Human Genetics* 53 (1993), 531.

Entine, Jon. *Taboo: Why Black Athletes Dominate Sports and Why We Are Afraid to Talk about It* (New York: Public Affairs, 2000), 246–271.

Eveleth, P. B. and J. M. Tanner. *Worldwide Variation in Human Growth*, second edition (Cambridge: Cambridge University, 1990).

Himes, John H. "Racial Variation in Physique and Body Composition," *Canadian Journal of Sports Science* 13 (1988), 117–126.

Hirata, K. I. *Selection of Olympic Champions, Vols. I and II* (Toyota, Japan: Chukyo University, 1979).

Hirata, K. I. and K. Kaku. *The Evaluating Method of Physique and Physical Fitness and Its Practical Application* (Gifu City, Japan: Hirata Institute of Health, 1968).

Levesque, M., M. R. Boulay, J. A. Simoneau. "Muscle Fiber Type Characteristics in Black African and White Males before and after 12 Weeks of Sprint Training," *Canadian Journal of Applied Physiology* 19 (1994), Supplement 25P.

Levesque, Martin, M. R. Boulay, G. Thériault, C. Bouchard, J. A. Simoneau. "Training-Induced Changes in Maximal Exercise of Short Duration and Skeletal Muscle Characteristics of Black

African and Caucasian Men," unpublished manuscript, 1995.

Martin, R. and K. Saller. *Lehrbuch der Anthropologie II* (Stuttgart: Fischer, 1959).

Overfield, Theresa. *Biologic Variation in Health and Illness,* second ed. (Boca Raton, Fla.: CRC, 1995), 1–2.

Saltin, Bengt. "Metabolic Fundamentals in Exercise," *Medicine and Science and Sports* 5, 3 (1973), 137–146.

Samson, Jacques and Magdeleine Yerlès. "Racial Differences in Sports Performance," *Canadian Journal of Sports Science* 13 (1988), 110–111.

Simoneau, J. A., C. K. Allah, M. Giroux, M. R. Boulay, P. Lagassé, G. Thériault, C. Bouchard. "Metabolic Plasticity of Skeletal Muscle in Black and White Males Subjected to High-Intensity Intermittent Training," *Medical Science Sports and Exercise* 23 (1991), S149.

Weber, Thomas K. et al. "Novel hMLH1 and hMSH2 Germline Mutations in African Americans with Colorectal Cancer," *Journal of the American Medical Association* 281 (June 23/30, 1999).

Race and Sports as Pseudoscience

MICHAEL SHERMER

In "An Essay on Man," the 19th century English poet and essayist Alexander Pope elucidated the pitfalls of speculating on ultimate causes derived from immediate events:

In vain the sage, with retrospective eye,
Would from th' apparent what
conclude the why,
Infer the motive from the deed, and show
That what we chanced was what we meant
to do.

Pope's wise words were in the back of my mind as I began writing this essay on March 5, 2000, a miserably cold and rainy Sunday morning, as I watched the elite runners in the Los Angeles Marathon—just a handful among the 23,000 weekend warriors who braved the elements—cross the finish line. Although I have run the L.A. Marathon, and even once completed a marathon after first swimming 2.4 miles in the open ocean and riding a bike 112 miles in the Hawaiian Ironman triathlon, I would not have given the results a second glance were it not for a book I had just read that called my attention to a characteristic shared by the top five finishers. They were: (1) Benson Mutisya Mbithi, 2:11:55, (2) Mark Yatich, 2:16:43, (3) Peter Ndirangu Nairobi, 2:17:42, (4) Simon Bor, 2:20:12, and (5) Christopher Cheboiboch, 2:20:41.

It was not the times of the top five finishers that stood out in this year's race, since they were well below both world and course records (understandable considering the conditions). What was startling was their country of origin. All were from Kenya. Coincidence? Hardly. Meaningful? To some, yes; to others, no; to science, maybe. That is the subject of the book I had just read, Jon Entine's controversial *Taboo: Why Blacks Dominate Sports and Why We're Afraid to Talk about It.*

I will not dissemble and pretend that I was not aware of the controversy surrounding claims that blacks are better athletes than whites due to heredity and being closer to the origin of humanity in Africa. I've been an athlete and sports fan all my life and recall the vitriolic reaction to Jimmy "the Greek" Snyder's 1988 off-the-cuff remarks at a restaurant about black slaves being bred for superior physicality (on Martin Luther King Day, no less, with a camera crew present): "The black is a better athlete because he's been bred to be that way. During slave trading, the slave owner would breed his big woman so that he would have a big black kid, see. That's where it all started." Blacks, Snyder explained, could "jump higher and run faster" because of their "high thighs and big size."

I even saw live the now-infamous 1987 ABC *Nightline* show (occasioned by a celebration of Jackie Robinson's shattering of the color barrier in baseball) when Ted Koppel asked Los Angeles Dodger baseball executive Al Campanis why there were no blacks in upper management. Campanis said that blacks "may not have some of the necessities" for such positions. "Do you really believe that?" Koppel rejoined. "Well, I don't say all of them," Campanis demurred, "but they certainly are short in some areas. How many quarterbacks do you have, how many pitchers do you have that are black?" After continuing with his folk lesson in sports physiology, Campanis noted why blacks do not compete in elite swimming: "because they don't have buoyancy." Whites are floaters, blacks are sinkers.

Campanis's attempts to explain himself opened the gates into the largely unspoken but pervasive attitudes held by many whites about blacks, even whites who would not consider themselves racist. "I have never said that blacks aren't intelligent, but they may not have the desire to be in the front office," Campanis continued. "I know that they have wanted to manage, and many of them have managed. But they are outstanding athletes, very God-gifted and they're very wonderful people. They are gifted with great musculature and various other things. They are fleet of foot, and this is why there are a number of black ballplayers in the major leagues." Blacks are fast around the bases, slow around the boardroom.

As University of Texas Professor John Hoberman explained in his 1998 book *Darwin's Athletes*, even many blacks embrace part of the thesis (to their cultural detriment, he believes). Dallas Cowboys all-star player Calvin Hill, a Yale graduate, opined: "On the plantation, a strong black man was mated with a strong black woman. [Blacks] were simply bred for physical qualities." San Francisco '49ers wide receiver Bernie Casey explained: "Think of what the African slaves were forced to endure in this country merely to survive. Black athletes are their descendants." Even the liberal champion of cultural determinism, Jesse Jackson, in a 1977 CBS *60 Minutes* segment on his P.U.S.H. program for black school kids, made a case for heredity over environment when he stated (in response to sociologists' environmental explanations for blacks' poorer school performances) that "If we [blacks] can run faster, jump higher, and shoot a basketball straighter [than whites] on those same inadequate diets . . ." then there is no excuse. It is time, Jackson argued, for blacks to start living up to their potentials in the classroom as well as the gym.

With such comments from both blacks and whites it is understandable why some blacks, such as the noted U.C. Berkeley sports sociologist Harry Edwards, respond so strongly, and usually wrongly, going to the opposite extreme of environmental determinism. On a March 8, 2000, radio show I hosted with Entine, Edwards, and Hoberman as guests, Edwards actually made the argument that the only reason blacks dominate NBA basketball, despite more than equal opportunity for whites to make it to the top, was that at this period of time the "black style" of basketball happens to be popular instead of the "white style" prominent in the 1950s, and that neither "style" was in any way superior. My co-host Larry Mantle and I, both enthusiastic L.A. Laker fans, gave each other a knowing glance of acknowledgement that this was, of course, utter nonsense.

Somewhere between Edwards's extreme environmental determinism and the Greek's radical biological determinism lies the truth about the cause and meaning of black-white differences in sports. But the Campanis episode was the most enlightening because these were not the remarks of a rabid bigot spewing racial epithets; rather, Campanis had spent decades in close proximity and in tight friendship with some of the greatest black ballplayers of the 20th century. So his comments were emblematic of the common attitudes shared by many, perhaps most, lay people and sports enthusiasts who know just enough to speculate in a social Darwinian mode about how and why blacks dominate in some fields but not others, and what these differences tell us about the human condition.

What do these differences mean? The answer depends on what it is you want to know. I shall address this subject neither to embrace the theory nor to debunk it; rather, the question itself raises a number of other questions and problems in this field of research that makes reaching grand and sweeping conclusions problematic at best.

From the Particular to the General: Do Black Athletes Dominate Sports?

If you are a basketball, football, or track-and-field fan, the black-white differences are obvious and real. You'd have to be blind not to see the gaping abyss any given day of the week on any one of the numerous 24-hour a day sports channels. Further, there are quantifiable within-race differences in some of these sports. Kenyans dominate marathon running, but you'll likely never see one line up for the 100-meter dash. On the other hand, blacks whose origins can be traced to West Africa own the 100-meter dash but will not likely soon be taking home the $35,000 automobile awarded to the L.A. Marathon winner. And it could be a long while before we see a white man on the winner's platform at either distance. As Entine carefully documents, at the moment "every men's world record at every commonly run track distance belongs to a runner of African descent," and the domination of particular distances are determined, it would seem, by the ancestral origin of the athlete, with West Africans reigning over distances from 100 meters to 400 meters, and East and North Africans prevailing in races from 800 meters to the marathon.

But my first quibble with the debate is how quickly it shifts from Kenyans winning marathons or West Africans monopolizing the 100-meter dash to, as stated in Entine's subtitle, "why black athletes dominate sports." I understand a publisher's desire to economize cover verbiage and maximize marketability (the actual text of *Taboo* is, appropriately, filled with qualifiers, caveats, and nuances), but the simple fact is that black athletes do not dominate sports. They do not dominate speed skating, figure skating, ice hockey, gymnastics, swimming, diving, archery, downhill skiing, cross-country skiing, biathlons, triathlons, ping pong, tennis, golf, wrestling, rugby, rowing, canoeing, fencing, strong-man competitions, auto racing, motorcycle racing, and on and on.

In my own sport of cycling, in which I competed at elite ultra-marathon distances (200 miles to 3,000 miles) for 10 years, there are almost no blacks to be found in the pack. Where are all those West African sprinters at velodrome track races? Where are all those Kenyans in long-distance road races or ultra-marathon events? They are almost nowhere to be found. In fact, in over a century of professional bicycle racing there has been only one undisputed black champion—Marshall W. "Major" Taylor. And Taylor's reign was a century ago! He started racing in 1896 and within three years he became only the second black athlete to win a world championship in any sport, and this was at a time when bicycle racing was as big as baseball and boxing. Since there were few automobiles and no airplanes, cyclists were the fastest humans on earth and were rewarded accordingly with lucrative winnings and more than 15 minutes of fame. Major Taylor was the first black athlete in any sport to be a member of an integrated team, the first to land a commercial sponsor, and the first to hold world records, including the prestigious mile record. He competed internationally and is still revered in France as one of the greatest sprint cyclists of all time. The fact that outside cycling circles he is completely unknown in America tells us something about the influence of culture on sports.

By the theory proffered by Entine and others, there is no reason blacks should not be prominent in cycling since the physical requirements are so similar to running. The reason they are not, in fact, is almost certainly cultural. Although there are no longer racial barriers (as witnessed by the wide range of colors and nationalities that fill out the pelotons throughout Europe and the Americas), the reason blacks are not in cycling is obvious, says Dr. Ed Burke, a sports physiologist at the University of Colorado in Boulder: "No money, no publicity, no grass roots program. Why would gifted American athletes, with so many lucrative opportunities in other sports, choose cycling?" In Europe working class fathers introduce their sons to the sport at an early age where they can be nursed through junior cycling programs until they turn professional and permanently bootstrap themselves into the middle classes. But there are not that many blacks in Europe, and in America no such social structure exists. Bottom line: in cycling culture trumps biology.

(After Major Taylor, many cite the black sprinter Nelson Vails, since he took the silver medal on the track in the 1984 Olympics. But this is problematic because the East Germans boycotted that Olympics, and they were dominating the sport in those years, having thoroughly trounced both Vails and the 1984 gold medalist, Mark Gorski, in the world championships the year before. After Vails, Scott Berryman was a national sprint champion, and 19-year old Gideon Massie recently won the Jr. Worlds on the track and is an Olympic hopeful for 2004. The few other isolated cases—Shaums March in downhill mountain biking and Josh Weir on the road—only further call our attention to the dearth of blacks in cycling.)

Would blacks dominate cycling ceteris paribus? The problem is that all other things are never equal so it is impossible to say until the natural experiment is actually run. There is no reason why they should not, by the arguments put forth by Entine, since track cycling is much like sprinting, and road cycling is similar to marathon running in terms of the physical demands on the athlete. But we simply do not know and thus it would be unwise to speculate. For that matter, the ceteris paribus assumption never holds true in the messy real world, so this whole question of race and sports is fraught with complications, making it exceptionally difficult to say with much confidence what these differences really mean.

The Hindsight Bias: Did Evolution Shape Black Bodies Best for Running?

Tiger Woods may very well be the greatest golfer of all time. Although he is not "pure" black, he is considered to be black by most people, especially the black community. Thus, he very well could inspire other blacks to go into the sport. What if this were to happen on such a scale that blacks came to dominate golf as they have football and basketball? Would the explanation for this dominance be role modeling coupled to cultural momentum, or would we hear about how blacks are naturally gifted as golfers because of their superior ability to swing a club and judge moving objects at a distance due to the fact that they are closer to the Environment of Evolutionary Adaptation (or EEA, as evolutionary psychologists call the Pleistocene period of human evolution)?

In cognitive psychology there is a fallacy of thought known as the hindsight bias, which states that however things turn out we tend to look back to justify that particular arrangement with a set of causal explanatory variables presumably applicable to all situations. Looking back it is easy to construct plausible scenarios for how matters turned out; rearrange the outcome and we are equally skilled at finding new reasons why that particular arrangement was also inevitable.

Consider professional basketball. At the moment blacks dominate the sport and it is tempting to slip into the adaptationist mode of Darwinian speculation and suggest that the reason is because blacks are naturally superior at running, jumping, twisting, turning, hang time, and all the rest that goes into the modern game. Then it is only a step removed from suggesting, as does Entine and others do, that the reason for their above average natural abilities is that since humans evolved in Africa where they became bipedal, populations that migrated to other areas of the globe traded off those pure abilities through adaptations to other environments—e.g., colder climates led to shorter, stockier torsos (Bergmann's Rule) and smaller arms and legs (Allen's Rule)—thereby compromising the ability to run and jump. African blacks, however, are closer to the EEA and thus their abilities are evolutionarily less modified.

For basketball, however, I would point out the remarkable range of skin tone one sees on the court. Are these black players all equally "black" in this racial sense? I grant that races may exist as fuzzy sets where the boundaries are blurred but the interiors represent a type we might at least provisionally agree represents a group we can label "black" or "white." But when I see a range of "black" skintone on the court—from Manute Bol's dark chocolate to Dennis Johnson's sandy beige—I cannot help but question the validity of allowing a single category to represent so many shades. The fuzzy boundaries of the "black" set are so wide and the overlap with the "white" set so great that it seems scientifically untenable to draw the same conclusions about basketball that are made for track and field.

I also find it interesting that individuals with a small percentage of "black" genes are always classified in the "black" set, whereas whites are not accorded an equally broad latitude. In other words, if we were to graph the range of skin tones in so-called blacks and whites as two bell curves, the overall width of the black curve would be much greater, and the standard deviation for the black curve would be considerably greater than it would be for the whites. Why is this? The answer is clearly cultural, I suspect, having to do with the eugenics notion of a "pure" white race being contaminated with the blood of other, lesser races. A fuzzy-logic solution to this problem is to have just one set with fractional numbers assigned. For example, just as we might label the early morning sky as .3 blue/.7 orange, the midday sky as .9 blue/.1 orange, and the sunset sky as .2 blue/.8 orange, we could label Manute Bol as .9 black/.1 white and Dennis Johnson as .2 black/.8 white. Better still, we could just not label people by skin color at all.

Finally, the step from racial group differences on a basketball court to racial evolutionary differences in the Paleolithic is a significant one, and it is here where the hindsight bias is especially obvious. Let's go back in time and see how—not to the Paleolithic, but just to the earlier part of the 20th century. It may come as a surprise, especially to younger readers, to hear that at one time Jews dominated basketball. What sorts of arguments were made for their "natural" abilities in this sport? In the 1920s, 1930s, and 1940s basketball was an east coast, inner-city, blue-collar immigrant game largely dominated by the oppressed ethnic group of that age, the Jews. Like blacks decades later, the Jews went into professions and sports open to them. As Entine so wonderfully tracks this history in *Taboo,* according to Harry Sitwack, star player of the South Philadelphia Hebrew Association (SPHA),

"The Jews never got much into football or baseball. They were too crowded [with other players] then. Every Jewish boy was playing basketball. Every phone pole had a peach basket on it. And every one of those Jewish kids dreamed of playing for the SPHA's."

The reason why is obvious, right? Cultural trends and socio-economic opportunities set within an autocatalytic feedback loop (where variables operate on each other to drive the system forward) led more and more Jews to go into the game until they came to dominate it. That is not what the scientific experts of the day said. As Entine shows, according to the wisdom of the time the Jews were just naturally superior basketball players.

Writers opined that Jews were genetically and culturally built to stand up under the strain and stamina of the hoop game. It was suggested that they had an advantage because short men have better balance and more foot speed. They were also thought to have sharper eyes, which of course cut against the other stereotype that they suffered from myopia and had to wear glasses. And it was said they were clever. "The reason, I suspect, that basketball appeals to the hebrew with his Oriental background," wrote Paul Gallico, sports editor of the *New York Daily News* and one of the premier sports writers of the 1930s, "is that the game places a premium on an alert, scheming mind, flashy trickiness, artful dodging and general smart aleckness."

By the late 1940s Jews moved into other professions and sports and, Entine notes, "the torch of urban athleticism was passed on to the newest immigrants, mostly blacks who had migrated north from dying southern plantations. . . . It would not be long before the stereotype of the 'scheming . . . trickiness' of the Jews was replaced by that of the 'natural athleticism' of Negroes." If Jews were dominating basketball today instead of blacks, what explanatory models, in hindsight, would we be constructing? If, in 30 years, Asians come to control the game would we offer some equally plausible "natural" reason for their governance?

Does this mean that blacks are not really better than whites in basketball? No. I would be shocked if it turned out that what we are witnessing is nothing more than a culturally dominant "black style" of play. But because of the hindsight bias I cannot be certain that we are not being fooled and that the reasons for the differences we witness today are far more complex than we understand.

The Confirmation Bias: Why Asians Dominate Ping Pong and Why No One Cares—Sports in Black and White

Why, it seems reasonable to ask, are we so interested in black-white differences in sports? Why not Asian-Caucasian differences? Why has no one written a book entitled *Why Asians Dominate Ping Pong and Why We're Afraid to Talk about It?* The reason is obvious: because no one cares that Asians are the masters at ping pong. This is America, and what Americans care about are black-white differences, especially within high visibility activities. By way of analogy, no first-century Egyptian would have wondered if Cleopatra was black, but 20th-century Americans have debated that very question.

The confirmation bias holds that we have a tendency to seek confirmatory data that support our already-held beliefs, and ignore disconfirmatory evidence that might counter those beliefs. We all do this. Liberals read the paper and see greedy Republicans trying to rig the system so that the rich can become richer. Conservatives read the same paper and see bleeding-heart liberals robbing the rich of their hard-earned dollars to support welfare queens on crack. Context is everything and the confirmation bias makes it very difficult for any of us to take an objective perspective on our own beliefs.

Yes, there are black-white difference in sports, and there may even be good physical reasons for some of these differences. But, as noted above, the vast majority of sports are not dominated by blacks. Why don't we hear about them? Because they don't interest us, or they do not support our preconceived notions about the importance of black-white race questions. Out of the literally hundreds of popular sports played in the world today, blacks dominate only three: basketball, football, and track-and-field. That's it. That's what all the fuss is about. (At 15 percent they don't even dominate baseball.) Why do we focus on those three? Because we live in America where the black-white issue has bedeviled our experiment in democracy from the beginning, and where basketball, football, and track and field are the big sports which pay the big bucks.

I am not arguing that it is scientifically untenable or morally corrupt to focus on these differences, but I am curious why those particular differences are of such interest to some people. Is it nothing more than some people like chocolate pudding and others tapioca? I doubt it. I suspect the confirmation bias directs our attention to differences most likely to support already held beliefs about race differences. This would explain why it is almost always the same people, regardless of the particular trait or characteristic under study, who are interested in looking at racial group differences, and why Americans are interested in black-white differences but not others, and why non-Americans have little or no interest in this difference question.

Let's consider another case of evolutionary adaptation for the ability to run, and of within-species differences in this ability—thoroughbred race horses. Here we find rather disconfirming evidence that the underlying genetic variability of thoroughbreds long ago ran out despite the vigilant efforts of highly motivated horse breeders with millions of dollars at stake for a horse who could knock off a second or two.

The Kentucky Derby is the most prestigious of all thoroughbred races and has been run since 1875 when, by the way, 13 of the 15 jockeys were blacks. In fact, black jockeys dominated the Derby for the first 30 years, winning half of all races. The first race was 1.5 miles and was won in 2:37. In 1896 the distance was lowered to its present length of 1.25 miles and was won by Ben Brush in a time of 2:07. As evident in the table below (given in five-year increments with variation mostly accounted for by track surfaces being either "fast" or "slow"), since 1950 the horses are just not getting any faster.

1900	Lt. Gibson	2:06
1905	Agile	2:10
1910	Donau	2:06
1915	Regret	2:05
1920	Paul Jones	2:09
1925	Flying Ebony	2:07
1930	Gallant Fox	2:07
1935	Omaha	2:07
1940	Gallahadion	2:05
1945	Hoop Jr.	2:07
1950	Middle Ground	2:01
1955	Swaps	2:01
1960	Ventian Way	2:02
1965	Lucky Debonair	2:01
1970	Dust Commander	2:03
1975	Foolish Pleasure	2:02
1980	Genuine Risk	2:02
1985	Spend a Buck	2:00
1990	Unbridled	2:02
1995	Thunder Gulch	2:02

The greatest thoroughbred race horse of all time, Secretariat, is the only horse to break the two minute barrier at 1:59.2. If million dollar purses and stud fees have not been able to break the bounds of genetic variability, one wonders just how much genetic variability there is or just how much hypothesized adaptations like changes in body build in response to climate change could be achieved.

Blood or Sweat? The Nature-Nurture Debate in Sports

In the middle of the 1985 3,000-mile nonstop transcontinental bicycle Race Across America I was pedaling my way across Arkansas when the ABC *Wide World of Sports* camera crew pulled up alongside to inquire how I felt about my third place position—way ahead of the main pack but too far behind to catch the leaders. I answered: "I should have picked better parents."

The quote comes from the renowned sports physiologist Per-olof Astrand and was made at a 1967 exercise symposium: "I am convinced that anyone interested in winning Olympic gold medals must select his or her parents very carefully." At the time I regretted repeating it because I meant no disrespect for my always-supportive parents. But it was an accurate self-assessment for I had done everything I could do to win the race, including training over 500 miles a week in the months before, observing a strict diet, employing weight training, utilizing massage therapists and trainers, and more. My body fat was 4.5 percent, and at age 31 I was as strong and fast as I had ever been or would be. Nevertheless it was apparent I was not going to win the race. Why? Because despite maximizing my environmental nurture, the upper ceiling of my physical nature had been reached and was still below that of the two riders ahead of me.

This vignette is symbolic of the larger discussion in sports physiology on the relative roles of heredity and environment. In 1971, the exercise physiologist V. Klissouras, for example, reported that 81–86 percent of the variance in aerobic capacity, as measured by VO^2 uptake, is accounted for by genetics. In 1973 he confirmed his findings in another study that showed that only 20–30 percent of the variance in aerobic capacity can be accounted for by the environment—i.e., training can only improve aerobic capacity by that amount.

Randy Ice, the sports physiologist who has been testing Race Across America cyclists for the past 18 years, estimates that 60–70 percent of the variability between cyclists in aerobic capacity is genetically determined. Others estimate similar percentages for anaerobic threshold, workload capacity, fast twitch/slow twitch muscle fiber ratio, maximum heart rate, and many other physiological parameters that determine athletic performance. In other words, the difference between Pee Wee Herman and Eddy Merckx (the greatest cyclist of all time) is largely due to heredity.

Now, let's be clear that no one—not Jon Entine on one end or, hopefully, Harry Edwards on the other—is arguing that athletic ability is determined entirely by either genetics or environment. Obviously it is a mixture of the two. The controversy arises over what the ratio is, the evidence for that ratio, and the possible evolutionary origins of the difference. What surprised me in reading Entine's book, and other arguments for evolutionary origins of biologically based racial group differences in athletic ability, was the dearth of hard evidence and the need to draw questionable inferences and make sizable leaps of logic.

Although Entine's book is promoted as if it were a polemic for the hereditary position, he confesses that even in his best case examples of the Kenyan marathon runners, we cannot say for certain if they are "great long distance runners because of a genetic advantage or because their high-altitude lifestyle serves as a lifelong training program." It's a chicken-and-egg dilemma, Entine admits: "Did the altitude reconfigure the lungs of Kenyan endurance runners or was a genetic predisposition induced by the altitude? Is that nature or nurture . . . or both?"

It is both. But proving a particular percentage of each is tricky business. "Most theories, including those in genetics, rely on circumstantial evidence tested against common sense, known science, and the course of history," Entine explains. "That scientists may yet not be able to identify the chromosomes that contribute to specific athletic skills does not mean that genes don't play a defining role. . . ." Clearly that is so. But the real debate is not if; it is how, and how much. It is here where the science is weak and our biases strong.

What do we really know, for example, about the genetic coding for running? On the one hand it can be argued that this is a very simple activity compared to, say, a complex gymnastics routine. Even so, running ability depends on a host of variables—fast twitch/slow twitch muscle fiber ratio, VO^2 uptake capacity, lung capacity, maximum heart rate, anaerobic threshold figures (that determine the level one can sustain work output), measures of strength versus endurance, etc. We can estimate that these variables are half or three-quarters determined by genes, but we haven't a clue as to how they are coded, or even how genes and environment interact in the development of the ability under question. Autocatalytic feedback loops are powerful mechanisms in physical, biological, and social systems, and we are discovering them in nature-nurture interactions as well. Some genes are turned on or turned off by environmental stimuli. It may be possible that some human populations with a genetically encoded ability to run fast never have these genes turned on by the proper environment, or during a critical period of development. And perhaps other groups, like the Kenyans, have both the genetic propensity plus the cultural drive, high-altitude training, and so forth. Further, we have no idea if different human groups code for such variables in different ways as they interact with their environment; thus their autocatalytic feedback loops may be different. We just do not know.

Finally, while we can agree that different human characteristics are coded by differing genomic complexes—from simple to complex—we do not know enough genetics to say with any confidence that, for example, the ability to run a 100-meter dash is coded by n genes, the ability to slam dunk in basketball is coded by 2n genes, and that the ability to negotiate a complex gymnastic routine is coded by 8n genes. And this is just for physical abilities. Cognitive skills are another subject entirely, and we have even less knowledge on, say, how spatial reasoning or verbal skills are genetically coded, or autocatalytically determined through gene-cultural co-development.

All of this makes conclusions drawn about racial differences in sports problematic. No doubt some black-white differences in some sports are heavily influenced by genetics and might possibly even have an evolutionary basis of origin. But proving that supposition is another matter entirely. As it is, to be fair, for the extreme environmental position. Harry Edwards, for example, argued on my radio show that Kenyans are tenacious trainers, rising at 5:00 A.M. every morning to run mountains at high altitude. But that's just the hindsight and confirmation biases at work again, where we examine the winner of a race to see what ingredients went into the winning formula. It ignores all the other hard-working jocks who also got up every morning at 5:00 A.M. (oh don't I remember it so painfully well?) but didn't take the gold. Or the other winners who slept in until 8:00 A.M. and went for a leisurely jog on the flats. Training alone won't get you to the finish line first. Neither will genetics. Neither will luck. To be a champion you need all three.

Master of My Fate

We are all products of an evolutionary history of biological descent. Paraphrasing Astrand, our parents have been very carefully chosen for us—by natural selection. Yet as philosopher Michael Ruse notes:

> We are what we are because of our biology in conjunction with the environment. Dogs are friendly; if you beat and starve them, they are vicious. Scotsmen are as tall as Englishmen; if you feed them simply on oats they are runts. As well-known, long-term study has shown . . . thanks to improved nutrition, the height and physique of the Scots has improved dramatically.

The philosopher Karl Schmitz-Moormann also explained that such statistical percentages as those used in describing the relative influence of heredity and environment are descriptive for large populations, not individuals. Even the most complete knowledge of a person will not allow us to predict the precise future of this individual, because the laws for making such predictions are built around populations. Schmitz-Moormann calls this thinking "conditionalism." He writes: "At all levels of the evolving universe statistics might be understood as the description of freely evolving elements within more or less narrowly defined ranges of possibilities created by past evolution. Instead of being determined, the universe appears only to be conditioned on all levels."

The key element here is the range of possibilities. Behavior geneticists call it the genetic reaction range, or the biological parameters within which environmental conditions may take effect. We all have a biological limit, for example, on how fast we can ride a 40k time trial or run a 10k. There is a range from lowest to highest that establishes the parameters of our performance. In the diagram on the left, athlete A has a higher genetic reaction range than athlete B. But there is overlap of the ranges, and this is the key to where such environmental factors as nutrition, training, coaching, and desire take effect. A may be more "gifted" than B, but this does not mean he will always or even ever beat B. If B performs at his best and A is only at 50 percent of his potential, then the genetic advantage is negated. Inheritability of talent does not mean inevitability of success, and vice versa.

Why do some black athletes dominate some sports? For the same reason that some white athletes dominate some other sports, and some Asian athletes dominate still other sports—a combination of biological factors and cultural influences. We do not know for sure how to tease apart these variables, but we've got some reasonably good indications and Entine's book is a good place to start, as is Hoberman's *Darwin's Athletes.* What do the differences really mean? My answer is a consilience of both positions: We are free to select the optimal environmental conditions that will allow us to rise to the height of our biological potentials.

In this sense athletic success is measured not just against others' performances, but against the upper ceiling of our own ability. To succeed is to have done one's absolute best as measured against the high mark of one's personal range of possibilities. To win is not just to have crossed the finish line first, but to cross the finish line in the fastest time possible within the allowable genetic reaction range. The poet William Ernest Henley expressed this concept well in his stirring Invictus:

Out of the night that covers me,
Black as the pit from pole to pole,
I thank whatever gods may be
For my unconquerable soul.
It matters not how strait the gate,
How charged with punishments the scroll,
I am the master of my fate:
I am the captain of my soul.

Science Is at an End

JOHN HORGAN

In 1989, Gustavus Adolphus College in Minnesota held a symposium with the provocative but misleading title "The End of Science?" The meeting's premise was that belief in science—rather than science itself—was coming to an end. As one organizer put it, "There is an increasing feeling that science as a unified, universal, objective endeavor is over" (in Selve, 1992). Most of the speakers were philosophers who had challenged the authority of science in one way or another. The meeting's great irony was that one scientist present, U.C. Berkeley biologist Gunther Stent, had for years promulgated a much more dramatic and persuasive scenario than the one posed by the organizers. Stent had asserted that science itself might be ending, and not because of the skepticism of a few academic sophists. Quite the contrary. Science might be ending because it worked so well.

Stent is hardly a fringe figure. He was a pioneer of molecular biology; he founded the first department dedicated to that field at Berkeley in the 1950s and performed experiments that helped to illuminate the machinery of genetic transmission. Later, after switching from genetics to the study of the brain, he was named chairman of the neurobiology department of the National Academy of Sciences. Stent is also the most astute analyst of the limits of science I have encountered (and by astute I mean of course that he articulates my own inchoate premonitions). In the late 1960s, while Berkeley was wracked with student protests, he wrote an astonishingly prescient book, now long out of print, called *The Coming of the Golden Age: A View of the End of Progress*. Published in 1969, it contended that science—as well as technology, the arts, and all progressive, cumulative enterprises—is coming to an end.

Most people, Stent acknowledged, consider the notion that science might soon cease to be absurd. How can science possibly be nearing an end when it has been advancing so rapidly throughout this century? Stent turned this inductive argument on its head. Initially, he granted, science advances exponentially through a positive feedback effect; knowledge begets more knowledge, and power begets more power. Stent credited the American historian Henry Adams with having foreseen this aspect of science at the turn of the century.

Adams's "law of acceleration," Stent pointed out, has an interesting corollary. If there are any limits to science, any barriers to further progress, then science may well be moving at unprecedented speed just before it crashes into them. When science seems most muscular, triumphant, potent, that may be when it is nearest death. "Indeed, the dizzy rate at which progress is now proceeding," Stent wrote in *Golden Age*, "makes it seem very likely that progress must come to a stop soon, perhaps in our lifetime, perhaps in a generation or two."

Certain fields of science, Stent argued, are limited simply by the boundedness of their subject matter. No one would consider human anatomy or geography, for example, to be infinite endeavors. Chemistry, too, is bounded. "[T]hough the total number of possible chemical reactions is very great and the variety of reactions they can undergo vast, the goal of chemistry of understanding the principles governing the behavior of such molecules is, like the goal of geography, clearly limited." (In fact, many chemists think that goal was achieved in the 1930s when the chemist Linus Pauling showed how all chemical interactions could be understood in terms of quantum mechanics.)

In his own field of biology, Stent asserted, the discovery of DNA's twin-corkscrew structure in 1953 and the subsequent deciphering of the genetic code had solved the profound problem of how genetic information is passed on from one generation to the next. Biologists had only three major questions left to explore: how life began, how a single fertilized cell develops into a multi-cellular organism and how the central nervous system processes information. When those goals are achieved, Stent said, the basic task of biology, pure biology, will be completed.

Stent acknowledged that biologists can in principle continue exploring specific phenomena and applying their knowledge forever. But according to Darwinian theory, science stems not from our desire for truth per se but from our compulsion to control our environment in order to increase the likelihood that our genes will propagate. When a given field of science begins to yield diminishing practical returns, scientists may have less incentive to pursue their research and society may be less inclined to pay for it. Moreover, just because biologists complete their empirical investigations, Stent asserted, does not mean that they will have answered all relevant questions. For example, no purely physiological theory can ever really explain consciousness, since the "processes responsible for this wholly private experience will be seen to degenerate into seemingly quite ordinary, workaday reactions, no more or less fascinating than those that occur in, say, the liver . . ."

Unlike biology, Stent said, the physical sciences seem to be open-ended. Physicists can always attempt to probe more deeply into matter by smashing particles against each other with greater force, and astronomers can always strive to see further into the universe. But in their efforts to gather data from ever-more-remote regimes, Stent contended, physicists will inevitably confront various physical, economic and even cognitive limits.

Over the course of this century, physics has become more and more difficult to comprehend; it has outrun our "Darwinian epistemology," our innate concepts for coping with the world. Stent rejected the old argument that "yesterday's nonsense is today's common sense." Society may be willing to support continued research in physics as long as it has the potential to generate powerful new technologies, such as nuclear weapons and nuclear power. But when physics becomes impractical as well as incomprehensible, Stent predicted, society will surely withdraw its support.

Stent's prognosis for the future was an odd mixture of optimism and pessimism. He predicted that science, before it ends, might help to solve many of civilization's most pressing problems. It would eliminate disease and poverty and provide society with cheap, pollution-free energy, perhaps through the harnessing of fusion reactions. As we gain more dominion over nature, however, we may lose what Nietzsche called our "will to power"; we may become less motivated to pursue further research—especially if such research has little chance of yielding tangible benefits.

As society becomes more affluent and comfortable, fewer young people may choose the increasingly difficult path of science or even of the arts. Many may turn to more hedonistic pursuits, perhaps even abandoning the "real world" for fantasies induced by drugs or electronic devices feeding directly into the brain. Sooner or later, Stent concluded, progress would "stop dead in its tracks," leaving the world in a largely static condition that he called "the new Polynesia." The advent of beatniks and hippies, he surmised, signaled the beginning of the end of progress and the dawn of the new Polynesia. He closed his book with the sardonic comment that "millennia of doing arts and science will finally transform the tragicomedy of life into a happening."

A Trip to Berkeley

In the spring of 1992 I traveled to Berkeley to see how Stent thought his predictions had held up over the years. Stent had moved to the U.S. from Germany as a youth, and his gruff voice and attire still bore traces of his origins. He wore wire-rimmed glasses, a blue, short-sleeved shirt with epaulets, dark slacks and shiny black shoes. Stent had obtained a doctorate in chemistry at the University of Illinois, but upon reading Erwin Schrödinger's book *What Is Life?*, he became entranced by the mystery of genetic transmission. After studying at the California Institute of Technology under Max Delbruck, Stent obtained a professorship at Berkeley in 1952. In these early years of molecular biology, Stent said, "none of us knew what we were doing. Then Watson and Crick found the double helix, and within a few weeks we realized we were doing molecular biology."

Stent began pondering the limits of science in the 1960s partly in reaction to Berkeley's free-speech movement, which had challenged the value of rationalism and technological progress and other aspects of civilization that Stent held dear. The university appointed him to a committee to "deal with this, to calm things down," by talking to students. Stent sought to fulfill this mandate—and to resolve his own inner conflicts over his role as a scientist—by delivering a series of lectures. These lectures became *The Coming of the Golden Age*.

I told Stent that I could not determine, after finishing *Golden Age*, whether he believed the new Polynesia, the era of social and intellectual stasis and universal leisure, would be an improvement over our present situation. "I could never decide this!" he exclaimed, looking genuinely distressed. "People called me a pessimist, but I thought I was an optimist." He certainly did not think such a society would be in any sense utopian. After the horrors wreaked by totalitarian states in this century, he explained, it was no longer possible to take the idea of utopia seriously.

Stent felt his predictions had held up reasonably well. Although hippies had vanished (except for the pitiful relics on Berkeley's streets), American culture had become increasingly materialistic and anti-intellectual; hippies had evolved into yuppies. The cold war had ended, although not through the gradual merging of communist and capitalist states Stent had envisioned. He admitted he did not anticipate the resurgence, in the wake of the cold war, of long-repressed ethnic and even tribal conflicts. "I'm very depressed at what's happening in the Balkans," he said. "I didn't think that would happen." Stent was also surprised by the persistence of poverty and of racial conflict in the U.S., but he thought these problems would eventually diminish in importance. (Aha, I thought. He was an optimist after all.)

Stent was convinced that science was showing signs of the closure he had predicted in *Golden Age*. Particle physicists were having difficulty convincing society to pay for their increasingly expensive experiments, such as the multi-billion-dollar Superconducting Supercollider. As for biologists, they still had much to learn about how, say, a fertilized cell is transformed into a complex, multi-cellular organism, like an elephant, and about the workings of the brain. "But I think the big picture is basically over," he said. Evolutionary biology in particular "was over when Darwin published *The Origin of Species*," Stent said.

Stent was still convinced, in spite of all the advances in neuroscience following the publication of *Golden Age*, that a purely physiological explanation of consciousness would not be as comprehensible or as meaningful as most people would like; nor would it help us to solve moral and ethical questions. In fact, Stent thought the progress of science might give religion a clearer role in the future rather than eliminating it entirely, as many scientists had once hoped. Although it cannot compete with science's far more compelling stories about the physical realm, religion still retains some value in offering moral guidance. "Humans are animals, but we're also moral subjects. The task of religion is more and more in the moral realm."

When I asked about the possibility that computers might become intelligent and create their own science, Stent snorted in derision. He had a dim view of artificial intelligence, and particularly its more visionary enthusiasts. Computers may excel at precisely defined tasks such as mathematics and chess, he pointed out, but they still perform abysmally when confronted with the kind of problems—recognizing a face or a voice or walking down a crowded sidewalk—that humans solve effortlessly. "They're full of it," Stent said of Marvin Minsky and others who have predicted that one day we humans will be able to "download" our personalities into computers. "I wouldn't rule out the possibility that in the 23rd century you might have an artificial brain," he added. "But it would need experience." One could design a computer to become an expert in restaurants, "but this machine would never know what a steak tastes like."

Stent was similarly skeptical of the claims of investigators of chaos and complexity that with computers and sophisticated mathematics they can transcend the science of the past. In *The Coming of the Golden Age*, Stent had discussed the work of one of the pioneers of chaos theory, Benoit Mandelbrot. Beginning in the early 1960s, Mandelbrot had shown that many phenomena are intrinsically "indeterministic"—they exhibit behavior that is unpredictable and apparently random. Scientists can only guess at the causes of individual events, and they cannot predict them with any accuracy.

Proponents of chaos and complexity were attempting to create effective, comprehensible theories of the same phenomena studied by Mandelbrot, Stent said. He had concluded in *Golden Age* that these indeterministic phenomena would resist scientific analysis, and he saw no reason to change that assessment. Quite the contrary. The work emerging from those fields demonstrated his point that science, when pushed too far, always culminates in incoherence. So Stent did not think that chaos and complexity will bring about the rebirth of science? "No," he said with a rakish grin. "It's the end of science."

What Science Has Accomplished

We obviously are nowhere near the new Polynesia that Stent envisioned, in part because applied science has not come nearly as far as Stent had hoped (feared?) when he wrote *The Coming of the Golden Age*. But I have come to the conclusion that Stent's prophecy has, in one very important sense, already come to pass. If one believes in science, one must accept the possibility—even the probability—that science has passed its peak. By science I mean not applied science but science at its purest and grandest, the primordial human quest to understand the universe and our place in it.

Further research may yield no more great revelations or revolutions but only incremental, diminishing returns.

These are trying times for truth-seekers. The scientific enterprise is threatened by technophobes, animal-rights activists, religious fundamentalists, and, most important of all, stingy politicians. Social, political, and economic constraints will surely make it more difficult to practice science, and pure science in particular, in the future. Moreover, science itself, as it advances, keeps imposing limits on its own power. Einstein's theory of special relativity prohibits the transmission of matter or even information at speeds faster than that of light; quantum mechanics dictates that our knowledge of the microrealm will always be uncertain; chaos theory confirms that even without quantum indeterminacy many phenomena would be impossible to predict; Kurt Gödel's incompleteness theorem denies us the possibility of constructing a complete, consistent mathematical description of reality. And evolutionary biology keeps reminding us that we are animals, designed by natural selection not for discovering deep truths of nature, but for breeding.

But by far the greatest barrier to future progress in pure science is its past success. Researchers have already mapped out physical reality, ranging from the microrealm of quarks and electrons to the macrorealm of planets, stars, and galaxies. Physicists have shown that all matter is ruled by a few basic forces: gravity, electromagnetism, and the strong and weak nuclear forces. Scientists have also stitched their knowledge into an impressive, if not terribly detailed, narrative of how we came to be. The universe exploded into existence 15 billion years ago, give or take five billion years (astronomers may never agree on an exact figure), and is still expanding outwards. Some 4.5 billion years ago, the detritus of an exploding star, a supernova, condensed into our solar system. Sometime during the next few hundred million years, for reasons that may never be known, single-celled organisms bearing an ingenious molecule called DNA emerged on the still-hellish earth. These Adamic microbes gave rise, by means of natural selection, to an extraordinary array of more complex creatures, including *Homo sapiens*.

My guess is that this narrative that scientists have woven from their knowledge, this modern myth of creation, will be as viable 100 or even 1,000 years from now as it is today. Why? Because it is true. Moreover, given how far science has already come, and given the physical, social, and cognitive limits constraining further research, science is unlikely to make any significant additions to the knowledge it has already generated. There will be no great revelations in the future comparable to those bestowed upon us by Darwin or Einstein or Watson and Crick.

The Anxiety of Scientific Influence

In trying to understand the mood of modern scientists, I have found that ideas from literary criticism can serve some purpose. In his influential 1973 essay, *The Anxiety of Influence*, the literary critic Harold Bloom of Yale University likened the modern poet to Satan in Milton's *Paradise Lost*. Just as Satan fought to assert his individuality by defying the perfection of God, so must the modern poet engage in an Oedipal struggle to define himself in relation to Shakespeare, Dante, and other masters. The effort is ultimately futile, Bloom said, because no poet can hope to approach, let alone surpass, the perfection of his forebears. Modern poets are all essentially tragic figures, late-comers.

Modern scientists, too, are late-comers, and their burden is much heavier than that of poets. Scientists must endure not merely Shakespeare's *King Lear* but Newton's laws of motion, Darwin's theory of natural selection,

Einstein's theory of relativity. These theories are not merely beautiful; they are also true, empirically true, in a way that no work of art can be. Most researchers simply concede their inability to supersede what Bloom called "the embarrassments of a tradition grown too wealthy to need anything more." They try to solve what the philosopher of science Thomas Kuhn has denigrated as "puzzles," problems whose solutions buttress the prevailing paradigm. They settle for refining and applying the brilliant, pioneering discoveries of their predecessors. They try to measure the mass of quarks more precisely or to determine how a given stretch of DNA guides the growth of the embryonic brain. Others become what Bloom derided as a "mere rebel, a childish inverter of conventional moral categories." The rebels denigrate the dominant theories of science as flimsy social fabrications rather than rigorously tested descriptions of nature.

Bloom's "strong poet" accepts the perfection of his predecessors and yet strives to transcend it through various subterfuges, including a subtle "misreading" of their work; only by so doing can a modern poet break free of the stultifying influence of the past. There are strong scientists, too, those who are seeking to misread and therefore to transcend quantum mechanics or the big bang theory or Darwinian evolution. For the most part strong scientists have only one option: to pursue science in a speculative, post-empirical mode that I call ironic science. Like art, philosophy, literary criticism, theology—the other ironic modes of discourse—ironic science can be neither definitively confirmed nor falsified. It offers not truth in the conventional sense but points of view, opinions which are, at best, "interesting," which provoke further comment. It cannot achieve empirically verifiable "surprises" that force scientists to make substantial revisions in their basic description of reality.

The most common strategy of the strong scientist is to point to all the shortcomings of current scientific knowledge, to all the questions left unanswered. But the questions tend to be ones that may never be definitively answered, given the limits of human science. How, exactly, was the universe created? Could our universe be just one of an infinite number of universes? Could quarks and electrons be composed of still smaller particles, ad infinitum? What does quantum mechanics really mean? (Most questions concerning meaning can only be answered ironically, as literary critics know.) Biology has its own slew of insoluble riddles. How, exactly, did life begin on earth? Just how inevitable was life's origin, and its subsequent history?

Superstring theory, which for more than a decade has been the leading contender for a unified theory of physics, is a particularly striking specimen of ironic science. Often called a "theory of everything," it posits that all the matter and energy in the universe and even space and time stem from infinitesimal, string-like particles wriggling in a hyperspace consisting of 10 (or more) dimensions. Unfortunately, the microrealm that superstrings allegedly inhabit is even less accessible to human experimenters than the quasars haunting the edge of the visible universe. A superstring is as small in comparison to a proton as a proton is in comparison to the solar system. Probing this realm directly would require an accelerator 1,000 light years around. That is why the physicist Sheldon Glashow, a Nobel laureate at Harvard University, once likened superstring theorists to "medieval theologians" (1986, 7).

The practitioner of ironic science enjoys one obvious advantage over the strong poet: the appetite of the reading public for scientific "revolutions." As empirical science ossifies, journalists like myself, who feed society's hunger, will come under more pressure to tout theories that supposedly transcend quantum mechanics or the big bang theory or natural selection. Journalists have, after all, helped superstring theory to win acceptance as a legitimate extension of nuclear physics rather than mathematical smoke and mirrors, as Glashow has put it. Journalists have also created the popular impression that fields such as chaos and complexity represent genuinely "new" sciences superior to the stodgy old reductionist methods of Newton, Einstein, and Darwin.

The *Star Trek* Factor

If my experience is any guide, even people with only a casual interest in science will find it hard to accept that science's days are numbered. It is easy to understand why. We are drenched in progress, real and artificial. Every year we have smaller, faster computers, sleeker cars, more channels on our televisions. Our views of the future are also distorted by what could be called the *Star Trek* factor. How can science be approaching a culmination when we haven't invented spaceships that travel at warp speed yet?

To be sure, applied science will continue for a long time to come. Scientists can keep developing versatile new materials; faster and more sophisticated computers; genetic-engineering techniques that make us healthier, stronger, longer-lived; perhaps even fusion reactors that can provide cheap energy with few environmental side effects (although given the drastic cutbacks in funding, fusion's prospects now seem dimmer than ever). The question is, will these advances in applied science bring about any "surprises," any revolutionary shifts in our basic knowledge? Will they force scientists to revise the map they have drawn of the universe or the narrative they have constructed of the universe's creation and history? Probably not. Applied science in this century has tended to reinforce rather than to challenge the prevailing theoretical paradigms. Lasers and transistors confirm the power of quantum mechanics, just as genetic engineering bolsters belief in the DNA-based model of evolution.

What constitutes a surprise? Einstein's discovery that time and space, the I-beams of reality, are made of rubber was a surprise. So was the observation by astronomers that the universe is expanding, evolving. Quantum mechanics, which unveiled a probabilistic element, a Lucretian swerve, at the bottom of things, was an enormous surprise; God does play dice (Einstein's disapproval notwithstanding). The later finding that protons and neutrons are made of smaller particles called quarks was a much lesser surprise, because it merely extended quantum theory to a deeper domain; the foundations of physics remained intact.

Learning that we humans were created not de novo by God but gradually, by the process of natural selection, was a big surprise. Most other aspects of human evolution—those concerning where, when and how, precisely, *Homo sapiens* evolved—are details. These details may be interesting, but they are not likely to be surprising unless they show that scientists' basic assumptions about evolution were wrong. We may learn, say, that our sudden surge in intelligence was catalyzed by the intervention of alien beings, as in the movie *2001*. That would be a very big surprise. In fact, any proof that life exists—or even once existed—beyond our little planet would constitute a huge surprise. Science, and all human thought, would be reborn. Speculation about the origin of life and its inevitability would be placed on a much more empirical basis.

But how likely is it that we will discover life elsewhere? In retrospect, the space programs of both the U.S. and the U.S.S.R. represented elaborate displays of saber-rattling rather than the opening of a new frontier for human knowledge. The prospects for space exploration on anything more than a trivial level seem increasingly unlikely. We no longer have the will or the money to indulge in technological muscle-flexing for its own sake. Humans, made of flesh and blood, may someday travel to other planets here in our solar system. But unless we find some way to transcend Einstein's prohibition against faster-than-light travel, chances are that we will never even attempt to visit another star, let alone another galaxy. A spaceship that can travel one million miles an hour, an order of magnitude faster than any current technology can attain, would still take almost 3,000 years to reach our nearest stellar neighbor, Alpha Centauri.

That's What They Thought 100 Years Ago

The most common response to the suggestion that science might be ending is the "that's-what-they-thought-at-the-end-of-the-last-century" argument. The argument goes like this: As the 19th century wound down, physicists thought they knew everything. But no sooner had the 20th century begun than Einstein and other physicists discovered—invented?—relativity theory and quantum mechanics. These theories eclipsed Newtonian physics and opened up vast new vistas for modern physics and other branches of science. Moral: Anyone who predicts that science is nearing its end will surely turn out to be as short-sighted as those 19th-century physicists were.

Those who believe science is finite have a standard retort for this argument: The earliest explorers, because they could not find the edge of the earth, might well have concluded that it is infinite, but they would have been wrong. Moreover, it is by no means a matter of historical record that late 19th-century physicists felt they had wrapped things up. The best evidence for a sense of completion is a speech given in 1894 by Albert Michelson, whose experiments on the velocity of light helped to inspire Einstein's theory of special relativity. Michelson stated (*Physics Today*, April 1968, 9):

> While it is never safe to say that the future of Physical Science has no marvels even more astonishing than those of the past, it seems probable that most of the grand underlying principles have been firmly established and that further advances are to be sought chiefly in the rigorous application of these principles to all the phenomena which come under our notice. It is here that the science of measurement shows its importance—where quantitative results are more to be desired than qualitative work. An eminent physicist has remarked that the future truths of Physical Science are to be looked for in the sixth place of decimals.

Michelson's remark about "the sixth place of decimals" has been so widely attributed to Lord Kelvin (after whom the Kelvin, a unit of temperature, is named) that some authors simply credit him with the quote. But historians have found no evidence that Kelvin made such a statement. Moreover, at the time of Michelson's remarks physicists were vigorously debating fundamental issues, such as the viability of the atomic theory of matter, according to the historian of science Stephen Brush of the University of Maryland. Michelson was so absorbed in his optics experiments, Brush suggested, that he was "oblivious to the violent controversies raging among theorists at the time." The alleged "Victorian calm in physics," Brush concluded, is a "myth" (1969, 9).

The Apocryphal Patent Official

Other historians, as is their wont, disagree. Questions concerning the "mood" of a given era can never be completely resolved. But the view that scientists in the last century were complacent about the state of their field has clearly been exaggerated. Historians have provided a definitive ruling, moreover, on another anecdote favored by those reluctant to accept that science might be mortal. The story alleges that in the mid-1800s, the head of the U.S. Patent Office quit his job and recommended that the office be shut down because there would soon be nothing left to invent.

In 1995, Daniel Koshland, editor of the prestigious journal *Science*, repeated this story in an introduction to a special section on science's future. In this section, leading scientists offered predictions about what their fields might accomplish over the next 20 years. Koshland, a biologist at the University of California at Berkeley, exulted that his prognosticators "clearly do not agree with that commissioner of patents of yesteryear. Great discoveries with great import for the future of science are in the offing. That we have come so far so fast is not an indication that we have saturated the discovery market, but rather that discoveries will come even faster" (1995).

There were two problems with Koshland's essay. First, the contributors to his special section envisioned not "great discoveries" but, for the most part, rather mundane applications of current knowledge, such as better methods for designing drugs, improved tests for genetic disorders, more discerning brain scans and the like. Some predictions, moreover, were negative in nature. "Anyone who expects any human-like intelligence from a computer in the next 50 years is doomed to disappointment," proclaimed the physicist and Nobel laureate Philip Anderson.

The second problem with Koshland's essay was that his story about the commissioner of patents is apocryphal. In 1940, a scholar named Eber Jeffry examined the patent-commissioner anecdote in an article titled "Nothing Left to Invent," published in the *Journal of the Patent Office Society*. Jeffry traced the story to Congressional testimony delivered in 1843 by Henry Ellsworth, then the Commissioner of Patents. Ellsworth remarked at one point: "The advancement of the arts, from year to year, taxes our credulity and seems to presage the arrival of that period when human improvement must end."

But Ellsworth, far from recommending that his office be shut down, asked for extra funds to cope with the flood of inventions he expected in agriculture, transportation, and communications. Ellsworth did indeed resign two years later, in 1845, but in his resignation letter he made no reference to closing the patent office; he only expressed pride at having expanded it. Jeffry concluded that Ellsworth's statement about "that period when human improvement must end" represented "a mere rhetorical flourish intended to emphasize the remarkable strides forward in inventions then current and to be expected in the future." But perhaps Jeffry was not giving Ellsworth enough credit. Ellsworth was, after all, anticipating the argument that Gunther Stent would make more than a century later: The faster that science moves, the faster it will reach its ultimate, inevitable limits.

Consider the implications of the alternative position, the one implicitly advanced by Daniel Koshland. He insists that because science has advanced so rapidly over the past century or so, it can and will continue to do so, possibly forever. But this inductive argument is deeply flawed. Science has only existed for a few hundred years, and its most spectacular achievements have occurred within the last century. Viewed from an historical perspective, the modern era of rapid scientific and technological progress appears to be not a permanent feature of reality but an aberration, a fluke, a product of a singular convergence of social, intellectual, and political factors.

The Rise and Fall of Progress

In his 1932 book, *The Idea of Progress*, the historian J. B. Bury stated (italics in the original):

> Science has been advancing without interruption during the last three or four hundred years; every new discovery has led to new problems and new methods of solution, and opened up new fields for exploration. Hitherto men of science have not been compelled to halt, they have always found means to advance further. But *what assurance have we that they will not come up against impassable barriers?*

Bury himself had demonstrated through his scholarship that the concept of progress is only a few hundred years old, at most. From the era of the Roman Empire through the Middle Ages, most truth-seekers had a degenerative view of history: the ancient Greeks had achieved the acme of mathematical and scientific knowledge, and civilization had gone downhill from there. Those who followed could only try to recapture some remnant of the wisdom epitomized by Plato and Aristotle. It was such founders of modern, empirical science as Isaac Newton, Francis Bacon, René Descartes, and Gottfried Leibniz who first set forth the idea that humans could systematically acquire and accumulate knowledge through investigations of nature. Most of these Ur-scientists believed that the process would be finite, that we could attain complete knowledge of the world and then construct a perfect society, a utopia, based on that knowledge. (The new Polynesia!)

Only with the advent of Darwin did certain intellectuals become so enamoured with progress that they insisted it might be, or should be, eternal. "In the wake of the publication of Darwin's *On the Origin of Species*," Gunther Stent wrote in his 1978 book *The Paradoxes of Progress*, "the idea of progress was raised to the level of a scientific religion. . . . This optimistic view came to be so widely embraced in the industrialized nations . . . that the claim that progress could presently come to an end is now widely regarded [to be] as outlandish a notion as was in earlier times the claim that the Earth moves around the sun" (27).

It was not surprising that modern nation states became fervent proponents of the science-is-infinite creed. Science spawned such marvels as The Bomb, nuclear power, jets, radar, computers, and missiles. In 1945 the physicist Vannevar Bush (a distant relative of former President George) proclaimed in *Science: The Endless Frontier* that science was "a largely unexplored hinterland" and an "essential key" to U.S. military and economic security. Bush's essay served as a blueprint for the construction of the National Science Foundation and other federal organizations that thereafter supported basic research on an unparalled scale. The Soviet Union was perhaps even more devoted than its capitalist rival to the concept of scientific and technological progress.

Of course, powerful social, political and economic forces now oppose this vision of boundless scientific and technological progress. The cold war, which was a major impetus for basic research in the U.S. and the Soviet Union, is over; the U.S. and the former Soviet republics have much less incentive to build space stations and gigantic accelerators simply to demonstrate their power. Society is also increasingly sensitive to the adverse consequences of science and technology—such as pollution, nuclear contamination, and weapons of mass destruction.

The disillusionment with science was foreseen early in this century by Oswald Spengler, a German schoolteacher who became the first great prophet of the end of science. In his massive tome *The Decline of the West*, published in 1918, Spengler argued that science proceeds in a cyclic fashion, with "romantic" periods of investigation of nature and the invention of new theories giving way to periods of consolidation in which scientific knowledge ossifies. As scientists become more arrogant and less tolerant of other belief systems, notably religious ones, Spengler declared, society will rebel against science and embrace religious fundamentalism and other irrational systems of belief. Spengler predicted that the decline of science and the resurgence of irrationality would begin at the end of this millennium.

Spengler's analysis was, if anything, too optimistic. His view of science as cyclic implied that science may one day be resurrected and undergo a new period of discovery. But science is not cyclic but linear; we can only discover the periodic table and the expansion of the universe and the structure of DNA once. The biggest obstacle to the resurrection of science—human science, the quest for knowledge about who we are and where we came from—is science's past success.

No More Endless Horizons

Scientists are understandably loath to state publicly that they have entered an era of diminishing returns. No one wants to be recalled as the equivalent of those allegedly shortsighted physicists of a century ago. There is always the danger, moreover, that such prophecies will become self-fulfilling. But Gunther Stent is hardly the only prominent scientist to violate the taboo against such prophecies. In 1971, *Science* published an essay entitled "Science: Endless Horizons or Golden Age?," by Bentley Glass, a prominent biologist and the president of *Science*'s publisher, the American Association for the Advancement of Science. Glass weighed the two scenarios for science's future posited by Vannevar Bush and Gunther Stent and reluctantly came down on the side of Stent. Not only was science finite, Glass argued, but the end was in sight. "We are like the explorers of a great continent," Glass proclaimed, "who have penetrated to its margins in most points of the compass and have mapped the major mountain chains and rivers. There are still innumerable details to fill in, but the endless horizons no longer exist" (23).

According to Glass, a close reading of Bush's *Endless Frontier* essay suggested that he, too, viewed science as a finite enterprise. Nowhere did Bush specifically state that any fields of science could continue generating new discoveries forever. In fact, Bush described scientific knowledge as an "edifice" whose form "is predestined by the laws of logic and the nature of human reasoning. It is almost as though it already existed." Bush's choice of this metaphor, Glass commented, reveals that he considered scientific knowledge to be finite in extent. Glass proposed that the "bold title" of Bush's essay was "never intended to be taken literally, but supposed merely to imply that from our present viewpoint so much yet remains before us to be discovered that the horizons seem virtually endless."

In 1979, in *The Quarterly Review of Biology*, Glass presented evidence to back up his view that science was approaching a culmination. Upon analyzing the rate of discoveries in biology, he found that they had not kept pace with the exponential increase in researchers and funding. "We have been so impressed by the undeniable acceleration in the rate of magnificent achievements that we have scarcely noticed that we are well into an era of diminishing returns," Glass commented. "That is, more and more scientific effort and expenditure of money must be allocated in order to sustain our progress. Sooner or later this will have to stop, because of the insuperable limits to scientific manpower and expenditure. So rapid has been the growth of science in our own century that we have been deluded into thinking that such a rate of progress can be maintained indefinitely."

When I spoke to him in 1994, Glass confessed that many of his colleagues had been dismayed that he had even raised the issue of science's limits, let alone prophesied its demise.

But Glass felt, then and now, that the topic is too important to ignore. Obviously science, as a social enterprise, has some limits, Glass said. If science had continued to grow at the same rate as it had earlier in this century, he pointed out, it would soon have consumed the entire budget of the industrialized world. "I think it's rather evident to everybody," he said, "that there must be brakes put on the amount of funding for science, pure science." This slowdown, he observed, was evident in the decision of the U.S. Congress in 1993 to cuts funds for the Superconducting Supercollider, the gargantuan particle accelerator that physicists hoped would propel them beyond quarks and electrons into a deeper realm of microspace.

Even if society were to devote all its resources to research, Glass added, science would one day still reach the point of diminishing returns. Why? Because science works; it solves its problems. After all, astronomers have already plumbed the farthest reaches of the universe; they cannot see what, if anything, lies beyond its borders. Moreover, most physicists think that the reduction of matter into smaller and smaller particles will eventually end, or may have already ended for all practical purposes. Even if physicists unearth particles buried beneath quarks and electrons, that knowledge will make little or no difference to biologists, who have learned that the most significant biological processes occur at the molecular level and above. "There's a limit to biology there," Glass explained, "that you don't expect to be able to ever break through just because of the nature of the constitution of matter and energy."

Hard Times Ahead for Physics

In 1992, the monthly journal *Physics Today* published an essay entitled "Hard Times," in which Leo Kadanoff, a prominent physicist at the University of Chicago, painted a bleak picture for the future of physics. "Nothing we do is likely to arrest our decline in numbers, support, or social value," Kadanoff declared. "Too much of our base depended on events that are now becoming ancient history: nuclear weapons and radar during World War II, silicon and laser technology thereafter, American optimism and industrial hegemony, socialist belief in rationality as a way of improving the world." Those conditions had largely vanished, Kadanoff contended; both physics and science as a whole are now besieged by environmentalists, animal-rights activists, and other antiscientific movements. "In recent decades, science has had high rewards and has been at the center of social interest and concern. We should not be surprised if this anomaly disappears" (9–11).

Kadanoff, when I spoke to him over the telephone two years later, sounded even gloomier than he had been when he wrote his essay. He laid out his worldview for me with a muffled melancholy, as if he were suffering from an existential head cold. But rather than discussing science's social and political problems, as he had in his essay, he focused on another obstacle to scientific progress: science's past achievements. The great task of modern science, Kadanoff explained, has been to show that the world conforms to certain basic physical laws. "That is an issue which has been explored at least since the Renaissance and maybe a much longer period of time. For me, that's a settled issue. That is, it seems to me that the world is explainable by law."

Of course, scientists still have much to learn about how the fundamental laws generate "the richness of the world as we see it." Kadanoff himself is a leader in the field of condensed-matter physics, which studies the behavior not of individual subatomic particles but of solids or liquids. Kadanoff has also been associated with the field of chaos, which addresses phenomena that unfold in predictably unpredictable ways. Some proponents of chaos—and of the closely related field called complexity—have suggested that with the help of powerful computers and new mathematical methods they will discover truths that surpass those revealed by the "reductionist" science of the past. Kadanoff had his doubts. Studying the consequences of fundamental laws is "in a way less interesting" and "less deep," he said, than showing that the world is lawful. "But now that we know the world is lawful," he added, "we have to go on to other things. And yes, it probably excites the imagination of the average human being less. Maybe with good reason." Is this state of affairs permanent? I asked. Kadanoff was silent for a moment. Then he sighed, as if trying to exhale all his world-weariness. "Once you have proven that the world is lawful," he replied, "to the satisfaction of many human beings, you can't do that again."

Whistling to Keep Our Courage Up

One of the few modern philosophers to devote serious thought to the limits of science is Nicholas Rescher of the University of Pittsburgh. In his 1978 book, *Scientific Progress*, Rescher deplored the fact that Stent, Glass, and other prominent scientists seemed to think that science might be approaching a cul de sac. Rescher intended to provide "an antidote to this currently pervasive tendency of thought" by demonstrating that science was at least potentially infinite. But the scenario he sketched out over the course of his book was hardly optimistic. He argued that science, as a fundamentally empirical, experimental discipline, faces economic constraints. As scientists try to extend their theories into more remote domains—seeing further into the universe, deeper into matter—their costs will inevitably escalate and their returns diminish.

"Scientific innovation is going to become more and more difficult as we push out further and further from our home base toward more remote frontiers. If the present perspective is even partly correct, the half-millennium commencing around 1650 will eventually come to be regarded among the great characteristic developmental transformations of human history, with the age of The Science Explosion as unique in its own historical structure as The Bronze Age or The Industrial Revolution or The Population Explosion."

Rescher tacked what he apparently thought was a happy coda onto his depressing scenario: Science will never end; it will just go slower and slower and slower, like Zeno's tortoise. Nor should scientists ever conclude that their research must degenerate into the mere filling in of details; it is always possible that one of their increasingly expensive experiments will have revolutionary import, comparable to that of quantum mechanics or Darwinian theory.

When I telephoned Rescher, he acknowledged that his analysis had been in most respects a grim one. "We can only investigate nature by interacting with it," he said. "To do that we must push into regions never investigated before, regions of higher density, lower temperature, or higher energy. In all these cases we are pushing fundamental limits, and that requires ever more elaborate and expensive apparatuses. So there is a limit imposed on science by the limits of human resources."

The End of History

In *Golden Age,* Stent suggested that science, before it ends, may at least deliver us from our most pressing social problems, such as poverty and disease and even conflict between states. The future will be peaceful and comfortable, if boring. Most humans will dedicate themselves to the pursuit of pleasure. In 1992, Francis

Fukuyama set forth a rather different vision of the future in *The End of History and the Last Man*. Fukuyama defined history as the human struggle to find the most sensible—or least noxious—political system. By the 20th century liberal democracy, which according to Fukuyama had always been the best choice, had only one serious contender: Marxist socialism. After the collapse of the Soviet Union in the late 1980s, liberal democracy stood alone in the ring, battered but victorious. History was over.

Fukuyama went on to consider the profound questions raised by his thesis. Now that the age of political struggle has ended, what will we do next? What are we here for? What is the point of humanity? Fukuyama did not supply an answer so much as a rhetorical shrug. Freedom and prosperity, he fretted, might not be enough to satisfy our Nietzschean "will to power" and our need for constant "self-overcoming." Without great ideological struggles to occupy us, we humans might manufacture wars simply to give ourselves something to do.

Fukuyama did not overlook the role of science in human history. Far from it. His thesis required that history have a direction, that it be progressive, and science, he argued, provided this direction. Science had been vital to the growth of modern nation states, which saw science as a means to military and economic power. But Fukuyama did not even consider the possibility that science might also provide post-historical humanity with a common purpose, a goal, one that would encourage cooperation rather than conflict.

Hoping to learn the reason for Fukuyama's omission, I called him at the Rand Corporation, where he had obtained a job after *The End of History* became a bestseller. He answered with the wariness of someone accustomed to, and not amused by, kooks. At first, he misunderstood my question; he thought I was asking whether science could help us make moral and political choices in the post-historical era rather than serving as an end in itself. The lesson of contemporary philosophy, Fukuyama lectured me sternly, is that science is morally neutral, at best. In fact, scientific progress, if unaccompanied by moral progress among societies or individuals, "can leave you worse off than you were without it."

When Fukuyama finally realized what I was suggesting—that science might provide a kind of unifying theme or purpose for civilization—his tone became even more condescending. Yes, a few people had written him letters addressing that theme. "I think they were space-travel buffs," he snickered. "They said, 'Well, you know, if we don't have ideological wars to fight we can always fight nature in a certain sense by pushing back the frontiers of knowledge and conquering the solar system.'"

He emitted another scornful little chuckle. So you don't take these predictions seriously? I asked. "No, not really," he said wearily. Trying to goad something further out of him, I revealed that many prominent scientists and philosophers—not just fans of "Star Trek"—believed that science, the quest for pure knowledge, represented the destiny of mankind. "Hunh," Fukuyama replied, as though he was no longer listening to me but had re-entered that delightful tract by Hegel he had been perusing before I called. I signed off.

Without even giving it much thought, Fukuyama had reached the same conclusion that Stent had in *The Coming of the Golden Age*. From very different perspectives, both saw that science is less a byproduct of our will to know than of our will to power. Fukuyama's bored rejection of a future dedicated to science spoke volumes. The vast majority of humans, including not only the ignorant masses but also highbrow types such as Fukuyama, find scientific knowledge mildly interesting, at best, and certainly not worthy of serving as the goal of all humankind. Whatever the long-term destiny of *Homo sapiens* turns out to be—Fukuyama's eternal warfare or Stent's eternal hedonism, or, more likely, some mixture of the two—it seems unlikely to be the pursuit of scientific knowledge.

Gunther Stent left several loopholes open in his end-of-science scenario. Society might become so wealthy that it will pay for even the most whimsical scientific experiments—particle accelerators that girdle the globe!—without regard for cost. Alternatively, science could achieve some enormous breakthrough, such as a faster-than-light transportation system or intelligence-enhancing genetic engineering techniques that would enable scientists to transcend their physical and cognitive limits. I would add two other possibilites to Stent's list. One is that scientists might discover that life exists elsewhere, creating a glorious new era in comparative biology.

The other possibility—which Stent rejects but a surprising number of other scientists find compelling—is that one day we humans will create intelligent machines that can transcend our physical, economic and cognitive limits and carry on the quest for knowledge without us. In my favorite version of this scenario, machines transform the entire cosmos into a vast, unified, information-processing network. All matter becomes mind. This proposal is not science, of course, but wishful thinking. It nonetheless raises some interesting questions, questions normally left to theologians. What would an all-powerful, cosmic computer do? What would it think about? I can imagine only one possibility. It would try to answer The Question, the one that lurks behind all other questions, like an actor playing all the parts of a play: Why is there something rather than nothing? In its effort to find The Answer to The Question, the universal mind may discover the ultimate limits of knowledge.

References:

Bloom, H. 1973. *The Anxiety of Influence.* New York: Oxford University Press.

Bury, J. 1932. *The Idea of Progress.* New York: Macmillan, New York.

Bush, V. 1945. *Science: The Endless Frontier.* Reissued by the National Science Foundation (1990).

Fukuyama, F. 1992. *The End of History and the Last Man.* New York: The Free Press.

Glashow, S., and P. Ginsparg. 1986. "Desperately Seeking Superstrings?" *Physics Today,* May.

Glass, B. 1971. "Science: Endless Horizons or Golden Age?" *Science,* January 8, pp. 23–29.

——. 1979. "Milestones and Rates of Growth in the Development of Biology." *The Quarterly Review of Biology,* March, pp. 31–53.

Holton, G. 1993. *Science and Anti-Science.* Cambridge: Harvard University Press.

Jeffrey, E. 1940. "Nothing Left to Invent." *Journal of the Patent Office Society,* July, pp. 479–481.

Kadanoff, L. 1992. "Hard Times." *Physics Today,* October, pp. 9–11.

Koshland, D. 1995. "The Crystal Ball and the Trumpet Call." *Science,* March 17.

Rescher, N. 1978. *Scientific Progress.* Oxford: Basil Blackwell.

Selve, R. Q. (ed.). 1992. *The End of Science? Attack and Defense.* Lanham, MD: University Press of America.

Stent, G. S. 1969. *The Coming of the Golden Age: A View of the End of Progress.* Garden City, NY: Natural History Press.

——. 1978. *The Paradoxes of Progress.* San Francisco: W. H. Freeman.

Science Is Just Beginning

JOHN CASTI

Questions about the origin of things—the universe, life, language, human beings—have always held a strong fascination for the intellectually inclined, perhaps because such one-time-only events are difficult to study, thus providing a playpen for unbridled speculation and almost limitless armchair philosophy. Equally fascinating, it seems, are the no-time-only events of how things will end. Recent Cassandras publicly airing their angst over the incipient demise of something beloved range from Steven Weinberg dreaming of a final theory in particle physics (*Dreams of a Final Theory*, Pantheon, New York, 1992) to Francis Fukuyama pondering the end of history (*The End of History and the Last Man*, Free Press, New York, 1992). The latest addition to this cast of doomsayers is journalist John Horgan, who ups the ante by trumpeting to the world the imminent end of all science in a recently published book (*The End of Science*, Addison-Wesley, Reading, MA, 1996). Now what could such a temerarious claim actually mean?

Contrary to many accounts, science is not a noun or adjective by which we carve up the landscape of knowledge, labeling areas like biology and chemistry "science," while denying that label to fields of enquiry such as art, history, and literature. Rather, science is a verb; it is a procedure of a very special type. What distinguishes it from religion, mysticism, poetry, and all the other players in the reality-generation game is the way science gets at the scheme of things. That way is to provide answers to questions about the world around us by invoking a set of rules (read: theory, formula, algorithm, program). But not just any old rule will do. A scientific rule possesses certain properties—public accessibility, clarity, brevity, bias-free—and is generated by following a very definite procedure, the so-called "scientific method." So if science is indeed coming to an end, the only interpretation of this claim that seems to make any sense whatsoever is that either there are no interesting questions left to answer, or that it is flat-out impossible to produce a set of scientific rules by which to answer any question that still piques our curiosity. It stretches the imagination to suppose that anyone would take either alternative seriously.

A few years ago, I published a book (*Paradigms Lost*, Morrow, New York, 1989) in which I looked at six of the major problems facing science today, trying to identify the competing answers, how they were generated, who held to them, and why. These Big Questions are:

1. How did life originate on Earth?
2. Are human social behavioral patterns determined by our genes?
3. How do humans acquire language?
4. Is it possible to build a computing machine that will think, just like you and me?

5. Do there exist intelligent, extraterrestrial life forms in the Milky Way galaxy?
6. Does there exist an objective reality independent of human observers?

I think that even Horgan, who states that science is part of the "primordial human quest to understand the universe and our place in it," would agree that these questions are an integral part of that quest, and that the wellspring of deep and important questions is far from having run dry.

Let me hasten to add that the last time I looked (about a week ago), science was not much closer to offering a knockdown, airtight set of scientific rules for answering any of these questions than it was when my book was first published. But that in no way implies that such a set of rules does not exist. An analogy with similar Big Questions in mathematics is helpful in elucidating this point.

By now it is a well-chronicled story how, in 1931, Kurt Gödel stamped paid to David Hilbert's cherished belief that any mathematical question could be definitively answered. Gödel's result demonstrated the existence of forever unanswerable questions about numbers. So unlike the real worlds of physics, biology, chemistry and all the rest, here we have an area for which we can state unequivocally that there exist questions that can never be answered by following the rules of mathematics. Yet, strangely enough, I cannot ever recall seeing a book or article suggesting that mathematicians are losing any sleep over the end of mathematics. In fact, until recently the undecidable propositions underwritten by Gödel's results were regarded mostly as curiosities by the mathematical community, although occasionally someone might start dreaming in print about one or another famous unsolved problem being one of them. In fact, the celebrated Fermat Conjecture was thought of in just these terms at one time, although we all know now that what it took for the Conjecture to be settled was just a little more genius—and a lot more hard work—on the part of Andrew Wiles in wielding the traditional rules of mathematical argumentation.

Even more philosophically interesting is the 1976 answer offered by Kenneth Appel and Wolfgang Haken to the famed Four-Color Conjecture. In contrast to conventional mathematical proofs, which are at least in principle surveyable by the human mind, the Appel and Haken result affirming that no more than four colors are needed to color any planar map was based upon the computational investigation of nearly 2,000 individual cases. This examination involved many hundreds of hours' worth of supercomputer calculations, and would require thousands of years of work by an army of mathematicians to thoroughly check every step. Many mathematicians rejected this "proof," as it did not play fair by the traditional rules of the mathematician's game. Twenty years later we find that this computational exercise was merely the tip of an iceberg that is now threatening to change the very rules of the games mathematicians play. The same evolution of the rules of the game is just as likely to occur in science as in mathematics. All that is needed is a Big Question requiring new concepts and new methods. Let me briefly outline one.

A large number of the systems constituting the warp and weft of everyday life—a stock market or a road-traffic network, for example—involve a medium-sized number of agents (traders or drivers) interacting on the basis of limited, local information. Moreover, these agents are intelligent and adaptive; their behavior and interactions with one another are determined by rules, just like those governing the behavior of planets or molecules. But unlike these lifeless objects, adaptive agents are ready to change their rules in accordance with new information that comes their way, thus continually adjusting to their environment so as to prolong their own survival. This is about as good a definition as any I know as to what constitutes a complex adaptive system (CAS). At present there exists nothing remotely close to a formalism (that is, a set of scientific rules) for even stating, let alone understanding, the questions surrounding the weird and wondrous ways of such processes.

A few years back, the Santa Fe Institute was formed to serve as a center for the scientific investigation of just these types of complex systems. But the methods of choice for these studies are as different from the methods used in ordinary science as the use of the computer was to resolve the Four-Color Conjecture. Science, Santa Fe style, is based largely on the use of detailed simulations that serve as silicon surrogates for real-world correlates like stock markets or the immune system. The purpose of these surrogates is to provide a laboratory for carrying out controlled, repeatable experiments of the sort that are too expensive, too impractical, too time-consuming, or just plain too dangerous to do on the real-world system itself. I have given a detailed account elsewhere (*Would-Be Worlds*, Wiley, New York, in press) of how this use of the computer-as-a-laboratory promises to change the frontiers of science in the coming century. So let me just say here that there is every reason to believe that computer laboratories will provide the same kind of insight into the workings of CASes that the invention of the microscope gave to cell biologists or the telescope offered to astronomers. And if history is any guide, this tool is going to generate a plethora of as-yet-unstated Big Questions that will in turn serve as the basis for the creation of a bona fide science of complex systems in the decades to come.

Perhaps not surprisingly, one of the principal targets of Horgan's broadsides against the survival of science is exactly this claim. In a recent electronic debate on the World Wide Web with the imaginative theoretical biologist Stuart Kauffman, Horgan argued that the belief of "chaoplexologists" like Kauffman in the emergence of fundamental new laws of complex systems is so much wishful thinking. Reading the transcripts of this debate is eerily reminiscent of an imagined science-fiction dialogue I once ran across between a human and a human-like alien just in from the far corners of Andromeda. In his intellectual ping-ponging with Horgan, Kauffman valiantly upholds (for the most part successfully, in my view) his belief in the endless levels of complexity one sees in the universe around us, complexities that are well-chronicled in his *At Home in the Universe* (Oxford, New York, 1995). Kauffman makes his case by employing standards and styles of argument familiar in the world of scientific discourse. Horgan's response, however, makes one wonder if there might not really be a second Earth out there in Andromeda, where people use terms like "law," "discovery," "fundamental," and even "science," more as they might be employed in a journal of deconstructionist literary criticism or, perhaps, as they would be propounded by certain continental philosophers whose names I shall pass over with the silence of the grave.

Unlike many of today's "endologists," who hint darkly at the end of some field or other from their perspective as active researchers in the area under scrutiny, journalistic members of the "end-of-X" crowd have a predilection for invoking outside authority figures to buttress their woolly-headed claims. For some unaccountable reason, Nobel-prize-winning physicists seem especially popular in this regard. I don't know about you, but I'm not sure that an eminent physicist, actively engaged in promoting his field, is the first person I'd consult if I were seeking a balanced, non-partisan view of the future of physics. Notwithstanding this fairly obvious point, Horgan, for example, cites with benign approval Richard Feynman's remark that, "[This] is the age in which we are discovering the fundamental laws of nature, and that day will never come again."

Let me appeal to the same shameless rhetorical trick in offering an antidote to Feynman's brand of misguided hubris in the words of Lord Kelvin, former President of the Royal Society, and one of the preeminent physicists of the late 19th century. When told of the discovery of X-rays Kelvin solemnly intoned, "X-rays will prove to be a hoax." My friendly neighborhood radiologist will no doubt ponder this point with considerable pleasure on his next trip to the bank. And on his way from the bank to his summer home in the Swiss Alps, perhaps he'll also ponder another of Lord Kelvin's pronouncements: "I can state flatly that heavier-than-air flying machines are impossible." (I wonder if Lord Kelvin ever saw a bird!) All this brings to mind the statement made by science-fiction writer Arthur C. Clarke, an observation so pregnant with relevancy that it's now enshrined in the literature as Clarke's First Law: "When a distinguished but elderly scientist states that something is possible, he is almost certainly right. When he states that something is impossible, he is very probably wrong."

Let me conclude by noting that there is one genuinely interesting point struggling to emerge from the debate between the Kauffmans and the Horgans of the world. And it is not whether science as we know it is coming to an end. I hope that by now you will agree that that question hardly deserves the attention of a disciplined mind. Rather, the issue that merits considerably more attention than it has thus far received is whether the real world may not be just too complex for the human mind to fully comprehend. In other words, are there limits to what we can ever hope to know by using the tools and techniques of what we call "science"? If such limits do indeed exist, I'm sure we'd all like to know about them. But unless these as-yet-unknown limits happen to encompass every Big Question that we can conceive of asking about life, the universe, and everything else, we would still be as far away from the end of science as we were at its beginning.

Just in case you haven't noticed, heavier-than-air flight is alive and well. Unfortunately, so, too, are lighter-than-air frothings about the end of science. After the philosophical smokescreens, pretentious blatherings, selective quotations, and rhetorical flourishes all fade away, like a trickle of water in the desert, what remains is little more than a shapeless bit of intellectual fluff, pure cotton candy for the mind.

The Science Wars

Deconstructing Science Is Good Science

DR. RICHARD OLSON

"Men of science are now writing a book as fallible and as infallible, as wise and as foolish, as learned and as greatly mistaken, as are the scriptures . . . Newton will come to be as old in science as Moses, and, like the last pundit philosopher, will be smiled at by posterity as a man who saw wonderful things, but was walking in the thick darkness of the eighteenth century."
–George Dawson, *Sermons on Disputed Points*, 1878

"Nay, it is come to this, that truth meets nowhere with stronger opposition, than from many of those that raise the loudest cry about it, and would be taken for no less than the only dispensers of the favors and oracles of heaven. If any has the firmness to touch the minutest thing that brings them gain or credit, he's presently pursued with the hue and cry of heresy."
–John Toland, *Christianity Not Mysterious*, 1696

I have been hearing about, reading about, or involved in a series of events during the past two years that have encouraged me to re-visit a set of issues that I had been centrally concerned about during the late 1960s and early 1970s—contemporary relationships between the scientific and technological communities and their critics. During the late 1960s, I was a newly minted Ph.D. in the history of science with an ABD in physics, teaching at U.C. Santa Cruz, one of the national centers of counter-culture (now, "New Age") thinking. I became disturbed at that time by what seemed to me an unwarranted tendency of some of my more radical colleagues to blame many of the ills of contemporary America—including the Vietnam War—on science and technology.

The big problem for me then, as it remains for me now, was that I could see a substantial kernel of legitimacy in the claims that certain notions of rationality and objectivity associated with modern science and technology did undermine important traditional values that I was and am unwilling to abandon. And it did seem that, for reasons I could not yet begin to understand, the destructive and exploitative potentials of new scientific knowledge often seemed far easier to realize than the constructive and liberating ones. At the same time, though I was aware of some of the limitations in the extent of scientific knowledge, I was convinced—as I remain convinced—that most scientists genuinely believe that they are engaged in the pursuit of value-neutral and universal knowledge of a nature which is oblivious to their interests, and that ultimately such knowledge will be more beneficial than harmful to humanity.

The initial occasion for my return to this topic (in one sense, I never left, because I have worked for years on historical attitudes toward science) was a conversation I had almost two years ago with Peter Degan, a historian of 20th century physics with special interests in the interactions between physics and religion. Peter had been asked to review two recent popular works by distinguished contemporary American Nobel Laureate physicists—*The God Particle: If the Universe Is the Answer, What Is the Question?* (1993) by Leon Lederman and *Dreams of a Final Theory* (1992) by Steven Weinberg. Both books seemed to be intended in substantial part to drum up support for the since-cancelled Superconducting Supercollider, and what particularly struck Degan was the authors' open appropriation of theological and spiritual language in defense of their funding appeals. Degan observed (1994): "They portray the high-energy physicist as the last hero of Western Civilization and the divinely inspired bearer of high culture who pursues humanity's search for transcendent truth and beauty. . . . Consequently, the high Spiritual value of this enterprise makes the supercollider an absolute funding priority and justifies whatever amount of money is needed for its construction" [*Isis*, 85 (1994): 738].

Weinberg's argument is particularly intriguing because while it tries to claim a unique epistemic status for *contemporary* attempts to discover unified theories, it adopts the stance which George Dawson predicted in the mid-19th century, explaining that Newtonian mechanics, of course, had to be superseded because it failed to meet the demand for "logical inevitability" which fundamental particle physicists *now* recognize as essential to any truly legitimate comprehensive theory. The capacity to hold simultaneously that scientific theories are fallible and transcendent is quite marvelous. Apparently, as Einstein and Stephen Hawking have argued, physicists really can see into "the Mind of God," giving high energy physics the religious purpose that Jesse Helms insisted upon as a price for his Senatorial support for the Supercollider. At the same time, they only get to see one small and misleading bit at a time, so that complete enlightenment always demands the further investment of the seeker's time and somebody else's money.

The second event to get me riled was a session of the annual History of Science Society held on October 26, 1995. Billed as a panel discussion of the audiences for the History of Science, the conversation got off to an odd and disturbing start when a faculty member from a well known Northeastern institution claimed that Paul Gross and Norman Levitt's *Higher Superstition: The Academic Left and Its Quarrels with Science* (see Jeffrey Shallit's review in *Skeptic*, 3, No. 1, 98–100) received such a favorable reception that the scientists from whom encouragement for a position in Science and Technology Studies (hereafter, STS) was expected had withdrawn their support. Another historian, who had served as a curator in a public museum devoted to science, then reported that he had not been allowed to include clips from the film *Hiroshima, Nagasaki* (which was compiled from footage shot by Japanese cameramen in the aftermath of the two explosions) in exhibits on the atomic bomb because it reflected too negatively on the scientists involved. Finally, another faculty member argued that as historians of science we had to recognize that our primary audience was science students and scientists, and that we should consequently pay less attention to meeting the intellectual demands of our peers and more to keeping our audience happy and supportive, lest our jobs disappear. (I am in complete agreement that science studies types should seek to serve audiences beyond themselves and that in order to do so we need to write in ways that are accessible to others and that are not intended to provoke hostility. What I seemed to hear that disturbed me was an implication that we should go out of our way to be uncritical.)

About a month after this event, I picked up my Winter, 1996, issue of *Science, Technology, and Human Values* (Vol. 21, #1) to read an acrimonious exchange between Ron Gieryn and Paul Gross over the character of the Smithsonian Institution's "Science in American Life" exhibit. Gieryn, representing the Social Studies of Science community on the Advisory Board for the exhibit, became irate that the exhibit did not adequately acknowledge the insights regarding the social construction of scientific knowledge which have been developed within Science Studies in recent years. Gross, on the other hand, was upset because he viewed the exhibit as unbalanced in its extensive emphasis on negative consequences of science and its failure to adequately represent either the unique cognitive content of science or the "unprecedented human adventure of science" (119). Gross cited as his own chemist M. C. Lafollette's complaint about the reason for the character of the exhibit: ". . . the lead curators seemed so fearful of building a 'pro-science' exhibit (which would have antagonized some of their colleagues) that they wound up creating a largely negative one" (118).

Finally, on December 15, 1995, Richard Sclove, whose Loka Institute sponsors FAST-net, an internet newsgroup oriented toward a more democratic politics of science and technology, initiated a fascinating and disturbing exchange of views when he posted a series of questions under the subject heading "Tech Criticism and Emotion." Sclove has for some time given talks and interviews regarding the social effects of particular technologies, and he notes that those who disagree with him increasingly claim that he is "anti-technology" and react "with great passion, anger, outrage, and/or defensiveness." One of his questions is why so many people react to the criticism of a particular technology or scientific claim by labeling the critic "anti-technology" or "anti-science," since in the parallel case, persons who criticize a particular law are not labeled "anti-law." Second, he wondered about the basis for the intensity of emotion associated with resistance to technology or science criticism. Both questions are of special importance to Sclove for very personal reasons, because, as he writes, "maybe there is little point in pursuing public technology criticism if I'm doing it in ways that are counterproductively pushing a lot of folks' emotional buttons. Perhaps if I understood the 'buttons' better, I could learn to reframe my talks to make them more effective." This posting elicited an outpouring of impassioned responses from scientists and engineers as well as STS scholars and science and technology policy activists, most of which serve better to illustrate the problems that Sclove raises than to provide answers to his questions.

What unifies all of these episodes, it seems to me, is that they are all symptoms of an increasing polarization between scientists, engineers, and the managers of technological enterprises on the one hand, and students and consumers of science and technology on the other. Furthermore, it seems to me that there are at least two major and closely intertwined sets of causes for the current tensions between these two groups.

First is the passing of the "Golden Age" of research and development associated with the Cold War. With a diminished military justification for R&D expenditures, we are seeing a substantial "downsizing" of both governmental and corporate R&D programs in the name of cost cutting. Whether these policies are wise in the long run, even from a purely economic perspective, is debatable. What is absolutely certain is that they have created a real short term threat to the economic health of the science and engineering communities, with physics being hit particularly hard. One symptom of the current hard times in the sciences and engineering is the huge number of applicants for science positions in four year colleges—positions which serious professionally oriented scientists or engineers formerly looked on with disdain. The last two such searches that I know of produced over 1000 and over 800 applicants, respectively, many from senior scientists willing to accept entry level rank and pay.

A second symptom is a tendency on the part of scientists in particular—illustrated by the works of Lederman and Weinberg—to try to justify scientific activities increasingly in non-military and non-economic terms, drawing on longstanding Neo-platonic traditions associated with mystical elements in Christianity and Judaism. (Margaret Wertheim has identified this Neo-platonic or Pythagorean tradition used by Western scientists in an interesting way in her *Pythagoras' Trousers: God, Physics, and the Gender Wars* [New York: Random House, 1995], but she has paid little attention to the social and economic conditions which have recently produced a renewed focus on this line of argument.)

A third symptom is the completely understandable tendency of many contemporary scientists and technologists to respond defensively and violently to any perceived attack on the credibility, authority, or beneficence of the scientific and/or technological enterprises. Any professional elite that perceives itself to be losing status and economic support—whether it be the Anglican clergy in the 17th century in the chaotic aftermath of the English Civil War, or the scientific and technical community in the late 20th century in the chaotic aftermath of the Cold War—is likely to respond defensively and with all of the cultural resources that it can muster to perceived attacks. Indeed, it would be irrational for its members to do otherwise, according to the notion of rationality prevailing among economists today, for to act rationally is merely to act in ways consistent with one's perceived interests.

The second cause is related to the first. At the same time that scientists and engineers are threatened by social and economic forces that are largely beyond their control, it is certainly true that *some* members of the STS community really are openly and admittedly hostile to science and technology (at least as they are currently practiced or implemented). The STS community provides one of the few identifiable and reachable targets for the anxiety, frustration, and anger which some scientists feel about the very real threats to the status and economic health of their disciplines. If one chooses to define science and/or technology sufficiently narrowly, it is possible to argue legitimately that some of its members really are "anti-science" and/or "anti-technology." In that case it can hardly be surprising that some scientists and engineers are inclined to blame current trends in academic STS for some of their woes and to launch counterattacks against such perceived slight.

None of what I have said so far is intended to trivialize the arguments between certain scientists and certain STS figures, or to suggest that there are not important intellectual issues at stake. Rather, it is intended to suggest why some of these issues have become matters of intense public concern very recently, and why the parties to debates seem to be becoming increasingly strident and uncivil toward one another. When we turn to the content of the conflicts between those who speak on behalf of the scientific/technological community and those who are often taken to be the spokespersons for the STS community, that content seems to hinge on a small number of basic foundational principles, assumptions, and values. Among these, one of the most fundamental seems to be the question of commitment to some form of philosophical realism versus commitment to some form of social constructionism.

Many, perhaps most, scientists believe—with Paul Gross—that scientific knowledge claims refer to some "real" natural world which exists independent of scientists, and that anyone who denies this claim is "anti-science." On the other hand, most students of STS believe that the objects of scientific claims are "representations" whose meanings are always negotiated within a specific social context. The most extreme of these see no way to link such representations to any independent "reality," so they conclude that there is no independent reality to be represented and that scientists who claim otherwise are claiming an authority which does not belong to them. It seems to me that any good skeptic must suspend belief with respect to this issue. While plausibility arguments may be developed on both sides, nothing since the time of David Hume has happened to guarantee that humans have access to any reality underlying their experiences or that experience itself is possible outside the domain of customs and habits which are acquired in social settings. By the same token, we can have no knowledge that warrants the denial of some reality underlying experience, so any insistence upon pure social constructivism seems as unwarranted as an insistence on pure realism. The trans-cultural applicability of many scientific knowledge claims suggests that there may at least be some species-common forms of experience and cognition. Historians and sociologists of science, however, have developed enough case studies that demonstrate the cultural specificity of many explanatory structures to suggest that socio-cultural factors often play a significant role in what representational systems we construct and therefore in what we count as legitimate science at any particular time and place. It would thus seem safest to either admit that both culture-transcendent and culture-dependent factors play a role in the generation of scientific knowledge or to act in ways that are neutral relative to realist/ social-constructionist claims.

The realist/social-constructionist dichotomy is related to a second issue connected with the definitions of "science" and "technology." At the heart of this issue is the question of how extensive we believe employment of the terms science and technology should be. Do science and technology include all the motives which underlie the creation of knowledge or artifacts and the uses to which they are put (whether by the creators' designs or otherwise)? Do they include all of the institutions within which knowledge and objects are made and used? Or do they include only the sequences of knowledge claims and artifacts or tools, without regard for whom they were produced, how they were used, and how they have differentially affected the lives of different groups of people?

Until relatively recently (the mid 20th century), most studies of the scientific and technological enterprises were done by scientists and engineers who tended to define science and technology as a special kind of knowledge and as a sequence of inventions, with little regard to any social dimensions or contexts. George Sarton, for example, a physical chemist turned historian of science who founded the History of Science Society, defined science as "systematized positive knowledge, or what has been taken as such at different ages and in different places" (1936, 5). In his famous *A History of Mechanical Inventions*, Albert Payson Usher argued that it was best "to separate the history of the inventions from the discussion of their significance" (1954, ix).

In fact, neither Sarton, Usher nor any of their fellow travellers really thought that science or technology could be completely separated from all human context. Instead, they tended to argue that the communities of scientists and inventor-entrepreneurs are relatively autonomous and that each is self-defined in terms of a set of commitments to unique constitutive values which are aimed at producing objective knowledge or increasing the efficiency of productive processes respectively. For those who define themselves as scientists in this way, the production of objective knowledge—or TRUTH—becomes the ultimate value, with commitment to such subordinate values as honesty, independence of authority, disinterestedness, openness of communications, etc., defining the moral worth of individuals.

With rare exceptions, most scientists and engineers continue to prefer to understand their activities in these narrowly construed ways, with the consequence that they can insulate themselves or deny responsibility for the social consequences of their activities, at least with those that might be considered negative. (They are frequently willing to take credit for the positive ones in a move that is psychologically understandable, but logically suspect.) After all, objective knowledge claims, being value-neutral, are available to all to use, and it is the users who must incur the blame for any misuse.

Modern STS is dominated by persons whose primary interests and commitments are to an understanding of the broad social contexts and consequences of science and technology. These include the social considerations that direct money and effort at certain problems rather than others, as well as the social and economic consequences that follow from the utilization of scientific knowledge or the implementation of technological systems to serve particular interests in society. With rare exceptions most of these persons have a strong commitment to social and economic equity and to participatory democracy, with a concomitant suspicion of expertise. Moreover, they are inclined to think that the search for a good life is a communal rather than an individual enterprise. Such people are, as Gross and Levitt quite rightly point out, largely members of the academic left, although contra Gross and Levitt, that fact does not mean either that they are wrong or that they are misguided. Above all, in their minds, it does not mean that they are anti-scientific. Many, such as the feminist philosopher Sandra Harding, are inclined to believe that a more egalitarian and inclusive scientific community would be capable of producing a more nearly universal and objective knowledge.

It is a serious mistake, I think, to try to ask which group is more nearly "correct" about the nature of science or technology. How one chooses to define these terms is to a substantial extent an ideological choice which is made largely because of commitments to certain values. That is, in the broadest sense of the word, the choice is made for political or ideological reasons. If it is a set of superstitions or an ideology that guides the leadership of the STS community in its interpretations of the scientific and technological enterprises—and it is—it is no less a set of superstitions or an ideology that guides the scientific community's vision of itself. By the same token, *each* of these sets of value commitments is likely to be equally "rational," in the sense that each is as likely as the other to be consistent with the preferred ends of its advocates.

PLEASE NOTE: I am not suggesting that the definitions of science and technology are arbitrary, any more than the claim that for some purposes light can be considered as exhibiting particle-like characteristics while for other purposes it can be considered to have wave-like characteristics means that the definition of light is arbitrary. What I am suggesting is that different purposes may be served by considering science and technology narrowly as systems of propositions and aggregates of artifacts respectively on the one hand, and as socially and culturally embedded human activities on the other.

What does all of this mean for those of us who seek to make intelligent decisions about issues on which scientists and technologists or members of the STS community have something to say, or who hope to say something worthwhile and not merely inflammatory about contemporary science and technology? One important answer is suggested by the feminist "point of view" theorists. According to the advocates of point of view theory, every party to every argument starts from some set of value orientations that emerge out of the life history of the participant. Moreover, all such sets are probably either equally rational, a-rational, or non-rational. Since to proceed without at least implicitly adopting some set of values is impossible, we might all be better off if we could "own," or become aware of, our own point of view and learn to respect—not necessarily agree with—the points of view taken by others. This stance was articulated in a particularly illuminating way in a "Response to Sclove" posted on FASTnet on January 2, 1996, by Lars Kluver, director of the Danish Board of Technology, which has developed a system of citizen-based "consensus conferences" to assess the potential impact of new technologies. Kluver reports the results of surveys done for the Danish Board of Technology on attitudes toward biotechnologies:

> A positive attitude to biotech is seen among people who believe in economic growth, competition, a strong army, and who generally think technology is of the good. A negative or skeptical attitude is found among people who believe in social equality, a healthy environment, and who generally question the benefits of technology. Our general attitudes towards a technology, in other words, do not come from rational thinking, but rather from the values we already have and try to live out—from our value-conservatism. To be short (and of course ignoring a lot of details) many technology debates [the same is true about science debates] have more to do with ideology (or religion, if you like) than they have with rationality. That may be why your right to open up a debate is not respected. Your opponents simply do not have the strength or ability to question their own values and as a result, they spoil the debate instead.
>
> What can you do about it? I cannot think of any fail safe tactics. In the end we are dealing with psychology here. It is very much up to your skill as a debater to clear away the defenses of your opponent. But respecting the rights of opponents to say what they want to is a prerequisite to getting the same respect back.
>
> There is one final point I would like to make: the reactions you meet can be seen from both sides of the technology debate. Many industrialists meet the same kind of reactions from green-party "believers," when they try to initiate a constructive technology debate (which many industrialists do). One type of reaction from the green people is: "Why should I listen to your arguments—you only want to make money anyway" (the "you-are stupid," or "you are left-wing" kind of argument again). Value-conservatism is a widespread phenomenon.

I am virtually certain that the only real possibility for carrying out constructive discussions about science and technology policies depends upon the growth of abilities among people of all persuasions to question their own values. And of all people who can do this it is, or at least it should be, the skeptics. I believe this ability is precisely what the skeptics should be promoting. What do you think?

The Science Wars

Deconstructing Science Is Pseudoscience

NORM LEVITT

In the process of answering my critics—particularly Richard Olson in his article on the science wars (see previous entry)—I wish to address the difference between knowledge and knowingness. It seems to me transparently obvious that acquiring and extending knowledge about the natural world is the real business of science, and that science has been astonishingly successful in doing this over the past few centuries. Yet even among highly educated people this fact often breeds discontent. Much of this is understandable. A technocratic civilization of global dimensions has been raised on the foundations laid down by science, and not all of its manifestations are admirable or reassuring. What is there to like about toxic waste or multi-megaton warheads? But while moral unease about the fruits of science makes some sense, it has been known to give rise to extravagant philosophical positions.

Specifically there are those who claim to have tamed the monster by declaring that somehow it is all a fake; science isn't "real" knowledge, it's just a "narrative." It's not abstractly preferable to other systems of belief—myth for instance—merely attached to a culture that is, for the moment, more powerful than others. To make this strange doctrine even marginally plausible would seem to require an intellectual engine at least comparably powerful to that deployed by the sciences. How could one hope to reveal the errors of a flawed knowledge-system without having some keener instrument at hand to dissect it?

Quite obviously, no such thing has been invented. What serves in its place, however, is a stubbornly entrenched species of knowingness, an attitude that gives itself permission to avoid the pain and difficulty of actually understanding science simply by declaring in advance that knowledge is futile or illusory.

Knowingness is usually intertwined with cynicism. But cynicism is only palatable when it makes itself one of its own targets. Knowingness has the annoying habit of letting itself off the hook. It functions selectively, casting a nasty shadow only in certain preferred directions. In fact, knowingness can sometimes be allied with the grossest credulity. The UFO buff who will swallow whole the most grotesque tales of alien abduction pulls a very knowing attitude when you try to point out that there is no evidence that a flying saucer crashed in Roswell, New Mexico, in 1947. You can't fool him! He just *knows* that those devious government mandarins will go to incredible lengths to keep the information hidden, just as the militia member knows that the BATF is trying to take away his fully automatic weapons so that the Zionist Occupation Government can impose its New World Order.

It isn't always the case, however, that knowingness is predicated on falsity or delusion. It may well be founded on a truth or a genuine insight. The real problem with knowingness is that it is fundamentally lazy. It looks for a Royal Road to deep understanding, a methodology that excuses one from having to look closely at details or take complexity and fine distinctions into account. Thus, it rapidly becomes formulaic, perfunctory, and extremely closed-minded. Genuine knowledge, suffice it to say, is a very different and vastly more demanding creature.

Let me offer my favorite example of the distinction between knowledge and knowingness—Mozart's great opera *Cosi fan tutte*. The plot is a shallow, brittle piece of fluff that has nothing to recommend it but its superficial knowingness. It regards the perplexities of love with a smirk and a sneer. The idea is that two young soldiers wager on the fidelity of their sweethearts with an embittered friend. They pretend to be called away to the battlefield. Then each dons a disguse and woos the other's mistress. Within a matter of hours the girls' vows of undying faithfuness wilt under a barrage of flattery and hormones. The soldiers then return as themselves to humiliate their lovers. For the sake of theatrical convention, the disenchanted men agree to take back their tarnished goddesses, for all women are the same and these no worse than any others.

Clearly this is a very silly affair. One 19th-century critic called it "too stupid for criticism." And yet, when Mozart's music infuses it, this nasty trifle is transformed into a compelling human story. By some magic no critic can quite account for, the cardboard cutout characters become fully realized human beings and their seemingly absurd plight becomes deeply moving. Not a word or action strays from the conventions of sex-farce, yet at the end we are neither amused nor titillated, but saddened and thoughtful. Mozart is not only a great musical craftsman; he is a great psychologist and a great dramatist. He understands people down to their core as only a very few artists—Shakespeare, for one—do. Through his uncanny alchemy, he allows us to know what he knows about the pain of self knowledge. The contrast between the superficial nonsense of the ostensible plot and the deep truth that is revealed through the music makes that revelation all the more poignant.

I'll now fast-forward to the late 1980s when a trendy young director named Peter Sellars mounted controversial productions of Mozart's three great buffa operas, updating their settings to contemporary New York. Sellars was the perfect incarnation of what was then coming to be known as the postmodern sensibility—in other words, a knowing smart-aleck determined to deconstruct the life out of everything he touched. For my money, his stagings were wretched; their musical inadequacy alone doomed them. But what really riled me was the director's self-indulgent display of superficial knowingness. This was seen at its worst in Sellars' *Cosi*, which was set, if you can believe it, in a suburban diner, with all characters depicted as borderline psychotics. In numerous public statements, the director smugly insisted that in seeing through the comic exterior to the bitter inner reality, he was the first to understand the work deeply. This was, of course, nonsensical as well as arrogant. Sellars was hardly the first commentator to perceive the opera's autumnal sadness, merely the most vulgar and trivial. His knowingness was self-defeating; in discarding the farce, he also threw away the exquisite subtlety and the shimmering mystery of the piece. By presenting himself as smarter than Mozart, he proved himself an uncomprehending ass.

I mentioned Sellars and his mugging of Mozart because it was through this disagreeable episode that I first became aware that there was such a thing as postmodernism afoot in the land. Only later did it dawn on me that the academy had been deeply drawn into this dreadful vortex, with faculty (senior and junior) and graduate students by the thousands clamoring frantically to be let into the club. It was especially horrifying to realize that among the articles of faith required of postulants was the dogma that only through this creed could one enlist in the struggle against the social and political evils of the world; only by getting right with Foucault, Derrida, Lyotard, and Kristeva could one truly oppose racism, sexism, homophobia, imperialism, ethnocentrism, and all the attendant evils wrought by the capitalist West. Since my politics are those of my granddad—which is to say Debsian socialist—I was disconsolate that at the tail end of this horrid century the grand tradition of the engagé intellectual had deliquesced into this slobber. To the extent that I could unkink the prevailing rhetoric to see how its practitioners thought they might accomplish something in terms of real-world politics, the master-plan seemed to be this: if enough professors committed themselves to using bizarre, woolly, and pretentious language in books, papers, and lectures, then the contours of the world would shift, expelling all evils and inaugurating the reign of the just. This idea seemed pretty comical to me, although the joke was bitter, but it took many supposedly humanistic fields by storm, particularly literary criticism and related subjects. Frank Lentricchia, a repentant Duke English professor who was, until recently, a highly placed courtier in this little empire, put it this way:

> I believe what is now called literary criticism is a form of Xeroxing. Tell me your theory and I'll tell you in advance what you'll say about any work of literature, especially those you haven't read. Texts are not read, they are preread. All of literature is x and nothing but x, and literary study is the naming (exposure) of x. For x, read imperialism, sexism, homophobia, and so on. All of literary history is said to be a display of x, because human history is nothing but the structure of x. By naming x, we supposedly name the social order (ordure) as it is, and always has been. An advanced literature department is the place where you can write a dissertation on Wittgenstein and never have to face an examiner from the philosophy department. An advanced literature department is the place where you may speak endlessly about gender and never have to face the scrutiny of a biologist, because gender is just a social construction, and nature doesn't exist.

This comment is gratifying in that it pretty much summarizes what I've long believed about the weird course taken by lit-crits and the like in recent years; it's nice to hear it from a consummate insider. However, from my point of view, the antics of avant-garde English professors would merely have been part of the passing scene, and really none of my business, had not the infection spread to what used to be a sober, intelligent, and valuable discipline: the history, philosophy, and sociology of science. What emerged from this contagion is now usually called "science studies." It hasn't by any means completely obliterated traditional scholarship in the area, but it has become the most aggressively self-promoting and publicly visible branch. It has risen to prominence on the same current of enthusiasm for "postmodernism" and for ostensible political rectitude that has overwhelmed literature departments. It shares many of the current dogmas of literary studies, and colludes closely with academic manifestations of identity politics such as women's studies. It overlaps what is nowadays called cultural studies, a tendency that has effaced traditional scholarship in a number of areas, and it has absorbed many of the radically relativistic attitudes that predominate in postmodern cultural anthropology. The central doctrine of science studies is that science is "socially constructed" in a way that disallows traditional notions of scientific validity and objectivity. On this view, scientific theories are merely narratives peculiar to this culture and this point in its history. Their chief function is to create stories about the world consonant with dominant social and political values. Thus, they are no more "true," or even more reliable, than the myths, legends, and just-so stories of other cultures. All are equally culture-specific.

I can't claim that every would-be scholar connected with the science studies movement accepts this doctrine wholeheartedly in its most radical form. Yet it constitutes the ineluctable background assumption of most theorizing and discussion. It is the ultimate trump card in debate, and such misgivings as may exist tend to be expressed with exaggerated caution. To object too strongly is to invite the charge of collusion with Western intellectual hegemony and with the impermissibly universalistic claims of Western science and Western rationalism. Here, inviting comparison with Lentricchia's remarks on postmodern literary studies, is a disillusioned assessment of postmodern science studies by Meera Nanda, a scholar in that area who is, by the way, a leftist and feminist of nonwestern background:

> Indeed, constructionists admonish us to give up such outmoded notions of truth as a correspondence with a mind-independent reality. Rather, they insist that truth and falsity of knowledge claims be treated "symmetrically," that is, true knowledge to be contingent on social factors to the same degree as falsehoods are. In this remarkable feat of cognitive egalitarianism, one cannot say that true knowledge is true and preferable because it transcends social interests and describes the world as it is, for that would refute what sociologists set out to prove, namely, that *all* knowledge and not just ideology is constituted by social interests and power.... One not completely unintended consequence of their epistemological antirealism is that constructionists have taken it upon themselves to try to wean working scientists and ordinary people from their commonsensical distinction between truth and falsity as a better or worse match with an independent reality, a distinction the constructionists believe is itself a western social construct.

No less than in the circle of alien-abduction believers, knowingness rules the day for science critics, with the curious corollary that knowledge—that infinitely more precious substance—is tossed on the trash heap. "You can't dupe us!" cry the social constructionists, thereby duping themselves beyond hope of redemption.

Richard Olson's essay is an attempt to defend this rather indefensible cult as embodying a kind of cracker-barrel, commonsensical skepticism. I sense that his heart isn't completely in the project, and that he's rather uneasily aware that some of what he proposes to defend can't be defended but must be camouflaged instead. However that may be, his essay reveals, in a number of ways, the intellectual constipation that results when mere knowingness takes the place of analysis and inquiry. Since one of his points touches me personally, more or less, I'll begin with that one. Olson insists—and here he has a lot of company within the science studies confraternity—that the reason his cartel has come under heavy criticism from scientists is this: the end of the Cold War has diminished both popular enthusiasm and government backing for science; the era of the carte blanche is over. Chagrined scientists are therefore looking for scapegoats, and their ire has fastened upon the innocent science studies community, a clan of fellow scholars who are just doing their job.

This has a certain plausibility if cheap cynicism is your only benchmark. The problem is that it's simply untrue. Olson has been generous enough to point out that *Higher Superstition,* the book I co-authored with Paul R. Gross, was of some significance in triggering the counter-reaction of the scientific community. Thus it follows from his thesis that Paul and I must have been particularly obsessed with the post–Cold War shortfall in science funding, and that this sent us hunting for scapegoats. But this isn't so! Frankly, if someone had bothered to ask me at the time I started writing on these issues what I thought the end of the Cold War implied for science funding, I'd have answered that I expected at least a modest "peace dividend" for pure research in the basic sciences, even under a Republican administration. Alas, that's not how things worked out, but it's what I thought.

The simple truth is that I became a critic of the radical science-studies movement because it seemed so intellectually shallow and indefensible, and because its leading figures, a number of whom I had taken care to hear out, seemed to embody all the misplaced self-regard and self-certainty that make postmodernism so unappealing. In other words, the bumptious Peter Sellars and his unholy mangling of Mozart were much more to the front of my mind than the funding policies of the Pentagon, NASA, National Science Foundation, National Institute of Health, the Department of Energy, and so forth. Moreover, I can speak with some authority about the motives of other people who have become involved on my side of the issue. The "post–Cold War" hypothesis doesn't fit them either—for one thing, it's pretty clear to us all that a coterie of leftish professors, however fervent, doesn't have a hell of a lot of direct influence on high government policy or on popular opinion. Nor does it fit the mathematicians and physicists at the Institute for Advanced Study in Princeton who clobbered the proposed appointment of Bruno Latour, a character deified by science studies trendoids. There, the issue was whether charlatanry ought to be rewarded by tenure at the most prestigious scholarly institution in the country. I'm perfectly happy to stipulate that disciples of science studies don't have very much real-world political power. But that should not immunize them from intellectual accountability.

While I'm on the subject of leftish politics and its connection with these issues, let me point out that Olson has it backwards on a related question. He specifically accuses my book of arguing that the practitioners of postmodern science-critique must be in error simply because they are on the left. This is not only a distortion, it's an absurdity. I'm pretty much on the left myself—I even have a couple of scars to prove it. What I really object to is the way a claim of left sympathies is used as a perpetual Get-Out-of-Jail-Free card, something that allows one to dismiss any criticism, however cogent, as the spite-work of diabolical reactionaries. If anyone bothers to check, it will be found that many of the points made by *Higher Superstition* have also been made by Noam Chomsky, clearly no "rightist" and clearly no fan of postmodern "theory" as it applies to science or anything else. Olson sheepishly acknowledges that most of the science studies gendarmerie has sort-of-leftish aims, but glosses this as merely implying an interest in socio-economic equality and increased democracy. That's not the problem. The problem is that the version of leftist thought that dominates is a sectarian offshoot, and a weird one at that. Peruse the literature, and you will easily find that "democracy," by these peculiar lights, is supposed to mean that all "ways of knowing" are to be accorded equal epistemic dignity, with the possible exception of scientific rationalism itself, which is naturally to be reviled as imperialistic, sexist, homophobic, and so forth. It seems to me that this view is not only silly, but of no particular use to progressive causes, as I understand them. It is, however, of some use to reactionary causes. The purveyors of biblical creationism, for instance, have their antennas up for useful bits of academic blather, and they have found a trove in the stock of catchphrases that science studies has coined to pooh-pooh actual science. In fact, they may have found actual allies, to judge by the statements one very prominent constructionist theorist has made within my hearing. To cite another instance, Meera Nanda has conclusively demonstrated that the impact of postmodernism, relativism, and anti-universalism on the Indian intellectual left was devastating. It paralyzed the fight against religious obscurantism and its attendant reactionary, misogynist politics. Simultaneously, it handed the Hindu fundamentalist movements a heap of useful slogans to deploy. One result has been the displacement of science and mathematics in many public schools by their "Vedantic" versions. The reaction of the science studies community has been telling, particularly in response to another outrage cited by Nanda. This concerns a powerful politician whose credulousness with respect to a superstitious practice called Vastu Shastra led directly to the destruction of a poor community. Nanda relates:

> I have tested this case on my social constructionist friends here in the U.S. While they do see the injustice of the situation, they do not see why I am so exercised by the irrationality that led to it. We have our superstitions in the West, they tell me. Did not Nancy Reagan consult astrologers? As for my suggestion that if we want justice, we must challenge the irrationality of the ideas that lead to injustice, I am told that there is no need for proving that Vastu Shastra is wrong and modern science correct. I am told that seeing the two culturally bound descriptions at par with each other is progressive in itself, for then neither can claim to know the absolute truth, and this tradition will lose its hold on people's minds. I am told that this desire to prove that the traditional knowledge is an incorrect representation of nature is a sign of a scientistic mindset, a hangover from my training in biology, that I must overcome it if I do not want to reengineer the society of my birth on technocratic lines. Finally, I am told that I am an incorrigible modernist if I believe that Western science has any democracy-enhancing potential in the world.

Could there be any more pointed instance of smug, insular, airtight, infinitely condescending knowingness? Here we have a picture of sanctimonious science studies arrogance in full bloom. In my experience, it is quite characteristic.

Olson cites Sandra Harding as someone who is intent on "democratizing" science in order to make it "capable of producing a more nearly universal and objective knowledge." Perhaps he hasn't read her with particular care. What she says pretty much accords with the constructionist dogmatics cited above; she is horrified by the notion of universally valid knowledge. For a view of what she actually has in mind when she speaks of "democracy" and "objectivity," I recommend her essay in *Social Text* (no. 46/47). To wit:

> Most models of the scientific future . . . imagine "one true science." They do not imagine as existing or desirable many different, and in some respects conflicting representations of nature. Yet this vision is beginning to emerge in the new Northern [i.e., what is usually called Western] science studies.

No less than the constructionists cited by Nanda, she enthusiastically recommends regarding all local knowledge systems, of which standard science is but one instance, as equally mature and equally valid. As to objectivity, she seems to equate it with anything that serves her political goals.

Quite appropriately, perhaps, in that same issue of *Social Text* the mathematical physicist Alan Sokal published his now-famous hoax. Sokal induced the postmodern luminaries who edit that journal to publish a heap of double-talk under the pretext that it was a real live scientist's genuflection to the wisdom of postmodern sages. This illustrates the kind of trouble a supposed intellectual can get into by letting mere knowingness do the work of careful, critical thought. The editors approved of Sokal's pretended sentiments (including his fulminations against Gross and Levitt); they didn't understand the math or the physics, but they liked the postmodern slogans that surrounded the technical stuff; they really didn't understand the paper as a whole (you can't—it makes no sense) but it sounded like the kind of thing they assume one is supposed to pretend to take seriously. They invited disgrace, and it descended on them in spades. Goody! But the whole affair makes an important political point. Sokal is yet another opponent of postmodern science-critique who is himself a principled leftist. His prank brought dozens of such people out of the woodwork. Articles appeared in adamantly left publications like the *Nation* (Katha Pollitt), *In These Times* (Tom Frank), and *Z Magazine* (Michael Albert), praising Sokal's stunt and largely siding with him (and perforce with me) in the resulting doctrinal catfight.

It is either hopelessly naive or hopelessly disingenuous on Olson's part to imply that the quarrel between the science critics and "their" critics follows the standard Right-Left cleavage line. It doesn't—not even close.

Let us also consider one of Olson's more abstract philosophical points. At some length Olson defends, at least provisionally, the notion of anti-realism. Here, philosophical muddle clouds his efforts; he has confused the epistemological with the ontological. As the philosopher John Searle pointed out, ontological realism is a position virtually everyone takes automatically, while anti-realism is incoherent. For realism is not so much a formal doctrine as it is the unspoken ground of all discourse, all attempts at communication. Any sincere declarative utterance is an attempt to give a true account of something assumed to be real. Olson himself, for instance, speaks of scientists, science studies scholars, and the relations between them. He assumes, ipso facto, that there are such things in the universe and that they may be meaningfully described. Thus, he is, malgre lui, as much an ontological realist as any physicist talking about quarks and leptons. We all are. Even a solipsist is a kind of straitened realist. This is not to say that we all agree on the same ontology or the same hierarchy of categories. Plainly we do not. The social constructionists, when they're not pretending to be anti-realist, hold that the socially real is really the really real, and that the scientist's reality is a figment. Thus they are realists after all, albeit screwy ones.

Olson does allude to real and perplexing philosophical questions. The ontological conundrum is a deep one: to what extent may we reify any of our theories about the world, even the most sophisticated, phenomenologically adequate theories? When, and with what justification, may we assert that the objects that seem natural in the context of these theories are the pristine entities underlying the real universe? This problem has been around for millennia, and it is surpassingly deep. In this connection, one may evoke names like Plato, Duns Scotus, William of Occam, Hume, Kant, Poincaré, Mach, Bohr, Carnap, Ayer, Quine, Bohm, Margenau, and even Penrose and Hawking. The problem largely stands apart, however, from problems of epistemology, especially those addressed by the social constructionists. Scientists, qua scientists, are basically interested in phenomenological adequacy and logical economy. Thus, a sensible theory of scientific epistemology must keep ontological questions pretty much in the background. They are not relevant to the "social construction" debate. In any case, despite their claims, the constructionists haven't made much of a contribution to the ontological problem—about the same, I'd say, as Barney the Purple Dinosaur. Talking about it is, however, a pretty good smokescreen for doubtful epistemology.

The besetting sin of social constructionism, and therefore of the science studies movement that blazons forth social construction on all its banners, is one of laziness. A few anemic truisms about how everything we do as human beings is "social" are cobbled together into a vague General Theorem. A fatal knowingness suffuses every corner of the enterprise. It licenses practitioners to talk endlessly about science without ever talking about science. Since one knows that scientific theories are mere transcriptions of social prejudices and social processes, all one has to do is tell a just-so story about social imagery or the like. One needn't bother with the inner logic of the theory, or with the evidence directly bearing on it, since these are, by assumption, mere illusions. This is a very forgiving methodology in practice; it seems to allow highly selective choice of evidence, procrustean treatment of such evidence as is cited, special pleading and, when all else fails, recourse to moralistic intimidation. Consequently, the "case studies" Olson alludes to as illustrating social construction in action are remarkably weak, and interesting only for what they tell us about the sovereign power of the bandwagon, even among supposed intellectuals with real Ph.D.'s.

The version of science studies Olson is trying to defend is really a changeling child. When the idea was first formulated about 20 years ago, the intention was to study the interaction between science and history, politics, social circumstances, philosophy, ethics, religion, and art. This was a worthy undertaking and a difficult one, requiring scholars at least moderately well versed in some branch of science in addition to whatever other specialized knowledge and methodology might be required. It wasn't intended to be slavishly admiring of each and every scientist, nor to disguise the difficult problems that a technology often more powerful than wise imposes on us. But neither was it intended to minimize the intellectual strength and integrity of science, nor to "contextualize" it into a culture-bound tissue of prejudices. In short, it was an enterprise that required grown-ups. A funny thing happened, however. The infant discipline was whisked out of sight while a phalanx of postmodernist wiseacres put in its place a bizarre, misshapen, and antic creature, one which exhibited all the deformities of its cousins in literary studies, cultural anthropology, ethnic studies, and so on, as well as some peculiar organs all its own.

It may well be possible to return to the original intent and to create a discipline intellectually sound and with something important to contribute to the political, ethical, and even the esthetic, vision of our culture. Many scholars (perhaps including Olson himself) wish this were so. Alas, the faddists are still in charge, thanks largely to the imputation of deviation from political rectitude that awaits anyone who too skeptically challenges constructionist dogma. But questions are being raised and reluctant dragons prodded into battle. The caustic response of scientists has something to do with this. Nothing deflates a windbag like a horselaugh (for which reason Alan Sokal's drollery may well accelerate the process considerably). The adjustment may be painful for some young researchers who have been conned into thinking that they are on the cutting edge of enormous intellectual revolution. Science studies—the responsible version thereof—will enlighten, inform, and clarify in many respects, but it almost certainly won't produce earth-shattering epiphanies or mind-bending paradoxes. I suspect that it will deflect the course of science itself only modestly (though benevolently, I hope). To a generation that hoped to turn the world upside down, and was taught that the right jargon intoned in the proper style could do so, this no doubt will come as a disappointment. But perhaps maturity, together with an appropriately Darwinian winnowing of the field and a desacralization of smug smartasserie, will cushion the transition. I certainly hope so.

Knowingness doesn't work for scientists, not, at least, when they are practicing their trade. Knowingness invites you to cut corners, and when you do so, reality exhibits a most remarkable tendency to step right up and kick you in the tail. Knowingness simply gets in the way when it is knowledge you are after. Knowingness won't work for science studies either, not if one takes the long view. Olson's essay, like a number of other items, including the infamous issue of *Social Text,* erupts into view right now because, after years of relative immunity, the science studies racket is under scrutiny by intellectuals in and out of science who won't be put off by the usual line of patter or soothed by the standard aphorisms. Like any Mafia family when the indictments come down, science studies gets in touch with its lawyers and protestations of affronted virtue pour forth. Sorry. I don't think that sort of thing will work here. These days, the Tree of the Hesperides does not thrive in the Groves of Academe, for they are choked by postmodernist smog; such pomology brings forth apples not golden, but variously crab, sour, and just plain rotten. The word is out and it's getting hard to unload the crop at any price.

Face it, guys, the jig is up.

Olson Replies

I was deeply saddened to read Norman Levitt's response to my article because it seems to me to illustrate precisely the intensifying pattern of demonizing those who do not share every one of our assumptions and values that Lars Kluver so effectively pointed out and which I sought to discourage. Usually this process involves considerations which Professor Levitt identifies with the term knowingness—a stance which "dispenses one from having to look too closely at details or take complexity and fine distinctions into account. Thus, it rapidly becomes formulaic, perfunctory, and extremely closed minded." It is not clear to me that anyone would want to disagree with Levitt's antagonism to knowingness, but just to be absolutely clear, I happen to share his irritation with those who do not look closely and carefully at details, who ignore complexity and fine distinctions, and who are proudly closed minded. Moreover, I am not really thrilled about those whose arguments contain fundamental logical fallacies either.

Because I do not pretend to the knowledge which Professor Levitt has about literary criticism and Mozart, I would like to focus on the concept of knowingness in connection with science studies in general and my position with respect to radical social constructionism in particular. Let me begin with an issue of logic which will move us onto broader concerns. Levitt asserts that because I suggested that the impact of *Higher Superstition* was symptomatic of trends in the relationship between the scientific community and the broader public which are related to the post–Cold War downturn in science funding, "it follows from [my] thesis that Paul and I must have been particularly obsessed with the post–Cold War shortfall in science funding. . . ." I very carefully did not say anything about the motives of the authors of *Higher Superstition,* nor do I believe for one minute that there is any legitimate logical strategy that can allow one to infer the motives of any author by considering the way in which readers use that author's words. Indeed, one major theme of my *Emergence of the Social Sciences* (Twayne, 1993) was that early works in the social sciences almost universally ended up serving interests diametrically opposed to those intended by their authors.

Certainly Levitt is not alone in seeing logical implications where there are none. Radical social constructionists are as likely—perhaps even more likely—to assume that because some argument is used for a given purpose it was intended for that purpose; and I do not applaud this tendency among social constructionists any more than I do among their antagonists. In both cases, it seems to me to arise from an unappealing kind of "knowingness." This brings me closer to a central claim of Levitt's, which is that my essay was "an attempt to defend this rather indefensible cult [presumably social constructionism, because that is the subject of the previous sentence]." At the risk of being boring, let me repeat just part of two sentences from my earlier essay: ". . . any insistence upon pure social constructivism seems as unwarranted as an insistence on pure realism. The transcultural applicability of many scientific knowledge claims suggests that there may at least be some species-common forms of experience and cognition." It seems to me that only by ignoring details and gross distinctions, let alone fine ones, can one claim that an argument which includes these lines is an attempt to defend radical social constructivism. (For a brief positive statement of my position on this issue, please refer to my response to letters from John Thaler and John Toomay in *Skeptic*, V. 4, #3, 23–24).

This failure to acknowledge or perceive complexity and fine distinctions, however, is not primarily important in connection with my position. It seems to me to be at the very heart of Levitt's strategy of lumping together a huge range of perspectives which he and Paul Gross openly admit share few characteristics except that they differ broadly from those of Gross and Levitt. Nearly all persons who accept the notion that cultures have any bearing on the content of science in any degree are caricatured by identifying their views with the most radical cultural constructionists, post-modernists, academic feminists, and ecologists. Then, the views of these groups are further distorted by taking passages out of context and interpreting them in ways that are at best uncharitable and at worst, intentionally perverse.

Consider, for example, Levitt's use of the passage from Sandra Harding's essay in *Social Text* (no. 46/47), in which Harding suggests that in her vision of the scientific future, there may be "many different, and in some respects, conflicting representations of nature." From this statement, Levitt infers that "she enthusiastically recommends regarding all local knowledge systems, of which standard science is but one instance, as equally mature and equally valid."

Not only does the Levitt statement not follow from the Harding passage which he cites, it is contrary to any position I have read in any of Sandra Harding's works or heard her express either in public or in private. I am certainly not prepared to agree with all positions that she might hold; but she is quite open in saying that it would be absurd to try to use any knowledge system other than that of the modern exact sciences if one's goal is, for example, to send a rocket to Mars. What she does insist upon—and here I am convinced that she is correct—is that local knowledge systems often incorporate knowledge of local environmental conditions which are important for the health and sustainability of the local community, even if that knowledge is not articulated in the same propositional form in which Western science expresses its knowledge claims. Equally to the point is the fact that though a few scientists may hope for some eventual theory of everything, most of my scientific colleagues are inclined, like Harding, to accept and often emphasize the existence of different representations of natural phenomena associated with different disciplines.

I am inclined to agree completely with Levitt that it is "hopelessly naive or hopelessly disingenuous . . . to imply that the quarrel between the science critics and their critics follows the standard Right-Left cleavage line . . . ," but then I was not the person who subtitled a book "The Academic Left and Its Quarrels with Science." I do believe that many, but not all, science studies students have left political leanings (many of which I share) and, like Levitt, I think that to imply that all left-leaning academics quarrel with science is absurd. The only disagreement we seem to have on this issue is over which of us has encouraged (note, I did not say taken) the naive or disingenuous stance, and whether most left-leaning science studies professionals can reasonably be said to quarrel with science because they approach it from a perspective not shared by Norman Levitt.

There is a final extremely serious issue which sometimes seems to get confused with the issue of the undoubted instrumental success of the modern Western sciences. This issue is raised by Levitt's discussion of Meera Nanda's critiques of science studies because of the aid and comfort they have supposedly given to "religious obscurantism and its attendant reactionary, misogynist politics." It brings us back to the fundamental question of the degree of respect we are to offer to persons whose basic values are different from our own. Do we really want a democratic world culture in which other persons are free to hold beliefs of which we do not approve? It seems to me that this is fundamentally a moral, rather than an epistemic, question, although it is clear that epistemology is deeply implicated in many subjects over which people are willing to fight to the very death to impose their wills on others.

Even if we are totally convinced that Western science offers a knowledge system which is more powerful in controlling the physical world than any alternative, and even if we are convinced of the undesirability of the gender politics which attends the practice of Vastu Shastra, or the stunted intellectual life associated with those who promote biblical creationism, it is clear to me that we have neither an obligation nor a right to deny people the opportunity to make the "wrong" choice or to denigrate them for doing so without trying to understand why they choose as they do. Legitimate issues other than control of the physical world, gender equity, or intellectual stimulation may be at the heart of their choices. And I am certainly convinced that I would not want to live in a world in which I was not free to choose "wrongly" by someone else's standards. On this issue it seems to me there is little to choose from between the extreme self-proclaimed defenders of science-as-we-know-it and the extreme proponents of such movements as eco-feminism or fundamentalist Christianity. That each group should try and make its best case seems completely appropriate to me, but that they should do so by demonizing their opponents and by distorting their views seems to me to decrease the likelihood that we can sustain a relatively free, open, and non-coercive society.

5
HISTORICAL DOCUMENTS

Creationism

"Mr. Bryan's Address to the Jury in the Scopes' Case. The Speech Which Was Never Delivered."

by William Jennings Bryan

Why Creationists Fear Evolution: An Introduction to William Jennings Bryan's Last Speech Showing Nothing Has Changed Since Scopes

MICHAEL SHERMER

In the movie version of *Inherit the Wind*, about the 1925 Scopes' "Monkey Trial," in the middle of William Jennings Bryan's final moving speech he dramatically keels over dead in the courtroom, to the gasps of his faithful followers and the chagrin of his evolutionary opponents. The reality was perhaps a bit less dramatic, but the real speech is much more poignant (in the movie he is reduced to reciting by heart the books of the Bible). William Jennings Bryan's last speech was never delivered, and he died two days later rather unceremoniously. Bryan College in Dayton, Tennessee, still stands as a monument to an age gone by. Or has it?

Recent legislation in Tennessee, fortunately defeated, proposed that evolution be taught as a "mere" theory, and not as a fact of science, opening the door for other "theories" to be discussed in public school biology classes, such as the "theory" of special creation, AKA "Scientific Creationism," AKA Genesis. A few other states are moving toward trying to pass similar legislative proposals and creationists continue with their bottom-up strategy of electing school board members and influencing teachers and parents.

In my book *Why People Believe Weird Things*, I provided a thorough refutation of creationist arguments. I thought we would allow William Jennings Bryan to be the champion of the "other side" that thinks belief in the theory of evolution can actually lead to immoral behavior, and that acceptance of the theory has led to social ills. Bryan's argument in this speech is not an antiquated belief. On the following page is an illustration of the "Evolution Tree," in which evolution is shown to lead to all manner of evil, including Communism, Nazism, Imperialism, Monopolism, Humanism, Atheism, Amoralism, Scientism, Racism, Pantheism, Behaviorism, and Materialism; and "Evil Practices" including Promiscuity, Pornography, Genocide, Slavery, Abortion, Euthanasia, Chauvinism, Infanticide, Homosexuality, Child Abuse, Bestiality, and Drug Culture. As a brief rebuttal to their creationist tactic I wish to provide a short history to the creationist history and a brief response to Bryan's address.

For those not familiar with the history of the trial, John T. Scopes was a substitute teacher

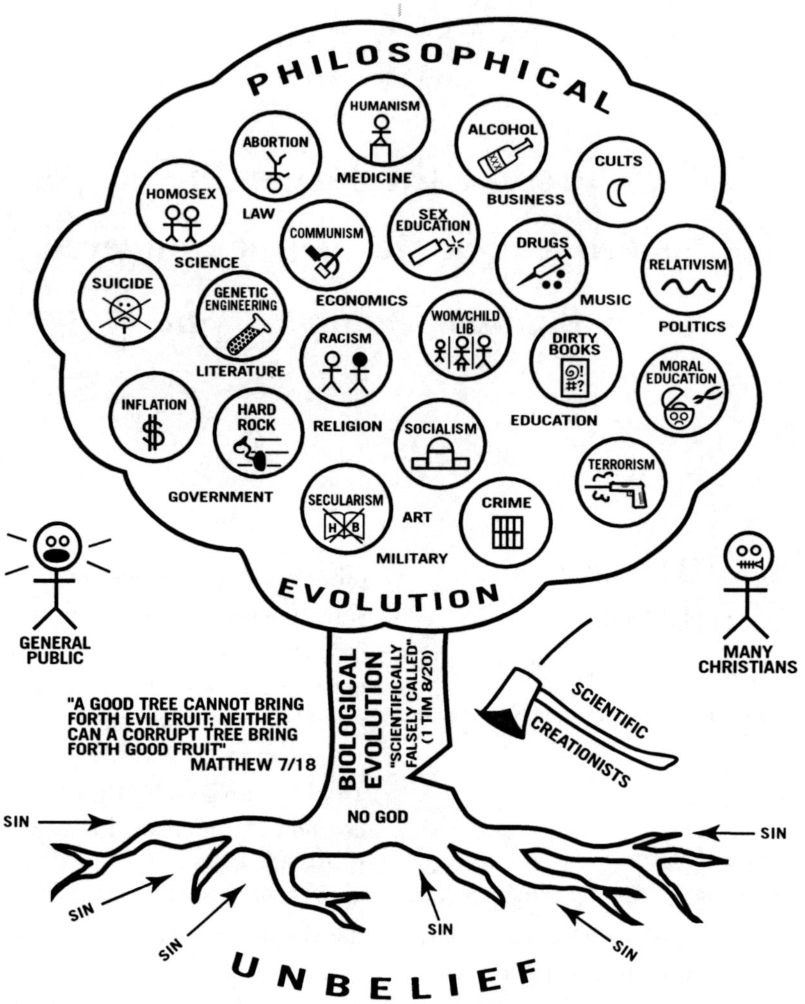

"Evolution Tree" from the Pittsburgh Creation Society. (R. G. Elmendorf)

who volunteered for the ACLU to be a test case to challenge Tennessee's "anti-evolution" law. It was the intention of the ACLU to take the case all the way to the U.S. Supreme Court. The most famous defense attorney of the day—Clarence Darrow—provided legal counsel for Scopes, and Bryan served as defender of the faith for the prosecution. Scopes was found guilty and fined $100 by Judge Raulston, but Tennessee law required that all fines above $50 must be set by a jury. Because of this technicality the defense was not able to appeal the case and it never was taken to the U.S. Supreme Court (though in 1987 an "equal time" law in Louisiana was challenged and defeated by a 7–2 vote of the justices; see my 1991 "Science Defended, Science Defined" in *Science, Technology, & Human Values*, V. 16, #4; reprinted in *Why People Believe Weird Things*).

Most people think that Scopes, Darrow, and the scientific community scored a great victory in Tennessee. H. L. Mencken, covering the trial for the *Baltimore Sun*, summarized it and Bryan this way: "Once he had one leg in the White House and the nation trembled under his roars. Now he is a tinpot pope in the Coca-Cola belt and a brother to the forlorn pastors who belabor half-wits in galvanized iron tabernacles behind the railroad yards. . . . It is a tragedy, indeed, to begin life as a hero and to end it as a buffoon."

But, in fact, this was no victory for evolution. Bryan died, but he had the last laugh, as the controversy stirred by the trial made others, particularly textbook publishers and state boards of education, reluctant to deal with the theory of evolution in any manner. Judith Grabiner and Peter Miller conducted a comparison study of high school textbooks before and after the trial, concluding: "Believing that they had won in the forum of public opinion, the evolutionists of the late 1920s in fact lost on their original battleground—teaching of evolution in the high schools—as judged by the content of the average high school biology textbooks [which] declined after the Scopes trial." A trial that seems comical in retrospect was really a tragedy, as Mencken concluded:

> Let no one mistake it for comedy, farcical though it may be in all its details. It serves notice on the country that Neanderthal man is organizing in these forlorn backwaters of the land, led by a fanatic, rid of sense and devoid of conscience. Tennessee, challenging him too timorously and too late, now sees its courts converted into camp meetings and its Bill of Rights made a mock of by its sworn officers of the law.

The speech that follows was vintage Bryan and should be read not just as a historical document and slice of fundamentalist Americana, it should toll a warning bell on the logic of faith and the power of rhetoric to move masses against reason and science. As Gould shows in "William Jennings Bryan's Last Campaign" (in *Bully for Brontosaurus,* Norton, 1991), Bryan's skepticism about evolution took a dramatic turn after the First World War when he became aware of the use of social Darwinism to justify militarism, imperialism, eugenics, and "paralyzing the hope of reform" through its program of "scientific breeding, a system under which a few supposedly superior intellects, self-appointed, would direct the mating and the movements of the mass of mankind" (to quote from the speech). Bryan feared for his faith and his country, and it was obvious to him who the enemy was: Darwinism and evolutionary theory.

Forget Duane Gish's demand for one transitional fossil, his obsession with the Bombardier Beetle, or his claim that evolution violates the second law of thermodynamics. These arguments are secondary matter. What really disturbs Gish and the creationists, as it did Bryan, is the implication of evolution for ethics and religion, and the following speech is an excellent summary of their fear that somehow a belief in evolution undermines morality. How do we answer this fear? The study of evolutionary ethics and the application of secular morality show how one can construct a meaningful existence and a moral life without religion. A brief response to Bryan and the creationists might include the following points:

1. The use or misuse of a theory does not negate the validity of the theory itself. Marx once claimed he was not a Marxist. Darwin would undoubtedly be spinning in his grave if he knew the uses of his theory in the 20th century to justify all manner of ideologies. The fact that Hitler implemented a eugenics program does not negate the theory of genetics. Theories are neutral; the use of theories is not. They are two different things.
2. The creationists' list of social problems—promiscuity, pornography, abortion, infanticide, racism, etc.—obviously existed long before Darwin and the theory of evolution. To blame Darwin for our own social and moral problems is to misdirect us from a deeper analysis and true understanding of these complex social issues.

3. The social evils that creationists fear have been with us since the birth of civilization. Organized religion has had thousands of years to solve these problems. To blame science and evolutionary theory for moral shortcomings is to admit that 6,000 years of religion has failed to do the job.
4. It is not the goal of science to replace faith and religion with evolutionary theory. The theory of evolution is a scientific theory, not a religious doctrine. It stands or falls on evidence alone. Religious faith, by definition, depends on belief when evidence is absent or unimportant. To fear the theory of evolution is an indication of a shortcoming in one's faith. If creationists had true faith in their religion it should not matter what scientists think or say. The fact that creationists have tethered themselves to science, even calling themselves "creation scientists," means that they feel their faith is not enough. They want proof. But proof of God is not possible, as the last 700 years of attempts to do so have shown (from Aquinas on).
5. The scientific attempt to understand human psychology and moral development, and the application of evolutionary theory to the origin and evolution of ethical behavior, are in their infancy. Religion has had 6,000 years, science less than 100. This 10 order-of-magnitude difference in time is significant. How much greater will our understanding of humanity be 6,000 years from now if science is applied to human affairs, no one can say, but given the relative difference in the rate of cumulative knowledge between science and religion we should be optimistic for the future. Religion may provide hope for some people. But only science has proven it can deliver the hopeful goods.

May It Please the Court, and Gentlemen of the Jury

William Jennings Bryan

Demosthenes, the greatest of ancient orators, in his "oration on the crown," the most famous of his speeches, began by supplicating the favor of all the gods and goddesses of Greece. If, in a case which involved only his own fame and fate, he felt justified in petitioning the heathen gods of his country, surely we, who deal with the momentous issues involved in this case, may well pray to the ruler of the universe for wisdom to guide us in the performance of our several parts in this historic trial.

Let me in the first place, congratulate our cause that circumstances have committed the trial to a community like this and entrusted the decision to a jury made up largely of the yeomanry of the state. The book in issue in this trial contains on its first page two pictures contrasting the disturbing noises of a great city with the calm serenity of the country. It is a tribute that rural life has fully earned.

I appreciate the sturdy honesty and independence of these who come into daily contact with the earth, who living near to nature, worship nature's god and who, dealing with the myriad mysteries of earth and air, seek to learn from revelation about the Bible's wonder working God. I admire the stern virtues, the vigilance and the patriotism of the class from which the jury is drawn, and am reminded of the lines of Scotland's immortal bard, which, when changed but slightly, describe your country's confidence in you:

O, Scotia, my dear, my native soil!

For whom my warmest wish to heaven is
sent,
Long may thy hardy sons of rustic toil
be blest with health, and peace, and sweet
content.
And, Oh, may heaven their simple lives
present
From luxury's contagion, weak and vile
Then, howe'er crowns and coronets be rent
A virtuous populace may rise the while,
And stand, a wall of fire, around their much
loved isle.

Let us now separate the issues from the misrepresentations, intentional and unintentional, that have obscured both the letter and the purpose of the law.

This is not an interference with freedom of conscience. A teacher can think as he pleases and worship God as he likes, or refuse to worship God at all. He can believe in the Bible or discard it; he can accept Christ or reject him. This law places no objections or restraints upon him. And so with freedom of speech, he can, so long as he acts as an individual, say anything he likes on any subject.

This law does not violate any rights guaranteed by any constitution to any individual. It deals with the defendant, not as an individual, but as an employee, an official or public servant, paid by the state, and therefore under instructions from the state.

The right of the state to control the public schools is affirmed in the recent decision in the Oregon case, which declares that the state can direct what shall be taught and also forbid the teaching of anything "manifestly inimical to the public welfare." The above decision goes even farther and declares that the parent not only has the right to guard the religious welfare of the child, but is in duty bound to guard it. That decision fits this case exactly. The state had a right to pass this law, and the law represents the determination of the parents to guard the religious welfare of their children.

It need hardly be added that this law did not have its origin in bigotry. It is not trying to force any form of religion on anybody. The majority is not trying to establish a religion or to teach it—it is trying to protect itself from the efforts of an insolent minority to force irreligion upon the children under the guise of teaching science. What right has a little irresponsible oligarchy of self-styled "intellectuals" to demand control of the schools of the United States, in which 25,000,000 of children are being educated at an annual expense of nearly $2,000,000,000?

Christians must, in every state of the union, build their own colleges in which to teach Christianity; it is only simple justice that atheists, agnostics and unbelievers should build their own colleges if they want to teach their own religious views or attack the religious views of others.

The statute is brief and free from ambiguity. It prohibits the teaching, in the public schools, of "any theology that denies the story of divine creation as taught in the Bible," and teaches, "instead, that man descended from a lower order of animals." The first sentence sets forth the purpose of those who passed the law. They forbid the teaching of any evolutionary theory that disputes the Bible record of man's creation and, to make sure that there shall be no misunderstanding, they place their own interpretation on their language and specifically forbid the teaching of any theory that makes man a descendant of any lower form of life.

The evidence shows that defendant taught, in his own language as well as from a book outlining the theory, that man descended from lower forms of life. Howard Morgan's testimony gives us a definition of evolution that will become known throughout the world as this case is discussed.

Howard, a 14-year-old boy, has translated the words of the teacher and the textbook into language that even a child can understand. As he recollects it, the defendant said "a little germ of one cell organism has formed in the sea; this kept evolving until it got to be a pretty good sized animal, then came on to be a land animal, and it kept evolving, and from this was man."

There is no room for difference of opinion here, and there is no need of expert testimony. Here are the facts, corroborated by another student, Harry Helton, and admitted to be true by counsel for defense. White, superintendent of schools, testified to the use of Hunter's civic biology, and to the fact that the defendant not only admitted teaching evolution, but declared that he could not teach it without violating the law. Robinson, the chairman of the school board, corroborated the testimony of Superintendent White in regard to the defendant's admissions and declaration. These are the facts; they are sufficient and undisputed; a verdict of guilty must follow.

But the importance of this case requires more. The facts and arguments presented to you must not only convince you of the justice of conviction in this case, but, while not necessary to a verdict of guilty, they should convince you of the righteousness of the purpose of the people of the state in the enactment of this law.

The state must speak through you to the outside world and repel the aspersions cast by the counsel for the defense upon the intelligence and the enlightenment of the citizens of Tennessee. The people of this state have a high appreciation of the value of education. The state constitution testifies to that in its demand that education shall be fostered and that science and literature shall be cherished. The continuing and increasing appropriations for public instruction furnish abundant proof that Tennessee places a just estimate upon the learning that is secured in its schools.

Religion is not hostile to learning; Christianity has been the greatest patron learning has ever had. But Christians know that "the fear of the Lord is the beginning of wisdom." Now, just as it has been in the past, and they therefore oppose the teaching of guesses that encourage Godlessness among the students.

Neither does Tennessee undervalue the service rendered by science. The Christian men and women of Tennessee know how deeply mankind is indebted to science for benefits conferred by the discovery of the laws of nature and by the designing of machinery for the utilization of these laws. Give science a fact and it is not only invincible, but it is of incalculable service to man.

If one is entitled to draw from society in proportion to the service that he renders to society, who is able to estimate the reward earned by those who have given to us the use of steam, the use of electricity, and enable us to utilize the weight of water that flows down the mountainside? Who will estimate the value of the service rendered by those who invented the radio? Or, to come more closely to our home life, how shall we recompense those who gave us the sewing machine, the tractor, the threshing machine, the tractor, the automobile and the method now employed in making artificial ice? The department of medicine also opens an unlimited field for invaluable service.

Typhoid and yellow fever are not feared as they once were. Diphtheria and pneumonia have been robbed of some of their terrors, and a high place on the scroll of fame still awaits the discoverer of remedies for arthritis, cancer, tuberculosis and other dread diseases to which mankind is heir.

Christianity welcomes truth from whatever source it comes, and is not afraid that any truth from any source can interfere with the divine truth that comes by inspiration from God Himself. It is not scientific truth to which Christians, therefore, can be scientific unless it is true.

Evolution is not truth; it is merely an hypothesis—is millions of guesses strung together.

It had not been proven in the day of Darwin; he expressed astonishment that with two or three million species, it had been impossible to trace any species to any other species. It had not been proven in the days of Huxley, and it has not been proven up to today. It is less than four years ago that Professor Bateson came all the way from London to Canada to tell the American scientists that every effort to trace one species to another had failed—every one.

He said he still had faith in evolution, but had doubts about the origin of species. But of what value is evolution, if it cannot explain the origin of species? While many scientists accept evolution as if it were a fact, they all admit, when questioned, that no explanation has been found as to how one species developed into another.

Darwin suggested two laws, sexual selection, and natural selection. Sexual selection has been laughed out of the class room, and natural selection is being abandoned, and no new explanation is satisfactory even to scientists. Some of the more rash advocates of evolution are wont to say that evolution is as firmly established as the law of gravitation, or the Copernican theory. The absurdity of such a claim is apparent when we remember that anyone can prove the law of gravitation by throwing a weight into the air, and that anyone can prove the roundness of the earth by going around it, while no one can prove evolution to be true in any way whatever.

Chemistry is an insurmountable obstacle in the path of evolution. It is one of the greatest of the sciences; it separates the atoms—isolates them and walks about them so to speak. If there were in nature a progressive force, an eternal urge, chemistry would find it. But it is not there.

All of the 92 original elements are separate and distinct; they combine in fixed and permanent proportions. Water is H_2O, as it has been from the beginning. It was here before life appeared and has never changed; neither can it be shown that anything else has materially changed.

Man a Special Creation

There is no more reason to believe that man descended from some inferior animal than there is to believe that a stately mansion had descended from a small cottage. Resemblances are not proof, they simply put us on inquiry.

As one fact, such as the absence of the accused from the scene of the murder, outweighs all resemblances that a thousand witnesses could swear to, so the inability of science to trace any one of the millions of species to another species, outweighs all the resemblances upon which evolutionists rely to establish man's blood relationship with the brutes.

But while the wisest scientists can not prove a pushing power, such as evolution is supposed to be, there is a lifting power that any child can understand. The plant lifts the mineral up into a higher world, and the animal lifts the plants up into a world still higher. So, it has been reasoned by analogy, man rises, not by a power within him, but only when drawn upward by a higher power.

There is a spiritual gravitation that draws all souls toward heaven, just as surely as there is a physical force that draws all matters on the surface of the earth towards the earth's center. Christ is our drawing power; he said, "I, if I be lifted from the earth, will draw all men unto Me," and his promise is being fulfilled daily all over the world.

It must be remembered that the law under consideration in this case does not prohibit the teaching of evolution up to the line that separates man from the lower form of animal. The law might well have gone farther than it does and prohibit the teaching of evolution in lower forms of life; the law is a very conservative statement of the people's opposition to an anti-Biblical hypothesis. The defendant was not content to teach what the law permitted; he, for reasons of his own, persisted in teaching that which was forbidden for reasons entirely satisfactory to the law makers.

Many of the people who believe in evolution do not know what evolution means. One of the science books taught in the Dayton high schools has a chapter on "The Evolution of Machinery." This is a very common misuse of the term. People speak of the evolution of the telephone, the automobile, and the musical instrument. But these are merely illustrations of man's power to deal intelligently with inanimate matter; there is no growth from within in the development of machinery.

Equally improper is the use of the word "evolution" to describe the growth of a plant from a seed, the growth of a chicken from an egg, or the development of any form of animal life from a single cell. All these give us a circle, not a change from one species to another.

Evolution—the evolution involved in this case, and the only evolution that is a matter of controversy anywhere—is the evolution taught by defendant, set forth in the books now prohibited by the new state law, and illustrated in the diagram printed on page 194 of *Hunter's Civic Biology*.

The author estimates the number of species in the animal kingdom at 518,900. These are then divided into 18 classes, and each class indicated on the diagram by a circle, proportioned in size to the number of species in each class and attached by a stem to the trunk of the tree. It begins at protozoa and ends with mammals.

Passing over the classes with which the average man is unfamiliar, let me call your attention to a few of the larger and better known groups. The insects are numbered at 360,000, over two-thirds of the total number of species in the animal world. The fishes are numbered at 13,000, the amphibians at 1,400, the reptiles at 3,500, and the birds at 13,000, while 3,500 mammals are crowded together in a little circle that is barely higher than the bird circle. No circle is reserved for man alone.

He is, according to the diagram, shut up in the little circle entitled "mammals," with 3,499 other species of mammals. Does it not seem a little unfair not to distinguish between man and lower forms of life? What shall we say of the intelligence, not to say religion of those who are so particular to distinguish between fishes and reptiles and birds, but put a man with an immortal soul in the same circle with the wolf, the hyena, and the skunk? What must be the impressions made upon children by such a degradation of man?

In the preface of this book, the author explains that it is for children, and adds that "the boy or girl of average ability upon admission to the secondary school is not a thinking individual." Whatever may be said in favor of teaching evolution to adults, it surely is not proper to teach it to children who are not yet able to think.

The evolutionist does not undertake to tell us how protozoa, moved by interior and resident forces, sent life up through all the various species, and can not prove that there was actually any such compelling power at all. And yet, the school children are asked to accept their guesses and build a philosophy of life upon them. If it were not so serious a matter, one might be tempted to speculate upon the various degrees of relationship that, according to evolutionists, exist between man and other forms of life.

It might require some very nice calculation to determine at what degree of relationship the killing of a relative ceases to be murder and the eating of one's kin ceases to be cannibalism. But it is not a laughing matter when one considers that evolution not only offers no suggestion as to a creator but tends to put the creative act so far away to cast doubt upon creation itself. And, while it is shaking faith in God as a beginning, it is also creating doubt as to heaven at the end of life.

Evolutionists do not feel that it is incumbent upon them to show how life began or at what point, in their long drawn out scheme of changing species man became endowed with hope and promise of immortal life.

God may be a matter of indifference to the evolutionists, and a life beyond may have no charm for them, but the mass of mankind will continue to worship their Creator and continue to find comfort in the promise of their Saviour that he has gone to prepare a place for them. Christ has made of death a narrow, starlit strip between the companionship of yesterday and the reunion of tomorrow, and evolution strikes out the stars and deepens the gloom that enshrouds the tomb.

If the results of evolution were unimportant, one might require less proof in support of the hypothesis, but before accepting a new philosophy of life, built upon a materialistic foundation, we have reason to demand something more than guess; "we may well suppose" is not a sufficient substitute for "thus saith the Lord."

If you, your honor, and you, gentlemen of the jury would have an understanding of the sentiment that lies back of the statute against the teaching of evolution, please consider these facts: First, as to the animals to which evolutionists would have us trace our ancestry. The following is Darwin's family tree, as you will find it set forth on pages 180–181 of his "Descent of Man."

> The most ancient progenitors in the kingdom of vertebrata, at which we are able to obtain an obscure glance, apparently consisted of a group of marine animals, resembling the larvae of existing asidians. These animals probably gave rise to a group of fishes, as lowly organized as the lancelot; and from these the canoids, and other fishes like the lepidosiren, must have been developed. From such fish a very small advance would carry us on to the amphibians. We have seen that birds and reptiles were once intimately connected together; and the monotrematata now connect mammals with reptiles in a slight degree. But no one can at present say by what line of descent the three higher and related classes, namely, mammals, birds and reptiles, were derived from the two lower vertebrate classes, namely, amphibians and fishes.
>
> In the class of mammals the steps are not difficult to conceive which led from the ancient monotremata to the ancient marsupials; and from these to the early progenitors of the placental mammals. We may thus ascend to the lemuridae; and the interval is not very wide from these to the simiadac. The simiadae then branched off into two great stems, the new world and the old world monkeys; and from the latter, at a remote period, man, the wonder and glory of the universe, proceeded. Thus we have given to man a pedigree of prodigious length, but not, it may be said, of noble quality.

Darwin, on page 171 of the same book, tries to locate his first man, that is, the first man to come down out of the trees, in Africa. After leaving man in company with gorillas and chimpanzees, he says: "But it is useless to speculate on this subject." If he had only thought of this earlier, the world might have been spared much of the speculation that his brute hypothesis has excited.

On page 79 Darwin gives some fanciful reasons for believing that man is more likely to have descended from the chimpanzee than from the gorilla. His speculations are an excellent illustration of the effect that the evolutionary hypothesis has in cultivating the imagination. Professor J. Arthur Thomson says that the "idea of evolution is the most potent thought economizing formula the world has yet known." It is more than that; it dispenses with thinking entirely and relies on the imagination.

On page 141 Darwin attempts to trace the mind of man back to the mind of lower animals. On pages 118 and 114 he endeavors to trace man's moral nature back to the animals. It is all animal, animal, animal, with never a thought of God or religion.

Our first indictment against evolution is that it disputes the truth of the Bible account of man's creation and shakes faith in the Bible as the word of God. This indictment we prove by comparing the process described as evolutionary with the text Genesis. It not only contradicts the Mosaic record as to the beginning of human life, but it disputes the Bible doctrines of reproduction according to kin—the greatest scientific principle known.

Evolution Incompatible with Faith

Our second indictment is that the evolutionary hypothesis carried to its logical conclusion, disputes every vital truth of the Bible. Its tendency, naturally, if not inevitably, is to lead those who really accept it, first to agnosticism and then to atheism. Evolutionists attack the truth of the Bible, not openly at first, but by using weasel-words like "poetical," "symbolical," and "allegorical" to search out the meaning of the inspired record of man's creation.

We call as our first witness Charles Darwin. He began life as a Christian. On page 39, volume l, of the life and letters of Charles Darwin, by his son, Francis Darwin, he says, speaking of the period of 1828 to 1831, "I did not then in the least doubt the strict and literal truth of every word in the Bible." On page 412 of volume 2, of the same publication, he says, "when I was collecting facts for 'The Origin' my belief in what is called a personal God was firm as that of Doctor Puzey himself."

It may be a surprise to your honor, and to you, gentlemen of the jury, as it was to me, to learn that Darwin spent three years at Cambridge studying for the ministry.

This was Darwin as a young man, before he came under the influence of doctrine that man was from a lower order of animals. The change wrought in his religious views will be found in a letter written to a German youth in 1879, and printed on page 277 of volume 1 of the life and letters above referred to. The letter begins:

> I am much engaged, an old man, and out of health, and I can not spare time to answer your questions fully, nor indeed can they be answered. Science has nothing to do with Christ, except insofar as the habit of scientific research makes a man cautious in admitting evidence. For myself, I do not believe that there ever has been any revelation. As for a future life, every man must judge for himself between conflicting vague probabilities.

Note that "science has nothing to do with Christ, except insofar as the habit of scientific research makes a man cautious in admitting evidence," stated plainly, that simply means that "the habit of scientific research" makes one cautious in accepting the only evidence that we have of Christ's existence, mission, teaching, crucifixion, and resurrection, namely the evidence found in the Bible.

To make this interpretation of his words the only possible one, he adds "for myself, I do not believe that there ever has been any revelation." In rejecting the Bible as a revelation from God he rejects the Bible's conception of God, and he rejects also the supernatural Christ of whom the Bible, and the Bible alone, tells. And, it will be observed, he refuses to express any opinion as to a future life.

Now let us follow with his son's exposition of his father's views as they are given in extracts from a biography written in 1876. Here is Darwin's language as quoted by his son:

> During these two years (October, 1838, to January, 1839) I was led to think much about religion. Whilst on board the Beagle I was quite orthodox, and I remember being heartily laughed at by several of the officers (though themselves orthodox) for quoting the Bible as an unanswerable authority on some point of morality. When thus reflecting I felt compelled to look for a first cause, having an intelligent mind, in some degree analogous to man; and I deserved to be called an atheist. This conclusion was strong in my mind about the time, as far as I can remember, when I wrote the "Origin of Species." It is since that time that it has very gradually, with many fluctuations, become weaker. Then arises the doubt, can the mind of man, which has, as I fully believe, been developed from a mind as low as that possessed by the lowest animals, be trusted when it draws such grand conclusions?
>
> I can not pretend to throw the least light on such abstruse problems. The mystery of the beginning of all things is insolvable by us; and I, for one, must be content to remain an agnostic.

When Darwin entered upon his scientific career he was "quite orthodox and quoted the Bible as an unanswerable authority on some point of morality." Even when he wrote "Origin of Species," the thought of "a first cause, having an intelligent mind, in some degree analogous to man," was strong in his mind. It was after that time that "very gradually, with many fluctuations, his belief in God became weaker." He traces this decline for us and concludes by telling us that he can not pretend to throw the least light on such abstruse problems—the religious problems above referred to. Then comes the flat statement that he "must be content to remain an agnostic," and, to make clear what he means by the word agnostic, he says that "the mystery of the beginning of all things is insolvable by us"—not by him alone but by everybody. Here we have the effect of evolution upon its most distinguished exponent; it led him from an orthodox Christian, believing every word of the Bible and in a personal God, down and down to helpless and hopeless agnosticism.

But there is one sentence upon which I reserve comment—it throws light upon its downward pathway: "Then arises the doubt, can the mind of man, which has, as I fully believe, been developed from a mind as low as that possessed by the lowest animals, be trusted when it draws such grand conclusions?"

Here is the explanation; he drags man down to the brute levels, and then, judging man by brute standards he questions "'whether man's mind can be trusted to deal with God and immortality.'"

How can any teacher tell his students that evolution does not tend to destroy his religious faith? How can an honest teacher conceal from his students the effect of evolution upon Darwin himself? And is it not stranger still that preachers who advocate evolution never speak of Darwin's loss of faith, due to his belief in evolution? The parents of Tennessee have reason enough to fear the effect of evolution upon the mind of their children. Belief in evolution can not bring to those who hold such a belief any compensation for the loss of faith in God, trust in the Bible and belief in the supernatural character of Christ. It is belief in evolution that has caused so many scientists and so many Christians to reject the miracles of the Bible, and then give up, one after another, every vital trust in Christianity. They finally cease to pray and sunder the tie that binds them to their Heavenly Father.

The miracle should not be a stumbling block to anyone. It raises but three questions: First, could God perform a miracle? Yes, the God who created the universe can do anything he wants to do with it. He can temporarily suspend any law that he has made or he may employ higher laws that we do not understand.

Second: Would God perform a miracle? To answer that question in the negative one would have to know more about God's plans and purposes than a finite mind can know and yet some are so wedded to evolution that they deny that God would perform a miracle merely because a miracle is inconsistent with evolution.

If we believe that God can perform a miracle and might desire to do so, we are prepared to consider with open mind the third question, namely: Did God perform the miracles recorded in the Bible? The same evidence that establishes the authority of the Bible establishes the truth of miracles performed.

Now let me read of one of the most pathetic confessions that has come to my notice. George John Romanes, a distinguished biologist, sometimes called the successor of Darwin, like Darwin, was reared in the orthodox faith, and like Darwin, was led away from it by evolution.

For 25 years he could not pray. Soon after he became an agnostic, he wrote a book entitled, "A Candid Examination of Theism," publishing it under the assumed name "Physicus." In his book he says:

> And for so much as I am far from being able to agree with those who affirm that the twilight doctrine in the "new faith" is a desirable substitute for the waning splendor of "the old" I am not ashamed to confess that with this virtual negation of God the universe to me has lost its soul of loveliness; and although from hence the precept "work while it is day" will doubtless but gain an intensified force from the terribly intensified meaning of the words that "the night cometh when no man can work," yet when at times I think, as think at times I must, of the appalling contrast between the hallowed glory of that creed which once was mine, and the lonely mystery of existence as now I find it—at such times I shall ever feel it impossible to avoid the sharpest pang of which my nature is susceptible.

Do these evolutionists stop to think of the crime they commit when they take faith out of the hearts of men and women and lead them out into a starless night? What pleasure can they find in robbing a human being of "the hallowed glory of that creed" that Romanes once cherished, and in substituting the "lonely mystery of existence" as he found it? Can the fathers and mothers of Tennessee be blamed for trying to protect their children from such a tragedy?

If any one has been led to complain of the severity of the punishment that hangs over the defendant, let him compare this crime and its mild punishment with the crimes for which a greater punishment is ascribed. What is the taking of a few dollars from one in day or night in comparison with the crime of leading one away from God and away from Christ?

He who spake as never man spake, thus describes the crimes that are committed against the young: "It is impossible but that offenses will come: but woe unto him through whom they come. It were better for him that a millstone were hanged about his neck and he be cast into the sea than he should offend one of these little ones."

Christ did not overdraw the picture. Who is able to set a price upon the life of a child—a child into whom a mother poured her life and for whom a father has labored? What may a noble life mean to the child itself, to the parents and to the world?

And, it must be remembered that we can measure the effect on only that part of life which is spent on earth; we have no way of calculating the effect on that infinite circle of life which existence here is but a small arc. The soul is immortal and religion deals with the soul; the logical effect of the evolutionary hypothesis is to undermine religion and thus affect the soul. I recently received a list of questions that were to be discussed in a prominent eastern school for women. The second question in the list read: "Is religion an obsolescent function that should be allowed to atrophy quietly, without arousing the passionate prejudice of outworn superstitions?" The real attack of evolution, it will be seen, is not upon orthodox Christianity or even upon Christianity, but upon religion—the most basic fact in man's existence and the most practical thing in life.

James H. Leuba, a professor of psychology at Bryn Mawr college, Pennsylvania, published a few years ago a book entitled, "Belief in God and Immortality." In this book he relates how he secured the opinions of scientists as to the existence of a personal God and a personal immortality. He issued a volume entitled, "American Men of Science," which he says, included the names of "practically every American who may properly be called a scientist."

There are 5,500 names in the book. He selected 1,000 names as representative of the 5,500, and addressed them personally. Most of them, he said, were teachers in schools of higher learning. The names were kept confidential. Upon the answer received, he asserts that over half of them doubt or deny the existence of a personal God and a personal immortality, and he asserts that unbelief being greatest among the most prominent. Among biologists, believers in a personal God numbered less than 31 per cent while unbelievers in a personal immortality numbered only 37 per cent.

He also questioned the students in nine colleges of high rank and from 1,000 answers received, 97 per cent of which were from students between 18 and 20, he found that unbelief increased from 15 per cent in the Freshman class up to 40 to 45 per cent among the men who graduated. On page 280 of this book, we read "the students' statistics show that young people enter college, possessed of the beliefs still accepted, more or less perfunctorily, in the average home of the land, and gradually abandon the cardinal Christian beliefs." This change from belief to unbelief he attributed to the influence of the persons "of high culture under whom they studied."

The people of Tennessee have been patient enough; they acted none too soon. How can they expect to protect society, and even the church, from the deadening influence of agnosticism and atheism if they permit the teachers employed by taxation to poison the mind of the youth with this destructive doctrine? And remember, that the law has not heretofore required the writing of the word "poison" on poisonous doctrines. The bodies of our people are so valuable that the druggists and physicians must be careful to properly label all poisons; why not be as careful to protect the spiritual life of our people from the poisons that kill the soul?

There is a test that is sometimes used to ascertain whether one suspected of mental infirmity is really insane. He is put into a tank of water and told to dip the tank dry while a stream of water flows into the tank. If he has not sense enough to turn off the stream he is adjudged insane. Can parents justify themselves if, knowing the effect of belief in evolution, they permit irreligious teachers to inject skepticism and infidelity in the minds of their children?

Do bad doctrines corrupt the morals of students? We have a case in point. Mr. Darrow, one of the most distinguished criminal lawyers in our land, was engaged about a year ago in defending two rich men's sons who were on trial for as dastardly a murder as was ever committed. The older one, "Babe" Leopold, was a brilliant student, 19 years old. He was an evolutionist and an atheist. He was also a follower of Nietzsche, whose books he had devoured and whose philosophy he had adopted. Mr. Darrow made a plea for him, based upon the influence that Nietzsche's philosophy had exerted on the boy's mind. Here are extracts from his speech:

> Babe took to philosophy. . . . He grew up in this way; he became enamored of the philosophy of Nietzsche. Your honor, I have read almost everything that Nietzsche ever wrote. A man of wonderful intellect; the most original philosopher of the last century. A man who made a deeper imprint on philosophy than any other man within a hundred years. In a way he has reached more people, and still he has been a philosopher of what we might call the intellectual cult.
>
> He wrote one book called "Beyond Good and Evil," which was a criticism of all moral precepts, as we understood them, and a treatise that the intelligent was beyond good and evil; that the laws for good and the laws for evil did not apply to anybody who approached the superman. He wrote on the will to power.
>
> I have just made a few short extracts from Nietzsche that show the things that he (Leopold) has read, and these are short and almost taken at random. It is not how this would affect you. It is not how it would affect me. The question is, how it would affect the impressionable, visionary, dreamy mind of a boy—a boy who should never have seen it—too early for him.
>
> Quotations from Nietzsche: "Why so soft, oh my brethren? Oh why so soft, so unresisting and yielding? Why is there so much disavowal and abnegation, in your heart? Why is there so little faith in your looks? For all creators are hard and it must seem blessedness unto you to press your hand upon millenniums and upon wax. This new table, ah, my brethren, I put over you; become hard. To be obsessed by moral consideration presupposes a very low grade of intellect. We should substitute for morality the will to our own end and consequently to the means to accomplish that. A great man, a man whom nature has built up and invented in a grand style, is colder, harder, less cautious and more free from the fear of public opinion. He does not possess the virtues which are compatible with respectability with being respected, nor any of these things which are counted among the virtues of the herd.

Mr. Darrow says: That the superman, a creation of Nietzsche, has permeated every college and university in the civilized world.

> There is not any university in the world where the professor is not familiar with Nietzsche, not one. . . . Some believe it and some do not believe it. Some read it as I do and take it as a theory, a dream, a vision, mixed with good and bad but not in any way related to human life. Some take it seriously. . . . There is not a university in the world of any high standing where the professors do not tell you about Nietzsche and discuss him or where the books are not there.
>
> If this boy is to blame for this, where did he get it? Is there any blame attached because somebody took Nietzsche's philosophy seriously and fashioned his life upon it? And there is no question in this case but what that is true. Then who is to blame? The university would be more to blame than he is; the scholars of the world would be more to blame than he is. The purposes of the world . . . are more to blame than he is. Your honor, it is hardly fair to hang a 19 year-old boy for the philosophy that was taught him at the university. It does not meet my ideas of justice and fairness to visit upon his head the philosophy that has been taught by university men for 25 years.

In fairness to Mr. Darrow, I think I ought to quote two more paragraphs. After this bold attempt to excuse the student on the ground that he was transformed from a well-meaning youth into a murderer by the philosophy of an atheist, and on the further ground that his philosophy was in the libraries of all the colleges and discussed by the professors—some adopting the philosophy and some rejecting it—on these two grounds, he denied that the boy should be held responsible for the taking of human life. He charges that the scholars in the universities were more responsible than the boy, and that the universities were more responsible than the boy, because they furnished such books to the students, and then he proceeds to exonerate the universities and scholars, leaving nobody responsible. Here is Mr. Darrow's language:

> Now I do not want to be misunderstood about this. Even for the sake of saving, the lives of my clients, I do not want to be dishonest and tell the court something that I do not honestly think in this case. I do not think that the universities are to blame. I do not think they should be held responsible. I do think however, that they are too large and that they should keep a closer watch, if possible, upon, the individual.
>
> But you can not destroy thought, because forsooth, some brain may be deranged by thought. It is the duty of the university as I conceive it, to be the greatest storehouse of the wisdom of the ages, and to have its students come there and learn, choose. I have no doubt but that it has meant the death of many, but that we can not help.

This is a damnable philosophy, and yet it is the flower that blossoms on the stalk of evolution. Mr. Darrow thinks the universities are in duty bound to feed out this poisonous stuff to their students, and when the students become stupefied by it and commit murder, neither they nor the universities are to blame. I protest against the adoption of any such a philosophy in the state of Tennessee. A criminal is not relieved from responsibility merely because he found Nietzsche's philosophy in a library which ought not to contain it. Neither is the university guiltless if it permits such corrupting nourishment to be fed to the souls that are entrusted to its care. But, go a step farther, would the state be blameless if it permitted the universities under its control to be turned into training schools for murder? When you get back to the root of this question, you will find that the legislature not only had a right to protect the students from the evolutionary hypothesis, but was in duty bound to do so.

While on this subject, let me call your attention to another proposition embodied in Mr. Darrow's speech. He said that Dickey Loeb, the younger boy, had read trashy novels, of the blood and thunder sort. He even went so far as to commend an Illinois statute which forbids minors reading stories of crime. Here is what Mr. Darrow said:

> We have a statute in this state, passed only last year, if I recall it, which forbids minors reading stories of crime. Why? There is only one reason; because the legislature in its wisdom, thought it would have a tendency to produce these thoughts and this life in the boys who read them.

If Illinois can protect her boys, why can not this state protect the boys of Tennessee? Are the boys of Illinois any more precious than yours?

But to return to the philosophy of an evolutionist, Mr. Darrow said:

> I say to you seriously that the parents of Dickey Loeb are more responsible than he, and yet few boys had better parents.

Again he says:

> I know that one of two things happened to this boy: That this terrible crime was inherent in his organism and came from an ancestor or that it came through his education and his training after he was born.

He thinks the boy was not responsible for anything; his guilt was due, according to this philosophy, either to heredity or environment. But let me complete Mr. Darrow's philosophy based on evolution. He says:

> I do not know what remote ancestor may have sent down the seed that corrupted him, and I do not know through how many ancestors it may have passed until it reached Dickey Loeb. All I know, it is true, and there is not a biologist in the world who will not say I am right.

Psychologists who build upon the evolutionary hypothesis teach that man is nothing but a bundle of characteristics inherited from brute ancestors. This is the philosophy which Darrow applied to his celebrated criminal case. "Some remote ancestor"—he does not know how remote—"sent down the seed that corrupted him." You can not punish the ancestor—he is not only dead, but, according to the evolutionists, he was a brute and may have lived 1,000,000 years ago. And he says that all the biologists agree with him—no wonder so small a percent of the biologists, according to Leuba, believe in a personal God.

This is the quintessence of evolution, distilled for us by one who follows that doctrine to its logical conclusion. Analyze this dogma of darkness and death. Evolutionists say that back in the twilight of life a beast, name and nature unknown, planted a murderous seed and that the impulse that originated in that seed throbs forever in the blood of the brute's descendants, inspiring killings innumerable, for which murderers are not responsible because coerced by a fate fixed by the laws of heredity. It is an insult to reason and shocks the heart. That doctrine is as deadly as leprosy; it may aid a lawyer in a criminal case, but it would, if generally adopted, destroy all sense of responsibility and menace the morals of the world. A brute, they say, can predestine a man to crime, and yet they deny that God-incarnated flesh can release a human being from his bondage or save him from ancestral sins. No more repulsive doctrine was ever proclaimed by man; if all the biologists of the world teach this doctrine—as Darrow says they do—then may Heaven defend the youth of our land from their impious babblings.

We Must Not Forget God

Our third indictment against evolution is that it diverts attention from pressing problems of great importance to trifling speculation. While one evolutionist is trying to imagine what happened in the dim past, another is trying to pry open the door of the distant future. One recently grew eloquent over ancient worms, and another predicted that 75,000 years hence everyone will he bald and toothless. But those who endeavor to clothe our remote ancestors with hair and those who endeavor to remove the hair from the heads of our remote descendants ignore the present with its imperative demands. The science of "how to live" is the most important of all the sciences. It is desirable to know the physical sciences, but it is necessary to know how to live. Christians desire that their children shall be taught all the sciences, but they do not want them to lose sight of the rock of ages while they study the age of the rocks; neither do they desire them to become so absorbed in measuring the distance between the stars that they will forget Him who holds the stars in his hand.

While not more than two per cent of our population are college graduates, these, because of enlarged powers, need a "heavenly vision," even more than those less learned, both for their own restraint and to assure society that their enlarged powers will be used for the benefit of society and not against the public welfare.

Evolution is deadening to spiritual life of a multitude of students. Christians do not desire less education, but they desire that religion shall be entwined with learning so that our boys and girls will return from college with their hearts aflame with love of God and love of fellow men, and prepared to lead in the altruistic work that the world so sorely needs. The cry in the business world, in the industrial, even in the religious world—is for consecrated talents—for ability plus a passion for service.

Our fourth indictment against the evolutionary hypothesis is that, by paralyzing the hope of reform, it discourages those who labor for the improvement of man's condition. Every upward-looking man or woman seeks to lift the level upon which mankind stands, and they trust that they will see beneficent changes during the brief span of their own lives. Evolution chills their enthusiasm by substituting aeons for years. It obscures all beginnings in the midst of endless ages. It is represented as a cold and heartless process, beginning with time and ending in eternity, and acting so slowly that even the rocks can not preserve a record of the imaginary changes through which it is credited with having carried an original germ of life that appeared sometime from somewhere. Its only program for man is scientific breeding, a system under which a few supposedly superior intellects, self-appointed, would direct the mating and the movements of the mass of mankind—an impossible system. Evolution, disputing the miracle and ignoring the spiritual in life, has no place for the regeneration of the individual. It recognizes no cry of repentance and scoffs at the doctrine that one can be born?

It is thus the tolerant and unrelenting enemy of the only process that can redeem society through the redemption of the individual. An evolutionist would never write such a story as the Prodigal Son; it contradicts the whole theory of evolution. The two sons inherited in the same parents, and through their parents from the same ancestors, proximate and remote. And these sons were reared at the same fireside and were surrounded by the same environment during all the days of their youth; and yet they were different.

If Mr. Darrow is correct in the theory applied to Loeb, namely, that his crime was due either to inheritance or to environment, how will he explain the difference between the elder brother and the wayward son? The evolutionist may understand from observation, if not by experience, even though he can not explain why one of these boys was guilty of every immorality, squandered the money that the father had laboriously earned, and brought disgrace upon the family name; but his theory does not explain why a wicked man underwent a change of heart, confessed his sins and begged forgiveness, and because the evolutionist can not understand this fact, one of the most important in the human life, he can not understand the infinite love of the Heavenly Father who stands ready to welcome home any repentant sinner, no matter how far he has wandered, how often he has failed, or how deep he has sunk in sin.

Your honor has quoted from a wonderful poem written by a great Tennessee poet, Walter Malone. I venture to quote another stanza which puts into exquisite language the new opportunity which a merciful God gives everyone who will turn from sin to righteousness:

Tho' deep in mire wring not your hands and weep,
I lend my arm to all who say "I can."
No shamefaced outcast ever sank so deep,
But he might rise and be a man.

There are no lines like these in all that evolutionists have ever written. Darwin says that science has nothing to do with the Christ who taught the spirit embodied in the words of Walter Malone, and yet this spirit is the only hope of human progress. A heart can be changed in the twinkling of an eye, and, a change in the life follows a change in the heart. If one heart can be changed, then a world can be born in a day. It is the fact that inspires all who labor for man's betterment. It is because Christians believe in individual regeneration and in the regeneration of society through the regeneration of individuals that they pray: "Thy Kingdom come, Thy will be done in earth as it is in heaven." Evolution makes a mockery of the Lord's prayer!

To interpret the words to mean that the improvement desired must come slowly through unfolding ages—a process with which each generation could have little to do—is to defer hope, and hope deferred makes the heart sick.

Evolution Demoralizing & Deadly

Our fifth indictment of the evolutionary hypothesis is that if taken seriously and made the basis of a philosophy of life, it would eliminate love and carry man back to a struggle of tooth and claw. The Christians who have allowed themselves to be deceived into believing that evolution is a beneficent, or even a rational, process have been associating with those who either do not understand its application or dare not avow their knowledge of these implicators. Let me give you some authority on this subject. I will begin with Darwin, the high priest of evolution, to whom all evolutionists bow.

On pages 149 and 150, in "The Descent of Man," already referred, he says:

> With savages, the weak in body or mind are soon eliminated and those that survive commonly exhibit a vigorous state of health. We civilized men, on the other hand, do our utmost to check the process of elimination, we build asylums for the imbecile, the maimed and the sick; we institute poor laws; and our medical men exert their utmost skill to save the life of every one to the last moment. There is reason to believe that vaccination has preserved thousands who from a weak constitution would formerly have succumbed to smallpox. Thus the weak members of civilized society propagate their kind. No one who has attended to the breeding of domestic animals will doubt that this must be highly injurious to the race of man. It is surprising how soon a want of care, or care wrongly directed, leads to the degeneration of a domestic race, but, excepting in the case of man himself, hardly a one is so ignorant as to allow his worst animals to breed.
>
> The aid which we feel impelled to give to the helpless is mainly an incidental result of the instinct of sympathy, which was originally acquired as part of the social instincts, but subsequently rendered in the manner previously indicated, more tender and more widely diffused. Nor could we check our sympathy, even at the urging of hard reason, without deterioration in the noblest part of nature. . . . We must therefore, bear the undoubtedly bad effects of the weak surviving and propagating their kind.

Darwin reveals the barbarous sentiment that runs through evolution and dwarfs the moral nature of those who become obsessed with it. Let us analyze the quotation just given. Darwin speaks with approval of the savage custom of eliminating the weak so that only the strong will survive and complains that "we civilized men do our utmost to check the process of elimination." How inhuman such a doctrine as this! He thinks it injurious to "build asylums for the imbecile, the maimed and the sick." Or to care for the poor. Even the medical men come in for criticism because they "exert their utmost skill to save the life of everyone to the last moment," and then note his hostility to vaccination because it has "preserved thousands who, from a weak constitution, would, but for vaccination, have succumbed to smallpox." All of the sympathetic activities of civilized society are condemned because they enable "the weak members to propagate their kind." Then he drags mankind down to the level of the brute and compares the freedom given to man unfavorably with the restraint that we put on barnyard beasts.

The second paragraph of the above quotation shows that his kindly heart rebelled against the cruelty of his doctrine. He says that we "feel impelled to give to the helpless," although he traces it to a sympathy which he thinks is developed by evolution; he even admits that we could not check this sympathy "even at the urging of hard reason, without deterioration of the noblest part of our nature." "We therefore bear," what he regards as "the undoubtedly bad effect of the weak surviving and propagating their kind." Could any doctrine be more destructive of civilization? And what a commentary on evolution! He wants us to believe that evolution develops a human sympathy that finally becomes so tender that it repudiates the law that created it and thus invites a return to a level where the extinguishing of pity and sympathy will permit the brutal instincts to again do their progressive work.

Let no one think that this acceptance of barbarism, as the basic principle of evolution, died with Darwin. Within three years a book has appeared whose author is even more frankly brutal than Darwin. The book is entitled "The New Decalogue of Science," and has attracted wide attention.

One of our most reputable magazines has recently printed an article by him defining the religion of a scientist. In his preface he acknowledges indebtedness to 21 prominent scientists and educators, "nearly all of them doctors" and "professors."

One of them who has recently been elevated to the head of a great state university read the manuscript over twice and made many valuable suggestions. The author describes Nietzsche, who, according to Mr. Darrow, made a murderer out of Babe Leopold, as the bravest soul since Jesus.

He admits Nietzsche was "gloriously wrong," but he affirms that Nietzsche was "gloriously right in his fearless questioning of the universe and of his own soul."

In another place the author says:

> Most of our morals today are jungle products.

And then he affirms that:

> It would be safer, biologically, if they were more so.

Now, after these two samples of his views, you will not be surprised when I read you the following:

> Evolution is a bloody business, but civilization tries to make it a pink tea. Barbarism is the only process by which man has ever organically progressed and civilization is the only process by which he has ever organically declined.
>
> Civilization is the most dangerous enterprise on which man ever set out. For when you take man out of the bloody, brutal, but beneficent hand of natural selection you place him at once in the soft, daintily gloved, but far more dangerous hand of artificial selection.
>
> And unless you call science to your assistance and make this artificial selection as efficient as the rude methods of nature, you bungle the whole task.

This aspect of evolution may amaze some of the ministers who have not been permitted to enter the inner circle of the iconoclasts whose theories menace all the ideals of civilized society. Do these ministers know that evolution is a "bloody business"? Do they know that barbarism is the only process by which man has ever organically progressed, and "that civilization is the only process by which he has ever organically declined"?

Do they know that the bloody, brutal hand of natural selection is beneficent and the artificial selection "found in civilization is dangerous"? What shall we think of the distinguished educators and scientists who read the manuscript before publication and did not protest against this pagan doctrine?

To show that this is a worldwide matter, I now quote from a book issued from the press in 1918, seven years ago. The title of the book is "The Science of the Power," and its author, Benjamin Kidd, being an Englishman, could not have any national prejudice against Darwin. On pages 46 and 47 we find Kidd's interpretation of evolution:

> Darwin's presentation of the evolution of the world as the product of natural selection in never-ceasing war, as a product that is to say, of a struggle in which the individual efficient in the fight for his own interests was always the winning type—touched the profoundest depths of the psychology of the west.
>
> The idea seemed to present the whole order of progress in the world as the result of a purely mechanical and materialistic process resting on force. In so doing it was a conception which reached the springs of that heredity born of the unmeasured ages of conquest out of which the western mind has come. Within half a century the "Origin of Species" had become the Bible of the doctrine of the omnipotence of force.

Kidd goes so far as to charge that "Nietzsche recited the interpretation of the popular Darwinism, delivered with the fury and intensity of genius." And yet Nietzsche denounced Christianity as the "doctrine of the degenerate," and mercy as "the refuge of weaklings."

Kidd says that Nietzsche gave Germany the doctrine of Darwin's efficient animal in the voice of his sermon, and that Bernhardi and the military textbooks in due time gave Germany the doctrine of the superman translated into the national policy of the superstate aiming at world power.

And what else but the spirit of evolution can account for the popularity of the selfish doctrine, "each one for himself, and the devil take the hindmost," that threatens the very existence of the doctrine of brotherhood?"

In 1900—25 years ago, while an international peace congress was in session at Paris, the following editorial appeared in L'Univers:

> The spirit of peace has fled the earth because evolution has taken possession of it. The plea for peace in past years has been inspired by faith in the divine nature and the divine origin of man; men were then looked upon as children of one father and war therefore was fratricide. But now that men are looked upon as children of apes, what matters it whether they were slaughtered or not?

When there is poison in the blood, no one knows on what part of the body it will break out, but we can be sure that it will break out unless the blood is purified.

One of the leading universities of the south (I love the state too well to mention its name) publishes a monthly magazine entitled, "Journal of Social Forces." In the January issue of this year a contributor has a lengthy article on "Zoology and Ethics," in the course of which he says:

> No attempt will be made to take up the matter of the good or evil of sexual intercourse among humans aside from the matter of conscious procreation, but as an historian it might be worth while to ask the exponents of the impurity complex to explain the fact that without exception the great herds of cultural afflorescence have been those characterized by a large amount of freedom in sex relations and that those of the greatest cultural degradation and decline have been accompanied with greater sex repression and purity.

No one charges or suspects that all or any large percentage of the advocates of evolution sympathize with this loathsome application of evolution to social life, but it is worth while to inquire why those in charge of a great institution of learning allow such filth to be poured out for the stirring of the passions of its students.

Just one more quotation: "The Southeastern Christian Advocate" of June 25, 1925, quotes five eminent college men of Great Britain as joining in answer to the question: "Will civilization survive?"

Their reply is that "Greatest danger to our civilization is the abuse of the achievements of science. Mastery over the forces of nature has endowed the twentieth century man with a power which he is not fit to exercise. Unless the development of morality catches up with the development of technique, humanity is bound to destroy itself."

Can any Christian remain indifferent? Science needs religion to direct its energies and to inspire with lofty purpose those who employ the forces that are unloosed by science. Evolution is at war with religion because religion is supernatural, it is therefore the relentless foe of Christianity which is a revealed religion.

Let us, then, hear the conclusion of the whole matter. Science is a magnificent material for force, but is not a teacher of morals. It is perfect machinery, but it adds no moral restraints to protect society from the misuse of the machine. It can also build gigantic intellectual ships, but it constructs no moral rudders for control of storm tossed human vessels.

It not only fails to supply the spiritual element needed, but some of its unproven hypotheses rob the ship of its compass and thus endanger its cargo.

In war, science has proven itself an evil genius, it has made war more terrible than it ever was before. Man used to be content to slaughter his fellowman on a single plane—the earth's surface.

Science has taught him to go down into the water and shoot up; to go up into the clouds and shoot down from above, thus making the battlefield three times as bloody as it was before; but science does not teach brotherly love.

Science has made war so hellish that civilization has but to commit suicide; and now we are told that newly discovered instruments of destruction will make the cruelties of the late war seem trivial in comparison with the cruelties of war that may come in the future.

If civilization is to be saved from the wreckage threatened by intelligence not consecrated by love, it must be saved by the moral code of the meek and lowly Nazarene. His teachings and His teachings alone can solve the problems that vex the heart and perplex the world.

The world needs a saviour more than it ever did, and His is the only name under Heaven whereby we must be saved. It is this name that evolution degrades, for carried to its logical conclusion, it robs Christ of the glory of a virgin birth, of the majesty of His deity and mission, and of the triumph of His resurrection. It also disputes the doctrine of the atonement.

It is for the jury to determine whether this attack upon the Christian religion shall be permitted in the public schools of Tennessee by teachers employed by the state and paid out of the public treasury.

This case is no longer local; the defendant ceases to play an important part.

The case assumes the proportions of a battle royal between unbelief that attempts to speak through so called science and the defenders of the Christian faith, speaking through the legislators of Tennessee.

It is again a choice between God and Baal.

It is a renewal of the issue in Pilate's court. In that historic trial—the greatest in history—force, impersonated by Pilate, occupied the throne.

Behind it was the Roman government, mistress of the world, and behind the Roman government were the legions of Rome.

Before Pilate stood Christ, the apostle of love.

Force triumphed, they nailed Him to the tree and those who stood around mocked and jeered and said, "Christ is dead." But from that day the power of Caesar waned and the power of Christ increased.

In a few centuries the Roman government was gone and its legions forgotten; while the crucified and risen Lord is the greatest fact in history and the growing figure of all time.

Again love and force meet face to face and again, "What Shall I Do With Jesus" must be answered. A bloody doctrine, evolution demands, as the rabble did 1,900 years ago, that He be crucified.

This can not be the answer of the jury representing a Christian state and sworn to uphold the laws of Tennessee. Your answer will be heard throughout the world; it is eagerly awaited by a praying multitude.

If the law is nullified there will be rejoicing where ever God is repudiated, the Saviour scoffed at and the Bible ridiculed. Every unbeliever of every kind and degree will be happy.

If, on the other hand, the law is upheld and the religion of the school children protected, millions of Christians will call you blessed, and with hearts full of gratitude to God will again sing that old song of triumph:

Faith of our fathers, living still,
In spite of dungeon, fire and sword,
Oh, how our hearts beat high with joy,
Whene'er we hear that glorious word:–
Faith of our fathers–holy faith,
We will be true to thee till death.

David Hume's "Of Miracles"

An Enquiry Concerning Human Understanding, 1758

Introduction

Michael Shermer

The importance of skeptical publications in this New Age resurgence of interest in miracles and various claims of the paranormal cannot be overstated. Yet it is equally important to remember our historical antecedents and how they analyzed and critiqued such claims in their own time. One of the greatest skeptics of the Modern Age is the Scottish philosopher David Hume (1711–1776), whose work, *An Enquiry Concerning Human Understanding*, is a classic in skeptical analysis. The book was originally published anonymously in London in 1739, as *A Treatise of Human Nature*, but, in Hume's words, "fell dead-born from the press, without reaching such distinction as even to excite a murmur among the zealots." (An author's biggest fear is not being panned; it is being ignored.)

Hume blamed his own writing style and reworked the manuscript into *An Abstract of a Treatise of Human Nature* in 1740, and again in 1748, as *Philosophical Essays Concerning the Human Understanding*. The work still gained Hume no recognition, so in 1758 he brought it out in a final version as *An Enquiry Concerning Human Understanding*, which comes down to us today as his greatest philosophical work. Ironically, when Hume finally did achieve fame and position, his critics often attacked his earlier works, a practice Hume found "very contrary to all rules of candour and fair-dealing, and a strong instance of those polemical artifices, which a bigotted zeal thinks itself authorized to employ," as he wrote in an "Advertisement" to the final publication!

In Section XII, "Of the Academical or Sceptical Philosophy," Hume distinguished between "antecedent skepticism," such as Descartes's method of doubting everything, that has no "antecedent" infallible criterion for belief; and "consequent skepticism," the method Hume employed that recognizes the "consequences" of our fallible senses, but corrects them through reason: "A wise man proportions his belief to the evidence." Wiser words could not be chosen for a skeptical motto.

For the modern skeptic, Hume's Section X, "Of Miracles," provides a generalized, when-all-else-fails analysis of miraculous claims. That is, when one is confronted by a true believer whose apparently supernatural or paranormal claim has no immediately apparent natural explanation, Hume gives us an argument that even he thought was so important (and Hume was not a modest man) that he placed his own words in quotes and called it a maxim. I think it is so useful an argument that it bears repetition, as *Hume's Maxim:*

> The plain consequence is (and it is a general maxim worthy of our attention), "That no testimony is sufficient to establish a miracle, un-

less the testimony be of such a kind, that its falsehood would be more miraculous than the fact which it endeavours to establish."

When anyone tells me that he saw a dead man restored to life, I immediately consider with myself whether it be more probable, that this person should either deceive or be deceived, or that the fact, which he relates, should really have happened. I weigh the one miracle against the other; and according to the superiority, which I discover, I pronounce my decision, and always reject the greater miracle. If the falsehood of his testimony would be more miraculous than the event which he relates; then, and not till then, can he pretend to command my belief or opinion.

So to honor *Hume's Maxim,* and to give the reader the full context of Hume's analysis, we present below the entirety of Section X "Of Miracles."

Section X. Of Miracles.

David Hume

Part I

THERE is, in Dr. Tillotson's writings, an argument against the *real presence,* which is as concise, and elegant, and strong as any argument can possibly be supposed against a doctrine, so little worthy of a serious refutation. It is acknowledged on all hands, says that learned prelate, that the authority, either of the scripture or of tradition, is founded merely in the testimony of the apostles, who were eyewitnesses to those miracles of our Saviour, by which he proved his divine mission. Our evidence, then, for the truth of the *Christian* religion is less than the evidence for the truth of our senses; because, even in the first authors of our religion, it was no greater; and it is evident it must diminish in passing from them to their disciples; nor can any one rest such confidence in their testimony, as in the immediate object of his senses. But a weaker evidence can never destroy a stronger; and therefore, were the doctrine of the real presence ever so clearly revealed in scripture, it were directly contrary to the rules of just reasoning to give our assent to it. It contradicts sense, though both the scripture and tradition, on which it is supposed to be built, carry not such evidence with them as sense; when they are considered merely as external evidences, and are not brought home to every one's breast, by the immediate operation of the Holy Spirit.

Nothing is so convenient as a decisive argument of this kind, which must at least *silence* the most arrogant bigotry and superstition, and free us from their impertinent solicitations. I flatter myself, that I have discovered an argument of a like nature, which, if just, will, with the wise and learned, be an everlasting check to all kinds of superstitious delusion, and consequently, will be useful as long as the world endures. For so long, I presume, will the accounts of miracles and prodigies be found in all history, sacred and profane.

Though experience be our only guide in reasoning concerning matters of fact; it must be acknowledged, that this guide is not altogether infallible, but in some cases is apt to lead us into errors. One, who in our climate, should expect better weather in any week of June than in one of December, would reason justly, and conformably to experience; but it is certain, that he may happen, in the event, to find himself mistaken. However, we may observe, that, in such a case, he would have no cause to complain of experience; because it commonly informs us beforehand of the uncertainty, by that contrariety of events, which we may learn from a diligent observation. All effects follow not with like certainty from their supposed causes. Some events are found, in all countries and all ages, to have been constantly conjoined together: Others are found to have been more variable, and sometimes to disappoint our expectations; so that, in our reasonings concerning matter of fact, there are all imaginable degrees of assurance, from the highest certainty to the lowest species of moral evidence.

A wise man, therefore, proportions his belief to the evidence. In such conclusions as are founded on an infallible experience, he expects the event with the last degree of assurance, and regards his past experience as a full *proof* of the future existence of that event. In other cases, he proceeds with more caution: He weighs the opposite experiments: He considers which side is supported by the greater number of experiments: to that side he inclines, with doubt and hesitation; and when at last he fixes his judgement, the evidence exceeds not what we properly call *probability*. All probability, then, supposes an opposition of experiments and observations, where the one side is found to overbalance the other, and to produce a degree of evidence, proportioned to the superiority. A hundred instances or experiments on one side, and fifty on another, afford a doubtful expectation of any event; though a hundred uniform experiments, with only one that is contradictory, reasonably beget a pretty strong degree of assurance. In all cases, we must balance the opposite experiments, where they are opposite, and deduct the smaller number from the greater, in order to know the exact force of the superior evidence.

To apply these principles to a particular instance; we may observe, that there is no species of reasoning more common, more useful, and even necessary to human life, than that which is derived from the testimony of men, and the reports of eye-witnesses and spectators. This species of reasoning, perhaps, one may deny to be founded on the relation of cause and effect. I shall not dispute about a word. It will be sufficient to observe that our assurance in any argument of this kind is derived from no other principle than our observation of the veracity of human testimony, and of the usual conformity of facts to the reports of witnesses. It being a general maxim, that no objects have any discoverable connexion together, and that all the inferences, which we can draw from one to another, are founded merely on our experience of their constant and regular conjunction; it is evident, that we ought not to make an exception to this maxim in favour of human testimony, whose connexion with any event seems, in itself, as little necessary as any other.

Were not the memory tenacious to a certain degree; had not men commonly an inclination to truth and a principle of probity; were they not sensible to shame, when detected in a falsehood: Were not these, I say, discovered by *experience* to be qualities, inherent in human nature, we should never repose the least confidence in human testimony. A man delirious, or noted for falsehood and villany, has no manner of authority with us.

And as the evidence, derived from witnesses and human testimony, is founded on past experience, so it varies with the experience, and is regarded either as a *proof* or a *probability*, according as the conjunction between any particular kind of report and any kind of object has been found to be constant or variable. There are a number of circumstances to be taken into consideration in all judgements of this kind; and the ultimate standard, by which we determine all disputes, that may arise concerning them, is always derived from experience and observation. Where this experience is not entirely uniform on any side, it is attended with an unavoidable contrariety in our judgements, and with the same opposition and mutual destruction of argument as in every other kind of evidence. We frequently hesitate concerning the reports of others. We balance the opposite circumstances, which cause any doubt or uncertainty; and when we discover a superiority on any side, we incline to it; but still with a diminution of assurance, in proportion to the force of its antagonist.

This contrariety of evidence, in the present case, may be derived from several different causes; from the opposition of contrary testimony; from the character or number of the witnesses; from the manner of their delivering their testimony; or from the union of all these circumstances. We entertain a suspicion concerning any matter of fact, when the witnesses contradict each other; when they are but few, or of a doubtful character; when they have an interest in what they affirm; when they deliver their testimony with hesitation, or on the contrary, with too violent asseverations. There are many other particulars of the same kind, which may diminish or destroy the force of any argument, derived from human testimony. Suppose, for instance, that the fact, which the testimony endeavours to establish, partakes of the extraordinary and the marvellous; in that case, the evidence, resulting from the testimony, admits of a diminution, greater or less, in proportion as the fact is more or less unusual. The reason why we place any credit in witnesses and historians, is not derived from any *connexion,* which we perceive *a priori,* between testimony and reality, but because we are accustomed to find a conformity between them. But when the fact attested is such a one as has seldom fallen under our observation, here is a contest of two opposite experiences; of which the one destroys the other, as far as its force goes, and the superior can only operate on the mind by the force, which remains. The very same principle of experience, which gives us a certain degree of assurance in the testimony of witnesses, gives us also, in this case, another degree of assurance against the fact, which they endeavour to establish; from which contradiction there necessarily arises a counterpoize, and mutual destruction of belief and authority.

I should not believe such a story were it told me by Cato, was a proverbial saying in Rome, even during the lifetime of that philosophical patriot. The incredibility of a fact, it was allowed, might invalidate so great an authority. The Indian prince, who refused to believe the first relations concerning the effects of frost, reasoned justly; and it naturally required very strong testimony to engage his assent to facts, that arose from a state of nature, with which he was unacquainted, and which bore so little analogy to those events, of which he had had constant and uniform experience. Though they were not contrary to his experience, they were not conformable to it.

But in order to encrease the probability against the testimony of witnesses, let us suppose, that the fact, which they affirm, instead of being only marvellous, is really miraculous; and suppose also, that the testimony considered apart and in itself, amounts to an entire proof; in that case, there is proof against proof, of which the strongest must prevail, but still with a diminution of its force, in proportion to that of its antagonist.

A miracle is a violation of the laws of nature; and as a firm and unalterable experience has established these laws, the proof against a miracle, from the very nature of the fact, is as entire as any argument from experience can possibly be imagined. Why is it more than probable, that all men must die; that lead cannot, of itself, remain suspended in the air; that fire consumes wood, and is extinguished by water; unless it be, that these events are found agreeable to the laws of nature, and there is required a violation of these laws, or in other words, a miracle to prevent them? Nothing is esteemed a miracle, if it ever happen in the common course of nature. It is no miracle that a man, seemingly in good health, should die on a sudden: because such a kind of death, though more unusual than any other, has yet been frequently observed to happen. But it is a miracle, that a dead man should come to life; because that has never been observed in any age or country. There must, therefore, be a uniform experience against every miraculous event, otherwise the event would not merit that appellation. And as a uniform experience amounts to a proof, there is here a direct and full *proof*, from the nature of the fact, against the existence of any miracle; nor can such a proof be destroyed, or the miracle rendered credible, but by an opposite proof, which is superior.

The plain consequence is (and it is a general maxim worthy of our attention), 'That no testimony is sufficient to establish a miracle, unless the testimony be of such a kind, that its falsehood would be more miraculous, than the fact, which it endeavours to establish; and even in that case there is a mutual destruction of arguments, and the superior only gives us an assurance suitable to that degree of force, which remains, after deducting the inferior.' When anyone tells me, that he saw a dead man restored to life, I immediately consider with myself, whether it be more probable, that this person should either deceive or be deceived, or that the fact, which he relates, should really have happened. I weigh the one miracle against the other; and according to the superiority, which I discover, I pronounce my decision, and always reject the greater miracle. If the falsehood of his testimony would be more miraculous, than the event which he relates; then, and not till then, can he pretend to command my belief or opinion.

Part II

In the foregoing reasoning we have supposed, that the testimony, upon which a miracle is founded, may possibly amount to an entire proof, and that the falsehood of that testimony would be a real prodigy: But it is easy to shew, that we have been a great deal too liberal in our concession, and that there never was a miraculous event established on so full an evidence. For *first*, there is not to be found, in all history, any miracle attested by a sufficient number of men, of such unquestioned good-sense, education, and learning, as to secure us against all delusion in themselves; of such undoubted integrity, as to place them beyond all suspicion of any design to deceive others; of such credit and reputation in the eyes of mankind, as to have a great deal to lose in case of their being detected in any falsehood; and at the same time, attesting facts performed in such a public manner and in so celebrated a part of the world, as to render the detection unavoidable: All which circumstances are requisite to give us a full assurance in the testimony of men.

Secondly. We may observe in human nature a principle which, if strictly examined, will be found to diminish extremely the assurance, which we might, from human testimony, have, in any kind of prodigy. The maxim, by which we commonly conduct ourselves in our reasonings, is, that the objects, of which we have no experience, resemble those, of which we have; that what we have found to be most usual is always most probable; and that where there is an opposition of arguments, we ought to give the preference to such as are founded on the greatest number of past observations. But though, in proceeding by this rule, we readily reject any fact which is unusual and incredible in an ordinary degree; yet in advancing farther, the mind observes not always the same rule; but when anything is affirmed utterly absurd and miraculous, it rather the more readily admits of such a fact, upon account of that very circumstance, which ought to destroy all its authority. The passion of *surprise* and *wonder*, arising from miracles, being an agreeable emotion, gives a sensible tendency towards the belief of those events, from which it is derived. And this goes so far, that even those who cannot enjoy this pleasure immediately, nor can believe those miraculous events, of which they are informed, yet love to partake of the satisfaction at second-hand or by rebound, and place a pride and delight in exciting the admiration of others.

With what greediness are the miraculous accounts of travellers received, their descriptions of sea and land monsters, their relations of wonderful adventures, strange men, and uncouth manners? But if the spirit of religion join itself to the love of wonder, there is an end of common sense; and human testimony, in these circumstances, loses all pretensions to authority. A religionist may be an enthusiast, and imagine he sees what has no reality: he may know his narrative to be false, and yet persevere in it, with the best intentions in the world, for the sake of promoting so holy a cause: or even where this delusion has not place, vanity, excited by so strong a temptation, operates on him more powerfully than on the rest of mankind in any other circumstances; and self-interest with equal force. His auditors may not have, and commonly have not, sufficient judgement to canvass his evidence: what judgement they have, they renounce by principle, in these sublime and mysterious subjects: or if they were ever so willing to employ it, passion and a heated imagination disturb the regularity of its operations. Their credulity increases his impudence: and his impudence overpowers their credulity.

Eloquence, when at its highest pitch, leaves little room for reason or reflection; but addressing itself entirely to the fancy or the affections, captivates the willing hearers, and subdues their understanding. Happily, this pitch it seldom attains. But what a Tully or a Demosthenes could scarcely effect over a Roman or Athenian audience, every *Capuchin*, every itinerant or stationary teacher can perform over the generality of mankind, and in a higher degree, by touching such gross and vulgar passions.

The many instances of forged miracles, and prophecies, and supernatural events, which, in all ages, have either been detected by contrary evidence, or which detect themselves by their absurdity, prove sufficiently the strong propensity of mankind to the extraordinary and the marvellous, and ought reasonably to beget a suspicion against all relations of this kind. This is our natural way of thinking, even with regard to the most common and most credible events. For instance: There is no kind of report which rises so easily, and spreads so quickly, especially in country places and provincial towns, as those concerning marriages; insomuch that two young persons of equal condition never see each other twice, but the whole neighbourhood immediately join them together. The pleasure of telling a piece of news so interesting, of propagating it, and of being the first reporters of it, spreads the intelligence. And this is so well known, that no man of sense gives attention to these reports, till he find them confirmed by some greater evidence. Do not the same passions, and others still stronger, incline the generality of mankind to believe and report, with the greatest vehemence and assurance, all religious miracles?

Thirdly. It forms a strong presumption against all supernatural and miraculous relations, that they are observed chiefly to abound among ignorant and barbarous nations; or if a civilized people has ever given admission to any of them, that people will be found to have received them from ignorant and barbarous ancestors, who transmitted them with that inviolable sanction and authority, which always attend received opinions. When we peruse the first histories of all nations, we are apt to imagine ourselves transported into some new world; where the whole frame of nature is disjointed, and every element performs its operations in a different manner, from what it does at present. Battles, revolutions, pestilence, famine and death, are never the effect of those natural causes, which we experience. Prodigies, omens, oracles, judgements, quite obscure the few natural events, that are intermingled with them.

But as the former grow thinner every page, in proportion as we advance nearer the enlightened ages, we soon learn, that there is nothing mysterious or supernatural in the case, but that all proceeds from the usual propensity of mankind towards the marvellous, and that, though this inclination may at intervals receive a check from sense and learning, it can never be thoroughly extirpated from human nature.

It is strange, a judicious reader is apt to say, upon the perusal of these wonderful historians, *that such prodigious events never happen in our days.* But it is nothing strange, I hope, that men should lie in all ages. You must surely have seen instances enough of that frailty. You have yourself heard many such marvellous relations started, which, being treated with scorn by all the wise and judicious, have at last been abandoned even by the vulgar. Be assured, that those renowned lies, which have spread and flourished to such a monstrous height, arose from like beginnings; but being sown in a more proper soil, shot up at last into prodigies almost equal to those which they relate.

It was a wise policy in that false prophet, Alexander, who though now forgotten, was once so famous, to lay the first scene of his impostures in Paphlagonia, where, as Lucian tells us, the people were extremely ignorant and stupid, and ready to swallow even the grossest delusion. People at a distance, who are weak enough to think the matter at all worth enquiry, have no opportunity of receiving better information. The stories come magnified to them by a hundred circumstances. Fools are industrious in propagating the imposture; while the wise and learned are contented, in general, to deride its absurdity, without informing themselves of the particular facts, by which it may be distinctly refuted. And thus the impostor above mentioned was enabled to proceed, from his ignorant Paphlagonians, to the enlisting of votaries, even among the Grecian philosophers, and men of the most eminent rank and distinction in Rome: nay, could engage the attention of that sage emperor Marcus Aurelius; so far as to make him trust the success of a military expedition to his delusive prophecies.

The advantages are so great, of starting an imposture among an ignorant people, that, even though the delusion should be too gross to impose on the generality of them (*which, though seldom, is sometimes the case*) it has a much better chance for succeeding in remote countries, than if the first scene had been laid in a city renowned for arts and knowledge. The most ignorant and barbarous of these barbarians carry the report abroad. None of their countrymen have a large correspondence, or sufficient credit and authority to contradict and beat down the delusion. Men's inclination to the marvellous has full opportunity to display itself. And thus a story, which is universally exploded in the place where it was first started, shall pass for certain at a thousand miles distance. But had Alexander fixed his residence at Athens, the philosophers of that renowned mart of learning had immediately spread, throughout the whole Roman empire, their sense of the matter; which, being supported by so great authority, and displayed by all the force of reason and eloquence, had entirely opened the eyes of mankind. It is true; Lucian, passing by chance through Paphlagonia, had an opportunity of performing this good office. But, though much to be wished, it does not always happen, that every Alexander meets with a Lucian, ready to expose and detect his impostures.

I may add as a *fourth* reason, which diminishes the authority of prodigies, that there is no testimony for any, even those which have not been expressly detected, that is not opposed by an infinite number of witnesses; so that not only the miracle destroys the credit of testimony, but the testimony destroys itself. To make this the better understood, let us consider, that, in matters of religion, whatever is different is contrary; and that it is impossible the religions of ancient Rome, of Turkey, of Siam, and of China should, all of them, be established on any solid foundation. Every miracle, therefore, pretended to have been wrought in any of these religions (and all of them abound in miracles), as its direct scope is to establish the particular system to which it is attributed; so has it the same force, though more indirectly, to overthrow every other system. In destroying a rival system, it likewise destroys the credit of those miracles, on which that system was established so that all the prodigies of different religions are to be regarded as contrary facts, and the evidences of these prodigies, whether weak or strong, as opposite to each other. According to this method of reasoning, when we believe any miracle of Mahomet or his successors, we have for our warrant the testimony of a few barbarous Arabians: And on the other hand, we are to regard the authority of Titus Livius, Plutarch, Tacitus, and, in short, of all the authors and witnesses, Grecian, Chinese, and Roman Catholic, who have related any miracle in their particular religion; I say, we are to regard their testimony in the same light as if they had mentioned that Mahometan miracle, and had in express terms contradicted it, with the same certainty as they have for the miracle they relate. This argument may appear over subtile and refined; but is not in reality different from the reasoning of a judge, who supposes, that the credit of two witnesses, maintaining a crime against any one, is destroyed by the testimony of two others, who affirm him to have been two hundred leagues distant, at the same instant when the crime is said to have been committed.

One of the best attested miracles in all profane history, is that which Tacitus reports of Vespasian, who cured a blind man in Alexandria, by means of his spittle, and a lame man by the mere touch of his foot; in obedience to a vision of the god Serapis, who had enjoined them to have recourse to the Emperor, for these miraculous cures. The story may be seen in that fine historian; where every circumstance seems to add weight to the testimony, and might be displayed at large with all the force of argument and eloquence, if any one were now concerned to enforce the evidence of that exploded and idolatrous superstition. The gravity, solidity, age, and probity of so great an emperor, who, through the whole course of his life, conversed in a familiar manner with his friends and courtiers, and never affected those extraordinary airs of divinity assumed by Alexander and Demetrius. The historian, a cotemporary writer, noted for candour and veracity, and withal, the greatest and most penetrating genius, perhaps, of all antiquity; and so free from any tendency to credulity, that he even lies under the contrary imputation, of atheism and profaneness: The persons, from whose authority he related the miracle, of established character for judgement and veracity, as we may well presume; eye-witnesses of the fact, and confirming their testimony, after the Flavian family was despoiled of the empire, and could no longer give any reward, as the price of a lie. *Utrumque, qui interfuere, nunc quoque memorant, postquam nullum mendacio pretium.* To which if we add the public nature of the facts, as related, it will appear, that no evidence can well be supposed stronger for so gross and so palpable a falsehood.

There is also a memorable story related by Cardinal de Retz, which may well deserve our consideration. When that intriguing politician fled into Spain, to avoid the persecution of his enemies, he passed through Saragossa, the capital of Arragon, where he was shewn, in the cathedral, a man, who had served seven years as a doorkeeper, and was well known to every body in town, that had ever paid his devotions at that church. He had been seen, for so long a time, wanting a leg; but recovered that limb by the rubbing of holy oil upon the stump; and the cardinal assures us that he saw him with two legs. This miracle was vouched by all the canons of the church; and the whole company in town were appealed to for a confirmation of the fact; whom the cardinal found, by their zealous devotion, to be thorough believers of the miracle. Here the relater was also cotemporary to the supposed prodigy, of an incredulous and libertine character, as well as of great genius; the miracle of so *singular* a nature as could scarcely admit of a counterfeit, and the witnesses very numerous, and all of them, in a manner, spectators of the fact, to which they gave their testimony. And what adds mightily to the force of the evidence, and may double our surprise on this occasion, is, that the cardinal himself, who relates the story, seems not to give any credit to it, and consequently cannot be suspected of any concurrence in the holy fraud. He considered justly, that it was not requisite, in order to reject a fact of this nature, to be able accurately to disprove the testimony, and to trace its falsehood, through all the circumstances of knavery and credulity which produced it. He knew, that, as this was commonly altogether impossible at any small distance of time and place; so was it extremely difficult, even where one was immediately present, by reason of the bigotry, ignorance, cunning, and roguery of a great part of mankind. He therefore concluded, like a just reasoner, that such an evidence carried falsehood upon the very face of it, and that a miracle, supported by any human testimony, was more properly a subject of derision than of argument.

There surely never was a greater number of miracles ascribed to one person, than those, which were lately said to have been wrought in France upon the tomb of Abbé Paris, the famous Jansenist, with whose sanctity the people were so long deluded. The curing of the sick, giving hearing to the deaf, and sight to the blind, were every where talked of as the usual effects of that holy sepulchre. But what is more extraordinary; many of the miracles were immediately proved upon the spot, before judges of unquestioned integrity, attested by witnesses of credit and distinction, in a learned age, and on the most eminent theatre that is now in the world. Nor is this all: a relation of them was published and dispersed every where; nor were the *Jesuits*, though a learned body, supported by the civil magistrate, and determined enemies to those opinions, in whose favour the miracles were said to have been wrought, ever able distinctly to refute or detect them. Where shall we find such a number of circumstances, agreeing to the corroboration of one fact? And what have we to oppose to such a cloud of witnesses, but the absolute impossibility or miraculous nature of the events, which they relate? And this surely, in the eyes of all reasonable people, will alone be regarded as a sufficient refutation.

Is the consequence just, because some human testimony has the utmost force and authority in some cases, when it relates the battle of Philippi or Pharsalia for instance; that therefore all kinds of testimony must, in all cases, have equal force and authority? Suppose that the Caesarean and Pompeian factions had, each of them, claimed the victory in these battles, and that the historians of each party had uniformly ascribed the advantage to their own side; how could mankind, at this distance, have been able to determine between them? The contrariety is equally strong between the miracles related by Herodotus or Plutarch, and those delivered by Mariana, Bede, or any monkish historian.

The wise lend a very academic faith to every report which favours the passion of the reporter; whether it magnifies his country, his family, or himself, or in any other way strikes in with his natural inclinations and propensities. But what greater temptation than to appear a missionary, a prophet, an ambassador from heaven? Who would not encounter many dangers and difficulties, in order to attain so sublime a character? Or if, by the help of vanity and a heated imagination, a man has first made a convert of himself, and entered seriously into the delusion; who ever scruples to make use of pious frauds, in support of so holy and meritorious a cause? The smallest spark may here kindle into the greatest flame; because the materials are always prepared for it. The *avidum genus auricularum*, the gazing populace, receive greedily, without examination, whatever sooths superstition, and promotes wonder.

How many stories of this nature have, in all ages, been detected and exploded in their infancy? How many more have been celebrated for a time, and have afterwards sunk into neglect and oblivion? Where such reports, therefore, fly about, the solution of the phenomenon is obvious; and we judge in conformity to regular experience and observation, when we account for it by the known and natural principles of credulity and delusion. And shall we, rather than have a recourse to so natural a solution, allow of a miraculous violation of the most established laws of nature?

I need not mention the difficulty of detecting a falsehood in any private or even public history, at the place, where it is said to happen; much more when the scene is removed to ever so small a distance. Even a court of judicature, with all the authority, accuracy, and judgement, which they can employ, find themselves often at a loss to distinguish between truth and falsehood in the most recent actions. But the matter never comes to any issue, if trusted to the common method of altercation and debate and flying rumours; especially when men's passions have taken part on either side.

In the infancy of new religions, the wise and learned commonly esteem the matter too inconsiderable to deserve their attention or regard. And when afterwards they would willingly detect the cheat, in order to undeceive the deluded multitude, the season is now past, and the records and witnesses, which might clear up the matter, have perished beyond recovery.

No means of detection remain, but those which must be drawn from the very testimony itself of the reporters: and these, though always sufficient with the judicious and knowing, are commonly too fine to fall under the comprehension of the vulgar.

Upon the whole, then, it appears, that no testimony for any kind of miracle has ever amounted to a probability, much less to a proof; and that, even supposing it amounted to a proof, it would be opposed by another proof; derived from the very nature of the fact, which it would endeavour to establish. It is experience only, which gives authority to human testimony; and it is the same experience, which assures us of the laws of nature. When, therefore, these two kinds of experience are contrary, we have nothing to do but subtract the one from the other, and embrace an opinion, either on one side or the other, with that assurance which arises from the remainder. But according to the principle here explained, this subtraction, with regard to all popular religions, amounts to an entire annihilation; and therefore we may establish it as a maxim, that no human testimony can have such force as to prove a miracle, and make it a just foundation for any such system of religion.

I beg the limitations here made may be remarked, when I say, that a miracle can never be proved, so as to be the foundation of a system of religion. For I own, that otherwise, there may possibly be miracles, or violations of the usual course of nature, of such a kind as to admit of proof from human testimony; though, perhaps, it will be impossible to find any such in all the records of history. Thus, suppose, all authors, in all languages, agree, that, from the first of January 1600, there was a total darkness over the whole earth for eight days: suppose that the tradition of this extraordinary event is still strong and lively among the people: that all travellers, who return from foreign countries, bring us accounts of the same tradition, without the least variation or contradiction: it is evident, that our present philosophers, instead of doubting the fact, ought to receive it as certain, and ought to search for the causes whence it might be derived. The decay, corruption, and dissolution of nature, is an event rendered probable by so many analogies, that any phenomenon, which seems to have a tendency towards that catastrophe, comes within the reach of human testimony, if that testimony be very extensive and uniform.

But suppose, that all the historians who treat of England, should agree, that, on the first of January 1600, Queen Elizabeth died; that both before and after her death she was seen by her physicians and the whole court, as is usual with persons of her rank; that her successor was acknowledged and proclaimed by the parliament; and that, after being interred a month, she again appeared, resumed the throne, and governed England for three years: I must confess that I should be surprised at the concurrence of so many odd circumstances, but should not have the least inclination to believe so miraculous an event. I should not doubt of her pretended death, and of those other public circumstances that followed it: I should only assert it to have been pretended, and that it neither was, nor possibly could be real. You would in vain object to me the difficulty, and almost impossibility of deceiving the world in an affair of such consequence; the wisdom and solid judgement of that renowned queen; with the little or no advantage which she could reap from so poor an artifice: All this might astonish me; but I would still reply, that the knavery and folly of men are such common phenomena, that I should rather believe the most extraordinary events to arise from their concurrence, than admit of so signal a violation of the laws of nature.

But should this miracle be ascribed to any new system of religion; men, in all ages, have been so much imposed on by ridiculous stories of that kind, that this very circumstance would be a full proof of a cheat, and sufficient, with all men of sense, not only to make them reject the fact, but even reject it without farther examination. Though the Being to whom the miracle is ascribed, be, in this case, Almighty, it does not, upon that account, become a whit more probable; since it is impossible for us to know the attributes or actions of such a Being, otherwise than from the experience which we have of his productions, in the usual course of nature. This still reduces us to past observation, and obliges us to compare the instances of the violation of truth in the testimony of men, with those of the violation of the laws of nature by miracles, in order to judge which of them is most likely and probable. As the violations of truth are more common in the testimony concerning religious miracles, than in that concerning any other matter of fact; this must diminish very much the authority of the former testimony, and make us form a general resolution, never to lend any attention to it, with whatever specious pretence it may be covered.

Lord Bacon seems to have embraced the same principles of reasoning. 'We ought,' says he, 'to make a collection or particular history of all monsters and prodigious births or productions, and in a word of every thing new, rare, and extraordinary in nature. But this must be done with the most severe scrutiny, lest we depart from truth. Above all, every relation must be considered as suspicious, which depends in any degree upon religion, as the prodigies of Livy: And no less so, every thing that is to be found in the writers of natural magic or alchimy, or such authors, who seem, all of them, to have an unconquerable appetite for falsehood and fable.'

I am the better pleased with the method of reasoning here delivered, as I think it may serve to confound those dangerous friends or disguised enemies to the *Christian Religion*, who have undertaken to defend it by the principles of human reason. Our most holy religion is founded on *Faith*, not on reason; and it is a sure method of exposing it to put it to such a trial as it is, by no means, fitted to endure. To make this more evident, let us examine those miracles, related in scripture; and not to lose ourselves in too wide a field, let us confine ourselves to such as we find in the *Pentateuch*, which we shall examine, according to the principles of these pretended Christians, not as the word or testimony of God himself, but as the production of a mere human writer and historian. Here then we are first to consider a book, presented to us by a barbarous and ignorant people, written in an age when they were still more barbarous, and in all probability long after the facts which it relates, corroborated by no concurring testimony, and resembling those fabulous accounts, which every nation gives of its origin. Upon reading this book, we find it full of prodigies and miracles. It gives an account of a state of the world and of human nature entirely different from the present: Of our fall from that state: Of the age of man, extended to near a thousand years: Of the destruction of the world by a deluge: Of the arbitrary choice of one people, as the favourites of heaven; and that people the countrymen of the author: Of their deliverance from bondage by prodigies the most astonishing imaginable: I desire any one to lay his hand upon his heart, and after a serious consideration declare, whether he thinks that the falsehood of such a book, supported by such a testimony, would be more extraordinary and miraculous than all the miracles it relates; which is, however, necessary to make it be received, according to the measures of probability above established.

What we have said of miracles may be applied, without any variation, to prophecies; and indeed, all prophecies are real miracles, and as such only, can be admitted as proofs of any revelation. If it did not exceed the capacity of nature to foretell future events, it would be absurd to employ any prophecy as an argument for a divine mission or authority from heaven. So that, upon the whole, we may conclude, that the *Christian Religion* not only was at first attended with miracles, but even at this day cannot be believed by any reasonable person without one. Mere reason is insufficient to convince us of its veracity: And whoever is moved by *Faith* to assent to it, is conscious of a continued miracle in his own person, which subverts all the principles of his understanding, and gives him a determination to believe what is most contrary to custom and experience.

Mesmerism

"Report of the Commissioners Charged by the King to Examine Animal Magnetism, Printed on the King's Order Number 4 in Paris from the Royal Printing House"

by Benjamin Franklin and Antoine Lavoisier

Introduction

Michael Shermer

In 1991, about the time we were creating and organizing the Skeptics Society and *Skeptic* magazine, I read an essay by Stephen Jay Gould entitled "The Chain of Reason Versus the Chain of Thumbs," in *Bully for Brontosaurus* (1991, W. W. Norton). It is the story of an 18th-century scientific investigation of an extraordinary claim—mesmerism—commissioned by King Louis XVI of France and conducted by such scientific luminaries as Benjamin Franklin and Antoine Lavoisier. The result of that investigation was the Report of the Commissioners Charged by the King to Examine Animal Magnetism, "Printed on the King's Order Number 4 in Paris from the Royal Printing House" in 1784, just five years before the demise of the ancien régime. Gould called the report "an enduring testimony to the power and beauty of reason," a "key document in the history of human reason," and said that "it should be rescued from its current obscurity, translated into all languages, and reprinted by organizations dedicated to the unmasking of quackery and the defense of rational thought" (188–189).

I kept that challenge in the back of my mind for the next five years, awaiting the time when we would have the space to allocate for the resurrection of this "key document" (it runs 18 pages, making it the third longest piece we have ever run). It is not a waste of space because the history of skepticism and the skeptical movement should be tracked and recorded as any field should be, and this is the first scientific investigation that we know of into what would today be considered a paranormal or pseudoscientific claim. No one else has taken up Gould's challenge, so in the pages to come we present you with this delightful piece of science and reasoning, with thanks to Steve Gould for providing a copy from the original in Harvard's Houghton Library, and to my friend and colleague Charles Salas and his wife Danielle for the translation; both write and speak fluent French (plus Charles is an intellectual historian of the period).

The historical context for the report is given in great detail by the renowned intellectual historian Robert Darnton, in his 1968 book *Mesmerism and the End of the Enlightenment in France* (Harvard University Press). The German physician Franz Anton Mesmer was the "discoverer" of animal magnetism, and he has ever since been remembered whenever we are "mesmerized" by something that seems to draw us to it like a magnet. The analogy is appropriate, for Mesmer reasoned that just as an invisible force of gravity binds the planets together, and an invisible force of electricity flows through various substances, and an invisible force of magnetism draws iron shavings to a lodestone, so an invisible force—animal magnetism—flows through living beings. To Mesmer these forces were actually manifestations of a single fluid flowing throughout the universe, the blockage of which can cause disease. Cure comes through releasing the blockage (similar to what is claimed for Chi power, acupuncture and acupressure, therapeutic touch, and other modern nostrums). Mesmer's technique involved facing the patient, touching fingers, and staring for prolonged periods into her eyes. By most accounts Mesmer was, well, rather mesmerizing, especially to his female patients, who would shake, groan, scream, and even faint (is this beginning to sound familiar to those who have ever witnessed a faith healing?).

Group healings involved everyone surrounding a "baquet," or vat, filled with "magnetized" water and placed in the center of the room. "Magnetized" rods protruding from the vat were grabbed by the patients who, with their other hand, held each other's thumbs between their thumb and forefinger and squeezed at the appropriate time to allow the magnetism to flow evenly through the group. To ensure proper conductivity in this "mesmeric chain," Mesmer looped a rope around them (without knots, for this might impede flow).

Mesmerism became all the rage, triggering a skeptical response by the medical establishment, which, along with other concerned scientists, persuaded King Louis XVI to establish a Royal Commission to test Mesmer's claims. (In the film *Jefferson in Paris*, the vat and rods are depicted, along with a skeptical Thomas Jefferson.) Franklin, the world's leading authority on electricity, was in Paris as a U.S. representative; Lavoisier, one of the founders of modern chemistry, lived there. The others on the Commission were respected scientists and medical doctors, including Dr. Guillotin, inventor of the device that would cut off Lavoisier's head, along with many others, over the course of the next decade of revolutionary mayhem.

The problem for the Commission, as the report reveals, is that animal magnetism is invisible. No problem, so is gravity. They would test its effects on objects, which was the basis for Mesmer's claims of curative power. (James Randi is fond of stating that it doesn't matter whether there is a scientific basis to astrology, ESP, and other psychic forces; the only thing that matters is if they actually work, which they don't.) The problem was that "cures" take too long for an experiment and may be caused by other conditions anyway (Franklin suspected that Mesmer's patients were cured by staying away from medical doctors!). Mesmer, however, did not take the test; his top student, Charles Deslon, took his place, which subsequently led to Mesmer disputing the findings. The experimenters began by trying to magnetize themselves—joined by rods, rope, and thumbs with Deslon giving proper instruction—to no effect. They then tried seven people from the lower classes and compared their results against seven people from the upper classes (recall the importance of class in prerevolutionary France). Only three, all from the lower classes, experienced anything significant, so the Commission concluded it was due to the power of suggestion.

To test the null hypothesis that magnetism is really just a placebo effect, Franklin and Lavoisier devised a test whereby some subjects would be deceived into thinking they were receiving the experimental treatment (magnetism) when they really were not, while others did receive the treatment and were told that they had not. The results were clear: the effects were due to the power of suggestion only.

To reinforce this conclusion, Franklin had

Deslon magnetize a tree in his garden. The experimental subject—allegedly "sensitive" to the magnetic effect but not told which tree was affected—then walked around the garden hugging trees until he declared he had sensed it. He collapsed in a fit in front of the fourth tree, but it was the fifth one that was "magnetized." Undaunted, Deslon claimed that all trees carry some magnetism and therefore the test was invalid (not unlike the excuses of failed water dowsers and other modern mystics). In test after test, Deslon failed. One woman was blindfolded and told that Deslon was "influencing" her, causing her to collapse in a mesmeric "crisis." He wasn't. Another woman could supposedly sense "magnetized" water. Lavoisier filled several cups with water, only one of which was "magnetized." After touching an unmagnetized cup she collapsed in a fit, upon which Lavoisier gave her the "magnetized" one, which "she drank quietly & said she felt relieved. Therefore the cup & magnetism missed their marks, because the crisis was quieted rather than exacerbated." Q.E.D.

The Commission concluded that "nothing proves the existence of Animal-magnetism fluid; that this fluid with no existence is therefore without utility; that the violent effects observed at the group treatment belong to touching, to the imagination set in action & to this involuntary imitation that brings us in spite of ourselves to repeat that which strikes our senses." In other words, the effect is mental, not magnetic.

The control of intervening variables and the testing of specific claims, without resort to unnecessary hypothesizing about what is behind the "power," is the lesson modern skeptics should take from this historical masterpiece. The other historical lesson is clear as well—true believers remain unaffected by contradictory evidence, in the 18th century as well as today. So why bother testing? Because the vast majority of people are neither true believers nor skeptics, but just intellectually curious and looking for a natural explanation for an apparently supernatural phenomenon.

Report of the Commissioners on Mesmerism

Translation by Danielle and Charles Salas

On March 12, 1784, the King appointed Physicians chosen from the Paris Faculté, *Messieurs* Borie, Sallin, d'Arcet, Guillotin, to examine & report on Animal magnetism practiced by *Monsieur* Deslon; & as requested by these four Physicians, His Majesty has appointed five of the Members of the Royal Academy of Sciences to conduct this examination with them: Messieurs Franklin, le Roy, Bailly, de Bory, Lavoisier. As M. Borie died at the beginning of the Commissioners' work, His Majesty chose M. Majault, a Doctor from the Faculté, to replace him.

The agent that M. Mesmer claims to have discovered, which he has made known under the name Animal magnetism, is, as he characterizes it himself & according to his own words,

> a universally spread fluid; it is the means of a mutual influence between celestial bodies, the earth, & living bodies; it is continuous so as not to permit any vacuum; it is incomparably subtle; it is capable of receiving, spreading, & communicating all the sensations of movement; it is sensitive to flux & reflux. The physical body feels the effects of this agent; &, when it insinuates itself into the substance of nerves, it affects them immediately. One recognizes particularly in the human body, properties similar to those of the magnet. One dis-

> tinguishes two diverse & opposed poles. The action & property of Animal magnetism may be transmitted from one body to another, animate & inanimate: This action operates from a distance, without the help of any intermediary body; it is increased when reflected by mirrors, communicated, spread, & increased by sound; this property may be accumulated, concentrated, transported. Although this fluid is universal, all animated bodies are not equally susceptible. There are some, albeit few, in whom the polar property is so strong that their mere presence destroys all the effects of this fluid in other bodies.
>
> Animal magnetism may itself cure nervous disorders & be a medium for curing others; it improves the action of medications; it induces & guides crises in such a way that disorders can be understood & mastered. In this way, the Physician knows the state of health of each individual & determines with certainty the origin, nature, & progress of even the most complicated of diseases; he prevents their spread & reaches a cure without ever exposing the patients to dangerous effects or unfortunate consequences, regardless of age, temperament & sex. Nature offers in Magnetism a universal means of healing and protecting people.

Such is the agent that the Commissioners have been charged to examine & whose properties are attested to by M. Deslon, who endorses all of M. Mesmer's principles. This theory is the basis of a paper read May 9 at the home of M. Deslon in the presence of the Lieutenant General of Police & the Commissioners. In the paper it is claimed that there is but one nature, one disease, one remedy; & this remedy is Animal magnetism. In instructing the Commissioners about the theory & action of magnetism, this Physician also taught them practical exercises, indicating where the poles are, how patients are to be touched & the manner in which this magnetic fluid is to be trained upon them.

M. Deslon pledged with the Commissioners, 1. to ascertain the existence of Animal magnetism; 2. to make known their findings; 3. to prove the usefulness of these findings & of Animal magnetism in the cure of diseases.

Having been introduced to the theory & techniques of Animal magnetism, it was time to learn about the effects. The Commissioners visited (& all of them more than once) the place where M. Deslon had his practice. In the middle of a large room they saw a circular vat, made of oak & raised a foot or a foot & a half, called a *baquet.* The covering of this vat has many holes from which protrude bent, flexible metal rods. The patients are arranged in rows around this vat, one rod to a person which because it is bent may be applied directly to the afflicted area of the body; the patients are chained together by a rope looped around their bodies; sometimes a second chain is created by touching hands, which is to say, the thumb is pressed between a neighbor's thumb & index finger, & squeezed; the sensation received from the left is sent through the right, & it circulates all around.

There is a pianoforte in the corner on which different tunes with various movements are played; sometimes the sounds of voice & singing are added.

All those who magnetize hold a metal rod ten to twelve inches long.

M. Deslon declared to the Commissioners, 1. that this rod conducts magnetism; this rod has the advantage of concentrating magnetism in the tip, & making the emanations more powerful. 2. Sound, in accordance with M. Mesmer's principle, is also a conductor of magnetism, & to communicate the fluid to the pianoforte, it is enough to bring the metal rod closer; the person in contact with the instrument also provides some fluid, & magnetism is transmitted through sound to near-by patients. 3. The rope wrapped around the patients is intended, like the chain of thumbs to augment the effects through communication.

4. The inside of the vat is made so as to concentrate magnetism. It is a large basin from which magnetism is spread through the metal rods dipped within it.

The Commissioners used an electrometer & a non-magnetic, metal needle to check that the vat did not contain any electrical or charged matter; & upon the declaration of M. Deslon regarding the composition of the inside of the vat, they agreed that no physical agent capable of contributing to the reported effect of magnetism was present.

A large number of patients arranged in several rings around the vat receive magnetism simultaneously therefore through these means: through the metal rods that transmit the magnetism from the vat; through the rope intertwined about the body, & by the union of thumbs communicating that of their neighbors; through the sound of the pianoforte, or through a pleasant voice that spreads it through the air. Patients are directly magnetized as well by passing the finger & the metal rod in front of the face, on top of or behind the head, & on afflicted areas, always maintaining the distinction of the poles; sight, staring at them, activates the effects. But above all patients are magnetized by the laying of hands & the pressure of fingers on the hypochondria & lower abdominal areas; the contact often maintained for a considerable time, sometimes a few hours.

Patients then display a variety of reactions depending on the different states they find themselves in. Some are calm, quiet, & feel nothing; others cough, spit, feel slight pain, a warmth either localized or all over, & perspire; others are agitated & tormented by convulsions. These convulsions are extraordinary in their number, duration, & strength. As soon as a convulsion begins, many others follow. The Commissioners have seen some lasting for more than three hours; convulsions are accompanied by murky & viscous expectorations drawn out by the violence of the exertions. Sometimes the expectorations contain streaks of blood; there is a young male patient, in particular, who spit out blood in abundance. These convulsions are characterized by quick, involuntary movements of limbs & the entire body, by a tightening of the throat, by the twitching of the hypochondria & epigastric area, by blurred & unfocused vision, by piercing shrieks, tears, hiccups & excessive laughter. They are preceded or followed by a state of languor & dreaminess, of a kind of prostration & even sleepiness. The slightest unexpected noise causes shivers; & it has been noticed that the change of tone & measure in the pieces played on the pianoforte had an influence on the patients—a faster movement, for example, agitated them more & renewed the intensity of their convulsions.

There is a padded room, intended primarily for patients racked by convulsions, a room named *des Crises;* but M. Deslon does not deem its usage necessary, & all patients, regardless of condition, are gathered together in the group treatment rooms. Nothing is more astonishing than the spectacle of these convulsions; without seeing it, it cannot be imagined: & in watching it, one is equally surprised by the profound repose of some of these patients & the agitation that animates others; the various reactions that are repeated, the fellow-feeling that sets in. One sees patients specifically searching for others & while rushing towards each other, smile, speak with affection & mutually soothe their crises. All submit to the magnetizer; even though they may appear to be asleep, his voice, a look, a signal pulls them out of it. Because of these constant effects, one cannot help but acknowledge the presence of a great power which moves & controls patients, & which resides in the magnetizer.

This convulsive state is improperly called Crisis in the theory of Animal magnetism: in this doctrine, the crisis is considered healthy, like those brought about by Nature or by the skillful physician to facilitate the cure of diseases. The Commissioners will adopt this term hence forward in this report, & when they make use of the word *crisis*, they will always mean the state of either the convulsions or the lethargy produced by the processes of Animal magnetism.

The Commissioners noticed that out of the number of patients in crisis, there were always many women & few men; that these crises took one to two hours to build; & that as soon as one was established, all the others would start successively soon after. This having been remarked upon, the Commissioners soon came to the conclusion that group treatment rooms could not be the setting for their experiments. The multiplicity of effects is a first obstacle; one sees too many things at once to see particular things clearly. Moreover, distinguished patients who come to the treatment for their health could be bothered by the questioning; being so carefully observed could inconvenience or displease them; the Commissioners themselves would be hindered by their concern for discretion. They then decided that their constant attendance not being necessary to the treatment, it sufficed that a few of them should come from time to time to confirm the preliminary general observations, to make new ones if necessary, & to report to the assembled commission.

The effect of group treatment having been observed, the next task was to unravel the causes & to search for proofs of the existence & the utility of magnetism. The question of existence is primary; the question of utility is not to be addressed until the first has been fully resolved. Animal magnetism may well exist without being useful but it cannot be useful if it does not exist.

In consequence, the principal purpose of the Commissioners' examination & the essential goal of their first experiments had to be to make certain of that existence. This purpose was still very broad & needed to be simplified. Animal magnetism embraces the whole of Nature; it is said to be the means by which celestial bodies influence us; the Commissioners thought that they should first set aside this mighty influence, to consider only the part of this fluid diffused upon the earth without bothering with whence it comes, & to ascertain the action it has upon us, around us & before our eyes, before considering its relations with the Universe.

The most reliable way to ascertain the existence of Animal-magnetism fluid would be to make its presence tangible; but it did not take long for the Commissioners to recognize that this fluid escapes detection by all the senses. Unlike electricity, it is neither luminescent nor visible. Its action does not manifest itself visibly as does the attraction of a magnet; it is without taste or smell; it spreads noiselessly & envelops or penetrates you without your sense of touch warning you of its presence. Therefore, if it exists in us & around us, it does so in an absolutely undetectable manner. Among those who profess magnetism, there are some who claim that it may occasionally be seen emanating from the tips of fingers serving as conductors or who believe that they feel its passage when the finger is moved back & forth in front of the face & over the hand. In the first instance, the visible emanation is only that of perspiration which becomes easily visible when magnified under a solar microscope; in the second, the feeling of cold or coolness that one feels, a feeling more noticeable the warmer one is, is caused by the finger disturbing the air which is always colder than body temperature. On the other hand, if the finger is brought close to the skin of the face, which is colder than the finger, & left there, one is made to feel a sensation of heat, which is communicated body heat.

It is also claimed that this fluid has an odor & that it is detectable when the finger or conducting rod is held under the nose; it is even said that these sensations are different under the two nostrils depending on the polar positioning of the finger or rod. M. Deslon has experimented upon several Commissioners; the Commissioners have repeated the experiment upon several subjects; none has felt this difference in sensation between one nostril & the other: & if by paying close attention, some odor is recognized, it is in the case of the iron rod, that is of the rod itself warmed & rubbed, & in the case of the finger, that of the emanation of perspiration, an odor often mixed with that of iron with which the finger is imprinted. These effects have been mistakenly attributed to magnetism, they all belong to known, natural causes.

In addition, M. Deslon never emphasized these fleeting sensations; he didn't think it necessary to have to produce them as proofs; &, on the contrary, he has expressly declared to the Commissioners that he could only prove the existence of magnetism through the action of this fluid, creating changes in animate bodies. This existence becomes even more difficult to ascertain through demonstrable effects whose causes are not unequivocal; through authenticated facts upon which mental circumstances have no influence; finally through proofs capable of impressing & convincing the mind, the only proofs that could satisfy enlightened Physicians.

The action of magnetism on animate bodies may be observed two different ways; either by prolonged action & its curative effects on the treatment of diseases, or by its temporary effects on the economy of the human body & by the observable changes it produces. M. Deslon insisted that the first of these methods be principally & almost exclusively used. The Commissioners did not believe they had to do so & here are their reasons:

Most diseases are seated inside the body. The long experience of a great many centuries has made the symptoms that precede & characterize these diseases well-known. That same experience has indicated their method of treatment. What is it in this method that is the goal of the Physician's effort? It is neither to oppose nor tame Nature, it is to help it in its operations. Nature heals the sick, said the Father of Medicine; but sometimes it meets obstacles that hinder its course, obstacles that needlessly consume its strength. The Physician is Nature's Minister; attentive observer, he studies its course. If that course is steady, sure, level & without deviations, the Physician observes it in silence & is careful not to disturb it with remedies at best useless; if this course is hampered, he facilitates it; if it is too slow or too fast, he accelerates it or slows it down. He sometimes limits himself to regulating diet to fulfill his goal; sometimes he uses medications. The action of medication in the human body is a new force that combines with the great force that sustains life: if the remedy follows the same paths already opened by this force, it is salutary & useful in expelling disease; if it tends to open contrary paths & divert this inner action, it is harmful. However, it must be agreed that this very real effect, salutary or harmful, may often escape common observation. The physical history of mankind offers very peculiar phenomena in this regard. We see that the most different diets have not prevented the attainment of old age. We see men seemingly stricken by the same disease who are healed while following opposite diets, & while taking entirely different remedies; Nature is therefore powerful enough to maintain life in spite of a bad diet & to triumph over both the disease & the remedy. If it has this power to resist remedies, all the more reason that it has the power to operate without them. The experience of their effectiveness, therefore, always carries some degree of uncertainty; in the case of magnetism, there is an extra degree of uncertainty: the question of its existence. For, how can one ascertain, by the treatment of diseases, the action of an agent the existence of which is in dispute when one can doubt the effect of medications the existence of which is not in question?

The cure cited the most in favor of the existence of magnetism is that of M. le Baron de

* * * , of which both the Court & the city have been informed. We will not enter herein to a discussion of the facts; we will not examine whether the remedies previously used may have contributed to that cure. On the other hand, we acknowledge that the state of the patient was grave &, on the other, the ineffectiveness of all the means of ordinary medicine; magnetism was used & M. le Baron de * * * fully recovered. But could not a natural occurrence alone have been responsible for this recovery? A woman of the people & very poor, living at Gros-caillou, was struck in 1779 by a malevolent fever of well known characteristics; she consistently refused any help, asking only that a water pitcher by her bedside be kept full. She stayed quietly on her bed of straw, drinking water all day & doing nothing else. The sickness progressed, passed successively through its different stages, & ended with complete recovery.

Mademoiselle G *** living at the Petite-sécuries of the King had two glands on the right breast that worried her very much; a surgeon advised her to use Painter's water, an excellent dissolving agent, stating that, if the remedy did not succeed within a month, the glands would have to be removed. The frightened young lady consulted M. Sallin who deemed the glands treatable. M. Bouvart, consulted later, gave the same opinion. She was encouraged to seek entertainment & distractions before beginning treatment; fifteen days later, she suffered a violent coughing crisis at the Opera & expectorated so abundantly that she had to be brought back home; in four hours she spit out three pints of phlegm; one hour later M. Sallin examined the breast & could no longer find any trace of the glands. M. Bouvart, who was called the next day, verified the felicitous effect of this natural crisis. If Mlle. G *** had taken Painter's water, then Painter's water would have had to be credited for the cure.

Observations over the centuries prove & Physicians themselves recognize, that Nature alone & without the help of medical treatment cures a great number of patients. If magnetism were inefficacious, using it to treat patients would be to leave them in the hands of Nature. In trying to ascertain the existence of this agent, it would be absurd to choose a method that, in attributing to the agent all of Nature's cures, would tend to prove that it has a useful & curative action, even though it would have none.

The Commissioners are in agreement on this with M. Mesmer. He rejected the cure of diseases when this way of proving magnetism was proposed to him by a Member of the Académie des Sciences: *it is,* said he, *a mistake to believe that this kind of proof is irrefutable; nothing conclusively proves that the Physician or Medicine heals the sick.*

The treatment of diseases, therefore can only furnish results that are always uncertain & often misleading; this uncertainty could not be evaded, & all cause of illusion offset, except by an infinity of cures & perhaps the experience of a few centuries. The purpose & importance of the Commission require means more prompt. The Commissioners have had to confine themselves to purely physical proofs, that is, to the temporary effects of the fluid on the Animal body, by stripping these effects of all illusions possibly mixed up with them, & making sure that they cannot be due to any cause other than Animal magnetism.

They set out to experiment on isolated subjects, who were willing to participate in a variety of experiments imagined by the Commissioners; & who, some through their naivete, others through their intelligence, would be able to give a truthful & exact account of what they experienced. These experiments will not be presented here chronologically but in the order of the facts that they ought to clarify.

The Commissioners resolved to begin by experimenting upon themselves, & to submit themselves to the action of magnetism. They were very curious to experience through their own senses the reported effects of this agent.

They therefore submitted themselves to these effects with the determination not to be angered by the injuries or upsets to their health known to be produced by magnetism, putting themselves in a position to resolve this important question on the spot by means of their own evidence. But in submitting themselves to magnetism in this way, the Commissioners had to take a necessary precaution. There is no individual, even in the best of health who, if he listened to himself attentively, would not feel within himself an infinity of the movements & variations of either warmth or very minor pain in various areas of the body; these variations which can occur at any time are independent from magnetism. It may not be inconsequential to bring & sustain attention upon oneself in this way. There are so many connections, by whatever means, between the will of the soul & body movements that it is impossible to gauge the effect of attentiveness, which seems only to be a sequence of intentions directed towards the same object with perseverance & without interruption. When one considers that the will moves the arm at pleasure, how can one be certain that the attention focused upon an interior part of the body cannot excite slight movements there, bring warmth there, & make modifications so as to produce new sensations there? The first concern of the Commissioners was necessarily not to pay too much attention to what was happening inside themselves. If magnetism is a real & powerful agent, it does not require to be thought about to be manifest; it must, so to speak, force itself upon the attention & make itself noticeable even by a mind disturbed by design.

But in deciding to make experiments upon themselves, the Commissioners unanimously agreed to make them amongst themselves without allowing any stranger other than M. Deslon to magnetize them or other persons of their own choosing; they also promised each other not to magnetize in group treatment, so that they could freely discuss their observations, & be in all cases the only, or at least the first, judges of what they would be observing.

In consequence, a separate room & particular vat were set aside for them at M. Deslon's, & once a week they sat there; they stayed for two to two & a half hours at a time, the iron rod resting on the left hypochondrium, & themselves surrounded by the rope of communication, & from time to time making the chain of thumbs. They were magnetized, either by M. Deslon or a disciple sent in his place, some for a longer time & more often than others, & these should have appeared to be the most sensitive; they were magnetized, sometimes with the finger & iron rod held & moved over various parts of the body, sometimes by applying hands & finger pressure to either the hypochondria or on the pit of the stomach.

None of them felt a thing, or at least, nothing that could be attributed to the action of magnetism. A few of the Commissioners have robust constitutions; others have weaker constitution & are subject to discomforts: one of these felt a slight pain in the pit of the stomach, following strong finger pressure there. This pain lasted all day & the next day, accompanied by a feeling of fatigue & uneasiness. A second felt a slight irritation of the nerves, which he is susceptible to, on the afternoon of one of the days he was touched. A third, endowed with a greater sensitivity, & especially an extreme instability in the nerves, felt more pain & more intense irritations; but these slight mishaps are the consequence of incessant & ordinary variations in the state of health &, consequently, foreign to magnetism, or they follow from the pressure exerted on the stomach. The Commissioners only mention these minor details out of a desire for scrupulous accuracy; they report them because they have imposed on themselves the rule of always telling the truth in all things.

The Commissioners could not help but be struck by the difference between group treatment & private treatment at the vat. Calm &

silence in one, movement & agitation in the other; there, multiple effects, violent crises, the normal state of body & spirit interrupted & troubled, Nature overstrung; here the body without pain, the spirit without trouble, Nature conserving its equilibrium & natural course, in a word, the absence of all effects; one cannot find this great power so astonishing in the group treatment; magnetism without energy appeared to be devoid of all sensible action.

The Commissioners, who at first went to the vat only once a week, wanted to test whether continuity might produce something; they went three days in a row, but their lack of sensibility was the same & they obtained no result whatsoever. This experiment, done & repeated on eight subjects at a time, a few of whom have habitual discomforts, suffices to conclude that magnetism has little or no effect on a state of health, & even on a state of slight infirmity. It was resolved to experiment on really sick subjects, & they were chosen from the class of commoners.

Seven patients were brought in Passy at the home of M. Franklin; they were magnetized in front of him & in front of the other Commissioners by M. Deslon.

The widow Saint-Amand, an asthmatic with swollen abdomen, thighs & legs; & the woman Anseaume, who had a lump on her thigh, felt nothing; little Claude Renard, a child of six years, scrofulous, almost emaciated, with a swollen knee & a crooked leg with an almost unmovable joint, an interesting child & more reasonable than his age would dictate, also felt nothing, & also Geneviève Leroux, nine years old, subject to convulsions & a disease somewhat similar to what is called *chorea sancti Viti*. François Grenet felt some effects; his eyes are diseased, especially the right one with which he can hardly see & where there is a large tumor. During the magnetization on the left eye, by bringing the thumb closer & moving it back & forth at close range & for a long time, he felt pain in the eyeball & tears appeared. When the right eye, the sicker of the two, was magnetized, he felt nothing; he felt the same pain in the left eye, & nothing elsewhere.

The woman Charpentier, knocked to the ground against a wooden beam by a cow two years ago, suffered various after effects: she lost her eyesight, then recovered it partially, but has stayed in a habitual state of infirmity; she claimed to have had two prolapses, & an abdomen of such sensitivity that she cannot bear to tie her skirt belts; this sensitivity is a matter of nerves being irritated and set into motion; the slightest pressure on the abdomen can get this motion underway &, by the correspondence of nerves, produce effects throughout the whole body.

This woman was magnetized like the others, by application & finger pressure; this pressure was painful to her; then as the finger was directed towards the area of prolapse, she complained of a headache; with the finger placed in front of her face, she said she was short of breath. With repeated movements of the finger from high to low, she had quick movements of the head & shoulders such as one has when feeling surprise mixed with fear, & similar to those of a person whose face has been splashed with drops of cold water. It seemed that she felt the same movements with her eyes closed. Fingers were placed under her nose while her eyes were closed & she said she thought that she was going to faint if that continued. The seventh patient, Joseph Ennuyé felt similar effects, but to a much lesser degree.

Out of these seven patients, four felt nothing & three felt some effects. These effects were worthy of the Commissioners' attention & warranted a scrupulous exam.

To enlighten themselves & fix their ideas on this matter, the Commissioners decided to experiment with patients from other circumstances, patients chosen from high society who could not be suspected of ulterior motives & whose intelligence would permit them to discuss their own sensations & report on them. Mmes. de B** & de V**, Ms. M** & R** were admitted to the Commissioners' private vat; they were asked to observe what they felt, but without giving it too much attention. M. M** & Mme. de V** were the only ones to feel something. M. M** has a cold tumor over the entire knee joint & his patella is painful. After having been magnetized, he declared he felt nothing anywhere in his body except when the finger was moved in front of the bad knee; he thought he then felt a slight warmth at the place where he usually has pain. Mme. de V**, suffering from a nervous condition, was many times on the point of falling asleep while being magnetized. Magnetized without interruption for one hour & nineteen minutes, most often by the laying of hands, she felt only some agitation & uneasiness. These two patients came only once to the vat. M. R** sick from an unresolved liver congestion, following from an obstruction improperly healed, came three times & felt nothing. Mme. de B** suffering obstructions sat constantly with the Commissioners, she felt nothing; & it must be said that she submitted to magnetism with perfect calm, which stemmed from a great incredulity.

Various patients were tested on other occasions but not around the vat. One of the Commissioners struck by migraine was magnetized by M. Deslon for half an hour; one of the symptoms of this migraine is excessive coldness in the feet. M. Deslon brought his foot close to that of the patient, the foot was not warmed, the migraine lasted its usual length, & the patient after sitting down by the fireplace felt the salutary effects that heat has always provided, without having felt during the day or the next night any of the effects of magnetism.

Even though inconveniences prevented M. Franklin from being in Paris & witnessing the experiments, he was himself magnetized by M. Deslon, who visited him at his home in Passy. The gathering there was numerous; all those present were magnetized. A few patients who had accompanied M. Deslon felt the effects of magnetism, as they usually did during group treatment, but Mme. de B**, M. Franklin, his two parents, his secretary, an American officer, felt nothing, even though one parent of M. Franklin was convalescing, & the American officer sick at the time with a low grade fever.

These different experiments furnish facts worthy of being collected & compared, & from which the Commissioners have been able to draw conclusions. Out of fourteen patients, five seemed to have felt effects, & nine none at all. The Commissioner who had the migraine & ice cold feet felt no relief from magnetism, & his feet were not warmed. Therefore this agent does not have the property, attributed to it, of communicating heat to the feet. Magnetism is also heralded as indicating the type & especially the seat of disease through the pain that the action of this fluid inevitably brings there. This advantage would be precious; the fluid, indicator of disease, would be a great tool in the hands of the physician, often confounded by equivocal symptoms; but François Grenet only had sensation & some pain in the eye that was less sick. Had the other eye not been red & swollen, one would have believed it to be undamaged judging by the effect of magnetism. M. R** & Mme. de B**, both sick with obstructions, & Mme. de B** quite seriously, having felt nothing, would not have been made aware of either the seat or the type of their disease. & yet, obstructions are diseases claimed to be especially susceptible to the action of magnetism; because according to the new theory, free & fast circulation of this fluid through the nerves is a way to clear up channels & destroy obstacles, that is to say, the blockages that it meets. At the same time it is said that magnetism is the cornerstone of health. If M. R** & Mme. de B** had not felt discomforts & suffering inseparable from the obstructions, they would have firmly believed that they were in the best state of health in the world. The same should be said of the American officer:

magnetism, heralded as an indicator of disease, has therefore entirely missed its mark.

The heat that M. M** felt on the patella is too subtle & too fleeting to lead to any conclusion. We may suspect that it comes from the cause described above, that is, from too much attention paid to observing oneself: the same attention would find similar feelings at any other moment when magnetism was not in use. The drowsiness felt by Mme. de V** probably comes from the invariability & boredom of the same situation; if she has had a certain light movement, we know that the nature of nervous conditions depends heavily on the attention paid to them; it is enough to think about them or to hear about them to regenerate them. It can be judged what will happen to a woman whose nerves are very jittery, & who is magnetized for an hour & nineteen minutes, during which time she has no other thought than that of her habitual ailments. It would have not been surprising had she suffered a more considerable nervous crisis.

Of the effects that could appear to have to do with magnetism, only those on the woman Charpentier, on François Grenet & on Joseph Ennuyé remain. But then in comparing these three particular cases to all the others, the Commissioners were surprised that these three patients from the lower class were the only ones who had felt something, while those of a higher class, more enlightened, more able to give account of their feelings, felt nothing at all. No doubt François Grenet felt pain in his eye & cried because the thumb was brought so close to it; the woman Charpentier complained that when her stomach was touched, the pressure corresponded to the prolapse; & this pressure may have produced a part of the effects that this woman felt; but the Commissioners suspected that these effects had been augmented by mental circumstances.

Let us take the standpoint of a commoner, for that reason ignorant, struck by disease & desiring to get well, brought with great show before a large assembly composed in part of physicians, where a new treatment is administered which the patient is persuaded will produce amazing results. Let us add that the patient's cooperation is paid for, & that he believes that it pleases us more when he says he feels effects, & we will have a natural explanation for these effects; at the least, we will have legitimate reasons to doubt that the real cause of these effects is magnetism.

Moreover, one can ask why magnetism had these effects on those people who knew what was done to them, who may have believed they had an interest in saying what they said, whereas it had no hold over little Claude Renard, over this delicate organization of childhood, so fickle & so sensitive! The reason & ingenuity of this child guarantees the truth of his testimony. Why did this agent produce no effect upon Geneviève Leroux, who was in a perpetual state of convulsions? Her nerves were certainly jittery, why did magnetism not manifest itself, either by augmenting or diminishing her convulsions? Her indifference & impassibility lead to the conclusion that she felt nothing, because the lack of reason did not permit her to judge that she should have felt nothing.

These facts permitted the Commissioners to observe that magnetism has seemed to be worthless for those patients who submitted to it with a measure of incredulity; that the Commissioners, even when those with jittery nerves deliberately focused their attention elsewhere, having been armed with philosophical doubt that ought to accompany every examination, did in no way feel the impressions felt by the three lower-class patients, & they must have suspected that these impressions, even supposing them all to be real, followed from an anticipated conviction, & could have been an effect of the imagination. From this has resulted another plan of experiment. From now on, their research is going to be directed toward a new object; it is a question of disproving or confirming this suspicion, of determining up to what point the imagination can influence feelings & establishing whether it can be the cause of all or part of the effects attributed to magnetism.

Next the Commissioners heard about the experiments done at the home of the Dean of the Faculté by M. Jumelin, Doctor of Medicine; they requested to see these experiments & they met with him at the home of one of the Commissioners, M. Majault. M. Jumelin declared that he was not a follower of M. Mesmer or of M. Deslon, that he had learned nothing from them about Animal magnetism; & from what he had heard said on the subject he conceived principles & carried out proceedings. His principles consist of regarding Magnetic Animal fluid as a fluid circulating in the body, & which emanates from it, but which is essentially the same as that which produces body heat; a fluid that like all others, tending toward equilibrium, passes from the body which has the most to the body which has the least. His methods are equally different from those of M. Mesmer & M. Deslon; he magnetizes as they do using the finger & the metal rod as conductors, & by the laying of hands, but without making any distinction between poles.

First, eight men & two women were magnetized & felt nothing; finally a woman who is the door-keeper at the home of M. Alphonse le Roy, Doctor of Medicine, having been magnetized on her forehead, but without contact, said she felt heat while M. Jumelin was moving his hand, & with the tips of his five fingers next to the woman's face, she said she felt as if a moving flame were coming from it; magnetized on the stomach, she said she felt heat there; magnetized on the back, she said she felt the same heat there: she declared furthermore that she felt warm all over & had a headache.

The Commissioners, seeing that out of eleven persons subjected to the experiment only one was sensitive to the magnetism of M. Jumelin, thought that this person felt something only because she was doubtless more impressionable; the occasion was favorable for shedding light on the matter. The sensitivity of the woman being well established, it was only a question of protecting her from her imagination, or at least of getting it out of the way. The Commissioners proposed to blindfold her so that they could observe the nature of her sensations while experimenting without her knowledge. She was blindfolded & magnetized, whereupon the phenomena no longer corresponded to the places where the magnetism was directed. Magnetized successively over the stomach & the back, the woman felt heat in her head, pain in her right leg, her left eye & left ear.

The blindfold was removed, & M. Jumelin having applied his hands on the hypochondria, she said she felt heat; then after a few minutes she said she was going to faint &, in fact, did. When she recovered, she was again used as a subject, she was blindfolded, M. Jumelin was moved aside, the room was made silent & the woman was made to believe that she was magnetized. The results were the same, even though nothing was done to her from near or afar; she felt the same heat, the same pain in her eyes & ears; she also felt heat in her back & loins.

After a quarter of an hour, M. Jumelin was signaled to magnetize her over her stomach, she felt nothing, the same thing with her back. Sensations diminished instead of increasing. The headache remained, the heat in the back & loins came to an end.

One sees that there have been effects produced & that these effects are similar to those felt by the three patients mentioned above. But the former & the latter were obtained by different methods. It follows that the methods of proceeding play no role whatsoever. The method of Ms. Mesmer & Deslon & an opposite method give the same results. The distinction between the poles, therefore, is chimerical.

One can observe that when the woman could see, she placed her sensations precisely on the magnetized area; whereas when she could not see, she placed them haphazardly & in areas far from those being magnetized. It was natural to conclude that these sensations, true or false, were determined by the imagination. We became convinced of this when we saw that this woman, having rested, not feeling anything & being blindfolded, felt all the same effects even though she was not magnetized; but the demonstration was completed when, after a fifteen minute experiment and her imagination probably tired & cooled off, the effects diminished instead of increasing at the very moment she was really being magnetized.

If she fainted, that is a mishap that happens frequently to women when they are bothered by clothes that are too tight. The laying of hands on the hypochondria may have produced the same effect in an excessively sensitive woman; but this cause is not even needed to explain what happened. It was very hot, the woman no doubt felt strong emotions in those first moments as she prepared to submit to a new, unknown experiment, & after such a prolonged effort, it is not out of the ordinary to feel weak.

This swooning, therefore, has a natural & known cause, but the sensations she experienced when not magnetized, can only be the effect of the imagination. The same results were obtained in similar experiments made by M. Jumelin at the same place, on the following day, in the presence of the Commissioners, on a blindfolded man & a woman with eyes uncovered; it was clear that their answers were determined by the questions that were posed. The question indicated where the sensation ought to be; instead of directing the magnetism towards them, it was only their imagination that was being heightened & directed. A child of five years, magnetized afterwards, felt only the heat generated beforehand in play.

These experiments appeared important enough to the Commissioners to be repeated in order to shed new light & M. Jumelin graciously agreed to participate. It would be pointless to object that M. Jumelin's method is bad; for at this moment it was not magnetism being put to the test but the imagination.

The Commissioners agreed to blindfold the subjects being tested, to not magnetize them most of the time, & to skillfully question them in such a way as to lead them to answers. The point was not to induce error, only to mislead their imaginations. Indeed, when not being magnetized, the sole response ought to be that they feel nothing; & when they are being magnetized, it is the heartfelt sensation that ought to dictate their response, & not the manner in which they are questioned.

The Commissioners, having accordingly moved to the home of M. Jumelin, began by putting his servant to the test. A specially designed blindfold, the same that was used in all subsequent experiments, was placed over his eyes. This blindfold was composed of two rubber crowns, the concave side of which was filled with eiderdown; all this was enclosed in two pieces of cloth sewn into a round shape. These two pieces were attached to one another; they had cords that tied behind. Placed over the eyes, they left a gap for the nose so that the subject could breathe freely without being able to see a thing, not even daylight, through, above, or under the blindfold. These precautions having been taken to secure the comfort of the subjects & the certainty of the results, M. Jumelin's servant was persuaded that he was magnetized. He then felt an almost overwhelming warmth, stirrings in his abdomen, his head became heavier; little by little he began to nod & appeared on the point of falling asleep. All of which proves, as we said earlier, that this effect is due to the situation, to boredom, & not to magnetism.

Magnetized next with eyes uncovered, he feels tingling in his forehead when the metal rod is brought close to it; blindfolded again, he feels no tingling when the rod is brought close; & when it is not, & he is questioned whether he does not feel something on his forehead, he declares he feels something there moving back & forth across it.

M. B**, an educated man, particularly in the field of medicine, blindfolded, offers the same spectacle; feeling effects when there is no action taking place, often feeling nothing when there is. These effects were such that even before being magnetized in any way, but believing he had been for ten minutes, he felt a warmth in his loins that he compared to the warmth of a stove. It is obvious that M. B** had a strong sensation because to describe it, he had to resort to such a comparison; & this sensation was entirely due to the imagination, which alone was acting upon him.

The Commissioners, especially the Physicians, conducted numerous experiments on different subjects whom they magnetized themselves, or whom they led to believe had been magnetized. The Commissioners magnetized randomly with opposite poles or like poles in either sense, & in every instance, they obtained the same results; there was not in all those experiments any variation other than that of the degree of imagination.

They were therefore convinced by facts that the imagination on its own can produce various sensations & make one feel pain, heat, even a substantial amount of heat in all parts of the body, & they have concluded that for many the imagination plays a necessary role in the effects attributed to Animal magnetism. But one must agree that the practice of magnetism produces in animated bodies changes more pronounced & upsets more substantial than the ones which have just been reported. So far none of the subjects who believed that they were magnetized were moved to the point of having convulsions; it therefore was a new type of experiment to test, if by shaking the imagination alone, one could produce crises similar to the ones taking place at the group treatment.

This idea then led to several experiments. When a tree has been touched following principles & methods of magnetism, anyone who stops beside it ought to feel the effect of this agent to some degree; there are some who even lose consciousness or feel convulsions. We spoke of this to M. Deslon who replied that the experiment ought to succeed so long as the subject was very sensitive, & we came to agreement with him to conduct this experiment in Passy, in the presence of M. Franklin. The necessity that the subject be sensitive made the Commissioners think that in order to make the experiment decisive & unquestionable, it must be made on a person chosen by M. Deslon, a person whose sensitivity to magnetism had already been proved. M. Deslon consequently brought with him a young man of about twelve; in the garden orchard, an isolated apricot tree, fit to conserve the magnetism that would be impressed upon it, was marked. M. Deslon was led to it by himself so he could magnetize it, the young man staying in the house in the presence of someone who did not leave his side. One would have wished that M. Deslon not be present during the experiment, but he declared that it could miss the mark if he did not direct his cane & his attention to that tree to amplify the action. It was reluctantly decided to keep M. Deslon as far away as possible & to place the Commissioners between him & the young man in order to ensure that he could make no signals & attest to the fact that no information was exchanged. These precautions, in an experiment that is to be authentic, are necessary without being offensive.

The young man was then brought in, blindfolded & made to stand in front of four trees that had not been magnetized, & asked to hug them each for two minutes as prescribed by M. Deslon himself.

M. Deslon, present & at some distance, pointed the cane at the tree that was really magnetized.

At the first tree, the young man, questioned after one minute, declared that he was perspiring profusely; he coughed, spit & said he felt a slight pain on the head; the distance to the magnetized tree was approximately twenty-seven feet.

At the second tree, he felt giddy with the same pain on the head; the distance was thirty-six feet.

At the third tree, the dizziness increases & the headache as well; he says he thinks he is getting closer to the magnetized tree; it was then about thirty-eight feet away.

Finally, at the fourth non-magnetized tree, & at about twenty-four feet from the magnetized one, the young man had a crisis; he lost consciousness, his limbs stiffened & he was carried to a nearby lawn where M. Deslon gave him first aid & revived him.

The result of this experiment is totally contrary to magnetism. M. Deslon tried to explain what happened by saying that all trees are naturally magnetized & that their own magnetism was strengthened by his presence. But in that case, anyone sensitive to magnetism could not chance going into a garden without incurring the risk of convulsions, an assertion contradicted by everyday experience. M. Deslon's presence did nothing more than it had in the coach in which he arrived with the young man, who sat across from him & felt nothing. Had the young man not felt anything, even under the magnetized tree, it could have been said that he was not sensitive enough, at least on that day: but the young man fell into a crisis under a non-magnetized tree; consequently, it is an effect which has no physical cause whatsoever, no outside cause, & which can have no cause other than the imagination. The experiment is therefore absolutely conclusive: the young man knew he was being led to a magnetized tree, his imagination was struck, successively heightened, & at the fourth tree it rose to the degree necessary to produce the crisis.

Other experiments support this one, & yield the same result. One day the Commissioners met in Passy at M. Franklin's with M. Deslon, having requested the latter to bring some patients with him & choose from amongst the poor being treated those who would be the most sensitive to magnetism. M. Deslon brought two women; & while he was busy magnetizing M. Franklin & several people in another apartment, these two women were separated & placed in two different rooms.

One of them, the woman P**, has leukoma; but as she is able to see a little, her eyes were covered with the blindfold described above. She was persuaded that M. Deslon had been brought in to magnetize her; silence was insisted upon, three Commissioners were present, one to question her, the other to take notes, the third to represent M. Deslon. They acted as if they were addressing M. Deslon, asking him to begin, but the woman was not magnetized at all; the three Commissioners remained quiet, occupied only in observing what was going to happen. After three minutes, the patient started to feel a nervous shiver; then in succession she felt pain in the back of her head, in her arms, pins & needles in her hands, that's the expression she used; she stiffened, clapped her hands, got up from her chair, tapped her feet: the crisis was well defined. Two other Commissioners in the next room with the door closed heard the clapping of hands & tapping of feet &, without seeing anything, were witnesses to this loud affair.

Those two Commissioners were with the other patient, a Mlle. B**, suffering from a nervous ailment. With her eyes left uncovered, her sight was unimpeded; she was seated in front of a closed door & persuaded that M. Deslon was on the other side in the process of magnetizing her. It was barely a minute of sitting there in front of that door before she began to feel shivers. A minute after that she started to chatter even though she felt generally warm; finally, after the third minute, she fell into a complete crisis. Her breathing was racing, she stretched both arms behind her back, twisting them strongly & bending her body forward; her whole body shook. The chatter of teeth was so loud that it could be heard from outside; she bit her hand hard enough to leave teeth marks.

It is well to observe that these two patients were not touched in any way; not even their pulses were felt so that it could not be said that magnetism had been communicated to them, & nonetheless the crises were full blown. The Commissioners, who wanted to know the effect of the workings of the imagination & appreciate what role it could have in the crises of magnetism, obtained all that they had wanted. It is impossible to see the effect of these workings more overtly or in a more evident way than in these two experiments. If the patients have claimed that their crises are stronger during treatment, it is because the shaking of nerves is catching & in general everyone's own individual emotion is increased by the spectacle of similar emotions.

We had an opportunity to test the woman P** a second time & to realize the extent to which she was ruled by her imagination. We wished to conduct the experiment of the magnetized cups: this experiment consists of choosing from amongst a number of cups one that is magnetized. The cups are presented one after the other to a patient sensitive to magnetism; he ought to have a crisis or at least sense some effect when the magnetized cup is presented; he ought to be indifferent to all the others that are not. It is only necessary that, as recommended by M. Deslon, the direct pole be presented so that the person handling the cup does not magnetize the patient, & that no effect other than the cup's magnetism be involved. The woman P** was summoned to M. Lavoisier's Arsenal where M. Deslon was present; she started falling into shock in the anteroom, before having seen either M. Deslon or the Commissioners; but she knew she should be seeing him, & that is a striking effect of the imagination.

After the crisis had abated, the woman was led to the site of the experiment. Several cups not at all magnetized were presented to her; the second cup started to affect her, & at the fourth, she fell completely into a crisis. It can be said that her actual state was that of a nervous crisis that had begun in the anteroom & began again on its own; but what is crucial is that having asked for a drink, it was given to her in the cup magnetized by M. Deslon himself; she drank quietly & said she felt relieved. Therefore the cup & magnetism missed their marks, because the crisis was quieted rather than exacerbated.

Sometime later, while M. Majault was examining her leukoma, the magnetized cup was brought close to the back of her head & held there for twelve minutes; she noticed nothing & felt no effect whatsoever, she was even calmer than at any other time because her imagination was distracted & occupied by the eye examination being made.

The Commissioners were told that this woman, left alone in the anteroom, suffered renewed convulsions when approached by several persons who had nothing to do with magnetism. It was pointed out to her that she was not being magnetized; but her imagination was so excited that she replied: if you were not doing anything to me, I would not be in the state I am in. She knew she had come to be the subject of experiment; someone's approach, the least noise drew her attention, awakening the idea of magnetism & renewing the convulsions.

In order to act powerfully, the imagination often needs to be stimulated in different ways simultaneously. The imagination responds to all the senses; its reaction must be proportional to the number of senses that move it & the feelings received: this is what the Commissioners realized following an experiment that they are about to describe. M. Jumelin had told them of a young lady, age 20, whose speech he had removed by the power of magnetism; the Commissioners repeated this experiment at his house, and the young lady agreed to it & agreed to be blindfolded.

First we tried to obtain the same result without magnetizing her; but whether she felt or believed she felt the effects of magnetism, we were unable to stimulate her imagination enough for the experiment to succeed. When she was really magnetized with eyes blindfolded, we were not more successful. The blindfold was removed; then the imagination was stimulated by sight as well as by hearing, the effects were more noticeable; but even though her head began to droop, even though she felt pressure at the base of the nose & many of the symptoms that she had felt the first time, she did not however lose her ability to speak. What she asked for was done, & in three fourths of a minute she became mute; only a few inarticulate sounds could be heard despite the visible efforts of the throat to push out sounds & those of the tongue & lips to enunciate. This state lasted only a minute: one can see that finding itself in precisely the same circumstances, the seduction of the mind & its effects on the organs of speech were the same. But it was not enough that the spoken word alerted her to the fact that she had been magnetized, it was necessary that the sense of sight make a stronger impression capable of stirring the imagination; it was necessary also that it be a known gesture to revive her ideas. It seems that this experiment shows wonderfully how the imagination works, being heightened by degree & requiring extra outside help in order to be stimulated more effectively.

This power that sight has over the imagination explains the effects that the doctrine of magnetism attributes to it. It is preeminently sight that has the power to magnetize; signs & gestures employed are ordinarily useless, the Commissioners were told, unless the subject has already been taken hold of by being glanced upon. The reason is simple; it is in the eyes where the most expressive traits of the passions are, & it is there that all that is most important & most seductive in character is unfolded. Therefore, the eyes must have a great power over us; but they have this power because they stir the imagination, & in a manner more or less exaggerated according to the strength of that imagination. It is therefore sight that gets all the work of magnetism underway; & the effect is so powerful, its origins so deep, that a woman newly arrived at M. Deslon's, coming out of a crisis & meeting the gaze of the disciple of Deslon who magnetized her, stared at him for three quarters of an hour. For a long time she was hounded by this look; she kept seeing before her that same eye intent on watching her; & she constantly carried it in her imagination for three days, whether asleep or awake. One sees all that can be produced by an imagination able to preserve the same impression for such a long time, the same impression, that is to say, able to revive by its own power the same feeling for three days.

The experiments just reported are consistent & also decisive; they authorize the conclusion that the imagination is the real cause of the effects attributed to magnetism. But the supporters of this new agent will perhaps reply that the identity of the effects does not always prove the identity of the causes. They will allow that the imagination may excite these impressions without magnetism; but they will maintain that magnetism can also excite them without the help of the imagination. The Commissioners could easily destroy this assertion by using reason & the principles of Physics: first & foremost, new causes are not to be postulated unless absolutely necessary. When the effects observed can have been produced by an existing cause, already manifested in other phenomena, sound Physics teaches that the effect observed must be attributed to it; & when one announces the discovery of a cause hitherto unknown, sound Physics also demands that it be established, demonstrated by effects that cannot be attributed to any known cause, & that can only be explained by the new cause. It would thus be up to the followers of magnetism to present other proofs & to look for effects that were entirely stripped of the illusion of the imagination. But as facts are more conclusive than reasoning & provide more striking evidence, the Commissioners wanted to put to the test what magnetism would be when the imagination was not at work.

An apartment with adjacent rooms & a communicating door was prepared. The door was removed & replaced by a frame, covered with two layers of paper. In one of these rooms was one of the Commissioners there to write down all that would happen, & a lady introduced as being from the provinces & in need of a seamstress. Mlle. B**, a seamstress who had already been used during the experiments in Passy & whose sensitivity to magnetism was known, was asked to come over. When she arrived, all was arranged so that there was only one chair where she could sit & this chair was situated in the embrasure of the communicating door where she found herself as in a nook.

The Commissioners were in the other room, & one of them, a Physician trained to magnetize & having already produced effects, was put in charge of magnetizing Mlle. B** through the paper frame. It is a principle of the theory of magnetism that this agent passes through wooden doors, walls, etc. A paper frame could not be an obstacle; moreover, M. Deslon has positively established that magnetism passes through paper; & Mlle. B** was magnetized as if she had been in the open & in his presence.

For a half-hour, from a distance of a foot & a half, she was magnetized with opposite poles, following all the procedures which had been taught by M. Deslon, & which the Commissioners saw practiced at his home. During all this time, Mlle. B** was conversing cheerfully; asked about her health, she answered freely that she felt quite well: in Passy she had fallen into a crisis after three minutes; here she endured magnetism for thirty minutes without any effect. It is just that here she did not know she was magnetized, & in Passy she believed that she was. One sees therefore that the imagination alone produces all the effects attributed to magnetism; & when the imagination does not act, there are no more effects.

But one objection can be made to this experiment; that Mlle. B** could have been ill disposed & found herself less sensitive at that time to magnetism. The Commissioners anticipated the objection & consequently conducted the following experiment. As soon as one ceased to magnetize through the paper, the same Physician-Commissioner moved to the other room; it was easy to induce Mlle. B** to be magnetized. He then commenced magnetizing her, being careful, as in the preceding experiment, to stand at a distance of one & a half feet from her, to use only the gestures & movements of the index finger & the metal rod, for had he applied his hands & touched her hypochondria, it could have been said that magnetism had acted through this closer contact. The only difference between these two experiments is that in the first, he magnetized with opposite poles, following the rules, whereas in the second, he magnetized with direct poles & backwards. Acting in this way, by the theory of magnetism, no effect at all should have been produced.

However after three minutes, Mlle. B** felt ill at ease & short of breath; then followed interspersed hiccups, chattering of the teeth, a tightening of the throat & a bad headache; she anxiously stirred on her chair; she complained about lower back pain; she occasionally tapped her feet rapidly on the floor; she then stretched her arms behind her back, twisting them strongly as in Passy; in a word, a complete & perfectly characteristic convulsive crisis. She suffered all this in twelve minutes whereas the same treatment employed for thirty minutes found her insensitive. The only thing added here is the imagination; it is therefore to it that these effects are due.

If the imagination started the crisis, it is also the imagination that made it stop. The Commissioner who magnetized her said it was time to finish; crossing his two index fingers, he presented them to her; & it is well to observe that by this he was magnetizing her with direct poles as he had done so far; nothing therefore had changed, the same treatment should have continued the same impressions. But the intention was enough to calm the crisis; the heat & headache dissipated. The areas that hurt were attended to one after the other, while announcing that the pain would disappear. In this way, the voice, by directing the imagination, caused the pain in the neck to stop, then in succession the irregularities in the chest, stomach & arms. It took only three minutes; after which Mlle. B** declared that she no longer felt anything & was absolutely back in her natural state.

These last experiments along with several done at the home of M. Jumelin have the double advantage of simultaneously demonstrating the power of the imagination & the nullity of magnetism in the effects produced.

If the effects are even more marked & crises seemingly more violent during group treatment, it is because several causes concur with the imagination to multiply & magnify the effects. The process begins with staring to take hold of the mind; touching & applying hands soon follow; & it is appropriate here to develop an exposition of the physical effects.

These effects are more or less substantial; the lesser are the hiccups, stomach upsets, purges; the more substantial are convulsions which are called crises. The place where touching occurs is the hypochondria, at the pit of the stomach, & sometimes on the ovaries when it is women who are touched. Hands, fingers press & more or less squeeze these different areas.

The colon, one of the large intestines, runs across both regions of the hypochondria & the epigastric area that separates them. It is placed directly under the tegument. It is therefore on this intestine that touching takes place, on this sensitive & very irritable intestine. Movement alone, repeated movements without any other agent, excite the muscular action of the intestine & sometimes results in evacuations. Nature seems to indicate, as by instinct, this maneuvering to hypochondriacs. The practice of magnetism is nothing more than this very maneuvering; & the purges which it can produce are facilitated further in the magnetic treatment by the frequent & almost habitual use of a real purgative, diluted cream of tartar.

But when this movement principally excites the irritability of the colon, this intestine presents other phenomena. It swells more or less, & sometimes to a considerable volume. It then communicates to the diaphragm such an irritation that this organ enters more or less into convulsions & this is what we call *crisis* in the treatment of Animal magnetism. One of the Commissioners has seen a lady subject to a kind of spasmodic vomiting repeated several times a day. The efforts produced only a cloudy & viscous fluid similar to that vomited by patients in crisis during the practice of magnetism. The convulsion had its seat in the diaphragm; & the region of the colon was so sensitive that the slightest touching of that area, a strong disturbance of the air, the surprise caused by an unexpected noise, sufficed to stimulate the convulsion. Thus this woman had crises without magnetism due solely to the irritability of the colon & the diaphragm, & women who are magnetized have their crises due to the same cause & by this irritability.

The laying of hands on the stomach has physical effects equally remarkable. The application is made directly upon this organ. Sometimes compression there is strong & continuous, sometimes light & repeated; sometimes vibrations are transmitted to this part by rotating the metal rod; lastly, thumbs are sometimes passed along there quickly & successively one after the other. These maneuvers quickly bring to the stomach an irritation strong & more or less lasting depending on whether the subject is more or less sensitive & irritable. Compressing the stomach predisposes it to this irritation.

This compression allows it to act on the diaphragm, & to communicate to it the impressions it receives. It cannot become irritated unless the diaphragm is irritated, & from there, as by the action of the colon, result the nervous symptoms we have just talked about.

With sensitive women, if one has put pressure on the two hypochondria without making any movement, the stomach tightens & these women faint. This is what happened to the woman magnetized by M. Jumelin; & what often happens without any other cause when the clothes of women are too tight; there is then no crisis, because the stomach is squeezed without being irritated, & because the diaphragm remains in its natural state. These same maneuvers practiced on the ovaries, aside from the effects that are particular to them, produce the same symptoms even more powerfully. The influence & the power of the uterus on animal economy is well known.

The intimate relation between the colon, the stomach & the uterus with the diaphragm is one of the causes of the effects attributed to magnetism. The lower abdominal regions, subjected to various touches, respond to a different plexus that constitutes a veritable nervous center, by means of which, aside from all other systems, it very likely excites a sympathy, a communication, a correspondence between all parts of the body, an action & a reaction such that the sensations excited in this center shake the other parts of the body; & vice versa such that a sensation felt in one part gets the nervous system going, which often transmits this impression to all the other parts.

This explains not only the effects of magnetic touching but also the physical effects of the imagination. It has always been observed that the affections of the soul make their first impression on this nervous center, which leads to the common saying that one has a weight on the stomach & that one feels suffocated. The diaphragm joins in, from which come sighs, tears & laughter. Next a reaction is felt on the viscera of the lower abdomen; & that is how we can make sense of physical disorders produced by the imagination. A sudden chill occasions colic, fear causes diarrhea, sadness gives rise to jaundice. The history of Medicine contains infinite examples of the power of the imagination & the influence of the soul. The fear of fire, a violent desire, a strong & lasting hope, a crisis of anger return the use of legs to a man crippled by gout, to a paralytic; an intense & unexpected joy dissipates a quartan fever two months old; a strong attentiveness brings a halt to hiccups; accidental mutes recover speech following a strong emotion of the soul. History shows that this emotion suffices to recover speech, & the Commissioners saw that striking the imagination was enough to cause its loss. The action & the reaction of the physical upon the mental & of the mental upon the physical have been demonstrated since observation has been part of Medicine, that is, from its origin. Crises arise from touching & from the imagination.

Tears, laughter, coughs, hiccups, & in general all the effects observed during what are called the crises of the group treatment arise therefore from either the functions of the diaphragm disturbed by physical means, such as touching & pressure, or from the power of the imagination so gifted for acting upon this organ & disturbing its functions.

If it were objected that touching is not always necessary for these effects, the reply would be that the imagination may possess enough resources to manufacture everything by itself—especially the imagination acting in a group treatment, doubly excited therefore by its own movement & that of the surrounding imaginations. We have seen what it produced in the experiments made by the Commissioners on isolated subjects; one can judge of its multiplied effects on patients brought together in the group treatment. These patients are assembled in a tight place, relative to their number: the air is warm, although care is taken to renew it; & it is always more or less laden with mephitic gas the action of which particularly affects the head & the nervous system. If there is music, it is another means of acting upon nerves & of stimulating them.

Several women are magnetized simultaneously & at first feel only effects similar to those noted by the Commissioners in several of their experiments. They have recognized that even during the group treatment, it is more often only after two hours that the crises begin. Little by little, impressions are communicated & reinforce each other, as one may notice at theatrical spectacles where the impressions are greater when there are many spectators, & especially in the places where one is at liberty to applaud. This indication of particular emotions establishes a general emotion which each shares to the extent to which he is susceptible. It is this that one observes also in armies on the day of battle, when the enthusiasm of courage as well as the panic of terror spread with so much rapidity. The sound of the drum & of the military music, the noise of the cannon, the musket fire, the cries, the disorder rattle the organs, give to the mind the same movement & heighten imaginations to the same degree. In this drunken unity, one impression manifested becomes universal; it encourages a charge or determines flight. The same cause gives birth to revolts; the imagination governs the multitude: men gathered in numbers are more taken by their senses, reason has less hold on them; & when fanaticism presides over these assemblies, it gives rise to the Tremblers of the Cevennes.

It is in order to stop such disturbances which can spread so easily that gatherings are forbidden in seditious towns. The mind is everywhere influenced by example. Mechanical imitation brings the physical into play: by isolating individuals, one can quiet their minds; by separating them, one can stop convulsions, naturally always contagious: we have a recent example of this in the young girls of Saint-Roch, who when separated were healed of the convulsions they suffered from when together.

Thus we meet again with magnetism, or rather with the theatrical play of the imagination, in the army, in large gatherings like that around the vat, acting by different means, but producing the same effects. The vat is surrounded by a new crowd of patients: sensations are continuously communicated & returned; in the end the exercise wears out the nerves; they become irritated & the woman who is most sensitive gives the signal. At that point the cords, all pulled to the same degree & in unison, respond, & the crises multiply; they mutually reinforce each other; they become violent. At the same time, the men witnessing these emotions share them to the degree of their nervous sensibility, & those whose sensibility is greater & more easily affected fall into a crisis themselves. This great affectability, in part natural & in part acquired, in men as well as women, becomes habitual. Having felt these sensations once or several times, it is only a question of recalling their memory to stimulate the imagination to the degree necessary to create the same effects. This is something always easy to do by placing the subject in the same circumstances. Then there is no need for group treatment, one has only to touch the hypochondria, to pass the finger & the metal rod in front of the face; the gestures are known. It is not even necessary that they be employed, it suffices that patients, eyes blindfolded, believe that the gestures are being repeated, that they are persuaded they are being magnetized; the ideas awake, the sensations reproduce themselves, the imagination employing familiar means, & taking the same paths, makes the same phenomena reappear. It is this that happens to the patients of M. Deslon, who fall into crisis without a vat, & without being excited by the spectacle of group treatment.

Touching, imagination, imitation, these then are the real causes of the effects attributed to this new agent, known under the name Animal magnetism, to this fluid said to circulate in the body & to spread from individual to individual; such is the result of the experiments by the Commissioners, & the observations that they made on the methods employed, & on the effects produced. This agent, this fluid does not exist, but as chimerical as it is, the idea of it is not new. A few authors, a few physicians from the last century have expressly dealt with it in several works. The curious & interesting researches of M. Thouret prove to the group that the theory, the processes, the effects of Animal magnetism, proposed in the last century, closely resembled those being taken up again in this one. Magnetism therefore is only an old error. This theory is being presented today with a more impressive apparatus, necessary in a more enlightened century; but it is not for that reason less false. Man seizes, abandons, takes up again the error that gratifies him. There are errors which will be eternally dear to humanity. How many times has astrology not reappeared upon the earth! Magnetism draws us to return to it. The desire has been to link it to celestial influences so as to make it more captivating & attract men with the double hopes that touch them most, the hope of knowing their futures, & the hope of prolonging their days.

There is reason to believe that the imagination is the most important of the three causes that we have just assigned to magnetism. We have seen by the experiments cited that it suffices on its own to produce crises. Pressure, touching appear therefore to serve it as preparations; it is through touching that the nerves are unsettled, imitation communicates & spreads the sensations. But the imagination is this terrible, active power that produces the great effects one observes with astonishment in the group treatment. These effects are astonishing in the eyes of everyone, while the cause is obscure & hidden. When it is considered that in the last centuries these effects have captivated men esteemed for their merit, their knowledge, & even genius, such as Paracelsus, Vanhelmont, Kirker, etc., it should not be surprising if today, persons who are educated, enlightened, if even a great number of Physicians have been taken in. The Commissioners admitted only to the group treatment where there is neither time nor the ability to conduct decisive experiments could themselves have been led into error. The freedom to isolate the effects was necessary in order to distinguish the causes; one must like them have seen the imagination work, partially in some way, to produce its effects separately & in detail, so as to conceive of the accumulation of these effects, to get an idea of its total power & take account of its wonders. But such examination requires a sacrifice of time, & much follow-up research which one does not always have the leisure to pursue for the purpose of instruction or satisfying one's own curiosity, or which one does not have even the right to undertake unless one is like the Commissioners charged by the King's orders, & honored with the group trust.

M. Deslon does not stray from his principles. He declared at the committee meeting held at the home of M. Franklin on June 19 that he believed he could in fact lay down the principle that the imagination had the greatest part in the effects of Animal magnetism; he said that this new agent may be only the imagination itself, the power of which is so great that it is little understood: at the same time he certifies that he has constantly been cognizant of this power in the treatment of his patients, & he certifies also that several have been healed or remarkably relieved. He has remarked to the Commissioners that the imagination directed in this way toward the relief of human suffering, would be a great blessing in the practice of Medicine; & persuaded of the truth of the imagination's power, he invited them to study its workings & effects at his home. If M. Deslon is still attached to the first idea that these effects are due to the action of a fluid that is communicated from person to person through touching or under the direction of a conducting agent, it will not take him long to recognize with the Commissioners that all that is needed is one cause for one effect, & that because the imagination is sufficient, the fluid is useless. No doubt we are surrounded by a fluid that belongs to us, imperceptible perspiration forms around us an atmosphere of vapors equally imperceptible; but this fluid acts only like the atmospheres, can only be communicated in infinitely small quantities through touching, is not directed either by conductors, or by sight, or by intention, is not at all spread by sound, nor reflected in mirrors, & is in no way admitting of the effects attributed to it.

It remains to examine whether the crises or the convulsions produced by the processes of this so-called magnetism, in the gathering around the vat, can be useful in healing or relieving the sick. No doubt the imagination of patients often has an influence upon the cure of their maladies. The effect is only known through a general experiment & was not determined by positive experiments but it does not appear that we can doubt it. It is a well-known adage that in medicine faith saves; this faith is the product of the imagination: the imagination therefore acts only through gentle means; through spreading calm through the senses, through reestablishing order in functions, in reanimating everything through hope. Hope is the life of man; what can give him the one contributes to him the other. But when the imagination produces convulsions, it acts through violent means; these means are almost always destructive. In a few very rare cases, they can be useful; there are some desperate cases where all must be disturbed in order to be put in order anew. These dangerous upsets may only be used in Medicine the way poisons are. It must be necessity that dictates their use & economy that controls it. This need is momentary, the upset must be unique. Far from repeating it, the wise physician busies himself with repairing the damage it has necessarily produced; but at the group treatment of magnetism, crises repeat themselves everyday, they are long, violent; the situation of these crises being harmful, making a habit of them can only be disastrous. How can one conceive that a woman whose chest is affected may without danger have bouts of convulsive coughing, of forced expectorations; & by violent & repeated efforts, tire & perhaps tear the lung where one has so much difficulty bringing balm & soothing? How can one imagine that a man, whatever his disease, in order to cure it must fall into crises where sight appears to be lost, where limbs stiffen, where with furious & involuntary movements he batters his own chest; crises that end with an abundant spitting up of mucus & blood! This blood is neither polluted nor corrupted; this blood comes from vessels torn by the efforts & from whence it comes contrary to the wish of Nature. These effects therefore are real afflictions & not curative ones; they are maladies added to the disease whatever it may be.

These crises still have another danger. Man is constantly controlled by habit; habit modifies Nature by successive degrees, but it disposes it so strongly that it often changes it entirely & makes it unrecognizable. Who can tell whether that crisis-state, at first impressed upon the will, will not become habitual? Whether this habit, thus acquired, would often reproduce the same incidents against one's will, & almost without the help of the imagination, which would be the lot of an individual subjected to these violent crises, physically & morally tormented by their unhappy impression, whose days would be divided between apprehensiveness & pain, & whose life would be only a lasting torture? These afflictions of the nerves, when they are natural, are the scourge of Physicians; it should not be the place of art to produce them. This art is disastrous, disturbing the functions of animal economy, pushing Nature to deviate, & multiplying the victims of its disordering. This art is especially dangerous in that not only does it aggravate nervous disorders by bringing the accidents back to mind, by making them degenerate into habits, but if this malady is contagious, as one may suspect, the practice of provoking nervous convulsions, & exciting them publicly during the treatments, is a means of spreading them in large cities; & even of afflicting the generations to come because the ills & habits of parents are transmitted to their posterity.

The Commissioners, having recognized that this Animal-magnetism fluid cannot be perceived by any of our senses, that it had no action whatsoever, neither on themselves, nor on patients submitted to it; having certified that pressure & touching occasion changes rarely favorable to animal economy & perturbations always distressing in the imagination; having finally demonstrated by decisive experiments that the imagination without magnetism produces convulsions, & that magnetism without imagination produces nothing; they have unanimously concluded, on the question of the existence & utility of magnetism, that nothing proves the existence of Animal-magnetism fluid; that this fluid with no existence is therefore without utility; that the violent effects observed at the group treatment belong to touching, to the imagination set in action & to this involuntary imitation that brings us in spite of ourselves to repeat that which strikes our senses, & at the same time, they feel obliged to add, as an important observation, that the touchings, the repeated action of the imagination in producing crises can be dangerous; that the witnessing of these crises is equally dangerous because of this imitation which Nature seems to have made a law; & that, consequently, all group treatment in which the means of magnetism will be used, can in the long run have only disastrous effects.

In Paris, this August eleven one thousand seven hundred & eighty four.

Signed B. Franklin, Majault, le Roy, Sallin, Bailly, d'Arcet, de Bory, Guillotin, Lavoisier

What Ever Happened to N-Rays?

Robert Wood's 1904 N-Ray Letter in *Nature*

Editor's note: This essay is the third in a series of classic historical pieces in skeptical and pseudoscience literature. Following William Jennings Bryan's never-delivered "Address to the Jury in the Scopes Case" on "The Most Powerful Argument against Evolution Ever Made" and Benjamin Franklin's and Antoine Lavoisier's investigation of Mesmerism for King Louis XVI of France. Here we republish Robert W. Wood's famous letter in *Nature* that blew apart the chimerical search for n-rays, with an introduction by psychologist and skeptical investigator Terence Hines.

A Classic in Skeptical History

TERENCE HINES

In early 1903, the news of the discovery of a new type of radiation in France spread through the international physics community. Rene Blondlot, one of the most famous physicists in the world, had made the discovery at the University of Nancy. He named the new radiation n-rays in honor of the university and city. The discovery of a new form of radiation was certainly not an unprecedented event at the start of the 20th century. Several other types of radiation had been reported in the dozen or so years previously (including x-rays). But none would be more controversial than n-rays.

N-rays were supposedly a form of radiation exhibited by any number of substances, with the bizarre exceptions of green wood and "anesthetized" metal (metal soaked in ether or chloroform). Within less than a year of its announced "discovery," no fewer than 30 papers were published confirming the existence of the new rays. Other laboratories, however, using more sophisticated methods were unable to replicate the findings. Blondlot's measuring instrument was a spectroscope with an aluminum-coated prism and thread on the inside. The n-rays were refracted by the prism and spread out into a spectrum. The only way to see the normally invisible n-rays was to cause them to hit a treated thread (e.g., one coated in calcium sulfide). Moving the thread across the gap between the prism and n-ray source caused the thread to become illuminated and this is what was reported as a "detection."

In 1903 *Nature* sent Johns Hopkins University physicist Robert W. Wood, who was attending a scientific conference in Britain, to

Nancy, France, to investigate. During a series of experiments, when the lights were out, Wood secretly removed the prism from the spectroscope, after which n-rays were still detected, clearly an impossible result since the prism was supposedly critical for refracting the rays. In short, what Wood's little experiment proved was that n-rays didn't exist. Blondlot's use of a purely subjective methodology, as opposed to an objective one, led him to believe in the reality of the new rays, as it did in several other laboratories, mostly in France. (There may have been some nationalistic bias here since the Germans had discovered x-rays).

Wood was an extraordinary individual whose wide-ranging areas of interest included many in physics, as well as non-traditional areas such as investigating spiritualistic mediums and the use of scientific methodology in crime detection. Following his visit to Blondlot's laboratory, Wood reported his findings in the September 29, 1904, issue of *Nature,* then, as it is today, one of the leading scientific publications in the world. This letter, reprinted here, is a classic in skeptical literature. After its appearance in *Nature,* it was quickly published in French in the *Revue Scientifique* (Vol. 2, Oct. 22, 1904, 536–538) and in German in the *Physikalische Zeitschrift* (Vol. 1, 1904, 789–791).

The letter seems to have had quite an effect. According to M. Nye, whose excellent history of the n-ray affair should be consulted for further details ("N-rays: An Episode in the History and Psychology of Science." *Historical Studies in the Physical Sciences,* 1980, 125–156), "only one confirming account of n-rays was presented to the [French] Academy" in the following years. Thus, Wood's letter signaled the beginning of the end of the n-ray episode. The debate would simmer on for a few more years and Blondlot, who retired in 1909, continued his n-ray quest, but to no avail.

It is worth noting that nowhere in Wood's letter did he specify at which laboratory it was that he made his observations. But everyone in the field knew.

The n-Rays

BY ROBERT W. WOOD

Nature, *September 29, 1904, pp. 530–531*

The inability of a large number of skillful experimental physicists to obtain any evidence whatever of the existence of the n-rays, and the continued publication of papers announcing new and still more remarkable properties of the rays, prompted me to pay a visit to one of the laboratories in which the apparently peculiar conditions necessary for the manifestation of this most elusive form of radiation appear to exist. I went, I must confess, in a doubting frame of mind, but with the hope that I might be convinced of the reality of the phenomena, the accounts of which have been read with so much scepticism.

After spending three hours or more in witnessing various experiments, I am not only unable to report a single observation which appeared to indicate the existence of the rays, but left with a very firm conviction that the few experimenters who have obtained positive results have been in some way deluded.

A somewhat detailed report of the experiments which was shown to me, together with my own observations, may be of interest to the many physicists who have spent days and weeks in fruitless efforts to repeat the remarkable experiments which have been described in the scientific journals of the past year.

The first experiment which it was my privilege to witness was the supposed brightening of a small electric spark when the n-rays were concentrated on it by means of an aluminum lens. The spark was placed behind a small screen of ground glass to diffuse the light, the luminosity of which was supposed to change when the hand was interposed between the spark and the source of the n-rays.

It was claimed that this was most distinctly noticeable, yet I was unable to detect the slightest change. This was explained as due to a lack of sensitiveness of my eyes, and to test the matter I suggested that the attempt be made to announce the exact moments at which I introduced my hand into the path of the rays, by observing the screen. In no case was a correct answer given, the screen being announced as bright and dark in alternation when my hand was held motionless in the path of the rays, while the fluctuations observed when I moved my hand bore no relation whatever to its movements.

I was shown a number of photographs which showed the brightening of the image, and a plate was exposed in my presence, but they were made, it seems to me, under conditions which admit of many sources of error. In the first place, the brilliancy of the spark fluctuates all the time by an amount which I estimated at 25 per cent, which alone would make accurate work impossible.

Secondly, the two images (with n-rays and without) are built of "installment exposures" of five seconds each, the plate holder being shifted back and forth by hand every five seconds. It appears to me that it is quite possible that the difference in the brilliancy of the images is due to a cumulative favoring of the exposure of one of the images, which may be quite unconscious, but may be governed by the previous knowledge of the disposition of the apparatus. The claim is made that all accidents of this nature are made impossible by changing the conditions, i.e., by shifting the positions of the screens; but it must be remembered that the experimenter is aware of the change, and may be unconsciously influenced to hold the plate holder a fraction of a second longer on one side than on the other. I feel very sure that if a series of experiments were made jointly in this laboratory by the originator of the photographic experiments and Profs. Rubens and Lummer, whose failure to repeat them is well known, the source of the error would be found.

I was next shown the experiment of the deviation of the rays by an aluminum prism. The aluminum lens was removed, and a screen of wet cardboard furnished with a vertical slit about 3 mm. wide put in its place. In front of the slit stood the prism, which was supposed not only to bend the sheet of rays, but to spread it out into a spectrum. The positions of the deviated rays were located by a narrow vertical line of phosphorescent paint, perhaps 0.5 mm. wide, on a piece of dry cardboard, which was moved along by means of a small driving engine. It was claimed that a movement of the screw corresponding to a motion of less than 0.1 of a millimeter was sufficient to cause the phosphorescent line to change in luminosity when it was moved across the n-ray spectrum, and this with a slit 2 or 3 mm. wide. I expressed surprise that a ray bundle 3 mm. in width could be split up into a spectrum with maxima and minima less than 0.1 of a millimeter apart, and was told that this was one of the inexplicable and astonishing properties of the rays. I was unable to see any change whatever in the brilliancy of the phosphorescent line as I moved it along, and I subsequently found that the removal of the prism (we were in a dark room) did not seem to interfere in any way with the location of the maxima and minima in the deviated (!) ray bundle.

I then suggested that an attempt be made to determine by means of the phosphorescent screen whether I had placed the prism with its refracting edge to the right or the left, but neither the experimenter nor his assistant determined the position correctly in a single case (three trials were made). This failure was attributed to fatigue.

I was next shown an experiment of a different nature. A small screen on which a number of circles had been painted with luminous paint was placed on the table in the dark room. The approach of a large steel file was supposed to alter the appearance of the spots, causing them to appear more distinct and less nebulous. I could see no change myself, though the phenomenon was described as open to no question, the change being very marked. Holding the file behind my back, I moved my arm slightly towards and away from the screen. The same changes were described by my colleague. A clock face in a dimly lighted room was believed to become much more distinct and brighter when the file was held before the eyes, owing to some peculiar effect which the rays emitted by the file exerted on the retina. I was unable to see the slightest change, though my colleague said that he could see the hands distinctly when he held the file near his eyes, while they were quite invisible when the file was removed. The room was dimly lit by a gas jet turned down low, which made blank experiments impossible. My colleague could see the change just as well when I held the file before his face, and the substitution of a piece of wood of the same size and shape as the file in no way interfered with the experiment. The substitution was of course unknown to the observer.

I am obliged to confess that I left the laboratory with a distinct feeling of depression, not only having failed to see a single experiment of a convincing nature, but with the almost certain conviction that all the changes in the luminosity or distinctness of sparks and phosphorescent screens (which furnish the only evidence of n-rays) are purely imaginary. It seems strange that after a year's work on the subject not a single experiment has been devised which can in any way convince a critical observer that the rays exist at all. To be sure the photographs are offered as an objective proof of the effect of the rays upon the luminosity of the spark. The spark, however, varies greatly in intensity from moment to moment, and the manner in which the exposures are made appears to me to be especially favourable to the introduction of errors in the total time of exposure which each image receives. I am unwilling also to believe that a change of intensity which the average eye cannot detect when the n-rays are flashed "on" and "off" will be brought out as distinctly in photographs as is the case on the plates exhibited.

Experiments could easily be devised which would settle the matter beyond all doubt; for example, the following: Let two screens be prepared, one composed of two sheets of thin aluminum with a few sheets of wet paper between, the whole hermetically sealed with wax along the edges. The other screen to be exactly the same, containing, however, dry paper.

Let a dozen or more photographs be taken with the two screens, the person exposing the plates being ignorant of which screen was used in each case. One of the screens being opaque to n-rays, the other transparent, the resulting photographs would tell the story. Two observers would be required, one to change the screens and keep a record of the one used in each case, the other to expose the plates.

The same screen should be used for two or three successive exposures, in one or more cases, and it should be made impossible for the person exposing the plates to know in any way whether a change had been made or not.

I feel very sure that a day spent on some such experiment as this would show that variations in the density on the photographic plate had no connection with the screen used.

Why cannot the experimenters who obtain results with n-rays and those who do not try a series of experiments together, as was done only last year by Cremieu and Pender, when doubt had been expressed about the reality of the Rowland effect?

R. W. Wood, Brussels, September 22

Scientific Study of Unidentified Flying Objects

Introduction

Michael Shermer

The *Scientific Study of Unidentified Flying Objects* was conducted at the University of Colorado between 1966 and 1968, with physics professor Edward U. Condon as its primary investigator. It is commonly known as the "Condon Report" or the "Colorado Project Report." The publication represents the largest single scientific project ever undertaken in relation to the UFO question. The *Scientific Study of Unidentified Flying Objects* was originally copyrighted in 1968 by the Regents of the University of Colorado, a body corporate. It was subsequently published in reports of the United States Air Force and other governmental agencies and was published commercially by Bantam Books, but is currently out of print.

Because of the historical importance of this document, the National Capital Area Skeptics, with the permission of the Regents of the University of Colorado, republished the *Scientific Study of Unidentified Flying Objects* on their web page. Under the direction of Jim Giglio, who worked for more than a year to bring this document to the web, and with the permission of the National Capital Area Skeptics, we present these excerpts—the first two sections of the publication—as a slice of twentieth-century history related to UFOs.

My own skepticism about the UFO phenomenon stems from a simple observation involving evolutionary biology: the extra-terrestrial inhabitants of UFOs are invariably described as remarkably similar to terrestrial primates—bilaterally symmetrical with two legs, two arms, two eyes, two ears, fingers and toes, a nose and a mouth. The probability of such creatures being anything like primates, let alone humans, is so remote as to not be worthy of further consideration. Of the hundreds of millions of species to have roamed the earth over the past three billion years, only gorillas, orangutans, chimps, bonobos, and humans have survived as living great apes, and only one species—us—has reached a level of intelligence and culture to achieve space flight. Is it really possible that the evolution of life on some other planet would so resemble ours as to produce another primate-like creature? No.

There is an additional problem, and that is the question of technological evolution. I first addressed this question in my January 2002 column in *Scientific American*, in an essay entitled "Shermer's Last Law." It is based on the famous three "laws" of the science fiction writer Arthur C. Clarke:

> *Clarke's First Law:* "When a distinguished but elderly scientist states that something is possible he is almost certainly right. When he states that something is impossible, he is very probably wrong."
>
> *Clarke's Second Law:* "The only way of discovering the limits of the possible is

to venture a little way past them into the impossible."

Clarke's Third Law: "Any sufficiently advanced technology is indistinguishable from magic."

This last observation stimulated me to think more on the impact the discovery of an Extra-Terrestrial Intelligence (ETI) would have on civilization. To that end I have immodestly proposed Shermer's Last Law (I don't believe in naming laws after oneself, so as the good book warns, the last shall be first and the first shall be last): *Any sufficiently advanced ETI is indistinguishable from God.*

God is typically described by Western religions as omniscient and omnipotent. Since we are far from the mark on these traits, how could we possibly distinguish a God who has them absolutely, from an ETI who has them in relatively (to us) copious amounts? Thus, we would be unable to distinguish between absolute and relative omniscience and omnipotence. But if God were only relatively more knowing and powerful than us, then by definition it *would* be an ETI! Consider two observations and one deduction:

1. Biological evolution operates at a snail's pace compared to technological evolution (the former is Darwinian and requires generations of differential reproductive success, the latter is Lamarckian and can be implemented within a single generation). 2. The cosmos is very big and space is very empty (*Voyager I,* our most distant spacecraft, hurtling along at over 38,000 mph, will not reach the distance of even our sun's nearest neighbor, the Alpha Centauri system that it is *not* even headed toward, for over 75,000 years). Ergo, the probability of an ETI who is only slightly more advanced than us and also makes contact is virtually nil. If we ever do find ETI it will be as if a million-year-old *Homo erectus* were dropped into the middle of Manhattan, given a computer and cell phone, and instructed to communicate with us. ETI would be to us as we would be to this early hominid—godlike.

Science and technology have changed our world more in the past century than it changed in the previous hundred centuries. It took 10,000 years to get from the cart to the airplane, but only 66 years to get from powered flight to a lunar landing. Moore's Law of computer power doubling every eighteen months continues unabated and is now down to about a year. Ray Kurzweil, in *The Age of Spiritual Machines,* calculates that there have been thirty-two doublings since World War II, and that the Singularity point may be upon us as early as 2030. The Singularity (as in the center of a black hole where matter is so dense that its gravity is infinite) is the point at which total computational power will rise to levels that are so far beyond anything that we can imagine that they will appear near infinite and thus, relatively speaking, be indistinguishable from omniscience (note the suffix!).

When this happens the world will change more in a decade than it did in the previous thousand decades. Extrapolate that out a hundred thousand years, or a million years (an eye blink on an evolutionary time scale and thus a realistic estimate of how far advanced ETI will be, unless we happen to be the first space-faring species, which is unlikely), and we get a gut-wrenching, mind-warping feel for just how godlike these creatures would seem.

In Clarke's 1953 novel *Childhood's End,* humanity reaches something like a Singularity (with help from ETIs) and must make the transition to a higher state of consciousness in order to grow out of childhood. One character early in the novel opines that "science can destroy religion by ignoring it as well as by disproving its tenets. No one ever demonstrated, so far as I am aware, the nonexistence of Zeus or Thor, but they have few followers now."

Although science has not even remotely destroyed religion, Shermer's Last Law predicts that the relationship between the two will be profoundly affected by contact with ETI. To find out how we must follow Clarke's Second Law, venturing courageously past the limits of the possible and into the unknown. Ad astra!

This is best done, in my opinion, through the SETI program, the Search for Extra-Terrestrial Intelligence using radio telescopes in the hopes of detecting a signal from an ETI, rather than a close encounter of the third kind. Thus, I agree with the final conclusion of the Condon report, as summarized in "Section I Conclusions and Recommendations":

> We believe that the existing record and the results of the Scientific Study of Unidentified Flying Objects of the University of Colorado, which are presented in detail in subsequent sections of this report, support the conclusions and recommendations which follow.
>
> As indicated by its title, the emphasis of this study has been on attempting to learn from UFO reports anything that could be considered as adding to scientific knowledge. Our general conclusion is that nothing has come from the study of UFOs in the past 21 years that has added to scientific knowledge. Careful consideration of the record as it is available to us leads us to conclude that further extensive study of UFOs probably cannot be justified in the expectation that science will be advanced thereby.
>
> It has been argued that this lack of contribution to science is due to the fact that very little scientific effort has been put on the subject. We do not agree. We feel that the reason that there has been very little scientific study of the subject is that those scientists who are most directly concerned, astronomers, atmospheric physicists, chemists, and psychologists, having had ample opportunity to look into the matter, have individually decided that UFO phenomena do not offer a fruitful field in which to look for major scientific discoveries.
>
> . . .
>
> The question remains as to what, if anything, the federal government should do about the UFO reports it receives from the general public. We are inclined to think that nothing should be done with them in the expectation that they are going to contribute to the advance of science.
>
> This question is inseparable from the question of the national defense interest of these reports. The history of the past 21 years has repeatedly led Air Force officers to the conclusion that none of the things seen, or thought to have been seen, which pass by the name of UFO reports, constituted any hazard or threat to national security.
>
> . . .
>
> It has been contended that the subject has been shrouded in official secrecy. We conclude otherwise. We have no evidence of secrecy concerning UFO reports. What has been miscalled secrecy has been no more than an intelligent policy of delay in releasing data so that the public does not become confused by premature publication of incomplete studies of reports.
>
> The subject of UFOs has been widely misrepresented to the public by a small number of individuals who have given sensationalized presentations in writings and public lectures. So far as we can judge, not many people have been misled by such irresponsible behavior, but whatever effect there has been has been bad.

Scientific Study of Unidentified Flying Objects

Dr. Edward U. Condon
SCIENTIFIC DIRECTOR

Conducted by the University of Colorado under contract No. 44620-67-C-0035 with the United States Air Force

Section II Summary of the Study

1. Origin of the Colorado Project. The decision to establish this project for the Scientific Study of Unidentified Flying Objects stems from recommendations in a report dated March 1966 of an Ad Hoc Committee of the Air Force Scientific Advisory Board set up under the chairmanship of Dr. Brian O'Brien to review the work of Project Blue Book. Details of the history of work on UFOs are set forth in Section V, Chapter 2. (See also Appendix A.)

The recommendation was:

It is the opinion of the Committee that the present Air Force program dealing with UFO sightings has been well organized, although the resources assigned to it (only one officer, a sergeant, and a secretary) have been quite limited. In 19 years and more than 10,000 sightings recorded and classified, there appears to be no verified and fully satisfactory evidence of any case that is clearly outside the framework of presently known science and technology. Nevertheless, there is always the possibility that analysis of new sightings may provide some additions to scientific knowledge of value to the Air Force. Moreover, some of the case records at which the Committee looked that were listed as "identified" were sightings where the evidence collected was too meager or too indefinite to permit positive listing in the identified category. Because of this the Committee recommends that the present program be strengthened to provide opportunity for scientific investigation of selected sightings in more detail than has been possible to date.

To accomplish this it is recommended that:

A. Contracts be negotiated with a few selected universities to provide scientific teams to investigate promptly and in depth certain selected sightings of UFO's. Each team should include at least one psychologist, preferably one interested in clinical psychology, and at least one physical scientist, preferably an astronomer or geophysicist familiar with atmospheric physics. The universities should be chosen to provide good geographical distribution, and should be within convenient distance of a base of the Air Force Systems Command (AFSC).

B. At each AFSC base an officer skilled in investigation (but not necessarily with scientific training) should be designated to work with the corresponding university team for that geographical section. The local representative of the Air Force Office of Special Investigations (OSI) might be a logical choice for this.

C. One university or one not-for-profit organization should be selected to coordinate the work of the teams mentioned under A above, and also to make certain of very close communication and coordination with the office of Project Blue Book.

It is thought that perhaps 100 sightings a year might be subjected to this close study, and that possibly an average of 10 man days might be required per sighting so studied. The information provided by such a program might bring to light new facts of scientific value, and would almost certainly provide a far better basis than we have today for decision on a long term UFO program.

These recommendations were referred by the Secretary of the Air Force to the Air Force Office of Scientific Research for implementation, which, after study, decided to combine recommendations A and C so as to have a single contracting university with authority to subcontract with other research groups as needed. Recommendation B was implemented by the issuance of Air Force Regulation 80-17 (Appendix B) which establishes procedures for handling UFO reports at the Air Force bases.

In setting up the Colorado project, as already stated in Section I, the emphasis was on whether deeper study of unidentified flying objects might provide some "additions to scientific knowledge."

After considering various possibilities, the AFOSR staff decided to ask the University of

Colorado to undertake the project (see Preface). Dr. J. Thomas Ratchford visited Boulder in late July 1966 to learn whether the University would be willing to undertake the task. A second meeting was held on 10 August 1966 in which the scope of the proposed study was outlined to an interested group of the administrative staff and faculty of the University by Dr. Ratchford and Dr. William Price, executive director of AFOSR. After due deliberation, University officials decided to undertake the project.

The contract provided that the planning, direction and conclusions of the Colorado project were to be conducted wholly independently of the Air Force. To avoid duplication of effort, the Air Force was ordered to furnish the project with the records of its own earlier work and to provide the support of personnel at AF bases when requested by our field teams.

We were assured that the federal government would withhold no information on the subject, and that all essential information about UFOs could be included in this report. Where UFO sightings involve classified missile launchings or involve the use of classified radar systems, this fact is merely stated as to do more would involve violation of security on these military subjects. In our actual experience these reservations have affected a negligible fraction of the total material and have not affected the conclusions (Section I) which we draw from our work.

The first research contract with AFOSR provided $313,000 for the first 15 months from 1 November 1966 to 31 January 1968. The contract was publicly announced on 7 October 1966. It then became our task to investigate those curious entities distinguished by lack of knowledge of what they are, rather than in terms of what they are known to be, namely, unidentified flying objects.

2. Definition of an UFO. An unidentified flying object (UFO, pronounced OOFO) is here defined as the stimulus for a report made by one or more individuals of something seen in the sky (or an object thought to be capable of flight but when landed on the earth) which the observer could not identify as having an ordinary natural origin, and which seemed to him sufficiently puzzling that he undertook to make a report of it to police, to government officials, to the press, or perhaps to a representative of a private organization devoted to the study of such objects.

Defined in this way, there is no question as to the existence of UFOs, because UFO reports exist in fairly large numbers, and the stimulus for each report is, by this definition, an UFO. The problem then becomes that of learning to recognize the various kinds of stimuli that give rise to UFO reports.

The UFO is "the stimulus for a report . . ." This language refrains from saying whether the reported object was a real, physical, material thing, or a visual impression of an ordinary physical thing distorted by atmospheric conditions or by faulty vision so as to be unrecognizable, or whether it was a purely mental delusion existing in the mind of the observer without an accompanying visual stimulus.

The definition includes insincere reports in which the alleged sighter undertakes for whatever reason to deceive. In the case of a delusion, the reporter is not aware of the lack of a visual stimulus. In the case of a deception, the reporter knows that he is not telling the truth about his alleged experience.

The words "which he could not identify" are of crucial importance. The stimulus gives rise to an UFO report precisely because the observer could not identify the thing seen. A woman and her husband reported a strange thing seen flying in the sky and reported quite correctly that she knew "it was unidentified because neither of us knew what it was."

The thing seen and reported may have been an object as commonplace as the planet Venus, but it became an UFO because the observer did not know what it was. With this usage it is clear that less well informed individuals are more likely to see an UFO than those who are more knowledgeable because the latter are better able to make direct identification of what they see. A related complication is that less well informed persons are often inaccurate observers who are unable to give an accurate account of what they believe that they have seen.

If additional study of a report later provides an ordinary interpretation of what was seen, some have suggested that we should change its name to IFO, for identified flying object. But we have elected to go on calling it an UFO because some identifications are tentative or controversial, due to lack of sufficient data on which to base a definite identification. A wide variety of ordinary objects have through misinterpretation given rise to UFO reports. This topic is discussed in detail in Section VI, Chapter 2. (The Air Force has published a pamphlet entitled, "Aids to Identification of Flying Objects" [USAF, 1968] which is a useful aid in the interpretation of something seen which might otherwise be an UFO.)

The words "sufficiently puzzling that they undertook to make a report" are essential. As a practical matter, we can not study something that is not reported, so a puzzling thing seen but not reported is not here classed as an UFO.

3. UFO Reports. In our experience, the persons making reports seem in nearly all cases to be normal, responsible individuals. In most cases they are quite calm, at least by the time they make a report. They are simply puzzled about what they saw and hope that they can be helped to a better understanding of it. Only a very few are obviously quite emotionally disturbed, their minds being filled with pseudo-scientific, pseudo-religious or other fantasies. Cases of this kind range from slight disturbance to those who are manifestly in need of psychiatric care. The latter form an extremely small minority of all the persons encountered in this study. While the existence of a few mentally unbalanced persons among UFO observers is part of the total situation, it is completely incorrect and unfair to imply that all who report UFOs are "crazy kooks," just as it is equally incorrect to ignore the fact that there are mentally disturbed persons among them.

Individuals differ greatly as to their tendency to make reports. Among the reasons for not reporting UFOs are apathy, lack of awareness of public interest, fear of ridicule, lack of knowledge as to where to report and the time and cost of making a report.

We found that reports are not useful unless they are made promptly. Even so, because of the short duration of most UFO stimuli, the report usually can not be made until after the UFO has disappeared. A few people telephoned to us from great distances to describe something seen a year or two earlier. Such reports are of little value.

Early in the study we tried to estimate the fraction of all of the sightings that are reported. In social conversations many persons could tell us about some remarkable and puzzling thing that they had seen at some time in the past which would sound just as remarkable as many of the things that are to be found in UFO report files. Then we would ask whether they had made a report and in most cases would be told that they had not. As a rough guess based on this uncontrolled sample, we estimate that perhaps 10% of the sightings that people are willing to talk about later are all that get reported at the time. This point was later covered in a more formal public attitude survey (Section III, Chapter 7) made for this study in which only 7% of those who said they had seen an UFO had reported it previously. Thus if all people reported sightings that are like those that some people do report, the number of reports that would be received would be at least ten times greater than the number actually received.

At first we thought it would be desirable to undertake an extensive publicity campaign to try to get more complete reporting from the public. It was decided not to do this, because about 90% of all UFO reports prove to be quite plausibly related to ordinary objects. A tenfold increase in the number of reports would have multiplied by ten the task of eliminating the ordinary cases which would have to be analyzed. Our available resources for field study enabled us to deal only with a small fraction of the reports coming in. No useful purpose would have been served under these circumstances by stimulating the receipt of an even greater number.

Study of records of some UFO reports from other parts of the world gave us the strong impression that these were made up of a mix of cases of similar kind to those being reported in the United States. For example, in August 1967 Prof. James McDonald of Arizona made a 20-day trip to Australia, Tasmania and New Zealand in the course of which he interviewed some 80 persons who had made UFO reports there at various times. On his return he gave us an account of these experiences that confirmed our impression that the reports from these other parts of the world were, as a class, similar to those being received in the United States. Therefore we decided to restrict our field studies to the United States and to one or two cases in Canada. (See Section III, Chapter 1.) This was done on the practical grounds of reducing travel expense and of avoiding diplomatic and language difficulties. The policy was decided on after preliminary study had indicated that in broad generality the spectrum of kinds of UFO reports being received in other countries was very similar to our own.

4. Prologue to the Project. Official interest in UFOs, or "flying saucers" as they were called at first dates from June 1947. On 24 June, Kenneth Arnold, a business man of Boise, Idaho was flying a private airplane near Mt. Rainier, Washington. He reported seeing a group of objects flying along in a line which he said looked "like pie plates skipping over the water." The newspaper reports called the things seen "flying saucers" and they have been so termed ever since, although not all UFOs are described as being of this shape.

Soon reports of flying saucers were coming in from various parts of the country. Many received prominent press coverage (Bloecher, 1967). UFOs were also reported from other countries; in fact, more than a thousand such reports were made in Sweden in 1946.

The details of reports vary so greatly that it is impossible to relate them all to any single explanation. The broad range of things reported is much the same in different countries. This means that a general explanation peculiar to any one country has to be ruled out, since it is utterly improbable that the secret military aircraft of any one country would be undergoing test flights in different countries. Similarly it is most unlikely that military forces of different countries would be testing similar developments all over the world at the same time in secrecy from each other.

Defense authorities had to reckon with the possibility that UFOs might represent flights of a novel military aircraft of some foreign power. Private citizens speculated that the UFOs were test flights of secret American aircraft. Cognizance of the UFO problem was naturally assumed by the Department of the Air Force in the then newly established Department of Defense. Early investigations were carried on in secrecy by the Air Force, and also by the governments of other nations.

Such studies in the period 1947–52 convinced the responsible authorities of the Air Force that the UFOs, as observed up to that time, do not constitute a threat to national security. In consequence, ever since that time, a minimal amount of attention has been given to them.

The year 1952 brought an unusually large

number of UFO reports, including many in the vicinity of the Washington National Airport, during a period of several days in July. Such a concentration of reports in a small region in a short time is called a "flap." The Washington flap of 1952 received a great deal of attention at the time (Section III, Chapters).

At times in 1952, UFO reports were coming in to the Air Force from the general public in such numbers as to produce some clogging of military communications channels. It was thought that an enemy planning a sneak attack might deliberately stimulate a great wave of UFO reports for the very purpose of clogging communication facilities. This consideration was in the forefront of a study that was made in January 1953 by a panel of scientists under the chairmanship of the late H. P. Robertson, professor of mathematical physics at the California Institute of Technology (Section V, Chapter 2). This panel recommended that efforts be made to remove the aura of mystery surrounding the subject and to conduct a campaign of public education designed to produce a better understanding of the situation. This group also concluded that there was no evidence in the available data of any real threat to national security.

Since 1953 the results of UFO study have been unclassified, except where tangential reasons exist for withholding details, as, for example, where sightings are related to launchings of classified missiles, or to the use of classified radar systems.

During the period from March 1952 to the present, the structure for handling UFO reports in the Air Force has been called Project Blue Book. As already mentioned the work of Project Blue Book was reviewed in early 1966 by the committee headed by Dr. Brian O'Brien. This review led to the reaffirmation that no security threat is posed by the existence of a few unexplained UFO reports, but the committee suggested a study of the possibility that something of scientific value might come from a more detailed study of some of the reports than was considered necessary from a strictly military viewpoint. This recommendation eventuated in the setting up of the Colorado project.

The story of Air Force interest, presented in Section V, Chapter 2, shows that from the beginning the possibility that some UFOs might be manned vehicles from outer space was considered, but naturally no publicity was given to this idea because of the total lack of evidence for it.

Paralleling the official government interest, was a burgeoning of amateur interest stimulated by newspaper and magazine reports. By 1950 popular books on the subject began to appear on the newsstands. In January 1950 the idea that UFOs were extraterrestrial vehicles was put forward as a reality in an article entitled "Flying Saucers are Real" in *True* magazine written by Donald B. Keyhoe, a retired Marine Corps major. Thereafter a steady stream of sensational writing about UFOs has aroused a considerable amount of interest among laymen in studying the subject.

Many amateur organizations exist, some of them rather transiently, so that it would be difficult to compile an accurate listing of them. Two such organizations in the United States have a national structure. These are the Aerial Phenomena Research Organization (APRO), with headquarters in Tucson, Arizona, claiming about 8000 members; and the National Investigations Committee for Aerial Phenomena (NICAP) with headquarters in Washington, D.C., and claiming some 12,000 members. James and Coral Lorenzen head APRO, while Keyhoe is the director of NICAP, which, despite the name and Washington address is not a government agency. Many other smaller groups exist, among them Saucers and Unexplained Celestial Events Research Society (SAUCERS) operated by James Moseley.

Of these organizations, NICAP devotes a considerable amount of its attention to attacking the Air Force and to trying to influence members of Congress to hold hearings and in other ways to join in these attacks. It maintained a friendly relation to the Colorado project during about the first year, while warning its members to be on guard lest the project turn out to have been "hired to whitewash the Air Force." During this period NICAP made several efforts to influence the course of our study. When it became clear that these would fail, NICAP attacked the Colorado project as "biased" and therefore without merit.

The organizations mentioned espouse a scientific approach to the study of the subject. In addition there are a number of others that have a primarily religious orientation.

From 1947 to 1966 almost no attention was paid to the UFO problem by well qualified scientists. Some of the reasons for this lack of interest have been clearly stated by Prof. Gerard P. Kuiper of the University of Arizona (Appendix C). Concerning the difficulty of establishing that some UFOs may come from outer space, he makes the following cogent observation: "The problem is more difficult than finding a needle in a haystack; it is finding a piece of extraterrestrial hay in a terrestrial haystack, often on the basis of reports of believers in extra-terrestrial hay."

5. Initial Planning. A scientific approach to the UFO phenomenon must embrace a wide range of disciplines. It involves such physical sciences as physics, chemistry, aerodynamics, and meteorology. Since the primary material consists mostly of reports of individual observers, the psychology of perception, the physiology of defects of vision, and the study of mental states are also involved.

Social psychology and social psychiatry are likewise involved in seeking to understand group motivations which act to induce belief in extraordinary hypotheses on the basis of what most scientists and indeed most laymen would regard as little or no evidence. These problems of medical and social psychology deserve more attention than we were able to give them. They fell distinctly outside of the field of expertise of our staff, which concentrated more on the study of the UFOs themselves than on the personal and social problems generated by them.

Among those who write and speak on the subject, some strongly espouse the view that the federal government really knows a great deal more about UFOs than is made public. Some have gone so far as to assert that the government has actually captured extraterrestrial flying saucers and has their crews in secret captivity, if not in the Pentagon, then at some secret military base. We believe that such teachings are fantastic nonsense, that it would be impossible to keep a secret of such enormity over two decades, and that no useful purpose would be served by engaging in such an alleged conspiracy of silence. One person with whom we have dealt actually maintains that the Air Force has nothing to do with UFOs, claiming that this super-secret matter is in the hands of the Central Intelligence Agency which, he says, installed one of its own agents as scientific director of the Colorado study. This story, if true, is indeed a well kept secret. These allegations of a conspiracy on the part of our own government to conceal knowledge of the existence of "flying saucers" have, so far as any evidence that has come to our attention, no factual basis whatever.

The project's first attention was given to becoming familiar with past work in the subject. This was more difficult than in more orthodox fields because almost none of the many books and magazine articles dealing with UFOs could be regarded as scientifically reliable. There were the two books of Donald H. Menzel, director emeritus of the Harvard College Observatory and now a member of the staff of the Smithsonian Astrophysical Observatory (Menzel, 1952 and Boyd, 1963). Two other useful books were *The UFO Evidence* (1964), a compilation of UFO cases by Richard Hall, and *The Report on Unidentified Flying Objects* by E. J. Ruppelt (1956), the first head of Project Blue Book. In this initial stage we were also helped by "briefings" given by Lt. Col. Hector Quintanilla, the present head of Project Blue Book, Dr. J. Allen Hynek, astronomical consultant to Project Blue Book, and by Donald Keyhoe and Richard Hall of NICAP.

Out of this preliminary study came the recognition of a variety of topics that would require detailed attention. These included the effects of optical mirages, the analogous anomalies of radio wave propagation as they affect radar, critical analysis of alleged UFO photographs, problems of statistical analysis of UFO reports, chemical analysis of alleged material from UFOs, and reports of disturbances to automobile ignition and to headlights from the presence of UFOs. Results of the project's study of these and other topics are presented in this section and in Sections III and VI of this report.

6. Field Investigations. Early attention was given to the question of investigation of individual cases, either by detailed critical study of old records or by field trip investigation of current cases. From this study we concluded that there was little to be gained from the study of old cases, except perhaps to get ideas on mistakes to be avoided in studies of new cases. We therefore decided not to make field trips to investigate cases that were more than a year old, although in a few cases we did do some work on such cases when their study could be combined with a field investigation of a new case.

At first we hoped that field teams could respond to early warning so quickly that they would be able to get to the site while the UFO was still there, and that our teams would not only get their own photographs, but even obtain spectrograms of the light of the UFO, and make radioactive, magnetic, and sound measurements while the UFO was still present.

Such expectations were found to be in vain. Nearly all UFO sightings are of very short duration, seldom lasting as long as an hour and usually lasting for a few minutes. The observers often become so excited that they do not report at all until the UFO has gone away. With communication and travel delays, the field team was unable to get to the scene until long after the UFO had vanished.

This was, of course, a highly unsatisfactory situation. We gave much thought to how it could be overcome and concluded that this could only be done by a great publicity campaign designed to get the public to report sightings much more promptly than it does, coupled with a nationwide scheme of having many trained field teams scattered at many points across the nation. These teams would have had to be ready to respond at a moment's notice. Even so, in the vast majority of the cases, they would not have arrived in time for direct observation of the reported UFO. Moreover, the national publicity designed to insure more prompt reporting would have had the effect of arousing exaggerated public concern over the subject, and certainly would have vastly increased the number of nonsense reports to which response would have had to be made. In recruiting the large number of field teams, great care would have had to be exercised to make sure that they were staffed with people of adequate scientific training, rather than with persons emotionally committed to extreme pro or con views on the subject.

Clearly this was quite beyond the means of our study. Such a program to cover the entire United States would cost many millions of dollars a year, and even then there would have been little likelihood that anything of importance would have been uncovered.

In a few cases some physical evidence could be gathered by examination of a site where an UFO was reported to have landed. In such a case it did not matter that the field team arrived after the UFO had gone. But in no case did we obtain any convincing evidence of this kind although every effort was made to do so. (See below and in Section III, Chapters 3 and 4.)

Thus most of the field investigation, as it turned out, consisted in the interviewing of persons who made the report. By all odds the most used piece of physical equipment was the tape recorder.

The question of a number of investigators on a field team was an important one. In most work done in the past by the Air Force, UFO observers were interviewed by a single Air Force officer, who usually had no special training and whose freedom to devote much time to the study was limited by the fact that he also had other responsibilities. When field studies are made by amateur organizations like APRO or NICAP, there are often several members present on a team, but usually they are persons without technical training, and often with a strong bias toward the sensational aspects of the subject.

Prof. Hynek strongly believes that the teams should have four or more members. He recommends giving each report what he calls the "FBI treatment," by which he means not only thorough interviewing of the persons who made the report, but in addition an active quest in the neighborhood where the sighting occurred to try to discover additional witnesses. Against such thoroughness must be balanced the consideration that the cost per case goes up proportionately to the number of persons in a team, so that the larger the team, the fewer the cases that can be studied.

The detailed discussions in Section III, Chapter 1 and in Section IV make it clear that the field work is associated with many frustrations. Many of the trips turn out to be wild goose chases and the team members often feel as if they are members of a fire department that mostly answers false alarms.

We found that it was always worthwhile to do a great deal of initial interviewing by long distance telephone. A great many reports that seem at first to be worthy of full field investigation could be disposed of in this way with comparatively little trouble and expense. Each case presented its own special problems. No hard-and-fast rule was found by which to decide in advance whether a particular report was worth the trouble of a field trip.

After careful consideration of these various factors, we decided to operate with two-man teams, composed whenever possible of one person with training in physical science and one with training in psychology. When the study became fully operational in 1967 we had three such teams. Dr. Roy Craig describes the work of these teams in Section III, Chapters 1, 3, and 4. Reports of field investigations are presented in Section IV.

7. Explaining UFO Reports. By definition UFOs exist because UFO reports exist. What makes the whole subject intriguing is the possibility that some of these reports cannot be reconciled with ordinary explanations, so that some extraordinarily sensational explanation for them might have to be invoked. A fuller discussion of some misinterpretations of ordinary events by Dr. W. K. Hartmann is given in Section VI, Chapter 2.

A great many reports are readily identified with ordinary phenomena seen under unusual circumstances, or noted by someone who is an inexperienced, inept, or unduly excited observer. Because such reports are vague and inaccurate, it is often impossible to make an identification with certainty.

This gives rise to controversy. In some cases, an identification that the UFO was "probably" an aircraft is all that can be made from the available data. After the event no amount of further interviewing of one or more witnesses can usually change such a probable into a certain identification. Field workers who would like to identify as many as possible are naturally disposed to claim certainty when this is at all possible, but others who desire to have a residue of unexplained cases in order to add mystery and importance to the UFO problem incline to set impossibly high standards of certainty in the evidence before they are willing to accept a simple explanation for a report.

This dilemma is nicely illustrated by a question asked in the House of Commons of Prime Minister Harold Wilson, as reported in *Hansard* for 19 December 1967:

> Unidentified Flying Objects. Question 14. Sir J. Langford-Holt asked the Prime Minister whether he is satisfied that all sightings of unidentified flying objects which are reported from service sources are explainable, what inquiries he has authorized into these objects outside the defense aspect, and whether he will now appoint one Minister to look into all aspects of reports.
>
> The Prime Minister: The answers are "Yes, except when the information given is insufficient," "None" and "No."

Obviously there is a nice bit of semantics here in that the definition of "when the information is sufficient" is that it is sufficient when an explanation can be given.

Discussions of whether a marginal case should be regarded for statistical purposes as having been explained or not have proved to be futile. Some investigators take the position that, where a plausible interpretation in terms of commonplace events can be made, then the UFO is regarded as having been identified. Others take the opposite view that an UFO cannot be regarded as having been given an ordinary identification unless there is complete and binding evidence amounting to certainty about the proposed identification.

For example, in January 1968 near Castle Rock, Colo., some 30 persons reported UFOs, including spacecraft with flashing lights, fantastic maneuverability, and even with occupants presumed to be from outer space. Two days later it was more modestly reported that two high school boys had launched a polyethylene hot-air balloon.

Locally that was the end of the story. But there is a sequel. A man in Florida makes a practice of collecting newspaper stories about UFOs and sending them out in a mimeographed UFO news letter which he mails to various UFO journals and local clubs. He gave currency to the Castle Rock reports but not to the explanation that followed. When he was chided for not having done so, he declared that no one could be *absolutely* sure that *all* the Castle Rock reports arose from sightings of the balloon. There might also have been an UFO from outer space among the sightings. No one would dispute his logic, but one may with propriety wonder why he neglected to tell his readers that at least *some* of the reports were actually misidentifications of a hot-air balloon.

As a practical matter, we take the position that if an UFO report can be plausibly explained in ordinary terms, then we accept that explanation even though not enough evidence may be available to prove it beyond all doubt. This point is so important that perhaps an analogy is needed to make it clear. Several centuries ago, the most generally accepted theory of human disease was that it was caused by the patient's being possessed or inhabited by a devil or evil spirit. Different diseases were supposed to be caused by different devils. The guiding principle for medical research was then the study and classification of different kinds of devils, and progress in therapy was sought in the search for and discovery of means for exorcising each kind of devil.

Gradually medical research discovered bacteria; toxins and viruses, and their causative relation to various diseases. More and more diseases came to be described by their causes.

Suppose now that instead, medicine had clung to the devil theory of disease. As long as there exists one human illness that is not yet fully understood in modern terms such a theory cannot be disproved. It is always possible, while granting that some diseases are caused by viruses, etc. to maintain that those that are not yet understood are the ones that are really caused by devils.

In some instances the same sort of UFO is observed night after night under similar circumstances. In our experience this has been a sure sign that the UFO could be correlated with some ordinary phenomenon.

For example, rather early in our work, a Colorado farmer reported seeing an UFO land west of his farm nearly every evening about 6:00 p.m. A field team went to see him and quickly and unambiguously identified the UFO as the planet Saturn. The nights on which he did not see it land were those in which the western sky was cloudy.

But the farmer did not easily accept our identification of his UFO as Saturn. He contended that, while his UFO had landed behind the mountains on the particular evening that we visited him, on most nights, he insisted, it landed in front of the mountains, and therefore could not be a planet. The identification with Saturn from the ephemeris was so precise that we did not visit his farm night after night in order to see for ourselves whether his UFO ever landed in front of the mountains. We did not regard it as part of our duty to persuade observers of the correctness of our interpretations. In most cases observers readily accepted our explanation, and some expressed relief at having an everyday explanation available to them.

We sought to hold to a minimum delays in arriving at the site of an UFO report, even where it was clear that it was going to be impossible to get there in time actually to see the reported UFO. Once an observer made a report, the fact of his having done so usually becomes known to friends and neighbors, local newspapermen, and local UFO enthusiasts. The witness becomes the center of attention and will usually have told his story over and over again to such listeners, before the field team can arrive. With each telling of the story it is apt to be varied and embellished a little. This need not be from dishonest motives. We all like to tell an interesting story. We would rather not bore our listeners if we can help it, so embellishment is sometimes added to maximize the interest value of the narration.

It is not easy to detect how a story has grown under retelling in this way. Listeners usually will have asked leading questions and the story will have developed in response to such suggestions, so that it soon becomes impossible for the field team to hear the witness's story as he told it the first time. In some cases when the witness had been interviewed in this way by local UFO enthusiasts, his story was larded with vivid language about visitors from outer space that was probably not there in the first telling.

Another kind of difficulty arises in interviewing multiple associated witnesses, that is, witnesses who were together at the time that all of them saw the UFO. Whenever several individuals go through an exciting experience together, they are apt to spend a good deal of time discussing it afterward among themselves, telling and retelling it to each other, unconsciously ironing out discrepancies between their various recollections, and gradually converging on a single uniform account of the experience. Dominant personalities will have contributed more to the final version than the less dominant. Thus the story told by a group of associated witnesses who have had ample opportunity to "compare notes" will be more uniform than the accounts these individuals would have given if interviewed separately before they had talked the matter over together.

One of the earliest of our field trips (December 1966) was made to Washington, D.C. to interview separately two air traffic control operators who had been involved in the great UFO flap there in the summer of 1952. Fourteen years later, these two men were still quite annoyed at the newspaper publicity they had received, because it had tended to ridicule their reports. Our conclusion from this trip was that these men were telling in 1966 stories that were thoroughly consistent with the main points of their stories as told in 1952. Possibly this was due to the fact that because of their strong emotional involvement they had recounted the incident to many persons at many times over the intervening years. Although it was true that the stories had not changed appreciably in 14 years, it was also true for this very reason that we acquired no new material by interviewing these men again. (See Section III, Chapter 5.)

On the basis of this experience we decided that it was not profitable to devote much effort to re-interviewing persons who had already been interviewed rather thoroughly at a previous time. We do not say that nothing can be gained in this way, but merely that it did not seem to us that this would be a profitable way to spend our effort in this study.

In our experience those who report UFOs are often very articulate, but not necessarily reliable. One evening in 1967 a most articulate gentleman told us with calm good manners all of the circumstances of a number of UFOs he had seen that had come from outer space, and in particular went into some detail about how his wife's grandfather had immigrated to America from the Andromeda nebula, a galaxy located 2,000,000 light years from the earth.

In a few cases study of old reports may give the investigator a clue to a possible interpretation that had not occurred to the original investigator. In such a case, a later interview of the witness may elicit new information that was not brought out in the earlier interview. But we found that such interviews need to be conducted with great care as it is easily possible that the "new" information may have been generated through the unconscious use of leading questions pointing toward the new interpretation, and so may not be reliable for that reason.

8. Sources of UFO Reports. Usually the first report of an UFO is made to a local police officer or to a local news reporter. In some cases, members of UFO study organizations are sufficiently well known in the community that reports are made directly to them. In spite of the very considerable publicity that has been given to this subject, a large part of the public still does not know of the official Air Force interest.

Even some policemen and newsmen do not know of it and so do not pass on the UFO report. In other cases, we found that the anti–Air Force publicity efforts of some UFO enthusiasts had persuaded observers, who would otherwise have done so, not to report to the Air Force. We have already commented on the fact that for a variety of reasons many persons who do have UFO experiences do not report promptly.

Ideally the entire public would have known that each Air Force base must, according to AFR 80-17, have an UFO officer and would have reported promptly any extraordinary thing seen in the sky. Or, if this were too much to expect, then all police and news agencies would ideally have known of Air Force interest and would have passed information along to the nearest Air Force base. But none of these ideal things were true, and as a result our collection of UFO reports is extremely haphazard and incomplete.

When a report is made to an Air Force base, it is handled by an UFO officer whose form of investigation and report is prescribed by APR 80-17 (Appendix A). If the explanation of the report is immediately obvious and trivial—some persons will telephone a base to report a contrail from a high-flying jet that is particularly bright in the light of the setting sun—the UFO officer tells the person what it was he saw, and there the matter ends. No permanent record of such calls is made. As a result there is no record of the total number of UFO reports made to AF bases. Only those that require more than cursory consideration are reported to Project Blue Book. Air Force officers are human, and therefore interpret their duty quite differently. Some went to great lengths not to submit a report. Others took special delight in reporting all of the "easy" ones out of a zealous loyalty to their service, because the more "identifieds" they turned in, the higher would be the over-all percentage of UFO reports explained. When in June 1967 Air Force UFO officers from the various bases convened in Boulder some of them quite vigorously debated the relative merits of these two different extreme views of their duty.

Many people have from time to time tried to learn something significant about UFOs by studying statistically the distribution of UFO reports geographically, in time, and both factors together. In our opinion these efforts have proved to be quite fruitless. The difficulties are discussed in Section VI, Chapter 10.

The geographical distribution of reports correlates roughly with population density of the non-urban population. Very few reports come from the densely populated urban areas. Whether this is due to urban sophistication or to the scattering of city lights is not known, but it is more probably the latter.

There apparently exists no single complete collection of UFO reports. The largest file is that maintained by Project Blue Book at Wright-Patterson Air Force Base, Ohio. Other files are maintained by APRO in Tucson and NICAP in Washington. The files of Project Blue Book are arranged by date and place of occurrence of the report, so that one must know these data in order to find a particular case. Proposals have been made from time to time for a computer-indexing of these reports by various categories but this has not been carried out. Two publications are available which partially supply this lack: one is *The UFO Evidence* (Hall, 1964) and the other is a collection of reports called *The Reference for Outstanding UFO Reports* (Olsen, n.d.).

We have already mentioned the existence of flaps, that is, the tendency of reports to come in clusters at certain times in certain areas. No quantitative study of this is available, but we believe that the clustering tendency is partly due to changing amounts of attention devoted to the subject by the news media. Publicity for some reports stimulates more reports, both because people pay more attention to the sky at such a time, and because they are more likely to make a report of something which attracts their attention.

In the summer of 1967 there was a large UFO flap in the neighborhood of Harrisburg, Pa. This may have been in part produced by the efforts of a local NICAP member working in close association with a reporter for the local afternoon newspaper who wrote an exciting UFO story for his paper almost daily. Curiously enough, the morning paper scarcely ever had an UFO story from which we conclude that one editor's news is another's filler. We stationed one of our investigators there during August with results that are described in Case 27.

Many UFO reports were made by the public to Olmsted Air Force Base a few miles south of Harrisburg, but when this base was deactivated during the summer UFO reports had to be made to McGuire Air Force Base near Trenton, N.J. This required a toll call, and the frequency of receipt of UFO reports from the Harrisburg area dropped abruptly.

For all of these various reasons, we feel that the fluctuations geographically and in time of UFO reports are so greatly influenced by sociological factors, that any variations due to changes in underlying physical phenomena are completely masked.

In sensational UFO journalism the statement is often made that UFOs show a marked tendency to be seen more often near military installations. There is no statistically significant evidence that this is true. For sensational writers, this alleged but unproven concentration of UFO sightings is taken as evidence that extraterrestrial visitors are reconnoitering our military defenses, preparatory to launching a military attack at some time in the future. Even if a slight effect of this kind were to be established by careful statistical studies, we feel that it could be easily accounted for by the fact that at every base men stand all night guard duty and so unusual things in the sky are more likely to be seen. Moreover civilians living near a military base are more likely to make a report to the base than those living at some distance from it.

AFR 80-17a directed UFO officers at each base to send to the Colorado project a duplicate of each report sent to Project Blue Book. This enabled us to keep track of the quality of the investigations and to be informed about puzzling uninterpreted cases. Such reporting was useful in cases whose study extended over a long period, but the slowness of receipt of such reports made this arrangement not completely satisfactory as a source of reports on the basis of which to direct the activity of our own field teams. A few reports that seemed quite interesting to Air Force personnel caused them to notify us by teletype or telephone. Some of our field studies arose from reports received in this way.

To supplement Air Force reporting, we set up our own Early Warning Network, a group of about 60 active volunteer field reporters, most of whom were connected with APRO or NICAP. They telephoned or telegraphed to us intelligence of UFO sightings in their own territory and conducted some preliminary investigation for us while our team was en route. Some of this cooperation was quite valuable. In the spring of 1968, Donald Keyhoe, director of NICAP, ordered discontinuation of this arrangement, but many NICAP field teams continued to cooperate.

All of these sources provided many more quickly reported, fresh cases than our field teams could study in detail. In consequence we had to develop criteria for quickly selecting which of the cases reported to us would be handled with a field trip (See Section III, Chapter 1.)

9. Extra-terrestrial Hypothesis. The idea that some UFOs may be spacecraft sent to Earth from another civilization, residing on another planet of the solar system, or on a planet associated with a more distant star than the Sun, is called the Extra-terrestrial Hypothesis (ETH). Some few persons profess to hold a stronger level of belief in the *actuality* of UFOs being visitors from outer space, controlled by intelligent beings, rather than merely of the *possibility*, not yet fully established as an observational fact. We shall call this level of belief ETA, for extraterrestrial actuality.

It is often difficult to be sure just what level of belief is held by various persons, because of the vagueness with which they state their ideas.

For example, addressing the American Society of Newspaper Editors in Washington on 22 April 1967, Dr. McDonald declared: "There is, in my present opinion, no sensible alternative to the utterly shocking hypothesis that the UFOs are extraterrestrial probes from somewhere else." Then in an Australian broadcast on 20 August 1967 McDonald said: ". . . you find yourself ending up with the seemingly absurd, seemingly improbable hypothesis that these things may come from somewhere else."

A number of other scientists have also expressed themselves as believers in ETH, if not ETA, but usually in more cautious terms.

The general idea of space travel by humans from Earth and visitors to Earth from other civilizations is an old one and has been the subject of many works of fiction. In the past 250 years the topic has been widely developed in science fiction. A fascinating account of the development of this literary form is given in *Pilgrims through Space and Time—Trends and Patterns in Scientific and Utopian Fiction* (Bailey, 1947).

The first published suggestion that some UFOs are visitors from other civilizations is contained in an article in *True*, entitled "Flying Saucers are Real" by Donald E. Keyhoe (1950).

Direct, convincing and unequivocal evidence of the truth of ETA would be the greatest single scientific discovery in the history of mankind. Going beyond its interest for science, it would undoubtedly have consequences of surpassing significance for every phase of human life. Some persons who have written speculatively on this subject, profess to believe that the supposed extraterrestrial visitors come with beneficent motives, to help humanity clean up the terrible mess that it has made. Others say they believe that the visitors are hostile. Whether their coming would be favorable or unfavorable to mankind, it is almost certain that they would make great changes in the conditions of human existence.

It is characteristic of most reports of actual visitors from outer space that there is no corroborating witness to the alleged incident, so that the story must be accepted, if at all, solely on the basis of belief in the veracity of the one person who claims to have had the experience. In the cases which we studied, there was only one in which the observer claimed to have had contact with a visitor from outer space. On the basis of our experience with that one, and our own unwillingness to believe the literal truth of the Villas-Boas incident, or the one from Truckee, Calif. reported by Prof. James Harder (see Section V, Chapter 2), we found that no direct evidence whatever of a convincing nature now exists for the claim that any UFOs represent spacecraft visiting Earth from another civilization.

Some persons are temperamentally ready, even eager, to accept ETA without clear observational evidence. One lady remarked, "It would be so wonderfully exciting if it were true!" It certainly would be exciting, but that does not make it true. When confronted with a proposition of such great import, responsible scientists adopt a cautiously critical attitude toward whatever evidence is adduced to support it. Persons without scientific training, often confuse this with basic opposition to the idea, with a biased desire or hope, or even of willingness to distort the evidence in order to conclude that ETA is not true.

The scientists' caution in such a situation does not represent opposition to the idea. It represents a determination not to accept the proposition as true in the absence of evidence that clearly, unambiguously and with certainty establishes its truth or falsity.

Scientifically it is not necessary—it is not even desirable—to adopt a position about the truth or falsity of ETA in order to investigate the question. There is a widespread misconception that scientific inquiry represents some kind of debate in which the truth is adjudged to be on the side of the team that has scored the most points. Scientists investigate an undecided proposition by seeking to find ways to get decisive observational material. Sometimes the ways to get such data are difficult to conceive, difficult to carry out, and so indirect that the rest of the scientific world remains uncertain of the probative value of the results for a long time. Progress in science can be painfully slow—at other times it can be sudden and dramatic. The question of ETA would be settled in a few minutes if a flying saucer were to land on the lawn of a hotel where a convention of the American Physical Society was in progress, and its occupants were to emerge and present a special paper to the assembled physicists, revealing where they came from, and the technology of how their craft operates. Searching questions from the audience would follow.

In saying that thus far no convincing evidence exists for the truth of ETA, no prediction is made about the future. If evidence appears soon after this report is published, that will not alter the truth of the statement that we do not *now* have such evidence. If new evidence appears later, this report can be appropriately revised in a second printing.

10. Intelligent Life Elsewhere. Whether there is intelligent life elsewhere (ILE) in the Universe is a question that has received a great deal of serious speculative attention in recent years. A good popular review of thinking on the subject is *We Are Not Alone* by Walter Sullivan (1964). More advanced discussions are *Interstellar Communications*, a collection of papers edited by A. G. W. Cameron (1963), and *Intelligent Life in the Universe* (Shklovskii and Sagan, 1966). Thus far we have no observational evidence whatever on the question, so therefore it remains open. An early unpublished discussion is a letter of 13 December 1948 of J. E. Lipp to Gen. Donald Putt (Appendix D). This letter is Appendix D of the Project Sign report dated February 1949 from Air Materiel Command Headquarters No. F-TR-2274-IA.

The ILE question has some relation to the ETH or ETA for UFOs as discussed in the preceding section. Clearly, if ETH is true, then ILE must also be true because some UFOs have then to come from some unearthly civilization. Conversely, if we could know conclusively that ILE does not exist, then ETH could not be true. But even if ILE exists, it does not follow that the ETH is true.

For it could be that the ILE, though existent, might not have reached a stage of development in which the beings have the technical capacity or the desire to visit the Earth's surface. Much speculative writing assumes implicitly that intelligent life progresses steadily both in intellectual and in its technological development. Life began on Earth more than a billion years ago, whereas the known geological age of the Earth is some five billion years, so that life in any form has only existed for the most recent one-fifth of the Earth's life as a solid ball orbiting the Sun. Man as an intelligent being has only lived on Earth for some 5,000 years, or about one-millionth of the Earth's age. Technological development is even more recent. Moreover the greater part of what we think of as advanced technology has only been developed in the last 100 years. Even today we do not yet have a technology capable of putting men on other planets of the solar system. Travel of men over interstellar distances in the foreseeable future seems now to be quite out of the question (Purcell, 1960; Markowitz, 1967).

The dimensions of the universe are hard for the mind of man to conceive. A light-year is the distance light travels in one year of 31.56 million seconds, at the rate of 186,000 miles per second, that is, a distance of 5.88 million million miles. The nearest known star is at a distance of 4.2 light-years.

Fifteen stars are known to be within 11.5 light-years of the Sun. Our own galaxy, the Milky Way, is a vast flattened distribution of some 10^{11} stars about 80,000 light-years in diameter, with the Sun located about 26,000 light-years from the center. To gain a little perspective on the meaning of such distances relative to human affairs, we may observe that the news of Christ's life on Earth could not yet have reached as much as a tenth of the distance from the Earth to the center of our galaxy.

Other galaxies are inconceivably remote. The faintest observable galaxies are at a distance of some two billion light-years. There are some 100 million such galaxies within that distance, the average distance between galaxies being some eight million light-years.

Authors of UFO fantasy literature casually set all of the laws of physics aside in order to try to evade this conclusion, but serious consideration of their ideas hardly belongs in a report on the scientific study of UFOs.

Even assuming that difficulties of this sort could be overcome, we have no right to assume that in life communities everywhere there is a steady evolution in the directions of both greater intelligence and greater technological competence. Human beings now know enough to destroy all life on Earth, and they may lack the intelligence to work out social controls to keep themselves from doing so. If other civilizations have the same limitation then it might be that they develop to the point where they destroy themselves utterly before they have developed the technology needed to enable them to make long space voyages.

Another possibility is that the growth of intelligence precedes the growth of technology in such a way that by the time a society would be technically capable of interstellar space travel, it would have reached a level of intelligence at which it had not the slightest interest in interstellar travel. We must not assume that we are capable of imagining now the scope and extent of future technological development of our own or any other civilization, and so we must guard against assuming that we have any capacity to imagine what a more advanced society would regard as intelligent conduct.

In addition to the great distances involved, and the difficulties which they present to interstellar space travel, there is still another problem: If we assume that civilizations annihilate themselves in such a way that their effective intelligent life span is less than, say, 100,000 years, then such a short time span also works against the likelihood of successful interstellar communication. The different civilizations would probably reach the culmination of their development at different epochs in cosmic history. Moreover, according to present views, stars are being formed constantly by the condensation of interstellar dust and gases. They exist for perhaps 10 billion years, of which a civilization lasting 100,000 years is only 1/100,000 of the life span of the star. It follows that there is an extremely small likelihood that two nearby civilizations would be in a state of high development at the same epoch.

Astronomers now generally agree that a fairly large number of all main-sequence stars are probably accompanied by planets at the right distance from their Sun to provide for habitable conditions for life as we know it. That is, where stars are, there are probably habitable planets. This belief favors the possibility of interstellar communication, but it must be remembered that even this view is entirely speculation: we are quite unable directly to observe any planets associated with stars other than the Sun.

In view of the foregoing, we consider that it is safe to assume that no ILE outside of our solar system has any possibility of visiting Earth in the next 10,000 years.

This conclusion does not rule out the possibility of the existence of ILE, as contrasted with the ability of such civilizations to visit Earth. It is estimated that 10^{21} stars can be seen using the 200-inch Hale telescope on Mount Palomar. Astronomers surmise that possibly as few as one in a million or as many as one in ten of these has a planet in which physical and chemical conditions are such as to make them habitable by life based on the same kind of biochemistry as the life we know on Earth. Even if the lower figure is taken, this would mean there are 10^{15} stars in the visible universe which have planets suitable for an abode of life. In our own galaxy there are 10^{11} stars, so perhaps as many as 10^{8} have habitable planets in orbit around them.

Biologists feel confident that wherever physical and chemical conditions are right, life will actually emerge. In short, astronomers tell us that there are a vast number of stars in the universe accompanied by planets where the physical and chemical conditions are suitable, and biologists tell us that habitable places are sure to become inhabited (Rush, 1957).

An important advance was made when Stanley L. Miller (1955) showed experimentally that electrical discharges such as those in natural lightning when passed through a mixture of methane and ammonia, such as may have been present in the Earth's primitive atmosphere, will initiate chemical reactions which yield various amino acids. These are the raw materials from which are constructed the proteins that are essential to life. Miller's work has been followed up and extended by many others, particularly P. H. Abelson of the Carnegie Institution of Washington.

The story is by no means fully worked out. The evidence in hand seems to convince biochemists that natural processes, such as lightning, or the absorption of solar ultraviolet light, could generate the necessary starting materials from which life could evolve. On this basis they generally hold the belief that where conditions make it possible that life could appear, there life actually will appear.

It is regarded by scientists today as essentially certain that ILE exists, but with essentially no possibility of contact between the communities on planets associated with different stars. We therefore conclude that there is no relation between ILE at other solar systems and the UFO phenomenon as observed on Earth.

There remains the question of ILE within our solar system. Here only the planets Venus and Mars need be given consideration as possible abodes of life.

Mercury, the planet nearest the Sun, is certainly too hot to support life. The side of Mercury that is turned toward the Sun has an average temperature of 660°F. (Mercury rotates in 59 days and the orbital period is 88 days, so there is a slow relative motion.) Since the orbit is rather eccentric this temperature becomes as high as 770°F, hot enough to melt lead, when Mercury is closest to the Sun. The opposite side is extremely cold, its temperature not being known. Gravity on Mercury is about one-fourth that on Earth. This fact combined with the high temperature makes it certain that Mercury has no atmosphere, which is consistent with observational data on this point. It is quite impossible that life as found on Earth could exist on Mercury.

Jupiter, Saturn, Uranus, Neptune and Pluto are so far from the Sun that they are too cold for life to exist there.

Although it has long been thought that Venus might provide a suitable abode for life, it is now known that the surface of Venus is also too hot for advanced forms of life, although it is possible that some primitive forms may exist. Some uncertainty and controversy exist about the interpretation of observations of Venus because the planet is always enveloped in dense clouds so that the solid surface is never seen. The absorption spectrum of sunlight coming from Venus indicates that the principal constituent of the atmosphere is carbon dioxide. There is no evidence of oxygen or water vapor. With so little oxygen in the atmosphere there could not be animal life there resembling that on Earth.

Although it is safe to conclude that there is no intelligent life on Venus, the contrary idea is held quite tenaciously by certain groups in America. There are small religious groups who maintain that Jesus Christ now sojourns on Venus, and that some of their members have traveled there by flying saucers supplied by the Venusians and have been greatly refreshed spiritually by visiting Him. There is no observational evidence in support of this teaching.

In the fantasy literature of believers in ETH, some attention is given to a purely hypothetical planet named Clarion. Not only is there no direct evidence for its existence, but there is conclusive indirect evidence for its non-existence. Those UFO writers who try not to be totally inconsistent with scientific findings, recognizing that Venus and Mars are unsuitable as abodes of life, have invented Clarion to meet the need for a home for the visitors who they believe come on some UFOs.

They postulate that Clarion moves in an orbit exactly like that of the Earth around the Sun, but with the orbit rotated through half a revolution in its plane so that the two orbits have the same line of apsides, but with Clarion's perihelion in the same direction from the Sun as the Earth's aphelion. The two planets, Earth and Clarion, are postulated to move in their orbits in such a way that they are always opposite each other, so that the line Earth-Sun-Clarion is a straight line. Thus persons on Earth would never see Clarion because it is permanently eclipsed by the Sun.

If the two orbits were exactly circular, the two planets would move along their common orbit at the same speed and so would remain exactly opposite each other. But even if the orbits are elliptical, so that the speed in the orbit is variable, the two planets would vary in speed during the year in just such a way as always to remain Opposite each other and thus continue to be permanently eclipsed.

However, this tidy arrangement would not occur in actuality because the motion of each of these two planets would be perturbed by the gravitational attractions between them and the other planets of the solar system, principally Venus and Mars. It is a quite complicated and difficult problem to calculate the way in which these perturbations would affect the motion of Earth and Clarion.

At the request of the Colorado project, Dr. R. L. Duncombe, director of the Nautical Almanac office at U.S. Naval Observatory in Washington, D.C., kindly arranged to calculate the effect of the introduction of the hypothetical planet Clarion into the solar system. The exact result depends to some extent on the location of the Earth-Sun-Clarion line relative to the line of apsides and the computations were carried out merely for one case (see Appendix E).

These calculations show that the effect of the perturbations would be to make Clarion become visible from Earth beyond the Sun's limb after about thirty years. In other words, Clarion would long since have become visible from Earth if many years ago it were started out in such a special way as has been postulated.

The computations revealed further that if Clarion were there it would reveal its presence indirectly in a much shorter time. Its attraction on Venus would cause Venus to move in a different way than if Clarion were not there. Calculation shows that Venus would pull away from its otherwise correct motion by about 1 second of arc in about three months' time. Venus is routinely kept under observation to this accuracy, and therefore if Clarion were there it would reveal its presence by its effect on the motion of Venus. No such effect is observed, that is, the motion of Venus as actually observed is accurately in accord with the absence of Clarion, so therefore we may safely conclude that Clarion is nonexistent. (These calculations assume Clarion's mass roughly equal to that of the Earth.)

In his letter of transmittal Dr. Duncombe comments "I feel this is definite proof that the presence of such a body could not remain undetected for long. However, I am afraid it will not change the minds of those people who believe in the existence of Clarion."

We first heard about Clarion from a lady who is prominent in American political life who was intrigued with the idea that this is where UFOS come from. When the results of the Naval Observatory computations were told to her she exclaimed, "That's what I don't like about computers! They are always dealing death blows to our fondest notions."

[So we need consider Clarion no further.]

Mars has long been considered as a possible abode of life in the solar system. There is still no direct evidence that life exists there, but the question is being actively studied in the space research programs of both the United States and Soviet Russia, so it may well be clarified within the coming decade.

At present all indications are that Mars could not be the habitation of an advanced civilization capable of sending spacecraft to visit the Earth. Conditions for life there are so harsh that it is generally believed that at best

Mars could only support the simpler forms of plant life.

An excellent recent survey of the rapidly increasing knowledge of Mars is *Handbook of the Physical Properties of the Planet Mars* compiled by C. M. Michaux (NASA publication SP-3030, 1967). A brief discussion of American research programs for study of life on Mars is given in *Biology and Exploration of Mars*, a 19-page pamphlet prepared by the Space Science Board of the National Academy of Sciences, published in April 1965.

The orbit of Mars is considerably more eccentric than that of the Earth. Consequently the distance of Mars from the Sun varies from 128 to 155 million miles during the year of 687 days. The synodic period, or mean time between successive oppositions, is 800 days.

The most favorable time for observation of Mars is at opposition, when Mars is opposite the Sun from Earth. These distances of closest approach of Mars and Earth vary from 35 to 60 million miles. The most recent favorable time of closest approach was the opposition of 10 September 1956, and the next favorable opposition will be that of 10 August 1971. At that time undoubtedly great efforts will be made to study Mars in the space programs of the U.S.S.R. and the United States.

Some of the UFO literature has contended that a larger than usual number of UFO reports occur at the times of Martian oppositions. The contention is that this indicates that some UFOs come from Mars at these particularly favorable times. The claimed correlation is quite unfounded; the idea is not supported by observational data (Vallee and Vallee, 1966, 138).

Mars is much smaller than Earth, having a diameter of 4,200 miles, in comparison with 8,000 miles. Mars' mass is about one-tenth the Earth's, and gravity at Mars' surface is about 0.38 that of Earth. The Martian escape velocity is 3.1 mile/sec.

At the favorable opposition of 1877, C. V. Schiaparelli, an Italian astronomer, observed and mapped some surface markings on Mars which he called "canali," meaning "channels" in Italian. The word was mistranslated as "canals" in English and the idea was put forward, particularly vigorously by Percival Lowell, founder of the Lowell Observatory of Flagstaff, Arizona, that the canals on Mars were evidence of a gigantic planetary irrigation scheme, developed by the supposed inhabitants of Mars (Lowell, 1908). These markings have been the subject of a great deal of study since their discovery. Astronomers generally now reject the idea that they afford any kind of indication that Mars is inhabited by intelligent beings.

Mars has two moons named Phobos and Deimos. These are exceedingly small, Phobos being estimated at ten miles in diameter and Deimos at five miles, based on their brightness, assuming the reflecting power of their material to be the same as that of the planet. The periods are 7h39m for Phobos and 30h18m for Deimos. They were discovered in August 1877 by Asaph Hall using the then new 26-inch refractor of the U.S. Naval Observatory in Washington. An unsuccessful search for moons of Mars was made with a 48-inch mirror during the opposition of 1862.

I. S. Shklovskii (1959) published a sensational suggestion in a Moscow newspaper that these moons were really artificial satellites which had been put up by supposed inhabitants of Mars as a place of refuge when the supposed oceans of several million years ago began to dry up (Sullivan, 1966, 169). There is no observational evidence to support this idea. Continuing the same line of speculation Salisbury (1962), after pointing out that the satellites were looked for in 1862 but not found until 1877, then asks, "Should we attribute the failure of 1862 to imperfections in existing telescopes, or may we imagine that the satellites were launched between 1862 and 1877?" This is a slender reed indeed with which to prop up so sensational an inference, and we reject it.

11. Light Propagation and Visual Perception. Most UFO reports refer to things seen by an observer. Seeing is a complicated process. It involves the emission or scattering of light by the thing seen, the propagation of that light through the atmosphere to the eye of the observer, the formation of an image on the retina of the eye by the lens of the eye, the generation there of a stimulus in the optic nerve, and the perceptual process in the brain which enables the mind to make judgments about the nature of the thing seen.

Under ordinary circumstances all of these steps are in fairly good working order with the result that our eyes give reasonably accurate information about the objects in their field of view. However, each step in the process is capable of malfunctioning, often in unsuspected ways. It is therefore essential to understand these physical and psychological processes in order to be able to interpret all things seen, including those reported as UFOs.

The study of propagation of light through the atmosphere is included in atmospheric optics or meteorological optics. Although a great deal is known about the physical principles involved, in practice it is usually difficult to make specific statements about an UFO report because not enough has been observed and recorded about the condition of the atmosphere at the time and place named in the report.

Application of the knowledge of atmospheric optics to the interpretation of UFO reports has been especially stressed by Menzel (1952; Menzel and Boyd, 1963). A valuable treatise on atmospheric effects on seeing is Middleton's *Vision through the Atmosphere* (1952). A survey of the literature of atmospheric optics with emphasis on topics relevant to understanding UFO reports was prepared for the Colorado project by Dr. William Viezee of the Stanford Research Institute (Section VI, Chapter 4).

Coming to the observer himself, Menzel stressed in consulting visits to the Colorado project that more ought to be known about defects of vision of the observer. He urged careful interviews to determine the observers' defects of vision, how well they are corrected, and whether spectacles were being worn at the time the UFO sighting was made. Besides the defects of vision that can be corrected by spectacles, inquiry ought to be made where relevant into the degree of color blindness of the observer, since this visual defect is more common than is generally appreciated.

Problems connected with the psychology of perception were studied for the Colorado project by Prof. Michael Wertheimer of the Department of Psychology of the University of Colorado. He prepared an elementary presentation of the main points of interest for the use of the project staff (Section VI, Chapter 1).

Perhaps the commonest difficulty is the lack of appreciation of size-distance relations in the description of an unknown object. When we see an airplane in the sky, especially if it is one of a particular model with which we are familiar, we know from prior experience approximately what its size really is. Then from its apparent size as we see it, we have some basis for estimating its distance. Conversely, when we know something about the distance of an unknown object, we can say something about its size. Although not usually expressed this way, what is really "seen" is the size of the image on the retina of the eye, which may be produced by a smaller object that is nearer or a larger object that is farther away. Despite this elementary fact, many people persist in saying that the full moon looks the same size as a quarter or as a washtub. The statement means nothing. Statements such as that an object looks to be of the same size as a coin *held at arm's length* do, however, convey some meaningful information.

Another limitation of normal vision that is often not appreciated is the color blindness of the dark-adapted eye. The human eye really has two different mechanisms in the retina for the conversion of light energy into nerve stimulus. Photopic vision is the kind that applies in the daytime or at moderate levels of artificial illumination. It involves the cones of the retina, and is involved in color vision. Scotopic vision is the kind that comes into play at low levels of illumination. It involves the rods of the retina which are unable to distinguish colors, hence the saying that in the dark all cats are gray. The transition from photopic to scotopic vision normally takes place at about the level of illumination that corresponds to the light of the full moon high in the sky. When one goes from a brightly lighted area into a dark room he is blind at first but gradually dark adaptation occurs and a transition is made from photopic to scotopic vision. The ability to see, but without color discrimination, then returns. Nyctalopia is the name of a deficiency of vision whereby dark adaptation does not occur and is often connected with a Vitamin A dietary deficiency.

If one stares directly at a bright light which is then turned off, an afterimage will be seen; that is, the image of the light, but less bright and usually out of focus, continues to be seen and gradually fades away. Positive afterimages are those in which the image looks bright like the original stimulus, but this may reverse to a negative afterimage which looks darker than the surrounding field of view. Afterimages have undoubtedly given rise to some UFO reports.

The afterimage is the result of a temporary change in the retina and so remains at a fixed point on the retina. When one then moves his eyes to look in a different direction, the afterimage seems to move relative to the surroundings. If it is believed by the observer to be a real object it will seem to him to have moved at an enormous velocity. A light going out will seem to shrink and move away from the observer as it does so. If one light goes on while another is going off, it may appear as if the light that is going off is moving to the place where the other light is going on.

Autokinesis is another property of the eye which needs to be understood by persons who are interested in looking for UFOs. A bright light in a field of view which has no reference objects in it, such as a single star in a part of the sky which has very few other stars in it, will appear to move when stared at, even though it is in reality stationary. This effect has given rise to UFO reports in which observers were looking at a bright star and believed that it was rapidly moving, usually in an erratic way.

12. Study of UFO Photographs. The popular UFO literature abounds with photographs of alleged strange objects in the sky, many of which are clearly in the form of flying saucers. Some of these have been published in magazines of wide circulation. The editors of *Look*, in collaboration with the editors of United Press International and Cowles Communications, Inc. published a *Look* "Special" in 1967 that is entirely devoted to "Flying Saucers," which contains many examples of UFO pictures.

Photographic evidence has a particularly strong appeal to many people. The Colorado study therefore undertook to look into the available photographs with great care. Chapter 2 of Section III gives the story of most of this work and Chapter 3 of Section IV gives the detailed reports on individual cases.

It is important to distinguish between photographic prints and the negatives from which they are made. There are many ways in which an image can be added to a print, for example, by double-printing from two negatives. Negatives, on the other hand, are somewhat more difficult to alter without leaving evidence of the fact. We therefore decided wherever possible to concentrate our study of photographic case upon the negatives. This was not, of course, possible in every instance examined.

A barber whose shop is in Zanesville, Ohio, but whose home is in the suburb of Roseville, has made a widely publicized pair of UFO photographs. He did not attempt to exploit them in a big way. He merely exhibited them for local interest (and stimulation of his barbering business) in the window of his shop. There they remained for more than two months until they were discovered by a big city newspaperman from Columbus, Ohio, who arranged to sell them to the Associated Press. They were distributed in February 1967 and have been often printed in various magazines after their original presentation in many newspapers.

Early in the project we became acquainted with Everitt Merritt, photogrammetrist on the staff of the Autometrics Division of the Raytheon Company of Alexandria, Virginia. He undertook to do an analysis of the photographs. A pair of prints was supplied to Merritt by NICAP.

Each of the pair shows the home of the photographer, a small bungalow, with a flying saucer flying over it. The flying saucer looks like it might be almost as large as the house in its horizontal dimension. The photographer says that he was leaving home with a camera when he chanced to look back and see the saucer flying over his home. He says he quickly snapped what we call picture A. Thinking the UFO was about to disappear behind a tree, he ran to the left about 30 feet and snapped picture B, having spoiled one exposure in between. He estimated that there was less than a two minute interval between the two pictures, with A followed by B.

Merritt studied the negatives themselves by quantitative photogrammetric methods, and also did some surveying in the front yard of the Roseville home, as a check on the calculations based on the photographs. From a study of the shadows appearing in the picture, he could show conclusively that actually picture B was taken earlier than picture A, and that the time interval between the two pictures was more than an hour, rather than being less than two minutes as claimed.

The photographic evidence contained in the negatives themselves is therefore in disagreement with the story told by the man who took the pictures. Two letters written to him by the Colorado project requesting his clarification of the discrepancy remain unanswered.

We made arrangements with Merritt for his services to be available for photogrammetric analysis of other cases. These methods require a pair of pictures showing substantially the same scene taken from two different camera locations. Unfortunately this condition is seldom met in UFO photographs. Only one other pair came to our attention which met this criterion. These were the much publicized pictures taken on 11 May 1950 near McMinnville, Ore. (Case 46). But in this case the UFO images turned out to be too fuzzy to allow worthwhile photogrammetric analysis.

Other photographic studies were made for the Colorado project by Dr. William K. Hartmann (Section III, Chapter 2).

Hartmann made a detailed study of 35 photographic cases (Section IV, Chapter 3) referring to the period 1966–68, and a selection of 18 older cases, some of which have been widely acclaimed in the UFO literature. This photographic study led to the identification of a number of widely publicized photographs as being ordinary objects, others as fabrications, and others as innocent misidentifications of things photographed under unusual conditions.

On p. 43 of the *Look* Special on "Flying Saucers" there is a picture of an allegedly "claw-shaped" marking on the dry sand of a beach. Some of the dark colored moist sand making up the "claw mark" was shipped to Wright-Patterson AFB and analyzed. The liquid was found to be urine. Some person or animal had performed an act of micturition there.

A report by Staff Sergeant Earl Schroeder which says "Being a native of this area and having spent a good share of my life hunting and fishing this area, I believe that the so-called 'monster' (if there was such) could very well have been a large black bear." His report also notes that "during the week of July 26 the local TV stations showed a program called *Lost in Space*. In this program there were two monsters fitting their description controlled by a human being."

Summarizing, the investigation report says, "There was food missing from the picnic table which leads to the belief that some animal was responsible for the black shape portion of the total sighting. There are numerous bears and raccoons in the area."

Another photograph presented in the *Look* Special is of a pentagonal image, though called hexagonal. Photographic images of this kind arise from a malfunctioning of the iris of the camera and are quite commonplace. It is hard to understand how the editors of a national illustrated magazine could be unfamiliar with this kind of camera defect.

13. Direct and Indirect Physical Evidence. A wide variety of physical effects of UFOs have been claimed in the UFO literature. The most direct physical evidence, of course, would be the actual discovery of a flying saucer, with or without occupants, living or dead. None were found. Claims which we studied as direct evidence are those of the finding of pieces of material which allegedly came from outer space because it is a product of a different technology, so it is said, than any known on earth. Another kind of direct evidence studied was allegations that disturbance of vegetation on the ground, or of the soil was due to an UFO having landed at the place in question.

The claimed indirect physical evidence of the presence of an UFO is of the nature of effects produced at a distance by the UFO. Accounts of sounds, or the lack of sounds, associated with UFOs, even though reports of visual observation indicated speeds of the UFO far in excess of the velocity of sound were common. Whenever a terrestrial solid object travels through the atmosphere faster than the speed of sound, a sonic boom is generated. The argument has been advanced that the absence of a sonic boom associated with UFOs moving faster than cutoff Mach (see Section VI, Chapter 6) is an indication of their being a product of a technology more advanced than our own because we do not know how to avoid the generation of sonic booms. Another category of indirect physical effects is those associated with claims that UFOs possess strong magnetic fields, vastly stronger than those that would be produced by the strongest magnets that we know how to make.

There are many UFO reports in which it is claimed that an automobile's ignition failed and the motor stopped, and in some cases that the headlights failed also, and that after this happened, an UFO was seen nearby. Usually such reports are discussed on the supposition that this is an indication that the UFO had been the source of a strong magnetic field.

Reports of both direct and indirect physical evidence were studied by various staff members of the Colorado project, principally by Dr. Roy Craig, whose account of these studies is contained in Chapters 3 and 4 of Section III.

These studies resulted mostly in lack of substantiation of the claims that have been made. Claims of terrestrial magnetic disturbances at various Antarctic bases were either unconfirmed or seemed to be closely related to a practical joke that was played on a base commander.

During the period of field study of this project only one case of automobile engine malfunction came to our attention. There was some ground for skepticism about the report in that it was made by a diabetic patient who had been drinking and was returning home alone from a party at 3:00 a.m.

Some laboratory tests showed that engine failure due to the action of an external magnetic field on the car's ignition coil would require fields in excess of 20,000 gauss, at the coil. Owing to the magnetic shielding action of the sheet steel in the car body, the strength of the field outside the car would have to be considerably greater than this. But magnetic fields of such intensity would alter the state of magnetization of the car itself.

The process of forming car bodies by cold-forming the sheet steel introduces some quasi-permanent magnetization into all car bodies. Since all of the bodies of a given make in a given year are usually made with the same molds on the same presses they are all magnetized in the same pattern.

In the case in question we found that the car body that had been subjected to the presence of the UFO was magnetized. The pattern of magnetization quite closely resembled that of a car of the same make and year that was found a thousand miles away in a used car lot in Boulder, Colo. From this we can infer that the car that was supposedly near the UFO, had not been subjected to a strong magnetic field, otherwise this would have permanently changed the state of magnetization of the body of the exposed car.

In the area of direct physical evidence, probably the most interesting result of investigation was the analysis of a piece of metallic magnesium which was alleged to have come from an UFO that exploded over a stretch of tidal water at Ubatuba, São Paulo, Brazil in 1957. This was one of several pieces of magnesium from the same source that had been sent to the society editor of a Rio de Janeiro newspaper at the time.

Later one of the pieces was subjected to elaborate chemical analysis in government laboratories in Brazil. The results of the analysis are given in great detail in the first of the Lorenzen books (1962), the full account occupying some forty pages. The claimed result of these studies was that the laboratory work showed the metallic magnesium to be purer than any ever made by man on Earth. Therefore it could not have been a product of earthly technology, therefore it came from an extraterrestrial source.

Mrs. Lorenzen kindly supplied one of the magnesium specimens to the Colorado project. We arranged to have it studied by the method of neutron activation analysis in a laboratory in Washington, D.C. The result, which is presented in detail in Chapter 3 of Section III, was that the magnesium metal was found to be much less pure than the regular commercial metal produced in 1957 by the Dow Chemical Company at Midland, Michigan. Therefore it need not have come from an extraterrestrial source, leaving us with no basis for rational belief that it did.

14. Radar Sightings of UFOs. The public became generally aware of radar at the end of World War II when the story of its important use in that war was told, after having been kept secret for some 12 years. A good non-technical account of this development is given in R. M. Page, *The Origin of Radar* (1962).

The word radar is an acronym for *RA*dio *D*etection *A*nd *R*anging. Basically, most radar systems operate in the following way. A transmitter sends out short pulses of electromagnetic energy at regular intervals. These are sent out through an antenna designed to radiate a narrow beam within a small angle of its main direction. This beam of pulses travels outward at the speed of light. If it encounters an obstacle, which may be a metallic object like an airplane, a rain storm, or a bird or a flock of birds, it is partially scattered in all directions from the obstacle. In particular a part of the beam is scattered back toward the transmitter. When it arrives back at the transmitter it is received and indicated or displayed in various ways, depending on the special purpose for which the system was designed. By the fact of there being a returned signal at all, the function of detection is accomplished. By the time delay involved between the transmission of the outgoing signal and the return of the back-scattered signal, the distance of the scattering object is inferred, thus accomplishing the function of ranging.

To get a beam of sufficiently narrow distribution in angle as to enable inferring from what direction the scattered signal was returned, the antenna must have a diameter of the order of ten times the wavelength of the radio waves which it uses.

In the period since 1945 the technology has had an enormous development so that nowadays there are elaborate networks of land and ship based radar systems, as well as radar systems carried by most airplanes, which have become vitally necessary to the safe operation of civil and military aircraft. In addition to the use of radar in connection with navigation, it has become a valuable tool in meteorological work in that distant rain storms can be detected by radar. Also the trails of ionized air left by meteors can be detected and studied by radar, providing for the first time the means for observing meteors in the daytime.

There are many popular misconceptions about radar. It is important at the outset to realize that the returned radar signal does *not* give a sharply focused image or picture of the obstacle that has been detected. What one gets when it is displayed on a cathode-ray screen is simply a diffuse blob of light indicating that *something* is there, in the direction the antenna is pointed (with some exceptions) and at the distance indicated by the time delay between transmission and reception of the back-scattered pulse. Of course, a large airplane gives a more intense signal than a flock of small birds at the same range, and skilled operators learn to make valid inferences about the nature of the object detected from other things that they know about the general situation together with the magnitude of the returned signal.

It is important also to recognize that the propagation of the outgoing and the back-scattered pulses is ordinarily assumed to be rectilinear and at the normal speed of light. But the actual propagation is affected by temperature and humidity difference in the air path along which the radio pulse travels. This can give rise to anomalous propagation that is analogous to but in detail not identical with the effects which give rise to mirages in the propagation of light through such an atmosphere. Usually the radar set operator does not know enough about the actual atmospheric conditions to make allowance for effects of this kind and, if they happen to be pronounced, can be led to make erroneous decisions. Another point is that, although the antenna sends out most of its energy in a single narrow beam, small amounts of energy go out in several other directions, known as sidelobes, so that a large or a nearby object in the direction of a sidelobe can give rise to a received signal that is indistinguishable from a small or distant object in the direction of the main beam.

The overall radar system is a rather complicated set of electronic equipment which can malfunction in various ways giving rise to internally generated signals which the operator will tend to regard as reflections made by outside obstacles which are in reality not there.

Usually the returned radar signals are displayed on the screen of a cathode ray tube and observed visually by the operator. On this account, subjective judgments of the operator enter into the final determination of what is seen, how it is interpreted and how it is reported. The data obtained from radar systems are thus not as completely objective as is often assumed. In some few instances subjectiveness is somewhat reduced by the fact that the cathode ray screen is photographed, but even when this is done there is a subjective element introduced at the stage where a human observer has to interpret the photograph of the radar screen.

Radar operators do report unidentified targets from time to time and so there exists a category of UFO cases in which the unidentified flying object was seen on a radar screen. In a few cases there is a close correlation between an unknown thing in the sky seen visually and something also displayed on radar.

However in view of the many difficulties associated with unambiguous interpretation of all blobs of light on a radar screen it does not follow directly and easily that the radar reports support or "prove" that UFOs exist as moving vehicles scattering the radio pulses as would a metallic object. The Colorado project engaged the services of the Stanford Research Institute to make a general study of the functioning of radar systems from the point of view of the relation of their indications to UFOs. The study which was carried out resulted in the production of Section VI, Chapter 5, by Dr. Roy H. Blackmer, Jr. and his associates, R. J. Allen, R. T. S. Collis, C. Herold and R. I. Presnell.

Studies of specific UFO radar reports and their interpretation are presented in Section III, Chapter 5 by Gordon Thayer. Thayer is a radio propagation specialist on the staff of the Environmental Science Services Administration in Boulder. In his chapter, Thayer presents a detailed analysis of some 35 cases, some of which are visual, others radar, and some are both. Both optical and radar phenomena are treated together because of the similarity in the wave propagation problems involved.

In his summary of results he says: "... there was no case where the meteorological data available tended to negate the anomalous propagation hypothesis ..." However, Thayer points out that adequate meteorological data for a thorough interpretation is often lacking so that a great deal more observational material of this kind would be needed in order to deal with a larger proportion of all of the reported UFO radar cases.

In view of the importance of radar to the safe operation of all aircraft, it is essential that further research be done leading to the more precise knowledge possible of anomalous propagation of radar signals. However, it is felt that this can best be done by a direct attack on the problem itself rather than by detailed field investigation of UFO cases.

15. Visual Observation made by U.S. Astronauts. The popular UFO literature makes occasional reference to UFOs seen by the U.S. astronauts in the space program operated by the National Aeronautics and Space Administration. We do not know of similar reports by Soviet astronauts but they may well have seen similar things.

In flights conducted between 12 April 1961 and 15 November 1966, thirty U.S. and Russian astronauts spent a total of 2,503 hours in orbit. The Colorado project was fortunate in that Dr. Franklin Roach, one of the principal investigators, has worked closely with the astronaut program in connection with their visual observations and so was already quite familiar with what they had seen and also was able to conduct further interviews with several of them on the basis of close personal acquaintances already established.

Roach presents a detailed account of what they saw as related to the UFO question in Section III, Chapter 6. Nothing was seen that could be construed as a "flying saucer" or manned vehicle from outer space. Some things were seen that were identified as debris from previous space experiments. Three sightings that are described in detail remain quite unidentified and are, Roach says, "a challenge to the analyst."

Roach emphasizes that the conditions for simple visual observation of objects near the satellite are not as good as might be naively supposed. As he describes them, "The conditions under which astronauts made their observations are similar to those which would be encountered by one or two persons in the front seat of a small car having no side or rear windows and a partially covered, very smudged windshield." Moreover, the astronauts were kept occupied with other observations and activities during their flight and so did not have extended periods of time in which to concentrate on visual observation of their surroundings. Most of the available visual observations therefore have to be regarded as a by product rather than a primary purpose of the program in which they were engaged.

The conclusion is that nothing definite relating to the ETH aspect of UFOs has been established as a result of these rather sporadic observations.

16. Public Attitudes Toward UFOs. Opinion polls are widely employed nowadays to measure public attitudes on various important and trivial issues. It is natural therefore to apply the same method to a determination of public attitudes toward various phases of the UFO question.

Studies of this sort are not studies of the UFOs themselves, but an attempt at determination of what the American public thinks about UFOs. Some UFOs either do or do not come from outer space, and the fact of the matter would not be determined by finding out what the opinion of the American people about it may be. Nevertheless we considered that public attitudes do play a role in policy formation in America, and therefore it was appropriate to carry on some work in this area.

In 1947, 1950 and 1966 brief surveys of public attitudes on UFOs or flying saucers were conducted by the American Institute of Public Opinion, popularly known as the Gallup poll. Arrangements were made by the Colorado project for a more detailed study to be made during the spring of 1968. This was done for us by the Opinion Research Corporation. Findings of the earlier studies and of the study made for us are presented in Chapter 7 of Section III.

The first two studies indicated respectively that 90% and 94% of the American adult public had heard of flying saucers. The first of these results, taken within months of the original June 1947 sightings at Mt. Rainier indicates the extraordinary interest which the subject aroused from the outset. The 1966 survey indicated that 96% of the adult public had heard of flying saucers.

In the 1966 poll people were asked,

"Have you, yourself, ever seen anything you thought was a 'flying saucer'?"

The result was that 5% of the 96% who had heard of them answered yes to this question. The sample was designed to be representative of the American population, 21 years of age and older, of whom there are some 100 million. This is the basis of the oft-quoted statistic that five million Americans have said that they think they have seen a flying saucer.

In the same 1966 poll, 48% said they thought the things called flying saucers were "something real," and 31% said that they were "just people's imagination." The question does not distinguish between various kinds of "real" things, such as weather balloons, aircraft, planets, mirages, etc., so the result by no means indicated that 48% believe they are visitors from outer space. That question was not included in the 1966 poll.

The 1966 poll asked whether the person interviewed thinks "there are people somewhat like ourselves living on other planets in the universe." The question thus bears solely on ILE, not on whether such intelligences do in fact visit the Earth. Of the 1,575 interviewed 34% thought yes, 45% thought no, and 21% had no opinion.

There were no statistically significant regional differences between East, Midwest, South and West with regard to the proportion of the population which had heard of, had seen, or believed in the reality of flying saucers. However, as to belief in ILE, the existence of people on other planets, this belief was held by only 27% of southerners, as compared with 36% of easterners, 37% of midwesterners and 36% of westerners. The lower proportion of southerners who believe in ILE is statistically significant, that is, outside the range of chance variation due to finite size of sample. Although statistically significant, it is causally unexplained.

Significant variation with age is shown in responses to belief in the reality of flying saucers, and to belief in intelligent life on other planets. About 50% of persons under 60 believe in the reality of flying saucers as compared with about 33% of persons over 60. On the other hand, a significantly smaller proportion of those under 50 believe in ILE, than do those over 50. On both of these points, the decline in the number of "believers" among older people is mostly due to the increase of those having "no opinion" rather than to an increase of the number of "non-believers." Here again the poll gives no basis for conclusions as to the reasons for these differences.

As to dependence on sex, 22% of men or women have no opinion as to the "reality" of flying saucers. Significantly more women than men believe in their reality:

	% Real	*% Imaginary*
Men	43	35
Women	52	26

The poll showed that increased amount of formal education is associated with an increased tendency to believe in the reality of flying saucers. Perhaps this result says something about how the school system trains students in critical thinking.

An interesting correlation is found between tendency to believe in UFO reality, and to believe in ILE with having had a personal experience of having seen an UFO. The results are:

	% Believing UFOs Are Real	*% Believing in ILE*
Sighters	76	51
Non-sighters	46	34

As before, causal relations are unexplored; we do not know whether seeing is believing, or believing is seeing.

In the 1968 study conducted for the Colorado project by the Opinion Research Corporation, 2,050 adults over 17 years of age, living in private households in the continental United States were interviewed. In addition teenagers in the same household with an adult who was interviewed were also interviewed to give a sample of their views. Separate studies of opinions held by college students were conducted. These are reported in Section III, Chapter 7.

In the 1968 survey, 3% of adults replied affirmatively to "Have you, yourself, ever seen an UFO?" This parallels the 5% who answered affirmatively in the 1966 Gallup poll to the similar question, "Have you ever seen anything that you thought was a 'flying saucer'?" One might think that the smaller number in 1968 could be explained by perhaps less familiarity of the public with the term UFO than with the term flying saucer. This seems hardly likely, however, in that the question was part of a total interview in which the meaning of the term UFO would have become clear from the general context of other questions in the interview. It seems to us therefore that this poll actually indicated a smaller percentage of sighters than the earlier one.

An important finding is that 87% of those who said that they had seen an UFO, also declared that they had reported it to no one, other than to family or friends, that is, to no one by which it would have received official attention. Thus only about one-eighth of sightings were reported anywhere, and not all of these were reported to the Air Force. Hence if all sightings were reported to the Air Force, this result indicates that the number of reports received would be more than eight times as many as are now being received. From the small fraction who did report to the Air Force, it seems a fair inference that most of these non-reporting sighters did not think that what they saw constituted a security hazard.

In contrast, 56% of the non-sighters declared that they would report it to the police if they saw an UFO. We find this rather large discrepancy between the promised reporting behavior of the non-sighters and the actual reporting behavior of the sighters quite puzzling.

17. Other Psychological Studies. Consideration was given to a variety of modes of conducting psychological and psychiatric research into the UFO phenomenon. The possibility that an "experimental UFO" might be launched and reports of its sighting studied was given serious consideration and rejected on three grounds: In view of the fact that this was a government-sponsored, university-based study, it was felt that experiments in which the public might regard itself as having been victimized by what amounted to a hoax were unwise. Such experiments also might give rise, we thought, to the erroneous notion that the study regarded UFO phenomena *solely* as the result of misinterpretation of natural or man-made phenomena. Finally, we were advised by some of our experts in the psychological disciplines, that a "mock-up" UFO would introduce unknown variables that would render inconclusive any results derived from the conduct of experiments with it (see Section VI, Chapter 10).

Turning to the realm of psychiatry, we decided to refrain from mounting a major effort in this area on the ground that such a study could not be given priority over other investigations. This decision was buttressed by the evidence that we rapidly gathered, pointing to the fact that only a very small proportion of sighters can be categorized as exhibiting psychopathology and that, therefore, there is no reason to consider them any more suitable for study than psychotic or psychoneurotic individuals who belong to any other statistical class of the population as a whole (see Section VI, Chapter 3).

18. Instrumentation for UFO Searches. As remarked earlier, the short duration of most UFO sightings, the delays in reporting them and the delays caused by communication and travel, make it essentially impossible that investigators can bring physical observing equipment to a report site quickly enough to make UFO observations in that way. There is another way that is often proposed for getting better observational data than is now available; namely, to set up a permanently manned network of observing stations at various places in the country to observe such UFOs as might come within their range.

Such a network of stations might be set up solely for the purpose of UFO study, or it might be established in conjunction with one of the networks of stations which exist for other astronomical or meteorological purposes. This latter alternative, of course, would be much less expensive than the former, or could give a greater coverage for the same expenditure.

We gave considerable attention to the possibilities and difficulties in this direction (Section VI, Chapter 9). At first we hoped that some definite results could be obtained by such cooperation with existing stations in a way that would make results available for this report.

An all-sky camera was operated during most of August 1967 at Harrisburg, Penna. during an UFO flap in that locality (Case 25) but no interesting results were found on some 9,000 photographs. It would be quite expensive to operate a network of such cameras on a routine basis all over the United States. The likelihood of interesting images being recorded would be very small. Because of the short duration of an UFO appearance a proper plan for use of the all-sky camera would involve frequent processing and examination of the film, otherwise the presence of an UFO would not be recognized until long after it had disappeared. This would greatly increase the cost of operation of such a network.

Another suggestion that is often made is to make UFO studies in connection with the radar networks operating in this country for air traffic control under auspices of the Federal Aviation Agency. Consideration was given to this possibility and it was concluded that it is quite out of the question to burden this network with additional duties of any kind. The air traffic control operators are now heavily burdened with the work of safely guiding civil and military aviation. During the summer of 1968 especially, the heavy overloads that sometimes exist on the system were emphasized by troublesome traffic delays in the neighborhood of several of the nation's major airports. It would be quite out of the question to ask the air traffic controllers to assume the responsibility of watching for UFOs in addition to their primary responsibilities. It would likewise be impracticable for a separate group of personnel to be installed at these stations to watch the same radars for UFOs.

The Prairie Network is a group of camera stations operated in the mid-west by the Smithsonian Institution in connection with the Harvard Meteor Program. Its primary purpose is to detect and record meteor trails in such a way as to guide a search for actual meteoritic bodies that strike the earth's surface. The field headquarters of this network is at Lincoln, Nebraska.

We prepared a listing of reported UFO sightings since 1965 that fell within the geographic limits of this network and through the kind cooperation of the Smithsonian Institution obtained the records of the network for the times and locations of these sightings. About half of the sightings were so lacking in specific information that, Frederick Ayer reports (1229) "even if an object had been recorded by the film it would have been impossible to correlate it with the sighting." About one-third of the sightings could not be traced on the film because of overcast skies. Some 18% of all the UFO sightings were identified on the network's records with a fair degree of probability. Nearly all of these were identified as astronomical objects. Some consideration was given to the costs and likelihood of success of adapting the Prairie Network instruments to UFO searches without interfering with their primary purpose. We think that something might be done along this line at reasonable expense, but we do not make a positive recommendation that such a program be undertaken because of the inconclusiveness of the information that we believe would be gathered.

Another existing program that was studied for unrecognized UFO records was that of scanning the night sky for study of air glow from the upper atmosphere, and of zodiacal light. Detailed study was made of two records obtained from a station on the Hawaiian Islands. One of these remains unidentified but is thought to be related to an artificial satellite for which no information is readily available. The other was definitely identified as a suborbital missile launched from Vandenberg AFB on the coast of southern California. Mr. Ayer concludes that "because of their relatively extensive sky coverage, scanning photometers can be considered useful instruments in the conduct of UFO searches." This, however, is not to be construed as a recommendation that a network of scanning photometer stations be established for this purpose.

Consideration was also given to the adaptability to UFO search purposes of radars of the type used by the Weather Bureau, and the radar station of the Radar Meteor Project of the Smithsonian Institution located near Havana, Illinois.

Although frequent claims are made in the UFO popular literature of magnetic disturbances due to the presence of UFOs, a consideration of various official magnetometer records produced no evidence of an effect of this kind that, in our judgment would warrant the setting up of an observational program to look for UFOs by their alleged magnetic effects.

19. Conclusion. In our study we gave consideration to every possibility that we could think of for getting objective scientific data about the kind of thing that is the subject of UFO reports. As the preceding summary shows, and as is fully documented in the detailed chapters which follow, all such efforts are beset with great difficulties. We place very little value for scientific purposes on the past accumulation of anecdotal records, most of which have been explained as arising from sightings of ordinary objects. Accordingly in Section I we have recommended against the mounting of a major effort for continuing UFO study for scientific reasons.

This conclusion is controversial. It will not be accepted without much dispute by the UFO amateurs, by the authors of popular UFO books and magazine articles, or even by a small number of academic scientists whose public statements indicate that they feel that this is a subject of great scientific promise.

We trust that out of the clash of opinions among scientists a policy decision will emerge. Current policy must be based on current knowledge and estimates of the probability that further efforts are likely to produce further additions to that knowledge. Additions to knowledge in the future may alter policy judgments either in the direction of greater, or of less attention being paid to UFO phenomena than is being done at present.

We hope that the critical analysis of the UFO situation among scientists and government officials that must precede the determination of official policy can be carried out on a strictly objective basis.

Attacks on the integrity of various individuals on either side of this controversy ought to be avoided. The question of an individual's integrity is wholly distinct from the issue of what science should do in the future about UFOs.

In the Congress of the United States concern about the UFO problem from a defense viewpoint is the province of the House Committee on Armed Services. Concern about it from the point of view of the nation's scientific research program comes under the House Committee on Science and Astronautics. Here there seems to be a valid situation of overlapping jurisdictions because the UFO problem can be approached from both viewpoints.

A particular interest in the UFO problem has been shown by Congressman J. Edward Roush of Indiana, who is a member of the House Committee on Science and Astronautics. He performed a valuable service by arranging for the holding of a "Symposium on Unidentified Flying Objects" in Washington on 29 July 1968 (see references). As pointed out by one of the symposium participants, Prof. Carl Sagan of the department of astronomy of Cornell University, the presentations made in that symposium incline rather strongly to the side of belief that large-scale investigations of the UFO phenomenon ought to be supported in the expectation that they would be justified by what some speakers called "scientific paydirt."

We studied the transcript of this symposium with great care to see whether we would be led thereby to any new material related to this study. We did not find any new data.

Several of the contributors to that symposium have become trenchant advocates in the past several years of a continuing major government investment in an UFO program. Several have long urged a greater degree of congressional interest in this subject. The symposium of 29 July afforded them an occasion on which with the utmost seriousness they could put before the Congress and the public the best possible data and the most favorable arguments for larger government activity in this field.

Hence it is fair to assume that the statements presented in that symposium represent the maximum case that this group feels could be made. We welcome the fact that this symposium is available to the public and expect that its data and arguments will be compared with those in their report of this study by those whose duty it is to make responsible decisions in this area.

We have studied this symposium record with great care and find nothing in it which requires that we alter the conclusions and recommendations that we have presented in Section I, nor that we modify any presentation of the specific data contained in other sections of this report.

References:

Bailey, J. O. *Pilgrims through Space and Time—Trends and Patterns in Scientific and Utopian Fiction*, New York: Argus Books, 1947.

Bloecher, T. E. *Report of the UFO Wave of 1947*, Washington (?), 1967.

Cameron, A. G. W. *Interstellar Communication*, New York: Benjamin, 1963.

Hall, Richard H. *The UFO Evidence*, Washington: NICAP, 1964.

Keyhoe, Donald E. "Flying Saucers Are Real," *True*, 1950.

Lorenzen, Coral B. *The Great Flying Saucer Hoax*, New York: William-Frederick Press, 1962.

Lowell, Percival H. *Mars and Its Canals*, New York: The Macmillan Company, 1908.

Markowitz, William. "The Physics and Metaphysics of Unidentified Flying Objects," *Science*, 157 (1967), 1274–79.

Menzel, Donald H. *Flying Saucers*, Cambridge: Harvard University Press, 1952.

Menzel, Donald H., and Lyle G. Boyd. *The World of Flying Saucers*, New York: Doubleday, 1963.

Miller, Stanley L. "Production of Organic Compounds under Possible Primitive Earth Conditions," *Journal American Chemical Society*, 77 (1955), 2351–61.

Olsen, T. *The Reference for Outstanding UFO Sighting Reports*, Ridenwood, Maryland: UFOIRC, Inc.

Page, R. M. *The Origin of Radar*, Garden City, New York: Doubleday, Anchor Books, 1962.

Purcell, Edwin. "Radioastronomy and Communication through Space," *Brookhaven Lecture Series No. 1*, Brookhaven National Laboratory, New York, 16 November 1960.

Ruppelt, B. J. *The Report on Unidentified Flying Objects*, New York: Doubleday and Company, Ace Books, 1956.

Rush, J. H. *The Dawn of Life*, New York: Doubleday & Co., Inc. 1957 (also Signet Library of Science, New American Library, N.Y. 1962).

Salisbury, Frank B. "Martian Biology," *Science*, 136 (1962), 17–26.

Shklovskii, I. S. *Artificial Satellites of Mars and Riddle of the Martian Satellites*, Moskow: Komsomal'skaya Pravda, 1 May and 31 May 1959, English translation, FTD-T[-62-488-1], Wright Patterson AFB, 18 May 1962.

Shklovskii, I. S., and Carl Sagan. *Intelligent Life in the Universe*, San Francisco: Holden-Day, 1966.

Sullivan, Walter. *We Are Not Alone*, New York: McGraw-Hill Book Co., 1964, New York: New American Library (paperback edition), 1966.

U.S. Ninetieth Congress, Second Session, Hearings before the Committee on Science and Astronautics, 29 July 1968. *Symposium on Unidentified Flying Objects*, Washington: Govt. Print. Off., 1968.

Vallee, Jacques, and Janine Vallee. *Challenge to Science—The UFO Enigma*, Chicago: Henry Regnery Co., 1966.

Epilogue

Let Us Reflect

Thoughtful Inquiry on Twenty-Five Years of Skepticism

MICHAEL SHERMER

Skepticism dates back to the ancient Greeks, well captured in Socrates' famous quip that all he knew was that he knew nothing. Skepticism as nihilism, however, gets us nowhere, and thankfully, almost no one embraces it. The word *skeptic*, in fact, comes from the Greek *skeptikos*, for "thoughtful"—far from modern misconceptions of the word as meaning "cynical" or "nihilistic." According to the *Oxford English Dictionary*, *skeptical* has also been used to mean "inquiring," "reflective," and, with variations in the ancient Greek, "watchman" or "mark to aim at." What a glorious meaning for what we skeptics do! We are thoughtful, inquiring, and reflective, and in a way, we are the watchers who guard against bad ideas—consumer advocates of good thinking who, through the guidelines of science, establish the mark to aim at.

Since the time of the Greeks, skepticism (in its various incarnations) has evolved along with other epistemologies and their accompanying social activists. The Enlightenment, on one level, was a century-long skeptical movement, for there were no beliefs or institutions that did not come under the critical scrutiny of such great thinkers as Voltaire, Denis Diderot, Jean-Jacques Rousseau, John Locke, Thomas Jefferson, and many others. Immanuel Kant in Germany and David Hume in Scotland were skeptics' skeptics in an age of skepticism, and their influence continues unwaned to this day (at least in academic philosophy and skepticism). Closer to our time, Charles Darwin and Thomas Huxley were skeptics par excellence, not only for the revolution they launched and carried on, respectively, against the dogma of creationism but also for their stand against the burgeoning Spiritualism movement that was sweeping across the United States, England, and the Continent. (Darwin worked quietly behind the scenes, whereas Huxley railed publicly against the movement, bemoaning it in one of the great one-liners in the history of skepticism: "Better live a crossing-sweeper than die and be made to talk twaddle by a 'medium' hired at a guinea a seance.") In the 1900s, Bertrand Russell and Harry Houdini stand out as representative of the skeptical thinkers and doers, respectively, of the century's first half, and in the first year of its second half, Martin Gardner's *Fads and Fallacies in the Name of Science* launched what we think of today as "the skeptics."

We are at an appropriate time for reflection with this two-volume encyclopedia on science, pseudoscience, and skepticism. I date the modern skeptical movement to 1950, with the publication of an essay by Gardner in the *Antioch Review* entitled "The Hermit Scientist." The essay is about what we would today call pseudoscientists, and it was Gardner's first-ever publication of a skeptical nature. It launched a lifetime of critical analysis of fringe claims, and in 1952 (at the urging of his literary agent, John T. Elliott), Gardner expanded the article into a book-length treatment of the subject under the title *In the Name of Science*, with the descriptive subtitle *An Entertaining Survey of the High Priests and Cultists of Science, Past and Present*. Published by Putnam, the book sold so poorly that it was quickly remaindered, and it lay dormant until 1957, when it was republished by Dover. It has come down to us as *Fads and Fallacies in the Name of Science*, still in print and arguably *the* skeptic classic of the past half century.

What caught the attention of a youthful Martin Gardner half a century ago? The "hermit scientist" who worked alone and was usually ignored by mainstream scientists: "Such neglect, of course, only strengthens the convictions of the self-declared genius," Gardner concluded in his original 1950 paper. "Thus it is that probably no scientist of importance will present the bewildered public with detailed proofs that the earth did not twice stop whirling in Old Testament times, or that neuroses bear no relation to the experiences of an embryo in the mother's womb" (referring to L. Ron Hubbard's dianetics theory that negative engrams are imprinted in the fetus's brain while in the womb).

Gardner was, however, half wrong in his prognostications: "The current flurry of discussion about (Immanuel) Velikovsky and Hubbard will soon subside, and their books will begin to gather dust on library shelves." While Velikovskians are a quaint few surviving in the interstices of fringe culture, Hubbard has been canonized by the Church of Scientology and deified as the founding saint of a world religion.

In the first chapter of *In the Name of Science*, Gardner picked up where he left off, noting that "tens of thousands of mentally ill people throughout the country entered 'dianetic reveries' in which they moved back along their 'time track' and tried to recall unpleasant experiences they had when they were embryos." More than fifty years later, Scientology has converted those reveries into a worldwide cult of personality surrounding L. Ron Hubbard that targets celebrities for membership and generates hundreds of millions of dollars in tax-free revenue as an IRS-approved "religion."

Today, UFOs are big business, but in 1950, Gardner could not have known that the nascent flying-saucer craze would turn into an alien industry, but it was off to a good start: "Since flying saucers were first reported in 1947, countless individuals have been convinced that the earth is under observation by visitors from another planet." Absence of evidence then was no more a barrier to belief than it is today, and believers proffered the same conspiratorial explanations for the dearth of proof, as Gardner explained: "I have heard many readers of the saucer books upbraid the government in no uncertain terms for its stubborn refusal to release the 'truth' about the elusive platters. The administration's 'hush-hush policy' is angrily cited as proof that our military and political leaders have lost all faith in the wisdom of the American people."

From his perspective in 1950, Gardner was even then bemoaning the fact that some beliefs never seem to go out of vogue, as he recalled H. L. Mencken's quip from the 1920s that "if you heave an egg out of a Pullman car window anywhere in the United States you are likely to hit a fundamentalist."

Gardner cautioned that when presumably

religious superstition should be on the wane, it is all too easy "to forget that thousands of high school teachers of biology, in many of our southern states, are still afraid to teach the theory of evolution for fear of losing their jobs." Today, Kansas and other states enjoin the fight as the creationist virus spreads northward.

I devote an entire chapter in my book *The Borderlands of Science* to Martin Gardner and his seminal work, but suffice it to say here that *Fads and Fallacies in the Name of Science* has been a cherished classic read by legions of skeptics and scientists, and it laid the foundation for a bona fide skeptical movement that found its roots in the early 1970s. There has been some debate (and much quibbling) about who gets what amount of credit for the founding of the Committee for the Scientific Investigation of Claims of the Paranormal (CSICOP) and its journal, *Skeptical Inquirer* (much of this played out in the pages of *Skeptic* magazine in our interviews with the major players). This is not the place to present a definitive history of the movement, but from what I have gleaned from first- and secondhand sources, Gardner, magician James Randi, psychologist Ray Hyman, and philosopher Paul Kurtz played key roles in the foundation and planning of the organization, with numerous others, such as Phil Klass and Marcello Truzzi, in important supporting roles.

The founding of the Skeptics Society by myself, Pat Linse, and Kim Ziel Shermer in 1992, then, was also not without precedent and historical roots, and though the history of this organization has yet to be written, it is clear that without the likes of Gardner, Randi, Hyman, and Kurtz, there would be no Skeptics Society and no *Skeptic* magazine. And what an experience it has been.

Twenty-five years ago, I was twenty years old and in my third year of college at Pepperdine University, a Church of Christ–based institution located in Malibu, California, and overlooking the Pacific Ocean. Although the site was certainly a motivating factor in my choice of a college, the primary reason I went there was that I was a born-again Christian who took his mission for Christ seriously. I thought I should attend a school where I could receive some serious theological training, and I did. I took courses in the Old and New Testaments, Jesus the Christ, and the writings of C. S. Lewis. I attended chapel twice a week (although, truth be told, attendance was required for all students). Dancing was not allowed on campus (the sexual suggestiveness might trigger already-inflamed hormone production to go into overdrive), and we were not allowed into the dorm rooms of members of the opposite sex.

Despite the restrictions, it was a good experience because I was a serious believer and thought that was the way we should behave anyway. But somewhere along the way, I found science, and that changed everything (although not overnight). I was thinking of majoring in theology, but then I discovered that a Ph.D. required proficiency in several dead languages (Hebrew, Greek, Aramaic, and Latin). Knowing that I was not especially good at learning live languages, let alone dead ones, I went into psychology and mastered one of the languages of science: statistics. There (and in research methodology courses), I discovered that many problems can be solved by establishing parameters to determine whether a hypothesis is probably right (i.e., rejecting the null hypothesis at the .01, or 99 percent, level of significance) or definitely wrong (i.e., not statistically significant). Instead of the rhetoric and disputation of theology, there were the logic and probabilities of science. What a difference this difference in thinking makes!

By the end of my first year of a graduate program in experimental psychology at the California State University, Fullerton, I had deconverted out of Christianity and removed my silver ichthus, replacing what was for me the stultifying dogmas of a 2,000-year-old religion with the worldview of an always changing, always fresh science. The passionate nature of this perspective was enthused most emphatically by my evolutionary biology professor, Bayard Brattstrom, particularly in his after-class discussions at a local bar that went into the wee hours of the morning. This is where the action was for me.

About that time (1975–1976), Uri Geller showed up on my radar screen. I recall *Psychology Today* and other popular magazines published glowing stories about him, and reports were afloat that experimental psychologists had tested the Israeli psychic and determined that he was genuine. My adviser—a strictly reductionistic Skinnerian behavioral psychologist named Doug Navarick—didn't believe a word of it, but I figured there might be something to the Geller phenomenon, especially in light of all the other interesting research being conducted on altered states of consciousness, hypnosis, dreams, sensory deprivation, dolphin communication (John C. Lilly), and the like. I took a course in anthropology from a woman who researched shamans of South America and their use of mind-altering plants. It all seemed entirely plausible to me, and, being personally interested in the paranormal (the Ouija board consistently blew my mind), I figured that this was rapidly becoming a legitimate subfield of psychological research. After all, Thelma Moss had a research laboratory devoted to studying the paranormal, and it was at the University of California, Los Angeles (UCLA), no less, which had one of the most highly regarded psychology programs in the country.

Enter James "the Amazing" Randi. I do not recall exactly when or where I first encountered him. I believe it was on the *Tonight Show* when he was demonstrating how to levitate tables, bend spoons, and perform psychic surgeries. He didn't convince me to become a full-fledged skeptic overnight, but he got me thinking that if some of the psychics were fakes, perhaps they all were (and if not fakes, at least self-deceived). Herein lies an important lesson. There is little to no chance that we can convince True Believers of the errors of their thinking. Our purpose is to reach that vast middle ground between hard-core skeptics and dogmatic believers—people like me who thought that there might be something to these claims but had simply never heard a good counterexplanation. There are many reasons why people believe weird things, but certainly one of the most pervasive is simply that most people have never heard a good explanation for the weird things they hear and read about. Short of a good explanation, they accept the bad explanation that is typically proffered. This fact alone justifies all the hard work performed by skeptics toward the cause of science and critical thinking. It does make a difference.

Fast-forward ten years. My first contact with organized skepticism came in the mid-1980s through the famed aeronautics engineer and human-powered flight inventor Paul MacCready. I originally met Paul through the International Human Powered Vehicle Association (IHPVA), as he was interested in designing these vehicles and I was interested in racing them (I had a ten-year career as an ultra-marathon cyclist). One day, he phoned to invite me to a lecture at the California Institute of Technology being hosted by a group called the Southern California Skeptics (SCS). This was an offshoot of CSICOP and one of many groups that had spontaneously self-organized around the country throughout the 1980s. The lectures were fascinating, and because of my affiliation with Paul, I got to meet some of the insiders in what was rapidly becoming the "skeptical movement." Paul was a friend of such science megastars as Richard Feynman, Stephen Jay Gould, and Murray Gell-Mann, and with the likes of Randi and the magicians Penn and Teller affiliated with the movement, it seemed like it was a happening place to be.

In 1987, CSICOP hosted a convention at the Pasadena Civic Center that featured Carl Sagan as the keynote speaker, and he was so inspiring that I decided to return to graduate school to complete my doctorate.

By the end of the 1980s, however, the Southern California Skeptics folded, and the skeptical movement came to a grinding halt in the very place that so desperately needed it. In 1991, I completed my Ph.D., was teaching part-time at Occidental College, and was nosing around for something different to do. I had just published a paper in a science history journal on the Louisiana creationism trial; it featured the activities of SCS members who had organized the amicus curiae brief that was signed by seventy-two Nobel laureates (Murray Gell-Mann encouraged his fellow Nobelists) and was submitted to (and read by) the U.S. Supreme Court. One of SCS's former volunteer staff members, Pat Linse, heard about the paper, tracked me down, and dropped by to pick up a reprint of my article.

During that visit, she expressed her frustration—and that of many others—that skepticism in southern California had gone the way of the Neanderthals. Subsequent meetings with her and others inspired Kim, Pat, and me to jump-start the skeptics movement again by launching a new group and inviting James Randi for our inaugural lecture in March 1992. The event was a smashing success, as well over 400 people crammed into a 300-seat hall to hear the amazing one astonish us all with his wit, wisdom, and magic.

With that successful event, we were off and running. I starting planning a newsletter, but when Pat saw a sample copy of a bicycle magazine I was publishing—*Ultra Cycling* magazine (the publication of the Ultra-Marathon Cycling Association and Race across America, which I had cofounded in the early 1980s), which was 64 pages long, perfect-bound, and with a duotone coated cover—she said that if we could splurge for a skeptical publication of that quality, she would provide the appropriate artwork and typography. Since Pat is a professional artist who was working for movie studios generating film posters, she was more than capable of backing up her offer, which I accepted.

Our original cover was to feature Randi, and Pat produced a striking portrait of him. But just before publication, Isaac Asimov died, so Pat generated a new cover portrait, and that became the cover of volume 1, number 1, of what we came to call *Skeptic* magazine. (My originally planned title—*The Journal of Rational Skepticism*—was voted down by Pat and my wife, Kim Ziel Shermer, who reasoned that shorter is better. They were right.)

Allow me to close this epilogue with a quote from one of my favorite skeptical books, Paul Kurtz's *The Transcendental Temptation* and his discussion of the meaning and goals of skepticism. It is an admonition we should all bear in mind, a passage to be read once a year:

> The skeptic is not passionately intent on converting mankind to his or her point of view and surely is not interested in imposing it on others, though he may be deeply concerned with raising the level of education and critical inquiry in society. Still, if there are any lessons to be learned from history, it is that we should be skeptical of all points of view, including those of the skeptics. No one is infallible, and no one can claim a monopoly on truth or virtue. It would be contradictory for skepticism to seek to translate itself into a new faith. One must view with caution the promises of any new secular priest who might emerge promising a brave new world—if only his path to clarity and truth is followed. Perhaps the best we can hope for is to temper the intemperate and to tame the perverse temptation that lurks within.

Amen, brother!

Contributors

John Adams (The Alien Archetype: The Origin of the "Grays") is a professor of history at Alabama A & M University. He has written an article titled "Outer Space and the New World in the Imagination of Eighteenth-Century Europeans." He teaches courses in ancient and modern world civilizations, British history, and Asian history, and has written a book on the dynamics between science fiction in early modern Europe and attitudes toward race and ethnicity among early Copernicans.

D. Alan Bensley (Pseudoscience and Science: A Primer in Critical Thinking) is an associate professor of psychology at Frostburg State University in Frostburg, Maryland. He is the author of the textbook *Critical Thinking in Psychology: A Unified Skills Approach* and has done extensive research on the improvement of thinking skills. His research and teaching interests are in research methods, sensation and perception, and cognition and critical thinking.

John Berger (Handwriting Analysis and Graphology) is an expert on handwriting analysis and graphology and a long-time investigator of unusual claims.

Susan J. Blackmore (Memes as Good Science, Out-of-Body Experiences, Near-Death Experiences) is a senior lecturer in psychology at the University of the West of England and a long-time researcher on the paranormal. She is the author of *The Adventures of a Parapsychologist, Beyond the Body, Dying to Live,* and *The Meme Machine.* She is one of the best known and most highly regarded skeptical investigators in the world today.

Jon Blumenfeld (Stock Market Pseudoscience) is Connecticut chapter chairman of the New England Skeptical Society, and is a regular contributor to its quarterly newsletter, *The New England Journal of Skepticism.* He has ten years of experience as a bond trader and commodity trading advisor and is currently an interest rate strategist for the United States' fixed income derivatives trading unit of a large European bank.

Chris Bonds (Synchronicity) teaches music at Wayne State College, Wayne, Nebraska. His interests include music, psychology, the paranormal, and the theory of evolution.

Rebecca Bradley (Tutankhamun's Curse) has a doctorate in archeology, specializing in the Nile Valley. Her other personas include fantasy writer (*The Gil Trilogy*), horror writer, and disaster voyeur.

Maarten Brys (Bermuda Triangle) studied the philosophy of science. He's a member of SKEPP and SKEPSIS, two skeptical organizations in Belgium and the Netherlands, and he is interested in finding scientific explanations for apparently paranormal phenomena.

Tim Callahan (Anastasia: A Case Study in the Myth of the Miraculous Survival) is the religion editor for *Skeptic* magazine and a long-time contributor to skeptical and humanist publications. He is the author of *Bible Prophecy: Failure or Fulfillment?* and his latest book is *The Secret Origins of the Bible*. His research interests are in comparative mythology, creationism, environmental issues, and biblical criticism.

Steuart Campbell (Ball Lightning) is a science writer from Edinburgh, Scotland, whose work has appeared in the *New Humanist* and *Skeptic*. He is also the author of *The UFO Mystery Solved* and *The Rise and Fall of Jesus*.

Al Carroll (Shamans and Shamanism) is in the Ph.D. program in American Indian history at Arizona State University. He received his master's degree from Purdue University. He is a Native American activist and co-founder of New Age Frauds Plastic Shamans (NAFPS), a watchdog organization devoted to warning the public about frauds impersonating as Native American medicine people. His research interests include Native American veterans, the modern practice of Native American beliefs, and Native American/New Age conflict.

John Casti (Science Is Just Beginning) is executive editor of the journal *Complexity*, a fellow of the Santa Fe Institute in Santa Fe, New Mexico, and a professor at the Technical University of Vienna. He is the author of *Searching for Certainty, Paradigms Lost, Complexification*, and other science books.

Drew Christie (Séance, Societies for Psychical Research) is a professor of philosophy at the University of New Hampshire, where he teaches environmental philosophy, logic, philosophy of law, and social and political philosophy. His research interests also encompass American pragmatism (old and new) and the philosophy of education. His recent work includes "Dewey and the Splintered Vision"; encyclopedia entries, "Richard Rorty" and "Chauncey Wright"; and reviews of websites devoted to the classical pragmatists Peirce, James, and Dewey. He is currently a member of the local school board and is working to meaningfully integrate computers into the K-12 curriculum.

Brad Clark (Spiritualism) has studied spiritualism and the spiritual movement in the United Kingdom and America.

Kevin Courcey (Prayer and Healing) is a registered nurse who has practiced for 22 years. He has spent 15 years as a psychiatric nurse and six years in phone triage answering questions about medical problems ranging from lacerations to fevers to chest pain. He has also chaired an 800-nurse bargaining unit and been a board member of the Oregon Nurses Association. He is a member of Freedom From Religion Foundation and is active in the Corvallis Secular Society in Oregon.

Chris Cunningham (The Shroud of Turin) is interested in applying science to miracles and to claims of the paranormal.

Geoffrey Dean (Astrology, Placebo Effect, Undeceiving Ourselves) coedited with Arthur *Mather Recent Advances in Natal Astrology: A Critical Review 1900–1976*, the first book-length critical review of scientific research into astrology. He and Mather have been collaborators since 1975 on critical articles, debates, surveys, and prize competitions for research in astrology. Dean is a freelance technical writer and editor in Perth, Western Australia.

Perry DeAngelis (Cults, Dowsing) is the executive director of the New England Skeptical Society. His interests are in field investigations of paranormal and pseudoscientific claims.

Tana Dineen (Psychotherapy as Pseudoscience) is a licensed psychologist with three decades of clinical experience. Early in her career she became concerned about the growing influence of mental health "experts" on society, which lead her to write her controversial book: *Manufacturing Victims: What the Psychology Industry Is Doing to People.*

Clayton J. Drees (Witches and Witchcraft) is associate professor of history and chair of the history department at Virginia Wesleyan College in Norfolk, Virginia. He teaches courses on the medieval and early modern periods in Europe and has published two books, *Authority and Dissent in the English Church* (1997) and *The Late Medieval Age of Crisis and Renewal* (2001).

Dan Dugan (Anthroposophy and Anthroposophical Medicine) is well known in the field of audio engineering as the inventor of the automatic microphone mixer. His patented equipment is used in thousands of churches, courtrooms (including the U.S. Supreme Court), and on television shows, including *The Late Show with David Letterman* and *Hollywood Squares.* In addition to engineering, Dan has a lively interest in philosophy, particularly skepticism, the philosophy of science, and current controversies about scientific paradigms and alternative medicine.

Chris Duva (Anomalous Psychological Experiences) is an assistant professor of biological psychology at Eastern Oregon University and an adjunct faculty member at Capella University. His research interests include false memory, brain damage-induced amnesia, and the physiological basis of drug addiction.

Jon Entine (Race and Sports as Good Science) wrote and produced a widely acclaimed 1989 NBC television special with Tom Brokaw on black athletes. This led to the publication of his controversial book *Taboo: Why Black Athletes Dominate Sports and Why We're Afraid to Talk about It.* His next two projects focused on genetics: *Jewish Genes* examines the effort to identify medical cures for diseases that disproportionately effect specific populations, including the social and political tempest stirred by such research; and *Creating Superboy (and Girl),* which looks at the impact of genetic engineering on human performance in athletics and elsewhere.

Garrett G. Fagan (Alternative Archaeology) is associate professor of classics and ancient Mediterranean studies and history at Penn State University. Aside from the phenomenon of pseudoarchaeology, his research interests lie mainly in Roman history. His first book *Bathing in Public in the Roman World* was published in 1999.

Kenneth Feder (Ancient Astronauts, The Mars Face: Extraterrestrial Archaeology, Pseudoarchaeology: Native American Myths as a Test Case) is a professor of anthropology at Central Connecticut State University. He is the director of the Farmington River Archaeological Project and the author of several books including: *The Past in Perspective: An Introduction to Human Prehistory; A Village of Outcasts: Historical Archaeology and Documentary Research at the Lighthouse Site;* and *Frauds, Myths, and Mysteries: Science and Pseudoscience in Archaeology.*

Robert A. Forde (Hypnosis) is a chartered psychologist and over the last 30 years has worked in prisons and police services. He is now in private practice in North Somerset, England, specializing in forensic and health psychology. His special interests are researching suggestibility (especially in police interviews), psychological trauma, and malingering.

Ronald Fritze (Pseudoarchaeology: Precolumbian Discoverers of America as a Test

Case) is a professor of history at Lamar University in Beaumont, Texas. He received his Ph.D. from Cambridge University and is one of the authors of *Reference Sources in History, Reflections on Western Civilization,* and *Reflections on World Civilization.* He is also editor of *The Historical Dictionary of Stuart England.*

Diego Golombek (Biorhythms) is professor of physiology and a researcher at the National University of Quilmes (where he heads the Chronobiology Lab), the University of Buenos Aires, and the National Research Council (Buenos Aires, Argentina). He has published extensively on circadian rhythms and general science topics.

Gina Green (Facilitated Communication) was director of research at the New England Center for Autism and Associate Scientists in the behavioral sciences division of the E. K. Shriver Center for Mental Retardation. She has written extensively on facilitated communication and was one of the first scientists to submit facilitated communication to scientific analysis. The book she most recently edited is *Behavioral Intervention for Young Children with Autism: A Manual for Parents and Professionals.*

Diane Halpern (Race and I.Q. as Pseudoscience) is director of the Berger Institute for Work, Family, and Children and professor of psychology at Claremont McKenna College. She is the author of several hundred journal articles and many books, including *Thought and Knowledge: An Introduction to Critical Thinking and Sex Differences in Cognitive Abilities.* Her teaching and research have been recognized with many awards including the 2002 Outstanding Teaching Award from Western Psychological Association, the American Psychological Foundation Award for Distinguished Teaching, the American Psychological Association Award for Distinguished Career Contributions to Education and Training, and the Outstanding Professor Award from California State University (state-wide).

Alan Harris (Tunguska) is a former scientist at NASA's Jet Propulsion Laboratory in Pasadena, California, and is now a senior research scientist with the Space Science Institute. His research specialty is studying small bodies of the solar system such as satellites, asteroids, comets, and meteors. His long-time interest lies in studying the risks of cosmic impacts on the Earth.

Steve B. Harris (Immortality: The Search for Everlasting Life) is an internist and experimental physiologist who is particularly interested in issues of life extension and the philosophy of science. He is a long-time member of the Skeptics Society's editorial board, and is chief of research at Critical Care Research, Inc., a southern California biotech company that is developing advanced resuscitation technologies.

Michael Heap (Ideomotor Effect [the "Ouija Board" Effect]) is a freelance clinical and forensic psychologist in Sheffield, England, and chairman of the Association for Skeptical Enquiry. He has a special interest in hypnosis and is the coauthor of *Hypnosis in Therapy* and *Hartland's Medical and Dental Hypnosis* and the coeditor of *Hypnosis: Current Clinical, Experimental and Forensic Practices,* and *Hypnotherapy: A Handbook and Hypnosis in Europe.*

John Hochman (Recovered Memory Therapy and False Memory Syndrome: A Psychiatrist's Perspective as a Test Case) is a practicing psychiatrist in Los Angeles, California. He specializes in the evaluation and treatment of victims of cultic entities and/or undue influence; the theoretical study of cult phenomena, psychiatry and the law; post-traumatic stress disorders, and multiple personality disorder, of which he is skeptical. He is a consultant and expert witness in courtroom cases involving abuse allegations, coercive persuasion, and psychotherapy cult involvement.

Samuel Homola (Chiropractic: Conventional or Alternative Healing?) retired in 1998 after 43 years of full-time practice as a chiropractor. He is the author of 12 books, including *Inside Chiropractic: A Patient's Guide*. He has written numerous articles for magazines and journals, including "Finding a Good Chiropractor," published in *Archives of Family Medicine*.

John Horgan (Science Is at an End) was a senior writer at *Scientific American* from 1986 to 1997 and is now a freelance writer whose work has appeared in the *New York Times*, *Time*, the *Washington Post*, *Science*, the *London Times*, the *New Republic*, *Discover*, *Slate*, and elsewhere. His books include *The End of Science* (1996), *The Undiscovered Mind* (1999), and *Rational Mysticism* (2003).

Satyam Jain (Magnetic Therapy) is a resident in psychiatry at Brown University and a research fellow in psychiatry at Creighton University Medical.

William Jarvis (Homeopathy) is a professor at Loma Linda University with dual appointments in the schools of medicine and public health and a secondary appointment in the school of dentistry. He is a consumer health education specialist and is involved in a wide variety of activities related to this field. He is president of the National Council Against Health Fraud, a nonprofit voluntary health agency that combats health misinformation, fraud, and quackery. He is also a member of the American Cancer Society's National Committee on Questionable Methods of Cancer Management and the California attorney general's Task Force on Health Fraud.

Simon Jones (Uri Geller) has written articles and interviews related to paranormal claims and fringe beliefs. His award-winning interview with Uri Geller can be found online at: www.simon-jones.org.uk.

Professor Ivan W. Kelly (Astrology, Placebo Effect, Undeceiving Ourselves) chairman of the Astrology Subcommittee of the United States-based Committee for the Scientific Investigation of Claims of the Paranormal, and author or coauthor of over one hundred scientific or philosophical articles. He is professor of educational psychology at the University of Saskatchewan, Saskatoon, Canada, and is especially interested in philosophical aspects of science.

Steven Korenstein (Electromagnetic Fields and Cell Phones) holds a master's degree in environmental and occupational health and works as a hazardous substances scientist for the California Environmental Protection Agency in the department of toxic substances control. He has recently published a study on the health of children in the *Journal of Environmental Health*. His research interests include the effects of exposure to environmental toxicants at the community level.

David J. W. Lauridsen Jr. (Fairies, Elves, Pixies, and Gnomes) is currently a technical analyst and web developer for Qwest Communications. He holds a bachelor's degree in communications from Ohio State University and is an increasingly ardent skeptic. An avid private pilot, David is excited by the philosophy of science and would like to expand the teaching of critical thinking skills to today's youth.

Bernard Leikind (Science and God) is a senior editor of *Skeptic* magazine. He lives in Encinitas, California, and works for a San Diego area software company. Previously, he worked as a physicist for General Atomics and as a plasma physics and fusion energy researcher at UCLA. He achieved notoriety through his investigation and explanation of firewalking, a stunt that he foolishly performed many times. In one imaginative and astonishing experiment, he demonstrated that raw steaks have the same mental powers as human firewalkers. His research interests also include unusual atmospheric phenomena and other unexplained anomalies.

Norman Levitt (The Science Wars: Deconstructing Science Is Pseudoscience) is a professor of mathematics at Rutgers University in New Brunswick, New Jersey. He specializes in geometric topology. He is the coauthor with Paul R. Gross of the controversial book *Higher Superstition: The Academic Left and Its Quarrels with Science,* and *The Flight from Science and Reason.*

Ricki Lewis (Dietary Supplements) is the author of *Human Genetics: Concepts and Applications,* coauthor of three other textbooks for McGraw-Hill Higher Education, and a contributing editor to *The Scientist.* She has a Ph.D. in genetics and is affiliated with the TIGR Center for the Advancement of Genomics.

Scott O. Lilienfeld (EMDR, Multiple Personality Disorder) is associate professor of psychology at Emory University, founder and editor-in-chief of the journal, *The Scientific Review of Mental Health Practice,* and past president of the Society for a Science of Clinical Psychology. His research interests include the causes and assessment of personality disorders, anxiety disorders, and dissociative disorders, and the problem of pseudoscience in clinical psychology and allied disciplines.

Andrew O. Lutes (Animal Mutilations) has a bachelor's degree in history from Northern Kentucky University and a master's in library science from Kent State University. He writes letters and essays constantly and reads every issue of both Skeptic and Skeptical Inquirer (which he indexes) cover to cover.

Steven Jay Lynn (Multiple Personality Disorder) is professor of psychology at the State University of New York at Binghamton. He edited a book series on trauma, memory, hypnosis, and dissociation for the American Psychological Association; was consulting editor for the *Journal of Abnormal Psychology;* and was guest editor for *Current Directions in Psychological Science.* He is on the editorial board of Scientific Review of Mental Health Practices and is a fellow of the American Psychological Association, the American Psychological Society, the American Association of Applied and Preventive Psychology, and the Society for Clinical and Experimental Hypnosis. He has published more than 200 books, articles, and chapters in the areas of hypnosis, memory, sexual abuse, risk prevention, dissociation and fantasy, and forensic psychology.

Kevin MacDonald (Psychoanalysis as Pseudoscience) received his Ph.D. in personality development at the University of Connecticut. He is a professor of psychology at California State University–Long Beach. His research emphasizes evolutionary perspectives in personality, child development, and ethnic relations. He is the author of *Social and Personality Development: An Evolutionary Synthesis, A People That Shall Dwell Alone, Separation and Its Discontents, Sociobiological Perspectives on Human Development,* and *Parent-Child Play.*

Barry Markovsky (UFOs) is professor and chair of the department of sociology at the University of South Carolina. In addition to his interest in how social factors influence paranormal beliefs, he has published research on social power, status processes, perceptions of justice, social influence, group solidarity, social networks, and methods for theory construction.

Juan Carlos Marvizon (Meditation) is assistant professor in the department of medicine, UCLA. He is an author of thirty research papers on neurotransmission and the physiology of pain, contributes to *Skeptic* magazine, and moderates the Skeptic Forum (http://forums.delphiforums.com/skepticforum/start). He has practiced different forms of meditation for 25 years. His interests are on topics related to the brain and the mind, including acupuncture, meditation and consciousness

Arthur Mather (Astrology, Undeceiving Ourselves) coedited with Geoffrey Dean *Recent Advances in Natal Astrology: A Critical Review 1900–1976*, the first book-length critical review of scientific research into astrology. He and Dean have been collaborators since 1975 on critical articles, debates, surveys, and prize competitions for research into astrology. Mather is in charge of technical training projects in Livingston, Scotland.

William F. McComas (Science and Its Myths) is an associate professor of science education at the Rossier School of Education of the University of Southern California where he is also the founding director of the Program to Advance Science Education. He teaches courses in educational research, issues in science education, and advanced science teaching methods. He maintains an active research program focusing on the improvement of laboratory instruction, evolution education, the impact of the philosophy of science on science teaching, and science learning in museums and field sites.

Richard J. McNally (EMDR) is professor of psychology in the department of psychology at Harvard University. He is the author of more than 165 publications, mostly concerning anxiety and traumatic stress disorders. McNally served with the American Psychiatric Association's subgroup on post-traumatic stress disorder for the 4th edition of the *Diagnostic and Statistical Manual.*

Jean Mercer (Attachment Therapy) has a Ph.D. in psychology from Brandeis University. She is professor of psychology at Richard Stockton College and president of the New Jersey Association for Infant Mental Health. Together with Larry Sarner and Linda Rosa, she recently completed a book on a death caused by attachment therapists, and has published several related journal articles.

Frank Miele (Evolutionary Psychology as Good Science) is senior editor of *Skeptic* magazine and author of *Jensenism and Skepticism.* His interviews of the major figures in and introductions to the IQ, evolutionary psychology, and environmental debates are featured regularly in *Skeptic* (and include E. O. Wilson, Richard Dawkins, Charles Murray, Jerry Brown, Lionel Tiger, and many others.) He is also a technical writer in the silicon valley.

Robert L. Miller (Christian Science as Pseudoscience) is a trial attorney in southern California, and runs a law firm called Robert Miller & Associates. He holds a Juris Doctor degree from Western State University College of Law and has written articles for various newspapers and magazines on a variety of subjects.

Richard Milner (Piltdown Man [Hoax]), contributing editor of *Natural History Magazine* at the American Museum of Natural History, is the author of *The Encyclopedia of Evolution: Humanity's Quest for Its Origins.* A historian of science and anthropology, he also wrote the acclaimed one-man musical Charles Darwin: Live & In Concert, which he performs all over the world.

Phil Molé (Carlos Castenada, Holistic Medicine: The Case of Caroline Myss) has a bachelor's degree in chemistry from DePaul University, where he minored in biology and mathematics. He earned a master's in public health from the University of Illinois at Chicago, and works at an environmental consulting company in Elmhurst, Illinois. He is a regular contributor to *Skeptic* magazine.

Douglas G. Mook (Observer Effects and Observer Bias) is professor of psychology emeritus at the University of Virginia, where he taught courses in research methodology for a number of years, and is now adjunct professor at Cooper Union College. He is author of *Motivation: The Organization of Action*, and *Psychological Research: The Ideas Behind the Methods.*

John L. Moore (Cryptozoology) is studying biology and geology at the University of Utah. He was formerly the associate editor of *The Cryptozoology Review*, and his research interests include invertebrate paleontology and archaeology.

David Morrison (Velikovsky: Cultures in Collision on the Fringes of Science) is a NASA research space scientist and was one of Carl Sagan's first doctoral students at Harvard University. His research interests include near-earth asteroids and how to detect them, as well as Immanuel Velikovsky and his influence on astronomy and the sciences.

John Mosley (Planetary Alignments) is an astronomer at the Griffith Observatory in Los Angeles where he supervises the educational programs. He has produced 50 public planetarium shows, including the *Star of Bethlehem.* He specializes in amateur astronomy and astronomical computer software.

Steve Novella (Cults, Dowsing) is an assistant professor of neurology at Yale University School of Medicine, president of the New England Skeptical Society and editor of the *New England Journal of Skepticism.* His interests are in medical fraud and skeptical philosophy.

Richard Olson (Witchcraft and The Origins of Science, The Science Wars: Deconstructing Science Is Good Science) is professor of history and Willard W. Keith Fellow in humanities at Harvey Mudd College and the 2002–2003 Hennebach Visiting Professor in humanities at the Colorado School of Mines. His publications include *Science Deified* (vol 1, 1982; vol. 2, 1990) and *The Emergence of the Social Sciences* (1993). He is currently working on a reference volume on science and religion in the Christian West: 1450–1900 and is series editor for a fourteen volume reference series on science and religion of which that work will be a part.

Laura Pasley (Recovered Memory Therapy and False Memory Syndrome: A Patient's Perspective as a Test Case) worked for the Dallas police department, retiring in January 1999 following a 25 year career. She was the first person to sue a therapist for inducing false memories and creating an unhealthy dependence. Her story was published in the book *True Stories of False Memories.*

Mark Pendergrast (Recovered Memory Therapy and False Memory Syndrome: A Father's Perspective as a Test Case) is an independent scholar and author living in Vermont. His books include *Victims of Memory*, *For God, Country and Coca-Cola,* and *Uncommon Grounds*. He is working on a history of mirrors.

Massimo Pigliucci (Science and Religion) is assistant professor in the departments of botany, ecology, and evolutionary biology at the University of Tennessee in Knoxville. His research focuses on the ecology and evolution of genotype-environment interactions. He is the author of several textbooks, as well as a book on creationism entitled *Evolution Denial.*

James W. Polichak (Memes as Pseudoscience) has a Ph.D. in experimental psychology at SUNY Stony Brook where his research focused on the comprehension of noun phrases and on auditory attention. He writes about science and education and their impacts on people's daily lives.

Gary P. Posner (Police Psychics: Noreen Renier as a Case Study) practiced internal medicine for 15 years. He founded the Tampa Bay Skeptics in 1988 and is editor of its newsletter. He is a consultant to the Committee for the Scientific Investigation of Claims of the Paranormal and is a member of the Board of Scientific and Policy advisers of the American Council on Science and Health.

Mark Pratarelli (Polygraph and Lie Detection) is associate professor of psychology at Colorado State University–Pueblo, the director of the Cognitive Neuroscience Laboratory, and author of *Niche Bandits.* His research interests include memory, language and the brain, as well as the evolution of human behavior.

Jon Puro (Feng Shui) is a computer programmer, freelance writer, and classical pianist and composer. He lived in Japan for over five years studying Japanese language and culture, and his research interests include Asian culture and history, philosophy, and the sociological and psychological implications of religious belief systems.

James Randi (The Liquefying "Blood" of St. Januarius, Pseudoscience and the Paranormal) 74, is a MacArthur Prize winner, a professional magician who has now turned his attention to the examination of paranormal, supernatural, and occult claims. He is president and founder of the James Randi Educational Foundation (www.randi.org) and he lectures all over the world. His books, such as *Flim Flam!* and *The Faith Healers*, have been published in English, Chinese, Japanese, Korean, French, German, Italian, Spanish, Polish, Hungarian, and Norwegian.

Todd C. Riniolo (Psi and Psi-Missing) is an assistant professor of psychology at Medaille College in Buffalo, New York. His research interests include quantitative issues in psychophysiological research, the history of psychology, and teaching of psychology. His work has appeared in publications such as *Psychophysiology, Infant Behavior and Development, Teaching of Psychology*, and *Skeptic.*

Lance Rivers (Alien Abductions) is an assistant professor of English at Lake Superior State University in Sault Sainte Marie, Michigan, where he chairs the School of English and Speech's Writing Studies Committee. He has presented conference papers nationally on the misuse of scientific theories in literary and composition studies, especially postmodern misreadings of chaos theory. His research interests include the intersection between science and the study of composition, the rhetoric of reports of the paranormal, and alien abduction narratives as folklore.

Russell Robinson (Earthquake Prediction) is a senior seismologist at the New Zealand Institute of Geological and Nuclear Sciences. He holds a Ph.D. in geophysics from Stanford University and has published over 40 scientific papers. His research interests are in earthquake forecasting, seismicity and geologic structure of the New Zealand region, and computer modeling of seismicity.

Ben S. Roesch (Cryptozoology) is a B.Sc (Hons.) marine biology student at the University of Guelph and was formerly the editor of *The Cryptozoology Review*. His research interests include deep-sea biology, shark biology, and environmental and comparative physiology and biochemistry.

Gerald M. Rosen (EMDR) practices clinical psychology in Seattle, Washington and holds a joint appointment as clinical associate professor in the departments of psychology and psychiatry at the University of Washington. He received Level I and Level II training in EMDR from Dr. Shapiro, and has been writing commentaries and reviews on the method ever since. He has authored more than 50 scientific publications.

Rebecca Rush (Subliminal Perception and Advertising) teaches world studies and geography at Lockport Township High School. In addition, she has worked as a film researcher on numerous historical documentaries and is also an advocate for media literacy.

Charles Salas and Danielle Salas (Mesmerism: "Report of the Commissioners Charged by the King to Examine Animal Magnetism," by Benjamin Franklin and Antonie Lavoisier). Charles is research project analyst at the Getty Research Institute for the History of Art and the Humanities. He received his Ph.D. in modern intellectual history from the Claremont Graduate University. Both he and his wife, Danielle Salas, are particularly interested in eighteenth-century French studies.

Vince Sarich (Race and I.Q. as Good Science) has been a faculty member in the department of anthropology at U.C. Berkeley since 1966. He is best known for his work in molecular dating in which he found that the accumulation of immunological differences among albumins occurred as a regular function of time. His current research centers on racial variation within the human species in which he suggests that while the species may be relatively old, races are young with most of the interpopulational variation having developed within the last 15,000 to 20,000 years.

Larry Sarner (Therapeutic Touch) is a mathematician, cryptographer, and voting-machine inventor who resides in Loveland, Colorado. He cofounded the Front Range Skeptics and the National Therapeutic Touch Study Group, is a member of the National Council Against Health Fraud, and is coauthor of *Your Very Last Chance: How Attachment Therapists Killed Candace Newmaker*.

Thomas F. Sawyer (Clever Hans) is professor of psychology at North Central College in Naperville, Illinois. He received his Ph.D. from Bowling Green State University. He teaches the following courses in psychology: science of behavior, statistics, research design and experimentation, drugs and behavior, sensation and perception, physiological psychology, and history and systems of psychology. His research interests include: time perception; drug legislation/policy in the US; and behavioral genetics.

Theodore Schick Jr. (Do Extraordinary Claims Require Extraordinary Evidence?: A Reappraisal of a Classic Skeptics' Axiom) is a professor of philosophy at Muhlenberg College in Allentown, Pennsylvania. He has published articles on epistemology, philosophy of science, and philosophy of mind. He is coauthor of *How to Think About Weird Things*.

Henry Schlinger Jr. (Evolutionary Psychology as Pseudoscience) is a professor of psychology at Western New England College in Springfield, Massachusetts. He is the author of *A Behavior Analytic View of Child Development* and coauthor of *Psychology: A Behavioral Overview*. He has also published numerous theoretical articles on intelligence and artificial intelligence, the role of verbal behavior in learning, and child development.

Louis A. Schmidt (Psi and Psi-Missing) is an assistant professor of psychology at McMaster University in Ontario, Canada. His research interests include socioemotional development in infants and children, and developmental psychophysiology.

Rudolf Smit (Astrology) is secretary of the late professor H J Eysenck's Committee for Objective Research into Astrology, founding editor of what is now *Astrologie in Onderzoek* [Astrology under Scrutiny], and editor 1992–1999 of *Correlation,* the journal of research into astrology. Until late 2000 he was editor and translator for one of the Netherlands' leading scientific institutes.

Jorge Soto (Crop Circles) is a chemist by training, and an astronomer hobbyist. He did his graduate studies at Caltech (1983) and postgraduate studies at UCLA, both in chemistry. Jorge started to work at Dow Chemical in 1988 and moved to Midland, Michigan in 1993. Throughout his career, he has been involved in various projects related to making plastics, which have lead to numerous patents.

Bob Steiner (Cold Reading) is a CPA and magician; was the national president of The Society of American Magicians, of which he is chair of the Occult Investigation Committee; a fellow of The Committee for the Scientific Investigation of Claims of the Paranormal (CSICOP); a member of Professionals Against Confidence Crime; an associate of the Inner Magic Circle (London); and served on the board of directors of The National Council Against Health Fraud. Among his nine published books is *Don't Get Taken!*

David X. Swenson (Thought Field Therapy) is a licensed psychologist in Wisconsin and Minnesota, and a diplomate in forensic psychology. He teaches management courses at the College of St. Scholastica in Duluth, Minnosota, and has a small private practice in forensic and organizational psychology.

George A. Ulett (Acupuncture) has an M.D. and Ph.D. and has held professorships at the Missouri Institute of Mental Health, Washington University School of Medicine, St. Louis University School of Medicine, and was Chairman of the Department of Psychiatry at the University of Missouri Medical School. He is the author of over 270 scientific articles and books. He studied traditional Chinese acupuncture in the 1960s, received the first NIH grant to study acupuncture (1972) and has used the neuroelectric acupuncture method in his medical practice for many years. His 1992 book *Beyond Yin and Yang: How Acupuncture Really Works,* has been revised and updated. His latest book is *The Biology of Acupuncture.*

John van Wyhe (Phrenology) wrote his doctoral dissertation on the role of phrenology in the creation and spread of naturalism in nineteenth-century Britain. He is currently a senior research fellow at the National University of Singapore and a researcher at the department of history and philosophy of science at Cambridge. His general research interests include eighteenth- and nineteenth-century British and German intellectual/cultural history, as well as evolutionary thought, understandings of brain functions, science and religion, nature, and philosophies of materialism. He has published extensively on phrenology.

Jeffrey S. Victor (Satanic Ritual Abuse) is a professor of sociology at a branch of the State University of New York. He has published a book on human sexuality and many articles on rumor-panics and other sociological topics. He is the author of *Satanic Panic: The Creation of a Contemporary Legend.* He also studies the repressed memory movements, as well as the alien abduction movement.

Daniel R. Wilson (Magnetic Therapy) is professor and chairman of psychiatry and professor of anthropology at Creighton University Medical Center in Omaha. His research interests include evolutionary epidemiology, psychopharmacology, and forensic psychiatry.

Bill Wisdom (Skepticism and Credulity: Finding the Balance between Type 1 and Type 2 Errors) is emeritus professor of philosophy at Temple University in Philadelphia, where for more than thirty years he has taught formal logic, the philosophy of science, the philosophy of religion, and topics in the history of philosophy. He is the coauthor of a textbook in formal logic and metatheory, and has given lectures and published articles in the theory of knowledge, philosophy of religion, skepticism, and formal logic.

Eric Wojciehowski (Ancient Astronauts: Zecharia Sitchin as a Case Study) is a probation agent with the Michigan department of corrections currently working in the city of Detroit. He earned a bachelor's degree in psychology and has spent many years studying mythology, religion, history, and the general basis of paranormal beliefs. He is also a member of the Ancient Astronaut Society.

Roahn H. Wynar (Faster-Than-Light Travel, Laundry Balls) is an atomic physics researcher in the physics department at the University of Washington in Seattle. His interests include laser cooling and trapping and fundamental symmetries.

Julie Yau (Witchcraft and Magic) is a freelance writer and artist residing in Toronto. Her research interests include cultures and behaviors of the masses, and issues pertaining to the body. She is currently compiling research for a book on the fear of death.

Harry Ziel (Alternative Medicine v. Scientific Medicine) is emeritus clinical associate professor, Department of Ob/Gyn at the University of Southern California School of Medicine. He served as the director of Ob/Gyn at Kaiser Hospital, Los Angeles, from 1980–1991, as residency director from 1970–1980, and as medical education director from 1962–1970. He is a graduate of Harvard and attended the University of Pennsylvania medical school.

Index

About the Editors

Dr. Michael Shermer is the founding publisher of *Skeptic* magazine, the director of the Skeptics Society, contributing editor and monthly columnist for *Scientific American*, and the host of the Skeptics Lecture Series at Caltech. He is the author of *In Darwin's Shadow: The Life and Science of Alfred Russel Wallace*; *The Borderlands of Science: Where Sense Meets Nonsense*; *Denying History*; *How We Believe: The Search for God in an Age of Science*; and *Why People Believe Weird Things*. He is also the author of *Teach Your Child Science* and coauthored *Teach Your Child Math and Mathemagics*. According to Stephen Jay Gould: "Michael Shermer, as head of one of America's leading skeptic organizations, and as a powerful activist and essayist in the service of this operational form of reason, is an important figure in American public life."

Dr. Shermer received his B.A. in psychology from Pepperdine University, M.A. in experimental psychology from California State Univesity–Fullerton, and his Ph.D. in the history of science from Claremont Graduate University. Since his creation of the Skeptics Society, *Skeptic* magazine, and the Skeptics Lecture Series at Caltech, he has appeared on such shows as *20/20, Dateline, Charlie Rose, Tom Snyder, Donahue, Oprah, Sally, Lezza, Unsolved Mysteries,* and other shows as a skeptic of weird and extraordinary claims, as well as on documentaries aired on A & E, Discovery, and The Learning Channel.

Pat Linse is an award winning illustrator specializing in film industry art. Long active in the skeptical movement, she is one of the founders of the Skeptics Society, *Skeptic* magazine, and *Jr. Skeptic* magazine, of which she is publisher and editor-in-chief. As *Skeptic*'s art director, she has created most of the illustrations for both *Skeptic* and *Jr. Skeptic.* She is coauthor of the first *Baloney Detection Kit* and is presently working on a *Baloney Detection Kit* series of books for teaching children science and critical thinking.

CPSIA information can be obtained at www.ICGtesting.com
Printed in the USA
BVOW02*2016020714

357936BV00007B/69/P

To Paul,

The *Skeptic* Encyclopedia of Pseudoscience

Santa Barbara, California   Denver, Colorado   Oxford, England

THE *SKEPTIC* ENCYCLOPEDIA of PSEUDOSCIENCE

Michael Shermer, Editor

Pat Linse, Contributing Editor

VOLUME ONE

Michael Shermer, Skeptics Society, *Skeptic* magazine, P.O. Box 338, Altadena, CA 91001
URL: http://www.skeptic.com, email: skepticmag@aol.com
(626) 794-3119 (phone), (626) 794-1301 (fax)

Library of Congress Cataloging-in-Publication Data
The *Skeptic* encyclopedia of pseudoscience / edited by Michael Shermer.
p. cm.
Includes index.
ISBN 1-57607-653-9 (set : hardcover : alk. paper) — ebook ISBN 1-57607-654-7
1. Pseudoscience—Encyclopedias. I. Shermer, Michael. II. Skeptic.
Q172.5.P77 S44 2002
503—dc21
2002009653

02 03 04 05 06 10 9 8 7 6 5 4 3 2 1

This book is also available on the World Wide Web as an e-book.
Visit abc-clio.com for details.

ABC-CLIO, Inc.
130 Cremona Drive, P.O. Box 1911
Santa Barbara, California 93116-1911

This book is printed on acid-free paper.
Manufactured in the United States of America

To James “the Amazing” Randi,
our hero, colleague, friend, and inspiration

Contents

Volume Two

Colorful Pebbles and Darwin's Dictum

An Introduction

In a session before the British Association for the Advancement of Science in 1861, less than two years after the publication of Charles Darwin's *Origin of Species*, a critic claimed that Darwin's book was too theoretical and that the author should have just "put his facts before us and let them rest." In a letter to his friend Henry Fawcett, who was in attendance in his defense, Darwin explained the proper relationship between facts and theory:

> About thirty years ago there was much talk that geologists ought only to observe and not theorize; and I well remember someone saying that at this rate a man might as well go into a gravel-pit and count the pebbles and describe the colours. How odd it is that anyone should not see that all observation must be for or against some view if it is to be of any service!

Few thinkers in Western history have had more profound insights into nature than Charles Darwin, and for my money, this quote is one of the deepest single statements ever made on the nature of science itself, particularly in the understated denouement. If scientific observations are to be of any use, they must be tested against a theory, hypothesis, or model. The facts never just speak for themselves. Rather, they must be viewed through the colored lenses of ideas—percepts need concepts.

When Louis and Mary Leakey went to Africa in search of our hominid ancestors, they did so based not on any existing data but on Darwin's theory of human descent and his argument that because we are so obviously close relatives of the great apes of Africa, it is there that the fossil remains of our forebears would most likely be found. In other words, the Leakeys went to Africa because of a concept, not a precept. The data followed and confirmed this theory, which is the very opposite of the way in which we usually think science works.

If there is an underlying theme in this encyclopedia—a substrate beneath the surface topography (to continue the geologic metaphor)—it is that science is an exquisite blend of data and theory, facts and hypotheses, observations and views. If we conceive of science as a fluid and dynamic way of thinking instead of a staid and dogmatic body of knowledge, it is clear that the data/theory stratum runs throughout the archaeology of human knowledge and is an inexorable part of the scientific process. We can no more expunge from ourselves all biases and preferences than we can find a truly objective Archimedean point—a god's-eye view—of the human condition. We are, after all, humans, not gods.

In the first half of the twentieth century, philosophers and historians of science (mostly professional scientists doing philosophy and history on the side) presented science as a progressive march toward a complete understanding of Reality—an asymptotic curve to Truth—with each participant adding a few bricks to the edifice of Knowledge. It was only a matter of time before physics and eventually even the social sciences would be rounding out their equations to the sixth decimal place. In the second half of the twentieth century, professional philosophers and historians took over the field and, swept up in a paroxysm of postmodern deconstruction, proffered a view of science as a relativistic game played by European white males in a reductionistic frenzy of hermeneutical hegemony, hell-bent on suppressing the masses beneath the thumb of dialectical scientism and technocracy. (Yes, some of them actually talk like that, and one really did call Isaac Newton's *Principia* a "rape manual.")

Thankfully, intellectual trends, like social movements, have a tendency to push both ends to the middle, and these two extremist views of science are now largely passé. Physics is nowhere near realizing that noble dream of explaining everything to six decimal places, and as for the social sciences, as a friend from New Jersey says, "Fuhgeddaboudit." Yet there is progress in science, and some views really are superior to others, regardless of the color, gender, or country of origin of the scientists holding those views. Despite the fact that scientific data are "theory laden," as philosophers like to say, science is truly different from art, music, religion, and other forms of human expression because it has a self-correcting mechanism built into it. If you don't catch the flaws in your theory, the slant in your bias, or the distortion in your preferences, someone else will. Think of N rays and E rays, polywater and the polygraph. The history of science is littered with the debris of downed theories.

Throughout this encyclopedia, we explore these borderlands of science where theory and data intersect. As we do so, let us continue to bear in mind what I call Darwin's dictum: "all observation must be for or against some view if it is to be of any service."

Using the Encyclopedia

One important tool in finding the right balance between theory and data or ideas and facts is a broad base of knowledge tempered with wisdom in making judgments about knowledge claims. Without the facts, you can't "judge for yourself" (as television documentaries often suggest viewers do) in any objective manner. What we hope to provide in this encyclopedia is a thorough, objective, and balanced analysis of the most prominent scientific and pseudoscientific controversies made in the name of science, mixing both facts and theory. The encyclopedia entries are written at a level appropriate for high school and college students conducting research in science and pseudoscience, members of the media looking for a balanced treatment of a subject, and those in the general public who desire a highly readable yet trustworthy resource to go to for the most reliable assessments of the most controversial and interesting mysteries involving our universe, our world, and ourselves.

As the subjects span all manner of claims from around the world, audiences and markets across the globe will be interested in reading these volumes. In addition, members of the media desperately need a reference resource in order to quickly get their minds around a subject, to book guests on both sides of an issue in order to properly set up a debate, and to get "just the facts" needed for the sound-bite story that is often demanded in the hectic world of journalism. Every newspaper, magazine, radio, and television producer and interviewer should keep a copy of this encyclopedia right between the dictionaries and reference works on contacting experts.

This two-volume encyclopedia encompasses claims from all fields of science, pseudoscience, and the paranormal, and it includes both classic historical works and modern analyses by the leading experts in the world who specialize in pseudoscience and the paranormal. The encyclopedia is heavily illustrated (these subjects lend themselves to both historical and contemporary images), and most entries offer a respectable bibliography of the best sources on that subject from both the skeptics' and the believers' perspectives, allowing readers to conduct additional research on their own after learning what the encyclopedia's expert author has had to say on the subject.

To make this encyclopedia original and different and to provide readers with a variety of subjects and analytic styles in order to properly follow Darwin's dictum of getting a healthy balance of data and theory, five categories of pseudoscience analyses are presented here:

1. A-to-Z listings. The *Encyclopedia of Pseudoscience* includes an A-to-Z section of subject analyses conducted by scientists and researchers, exploring phenomena such as alternative medicine, astrology, crop circles, handwriting analysis, hypnosis, near-death experiences, reincarnation, séances, spiritualism, subliminal perception, UFOs, witchcraft, and much more. These fifty-nine entries are written in a straightforward manner and are of moderate length and depth, offering some theoretical foundation but not to the same extent as the articles in subsequent sections.

2. Investigations. Articles in this section consist of research investigations carried out by scientists and scholars as originally published in the pages of *Skeptic* magazine, republished and repackaged here for the first time. These twenty-three articles are more than brief summaries of subjects as presented in the A-to-Z section; they are also skeptical analyses and include much more extensive research and bibliographies. Such analyses are applied to acupuncture, Atlantis, chiropractic, facilitated communication, homeopathy, immortality, and many other topics, and there are several critical pieces on the pseudoscience often found in psychology and psychotherapy. These latter pieces are especially important: although some forms of pseudoscience are seemingly harmless—astrology and crop circles come to mind—other forms can be exceptionally dangerous, particularly those dealing with the mind and behavior.

3. Case studies. The *Encyclopedia of Pseudoscience* includes a special section comprising thirteen in-depth analyses of very specific case studies originally conducted for *Skeptic* magazine and used here as part of the larger phenomena under investigation. For example, three special articles are devoted to recovered memory therapy and false memory syndrome—one from a psychiatrist's perspective, one from a patient's perspective, and one from a father's perspective. Through these case studies, the reader will be given a complete analysis of a subject. The cases will interest both amateurs and professionals in a field, and they are ideal for research papers by students or background research by scientists and professionals. Journalists and interested readers wanting details on a case study need go no further than this section of the encyclopedia.

4. For-and-against debates. The *Encyclopedia of Pseudoscience* includes the most original section ever compiled in an encyclopedia in the form of a "pro and con" debate between experts, allowing readers to judge for themselves by hearing both sides of an issue. Thus, for instance, "Memes as Good Science," by experimental psychologist Susan Blackmore, is contrasted with "Memes as Pseudoscience," by cognitive psychologist James W. Polichak. Even more controversially, the study of "Race and

Sports as Good Science," by author Jon Entine, is contrasted with the study of "Race and Sports as Pseudoscience," which I authored. Also included are debates on evolutionary psychology, on the question of whether science is at an end, and on the science wars. These twelve articles, originally published in *Skeptic* magazine, have been used extensively by high school teachers and college professors around the world as supplemental reading material for students in search of the terms of a debate on one or more of these vital and controversial issues.

5. Historical documents. The encyclopedia includes five classic works in the history of science and pseudoscience. For example, the first scientific and skeptical investigation of a paranormal/spiritual phenomenon—mesmerism—is offered in the "Report of the Commissioners Charged by the King to Examine Animal Magnetism, Printed on the King's Order Number 4 in Paris from the Royal Printing House." Published in 1784, five years before the French Revolution, this piece was the first attempt to put to the test (including under controlled conditions) a quasi-scientific phenomenon. What made this report so special was that the test was conducted by none other than Benjamin Franklin and Antoine Lavoisier.

So, as you work your way through this encyclopedia—either moving from start to finish or, more appropriately for this genre, skimming and scanning and plucking out what is needed or wanted—remember Darwin's dictum that every observation must be for or against some view if it is to be of any service. Remember, as well, the words of wisdom offered by the Harvard paleontologist who inherited Darwin's mantle, Stephen Jay Gould, in a 1998 essay entitled "The Sharp-Eyed Lynx, Outfoxed by Nature":

> The idea that observation can be pure and unsullied (and therefore beyond dispute)—and that great scientists are, by implication, people who can free their minds from the constraints of surrounding culture and reach conclusions strictly by untrammeled experiment and observation, joined with clear and universal logical reasoning—has often harmed science by turning the empiricist method into a shibboleth. The irony of this situation fills me with a mixture of pain for a derailed (if impossible) ideal and amusement for human foibles—as a method devised to undermine proof by authority becomes, in its turn, a species of dogma itself. Thus, if only to honor the truism that liberty requires eternal vigilance, we must also act as watchdogs to debunk the authoritarian form of the empiricist myth—and to reassert the quintessentially human theme that scientists can work only within their social and psychological contexts. Such an assertion does not debase the institution of science, but rather enriches our view of the greatest dialectic in human history: the transformation of society by scientific progress, which can only arise within a matrix set, constrained, and facilitated by society.

It is my fondest hope that this encyclopedia will facilitate a deeper understanding of pseudoscience and in the process illuminate the process of science itself.

Michael Shermer
General Editor

The *Skeptic* Encyclopedia of Pseudoscience

1
IMPORTANT PSEUDOSCIENTIFIC CONCEPTS

Alien Abductions

LANCE RIVERS

An alien abduction involves the removal of a human being by an extraterrestrial species for the purpose of medical experimentation, crossbreeding, or spiritual enlightenment. Many skeptics believe that abduction narratives are related to a variety of experiences such as sleep paralysis and dreams, rather than actual events in the physical world. Persons who claim to have experienced alien abduction can be divided into two categories: abductees (subjects of alien experiments who suffer traumatic scarring from the abduction) and experiencers (subjects of alien experiments who derive spiritual enlightenment as a result).

An individual's placement within one of these categories shows a correlation with the abduction researcher(s) with whom he or she has had primary contact. Given that the tool of choice used by abduction researchers is hypnosis, even in cases in which the abductee consciously remembers the experience, this categorization of abduction experiences suggests to skeptics that researcher bias is the driving force behind the phenomenon. Most abduction researchers respond to such criti-

Children looking at a flying saucer. (The Image Bank)

cism with the assertion that hypnosis, when properly and cautiously used, is a powerful tool for uncovering repressed memories. Other researchers point to scars, implantations, and terminated pregnancies as objective evidence of abduction, but they remain unable to provide the medical records necessary to corroborate these claims. Whether the skeptics or the proponents of alien abduction are correct, what remains certain is that those men and women who report experiencing it have been subjected to something deeply and personally traumatic.

The use of hypnosis as a tool in abduction research dates to the first well-publicized case, that of Betty and Barney Hill. In the early morning hours of September 20, 1961, the Hills were traveling on U.S. Route 3 near Lincoln, New Hampshire, when they noticed a bright light moving rapidly across the sky. Frequently stopping to observe the object, they became increasingly agitated as it changed course and eventually hovered about 100 feet from their car. Barney, who had been standing in the road watching the craft when it approached, immediately returned to the car in fear that he and his wife were going to be captured. As they drove away, the Hills heard several beeping noises from the rear of the car, though they did not see the object again. Later, they noticed that they were unable to account for about two hours of their trip (though their frequent stops might explain this). The following morning, the Hills reported the sighting to their local air force base, and after reading the book *The Flying Saucer Conspiracy,* they notified the National Investigations Committee on Aerial Phenomena (NICAP). Over the course of the next several months, Betty Hill began dreaming about contact with the craft's inhabitants. Finally, in December 1963, the Hills entered therapy with Dr. Benjamin Simon for treatment of Barney's increasing anxiety. Under hypnosis, both Hills recounted the story of the unidentified flying object (UFO) and their subsequent abduction.

In *The Interrupted Journey,* John G. Fuller emphasizes the fact that the Hills underwent separate hypnotic regressions, independently verifying each other's stories. Fuller is quick to point out that Barney's story in particular explains physical evidence in the case, such as the scuff marks on the top of his shoes that resulted from being dragged into the flying saucer. Yet the details of Betty's physical examination by the aliens suggest that her dreams are the more likely source of the abduction scenario. Further, Simon believed that the underlying source of the story might have been anxiety over the interracial aspects of the Hills' marriage. Indeed, the removal of skin and hair (sites of racial difference) samples from Betty Hill by the aliens suggests this. Coupled with discrepancies in the story, such as Betty Hill's failure to notice the missing lock of hair after the abduction and the Hills' inability to find the site of their abduction, these factors indicate that the Hill case is more likely the result of Betty and Barney discussing her dreams and confabulation, rather than the result of an actual abduction.

Despite the publicity the Hill case received, abduction reports remained unusual until the mid-1970s, when several questionable abduction stories began to emerge (such as the Pascagoula and Travis Walton cases). Then, in 1979, Raymond E. Fowler published *The Andreasson Affair,* the account of Betty Andreasson's encounter with an alien named Quazgaa. Andreasson's story emerged under hypnosis some ten years after the events described.

According to that story, the Andreasson family noticed a bright light coming from their backyard during a power outage on January 25, 1967. Suddenly, everyone present except for Betty Andreasson was paralyzed, and several humanoid figures entered the kitchen through a closed door. Initially identifying the beings as "angels," Andreasson agreed to accompany them to their ship, where she was escorted to an "upper room." After a brief "cleansing," Andreasson changed into a "white garment" and underwent a battery of examinations, finally being shown into a small room with several seats. Sitting in one of these, she was subsequently enclosed in formfitting plastic and immersed in a gray liquid. When the liquid was drained away, she was taken on a tour of an alien realm, culminating in an encounter with a "phoenix" and a feeling of religious ecstasy, before being returned (Fowler 1979, 24–104).

Although Fowler admits that the case is difficult given Betty's devout Christianity and the religious symbolism of the phoenix (as a Christ figure), he fails to connect Andreasson's Pentecostalism with several other religious symbols in the story. For example, Betty's mention of an upper room is a reference to the location of the Last Supper (Matt.14:15) and traditionally the location of the First Pentecost and the descent of the Holy Spirit (Acts 2:1–2). The white garment is reminiscent of the white robes worn by the elect (Rev. 7:9), with immersion while wearing this garment being suggestive of baptism. Indeed, the Andreasson story closely follows the structure of apocalyptic literature—cleansing, testing, baptism, tour of heaven, and revelation. It was predictable, then, that in a later hypnotic session, Betty would begin speaking in a strange tongue (Fowler 1979, 138) and prophesying ecological disaster. Thus, as a whole, the Andreasson case argues not for alien abduction but for a religious experience of the type frequently reported by Pentecostal Christians.

Before the 1981 publication of artist Budd Hopkins's book *Missing Time,* stories of alien abductions more closely resembled first-contact scenarios. Medical examinations were nonintrusive, involving surface observation and scanning with some sort of machinery. The aliens themselves were benevolent, usually speaking with the abductee and imparting some sort of wisdom. *Missing Time,* by contrast, presents intrusive examinations: anal probing, deep-tissue sampling that leaves scars, and surgical implantation of unknown devices using the most gruesome methods and no anesthesia. The aliens themselves emerge as coldhearted beings without regard for the trauma they cause abductees. Most disturbingly, the aliens seem to have been engaged in a longitudinal study of humans, abducting them multiple times throughout their lives, which suggests some sort of tracking and monitoring of their subjects.

Hopkins's first involvement with an abduction case, that of Steven Kilburn, began when the two men were introduced during a meeting of UFO researchers in Hopkins's home. Like Barney Hill, Kilburn could not account for a period of time following a UFO sighting and had been experiencing increasing anxiety ever since. At Hopkins's urging, Kilburn underwent hypnotic regression to uncover the events of this missing time, only to discover that he had been the subject of a painful alien medical examination, the exact nature of which is unclear. Although Kilburn's hypnosis sessions seem to have been relatively free from leading questions and hypnotist influence, there are several problems with Hopkins's interpretation of them, most noticeably in his reconstruction of the narrative. As Hopkins himself points out, Kilburn's story emerged not as a full-fledged narrative with a continuous logical stream but rather as a set of discrete episodes in no particular order, necessitating reconstruction to make sense of them (Hopkins 1981, 61). Without this rearrangement, Kilburn's story has the logical structure of a dream, and it is this point that Hopkins misses throughout his work. For example, Howard Rich, another abductee discussed in Hopkins's book, states during one regression, "It's really just a dream . . . it's not happening" (p. 88).

Hopkins interprets such statements as denial of reality, rather than accepting them at face value. However, the main problem with his research is his selective use of evidence to create a typical abduction scenario against which all other cases are judged. In other words, abduction stories that fit Hopkins's idea of abduction are accepted as real, and those that do not are rejected or are filtered to create an acceptable story. Given that four of the six cases discussed in *Missing Time* are identified by the abductees as dreams, what the book presents are not so much actual abduction stories as Hopkins's own retelling and reshaping of these stories to create an abduction scenario. Science-fiction and fantasy novelist Whitley Strieber came forward with his own story of multiple abductions in 1987 with his book *Communion: A True Story*. Compiled from transcripts and tapes of hypnotic regressions, personal journals, conversations with friends, and his own conscious memories, *Communion* is perhaps the most detailed account of a single abductee's experiences, and, as with *Missing Time* (Strieber had a close association with Hopkins at the time), dreams seem to play a significant part in the experience.

On December 26, 1985, Strieber relates, he was awakened by a loud noise and then rushed by a small figure. Immediately paralyzed, Strieber was levitated into a ship, where he was examined by aliens and implanted with an unknown device before being returned to his bed. Strieber reports that these are conscious memories and were not recovered by hypnosis. Indeed, he is careful to point out that he was awake throughout this ordeal, thus eliminating dreams as source material. Yet the incident is typical of sleep paralysis incurred during hypnagogic or hypnopompic sleep states. Such states are characterized by a sense of being awake, a painful tingling paralysis throughout the body, and a sense of a malevolent presence. It is not unusual for such states to be accompanied by nightmares that provide details to explain these sensations. Most important, dreams occurring in hypnagogic or hypnopompic sleep are always consciously remembered, as are the states themselves. Hypnagogic and hypnopompic sleep states are terrifying events, and chronic sufferers can easily find themselves obsessively trying to figure out what is happening.

It is in keeping with such anxiety that Strieber sought psychological help in sorting out his experience, eventually submitting to hypnotic regression as part of his treatment. During these sessions, he began to recall other incidents in which he had been abducted—incidents that had been taking place from his childhood. In each case, the abductions he relates have the logical structure of dreams, rather than reality: locations shift rapidly and without transition, beings shift identities (including a feminine figure in white, Carl Jung's anima), and time is distorted.

It is the level of detail provided by Strieber's narrative that causes skeptics difficulty in accepting alien abduction as the explanation for his experiences. For example, Strieber reports that in August 1967, he experienced a period of missing time lasting approximately twenty-four hours. However, his account suggests that he was actually abducted three times within this twenty-four-hour period. Logically, one has to wonder why his abductors felt it necessary to return Strieber to his home after each abduction only to abduct him again moments later. Nor is Strieber able to explain the reasons for his multiple abductions, calling into question the motives of the alleged aliens. Human scientists, even those engaged in longitudinal studies, rarely find it necessary to conduct monitoring sessions of their subjects with the frequency Strieber reports. Although abduction researchers are quick to point out that alien science cannot be compared to human science and that the frequency of multiple abductions may be necessary for alien purposes, skeptics continue to argue that this discredits alien science, since our own science could learn more from less frequent contact.

Of course, Strieber is ultimately more concerned with the impact of his experiences on his personal growth than such questions. Over the course of several books, especially *Transformation* and *Breakthrough*, he slowly detaches himself from Hopkins's view of alien abduction as a negative experience and comes to the realization that he has grown in his understanding of himself. So, although acknowledging that his initial terror and anxiety were probably justified on some level, he concludes that his increased concern for the environment, as well as his deepened spirituality and connection with the world at large, are the ultimate goals of alien abduction. In this sense, Strieber remains unconvinced of the physical reality of his experience but rather recognizes that his abductors may have only a spiritual reality. In the final analysis, then, the sources and causes of Strieber's abduction remain less important for him than the results they yield, even if they are the results of dreams.

David M. Jacobs, an associate professor of history at Temple University, disagrees. Like Budd Hopkins, Jacobs asserts that the alien abductors have a physical reality and are probably the inhabitants of a planet within our own plane of existence. And like Hopkins, he sees the abduction phenomenon as negative, rejecting positive views of the experience as the result of the abductees' need for peace of mind and the infiltration of New Age thought into abduction research (Jacobs 1998, 208–219). In the case of some experiencers, Jacobs points out, further hypnotic regression (with himself as hypnotist) can reveal a darker and more threatening alien agenda than spiritual enlightenment, converting the experiencer to an abductee, as in the case of Pam Martin (Jacobs 1998, 24–25). In short, no matter what the hypnotic subject feels about his or her abduction, a negative picture emerges under hypnosis with Jacobs. For this reason, his methods are among the most criticized in abduction research. And, as in Strieber's case, the primitive nature of alien science revealed by his research (especially with regard to genetic experimentation using techniques known to humans for over a century, such as amniocentesis and artificial insemination) suggests that the source of abduction narratives is scientifically unsophisticated human minds, rather than sophisticated alien science. Thus, internally conflicting stories and researcher bias result in many skeptics and abduction researchers rejecting Jacobs's arguments that the alien agenda is a hybridization program and invasion.

John E. Mack bears impressive credentials as a professor of psychiatry at Harvard Medical School and may have the widest influence of any abduction researcher. In *Abduction*, he presents a well-reasoned approach to the phenomenon, acknowledging that abduction narratives are potentially the result of his own interpretation rather than reported events. However, Mack is convinced of the reality of alien abduction, though he suspects the abductors are from a spiritual plane. Thus, he ignores mundane explanations of the phenomenon, such as cultural contamination (incorporation of images and stories from the mass media), hypnagogic and hypnopompic sleep states, sexual abuse or aberration, and dreams. Even more problematic is his focus on the spiritual nature of the aliens, suggesting as he does that their primary agenda is enlightenment; this raises the question of the necessity of medical examinations, since aliens seem to derive no benefit from them beyond the torture of their subjects. Further, because most abductees reporting positive experiences point to fear of the unknown as the source of their spiritual development, they also derive no benefits from this part of the abduction. Again, the logic of dreams is at work in the abductions Mack presents.

Mack is not unaware of this logical conundrum, nor does he ignore those raised by the impossible physical acts (such as walking through walls and closed doors) that are common in abduction accounts. However, because of his emphasis on the spiritual aspects of the phenomenon, Mack concludes not that these problems indicate dreams as a source of the phenomenon but rather that dualism and materialism render the Western scientific model inadequate for the study of alien abduction (Pritchard et al. 1994, 565–567). In this way, Mack is able to have his cake and eat it too. For example, he points to scooplike scars on abductees as evidence of abduction, but when he is asked to provide medical records showing that the scars did not exist prior to the abduction experience, he responds by pointing out that abduction research should not seek objective proof or evidence. It is this confusion of science and philosophy that led Mack's peers at Harvard to conduct an investigation into his methods (to which he responded, through lawyers, that his academic freedom allowed him to pursue this line of inquiry).

Few publications provide a skeptical look at the abduction phenomenon, but there are two that bear mention. Terry Matheson's *Alien Abductions: Creating a Modern Phenomenon* examines the major abduction cases from a narrative viewpoint, arguing that the phenomenon has grown as the result of a developing secular mythology describing a variety of experiences. Kevin D. Randle, Russ Estes, and William P. Cone's *Abduction Enigma* adopts a scientific approach in its study of alien abduction to conclude that a combination of factors is at work: sleep paralysis, a lack of personal boundaries, sexual-identity problems, and researcher bias. Although the source of the phenomenon remains controversial, both books present viable and more prosaic explanations than those offered by most abduction researchers.

References:

Fowler, Raymond E. 1979. *The Andreasson Affair: The Documented Investigation of a Woman's Abduction aboard a UFO.* Columbus, NC: Wild Flower Press.

Fuller, John G. 1966. *The Interrupted Journey.* New York: Berkley Medallion Books.

Hopkins, Budd. 1981. *Missing Time.* New York: Ballantine Books.

Jacobs, David M. 1998. *The Threat: Revealing the Secret Alien Agenda.* New York: Simon and Schuster.

Mack, John E. 1994. *Abduction: Human Encounters with Aliens.* New York: Macmillan Publishing.

Matheson, Terry. 1998. *Alien Abductions: Creating a Modern Phenomenon.* New York: Prometheus Books.

Pritchard, Andrea, David E. Pritchard, John E. Mack, Pam Kasey, and Claudia Yapp, eds. 1994. *Alien Discussions: Proceedings of the Abduction Study Conference.* Cambridge, MA: North Cambridge Press.

Randle, Kevin D., Russ Estes, and William P. Cone. 1999. *The Abduction Enigma.* New York: Tom Doherty Associates.

Strieber, Whitley. 1987. *Communion: A True Story.* New York: Avon Books.

Alternative Archaeology

GARRETT G. FAGAN

The term *alternative archaeology* (or *pseudoarchaeology*) is used to denote various approaches to investigating the ancient past using paranormal methods and/or pseudoscientific standards of argument. The genre has been around at least since the late nineteenth century—appearing in the wake of real archaeology's great achievements at that time—although its roots can be traced back further into the lore of such movements as Freemasonry and Rosicrucianism. Its practitioners stand apart from archaeologists—whom they deride—in that they argue for an essentially simplistic picture of antiquity populated with such wonders as lost supercivilizations, sunken continents, cyclical cataclysmic upheavals, extraterrestrial civilizers, or psychically attuned priestly ruling elites. Alternative authors can also claim to have uncovered hints of heretofore lost "ancient wisdom" that, in its pessimistic form, warns of dire catastrophes threatening humanity or, in a more cheerful manifestation, promises worldwide spiritual renewal.

Alternative writers usually have no qualifications in the field of archaeology, yet they claim to be revolutionaries who will transform our view of the deep past. They spurn the broadly scientific method by which real archaeology has painstakingly built up a picture of that past: namely, the rational analysis of associated artifacts, of whatever size and nature, found and studied in clearly defined and repeated contexts. Instead, the alternative camp relies on a variety of dubious "methods" (which will be outlined) in support of its contentions. In such works, the mode of argument is often legalistic and associative, with an abundance of rhetorical questions and innuendo, rather than scholarly and probative, that is, mounting a sustained and coherent argument based on verifiable data. Upon such weak foundations, alternative archaeologists confidently raise their towering, yet by no means mutually consistent, edifices of fantastic possibility.

The Distinction between Archaeology and Pseudoarchaeology

Real archaeology is both a practical and an interpretive discipline. The practical wing involves the gathering of physical evidence from the past through careful procedures (field surveys, excavation, core sampling, and so on). Once gathered, however, that evidence remains mute and only speaks when interpreted with reference to the physical and cultural environment that produced it. A wall is just a wall until it is identified as Hadrian's Wall, when analysis of it reveals much about the organization and deployment of the Roman army, ancient military construction techniques, Roman frontier policies in the time of Hadrian, and so forth. Since writing has only been with us for the past 5,000 years and many advanced civilizations never developed it at all, archaeology must, for the most part, proceed without the insights gained from what the culture under study has to say about itself. This makes the job of archaeological interpretation all the more painstaking and gradualist, as the new evidence that is constantly being uncovered often forces old views to be substantially revised or abandoned altogether. Any new and ambitious archaeological hypothesis, in order to be convincing, has to do a good job of explaining as much of the evidence as it can; more important, it must also do a better job of explanation than the hypothesis it seeks to replace (Renfrew and Bahn 2000).

Nevertheless, the interpretive nature of archaeology leaves it open to charges of not being "fact" but "mere conjecture." This is a very common indictment among pseudoarchaeologists and their supporters (e.g., Hancock 1995, 1998, 2002). By invoking it, debates between established and alternative views are presented as matters of mere opinion. In this way, the pseudoarchaeologists can portray their speculations as just another batch of opinions about the past, no more or less provable than the conventional take on such matters. (Funnily enough, alternative writers also hold the view that academic archaeology is a dogmatic orthodoxy; but if that is so, the dogma is supposedly composed of subjective conjectures—by what criteria and by what procedures are the conjectures selected either to join the official archaeological catechism or to be rejected as heresy? This is never explained.)

As with any science and its pseudo counterpart, the real distinction between archaeology and pseudoarchaeology lies in matters of method. For even though archaeological hypotheses must respect the evidence and, to gain acceptance, succeed in making good sense of it through a rigorously rational analysis, alternative archaeology suffers no such constraints and can range freely and widely. It can pick and choose what to present to its readers; eschew requirements of logic, consistency, and (in many cases) prima facie plausibility; and speculate endlessly about conceivable possibilities soon to be vindicated by as yet uncovered spectacular finds—finds that, disappointingly, never quite seem to materialize. Conclusions reached by flawed methods are themselves flawed, and it is this straightforward observation that invalidates the entire genre of alternative archaeology, the methods of which are so deeply faulty as to be entirely useless. Therefore, the crux of the issue lies with procedure and method, and so the rest of this entry focuses on these issues.

Pseudopractice

Pseudoarchaeologists are rarely, if ever, engaged in the practical side of archaeology in the field, but occasionally, they do rely on their own methods. Use of psychic visions to explicate the past is a prominent example. In the 1920s to the 1940s, the U.S. psychic Edgar Cayce (1877–1945) had some 700 visions concerning human evolution and Atlantis. For instance, he saw that humanity was originally composed entirely of thought, with no corporeal presence—a claim notably difficult to check in the physical record. He declared the remains of Piltdown man, unearthed in England between 1908 and 1912, to be those of an Atlantean colonizer who had found his way to Britain. His psychic sources, apparently, were unaware that Piltdown man was a man-made hoax, as was revealed conclusively in 1953 (see the "Piltdown Man [Hoax]" entry in this encyclopedia). (The uncovering of the hoax, by the way, is a testament to the self-correcting nature of archaeology.) Cayce further predicted that Atlantis would rise again in 1968 or 1969 and that an Atlantean Hall of Records would be opened in the 1990s at Giza in Egypt. Cayce also asserted that in a previous

Prophet and psychic Edgar Cayce. (Bettmann/CORBIS)

life, he himself had been an Atlantean priest named Ra Ta, in which role, rather impressively, he had founded ancient Egyptian culture in 10,500 B.C.

Other psychics (notably, "Madame" Helena Blavatsky) similarly filled antiquity with extraordinary events and marvelous beings. Blavatsky, on the testimony of obscure tablets from Tibet that only she got to see, claimed humanity had once been astral jellyfish who came to Earth and founded Atlantis and Lemuria (another lost continent, older than Atlantis); the latter was populated by four-armed, apelike hermaphrodites who laid eggs (for all this, see Feder 2002; James and Thorpe 1999; Jordan 2001; Steibing 1984). Needless to say, none of these claims have been substantiated convincingly (though there has been much special pleading by the faithful), many run contrary to the observable evidence, and some (such as Cayce's claims about Piltdown man) have been shown to be conclusively wrong. Psychic visions appear an unsafe basis on which to investigate the past. A related psychic archaeological "method," apparently more reasonable than visions, is dowsing. Extravagant claims have been made for the efficacy of dowsing in locating archaeological sites, but the technique has yet to prove itself as consistently effective as the standard methods of field survey and excavation (Von Leusen 1999).

Pseudointerpretation

Unlike archaeological practice (which produces very clear-cut results—or not, as the case may be), the process of archaeological interpretation is more open to "alternative" possibilities. The picture of the past painted primarily from archaeological sources can fairly be characterized as an amalgam of interpretations or explanatory hypotheses formulated by archaeologists and ancient historians to make sense of the evidence they have uncovered. Given this, we have two choices. We can consider all explanatory hypotheses equally valid, or we can consider some more valid than others. The former approach leads to a sort of intellectual anarchy, in which there is no way to distinguish good explanations from bad ones. In this universe, the claim that unicorns built the pyramids using magical powers would carry as much credence as the assertion that the ancient Egyptians built them through hard labor.

Clearly, then, all claims about the past are not equally valid. The real question therefore is, How do we distinguish good from bad explanatory hypotheses? Genuine archaeological hypotheses are fixed by two anchors: accountability to the evidence (which is to say, all the pertinent evidence, not just convenient parts of it) and rational analysis. Propositions in archaeology, as in any intellectual discipline, cannot be verified by a selective analysis any more than they can rely on inherently unverifiable assertions (such as psychic visions). In this respect, archaeology follows a largely scientific method: hypotheses are formulated and then checked for logical consistency and accountability against the available evidence; conclusions are provisional on continued support by the evidence; conclusions are constantly reviewed in light of new evidence or more sophisticated modes of analysis; and so on. Hypotheses that withstand such constant scrutiny come to be widely held (and don the mantle of "fact"); those that do not are discarded or shelved for possible future consideration. This process of interpretation is complicated by the substantial gaps in our evidence from antiquity. The further back we go, the greater the gaps. And the greater the gaps, the wider the latitude for competing explanatory hypotheses. For instance, far less can be reliably deduced from cultures that have left us no written evidence than from those that have, and far less can be said about human cultures of 20,000 B.C. than those of 2000 B.C. But such is the nature of the beast. Ancient historians and archaeologists are accustomed to living with holes in their evidence, with things that cannot be fully explained, with—for want of a better term—*ancient mysteries*. Their attitude toward such gaps is to put them aside until such time as new evidence or improved hypotheses can throw light on them. In the interim, they tend to focus on more productive lines of inquiry rather than speculate endlessly about that which cannot (as yet) be checked. There is, then, a built-in uncertainty at the heart of archaeological interpretation that leaves the door open to the notions of alternative archaeologists, who attempt to bypass appropriate methods in peddling their pet theories. Instead, they have developed a battery of "methods" and approaches to do so. It would be impossible to survey all of these here, so a sample of the most common of the genre's characteristic procedures will have to suffice (examples can be found throughout the likes of Hancock 1995 and 1998 and von Däniken 1970).

Selective and/or Distorted Presentation of Established Knowledge

Pseudoarchaeological claims are habitually based on a systematically selective presentation of ancient evidence and on outdated, disproven, or long-discredited modern "theories." They present specific items from the ancient past that seem to suit their claims and ignore the rest. They will often assert that mysteries remain unsolved when, in fact, they have long been solved, or they will present as mysterious and poorly understood sites or artifacts that have long been studied. In addition, such works often range widely over the canon of ancient cultures, from the Egyptians to the Maya to the Khmers to Easter Island (Hancock 1995, 1998; von Däniken 1970). Since general readers cannot possibly be informed about the recent developments in all of these specialized fields of study, they will be readily convinced by such an apparently impressive body of alternative "evidence." Invariably, when the claims are investigated further, more reasonable explanations quickly emerge (James and Thorpe 1999).

Reliance on Supposed "Anomalies"

Alternative archaeologists display an obsessive focus on odd finds and sites. For this reason, their books usually survey the same material ad nauseam, and their fantastic explanations fail to throw light on the mass of evidence from the ancient past, since they are built on the wrong data set (that is, the anomalous exceptions rather than the majority of the evidence). Since their explanations are based on such flimsy foundations, their work is further characterized by rampant speculation and wild possibilities offered to fill the yawning gaps in their scenarios. A classic example relates to the huge animal figures and complex patterns inscribed on the surface of the Nazca Desert in Peru. We have very little evidence from the culture that created them (though their construction dates are fixed by pottery finds to between c. 200 B.C. and A.D. 600), so the intended function of the lines remains a genuine mystery to real archaeologists, although several plausible hypotheses have been advanced (Aveni 1990). For alternative archaeologists, however, the Nazca depictions are landing sites for aliens or encoded messages from lost Atlantis. Unlike the real archaeologists, the pseudoarchaeologists show no desire to relate the Nazca lines to the local culture that produced them or to seek obvious or rational explanations for their existence. Instead, the lines are treated as if they defy any rational analysis (which they do not), and wild speculations are offered up as if they were reasonable explanations.

Reinterpretation of Specific Artifacts or Entire Sites without Regard for Their Context

Pseudoarchaeologists will often present artifacts or entire sites out of context to support their claims (as just noted with the Nazca lines). Thus, in recent works, the Sphinx in Egypt (built c. 2500 B.C.) and the city of Tiwanaku in Bolivia (fl. c. A.D. 100–900) are divorced from their firmly established historical contexts and presented as evidence of a sophisticated, "seeding" supercivilization of the era of 10,500 B.C. (though the upper and lower dates for this great civilization vary in alternative works by as much as 10,000 years). Proponents are unperturbed by the complete lack of any archaeological context for the Sphinx at this early date or at any time before 2500 B.C. or by the stratigraphically established radiocarbon dates for Tiwanaku that fix its earliest possible habitation at 1500 B.C. In the face of such objections, a battery of special pleas and conceivable possibilities is offered—but no corroborating evidence for the extravagant claims themselves.

"Decoding" of Traces of Hidden History or Advanced Knowledge in Ancient Myths or Iconography

Pseudoarchaeologists often lay claim to special knowledge that has been "encoded" in ancient myth cycles and/or imagery. They then claim to "decode" this knowledge, usually by means of literal-minded and subjective interpretation but sometimes (as in the Bible Code) by apparently sophisticated means. Naturally, the immense complexities in studying and making sense of the wide range of human mythologies are not addressed. The more basic methodological question is, of course, How do we know that this vital information was intentionally "encoded" in myths by the ancients and not placed there by the imaginations of the modern writers? (Steibing 1984).

Substitution of Speculation for Traditional Archaeological Evidence

Real archaeology relies on artifacts of all shapes and sizes (from pollen grains to entire cities) in forming its reconstruction of the past. Alternative archaeologists habitually cannot offer a single scrap of such evidence in support of their claims. They therefore resort to manufacturing evidence of their own. One way of doing so is to reassign monuments from established cultures to their alternative ones (see the previous discussion on the Sphinx and Tiwanaku). Another is to offer what are, in essence, speculations as if they were evidence. For instance, perceived correlations between the position of monuments on the ground and certain constellations in the sky have been used to redate familiar sites to vastly earlier eras on the basis of the precession of the Earth's axis (that is, the apparent movement of the stars across the sky in a 26,000-year cycle). The claim is that the stars and monuments lined up most closely at a very early date; therefore, either the monuments were built at that date or they intentionally "commemorate" it. Either way, the alignments seem to offer hard evidence of an early, advanced civilization. This so-called method has been applied to the pyramids at Giza, the Khmer monuments at Angkor in Cambodia, and some Central or South American sites (notably, Tiwanaku). The use of the star-alignment argument is made more effective when combined with selectivity of presentation (see the earlier discussion), so that only the monuments that "fit" are included—the rest are ignored. More worrying, all other pertinent dating evidence from the site under investigation is either neglected or dismissed. But when there is no indication that the ancient culture in question knew of precession or, in general, sited their monuments to map constellations, how can the researcher be sure that his or her perceived pattern is historically meaningful and reflects the ancient builders' intentions rather than the modern writers' skill in discerning patterns? In fact, star-map redating is really the substitution of modern speculation (that is, the selective star-monument correlations) for hard dating evidence.

Attempts to Deploy the Authority of Nonhistorical Sciences to Establish Historical Hypotheses

The star-alignment argument is an example of this method also. It attempts to use the hard science of astronomy and the fact of precession as the core elements of an "alternate" picture of the human past. Similarly, the use of often tortured mathematics to find "significant" numbers (pi, or the earth's circumference, for example) supposedly encoded in the proportions of ancient monuments, such as the pyramids of Giza, is really an appeal to the authority of numbers in modern culture—their "discovery" in the monuments seems an incontrovertible fact to the modern mind. But that the numbers were intentionally placed there by the ancient builders is not established by argument or evidence; it is just assumed as a given. The proposition is not tested against comparable monuments to see if the alleged encoding of significant numbers is found in all pyramids, for instance; rather, only a select few monuments are examined (see the previous discussion on selective presentation). Recently, the Sphinx has been redated to preposterously early epochs on the basis of one scientist's geologic opinion that it was weathered by water and not by wind and sand (Schoch 1999). Since Egypt has been arid at least since 5000 B.C., the argument goes, the Sphinx must predate the Egyptians and the traditional date of 2500 B.C. This, too, is an attempt to co-opt the authority of a hard science (geology) in support of alternative historical claims. In fact, geology is singularly unsuited as a historical dating tool, since its chronological perspective is vastly deeper than that of human history, and the rate at which rocks erode has not been sufficiently established for this method to be used as a "clock" for man-made monuments. A myriad of other explanations for the erosion patterns on the Sphinx are available that accommodate the traditional, archaeologically established date for the monument. The so-called water-erosion redating is not necessary and is yet another example of modern speculation being offered up as hard evidence (see the preceding section) (Jordan 1998).

Use of Innuendo to Undermine the Authority of Academic Archaeology

The flip side of the preceding method involves the use of innuendo, which is a cardinal feature of pseudoarchaeological presentation. Alternative writers exploit gaps in parts of what they call the "orthodox" reconstruction of events to undermine the credibility of the whole: because we don't know exactly how the pyramids were built, everything about them and Egyptology is up for debate; because no bodies were found in the pyramids, their status as tombs is open to question; and so on. The detailed evidence that has led archaeologists to conclude that the pyramids were built by the Egyptians as tombs for their pharoahs either is not presented or is dismissed summarily as "orthodoxy" or "opinion." The general reader, therefore, is left with a very one-sided view of the situation, one that favors the alternative possibilities over the established view founded in evidence.

Use of Rhetorical Tricks to Mask a Weak Case

Although all good writing uses rhetorical skill in its presentation, pseudoarchaeological works employ a battery of rhetorical strategies not in the service of a coherent argument but as a replacement for it. A possibility is raised on one page and resurrected as established fact a few pages later. Rhetorical questions are used to plant suggestions in the reader's head that cannot be sustained from the evidence or by detailed argument. Much rhetoric is also devoted to upbraiding academic archaeologists as arrogant egoists who are not only closed to any "new thinking" but also actively seeking to suppress it by means of a sort of inquisition. What the value to archaeologists might be in systematically suppressing new evidence is never explicated.

Use of Legal, Not Scientific, Standards of Argument

The founder of modern alternative archaeology, Ignatius Donnelly (1832–1901), was a lawyer by training; one of its most successful modern exponents, Graham Hancock, expressly likens his task to that of a defense attorney making the best case he or she can. Lawyers, of course, seek not to find out what actually happened in any given instance but to make the best case they can to benefit their client. It is habitual for them to seek to limit damaging evidence presented to a jury (or downplay its significance), to use any rhetorical tricks they can to enhance a case, and to undermine the credibility of powerful opposing witnesses to discredit their testimony. Archaeologists, by contrast, are less interested in winning rhetorical points and more interested in interpreting all the evidence to find out what happened in the past.

Conclusion

The methods outlined here can be considered diagnostic of "alternative" archaeological works. In surveying the history of pseudoarchaeology, what is most noteworthy is the complete lack of progress toward a fuller understanding since Ignatius Donnelly founded the genre in 1882. Despite millions of pages of alternative arguments, we are still not one millimeter closer to finding (never mind studying) Atlantis, the great but nameless supercivilization, or the alien civilizers. Instead, alternative propositions can only be lined up alongside each other, none any more verifiable than the next. This is because pseudoarchaeology offers no hard evidence to work with and thus has no data against which to check its claims. To date, over forty locations have been proposed for Atlantis, covering most corners of the planet. Future "theories" will only inflate that total, not solve the mystery. Over this same period of alternative stagnation, huge strides have been taken in our understanding of antiquity by real archaeology, and the picture is constantly being refined by new discoveries.

There could be no clearer demonstration of

which approach offers the more promising results. It used to be thought that high civilization arose in one place (usually Egypt or Mesopotamia) and diffused outward. Since World War II and the systematic application of carbon dating on a global scale, this view has become untenable. Rather, the evidence now tells us that the great civilizations of the past arose at different times and in different places, under intriguingly similar circumstances. Pseudoarchaeologists refuse to accept this scenario. They are all relentlessly diffusionist in their reconstructions, whether the civilizing source be identified as Atlantis, a nameless supercivilization, or outer space. Thus, ironically, their claim to be revolutionizing our view of antiquity with "new thinking" is, in reality, a call to regress to a long-outdated model unsupported by the evidence. Alternative archaeology does not point the way forward; rather, it directs us backward.

References:

Aveni, A. 1990. *The Lines of Nazca.* Philadelphia: American Philosophical Society.

*Collins, A. 2000. *Gateway to Atlantis.* London: Headline.

*Donnelly, I. [1882] 1949. *Atlantis: The Antediluvian World.* Rev. ed. New York: Harper.

Feder, K. 2002. *Frauds, Myths, and Mysteries: Science and Pseudoscience in Archaeology.* 4th ed. Boston: McGraw-Hill.

*Hancock, G. 1995. *Fingerprints of the Gods.* New York: Doubleday. (Second edition in 2001 has unchanged text but a new introduction).

*——. 1998. *Heaven's Mirror.* New York: Crown.

*——. 2002. *Underworld: Flooded Kingdom of the Ice Age.* London: Penguin.

James, P., and N. Thorpe. 1999. *Ancient Mysteries.* New York: Ballantine.

Jordan, P. 1998. *Riddles of the Sphinx.* New York: New York University Press.

——. 2001. *The Atlantis Syndrome.* Phoenix Mill, United Kingdom: Sutton.

Renfrew, C., and P. Bahn. 2000. *Archaeology: Theory, Methods and Practice.* 3d ed. London: Thames and Hudson.

*Schoch, R. M. 1999. *Voices of the Rocks.* New York: Crown.

Steibing, W. H. 1984. *Ancient Astronauts, Cosmic Collisions, and Other Popular Theories about Man's Past.* Amherst, NY: Prometheus.

*von Däniken, E. 1970. *Chariots of the Gods.* New York: Putnam.

Von Leusen, M. 1999. "Dowsing in Archaeology: Is There Anything Underneath?" *Skeptical Inquirer* 23, no. 2: 33–41.

*Asterisked items are examples of alternative archaeological works.

Ancient Astronauts

KENNETH L. FEDER

The notion that human antiquity might best be explained by reference to the intervention of extraterrestrial aliens has been called "the ancient astronaut hypothesis." This hypothesis exploded into public consciousness in the late 1960s, soon after Swiss author Erich von Däniken began circulating a manuscript for a book that would ultimately bear the English title *Chariots of the Gods? Unsolved Mysteries of the Past.* Neither an archaeologist nor a historian, von Däniken was a hotel manager with a long and checkered criminal past including convictions for repeated acts of embezzlement, fraud, and forgery.

Von Däniken's manuscript presented a detailed argument for the ancient astronaut hypothesis, consisting of examples of great technological leaps evidenced in the archaeological record that the author argued were inspired by human contact with extraterrestrials. Early in 1968, the German publisher Econ-Verlag published his book (titled *Erinnerungen an die Zukunft,* or *Recollections of the Future*), with an unenthusiastically small initial print run. Surprising many, the book took off, captivating the public's imagination, selling more than 170,000 copies by December 1968, and spurring a plethora of von Däniken sequels, not to mention copycat books that essentially repeated the arguments of the original. By 1989, *Chariots of the Gods?* had become one of the best-selling paperback books of all time; the cover of the 1999 reprint boasted sales of more than 7 million copies. To date, von Däniken has written twenty-five books on the subject of ancient astronauts that have sold a combined total of more than 60 million copies in twenty-eight different languages (according to his homepage, http://www.daniken.com). His most recently published work is *Odyssey of the Gods: The Alien History of Ancient Greece* (2000), presenting the same essential argument as his previous twenty-four books. Currently, a theme park based on the "mysteries" of the ancient world that von Däniken explored in his books is being planned for Interlaken, Switzerland; it is scheduled to open in late 2002 (http://www.worldmysteries.ch/).

Interestingly, though the name von Däniken and the ancient astronaut hypothesis became synonymous, making the author a fixture on the lecture and television talk-show circuit in the 1970s, he was not the originator of the idea that the archaeological record might evidence the visitation of extraterrestrials to Earth early in the history of the human species. The actual source for that idea is as surprising as it is revealing. A version of the ancient astronaut hypothesis was proposed in an article published in a small technical journal, *Planetary Space Science,* in 1963, five years before the publication of *Chariots.* In discussing the possibility of intelligent life on other planets, the author of that article addressed the so-called Fermi Paradox, attributed to the physicist Enrico Fermi. Fermi argued that intelligent life on other planets must be either rare or nonexistent, for otherwise, Earth would have already been visited or contacted by these extraterrestrials.

The 1963 article's author responded by agreeing that extraterrestrials were, in all likelihood, not currently visiting Earth, but he added that one had to assess the possibility of such visitations within the context of the entirety of Earth's history. If such a visit had occurred at all, it was statistically more likely to have happened sometime in the past, which is very long, rather than the historical present, which is very short. The author went on to suggest, referring to possible ancient contact, that "it is not out of the question that artifacts of these visits still exist." In other words, the author was suggesting that there might be archaeological evidence in the form of physical remnants of an alien technology ensconced in the archaeological or geologic records. The source for this audacious suggestion—essentially the ancient astronaut hypothesis—was none other than Carl Sagan (1963, 496).

Whether von Däniken was aware of Sagan's article is unknown (he did not cite it in *Chariots of the Gods?*), but certainly he took the idea that extraterrestrials might have visited Earth in the ancient past and inspired a virtual ancient astronaut industry with books, lectures, movies, and television documentaries all based on this possibility.

Contained within the general ancient astronaut hypothesis as presented by von Däniken are three specific assertions:

1. Human biological evolution is the result of the direct intervention of extraterrestrial aliens.
2. All over the world, there are ancient works of art—including cave paintings, images on pottery, and so on—and written accounts among the earliest literate peoples that can best be interpreted as artistic depictions and written descriptions by human beings of extraterrestrial aliens who visited Earth.
3. Great intellectual leaps are indicated in the archaeological record, particularly in the form of technological advancements that could not have been the result of simple human ingenuity. These jumps are the result of the introduction of new technologies by extraterrestrial aliens.

The first of these assertions might well be called the "amorous astronaut hypothesis" (Feder 2001). In *Chariots of the Gods?* von Däniken implied that the biological evolution of the human species resulted from actual interbreeding between extraterrestrial aliens and ancient hominids, producing an evolutionary leap by what amounts to interstellar hybridization. "A few specially selected women," he asserted, "would be fertilized by the astronauts" (von Däniken 1970, 11). This is a fascinating hypothesis, but as Sagan pointed out in an interview conducted for the Horizon/Nova documentary *The Case of the Ancient Astronauts*, a human ancestor could more likely have successfully mated with a petunia than with an extraterrestrial; at least the hominid and the petunia evolved on the same planet and therefore shared a biological connection, however remote. The likelihood of two species that evolved on different planets having combinable DNA is so minute it is hardly worth considering.

The second assertion of the ancient astronaut view might be called the "inkblot hypothesis" (Feder 2001). The interpretation of ancient art or even early historical writings as representations or descriptions of extraterrestrial visitors, produced by ancient human beings, is little more than a sort of "Rorschach archaeology": extraterrestrial aliens are seen in images or written descriptions much in the way experimental subjects see clouds, butterflies, or elephants in inkblots.

For instance, as discussed in *The Case of the*

Ancient Astronauts, von Däniken saw airport runways in the Nazca lines—lengthy, rectilinear features produced in the high-altitude desert of western South America beginning as much as 2,000 years ago and ending by about A.D. 700. According to von Däniken, "Seen from the air, the clear-cut impression that the 37-mile long plain of Nazca made on me was that of an airfield!" (1970, 17). In other words, to him, the lines resembled airport runways, so that is what they must be.

It is more than just a little problematic for von Däniken's interpretation that the so-called lines are not wide swaths produced with an extraterrestrial version of concrete but merely patterns that reflect a color difference produced by sweeping dark surface pebbles off the underlying, lighter-colored sandy soil. No aircraft, extraterrestrial or otherwise, could land on such a soft surface. The lines are interpreted by archaeologists as ceremonial pathways of the ancient Nazca people; they were used precisely in this way in the fairly recent past.

One of von Däniken's most notorious examples of the inkblot hypothesis concerns the sarcophagus lid of the Maya king Pacal (see illustration), arguably one of the best-known rulers in Maya history. We know that Pacal ascended to the throne of the Maya city-state of Palenque on July 29 in A.D. 615 when he was only twelve years old. We know that he ruled for sixty-eight years and oversaw a period of vigorous construction of temples, palaces, and pyramids at Palenque. And we also know that he died on August 31, 683. Yet, eschewing the detailed historical narrative for Pacal provided by the Maya themselves in their written language, von Däniken applied his best inkblot analysis and interpreted the bas-relief on the surface of the coffin lid as a spaceman working the controls of an alien craft. In fact, the coffin lid contains myriad elements of iconography seen throughout the Maya world: the quetzal bird, bearded dragons, an earth monster, and a stylized ceiba tree (a species revered by the Maya). Very few of these elements would be expected in an extraterrestrial spacecraft. And, of course, Pacal's mortal remains were found in the coffin; not surprisingly, the bones are those of a human being and not an extraterrestrial alien.

Pacal's sarcophagus lid at Palenque, Mexico, 1976. (Copyright Dr. Merle Greene Robertson)

Finally, the third assertion listed earlier has been labeled the "our ancestors, the dummies" perspective by anthropologist John Omunhundro (1976). The philosophical underpinnings of this part of the ancient astronaut hypothesis ignore or minimize the intellectual abilities of ancient human beings, essentially denying the possibility that they were able, by the application of their own intelligence and by the sweat of their own brows, to produce the marvelous achievements in engineering, architecture, mathematics, calendrics, agronomy, and the

The Collapsed Pyramid at Meidum. (M. H. Feder)

like that are so clearly reflected in the archaeological record. Instead, von Däniken and his followers interpret these intellectual achievements as having been inspired by contact with a superior extraterrestrial intelligence.

One of the most egregious and ill-informed examples of this line of reasoning concerns von Däniken's interpretation of the history of ancient Egypt. He argued that ancient Egypt's monumental architecture appeared in the region thousands of years ago without any evidence of the development that would be expected if it were the result of human technological progression. He also characterized Egyptian civilization as "ready-made" and as appearing "without transition." The pyramids, the Great Sphinx, and the spectacular temples are, in von Däniken's opinion, "genuine miracles in a country that is suddenly capable of such achievements without recognizable prehistory" (1970, 74).

Such characterizations were breathtakingly ignorant of Egyptian history even when von Däniken wrote *Chariots of the Gods?* Archaeology has revealed a long evolutionary sequence of cultural development in Egypt, including the slow adoption of agriculture beginning 8,000 years ago; increasing sedentism and a growth in village size along the Nile; intervillage competition for land, resources, and people; the concentration of wealth in the hands of a few families; an increase in tomb size for the leaders of these wealthy and increasingly powerful families; and the consolidation of social, economic, religious, and political power in the hands of a single leader (the first pharaoh, whom we know as Narmer) by about 3100 B.C. (Clayton 1994). And what about the single most diagnostic architectural feature of ancient Egypt, the pyramid? Here, too, there is clear evidence of a developmental sequence by which Egyptians perfected the pyramid-making craft over a period of several hundred years (Lehner 1997). Beginning at Hierakonpolis, powerful "pottery barons" were buried in large, impressive tombs cut into the earth, which were then capped by single-story burial structures called mastabas.

Mastabas became larger and more ornate until Djoser, the second pharaoh of the Third Dynasty (2668–2649 B.C.), produced an impressive elaboration of this theme, with a series of five such structures of decreasing size superimposed one on top of the other.

Sneferu (2613–2589 B.C.), the first pharaoh of the Fourth Dynasty, began construction at Meidum of a burial monument that was planned as another step pyramid. Toward the end of the project, however, a decision was made to fill in the steps in an effort to produce a true pyramid with four flat, triangular faces. The pyramid engineers confronted a major problem in doing this: Egyptian step pyramids were far too steep to be readily converted to true pyramids, and the attempt to do so at Meidum was plagued by problems. Severe cracks developed in the pyramid's stone facing as a result of subsidence at its base, and it likely was abandoned because of this. The ruin of this project is called the Collapsed Pyramid, though there is no evidence of a catastrophic collapse (see illustration); it is probable that the engineers simply walked away from the project and that much of the polished stone casing was recycled in other structures. Sneferu next initiated another true pyramid project 40 kilometers north of Meidum, at Dashur. That pyramid again was begun at a very steep angle (about 60°), and subsidence problems arose once more, jeopardizing the project. Rather than abandoning this attempt, the builders actually adjusted the angle of the pyramid surfaces in the middle of the job, building the upper part of the pyramid at a shallower angle (between 43° and 44°). This project was successfully completed, but because of the change in angle, the pyramid has a decidedly odd appearance and is accurately called the Bent Pyramid.

Apparently unsatisfied by this midcourse correction, Sneferu began yet another pyramid project in the thirtieth year of his reign. This time, his architects designed and built the monument from the beginning with a more gentle slope and for the first time successfully constructed a large, standard pyramid—this is the so-called North or Red Pyramid. It was during the reign of the pharaoh who succeeded Sneferu—Khufu (2589–2566 B.C., often called by his Greek name, Cheops)—that pyramid building reached its zenith with the construction of the Great Pyramid, nearly 500 feet in height and consisting of more than 2.5 million stone blocks.

Clearly, the history of Egyptian pyramid construction reflects an evolutionary process, with a sequence of projects revealing false starts, trial and error, and on-the-fly problem solving typical of the imperfect process by which human technology progresses (see Pyramid entry in section 2). Mistakes in pyramid building like those exhibited in the archaeological record of Egypt would be surprising indeed if these structures were built or supervised by extraterrestrials capable of building starships that could cruise the galaxy. This slow process of technological development characterizes the archaeological record of the other early civilizations as well, contradicting any hypothesis that relies on the sudden appearance of sophisticated technologies in the ancient world.

In underestimating the intelligence and capabilities of ancient people, von Däniken was not being original: he was merely updating the long-discredited, extreme diffusionist view that characterized anthropology in the late nineteenth and early twentieth centuries. Such hypotheses are meritless; the archaeological record clearly reflects the abilities of all ancient people to progress by their own efforts. There is no evidence for and no need to speculate about an extraterrestrial source for human cultural development.

References:

The Case of the Ancient Astronauts (television documentary). 1971. London and Boston: Horizon and Nova.

Clayton, Peter A. 1994. *Chronicle of the Pharaohs: The Reign-by-Reign Record of the Rulers and Dynasties of Ancient Egypt.* London: Thames and Hudson.

Feder, Kenneth L. 2001. *Frauds, Myths, and Mysteries: Science and Pseudoscience in Archaeology.* 4th ed. Mountain View, CA: Mayfield.

Lehner, Mark. 1997. *The Complete Pyramids: Solving the Ancient Mysteries.* London: Thames and Hudson.

Omunhundro, J. T. 1976. "Von Däniken's Chariots: A Primer in Crooked Science." *Zetetic* 1, no. 1: 58–67.

Sagan, C. 1963. "Direct Contact among Galactic Civilizations by Relativistic Interstellar Spaceflight." *Planetary Space Science* 11: 485–498.

von Däniken, Erich. 1970. *Chariots of the Gods? Unsolved Mysteries of the Past.* New York: Bantam Books.

——. 2000. *Odyssey of the Gods: The Alien History of Ancient Greece.* New York and London: Element Books.

Animal Mutilations

ANDREW O. LUTES

Domestic livestock have sometimes been found with seemingly unexplainable fatal wounds. Often, the animals' eyes and genitals have been removed. Claims have been made that they were killed and mutilated by unseen malevolent forces. Extraterrestrials gathering specimens, Satanists making sacrifices, government agencies up to no good, and strange cryptozoological animals on a rampage (such as the "goatsucker," also called the "Chupacabra") have all been blamed. Claims have also been made that the dead bodies show unusual signs—surgical precision cuts or laser precision cuts (this claim has been made by a scientist, John Altschuler)—and that their blood has been entirely drained.

These phenomena have normal explanations. For instance, if the animals were killed by predators or if they froze to death, their bodies may have been eaten by scavengers. The so-called precision and laser cuts could have been caused by bites from small teeth and beaks, and when the dead animals' bodies swelled and bloated, which is a normal event in decomposition, the cut lines may have stretched, making them look precise and rounded. The maggots of blowflies, eating at the wounds, could also make them look like precision cuts. For these reasons, Altschuler's claims have not been accepted by mainstream scientists. Beyond that, the claims of bodies being drained of blood are simply not true. Blood dries at body surfaces, and if bodies are examined in autopsies, blood remains are found in them. Animal predators and scavengers, not having our sensibilities, eat soft body parts first, such as eyes, teats, and genitals.

When investigators examined the bodies of some mutilated animals, they found ordinary explanations. In 1975, the Colorado State University Diagnostic Laboratory examined tissue samples from mutilated cattle and concluded that the animals died from natural causes. Federal Bureau of Investigation (FBI) pathologist Kenneth Rommel, studying New Mexican mutilations in 1980, concluded that predators were the cause. Zoologist Ron Magill examined mutilated creatures in Sweetwater, Florida, in 1996 and found that dogs had done the deed. In the same year, University of Miami veterinary professor Alan Herron cut open a dead goat and showed it had not been drained of blood; Herron said that bites on the animal revealed that wild dogs had killed it. Also in 1996, the Puerto Rico Agriculture Department found normal causes for the mutilation of dead stock.

Extraordinary claims require extraordinary proof, and the burden of proof is on the claimant. Yet those who have alleged extraordinary causes for animal mutilations have produced no extraordinary proofs at all, just ordinary phenomena that can be explained by known causes. Furthermore, the broadcast programs and publications that make extraordinary claims about animal mutilations are not held accountable. By calling themselves entertainment shows or by invoking First Amendment protections, they can say, quite legally, whatever they like. Consequently, it is up to the individual to determine if ordinary explanations are available, rather than immediately believing repeated sensational claims. It is the hype of these sensational shows and publications that keeps the alleged mystery of animal mutilations going.

References:

Ayers, Kathleen. "Review of *Mute Evidence*, by Ian Summer." *Skeptical Inquirer* 9, no. 4: 374–375.

Frazier, Kendrick. "Cattle Mutilations in New Mexico." *Skeptical Inquirer* 4, no. 1: 11–12.

———. "Cattle Mutilations: Mystery Deflated." *Skeptical Inquirer* 5, no. 1: 2–6.

Genoni, Thomas C., Jr. "Long Live the Goatsucker." *Skeptical Briefs* 6, no. 4: 6.

Harwell, Ray. "Natural Mutilations." *Skeptical Inquirer* 5, no. 1: 88.

Howe, Linda Moulton. 1998. *Glimpses of Other Realities*. Vol. 1, *Facts and Eyewitnesses*, and vol. 2, *High Strangeness*. New Orleans, LA: Paper Chase Press.

Nickell, Joe. "Goatsucker Hysteria." *Skeptical Inquirer* 20, no. 5: 12.

———. 1998. Interview by Art Bell. *Art Bell Show* (syndicated). November 19.

Sheaffer, Robert. "Psychic Vibrations." *Skeptical Inquirer* 6, no. 4: 16.

Stewart, James R. "Cattle Mutilations: An Episode of Collective Delusion." *Zetetic* 1, no. 2: 55–66.

Anomalous Psychological Experiences

CHRIS DUVA

Anomalous psychological experiences are any of a variety of atypical mental phenomena with an apparently unspecific or unknown origin, including hallucinations; distortions of time; heightened emotional arousal; and alterations of sensation, perception, memory, or attention (Reed 1988). As these experiences appear to fall outside the range of normal sensory and perceptual processes, they are frequently interpreted as being supernatural or paranormal in origin and, as such, are often given great meaning and personal significance by those who have them.

Interpretations of such psychological experiences vary widely, but common forms include alien visitations and abductions, near-death experiences (NDEs), out-of-body experiences (OBEs), anomalous healings, past-life experiences, mystical experiences, and psi-related phenomena such as psychokinesis, extrasensory perception (ESP), déjà vu, and precognition. The cultural context and religious beliefs of an individual who has an anomalous experience may significantly influence the interpretation. Synesthesia (the transmutation of the senses—e.g., being able to taste colors) and lucid dreaming (being aware that one is dreaming while a dream is occurring) are also characterized as anomalous psychological experiences but are typically not interpreted by experiencers as being supernatural or paranormal in origin (Cardena, Lynn, and Krippner 2000).

Anomalous psychological experiences have been documented throughout recorded history, and their occurrence crosses cultural, racial, ethnic, gender, and socioeconomic boundaries. The actual prevalence of these experiences is difficult to determine, as many may go unreported due to their unique and highly personal nature. Although their occurrence is generally not challenged, the cause, meaning, and interpretation of these experiences have generated considerable controversy both among the general public and within the scientific community itself. A variety of theories that emphasize cultural factors, religious socialization, personality characteristics, psychological process, and brain dysfunction have been proposed as nonsupernatural explanations for anomalous psychological experiences.

Cultural theories draw heavily on the fact that individuals who have had an anomalous experience tend to adopt supernatural explanations for it that are in accordance with the norms of their society. The profound impact of cultural expectations on the interpretation of such experiences can be appreciated in the context of the frequent reports of strange visitors in the night, a phenomenon that has been described for hundreds of years and continues today. In the past, these encounters were variously said to involve spirits, demons, the Devil, saints, or angels. Not until recently, with the advent of flight and the exploration of space, did encounters with extraterrestrial beings become widely reported as a means to explain these suspected visitors. Thus, individuals in preindustrialized or less technologically advanced societies have tended to interpret anomalous experiences as mystical or religious in nature, in accordance with their culturally derived belief systems. In recent times, particularly in Western culture, technological advances and exposure to popular media depictions of alien abductions, NDEs, and psychic phenomena may provide a framework for individuals to interpret anomalous experiences when other alternatives prove unsatisfactory.

Religious socialization is another cultural factor that appears to play an important role in the psychology of anomalous experience. Those whose religious practices encourage ritualistic prayer, chanting, or meditation are more likely to have anomalous experiences and are also likely to invoke a paranormal or supernatural explanation as their cause. By contrast, those with strong religious beliefs in Protestantism or Catholicism, for example, tend not to accept paranormal explanations for anomalous experiences if they require invoking claims of mystical powers, such as precognition, ESP, or psychokinesis. They are, however, likely to accept supernatural explanations for religious-related anomalous experiences, such as apparent miraculous healings, hallucinations, and NDEs. In those without strong religious beliefs, the acceptance of paranormal interpretations for anomalous experiences tends to be considerably greater, particularly with respect to psi-related phenomena. In some individuals, belief in the paranormal or supernatural may function in a manner similar to religious faith by reducing the anxiety associated with the unknown or unexplainable. Scientific training may mediate some of the influence that religious and cultural factors have on the occurrence and interpretation of anomalous experiences. In general, those with a high degree of scientific training are less likely to have had an anomalous experience and are less likely to believe in paranormal explanations for such phenomena than those with limited scientific background.

Scientific interpretations of anomalous experiences have focused on personality factors, known psychological processes, and an understanding of the nervous system as potential explanatory factors. Traditionally, the scientific investigation of anomalous experiences has been confined to the margins of psychology. More recently, however, the study of these phenomena has begun to make its way into the mainstream, scientific establishment. Despite this development, however, anomalous psychological experiences are notoriously difficult to study empirically because of their infrequent and unpredictable occurrence and their internal, subjective nature (Blackmore 1996). The Anomalous Experiences Inventory and the Paranormal Beliefs Inventory are two instruments that have been developed to aid researchers in identifying and classifying the various types of anomalous experiences and in evaluating an individual's interpretation of them. A variety of other standard psychological tests, such as the Keirsey Temperament Sorter, the Minnesota Multiphasic Personality Inventory (MMPI), the Rorschach Inkblot Test, and the Meyers-Briggs Type Indicator, have also been used to identify individual variables that may be related to the perception and interpretation of anomalous experiences. Modern brain-imaging techniques, including positron emission tomography (PET) and functional magnetic resonance imaging (fMRI), have also begun to be utilized and are on the cutting edge of advancing biological explanations for anomalous psychological experiences.

Researchers have identified several personality characteristics that appear to be related to an individual's propensity for anomalous experiences. Although the findings vary slightly depending on the specific nature of the experience, three factors have emerged consistently.

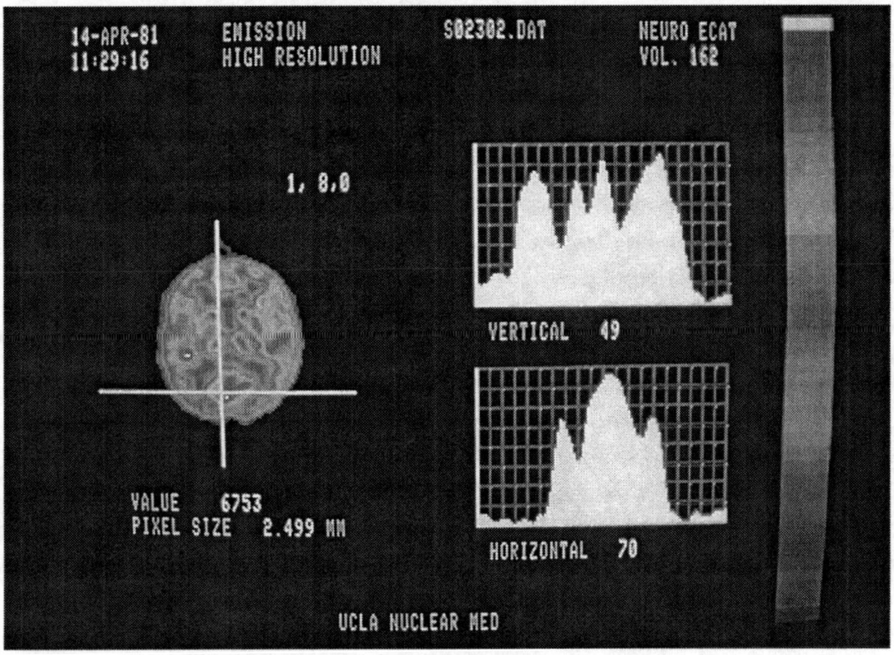

PET scans of a brain. (Roger Ressmeyer/CORBIS)

Individuals who report anomalous experiences tend to be highly prone to fantasy and are more susceptible to being hypnotized than nonreporters. These factors may be related to the third prominent factor—dissociative tendencies (defined as recurrent feelings of being detached from one's surroundings), another personality characteristic on which those who report anomalous experiences score high. Other personality traits, such as neuroticism, openness to experience, introversion/extraversion, and sensation seeking, have also been shown to correlate with the propensity for anomalous experiences but are somewhat less reliable markers. In general, psychopathology itself does not appear to be a major factor predisposing an individual to anomalous experiences or a belief in paranormal explanations for them. Although most studies have failed to find a firm link between psychopathology and anomalous experiences, it has been noted that many of the individuals reporting such experiences were more suspicious, distrustful, and lonely than nonreporters, as well as less happy and far more likely to have attempted suicide. There is also a higher than normal incidence of childhood abuse, broken homes, and familial alcoholism among those reporting anomalous experiences (Cardena, Lynn, and Krippner 2000).

A variety of psychological processes may play a role in the origin and interpretation of anomalous experiences as well. Prominent in many psychological explanations of such experiences is the phenomenon of sleep paralysis, a condition in which a person finds him- or herself trapped between a state of sleep and wakefulness, temporarily unable to move. Such a state may be accompanied by vivid hallucinations and may occur when the individual is falling asleep (hypnagogic hallucinations) or as he or she is waking up (hypnopompic hallucinations). These images appear quite real and can often be extremely terrifying. Essentially, the individual is dreaming while awake and is unable to move because the major motor systems of the body are temporarily paralyzed, possibly to keep the person from acting out his or her dreams. Hypnagogic and hypnopompic hallucinations have been suggested as a source for many anomalous experiences, particularly those involving nighttime visitations from strange beings. In addition to night paralysis, individuals subject to high levels of stress may experience psychological anomalies. In fact, stress has been shown to be a predisposing factor for OBEs, NDEs, lucid dreaming, and psi-related phenomena. Fatigue may also play a part in the generation of anomalous experiences, and it is well documented that prolonged sleep deprivation can lead to visual and auditory hallucinations. In this regard, it should be noted that numerous cultures engage in religious rituals that use stress and fatigue as mechanisms to induce an altered state of consciousness in an attempt to experience visions or make contact with the spirit world.

Abnormalities in brain function have also been suggested as a possible source for anomalous psychological experiences. Aspects of some such experiences (hallucinations, out-of-body experiences, tunneling, alterations in sensory and perceptual processes) can be induced by a variety of psychoactive drugs, suggesting that sporadic alterations in brain chemistry in the absence of drugs may occur and result in the production of atypical mental phenomena. Additionally, it has been known since the 1950s that OBE-like experiences can be induced by electrical stimulation of the temporal lobe, and temporal lobe dysfunction has been implicated in a wide range of anomalous experiences, including psi-related phenomena, NDEs, and hallucinations. NDEs themselves have received considerable attention from a biological perspective, and a variety of theories positing hypothetical neurochemicals, hypoxia, and known neurotransmitters have been put forth to explain the phenomenon. Yet, despite a considerable amount of speculation, there is little solid evidence linking most anomalous experiences to an underlying biological process or functional abnormality.

Two exceptions to this rule, however, are synesthesia and lucid dreaming. The brain-imaging technique PET, which is capable of measuring activity in localized brain areas, has shown that the brain of a synesthete (one who is capable of experiencing the phenomenon of synesthesia) actually functions differently than a typical brain. Thus, when a synesthete who claims to be able to see sounds actually hears something, not only does the area of the brain responsible for processing auditory information become active but so too does the visual processing area (Baron-Cohen and Harrison 1996). Biological indicators of lucid dreaming are slightly more subtle. Although no unique physiological states have been determined, lucid dreaming is associated with higher levels of activation in the cerebral cortex. Additionally, those highly skilled in the process can move their eyes in a predetermined direction while dreaming, as a way to signal to researchers measuring their eye movements electronically that they are experiencing a lucid dream. It was this experimental procedure that actually confirmed the existence of lucid dreaming, a phenomenon that had been reported for decades but whose existence remained questionable. The ability to experience a lucid dream does not appear to result from any unique properties of the individual, and a variety of techniques have been developed to train people to become lucid dreamers (Green and McCreery 1995). The recent findings with synesthesia and lucid dreaming may provide a model for the future understanding of other atypical mental phenomena. As techniques for measuring them become more sophisticated, they, like synesthesia and lucid dreaming, may also become less "anomalous" and may begin to be understood as normal, although rare, parts of the continuum of human psychological functioning.

A significant portion of the general public and a small minority of scientists believe that anomalous psychological experiences represent actual paranormal or supernatural phenomena. NDEs are taken as existence of an afterlife, past-life experiences are taken as evidence of reincarnation, anomalous healings are thought to be indicative of the power of prayer and faith, and psi-related phenomena are believed to represent the untapped powers of the human mind. Although empirical evidence is lacking, adherents offer several lines of reasoning to support their claims.

The perceived similarities in the nature and content of some anomalous experiences among individuals from vastly different cultures and with different religious beliefs figure prominently into many supernatural explanations. For example, the tunneling/bright-light phenomenon associated with NDEs is quite universal and invariant, irrespective of cultural variables. Skeptics might explain this observation by the fact that a lack of oxygen to the brain can produce similar types of hallucinatory experiences among people from different cultures. But the universality of the phenomena is often cited as a key piece of evidence in support of a supernatural explanation for NDEs, the logic being that if NDEs were culturally determined, then such similarities should not exist. Proponents of the supernatural also point to the positive transformational nature that many anomalous experiences have on the lives of those who report them. The argument here is that mere "hallucinations" or "perceptual anomalies" could not produce such life-changing responses and that a power of this type could only come from something that is beyond this world.

The inability of science to completely and unambiguously explain all aspects of anomalous experiences under all conditions is perhaps the most frequently cited piece of evidence in support of their supernatural origins. Indeed, there are some unexplainable phenomena, such as xenoglossy, in which a hypnotized individual appears to understand a foreign language to which he or she has had no known exposure. In addition, some individuals claim that they left their bodies after being pronounced clinically dead, and they are able to recount events that had transpired in the hospital room during that period with uncanny accuracy and detail. There are also cases of psi-related phenomena (such as ESP), some of which were generated under laboratory conditions, that supporters claim cannot be accounted for by scientific explanations. Proponents cite these rare cases, which appear to defy all scientific interpretations, as evidence for an afterlife, reincarnation, and the existence of psychic powers. However, the fact that something is not currently understood by science should not necessarily lead to the conclusion that it is supernatural in origin.

Science is replete with examples of phenomena that were at one time not understood but have since come to be explained. Synesthesia and lucid dreaming, for example, were previously unexplained and their existence questioned, but both are now known to be real phenomena whose biological mechanisms are beginning to be understood. However, proponents of supernatural explanations claim that even if brain mechanisms associated with anomalous experiences are identified, it will be impossible to infer causality. The crux of the issue is this: does an altered brain state cause an anomalous experience, or does an anomalous experience cause an alteration in brain function? Believers argue that, far from producing sensory and perceptual hallucinations that are mistakenly identified as paranormal, alterations in brain states may actually allow for the perception of real paranormal information or increase an individual's ability to come into contact with the supernatural world.

References:

Baron-Cohen, Simon, and John E. Harrison, eds. 1996. *Synaesthesia: Classic and Contemporary Readings*. Cambridge, MA: Blackwell.

Blackmore, Susan. 1996. *In Search of the Light.* Amherst, NY: Prometheus.

Cardena, Etzel, Steven Jay Lynn, and Stanley Krippner, eds. 2000. *Varieties of Anomalous Experience: Examining the Scientific Evidence.* Washington, DC: American Psychological Association Press.

Frazier, Kendrick, ed. 1999. *Encounters with the Paranormal.* Amherst, NY: Prometheus Books.

Green, Celia Elizabeth, and Charles McCreery. 1995. *Lucid Dreaming: The Paradox of Consciousness during Sleep.* London: Routledge.

Reed, Graham. 1988. *The Psychology of Anomalous Experience.* Amherst, NY: Prometheus Books.

Anthroposophy and Anthroposophical Medicine

DAN DUGAN

Anthroposophy, or Spiritual Science, is the philosophy of the followers of Rudolf Steiner (1861–1925). Steiner was the leader of the German section of Theosophy from 1902 to 1912, when disagreement over whether Krishnamurti was a reincarnation of Christ led him, followed by most of the German Theosophists, to form his own group. Anthroposophy combines elements of Buddhism and Hinduism (reincarnation and karma), Zoroastrianism (light and dark gods), Manichaean and Gnostic Christianity, and European esoteric traditions including Freemasonry, Rosicrucianism, and herbalism. It has been described by critics as a cultlike religious sect (PLANS 2001).

In addition to the study of the writings and lectures of Steiner, Anthroposophy's activities include the worldwide Waldorf education program; eurythmy, a dance form; Anthroposophical medicine, which claims to have ten hospitals in Europe; biodynamic agriculture; Camphill institutions for developmentally disabled children and adults; several busy publishing houses; politics (the Threefold Social Order); the manufacture of cosmetics and pharmaceuticals under the brands Weleda and Wala; a church called the Christian Community; residential care for elderly people; art schools; financial services; and architecture (ASA 2000).

Steiner claimed to be able to make scientific observations in the spirit world. His cosmology elaborated Helena Blavatsky's numerologically ordered succession of past and future epochs, the current time period being the fifth post-Atlantean epoch. In physics, Steiner championed Johann Wolfgang von Goethe's color theory over Isaac Newton, and he called relativity "brilliant nonsense." In astronomy, he taught that the motions of the planets were caused by the relationships of the spiritual beings that inhabited them. In biology, he preached vitalism and doubted germ theory. His physiology was based on the idea of the "threefold man," who contained (1) a nerve-sense system, (2) a metabolic-muscular system, and (3) a rhythmic system in which the heart served only as a regulator, not a pump.

Anthroposophical Medicine

Anthroposophical medicine is a system of medicine that extends medical science into the realm of the spiritual. Anthroposophical physicians attend special training in Switzerland or Germany after obtaining their regular medical degrees. The physician's primary focus is the soul life of the patient. Homeopathic "potentization" is used in the making of remedies, though homeopathic "proving" is not. (See "Homeopathy," this volume.) In the Anthroposophical analysis, methods include a capillary dynamolysis and sensitive crystallization, in which the physician or researcher interprets the picture formed by the crystallization of salts dissolved in a fluid such as blood: "If we let a salt solution crystallize out after having added to it some drops of an extract from a living plant or tissue, we shall see the crystals arrange themselves to give an image of the etheric forces of the living substance studied" (Bott 1984, 23).

Mineral and plant remedies are selected by occult correspondences with the planets and European traditions based on the form of the plant. The Anthroposophical cancer remedy Iscador, a mistletoe extract, is manufactured in an elaborate process at the Weleda factory in Switzerland. In the year 2000, there were fifteen Anthroposophical medical practices in the United States (ASA 2000, 57). Waldorf schools are primary recruiters of new patients.

Biodynamic Agriculture (Biodynamics)

Biodynamic agriculture, or biodynamics, involves an application of Anthroposophy to farming, combining organic practices with magical rituals and soil treatments prescribed by Rudolf Steiner. Close attention is paid to astrological conditions for soil preparation, planting, and harvesting. Developing healthy soil is a main concern, using manure from animals raised on the farm and carefully prepared compost. Rudolf Steiner pointed out that a new science of cosmic influences would have to replace old, instinctive wisdom and superstition. Based on his own insight, he introduced what are known as "biodynamic preparations":

> Naturally occurring plant and animal materials are combined in specific recipes in certain seasons of the year and then placed in compost piles. These preparations bear concentrated forces within them and are used to organize the chaotic elements within the compost piles. When the process is complete, the resulting Preparations are medicines for the Earth which draw new life forces from the cosmos. Two of the Preparations are used directly in the field, one on the earth before planting, to stimulate soil life, and one on the leaves of growing plants to enhance their capacity to receive the light. Effects of the Preparations have been verified scientifically. (Wildfeur n.d.)

Goethean Science

Goethean science is a philosophy of science practiced in Anthroposophy, taking inspiration from the scientific works of Goethe as interpreted by Steiner. Steiner built on the holistic and antimaterialistic sentiments of Weimar Germany to define a theory-free phenomenology. This method is intended to correct the errors of reductionist, materialistic science by penetrating to the primal phenomena (Urphaenomen) in nature by thoughtful observation.

> We can and must agree with Goethe's view: there can be no doubt that the method of exact intuitive perception leads to valid scientific knowledge, and this within a realm hardly accessible to the analytical mode of thought—the realm of qualities and relationships between forms. Above all we must agree with Goethe that the "archetypal images"—the formative principles behind phenomena—are spiritual realities accessible to our cognition, and we must view them as a part of the spiritual content of nature. (Heitler 1998, 65–66)

The method of Goethean science is applied in Anthroposophical medicine and defines the principles of the science curriculum taught in

Waldorf education, an international movement that grew from a school established by Rudolf Steiner in 1919 for the children of workers at the Waldorf-Astoria cigarette factory in Stuttgart, Germany. Today, there are over 500 Waldorf schools across the world, including 122 in the United States (ASA 2000, 55–57). Proponents believe the Waldorf program is a paragon of holistic education. Critics characterize it as the principal missionary activity of Steiner's Anthroposophy. The science curriculum follows the principles of Goethean science and Steiner's phenomenological approach that eschews all theory in favor of teaching only. Steiner taught that abstract reasoning before the age of fourteen would lead to illness later in life.

Anthroposophical beliefs that surface in Waldorf schools include the evolution of animals out of humanity; a physiology based on Steiner's threefold man (with nerve-sense, metabolic-muscular, and rhythmic systems); the notion that Goethe was right and Newton was wrong about color; and the idea that the heart does not pump blood. Waldorf schools have become controversial since 1991, when Waldorf-method public schools and Waldorf-inspired charter schools first were established in the United States and critics began alleging violations of the establishment clause of the U.S. Constitution.

The physical environment of the schools is carefully artistic. Angles and corners are softened with draperies. Walls are painted with transparent washes of pastel colors. Subjects are taught in two-week blocks. There are no textbooks. Students make their own books for each lesson block, copying text and illustrations from the blackboard. All modern media are strictly avoided. Some schools require families to eliminate radio, television, and movies. Artistic media are prescribed by age; in the early grades, wet-on-wet watercolor and block crayons are used with the intent of preventing the use of lines. Children graduate to using colored pencils in later grades. Books are rarely seen in kindergarten, and reading is approached slowly in the first three grades.

Nature altar in a California public Waldorf school. (Courtesy of the author)

The philosophical subtext of the Waldorf schools is Steiner's Anthroposophy. Proponents say that Anthroposophy informs Waldorf education but isn't taught directly. Critics point to numerous examples of Anthroposophical doctrine in student work and the exclusively Anthroposophical teacher training, a two-year program. The foundation year consists entirely of the study of Anthroposophy. The second year applies the Anthroposophical theory of child development to teaching. Occultism makes heavy use of numerology, and Steiner viewed human life as being organized in seven-year periods. Accordingly, he said, in the first seven years, only the physical body is fully incarnated; at age seven, the etheric body is born, at age fourteen the astral body, and at age twenty-one the I.

References:

Anthroposophical Society in America (ASA). 2000. *Directory of Initiatives 2000*. Ann Arbor, MI: ASA.

Association of Waldorf Schools of North America (AWSNA). URL: http://www.awsna.org. (Accessed on June 3, 2001).

Bott, Victor. 1984. *Anthroposophical Medicine: Spiritual Science and the Art of Healing*. Rochester, VT: Healing Arts Press.

Childs, Gilbert. 1991. *Steiner Education: In Theory and Practice*. Edinburgh: Floris Books.

The Goetheanum. URL: http://www.Goetheanum.ch. (Accessed on June 3, 2001).

Heitler, Walter. 1998. "Goethean Science." In *Goethe's Way of Science: A Phenomenology of Nature,* edited by David Seamon and Arthur Zajonc, 55–69. Albany: State University of New York Press.

Lovel, Hugh. 1994. *A Biodynamic Farm for Growing Wholesome Food.* Kansas City, MO: Acres, USA.

"Natural Science Section." *The Goetheanum.* URL: http://www.goetheanum.ch/sektion/nws/nws.htm. (Accessed on June 3, 2001).

Physicians' Association for Anthroposophical Medicine. URL: http://www.paam.net. (Accessed on June 3, 2001).

PLANS, Inc. URL: http://www.waldorfcritics.org. (Accessed on June 3, 2001).

——. "Our Concerns about Waldorf Schools." URL: http://www.waldorfcritics.org. (Accessed on June 3, 2001).

Science Group of the Anthroposophical Society in Great Britain. URL: http://www.anth.org.uk/Science. (Accessed on June 3, 2001).

Washington, Peter. 1995. *Madame Blavatsky's Baboon: A History of the Mystics, Mediums, and Misfits Who Brought Spiritualism to America.* New York: Schocken Books.

Wildfeur, Sherry. N.d. "What Is Biodynamics?" *Biodynamic Farming and Gardening Association.* URL: http://www.biodynamics.com/biodynamics.html. (Accessed on June 3, 2001).

Astrology

GEOFFREY DEAN, IVAN W. KELLY,
ARTHUR MATHER, AND RUDOLF SMIT

Astrology is the study (generally nonscientific) of supposed relationships between the heavens and human affairs; it has nothing to do with astronomy, which is the scientific study of celestial objects. Astrology takes two forms: the serious astrology of journals and consulting rooms, considered here, and the popular but trivial entertainment of sun sign columns (see the "Sun Sign Astrology" entry in this encyclopedia).

Astrologers say the heavens reflect our destiny from the cradle to the grave. During much of history, such an idea was compatible with the best available knowledge about the world. But around 1650, the scientific revolution changed everything. Today, scientists and philosophers reject astrology for the best of reasons: it has not contributed to human knowledge, it has no acceptable mechanism, its principles are known to be invalid, users disagree on almost everything (even on fundamentals such as which zodiac to use), and it has failed hundreds of tests. Furthermore, astrology is easy to explain: users do not guard against hidden persuaders (perceptual and reasoning errors), which is why an actually invalid astrology seems to work and why users can accept erroneous conclusions as true. Yet astrology still pervades modern culture. Regardless of its truth or falsity, astrology offers emotional comfort, spiritual support, and interesting ideas to stimulate self-examination. People seem to want it. But it faces strong competition from a barrage of self-help psychologies and philosophies.

Early History

Since the dawn of time, people everywhere have tried to decipher the sky. Our ancestors believed it had control over their lives. They loaded it with symbols, myths, and legends that showed its earthly connections, and these connections became astrology. Their astrology varied according to culture, but it was an important part of their history. The astrology that survives today in the West arose before 2000 B.C. in southern Iraq. It began with celestial omens, for example, "If the Moon can be seen on the first night of the month, the country will be peaceful." Around 500 B.C., the Greeks added their own ideas, such as number symbolism, planetary gods, and the oneness of nature. By A.D. 1, the basics of modern Western astrology (planets, signs, houses) had been established.

Recent History

Until 1600, an educated person could accept astrology because it was compatible with the best knowledge of the world then available.

But by 1700, the scientific revolution had destroyed the worldview on which astrology depended. Gone were the crystalline spheres on which the heavens revolved, the bodily humors needed for planets to act on, the planetary links with metals, the belief in number symbolism and planetary gods, and the undisputed authority of ancient Greek and Roman ideas. Among educated people, serious astrology was effectively dead, but at a popular level, it lived on in almanacs, which were the most widely read literature after the Bible at the time. Around 1850, when population and literacy were rapidly increasing, a handful of British enthusiasts began to revive serious astrology; around 1900, it reemerged in Europe and the New World, riding on a fascination with the occult and a general crisis of religious faith. A new breed of astrologers swept aside the complications of the Greeks and developed new marketing ploys, such as assembly-line horoscopes, sun sign astrology, and counseling. By the 1950s, astrology was becoming a boom industry, but the skepticism of educated people remained.

The Basics of Astrology

Today, astrology holds that the macrocosm (universe) and microcosm (man) are related; just as man is contained within the universe, so he contains a universe within himself, the one a reflection of the other. The idea is traditionally expressed as "as above so below" or more cautiously as "the stars incline but do not compel." It says we can learn about ourselves by looking at the stars. The starting point is the birth chart, a simplified map of the heavens at the moment of birth, whether for a person, a company, or a country. The birth chart is said to be a blueprint for life, and learning how to interpret it can take a year or more of hard study. The meanings of the individual parts, such as signs and planets, are derived not from observation, as some astrologers claim, but from symbolism and analogy, the assumption being that things similar in some respects are also similar in others. Thus, the number four and the fourth planet have the same qualities. Aries indicates ramlike impulsivity and willfulness. Mars, the red planet, indicates blood, anger, and war.

The assumption of similarity (also known as the doctrine of correspondences) was accepted without question in the days when nature was a mystery. And by 1500, planetary symbolism had passed into our vocabulary in such adjectives as *solitary*, *lunatic*, *mercurial*, *venereal*, *martial*, *jovial*, and *saturnine*. But the underlying assumption is wrong. We now know that things similar in some respects are rarely similar in other respects, so the assumption can lead to absurd conclusions—e.g., John Smith is tall, therefore John Brown is tall; Australian skies are blue, therefore Australians are melancholy. The assumption can also lead to unresolvable contradictions—e.g., the Moon was male to the Babylonians but female to the Greeks; to astrologers, Venus is symbolic of loving harmony, but its actual surface conditions, a searing 450°C under a crushing 90 atmospheres of carbon dioxide and a fog of sulphuric acid, are more like hell. Today, the way astrology uses symbolism cannot be taken seriously. Or, as one researcher put it, there are lies, damned lies, and symbols.

Disagreement Is Normal

Equally suspect is astrology's serious literature, which in the West encompasses nearly 1,000 shelf-feet of books and periodicals, of which about 3,000 books are in print (one-third in English). Astrology books typically offer speculation and strange ideas but rarely facts and critical thinking. Disagreement is normal; as-

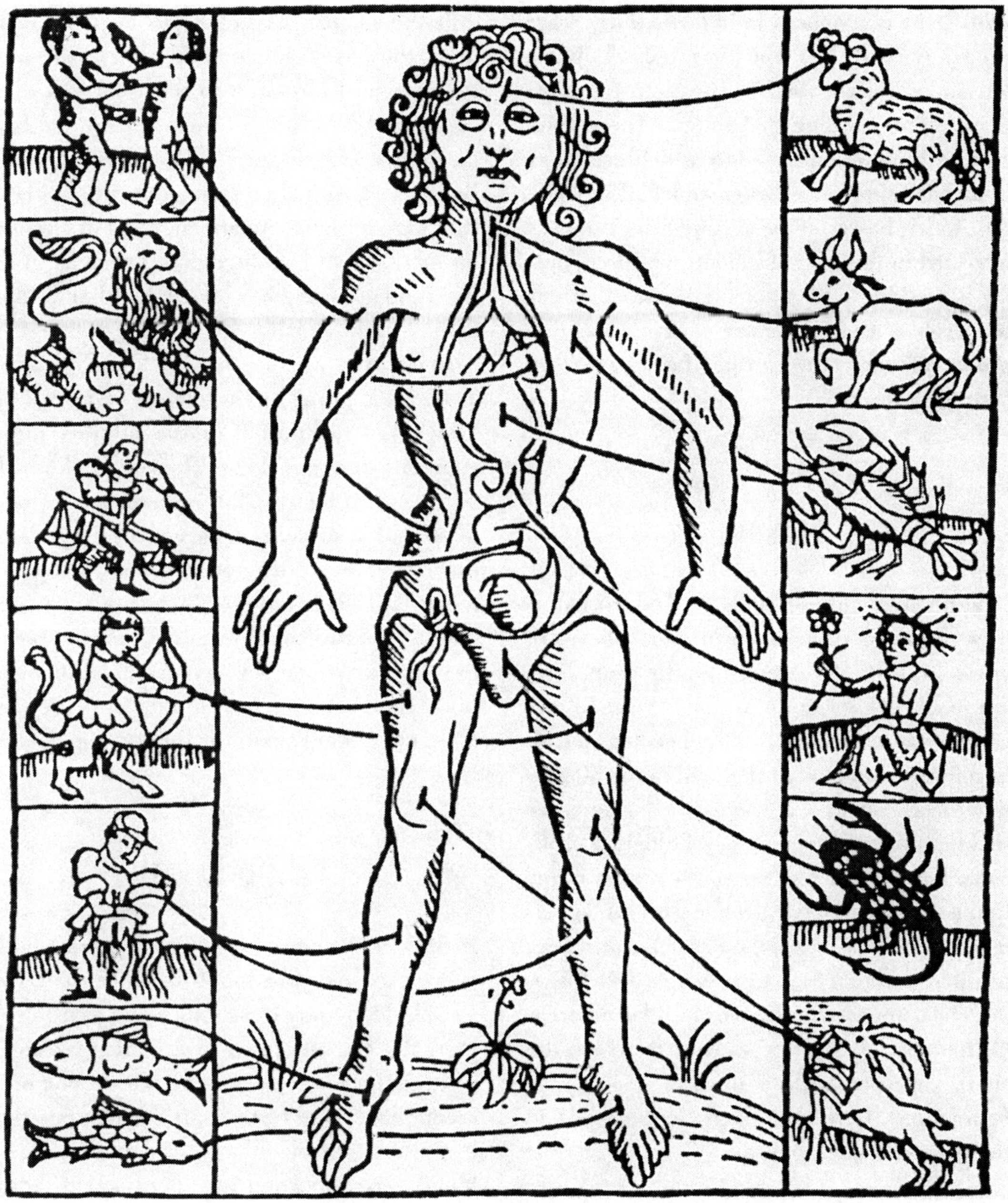

Sixteenth-century woodcut of *The Zodiacal Man*.

trologers tend to agree on the importance and meaning of the known planets, but there is general disagreement about everything else. They disagree on what a birth chart should include (signs, houses, aspects, midpoints, nodes, parts, harmonics, hypothetical planets, asteroids, progressions, directions; the list is seemingly endless), and they disagree on the details (there is more than one system of signs, of houses, of aspects, of progressions, and so on). They also disagree on the methods of interpretation, which range from using strict rules to ignoring rules in favor of psychic flashes.

Astrologers do not even agree on what a

birth chart is supposed to indicate. Fifty years ago, they tended to opt for minds, feelings, physique, health, wealth, vocation, relationships, events, destiny, and so on. "There is no area of human existence to which astrology cannot be applied" (Parker and Parker 1975, 60). Today, however, the shrewd astrologer has retreated in the face of modern research, opting instead for hidden potentials and other unobservables that are more secure from disproof and for the unspecific claim that "astrology works."

Popularity

Nevertheless, opinion polls in Western countries show that one person in four believes in astrology. In Eastern countries, the proportion approaches four in four. Each year in the United States, roughly 1 million people consult astrologers—not many compared to the 50 million Americans who, at any particular time, are seeking answers to their psychological problems, but still a lot of users. So who is right? Could the stars really correlate with human affairs? Before the 1970s, no conclusions were possible due to a lack of research. But this is no longer the case. Advances in relevant areas (astronomy, psychology, statistics, research design), together with a decisive technology (computers), have now given clear answers to these age-old questions.

Facts and Feelings

To start with, everything depends on what is meant by "astrology." Like most social beliefs, astrology covers areas where feelings matter but not facts, as in religion, and other areas where facts matter but not feelings, as in science. So astrology can mean different things to different people. Users tend to focus on feelings: they seek spiritual insight, emotional support, and direction to life, so when they claim that "astrology works," they tend to mean that "it feels good" or "it is meaningful." But many things are meaningful without being true (Santa Claus, Superman, faces in clouds). So this kind of astrology does not need to be true, and attacking it is on a par with attacking Santa Claus.

In contrast, critics tend to focus on facts. They seek proof of the claims of astrology. They want to know if Leos really are more generous than non-Leos. Their kind of astrology needs to be true, but research has consistently failed to deliver. So when critics claim that "astrology does not work," they tend to mean that "it is not true" or "any success is due to nonastrological factors." This difference in viewpoint explains why users and critics can disagree so completely over astrology—they are often not talking about the same thing.

Research Findings

Before 1950, almost no empirical studies of astrology existed. But by 1975, more than 100 were hidden away in astrology and psychology journals. The same year saw the first astrology software for home computers, which took only seconds to make birth chart calculations that had previously taken hours or days to make by hand. Astrology soon lost its mystery, Today, the number of empirical studies exceeds 500, and they have shown that it is meaningless to ask questions such as "Is astrology true?" because they are too vague (the answer could be yes or no depending on what is meant by "astrology"). Instead, the questions should be more specific, such as "Does astrology deliver anything not explained by nonastrological factors?" (the answer is no).

Of course, testing astrology is not easy. Test-

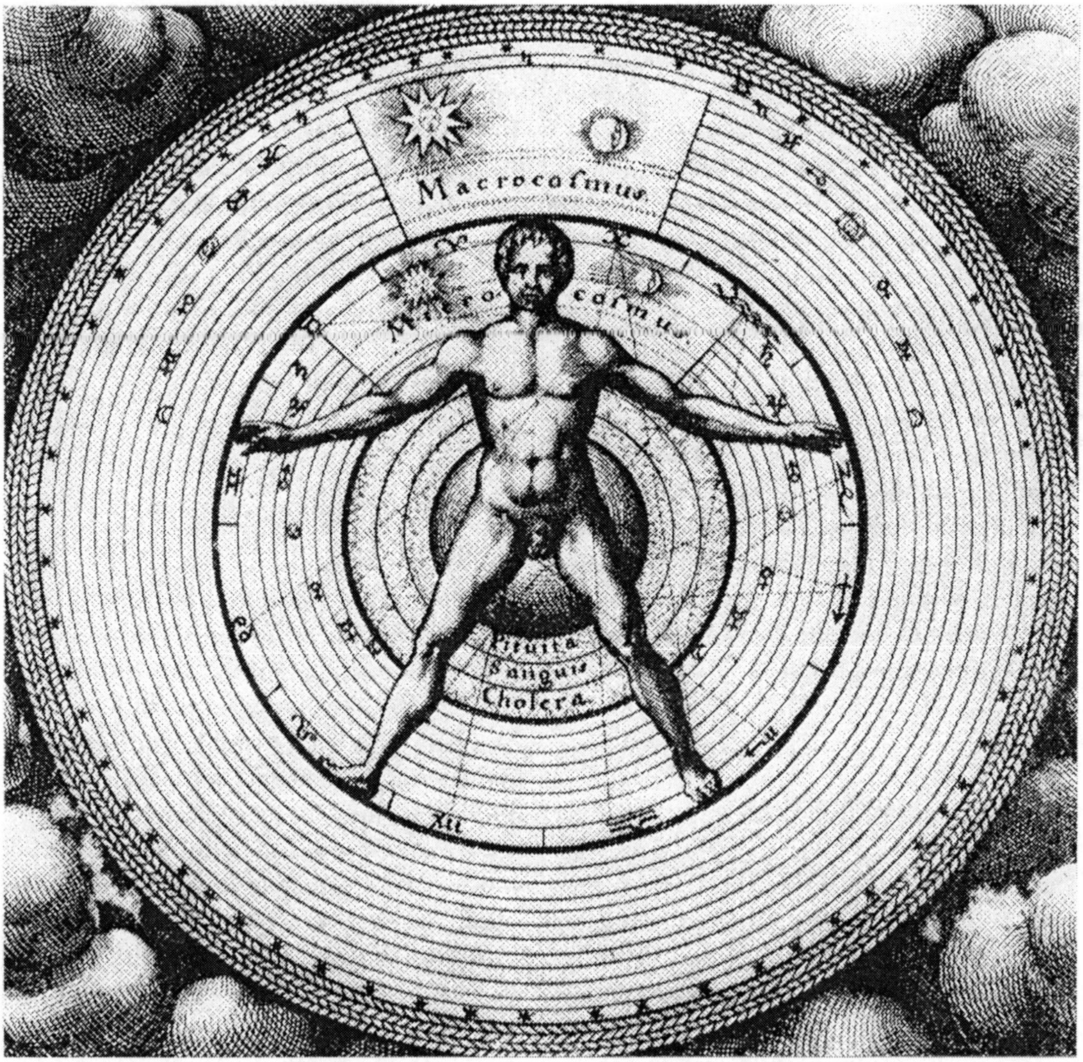

Detail from title page of Robert Fludd's cosmic *Historia*, 1716.

ing requires expertise in relevant areas such as psychology and statistics and experience in avoiding pitfalls. For example, astrology is said to arise from nothing we know about, so it is easy to misinterpret natural irregularities in the data as being due to astrology. But the picture emerging from these hundreds of studies is clear and consistent: astrology does not deliver factual truth, at least not truth commensurate with its claims. It contributes nothing to our knowledge of the world. Orthodox approaches are vastly better, which is why scientists and philosophers see astrology as unfruitful and (except for its historical and social implications) not worth serious study.

Hidden Persuaders

Astrologers generally ignore research findings. Some argue that astrology is symbolic or soul stuff or merely a language, so it cannot be tested. Others argue that astrology has no scientific explanation, so proper tests cannot be devised. And still others argue that their experience demonstrates the truth of astrology every day, so tests are superfluous. But astrology has to be testable; otherwise, astrologers could never know anything about it. And astrology does have a scientific explanation—human judgment errors. Normally hidden, these errors persuade users that astrology works even though it is actually invalid. They are the same hidden persuaders that have led millions of people to believe in pseudosciences such as phrenology and biorhythms, which we now know are completely invalid. Hidden persuaders have been the subject of thousands of scientific studies and dozens of scientific books, but they are rarely mentioned in astrology books. When these errors are prevented, astrology (like phrenology and biorhythms) suddenly fails to work. So the supposed veracity of experience is a delusion.

The Mars Effect

During forty years of immense labor, the most famous in modern astrology, the late Michel Ganquelin (1926–1991) and his wife, Françoise, tested numerous astrological claims. Their findings were almost entirely negative. Gauquelin (1991, 60) concluded, "Having collected half a million dates of birth from the most diverse people, I have been able to observe that the majority of the elements in a horoscope seem not to possess any of the influences which have been attributed to them." But one finding seemed positive: professional people tended to be born with a surplus or deficit of certain planets in the areas just past rise or culmination—but only if they were eminent and were born naturally. The tendency was later called the Mars effect, but depending on the occupation it could equally well have been called the Moon, Venus, Jupiter, or Saturn effect. Astrologers claimed the effect supported astrology, but it was extremely weak and could be detected only in samples of a thousand or more. Furthermore, contrary to astrological claims, there was no effect for half the planets, for signs or aspects, or (on Gauquelin's figures) for the 99.994 percent of the population who were not eminent. And it has recently been found that some parents were altering birth data before reporting them to the registry office, making them fit astrology in the same way as they might make them avoid Friday the Thirteenth (Dean 2002). So any support for astrology disappears.

Is Astrology Helpful?

Astrologers are generally nice people who want to help others. To them, the question is not "Does astrology work?" but "Is astrology helpful?" or "Does astrology produce change?" The answer to both is clearly yes, if only because astrology indirectly puts clients in touch with someone they can talk to. Astrologers tend to dismiss problems of how astrology might work, saying their concern lies with the client, not the means. But note the dilemma—to receive therapy by conversation or to produce change, the client has to believe in something that is untrue. Otherwise, why bother with birth charts? The same dilemma can apply elsewhere, as in palmistry, psychotherapy, and even religion, so it is not unique to astrology. Nevertheless, it presents an ethical problem that astrologers have generally failed to recognize, let alone resolve.

Astrology's Dark Side

Anything that fools people is potentially dangerous. In 1995, a survey of 509 British schoolchildren aged fourteen and fifteen found that most saw astrology as harmless fun. But more than one-third actually believed their stars. And a minority had been led to other occult practices that ended in trauma, so for them, astrology was not harmless fun (Boyd 1996). Nor is it harmless fun in Japan. According to Japanese astrology, women born in a fire-horse year will have unhappy marriages. The year 1966 was a fire-horse year, and the then annual total of 2 million births dropped by 25 percent due to an extra half million abortions, which in Japan is the principal means of birth control. People did not want to risk having girls who would be hard to marry off. Thus, in half a million cases, even this simplistic astrology was anything but harmless fun (Kaku 1975).

For and Against

The case *against* astrology is that it is untrue and falsely described. It does not deliver benefits beyond those produced by nonastrological factors, and astrologers ignore unwelcome evidence—defects that are never disclosed. So people are being misled. The case *for* astrology is that a warm and sympathetic astrologer provides low-cost, nonthreatening therapy that is otherwise hard to come by, especially as astrology implies no physical, mental, or moral weakness as might apply when visiting a doctor, psychiatrist, or priest. But whatever our verdict, for or against, if astrology encourages people to explore the problems of humanity's existence and to express spiritual values and if it provides a bridge between a person of wisdom and a person in need, then these qualities deserve study as much as any objective claim would. There is more to astrology than being true or false.

What of the Future?

Beliefs survive best when, as in astrology, believers create a close-knit group in which the sense of belonging makes their beliefs impervious to criticism. Nobody who derives benefit from astrology is going to believe evidence for its invalidity. So until a more rewarding belief comes along, astrology is unlikely to go away. Nevertheless, it faces the same scientific opposition and the same taint of pop that helped the downfall of phrenology. It also faces strong competition from a barrage of self-help psychologies and philosophies (see any New Age bookstore). Extrapolation of the growth in British astrology books suggests that an end may be near around 2100, but whether this is a genuine end or merely the result of books being replaced by the Internet is hard to say (Dean and Kelly 2001, 202). Only time will tell if astrology can survive.

References:

Boyd, A. 1996. *Dangerous Obsessions: Teenagers and the Occult.* London: Marshall Pickering.

Dean, G. 2002. "Is the Mars Effect a Social Effect?" *Skeptical Inquirer* 26, no. 3: 33–38. A comprehensive abstract is at URL: http://www.astrology-and-science.com/.

Gauquelin, M. 1991. *Neo-Astrology: A Copernican Revolution.* London: Arkana.

Kaku, K. 1975. "Increased Abortion Rate in 1966: An Aspect of Japanese Folk Superstition." *Annals of Human Biology* 2: 111–115.

Parker, D., and J. Parker. 1975. *The Compleat Astrologer.* London: Mitchell Beazley.

Descriptive books on astrology are numerous and easy to find—just visit any public library. By contrast, critical books on astrology are hard to find; they also tend to be incomplete because most of the important studies are not retrievable via computerized databases such as PsychInfo. For ordinary readers, good sources of information are:

Brau, J. L., H. Weaver, and A. Edmands. 1980. *Larousse Encyclopedia of Astrology.* New York: McGraw-Hill. Good general-purpose coverage of history, methods, and biographies but has almost nothing on research. Includes instructions for calculating and reading birth charts.

Phillipson, G., ed. 2000. *Astrology in the Year Zero.* London: Flare Publications. See URL: http://

www.flareuk.com. Explores key issues via interviews with astrologers and researchers. Includes a review of recent research findings, a look at human judgment errors, and a critical bibliography, all with expanded versions at URL: http://www.astrology-and-science.com/.

For academics, a comprehensive critique of astrological ideas is:

Kelly, I. W. 1997. "Modern Astrology: A Critique." *Psychological Reports* 81: 1035–1066. Has 131 references, including major critical works. Concludes that astrology does not have the means to solve its internal problems. An expanded version is at URL: http://www.astrology-and-science.com/.

The following titles address more specific issues:

Ankerberg, J., and J. Weldon. 1989. *Astrology: Do the Heavens Rule Our Destiny?* Eugene, OR: Harvest House. Very readable critique of astrology's claims and the personal consequences of its use. Many useful quotes, all referenced to the exact page. The answer to the title question is no.

Dean, G. 1992. "Does Astrology Need to Be True?" In *The Hundredth Monkey and Other Paradigms of the Paranormal,* edited by K. Frazier, 279–319. Buffalo, NY: Prometheus Books. Goes beyond the popular astrology of newspaper columns to examine the serious astrology of consulting rooms and learned journals. The answer to the title question is no.

Dean, G., and I. W. Kelly. 2001. "Does Astrology Work? Astrology and Skepticism, 1975–2000." In *Skeptical Odysseys,* edited by P. Kurtz, 191–207. Amherst, NY: Prometheus Books. Reviews the progress of research into astrology since 1975.

Dean, G., A. Mather, and I. W. Kelly. 1996. "Astrology." In *The Encyclopedia of the Paranormal,* edited by G. Stein, 47–99. Amherst, NY: Prometheus Books. An extended (28,000-word) scientific survey covering history, popularity, arguments for and against, conceptual problems, controlled tests, effect size comparisons, problems of birth chart interpretation, how belief in astrology arises, the role of human judgment errors, and the future of astrology.

Gambrill, E. 1990. *Critical Thinking in Clinical Practice: Improving the Accuracy of Judgments and Decisions about Clients.* San Francisco: Jossey-Bass. A survey of human judgment errors made by professionals in psychology, medicine, and the helping professions, showing how to reduce their influence on decision making. Equally applicable to astrology.

Gettings, F. 1990. *The Arkana Dictionary of Astrology.* London: Arkana. A scholarly work covering every aspect of astrology from its origins to the present day. Has nearly 4,000 entries from Ab and Abbreviations to Zubrah and Zugos. A rich source of information for academics but has little on research.

Holden, J. H. 1996. *A History of Horoscopic Astrology: From the Babylonian Period to the Modern Age.* Tempe, AZ: American Federation of Astrologers. Has a bibliography of 27 titles and a detailed index. Clear, readable, the only history to focus on technical developments. Includes brief biographies of about 1,000 individual astrologers.

Attachment Therapy

JEAN MERCER

Attachment therapy (AT) is a treatment used with the intention of creating emotional change in children and adolescents, but its theoretical background, empirical support, and safety remain questionable. The practice is used with the goal of causing children to feel a strong and positive emotional connection to the adults who care for them (their biological, foster, or adoptive parents). AT practitioners refer to this positive feeling toward the adults as "attachment" or "bonding," and they consider it to be the basis for obedience, desirable behavior, and the development of conscience and character (Levy and Orlans 2000a, 2000b).

Writings about AT attribute the violent, aggressive behavior of children and adolescents to a lack of attachment, and AT practitioners consider aggression, disobedience, and lack of affection to be part of a syndrome they call attachment disorder (AD). Many children brought into AT are adopted or have backgrounds of neglect or abuse. The ideas behind AT are shared to some extent by many types of psychotherapy, but AT is unusual in its practitioners' assumption that emotional change can be brought about by physical treatment rather than by talking, play, or other forms of communication. Various specific techniques are used by different AT practitioners, but all involve some type of physical restraint and some degree of discomfort or fear (Cline 1992; Levy and Orlans 2000a, 2000b).

Techniques of AT

The restraint used in AT ranges from an extreme form called "rebirthing" (Crowder 2000) to a less coercive approach that is often called "holding therapy." Rebirthing is a version of a practice in which adults sometimes participate because they expect psychological benefits from the experience. When done with a child, rebirthing involves wrapping the child tightly in a blanket and carrying out a reenactment of a birth. Several adults restrain the wrapped child and press against him or her in imitation of the contractions of the uterus during the birth process. The child is told to struggle, not just to escape but to be "reborn" as the child of an adoptive or other parent, who may be present during the process (Crowder 2000).

In holding therapy, the child is not wrapped but is restrained by adults, who may try to establish prolonged eye contact while holding the struggling child over a period of hours. Repeated sessions of restraint occur over days or even weeks (Levy and Orlans 2000a, 2000b).

Physical restraint is not the only tool used by AT practitioners, although it is at the heart of the treatment procedure. Children who are receiving AT also spend time with "therapeutic foster parents" who are trained in AT techniques and philosophies and whose goal is to stress adult authority and control. Obedience and an affectionate attitude are said to be created as a result of the foster parents' provision or denial of food, play, and approval; the rules by which the foster parents do this do not resemble the reinforcement techniques they suggest but are deliberately arbitrary and unpredictable (Thomas 2000). The child's biological or adoptive parents are also instructed to use these techniques, as well as to employ physical restraint, and support groups are available to encourage this treatment.

Outcomes

Both rebirthing and other holding techniques have been associated with the deaths of children (Crowder 2000; Horn 2000; Smith 1996). Public attention was drawn to AT after the suffocation death of ten-year-old Candace Newmaker during a rebirthing session conducted by AT practitioners in Colorado in April 2000 (Crosson 2000). Several years earlier, three-year-old Krystal Tibbets of Utah had died when her foster father lay on top of her in a "therapeutic" effort advised by AT practitioners (Smith 1996). Evidence in a trial connected with Candace Newmaker's death revealed the AT practitioners' belief that the child's pleas for help showed her resistance to treatment rather than real distress. Positive results for children treated by AT have also been claimed (Randolph 2001), but as will be noted, these claims are based on questionable research methodology.

Research on the outcome of therapy for emotional problems is quite difficult to carry out and to interpret, but the American Psychological Association has suggested some important criteria that should be met before a treatment is claimed to be effective. Published research on AT has not complied with those guidelines and is not adequate to support the claims made by some AT practitioners.

The major problem with the research asserted to be evidence for the efficacy of AT is that it has not employed a random assignment of children to groups that would or would not receive AT (Randolph 2001; Myeroff, Mertlich, and Gross 1999). The random assignment to groups is the only legitimate way to know that a difference between AT-treated and untreated children has resulted from the treatment rather than from other, preexisting differences between the groups. For example, one study (Myeroff, Mertlich, and Gross 1999) compared children whose parents brought them for treatment with other children whose parents applied to bring them but were unable to make the arrangements to do so. Later differences between the two groups of children could have resulted from systematic differences between the two sets of families (for instance, numbers of other children to be cared for) rather than from the experience of treatment.

Theoretical Background

Why would AT practitioners expect their treatment to cause positive emotional changes in children? The answer to this question requires an examination both of the theories on which AT writers claim to base their work and of other theories and practices that seem much more strongly connected to AT than the claimed sources.

AT practitioners (Levy and Orlans 2000a, 2000b) say they base much of their approach on the work of psychologists such as John Bowlby (1982), focusing on early emotional development. Bowlby and his colleagues used both clinical and experimental methods to examine children's development of strong positive feelings toward their parents or other familiar caregivers. These strong positive feelings of a child for an adult, called "attachment," normally develop between about six and eighteen months of age. Attachment occurs as a result of

Illustration of child holding hands with adult. (Bonnie Timmons/The Image Bank)

repeated pleasant social interactions, such as play between the child and a small number of consistent, responsive caregivers. The child's behavior shows evidence of attachment when he or she strongly prefers to be with a familiar person, cries or shows distress when separated from that person, avoids strangers, and turns to the familiar person for security when in an unfamiliar place.

When their caregivers are less responsive than is appropriate, some children form attachments that seem less secure and comforting for them than what is typically seen. Infants and toddlers who do not have consistent, responsive caregivers do not form attachments but instead show indiscriminate friendliness toward adults. Toddlers who have formed secure attachments suffer deep grief and distress if they undergo an abrupt, prolonged separation from the familiar person.

The long-term importance of attachment stems from its role as a foundation for later relationships. Children who have formed less secure and less comforting attachments may later have poor emotional relationships with friends, spouses, and their own children. Those who have formed no attachment tend to exploit other people and to have no sense of guilt about harming others. Those who have formed an attachment and been separated may be overly sensitive to later losses of this type (Bowlby 1982). Of course, later experiences and attempts to understand relationships can alter the individual's point of view; the early attachment experience is only one factor in an adult's attitude toward others (van IJzendoorn 1995).

AT practitioners stress Bowlby's concepts of attachment and separation and the roles they play in early emotional life, and they are particularly concerned with the exploitativeness and possible violence connected with a lack of attachment. However, the AT view of the timing and causes of attachment events is quite different from Bowlby's. AT practitioners believe that the emotional connection the child feels for the parent begins before birth and that separation at birth or in the first few months of life will cause serious grief to the child. (In Bowlby's theory, the child must be at least six months old before separation will have such an emotional effect.) AT practitioners stress the relationship between the child and the birth mother as being the most important. (Bowlby considers any consistent, responsive caregiver as a potential focus of attachment.) AT practitioners believe that postnatal attachment is strengthened by a set of experiences they call the "bonding cycle"; these are the child's experiences of pain, need, fear, or distress that are relieved by the adult caregiver on many occasions between birth and age two. (In Bowlby's approach, attachment is considered to occur between about six and eighteen months as a result of pleasurable social interactions between the child and one of a few consistent, responsive adult caregivers. Attachment occurs easily during this period because the child is especially ready for it developmentally.)

Because AT practitioners believe that attachment follows a set of experiences of suffering and relief, they assume that subjecting the child to repeated instances of distressing physical restraint followed by release can create attachment where none exists (Cline 1992). The AT stress on the process rather than the developmental needs or abilities of the child leads to the assumption that attachment can be formed in the same way at any time during childhood or adolescence. AT practitioners view attachment as an intrapersonal event whose occurrence prepares the way for filial affection and obedience, rather than as part of the ongoing, lifelong development of attitudes toward and relationships with specific other people.

It would appear that AT is actually derived not from Bowlby's attachment theory but from the writings of other therapists such as Robert Zaslow (Zaslow and Menta 1975), who described the bonding cycle, and others in the 1940s and after who suggested physical restraint and discomfort as forms of therapy. It should be noted that these latter writers are of little relevance to the work of most modern psychotherapists in the United States, who generally employ a more cognitive and communicative approach, both with adults and with children.

References:

Bowlby, John. 1982. *Attachment.* New York: Basic Books.

Cline, Foster. 1992. *Hope for High Risk and Rage Filled Children.* Evergreen, CO: EC Publications.

Crosson, J. 2000. "'Rebirthing' Halted after a Death." *Philadelphia Inquirer,* May 22, A5.

Crowder, Carla. "Prosecutors Add Charges for Rebirthing Therapist." *Denver Rocky Mountain News.* URL: http://www.insidedenver.com/news/0729char0.shtml. (Accessed on July 29, 2000).

Horn, M. "A Dead Child, a Troubling Defense." *US News Online.* URL: http://www.usnews/issue/970714/14atta.htm. (Accessed on July 21, 2000).

Levy, Terry, and Michael Orlans. 2000a. "Attachment Disorders as an Antecedent to Violent and Antisocial Patterns in Children." In *Handbook of Attachment Interventions,* edited by Terry Levy, 1–16. San Diego, CA: Academic Press.

———. 2000b. "Attachment Disorder and the Adoptive Family." In *Handbook of Attachment Interventions,* edited by Terry Levy, 244–260. San Diego, CA: Academic Press.

Myeroff, Robin, Gary Mertlich, and Jim Gross. 1999. "Comparative Effectiveness of Holding Therapy with Aggressive Children." *Child Psychiatry and Human Development* 29, no. 4: 303–313.

Randolph, Elizabeth. "Attachment Therapy Does Work!" *The Attachment Center at Evergreen.* URL: http://www.attachmentcenter.org/articles/article015.htm. (Accessed on January 20, 2001).

Smith, L. 1996. "Full of Woe." *Los Angeles Times,* July 12, E1.

Thomas, Nancy. 2000. "Parenting Children with Attachment Disorders." In *Handbook of Attachment Interventions*, edited by Terry Levy, 67–111. San Diego, CA: Academic Press.

Van IJzendoorn, Martinus. 1995. "Adult Attachment Representations, Parental Responsiveness, and Infant Attachment: A Meta-analysis on the Predictive Validity of the Adult Attachment Interview." *Psychological Bulletin* 117: 387–403.

Zaslow, Robert, and Marilyn Menta. 1975. *The Psychology of the Z-Process: Attachment and Activity*. San Jose, CA: San Jose State University Press.

Ball Lightning

STEUART CAMPBELL

Ball lightning (BL) is popularly described as a slow-moving luminous ball not more than 12 inches (30 centimeters) in diameter occasionally seen at ground level during a thunderstorm. Scientists usually understand it as an electrical discharge phenomenon somehow associated with normal lightning.

The existence of BL is controversial, with opinions and explanations changing over time. Although many theories have been advanced to explain it, none of them account for all the reported characteristics of ball lightning. Further, BL has not been created in laboratory conditions with all its characteristics, and reliable accounts of it are rare and often suspect. Because of perceptual and memory problems, anecdotal evidence is of doubtful value. There is no photograph, film, or video recording that can be accepted unreservedly as showing BL. Many forget the null hypothesis, which has explained many postulated phenomena (such as phlogiston and the ether) that turn out to be nonexistent. The null hypothesis may also explain BL, which could be a chimera or a pseudophenomenon.

Skepticism regarding the existence of BL goes back at least to Michael Faraday and François Arago in the nineteenth century. In 1839, Faraday allowed that balls of fire might appear in the atmosphere but doubted that they had anything to do with lightning or atmospheric electricity (Barry 1980, 133). More recently, Karl Berger reported that, in over twenty years of study as a meteorologist and lightning investigator, he had never observed BL. He concluded that it did not exist (Barry 1980, 133). Other scientists have reached the same conclusion. James Lovelock put tales of BL in the same category as those of spontaneous human combustion and crop circles (Lovelock 2000, 86). Even James Barry (1980, 134) allowed that the unbiased examination of reports leads to the conclusion that a great percentage of them are highly questionable and could be interpreted in several ways. Among those ways is the persistence of the vision theory proposed by Lord Kelvin (William Thomson) in 1888. He claimed that the uniform size of the ball lightning reported in many cases was ascribed to an illusion associated with the blind spot in the eye (Singer 1971, 19). Lovelock (2000) reported such a case after a lightning flash. Other sources of deception that have been proposed are the will-o'-the-wisp and owls with luminous wings, but the existence of either of these entities is itself doubtful. In recent years, some scientists have accepted the existence of ball lightning, albeit with little evidence.

Reports of BL suffer from defects inherent in the human perceptual and memory systems. Because both perception and memory are reconstructive processes, what we perceive is not necessarily what the sense organs receive. This fact is demonstrated by various well-known optical illusions, such as the moon illusion. Distant stationary lights are subject to several movement illusions, all of which attribute movement to the light. The most famous example is the autokinetic illusion, in which a stationary light (usually a star) will appear to move about at random.

The size or distance of an unknown object cannot be determined by observers without additional information. Observers usually make a guess about either the size or the distance of an object and then determine the other parameter from their guess. In fact, both can be wrong. The size of distant objects seen near the horizon can be exaggerated (the moon illusion), as can an object's altitude (angle above the horizon). Nor can observers usually distinguish between a change in size of an object and a change in its distance, and they usually interpret a change in size as a change in distance. A phenomenon called size constancy can interfere with size perception. Even estimates of time span are unreliable; intense interest tends to shorten the estimation of time elapsed. Estimates of brightness are meaningless (brightness is a relative term), and observers tend to make false associations, drawing unwarranted conclusions from what they perceive. They may associate effects with the wrong cause. In the case of anomalous luminous phenomena, observers try to identify them by reference to the models they carry in their minds. Clearly, they can only identify a phenomenon as BL if they have heard of it. Conversely, they are likely to identify an anomalous object as BL simply *because* they have heard of it.

Memory is not much more reliable than perception. People who report BL and who have heard about other reports may, inadvertently, draw on these previous reports for their own report. Tests show that reliability decreases with time, and it is strongly suspected that observers attempt to make facts fit theory.

Consequently, genuine anecdotal reports of BL must be regarded with suspicion. Observers are mostly unaware of the distortions involved in perception and memory. Worse still, asking people if they have seen BL begs the question of its existence and ignores their inability to distinguish it (if it exists) from other phenomena. The question plants a concept in the mind that will distort the memory of any genuine perception. Such a question should not be asked, and surveys based on this question are worthless.

The contradictory results obtained from reports were noted by an early investigator, F. von Lepel (Singer 1971, 62). According to reports, BL occurs in any type of weather, not just storms; it can be any color; it can be motionless or moving at any speed, often against the wind; it can disappear violently or silently; it may follow wires or edges or travel independently; it may be outside or inside; its lifetime varies from a fraction of a second to several minutes; it can be spherical or pear-shaped; it is either silent or noisy; and so forth.

In other words, the phenomenon exhibits no consistent characteristics and appears to be all things to all observers. And as one investigator commented, there are very few natural phenomena that observation makes more difficult to explain (Singer 1971, 62). However, such contradictions might be explained if the observers are reporting many different phenomena, none of which are actually BL. Among the objects mistaken for BL are bright astronomical objects at low altitude, sometimes seen in mirage (Campbell 1988a).

Anecdotal reports are unreliable, and so are illustrations based on these reports. However, it is more difficult to explain reports of physical damage and photographic evidence. It is sometimes alleged that BL can penetrate closed windows, and the literature contains several alleged examples. When a mysterious hole appeared in a window of his department during a storm, a professor of meteorology in Edinburgh concluded that BL was the cause. But later investigation showed a simpler explanation—mechanical damage (Campbell 1981a).

And almost circular cracks can appear in sheet glass when it is subjected to the appropriate sudden stress or impact. Reports of extensive damage, such as fires or explosions, may more easily be explained as the result of ordinary lightning strikes. Such reports are not clarified by the popular belief that lightning strikes are the result of something called a "thunderbolt."

Barry (1980, 108) demonstrated that a long-lived luminous ball phenomenon can be produced by a spark-initiated combustion of low-density hydrocarbon gas at atmospheric pressure. This phenomenon may explain the 1975 report from a housewife in Smethwick, England, that BL appeared over her gas range (Campbell 1988b). Normal lightning may ignite hydrocarbon gases in the atmosphere, producing similar phenomena, but this is not what is understood as BL.

Photographs alleged to show BL are as suspect as anecdotal reports and sketches. The camera cannot lie, but what it shows can be misinterpreted—and the *photographer* can lie. Until the early 1970s, a photograph taken in 1961 at Castleford, England, had been interpreted as showing the path of BL. Even *New Scientist* magazine described the image as the "Path of a Thunderbolt." But a decade later, it was claimed that the photograph showed the pulsed trace from a street lamp (Davies and Standler 1972), and a decade after that, it was demonstrated that this assessment was correct (Campbell 1981b): the photographer incautiously moved the camera while the shutter was still open. A Russian photograph taken in 1957 was explained in the same way—but not before a member of the Soviet Academy of Sciences endorsed the photo on the basis of similar pictures he had seen in a 1939 U.S. journal (Campbell 1987). He did not know that the pictures were all produced by lamps, presumably as hoaxes.

Many alleged pictures of BL are deliberate fakes. One example seems to be a picture produced in 1966 by a former Canadian air force pilot, which misled the U.S. editor of *Aviation Week and Space Technology;* he used it on the cover of his skeptical books on unidentified flying objects (UFOs) (Campbell 1988c).

Although it is fairly easy to take a photograph—or to fake one—that many mistakenly interpret as showing BL, it should be less easy to produce a film or video sequence that could fool anyone. However, in 1973, a film appeared that allegedly depicted BL traveling slowly across the horizon near Aylesbury, England. It showed a bright ball of light moving on a steady horizontal course for 23 seconds, until it suddenly vanished. Because the object was reported initially as a UFO, the film has been shown many times at UFO conferences and has been featured in a British Broadcasting Corporation (BBC) television program about UFOs. But it was also thought that the film showed BL. Later, it was demonstrated that the "ball" was actually burning fuel being dumped from a damaged U.S. fighter-bomber; the aircraft itself, nearly 4 miles (6 kilometers) away, was not visible beside the fireball and was too far away to be heard (Campbell 1991).

In 1989, a TV station in southeast England screened a video of a smudgy spherical object with a hole that was captured accidentally as the videographer attempted to film normal lightning; he had not seen anything unusual during the recording. The videographer thought it might show BL, and this explanation was initially endorsed by Roger Jennison, a professor at the University of Kent (who has himself reported seeing BL). However, it was later demonstrated that the object in the sequence was a combination of an artifact of the camera itself and a distant streetlight (Bergstrom and Campbell 1991).

References:

Barry, James Dale. 1980. *Ball Lightning and Bead Lightning: Extreme Forms of Atmospheric Electricity*. New York: Plenum Press.

Bergstrom, Arne, and Steuart Campbell. 1991. "The Ashford 'Ball Lightning' Video Explained." *Journal of Meteorology, UK* 16, no. 160: 185–190.

Campbell, Steuart. 1981a. "Not Lightning Damage." *Weather* 36, no. 3: 66–71.

——. 1981b. "How Not to Photograph Ball Lightning." *British Journal of Photography* 128, no. 6326: 1096–1097, 1105.

——. 1987. "Ball Lightning Exposed! Another Picture Puzzle . . ." *British Journal of Photography* 134, no. 6645: 1537–1538.

——. 1988a. "Russian Accounts of Ball Lightning." *Journal of Meteorology, UK* 13, no. 128: 126–128.

——. 1988b. "The Smethwick Ball Lightning Report." *Journal of Meteorology, UK* 13, no. 134: 391–393.

——. 1988c. "The Childerhose UFO: Fact or Fiction?" *British Journal of Photography* 135, no. 6686: 72.

——. 1991. "Fireball by Day." *British Journal of Photography* 138, no. 6814: 22–23.

Davies, D. W., and R. B. Standler. 1972. "Ball Lightning." *Nature* 240 (November 17): 144.

Lovelock, J. 2000. *Homage to Gaia*. New York: Oxford University Press.

Singer, Stanley. 1971. *The Nature of Ball Lightning*. New York: Plenum Press.

Bermuda Triangle

MAARTEN BRYS

The Bermuda Triangle is the triangular area in the Atlantic Ocean between the Bahamas, Bermuda, and the East Coast of the United States in which ships and airplanes are said to mysteriously disappear. The absolute peak in cultural interest in the Bermuda Triangle followed the 1974 publication of the best-selling book *The Bermuda Triangle*, by Charles Berlitz and J. Manson Valentine, of which millions of copies were sold.

Some of the more imaginative explanations for the disappearances are kidnappings by unidentified flying objects (UFOs) and dangerous force fields originating in the lost continent of Atlantis below the ocean's surface. The truth is that there is absolutely nothing unusual with that area.

The exact position and size of the Bermuda Triangle are somewhat disputed: certain authors say that it has a surface of 500,000 square kilometers, whereas others mention figures three times as high and also consider the Azores and parts of the West Indies as being part of the triangle. The rumors about mysterious disappearances in that part of the Atlantic Ocean already existed in the era of Columbus, but the triangle craze reached its peak in the 1970s.

What were the claims? All the stories about the Bermuda Triangle contain certain similarities: they are always about ships or airplanes that, although in the hands of experienced pilots or sailors, mysteriously disappear in a calm sea and in bright weather conditions. Usually, strange radio messages are mentioned to liven up the story. But those who truly investigate the facts will find out that these stories often are transferred from one book to another, with each author adding a number of juicy details. As such, an unseaworthy ship that sank during a heavy storm is slowly turned into an unsinkable ship that mysteriously disappears in a calm sea.

The most famous example is the story of Flight 19, the crew of which is brought home by a UFO in Steven Spielberg's 1977 box-office hit *Close Encounters of the Third Kind*. Bermuda Triangle books tell the story of experienced pilots flying out to sea and sending odd radio messages just before disappearing. The facts about this case, however, make explanations of the incidents rather mundane: inexperienced pilots, inaccurate navigation, broken compasses, bad weather conditions, and poor radio connections. The pilots got lost, ran out of fuel, and crash-landed in the sea. The heavy airplanes sank to the bottom within minutes.

A year after the book by Berlitz and Valentine was published, the complete and partial lies that had been copied from book to book over the years until they ended up in that 1974 publication were finally exposed in *The Bermuda Triangle Mystery—Solved* by Lawrence Kusche (1975). Kusche demonstrated that there is nothing wrong with that part of the sea. He indicated that there are no more

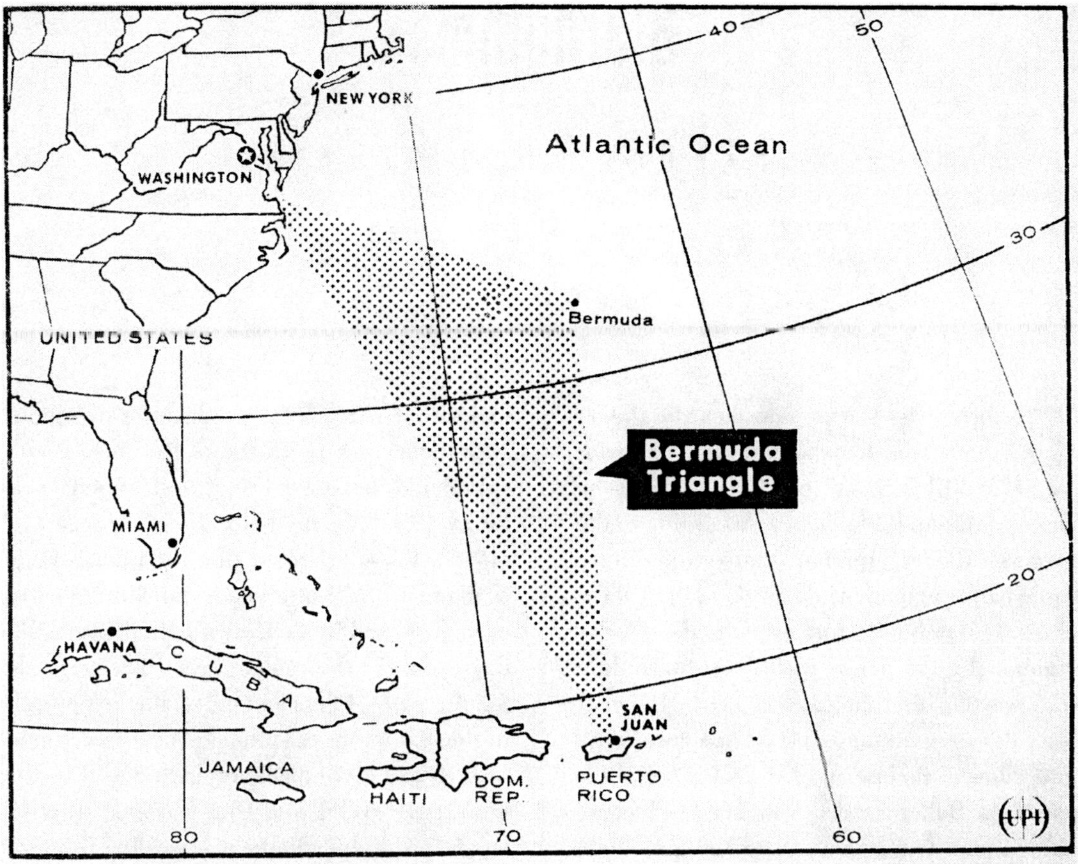

Map of the Bermuda Triangle. (Bettmann/CORBIS)

accidents there than in other heavily used sea routes and that all the exaggerated stories about mysterious disappearances were no more than products of the imaginations of a number of writers. Kusche's book is still held up as a classic in skeptical research.

Slowly, the subject was forgotten. Today, only occasionally does one hear about the Bermuda Triangle, even though ships and planes still encounter disasters in that region in the normal course of traversing the stormy Atlantic. In 1980, Berlitz presented information on some new "unexplainable" accidents, but it turned out that they were not so unexplainable at all, and only three of them occurred in the famous triangle. And in 1991, there was a stir when one of the hundred airplane wrecks near Fort Lauderdale, Florida, was thought to be the infamous 1945 Flight 19. But like so much that is said about the Bermuda Triangle, that report turned out to be false.

References:

Dennett, M. R. 1981. "Bermuda Triangle, 1981 Model." *Skeptical Inquirer* 6, no. 1: 42–52.

Kusche, Larry. 1975. *Bermuda Triangle Mystery—Solved.* Amherst, NY: Prometheus Books.

Randi, James. 1982. *Flim-Flam!* Amherst, NY: Prometheus Books.

Stein, G., ed. 1996. *The Encyclopedia of the Paranormal.* Amherst, NY: Prometheus Books.

Biorhythms

DIEGO GOLOMBEK

Biorhythms is a pseudoscientific theory that claims to explain human behavior and physiology according to exact cycles of different periodicities. According to the original theory, human biorhythms have three basic components: a 23-day physical cycle, a 33-day intellectual cycle, and a 28-day emotional cycle. These cycles constitute sine-like waveforms that cross a central "zero" line, thus generating "up" (above zero) days and "down" (below zero) days for the three variables. Believers of the theory say that we should plan our activities according to these different cycles, making sure that important activities are carried out on the so-called up days.

Biorhythms theory has no scientific value. It has been rigorously tested by different studies, none of which support any of its claims. Those studies that do support biorhythms have not been published in scientific journals, nor have they been conducted in ways that can be reproduced or contrasted scientifically, since they incur numerous methodological mistakes and/or fallacies. The most classical biorhythms theory combines the three basic cycles, but more recently, other periodical variables have been added, including a 38-day intuition cycle and a 53-day spiritual cycle.

Biorhythms theory was created almost simultaneously by two physicians. One of them was Wilhelm Fliess, a nose and throat specialist from Berlin who was a contemporary and close friend of Sigmund Freud (some scholars claim that part of the origins of psychoanalytic theory can be traced to the early correspondence between Freud and Fliess) (Sulloway 1993). In his book *The Course of Life* (1909), Fliess proposed that the human body is made up of both male and female cells, each governed by their own periodicity: a 28-day cycle for the female and a 23-day cycle for the male. Fliess expanded this hypothesis to the claim that emotional processes respond to a 28-day cycle (probably influenced by the menstrual cycle) and that physical (that is, "male") abilities fluctuate according to a 33-day cycle. Fliess came up with the idea that every number could be constructed from 28 and 33 according to a simple formula: $23x + 28y$, where x and y are positive or negative integers. However, Martin Gardner showed that "if any two positive integers that have no common divisor are substituted for 23 and 28 in his basic formula, it is possible to express any integer whatever" (Gardner 1981). At the same time, Hermann Swoboda, a professor of psychology at the University of Vienna, started to record the behavior of his patients and proposed that the same 23- and 28-day cycles ruled human activities. Swoboda published his findings in several books that were enormously popular at the beginning of the twentieth century, among them *Periods of Human Life* (1904) and *The Critical Days of Man* (1909), which included a slide rule to calculate critical days.

The 33-day intellectual cycle was proposed

by Alfred Teltscher, an Austrian engineer from the University of Innsbruck who gathered information about his students' academic performance in written tests. He claimed that good and bad notes taken by students in class corresponded to an exact cycle, suggesting that mental activity undergoes cyclic changes of this exact periodicity. According to the theory, the three rhythms start at birth and continue exactly throughout the entire life span of the individual. Simple mathematical programs allow one to chart the three cycles simultaneously, just by entering the birth date and the time span for which the biorhythm will be calculated. The graph below is an example of a biorhythm chart for an individual born on January 1, 2000.

The graph shows the simultaneous progression of the emotional, physical, and intellectual biorhythms. The middle "zero" line defines the "up" and "down" days of the different variables. The time points where the plots cross the zero line are "critical" days, with supposedly poor performance ratings for the variable under consideration. According to biorhythms theory, important activities (such as a physical or intellectual test or a marriage proposal) should not be scheduled for these critical days, since the odds of failure are high. The most critical day of all is the "triple critical"—the day when all three cycles are on the cross-zero line. "Double critical" days—when any two variables cross the zero line on the same day—are also to be avoided for important tasks or decision making.

Advocates of biorhythms theory claim that enormous amounts of data validate their findings. Among the most popular "proofs" of biorhythms are charts that demonstrate that Clark Gable had a heart attack on a "negative" physical day, Marilyn Monroe committed suicide on a "negative" emotional day, and Mark Spitz won seven Olympic medals during a period characterized by "high" emotional and physical days. Another concept incorporated in biorhythms theory is that of "compatibility," which represents the likelihood that two individual's rhythms will correspond or match. Compatibility ranges from 0–19 (very low) to 80–100 (very high), and it is offered as an explanation for how two people might get along.

None of the studies mentioned earlier in-

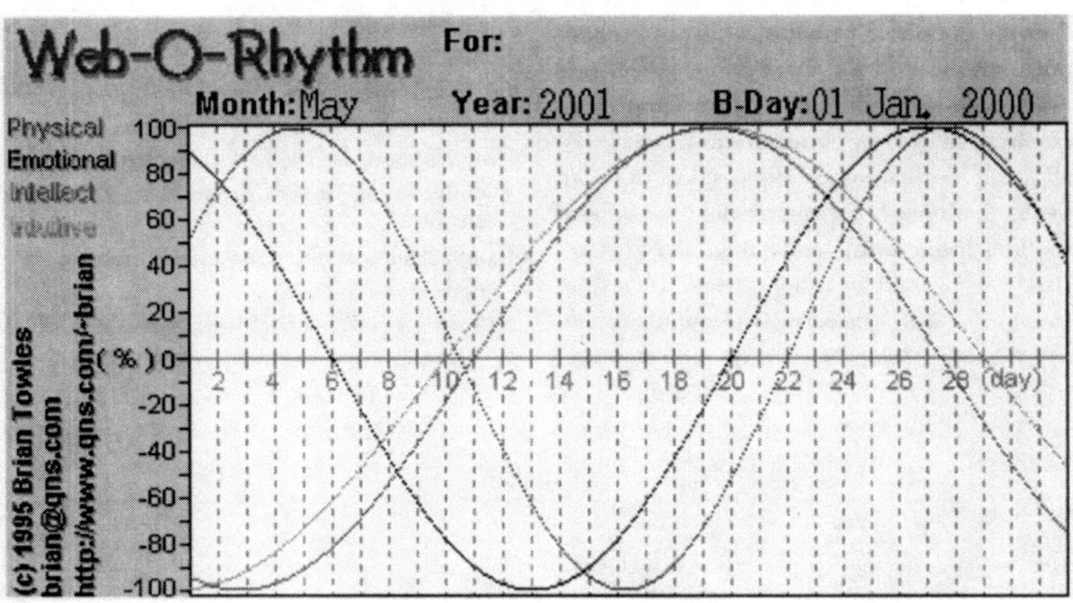

Biorhythm chart.

clude a rigorous and testable statistical analysis that can rule out subjective appraisal of behavior of simple probabilities. Metanalyses of biorhythms studies have demonstrated that the theory is not valid and is not supported by rigorous statistical testing (Hines 1998). Blind tests have failed completely to support biorhythms theory: volunteers have been found to report complete accordance with biorhythmic charts that they believed were their own when, in fact, the charts had been calculated for different persons (Randi 1982).

Biorhythm supporters sometimes confuse the basis of the theory with the scientific foundations of chronobiology, which is the study of biological clocks and rhythms (Aschoff 1981). Here, we will use the term *biological rhythms* for the cycles studied scientifically by chronobiology, not to be confounded with the three basic *biorhythms* discussed earlier. Endogenous clocks generate rhythms that are about 24 hours in duration (termed *circadian,* from the Latin *circa,* meaning "about," and *diem,* meaning "day"), which are entrained to exact 24-hour periods by environmental cycles such as the daily photoperiod. Other biological rhythms have circa-tidal periodicities (that is, they are entrained by wave cycles) or circannual periodocities (of about 365 days). However, biological rhythms differ from biorhythms in several fundamental ways: (1) although biorhythms are supposed to begin at birth, biological (in particular circadian) rhythms begin during embryonic life; (2) biorhythms respond to exact periodicities that cannot be easily traced to environmental cycles, whereas biological rhythms are not exact, vary between and within individuals, and tend to correlate with environmental changes; (3) biorhythms are based upon individual and subjective recordings of human behavior, but biological rhythms are universal in nature and are subject to objective recording under a variety of experimental conditions; and (4) although most biological rhythms (in particular, circadian cycles) have a clear anatomical site of origin, as well as physiological explanations, there are no indications of a possible endogenous correlate of biorhythms.

Biorhythms theory should continue to be considered a pseudoscientific theory because it cannot be tested scientifically and because all of its claims have been refuted by rigorous analysis. As is the case with astrology and other pseudoscientific approaches to human behavior, the relative strength of biorhythms theory is probably based on the subjective validation of advocates who rule their lives according to the ups and downs of their computer-generated graphs.

References:

Aschoff, Jurgen, ed. 1981. "Biological Rhythms." In *Handbook of Behavioral Neurobiology.* Vol. 4. New York: Plenum Press.

Gardner, Martin. 1981. *Science: Good, Bad and Bogus.* Buffalo, NY: Prometheus Books.

Hines, Terence M. 1998. "Comprehensive Review of Biorhythm Theory." *Psychological Reports* 83: 19–64.

Randi, James. 1982. *Flim-Flam!* Amherst, NY: Prometheus Books.

Sulloway, Frank. 1993. *Freud: Biologist of Mind.* Cambridge, MA: Harvard University Press.

Castaneda, Carlos

PHIL MOLE

Carlos Castaneda was one of the most popular authors in the alternative religious movement known as the New Age. Like many New Age authors of his generation, he combined ancient religious rites and beliefs with the countercultural attitudes of the late 1960s, especially regarding psychedelic drug use as a pathway to transcendence. Castaneda's books, now translated into twenty languages, teach the reality of an extraordinary world of sorcery and mysticism lying just beyond the borders of our everyday perceptions. The overwhelming success of these books suggests that many people find consolation in this message.

Not much is accurately known about Castaneda because he often falsified important details about his life. In the late 1960s and early 1970s, he repeatedly informed interviewers that he was born in 1935 in Brazil and that his uncle was Oswaldo Aranha, who was president of the UN General Assembly and an ambassador to the United States. He also claimed that he studied art in Italy, graduated from Hollywood High School, and served in the U.S. Army in Spain during wartime. However, Castaneda was actually born in 1925 in Peru and was unrelated to Aranha (Lindskoog 1993). He attended high school and art school in Lima and earned a reputation for telling tall tales. He married in 1951 but soon deserted his wife and child and moved to Los Angeles. Castaneda enrolled as a psychology major at Los Angeles City College and took courses in creative writing and journalism. He married a woman named Margaret Runyon while living in Los Angeles, but he lived with her for only six months and often disappeared for months at a time without an explanation (Gardner 1999, 13).

Castaneda entered the University of California at Los Angeles (UCLA) in 1959 and began research in the southwestern United States classifying the psychotropic plants used by sorcerers. According to Castaneda, he met an elderly Yaqui sorcerer while waiting at a bus stop in Arizona in 1960. This Yaqui, whom he referred to as "Don Juan," quickly befriended the young student. Apparently, Don Juan decided to share his secret Yaqui philosophical knowledge with his new friend after witnessing Castaneda and a dog urinate on each other. This event, to Don Juan, was a positive omen that Castaneda was worthy of the sacred Yaqui teachings (Castaneda 1968, 39–41). Don Juan then reportedly accepted him as an apprentice and began training him in the preparation and proper consumption of drugs such as peyote, jimsonweed, and hallucinogenic mushrooms.

In his 1968 book, *The Teachings of Don Juan: A Yaqui Way of Knowledge*, Castaneda described his thrilling, reluctant entry into the world of Yaqui sorcery. He revealed that he had used Yaqui sorcery to become a crow and talk to lizards and that he had encountered evil spirits in the desert. Although he concluded *The Teachings of Don Juan* with a confession that he was too terrified to pursue Yaqui mysticism further, Castaneda wrote nine other books over the next three decades. All of them contained more alleged descriptions of ancient Yaqui wisdom and argued that the human experience of reality is mostly a product of cultural consensus. Castaneda maintained that people who allowed a shaman such as Don Juan to shape their perceptions differently would see all of the magical forces at work behind the veil of ordinary reality and could become sorcerers themselves.

In 1972, UCLA awarded a doctorate in anthropology to Castaneda for a thesis based on his third book, *Journey to Ixtlan.* Soon after, other anthropologists carefully investigated his work and found it to contain an overwhelming number of inconsistencies and factual errors (Gardner 1999, 14). Specialists in Native American culture demonstrated that Castaneda's books contained no valid information about Yaqui beliefs and practices and that the author did not seem to know even one Yaqui name for native plants and animals after nine years of alleged research. He also claimed to find hallucinogenic mushrooms in areas where none grow and described the sensation of being drenched by warm winter rains in a desert where winter rains are extremely cold. These investigations proved Castaneda's books to be mostly fiction, although authors as respected as Joyce Carol Oates considered them fiction of reasonably high literary quality. Castaneda's only genuine act of sorcery, as one critic noted, was turning UCLA into an ass (Lindskoog 1993).

Some of Castaneda's defenders fondly argue that he never meant to promote his books as factual anthropological accounts but intended them to be taken as religious allegories disclosing a different and higher kind of truth. However, Castaneda very clearly defended the literal truth of his books in a 1972 interview. When asked about the factual basis of the Don Juan books, he replied, "The idea that I concocted a person like Don Juan is inconceivable. He is hardly the kind of figure my European intellectual tradition would have led me to invent. The truth is much stranger. I didn't create anything. I am only a reporter" (Noel 1976, 77).

Despite Castaneda's assertions to the contrary, an educated reader will be very impressed with how much Don Juan and Carlos Castaneda really are products of the Western intellectual tradition. Don Juan's insistence that the quest for power underlies all human action virtually quotes Friedrich Nietzsche, and his teachings about different cultural patterns of experiencing reality echo the cultural relativism of social scientists such as George Boas, Ruth Benedict, and Margaret Mead. The "structural analysis" of the logic of Don Juan's teachings in *The Teachings of Don Juan* also reveals a familiarity with the work of anthropologist Claude Levi-Strauss.

In 1993, Castaneda introduced a system of meditation and bodily movements known as Tensegrity and claimed it was the culmination of centuries of Native American mystical wisdom. Unfortunately, the magical powers of Tensegrity could not prevent Castaneda's death in 1998, at the age of seventy-two. Even in death, Castaneda's influence on the New Age movement remains considerable, and skeptics will most likely contend with his legacy for many years to come.

References:

Brown, Mick. 1998. "Shaman or Sham?" *London Daily Telegraph,* August 1.

Castaneda, Carlos. 1968. *The Teachings of Don Juan: A Yaqui Way of Knowledge.* New York: Ballantine Books.

———. 1971. *A Separate Reality: Further Conversations with Don Juan.* New York: Pocket Books.

———. 1972. *Journey to Ixtlan: The Lessons of Don Juan.* New York: Pocket Books.

Cohen, Hal. 1998. "Shaman or Sham?" *Lingua Franca* 8, no. 6 (September): 22–24.

De Mille, Richard. 1976. *Castaneda's Journey: The Power and the Allegory*. Santa Barbara, CA: Capra Press.

De Mille, Richard, ed. 1980. *The Don Juan Papers: Further Castaneda Controversies*. Santa Barbara, CA: Ross-Erickson.

Fikes, Jay C. 1996. *Carlos Castaneda, Academic Opportunism, and the Psychedelic Sixties*. Victoria, British Columbia: Millennia Press.

Gardner, Martin. 1999. "Carlos Castaneda and New Age Anthropology." *Skeptical Inquirer* 23, no. 5: 13–15.

Lindskoog, Kathryn. 1993. *Fakes, Frauds & Other Malarkey: 301 Amazing Stories and How Not to Be Fooled*. Grand Rapids, MI: Zondervan Publishing House.

Noel, Daniel C., ed. 1976. *Seeing Castaneda: Reactions to the "Don Juan" Writings of Carlos Castaneda*. New York: G. P. Putnam's Sons.

Clever Hans

THOMAS F. SAWYER

Clever Hans was a horse that purportedly could do mathematics and perform other intelligent acts. Hans was owned by Wilhelm von Osten, a former mathematics teacher who firmly believed horses to be capable of feats of high intelligence. However, skeptics discovered the phenomenon of Clever Hans and his accomplishments was an interesting case of observer bias affecting the effect being examined.

During the early part of the 1900s, von Osten exhibited Hans throughout Germany to large and enthusiastic audiences. Never charging a fee for such displays, he would ask the horse questions such as, "If the first of the month is a Friday, what is the date of the following Tuesday?" and in answer, Hans would tap the appropriate response with his hoof.

Skeptical of the true nature of Hans's abilities, the German Board of Education appointed a commission to investigate the animal. The commission was headed by German psychologist Carl Stumpf and included, among others, the director of the Berlin Zoo, a veterinarian, a cavalry officer, and a circus manager. The preliminary report concluded that Hans's achievements were not the result of deliberate trickery or influence by his owner.

Oskar Pfungst, an assistant to Stumpf, continued the investigation of Clever Hans in a more controlled environment. One variable Pfungst manipulated was whether the questioner was aware of the correct answer to the questions posed to the horse. When the questioner did know the answer, the horse usually responded correctly, but when the correct response was unknown, Hans almost always failed. In addition, Pfungst discovered the importance of visual cues, as in order to answer correctly, Hans had to be able to see the questioner as he tapped his answer. If the questioner moved out of the animal's visual field, Hans would struggle to regain sight of him. Further observation by Pfungst suggested that Hans was detecting and responding to unintentional movements of the questioner, such as slight postural adjustments or upward movements of the eyebrows and head as Hans approached the correct number of taps.

Once this was discovered, Pfungst was able to ask a question and influence Hans to make almost any response by voluntarily producing the appropriate movements himself. To further confirm the impact of such observer bias, Pfungst put himself in the place of Hans. He brought several persons into the laboratory, hooked them up to an apparatus that would monitor head movements, and had them put questions to him in a manner comparable to the way they had been posed to Hans. As Pfungst tapped out the answers to a question, he watched for movements comparable to those made by persons questioning Clever Hans. In the vast majority of cases, the participants made involuntary movements that coincided with the point at which Pfungst's tapping was supposed to cease.

The century-old case of Clever Hans illus-

Woodcut shows horse performing from brochure published in 1595. (Fortean Picture Library)

trates the fact that unintentional actions by observers may bias or affect the behavior of participants in psychological research. More recently, it has been suggested that the phenomenon of facilitated communication (FC) results from similar observer effects. FC is performed with autistic and other developmentally disabled children, and it consists of a child typing his or her thoughts or responses to questions on a keyboard while the "facilitator" touches the child's hand, arm, or shoulder. Such children, previously thought to be noncommunicative, seem to function at normal or above-normal levels in such situations, doing high-level mathematics and even typing poetry. However, in controlled situations, FC was found to result from unintentional facilitator control of the typing, which is a form of observer bias. Just as Pfungst found with Clever Hans, researchers have demonstrated that children are unable to answer very simple questions or identify objects shown to them via FC unless their facilitators are aware of the correct response.

References:

Hothersall, D. 1995. *History of Psychology*. New York: McGraw-Hill.

Montee, B. B., R. G. Miltenberger, and D. Wittrock. 1995. "An Experimental Analysis of Facilitated Communication." *Journal of Applied Behavior Analysis* 28: 189–200.

"Oscar Pfungst: Clever Hans (The Horse of Mr. Von Osten)." *Thoemmes Press*. URL: http://www.thoemmes.com/psych/psy/pfungst.htm. (Accessed on May 1, 2001).

Pfungst, O. 1911. *Clever Hans*. New York: Henry Holt.

Cold Reading

BOB STEINER

Cold reading is the method used by mystical readers to talk to customers whom they have never met and about whom they have no advanced information. Part of the readers' job is to dazzle their customers with how much they apparently know about them. That skill lends enormous credibility to the advice given by the readers. Cold reading has many manifestations, including (but not limited to) channeling (talking with the dead or other spirits), astrology, psychic reading, tarot reading, crystal ball gazing, palmistry, and tea leaf reading.

Generally, the explanation of the cold readers' methods analyzes the techniques used from the perspective of an outside investigator looking in, in order to determine the what and how of the readings. Here, the claims will be examined from the inside looking out.

Imagine that you are a reader. Choose your discipline: you can be a crystal ball gazer, psychic, astrologer, Ouija board expert, or whatever you like. You are bright, with a good sense of humor. And you must appear to be self-assured, for your customer must have confidence in you. "Confidence" is the name of the game, whence comes the term *confidence game*, frequently shortened to *con game*. Unless you appear self-assured and self-confident, you will not be able to sell your advice.

Next, add a liberal sprinkling of charisma. And, oh yes, you are unethical, virtually without conscience or morals. Why do people come to you and pay for your conversation? Sometimes, customers (call them "clients"—it sounds much more professional) come for advice. Often, they just want someone to talk to. And many times, they simply want your permission to do what they want to do themselves. In all cases, they want you to impress them with your insight. There are three basic reasons why clients come to you:

1. They seek advice. Some people really do want your input. But often, they just want you to agree with them, in the guise of giving advice.

2. They want someone to talk to. Many people simply want someone to listen to them. You are a trusted, "unbiased," intelligent listener. They lack that at home. Perhaps the discussion concerns the client's spouse. The client cannot dump this stuff on coworkers—that could backfire in the future. Where can your client find someone to trust, someone who is impartial, intelligent, compassionate, understanding, and more? The psychic!

3. They are looking for permission. Take, for example, Mary, a woman in her midfifties. She was recently widowed, after thirty-three years of marriage. She and her husband truly loved each other. They loved their children, and that love was returned in kind. They had a wonderful family. Friends said that theirs was a marriage made in Heaven. Mary tells you about the sickness and death of her husband, Joe. It was a tough time. As you listen and watch, you observe that the pupils in Mary's eyes contract as she talks about Joe's

death. That subtle change, augmented by a slight squinting, strikes you as being considerably different from the joyous look in her eyes when she was telling about the good times with Joe. Mary's children are grown, married, and have their own lives to live. Yes, they love their mother, and they keep in touch. But they have young children, job and money pressures, and the goal of keeping marriage and family together. Mary has been living alone since Joe died. She has been dreadfully lonely. You observe that it seems to be a relief for her to have someone on whom to unload this entire story. You ask how long she has been widowed. She looks at her watch and replies, "Ten months, four days, three hours, and fourteen minutes." She smiles faintly as she adds, "But who's counting?" It is clear that Mary's thoughts dwell on her beloved late husband—all day and night, every day and night. You remark that she has trouble occupying her time with any degree of comfort. Mary is amazed by your insight! She tells you that she has a job. She has many good friends with whom she visits. But when you ask if she dates, she averts her eyes and pulls away from you. Before she has a chance to answer, you jump in with, "Of course you have not yet dated." Again, Mary is astonished by your "extrasensory" perception. She coughs and tries to calm herself. She has trouble with the next few words, and then her more recent story comes pouring out. Pay attention! You will soon learn why Mary came to see you.

Last Saturday night, Mary went to a party at the home of dear friends. There she met George. He is fifty-eight years old. She has trouble saying these words about a man other than her late husband, but she tells you that George is attractive, bright, alert, and sparkling. He has been widowed for fourteen months. He loved his late wife. At this, Mary chokes up. You ask if she would like something to drink. You have learned that alcohol often loosens up your clients and allows them to talk more freely. "Just orange juice, if you have any," Mary replies. Between sips of orange juice, she opens up some more. It is obvious that she likes you and trusts you. She tells you that the rapport with George was wonderful. And instantaneous! They even laughed together! It has been a long time since Mary had someone to laugh with. You think you detect a blush as Mary says, "George asked me out to dinner. I told him that I would have to think about it." As you listen and watch, you see that the pupils in Mary's eyes have dilated as she talks about George. Translation: George is a happy topic for Mary.

Why did Mary come to see a psychic? What does she want to hear? Her meeting with George gave her a warm feeling, the likes of which she has not felt in over ten months. But she also feels a tinge of guilt. Would Joe approve? Would she be disrespecting his memory if she were to date another man? Would it lead to sex? Would that be all right? And now we come to the number one rule for a psychic reader: tell 'em what they want to hear. Mary has come to you, a psychic, for permission. She wants your validation, approval, and encouragement to go on a date with George. She doesn't realize that is why she came to you. But you know that is true. Try this: "It is clear that you and Joe truly loved each other. Let us assume, just for a moment, that you had died and that Joe remained alive. You are looking down from Heaven, and you see how lonely he is. He is still being faithful to you." The mere mention of God or Heaven lends a supernatural aura to your reading. It makes your words seem to be larger than life, even holy. You continue: "Surely Joe would not go out on a date one week after you passed away. But now you look down and see that, almost a year after you have passed on [avoid the words *died* and *death*], Joe has met a woman who is really nice. He could actually find some happiness with her." You pause, then add: "While you were alive, Joe was true and faithful to you.

What would you want him to do now, in the example I just gave you?" As soon as you started your hypothetical example, tears came to Mary's eyes. Now the tears are streaming down her cheeks. You pass a box of tissues across the table. She wipes her eyes, blows her nose, and replies, "I would want him to ask this new woman out on a date."

You immediately state, "And that is exactly what Joe wants for you and George!" Mary's tears become a waterfall. But this time, they are tears of relief. You have confidently stated as fact what Mary came to hear! Sure that you have read the situation correctly, you continue, "Mary, as soon as you get home, call George and accept his invitation." Yes, Mary consulted a psychic in order to ask for and receive permission to go out to dinner with George.

Before you get all choked up and start to argue that you did a real service for Mary and that you gave her correct advice—and what was so wrong with the psychic in this case?—allow me to show you what is wrong with this picture by describing another example.

Your next client is Charles. He is moderately happy in his marriage. Soon, he will be going to a business conference, accompanied by an attractive female coworker named Darlin. She is also married. Charles tells you that he and Darlin have excellent rapport and easily laugh with each other at work. They enjoy each other's company when they go out to lunch together—usually two or three times a week. This is the first time that they will be alone together. Coincidentally, they have adjoining rooms. Who arranged that coincidence? you wonder to yourself. Your client assures you that if he were to sleep with Darlin on the trip, neither his wife nor her husband would ever find out. He tells you that he believes that the variety and change of pace would actually strengthen his marriage. Every time he says the name Darlin, he smiles. Remember the rule: tell 'em what they want to hear.

Why did Charles come to you? Do you want to keep him as a client? Remember the character traits you possess: you are unethical, virtually without conscience or morals. You are certain that Charles came to you for permission. What do you tell him?

We now come to realize that the fact that you gave Mary good, wholesome, correct advice was merely a coincidence! The rule is not that you should give correct advice. Rather, the rule is to tell 'em what they want to hear. And that, dear friends, is the difference between a good friend, a good psychologist, or a good adviser of any kind and an unethical charlatan.

And now for the crucial question: why can't a psychic just give what he or she believes to be the correct advice all the time? A further example will demonstrate the problem inherent in that scenario.

Now you are a palm reader. A young man comes to you. Although he appears to be happy and healthy, the heart line in his hand is unusually short, indicating that he will die young. What do you tell him? If you tell him that he will die prematurely, this will doubtless be the last time you will see him; he will find a more understanding palm reader. If you tell him that he will have a long, healthy life, you are going against what his palm tells you. Oh! What's that? You don't believe that the palm really tells such things. Then you are a liar and a fraud. But then, so is the self-proclaimed psychic who says but does not believe that the advice he or she offers comes from vibrations (vibes). Welcome to the mystical world of successful readers!

You must be prepared to draw out the person who comes to you for psychic advice. A simple statement such as "I see a disagreement with a woman or girl in your life" will cause your male client to search for and find a fit. It might involve his wife, mother, sister, daughter, coworker, bank teller, or even a disagreeable clerk at a store. Your statement was broad enough that he will probably find someone who fits the statement. What topics might you introduce? The following mnemonic will serve as a memory jogger. Think of a hemmed skirt with a slit down the side. SLIT HEMMED:

Sex
Love (relationships)
Interests (sports, hobbies, politics, religion)
Travel

Health (careful—don't make medical diagnoses or give medical advice)
Exercise
Money
Move (I see a move in your future)
Employment (job, change of jobs, promotion, your boss doesn't appreciate you)
Diet

Study your client. If the person is slightly overweight, try: "I see that you started a wonderful exercise program. But you didn't follow through, as you know you should have." That statement applies to almost everyone, of course, but your client will give you credit for having picked that up from him or her.

Does this technique work? Many claim that it does. The words of clinical psychologist Terry Sandbek are appropriate:

> We have a discipline that can test the validity of claims. That discipline is called *science.* Through science we can devise tests and standards. Physicians must pass medical boards; psychologists must pass written and oral examinations; hospitals are regulated by strict standards of the Joint Commission on Hospital Accreditation; drug companies can offer chemicals to the public that have passed rigid scientific scrutiny. It would not be difficult to devise scientific procedures and standards to test the validity of the claims of self-proclaimed psychics, astrologers, and the like. If the person cannot pass the test, or refuses to be tested, the business falls into the category of *profiteering from the suffering of others.*

Crop Circles

JORGE SOTO

The term *crop circle* refers to a geometric or otherwise orderly design that has been formed when wheat or other crops have been crushed by swirling or stumping motions. Such designs tend to appear overnight, and therefore, how they are formed is a matter of controversy. Crop circles have occurred in Canada, the United States, Japan, India, Australia, Denmark, Holland, Siberia, the Czech Republic, Finland, England, and other parts of the world. Crop circles have been reported since the 1960s, but their numbers rose dramatically in the 1980s and the 1990s. However, they seem to prevail in the Wiltshire district in southern England. Since the late 1970s, approximately 8,000 crop circles have been reported, the majority of which are in England. People who attempt to study these patterns (which are usually circular) have coined a name for themselves—*cereologists,* from the name of Ceres, the Roman goddess of vegetation. Various theories (Oxlade 1999) have been advanced to explain the formation of crop circles, ranging from plasma vortex phenomena to unidentified flying objects (UFOs) to simple hoaxes.

The plasma vortex hypothesis proposes that some sort of swirling air current, similar to a tornado, leaves an imprint on these crops as the rotation of the vortex bends the stems to the ground. It is alleged that a spinning mass of air, which has accumulated a significant fraction of electrically charged matter, slams on the crop, creating a circle as it crushes the stems of the plants. This proposal is weak, since it is hard to conceive that a swirling air current will "drop" at the precise points necessary to generate an orderly geometric design on the ground. (Only the very first crop circles were simple circles. Over time, they became very complicated, with a multitude of geometric patterns.) Moreover, many other patterns, such as the crop "star," could not be originated by this process (see figure below).

A second theory assumes that crop circles are created by UFOs. Proponents of this theory note that crop circles occasionally seem to appear in conjunction with a UFO sighting. Some of the early, simple (single) crop circles certainly do suggest that the crop stems could have been flattened by a flying saucer landing on the field. The "rotating" pattern of the crop circle could be explained by the spinning of the saucer. As the circles have become

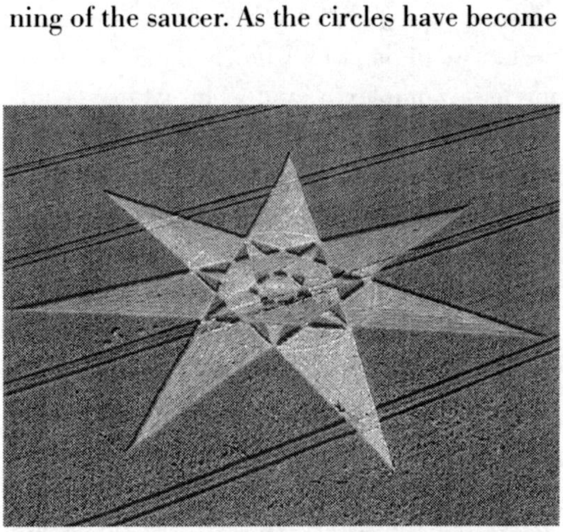

A crop "star." (Copyright Steve Alexander, 1999)

Videotape image of a complex crop circle and pi. (Copyright Peter Sørensen, 1999)

more complex in shape, though, proponents of the UFO theory have had to modify their ideas, suggesting that the marks left are due to a strange effect of the craft's drive force on the plants. Others even argue that the shapes are messages purposefully left by the saucer's crew. The photograph shown above is intriguing in its complexity, and a clue of its origin is seen on the bottom right of the photograph, where the Greek letter π (pi) is "imprinted" on the ground (Sørensen 1999).

Many of these crop circles are complex mathematical descriptions. Because the term *crop circle* seems rather inappropriate for the majority of the latest discoveries, anything that is more than a single circle has been generically named a *crop circle formation*. This term and other terms such as *pictogram* (a picture or symbol that represents a word or phrase) and *agriglyph* (a complex shape or form of flattened crops) are used to try to convey tangible descriptors of the physical appearance of the phenomena that can be found in so many forms.

Some people suggest that crop circles and crop circle formations are an indication of supernatural forces and that the areas where they are located show particularly high levels of electromagnetic fields (see *Crop Circle Central* Web site). However, a number of people have claimed to be the creators of such designs (see *CircleMakers* Web site). Indeed, some creators of these designs have even been prosecuted in recent years (see *Wiltshire Archive Files* Web site). And with the technological advances in computer imaging, pictures of crop circles can even be "virtually" (digitally) faked (see *Eye Wire* Web site).

Yet another claim (see *The Crop Circular* Web site) is that in genuine formations, the

A circle on ice. (Copyright Lyn Winer, 2001)

stems are not broken but are bent, normally about an inch off the ground at the plants' first nodes. According to the source of this claim, the only method capable of producing such an effect would employ microwaves or ultrasound. No further discussion is offered as to the validity of this claim and no experimental proof is given, so one is forced to accept it on faith alone. This source also proposes, with no specific information about the claim, that there is "an increase of infrared output within and around a new formation, indicating that both the heat content of the plants and the watershed have been affected" (*The Crop Circular* Web site). The source even goes so far as to claim that there is evidence of four non–naturally occurring, short-life (few hours) radioactive isotopes, but these isotopes are not further described in any way.

An unusual twist was reported (see *Crop Circle Research* Web site) about a circle on ice, as shown above. The formation was reportedly found in Churchville, Maryland, in February 2001. The ring was approximately 30 feet in diameter, making it also somewhat larger than the average size of most other ice rings reported to date (which range from 15 to 20 feet). Such phenomena obviously require a much different explanation than that of the classical crop circles.

There is no way that anyone can definitively differentiate between a hoaxed and a "genuine" circle. However, several pieces of information support the complete hoax theory. To begin with, there is a lack of historical precedent for crop circles. Curiously, the number and complexity of the circles have grown in proportion to the media coverage of the phenomenon, which strongly suggests that the originator is more interested in making circles if they make the news. Further, the increased complexity of the crop circles over time implies a "thinking behind the machine." An intriguing characteristic of the phenomenon is that all of the circles are formed at night without detection. In fact, no credible reports have been filed of a circle being made in the presence of any witnesses. Why is it that this is a nocturnal phenomenon? Taking into consideration these facts, it is more sensible to explain the phenomenon as a work of hoaxers. Of course, if solid evidence that the formation of the circles occurs in some other manner, it should be considered and analyzed to determine the real nature of this amusing phenomenon.

References:

CircleMakers. URL: http://www.circlemakers.org. (Accessed in January 2002). (Formation made by Matthew Williams.)

Crop Circle Central. URL: http://www.paradixmshift.com/ccc.html. (Accessed in January 2002).

Crop Circle Research. URL: http://www.cropcircleresearch.com/millennium/icering.html. (Accessed in January 2002).

"Crop Circles Revealed." *Eye Wire*. URL: http://www.evewire.com/magazine/features/crop secrets/. (Accessed in January 2002).

The Crop Circular. URL: http://www.home.clara.net/lovely/education.html. (Accessed in January 2002).

Oxlade, Chris. 1999. *The Mystery of Crop Circles*. Hong Kong: Heinemann Library.

Sorensen, Peter. "The Last Crop Formation of the Millennium." URL: http://www.greatdreams.com/southfld.htm. (Accessed in January 2002).

Wiltshire-Archive Files. URL: http://www.thisiswiltshire.co.uk/wiltshire/archive/2000/11/07/wilts_news 1 ZM.html. (Accessed in January 2002).

Cryptozoology

BEN S. ROESCH AND JOHN L. MOORE

Cryptozoology is the study of animals whose existence has not been proven. Its practitioners, including amateurs and professional zoologists, search for both well-known monsters, such as the Loch Ness monster and bigfoot, and smaller, lesser-known animals. Cryptozoology ranges from pseudoscientific to useful and interesting, depending on how it is practiced.

What Is Cryptozoology?

The word *cryptozoology* was coined by Belgian zoologist Bernard Heuvelmans in the late 1950s and literally means the study of hidden animals. The evidence for these "hidden animals," or cryptids, is not sufficient to allow all (or even most) zoologists to accept their existence, but there is some evidence (such as eyewitness accounts, legends, or footprints) that suggests they do exist. Furthermore, Heuvelmans said that, to be a cryptid, an animal must potentially be either a new species or a new subspecies. Others have extended the meaning of *cryptid* to include supposedly extinct animals that may still survive and animals that possibly exist far outside their known ranges. According to Heuvelmans, the task of cryptozoology is to derive a thorough account of a cryptid's appearance and habits from whatever evidence exists for it, allowing anyone to visit its habitat and verify the animal's existence. With proof that it exists, the organism ceases to be a cryptid, and the animal is moved from the world of cryptozoology into zoological knowledge.

People have always been interested in reports of unusual animals, and sometimes these were studied as part of natural history or zoology, notably by such scholars as A. C. Oudemans, Willy Ley, Ingo Krumbiegel, and Ivan T. Sanderson. Cryptozoology was not organized as a separate field until the 1950s, when Heuvelmans coined the word and wrote an important book on terrestrial cryptids, *On the Track of Unknown Animals;* he later covered marine cryptids in *In the Wake of the Sea-Serpents*. In 1982, a group of scientists founded the International Society of Cryptozoology, with Heuvelmans as president and Roy P. Mackal as vice-president; it publishes the journal *Cryptozoology*. Other contemporary cryptozoologists include Loren Coleman, Michel Raynal, and Karl P. N. Shuker.

Although Heuvelmans and many of the individuals associated with the International Society of Cryptozoology are professional scientists and view their work as science, many other scientists think that the field is a pseudoscience. Papers on the topic are rarely published in scientific journals, no formal education on the subject is available, and no scientists are employed to study cryptozoology. Consequently, much cryptozoological work is done by amateurs. They vary from those who accept every claim, regardless of its

Cornish sea monster "Morgawr," allegedly photographed during early February 1976 from Rosemullion Head near Falmouth, Cornwall. (Fortean Picture Library)

improbability, to those who carefully and skeptically investigate cryptozoological assertions.

Cryptids

Cryptozoologists have described at least 250 cryptids from around the world, ranging from the famous "monsters" that receive much attention to lesser-known and sometimes more zoologically feasible animals. Accounts of some cryptids, both well-known and unfamiliar, follow.

Lake Monsters

Lake-monster legends are widespread, and reports of unusual creatures are relatively common not only from such places as Loch Ness, Lake Champlain, and Lake Okanagan (the purported homes of Nessie, Champ, and Ogopogo, respectively) but also in many other lakes and rivers worldwide. Lake monsters are often described as serpentine animals with humps, long necks, and horselike heads; some are said to be similar to overturned boats. Among the identifications that have been offered by cryptozoologists are surviving plesiosaurs, surviving basilosaurids (early long-bodied whales), long-necked seals, giant eels, and giant amphibians. Skeptics argue that sightings of lake monsters are the result of misidentifications of known animals (e.g., deer, waterbirds, eels, sturgeons, and seals) and physical phenomena (e.g., odd waves and optical effects). Some photographs and films supposedly show lake monsters, but though sometimes interesting, these are usually inconclusive.

Sea Serpents

Tales of sea serpents have long been popular, and common cryptozoological explanations for the mystery include surviving plesiosaurs, giant eels, and huge long-necked seals. In his 1968 book *In the Wake of the Sea-Serpents,* Heuvelmans collected nearly 600 sightings of unidentified marine animals since the seventeenth century. He proposed that there are nine types of sea serpents, including living basilosaurids, giant eels, and a long-necked seal. Some have criticized his classification as too speculative, but it is embraced with reservations by other cryptozoologists. Skeptics suggest that sea serpent reports are the result of misidentifications of known animals, including whales, seals, oarfish, and giant squid. Carcasses resembling sea serpents have frequently washed up on beaches around the world, but when they are investigated, they invariably turn out to be the decaying remains of sharks, whales, or other known animals. Nevertheless, the oceans are underexplored, and of all major cryptids, perhaps the most plausible are sea serpents.

Ape photographed by Francis de Loys in Venezuela/Colombia area, 1929. (Fortean Picture Library)

Bigfoot and Similar Creatures

Stories of hairy, apelike humanoids are found across the globe. North America's bigfoot is the most famous, and others include yeti in the Himalayas, almas in southern Russia, wildman in China, orang-pendek in Malaysia, agogwe in Zimbabwe, and didi in Guyana. Based on apparent physical differences among these creatures, some cryptozoologists have proposed that several species of large, bipedal, nonhuman primates exist. Various fossil taxa have been resurrected to explain these reports, including *Gigantopithecus* (a giant ape that lived several hundred thousand years ago), *Australopithecus,* and Neanderthals (*Homo neanderthalensis*). Skeptics, by contrast, argue that these cryptids are based on misidentifications and exaggerated tales of known animals such as bears and known primates. Tracks and hair samples from supposed "ape-men" have been collected, but they are inconclusive.

Living Dinosaurs

Living dinosaurs—or at least animals that resemble living dinosaurs—are another popular quarry of cryptozoologists and are reported most commonly from tropical areas, particularly in Africa. A famous example is mokele-mbembe, which is purported to haunt swamps and lakes in tropical West Africa. It is described by local people as an amphibious vegetarian (whose favorite food is the fruit of the malombo vine) about the size of a hippopotamus but with a long neck and tail. When shown illustrations of various extinct and living animals, locals reportedly usually pick out sauropod dinosaurs as bearing the closest resemblance to mokele-mbembe. Numerous expeditions have been mounted in search of this cryptid, but little more than anecdotal evidence has been obtained.

Giant Octopus

In 1896, a large, decaying carcass washed ashore near Saint Augustine, Florida, which a local physician thought was the remains of a giant octopus. When zoologist A. E. Verrill heard of the find, he named the creature *Octopus giganteus.* Upon examination of tissue samples from the animal, however, he decided it was actually the remains of the head of a sperm whale (*Physeter macrocephalus*). In 1971, a study of the microscopic structure of preserved samples from the carcass concluded that it probably was part of an octopus. A later biochemical analysis supported this identification, and proponents also cited reports and legends from the Caribbean of a similar animal known as the lusca. In 1995, a new study of the samples via both microscopic and biochemical analyses concluded that they probably came from a whale. Others argued that the matter was inconclusive, pending further studies.

Cryptophidion

In 1994, zoologists Van Wallach and Gwilym S. Jones published three photographs of a snake that was collected in central Vietnam in 1968 by U.S. Naval Medical Research Unit No. 2. Unfortunately, the specimen was later lost. It appears to have been a small, burrowing snake, probably belonging to the family Colubridae (a large group comprising over 60 percent of all known snakes, including such familiar kinds as garter snakes and king snakes). The characteristics of the snake in the photos did not match those of any known species, and Wallach and Jones named it the Vietnamese sharp-nosed snake, *Cryptophidion annamense.*

Leguatia

In 1708, the French traveler François Leguat published a description of a large, long-necked bird he had seen on Mauritius (an island in the southwestern Indian Ocean famous for its extinct birds, including the dodo) and included an illustration of a bizarre rail-like bird. This account was used by later scholars to support the idea that a giant rail, which was named *Leguatia gigantea,* had formerly lived there. The bird Leguat saw was probably a flamingo (*Phoenicopterus ruber*), however; later research showed that he had copied the illustration from a print made by Dutch artist Adriaen Collaert over 100 years earlier. Collaert's bird remains a mystery; similar birds are included in the work of a seventeenth-century English artist and a nineteenth-century Japanese artist. It has been suggested that the animal inhabited China (there is no reason to think it actually lived on Mauritius).

Cryptozoological Evidence

Few will dispute that most cryptozoological evidence consists of eyewitness testimony and folklore. Unfortunately, traditional accounts can be inaccurate, and eyewitness testimony is poor evidence. Humans are imperfect observers who can easily mistake known animals or physical phenomena for supposed cryptids or poorly recall the details of a sighting. This reality is especially relevant to cryptozoology, for sightings are often of short duration, occur unpredictably in uncontrolled settings, and are made by untrained observers. Also, the credence one puts in an eyewitness is arbitrary, and the possibility of a hoax is always present. These points indicate that cryptozoologists must be careful in their interpretations of cryptozoological evidence; unfortunately, as will be discussed later, this is not always the case.

Other than anecdotal evidence, there is occasional physical evidence for some cryptids, such as hair samples and tracks, but the analysis of such data is usually inconclusive. Film and photographs of alleged cryptids also exist, but they are typically not clear enough to be good evidence. Another problem with images of cryptids is that they may be hoaxes. For example, a photo of an unidentified primate killed in Venezuela was used by French anthropologist George Montandon in 1929 to coin the name *Ameranthropoides loysi* for a new species of ape (the first from the New World), but it has since been identified as a spider monkey (*Ateles belzebuth*). Today, with the availability of advanced photoimaging computer programs and digital effects, the ability to create convincing hoaxes is even easier. Images of cryptids will probably never be able to prove their existence, perhaps excepting remarkably detailed video or photographs.

Successes and New Species

A major problem for cryptozoology is its general lack of success. None of the cryptids discussed in Heuvelmans's 1958 book, for example, have yet been confirmed, and none of the major mysteries—bigfoot, sea serpents, and lake monsters—have been resolved. Many new animals are discovered every year, however. Most of these are small and unspectacular—and not cryptozoological, since they usually weren't known from eyewitness testimony, folklore, or physical traces. Nevertheless, large or unusual animals are discovered fairly often. These include the following:

Okapi

The okapi (*Okapia johnstoni*), a short-necked relative of the giraffe, was named as a new species in 1901. Several travelers had previously reported that a donkeylike animal was known to people in central Africa, and a French army officer had actually seen a live okapi. Aware of some of these reports, Sir Harry Johnston decided to investigate, and he was soon sent a skull and the complete skin of an okapi.

Coelacanths

Although it had been believed that coelacanths were long extinct (fossils are found until the late Cretaceous, with only a few less certain records since), a live individual was collected by fishermen off the coast of South Africa in 1938. Named *Latimeria chalumnae*, it is a blue, deepwater fish up to 6 feet long primarily found around the Comoros. In 1998, a second (although quite similar) species, now named *Latimeria menadoensis*, was found off Sulawesi in Indonesia. Both populations were known to some local fishermen. On the basis of reported captures and purported artistic representations, cryptozoologists have argued that coelacanths are found elsewhere in the world, notably the Gulf of Mexico.

Megamouth Shark

The first specimen of megamouth shark (*Megachasma pelagios*) was discovered entangled in a U.S. Navy ship's anchor off Hawaii in 1976. This deep-sea, planktivorous shark, which can grow to over 16 feet in length, was a completely unexpected find. Since the initial discovery, seventeen more megamouths have been found around the world.

Giant Gecko

Although *Hoplodactylus delcourti* is the largest known gecko, at 2 feet in length, it was not recognized until the 1980s, when the only known specimen was found in a natural history museum in Marseille. Unfortunately, no records indicated when or where it was obtained, but it was probably collected in the

1800s in New Zealand, where all its close relatives are found. Additionally, the native Maori described a similar lizard, the kawekaweau, to early explorers. Although some suggest the giant gecko is still alive, searches for it have been unsuccessful.

New Beaked Whales

The beaked whales, a group of deep-diving toothed whales, are probably the least-known large mammals, and two new species were described recently. The pygmy beaked whale (*Mesoplodon peruvianus*) was named in 1991 and is known from a few specimens (up to 13 feet long) stranded or captured off Peru, Mexico, and New Zealand. Bahamonde's beaked whale (*Mesoplodon bahamondi*) was named in 1997 on the basis of a single skull, collected on an island off Chile, from an animal estimated to be 16 to 18 feet long. Another skull from this species has been found in New Zealand. An unidentified beaked whale seen many times in the eastern tropical Pacific may be a pygmy beaked whale. An additional new species of beaked whales is known but has not yet been described scientifically.

Saola and New Muntjacs

The saola or Vu Quang ox (*Pseudoryx nghetinhensis*) was discovered in Vietnam in 1992. It is a member of the group that includes cattle, buffaloes, and bison, although it is smaller and has long, spindle-shaped horns. In the same poorly explored region along the border of Vietnam and Laos, two species of muntjac (a kind of small deer with simple horns) have recently been discovered: the giant muntjac (*Megamuntiacus vuquangensis*) and the Truongson muntjac (*Muntiacus truongsonensis*). The leaf muntjac (*Muntiacus putaoensis*), the smallest known species of deer, was discovered in 1997 in northern Myanmar. Another supposed new mammal from Southeast Asia remains controversial: at least some of the horns attributed to a buffalo- or goatlike animal named *Pseudonovibos spiralis* were actually manufactured from the horns of domestic cattle (*Bos taurus*), although others may be genuine. Local people were aware of all of these animals before their scientific discovery, and a photograph of the horns of the giant muntjac was published in a zoology journal in 1899.

Although few of these new animals were actually predicted by cryptozoologists, many of them conceivably could have been—anecdotal evidence or physical traces, often provided by locals, suggested the existence of an animal later confirmed by scientists. This shows that eyewitness testimony and local folklore are not completely worthless, and they can be especially useful if they are detailed and provided by knowledgeable observers. Nevertheless, the discovery of a new deer, for instance, clearly does not imply that more sensational cryptids such as bigfoot are equally likely to be found. Similarly, unlike bigfoot and other spectacular cryptids, actual new species were usually discovered shortly after scientists heard rumors of their existence and started searching for them.

Creationism and the Paranormal in Cryptozoology

An unfortunate problem for cryptozoology is the occasional presence of creationist and paranormal claims in this field. Creationists are attracted to cryptozoology because of ideas regarding prehistoric survivors, which they take as support for a young Earth and other creationist myths. Paranormalists may get involved with famous monsters, such as bigfoot, which they sometimes declare are extraterrestrials or have teleportation and extrasensory perception (ESP) powers. These two groups of individuals occasionally are tolerated and even accepted by some cryptozoologists, particularly popularizers, which only damages cryptozoology's image among scientists. More serious cryptozoologists, however, reject paranormalists and creationists as the "lunatic fringe," noting that their own field was founded on zoological principles, including evolution.

Is Cryptozoology a Science?

Because much of the evidence they have to work with is undetailed and anecdotal, cryptozoologists may interpret data rather freely. The cryptozoological literature is full of rampant speculation on the identity of cryptids, with little skepticism or critical examination being applied to the evidence. For example, at Loch Ness, some researchers attribute any strange water disturbance to the supposed monster, without considering other explanations. Perhaps the most common form of cryptozoological speculation is the notion that various cryptids are surviving forms of prehistoric animals. Many different kinds of large extinct vertebrates (and a few invertebrates) have been resurrected by cryptozoologists, including dinosaurs, plesiosaurs, pterosaurs, extinct hominids, giant ground sloths, and mammoths. As paleontologist Darren Naish has noted, the fossil animals favored by cryptozoologists are usually those widely known to the public, rather than lesser-known animals that may fit a certain cryptid's description better. This tendency indicates a misinformed, rather arbitrary method of selecting extinct animals to identify cryptids. Moreover, well-established scientific theories often counter cryptozoological resurrections. For example, the fact that the fossil record for plesiosaurs and dinosaurs (except birds) ends sixty-five million years ago indicates they are extinct, yet they are constantly touted as possible survivors in Loch Ness and West Africa, respectively. Cryptozoologists reply that the discovery of the coelacanth, believed extinct for millions of years, demonstrates that prehistoric animals can survive to the present without leaving a fossil record. The coelacanth is a poor model for dinosaur or plesiosaur survival, however, since it is much smaller and has more fragile bones that fossilize less reliably and are harder to identify. The evidence available is not sufficient to support prehistoric resurrections and other such speculations.

Pointing to this rampant speculation and ignorance of established scientific theories in cryptozoology, as well as the field's poor record of success and its reliance on unsystematic, anecdotal evidence, many scientists and skeptics classify cryptozoology as a pseudoscience. Some cryptozoologists respond that anecdotal evidence is valid (or they concentrate on the available physical evidence), and they sometimes appeal to the philosophical argument that there is no fixed scientific method. Some also suggest cryptozoology offers new paradigms that challenge accepted scientific theories (despite the fact that the latter are actually based on more reliable evidence) and argue that the discipline is a multidisciplinary approach that defies classification as a typical science. Perhaps, others suggest, cryptozoology is natural history, being focused on rough and sometimes inconclusive descriptive data. In the end, the argument of whether cryptozoology is a science or not may come down to one's particular view of science.

It has been argued that many of the problems cryptozoology experiences are committed most often by popularizers, who may concentrate on Nessie, bigfoot, and other monsters. Cryptids large and small can usually be discussed rationally, however, without wild speculation that flagrantly disagrees with scientific theories. When this method is used, it seems that cryptozoology can be a useful and interesting field of inquiry, even if not strictly a science or always conclusive in its answers. Some cryptozoologists follow such an approach, and the discipline's problems may lie more with others who ignore such guidelines and put too much weight on weak evidence. If these individuals were to consider the quality of the evidence more critically and be more reserved in their judgments and hypotheses, cryptozoology might become more appreciated by mainstream science.

References:

Bille, Matthew A. 1995. *Rumors of Existence: Newly Discovered, Supposedly Extinct, and Unconfirmed Inhabitants of the Animal Kingdom.* Surrey, British Columbia: Hancock House.

Coleman, Loren, and Jerome Clark. 1999. *Cryptozoology A to Z: The Encyclopedia of Loch Monsters, Sasquatch, Chupacabras, and Other Authentic Mysteries of Nature.* New York: Fireside.

Ellis, Richard. 1994. *Monsters of the Sea.* New York: Alfred A. Knopf.

Heuvelmans, Bernard. 1958. *On the Track of Unknown Animals.* New York: Hill and Wang.

———. 1968. *In the Wake of the Sea-Serpents.* New York: Hill and Wang.

Mackal, Roy P. 1980. *Searching for Hidden Animals.* Garden City, NY: Doubleday.

Naish, Darren. 2001. "Sea Serpents, Seals and Coelacanths: An Attempt at a Holistic Approach to the Identity of Large Aquatic Mystery Creatures." In *Fortean Studies,* vol. 7, edited by Ian Simmons and Melanie Quin, 75–94. London: John Brown Publishing.

Shuker, Karl P. N. 1995. *In Search of Prehistoric Survivors: Do Giant "Extinct" Creatures Still Exist?* London: Blandford.

Cults

STEVE NOVELLA AND

PERRY DEANGELIS

Ever since the Bronze Age, charismatic leaders that have sought, for their own selfish desires, to control people drawn to them. Psychological and physical coercion have been used by those who would compel the will of others—but always in the guise of beneficence. Masking the actual beliefs and goals of the group is common to almost all cults. This is an insidious and deliberate use of deception. The details of the cult's belief system are revealed to recruits only in stages, which is a move calculated to draw them in, step by step, without scaring them off early on. In contrast, mainstream belief systems are up front with their principles and make no attempt to conceal them.

Cult leaders commonly seek to eradicate their members' ability to think critically and make life decisions. They retrain their victims to think in their own highly defined and constricting manner, so that they become pawns subject to the will of the leader. This process is not accomplished through argument or force, as is often thought, but rather with subtle persuasion, flattery, guilt, and always deception.

Due to the large variety of cultic organizations, there is no simple method to determine whether a particular group can be defined as a cult. One must evaluate both the group's belief structure and its behavior in order to determine if it adheres to the hallmarks of a cult. Robert Lifton defined five tried-and-true methods destructive groups use to ensnare their flocks and keep them corralled (Lifton 1961):

1. Totalism—Totalism is an us-against-them philosophy used to achieve complete separation from the past, which is portrayed as filled with evil forces or unenlightened individuals.
2. Environmental control—Everything that perspective recruits see, eat, and do in their waking hours is carefully manipulated.
3. Loading the language—The jargon of the cult features quick and easy phrases and statements that only have meaning to the cultists. Such jargon encourages isolationism and psychological cloning.
4. Demand for purity—All actions are judged by the cult's definition of purity, which is crafted by the leaders to suit their needs. Such definitions are applied in an absolute, black-and-white manner. Anything is acceptable in the pursuit of this purity.
5. Mystical leadership—The cult leader endows him- or herself with a mystical mantle, often supposedly as an agent of divine powers on Earth. Confession to and denunciation by the leader are ingrained. The victim acquires a pawnlike attitude, wherein devotion and obedience to the leader supersede standards of morality or self-preservation.

The more obvious and pervasive these philosophies are in a group, the more group members will adhere to ideological totalism and the more these devices will be used to corrupt an individual's will, making the group more of a cult.

How Cults Recruit

Historically, cults have thrived during times of societal instability. When people are at a loss to make sense of the rapid changes around them and are forced to rethink much of what they once held as true, they are ripe for cult membership.

For example, after the fall of Rome or during the French and Industrial Revolutions, cults sprang up in theretofore unprecedented numbers. The siren song of the recruiters promising to wash away the fear and uncertainty of the time was simply too alluring for many to deny, and cults flourished. The black-and-white philosophies of cults are much easier to digest than the complex and dynamically changing realities of society.

It has been argued that cults are not destructive because people who join them are "seekers," that is, individuals who seek answers to grand questions. Yet this statement cannot be true because cult recruiters reach out with malign intent and trap their victims with deception. The true nature of the cult is never discussed at the outset. One cannot join a group to aid its members in their search if one does not know what the group itself is actually about. Cults recruit people from all socioeconomic strata and all age groups. In fact, they actively seek out older people who have accumulated estates that can be willed to the group.

Two main aspects are predominant in making individuals vulnerable to cult recruitment. The first is that they feel unconnected to those around them. This feeling is likely to occur at certain times in life, such as between completing college and taking a job, when traveling for an extended period, on arriving at a new location, or soon after being rejected, fired, or divorced. Anytime that people do not have compelling connections in their lives, they are extremely vulnerable to the seeming familialness of cultic recruitment. The second aspect is depression. A person suffering from a depression that is not completely debilitating is very malleable and is easily soothed by the honey coating of cultic deception. Cults seem to offer nearly instant and often simplistic and focused solutions to the myriad problems daunting vulnerable individuals.

Recruiters for cultic groups are trained to spot susceptible people, and they know well the signs of vulnerability. They will often strike up a conversation with a potential recruit and perform a purposeful "cold reading" of the victim. They are trained to assess the individual's needs very rapidly and will then speak directly to these needs. To a person suffering from anxiety and want of affiliation and affection, these soothing words from a stranger, whose only apparent motivation is to help, are very persuasive. In addition, the recruiter carefully scripts the physicality of the first contact with a recruit. He or she knows what posture to hold and at what distance to sit from a mark so as to not be too threatening, while still maintaining control of the conversation. Keeping direct eye contact is always emphasized, as this relays a sense of both empathy and sincerity. These are highly intelligent practices, and they speak directly to the insidious nature of destructive cults.

After the initial contact, the first crucial step in absorbing the victim is to proffer an invitation. This invitation is anything that the recruiter believes the victim will acquiesce to after the first assessment. This step is essential, for it is at this time that the victim will be moving out of his or her familiar world and entering into the nebulous world of the cult. What the victim who accepts the invitation will usually find on attending the first gathering is the cult's "front group." A front group is a cadre of select individuals from within the cult who act to mask the cult's real agenda. Often, cults will have several different front groups that can appeal to a wide variety of interests and needs. At this initial meeting, the victim is swarmed by the front group, and affection and attention are lavished. The primary purpose of this step is to get the recruit to agree to a longer-term commitment at the cult's facility.

Persuading the victim to accept an invitation into the cult's facility is the second crucial step, for once he or she is there, the actual separation from the outside world is effectuated. Then, the process of thought reform begins in earnest. The recruits are surrounded by veteran members of the cult who sing its praises ceaselessly, going on and on about the inherent strength of whatever new belief system is being advocated. The leader is praised without end as his or her uniqueness is revealed and claimed to be the savior of humanity via whatever method he or she has chosen—revealed knowledge, perfect social paradigms, ancient or alien wisdom, and so forth. The fact that there is little or no objective evidence to support these claims is glossed over with the group's jargon. Again, the recruits feel that they are somehow not as "good" as the other members because they do not understand the specific language and nonsense words of the cult. Only through a parroting of this jargon do they get approval.

Veteran cult members immediately begin to direct the recruits' actions, keeping their time carefully filled with meetings, exercise, reading cult propaganda, chanting, and whatever else the particular group has found that will occupy the recruits' time. This oppressive atmosphere does not allow for reflection and negative feelings, and questions are suppressed, as these are only the victims' old and unclean ways rising to surface. All connections with the past are severed. The recruits' families and friends are painted as unenlightened individuals who need to be shunned until they have seen the way. Victims are made to feel that they, too, were bad in their old lives, and this guilt is reinforced by the denunciation of the past. The guilt is embedded after the initial waves of love that the group showered on the recruits, and it is very confusing and causes much anxiety. Recruits are never allowed to speak with other recruits who might share their initial doubts and hesitations. They are made to believe that if they have doubts, they are the only ones with such qualms and should be ashamed of them. Their critical faculties are derided at every turn.

In addition to psychological conditioning, a careful program of physiological control is instituted. Recruits are often kept so busy by the cult that they become sleep-deprived. Prospective members can also be made to hyperventilate by loud repetitive chanting, an activity that reduces the level of carbon dioxide in the blood, causing it to become too alkaline and leading to respiratory alkalosis. This, in turn, makes the victims light-headed and woozy, further diminishing their critical processes. Special and restrictive diets are enforced to make the recruits weak and uncomfortable. Drugs and even sugar can be used to induce an artificial high so the cult's activities and propaganda will temporarily excite the recruits. Purging and colonics may also be used, as well as dehydration, all to make the subjects more confused, disoriented, and dependent.

The recruits' appearance is often altered to suit the cult's standards. This can involve anything from wearing a special uniform to cutting hair a certain way to constantly wearing cult paraphernalia. Changing a person's long-held appearance can have a profound effect on his or her sense of continuity, and it reinforces the notion that an entirely new life has begun. Sometimes, the recruits are even required to take on new names.

A pattern of "antagonism, apathy, and acceptance" is classic in brainwashing. The antagonism is any resistance that the victims might have to the indoctrination process when first inducted. This is quickly quelled via the previously described methodologies, and the recruits move into the apathy stage. In this state, it is simply easier to just go along, drop any excoriated resistance, and fall into the reassuring conformity of the encompassing group—which, of course, leads directly to acceptance and the final attenuation of individuality and self-preservation.

Exiting a Cult

It is much more difficult to exit a cult than to enter one. On the way in, all is sweetness and light, the courting process has just begun, and recruits still feel that the cult is enhancing their personalities. It is only during the exiting process that they learn that their personalities have, in fact, been stolen. The damaging methods a cult uses to beguile its victims leave mental wounds in former members that often take years to heal. The reduction of one's will to resist and the degradation of one's critical faculties make the transition back to freedom very difficult. Several key characteristics typify almost all ex-cult members. The most predominant is fear. Many cults use fear to maintain loyalty to the group; everything from denunciation to claims of damnation to physical force is used to both retain and return members. Ex-members are often encouraged, if not forced, to change locations, telephone numbers, and even their names to escape the harassment of their former groups. Of course, this fear is always much more acute when a family member, particularly a child, is left in the group. The group can threaten the child with sanctions unless the member returns; at the very least, it will almost always sever all contact between the child and the "traitor."

Another aftereffect that ex-cult members must deal with is "flashback." Not unlike shell shock (wherein combat veterans react with inappropriate emotion and fear to any loud noise), ex-cult members will sometimes find themselves wandering back to the trancelike state they were ensconced in during their days in the group. These times of "floating" are triggered by stress or deep depressions or even when the cult's jargon is heard in completely unrelated contexts. These flashbacks decrease in frequency over time, but they can last for months (Singer 1995).

The attack on one's mentality when in a cult leaves the victim's cognitive skills dulled. It takes time to retrain the mind to evaluate and perceive in real time. The outside world is a busy and complex place. The empty simplicity of the cultic core is gone, and the sensory input can sometimes be daunting to one who has languished in zombielike obedience for an extended period of time. For this reason, the ex-member should attempt tasks in an ascending level of difficulty and complexity, as one would do when training to do these things for the first time.

Many ex-members report that they are consumed with guilt, a guilt that may take many forms. Within the cult, members are often forced to perform illegal activities, learning to con, trick, and steal from others. They compel donations in a variety of dishonest and coercive ways. And they suppress personal morality to the will of the cult. Such actions leave them deeply ashamed once they separate from the group. They are uncertain how they can face up to these actions and how they can repay those they themselves victimized. Further, ex-members may feel very troubled about close friends and family members who were left behind in the cult. This makes the dismissal of the cult extraordinarily difficult. When their feelings for those still within the cult call to them most strongly, they may even begin to doubt their wisdom in leaving the cult. This miasma of doubt and confusion can be debilitating and slow recovery to a crawl. Finally, ex-members must come to face those in the outside world with whom they suddenly severed ties when they were absorbed by the cult. When confronted with the compassion and concern that their loved ones have maintained for them, even as they were chanting about their loved one's evils at the leader's behest, the guilt can be overwhelming.

This shame leads directly to another problem faced by ex–cult members—the continual bombardment of questions from others and the sense that they have an obligation to explain what happened to them. It is exceptionally difficult to talk to those never victimized by a recruiter and thence a cult about the subtleties of manipulation and coercion that ensnared them. To describe the charisma of the leader in full plume and the atmosphere of euphoria that the masterful manipulation of the cult could cause is all but impossible. As a result, ex–cult members sense that no one on the outside understands what they went through, and they feel pitied. Further, family and friends often put the emerged individual under a microscope, watching for any signs of weakness that may be indications that he or she may again become the mark of the old (or even a new) group. This situation often leads to encounters in which both the watchers and the watched grow concerned but fail to communicate that concern effectively. Tensions can quickly rise under such circumstances, and the ex-member's sense of self-worth can be eroded by the perception that loved ones do not believe he or she can manage things properly.

The entire sense of self that was so artificially inflated at times within the cult must be reassessed in a realistic state. No longer can the victims consider themselves the "chosen." They are suddenly just like everyone else—still searching, still hoping, and still struggling. These ex-members are left feeling that perhaps they are not only no longer chosen but also valueless. They have a very difficult time learning to trust again. Fear of being victimized anew can make them cynical and distant (Lifton 1961).

Helping Victims Cope

The primary way to help ex–cult members reemerge as healthy persons is through understanding—understanding their plight and helping them to understand what happened to them. It must be explained to them with firm compassion that they were the victims of time-tested, cohesive, and insidious methods of manipulation that have trapped countless thousands through the ages. This will allow them to talk openly about their fears, both past and current. Once the victims begin to see that they were, in fact, victimized, the process of rebuilding and reawakening their atrophied critical faculties can begin in earnest. They must be made to see how and why they were ensnared and be given the tools to avoid such an outcome in the future.

Ex-members must be reoriented to reality. This process can be accomplished by simple tasks that help them to rebuild a fulfilling connection with the outside world. Anything that might bewilder or entrance them must be meticulously avoided. No drugs or alcohol should be consumed during this tenuous time. Anything that might cause a state of sensory overload should be avoided (loud music, crowds, a large urban environment, and the like). The maintenance of routines in the early recovery stages is a good idea. Making checklists of activities and following a schedule are important, as is planning out purchases and projects well beforehand. The reorientation to reality can be accomplished by keeping apprised of current events via newspapers, television news and talk shows, and talk radio (Ryan 1999).

For information on specific cults as well as anticult and support groups, contact the Leo J. Ryan Society at http://www.cultinfo.com.

References:

Appel, Willa. 1983. *Cults in America: Programmed for Paradise.* New York: Holt, Rinehart and Winston.

Heinsohn, Gunnar. URL: http://www.teleport.com/~kronia/journals/sacrfice.txt. (Accessed in 1998).

Langone, Michael D. URL: http://www.mhsource.com/edu/psytimes/p960714.html. (Accessed in 1998); URL: http://www.csj.org/studyindex/studyrecovery/study_trance.htm. (Accessed in January 1999).

Lifton, Robert Jay, M.D. 1961 (1989). *Thought Reform and the Psychology of Totalism.* Chapel Hill and London: University of North Carolina Press.

MacCollam, Joel A. 1979. *Carnival of Souls: Religious Cults and Young People.* New York: Seabury Press.

Ryan, Patrick L. 1999. *Coping with Trance States: The Aftermath of Leaving a Cult.*

Singer, Margaret T. 1995. *Cults in Our Midst.* San Francisco: Jossey-Bass Publishers.

Tobias, Madeleine, and Janja Lalich. 1994. *Captive Hearts, Captive Minds: Freedom and Recovery from Cults and Abusive Relationships.* New York: Hunter House Publishers.

Dietary Supplements

RICKI LEWIS

Nestled among the vitamins and minerals on supermarket shelves are many apparently natural remedies sold as "dietary supplements." The labels state that these products are not intended to treat anything but are designed to promote health. For example, kava, lemon balm, German chamomile, lavender, passionflower, and valerian root all purportedly provide restful sleep.

Elsewhere in the supermarket, cough drops containing echinacea promise to boost immunity, juice drinks offer ginkgo to rev up memory, and St. John's wort is said to elevate mood. But it would take hundreds of cough drops to shorten the course of a respiratory infection, and forty-five drinks to administer enough ginkgo or St. John's wort to exert an effect. The Internet offers dietary supplements, too. One proprietary blend "supports the healthy functioning of the heart muscle." It includes pig intestine, sheep spleen, assorted bovine parts, several grains, pea juice, soy, and mushrooms—everything but eye of newt.

Dietary supplements are a multibillion-dollar industry in the United States, where a third of the population has tried them (Blendon et al. 2001). St. John's wort alone racked up $195 million in sales in 2000. Consumers key in on the "organic," "herbal," "natural," and even the nonsensical "chemical-free" claims on product labels, often believing that a chemical synthesized in a laboratory is somehow different from the same chemical occurring in a living organism. This isn't the case. Salicylic acid is salicylic acid, whether it comes from a bottle of aspirin tablets or willow bark. The popularity of these products is also due to the ease of obtaining them, thanks to loose government regulations that are, in part, a response to consumer demand.

Drugs versus Dietary Supplements

People have used medicines derived from organisms for millennia. Many prescription drugs are modeled after such "natural products." Two commonly used cancer drugs, for example, are alkaloids that come from the rosy periwinkle, *Vinca rosea*.

Drugs differ from dietary supplements in purity and method of preparation. The development of a drug often takes many years. The probability that a chemical derived from an organism will make its way to drugstore shelves as a prescription or over-the-counter medication is 1 in 10,000.

To develop a drug, researchers identify and isolate active ingredients from organisms, then either use these compounds or synthesize related ones that are more effective, less toxic, or both. For example, researchers first reported the ability of mayapple (*Podophyllum peltatum*) extract to shrink tumors in mice in 1942. In the 1950s, investigators funded by the National Cancer Institute's Drug Research and Development program isolated and described the active ingredient podophyllotoxin. But in clinical trials, the compound proved to be toxic, so human testing was stopped. Then, in the 1960s, a pharmaceutical company created a safer, semisynthetic version of podophyllotoxin, which the Food and Drug Administration (FDA) approved in 1983 to treat four types of cancer.

In contrast to the strict requirements of purity applied in drug development, herbal remedies and other dietary supplements can include many parts of the source organism. Consequently, along with the desired substance, a consumer may ingest microorganisms, fungal spores, heavy metals, toxins, herbicides, and pesticides. Labels don't always tell the full story. ConsumerLab.com, for instance, detected 30 milligrams of manganese per dose in a glucosamine and chondroitin preparation, a "joint health" product. But the Institute of Medicine recommends a maximal daily intake of 11 milligrams of manganese, and a normal diet supplies 2 milligrams daily! Manganese overdose causes symptoms similar to those of Parkinson's disease.

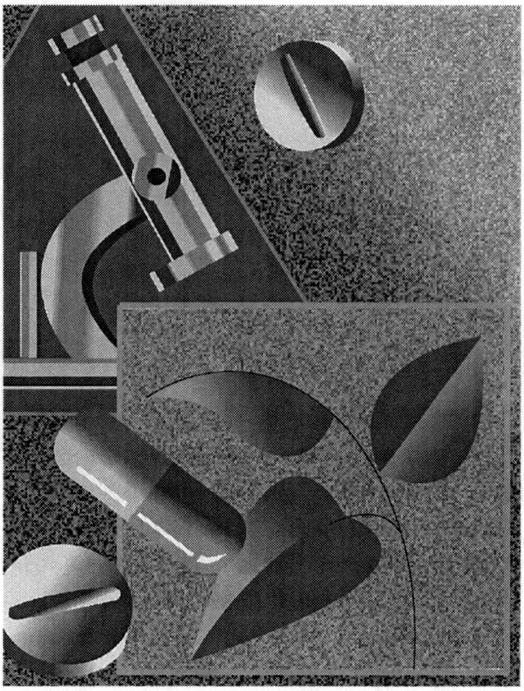

Illustration of microscope and medical research. (Shakirov/Artville)

The Dietary Supplements Health and Education Act (DSHEA)

By definition, dietary supplements are neither food nor drug, thanks to the FDA's Dietary Supplements Health and Education Act (DSHEA) of 1994, which amended the Federal Food, Drug and Cosmetic Act. This earlier regulation classified substances as being either food additives, which require premarket approval based on a demonstration of safety, or "generally recognized as safe" (GRAS). Before 1994, added botanicals or other biological materials were deemed an adulteration. The Food, Drug and Cosmetic Act also stipulated that the FDA must approve health claims after evaluating substantial scientific evidence. This happens rarely (it occurred, for example, with calcium to prevent osteoporosis and folic acid to prevent neural tube defects).

DSHEA expanded definitions of "dietary supplement," applying the label not only to essential nutrients but also to "a product (other than tobacco) that is intended to supplement the diet that bears or contains one or more of the following dietary ingredients: a vitamin, a mineral, an herb or other botanical, an amino acid, a dietary substance for use by man to supplement the diet by increasing the total daily intake, or a concentrate, metabolite, constituent, extract, or combinations of these ingredients."

A dietary supplement need not be tested for safety or efficacy. All that the FDA requires is that labels include a disclaimer: "This statement has not been evaluated by the FDA. This product is not intended to diagnose, treat, cure or prevent any disease." Should a dietary supplement label claim to treat anything, the content becomes a drug and must be evaluated for safety and efficacy. Only after many consumers file reports of adverse effects of a particular dietary supplement does the FDA issue a warning or withdraw the product, pending investigation.

Because dietary supplements don't have to undergo much, if any, testing, how they work is often shrouded in mystery or scientific-sounding jargon. For example, makers of some animal-based products seem to believe that if a human organ needs help, the person need only swallow an extract of said organ from another type of animal. Marketers of one product advertised on the Internet claim that "bovine hypothalamus extract helps maintain the hypothalamus in a good state of repair to support healthy hypothalamic function" (http://www.standardprocess.com/sp_catalog_product_detail.asp). It's hardly that simple—the product is first digested into its constituent nutrients, not shuttled to where it's needed intact. The shark cartilage craze fell into this category of like helping like. The idea that shark cartilage prevents cancer arose from initial studies that suggested that sharks do not get cancer. Since sharks have cartilaginous skeletons, the thinking went, their cartilage may protect against cancer. Further investigation revealed that sharks actually do get cancer—perhaps sick sharks simply die and sink and therefore aren't studied!

Shark liver capsules. The outer packaging advertises the dietary supplement as a "healthy food for middle and high age people." (Jeffrey L. Rotman/CORBIS)

The like helps like approach is just one way that marketers delude consumers who have forgotten Biology 101. There's simply no rhyme or reason to take some products. Consider the genetic material deoxyribonucleic acid (DNA) and ribonucleic acid (RNA), which enables the cell to use the genetic information to synthesize protein. There are products claiming to contain DNA and RNA that sell for more than $20 a bottle and are just brewer's yeast. RNA, DNA, and their building blocks are abundant in food and are synthesized constantly in nearly all cells. Yet the bottle claims that depletion of RNA in the body can result from lack of exercise, pollution, and stress. Pyruvic acid is another unnecessary dietary supplement. It is an intermediate in the breakdown of glucose. Again, basic biology. Yet some health-food stores actually lock the stuff up as if it were precious.

Beware!

Consuming desiccated pig spleen, shark cartilage, or yeast can't harm much more than a person's wallet. But some supplements can threaten health if they are taken in too high doses, if they are taken instead of conventional therapeutic drugs, or if they interact with drugs. Despite the FDA's definition of dietary supplements, abundant scientific evidence indicates that these products can indeed contain chemicals that act as drugs in the human body. Following are some frightening examples of the dangers of certain dietary supplements.

A Weight Lifter and Creatine

A previously healthy twenty-four-year-old weight lifter awoke one morning with intense pain in his thighs (Robinson 2000). In the hospital, he passed blood and protein in his urine and had difficulty breathing; he also had an enlarged heart and lungs. Water was rushing into his skeletal muscle cells, upsetting his body's water distribution in a way that swelled and threatened his vital organs. His enlarged muscles were not stronger but were actually disintegrating, a condition called rhabdomyolysis. The diagnosis: acute quadriceps compartment syndrome and rhabdomyolysis. The likely cause: creatine supplements.

Cells synthesize creatine from the amino acids arginine, glycine, and methionine. Creatine binds phosphate, which it can then transfer to adenosine diphosphate (ADP) to form adenosine triphosphate (ATP), the body's primary energy molecule. Creatine is then metabolized to form creatinine, which built to sky-high levels in the weight lifter's blood and urine. He had overdosed on creatine. The product label advises that to increase muscle mass and strength, one should take a loading dose of 20 to 25 grams a day for 5 to 7 days, then take 2 to 5 grams a day thereafter. No study had ever monitored the effects for longer than 10 weeks. The recommended daily allowance for non–weight lifters is a mere 2 grams. But the weight lifter took 25 grams a day for a year! After 22 days in the hospital and months of physical therapy, he recovered.

The FDA and the National Collegiate Athletic Conference are concerned at the lack of long-term studies on the safety and efficacy of creatine, as well as the fact that most studies that have been conducted do not demonstrate athletic enhancement. Indeed, many adverse-event reports to the FDA document muscle cramps, diarrhea, loss of appetite, seizures, strains, and dehydration as side effects. In 1997, for instance, three college wrestlers died from dehydration associated with creatine use. Yet this product is still widely used. A recent survey of high school athletes revealed that of 328 participants, 26 males and 1 female—8.2 percent—took creatine supplements (Smith and Dahm 2000, 1257). Most reported that they either did not know the recommended dose or intentionally exceeded it. Among professional football players, 25 percent to 75 percent use creatine. Many of them claim that creatine is a safe alternative to steroids for bulking up. The U.S. market for creatine is $200 million a year.

Celebrity Power—The Iscador Story

Another way that dietary supplements attain credibility without scientific evidence is through celebrity endorsement, even when it is unintentional. Consider the case of Suzanne Somers and her use of the mistletoe extract Iscador to treat breast cancer.

Many people regard the bubbly Somers, former star of *Three's Company* and promoter of the ThighMaster, as a health expert. When claiming that she had liposuction to correct damage from breast cancer surgery, she revealed that she injects herself daily with Iscador.

The use of mistletoe was originally part of a movement called Anthroposophy that was founded by Rudolf Steiner (1861–1925) (see the "Anthroposophy and Anthroposophical Medicine" entry in this section). A self-described "spiritual scientist," Steiner thought cancer to be an imbalance of the forces that control cell growth and division. He noted unusual properties of mistletoe that suggested it would combat cancer. The plant's spherical shape, lack of roots, and parasitic growth on other plants, he thought, suggested a defiance of gravity, and the fact that it flowers in winter and has berries all year long indicated the plant ignored seasonal changes. With such characteristics, Anthroposophists thought, perhaps mistletoe extract could also correct whatever imbalances underlie cancer. Adding trace amounts of certain metals to the extract can supposedly target the treatment to certain types of cancer, and activities such as dancing or sculpting with modeling clay are said to add therapeutic benefit. The goal is to strengthen a patient's "organic self-supportive systems," whatever they are (Office of Technology Assessment 1990).

The source of Iscador is the European mistletoe *Viscum album*. Usually, the entire plant is ground up and injected. Mistletoe contains several compounds that are toxins in high doses but do affect human cells growing in culture when administered in small doses. These chemicals stimulate the proliferation of white blood cells (an immune response), yet they dampen the division of other cell types—hence, the idea that the plant can treat cancer arose. The possible pharmacologically active molecules in mistletoe are:

- Lectins, which are complex carbohydrates bound to proteins. Lectins bind certain cell types.
- Viscotoxins, which are proteins that kill certain dividing cells.
- Alkaloids, which inhibit cell division.

The active ingredients of European mistletoe have not been adequately studied in the United States. Investigations conducted in other countries use either the whole plant in extract form or consider evidence of immune stimulation as a clinical end point, rather than efficacy in treating cancer.

Drug Interactions

Herbal and other remedies that might not have been dangerous in the past are potentially threatening now because people today take many more medications. The active ingredients in some dietary supplements, whether the FDA calls them drugs or not, interact with drugs. An unfolding disaster in this area involves St. John's wort, whose active ingredient, hypericin, is widely regarded as an antidepressant. But hypericin does more than possibly elevate mood—it also lowers the blood levels of nearly half of all prescription drugs by interfering with the enzyme system in the liver that metabolizes drugs. Affected drugs include blood thinners, antibiotics, oral contraceptives, antirejection drugs, heart medications, and protease inhibitors used to treat HIV infection.

A patient who does not inform a physician that he or she is taking St. John's wort could be in for trouble. This happened to a man who had recovered well from a heart transplant and then began to reject the organ (Ruschitzka 2000). Blood tests indicated low levels of the antirejection drugs that he had been taking daily. The patient had not told his physician that he was taking St. John's wort because he did not think it was a drug. Once he stopped taking it, the heart rejection ceased.

Studies to Evaluate Dietary Supplements as Drugs

It is clear that the way to improve the safety of dietary supplements is to conduct more research, but this is easier said than done. Unfortunately, few studies of dietary supplements meet all of the criteria of a well-planned experiment. The most meaningful type of study has a large sample and control groups; it should also be a double-blind protocol, meaning that neither researchers nor participants know who is receiving the treatment or a placebo, and it should be as unbiased as possible. Following are a few examples of the variety of scientific investigations conducted into the efficacy of dietary supplements.

Ephedra

Ephedra (also known as ma huang), a component of a product called herbal ecstasy, is a stimulant that has been associated anecdotally with hypertension, arrhythmia, tremors, headache, seizures, heart attack, stroke, and death. In one evaluation of 140 adverse-event reports, researchers concluded that 43 (31 percent) of the cases could be definitely or probably related to ephedra (Haller and Benowitz 2000). The problem with evaluating adverse-event reports is that they rely on volunteers and will likely include some people whose symptoms occurred coincidentally with taking the dietary supplement but were not necessarily caused by the product. Conversely, some people who react to the dietary supplement may not realize the link or contact the FDA. The agency has not taken further action against ephedra but is continuing to monitor reports of adverse events.

Gingko Biloba

A study of the effect of *Gingko biloba* extract on tinnitus (ringing in the ears and persistent sound) had large enough numbers—978 individuals were studied—but each participant was given either *Gingko biloba* or a placebo for twelve weeks, then asked to evaluate symptoms using a rating scale (Drew and Davies 2001). Such a study cannot account for individual differences in perception. A crossover design might have been more meaningful, with each participant experiencing one trial of *Gingko biloba* and one of the placebo. (The study found no difference between the groups.)

Glucosamine

Well-done studies, by contrast, can validate anecdotal reports of the efficacy of a dietary supplement. This was the case for glucosamine, a component of cartilage and the synovial fluid that bathes joints. Because glucosamine is found in the joint space, the theory goes, supplying it exogenously should help maintain joint structure—a variation of like treats like thesis. Debate centered on whether digestion would dismantle glucosamine pills. Although that question has yet to be answered, a clinical trial provided compelling evidence that this supplement actually performs as people say it does (Reginster et al. 2001).

In the double-blind trial, 106 people with osteoarthritis of the knee took glucosamine and 106 took a placebo for three years. The researchers assessed the outcome through patient reports of pain improvement and X rays of the joint space taken at the start of the trial, at one year, and at three years. Results were remarkably clear. In the patients receiving a placebo, the joint space progressively narrowed, and their symptoms slightly worsened. But those receiving glucosamine had no loss of joint space, and their symptoms improved. The researchers concluded that glucosamine might not only relieve symptoms but also actually modify the course of the disease.

St. John's Wort

When studies contradict each other, researchers scramble for explanations, and journalists wonder what to report. This is the case for St. John's wort. In September 2000, the *British Medical Journal* published results of a randomized, double-blind trial (Woelk 2000). Of 324 people with mild to moderate depression, 157 received St. John's wort and 167 received imipramine, an older antidepressant. Patients assessed their moods six weeks later, and the treatments were deemed "therapeutically equivalent." However, criticism of the study came quickly. Critics charged that comparison with a newer selective serotonin-reuptake inhibitor (SSRI) would have been more useful. They also said that an inert placebo should have been included to assess whether symptoms resolved over time without treatment. And finally, they pointed out, six weeks is not nearly long enough to detect recovery.

A few months later, the *Journal of the American Medical Association* published the results of another clinical trial (Shelton et al. 2001). In the study, 98 people with major depression were assigned St. John's wort for an eight-week trial and 102 individuals received an inert placebo. All participants were given a placebo for one week prior to the eight-week period. The investigators used standard scales of depression to assess outcome. The passage of time improved symptoms in all participants, but the difference in efficacy between the two treatments was not statistically significant.

Why did one study find that St. John's wort worked and the other find that it did not? Perhaps this outcome can be traced to the fact that the investigations involved different types of patients. St. John's wort appeared to be effective in individuals with mild to moderate depression but not in people suffering from major depression. Ongoing trials sponsored by the National Institutes of Health are comparing St. John's wort with the SSRIs and with the placebo in the treatment of different degrees of depression.

Integrating Dietary Supplements into Health Care

The idea that dietary supplements are not drugs and do not contain drugs has spilled over into the doctor-patient relationship. According to a Harvard School of Public Health analysis, many people do not reveal their use of dietary supplements to their physicians. The primary reason is that people think doctors do not know enough about these products or are biased against them. However, more and more physicians and medical history forms are asking patients about their use of alternative or complementary medical therapies in general and about dietary supplements in particular. The reason is that the medical community hopes to prevent the sort of tragedy that happened to a seven-year-old with HIV infection being treated at Columbus Children's Hospital in Ohio. The child's mother became convinced that bovine colostrum—a cow's first milk, which is rich in antibodies—could treat the condition better than the standard protease inhibitors. She started her son on the supplement and discontinued the prescribed medications without telling the doctors. Several months later, the boy's infection was no longer under control. Yet another patient trying the same supplement shared the information with health-care providers, who saw that he took the protease inhibitors when his viral load increased.

Whether the FDA considers dietary supplements to be drugs or not, two facts are apparent: these products can indeed *act* as drugs, and many people are taking them. In the future, patients and physicians will have to communicate more effectively, and the FDA will have to catch up with medical science to ensure that people can take dietary supplements safely.

References:

Blendon, Robert J., et al. 2001. "Americans' Views on the Use and Regulation of Dietary Supplements." *Archives of Internal Medicine* 151, no. 6 (March 26): 805.

Drew, Shelley, and Ewart Davies. 2001. "Effectiveness of Ginkgo Biloba in Treating Tinnitus: Double Blind, Placebo Controlled Trial." *British Medical Journal* 322 (January 13): 73.

Haller, Christine A., and Neal L. Benowitz. 2000. "Adverse Cardiovascular and Central Nervous System Events Associated with Dietary Supplements Containing Ephedra Alkaloids." *New England Journal of Medicine* 343, no. 25 (December 21): 1833.

Office of Technology Assessment. 1990. *Unconventional Cancer Treatments.* Washington, DC: U.S. Government Publications.

Reginster, Jean Yves, et al. 2001. "Long-Term Effects of Glucosamine Sulphate on Osteoarthritis Progression: A Randomized, Placebo-Controlled Clinical Trial." *Lancet* 357 (January 27): 251.

Robinson, Stacey J. 2000. "Acute Quadriceps Compartment Syndrome and Rhabdomyolysis in a Weight Lifter Using High-Dose Creatine Supplementation." *Journal of the Air Force Family Practice* 13, no. 2 (March-April): 134.

Ruschitzka, F., et al. 2000. "Acute Heart Transplant Rejection Due to St. John's Wort." *Lancet* 355 (February 12): 548.

Shelton, Richard C., et al. 2001. "Effectiveness of St. John's Wort in Major Depression." *Journal of the American Medical Association* 285, no. 15 (April 18): 1978.

Smith, Jay, and Diane L. Dahm. 2000. "Creatine Use among a Select Population of High School Athletes." *Mayo Clinic Proceedings* 75, no. 12 (December): 1257.

Woelk, Helmut. 2000. "Comparison of St. John's Wort and Imipramine for Treating Depression: Randomized Controlled Trial." *British Medical Journal* 321 (September 2): 536.

Dowsing

STEVE NOVELLA AND
PERRY DEANGELIS

Dowsing is the alleged paranormal ability to garner a simple answer from a person's surroundings. This feat is supposedly accomplished with the use of a focus and any question that can be answered with either a yes or a no. Although water witching (the locating of subterranean water) is the most widely known application of dowsing, the technique has also been used to attempt to find lost articles and people and even as a guide to financial decisions.

Dowsing can be traced back to our most primitive dwellings. On cave paintings near Tassili, Algeria, there are depictions of herders holding divining sticks pointed toward the skies. Egyptian priest carvings show similar rods, and there are statues from 2200 B.C. of the Chinese emperor Hwang-Yu holding such devices. Even the great philosopher Confucius spoke of the practice.

However, it was not until 1556 that a book was written that included a description of dowsing that was taken seriously. The tome is known as the *De Re Metallica* and was written by Georgius Agricola, a German. This book was not about water dowsing but about using the dowsing technique to find precious metals. It was designed with miners in mind and was widely distributed in its time.

The instrument a dowser uses is called a dowsing rod, dowsing stick, doodlebug (when used to locate oil), or divining rod. Almost any item can be used for this purpose: a birch twig, a whalebone, and even a hanger. The focus is held loosely in the hands and perpendicular to the ground while the dowser walks about until it begins pointing downward. Sometimes, two foci are used, one held in each hand. The location point is then said to be where the foci cross one another. In the case of a dowser using a map to locate something (or the stock market listings to choose a stock), a pendulum is often employed. It is swung gently above the map until it stops over the location where the sought object is said to be.

Dowsers typically give one of three explanations for the mechanism behind dowsing. The first is "physical," involving a force that emanates from the object an individual is attempting to locate. Supposedly, the dowser is simply attuned to this force and thus detects it. Different dowsers are said to be attuned to different stimuli; one attuned to gold may not be attuned to other metals. The second explanation is "psychical," involving the power of the conscious self. This power is said to somehow reach forth and connect with the item being sought, thus revealing the object. This is how map dowsers claim to work. A specialized dowser can allegedly dowse a map in California and locate a missing person in New York. The third explanation involves a combination of the first two methodologies. That is, it presupposes that there are emanations from all objects and that the dowser detects them with his or her consciousness. Many dowsers believe that the subconscious knows the answers to all questions and that one must cultivate an intimate rapport with the nonconscious mind to wrest these answers from within.

Despite widespread belief, careful investigation has demonstrated that the technique of dowsing simply does not work. No researcher has been able to prove under controlled conditions that dowsing has any genuine divining power. In the so-called Munich experiments, the most extensive and celebrated investigation of the technique was conducted by Professor H.-D. Betz and his colleagues in Germany. Betz concluded from his data that "a real core of dowser-phenomena can be regarded as empirically proven" (1997, 55). However, a review of his data (Enright 1999) revealed that Betz came to this conclusion by counting only the best "skilled" dowsers. In effect, he selected for the positive data and rejected the negative data, a clear violation of standard scientific protocol. An analysis of all the data revealed no statistical phenomenon—no effect of dowsing. The same was true when the dowsers who performed the best on initial testing were retested—no effect. In short, despite the claims of some researchers, the largest and best study of dowsing to date demonstrated no dowsing ability.

Likewise with the proposed mechanisms of dowsing, no studies done under proper conditions have been able to demonstrate in a meaningful manner either the physical or psychical explanations advanced by dowsing's proponents. A more likely explanation for the movement of a dowser's focus is the ideomotor effect, which entails involuntary and unconscious motor behavior. In 1852, William Carpenter gave a lecture, reprinted in the *Proceedings of the Royal Institution,* in which, for the first time, ideomotor activity was identified as a third category of unconscious, instinctive behavior. (The other categories are excitomotor [breathing and swallowing] and sensorimotor [startle reactions] activity.) Ideomotor movement is secondary to thought, and it begins in the cerebrum. Any body movement without volition can be attributed to the ideomoter effect. According to the ideomotor explanation, it is the dowsers' expectations that cause the subconscious movement of the dowsing rods. This is also the best explanation for the movement of the planchette on a Ouija board or the actions taken by those engaged with "facilitated communication" (see entry in section 2).

References:

Betz, H.-D. 1997. "Neue Ergebnisse der Rutengängerforschung: Wetter-Boden-Mensch." *Zeitschrift für Geobiologie* 5: 55–59.

DeAngelis, Perry. 1996. *The Connecticut Skeptic* 1, no. 4.

Enright, J. T. 1999. "The Failure of the Munich Experiments." *Skeptical Inquirer,* January-February.

Gardner, Martin. 1957. *Fads and Fallacies in the Name of Science.* New York: Dover Publications.

URL: http://www.barrettdorko.com/articles/ideomotor.htm. URL: *http://www.skeptic.com/contents.html#D.*

Vogt, Evon, and Ray Hyman. 2000. *Water Witching USA.* 2d ed. Chicago: University of Chicago Press.

Earthquake Prediction

RUSSELL ROBINSON

The ability to accurately forecast the time, place, and size of large earthquakes would clearly be useful. Civil defense authorities could be alerted for a rapid response, critical facilities could be secured, and perhaps lives could be saved by evacuating unsafe structures. Consequently, much work, both scientific and pseudoscientific, has been dedicated to earthquake prediction. Many types of observations have been suggested as possible forecasting tools, along with a few examples from the past that look convincing at first sight. Further rigorous testing is rarely undertaken, however, and no scientifically accepted forecasting method has yet been identified. Most proposals slowly fade from attention when experience shows they are invalid or unreliable. But some people remain convinced they can forecast earthquakes by pseudoscientific means. Because the statistics of earthquake occurrence are not understood, such individuals take inevitable random coincidences as proof that their proposed methods are valid. Also, failures and false alarms are often forgotten. Reasonable physical mechanisms explaining why the proposed methods work are usually lacking. This is not to say that reliable and useful earthquake forecasting will never be achieved, whether by scientific or pseudoscientific means, but much work remains to be done.

It is useful to be aware of some basic statistics of earthquakes when faced with a proposed method of forecasting that, on the face of it, seems scientifically implausible but has been "successful" in some circumstances. On average, there are seventeen earthquakes of magnitude 4 or more somewhere in the world every day of the year. In populated regions, most such events are felt. The numbers decrease if only bigger events are considered. So, for example, a forecast that doesn't specify a particular location or magnitude is almost certain to be correct but useless. If a large-magnitude event is forecast but no position is given, the chance of a coincidence is lessened but is still reasonably large. Most people would consider a magnitude 6 or more earthquake to be large. Such an event will occur somewhere on Earth 120 times a year on average. If the forecast is considered correct when the actual date is within a week of the forecast date, then the probability of a random coincidence is about one in two, the same as flipping a coin.

Another point to remember is that large earthquakes don't occur randomly over Earth's surface but mainly in well-defined zones as mapped out by previous events and geologic investigations. For example, large earthquakes are rare in the eastern United States but much more common along the San Andreas and associated faults in California (earthquakes are caused by rapid slip along a fault). This observation is, in fact, the basis of long-term statistical forecasts of earthquake hazard; seismologists know more or less *where* large earthquakes will happen in the future but not *when*.

Some of the more common pseudoscientific methods proposed for forecasting earthquakes involve earthquake weather, unusual animal activity, Earth tides, eclipses, planetary alignments, unusual sounds, astrology, and just plain psychic ability. Most people using these methods probably truly believe they are correct. But fraud has been suggested (if not proved) in a few cases. For example, predictions made *after* an earthquake may sometimes be presented as true forecasts.

Of the many proposed pseudoscientific earthquake forecasting methods, perhaps only those referencing Earth tides (produced in the solid Earth by the Moon and Sun) and unusual animal behavior (see the later discussion) have any plausible physical mechanism behind them, at least as physics is presently understood. And even then, the explanations are questionable. A recent study on Earth tides as a trigger for earthquakes shows a very small statistical effect for small earthquakes along the San Andreas fault in central California. The researchers had to use thousands of events to make the effect even barely visible. This accords with calculations of the size of the stress caused by the tides, which is very small compared to the stress released by the earthquakes themselves. So the tides may conceivably trigger an event that was "almost ready to go" anyway and would have occurred shortly even if there were no tidal effect. It should also be noted that tidal stresses vary throughout the day and at some times are likely to inhibit earthquakes rather than trigger them.

Related to possible tidal triggering are proposals that planetary alignments will cause large earthquakes (or worse). An alignment occurs when several planets, rotating about the Sun at different speeds, all fall on (or close to) a line from Earth. The idea is that at such times, the gravitational effect of the planets on

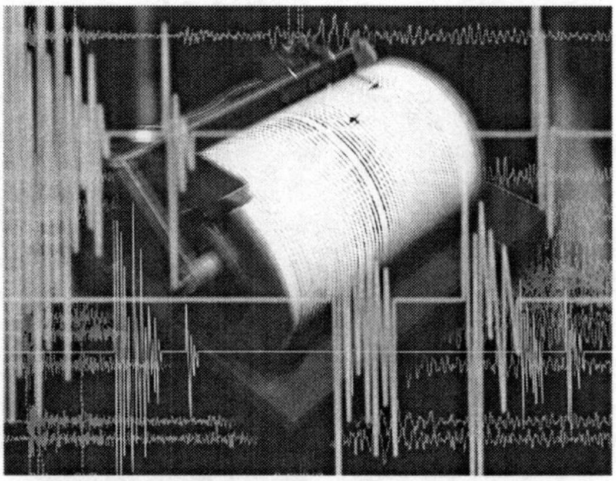

Composite image of a chart and a Richter scale. (Jason Reed/Photodisc)

Earth all add up, thereby causing earthquakes. In 1974, a book appeared forecasting doom during one such alignment that would occur in 1982. The book reportedly sold a large number of copies, especially in southern California (where people should be worried about earthquakes anyway). But there were no large earthquakes near the predicted time. More recently, there was the so-called Grand Alignment of May 5, 2000, when Earth and five other planets, plus the Sun and Moon, all came close to falling on a line (and it was a new millennium to boot). There were similar forecasts of earthquake disasters, but none occurred. It is easy to calculate the tides due to the planets and show that the effect of the planets involved in these predictions is very small compared to that of the Sun and Moon (as was discussed earlier). In fact, because the Moon's orbit is not exactly circular, its tidal effect varies throughout a month by a much larger amount than any possible tidal effect of the more distant planets.

Animals may be sensitive to changes in the environment that people cannot sense and for which instruments are not in place. Unusual animal behavior was apparently one factor that led Chinese officials to predict the 1975 Haicheng earthquake (magnitude 7.3) a day in advance. However, there were other signs as well, such as ground tilting and foreshocks. (The Chinese failed to predict the larger, more deadly Tangshan earthquake one year later.) Some large earthquakes are preceded (by a few days) by a small number of much less intense foreshocks in the same region. These may be too small to be felt by people, although nearby seismographs would record them, and animals sensitive to small vibrations might sense them. Foreshocks may be part of a preearthquake process in which the fault involved begins to slip very slowly for perhaps a few days before escalating into the much faster slip of the earthquake itself. This slow preslip may have effects on concentrations of soil gases near the surface that some burrowing animals could sense. On a shorter time scale, some people have reported that their pets act oddly for ten seconds or so before they themselves feel an earthquake. This may be because of the two main types of waves traveling out from an earthquake focus. The P (primary) waves travel faster than the S (secondary or shear) waves but are usually weaker. Sensitive animals may sense the P waves, whereas people only feel the stronger but slower and thus later S waves.

What would seismologists like to see erstwhile earthquake predictors do in order to test their proposals? First, they should set down the specifics of the method they advocate so that others can reproduce the predictions using the same input data. These specifics should include uncertainties—in other words, how much error is allowed in the place, time, and magnitude. Second, they should set up a formal system in which predictions are made in real time (not after the fact). Third, after a certain period, they should evaluate the number of successes, failures, and false alarms their method has produced as compared to random coincidence.

Would-be earthquake predictors can, in the United States, submit specific predictions to the National Earthquake Prediction Evaluation Council. In additon, the International Association of Seismology and Physics of the Earth's Interior (IASPEI) periodically evaluates prediction methods, as opposed to specific predictions.

References:

An Internet search for "earthquake prediction" (http://www.google.com is recommended) will produce many interesting sites, presenting both the scientific and pseudoscientific viewpoints. Those included in this list of references may be particularly informative.

Bolt, B. A. 1992. *Earthquakes.* New York: W. H. Freeman.

Geller, R. J., D. D. Jackson, Y. Y. Kagan, and F. Mulgaria. "Earthquakes Cannot Be Predicted." *Science* 275: 1616–1617.

Phil Plait's Bad Astronomy Misconceptions. URL: http://www.badastronomy.com/bad/misc/planets.html. Gives a discussion of planetary alignments.

USGS Earthquake Hazards Program–Northern California. URL: http://www.quake.wr.usgs.gov. Provides a great deal of earthquake-related information, including material on prediction.

Wesson, R. L., and R. E. Wallace. 1985. "Predicting the Next Great Earthquake in California." *Scientific American* 252: 35–43.

Electromagnetic Fields and Cell Phones

STEVEN KORENSTEIN

Since at least the 1980s, public concern has grown in response to reports of possible negative health effects from exposure to electromagnetic fields (EMFs). Produced from power transmission and distribution lines, as well as from household appliances, EMFs have reportedly been linked to cancers of the nervous system and blood, immune system dysfunction, and a variety of other, ubiquitous illnesses. Although some studies have shown slight increases in risk, most reputable, large-scale investigations have drawn no clear or conclusive link between exposure to environmental levels of EMF and human disease (Jackson 1992).

Electromagnetic fields are defined by their ability to cause changes in objects around them, such as the pull a magnet has on an iron rod or iron filings. The movement of electrons in, for instance, a wire of a transmission line causes EMFs. EMFs are a combination of two types of waves: electrical waves (sometimes called voltage) and magnetic waves.

Since the nineteenth century, scientists have known that electricity and magnetism are closely related phenomena. It was James Clerk Maxwell whose equation described how a moving electrical current produced magnetic fields and, conversely, a moving magnetic field produced an electrical current. Radio waves, microwaves, and X rays are all familiar examples of the forms of waves that cause EMFs.

All part of the electromagnetic spectrum, energy waves share several general properties. For example, all electromagnetic waves can travel through empty space at the speed of light (although they are much slower when passing through something like a wire). Also, every electromagnetic wave can be described by its wavelength and frequency. When visualized by using special devices, electromagnetic waves have peaks and valleys, just like the waves of an ocean. The distance between peaks is the wavelength and the number of times each wave passes some point in one second is the frequency.

For electromagnetic waves, there is a simple relationship between frequency and wavelength. As the wavelength increases, the frequency decreases. By knowing the wavelength and frequency, we can determine how much energy is associated with the wave. A wave with a high frequency and a small wavelength will have more energy than a wave with a low frequency and a long wavelength. Thus, the higher the frequency of a wave, the greater the damage we can expect to occur from exposure.

Although there are several types of electromagnetic waves, they can be broadly classified into two categories: ionizing and nonionizing radiation. Ionizing radiation has more energy. X rays and gamma rays are examples of ionizing radiation. Ionizing radiation also has a high frequency and, because it has a lot of energy, is able to penetrate living cells and damage genetic material. Nonionizing radiation has less energy. The energy level of nonionizing radiation is not great enough to break the bonds in genetic material. Nonionizing radiation, therefore, is much less damaging to living things than ionizing forms of radiation. Light waves, radio waves, and microwaves are all examples of nonionizing radiation.

Questions concerning the potential health effects from human exposure to EMFs first arose in 1979 with the findings of two epidemiologists, Nancy Wertheimer and Ed Leeper (1979). They observed that children living near electrical transmission lines in Colorado seemed to have a higher incidence of cancer than children living farther away. The researchers stated that the reason for the higher incidence was uncertain, and, in fact, took no actual measurements of EMFs. Their findings, however, were published in the *Journal of Epidemiology* and set off a firestorm of reports in the popular media on the negative health effects associated with exposure to electromagnetic fields.

Researchers involved in the Wertheimer and Leeper study and others like it have been unable to demonstrate any scientifically sound mechanism that would support their claims about the type of diseases purportedly caused by environmental exposure to EMFs. To have a potential for biological damage, electromagnetic fields must be in the ionizing-radiation category, or if they are in the nonionizing category, they must be of such intensity that the waves can injure cells through the physical process of heating. Thus, environmental EMFs must be of greater intensity than the normal fields generated within the human body.

The human body naturally produces electromagnetic impulses, generated by such familiar activities as a beating heart. These naturally occurring impulses have been shown to be ten times greater than a typical exposure from the energy in a transmission line, which commonly has a 60 hertz, 5 milligauss field. The electrical field produced in the human body from this level of exposure would be about ten-millionths of a volt per meter, not enough to cause a significant energy transfer to cells (Adair 1991).

Although the magnetic forces emanating from electrical lines could, in theory, cause biological changes in electrically sensitive cells such as neurons, the field produced from transmission lines is too weak to do so. The field produced from these lines is several orders of magnitude weaker than Earth's magnetic field (Lee, Astumian, and Weaver 1996). Accordingly, man-made environmental fields are too small to produce damaging biological effects (Astumian, Weaver, and Adair 1995). Results from recent studies (Linet et al. 1997) have confirmed that environmental exposure to EMFs does not lead to such diseases as leukemia.

Cell Phones

As with electrical transmission lines, reports of negative health effects have surrounded the use of cell phones since their introduction in the early 1980s. Yet today, nearly 500 million people use cell phones across the world, without any proven illness (Moulder et al. 1999).

Handheld cellular phones use a portion of the electromagnetic spectrum known as microwaves. Microwaves are nonionizing radiation, close to radio waves in their frequency and energy. Accordingly, nonionizing radiation is unable to break molecular bonds within the cell. However, as with a microwave oven, if the microwave energy is intense enough it can cause heating effects. The question becomes whether cell phones produce enough energy to cause biological damage.

As with all electromagnetic waves, microwaves lose intensity rapidly with distance. As the distance from the microwave source is doubled, the strength of the field is reduced by a factor of four. Thus, the energy emitted by a cell phone is much higher than the energy actually absorbed by the user's head. Research published in the *New England Journal of Medicine* (Inskip et al. 2001) did not demonstrate that diseases such as brain cancer would result from such low-intensity exposure. These results agree with other large-scale studies designed to determine if cell phones produce disease. For example, scientists in Europe examined over 400,000 cell phone users in Denmark. The results of the study, the first to be conducted on a nationwide basis, found no correlation between cell phone usage and disease (Johansen et al. 2001).

References:

Adair, Robert K. 1991. "Constraints on Biological Effects of Weak Extremely-Low-Frequency Electromagnetic Fields." *Physical Review* 43: 1039.

Astumian, R. D., J. C. Weaver, and R. K. Adair. 1995. "Rectification and Signal Averaging of Weak Electric Fields by Biological Cells." *Proceeding of the National Academy of Science* 92 (April): 3740–3743.

Inskip, P. D., R. E. Tarone, E. E. Hatch, T. C. Wilcosky, W. R. Shapiro, R. G. Selker, H. A. Fine, P. M. Black, J. S. Loeffler, and M. S. Linet. 2001. "Cellular-Telephone Use and Brain Tumors." *New England Journal of Medicine* 344, no. 2 (January 11): 79–86.

Jackson, J. D. 1992. "Are the Stray 60-Hz Electromagnetic Fields Associated with the Distribution and Use of Electric Power a Significant Cause of Cancer?" *Proceeding of the National Academy of Science* 89 (April): 3508–3510.

Johansen, C., J. D. Boice Jr., J. K. McLaughlin, and J. H. Olsen. 2001. "Cellular Telephones and Cancer: A Nationwide Cohort Study in Denmark." *Journal of the National Cancer Institute* 93, no. 3 (February 7): 203–207.

Lee, R. C., R. D. Astumian, and J. C. Weaver. 1996. "The Salvatore/Weitberg/Wehta Article Reviewed." *Oncology* 10, no. 4 (April): 577–579.

Linet, M. L., E. E. Hatch, R. A. Kleinerman, L. L. Robison, W. T. Kaune, D. R. Friedman, R. K. Severson, C. M. Haines, C. T. Hartsock, S. Niwa, S. Wacholder, and R. E. Tarone. 1997. "Residential Exposure to Magnetic Fields and Acute Lymphoblastic Leukemia in Children." *New England Journal of Medicine* 337, no. 1 (July 3).

Moulder, J. E., L. S. Erdreich, R. S. Malyapa, J. Merritt, W. F. Pickard, and Vijayalaxmi. 1999. "Cell Phones and Cancer: What Is the Evidence for a Connection?" *Radiation Research* 151: 513–531.

Wertheimer, N., and E. Leeper. 1979. "Electrical Wiring Configurations and Childhood Cancer." *American Journal of Epidemiology* 109, no. 3 (March): 273–284.

Fairies, Elves, Pixies, and Gnomes

DAVID J. W. LAURIDSEN JR.

Fairies, pixies, elves, and other related creatures are typically characterized as being diminutive, sometimes winged beings of more or less human form, possessing magical or supernatural powers and living in forest glades, gardens, or other watered and green spaces. Gnomes are historically described as diminutive as well but are generally portrayed as being deformed, crippled, or grotesque; they are sometimes subterranean creatures and are usually said to guard and protect secrets or treasure. These and similar creatures may be good or evil, benevolent or malicious, depending on the story or account. These mythical beings are often said to generally avoid human contact. References to these types of creatures date back several centuries and are most prominent in northern European folklore and literature.

Although no objective, modern scientific evidence concretely documenting these creatures exists, there are true believers. Personal recollections of firsthand encounters with such creatures have been around for as long as the folklore and literature describing them. The explanation given for the lack of credible evidence usually revolves around the description of these beings and their inherent properties. Generally speaking, these creatures are said to only be seen by:

- Certain types of people (the open-minded and willing, the pure of heart, innocent children, and so forth)
- Those having some predisposition to the special ability or skill required to witness or even summon the creatures
- Accident, surprise, or stealth (as by an individual upon waking from sleep or surreptitiously stumbling upon creatures, while remaining undetected by them)
- Those possessing some thing or some quality that makes or forces the creatures to reveal themselves to witnesses

In some explanations, it is forbidden or dangerous for the creatures to reveal themselves to humans, which helps to account for their severe reluctance to be discovered. Some reports use the presence of "telltale" markings or manifestations such as fairy rings, stone circles, hollow hills, or other earthly indications as proof of their existence. Upon realizing they have been discovered, the creatures are sometimes said to disappear, leaving no trace or evidence aside from the witnesses' accounts. This inherent inability to conclusively prove or disprove the existence of such beings reinforces the mystery and enchantment surrounding the stories.

The most documented, highly touted, and widespread account relating the supposed existence of fairies began in 1917 in the Yorkshire village of Cottingley, England. Photographs taken by two young girls (Elsie Wright, age sixteen, and her cousin Frances Griffiths, age ten) became the center of a con-

Elsie Wright with her photograph of the Cottingley fairies, which she and Frances Griffiths faked in 1917–1920. (Fortean Picture Library)

troversy that lasted for over sixty years. Two photographs taken in the glen behind their home in 1917 appeared to show the girls cavorting with a number of small, winged, humanoid creatures that the children insisted were fairies and a gnome.

The girls vehemently denied any impropriety whatsoever in the creation of the photographs, sparking both the interest and the incredulity of their friends and neighbors. It wasn't until three years later, however, after being approached by Spiritualists to further document the existence of these fairies, that the girls achieved international fame regarding their tale. That year, they produced three more photographs of the apparently supernatural beings.

The photographs were published in *Strand Magazine*, with a supporting article written by a seemingly unlikely author—Sir Arthur Conan Doyle, the creator of the analytical detective Sherlock Holmes. Although Doyle's fictional detective was steadfast in his logical approach to the unusual or unexplained, Doyle himself had a penchant for the mystical, and he was a fervent Spiritualist. He eagerly became a believer in the "fairy tale" after various photography experts at the time were unable or unwilling to declare that the photographs were fakes. Without conclusive evidence of tampering and with the firm resolve of the two girls to stick to their story, the Theosophists and Spiritualists of the day heralded the pictures as proof of the existence of fairies. This claim, of course, was the focus of some controversy. Among the most vocal critics was Harry Houdini, the acclaimed magician, illusionist, and friend to Doyle. Houdini quickly and adamantly criticized his friend for his all-too-eager acceptance of the photos and the explanation given for them.

The Cottingley fairy photos (as they came to be called) drifted in and out of controversy until 1982, when their true origins were finally revealed. In an interview with Joe Cooper, Wright and Griffiths, who were in their seventies and eighties at the time, finally admitted that at least four of the photos were complete frauds. They were divided on whether the final picture had been faked, however. Elsie stated quite firmly that it too was a hoax, but Frances resolutely repeated her claim that they had accidentally photographed the fairies she and Elsie said they had seen in the Cottingley glen when they were children. Both women died professing their belief in the existence of fairies, despite their admission that the photographs were fabrications of their childhood imaginations.

References:

Briggs, Katherine. 1976. A *Dictionary of Fairies.* Boston: Allen Lane.

Cooper, Joe. 1990. *The Case of the Cottingley Fairies.* London: Robert Hale.

Doyle, Arthur Conan. 1920. "Fairies Photographed: An Epoch-Making Event." *Strand Magazine.*

——. 1921. *The Coming of the Fairies.* Hodder and Stoughton.

Ernst, Bernard M. L., and Hereward Carrington. 1972. *Houdini and Conan Doyle: The Story of a Strange Friendship.* New York: Benjamin Blom.

"Hidden Ireland: A Guide to Irish Fairies." *Irelands Eye.com.* URL: http://www.irelandseye.com/animation/intro.html. (Accessed on April 13, 2001).

"Library: The Case of the Cottingly Fairies." *James Randi Educational Foundation.* URL: http://www.randi.org/library/cottingley/. (Accessed on April 13, 2001).

Randi, James. 1982. *Flim-Flam!* Amherst, NY: Prometheus Books.

Trubshaw, Bob. "Fairies and Their Kin." *At the Edge,* no. 10. URL: http://www.indigogroup.co.uk/edge/fairies.htm. (Accessed on April 13, 2001).

Van Gelder, Dora. 1977. *The Real World of Fairies.* Theosophical Publishing House.

Wentz, W. Y. Evans. [1911] 1973. *The Fairy Faith in Celtic Countries.* Reprint, New York: Lemma.

Faster-Than-Light Travel

ROAHN H. WYNAR

Faster-than-light travel involves a material object or communication signal moving from one point to another faster than a beam of light can travel between the same two points—that is, faster than 299,792.458 kilometers per second in empty space.

The possibility of faster-than-light travel was never in dispute until Albert Einstein proposed the phenomenally successful Special Theory of Relativity in 1905. That theory states that, for very fundamental reasons, the speed of light is the ultimate speed limit for any object or signal. This limitation applies to immaterial notions as well, such as thoughts, messages, or feelings. The Special Theory of Relativity asserts that the concept of faster-than-light travel is a violation of basic physical principles and therefore an impossibility.

Perhaps due to the fact that science fiction captivated twentieth-century popular culture, the notion of faster-than-light travel has become deeply entwined with our expectation of how the future will unfold. Space travel to distant planets orbiting distant stars is often viewed as being merely a few generations of technology away. However, the prospect of ever moving through space at speeds that make such rapid interstellar trips possible is as unlikely today as it was at the turn of the twentieth century. This is because special relativity is not a technical constraint that can be defeated by clever engineering or yet-to-be-invented engines. Rather, it is a fundamental principle of physics that is, as far as we know, inviolate.

The notion that there is one immutable set of physical laws that operate throughout the universe for all physicists no matter who they are or how fast they are moving is a fundamental assumption of Einstein's theory. The science of physics would simply not be possible if each physicist discovered entirely different laws. This idea is closely related to the concept of causality, which dictates that all occurrences in the physical universe must have a cause and that this cause must precede the effect in time. Causality is also a fundamental assumption of special relativity.

Another basic assumption of special relativity is the idea that empty space is truly empty. Although this notion may seem obvious to the modern mind, the exact nature of so-called empty space and what might be filling it was a question of great scientific interest at the turn of the twentieth century. Consider the example of a submarine. When a physicist travels in a submarine, it is possible to determine whether he or she is moving through the water by several methods. For instance, one could place a small pinwheel on the hull of the sub and infer the motion through the water by watching it spin. Or one could measure the difference in the pressures on the bow and the stern of the sub. But now consider the same submarine in empty space. If space is

Albert Einstein. (Hulton-Deutsch Collection/ CORBIS)

truly empty, then there is no medium to spin a pinwheel or to push against the hull. In fact, if space is really "nothing," then it does not even make sense to talk about moving "through" space at all. All the physicist can say is that he or she is in motion relative to other objects, such as planets, asteroids or other spaceships.

The ultimate implication of this is that all space travelers, no matter how they are moving relative to one another, are equally unable to determine their motion relative to empty space. However, consider the alternate hypothesis: that space is not really empty but instead is filled with a special substance called "ether." If there was some way to detect the ether, then a space pilot could conclude that he or she was moving, say, 100 kilometers per hour relative to the ether. In addition, one could imagine traveling "with" or "against" the ether, and, of course, ether "crosswinds" would also make perfect sense. But more important, there is a special group of physicists who might determine that they are not moving at all in the ether; in fact, members of this group would consider themselves to be at "absolute rest" in the universe. The idea of ether is quite compelling because with it, we can understand light in much the same way we understand sound: light is vibration of the ether just like sound is vibration of the air. The ether hypothesis raises the following obvious question: shouldn't the observer's movement through the ether affect his or her determination of the speed of light? If the ether is moving relative to Earth, then our measurement of the speed of light should depend on which direction the light is traveling. The problem is analogous to the idea of "airspeed" and "groundspeed" for airplanes; if one travels with the wind, one's groundspeed is faster than if one travels against the wind. Despite many years of effort, no physicist has ever observed light moving through empty space at any speed other than 299,792 kilometers per second. No attempt to observe the ether or any of the ether's effects has ever been successful. Einstein proceeded to develop a theory of physics assuming that space is truly empty, that ether does not exist, and that there is no such thing as absolute rest.

Now we can begin to see why light plays such a special role in physics. Light propagates through "nothing," and one cannot determine one's velocity relative to nothing. Therefore, it is not possible, even in principle, to correct the propagation of light for the motion of the observer. This leads to the following amazing conclusion: all physicists measure the speed of light at 299,792 kilometers per second no matter where they are or how fast they are moving. To help visualize this point, imagine a spaceship moving at 99 percent of the speed of light relative to a space station. The pilot of the spaceship turns on a headlight. The spaceship crew will see the light beam zip off ahead of the ship at a speed of 299,792 kilometers per second. The crew of the space station, however, also sees the same light ray move at 299,792 kilometers per second. This means the space station crew sees the light just barely creep ahead of the spaceship. Obviously, these two observations contradict one another. Does the light fly away from the spaceship at 299,792 kilometers per second, or does it just creep ahead of the ship at a mere 2,997 kilometers per second? The answer is that it depends on who you are. But both observers measure the exact same speed for the light ray, namely, 299,792 kilometers per second. The fascinating implication of this is that distances and time intervals for various moving space travelers must be different. This observation leads to the celebrated notions of "time dilation" and "length contraction" (Griffiths 1989). These adjustments in time and length exactly compensate for the relative motion between the observers so that all the laws of physics are the same for the spaceship crew and the station crew. Interestingly, their observations are different, but the laws that describe these observations are the same.

At this point, you should be impressed by the tremendous generality of special relativity. The speed of light is not a technical constraint. It is a consequence of our most deeply held assumptions about the way physics works: causality, empty space, and the existence of universally applicable laws of physics.

Two important conclusions of the Special Theory of Relativity are that all forms of energy that are able to self-propagate in empty space must share the same fixed speed and that nothing can ever move faster than this speed. Only light and gravity are currently known to self-propagate through empty space. All other forms of matter (for this discussion, light and gravity are themselves forms of matter) must be pushed.

In the years since 1905, special relativity has become one of the most well-verified theories in all of science, and its credibility has never once been questioned by sound experimental evidence. Because of this extreme credibility, any observation of faster-than-light travel would be a shocking discovery.

Nonetheless, faster-than-light travel is an important part of many pseudoscientific models, especially models of alien visitation, astral projection, and channeling. Aliens visiting Earth would certainly have come from places outside the solar system. If we take seriously the notion that they travel here in material form, they certainly would travel faster than light; otherwise, their trip would have taken thousands, if not millions, of years. Savvy promoters of alien visitation theories might point to the notion of time dilation in order to explain such trips. Special relativity predicts that objects moving near the speed of light experience time moving at a slower rate, and thus, a thousand-year voyage to an earthbound observer might be over in a few days for the alien traveler. However, special relativity makes it extremely difficult to even approach the speed of light, and it could easily require more energy than the total of several years' output of all the power plants on Earth to reach speeds at which exploiting time dilation is plausible. But even if such energy could be harnessed, the fact remains that this form of travel is not faster than light.

Pseudoscientific models of astral projection and channeling attempt to get around the constraints of special relativity by claiming that the physical body does not actually travel anywhere. Instead, only some ethereal body or even just the thoughts of people or aliens living near faraway stars or galaxies actually cross the vast expanse of space faster than the speed of light. Despite the fact that the properties of astral bodies and channeled thoughts are completely speculative and unknown, the foundation of special relativity is so fundamental that any appeal to incorporeal communication or transfer is still expressly forbidden. This is because special relativity forbids the transmission of information, in any form, at a speed faster than light. As previously mentioned, the basis of this restriction is the scientific commitment to the notion of cause and effect.

Are there any loopholes that may allow for faster-than-light travel? First, we could abandon the idea of causality, but no observation ever made supports that option. Second, we might try to appeal to the General Theory of Relativity, also devised by Albert Einstein, which connects the presence of matter and energy with the structure of space and time. Certain extreme manipulations of general relativity combined with some speculative assumptions allow for the possibility of manipulating the very structure of space in a way that permits faster-than-light travel. Among these ideas are the so-called wormholes said to connect vastly separated regions of space-time and the Alcubierre Warp Drive (Alcubierre 1994). Although it is risky to predict what the future will bring, it seems clear that the energy requirements needed to build wormholes are vastly more than a terrestrial civilization could ever hope to generate. Indeed, any attempts to achieve faster-than-light travel using general relativity are currently plagued by notions of exotic matter, extreme energies, and heavy speculation.

The third possibility that many turn to is that another great theory of physics—quantum mechanics—might defeat the light barrier. The famous Einstein-Rosen-Poldovsky (ERP) paradox seeks to demonstrate that quantum mechanics is inconsistent with special relativity (Sakurai 1994). Thus, the ERP paradox pits the two most highly regarded theories of physics against each other in an attempt to demonstrate that they cannot both be true. Experimental investigation into the question has clearly indicated that quantum mechanics does indeed exhibit certain forms of faster-than-light influences; however, it is not difficult to show that these influences can never be used to transmit information and thus do not threaten the assumption of causality. It appears that the two theories do not, in fact, contradict each other.

Lastly, a recent experiment has discovered that certain laser pulses propagating through a specially prepared gas have traversed that gas faster than light. However, in the abstract of their report, the researchers who conducted this experiment included this statement: "The observed superluminal light pulse propagation is not at odds with causality" (Wang, Kuzmich, and Dogariu 2000). This statement means that it will never be possible to use their technique as a signaling device and, therefore, that no violation of special relativity is involved.

References:

Alcubierre, Miguel. 1994. "The Warp Drive: Hyper-Fast Travel within General Relativity." *Classical and Quantum Gravity* 11: L73–L77.

Griffiths, David J. 1989. *Introduction to Electrodynamics.* Englewood Cliffs, NJ: Prentice-Hall.

Sakurai, J. J. 1994. *Modern Quantum Mechanincs.* Rev. ed. New York: Addison-Wesley.

Wang, L. J., A. Kuzmich, and A. Dogariu. 2000. "Gain-Assisted Superluminal Light Propagation." *Nature* 406 (July 20): 277.

Wynar, Roahn. 2000. Ph.D. diss. University of Washington.

Feng Shui

JON PURO

Feng shui (pronounced "fung shway") is the Chinese tradition of attempting to control one's health, fortune, and future by placing and arranging living quarters, gravesites, physical structures, and interior objects to be in harmony with ancient beliefs about how humans and their environments interact. Based on ancient Chinese philosophical traditions, feng shui has developed for over two millennia to include knowledge, rituals, aphorisms, and superstitions from throughout China. As such, it is central to any understanding of Chinese cultural history, life, and psychology, as well as that of many other East Asian cultures that also practice Chinese feng shui.

Chinese Philosophy and Naturalism

Feng shui has its roots in the Chinese philosophical school of thought called the Yin-Yang School. During the Period of Warring States (403–221 B.C.E.), a turbulent period of Chinese history in the late Zhou dynasty (1122–211 B.C.E.), the principles of the six major schools of Chinese thought were established: Confucianism, Mohism, the Legalist School, the School of Names, Taoism, and the Yin-Yang School. The naturalistic Yin-Yang School, with its emphasis on the interdependence of humankind with nature, became one of the most influential schools of thought in Chinese culture. The school later formed the philosophical basis for feng shui as well as other aspects of Chinese culture, including art, marriage, politics, medicine (e.g., acupuncture), and other practices of divination (e.g., astrology and numerology).

Chinese philosophy is primarily rural in nature. Most Chinese were and still are farmers and thus have always been dependent on nature for their livelihood and preservation of their way of life. The agrarian life has been idealized by most Chinese schools of thought; it was viewed as simple, pure, and innocent, as opposed to the life of the urban merchant, who was frequently characterized as self-centered, greedy, and antisocial. Feng shui draws from this tradition and idealization of rural naturalism.

Reverence and respect for nature can also be attributed to the topology and climate of China. Farms were vulnerable to bitter cold winds and storms from the north, and frequent flooding of the country's major rivers meant the people and their farms were vulnerable to water. Farms had to be located near rivers for supplies of water (particularly for rice farmers), but flooding could quickly destroy a farm. In this way, the majority of people in China felt dependent on nature to bring either prosperity or devastation and thus sought means to control it.

The Yin-Yang School

The Yin-Yang School, as its name implies, was based on the concepts of Yin and Yang, which are the two complementary opposites believed to underlie all of nature. Yin is negative, female, dark, cold, or passive; Yang is positive, male, light, hot, or active. All materials, vegetation, and animals have Yin and Yang, according to this philosophy, but each has more of one than the other and hence tends to be either more Yin or more Yang. These opposing characteristics are not statements of value or worth; rather, they describe the dualistic nature of reality, much like a scientist would speak of the positively and negatively charged poles of a magnet. Yang is not better than Yin or vice versa; they are to be understood as complementary and necessary properties of nature.

According to the Yin-Yang School, the complementary opposites of Yin and Yang originated in the tai chi, or the ultimate "oneness." The unity of the tai chi is composed of the duality of the Yin-Yang, the duality of the Yin-Yang leads to "the four secondary forms," the four secondary forms give rise to "the eight elements," and these eight elements lead ultimately to the complexity of all reality.

The Tai Chi ("Yin Yang") surrounded by the Eight Trigrams. (Courtesy of author)

The philosophy of Yin-Yang is well over 2,000 years old. It plays a central role not only in feng shui but also in all aspects of Chinese life and the development of Chinese culture and science. The concepts of Yin and Yang were first used in Chinese astronomy to understand the movements and relationships of celestial objects (for example, representing Earth as Yang and the Moon as Yin). It was also used to explain natural phenomena, such as earthquakes. The pairing and interplay of opposites in the Yin-Yang School contributed to the underlying Chinese philosophy of harmony in all things and the moral teaching of moderation over extremism.

In the second and third centuries B.C.E., the Yin-Yang School incorporated the Theory of the Five Agents. This theory held that all changes in nature are predicated on the interaction of five "forces" or "agents" that compose all matter. These agents are not materials but are instead processes or properties of nature, similar to the "four elements" proposed by the Greek philosopher Anaximander. The agents (metal, wood, earth, fire, and water) interact in a way to produce change, and thus, changes to any substance can be predicted by its underlying dominant agent. As the five agents interact, more complex items are created, such as trees, mountains, and rivers. Change is always occurring as the five agents destroy and create each other in a cycle.

Coupled with this recognition of change as the only universal constant, a key ancient text of the Yin-Yang School, the *I Ching*, or *Book of Changes*, stated that these changes follow a pattern. The *I Ching* taught that eight trigrams, each made up of three solid or dashed lines, could be used to predict change. This became a very popular method of fortune-telling and is still used today. The eight trigrams originated in the Shang dynasty (approximately 1766–1123 B.C.E.) practice of divination using tortoiseshells and bones. Shells or bones were heated until cracks began to appear, then the fortune-teller would "read" the cracks to see the future. The trigrams were an attempt to copy this practice, and they are used to assist the feng shui expert in divining a person's future.

Despite the occultist practices that grew out of the Yin-Yang School, this school of thought was central to the later development of Chinese science. The naturalism of the Yin-Yang School and feng shui, when they were developed over 2,000 years ago, was a kind of protoscience. Humankind and nature were viewed as interdependent agents, each affecting the welfare of the other. As such, humans could attempt to improve their lives by understanding and controlling their environment. Of course, the Yin-Yang School and feng shui were not true science, since there was no reliance on physical evidence to prove their contentions and especially since feng shui later came to employ occult rituals, supernatural forces, and superstitions in its practices, all of which are quite contrary to modern science. But the view that humankind could control its destiny by interacting with and using nature to its advantage was significant in its time. Due to the development of philosophy and naturalism, the ancient Chinese did not need to employ any divinities to explain the universe or for moral guidance, so the Chinese people never developed any dependence on gods or religion. For them, philosophy and naturalism filled all spiritual and moral needs.

風水

The Chinese characters "Feng Shui" ("Wind and Water").

The Practice of Feng Shui

Feng shui is an eclectic mix of naturalistic philosophy, environmental awareness, ancient astronomy and astrology, fortune-telling, magic, and folk traditions. The term *feng shui* literally means "wind and water," and the emphasis on living harmoniously with nature is evident in many feng shui principles, some of which are quite rational. For example, feng shui teaches that building a home on the south side of a hill is optimal. This is likely due to the fact that China is subject to bitterly cold north winds, so a home built on the south side of a hill would have natural insulation from those winds. Feng shui also teaches that a home should be placed midway up a hill, not at the base or the top. This is also logical given China's topography: building one's home at the top of a mountain often would expose it to the same frigid northerly winds, and building it at the base of a hill could bring disaster because of the oft-flooding rivers in China. From these logical foundations, however, feng shui has grown into a vast and complex tapestry of protoscientific or pseudoscientific theories, fortune-telling, and superstition.

The current practice of feng shui is the result of the fusion of the two primary feng shui schools around the third century A.D. One school, developed in Fukien Province, stresses the importance of direction. This so-called Fukien or Compass School of feng shui uses the ancient book of divination, the *I Ching*, to determine optimal geometric balance and placement. Building orientations may be classified by the Compass School as conforming to one of the eight trigrams (discussed earlier). The eight trigrams pertain to eight different directions on the compass (north, northeast, east, and so on). Each of the eight directions is said to possess characteristics that make certain activities in that location more or less favorable as compared with other locations.

The Compass School of feng shui also incorporated ancient Chinese astronomical knowledge. Thousands of years ago, Chinese astrologers developed techniques to view patterns and messages in the stars, Sun, Moon, and planets. The feng shui terms for the four directions of right, front, left, and back are taken directly from the Chinese astronomical terms for east, south, west, and north and are, respectively, the dragon, bird, tiger, and tortoise. The need for accurate identification of direction in feng shui was one of the primary reasons for the development of the compass in China.

The second school of feng shui, from Kiangsi Province, was primarily concerned with shapes of landmasses and bodies of water. Associated with other practices that grew out of the Yin-Yang School, such as astrology, physiognomy, numerology, and acupuncture, this school of feng shui, often called the Form School, was a type of geomancy, which is the reading and interpretation of meanings from shapes and patterns in the physical environment. According to this school, different shapes and contours in the Earth are taken to mean different things. For example, a feng shui practitioner working on a home design or planning a gravesite may look at the surrounding hills to determine what animal or beast they resemble. A hill or a combination of hills might be seen to resemble a tortoise, tiger, dragon, snake, or phoenix. A dragon is usually considered optimal, since the dragon is seen as a fierce protector. However, a dragon could also be bad if a home or grave is placed near the dragon's mouth or tail. Near its mouth, the structure might get eaten; near its tail, it might be destroyed as the tail swings. The family inhabiting the house or the relatives of the deceased could experience bad fortune or even death because of the location of the house or the grave. A river nearby is usually viewed as good, since the river flow brings ch'i, but a river might also be seen as resembling a dragon or serpent; placement of a structure near a river must also be done carefully so as not to harm the beast and not to place the structure near its tail or mouth. In *The Golden Bough,* James Frazer called this association of the properties of separate objects based on their similar appearance the "Law of Similarity." This law is common the world over and is the basis for many Western and Eastern traditions and folklore.

Another use of the Law of Similarity in feng shui is in the perceived connection between the appearance of buildings and any of the five elements. (See the earlier discussion of the five elements.) A building that is tall and thin (such as a tower), no matter what material it is made from, is called a "wood" type of building because of its resemblance to a tree. Such a building is said to possess the properties of the wood agent. A building that is flat and square is an "earth" type of building. There are also metal, fire, and water types of buildings. Each of these is said to react differently if placed within environments of different types, which may also be wood, earth, metal, fire, or water types. For example, a wood building placed in a fire environment will give more than it receives, as the environment takes from the building in the same way that fire takes from wood. Such a building would be deemed a poor place for a business because the business might lose money, and it would be better used as a school, hospital, or some other such function that gives to the surrounding community. Two adverse elements may also be neutralized with the use of a third "controlling" element.

The concept of ch'i is also central to feng shui. Ch'i refers to a hypothetical life force or energy that permeates all living and natural bodies: all animals, including people, have ch'i, as do plants, mountains, rivers, wind, the Earth, the Sun, the Moon, and the planets. Ch'i is viewed as a force that flows through the universe, thus connecting all living and non-living objects. Ch'i, it is said, can be used for one's benefit, but it can never be controlled. The goal of feng shui is to use the Earth's ch'i to one's advantage. It is also the goal of feng shui to minimize the effects of *sha*, which is the term for the negative current that carries bad fortune and is seen as the opposite of ch'i.

The flow of ch'i is viewed in feng shui as vital to the well-being of one's home. Doors, hallways, gardens, and furniture all must be placed to provide for its optimal flow in order to prevent the disharmony that would result from holding ch'i in one place. According to Yin-Yang philosophy, blocking or preventing the smooth flow of ch'i is bad because movement and change are viewed as fundamental properties of the universe. Thus, interfering with this natural flow can be disruptive of the natural order and potentially disastrous. For example, feng shui teaches that a bed should be placed in a room so that it is not directly in front of a door and thus blocking the incoming ch'i energy. Mirrors are also viewed as being potentially powerful reflectors of this energy and thus must be placed so as not to concentrate ch'i into one area. There are many other such prescriptions for interior design and arrangement. Recommendations will often vary according to the feng shui practitioner, as the rules tend to be very general and subject to personal interpretation, but the goal is always the comfort, security, and prosperity of the dweller through the proper management of ch'i.

Finally, the feng shui practitioner might also employ any of the myriad Chinese folk rituals or traditional healing methods. For example, a lucky charm or a bamboo flute may be placed in a particular location to ward off lurking evil spirits. Any of a number of folk adages might be quoted to justify this or some other recommendation. Various traditions, folk remedies, and superstitions from throughout Chinese culture have also been incorporated into feng shui's eclectic mix.

Feng shui is widely practiced today. Still popular in China, it has made its way throughout East Asia and is practiced in Singapore, Korea, Laos, Thailand, Vietnam, the Philippines, Malaysia, and Japan. Recently, feng shui has made its way to the West and has become very popular among New Age enthusiasts. Books, TV shows, and Web sites attest to the efficacy of feng shui, expound its teachings, and sell an assortment of products and services purported to ward off bad luck and improve one's fortune.

References:

Frazer, James George. 1922. *The Golden Bough*. New York: Macmillan.

Lin, Henry B. 2000. *The Art and Science of Feng Shui*. St. Paul, MN: Llewellyn Publications.

Rossbach, Sarah. 1983. *Feng Shui: The Chinese Art of Placement*. New York: Penguin Books.

Walters, Derek. 1991. *The Feng-Shui Handbook*. London: Aquarian Press.

Wilhelm, Richard, trans. 1950. *The I Ching or Book of Changes*. New York: Bollingen Foundation.

Wu, Lawrence. 1986. *Fundamentals of Chinese Philosophy*. Lanham, MD: University Press of America.

Yu-Lan, Fung. 1948. *A Short History of Chinese Philosophy*. New York: Macmillan.

Geller, Uri

SIMON JONES

Uri Geller is the Israeli metal bender and psychic illusionist who became a sensation in the early 1970s. Geller has convinced many people, including several scientists who have tested his abilities, that he possesses genuine psychic powers. Skeptics point out that skilled conjurors can replicate all of Geller's feats using trickery and that nonpsychic explanations must be eliminated before one assumes that the laws of nature have been broken.

Uri Geller is chiefly known for being able to bend or break small metallic objects such as spoons. His reputation also rests on his ability to read the contents of sealed envelopes (which usually contain drawings allegedly prepared out of his sight) and restart watches that appear to have stopped working. Geller maintains that he has never used trickery to achieve his effects. However, conjurors have produced similar feats using sleight-of-hand and misdirection techniques. In addition, some observers claim to have caught Geller in the act of bending cutlery with his hands (see Emery 1987).

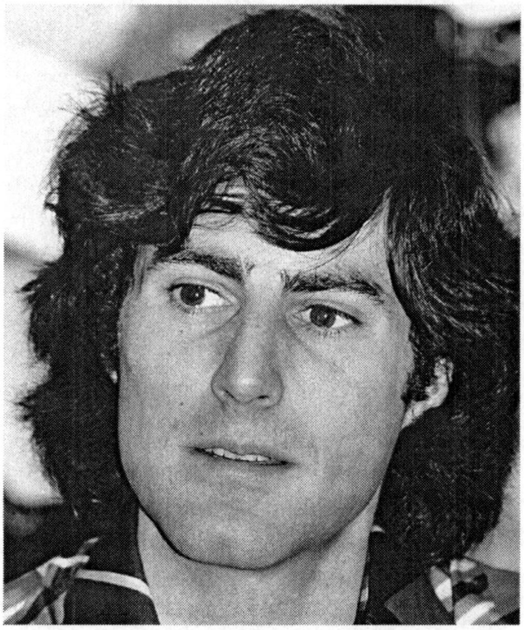

Uri Geller, the psychic performer who claims he can bend cutlery and perform other feats using only the power of his mind, c. 1978. (Hulton-Deutsch Collection/CORBIS)

The Spoon Bend

A conjuror can create the appearance of a spoon bending while it is gently stroked. A momentary misdirection by the conjuror, such as moving position to show the spoon to other people, allows the conjuror to bend the spoon physically. He or she can then disguise the bend with a hand and slowly reveal it at the appropriate moment. The effect is so convincing to most observers that they believe the spoon is bending before their eyes. There are many ways by which an observer can be misdirected. For instance, when Geller performs metal bending, he often moves the item toward other metallic objects in the room, which he claims enhances the effect. He also frequently fails in his initial attempts to bend the metal but returns to the object a short time later (after trying other psychic effects) and achieves the bend. This again provides the opportunity for misdirection.

An alternative nonpsychic technique, which often causes the spoon to break, requires the conjuror or an accomplice to have prior access to the cutlery. The conjuror or another individual prestresses the spoon by carefully bending it back and forth until it reaches the point of breakage. The stress point is not readily visible. The conjuror then picks this item apparently at random and subjects it to gentle rubbing, causing "plasticity" followed by complete fracture. Prior to some of his television appearances, Geller has been known to have had access to the cutlery that he later broke in front of the cameras. Some critics have accused Geller of using chemicals on his hands to soften the metal, but since there is no such chemical that could be used safely, this explanation can be discounted.

Reproducing a Drawing in a Sealed Envelope

Over the years, magicians have developed many different techniques for divining the contents of a sealed envelope. Some of these techniques (such as gimmicked notepads on which the drawing or message is made) are available on the market; others are still used by professional magicians. The methods can be as simple as peeking through one's fingers to see the drawing being made, holding the envelope up to the light, or even opening the envelope when the viewer's attention is distracted. A confederate may also be able to assist by conveying information about the drawing. Skeptics allege that Geller's manager and brother-in-law, Shipi Shtrang, has acted as a confederate, and Shtrang *has* been present at many of Geller's successful demonstrations. Observers often forget the presence of an accomplice, particularly if that person appears to have no active role in the proceedings.

Restarting "Broken" Watches

The restarting of apparently broken watches has persuaded many observers of the reality of psychic phenomena. However, neither trickery nor psychic powers are required to achieve this effect. According to researchers David Marks and Richard Kammann (1977), jewelers estimate that "over 50 per cent of watches brought in for repair are not mechanically broken, but have stopped because of dust, dirt, gummed oil, or badly distributed oil." When such a watch is bumped or held between one's hands to warm up the oil, it may start working for a short period. Marks and Kammann have demonstrated this effect in more than half of a random selection of "broken" watches. In his appearances on television and radio, Geller can also rely on the statistics of a large pool of viewers or listeners. Among such a sizable group, there will inevitably be a few people who claim to find their watches and clocks working after years of apparent inactivity. Skeptics have demonstrated that this effect can be produced by nonpsychics in the same circumstances.

Testing by Scientists

Uri Geller has repeatedly claimed that science has proven the existence of his alleged psychic powers. The principal experiments on which he bases his assertion are those conducted at the Stanford Research Institute in California. The results of these experiments were published in the science journal *Nature* in 1974, where it was suggested that under controlled laboratory conditions, Geller had demonstrated extrasensory perception but not paranormal metal bending. The paper was accompanied by an extensive editorial that explained that the paper's referees had expressed serious reservations about its scientific merit. Others have also condemned the protocols used in these experiments as lax and unscientific. Moreover, Geller has never participated in repeatable experiments under conditions that would preclude fraud. Magicians have also pointed out that scientists are rarely experienced in detecting legerdemain.

Personality

Skeptics argue that Uri Geller's personality is a powerful factor in his ability to convince people that he has genuine psychic powers. Those who have met him have attested to his engaging warmth and youthful enthusiasm at the effects he produces. However, Geller's charm is not extended to skeptics, and he has threatened or pursued numerous legal actions against his detractors. In 1991, he filed a $15 million lawsuit against magician James Randi and the skeptics organization Committee for the Scientific Investigation of Claims of the Paranormal (CSICOP) after Randi told a newspaper that Geller had "tricked even reputable scientists" with tricks that "are the kind that used to be on the back of cereal boxes when I was a kid" (interview in the *International Herald Tribune*, April 9, 1991). The court found against Geller, who eventually settled the case at a cost of $120,000.

References:

Emery, C. Eugene. 1987. "Catching Geller in the Act." *Skeptical Inquirer* 12, no. 1: 75–80.

Frazier, Kendrick, ed. 1986. *Science Confronts the Paranormal.* New York: Prometheus Books.

Hyman, Ray. 1977. "Review of *The Geller Papers*, edited by Charles Panati." *Zetetic* 1, no. 2: 73–80.

Margolis, Jonathan. 1999. *Uri Geller: Magician or Mystic?* New York: Welcome Rain.

Marks, D., and R. Kammann. 1977. "The Nonpsychic Powers of Uri Geller." *Zetetic* 1, no. 2: 9–17.

Randi, James. 1982. *The Truth about Uri Geller.* New York: Prometheus Books.

Handwriting Analysis and Graphology

JOHN BERGER

Graphology is the method of interpreting personality through examination of an individual's handwriting. It is not to be confused with document examination, which is the inspection of handwriting, ink, and paper for clues to origin and circumstances of the message. Document examiners use microscopes, chemical analyses, and electrostatic machines to carry out their investigations; graphologists use magnifying glasses, onionskin paper, and slant gauges to do their studies.

Handwriting analysis is based on the premise that the appearance of one's handwriting is influenced by the writer's subconscious and that proper interpretation of handwriting will lead to analysis of the writer's personality. There is little mention of external or environmental influences on the appearance of one's handwriting, which is why most graphologists prefer to look at handwriting samples made under normal conditions over a period of time. This is also why skeptics criticize the reliability of handwriting analysis. Graphologists look at several trends and traits in a person's handwriting. Some of the more important tendencies are slant of writing, page layout, preference for a particular zone, distortion of words or letters, speed of writing, and pen pressure.

Much of the analysis and many of the conclusions drawn by handwriting analysts are common sense: quick thinkers write quickly, messy people have messy writing, and uneducated folks make many spelling errors. The interpretations get more interesting when the script is construed as a series of doodles. In handwriting, we often stylize letters subtly, sometimes for fun. The letter *B* becomes the profile of a woman's torso, the letter *S* becomes a dollar sign, and the letter *T* becomes one's spine and shoulders. Of course, these letters may be interpreted differently; the *T* might bring to mind a table or a telephone pole, and the *S* can suggest a snake or a roadway. The handwriting analyst then notes any fugue in the art that may be a theme in the writer's personality.

According to analysts, the way in which people write their words varies according to the way they feel about them. For example, one woman's hand trembled as she wrote *we*, *us*, and *our* (she was going through a divorce). A man made *Mom* illegible because he didn't want to talk about his mother. Some people underline their signature as if to say, "Look at me, I'm important" or "I worry that I am not important." For the most part, the graphologist looks at the hidden message; that is, the overall appearance of the note is more important than the words. The appearance reveals the topics that a writer wants to emphasize and neglect, thus revealing several aspects of the personality. It is not unlike interpreting the mind of an artist through examination of the preference of color and emphasis of subjects in his or her paintings.

Despite the simplicity of the basis of

graphology, there are problems with this method. Primarily, each handwriting trait has multiple explanations. That is, there is no one-to-one correlation between a personality attribute and a handwriting peculiarity. For example, a loud and energetic man may write in a large script. But there could also be another reason for his script: if he has bad eyesight, he may need to write in large letters in order to read what he has written.

More to the point, when a graphologist sees a message written by a quivering pen and presumes the author has had a stroke or a similar health problem, how do we know that it wasn't the desk that was shaking or that the note wasn't written while its author was riding on a city bus moving along a roughly paved road? Look at the implications. Let's assume that every standard graphological trait can appear for two reasons. Now let's assume that we are looking at a handwriting sample that contains five significant traits. If a graphologist attaches only one reason for each trait, then he can make one conclusion, based on the five traits, about the author. If there are two reasons for each trait, then the number of interpretations rises to five squared. In other words, we can have as many as twenty-five explanations of that handwriting sample. In real life, we reduce the number of interpretations to a manageable few by assuming the traits that share similar origins will be the ones to include in the report that will be prepared. For example, a page written in tiny script with excellent layout and alignment is more likely made by a person who has great attention to detail than by a person who is secretive. Nevertheless, the number of interpretations is still greater than one, and that fact erodes the validity of graphology.

Another area of handwriting analysis that is not germane for all is the association of certain letters with personality traits. For example, the letter *k* is associated with feelings of aggression, and any person who emphasizes their *k*s is said to have violent tendencies. We are told the letter *k* is the initial letter of such cruel words as *killing, karate, knifing, kicking,* and *kamikaze*. However, a quick look in the dictionary reveals a large number of nice words that start with the letter *k, kisses, kitten, kindness, kinship,* and *Kamasutra* among them. As well, if the writer's name starts with the same letter, he or she may emphasize it out of habit. Therefore, the association of the *k* with violent tendencies cannot be true for all people.

A major trait in handwriting is speed. Although we are told that a quick writer is a fast thinker, we must realize that a person who has consumed excessive caffeine will also write quickly or that a sleepy person will write very slowly. Although these last two reasons support the idea that speed of writing is related to speed of thinking, the caffeine high and the sleepy condition are temporary and may not be typical of the person's normal condition. Thus, there is a question of validity when relating speed of writing to quickness of thought. Another key point in graphology is linking pen pressure to anger or energy. Angry writers tend to press hard with a pen and will sometimes tear the paper. Normal people will also press hard when they are using a ballpoint pen that rolls with difficulty, or they may tear the paper if writing on a soft surface. Either state of affairs will make the pen pressure appear excessive, and the graphologist may conclude the writer is angry, yet both situations have external influences acting on the process of writing that have nothing to do with the writer's state of mind.

The slant of a person's writing is supposed to reveal his level of extraversion or introversion. The International Graphoanalysis Society of Chicago provides a clear-plastic handwriting slant gauge to their students to enable them to determine the slope of handwriting on a scale of A to F, with A being introverted (backhanded writing) and F being very extraverted (slanted far to the right). One can

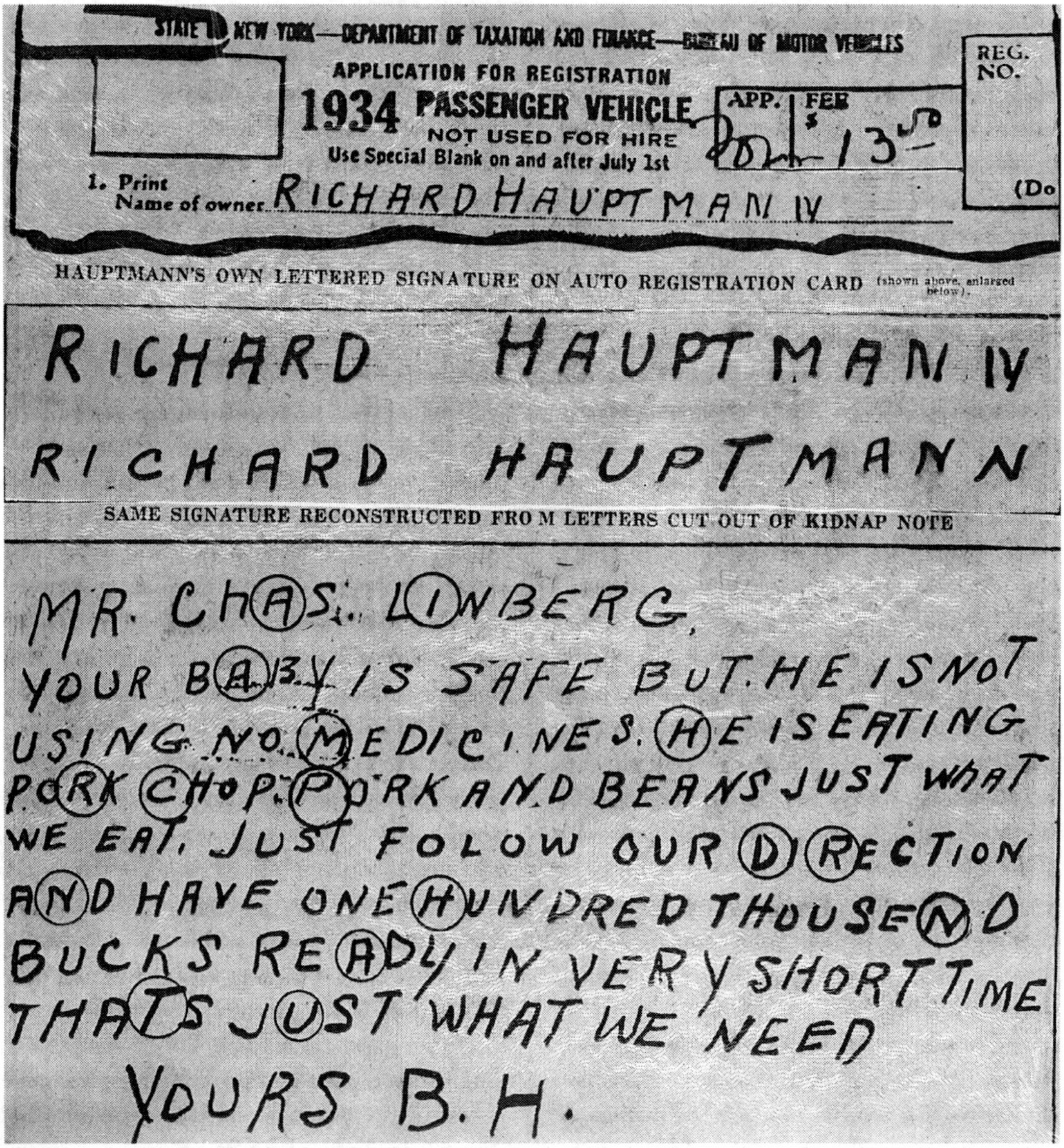

STATE OF NEW YORK—DEPARTMENT OF TAXATION AND FINANCE—BUREAU OF MOTOR VEHICLES

APPLICATION FOR REGISTRATION

1934 PASSENGER VEHICLE

NOT USED FOR HIRE

Use Special Blank on and after July 1st

APP. | FEE $ 13.50

REG. NO.

1. Print Name of owner RICHARD HAUPTMAN IV

(Do

HAUPTMANN'S OWN LETTERED SIGNATURE ON AUTO REGISTRATION CARD (shown above, enlarged below).

RICHARD HAUPTMAN IV

RICHARD HAUPTMANN

SAME SIGNATURE RECONSTRUCTED FROM LETTERS CUT OUT OF KIDNAP NOTE

MR. CHAS. LINBERG,
YOUR BABY IS SAFE BUT. HE IS NOT
USING NO MEDICINES. HE IS EATING
PORK CHOP. PORK AND BEANS JUST WHAT
WE EAT. JUST FOLOW OUR DIRECTION
AND HAVE ONE HUNDRED THOUSEND
BUCKS READY IN VERY SHORT TIME
THATS JUST WHAT WE NEED
YOURS B. H.

Handwriting evidence from the Lindbergh Case, 1934. (Bettmann, CORBIS)

conduct a simple experiment to show slant is not a reliable aspect of handwriting analysis. Ask a man to sit comfortably and start writing. After he has written a few sentences, ask him to stop and turn the paper counterclockwise a bit. He will continue to write comfortably, but his writing will now display a greater rightward slant (the author conducted this experiment). His emotional state has not changed; only the position of the paper has shifted. Thus, writing slant is shown to be an unreliable approach to determining extraversion.

The appearance of a particular trait does not always mean the author has the matching mannerism at all times. For example, one university professor had appalling handwriting, from which a graphologist might have correctly predicted a disheveled appearance.

However, this instructor was also an expert in molecular structure, and his work in that field was impeccable. The graphological conclusion for this professor missed an important facet of his life, and therefore was not relevant. A person who prefers to print sentences instead of writing in copybook fashion is said to be "construction minded." This is a common trait for men or women in the building trades, architecture, and engineering who sketch boxlike structures and sketch legible plans for a living. This statement is a generalization but not a rule. Consider the case of a man who wrote and spoke Chinese, which has a picture-alphabet. That is, the Chinese word for house is a symbol that is derived from a sketch of a house. When the man wrote in English, he sketched his letters instead of writing in a script. Yet he was not in the construction business; he was the manager of a garage. Thus, the graphological trait of the "construction-minded" personality is not interpreted the same across all nationalities.

Handwriting experts are often used to select the best potential employees from a stack of handwritten job applications. In a series of studies reported by Abraham Jansen (1973, 126), when graphologists examined several handwritten résumés and the results were compared to business personnel ratings, the graphological judgment showed a positive but very slight agreement. Although a handwriting analyst can be useful in choosing a few good applicants from a large number of applications, there are some potential problems. For example, if job seekers know that a company's decision to interview and hire is based on handwriting, they may ask honest, reliable, and intelligent acquaintances to write the résumés. Alternatively, the job seekers could pick up a book on graphology at the library and learn the handwriting style necessary to get hired. Thus, a graphologist may be tricked into recommending an unsuitable person for a job.

Graphotherapy is like handwriting analysis in reverse. Graphotherapists believe that a change in handwriting causes a change in personality. They cite instances where a graphotherapist has been able to cure emotional illnesses by teaching the subject to write better. They defy credibility by claiming the current handwriting taught in public schools in the United States and Canada is creating mental illness in our children. Of course, anybody can practice inflating their upper loops and making Greek-style *e*s (two traits associated with higher intelligence) to impress handwriting analysts, but such changes in handwriting style won't increase the number of neurons in one's brain, nor will it erase blocks to memory, improve oxygen transfer across arterial walls, or alter one's IQ score. At most, it can create a new persona that pretends to be intelligent and organized. Another issue in job selection by handwriting analysis is correlation. For example, in graphology, the letter *f* represents organizational ability. An inflated upper loop suggests management potential; an inflated lower loop is a sign of a person who follows orders well. The letter *f* may also have a personal meaning; perhaps it is the first letter of an applicant's girlfriend's name or a reminder of the time he was beaten for uttering the *f*-word. Thus, although the lowercase *f* is an indicator of organizational position, there are other factors that can reduce the correlation of management ability to handwriting.

Handwriting is not the only external indicator of personality. Body language, speech, and choice of colors, cars, and dress are outward traits that reveal certain things about our character. They are studied by salespeople to help select the right product for the customer. For example, people who are conforming and reliable often speak in a predictable tone, wear blue clothing, drive conservative cars, and write in copybook style. However, just as not all owners of blue sedans are conformists, not all outward signs correlate to personality traits.

Someone can buy a blue car because of its price, thus showing they are more motivated to acquire a bargain than to travel in a vehicle having the correct color for their personality. We can say there is a correlation between observed behavior and personality, but this correlation is never perfect.

In conclusion, graphology is based on a relationship between personality and handwriting, but it does not accommodate external factors in the analysis. The relation between handwriting and personality is neither reliable nor valid enough to call handwriting analysis a science. It is an art with many interpretations, even among professional graphologists who seldom agree on the same interpretation of a handwriting sample (Jansen 1973, 126).

References:

Hill, Barbara. 1981. *Graphology*. London: Robert Hall.

Jacoby, H. J. 1939. *Analysis of Handwriting*. London: George Allen and Unwin.

Jansen, Abraham. 1973. *Validation of Graphological Judgments*. The Hague: Mouton Publishers.

Lester, David. 1981. *The Psychological Basis of Handwriting Analysis*. Chicago: Nelson Hall.

Hypnosis

ROBERT A. FORD

Hypnosis is a deeply relaxed state in which one person (the subject) becomes unusually receptive to suggestions made by another (the hypnotist). Scientists are not agreed on whether hypnosis is an altered state of consciousness (trance), different from both sleep and wakefulness, or whether its effects can be explained by the expectations of the subject and others present. Its use as a form of entertainment on the stage and on television has deterred many from studying it seriously, but it has also been used in both medical and psychotherapeutic practice and research.

A session of hypnosis begins with the subject in relaxed pose and the hypnotist making a series of suggestions, referred to as "hypnotic induction." These suggestions are aimed at causing the subject to relax and become absorbed in thoughts and images mentioned by the hypnotist. Sometimes, a subject may be asked to fix his or her gaze on a small object, light, or pattern while listening to the hypnotist. The monotony of this activity helps to induce relaxation and openness to suggestion. When the induction is complete, the hypnotist may make various suggestions to test the depth of the hypnotic state. For example, the subject may be told that a limb has become rigid and immobile or that a part of the body will move by itself (ideomotor suggestions). Perceptual suggestions induce experiences in the subject, such as the feeling that an arm has become very heavy or very cold. Complex experiences may also be suggested, for instance, that the subject is taking part in some social event or reliving a past experience. Subjects may respond freely to these situations, or they may be directed to adopt a particular response; thus, they could be told, "Your foot is becoming hotter and hotter, and to relieve this you must take off your shoe and put it on the table." Any response may be directed to take place after the hypnotic session is finished (posthypnotic suggestion). People can also be instructed to forget the instruction was given, so that they later act upon it without knowing why.

Stage hypnotists often select the most suggestible subjects by making suggestions to an entire audience and seeing which people respond best. These individuals will be the subjects who can be induced to perform the most bizarre acts with the greatest entertainment value. A favorite such "test" is arm levitation, in which an entire audience is asked to stand with their eyes closed while the hypnotist suggests that their right arms are getting lighter and lighter and floating upward. A few will respond with highly raised arms, showing that they are very receptive to suggestion. The U.S. psychologist Ernest Hilgard (1986) estimated that about 15 percent of people are highly susceptible to hypnosis and that 5 to 10 percent highly resistant. He also showed that susceptibility decreases with age, a fact that has led some researchers to believe that it reflects a biological characteristic.

It is generally maintained by hypnotists that no one can be hypnotized to do anything that is "against their will." In practice, this seems to mean anything that conflicts with their moral code or personal values, but attempting to demonstrate this can backfire. One college lecturer used to demonstrate hypnosis to his psychology classes and finish with a suggestion that the subjects would take off their clothes. The students would invariably refuse, but one day, a young woman began to strip without hesitation. The lecturer hastily terminated the session and asked her why she hadn't refused like the others. She replied that she was paying her way through college by working weekends as an exotic dancer. Taking her clothes off in front of a group of strangers was nothing unusual for her.

The History of Hypnosis

The power of suggestion in effecting "miracle cures" may have been known as far back as biblical times and was mentioned by Aesculapius in 400 B.C.E. However, modern interest in the phenomenon is generally traced back to Franz Anton Mesmer, an Austrian physician practicing in Paris in the 1780s. Mesmer believed that everyone possessed magnetic fields, which he termed "animal magnetism," and that illness resulted when the balance of these fields was disturbed. Since he believed his own magnetism to be unusually abundant, he thought he could cure disease by channeling some of his surplus into patients to restore the balance in their own fields.

Mesmer had patients sit in a darkened room while soft music was played. Around them were barrels of water, ground glass, and iron filings, which were supposed to influence the magnetic fields. Mesmer donned a theatrical robe and carried an iron bar, with which he tapped patients lightly in passing. Some were so affected that they had fits or convulsions and had to be taken into another room to recover. The whole setting was designed to impress with suggestions of mystical power, and some patients did report cures. However, a committee set up by King Louis XVI to investigate Mesmer concluded that the cures were due not to animal magnetism but to suggestion, and Mesmer's work fell into disrepute (see Mesmerism entry in section 5). His name survives in the word *mesmerized.*

In the 1840s, James Eskdale, a British physician working in India, reported conducting painless surgical operations without anesthetic using *hypnosis,* a term coined by a colleague from the Greek word for sleep. According to Eskdale, patients reported no pain during surgery, and they had no memory of pain afterward. Anesthetics were just being developed, however, and little attention was paid to his work.

Hypnosis as Entertainment

Since hypnotized subjects can be induced to perform suggestive or outrageous acts in front of others, hypnosis has been used for decades as a form of entertainment on the stage and television. Sometimes, simple demonstrations are used, such as the human plank trick. In this trick, the subject usually lies across three chairs and is told that his or her body has become totally rigid. The hypnotist then removes the middle chair, leaving the subject suspended from the other two chairs by head and heels.

More complex perceptual suggestions are often made, so that a whole group of subjects act out different responses to a particular word from the hypnotist or some other trigger stimulus. These responses might be shouting, singing, or doing an animal impression. People can also be persuaded that they have special powers, such as X-ray vision to see through people's clothing and reveal the audience naked. Such effects are often referred to as "hallucinations," but when questioned after being part of such an experience, subjects clearly indicate they did not actually see the effects suggested (such as a naked audience), although at the time they behaved as if they did. People can also experience "negative hallucinations" (believing something is absent when it is not). However, although they will say that they cannot see the object in question, they will still walk around it instead of bumping into it, rather as sleepwalkers do. People who are merely pretending to be hypnotized usually bump into such objects.

Since show business hypnotists have sometimes led subjects to perform indecent or otherwise unacceptable acts, most Western countries now have some system for the regulation or licensing of hypnosis when used for entertainment. A few people have claimed to suffer psychological damage after participating in stage hypnosis, but recent reviews suggest that there is little evidence to support their claims (Heap 2000). It seems likely that subjects experiencing problems after the event may attribute them to having been hypnotized, and others may feel in retrospect that they were humiliated by being made a public show.

The Therapeutic Use of Hypnosis

Since Mesmer, hypnosis has been used by many physicians, psychiatrists, and psychotherapists for therapeutic purposes. As well as general relaxation, therapeutic uses include suggestions intended to encourage change in attitudes and behavior—for example, to improve self-esteem and confidence or to reduce the craving for tobacco. Naturally, these changes should be agreed upon in advance between therapist and patient.

As well as psychological or behavioral problems, psychosomatic disorders (physical ailments with a psychological cause) are reported to be amenable to hypnotic treatment, and a number of scientific studies confirm that this may be so. These disorders include migraine headaches, some intestinal problems such as irritable bowel syndrome, asthma, and several skin complaints such as eczema, psoriasis, and even warts (Agras 1984). Since physicians generally accept that there is a psychological element in much physical disease, it is not surprising that hypnosis has also been used in the treatment of "purely" physical illness. Thus, it has been used to alleviate pain in both medical and dental practice, as well as in natural processes such as childbirth.

It is very difficult to evaluate many claims for the therapeutic effectiveness of hypnosis, as the whole field is plagued by problems of definition. For example, many therapists use elements of hypnosis in their work, such as relaxation and suggestion, without calling them by that name. It is also certain that suggestion can have significant effects outside of hypnosis, such as the well-known placebo effect—the tendency of people to report improvement when given any treatment, even one that has no treatment value. Sorting out these influences takes very careful experimental design and statistical evaluation of the results. Nonetheless, in a recent report, the British Psychological Society (2001) accepted that there can be therapeutic value in hypnosis for both physical and psychological problems.

Hypnosis and Memory

The use of hypnosis to enhance memory dates back to at least 1895, when the Viennese psychoanalyst Sigmund Freud reported its use in the case of a female hysteric patient. He claimed to have traced the cause of her problem to sexual abuse in childhood, perpetrated

Illustration of hypnotic pendulum. (Bruce T. Brown/Stone Images)

by her father, and he suggested this as a general cause of hysteria. The woman's memory of the abuse was supposed to have been repressed (removed from consciousness and hidden in the unconscious mind). Freud was ridiculed, however, as hysteria is a common condition and his theory implied the existence of a great many abusive fathers in middle-class Vienna. Freud eventually accepted that what he had uncovered was a fantasy rather than a repressed memory, and he subsequently dropped the use of hypnosis to recover lost memories.

Despite Freud's experience, some therapists continue to use hypnosis—or elements of it—to regress patients to an earlier age in an attempt to recover lost memories. In regression, patients are hypnotized, encouraged to imagine that they are still children, and asked questions about what they experience at that age. Some therapists even claim to be able to regress people beyond birth to previous lives. There is no scientific evidence that this process has any validity, nor any evidence for its effectiveness in therapy.

Hypnosis has also been used in attempts to enhance the memory of witnesses in police investigations, perhaps to improve the descriptions of suspects or to confirm car license numbers that were only briefly glimpsed and imperfectly remembered. Supposedly, by being hypnotized, people can be encouraged to return mentally to the scene that they witnessed and observe it more carefully than they did at the time. Unfortunately, the only material they can work with is what they observed at the time, and this procedure has often succumbed to the same problems encountered by Freud. One British police force hired a forensic psychologist to hypnotize a witness who had seen a suspect car but could not remember the license number. Under hypnosis, the witness reported a full number, and the owner was duly traced and raided. Unfortunately, the owner turned out to be the witness's former girlfriend. She had recently broken off their relationship, to his great distress, and her car number had considerable emotional significance for him at the time.

Examples like this highlight the main problem with regression and hypnotically recovered memory—the memories "recovered" are often very inaccurate. In recent years, the problem has come to prominence in cases of so-called false memory syndrome, although the term *syndrome* is inappropriate in this context, as it suggests a disease pattern. In fact, it is completely normal for hypnotically induced memories to be held with great confidence by subjects, however inaccurate they may be. This may be why allegedly recovered memories can lead to protracted court cases brought by the supposed victims of childhood abuse and reinforce fixed ideas in others who believe they have been abducted by aliens. Studies have shown that, although people remember more under hypnosis than they otherwise would, they recall more inaccurate information as well as more accurate information. Under less controlled conditions than those of a scientific experiment, it is often impossible to distinguish between the two. The situation can be complicated still further by "source amnesia," in which the subject remembers the information (accurate or not) but forgets that it was recalled under hypnosis. Finally, it has not yet been demonstrated that memories can be repressed while remaining accessible through hypnosis; the existence of false memories, however, is not in doubt (Conway 1997). Consequently, in both the United States and Britain, the authorities have laid down guidelines for the forensic use of hypnosis, and in Britain specifically, it is actively discouraged, especially with witnesses who may later give evidence in court. The use of hypnotic regression in therapy is much less well regulated.

The Nature of Hypnosis

Strangely enough, although the phenomenon has been known and used for 200 years, the nature of hypnosis has received little scientific attention until the last few decades. Inevitably, there is disagreement among researchers about what hypnosis actually is, and two schools of thought have emerged. One takes the traditional view that hypnosis is a trance state—a state of consciousness different from both sleep and wakefulness. The other believes that hypnosis is nothing more than social role-playing, induced by the expectations of the subject and perhaps the audience, if there is one. Both of these theories have problems in terms of explaining all the phenomena associated with hypnosis.

Prominent among the "state" theorists is Ernest Hilgard (1986), who proposed a "neodissociation" theory. According to this theory, hypnosis divides consciousness into separate and parallel channels of mental activity, so that the subject can attend to the hypnotist and other events simultaneously. Thus, when told under hypnosis that he was deaf, a subject appeared to be so. However, when asked to give a signal if there was some part of him that could still hear, he gave the signal. Similarly, when told that one of their hands is anesthetized, subjects typically say that they cannot feel when that hand has been touched. However, when both hands are touched several times and the subjects are asked to count the number of touches, they report the total number of touches to both hands, including the "anesthetized" one. Like the "deaf" man who could still hear, it seems there is some part of them that can still feel the anesthetized limb.

T. X. Barber (2000) argued that hypnosis is nothing more than subjects acting out a role in accordance with what they feel is expected of them. What they do, Barber suggested, is suspend the normal rules of self-control, enabling them to carry out acts that are normally inhibited. He has shown that many hypnotic demonstrations—including the human plank trick—can be performed by nonhypnotized people. Barber also showed that even skilled hypnotists are unable to tell whether someone is genuinely hypnotized or just pretending.

The controversy and the research continue. The most recent review of research and practice (British Psychological Society 2001) reported that, 200 years after Mesmer, no firm conclusion could yet be drawn about what hypnosis actually is.

References:

Agras, W. S. 1984. "The Behavioral Treatment of Somatic Disorders." In *Handbook of Behavioral Medicine*, edited by W. D. Gentry. New York: W. H. Freeman.

Barber, T. X. 2000. "A Deeper Understanding of Hypnosis: Its Secrets, Its Nature, Its Essence." *American Journal of Clinical Hypnosis* 42: 208–272.

British Psychological Society. 2001. *The Nature of Hypnosis*. Leicester: British Psychological Society.

Conway, M. A., ed. 1997. *Recovered Memories and False Memories*. Oxford: Oxford University Press.

Heap, M. 2000. "The Alleged Dangers of Stage Hypnosis." *Contemporary Hypnosis* 17: 117–126.

Hilgard, E. R. 1986. *Divided Consciousness: Multiple Controls in Human Thought and Action*. New York: Wiley.

Ideomotor Effect (the "Ouija Board" Effect)

MICHAEL HEAP

The ideomotor effect is the phenomenon whereby a seemingly involuntary movement occurs in response to the suggestion or expectation of that movement. Ideomotor responding is thought to underlie certain activities or phenomena that are accorded paranormal status, such as the Ouija board (or planchette) and dowsing, as well as other extraordinary practices such as facilitated communication.

A common illustration of ideomotor responding is postural sway. One person is instructed by another to stand upright and to focus on the slight tendency for his or her body to sway backward and forward to maintain balance. The suggestion is continually repeated that these movements are becoming more and more pronounced. Individuals vary in their responsiveness to this suggestion; some do not respond at all, whereas some respond so well that they have to be kept from falling over.

Another example is arm levitation in response to repeated suggestions that the subject's arm is feeling light and automatically rising; similar responses are seen in finger levitation and as a reaction to suggestions that the participant's outstretched arms are being drawn together by an invisible force. Often, an apposite image is introduced; for example, imaginary magnets may be held in each hand in the case of the last example mentioned here.

The ideomotor effect exemplifies what is sometimes called the "classic suggestion effect," which also includes changes in perceptual experiences (e.g., warmth or coolness of the hand). The key characteristic is that the response is experienced as involuntary. Individuals differ in their responsiveness to ideomotor suggestions. This responsiveness appears to be a stable characteristic and is sometimes termed "primary suggestibility."

The ideomotor effect is most likely to be the basis of certain unusual phenomena whereby observed movements in humans are ascribed to some paranormal entity or force, thus contravening Occam's razor. (This is the principle that hypothetical constructs should not be used when the phenomenon can be explained by existing knowledge.) An example is the Ouija board, or planchette, and its variations (e.g., turning tables). One or more participants place a finger or fingers on a movable object such as a small platform on wheels. The participants then ask questions, and the object moves, apparently automatically, toward one of two written replies, "Yes" or "No." A popular variant is to use an inverted tumbler surrounded by a circle of cards, each bearing a letter of the alphabet, with the answers spelled out as the tumbler moves toward different letters in sequence. Some people believe that this arrangement allows a dialogue between the participants and the spirits of people who have died. However, the effect may be more parsimoniously explained by the net result of the participants' ideomotor responding due to expectation.

Another example is sometimes known as Chevreul's pendulum. One person holds the string of a pendulum at the top and fixates the bob, which initially is in the resting position. The experimenter suggests that when the other person thinks that the pendulum is moving in a particular direction (backward and forward, side to side, clockwise, counterclockwise, and so on), it will gradually start to do so, without any deliberate effort on his or her part.

Hypnotherapists have employed pendulums in this way on the assumption that the ideomotor reponses allow a dialogue with a patient's unconscious mind. A particular movement of the pendulum, say, backward and forward, is identified as the unconscious communication "Yes," whereas a different movement, say, side to side, denotes "No." The therapist may then address questions to the patient's "unconscious mind," such as, "Is there any particular memory that may be still troubling you?" Nowadays, most hypnotists prefer signals by ideomotor finger movements.

The idea that the communications come from the patient's unconscious mind can best be regarded as a metaphor. There is no reason to suppose that the answers elicited have any special validity, but there may be circumstances in psychotherapy in which this procedure has some advantage over direct verbal communication when broaching sensitive and potentially distressing issues.

Simple pendulums are also employed by dowsers, some of whom claim to be able to use them to locate missing objects or people by holding them over maps. There is also a tradition of foretelling the sex of an unborn child by the direction of the swing of a pendulum over the expectant mother's abdomen.

Another dowsing technique uses rods. A dowsing rod can be manufactured by straightening out a wire coat hanger, cutting it down to an appropriate size, and bending it at a right angle near one end so the small shaft fits comfortably into the hand. One such rod is held in each hand at about shoulder height with the long shafts parallel and horizontal, pointing ahead. At some stage, as one processes around an area, the long shafts will swing toward (or sometimes away from) one another. If just one of these rods is used, the same movement will be observed.

The relevant ideomotor response in this case is the raising of the hand, thus changing the position of the rod's center of gravity in relation to the fulcrum at the hand. (Some dowsers claim that putting the short end of the rod in a sleeve, such as the empty stem of a ballpoint pen, eliminates the influence of the ideomotor effect, but this is clearly fallacious.) Therefore, a plausible explanation of dowsing is the ideomotor effect in response to the expectations of the dowser.

The ideomotor effect appears to underlie facilitated communication, whereby children with learning disabilities or autism seem able to type out complex messages on a keyboard even when there is no prior evidence of any degree of literacy. This phenomenon only occurs when the child's hand is supported by a trained facilitator. In this case, it is the expectations and ideomotor responding of the latter that are responsible for the message produced on the screen or paper.

One explanation for the occurrence of the ideomotor effect is that imagining or expecting the movement generates equivalent neuromuscular activity that is too slight to be consciously experienced. A gross movement, such as arm levitation, may be the result of a sequence of small ideomotor responses. However, a second mechanism—dissociation—may underlie more complex activity, as in automatic writing. It is hypothesized that some individuals may have a well-developed capacity for suppressing from conscious awareness activity that would normally be experienced at a conscious level. In the case of the ideomotor effect, this would be awareness of the intentional effort normally associated with the movement; hence, say, the hand appears to be moving on its own. A third mechanism at work may be attribution; the participant is more inclined to experience the movement as involuntary because the context (the instructions of the experimenter) have defined it thus.

References:

Hilgard, Ernest R. 1986. *Divided Consciousness: Multiple Controls in Human Thought and Action.* New York: Wiley.

Hoggart, Simon, and Michael Hutchinson. 1995. *Bizarre Beliefs.* London: Richard Cohen Books.

Lynn, Steven J., and Judith W. Rhue, eds. 1994. *Dissociation: Clinical and Theoretical Perspectives.* New York: Guilford Press.

Lynn, Steven J., Judith W. Rhue, and John R. Weekes. 1989. "Hypnosis and Experienced Volition." In *Hypnosis: The Cognitive-Behavioural Perspective*, edited by Nicholas P. Spanos and John F. Chaves, 78–109. Buffalo, NY: Prometheus.

Laundry Balls

ROAHN H. WYNAR

Laundry balls are spherical or toroidal (doughnut-shaped) objects intended to be inserted into a washing machine instead of soap. The basic claim for laundry balls is that they are as effective as soap at cleaning clothes but have no environmental impact, since they release no chemicals and can be reused indefinitely. Laundry balls are additionally claimed to deodorize, sterilize, bleach, and soften clothes. They are typically sold via multilevel marketing but can often be found in retail stores that cater to the environmentally conscious. Most are plastic, but some are ceramic. Some contain a colored fluid that never escapes the ball but is supposedly connected with its mechanism of action.

Laundry ball manufacturers usually make one of two claims regarding how the balls work. Some claim that they modify the nature of the water in the washing machine. The water is said to become "structured," "ionized," or "clustered," depending on the manufacturer. This special, modified water is supposedly able to more deeply penetrate the fabric of clothing and carry away dirt. The fact that no two manufacturers make precisely the same claims about how the laundry balls work is a good clue that the entire concept is invented to exploit the inclination of people to be environmentally conscious. Liquid water has no structure, and water in general forms structure only in the solid state, namely, ice. The idea that liquid water contains some sort of complex internal organizing structure is similar to the claims of homeopathy and has no scientific basis. Since the laundry balls add no chemicals to the water, it is not clear how they can affect the water at all. Other claims involve supposed infrared rays that emanate from the laundry ball and affect either the clothes directly or the water around the clothes. However, there is no reason to believe that infrared light has any special effect on water or that laundry balls emit such light in unusual quantities.

The second common claim is that "liquid magnetism" emanates from the laundry ball. Suppliers making this claim use words from the science of magnetohydrodynamics to string together unsubstantiated claims about how the special magnetism emanating from the balls cleans clothes and helps to prevent disease. However, no details are ever provided, and the entire notion of liquid magnetism appears to have been invented in the manufacturer's imagination.

The laundry ball is completely inert and has no effect on the washing of clothes. Instead, the belief that they work is a delusion reinforced by the simple fact that warm water and agitation will clean clothes to a considerable degree as long as the amount of organic staining is low. When using laundry balls on normally soiled clothing, consumers are discovering that detergent is only marginally beneficial. But heavily soiled or greasy laundry will not become clean with a laundry ball. With the success of laundry ball promotions, the uses began to expand into every conceivable cleaning niche, including washing cars, people, and food.

In both Utah and Oregon, state agencies have forced laundry ball manufacturers to stop making false claims about their product. However, it must be emphasized that marketing of laundry balls is usually not illegal. By using certain vague language and limiting details regarding claims, companies can sell laundry balls free from the fear of prosecution.

References:

The best source for information about laundry balls is the Internet. See:

"Laundry Balls Online." URL: http://www.syntac.net/hoax/Laundry/index.php.

State of Oregon Department of Justice. URL: http://www.doj.state.or.us/FinFraud/97C14017.htm.

World Wide Scam. URL: http://www.worldwidescam.com/show.htm.

Magnetic Therapy

SATYAM JAIN
AND DANIEL R. WILSON

Magnetic therapy involves the application of magnetic fields on parts of the body to speed healing, relieve pain and inflammation, or improve bodily functions. The use of magnets can be traced to ancient Greek, Egyptian, Chinese, and Indian physicians. Modern magnetic therapy begins with the Viennese physician Franz Anton Mesmer.

Magnetic therapy is becoming a more visible part of the alternative-medicine boom in the United States and Europe. Some claim that properly designed magnetic products are not simply useful for therapy but are also essential for proper health; they suggest that people are healthier in parts of the world where magnetic fields are stronger. And millions of people visit Lourdes, France, where greater magnetic fields allegedly prevail. Is it all just hokum, as many previously assumed, or is magnetic therapy becoming scientifically respectable?

Therapeutic Claims

The intimate physiology underlying such a wide range of claimed benefits is not completely understood—if, indeed, either the benefits or the physiology exists at all. A common theme is that magnetic fields increase blood circulation to and from bodily tissues, which, it is claimed, enriches the supply of oxygen and nutrients while also enhancing the exfiltration of contaminants, toxins, and inflammatory mediators.

Other studies suggest magnetic fields may modulate pain receptors to induce a slight anesthetic effect. Other prototheses abound. One is that magnetic field effects are directly transmitted to the brain via blood vessels' "circuitry," with a subsequent release of endorphins (chemicals that act as natural pain relievers). Another is that magnetic fields attract positive ions to enhance the body's reaction to the Earth's magnetic fields, with positive benefits as a presumed consequence (via a rather obscure mechanism). The explanation accepted by most medical experts is more mundane—magnets simply harness the power of placebo. In other words, they work because people think they work.

Explanations that magnetic fields increase circulation, reduce inflammation, or speed recovery from injuries are simplistic and are not supported by the weight of experimental evidence, which, it must be said, is itself quite limited. The effects of magnetic fields on body tissues are complex and appear to vary from tissue to tissue and with different intensities and durations of the magnetic field applied. The nature of magnetic devices does not always make them amenable to randomized, controlled, double-blind studies (which again are few in number).

Research

Such research as exists is largely divided into two distinct areas: pulsed bioelectric magnetotherapy and therapy via fixed magnets. Some 85 to 90 percent of the scientific citations relate to the former, largely as pulsed bioelectric biomagnetic therapy; the remainder are based on therapy with fixed solid magnets.

Fixed Solid Magnets

A small scientific study at the Baylor College of Medicine suggested magnets may ease pain (Vallbona, Hazelwood, and Jurida 1997). More specifically, the study showed that magnets reduced muscular and osteoarthritis pain in a small group of postpolio patients.

In this double-blind study, 29 of 50 patients enrolled had magnets strapped to their most tender spots for 45 minutes. The other 21 patients had sham magnets that looked exactly the same placed on painful areas. The results revealed that 76 percent of those with real magnets said their pain decreased, but only 19 percent of those with fake magnets felt any improvement.

This study involved only one 45-minute treatment, did not compare the magnets to other treatments, and did not evaluate how long the reported pain relief lasted. Moreover, scientific dictates require positive results to be repeated before they can be considered reliable. Throughout medical history, numerous theories that looked promising based on initial studies failed when subjected to repeated examination.

Apart from that, several other "studies" listed on commercial Web sites for diverse magnetic products claim similarly diverse effects. These include reduced foot pain in diabetics (magnetic insoles); "clinically relevant" pain relief and sleep improvement in fibromyalgia sufferers (magnetic mattress pads); improvement in visual memory, cognitive function, drawing performance, and social interactions in Alzheimer's patients (external application of electromagnetic fields ranging from 5 to 8 hertz); and remission of depressive symptoms (transcranial magnetic stimulation). Further claims attributed to unspecified magnetotherapeutic products include improvement in everyday performance in children with attention deficit disorder, relief of pain from sports injury, edema reduction in ankle sprain, and much more.

Without exception, these investigations have not been replicated and/or they lack scientific controls. Rigorous scientific studies sufficient for acceptance by mainstream medicine are in short supply. One rigorous study of low-back pain treated by magnets versus sham magnets was published in the *Journal of the American Medical Association* in 2000, but no pain or mobility differences were evidenced. Hence, most physicians do not recommend magnets for pain relief or other uses. This dearth of data is also why the U.S. Food and Drug Administration (FDA) has not approved any health product with magnets and why the Federal Trade Commission is cracking down on companies that claim magnets can treat or cure illnesses.

However, the results of several studies were intriguing enough that the National Institutes of Health's Office of Alternative Medicine (NIH-OAM) has commissioned two studies of magnetotherapy, but results are not yet published.

Pulsating Electromagnetic Field Therapy

The most widely studied application of electromagnetic field therapy in human medicine is in fracture therapy. Although the mechanisms remain undetermined, several studies report electrical fields generated by pulsatile electromagnetic field therapy stimulate biological processes pertinent to osteogenesis and bone-graft incorporation. This form of therapy is approved for the treatment of delayed and non-union fractures in humans in the United States by the FDA. Pulsating electromagnetic field therapy, however, delays the healing of fresh, experimentally induced fractures in rabbits.

Pulsating electromagnetic field therapy has also been evaluated in the treatment of soft-tissue injuries, with the results of some studies providing evidence that this form of therapy may be of value in promoting the healing of chronic wounds (such as bedsores), in neuronal regeneration, and in many other soft-tissue injuries.

In contrast, a number of investigators have been unable to show any effect of low-level electromagnetic fields on tissue healing. One study, for example, failed to identify any beneficial effect of applying a magnetic field to a nonhealing fracture and concluded that the long periods of immobilization and inactivity required for the application of the magnetic field therapy were just as likely to be responsible for tissue healing.

Criticisms of pulsating electromagnetic field studies include several points: some of the studies are poorly designed, independent trials have not been conducted to confirm positive results, and the electrical fields induced by the machines are several orders of magnitude lower than required to alter the naturally occurring electrical fields that exist across biological membranes. Even proponents of the therapy concede that much work needs to be done to optimize such variables as signal configuration and duration of treatment before pulsating electromagnetic field therapy can be generally recommended.

Magnetic therapy advertisement for Magnetic Shields, a sort of body armor to enhance the magnetic power of the body, 1893. (The Fortean Picture Library)

Marketing Magnetism

Magnetic therapy appeals to those who want to relieve chronic pain without drugs and invasive intervention. Both athletes as well as ordinary folks are strapping magnets onto sore spots in order to find pain relief and faster healing from sprains, strains, cramps, and mangled muscles. As a result, magnetic therapy is becoming increasingly popular, particularly among osteopaths, physiotherapists, and chiropractors as well as various holistic practitioners.

About $150 million worth of magnets are sold as medical products each year in the United States alone. They come under more than a dozen different brands and in various shapes and sizes. The range goes from coin-sized patches that cost a few dollars to king-sized mattress pads that sell for up to $1,000. One can even buy magnet-studded facial masks, car seats, shoe insoles, and pet collars.

Vendors selling magnets claim that magnetic therapy does not have any side effects, but its effects are not warranted in pregnant women, people with pacemakers and defibrillators, and people suffering from open sores, tuberculosis, viral infections, and mycoses. Currently, no study is available to document the long-term effects of magnetic therapy.

Conclusions

The status of magnet therapy is typical of other largely unconfirmed claims rampant in alternative medicine. It is certainly possible and even likely that magnets have significant physiological properties and so may well have therapeutic applications. Still, very few data have been derived from well-designed scientific studies testing the efficacy of magnetic therapy in the treatment of specific medical syndromes. Moreover, the fact that magnetic therapies appear harmless does not mean they are safe. Thus, comprehensive issues of safety and efficacy persist with respect to the medical use of magnets. Indeed, there is such a dearth of systematic data as to either the safety or effectiveness of magnetotherapy that, for now, the best advice is caveat emptor.

References:

Burcum, J. 1999. "Magnets as Pain Therapy Attract Skeptics and Believers." *Minneapolis Star Tribune*, August 25.

Collacott, E. A. 2000. "Bipolar Permanent Magnets for the Treatment of Chronic Low Back Pain." *Journal of the American Medical Association* 283: 1322–1325.

Conca, A., S. Koppi, P. Konig, E. Swoboda, and N. Krecke. 1996. "Transcranial Magnetic Stimulation: A Novel Antidepressive Strategy?" *Neuropsychobiology* 34, no. 4: 204–207.

Pilla, A. A., and L. Kloth. 1997. "Effect of Pulsed Radio Frequency Therapy on Edema in Ankle Sprains: A Multisite Double-Blind Clinical Study." In *Abstracts of the Second World Congress for Electricity and Magnetism in Biology and Medicine*, 300. Bologna, Italy.

Ramey, D. W. 1998. "Magnetic and Electromagnetic Therapy. *Scientific Review of Alternative Medicine* 2, no. 1: 13–19.

Sandyk, R. 1994. "Alzheimer's Disease: Improvement of Visual Memory and Visuoconstructive Performance Treatment with Picotesla Range Magnetic Fields." *International Journal of Neuroscience* 76, nos. 3–4: 185–225.

Vallbona, C., C. F. Hazelwood, and G. Jurida. 1997. "Response of Pain to Static Magnetic Fields in Postpolio Patients: A Double-Blind Pilot Study." *Archives of Physical and Rehabilitative Medicine* 78: 1200–1203.

The Mars Face

Extraterrestrial Archaeology

KENNETH L. FEDER

In a general historical perspective, the ultimate source of the Mars Face controversy can be traced to the musings of Italian astronomer Giovanni Schiaparelli, who, in 1877, observed linear markings on the face of Mars. He called them *canali* (the Italian word for channels), which he identified as natural features of the Martian landscape. U.S. astronomer Percival Lowell became the chief proponent of the hypothesis that Schiaparelli's canali were, in fact, "canals" that were the result of intelligent activity, designed to distribute water to the agricultural fields and settlements of Martians. Speculation about the possibility of intelligent life on our planetary neighbor has burgeoned ever since.

The proximate source of the Mars Face controversy was a grainy photograph (see figure) taken on July 25, 1976, by a camera mounted on the *Viking* orbiter spacecraft. Searching for potential landing areas on the Martian surface while flying at a little more than 1,860 kilometers (1,162 miles) above a region of the Red Planet called Cydonia, the *Viking* camera caught sight of a surface feature that appeared to bear a striking resemblance to a human face (Malin Space Science Systems 2002).

Though conspiracy theories abound surrounding this feature of the Martian landscape, it cannot be said that the National Aeronautics and Space Administration (NASA) made any attempt to hide or cover it up. In fact, on July 31, just six days after the photograph was radioed back to Earth and processed, NASA released the image to the public and included this caption: "The huge rock formation in the center, which resembles a human head, is formed by shadows giving the illusion of eyes, a nose and a mouth" (NASA press release). Ultimately, nine images of the so-called face were captured by the *Viking* spacecraft, but none was as evocative as the original (to view a selection of these images, see Malin Space Science Systems 2002).

Soon thereafter, some were suggesting that the facelike feature on Mars, which was 3 kilometers (nearly 2 miles) long and 240 meters (about 800 feet) high, was neither an illusion of light and shadow nor a fortuitously shaped natural feature. Rather, they said, it was actually an artistic depiction of a face—a monumentally scaled archaeological artifact of an ancient and now most likely extinct Martian civilization (DiPietro and Molenar 1982). Some looked beyond the face, seeing in its proximity the archaeological ruins of a great city replete with a five-sided pyramid, a fortress, transportation arteries, and an artificial mound, or "tholus," surrounded by a moat (Hoagland 1987). Richard Hoagland, who has been a lightning rod in this debate (see Posner 2000 for some interesting insights into Hoagland), has argued that the Cydonia images show the remains of a complex settlement—not of indigenous Martians but of aliens who colonized Mars as much as half a million years ago. In Hoagland's speculation, these extraterrestrials next colonized Earth, and we are their descendants, explaining the fact that the Mars Face is recognized as a face precisely because it looks human.

Much of the Mars Face argument seems to be based on an iteration of the kind of "Rorschach archaeology" applied by Erich von Däniken (see the "Ancient Astronauts" entry in this encyclopedia). In other words, those who support the hypthesis that the face and associated features are artificial have come to their conclusions subjectively, by eyeballing low-resolution photographs taken of the Martian surface. They argue that the Cydonia feature must actually be a carved human face because it seems to resemble one.

A fundamental problem in this line of reasoning is the fact that it is not at all uncommon for natural features of the landscape here on Earth to suggest artificial images, again in the way inkblots suggest specific pictures. For example, the "Old Man in the Mountain," so emblematic of New Hampshire that its profile is on that state's license plates, indeed looks like an old man but is obviously an entirely natural feature. Wisconsin has its remarkable—and entirely natural—profile of the Indian leader Black Hawk. Virtually all solution caverns have rock formations that suggest a variety of recognized artifacts, such as the Statue of Liberty, the U.S. Capitol dome, Abraham Lincoln, two eggs sunnyside up, and so forth. NASA image researchers have found other remarkable elements of the Martian landscape that bear a striking resemblance to seemingly nonnatural features. For example, even a cursory glance at a Martian meteor-impact crater that is 8 kilometers in diameter indicates that it deserves its designation as the "largest happy face" in the known universe (see http://www.msss.com/education/happy_face/happy_face.html). Another feature produced by two intersecting craters is readily recognizable as a Valentine's Day heart (see http://mpfwww.arc.nasa.gov/mgs/msss/camera/images/6_17_99_heart/index.html). And a Martian lava flow bears an uncanny resemblance to the Muppet character Kermit the Frog.

More detailed—and, ostensibly, more objective—claims concerning the possible artificial source of the Cydonia features rely on mathematical arguments concerning their alignments and locations. It is maintained that these alignments are not random, as would be expected if the features were natural, but instead reflect precise mathematical relationships that only a highly sophisticated society could incorporate into their design (Hoagland 1987).

NASA scientists are quite skeptical of any such conclusions based on measurements taken from the *Viking* photographs of Mars. These photographs are extremely low-resolution pieces—each pixel in the original face photograph corresponds to 43 meters (141 feet) on the ground—and are quite imprecise and, therefore, highly inaccurate. For example, the so-called pyramid's location on the face photograph varies by as much as 17 kilometers, depending on which positioning system is employed when analyzing the photograph (M. Malin 2002).

New photographs of the Cydonia region have been taken by the *Mars Global Surveyor* (MGS) satellite, which has been in orbit around Mars since September 1997. The MGS camera is able to take photographs of a far higher resolution than the *Viking* orbiter and in April 1998 snapped an image of the Mars Face feature (see http://www.msss.com/mars_images/moc/4_6_98_face_release/compare.gif). Each pixel in this photograph represents only 4.3 meters (14.1 feet) of the Martian surface; this resolution is ten times higher than the best images of the face taken by *Viking*.

Viking orbiter images acquired in 1976 showed that one of thousands of buttes, mesas, ridges, and knobs in the transition zone between the cratered uplands of western Arabia Terra and the low, northern plains of Mars looked somewhat like a human face. (Image 35A72, NASA/JPL/Malin Space Science Systems)

One would be hard-pressed to coax any kind of a facelike image from the 1998 photograph. But there was apparently enough ambiguity left (the photograph was taken at a substantial, 45° angle, and there was significant cloud cover) that supporters of the face hypothesis held out hope that subsequent images would provide clearer evidence. In one sense, they were correct. On April 8, 2001, the MGS was rolled to an angle of 24.8° to its left and photographed the face from a distance of only about 450 kilometers (280.5 miles). The Martian atmosphere was quite clear when this photograph was taken, and the image has an even higher resolution than the April 1998 image, with each pixel representing only about 1.56 meters (about 5 feet, the maximum resolution possible with this camera). According to NASA (press release), an object the size of a small building would be discernible in this image; genuine cultural features, especially monumental features such as pyramids or temples, would be easily recognized. However, no such features are present in the new photograph, and the fabled Mars Face has disappeared completely; all that is left is an eroded mesa with a rather nondescript depression where the supposed eye was located and a linear valley where people saw a mouth (see next Figure).

The face mesa is only one of many on the Cydonia plain. The geologic processes that produced these landforms are uncertain; the mesas may have been eroded by wind, water, or even glacial activity sometime in Mars's past. Measurements taken with the Mars Orbiter Laser Altimeter (MOLA) provide precise measurements of the heights, proportions, and volumes of these mesas. The face mesa is in no way unique or unlike the other mesas that dot the plain, except for the fact that with a low-

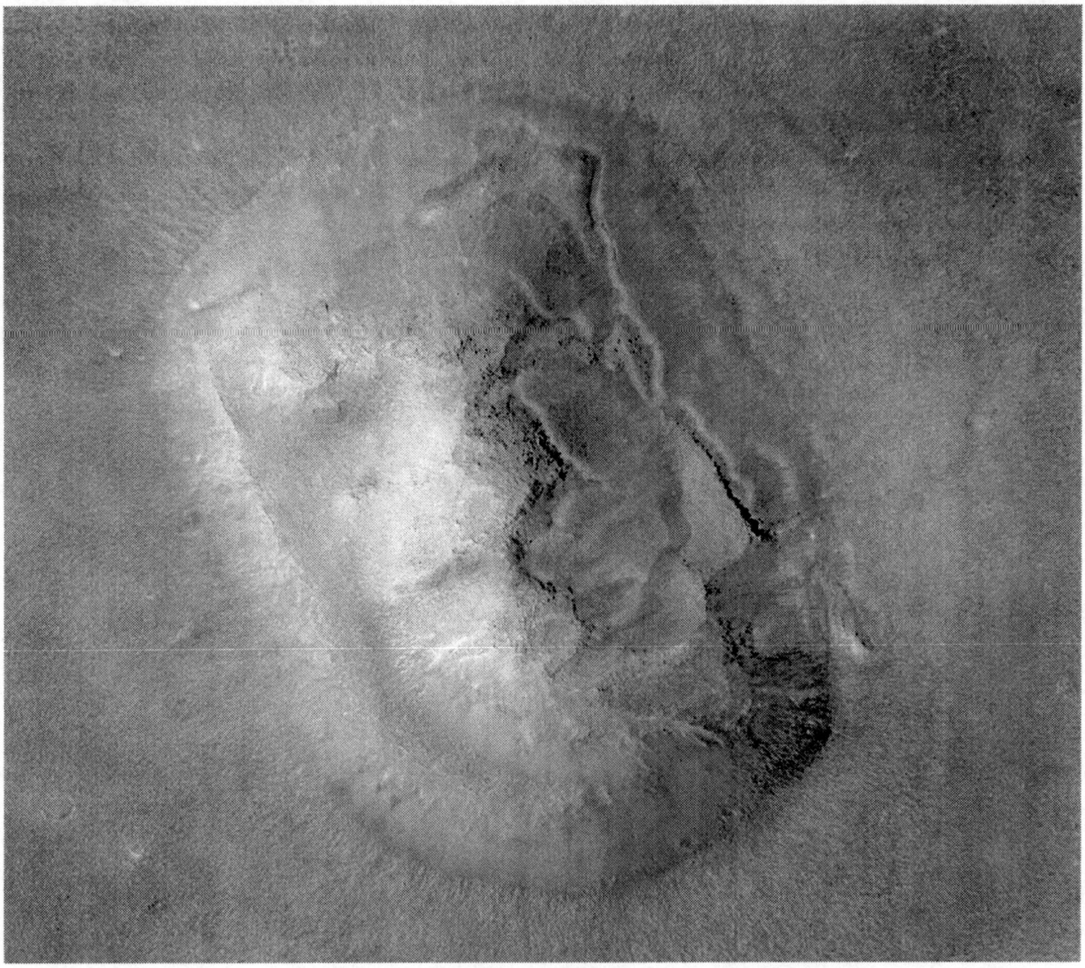

On April 8, 2001, the Mars orbiter camera on the *Mars Global Surveyor* spacecraft obtained a new high-resolution view of the "face." (MOC image E03-00824, NASA/JPL/Malin Space Science Systems)

resolution camera and the right lighting conditions, it looks like a human face. The measurements are so precise—the MOLA has a remarkably accurate vertical precision of between 20 and 30 centimeters (8 and 12 inches)—that NASA scientist Jim Garvin has even produced a trail map leading to the top of the mesa, should astronauts ever reach the Cydonia region of Mars (NASA press release). He estimates it will be a two-hour hike.

With the application of high-resolution photography and instrumentation such as the laser altimeter, the face on Mars has disappeared, feature by feature, like the Cheshire cat in *Alice in Wonderland.* The Cydonia mesas provide no proof that intelligent life once existed on Mars.

References:

DiPietro, V., and G. Molenar. 1982. *Unusual Martian Surface Features.* Glendale, MD: Mars Research.

Hoagland, Richard. 1987. *The Monuments of Mars: City on the Edge of Forever.* Berkeley, CA: North Atlantic Books.

Malin, Michael. "Observations of the 'Face on Mars' and Similar Features by the Mars Global Surveyor Orbiter Camera." URL: http://barsoom.

MSSS.com/education/facepage/face.html. (Accessed in February 2002).

Malin Space Science Systems. "The 'Face on Mars.'" URL: http://barsoom.MSSS.com/education/facepage/face.html. (Accessed in February 2002).

NASA. Press Release P-17384. URL: http://barsoom.MSSS.com/education/facepage/pio.html. (Accessed in February 2002).

——. 2001. "Unmasking the Face on Mars." URL: http://science.nasa.gov/headlines/y2001/ast24may_1.htm?list540155. (Accessed in February 2002).

Posner, Gary P. 2000. "The Face Behind the 'Face' on Mars." *Skeptical Inquirer* 24, no. 6: 20–26.

Meditation

JUAN CARLOS MARVIZON

The word *meditation* refers to religious practices aimed at producing transcendental experiences and changes in consciousness by quieting or focusing the mind. Meditation is present in nearly all religions, but its practice is central in Buddhism, Hinduism, and Taoism. In Christianity, the practice of meditation with the aim of producing mystical states is often discouraged (White 1974, 149–158). In any event, in Christianity and the other monotheistic religions, meditation is primarily a form of prayer, whereas in Eastern religions, meditation consists of elaborate techniques designed to produce states of mind in which truth can be directly perceived. Over the last hundred years or so, these Eastern meditation techniques have been imported to the West, where they now exert a considerable cultural influence. Various claims have been made about the effects of meditation, ranging from stress relief to psychic powers, and a fair number of scientific studies have been conducted to examine these claims, which are classified here as paranormal, extraordinary, and ordinary effects.

Paranormal Effects

These effects are events that would be in direct violation of the known laws of physics, and they run the whole gamut of paranormal phenomena: clairvoyance, telepathy, levitation, teleportation, astral voyages, seeing auras, and so forth. Many examples of these psychic wonders can be found in the autobiography of Swami Yogananda (1968) and in stories of other yogis and Tibetan mystics (White 1974, 209–224), but none of them have been proved. In the 1970s, the Transcendental Meditation (TM) program claimed to produce "sidhis," or psychic powers, in its practitioners, including levitation, invisibility, and omniscience. The Maharishi European University published "scientific research" that allegedly showed that psychic powers are bestowed by the "higher states of consciousness" achieved during the practice of TM (Orme-Johnson and Farrow 1977, 209–210). This publication is a prime example of pseudoscience, being seriously flawed at best and an outright fraud at worst. The issue of whether paranormal phenomena actually occurred is never directly addressed. Instead, the authors repeatedly confuse the subjective experiences of the meditators with objective events. The book in which this research and other papers are compiled does show pictures of people levitating one foot above the ground in the lotus position (see figure). In fact, the position of their clothing and hair indicates that these people were just jumping with their legs crossed. The authors complicate matters by stating that levitation starts with "hopping" and develops as "a feeling of suspension in the air for a few seconds" (Orme-Johnson and

Farrow 1977, 708). Two follow-up papers published in the *International Journal of Neuroscience* dropped the levitation claims but continued to refer to this hopping activity as "Yogic Flying."

Extraordinary Effects

These effects relate to claims that, although not in violation of the laws of nature, would certainly revolutionize our view of ourselves and the world if they were to be proven. People generally do not meditate to acquire supernatural powers; they do it because they hope to achieve a state of mind that transcends their personal limitations. How that transcendent state is described is intimately linked to their particular religious beliefs. For example, Buddhism strives to transcend suffering in nirvana, described as a state of profound insight accompanied by complete detachment and loss of the ego (Kapleau 1989; Austin 1998). In modern Hinduism, the transcendental state is called "Samadhi" or "Moksa" and consists in the realization that our innermost soul, or "Atman," is identical with "Brahman," the primordial God. Similarly, for the mystics of monotheistic religions, the transcendental state is a direct vision or communication with God. It is hard to separate the supernatural from the extraordinary when it comes to these religious experiences. In Buddhism, nirvana is sometimes considered a supernatural state that provides direct knowledge of the ultimate reality, whereas agnostic Buddhists (Batchelor 1997) propose a more modest interpretation as a change in consciousness that solves the problem of suffering. One idea often encountered is that all religions lead to the same transcendental state. Once that state is achieved, all apparent differences between religions disappear. Accordingly, it may be possible to classify the "alternate" states of consciousness produced by meditation in a hierarchical order, from ordinary wakefulness to a profound and permanent state of illumination (Austin 1998, 298–305). However, there seems to be little factual support for this idea.

A woman in Tokyo meditates inside a portable meditation room. (Reuters NewMedia Inc./ CORBIS)

Ordinary Effects

Hindu and Buddhist meditation techniques have been introduced in the West as secular practices with the more modest goal of improving physical and psychological health. The main proponent of this effort has been Maharishi Mahesh Yogi and his TM program (White 1974, 85–109). However, despite its claims of secularity, TM retains many philosophical elements and rituals of Hinduism. The various organizations linked to TM have encouraged scientific research to provide evidence for their claims about the beneficial effects of medita-

Hatcha-Yogi L. S. Rao sits comfortably on a heavily spiked fence, 1952, New York. (Bettmann/CORBIS)

tion. A compilation of this research was published by David Orme-Johnson and John T. Farrow (1977), including some articles that appeared in mainstream scientific journals and many that were not peer-reviewed.

Scientific Research on Meditation

Numerous scientific studies have tested the idea that meditation produces measurable physiological changes indicating increased relaxation and reduced stress (ordinary claims) or even "transcendental" states of consciousness (extraordinary claims). Most of these studies were done in the 1960s and 1970s using electroencephalogram (EEG) readings to measure brain activity, electrocardiogram (EKG) readings to measure heart rate, electromyographs (EMG) readings for recording muscle activity, and measures of respiration rate and skin resistance. Although modern imaging techniques (such as positron emission tomography [PET] and functional magnetic resonance imaging [fMRI]) are far superior to

EEG in probing the living human brain, there are very few studies on meditation using them.

EEG studies have provided quite intriguing results. In this technique, electrodes are placed on the skin on the surface of the skull to measure minute variations in the electric field resulting from the activity of large populations of neurons in the cerebral cortex. Although the relationship between the function of brain cells and the EEG remains obscure, this is a fairly old technique, and there is a vast literature on the association of various EEG features with distinct physiological and pathological states. Two measures are made from the EEG trace: amplitude (vertical size of the wave) and frequency (number of waves per unit of time). Brain wave patterns have been classified in four types according to their frequency: beta waves (more than 14 hertz or waves per second) are associated with wakefulness; alpha waves (8–13 hertz) occur in relaxed wakeful states, particularly with eyes closed; theta waves (4–7 hertz) are quite rare, being found during periods of intense emotion and other anomalous states; and finally, delta waves (below 3.5 hertz) are observed during sleep and coma. A flat EEG is interpreted as a dead brain. One interesting property of EEG recordings that has been applied to the study of meditation is the "alpha block." If a person is relaxed with eyes closed, his or her EEG will show alpha activity. However, a sudden noise or other unexpected stimulus will produce an abrupt cessation of the alpha waves and the onset of beta waves. After a few seconds, the alpha activity returns. Alpha block is subject to habituation: a stimulus will stop producing alpha block after a few repetitions.

In regard to meditation research, it must be kept in mind that meditation techniques are far from homogeneous. Some techniques stress concentration: the attention is focused on a single object while the rest of the activity of the mind is suppressed. Another meditation approach is "mindfulness," in which the activity of the mind is not to be suppressed but only directed so that all experiences (sensations, thoughts, emotions, and the like) are given the same importance, neither being pursued nor rejected (White 1974, 56–57, 127–128). Other types include devotional meditations that cultivate particular emotional states and discursive meditations encountered in Western religions. Results of different studies show that the meditation technique determines changes in the EEG pattern observed.

One of the most comprehensive and well-designed studies on meditation was conducted by Tomio Hirai and collaborators (1989) on Zen monks. The study included 48 monks with extensive meditation experience (25 to 55 years), 98 Zen trainees classified in three groups of different proficiency, and 15 control individuals with no meditation experience. In experienced monks, meditation produced a distinctive, reproducible EEG pattern with four phases: (1) alpha waves at the beginning of meditation, (2) an increase in the amplitude of the alpha waves as the meditation progressed, (3) a slowing of the frequency of the alpha waves, and (4) a rhythmical theta wave alternated with alpha waves in the final stage of meditation. The high-amplitude alpha waves and the intermittent theta waves are unique, especially considering that Zen meditation is done with eyes open, which normally disrupts alpha waves. Less experienced Zen trainees showed some of the EEG changes detected in the more experienced meditators, and the extent of these changes correlated well with the length of their training and with their proficiency in meditation as evaluated by their teachers. The EEG pattern produced by Zen meditation was different from that produced by sleep, hypnosis, and psychotropic drugs such as LSD, mescaline, or marijuana. One of the most remarkable findings of this study was the lack of habituation to repeated stimuli. A clicking sound produced the alpha block phenomenon described earlier. However, repetition of the clicking sound continued to produce the alpha block in meditators almost indefinitely, whereas in the control individuals, the alpha block disappeared after a few repetitions. This finding is consistent with the "mindful" nature of Zen meditation, which tries to achieve a sustained awareness of all types of stimuli.

Studies of other types of meditation produced different results (White 1974, 225–243). Yoga meditation also produced alpha waves, but there was no alpha block in response to external stimuli. This result could be due to the fact that yoga meditation is concentrative: its aim is to block all external stimuli and to reduce the activity of the mind. A third type of meditation, called "krya yoga," consists of focusing attention on an unfolding vision of the *Kundalini*, or "serpent energy," traveling from its resting place at the base of the spine to the peak of the head. This type of meditation produced a strikingly different EEG pattern, consisting of extremely fast beta activity with high-amplitude waves, accompanied by an acceleration of the heart rate. External stimuli applied during meditation had no effect on the EEG. All these findings are indicative of a state of high arousal isolated from the external environment, which again coincides with the subjective description of this meditative experience. Studies on EEG changes during TM (Orme-Johnson and Farrow 1977, 151–186) produced results that seem to be a mixture of those observed with Zen, yoga, and krya meditation. These results include high-amplitude alpha waves with little alpha block and also theta, delta, and beta waves.

In conclusion, there is evidence that meditation produces states of consciousness different from wakefulness, sleep, hypnosis, and those states induced by psychotropic drugs. Different forms of meditation have different physiological signatures, and these appear to be consistent with the stated goals of the meditation. However, the existence of particular states of consciousness during meditation does not prove that these states are extraordinary or transcendent in any way. It would be interesting to see whether it is possible to provide objective evidence of transcendent experiences.

References:

Austin, James H. 1998. *Zen and the Brain: Toward an Understanding of Meditation and Consciousness.* Cambridge, MA: MIT Press.

Batchelor, Stephen. 1997. *Buddhism without Beliefs: A Contemporary Guide to Awakening.* New York: Riverhead Books.

Hirai, Tomio. 1989. *Zen Meditation and Psychotherapy.* Tokyo and New York: Japan Publications.

Kapleau, Philip. 1989. *The Three Pillars of Zen: Teaching, Practice, and Enlightenment.* New York: Anchor Books.

Orme-Johnson, David W., and John T. Farrow. 1977. *Scientific Research on the Transcendental Meditation Program: Collected Papers.* Weggin, Switzerland: Maharishi European Research University Press.

White, John W. 1974. *What Is Meditation?* New York: Anchor Press/Doubleday.

Yogananda, Paramahansa. 1968. *Autobiography of a Yogi.* Los Angeles: Self-Realization Fellowship.

Multiple Personality Disorder

SCOTT O. LILIENFELD AND
STEVEN JAY LYNN

Multiple personality disorder (MPD), currently known as dissociative identity disorder in the American Psychiatric Association's official diagnostic manual of mental disorders, has long been among the most controversial diagnoses in psychology and psychiatry. Although the precise diagnostic criteria for MPD have shifted somewhat over the years, it is now defined as a condition characterized by the presence of two or more distinct personalities or personality "states" (that is, temporary patterns of behavior) that recurrently assume control of the individual's behavior. Such alternate personalities or personality states, which are called "alters," frequently exhibit personality features that differ markedly from those of the primary, or "host," personality. For example, if the primary personality is shy and retiring, one or more of the alters are often outgoing or flamboyant. In addition, individuals with MPD report significant episodes of amnesia (memory loss) in regard to important personal information. For example, they may report frequent hours or days of "lost time": they cannot recall where they were or what they were doing in those periods. This amnesia is commonly reported to be asymmetrical, whereby the primary personality knows little about the behaviors of the alters but not vice versa (American Psychiatric Association 1994). Most MPD patients are women, although the reason for this imbalanced sex ratio is unknown.

The nature and features of MPD alters are highly variable both across and within individuals. The number of alters has been reported to range from one (in the so-called split personality) to hundreds or even thousands, with one clinician reporting a case of an MPD patient with 4,500 alters (Acocella 1999). These alters are not uncommonly of different sexes, ages, and even races. There have even been reported alters of Mr. Spock, lobsters, chickens, gorillas, unicorns, God, and the bride of Satan (Acocella 1999). In addition, alters have been purported to differ in their allergies, handwriting, voice patterns, and eyeglass prescriptions.

Reports of MPD in both popular and clinical literature date back at least to the nineteenth century. Robert Louis Stevenson's classic 1885 book, *The Strange Case of Dr. Jekyll and Mr. Hyde*, which describes a scientist who ingests a mysterious potion that transforms him into an entirely different individual, is among the first tales reminiscent of the modern-day notion of MPD. Although the remarkable symptoms of MPD captured the imagination of authors and researchers throughout the nineteenth and twentieth centuries, reports of this condition were exceedingly rare until the late twentieth century. As of 1970, there were a total of 79 cases of clear-cut

John Barrymore starring in *Dr. Jekyll and Mr. Hyde.* (Underwood & Underwood/CORBIS)

MPD in the world literature. One of these was the celebrated case of Chris Sizemore, which formed the basis of the book (and later the Hollywood film) *The Three Faces of Eve* (Thigpen and Cleckley 1954). As of 1986, however, the number of reported MPD cases had swelled to approximately 6,000. This massive increase coincided closely with the release of the popular nonfiction book *Sybil* (Schreiber 1974), which told the story of a young woman with sixteen personalities who reported a history of severe and sadistic child abuse. This best-selling book was turned into a widely viewed television film in 1977. The number of reported cases of MPD at the turn of the twenty-first century is difficult to estimate, although it appears to be in the tens of thousands (Acocella 1999). The reasons for this apparent "epidemic" in the number of MPD cases is unknown, and, as we will see shortly, it remains a point of considerable debate.

It is also worth noting that the number of MPD alters has increased markedly over time. Whereas most cases of MPD prior to the 1970s were characterized by only one or two alters, recent cases are generally characterized by considerably more (North et al. 1993). A 1989 study, for example, reported that the mean number of MPD personalities was sixteen, which was precisely the number reported by Sybil (Acocella 1999).

The causes of MPD have been a source of heated and at times acrimonious disagreement among researchers. Colin Ross (1997) and others have proposed that MPD is a "posttraumatic" condition that arises primarily from a history of severe physical and/or sexual child abuse. Ross maintained that individuals who experience horrific abuse in early life often "dissociate" (hence, the term *dissociative identity disorder*) or compartmentalize their personalities into distinct alters as a means of coping with the profound emotional pain of this trauma. According to Ross (1997, 59), "MPD is a little girl imagining that the abuse is happening to someone else." In support of this assertion, proponents of the posttraumatic model cite data suggesting that a large proportion (perhaps over 90 percent) of individuals with MPD report a history of child abuse (Gleaves 1996).

Proponents of the posttraumatic model attribute the dramatic recent increase in the reported prevalence of MPD to the heightened awareness and recognition of this condition by psychotherapists. Specifically, they argue that clinicians have only recently become attuned to the presence of possible MPD in their clients and, as a consequence, now inquire more actively about potential symptoms of this condition (Gleaves 1996). In many cases, these proponents advocate the use of hypnosis, sodium amytal (so-called truth serum), guided imagery, and other suggestive therapeutic techniques to "call forth" alters that are otherwise inaccessible, as well as to recover apparent memories of child abuse that have seemingly been repressed.

Although the posttraumatic model of MPD remains popular among many psychotherapists, numerous critics have called its core assumptions into question. The most influential alternative model of MPD is the sociocognitive model advanced by Nicholas Spanos (1994). According to this model, MPD is largely a socially constructed condition that results from inadvertent therapist cueing (e.g., suggestive questions regarding the existence of possible alters), media influences (e.g., film and television portrayals of MPD), and broader sociocultural expectations regarding the presumed features of MPD. Specifically, Spanos and his colleagues contended that individuals with MPD are engaged in a form of role-playing that is similar in some ways to the intense sense of imaginative involvement that some actors report when performing in a part. Because individuals who engage in role-playing "lose themselves" in the enacted part, this phenomenon should not be confused with simulation or conscious deception. Advocates of the sociocognitive model do not believe that most individuals with MPD are consciously faking this condition, although there is compelling evidence that at least a few well-publicized criminals (e.g., serial murderer Kenneth Bianchi) have attempted to fake MPD. Instead, according to the sociocognitive model, the symptoms of MPD are almost always genuine, but they are induced primarily by suggestive therapeutic practices and expectations regarding the features of the disorder. Moreover, according to this model, the dramatic "epidemic" in MPD cases since the 1970s stems not from improved diagnostic and assessment practices but rather from iatrogenic (therapist-induced) influences and the increased media attention accorded MPD.

Advocates of the sociocognitive model invoke a variety of pieces of evidence in support of this position (see Lilienfeld et al. 1999; Spanos 1994). For example, a large proportion and perhaps even a substantial majority of MPD patients exhibit few or no unambiguous signs of this condition (e.g., alters) prior to psychotherapy. Moreover, patients with MPD are in psychotherapy an average of six to seven years before being diagnosed with this condition (Gleaves 1996). These pieces of evidence raise the possibility that such patients developed unambiguous MPD symptoms only after receiving treatment.

In addition, the distribution of MPD cases across therapists appears to be strikingly nonrandom. For example, a 1992 study in Switzerland revealed that 66 percent of MPD diagnoses were made by .09 percent of clinicians (Spanos 1994; see Lilienfeld et al. 1999, for other examples). Such findings could perhaps be explained by positing that patients with actual or possible MPD are selectively referred to MPD experts. Nevertheless, they are also consistent with the suggestion that only a handful of clinicians are diagnosing and perhaps eliciting MPD in their patients.

Although the MPD epidemic is relatively recent, a number of other psychopathological "epidemics" have been observed throughout history. Indeed, there is suggestive evidence that the overt manifestations of so-called hysteria have shifted substantially over time in accord with prevailing sociocultural beliefs and expectations. For example, Victorian England in the nineteenth century witnessed a dramatic upsurge in the incidence of certain unexplained somatic symptoms, such as paralyses and aphasias (inability to speak) that lacked a demonstrable physical basis. These symptoms were subsequently displaced by less florid symptoms of fainting ("the vapors"; see Veith 1965). Proponents of the sociocognitive model contend that the recent epidemic of MPD is merely one manifestation of the capacity of sociocultural expectations to induce the large-scale transmission of certain conditions (see also Showalter 1997). This hypothesis is difficult to test but merits careful scientific consideration.

Still other evidence provides indirect support for the sociocognitive model. For example, many standard therapeutic practices for MPD appear to reward patients' displays of multiplicity and encourage the emergence of alters (Lilienfeld et al. 1999; Piper 1997). Some of these practices appear to be highly suggestive. Frank Putnam (1989), for example, recommended using a technique known as the "bulletin board," which encourages MPD alters to post notes to one another. Ross advocated giving names to alters to affirm their existence, and he used a technique known as the "inner board meeting" to "map" the system of alters and recover repressed memories: "The patient relaxes with a brief hypnotic induction, and the host personality walks into the boardroom. The patient is instructed that there will be one chair for every personality in the system. . . . Often there are empty chairs because some alters are not ready to enter therapy. The empty chairs provide useful information, and those present can be asked what they know about the missing people" (Ross 1997, 351). These and other techniques may inadvertently "reify" alters and encourage patients to view different aspects of their personalities as entirely distinct entities.

Moreover, some psychotherapists regularly use hypnosis and other suggestive techniques in efforts to unearth presumed latent alters and memories of past abuse. Nevertheless, there is consistent evidence that hypnosis does not enhance memory and instead may often lead to false memories (that is, memories of events that never occurred), although it typically increases individuals' subjective confidence in their memories (Lynn et al. 1997).

It is important to emphasize that the sociocognitive model does not assert that MPD symptoms arise in a vacuum. There is compelling evidence that many individuals with MPD enter psychotherapy with a host of psychological difficulties, including depression, anxiety, interpersonal difficulties, and symptoms of eating disorders. In particular, a large proportion (perhaps 50 percent or more; see Lilienfeld et al. 1999) of individuals with MPD in clinical samples meet diagnostic criteria for borderline personality disorder (BPD), a condition characterized by such symptoms as unstable identity, dramatic and seemingly inexplicable mood swings, impulsive and self-damaging behaviors (e.g., cutting oneself), and dramatic shifts in one's attitudes toward people (e.g., alternating between worshiping and devaluing the same individual within a short span of time). Advocates of the posttraumatic model typically maintain that the extensive overlap between MPD and BPD actually reflects the fact that many BPD patients suffer from "latent" (undiagnosed) MPD. But another and perhaps more plausible interpretation is possible. Specifically, individuals who enter psychotherapy with multiple BPD features may be seeking an explanation for their seemingly inexplicable instability in self-concept, mood, and impulse control. A therapist who repeatedly prompts such individuals with suggestive questions (e.g., "Might there be a part of you to whom I haven't yet spoken?") or encourages clients to search for dissociated alters may be especially likely to elicit reports of indwelling "identities" that can ostensibly account for these individuals' puzzling behaviors (see Ganaway 1995).

A final important source of evidence in support of the sociocognitive model is the fact that MPD is largely a culture-bound syndrome. Until quite recently, MPD was diagnosed almost exclusively in North America (Spanos 1994). Interestingly, however, MPD has recently begun to be diagnosed with considerable frequency in certain European countries, such as the Netherlands, where several prominent MPD proponents have sparked public interest in the condition. In addition, several psychiatric disorders bearing intriguing similarities to MPD have been observed in non-Western countries. For example, some women in Malaysia and Indonesia suffer from a condition known as "latah," which is marked by sudden and short-lived episodes of profanity, trancelike states, and command obedience (responding automatically to others' suggestions), followed by amnesia for the episode. Some men in Malaysia and certain other countries (e.g., Laos) exhibit a condition known as "amok," which is marked by a period of intense brooding in response to a perceived insult, followed by a dramatic outburst of uncontrolled and extremely aggressive behavior toward others (hence, the phrase *running amok*). The episode is often followed by amnesia for what transpired. Although the underlying commonalities among MPD, latah, amok, and similar conditions (see American Psychiatric Association 1994) remain to be clarified, the possibility that these conditions are superficially different cross-cultural manifestations of the same underlying disorder is intriguing.

Finally, Spanos (1994) and other critics (e.g., Lilienfeld et al. 1999) have challenged a core assumption of the posttraumatic model, namely, the presumed relation between child abuse and subsequent MPD. Almost all of the findings on the child abuse–MPD association are based on retrospective memory reports that have not been corroborated by objective evidence (e.g., documented records of abuse). Because psychologists have long known that retrospective reports are often subject to memory distortion, such reports must be interpreted with considerable caution. Rendering these reports even more problematic is the fact that they typically derive from MPD patients who have been in psychotherapy for years. Because many of these patients have been subjected to hypnosis and other suggestive procedures that are known to increase the occurrence of false memories, their reports of past child abuse should not be accepted without external corroboration. These methodological limitations do not refute the claim that child abuse is associated with and perhaps even causally related to MPD in certain cases. But they indicate that the data supporting this claim are less convincing than has often been claimed. Interestingly, prior to the publication of *Sybil* in 1974, reports of past child abuse among MPD patients were very rare. This intriguing but often overlooked fact is consistent with the possibility that the marked increase in child abuse reports among MPD patients is largely a function of therapists' use of suggestive procedures to recover memories of abuse.

At the present time, the data do not allow an impartial observer to definitively choose between the posttraumatic and sociocognitive models of MPD. Nevertheless, the powerful convergence across several independent lines of evidence provides compelling support for many aspects of the sociocognitive model. Indeed, even many proponents of the posttraumatic model now acknowledge that a certain number of MPD cases are likely to be largely iatrogenic in origin. It is conceivable that both models are at least partly correct. For example, perhaps an early history of child abuse leads certain individuals to adopt a fantasy-prone personality style as a means of coping with this trauma. This personality style may, in turn, increase individuals' susceptibility to suggestive therapeutic procedures, leading to the induction of alters. This and even more sophisticated models of MPD have yet to be subjected to direct empirical tests.

Nevetheless, the recent MPD epidemic imparts one clear and crucial lesson: beliefs can help to shape reality. Psychotherapists must therefore remain cognizant of the possibility that their therapeutic practices can inadvertently exacerbate and perhaps even cause psychopathology.

References:

Acocella, Joan. 1999. *Creating Hysteria: Women and Multiple Personality Disorder.* San Francisco: Jossey-Bass.

American Psychiatric Association. 1994. *Diagnostic and Statistical Manual of Mental Disorders.* 4th ed. Washington, DC: American Psychiatric Association.

Ganaway, George K. 1995. "Hypnosis, Childhood Trauma, and Dissociative Identity Disorder: Toward an Integrative Theory." *International Journal of Clinical and Experimental Hypnosis* 43: 127–144.

Gleaves, David H. 1996. "The Sociocognitive Model of Dissociative Identity Disorder: A Reexamination of the Evidence." *Psychological Bulletin* 120: 42–59.

Lilienfeld, Scott O., Steven J. Lynn, Irving Kirsch, John F. Chaves, Theodore R. Sarbin, and George Ganaway. 1999. "Dissociative Identity Disorder and the Sociocognitive Model: Recalling the Lessons of the Past." *Psychological Bulletin* 125: 507–523.

Lynn, Steven J., Timothy G. Lock, Bryan Myers, and David G. Payne. 1997. "Recalling the Unrecallable: Should Hypnosis Be Used to Recover Memories in Psychotherapy?" *Current Directions in Psychological Science* 6: 79–83.

McHugh, Paul R. 1993. "Multiple Personality Disorder." *Harvard Mental Health Newsletter* 10: 4–6.

North, Carol S., JoEllyn N. Ryall, Daniel A. Ricci, and Richard D. Wetzel. 1993. *Multiple Personalities, Multiple Disorders: Psychiatric Classification and Media Influence.* Oxford: Oxford University Press.

Piper, August. 1997. *Hoax and Reality: The Bizarre World of Multiple Personality Disorder.* Northvale, NJ: Jason Aronson.

Putnam, Frank. 1989. *Diagnosis and Treatment of Multiple Personality Disorder.* New York: Guilford Press.

Ross, Colin A. 1997. *Dissociative Identity Disorder: Diagnosis, Clinical Features, and Treatment of Multiple Personality Disorder.* New York: Wiley.

Schreiber, Flora R. 1974. *Sybil.* New York: Warner Books.

Showalter, Elaine. 1997. *Hystories: Hysterical Epidemics and Modern Culture.* New York: Columbia University Press.

Spanos, Nicholas P. 1994. "Multiple Identity Enactments and Multiple Personality Disorder: A Sociocognitive Perspective." *Psychological Bulletin* 116: 143–165.

Thigpen, Corbett H., and Hervey M. Cleckley. 1954. "A Case of Multiple Personality." *Journal of Abnormal and Social Psychology* 49: 135–151.

Veith, Ilza. 1965. *Hysteria: The History of a Disease.* Chicago: University of Chicago Press.

Near-Death Experiences

SUSAN J. BLACKMORE

What is it like to come very close to death and survive? In 1975, physician Raymond Moody hit the best-seller lists with *Life after Life*, claiming that hundreds of near-death survivors had reported overwhelmingly pleasant experiences. During these experiences, he noted, they seemed to leave their bodies and view resuscitation attempts from above; then they passed down a dark tunnel toward a brilliant light, met a "being of light" who helped them to evaluate and judge their own lives, and finally decided to return to life rather than go on into the peace and bliss of death (Moody 1975). The near-death experiences (NDEs) were difficult to talk about for the survivors but left them changed for the better—reportedly less materialistic and with a reduced fear of death. Reactions to these claims ranged from the popular view that these experiences must be evidence for life after death to outright rejection of the experiences as, at best, drug-induced hallucinations or, at worse, pure invention.

The State of the Evidence on Near-Death Experiences

Twenty years and much research later, it is clear that neither extreme is correct. On the one hand, the claim that the experiences are evidence for survival after death is untenable. Even though the boundary between life and death is pushed back by improved techniques, it is always possible to argue that the person did not actually die and that the experiences were part of life and not death. Of course, if there is life after death, these experiences may give a clue as to what it is like, but they can never be definitive evidence that there is.

On the other hand, the experiences cannot be dismissed as either totally invented or hallucinations caused by medical intervention or drugs. Moody simply collected cases as they came along, but research by Kenneth Ring, conducted on 101 randomly selected survivors, soon confirmed that such reports are common. In that research, about 60 percent of the participants reported peace, one-third mentioned out-of-body experiences (OBEs), one-quarter said they had entered the darkness (or a tunnel), and rather fewer reported experiences such as life review and the decision to return (Ring 1980). Near-death experiences (NDEs) also appear to be widespread through many ages and cultures. Long before Moody, there were similar descriptions of deathbed experiences (when the patients did go on to die) in the psychical research literature (Barrett 1926; Osis and Haraldsson 1977), as well as isolated reports in the medical literature (Dlin, Stern, and Poliakoff 1974; Dobson et al. 1971; Druss and Kornfeld 1967; MacMillan and Brown 1971). In addition, there are both historical and contemporary accounts from many different cultures

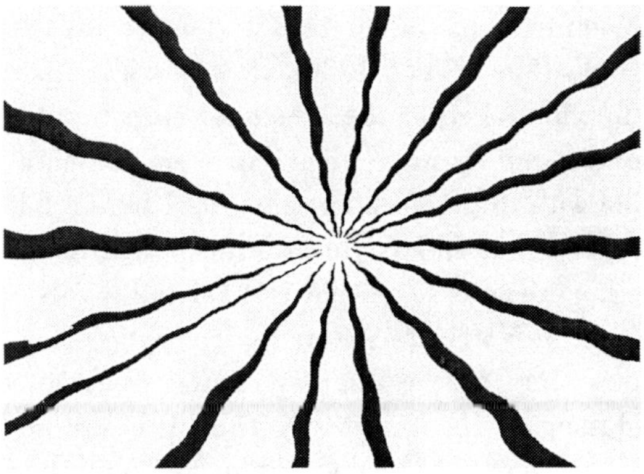

Drawing of tunnel with light at the end, as seen by those in near-death experiences. (Dr. Susan Blackmore/Fortean Picture Library)

(Blackmore 1993), and in our own culture, children also report similar experiences, although their reports tend to be fragmentary compared with those of adults (Morse et al. 1986; Morse 1990).

Explanations for the Near-Death Experience

Although some modern stories may be inventions based on the widespread publicity about the phenomenon, it seems unlikely that people across so many other ages and cultures would have invented similar stories. The question then becomes why the features are so often the same. Common theories include the effects of (1) expectation, (2) administered drugs, (3) endorphins, (4) anoxia (oxygen depletion) or hypercarbia (excess carbon dioxide), (5) temporal lobe stimulation, and (6) life after death. Each will be considered in turn.

Expectation

Expectation clearly has an effect on NDEs, though there are two different aspects to this factor. First, NDEs often happen to people who think they are dying when, in fact, there is no serious clinical emergency. This adds to the general conclusion that you do not have to be physically near death to have an NDE (Gabbard, Twemlow, and Jones 1981; Owens, Cook, and Stevenson 1990). Indeed, some aspects of the NDE, such as the out-of-body experience (see the "Out-of-Body Experiences" entry in this encyclopedia) can occur at any time and to perfectly healthy people (Blackmore 1982; Gabbard and Twemlow 1984; Irwin 1985). There are some differences between the NDEs of those who are and are not close to death, but they are small compared to the similarities (Owens, Cook, and Stevenson 1990).

Second, the details of the NDE may vary with expectations about death. For example, Christians tend to see Jesus in the light, and Hindus see the messengers of Yamraj coming to take them away—and they often refuse to go! (Osis and Haraldsson 1977). However, the general pattern seems to be similar across cultures, suggesting that religious expectations are not responsible for the entire experience or for most of its common features. If they were, we might expect more pearly gates and fewer tunnels. We might also expect those who attempt suicide to have more hellish experiences, but they do not (Greyson and Stevenson 1980; Ring and Franklin 1981–1982; Rosen 1975). Their NDEs are much like others and tend to reduce future attempts at suicide.

All this suggests that, although expectation may change the details of NDEs, it cannot be used to explain their occurrence entirely or even to account for the similarities across ages and cultures.

Administered Drugs

The suggestion that the NDEs are created by drugs administered to dying patients does not hold up either. Many classic cases have been reported from drug-free patients and from people who were falling from mountains (Noyes and Kletti 1972) or involved in other accidents in which no drugs were involved. More specifically, research shows that patients given anesthetics or painkillers have fewer or more muted and less detailed NDEs than others (Greyson and Stevenson 1980; Osis and Haraldsson 1977; Ring 1980). It seems likely that it is the brain's own drugs that are more important for the NDE than drugs administered from outside.

Endorphins

Daniel Carr (1981, 1982) first suggested that endorphins could account for the NDE. Endorphins are released under stress (including both actual physical trauma and extreme fear—such as the fear of dying). They are known to block pain and to induce feelings of well-being, acceptance, and even intense pleasure, which might suggest they are responsible for the positive emotional tone of most NDEs. There is much controversy over the occurrence of "hellish" NDEs, with some researchers arguing that they are far more common than previously suspected (Atwater 1992; Greyson and Bush 1992; Rawlings 1978). Occasionally, NDEs change from pleasant to hellish, as occurred in one seventy-two-year-old cancer patient who was administered naloxone. His pleasant NDE turned to horror and despair as the friendly creatures morphed into the doctors treating him—suggesting that the naloxone (a morphine antagonist) had blocked the endorphins that were providing the pleasant feelings (Judson and Wiltshaw 1983). This is circumstantial, though, and Melvyn Morse has argued that endorphins are not responsible, suggesting that the neurotransmitter serotonin plays a more important role. Of eleven children who had survived critical illnesses, including coma and cardiac arrest, seven reported NDEs, while twenty-nine age-matched controls, who had had similar treatments including the use of narcotics, did not report any NDEs (Morse et al. 1986). However, it is questionable whether the effects of narcotics administered during critical illness are comparable with those of endorphins. Karl Jansen has argued that endorphins are not potent hallucinogens and suggested instead the involvement of NMDA receptors (postsynaptic receptors for the excitatory neurotransmitter glutamate) (Jansen 1989). Thus, it is still not known just how far endorphins are implicated in the NDE.

Anoxia or Hypercarbia

The argument over the role of anoxia has been complex. Some attribute to anoxia all the features of the NDE, though this reasoning is implausible, since so many NDEs clearly occur in the absence of anoxia (e.g., when the person only *thinks* he or she is going to die).

Others have argued that the cortical disinhibition associated with anoxia may be responsible for the tunnel and light experiences. Since the visual cortex is organized with many cells devoted to the center of the visual field and few to the periphery, random excitation will produce the effect of a bright light in the center fading out toward darkness—in other words, a tunnel effect (Blackmore and Troscianko 1988). More generally, it has been suggested that it is the disinhibition (not the anoxia per se) that is responsible for much of the NDE (Blackmore 1993).

Anoxia in non-life-threatening situations does cause odd experiences, such as the visions and out-of-body experiences reported by pilots trained in gravity-induced loss of consciousness (Whinnery 1990). There are also suggestions of NDE-like experiences in children suffering from reflex anoxic seizures, though most of these children are too young to describe their experiences (Appleton 1993; Blackmore 1998).

Against all this, others argue that the effects

of anoxia are not like those of NDEs (for example, producing confusion rather than the clear thinking of a typical NDE), though this is complicated by the fact that different types and speeds of anoxia cause different effects. There is also one case of an NDE in a patient with measured, normal blood gases (Sabom 1982), although it has been argued that his blood was taken from the femoral artery and that peripheral blood bases are not a reliable indicator of cortical blood gases (Gliksman and Kellehear 1990).

There may also be a role for hypercarbia, which has long been known to induce strange experiences such as lights, visions, and out-of-body and mystical experiences (Meduna 1958).

Temporal Lobe Stimulation

The temporal lobe is likely to be crucial in NDEs, since it is sensitive to anoxia and its stimulation is known to induce hallucinations, memory flashbacks, body distortions, and out-of-body experiences (Halgren et al. 1978; Penfield 1955). The limbic system is also sensitive to anoxia and involved in the organization of emotions and memory, suggesting a possible link with the life review that sometimes occurs during NDEs. An interesting effect of endorphins is that they lower the seizure threshold in the temporal lobe and limbic system (Frenk, McCarty, and Liebeskind 1978), so they might produce the same effects as anoxia. One neurobiological model of the NDE is based almost entirely on the notion of abnormal firing in the temporal lobe and associated parts of the brain (Saavedra-Aguilar and Gomez-Jeria 1989). Also, research looking for an "NDE-prone personality" has led to the conclusion that those most likely to have NDEs may have more unstable temporal lobes and show more "temporal lobe signs" than others (Ring 1984), though it is not clear how much of this association is a cause or an effect of the NDE.

Life after Death

None of the previous mechanisms can account entirely for the NDE, and many theorists argue that something beyond the brain is involved—for example, that there is a soul or something else that leaves the body at death and that the NDE is a glimpse of what follows. Direct evidence for this explanation is impossible to obtain. However, there are claims that during NDEs, people have been able to hear conversations and see the actions of people around them and even observe things such as the behavior of needles on dials, all of which they could not possibly have known about while in a comatose state (Sabom 1982). If such paranormal acquisition of information really occurs, it is evidence that any naturalistic account of NDEs must be incomplete. But does it occur? Many of these claims are based purely on anecdotal evidence, and very few have any independent corroboration.

For example, the most famous case involves a woman named Maria, who was taken to a Seattle hospital after a severe heart attack and then suffered a cardiac arrest. She later told her social worker, Kimberley Clark, that as she was being driven into the hospital in an ambulance, she had looked down from above and seen a tennis shoe on an inaccessible window ledge. Clark then searched for the shoe and apparently found it, just as Maria had described it. The problem with this case is that we have only Clark's description to go on. Neither Maria nor anyone else involved gave an independent account of the original experience or of the existence of the shoe, and Maria herself is now untraceable and presumed dead. Like so many other cases, this one does not stand up under scrutiny. There are other similar cases (Ring and Lawrence 1993). Yet skeptics tend to reject the evidence as inadequate, whereas proponents think it is conclusive. Perhaps the matter might be resolved by appropriate experiments, such as those using concealed targets in operating theaters and recovery rooms. Some are presently under way, but no successful results have yet been published.

The transformations reported in the lives of some individuals after near-death experiences are also taken as evidence of the NDE's heavenly nature. However, simply facing up to death can bring about a change in personal values, and there is conflicting evidence about whether an NDE is necessary for such an outcome (Greyson 1990; Pope 1994). It has also been argued that during the NDE, the usual model of self breaks down, and this brief experience of selflessness may bring about personal changes (Blackmore 1993).

In the end, it is probably a matter of personal preference whether to interpret the NDE as a glimpse of the life beyond or the product of the dying brain. In either case, the NDE deserves serious research, and the dying, the recovering, and their relatives deserve to know what we have learned. As Morse (1994) put it, these experiences can help us to restore dignity and control to the dying process. Just as NDEs reduce the fear of death in the people who have them, so they can help all of us to accept death as a positive aspect of life. Indeed, the study of life at its last limits may tell us more about ourselves and our lives than it does about death.

Originally published as an invited editorial review in the Journal of the Royal Society of Medicine, *1996, 89; 73–76. Copyright for original article by the Royal Society of Medicine.*

References:

Appleton, R. E. 1993. "Reflex Anoxic Seizures." *British Medical Journal* 307: 214–215.

Atwater, P. M. H. 1992. "Is There a Hell? Surprising Observations about the Near-Death Experience." *Journal of Near-Death Studies* 10: 149–160.

Barrett, W. 1926. *Death-Bed Visions*. London: Methuen.

Blackmore, S. J. 1982. *Beyond the Body*. London: Heinemann.

———. 1993. *Dying to Live: Science and the Near-Death Experience*. London: Grafton.

———. 1998. "Experiences of Anoxia: Do Reflex Anoxic Seizures Resemble Near-Death Experiences?" *Journal of Near-Death Studies* 17: 111–120.

Blackmore, S. J., and T. Troscianko. 1988. "The Physiology of the Tunnel." *Journal of Near-Death Studies* 8: 15–28.

Carr, D. 1981. "Endorphins at the Approach of Death." *Lancet* (February 14): 390.

———. 1982. "Pathophysiology of Stress-Induced Limbic Lobe Dysfunction: A Hypothesis Relevant to Near-Death Experiences." *Anabiosis: The Journal of Near-Death Studies* 2: 75–89.

Clark, K. 1984. "Clinical Interventions with Near-Death Experiencers." In *The Near-Death Experience: Problems, Prospects, Perspectives*, edited by B. Greyson and C. P. Flynn, 242–255. Springfield, IL: Charles C. Thomas.

Dlin, B. M., M. D. Stern, and S. J. Poliakoff. 1974. "Survivors of Cardiac Arrest: The First Few Days." *Psychosomatics* 15: 61–67.

Dobson, M., A. E. Tattersfield, M. W. Adler, and M. W. McNicol. 1971. "Attitudes and Long-Term Adjustment of Patients Surviving Cardiac Arrest." *British Medical Journal* 3: 207–212.

Druss, R. G., and D. S. Kornfeld. 1967. "The Survivors of Cardiac Arrest: A Psychiatric Study." *Journal of the American Medical Association* 201: 291–296.

Frenk, H., B. C. McCarty, and J. C. Liebeskind. 1978. "Different Brain Areas Mediate the Analgesic and Epileptic Properties of Enkephalin." *Science* 200: 335–337.

Gabbard, G. O., and S. W. Twemlow. 1984. *With the Eyes of the Mind*. New York: Praeger.

Gabbard, G. O., S. W. Twemlow, and F. C. Jones. 1981. "Do 'Near Death Experiences' Only Occur Near Death?" *Journal of Nervous and Mental Disease* 169: 374–377.

Gliksman, M. P. H., and A. Kellehear. 1990. "Near-Death Experiences and the Measurement of Blood Gases." *Journal of Near-Death Studies* 9: 41–43.

Greyson, B. 1990. "Near-Death Encounters with and without Near-Death Experiences: Comparative NDE Scale Profiles." *Journal of Near-Death Studies* 8: 151–161.

Greyson, B., and N. E. Bush. 1992. "Distressing Near-Death Experiences." *Psychiatry* 55: 95–110.

Greyson, B., and I. Stevenson. 1980. "The Phenomenology of Near-Death Experiences." *American Journal of Psychiatry* 137: 1193–1196.

Halgren, E., R. D. Walter, D. G. Cherlow, and P. H. Crandall. 1978. "Mental Phenomena Evoked by Electrical Stimulation of the Human Hippocampal Formation and Amygdala." *Brain* 101: 83–117.

Irwin, H. J. 1985. *Flight of Mind: A Psychological Study of the Out-of-Body Experience.* Metuchen, NJ: Scarecrow Press.

Jansen, K. 1989. "Near Death Experience and the NMDA Receptor." *British Medical Journal* 298: 1708.

Judson, I. R., and E. Wiltshaw. 1983. "A Near-Death Experience." *Lancet* (September 3): 561–562.

MacMillan, R. L., and L. Brown. 1971. "Cardiac Arrest Remembered." *Canadian Medical Association Journal* 104: 889–890.

Meduna, L. J. 1958. *Carbon Dioxide Therapy: A Neurophysiological Treatment of Nervous Disorders.* 2d ed. Springfield, IL: Charles C. Thomas.

Moody, R. A. 1975. *Life after Life.* Atlanta, GA: Mockingbird.

Morse, M. 1990. *Closer to the Light.* London: Souvenir.

Morse, M., P. Castillo, D. Venecia, J. Milstein, and D. C. Tyler. 1986. "Childhood Near-Death Experiences." *American Journal of Diseases of Children* 140: 1110–1114.

Morse, M. L. 1994. "Near Death Experiences and Death-Related Visions in Children: Implications for the Clinician." *Current Problems in Pediatrics* 24: 55–83.

Noyes, R., and R. Kletti. 1972. "The Experience of Dying from Falls." *Omega* 3: 45–52.

Osis, K., and E. Haraldsson. 1977. *At the Hour of Death.* New York: Avon.

Owens, J. E., E. W. Cook, and I. Stevenson. 1990. "Features of 'Near-Death Experience' in Relation to Whether or Not Patients Were Near Death." *Lancet* 336: 1175–1177.

Penfield, W. 1955. "The Role of the Temporal Cortex in Certain Psychical Phenomena." *Journal of Mental Science* 101: 451–465.

Pope, J. 1994. "Near-Death Experiences and Attitudes Towards Life, Death and Suicide." *Australian Parapsychological Review* 19: 23–26.

Rawlings, M. 1978. *Beyond Death's Door.* Nashville, TN: Thomas Nelson.

Ring, K. 1980. *Life at Death: A Scientific Investigation of the Near-Death Experience.* New York: Coward, McCann and Geoghegan.

——. 1984. *Heading toward Omega: In Search of the Meaning of the Near-Death Experience.* New York: Quill.

Ring, K., and S. Franklin. 1981–1982. "Do Suicide Survivors Report Near-Death Experiences?" *Omega* 12: 191–208.

Ring, K., and M. Lawrence. 1993. "Further Evidence for Veridical Perception during Near-Death Experiences." *Journal of Near-Death Studies* 11: 223–229.

Rosen, D. H. 1975. "Suicide Survivors." *Western Journal of Medicine* 122: 289–294.

Saavedra-Aguilar, J. C., and J. S. Gomez-Jeria. 1989. "A Neurobiological Model for Near-Death Experiences." *Journal of Near-Death Studies* 7: 205–222.

Sabom, M. B. 1982. *Recollections of Death.* London: Corgi.

Sabom, M. B. 1982. *Recollections of Death.* London: Corgi.

Whinnery, J. E. 1990. "Acceleration-Induced Loss of Consciousness: A Review of 500 Episodes." *Archives of Neurology* 47: 764–776.

Observer Effects and Observer Bias

DOUGLAS G. MOOK

Fundamental to the skeptical outlook is a simple and familiar idea: we can and do delude ourselves. Even if our conclusions are based on direct observations, we may still distort our observations so that they lead us to form or maintain conclusions that are simply wrong. There are many ways this can happen, but two in particular occur so often that they rate special treatment, and this article is devoted to them. First, we may distort what happens: the problem of observer effects. Second, we may distort how we see it happening: the problem of observer bias. To better understand what these ideas mean, consider an instance that involves both types of distortions.

The Double-Blind Experiment

The double-blind experiment is a staple of medical research. Suppose a team of physicians wants to evaluate a new way of treating some disorder. The researchers treat a group of patients with the new method, comparing that group with another group of patients treated the old way. (There are many variations on this theme, but this one will do.) If at all possible, they will conduct the experiment as a "double-blind" study, which simply means that neither the patients nor the researchers who interact with them and evaluate them will know which patient is in which group until the experiment is over.

Why is the double-blind method valuable? First, if the physicians and attendants knew which patients were receiving which treatments, they might—perhaps quite unconsciously—give more care and attention to the patients receiving the new, experimental treatment than to the patients receiving the old one. If so, then the greater care and attention might in themselves produce more improvement in the experimental group, making the new treatment look more effective than it is. This would be an example of observer effects. The clinicians, as observers of the patients, are also having unintended effects on them.

Second, knowledge of which patients are receiving which treatments could lead to biased evaluations of the patients. This could happen in either or both of two ways. If the clinicians expect the new treatment to be more effective, they could "see" what they expect to see and rate the experimental patients as more improved even if they really are not. Then, too, the patients could have a similar bias in their observations of themselves, so that they also might "see" themselves as more improved than they really are. Both of these outcomes would be cases of observer bias: the clinicians and/or the patients themselves are seeing what they wish or expect to see, rather than what is really there to be seen.

For all these reasons, the evaluation of a new treatment procedure will routinely be conducted in a double-blind manner if possible. The origin of the term is easy to see: the patient will not know which group he or she is in (single-blind), and the physicians and other personnel who interact with and evaluate the patient also will not know this (double-blind).

Medical research is not the only arena in which observer effects and observer bias are problematic. They may operate in other kinds of scientific research, too, and they may operate in our everyday observations—the informal "research projects" that we all use to form or support our beliefs. They can be sneaky and fertile sources of self-delusion. So let us look at some further examples of how the problems may arise and of how we can combat them.

Observer Effects

We may influence events to make them fit our preconceptions—whether or not they should and whether or not we intend it. As an example, the story of Clever Hans, the mathematical horse (Spitz 1997; Stanovich 2001), is a classic cautionary tale known to every researcher (see the "Clever Hans" entry in this encyclopedia).

Clever Hans

The story takes place in Germany in the late 1800s. A certain schoolteacher had a horse named Hans, which had remarkable capabilities: Hans could do arithmetic! Ask him to add, say, 7 and 4, and he would tap 11 times with his hoof and then stop. He was just as gifted at subtraction and even multiplication and division.

This schoolteacher was not a faker. He could have made a pretty pfennig exhibiting his horse in public, but he did not do so. Rather, he invited scientists to study Clever Hans. They did, and they came away convinced that Hans was genuine. The schoolteacher was not feeding him cues; in fact, the man could be out of sight altogether, and still Clever Hans could solve problems that were put to him by someone else.

But one scientist, an experimental psychologist named Oskar Pfungst, said, "Let's do one more check." Accordingly, he had one of the onlookers whisper into Hans's ear, for example, "Seven." Then another, different observer would whisper into the horse's ear, "Plus four." Neither of the whisperers could hear the other. Therefore, none of the observers knew the correct answer. Not until Hans told them—if he could. And he couldn't. Under these conditions, after Hans was asked the question, he would go on tapping and tapping indefinitely.

It seems that Hans had depended on unconscious cuing by the observers. As long as the audience knew what the correct answer was—$7 + 4 = 11$—they would wait until Hans's eleventh tap. Then they would lean forward alertly as if to say: "That's the answer. Will he stop now?" Those subtle movements were Hans's cue to stop tapping.

So, yes, Hans was a very clever horse—but he was not clever at arithmetic. He was clever at reading humans and the unconscious cues their movements gave him. The schoolteacher, too, must have been giving him such unconscious cues all along, though he neither knew nor intended it.

Since his day, the "Clever Hans effect" has been a standard phrase used to remind researchers—and the rest of us—of the danger of observer effects. An observer can give unintended cues that can affect what his or her subjects do, and this can lead the observer, in all honesty, to self-delusion.

Facilitated Communication

Hans taught us a valuable lesson. Unfortunately, not all of us have learned it. Unintended effects on what we observe, resulting from unconscious movements or other unconscious cuing, can cause our observations to match our preconceptions even when otherwise they would not. Examples range from Ouija boards and dowsing rods to the tragic story of "facilitated communication" (FC) in autistic children (see Facilitated Communication entry in section 2).

Autism is a severe developmental disorder, apparent in early childhood. It is characterized by a profound difficulty—indeed, what seems a profound disinterest—in communicating with other human beings. In the late 1980s, an exciting breakthrough was announced. It was reported that in autistic children, a technique called facilitated communication could reveal a hidden ability to communicate. Autistic children could type coherent messages on a keyboard if their hands and arms were supported over the keyboard by a trained and sympathetic facilitator. In that way, seemingly uncommunicative children could tell us much about themselves.

The reports of success with this technique were widely publicized, of course, and they led to a great surge of optimism among both therapists and parents. But certain disturbing things came to light. Some autistic children would type messages that reported incest or abuse occurring in their homes. In some of these cases, children were actually removed from their homes because of the abuses "uncovered" by FC.

Fortunately, these dramatic claims were followed up by more careful experiments. These studies made it clear that FC is really the Clever Hans effect all over again. The facilitators who supported the children's hands were unconsciously cuing the children as to what keys to hit. How do we know this? By the same sorts of checks that exposed Clever Hans. In one experiment, for example, questions were read to the patient and to his or her facilitator through headphones. If the questions were the same for both, then the child would type out the correct answers. But if the child heard one question and the facilitator heard a different one (without knowing it), then the child would type answers to the questions that the facilitator had received.

These findings are exactly parallel to the Clever Hans case. If the observers did not know the answer to the arithmetic problem, Hans couldn't tap it. If the facilitator did not know what the child was hearing, the child couldn't type it. Both Clever Hans and the autistic children were performing in accordance with unconscious and unintentional cues given by the onlookers in one case, the facilitators in the other.

And if that is so, then it is also likely that the stories of abuse, incest, and the like were not reports by the children but unconscious inventions by the therapists. At this writing, the useless and dangerous FC fad is mercifully fading away. But in the meantime, families were shattered by false accusations of parental abuse.

Why did this happen? Like Clever Hans's owner, the FC therapists were sincere and well-meaning. But they were also untrained in research, and so they were unaware of how subtly and unintentionally we can influence what other humans do. Professionals who were adequately trained in research would have seen that the Clever Hans effect was a danger, and they would have checked for it.

Observer Bias

We don't want to distort what happens, which can occur with observer effects. But we also don't want to distort how we see it happening, which introduces the problem of observer bias. Most of us are aware that we may "see" what we expect to see and perhaps what we want to see rather than what is "out there" to be seen. Despite this, we often underestimate the danger inherent in observer bias, and we may overlook this factor when we should not.

Seeing what we expect to see can lead to distortions that would be funny if they weren't so dangerous. Here's a story told by James "The Amazing" Randi, a stage magician and a well-known skeptic. He once appeared, in disguise, as a guest on a talk show. There, he informed his audience that while driving from New Jersey into New York, he had seen a V formation of objects flying toward the north overhead. Within seconds, Randi said, the "station switchboard lit up like an electronic Christmas tree" (quoted in Stanovich 2001, 69). One "eyewitness" after another called in to confirm Randi's "sighting"—which was purely imaginary. He had made it all up. Nevertheless, observers looked at the skies and saw what they expected to see—the little unidentified flying objects (UFOs) that were not there.

N Rays

Practicing scientists, too, are subject to observer bias. The classic example is the scientist who has a pet theory and "sees" events that confirm it. The danger is that the scientist may "see" events or objects that simply are not there, like Randi's made-up UFOs. That this can and does happen is shown by the following cautionary tale.

In 1903, a distinguished physicist, René Blondlot, announced a new form of radiation that he had discovered, naming the new rays N rays for the University of Nancy, where he taught. The problem was that other scientists could not verify his findings. Something clearly was wrong. Another physicist visited Blondlot's lab, where, sure enough, he could see no evidence of N rays—though Blondlot and his coworkers insisted that they could. Finally, the visitor secretly jiggered the apparatus so that it could not possibly produce any rays at all. But his hosts, not knowing this, continued to see the rays that simply were not there. They saw what they expected to see (Hyman 1964).

Patient or Applicant?

Observer bias has been demonstrated many times by direct experiment. In one such experiment, panels of mental health professionals were shown a videotape of a younger man talking to an older man about his feelings and experiences. Some of the clinicians were told that the young man was a patient; others were told that he was a job applicant. After seeing the videotape, the viewers were asked what they thought about the young man: what kind of person was he?

The videotape was the same for all observers, but the judges' reactions were very different, depending on what they thought they were seeing. Those who saw the "job applicant" tended to describe him as "attractive, candid, innovative, and realistic." But those who thought the young man was a patient saw him as a "tight, defensive person," whom they described as "frightened of his own aggressive impulses" (Langer and Abelson 1974, 7). Since the tape was the same, these striking differences can mean only that the judges' observations were biased by their ideas as to what patients and job applicants are like or, perhaps, how they ought to behave.

This, of course, was done as an experiment, to demonstrate the effect. Consider, though, that these professionals were seeing real patients on a daily basis. There is no reason to doubt that their perceptions of these very real people were similarly affected by preconceptions. (For a dramatic experimental demonstration of this, see Rosenhan 1973.)

Subliminal Tapes

Subliminal tapes are said to have messages embedded in them that are too faint to hear consciously because they are masked by noise or soft music. The claim is that the subliminal messages are registered by the unconscious, so that their messages can affect us without our awareness.

Indeed, some years back, there was a considerable flap about so-called subliminal advertising. In a movie theater, very brief messages—"Eat X" or "Drink Y"—were flashed on the screen so briefly that people in the theater were unaware of them—and sales of X and Y soared. The only trouble is that this scenario never happened. The "researcher" who conducted these studies later admitted publicly that he had made up the data in order to get research funding.

But in our own time, such subliminal stimuli have come back, for example, as aids that can teach foreign languages or improve memory or self-esteem, all without our awareness and without effort. The claims are backed up with glowing testimonials by those who have tried these tapes.

But do the claims hold up? A number of research teams (e.g., Greenwald et al. 1991; Pratkanis 1992) have checked the effectiveness of subliminal tapes, and they have checked something else as well: our capacity for self-delusion. In one such experiment, subject volunteers were given tapes claimed to enhance either self-esteem (for half the subjects) or memory (for the other half). These were commercially available subliminal tapes, and subjects listened to them every day for a few weeks, as instructed by the manufacturer. Memory and self-esteem were measured before and after this period.

What happened? Results showed that there was no consistent effect of either tape on memory or self-esteem, as measured by objective tests. But the subjects thought there was. They told the experimenters that their self-esteem or their memories had, in fact, improved over the course of listening to the tapes.

However, there was a refinement to this experiment. For half the subjects in each group, the tapes were mislabeled. Thus, half of those who thought they were hearing memory tapes were actually hearing self-esteem tapes, and half who thought they were hearing self-esteem tapes were listening to memory tapes. At the conclusion of the listening period, subjects who thought they had been listening to self-esteem tapes reported enhanced self-esteem—even if they had really been listening to memory tapes. And subjects who thought they had been listening to memory tapes reported better memories—even if they had really been listening to self-esteem tapes.

In short, the subjects were deluding themselves twice. First, they thought they had improved when they hadn't. And second, they thought the tapes had done it, but the imaginary improvement was present where they expected it to be, not where the tapes they had heard should have caused it to be. The subjects were observing themselves with biased eyes and finding change where they expected to find it.

What is common to all of the cases we've considered is just this: perfectly sensible people deluded themselves. Their observations purportedly showed that Hans could do arithmetic, that autistic children could communicate by keyboard, that N rays existed, and that subliminal tapes had improved memory or self-esteem—when none of these was true. In some cases, observers unintentionally influenced what happened (observer effects). In others, they distorted what they saw happening (observer bias).

But remember that we all have preconceptions. Like Professor Blondlot, we have our pet theories, and, as in each of the cases mentioned, we may feel that our observations confirm them. We may do well to consider the possibility that we are deluding ourselves and check it out.

But check it out how? The first thing to realize is that collecting testimonials is not enough. There will be no shortage of people who will tell us, quite sincerely: "I saw that horse tap out a sum [or an autistic child type out a message] with my own eyes!" Or "I saw for myself how much my memory improved!" This, as we've seen, does not rule out the possibility that our excited informants are simply deluding themselves, seeing what they expect to see, or perhaps even causing to happen what they expect to happen. But just such testimonials are what blare at us from newsstands and media screens. We should take them with a healthy, salty dose of skepticism.

However, running through the prior examples, we also notice that tests for self-delusion need not be elaborate. It may not be at all difficult to make the observer "blind" as to what the answer ought to be, as in the case of Clever Hans. Similarly, if people had listened to subliminal tapes not knowing whether they were memory or self-esteem tapes, their expectations could not have biased their evaluations. If Professor Blondlot had checked for his N rays, not knowing whether the apparatus was working or not, he would have saved himself embarrassment. And if people are asked simply, "What do you see in the sky?"—not knowing what they "ought" to be seeing—it's unlikely that Randi's UFOs over New Jersey would have been confirmed by so many radio listeners.

Now, things are not always that easy. Sometimes it takes special conditions to check out the possibility of observer effects or observer bias—as, for instance, by mislabeling some tapes or by arranging for a child to hear one thing and an adult facilitator another. These studies required full-scale experiments. And each one set up an artificial set of conditions that one would not encounter in ordinary life—which also means that no amount of ordinary-life observation would have shown what was really going on. We ourselves, busy with other things, may not always be able to set up the conditions for such an experiment.

What we can do, however, is (1) be aware of the two dangers, (2) check them out when we can, and (3) remember that if we cannot easily check them out ourselves, others may be able to do so. We can learn where to find the investigations that do set up the necessary conditions for evaluating claims. The references that follow are a good place to start.

References:

Gilovich, T. 1991. *How We Know What Isn't So.* New York: Free Press.

Greenwald, A. G., E. Spangenberg, A. Pratkanis, and J. Eskenazi. 1991. "Double-Blind Tests of Subliminal Self-Help Audiotapes." *Psychological Science* 2: 119–122.

Hyman, R. 1964. *The Nature of Psychological Inquiry.* Englewood Cliffs, NJ: Prentice-Hall.

Langer, E. J., and R. P. Abelson. 1974. "A Patient by Any Other Name . . . Clinician Group Differences in Labeling Bias." *Journal of Consulting and Clinical Psychology* 42: 4–9.

Mook, D. G. 2001. *Psychological Research: The Ideas behind the Methods.* New York: W. W. Norton.

Nisbett, R. E., and L. Ross. 1980. *Human Inference: Strategies and Shortcomings of Social Judgment.* Englewood Cliffs, NJ: Prentice-Hall.

Pratkanis, A. R. 1992. "The Cargo-Cult Science of Subliminal Persuasion." *Skeptical Inquirer* 16: 260–272.

Rosenhan, D. L. 1973. "On Being Sane in Insane Places." *Science* 179: 250–258.

Shermer, M. 1997. *Why People Believe Weird Things: Pseudoscience, Superstition, and Other Confusions of Our Time.* New York: W. H. Freeman.

Spitz, H. H. 1997. *Unconscious Movements: From Mystical Messages to Facilitated Communication.* Mahwah, NJ: Erlbaum.

Stanovich, K. E. 2001. *How to Think Straight about Psychology.* 6th ed. New York: Longmans.

Acknowledgments

Portions of this chapter are adapted from D. G. Mook, *Psychological Research: The Ideas behind the Methods* (New York: W. W. Norton, 2001). The author thanks Melody Browning-Mook, Dale Marshall, and Shane Parkhill for helpful comments on a preliminary version of the manuscript.

Out-of-Body Experiences

SUSAN J. BLACKMORE

An out-of-body experience (or OBE) is an experience in which a person seems to see the world from a location outside of the physical body. In other words, when you have an OBE, you feel as though you have left the body and are able to see, feel, and move around without it. Note that this definition treats the OBE as an *experience* only. So, if you *feel* as though you are out of your body, then you are, by definition, having an OBE, whether or not anything has actually left the body. As John Palmer (1978, 19) pointed out, "The OBE is neither potentially nor actually a psychic phenomenon. It is an experience or mental state, like a dream or any other altered state of consciousness. It may be associated with psi [a general term covering all four forms of psychic phenomena, i.e., telepathy, clairvoyance, precognition, and psychokinesis, or the supposed mechanism underlying them], but it is not a psychic phenomenon itself." This broad definition allows researchers to study the experience without committing themselves to any particular theory of the OBE.

Surveys show that about 15 to 20 percent of the population have had an OBE at some time during their lives (Blackmore 1982, 1996). Most of these people have had only one or a very few OBEs, although a small percentage have many. Some people find the experience frightening, but others value the pleasant sensations and visions it entails; some even learn to induce the experience at will.

Spontaneous OBEs most often occur while resting, just before sleep, or when meditating. However, they can occur at almost any time, and occasionally, the person experiencing an OBE carries on with what he or she was doing (such as walking, driving, or even speaking) apparently without interruption. Common factors that provoke OBEs include relaxation, loss or disruption of the body image, and reduced sensory input. Most spontaneous OBEs are very brief, lasting only a few seconds. Some begin with the experience of traveling down a dark tunnel, often with a bright white or golden light at the end. Others begin with rushing or whirring noises, odd vibrations, or simply brief periods of blackout. The return from an OBE is usually gradual, but occasionally, it is accompanied by a sensation of shock or disorientation (Muldoon and Carrington 1929; Alvarado and Zingrone 1997).

People who have OBEs (OBErs) often feel as though they can travel anywhere and see anything they wish. According to various occult traditions, the OBEr has another body or a double, sometimes referred to as the "astral body." This double is usually something like a replica of the physical body, though less distinct. At times, it is said to be ghostly or transparent and described as whitish or pale gray. In rare cases, the double is perceived to be connected to the physical body by a silver cord. Experiences with a second body have been called "parasomatic" by Celia Green (1968) and are contrasted with the more common "asomatic" experiences in which the person feels as though he or she is just disembodied awareness or a point of consciousness. Occasionally, the OBEr reports feeling more like a bubble, a spot of light, or a patch of mist.

During an OBE, vision and hearing are said to be more powerful and clearer than normal. Some people even get the impression that they could see all around at once or hear anything anywhere if they wished to; there is a sense of limitlessness. OBEs are like dreams in some ways—for example, the scenery and lighting can be very strange, and the ordinary constraints of the physical world do not seem to apply. However, unlike ordinary dreams, OBEs feel very real, consciousness is clear, and the experience is usually remembered very vividly afterward. In some ways, OBEs are more like lucid dreams, that is, dreams in which you know *during the dream* that you are dreaming. OBEs can also merge into mystical experiences, and they form a central part of the near-death experience (NDE). After OBEs, people say that their attitudes and beliefs are changed, usually in positive ways. A psychiatric analysis of hundreds of cases of OBEs showed that for many people, their fear of death was reduced and their belief in survival increased (Gabbard and Twemlow 1984).

OBErs frequently interpret their experiences as psychic, paranormal, or mystical. Some claim that they could see not only their own bodies but also distant scenes, although the experimental evidence to support these claims is weak. More rarely, they also claim to be able to influence distant events, but the frustration of being *unable* to influence things is more commonly reported. For example, OBErs at times try to touch people they see, only to find that the people do not notice them at all; similarly, they may try to switch on lights or move objects, only to discover that their hands go right through these objects.

The occurrence of OBEs is not related to age, sex, educational level, or religion (Green 1968; Gabbard and Twemlow 1984). However, OBErs score higher on measures of hypnotizability and absorption—that is, they can more easily become absorbed in films, books, or fantasies (Irwin 1985). OBErs are also more likely to believe in the paranormal, to have various kinds of psychic experiences, and to report frequent dream recall and lucid dreams.

Many popular techniques are available for inducing the experience, most of which use imagery and relaxation as key components (Blackmore 1982; Rogo 1983). Experimental techniques have also been developed using special sounds and visual displays, as well as imagery exercises and relaxation. From the early days of psychical research, hypnosis has been used to induce OBEs or "traveling clairvoyance" (see Blackmore 1982; Alvarado 1992). Drugs have been known to produce OBEs, especially the psychedelics LSD, psilocybin, DMT and mescaline, and the dissociative anesthetic ketamine, which often induces feelings of body separation, floating, and even dying. However, there is no known drug that can reliably induce an OBE.

There is no evidence to suggest that people who have OBEs are mentally ill in any way. In their study of over 300 OBErs, Glen Gabbard and Stuart Twemlow (1984) found that the participants were generally well adjusted, with low levels of alcohol and drug abuse and no signs of psychotic thinking. Several other studies have failed to detect any differences in various measures of psychopathology between individuals who do and do not have OBEs. Some people who have OBEs fear that they are ill or even going mad. They can often be helped simply by giving them information about how common OBEs are and the circumstances under which they are most likely to happen. Others fear that they will go out of their bodies and not be able to get back. This fear does not seem well founded. In both the occult and the scientific literature, it is said that if the physical body is touched or if attention is demanded

Out-of-body experience: astral body lying above physical body, connected to it by a cord. (Fortean Picture Library)

in some way, the out-of-body experience will end. Indeed, it is generally hard to induce OBEs, and it takes much practice to keep them going for long periods.

To date, most experiments on OBEs have fallen into one of three types. The first type, involving attempts to detect the double during OBEs, began early in the twentieth century. Photographs were taken of projected doubles, and efforts were made to weigh the souls as they left the bodies of people dying of tuberculosis; however, the studies were not well controlled, and the perceived effects disappeared when better methods were developed (for a review of these studies, see Blackmore 1982). Recent studies have used magnetometers, thermistors, and ultraviolet and infrared detectors, as well as both human and animal subjects. Although there has been limited success with animals, no reliable detector of an out-of-body presence has been discovered thus far (Morris et al. 1978).

Experiments of the second type have tried to detect paranormal vision or hearing during OBEs—that is, to determine whether people can really see and hear what is going on at a distance. In early experiments, mediums were asked to exteriorize their doubles and to smell scents or view the actions of people at a distance, but the mediums usually could have seen what was going on around them. More recently, target letters, numbers, or objects have been concealed from view in the laboratory, and people who can have OBEs at will have been asked to try to see them (for a review of these experiments, see Alvarado 1982). In one well-known experiment, a subject correctly saw a five-digit number, but this type of success has never been repeated, and most other experiments have had equivocal results. Case studies include many claims that people can really see at a distance during OBEs, but the experimental evidence does not substantiate them.

The third type of experiment has involved physiological monitoring of OBErs. No unique physiological state seems to be involved in the out-of-body experience. In the few studies done, the subjects were found to be in very relaxed waking states or in states resembling that on the very edge of sleep. It is certainly clear that OBErs were not in rapid-eye-movement (REM), or dreaming, sleep. Therefore, OBEs cannot be considered to be a kind of dream.

Three main theories have been advanced in the attempt to explain the out-of-body experience. The first theory is based on the idea that something leaves the body during the OBE. Most people who have an OBE find the experience so compelling and realistic that they jump to the obvious conclusion that "they" have left their physical bodies. Many further conclude that this "something" that has left does not depend on the body and can therefore continue after physical death (although this conclusion does not necessarily follow, since the physical body was alive at the time).

The idea that we have doubles can be traced back to ancient Egypt and to Greek philosophy, and it can be found in folklore and mythology from many cultures. It is also a part of many religious doctrines. A popular modern version, based in the teachings of Theosophy, involves astral projection. It is suggested that we all have several bodies, of which the first three are the physical, the etheric, and the astral. The astral body is said to be able to separate from the physical body and move about on the astral plane without it. In life, the two bodies are connected by a silver cord, but at death, this cord is broken, freeing the consciousness.

There are numerous problems with this and all similar theories. They cannot specify what the astral body consists of, in what sense it is conscious, or how this consciousness is related to the obvious sensory and memory functions of the brain. Further, they cannot identify how the astral body perceives the world without using any sensory apparatus and without being detected or why the astral world appears the way it does. Also, studies have found that experiences involving astral doubles are far from common and that reports of the silver cord are extremely rare.

A second kind of theory suggests that OBEs are imagination plus extrasensory perception (ESP). In principle, this theory might account for the claims of paranormal perception during OBEs without involving all the problems of other worlds and other bodies. However, this is the weakest possible kind of theory, since imagination is such a broad concept and ESP is not well understood. In addition, it is not easy to see how this theory could be tested, which makes it useless as a scientific theory.

The final type of theory holds that the OBE is a purely psychological phenomenon, involving no self, soul, spirit, or astral body that leaves the physical body. Several theories of this kind have been proposed. They do not necessarily preclude the possibility of ESP, but they make no special provision for it.

Psychoanalytic interpretations treat the OBE as a dramatization of the fear of death, an uncoupling of the bodily ego from the mental ego, regression of the ego, or a reliving of the trauma of birth. Carl Jung himself saw OBEs as part of the process of individuation, and others have used his ideas of archetypes in trying to understand the OBE (see Alvarado 1992 for a review of this approach). In 1978, Palmer suggested that a loss of or change in the body image threatened the self-image and that the OBE then occurred, through unconscious mechanisms, as one of several possible ways of reestablishing personal identity. The ego strives to reestablish the normal body image, and when this effort succeeds, the OBE ends. Palmer added that some people could gain ego control over the process and so learn to have OBEs at will, providing access to deep, unconscious material and latent psychic abilities. This theory relies heavily on psychoanalytic concepts that have not stood up well to psychological research, generally being found to be either untestable or false.

One popular idea likens the OBE to birth. Superficially, there may be similarities between the tunnel and the birth canal or between the silver cord and the umbilicus. However, the birth canal would look nothing like a tunnel to a fetus being pushed through it, the light would more likely be red with blood rather than bright white or yellow, and silver cords are, in any case, very rarely reported. Memory from childhood is also extremely limited and unlike adult memory. Birth theories predict that people born by cesarean section should not have either tunnel experiences or OBEs, but in one study comparing people born normally with those born by cesarean, the same proportion of each group had had tunnel experiences and the cesareans reported more OBEs.

Harvey Irwin (1985) suggested that the OBE begins with a disruption of the normal body sense, leading to somesthetic (bodily) sensations of floating or flying. This is then translated, by synesthesia, into a complete experience of leaving the body, with visual, tactile, auditory, and other senses all being transformed. The process requires attention to or absorption in the new experience and loss of contact with somatic sensations. This concept would explain not only the conditions under which the OBE occurs but also the tendency for OBErs to score higher in tests of absorption.

Susan Blackmore suggested a theory based on a change in perceptual perspective. We normally feel as though "we" are behind the eyes looking out, but this viewpoint requires good sensory input to be sustained. OBEs occur when the normal self-image breaks down and the brain attempts to reconstruct it from memory and imagination. Memory images are often built in a bird's-eye view (you remember scenes as though looking down from above). If such a memory image takes over as the current "model of reality," an OBE occurs. Sounds might be incorporated relatively easily into the bird's-eye view, but anything that is likely to rebuild the normal body image, such as movement or a touch on the body, will end the OBE. According to this theory, the OBE is entirely imagined, but because the new viewpoint has taken over completely, the experience feels totally realistic.

These psychological theories account for the conditions under which OBEs occur and explain why the out-of-body world is rather like the world of imagination, with transparent walls and the ability to move around at will and to see in all directions. They also explain why apparently correct details are often mixed with false ones, since the brain has simply put together the best information it has.

When psychological theories of the OBE were first proposed, it was predicted that people with vivid imagery would be more likely to have OBEs. This expected result was not found, but since that time, OBErs have been shown to be better at spatial imagery and at switching viewpoints in imagery; they also have superior dream-control skills and a greater tendency to dream in bird's-eye view.

Theories like these can potentially explain all the phenomenology of the OBE without recourse to any other bodies or alternative realities. However, claims of paranormal events during OBEs continue to be made, and though they are not incompatible with the psychological theories, they provide a challenge to any purely materialist theory of human nature. Only further research will determine whether these claims are valid.

This article was prepared for publication in *A–Z Parapsychology*, edited by Jane Henry, Routledge 2003.

References:

Alvarado, C. S. 1982. "ESP during Out-of-Body Experiences: A Review of Experimental Studies." *Journal of Parapsychology* 46: 209–230.

———. 1992. "The Psychological Approach to Out-of-Body Experiences: A Review of Early and Modern Developments." *Journal of Psychology* 126: 237–250.

Alvarado, C. S., and N. L. Zingrone. 1997. "Out-of-Body Experiences and Sensations of 'Shocks' to the Body." *Journal of the Society for Psychical Research* 61: 304–313.

Blackmore, S. J. 1982. *Beyond the Body*. London: Heinemann. Overview of cases, theories, and scientific research.

———. 1996. "Out-of-Body Experiences." In *Encyclopedia of the Paranormal*, edited by G. Stein, 471–483. Buffalo, NY: Prometheus.

Crookall, R. 1961. *The Study and Practice of Astral Projection*. London: Aquarian Press. A classic description from an occultist's point of view.

Fox, O. 1962. *Astral Projection*. New York: University Books. Firsthand descriptions of extraordinary experiences.

Gabbard, G. O., and S. W. Twemlow. 1984. *With the Eyes of the Mind*. New York: Praeger. A study of cases, with psychoanalytic interpretations.

Green, C. E. 1968. *Out-of-the-Body Experiences*. London: Hamish Hamilton. A large collection of cases, with analysis of different types.

Irwin, H. J. 1985. *Flight of Mind: A Psychological Study of the Out-of-Body Experience*. Metuchen, NJ: Scarecrow Press. A scientific and psychological overview of research and theories.

Mitchell, J. L. 1981. *Out-of-Body Experiences*. Northamptonshire, England: Turnstone. Simple coverage of the research and theories.

Morris, R. L., S. B. Harary, J. Janis, J. Hartwell, and W. G. Roll. 1978. "Studies of Communication during Out-of-Body Experiences." *Journal of the American Society for Psychical Research* 72: 1–22.

Muldoon, S., and H. Carrington. 1929. *The Projection of the Astral Body*. London: Rider and Co. Classic account, full of descriptions of experiences and the authors' own theories.

Palmer, J. 1978. "The Out-of-Body Experience: A Psychological Theory." *Parapsychology Review* 9: 19–22.

Rogo, D. S. 1983. *Leaving the Body: A Practical Guide to Astral Projection*. Englewood Cliffs, NJ: Prentice-Hall. Among the best of the many "how-to-do-it" books.

Rogo, D. S., ed. 1978. *Mind beyond the Body: The Mystery of ESP Projection*. Harmondsworth, England: Penguin. A compilation of chapters by various experts, including some scientists.

Phrenology

JOHN VAN WYHE

Phrenology was considered a science of mind and character divination based on the shape of the head or skull. Phrenologists believed that the brain was the organ of the mind and that the mind consisted of a number of distinct, inborn departments or faculties. Because they were considered distinct, the faculties of the mind were thought to have their own discrete locations or "organs" in the brain, and the size of an organ, other things being equal, was considered a measure of its power or activity. As the skull was believed to take its shape from the underlying brain, the surface of the skull could be read as an accurate mold of the shape of the brain and therefore of its mental faculties.

The sizes of the phrenological organs were usually determined by an examination with the hands and sometimes with calipers or a measuring tape. So, for example, if a person's head showed a large organ of Benevolence, the phrenologist would conclude that this person was highly benevolent. If the area of the head marked as Benevolence was comparatively small or if it presented an indentation, the phrenologist would conclude that the person was "sordid, avaricious, mean, and totally insensible to charity" (Lundie 1844, 8). The surface of the head that covers the brain was mapped out to help the phrenologist locate the relevant areas of brain. Therefore, phrenology was more about determining the shape of the brain rather than just reading the head. There were about thirty organs/faculties in phrenology; the number varied by phrenologist and generally tended to increase over time as new organs were "discovered" or added.

Phrenologists believed that theirs was the only true "science of mind," which marked a difference between phrenology and other forms of divining character and personality. The phrenologist claimed privileged access to otherwise hidden knowledge about people, but these claims were linked to the growing authority of the natural sciences. Phrenological character readings and predictions based on them could be seen to be "true" and may still seem so by their vagueness and the flexibility of the phrenological system to adapt to individual cases. For example, if a phrenologist were presented with a convicted murderer whose organ of Destructiveness (earlier called Murder) was exceedingly small, the murderous behavior could easily be explained by appealing to the influence of other organs. However, with such a lax treatment of evidence and such a cavalier application of theoretical explanation, it was possible to claim absolutely anything with phrenology. All phrenologists shared the all-too-common practice of trumpeting confirmatory evidence as "proof" and explaining away contradictory evidence. If the difference between science and pseudoscience is the degree to which this practice is cultivated, then phrenology could rank along with palmistry or astrology as one

THE PHRENOLOGICAL BUST.

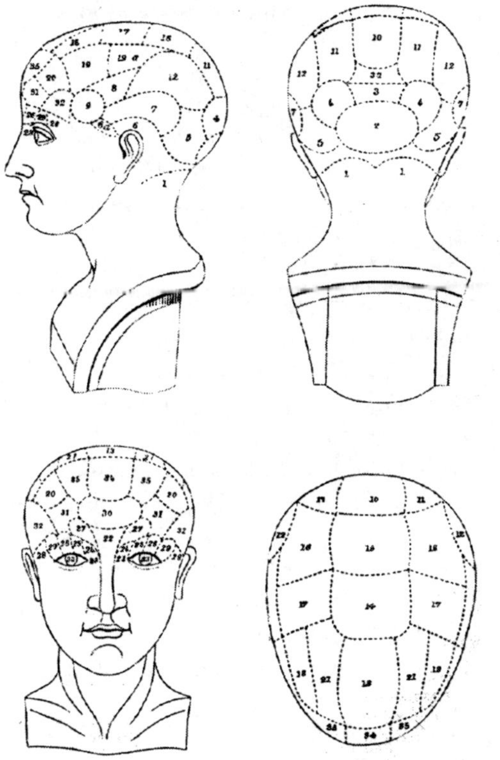

Frontispiece to George Combe's System of Phrenology, 1853. (Courtesy of author)

of the most visible examples. In this sense, phrenology could be compared to so-called parapsychology today.

Despite phrenology's reputation as the epitome of pseudoscience, most of its basic premises have been vindicated, though character reading from the outside of the head has not. For example, the principle that many functions are localized in the brain is now commonplace (although many other functions are distributed). Also, areas of the brain that are more frequently used may become enlarged with use, just as phrenologists asserted. This was recently seen in a study of the hippocampi of London taxi drivers; the hippocampus is involved in remembering routes and pathways (Maguire, Frackowiak, and Frith 1997). In addition, some personality disorders correlate to specific diseased or damaged regions of the brain. And paleontologists who have made endocasts of the skulls of early hominids to determine the shapes of their brains have suggested that an enlarged node at Broca's area is evidence of language use. This is essentially phrenology in a new guise. Size is taken as evidence for power, and functions are believed to reside in specifically bounded regions. Finally, the once controversial belief that the mind is in the brain is now common sense.

The science of phrenology was created by the Viennese physician Franz Joseph Gall (1757–1828) in the 1790s. Gall called his system *Schädellehre* (doctrine of the skull) or organology, and later, it was simply known as "the physiology of the brain." Contrary to common legend, the system was not created by Gall with the assistance of his German disciple Johann Gaspar Spurzheim (1776–1832) but by Gall alone. Spurzheim later took Gall's system to Britain in 1814, where the word *phrenology* was coined the following year. Spurzheim increased Gall's twenty-seven faculties/organs to thirty-three, and British phrenologists soon added a few more. Spurzheim also changed the character of Gall's mental and physiological science, making it a moral science and a philosophy of nature, which Gall, a respected physician and anatomist, never intended. For Spurzheim and later phrenologists, each faculty/organ had its proper area of function and abuses. Hence, the proper function of what Gall called the organ of Murder was, according to Spurzheim, morally acceptable self-defense. Spurzheim also arranged the phrenological faculties/organs into a more scientific-sounding taxonomy of orders and genera.

In Britain, Spurzheim's phrenology was added to a native culture of scientific, gentlemanly societies, complete with meeting halls, journals, and museums. These first phrenologists produced the now familiar phrenological plaster bust. The British phrenological bust differed from earlier diagrams of the organs, which had represented actual specimens—and usually skulls—not lifelike heads. The phrenological bust, by contrast, was a generalized model that represented an ideal "development" of cerebral organs.

The first phrenological society was founded in Edinburgh in 1820, and throughout Britain, scores more appeared over the following two decades. From Britain, the science spread to the United States, where the first society was founded in Philadelphia in 1822. The most important popularizer of the science after Spurzheim was the Edinburgh lawyer George Combe (1788–1858). This first wave of phrenological popularity had died away in Britain by the 1850s. However, a new U.S.-based revival of the science began in New York in the 1840s under the leadership of the Fowler family. Although essentially all of the Fowlers' phrenology and philosophy was borrowed from Combe, their science was more earthy and profitable and less scientifically pretentious than the earlier phrenology. The Fowlers took phrenology back to Britain in the 1860s and 1870s. It is this latter phrenology that lingered into the twentieth century and whose vestiges are sometimes seen today.

References:

Cooter, R. 1984. *The Cultural Meaning of Popular Science: Phrenology and the Organization of Consent in Nineteenth-Century Britain.* Cambridge: Cambridge University Press.

——. 1989. *Phrenology in the British Isles: An Annotated, Historical Biobibliography and Index.* Metuchen, NJ, and London: Scarecrow Press.

de Giustino, David. 1975. *Conquest of Mind: Phrenology and Victorian Social Thought.* London: Croom Helm.

Maguire, Eleanor A., Richard S. J. Frackowiak, and Christopher D. Frith. 1997. "Recalling Routes around London: Activation of the Right Hippocampus in Taxi Drivers." *Journal of Neuroscience* 17: 7103–7110.

Stern, M. 1971. *Heads and Headlines: The Phrenological Fowlers.* Norman: University of Oklahoma Press.

van Wyhe, John. "The History of Phrenology on the Web." URL: http://www.jmvanwyhe.freeserve.co.uk.

Piltdown Man (Hoax)

Famous Fossil Forgery

RICHARD MILNER

On December 18, 1912, newspapers throughout the world blared sensational headlines: "MISSING LINK FOUND—DARWIN'S THEORY PROVED." On that date, at a geologic society meeting in southwestern England, an amateur archaeologist named Charles Dawson announced that he had found the skull and jaw of a fossil ape-man in a Sussex gravel pit—the long-sought "missing link" between apes and humans. Later, he reported finding other bones and teeth in the same area.

A lawyer and "antiquarian" hobbyist from the country town of Lewes, Dawson had spent years searching for fossils and stone tools in the countryside near his home. Assisted by workmen and later by the young French Jesuit priest Pierre Teilhard de Chardin, he claimed to have found pieces of a skull (cranium) and jaw and stone and bone tools along with remains of various extinct mammals, including the giant beaver, mastodon, and hippopotamus. Proudly, he proclaimed the bones were from a new species of human ancestor, which experts from the British Museum obligingly named *Eoanthropus dawsonii*, "Dawson's Dawn Man."

Remarkably, the skull appeared entirely human, but the jawbone fragment looked apelike. A good deal of interpretation was necessary, as the face, along with chunks of the jaw hinges, was missing. A few scientists questioned whether the skull and jaw really belonged together, but most came around to accepting Piltdown as authentic. To the public, Piltdown became famous as "the Earliest Englishman."

Artists sketched Piltdown man's brutish but intelligent face, and statues of his presumed physique began to appear in museums. The local pub, The Lamb's Inn, was renamed The Piltdown Man. In the United States some years later, a popular comic strip called "Peter Piltdown" appeared regularly in the Sunday funnies—the cartoon ancestor of "Alley Oop" and "Fred Flintstone."

Several prestigious British anthropologists put their names and reputations on the line in authenticating Piltdown. When Sir Arthur Keith, curator of the Hunterian Museum, was challenged on his reconstruction, he gave a dramatic demonstration: a known skull was smashed to pieces, and Keith correctly reconstructed its shape and cranial capacity from some of the fragments. Among others who championed the authenticity of the "great discovery" were the anatomists Arthur Smith-Woodward and Sir Ray Lankester, both of the British Museum of Natural History.

Forty years later, in 1953, the famous bones again made world headlines: "PILTDOWN APE-MAN A FAKE—FOSSIL HOAX MAKES MONKEYS OUT

OF SCIENTISTS!" A geologist and chemist named Kenneth Oakley and his colleagues (including J. S. Weiner) at the British Museum had applied new technology to the relative dating of the bones, and the fraud was revealed. Levels of radioactive potassium fluoride at the site did not match those of the bones, which were much younger. And the jaw was much younger than the skull with which it was supposed to belong. The skull and jaw could not possibly have come from the same individual unless, as one scientist put it, the man died but his jaw lingered on for a few thousand years.

Oakley and his colleagues concluded that someone had taken an old human cranium and planted it at the gravel excavation together with a doctored mandible from a modern ape. The jaw had come from an orangutan and was deliberately broken at the hinge to obscure its fit with the skull. All the bones had been stained brown with potassium bichromate to make them appear equally old.

Why was the fraud so eagerly and readily accepted? According to hoax expert Nick Yapp, a successful hoax usually props up cherished but questionable beliefs and expectations, and it often bolsters local pride or patriotism. Piltdown man brilliantly met these criteria. The skull appeared at a time when fossil humans or near humans were very few and far between,

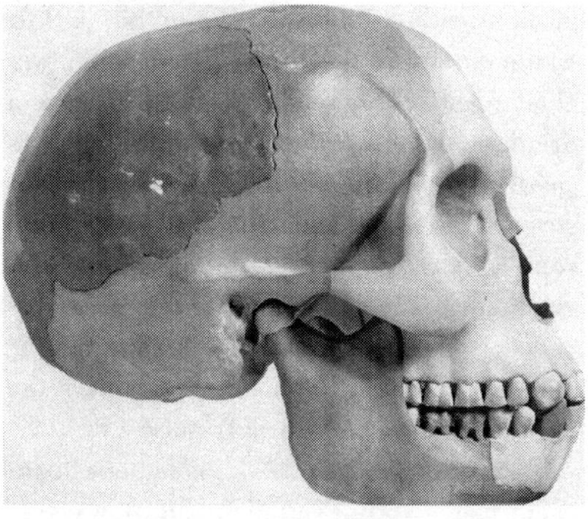

The Piltdown skull. (Author's collection)

and it provided the sought-after missing link that Darwin's theory seemed to require. As for patriotism, British scientists had long been jealous of the sensational fossil men found in Germany and France, and they strongly craved an English ancestor of comparable age.

In addition, British anthropologist Sir Arthur Keith had theorized that a "big brain" came first in human evolution and was the hallmark of humanity. Piltdown filled that bill, too. Because of that bias, Keith had scoffed at Raymond Dart's discovery of a "man-ape" or australopithecine in South Africa because it had an ape-sized brain and humanlike teeth and jaws—exactly the opposite of Piltdown, with its large braincase and apish teeth. Of course, Dart's "Taung child" was to be the first of many African hominid fossils that eventually revolutionized paleoanthropology.

In their enthusiasm at the time, few scientists thought it strange that the ancestral apeman should have been found about 30 miles from the home of Charles Darwin or that an odd, paddlelike bone implement found in the pit resembled nothing so much as a prehistoric cricket bat—a bit *too* appropriate an artifact for the "first Englishman." The hoaxer could not resist going over the top and tipping his hand, but the scientists had swallowed the story so completely that they did not even choke at a crude cricket bat carved from a mammoth bone.

When Oakley's group finally unraveled the hoax during the early 1950s, there were cheers and jeers. Creationists proclaimed that it proved all evolutionary science is phony. Anthropologists said the exposure proved their discipline is self-correcting and eventually roots out frauds. Advocates of the African australopithecine fossils felt vindicated in their view that hominids had small canines and walked upright very early on and only developed expanded brains much later. A member of the British Parliament proposed a vote of no confidence in the scientific leadership of the

Sir Arthur Keith demonstrates his skill by correctly restoring a smashed skull whose cranial capacity was known. A portrait of Charles Darwin gazes down from the wall. (Author's collection)

British Museum. The motion failed to carry amid laughter, when another member of Parliament reminded his colleagues that politicians had "enough skeletons in their own closets" to worry about.

The Piltdown hoax remains one of the most intriguing mysteries in the history of science. Who was the culprit, and what was his motive? For years, the finger of accusation pointed only at Dawson. His motive would have been the fame and attention he gained as discoverer of England's most ancient inhabitant. As the years went by, however, Piltdown has become a perennial whodunit, with an increasing number of suspects added to the list of possible perpetrators.

Harvard paleontologist and historian of science Stephen Jay Gould, for instance, believed Dawson initiated the hoax as a prank, then enlisted the cooperation of his sometimes assistant Pierre Teilhard de Chardin, who was then a young seminarian. Fans of Father Teilhard, of whom there are still many, were outraged. The French paleontologist and evolutionist suffered greatly at the hands of his order for his forthright honesty. When he attempted to reconcile his two most deeply held beliefs, evolution and Catholic doctrine, the church sent him as a missionary to China and forbade him to publish or teach his unorthodox ideas. (Of course, during his exile, Teilhard was one of the first to study and publish on the *Homo erectus* fossils that were discovered near Beijing, then Peking, and home of the famous "Peking man.") To this day, Teilhard's supporters believe he was constitutionally incapable of promoting a deliberate lie.

During the 1980s, anthropologist John Winslow accused Sherlock Holmes's creator, Sir Arthur Conan Doyle, who lived only a few miles from the Piltdown site, of being the hoaxer. A fanatic believer in Spiritualism,

Doyle might have sought revenge by fooling the scientists who ridiculed his fervent belief in spirit-mediums and accused him of credulity in assessing evidence.

According to this theory, E. Ray Lankester, who later became director of London's Natural History Museum, incurred Doyle's lasting wrath for having been the first scientist to prosecute a spirit-medium for fraud in 1876, an incident that Doyle always referred to as a "persecution" rather than a "prosecution." Lankester, the nemesis of hoaxers, was himself completely taken in by Piltdown. Winslow has also suggested that perhaps Doyle wanted to demonstrate how easily scientific experts would uncritically embrace flimsy evidence that supported their own beliefs, even as they scoffed at the authenticity of his so-called spirit photographs. Doyle played golf frequently near the Piltdown site and visited it several times in the company of the scientists who were involved, even offering to drive them around the neighborhood in his motorcar. Like Teilhard, however, Doyle has many admirers who continue to maintain that his involvement in the hoax is unthinkably inconsistent with his character, which was one of extraordinary integrity.

Nevertheless, Doyle did have a prankster side. His classic book *The Lost World* (the granddaddy of all dinosaur adventures) is full of tantalizing references to faked bones and photographs, and the frontispiece of the book features the author himself in stage makeup as the "gorilloid" Professor Challenger. But the case against the great storyteller has never been proved. Sometimes, Doyle deceived people for the fun of it, but (as far as we know) he always revealed the joke afterward. Perhaps, as some have suggested, this is a mystery only Sherlock Holmes could solve!

Another candidate for "perp," proposed by paleontologists Brian Gardiner and Andrew Curran of the British Museum, was the geologist Martin Hinton. Curran found a satchel that had belonged to Hinton stashed in the bowels of the museum. It contained dozens of rodent teeth that had been subjected to various stains and chemical treatments. At the bottom of the trunk were ten carved pieces of hippopotamus and elephant teeth, colored in a similar manner to the Piltdown remains and the alleged cricket bat. Gardiner is convinced these were the products of tests Hinton had performed before treating the Piltdown bones to make them appear old. According to Gardiner, Hinton, a great expert on Pleistocene gravels, was then the only person on the scene who was sufficiently knowledgeable to plant the convincing assemblage of extinct beavers and mammoths along with the skull. "Dawson wasn't clever enough to have correctly seeded the gravels, which were totally unfossiliferous," says Gardiner. Hinton's motive, he believes, would have been to get revenge on his boss, Smith-Woodward, against whom he had a smoldering grievance.

So far, there is still no proven solution to the mystery. Many other possible candidates have been offered, as have stories of mysterious tape-recorded accusations by aged survivors, apocryphal anecdotes about family recollections, and allegations of gross inaccuracy in accepted accounts of the events. Some years ago, Charles Blinderman, in his 1986 *The Piltdown Inquest*, furnished a detailed review of a roster of ten of the usual suspects. More recently, John E. Walsh has reviewed the case in his 1996 *Unravelling Piltdown: The Scientific Fraud of the Century and Its Solution*, in which he opts for a lone conspirator.

After extensive sleuthing at almost a century's distance, Walsh brings the case back to its original prime suspect, Charles Dawson. According to Walsh, Dawson left a trail of hoaxes over thirty-five years, along with some legitimate scientific work. At various times, Dawson had reported finding in Sussex a cast-iron Roman figurine, a prehistoric flint weapon hafted to a crumbling wooden handle, the remains of a very ancient boat, a strangely shaped "Roman horseshoe," and a clock face ostensibly from the Middle Ages—none of which were ever authenticated. If he was indeed the Piltdown mastermind, perhaps he took an accomplice into his confidence; there is certainly no shortage of suspects. But Dawson also has supporters (though few) who believe that he was duped himself and was not the perpetrator.

Though some might find it an amusing and delightful historical puzzle, the Piltdown hoax was no harmless prank. It consumed the energies of dozens of able men, destroyed the careers of a few, and put science on the wrong track for decades in its quest to understand human evolution. Charles Darwin believed that accepting shoddy evidence is much more dangerous than adopting incorrect theories. In his 1871 *Descent of Man,* he wrote: "False facts are highly injurious to the progress of science, for they often endure long; but false views, if supported by some evidence, do little harm, for everyone takes a salutary pleasure in proving their falseness, and when this is done, one path toward error is closed and the road to truth is often at the same time opened" (Darwin 1883, 606).

References:

Blinderman, Charles. 1986. *The Piltdown Inquest.* Buffalo, NY: Prometheus Books.

Darwin, Charles. 1883. *The Descent of Man and Selection in Relation to Sex.* 2d ed. New York: Appleton.

Spencer, Frank. 1990. *Piltdown: A Scientific Forgery.* London: Oxford University Press.

——. 1990. *The Piltdown Papers, 1908–1955.* London: Oxford University Press.

Walsh, John Evangelist. 1996. *Unravelling Piltdown: The Science Fraud of the Century and Its Solution.* New York: Random House.

Weiner, J. S. 1955. *The Piltdown Forgery.* Oxford: Oxford University Press.

Yapp, Nick. 1992. *Hoaxers and Their Victims.* London: Robson Books.

Placebo Effect

GEOFFREY DEAN AND IVAN W. KELLY

Placebo (pronounced plaSEEbo) is Latin for "I will please." The placebo effect is the positive effect produced by treatments that should have no effect; for example, a gelatin capsule filled with table salt and given with the assurance that it will bring sleep will actually do so for about one person in three. Interest in the placebo effect began around 1955, and by 1980, it was the subject of nearly a thousand articles and books. Many medical treatments, especially old ones, have been found to derive their benefits from the placebo effect. In fact, when testing new drugs, the effect is so pervasive that special strategies are needed to cope with it.

A placebo treatment gives positive expectations and a sense of control, and it is quite different from no treatment. No actual understanding of the treatment is required, only a belief that it will work. When a witch doctor asks the spirits for help, clients are given hope and support even though the whole process may be a fraud. This is the placebo effect. When an astrologer reads a birth chart, the trappings are different but the effect on the client is the same. In each case, what matters is not validity but faith. Faith is more than just believing in something; it means having no doubts at all. When doubts are absent, the client can immediately expect positive outcomes (Plotkin 1985).

Much about placebos is controversial. The actual mechanism is not fully understood, although endorphins seem to be involved in any suppression of pain. Nor is there a single, agreed-upon definition of the placebo effect; in fact, discussion of the various definitions has filled an entire chapter (Shapiro and Shapiro 1997). It is sometimes not even clear whether a placebo effect has occurred at all—some medical conditions get better spontaneously, participants tend to be selected at low points in their well-being and thus seem worse than they really are, or they may report improvements just to be well mannered, and so on. Tests that depend on participants not knowing whether a placebo is used can be suspect; for example, in 27 studies involving a total of over 13,000 people, about two-thirds guessed correctly versus the 50 percent expected by chance (one explanation is that side effects from the genuine pills gave the game away). But it is generally accepted that the expectations of therapist and client are important and that the response to the placebo can vary from 0 percent to 100 percent depending on the situation (Wall 1999). Other things being equal, the more faith the patient has in the treatment, the better the chance that the therapy will work.

The placebo effect is not clearly related to individual variables such as gender, age, intelligence, and personality (with the possible exception of anxiety), probably because the effect depends on too many variables to have a simple relationship with any one. A link with suggestibility seems inevitable but has yet to be formally established (tests of suggestibility usually measure only gullibility, which is not necessarily related to the kind of needs that lead to placebo effects).

But the effect does depend on the situation, that is, on our need for therapy and our perceptions; for example, war wounds cause less pain than the same wounds resulting from surgery. The effect is increased by positive staff attitudes and appropriate surroundings and decreased by negative attitudes or unpleasantness. Furthermore, the effect is increased for actual sufferers (who need relief) compared with volunteers (who do not). In general, two pills have more effect than one pill, and red pills are more stimulating than blue pills, but in individual cases, the effect of pill shape, size, and color depends on individual likes and dislikes. The same applies to the actual treatment procedure. Thus, people unwilling to accept emotional support directly may accept it when suitably disguised, which is a point in favor of otherwise untenable systems such as tea leaf reading. Many other variables that have not been fully explored, such as cost and perceived efficiency, may also play a role.

The active ingredients in a placebo treatment are no more than those shared by any therapy—a person in need, a consultant, an explanation for the condition, a healing ritual, and an expectation of improvement. In other words, warmth, attention, and being told you will feel better. To boost the placebo effect, choose a practitioner who is optimistic, has your confidence, and involves you in decision making; if there is a choice of decisions, choose whatever you think will work best; and if something worked in the past, use it again unless there are good reasons not to (Brown 1997, 80).

What makes a good therapist? For many therapies but not all (e.g., not behavior modification), the single most important factor may be whether the therapist and client like each other. Good therapists tend to be warm, optimistic, and empathetic, all of which determine the liking by clients much more than competence does. When the liking is mutual, four times as many clients report an improvement compared to when they dislike the therapist (this could also mean that clients like their therapists because they are improving).

What makes a good client? In psychotherapy (the treatment of psychological problems by psychological methods), the ideal client "must be suggestible. He should be able to easily absorb dogmas and ideas of the most abstract, even outlandish dimension. He should be philosophically adaptable and able to ape the therapist's value system and biases. The more he agrees with the therapist, the better his chances of being helped. This conditioning process is at the core of all faith healing, magic and religion" (Gross 1978, 48). In short, a good client is one who shares the therapist's beliefs, which clearly has some importance for pseudoscience—if therapists take care to accept only good clients, their beliefs (whether true or false) cannot fail to be reinforced.

An important factor for maintaining improvement after therapy is personal effort. People who merely swallow placebo pills do worse than those who swallow pills and do exercises. Placebos also work if they promise not getting better but getting worse; for example, people given pills correctly described as "inactive" generally feel worse. Such placebos are sometimes called "nocebos" (Latin for "I will harm"). An extreme example of a nocebo is the voodoo hex, where the victim allegedly dies of fright. In more general terms, when people think sick, they get sick; thus, women who believe they are prone to heart disease are four times as likely to die as women equally at risk but without such fatalistic beliefs, and people are several times less likely to experience side effects from real drugs if they don't know about them. Interestingly, side effects from placebo pills are reported by about one person in five, with the side effects being similar to those reported for the supposed medication; for example, supposed antispasmodics produce nausea, and supposed tranquilizers produce sleepiness. Side effects also increase with client hostility, so they may indicate dissatisfaction with the treatment, or they may simply indicate that the treatment is working (meaning the placebo is having the same effect, and the same side effects, as a real treatment). Not much is known about the long-term effects of placebos versus real treatments.

The placebo effect suggests ways of turning to advantage approaches that might otherwise be dismissed as pseudoscience, for instance, by alternately taking placebo pills and drug-containing pills. For mild disorders, the effect should be the same, but the cost would be less, as would the dependence on drugs, doctors, and hospitals (Brown 1997, 60). In fact, many psychotherapies have appeared since the 1970s that are little more than the placebo effect in disguise, for example, imagery, relaxation, and self-hypnosis (Plotkin 1985, 251). In medicine, such an approach by doctors might be seen to imply deception or to involve real dangers if the placebo effect happened not to work. Nevertheless, it seems indefensible to use a negative tone ("we have no idea if a problem exists") when a positive tone ("you should soon feel better because we can find no problems") produces notably better outcomes (see Thomas 1987).

Alternative therapists are better able to exploit the placebo effect because they tend to have great faith in their therapy, whether justified or not, whereas conventional therapists are more cautious and are professionally bound to reveal any limitations. But how to encourage the faith on which everything depends if genuine effects are disclaimed? This is the fundamental dilemma from which there is no obvious escape. Of course, it could be argued that the faith involved would be of a new kind, namely, faith in faith itself rather than faith in the existence of genuine effects. To psychologists, this is clearly a more acceptable faith, but whether it can invoke as strong a placebo effect is a matter for research. It seems unlikely to invoke as strong a following.

References:

Brown, W. A. 1997. "The Best Medicine." *Psychology Today* 30, no. 5: 57–60, 80, 82. A very readable account. The author's follow-up article, "The Placebo Effect," *Scientific American* 278 (January 1998): 68–73, offers much sensible advice to doctors about maximizing placebo effects in an ethically acceptable way.

Gross, M. L. 1978. *The Psychological Society: A Critical Analysis of Psychiatry, Psychotherapy, Psychoanalysis and the Psychological Revolution.* New York: Random House. Concludes that much of the field borders on pseudoscience.

Harrington, A., ed. 1997. *The Placebo Effect: An Interdisciplinary Exploration.* Cambridge, MA: Harvard University Press. An anthology by experts across many fields of interest.

Plotkin, W. B. 1985. "A Psychological Approach to Placebo: The Role of Faith in Therapy and Treatment." In *Placebo: Theory, Research, and Mechanisms,* edited by L. White, B. Tursky, and G. E. Schwartz, 237–254. New York: Guilford.

Shapiro, A. K., and E. Shapiro. 1997. *The Powerful Placebo: From Ancient Priest to Modern Physician.* Baltimore, MD: Johns Hopkins University Press. Focus is on medicine rather than psychotherapy.

Thomas, K. B. 1987. "General Practice Consultations: Is There Any Point in Being Positive?" *British Medical Journal* 294: 1200–1202. Finds improvement rates of 64 percent for positive consultations versus 39 percent for negative consultations.

Wall, P. D. 1999. "The Placebo and the Placebo Response." In *Textbook of Pain,* 4th ed., edited by P. D. Wall and R. Melzack, 1419–1430. St. Louis, MO: W. B. Saunders. Dispels myths about the placebo effect.

Planetary Alignments

JOHN MOSLEY

Planetary alignments occur when two or more planets and the Sun and Moon occasionally line up as seen from Earth. Some wonder if the planets' gravitational forces are enhanced at such times and if Earth is affected. In reality, the planets are so distant that their tidal forces on Earth are negligible. Planetary alignments can be predicted in advance and are curiosities that can be aesthetically pleasing when observable, but they have no effect on Earth or its inhabitants.

The solar system is flat, with the Sun at its center and the planets circling it on concentric orbits. Earth lies in the plane of the solar system, and we see the planets' orbits edge-on. As we look across space, the Sun seems to travel around the sky through the constellations of the zodiac on a path called the ecliptic, while the planets (and the Moon) travel near the Sun's path. They travel at different speeds, so inevitably, the Sun, Moon, and planets must pass each other. When one planet passes another (or the Sun or Moon), the two are in "conjunction." Occasionally, three or more planets gather together in what astronomers informally call a "massing" or "grouping." The public calls all such gatherings "alignments."

Historically, the sky has been the ultimate source of mystery for humankind, and powerful forces have long been believed to dwell there. Chief among those forces is the Sun, which was worshiped as a god and which, at a minimum, is the source of our light and warmth. In former times, when the planets were unknowable lights in the sky that moved in mysterious ways, it seemed reasonable to suppose that there was a link between their motions and events on Earth. We humans are pattern-seeking animals, and our brains are wired to find relationships—even when there are none. Astrology was developed to give meaning to such imagined correspondences. Jupiter appeared near Venus, and an earthquake rocked the land only a week later—was it only coincidence? A miscarriage followed a solar eclipse—was there a connection? Associating astronomical occurrences with terrestrial events seemed to help us understand the vagaries of nature. Eventually, we discovered that those puzzling moving lights in the sky are worlds that move according to the laws of gravity, not through the intentions of sometimes malevolent gods, and thus that there is no reason to fear their comings and goings.

Today, amateur astronomers enjoy watching planetary alignments for their novelty and, if the two objects are close enough, for the beauty. The public imagines that the planets are much closer to each other than they are in reality and that they exert significant gravitational forces on each other when they are aligned. Sensationalist books and films foster the misconception that enhanced tidal forces and unspecified "energy fields" trigger earthquakes or can even cause Earth to fall over on its axis. Astrologers, psychics, pyramid experts, and prophesiers of apocalypse link planetary alignments with historical mysteries such as the collapse of the Maya civilization or Noah's flood, and they predict disaster of cosmic proportions when the planets align. The broadcast media have a strong sensationalist element and often claim to present an "unbiased view so the viewer can decide," spreading unfiltered cries of doom and gloom quickly and widely.

Recent Planetary Alignments

How often do the planets align, and what really happens when they do? To explore these questions, we can look at recent planetary alignments, all of which were accompanied by predictions of cosmic catastrophe.

February 5, 1962

The best planetary grouping in modern times happened on February 5, 1962. All five naked-eye planets plus the Sun and Moon were massed within a circle 17° in diameter. The Moon passed directly in front of the Sun as seen from New Guinea and part of the Pacific Ocean, where there was a total eclipse of the Sun! The end of the world was predicted, but the news media were not so sensational in 1962, and the alignment received much less press than it would today. The alignment was especially feared in India, but the day turned out to be like any other.

March 10, 1982

The Jupiter Effect, a book published in 1974, claimed that a circuitous sequence of events would trigger earthquakes, especially in California, in 1982. The book's thesis was that a rare alignment of planets (as seen from the Sun) that occurs once every 179 years would exert a strong tidal effect on the Sun, which would increase solar activity, which would cause more sunspots, which would propel more atomic particles toward Earth, which would disturb the normal circulation of Earth's atmosphere, which would cause sudden major storms, which would cause abrupt changes in Earth's rotation, which would trigger major quakes along faults (specifically, the San Andreas) already subject to strain. The reality was somewhat different. At their closest—on March 10, 1982, as seen from the Sun—the planets spanned 95°, which is more than a quarter of the sky and far from an alignment. No chain is stronger than its weakest link, and the "Jupiter effect" chain had several links that were suppositions at best. Nothing unusual happened on the predicted date.

May 5, 2000

In the book *5/5/2000,* Richard Noone (1982, 53) claimed that on that date, "for the first time in 6,000 years all the planets of our solar system will be arrayed in practically a straight line in space"—an alignment that would cause the ice accumulated at the South Pole to upset Earth's axis and thereby initiate sudden and catastrophic floods and earthquakes. The author argued that the ancient Egyptians had warned us about this 6,000 years ago, adding that the same phenomenon happened to them, too, which is why they built the Great Pyramid. Noone did not hint at how the planetary alignment would cause Earth to self-destruct in this book, which was really about "secrets of the ancients."

In contrast to the predictions advanced in *The Jupiter Effect,* when there was no actual planetary alignment, the planets did gather together in May 2000, as was first predicted by the Belgian astronomer Jean Meeus in 1961.

Illustration of aligned planets. (Antonio M. Rosario/The Image Bank)

On May 5, the five naked-eye planets plus the Sun and Moon spanned just 25.9°. The Sun was part of the alignment, and the Moon and planets could not be seen. Twelve days later, on May 17, the five planets and the Sun (but not the Moon) spanned only 19.5°. There was widespread concern that such an alignment would trigger a cosmic disaster, but, as always, Earth emerged unscathed.

How Often Do the Planets Align?

Inevitably when discussing a planetary alignment, someone asks, "How often does this happen?" or "When will it happen again?" Alignments between two planets or a planet and the Sun and Moon (which, as mentioned, are properly called conjunctions) happen roughly weekly. Loose groupings of three objects are also common and happen almost monthly. Groupings of four or more objects (or tight groupings of three) are comparatively rare (De Meis and Meeus 1994). The Belgian astronomer Jean Meeus found that there are 103 groupings between the years 3100 B.C. and A.D. 2735 when the five naked-eye planets fit within a circle 25° or less in diameter, which is an average of once every 57 years. Of these 103 groupings, the five planets fit within a circle 10° in diameter on ten occasions (an average of once every 584 years). The minimum separation of the naked-eye planets within this long time span was 4.3° on February 27, 1953 B.C. (Apparently, the Chinese calendar was reset to begin with the following new Moon on March 5.) The last three close groupings were on A.D. April 30, 1821 (19.7°), A.D. February 5, 1962 (15.8°), and A.D. May 17, 2000 (19.5°). The next will be on September 8, 2040 (9.3°). Most of these groupings include the Sun and are not observable, but the 2040 grouping, which will also include the crescent

Moon, will occur well to the east of the Sun and will be spectacular at 7:30 P.M.

Planetary Alignments and Earthquakes

Many people believe that when the planets align, they have an effect on Earth. The assumption is that their gravity is focused and magnified, increasing their tidal forces and triggering earthquakes. An understanding of gravity and tides shows that this is not so.

Usually, the closeness of the alignment is grossly overstated. The covers of both *The Jupiter Effect* and *5/5/2000* show the planets perfectly aligned and as close to each other as billiard balls on a table. This common imagery shapes the public perception. Generally, the alignment is far less dramatic, but regardless of how precisely the planets line up, we can evaluate the idea that planets cause earthquakes. We can take two separate approaches in this regard. First, it is supposedly gravitational tidal forces that trigger earthquakes. The accompanying list shows the relative tidal force of each planet on Earth when each planet is at its closest. The Sun has 1 unit of tidal force on Earth; the Moon has a little more than twice the effect of the Sun; and the other nine planets together with all their moons add only another 1/5000 as much. If all the planets were to align perfectly, their gravity would raise a 2-meter ocean tide by an additional 1/25 of a millimeter. The tidal forces of the planets are entirely negligible, and it makes no difference to Earth whether they are aligned or not.

Tidal Forces of the Sun, Moon, and Planets on Earth
(derived from Thompson 1981, 220)

(The Sun's tidal force equals 1.00)
Moon: 2.21
Sun: 1.00
Venus: 0.000113
Jupiter: 0.0000131
Mars: 0.00000230
Mercury: 0.000000723
Saturn: 0.000000462
Uranus: 0.00000000735
Neptune: 0.00000000213
Pluto: 0.000000000000139

The tidal force of one object on another is proportional to its mass and inversely proportional to the cube of the distance between them. The Moon has only 1/81 the mass of Earth, but it exerts more than twice as much tidal force as the Sun, which has 333,000 times the mass of Earth but is about 400 times farther away than the Moon. Venus, which has the same mass as Earth, exerts almost 10 times as much tidal force on Earth as Jupiter, which has 318 times the mass of Earth but is 15 times farther away than Venus. The book you are holding in your hands exerts 1 billion times as much tidal force on your body as the planet Mars when Mars is at its closest.

Second, one could make lists of past earthquakes and planetary alignments and compare them to see if there is a correlation. This would be, in principle, a simple task requiring no theory and almost no knowledge—just pads of paper and lots of time. Seismologists record tens of thousands of earthquakes each year, which is more than enough to do a proper statistical sample. Anyone with time and access to a research library could look for a correlation between tides and earthquakes, and the first person to find such a correlation would be famous. Yet no one has yet found a convincing relationship—probably because there is none.

Earthquakes are caused by motions within Earth. We would like to predict them for obvious reasons, but an appeal to the other planets or to astrology won't help.

Times and circumstances of upcoming conjunctions are printed in popular astronomy magazines and can be downloaded from astronomy sites on the World Wide Web. Be sure to go outside and watch those that are visible and enjoy one of nature's more sublime spectacles as the grand gearwork of the cosmos occasionally brings the planets into and out of alignment.

References:

De Meis, Salvo, and Jean Meeus. 1994. "Quintuple Planetary Groupings—Rarity, Historical Events, and Popular Beliefs." *Journal of the British Astronomical Association* 104, no. 6 (December): 293–297.

Gribbin, John, and Stephen Plagemann. 1974. *The Jupiter Effect*. New York: Walker.

Meeus, Jean. 1961. "Compact Planetary Groupings." *Sky & Telescope* 22: 320–321.

Noone, Richard. 1982. *5/5/2000 Ice: The Ultimate Disaster*. New York: Harmony Books.

Schaefer, Bradley E. 2000. "Conjunctions That Changed the World." *Sky & Telescope* 99: 28–34.

Thompson, L. C. 1981. "On the Trail of the Jupiter Effect." *Sky & Telescope* 62: 220.

Polygraph and Lie Detection

MARC E. PRATARELLI

Using the polygraph to tell us who is lying and who is being truthful is one of the most controversial yet challenging enterprises facing social scientists, criminal investigators, security personnel, the justice system, and lawmakers. The controversy pits proponents of the polygraph, who believe they need every means available to detect deceptive individuals, based solely on its functionality, against scientists or policymakers, who argue that any technology that supports our national security and influences judgments about who did or didn't commit a crime or who can or can't be trusted with sensitive information must be theoretically and experimentally sound before it is implemented. There is general agreement among both the proponents and detractors of the polygraph that, in practice, it is both science and art. The problem is that science is inherently an objective enterprise bent on making logical and rational interpretations of available data. In contrast, art and clinical skills involving human beings are inherently fuzzy or subjective enterprises clouded by subtle nuances and intuitive or special, indescribable insights acquired through much prior training and experience; some refer to such special skills as a gift. In either case, the ideal "lie detector" would be one that commands the same kind of scientific respectability that DNA technology has earned since the early 1990s.

During the spring of 2001, the U.S. Federal Bureau of Investigation (FBI) discovered that Special Agent Robert Hanssen, a veteran of more than two decades of service and a member of the FBI's elite counterintelligence division, had been secretly sharing highly classified information with the Russian KGB for over fifteen years. The FBI alleged that other intelligence agents had lost their lives because of the information Hanssen had provided, and more important, that U.S. security had been compromised so extensively that the results were incalculable. The criminal investigation of Hanssen quickly led to an April 5, 2001, hearing of the Senate Judiciary Committee, whose members wanted to know not only why Hanssen had gone undetected for so long but also what internal security procedures were being relied upon by one of the most prestigious criminal investigation agencies of the federal government. The most notorious technique in the FBI's arsenal—as it is in many other federal agencies, including the army, navy, and air force—was the polygraph.

Less than a year before the Hanssen event, Wen Ho Lee, a scientist of Chinese descent working at the Department of Energy's Los Alamos National Laboratory in New Mexico, had been jailed, investigated, and released after many months of incarceration for allegedly leaking information to China about details of the U.S. nuclear program. Administrators reacted by increasing security at nuclear facilities and announcing they would polygraph some 13,000 employees. The resulting public outcry led to a million-dollar

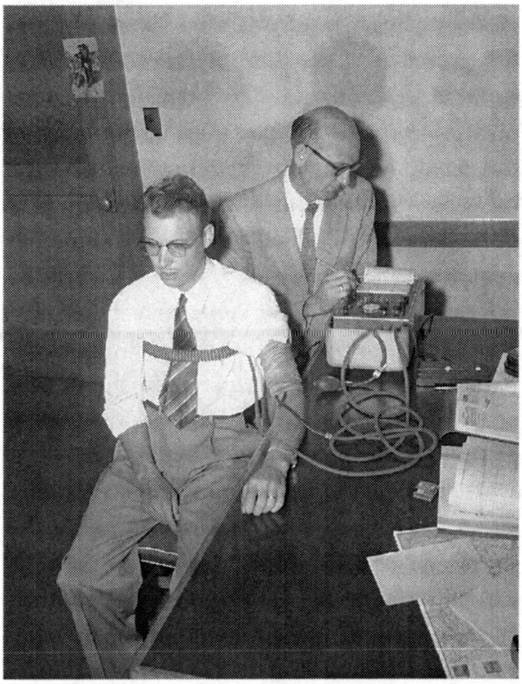

Suspect getting lie detector test, 1955. (Bettmann/CORBIS)

investigation by the National Academy of Sciences to evaluate the legitimacy of the polygraph technique as a means of detecting spies and their espionage.

The academy's investigation, like the Senate Judiciary Committee hearing and other such investigations, was a predictable reaction given both the history and the current state of polygraph technology and the behavioral technique used to elicit confessions and assess guilt. Although such hearings and investigations continue at present, one important issue has resurfaced in the ongoing polygraph controversy: how scientific is the technique, and in particular, how much of that technique is dependent on the subjective clinical skills of the polygraph examiner?

Problems with the polygraph test were judged significant enough to warrant passage of the federal Employee Polygraph Protection Act (EPPA) in 1988, designed to protect U.S. citizens from its use for employment purposes. Those who drafted and endorsed this measure pointed to research that demonstrated the unreliability of the polygraph, including the fact that it failed to detect deceivers (known as false negatives) almost as often as it falsely implicated innocent people (known as false positives). Not surprisingly, results of polygraph examinations are also not admissible in court.

The EPPA, however, places limits only on the private sector. Agencies and departments of local, state, and federal governments and certain federally contracted businesses have license to use it as needed. The principal areas in which it is used include criminal investigations, routine personnel security screenings, and preemployment screenings; thus, the FBI, the Central Intelligence Agency (CIA), and other federal agencies require preemployment polygraph exams as part of the battery of honesty/integrity checks prior to being hired. Various statistics suggest that better than 90 percent of all polygraph examinations given are for noncriminal investigations. The reason the EPPA exempts government agencies is the important practical role the polygraph plays in enhancing national security, in forensic investigations, and as a purported deterrent to criminal activity. Notwithstanding its pseudoscientific nature (which will be discussed), nearly all professional polygraph examiners, as well as many government personnel working in national security, criminal investigations, and other domains, will attest to its demonstrated usefulness in rooting out the occasional spy. (For example, Harold Nicholson of the CIA produced a 97 percent probability of lying on two critical questions in his polygraph examination in 1995; subsequent FBI investigations of Nicholson uncovered his regular pattern of foreign travel and large, unaccounted-for payments to his personal bank account.) There is little dispute that the polygraph is a useful tool in the criminal investigator's arsenal for uncovering criminal behavior. In addition, with less sophisticated individuals not trained in counterintelligence and countermeasures (which are designed to beat the polygraph), the mere mention that the polygraph will be used has helped to extract confessions. In a similar manner, the polygraph may also act as a deterrent to future criminal behavior. Employees are likely to be more honest given their awareness that they will be screened periodically. Moreover, awareness that a polygraph examination is a routine part of the preemployment process, as is customary with the CIA, FBI, and other agencies, helps guarantee that the candidate pool has high integrity. But it is important to note that polygraph exams are never the only tool used to validate employee integrity.

What is the basis for claims that the polygraph is nothing more than junk science and that a myth surrounding its ability to ferret out liars pervades much of society? The answer is complex and multifaceted because it involves a combination of the technological, physiological, and psychological underpinnings of the polygraph examination process. The earliest recognizable version of the polygraph was developed by William Marston, a psychologist in the early 1900s (Marston 1917). Over the next few years, as the crude technology of electrophysiological recording advanced, the first version of the present polygraph appeared. It measured three essential components of a person's physiological reactivity. According to theory, this reactivity was derived from some degree of psychological anxiety arising out of guilt because the examinee was knowingly and intentionally deceiving or concealing information from the examiner (Abrams 1989).

The conventional polygraph records respiration and heart rate, blood pressure, and skin conductance (the sweating response). From these three autonomic indicators of bodily arousal, dozens of refined measures can be extracted—and therein lies one of the major, fundamental problems with the physiological element of the polygraph. Changes in breathing patterns, heart rate, and blood pressure and the presence of sweating are secondary measures in that they must be preceded by some mental or emotional operation in the brain of the examinee. Because the physiological activity occurs subsequent to mental activity, it is not a surprise to scientists that no unique lie-response pattern has ever emerged. Clearly, an informed examinee can compromise the polygraph test by any one of several techniques designed to modify one's thinking during the test or by adding a physical reaction to the mental operations during the test.

In practice, the trained polygraph examiner attempts to make a determination about the examinee's truthfulness during the test by carefully comparing the responses to "control" or "comparison" questions with those to "relevant" questions (Reid 1947). Relevant questions are defined by the critical interests of the examiner or whoever is requiring the test results. For example, if an examinee is a suspect in a kidnapping or, alternatively, is being considered for a position in the federal government that would entail access to highly secret information, relevant questions to these two different scenarios might be, respectively: "Did you remove the child from the Johnsons' home? or "Have you ever had contact with an agent of a foreign government?" Guilty or deceptive individuals would, in theory, produce greater physiological reactivity in one or more of the three indicators mentioned earlier if they responded "No" to a relevant question because they might perceive a sense of guilt in their dishonest response. But one of the major criticisms of the polygraph is that innocent persons who are suspected of the same crimes might produce significant responses to the relevant question(s) because they are nervous or petrified that their reaction might suggest they are guilty or being deceptive when, in fact, they are not. For this reason, comparison (or control) questions were developed. Individuals are instructed to lie intentionally to benign questions so that their deceptive or nervous physiological reactions can be compared against their reactions to relevant questions.

Comparison questions in the Control Question Test (CQT) are questions about behaviors that most, if not all, individuals in society have committed at one time or another—for example, "Have you ever lied to anyone?" or "Have you ever failed to return something that wasn't yours?" or "Have you ever exceeded the speed limit?" Polygraph researchers have suggested that, in theory, the innocent person should be more concerned with their responses to these comparison questions than to the relevant questions. But therein lies yet another problem with the CQT approach. It is assumed that certain questions will always elicit a deceptive response. The assumption is that during an interrogation, the subject will be fearful of admitting to any criminal or immoral act and therefore will lie to the investigator, even though virtually every person has committed the act in question (Ford 1995).

In response to the criticisms leveled at the CQT, a somewhat more sophisticated version of the test was elaborated by Charles Honts and David Raskin (1988). Under their Directed Lie Control Test (DLCT), the examinee is instructed not only to lie in response to control or comparison questions but also to actually think of a particular case when they had violated the issue being asked in the comparison question. In this way, an examinee should be even more concerned about their lie response to the comparison question than to their truthful response to the relevant questions. The benefit to using either the CQT or DLCT is that the polygraph examiner needn't have any factual information or evidence relating to a particular crime. This is why the most modern version of the CQT is currently the most popular questioning technique in use both for criminal investigations and, more important, for the routine periodic security screening exams given to many government employees and for the preemployment screening exams given to job applicants.

Another exam that was popular in the rare cases when the criminal investigator could furnish reliable factual information to the polygraph examiner was called the Guilty Knowledge Test (GKT) (Lykken 1959). The essence of the questioning in the GKT was a multiple-choice exam with questions much like the kidnapping question asked earlier. In the GKT, examinees were asked crime-relevant questions, and the physiological signatures of the responses were later compared to responses to irrelevant truthful response questions such as "Is your name John Doe?" or "Were you born on September 12, 1971?" A person who produced more physiological reactivity to the crime-relevant questions as compared to the irrelevant questions was judged to be deceptive.

References:

Abrams, S. 1989. *The Complete Polygraph Handbook.* Lexington, MA: Lexington Books.

Ford, C. V. 1995. *Lies, Lies, Lies: The Psychology of Deceit.* Washington, DC: American Psychiatric.

Honts, C. R., and D. C. Raskin. 1988. "A Field Study of the Validity of the Directed Lie Control Question." *Journal of Police Science and Administration* 16: 56–61.

Lykken, D. T. 1959. "The GSR in the Detection of Guilty Knowledge." *Journal of Applied Psychology* 43: 385–388.

Marston, W. M. 1917. "Systolic Blood Pressure Symptoms of Deception." *Journal of Experimental Psychology* 2: 117–163.

Reid, J. E. 1947. "A Revised Questioning Technique in Lie Detection Tests." *Journal of Criminal Law, Criminology, and Police Science* 37: 542–547.

Prayer and Healing

KEVIN COURCEY

When a person becomes ill, it is virtually inevitable that a friend, family member, or loved one will offer prayers for a speedy recovery—but do such prayers really have any effect? Do those who are prayed for recover more quickly and have fewer complications? Although most skeptics have their doubts, believers have begun using the scientific method to test the efficacy of the power of prayer, and they are claiming they now have proof that prayer heals.

One of the most significant achievements of the last millennium has been the shift from a religious- and folklore-based system of medicine to a more scientific, evidence-based model. For example, what once was believed to be the result of possession by demons is now recognized as a brain disorder called epilepsy and can be treated with medication. The average U.S. citizen born in 1900 had a life expectancy of only forty-seven years; life expectancy today exceeds seventy-seven years. Smallpox, once responsible for 2 million deaths a year, has been eradicated (excepting possible terrorist sources). And the last crippling case of polio in the United States occurred in 1979. Yet despite the obvious success of these science-based medical advances, the belief that prayer alone can heal the sick persists.

The Christian Science religion, for example, teaches that illness is an illusion and that prayer, by invoking natural spiritual laws, can dispel illness. Thus, Christian Scientists avoid medical doctors when they are ill and choose prayer instead, either by themselves or with a Christian Science "practitioner"—someone who has a minimum of two weeks of instruction in the use of prayer to conquer disease.

If prayer is effective at curing illness, one would think that the life expectancy of Christian Scientists would be at least equal to that of their non–Christian Science peers who resort to traditional Western medicine. Although the Christian Science church does not publish (or even collect) any statistics about their success rates, a clever study done in 1989 was able to shed some light on this matter by comparing the mortality rates of those who had graduated from Principia College, a Christian Science college in Illinois, between 1934 and 1948 and those who graduated from the University of Kansas during the same years. The study found that even though Christian Science tenets forbid the use of alcohol or tobacco, factors that should improve mortality rates, the male death rate was 25 percent higher for Christian Scientists than for their peers at the University of Kansas. The female death rate was 15 percent higher (Simpson 1989).

One of the first people to use rational, scientific inquiry to determine the effectiveness of prayer was Sir Francis Galton of Britain, in 1872. Galton framed his inquiry as a simple statistical question: are prayers answered, or are they not? "There are two lines of research, by either of which we may pursue this inquiry," he wrote. "The one that promises the most trustworthy results is to examine large classes of cases, and to be guided by broad averages; the other, which I will not employ in these pages, is to deal with isolated instances" (Galton 1872, 125).

Galton compared the longevity of several different groups to arrive at his conclusions. Since every person who attended the Church of England at that time prayed for the well-being of the royal family, he compared the royals' life spans with those of members of other well-to-do classes and found the sovereigns to have the shortest lives of the group. He then went on to compare the life spans of distinguished members of the clergy, whom he noted would be among the most prayerful people, with those of lawyers and medical doctors. He found the clergy to have the shorter life spans. Galton also looked at infant mortality rates, rates of psychiatric disturbance, and even whether ships carrying missionaries, for whom many would be praying, had better outcomes than ships carrying slave traders. In every instance, he found that the prayers of the faithful had no statistical effect on the outcomes (Galton 1872).

One of the most significant developments in research techniques over the past century has been the use of the double-blind, placebo-controlled study. In such a study, neither the subjects nor the researchers know who is receiving the test treatment and who is receiving the placebo, or control, treatment until after the study is complete. This research method significantly reduces the chance that a normal placebo response will be mistaken for a therapeutic response; it also reduces the likelihood of researcher bias affecting the results.

One of the first uses of the double-blind method to investigate intercessory prayer (prayer said on behalf of others) was conducted at the London Hospital Medical College in 1965. Researchers assigned patients to two groups: one was prayed for by volunteers, and the other was not. The patients did not know they were part of a study, and the examining physicians did not know to which group the patients were assigned. When the patients' progress was evaluated after several months, no significant difference between the groups could be found (Joyce and Welldon1965).

Probably the most cited study in this field was done by Dr. Randolph Byrd and published in 1988. Byrd divided patients on a cardiac unit of a major San Francisco hospital into two groups and had Christian volunteers pray for half of them. He tracked twenty-six different "problem" events during the study, such as the need for medication for chest pain, the development of pneumonia, or the need for a pacemaker. Byrd claimed that prayer was effective in twenty-one of the twenty-six measured categories (Byrd 1988). But was it really?

Byrd admitted that he studied so many interrelated variables that his statistical analysis was of limited value; in reality, only three of the twenty-six variables showed a significant result. The Byrd study also failed to adequately control for preexisting conditions. That is, the control group had more admission diagnoses of acute heart attacks, more cases of irregular heart rhythm, more heart valve disease, and even more patients admitted with cardiac arrest. Yet despite their advantage over the control group, the prayed-for patients still needed more medication for heart pain, had more unstable heart pain, had a higher percentage of readmissions to the coronary care unit, and needed four times the number of temporary pacemakers and three times the number of permanent pacemakers as did the control group. Even though Byrd asked his volunteers to pray specifically for a "rapid recovery and for prevention of complications and death" for their patients, his study found no significant difference between the groups in the length of stay in the cardiac unit, the total days spent in the hospital, or the number of deaths during the study.

At least one subsequent study attempted to replicate Byrd's experiment. On October 25, 1999, the *Archives of Internal Medicine*, a peer-reviewed journal of the American Medical Association, published a study on the efficacy of remote, intercessory prayer on a group of nearly 1,000 patients admitted to the cardiac unit of a major hospital (Harris et al. 1999). The patients were divided into two groups (depending upon whether their medical record number was odd or even), with one group receiving prayer and the other acting as a control. The praying was done by Christian volunteers in their own homes who knew only the first names of the patients for whom they were praying. Neither the patients nor their doctors were informed that a study was occurring. The authors claimed that the patients who were prayed for had a significantly (11 percent) better outcome than those in the control group.

The authors of this study measured thirty-three different variables during the course of the patients' hospitalization but found no significant differences between the prayer group and the control group on any of these variables. But in fact, the prayed-for patients had a higher rate of readmission to the coronary care unit, a higher rate of pneumonia, longer hospital stays, and even a higher mortality rate. It was only after the researchers imposed what they described as a global "hospital course" rating scale on the data that they were able to discover a positive response in the prayed-for group. Also, a subsequent analysis of this study found that five serious preexisting medical conditions were overrepresented in the control group in this study (Courcey 2000). Consequently, the control group started off with a 62 percent higher rate of patients with acute pulmonary edema (which causes the lungs to fill with fluid), a 31 percent higher rate of patients with heart valve disease, an 18 percent higher rate of patients who had a history of previous heart attacks, a 10 percent higher rate of diabetics, and a 10 percent higher rate of patients with chronic kidney failure. No preexisting conditions were similarly overrepresented in the prayed-for group. The increased chance of complications in the control group due to these preexisting conditions could easily account for the small difference noted between the groups on the "hospital course" rating scale.

Monk praying, Burma. (Eric Meola/The Image Bank)

In yet another attempt to prove the power of intercessory prayer, Scott Walker, a physician and an assistant professor of psychiatry at the University of New Mexico, tracked whether patients in his alcohol treatment program who were being prayed for had better outcomes than those who reported no one praying for them. Walker concluded that "compared with a normative group of patients treated at the same facility, participants in the prayer study experienced a delay in drinking reduction.

Those who reported at baseline that a family member or friend was already praying for them were found to be drinking significantly more at 6 months than were those who reported being unaware of anyone praying for them" (Walker et al. 1997, 85).

Another thrust of recent prayer research has been to document the health effects of self-prayer. Much of this research has been funded by the Templeton Foundation, a Christian organization whose stated goal is to encourage religious faith by using scientific research to show the positive effects of spiritual practice.

When the funding organization has such an overt agenda, one must examine the research very critically. For example, one of Templeton's primary sponsored researchers, Harold Koenig of Duke University, conducted a study of hospitalized, medically ill, elderly men and focused on the coping strategies these men used for the depression that can arise from being diagnosed with a serious illness. Koenig and colleagues (1992) reported that those who used religious coping, including prayer, were less depressed on a subsequent admission to the hospital. A more objective review of the data from this study, however, would indicate that those who expressed strong religious beliefs and used prayer as a coping mechanism were being hospitalized with acute illnesses (such as cancer, heart disease, kidney disease, respiratory disease, and neurological dysfunction) at rates two to four times higher than those who expressed "no religious preference." By focusing on whether the subjects' prayers simply made them less depressed about their physical illness, Koenig was able to avoid the more penetrating question of whether this group was actually more prone to physical and psychiatric illness. Furthermore, he was unable to replicate his finding of decreased depression in a similar study done in 1998.

Much of the research attempting to establish a link between prayer and health is similarly flawed. In a review of the research published in the prestigious British medical journal *Lancet*, Richard Sloan of Columbia University noted that the research linking prayer and health "is weak, with significant methodologic flaws, conflicting findings, and a lack of clarity and specificity" (Sloan, Bagiella, and Powell 1999, 664). He concluded that there is no evidence at all that "religious activities, such as prayer or reading the Bible, play a role in improving health, despite their importance in people's spiritual lives" (Sloan, Bagiella, and Powell 2000).

As former editor of the *Journal of the American Medical Association*, George Lundberg has had the opportunity to review reams of research purporting to document the health benefits of spirituality, faith, and prayer, yet he remains thoroughly skeptical: "Evidence of religious faith producing healing is anecdotal only," Lundberg noted. "In the past 15 years, not one of the articles submitted to the journal describing the direct effects of spirituality, prayer or church attendance on staying well or getting well has survived the journal's peer review process" (Rubin 1998).

References:

Byrd, R. C. 1988. "Positive Therapeutic Effects of Intercessary Prayer in a Coronary Care Unit Population." *Southern Medical Journal* 81, no. 7: 826–829.

Courcey, K. M. 2000. "Medical Claims for Intercessory Prayer Remain Elusive." *Scientific Review of Alternative Medicine* 4, no. 2: 9–11.

Galton, F. 1872. "Statistical Inquiries into the Efficacy of Prayer." *Fortnightly Review* 12: 125–135.

Harris, W. S., M. Gowda, J. W. Kolb, C. P. Strychacz, J. L. Vacek, P. G. Jones, A. Forker, J. H. O'Keefe, and B. D. McCallister. 1999. "A Randomized, Controlled Trial of the Effects of Remote, Intercessory Prayer on Outcomes in Patients Admitted to the Coronary Care Unit." *Archives of Internal Medicine* 159, no. 19: 2273–2278.

Joyce, C. R. B., and R. M. C. Welldon. 1965. "The Efficacy of Prayer: A Double-Blind Clinical Trial." *Journal of Chronic Disease* 18: 367–377.

Koenig, H. G., H. J. Cohen, D. G. Blazer, C. Pieper, K. G. Meador, F. Shelp, V. Goli, and B. DiPasquale. 1992. "Religious Coping and Depression among Elderly, Hospitalized Medically Ill Men." *American Journal of Psychiatry* 149: 1693–1700.

Rubin, A. J. 1998. "Pills and Prayer." *Washington Post,* January 11.

Simpson, W. F. 1989. "Comparative Longevity in a College Cohort of Christian Scientists." *Journal of the American Medical Association* 262: 1657–1658.

Sloan, R. P., E. Bagiella, and T. Powell. 1999. "Religion, Spirituality, and Medicine." *Lancet* 353: 664–667.

Sloan, R. P., and L. VandeCreek. 2000. "Religion and Medicine: Why Faith Should Not Be Mixed with Science." *Medscape,* August 2.

Walker, S. R., J. S. Tonigan, W. R. Miller, S. Corner, and L. Kahlich. 1997. "Intercessory Prayer in the Treatment of Alcohol Abuse and Dependence: A Pilot Investigation." *Alternative Therapies in Health and Medicine* 3: 79–86.

Pseudoscience and Science

A Primer in Critical Thinking

D. ALAN BENSLEY

In a commonly cited definition, *critical thinking* has been described as "reasonable, reflective thinking that is focused on deciding what to believe or do" (Ennis 1987, 9). As such, critical thinking is an approach to knowledge that emphasizes the importance of reflecting on or thinking about the basis of one's beliefs. The critical thinker comes to believe that which is supported by good reasons or strong evidence. If people do not think critically, they may fall prey to many exaggerated, unfounded, and dubious claims made by those who practice pseudoscience. Pseudoscience is an approach that only *appears* to be scientific; it is thought to differ from science in important ways. Although critical thinking is often associated with those taking a scientific approach and not with those taking a pseudoscientific approach, it has sometimes proven difficult to differentiate science from pseudoscience in practice.

Using the definition of critical thinking just given, scientific thinking can be framed as a kind of critical thinking and can be contrasted with the approach of pseudoscience. Like the critical thinker, the scientist should seek good reasons supported by strong evidence in evaluating claims and hypotheses. Science especially values empirical evidence, that is, evidence based on carefully made observations. Also, both critical and scientific thinkers recognize the need to examine and evaluate all of the relevant evidence—not just the positive evidence that supports some favored claim or theory but also the negative evidence that could refute it.

Those who practice a pseudoscience, such as astrology or creationism, do not consistently follow the rules of good reasoning and are not careful in their use of evidence. They often accept low-quality evidence in support of their claims and ignore evidence that does not support the claims they advance. For example, many people who believe that we are visited by alien beings from other planets base their belief almost totally upon the informal observations of people who claim to have seen aliens and their spaceships. These anecdotes or descriptions of personal experiences are a weak kind of evidence because individual experiences are unique, unrepeatable events that are subject to many kinds of error. At the same time, these believers tend to ignore the almost complete lack of physical evidence for alien visitation. Using careful, systematic methods of observation, scientists have been unable to verify alien visitation. Believers tend to ignore this negative, higher-quality evidence.

A core assumption in the definition of critical thinking is that it involves the use of criteria for deciding what to believe or do. A criterion is a standard, benchmark, or condition that is used to weigh the truth or value of some claim. The rules of logic may be thought of as criteria applied to evaluating the soundness of an argument. One such rule used by critical thinkers is that a conclusion should be consistent with evidence. Scientists often use a similar rule in assuming that a theory must be consistent with observations relevant to it. In contrast, pseudoscience does not consistently rely upon the criterion that a theory be consistent with carefully and systematically obtained observations.

A scientific community also develops its own criteria for deciding what makes a conclusion sound and how to handle evidence. For example, geologists use a method from physics known as radiocarbon dating to decide the age of certain fossils. A radioactive form of carbon called carbon 14 is found in living things and is part of their fossilized remains after they have died. By tracking the constant rate of decay in the amount of carbon 14 in a fossil, scientists can measure its age back to about 70,000 years. By comparing their observations using carbon dating with the results from other methods for dating old objects, scientists have determined that carbon-dating estimates of age have a margin of error. So, for example, scientists might estimate a fossilized bone from an early modern human to be 30,000 years old, give or take up to 8,000 years under ordinary conditions. Some creationists who support a literal interpretation of the Bible and believe the earth is only 6,000 years old object to this margin of error. Instead, they say that the method produces errors on the order of tens of thousands of years and that, as a result, this method does not disprove their belief that humans originated only 6,000 years ago. When they do this, creationists have rejected the use of a scientific criterion that a community of scientists has developed from physical evidence. In general, those who take a pseudoscientific approach often fail to accept and abide by the conventional rules and methods that science establishes.

Astrology is often taken as the classic example of a pseudoscience. Astrologers use mathematics (rules for handling numbers) to calculate horoscopes and so may appear to be doing science. In this case, they follow the rules of mathematics and thus are using the criterion that a good horoscope is calculated accurately. The problem, however, is that these complex calculations are made using incorrect assumptions about the relationship between patterns in the planets and stars and a person's time of birth. There is no good evidence to support the astrological claim that one's personality characteristics or future can be accurately predicted from the position of the stars and planets (Kelly 1997).

Scientists also sometimes make predictions from incorrect premises, but there is a big difference in the two approaches. Scientists systematically check the outcome of their predictions and by doing so find out when those predictions are in error. And when their predictions are not supported by observations, scientists eventually reject a theory or change it to be consistent with observations. Thus, science is a dynamic approach to knowledge that is self-correcting. In contrast, astrology has remained mostly unchanged for centuries despite evidence that its predictions are often in error.

Critical thinkers and scientists also use other criteria, such as plausibility, to evaluate claims and to guide their inquiries. A claim is plausible if it seems reasonable, given other things that are known. For example, given what is known about mechanics and the law of thermodynamics, it is implausible that someone could build a perpetual-motion machine that would not create friction and run down. But despite the implausibility of this notion, people have claimed they have built such machines, though none have been able to convincingly demonstrate that their machines do not run down. It is important to note, however, that applying the criterion of plausibility may not always lead to a sound conclusion. In the 1800s, landing a person on the moon must have seemed very implausible, but that very feat was accomplished in 1969. This example illustrates how critical thinkers need to take into account what is known in deciding what to believe or do. In contrast to pseudoscience, critical thinkers and scientists treat their beliefs, hypotheses, and theories as tentative and capable of revision based on the evidence.

Many people persist in believing paranormal and pseudoscientific claims for which there is no good evidence. A recent study of college students' beliefs has shown that 99.4 percent of those surveyed expressed some belief in one or more of the following: astrology, biorhythms, extrasensory perception (ESP)/ psychokinesis, extraterrestrial visitation, fire walking, out-of-body experiences, precognition in dreams, reincarnation, spiritual communication with the dead through mediums, and tarot cards (Messer and Griggs 1989). As noted in many of the articles in this volume, these phenomena have not been substantiated by scientific evidence. People also adopt many commonsense or folk psychological beliefs that are at odds with the findings of science. For example, a good number believe that the phases of the moon cause people to go crazy or behave abnormally, as reflected in the common term *lunatic*. Yet a review of research on the connection between changes in the moon and various indicators of abnormal/deviant behavior (such as admissions to mental hospitals and fights at hockey games) consistently show no correlation (Rotton and Kelly 1985). Therefore, the critical thinker should reject the popular but unsupported belief in this relationship between the moon and human behavior (Bensley 1998).

Perhaps surprisingly, scientists and other scientifically trained individuals sometimes fail to think critically. A striking example in this regard was Sir Arthur Conan Doyle who, before writing the Sherlock Holmes books, was trained as a physician. Although his famous hero was known for his great powers of reasoning, Doyle himself was taken in by spirit mediums who claimed they could contact his dead relatives. He also believed in the Cottingley fairies shown in the next figure. In 1917, two young English girls, Frances Griffiths and Elsie Wright, reported that they had taken pictures of fairies that visited them; Doyle tended to believe the statements of Theosophists (members of a group that believed in astrology and Spiritualism) when they endorsed these supposed sightings.

Doyle turned to photography experts to check out the photos. When he asked workers at Eastman Kodak to examine the first photo, they reported that it did not appear to be a hoax but added that they could produce such a photo with their advanced techniques. Doyle did not seem to be overly troubled by this report or the fact that one of the girls, sixteen-year-old Elsie, had artistic skill and worked in a photography shop. More recent expert analysis of the first photo suggests that it was contrived by Elsie using cutouts of fairy figures (Randi 1982).

That working scientists sometimes fail to think critically was shown in a study by Michael Mahoney (1977), who asked reviewers at a scientific journal to evaluate manuscripts of articles that were identical except for their results. He found that the reviewers gave higher ratings to those articles with results that supported their own favored theoretical position as opposed to identical submissions that had contrary results. This suggests that scientists who lack objectivity may fail to think critically about the evidence presented to them.

Scientists sometimes even come to believe in pseudoscientific theories. An important example of this involved the theory known as phrenology, which was developed in the nineteenth century by the anatomists Franz Joseph

Frances Griffiths with fairies, photographed by Elsie Wright at Cottingley Glen, West Yorkshire, July 1917. Later admitted as hoax by Wright. (Fortean Picture Library)

Gall and Johann Spurzheim. Phrenology was the mistaken notion that bumps and indentations on a person's skull indicated that individual's specific characteristics and abilities. Gall and his student Spurzheim were originally in the mainstream of conventional science, and they wrote a significant book on anatomy that was published in 1813. However, based on the informal observation of a schoolmate who had good verbal memory and also protruding eyes, Gall developed the more controversial phrenology hypothesis. Gall's theory incorporated elements from the more commonly accepted philosophical view that the mind was composed of faculties with the developing view that the brain was the site of the mind (Leahey and Leahey 1983). He initiated a careful research program using behaviors as indicators of mental faculties to identify those that were supported by observations, much as scientific psychologists do today. He also began with the working assumption that the structures in the brain accounting for these faculties could be revealed in the surface features of the skull. Thus, Gall began with a scientific approach.

Phrenology, however, evolved into a pseudoscientific movement when Spurzheim began to part ways from his more scientifically rigorous teacher. He announced that he intended to take a philosophical approach to faculties. In the years that followed, Spurzheim and other phrenologists added and subdivided faculties to make a long list of characteristics that allegedly correlated with the features of the skull, based on little or no empirical support. The figure opposite shows a phrenological map of the skull divided into many areas, each of which is associated with a specific characteristic. Spurzheim and other phrenologists also began to accept as a principle the idea that the skull reflected underlying faculties, which was apparently only a working assumption for Gall (Leahey and Leahey 1983).

The phrenologists used sophisticated-looking equipment to measure the bumps and indentations on a person's skull. From these observations and information about the person, they would connect the bumps and indentations to characteristics such as benevolence and self-esteem.

The problem with the phrenology that arose after Gall was that it was pseudoscience. Although the equipment looks sophisticated, there is no plausible reason to think that having measurements of the relatively small bumps and indentations on the surface of the skull would have anything to do with the contour of the brain underneath. The brain is a gelatinous mass that can take on many forms. Thus, the skull need not have bumps or indentations to accommodate the rather small and specific brain structures underneath, which the phrenologists thought corresponded to the many specific characteristics. In fact, the inside surface of the skull is relatively smooth and rounded to make room for relatively large areas of brain. These facts were known at the time of the phrenologists but were apparently ignored.

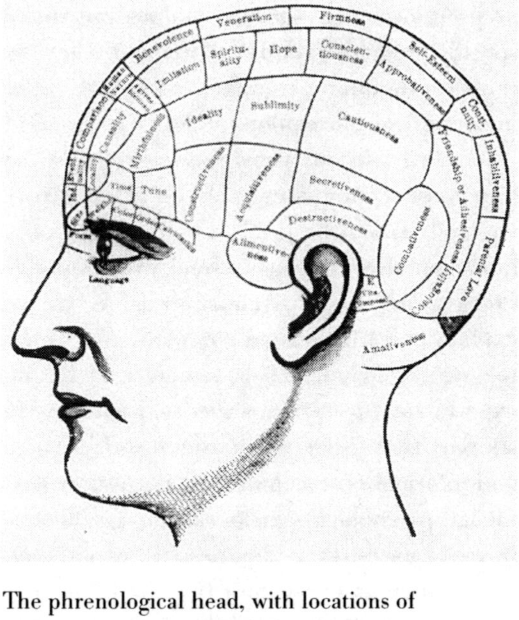

The phrenological head, with locations of propensities. From Nelson Sizer's *Heads and Faces, and How to Study Them*, 1892. (Fortean Picture Library)

From the perspective of psychological science, phrenology was a very complex theory with many untested assumptions. Psychology as a science was not founded formally until 1879, and there was no comprehensive, scientifically based theory of personality until the twentieth century. Nevertheless, phrenologists proposed that many characteristics, such as cautiousness and secretiveness, were related to skull features, even though these characteristics had themselves never been studied scientifically.

Science is a very careful, deliberate approach to knowledge and explanation. As such, it does not propose complex, untested ideas to explain phenomena that may have simpler and more empirically justified explanations. In contrast, practitioners of pseudoscience often propose notions for which simpler explanations based on what is known would suffice. For example, advocates of the existence of ESP propose psychic ability as an explanation of occasions when people seem to know things they wouldn't ordinarily be expected to know. In many of these cases, the advocates fail to offer the simpler, more plausible explanations that are available. Thus, a person thought to be showing psychic ability may actually be receiving information from someone else, be engaging in trickery, or may just be making some lucky guesses. All three of these explanations are well documented, whereas unseen, nonphysical psychic abilities such as extrasensory perception are not. Similarly, the phrenologists accounted for human behavior based on elaborate sets of traits that were not supported by observation.

Another problem with phrenology as pseudoscience was that the phrenologists did not carefully measure individual characteristics. Rather, they often identified the characteristics that were thought to go with the bumps and indentations after the fact. Consequently, they could not make accurate predictions about a person's characteristics from the shape of that person's skull. This is illustrated by an anecdote of how François Magendie, the great nineteenth-century physiologist, tested Spurzheim's use of phrenological theory (Krech 1962). Magendie invited Spurzheim to his home to examine the preserved brain of the brilliant French philosopher and mathematician Pierre Laplace. Unbeknownst to Spurzheim, Magendie had substituted the brain of a mentally retarded man. Spurzheim admired the brain of the retarded person as if it had belonged to Laplace, which clearly suggests that he was unable to use phrenology to make accurate predictions or diagnoses. Spurzheim was unable to critically evaluate the surface features of the retarded person's brain to recognize that they were inconsistent with the features of a highly intelligent person as identified by phrenological theory.

Another problem with pseudoscience is the tendency to form irrefutable or unfalsifiable hypotheses. For example, when proponents of ESP are confronted with research evidence that does not support its existence, they often claim that the presence of an experimenter making observations causes "negative energy" to be transmitted. The skepticism of the scientists is thought to interfere with the sensitive psychic abilities of the subjects. As noted by Keith Stanovich (1998), if scientists are not allowed to make observations, it is impossible to scientifically study ESP because the hypothesis cannot be tested and possibly disconfirmed. Similarly, when confronted with geologic evidence suggesting that fossils of living things are millions and not thousands of years old, creationists sometimes counter with the proposal that God has made the world appear to be much older than it actually is in order to test the faith of people. Another variation of this rebuttal is that Satan is tempting people with false evidence. When, however, creationists propose that some unobservable entity such as God or Satan has made it so that no good observations can be made to study creation, then they have made it impossible to disconfirm or refute the hypothesis (Hines 1988). The conduct of science depends on being able to make observations, but creationists have rendered the question into a nonscientific form.

Yet another problem with pseudoscientific claims and predictions is their vagueness. Critical thinkers and scientists strive for clarity in their use of language. They make their predictions specific so that they can be tested, yielding one outcome or another. In contrast, the predictions of psychics are notorious for their vagueness. As a result, psychic predictions are susceptible to post hoc, or after-the-fact, explanations when they do not turn out to be correct. Scientists, too, must be careful to make specific predictions that can be disconfirmed. For example, philosophers and psychologists have criticized the psychoanalytic approach of Sigmund Freud and his followers as pseudoscience because it does not make specific predictions and stick to them. They say that psychoanalysis makes complex, often metaphorical assumptions about how the mind works but without providing clear rules for how those assumptions apply. As a result, psychoanalysis can be used to explain any phenomenon after it occurs. Freud wrote voluminously and used psychoanalysis to try to explain everything from common slips of the tongue to religious beliefs. However, he did no experimental tests of his theory, preferring to support that theory with after-the-fact cases and informal observations. In contrast, experimental psychologists make specific predictions in order to test their theories and hypotheses. Psychoanalysts might tout the ability of their theory to explain any event as a virtue, but philosopher Karl Popper has said that this is actually a weakness. When scientists make specific predictions that can be disconfirmed, they add strength to a theory if the predictions turn out to support it (Popper 1959). For example, before the solar eclipse of 1919, Albert Einstein used his new theory of relativity to make the specific and risky prediction that the immense gravity of the Sun would bend the light coming from background stars. Although the results of testing this risky prediction could have disconfirmed the theory of relativity, many physicists strengthened their belief in the theory when it was observed that the Sun did, in fact, bend the light of the background stars.

In contrast to pseudoscience, science involves thinking critically about the meaning of systematic observations. But some philosophers of science have challenged this view, pointing to the irrational origins of modern science. For instance, Sir Isaac Newton, who did much to champion a rational approach to experimental physics, also did many experiments on alchemy and wrote on that subject. Alchemy—a very old mystical approach in which various substances were mixed together with the goal of producing gold—is inconsistent with the basic principles of physical science and never actually yielded gold. The fact that Newton did not merely dabble in alchemy but conducted much alchemical research suggests that he was not taking a rational approach in developing physical theory.

Another criticism of science is the fact that, in practice, scientists do not appear to think critically when they obtain negative evidence for a theory but do not reject it. For example, when they conduct an experiment to test a prediction from a theory and observations do not support that theory, they often fail to reject the theory even though it has been shown to be inconsistent with observations. Instead, they are likely to blame the failure on poor experimental method or some other problem in observation. They only reject the initial theory after many failed attempts to obtain observations consistent with it and often only after a better theory comes along. Philosopher of science Imre Lakatos (1970) argued that although scientists may appear to be uncritical and irrational in the short run, they no longer seem so when we look at how they change their theories over the long run. When scientists first obtain the negative evidence, they are not sure what these observations mean. Because theories are general principles that are consistent with many observations, they take a conservative approach and do not reject the entire theory at first but instead reject more peripheral assumptions. Over time, scientists are rational in that the theories they move toward are more consistent with all of the evidence. For Lakatos, the failure to quickly reject a theory with negative evidence serves as a practical strategy that helps scientists move toward a sound theory over time. It is not a sign of dogmatism or of holding fast to a favored theory regardless of the evidence.

Authorities in the area of critical thinking have often argued that the ability to think critically involves acquiring the skills of reasoning and good thinking along with the dispositions and attitudes required to make use of those skills. Examples of critical-thinking skills are the abilities to identify claims, to evaluate different kinds of evidence, to identify assumptions, and to draw a sound conclusion from evidence. Examples of critical-thinking dispositions and attitudes are the tendency to be fair-minded, the tendency to be reflective, and an attitude of skepticism toward claims made. It seems clear that although someone might possess the critical-thinking skills necessary for coming to a sound conclusion, that same person might not be disposed to use those skills. For example, a person might be able to reason and understand evidence but not be skeptical, that is, might not be inclined to question evidence presented in support of a claim. Consequently, someone may fail to think critically because he or she lacks necessary skills, is not inclined to use those skills, or both.

This observation raises the question of whether people who believe in paranormal and pseudoscientific claims might be deficient in critical-thinking skills or dispositions as compared to nonbelievers. Research by Susan Blackmore and Tom Troscianko (1985) showed that believers in ESP may be poorer at estimating probabilities than nonbelievers. People who are inaccurate in estimating the probability of events may tend to overestimate the likelihood of a coincidence or an unusual event. This, in turn, may lead them to attribute a given event to ESP or some paranormal cause instead of to chance. Other research suggests that people with poor critical-thinking skills may come to believe paranormal claims despite the considerable amount of good evidence against such claims. James Alcock and Laura Otis (1980) found that compared with believers in paranormal phenomena, nonbelievers had significantly higher scores on a test of critical-thinking ability and significantly lower scores on a measure of dogmatism. In this context, dogmatism is a disposition or attitude by which people hold strongly to their favored opinion and are resistant to opposing views even when the evidence supports one.

A recent study by Chris Roe (1999) further emphasized the importance of critical-thinking dispositions in approaching questions concerning the paranormal. Roe found no difference in the critical-thinking abilities of believers versus nonbelievers on a task in which they evaluated evidence for and against ESP. However, he found that both believers and nonbelievers tended to rate the studies offered against their position as having lower quality than those that were favorable to their position, even though they were equal in quality. This result suggests that people's prior beliefs and dispositions may be more important in the way they evaluate claims than their critical-thinking ability.

Given that individuals such as college students often show deficiencies in critical-thinking skills and dispositions that persist even after considerable education, special kinds of instruction designed to address these problems may be required. Research on these instructional approaches has shown that some of them do help students improve their ability to think critically about scientific and pseudoscientific claims and to reduce their belief in unsupported claims. Alan Bensley (1998) developed a method for teaching students how to think critically about scientific and paranormal claims designed to achieve these ends. He and Cheryl Haynes (1995) found that students in a class using the method significantly improved their ability to critically analyze a scientific discussion of a psychological question and increased their use of the critical-thinking language experts employ as compared to students in a similar class that was not getting the critical-thinking instruction. In a study of the beliefs of students in other classes, Bensley and Tanya De Both (1998) found that students in courses using the method changed their commonsense beliefs about the mind and behavior to be more in line with scientific psychology. Compared with others who received ordinary instruction, students who got the critical-thinking instruction were more likely to change their minds about questions regarding the effects of the moon on abnormal behavior and the ability of hypnosis to bring back forgotten memories; however, this research was not specifically designed to examine changes in belief in the paranormal.

Other research studies have more directly tested the effects of critical-thinking instruction in terms of belief in the paranormal. Jerome Tobacyk (1983) reported that there was a reduction in belief in paranormal phenomena among students who had taken a course especially designed to help them examine claims of the paranormal. Similarly, Davina Mill, Thomas Gray, and David Mandel (1994) found that students taking courses on research methods and statistics only improved their ability to think critically when they were given special instruction in applying the methods of science and in analyzing paranormal claims. However, neither the traditional course work nor this special instruction reduced students' beliefs about the paranormal. Research on the teaching of thinking has found that people who acquire thinking skills in one area often fail to use those skills in a different area. This and other research not reviewed here suggest that it may be difficult for people to acquire the thinking skills and dispositions needed to critically examine paranormal and pseudoscientific claims. Furthermore, it may be necessary to directly teach critical and scientific thinking to help students apply these skills and to acquire important critical-thinking dispositions such as having a questioning attitude toward claims.

References:

Alcock, James E., and Laura P. Otis. 1980. "Critical Thinking and Belief in the Paranormal." *Psychological Reports* 46: 479–482.

Bensley, D. Alan. 1998. *Critical Thinking in Psychology: A Unified Skills Approach.* Pacific Grove, CA: Brooks/Cole.

Bensley, D. Alan, and Tanya De Both. 1998. "The Influence of Instructional Methods on Adherence to Commonsense Theories of Psychology." Poster presented at the Tenth Annual Meeting of the American Psychological Society, May 23, Washington, DC.

Bensley, D. Alan, and Cheryl Haynes. 1995. "The Acquisition of General Purpose Strategic Knowledge for Argumentation." *Teaching of Psychology* 22: 41–45.

Blackmore, Susan, and Tom Troscianko. 1985. "Belief in the Paranormal: Probability Judgments, Illusory Control, and the 'Chance Baseline Shift'." *British Journal of Psychology* 76: 459–468.

Ennis, Robert H. 1987. "A Taxonomy of Critical Thinking: Dispositions and Abilities." In *Teaching Thinking Skills,* edited by J. B. Baron and R. J. Sternberg, 9–37. New York: W. H. Freeman.

Hines, Terence. 1988. *Pseudoscience and the Paranormal: A Critical Examination of the Evidence.* Buffalo, NY: Prometheus Books.

Kelly, Ivan W. 1997. "Modern Astrology: A Critique." *Psychological Reports* 81: 1035–1066.

Krech, David. 1962. "Cortical Localization of Function." In *Psychology in the Making,* edited by L. Postman. New York: Knopf.

Lakatos, Imre. 1970. "Falsification and the Methodology of Scientific Research Programmes." In *Criticism and the Growth of Knowledge,* edited by I. Lakatos and A. Musgrave, 91–95. Cambridge: Cambridge University Press.

Leahey, Thomas H., and Grace E. Leahey. 1983. *Psychology's Occult Doubles: Psychology and the Problem of Pseudoscience.* Chicago: Nelson Hall.

Mahoney, Michael J. 1977. "Publication Prejudices: An Experimental Study of Confirmatory Bias in the Peer Review System." *Cognitive Therapy and Research* 1: 161–175.

Messer, Wayne S., and Richard A. Griggs. 1989. "Student Belief in the Paranormal and Performance in Introductory Psychology." *Teaching of Psychology* 16: 187–191.

Mill, Davina, Thomas Gray, and David Mandel. 1994. *Canadian Journal of Behavioral Science* 26: 246–258.

Paul, Richard W. 1993. *Critical Thinking: What Every Person Needs to Survive in a Rapidly Changing World.* 3d ed. Santa Rosa, CA: Foundation for Critical Thinking.

Popper, Karl R. 1959. *The Logic of Scientific Discovery.* New York: Basic Books.

Randi, James. 1982. *Flim-Flam!* Amherst, NY: Prometheus Books.

Roe, Chris A. 1999. "Critical Thinking and Belief in the Paranormal: A Re-evaluation." *British Journal of Psychology* 90: 85–98.

Rotton, James, and Ivan Kelly. 1985. "Much Ado about the Full Moon: A Meta-analysis of Lunar-Lunacy Research." *Psychological Bulletin* 97: 286–306.

Stanovich, Keith E. 1998. *How to Think Straight about Psychology.* 5th ed. New York: Longman.

Tobacyk, Jerome J. 1983. "Reduction in Paranormal Belief among Participants in a College Course." *Skeptical Inquirer* 8: 57–61.

Reincarnation

PHIL MOLÉ

Reincarnation is the belief that the souls of human beings inhabit a succession of physical bodies during their existence. According to this doctrine, physical death is a transitional period in which a soul ends its lifetime in one body and prepares to begin a new life in another body.

The concept of reincarnation existed to a limited extent in ancient Greek and Egyptian cultures, but it did not become an essential component of a philosophical system until the development of Eastern traditions such as Buddhism and Hinduism. Many historians of religion trace the first fully articulated reincarnation doctrines to the *sramanas*, or wandering ascetics, present in India and South Asia in the fifth and sixth centuries B.C. (Smart 1998, 56). Emphasis on the ascetics and their teachings of samsara, or rebirth, entered the philosophies of Buddhist and Jainist religious movements. The influence of these religions integrated reincarnation into the priestly religion that eventually became Hinduism, especially after the composition of some of the later philosophical documents known as the Upanishads (Olivelle 1998, xxxiii). The concept is now widespread throughout South Asia, and reincarnation is integral to classic Eastern religious texts such as the *Tibetan Book of the Dead* and the *Bhagavad Gita*.

Reincarnation is sometimes but not always associated with the concept of karma, or the spiritual effect of past actions. The reincarnationist who believes in karma considers the experiences of this lifetime to be the result of actions from previous lifetimes. In classic Hindu tradition, a person strives to free him- or herself from the restraints of past actions through meditation and self-denial. Liberation from the effects of karma will lead to personal salvation and escape from the cycle of birth and rebirth (Flood 1996, 76).

Although Western supporters of reincarnation frequently cite historical figures such as Benjamin Franklin, Voltaire, Johann Wolfgang von Goethe, David Hume, and Thomas Henry Huxley as fellow believers, very few distinguished Western thinkers have accepted the idea. Franklin, Voltaire, Hume, and Huxley were skeptical of the existence of a soul independent of the human body, and belief in a soul is a prerequisite for belief in reincarnation. Goethe seemingly expressed sympathy for reincarnation in some of his writings, but he also expressed contrary opinions on many occasions and cannot properly be considered a believer. Arthur Schopenhauer, the prominent German philosopher and scholar of Eastern philosophies, was one of the few notable supporters of reincarnation in the West. Still, the truth of a doctrine cannot be determined by simply listing its most famous supporters, and proponents of reincarnation often resort to such lists when their philosophical arguments are weakest.

Since the late nineteenth century, reincarnation has earned fairly significant popular acceptance as a component of alternative religious movements. Edgar Cayce, the infamous "sleeping prophet" who attracted attention for his reputed clairvoyant abilities, included reincarnation on the long list of paranormal phenomena in which he passionately believed. Reincarnation also earned the support of Madame Helena Blavatsky, the founder of a fringe religion known as Theosophy. An eclectic mixture of mystical traditions from all over the world, Theosophy taught that consciousness pervades all matter in the universe. Blavatsky and her followers embraced reincarnation as an important argument against contemporary materialist philosophies that questioned the existence of souls and other supernatural entities (Washington 1998, 45). Despite numerous factual and philosophical errors in the writings of Cayce and Blavatsky, both figures continue to be important influences on contemporary reincarnationists such as Elizabeth Kubler-Ross and Raymond Moody.

Reincarnationists cite various types of evidence as "proof" of their belief. A common argument involves cases of alleged child prodigies who show an unusual amount of talent or intellectual ability at a very early age. One of the most frequently cited examples is William Hamilton (1805–1865), who acquired vast mathematical expertise and the ability to speak thirteen languages before his adolescence. Other common examples include composers such as Felix Mendelssohn, Wolfgang Amadeus Mozart, and Franz Schubert, who produced sophisticated music while very young (Edwards 1996, 48–49). Supporters of reincarnation claim that traditional genetic and cultural explanations of human learning abilities cannot account for the talents of these prodigies, especially since they often display abilities absent in both parents. However, this claim has serious shortcomings. First, the premise of the genetic argument rests on the false assumption that all features of an offspring must be present in one or both parents if they were acquired through normal heredity. In reality, many genes are recessive and can be passed from parents to children without being activated. Genes also do not function independently but are stimulated or repressed by environmental influences. As studies of identical twins have shown, people with the same sets of genes can develop talents to very different degrees if they are raised in different environments (Segal 1999, 314).

Second, the attempt to explain novel or extraordinary data with theories such as reincarnation is inherently misguided. The fact that scientists currently do not fully understand the cognitive or physiological basis for intellectual talent does not justify paranormal explanations. Since the human mind is extraordinarily complicated and powerful, there is no reason to consider anything other than strictly biological and cultural factors to explain the abilities of child prodigies. The reincarnationist claims are merely "god of the gaps" arguments advanced by those seeking to fill the holes in human knowledge with fantastic and mostly arbitrary explanations. As critics argue, a reincarnationist could just as feasibly apply his or her argument to anyone with any type of special ability, be it Albert Einstein, Paul McCartney, William Faulkner, or a talented teenage track-and-field runner. Generally, skeptics also contend that to suggest talent can only be explained through appeals to the supernatural is both unwarranted and demeaning to human potential.

Strange birthmarks on a person's body are another commonly cited "proof" of reincarnation. Ian Stephenson, one of the most prominent contemporary reincarnationists, considers birthmarks to be the strongest evidence in favor of the doctrine. He and many of his colleagues find a correspondence between birthmarks on living individuals and wounds or other markings on the bodies of deceased persons, and they claim the similarity of these marks is too strong to result from chance alone. The only sensible explanation, in their view, is that the deceased person has been reincarnated in a new body, with the previous bodily markings intact. However, most of these alleged cases of physical similarities are based on anecdotal evidence, since it is usually impossible to inspect the body of the deceased person or to analyze a detailed photograph of the body. Many of the alleged correlations are invented retrospectively by family members who already believe in reincarnation. After a child is born, family members believing in reincarnation look for birthmarks on the child and then try to recall a dead friend or relative who had similar marks. This method of selectively reviewing data to verify preconceived ideas virtually guarantees errors in judgment and reasoning. Aside from these difficulties, reincarnationists must also explain how the presumably immaterial soul of a deceased person can transmit physical characteristics to a new body. Since there is no logical way that a nonphysical entity can cause changes on physical bodies, such a transmission of characteristics must be extremely improbable, if not impossible. This modus operandi problem of conserving the physical traits of the dead continues to defeat the best arguments of reincarnationists (Edwards 1996, 135).

Another category of evidence used to support reincarnation concerns déjà vu, or the inexplicably strong feeling that a current event has been experienced previously. Believers in reincarnation consider déjà vu experiences to be spontaneous memories of events from past lives, and they maintain that science will never adequately account for them. Few reincarnationists appear to have actually explored scientific explanations of déjà vu, since viable theories have been available since the nineteenth century. Philosopher and psychologist William James, for example, suggested two possible explanations for déjà vu in his classic text *Principles of Psychology* (1890). The first explanation involves the inability of a person to distinguish between a current experience that resembles a past experience in some important aspects. The uncanny feeling associated with the déjà vu experience fades as soon as the uniqueness of the current experience becomes more apparent. The second explanation is that the two hemispheres of the brain sometimes process sensory information at slightly different rates. A neural short circuit results, causing the general impression of an experience to register in the memory before the conscious mind has fully analyzed it. Modern cognitive researchers have found significant evidence that this theory explains a large number of déjà vu experiences. For instance, psychologist Arthur Reber noted that patients with certain types of brain damage frequently have déjà vu experiences (Reber 1985, 183). This evidence strongly suggests that these experiences are physiological and psychological phenomena. Cognitive researchers consider déjà vu to be fully explicable in scientific terms and do not endorse mystical explanations such as reincarnation.

Since the 1950s, hypnotically induced memories of past lives have been the most widely discussed evidence for reincarnation. The process of using hypnosis to recover alleged memories of previous lives is known as past-life regression. While hypnotized, a subject answers a series of questions and gradually reveals the identity and nature of past lives. This methodology is similar to the techniques used by researchers in the recovered-memory movement, in which therapists apparently retrieve details of long-repressed memories from hypnotized subjects. Past-life regression and other recovered-memory therapists falsely consider human memory to be a faithful record of actual events, requiring only the prompting of a skilled hypnotist to accurately reveal the details of past experiences. But researchers such as Elizabeth Loftus have demonstrated that memories are constructed rather than simply retrieved, and memories recalled through hypnosis are especially prone to inaccuracies. Suggestive questions asked by the therapist can cause a hypnotic subject to hold distorted or completely false memories of past events (Loftus 1997, 72). In the 1990s, documented cases involving false accusations of sexual and physical abuse resulting from recovered memories further proved the unreliability of hypnotherapy for accurate memory retrieval.

The most famous case of hypnotically induced past-life regression concerned the case of a young housewife named Virginia Tighe. An amateur hypnotist named Morey Bernstein conducted six hypnotic sessions with Tighe between November 1952 and October 1953 and allegedly regressed her to a previous life as a nineteenth-century Irish woman named Bridey Murphy. While under hypnosis, Tighe described many details of her life as Murphy, including descriptions of her birth in the small town of Cork in 1798, her marriage to a young Protestant man named Joseph MacCarthy, their life together in Belfast, and her death in 1864 (Bernstein 1956, 108–163). She also spoke in an Irish brogue, captured on an audio recording of the sessions that was later released as a best-selling album. Bernstein published a serialized account of the case in the *Denver Post*'s Sunday supplement in September 1954 before releasing his book *The Search for Bridey Murphy* in 1956. The book was an enormous success, and public interest in reincarnation in the United States immediately increased. The popularity of Bridey Murphy subsided after a chain of newspapers owned by William Randolph Hearst ran an exposé of the case, claiming to debunk Bernstein's conclusions in *The Search for Bridey Murphy*. Unfortunately, editors at the Hearst papers were motivated by factors other than a fondness for truth. They were mainly interested in discrediting newspapers such as the *Chicago Daily News*, which had obtained the enviable syndication rights to the Bridey Murphy story. Reporters for Hearst's *Chicago American* unscrupulously fabricated most of the details of their "debunking" and opened the door for later reincarnationists to uphold the validity of the Bridey Murphy case (Gardner 1957, 317–318). Subsequent investigators have shown Tighe's descriptions of persons and places in nineteenth-century Belfast to be incorrect, and many of the supposed anecdotes about Bridey Murphy's life probably resulted from subconscious recollection of stories told by Tighe's Irish friends and neighbors. These investigations have thoroughly disproved the Bridey Murphy case, although ardent reincarnationists still cite it as incontrovertible evidence of their doctrine.

Several important philosophical problems also undermine the theory of reincarnation. The "population growth" objection, first found in *Treatise of the Soul* by the early Christian thinker Tertullian, points to a discrepancy between the number of living souls and the number of souls in early human history. Reincarnationists are committed to the notion that each human soul is eternal and has lived countless lives as it has traveled from one human body to the next. However, the total population of people alive today is now greater than it has been at any previous time in history. In the first century C.E., only 200 million people were living on the planet, whereas there are over 6 billion people alive today. Therefore, the overwhelming majority of people living now could not be reincarnations of people from the past, since the earliest populations of humans were much smaller than the current population. Many souls of the living are simply not accountable through the theory of reincarnation.

Other important objections concern the nature of the soul itself. Many reincarnationists insist that the soul is a replica of a human personality and is capable of learning and changing in analogous ways. However, if the soul really does change correspondingly with our conscious personality, it follows that any good or bad effects on the personality will also affect the soul. In practice, reincarnationists hold the arbitrary and indefensible belief that only positive changes in a person's personality are transmitted to his or her eternal soul. Few reincarnationists would maintain that brain damage that adversely affects a person's conscious thought and personality also damages the health of his or her soul, but that is exactly what they must maintain if they apply their doctrine consistently. They cannot simultaneously claim that the soul is unchanging and changeable in order to save their theory from its unpleasant consequences. There is also the troubling fact that people do not consciously remember any of the details of past lives. This implies a less-than-perfect continuity between the identities of a soul from one lifetime to the next.

Logical considerations have prevented reincarnation from earning the assent of most people trained in critical thinking. Even some Eastern religious thinkers, such as the Hindu reformer Ram Mohan Roy (1772–1833), have considered reincarnation incompatible with a system of rational ethics (Flood 1996, 252–253). However, the doctrine has survived for millennia, and the appeal of its simplistic view of life is not likely to disappear anytime soon.

References:

Bernstein, Morey. 1956. *The Search for Bridey Murphy*. New York: Doubleday.

Edwards, Paul. 1996. *Reincarnation: A Critical Examination*. Buffalo, NY: Prometheus Books.

Flood, Gavin. 1996. *An Introduction to Hinduism*. New York: Cambridge University Press.

Gardner, Martin. 1957. *Fads and Fallacies in the Name of Science*. New York: Dover.

Hick, John. 1976. *Death and Eternal Life*. New York: Harper and Row.

James, William. 1890. *Principles of Psychology*. Cambridge, MA: Harvard University Press.

Lamont, Corliss. 1990. *The Illusion of Immortality*. New York: Continuum.

Loftus, Elizabeth. 1997. "Creating False Memories." *Scientific American* 277, no. 3: 70–75.

Olivelle, Patrick, trans. 1998. *Upanishads*. Oxford: Oxford University Press.

Reber, Arthur. 1985. *Dictionary of Psychology*. New York: Penguin Books.

Segal, Nancy. 1999. *Entwined Lives: Twins and What They Tell Us about Human Behavior*. New York: Pantheon.

Smart, Ninian. 1998. *The World's Religions*. New York: Cambridge University Press.

Washington, Peter. 1998. *Madame Blavatsky's Baboon*. New York: Schocken.

Séance

DREW CHRISTIE

A séance (pronounced SAY-ahnce) is a gathering to communicate with the dead. A small party sits around a table in a darkened room with a medium who calls the spirits. After a sometimes considerable period of waiting, mysterious knocks are heard, a nearby musical instrument sounds, the table turns and rises, spirits appear, or the medium goes into a trance and gives voice to messages from the dead. Ouija boards were introduced in 1889 and rapidly became a popular part of séances.

There have always been those who claimed to communicate with the dead. However, the rituals and conventions specific to séances stem from the wide publicity that table tapping received starting in 1848 with the Fox sisters, Katie and Margaret, of Rochester, New York. Word of their story quickly spread to England and throughout Europe. Séances became the rage. They were a form of popular entertainment before movies, radio, and television, and skeptics were invited to enjoy the show (Oppenheim 1985).

Séances remained popular even though most everyone knew that fraud was frequently uncovered. Newspapers carried stories about secret compartments found in a medium's cabinet, "spirits" that had been trapped scampering across the floor, and mediums who were caught in the act of changing costumes. The Fox sisters eventually confessed that they had produced the mysterious tappings heard in their séances by popping their toe joints. Both belief and skepticism were widespread. A common complaint among skeptics was how trivial the content of alleged communication from the dead was. Yet several mediums became household names: Daniel Home, Florence Cook, Leonora Piper, Eusapia Palladino.

Séances became so much a part of the culture of all social classes that Victorians inevitably had strong opinions about them. The novelist George Eliot was contemptuous and described séances as "either degrading folly, imbecile in the estimate of evidence, or else as imprudent imposture" (in Haight 1955). The poet Robert Browning penned a satire, "Mr. Sludge the Medium," and expressed concern that his poet-wife, Elizabeth Barrett Browning, believed in such nonsense. Others gave Spiritualism sufficient credibility to form the American and British Societies for Psychical Research. The influence of séances should not be underestimated; contemporary literary critics note the tremendous impact of the occult on many important writers (Surette 1993).

Cultural historians are intrigued by the popularity and credulity that séances commanded for nearly 100 years even among the well educated (Brandon 1983). Some have noted that, during the heyday of séances, the medium was often a young woman and that many séances had an erotic cast (Owen 1990). During a time of rigid sex roles, the séance was, in part, an expression of Victorian ambivalence about sexuality, and becoming a medium was an appealing job option for some women.

References:

Brandon, R. 1983. *The Spiritualists: The Passion for the Occult in the Nineteenth and Twentieth Centuries.* New York: Alfred A. Knopf.

Haight, G. S., ed. 1955. *The George Eliot Letters.* Vol. 5. New Haven, CT: Yale University Press, 49.

Oppenheim, J. 1985. *The Other World: Spiritualism and Psychical Research in England, 1850–1914.* New York: Cambridge University Press.

Owen, A. 1990. *The Darkened Room: Women, Power, and Spiritualism in Late Victorian England.* Philadelphia: University of Pennsylvania Press.

Surette, L. 1993. *The Birth of Modernism: Ezra Pound, W. B. Yeats, T. S. Eliot, and the Occult.* Montreal: McGill-Queen's University Press.

Shamans and Shamanism

AL CARROLL

As used by anthropologists, the term *shamanism* means any tribal or earth-based religion or any religion that is not part of the world's "major" religions. But as used by the New Agers, the concept of shamanism is a troubling mix of marketing angles, cultural biases, and outright fraud. That's why anyone calling him- or herself a "shaman" is commonly referred to as a "shame-on" by American Indians.

The modern movement of would-be shamans got its start in 1980 when Michael Harner published *The Way of a Shaman*. Harner was seeking to avoid many of the pitfalls the New Age movement had fallen into, such as exploitative leaders; unclear and unrealistic goals; incoherent, contradictory, or nonsensical beliefs that were widely mocked by most of the public; blatant abuse and exploitation of tribal peoples and beliefs; and a complete lack of credibility with either academia or the public. In all of these goals, Harner and the rest of the shamanism movement have utterly failed. Many of the most disreputable New Age leaders, such as Lynn Andrews and Ed McGaa, sensed the marketing potential and simply adopted the shaman pose. Harner's methods were little different from the New Agers' in his assumptions that one could easily and quickly learn methods that actually take decades to master among tribal traditionalists. Even his "advanced" seminars only last three days, and he is clearly engaged in a highly profitable enterprise as much as an attempt to form a new spirituality, thereby fitting well into the New Age.

Harner and the other would-be shamans also make the same mistakes as New Agers in trying to homogenize tribal traditions worldwide and deny their diversity and important differences by lumping several thousand belief systems together. Harner pretends one can master elements that are supposedly common or universal to all ("core" shamanism, in his lingo). In fact, the supposed commonalities of shamanism are largely superficial or even self-delusional. For example, many would-be shamans falsely claim the sweat lodges used by some American Indian groups are a "core universal shamanic" practice. They allege the Romans and Celts also used sweat lodges. In fact, both those groups used saunas with no spiritual aim or practice involved. Not even all American Indian groups use the sweat lodge.

Finally, Harner and the rest of the would-be shamans are no different in exploiting both tribal peoples and Western seekers of spiritual truths. Shame-ons exploit the former by deceptively misrepresenting traditional beliefs and trying to subjugate native, community-oriented beliefs to Western, egoistic individual needs. And they exploit the latter for purposes of obtaining cash, boosting their own egos, and in some cases, sexual exploitation. Anyone seeking to understand the beliefs of tribal peoples would be far better off reading the writings of respected native authors such as

Vine Deloria Jr. and Wilma Mankiller rather than turning to the works of opportunists.

References:

Churchill, Ward. "Indigenous Knowledge Not for Sale: Spiritual Hucksterism." *NativeNet.* URL: http://cs.fdl.cc.mn.us/natnet/h-storm.html. (Accessed on March 11, 2001).

Cubbins, Elaine. "Native Web Site Evaluation." *University of Arizona.* URL: http://www.arizona.edu/~ecubbins/webcrit.html. (Accessed on September 29, 2001).

Guarding Indian Cultural Spiritual Beliefs. URL: http://www.iktome.freewebsites.com. (Accessed on March 11, 2001).

Smith, Andy. "Readings on Cultural Respect." URL: http://www.alphadc.com/treaty/r-explt.html. (Accessed on March 11, 2001).

Wallis, Robert J. "Journeying the Politics of Ecstasy: Anthropological Perspectives on Neoshamanism." URL: http://www.interchange.ubc.ca/fmuntean/POM6a2.html. (Accessed on March 11, 2001).

The Shroud of Turin

CHRIS CUNNINGHAM

The Shroud of Lirey-Chambéry-Turin has undergone more scientific scrutiny than any other religious artifact in history. It has been subjected to batteries of sophisticated tests and intense examinations by highly respected academics and scientists, and it has become a symbol both to those who seek to unify religion and science and to those who see these two concepts as diametrically opposed. Whatever tests are performed on the shroud, however, the research ultimately proves just one thing—that even the most respected scientists can be guilty of seeing only what they want to see when matters of faith come into play.

The shroud is a linen cloth some 14 $^1/_2$ feet long by 3 $^1/_2$ feet, made with an unusual type of weave called a 3-to-1 herringbone twill. Indeed, very few samples of this pattern from any time period exist, and most of those that survive are made of silk. It possesses what appears to be both the front and back images of an adult male with long hair and beard, lying down with his hands crossed. The features on the image are similar in style to medieval Gothic art. Vivid red bloodstains appear in various places, corresponding to accepted accounts of the wounds of Jesus. The image is a sepia-yellow in color, but it does not have a sharp outline. To add to the odd effect, the image on the shroud, as was discovered by an early photographer, is actually very similar to a photographic negative.

The shroud has been definitively traced to the mid-fourteenth century by historians. It was owned by a French knight, Geoffrey de Charny, who built a special chapel in Lirey, France, to house it. The first known exhibition of the shroud was in this chapel in 1357, and it attracted many pilgrims (and their money).

By 1389, the first of many scandals involving the shroud erupted. Pierre d'Arcis, the bishop under whose authority the Lirey chapel fell, claimed in a letter to the pope that the dean of the church had "procured for his church a certain cloth cunningly painted, upon which by a clever sleight of hand was depicted the twofold image of one man" (Nickell 1998, 17). The bishop went on to state that his predecessor had even received a confession from the artist who allegedly painted the cloth. Despite various orders and protestations from the bishop and even from the king of France, the shroud continued to be exhibited as the genuine burial cloth of Jesus.

In 1502, the shroud was transferred to the Royal Chapel in the castle at Chambéry. Thirty years later, it was nearly destroyed when the chapel burned down. The silver reliquary in which it was stored melted, and only the efforts of the clergy and a local blacksmith saved the shroud. Melted silver burned a hole in one corner of the folded cloth but fortunately did not damage the image. Thereafter, the shroud was periodically exhibited and moved about, finally coming to rest in the town of Turin, Italy, in 1578. It has remained there ever since, except for a period when it was kept in a remote abbey for safekeeping during World War II.

Occasional accusations of forgery have dogged the shroud, the most contentious of which began at the dawn of the twentieth century. In 1898, an amateur photographer named Secondo Pia was granted permission to photograph the shroud. When he developed the negative, he discovered that, instead of the usual oddly shaded inverted image, it showed a very lifelike image of a man. The only way this could have occurred, he believed, was if the image on the shroud was, in fact, a negative itself. The challenges started immediately. Pia was accused of having overexposed the image during development; others alleged that the effect was the result of backlighting. It was not until 1931 that Pia's work was verified by additional photography.

Major scientific examination of the shroud began in 1979. The Holy Shroud Guild, a Catholic organization that "promotes study and devotion of the Shroud of Turin" (Holy Shroud Guild 2001), arranged for a group of scientists to study the shroud using modern equipment. The scientists, led and organized by two members of the Executive Council of the guild, became known as the Shroud of Turin Research Project (STURP).

STURP members examined the shroud for five days with all means at their disposal. Samples of the shroud's fibers and any contaminants that may have been present were collected using a special type of sticky tape, to be examined later in a laboratory setting. These tapes were to become the source of much information regarding the shroud. A large number of these tapes were sent to microanalyst Walter McCrone, who, upon examination, discovered the presence of iron oxide and various other pigments common to the mid-fourteenth century. These pigments were present only on the areas of the shroud where there was either blood or part of an image.

Most tests using extremely sensitive equipment and conducted by forensics experts detected no trace of blood on the shroud image areas. A series of tests performed by chemist Alan Adler and biophysicist John Heller are perhaps the most referenced, having been among the few that found positive results. However, none of the tests performed by this pair were specific for blood, and none could provide similar results in the presence of the same pigments detected by McCrone.

Much was made by the STURP members about the apparent realism of the blood flows and the anatomy. Some who believe the blood came from a body wrapped in the shroud point to this as evidence. No explanation has been given, however, as to how such artistically perfect blood patterns were transferred to the cloth without smearing or blotching, particularly given that Jewish custom required that the body be washed prior to burial. Another difficulty with explaining the blood as having come from a genuine crucified body is that the blood itself is a very bright red, not the dark black that would be expected of old blood.

The image itself has little distortion to it. Simple experiments can be done by placing some form of pigment on a bust or statue's face and laying a cloth over it like a shroud. When the cloth is pulled up, the image is badly distorted, with eyes appearing much wider than they actually are and a wide, flattened nose, among other problems. If the shroud had been lying loosely on a body and had an image imprinted from contact, it would have such distortions. The image on the shroud appears as if the cloth were relatively flat when the image was made. Shroud investigator Joe Nickell has shown through experimentation that such an image could have been produced using a rubbing technique common in the mid-fourteenth century, and Nicolas Allen has produced actual photographic negatives using materials and knowledge available in the 1300s.

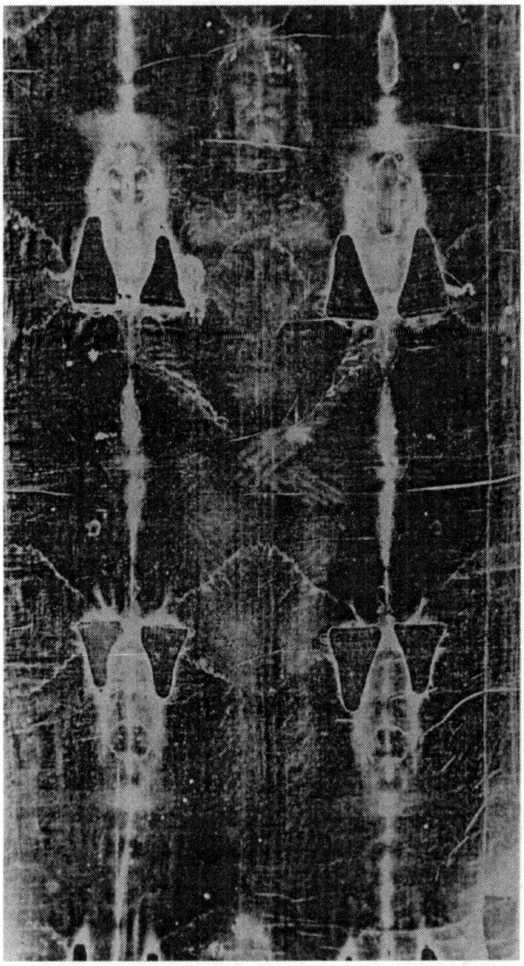

Frontal view of the Turin Shroud image (reversed to negative as to appear positive; enhanced contrast). (Fortean Picture Library)

Alan Whanger developed a technique he termed the polarized image overlay technique (PIOT) for detecting previously undetected images on the shroud. With this method, Whanger found what he believed to be coins on the eyes and even the instruments of crucifixion, including the spear, the sponge, the scourge, and various other items generally associated with the crucifixion. The images are detected by overlaying the slides of two images, one the "target" image and the other the image that is being sought. If the detection of images of the crucifixion instruments is accurate, that finding lends further support to the forgery hypothesis. If images of crucifixion instruments actually have been detected, the forgery hypothesis would be further supported, for the middle of the fourteenth century saw an interest in painting images of a crucified Jesus with the instruments about him. Some argue that this is evidence of the influence of the shroud on modern art. However, this argument fails to take into account that the images on the shroud are so faint as to be undetectable without modern techniques. If the images were as relatively faded then as they are now, as happens with the formation methods suggested by those who support the 2,000-year age of the shroud, medieval artists would have been unable to see the images in order to imitate them. If, however, the image has faded with time, as is suggested by those who believe the shroud is a painting, it is possible that the items were once painted on in keeping with the contemporary style but have since faded to the point of being nearly undetectable.

The most conclusive tests were performed after much discussion between scientists and the shroud's caretakers. Radiocarbon dating was agreed to and conducted in 1988. Samples of the shroud were sent to three laboratories to have their carbon 14 content examined. By comparing the amount of this form of mildly radioactive carbon with the amount present in modern-day materials, scientists could generally determine (within 150 years) the age of any formerly living material.

The results of the tests all came back in close agreement: the linen used to make the shroud had been harvested somewhere between A.D. 1260 and 1390. These results were in line with both the historical and the artistic information. However, the results of the tests eventually came under attack. Several conspiracy theories arose, for which no evidence was ever provided. One Russian scientist attempted to explain the results as having been skewed as a result of the fire of 1532, but his own results could never be corroborated by other scientists, and the veracity of his work was later brought into question.

In 1993, a researcher from the University of Texas, Leoncio Garza-Valdez, showed that a coating of difficult-to-remove material was likely on the threads of the shroud. This material would have grown there over the years because of bacteria and fungi, which leave substances behind as they die off. Because the living organisms contain carbon, their presence could have affected the radiocarbon test and suggested that the material was newer than it actually is. But, though it has been generally accepted that this material may be present, several other researchers have shown that it is extremely unlikely that there was enough of the material to skew the results by over 1,200 years. Physicist Thomas Pickett showed by calculations that the weight of the "varnish" would have to be twice the weight of the sample if it were to throw the data off that far. Meanwhile, the microscopic analysis of the material indicated that there is at most 57 percent contamination.

References:

Case, T. W. 1998. *The Shroud of Turin and the C-14 Dating Fiasco: A Detective Story.* Cincinnati, OH: White Horse Press.

Holy Shroud Guild. URL: http://www.shroud.org. (Accessed on May 20, 2001).

McCrone, Walter C. 1999. *Judgement Day for the Shroud of Turin.* Amherst, NY: Prometheus Books.

Nickell, Joe. 1998. *Inquest on the Shroud of Turin.* Amherst, NY: Prometheus Books.

Whanger, Alan, and Mary Whanger. 1998. *The Shroud of Turin: An Adventure of Discovery.* Franklin, TN: Providence House Publishers.

Wilson, Ian. 1998. *The Blood and the Shroud: New Evidence That the World's Most Sacred Relic Is Real.* New York: Simon and Schuster.

Societies for Psychical Research

DREW CHRISTIE

A distinguished group of Cambridge University scholars founded the first Society for Psychical Research in 1882. Its purpose was to scientifically examine the séances, apparitions, table tapping, fairy photographs, clairvoyance, telepathy, automatic writing, and trance states that were mainstays of Victorian culture. In its early years, the society had a number of distinguished presidents: Prime Minister A. J. Balfour; physicists Sir William Crookes, Sir Oliver Lodge, and John William Strutt, Lord Rayleigh; philosophers Henry Sidgwick, C. D. Broad, and Henri Bergson; psychologist William James; physiologist and Nobel laureate Charles Richet; and zoologist Sir Alister Hardy (Oppenheim 1985). The American Society for Psychical Research was founded three years later, in 1885. The most prominent scientist in the United States at the time, Johns Hopkins astronomer Simon Newcomb, was the first president of the American society.

Today, the American and British societies continue to occupy handsome Victorian buildings in New York and London, to sponsor research and lecture series, to publish journals, to run Web sites (http://www.aspr.com, http://www.spr.ac.uk), and to maintain library collections. Although they no longer include among their members the most prominent intellectuals and scientists of the day, the societies do continue to attract the occasional scientist with impressive academic credentials. The British and American societies parented additional societies around the world and are the grandparents of the numerous psi institutes, associations, and centers currently found on every continent.

People have always been intrigued by ghosts, spirits, and mystical powers, but cultural historians have sought to explain the explosion of interest in the paranormal during the second half of the nineteenth century. The majority of Victorian scientists and intellectuals scorned the paranormal, yet a significant number of them were open-minded (if often skeptical) and considered the alleged phenomena worthy of serious investigation. One factor was a negative reaction to Charles Darwin's theory of evolution, which many considered unproven and dangerous because it claimed to remove God's design from the universe. There was widespread concern that a cold materialism was coming to dominate not only science but also culture. Telepathy and the spirit world, if true, would show how little scientists understood.

Developments in nineteenth-century physics were another factor. James Clerk Maxwell's 1873 unification into a single set of equations the laws of seemingly disparate phenomena—heat, light, motion, electricity, and magnetism—may have encouraged thoughts of relating even more distinct phenomena, namely, mind and matter. In other words, if motion changed into light and magnetism, why couldn't matter transform into spirit? Additionally, the certainty of mechanical, Newtonian physics was challenged toward the end of the nineteenth century. A mysterious and unverified ether, considered necessary for light to travel through, was said by physicists to permeate space. The ether provided a readily available example of a puzzling substance that played a fundamental role in the nature of things. The discovery of non-Euclidean geometries, like evolutionary theory, cast doubt on ancient certainties. Some speculated that the paranormal only appeared abnormal because humans lacked access to an unseen fourth dimension. The disruption of nineteenth-century physics by the discovery of radiation (in 1895) and the electron (in 1897) further heartened psychics. Within the context of times, it is not startling that Marie Curie, who received a Nobel Prize for her discovery of radium, and Sigmund Freud, who tantalized society with his theory of the unconscious, became members of the British Society for Psychical Research.

As prominent as any convert to Spiritualism was Alfred Russel Wallace (1823–1913), the brilliant naturalist who codiscovered natural selection in the 1850s with Darwin. Within a decade, Wallace started attending séances and quickly became a fervent believer in the existence of a spirit world. Darwin was greatly disappointed when Wallace published an article in 1869 arguing that evolutionary theory could not account for human consciousness. Wallace developed the striking view that evolution governed bodies but not the uniquely human form of consciousness. Writing decades before the discovery of genetics, he rightly understood that evolutionary theory was far from complete. He attempted to close genuine gaps in evolutionary theory with what he thought he saw at séances. Whatever Wallace's reasons were for denying that the human mind evolved biologically, his espousal of Spiritualism was well-known and widely discussed. Although he did not share all the convictions concerning the existence of spirits held by many members of the British Society for Psychical Research, he was a longtime honorary member. Wallace declined invitations to become president of the British society.

The research conducted by the societies at the turn of the twentieth century was exhaustive, though relatively crude as judged by contemporary standards of psychological research. In the late nineteenth century, neither the ease with which false memories are implanted nor the psychology of belief perseverance had been explored. In other words, researchers gave sincere, earnest testimony more credence than do modern research psychologists. And some psychical research was supportive of paranormal claims. In *Phantasms of the Living*, 1,300 cases of allegedly veridical apparitions or hallucinations (often of someone's death) were investigated. Death records were checked, and interviews were conducted. The researchers claimed, based on a crude statistical analysis, that the chance of so many firsthand, well-attested veridical visual phantasms was trillions to one (Gurney et al. 1886). Numerous séances were attended and reported on (only some favorably), but researchers did not go beyond inspecting the medium's cabinet, holding his or her hands during the proceedings, and watching the door. A typical investigation took place at a location chosen by the medium and with the medium's selection of props and lighting. Moreover, there was little concern that phenomena be repeatable because investigators were all too willing to accept that genuine psychic capabilities were rare and unpredictable.

Only some psychical research supported paranormal claims. After reluctantly taking office as the first president of the American Society for Psychical Research, Simon Newcomb assiduously pursued investigations of the paranormal. Following two years of studying the literature and attending séances, he concluded that psychical research was a scientific dead end (Moyer 1998). The most famous investigation by the British society was its study of Helena Petrovna Blavatsky, founder of Theosophy (wisdom of the gods). Claiming that there was a close affiliation between mysticism and Buddhism, Blavatsky initiated a wave of interest in Eastern religions among Westerners. She contended that the Mahatmas dictated her first book, *Isis Unveiled*, to her in 1877 through automatic writing. An international sensation, Blavatsky was unusual in claiming paranormal powers while counseling against séances. The trouble with séances, according to her elaborate cosmology, was that communication with the living might well hinder a soul's migration to the next level of existence and eventual reincarnation. In 1885, the British society funded a trip by Richard Hodgson to a site of alleged miracles in India. Over the course of several months, Hodgson grew increasingly skeptical. He exposed hidden sliding panels in the temple and documented the path of supposedly mysterious telegrams. Hodgson revealed the trick behind letters that appeared to float down from the ceiling, and sources were found from which Theosophical wisdom had been copied. But then as now, followers were quite willing to believe their leader was set up or merely caught on a bad day. Though the society's exposure of Blavatsky was widely publicized, Theosophy continued to thrive and currently has a significant influence on the New Age movement (Oppenheim 1985, 174–178).

Was and is psychical research a pseudoscience? In the late nineteenth century, work in that field was not mainstream science, but it was not beyond the pale. Though the research methods employed in that era now appear flawed, they included serious efforts at verifying claims and controlling for fraud. On a continuum in which physics is the extreme of pure science and alien abduction theory is at the pseudoscience extreme, the work of the early psychical societies is closer to pseudoscience than it is to physics, but in historical context, it was far from the extreme.

References:

Gurney, E., F. W. H. Myers, F. Podmore, and Society for Psychical Research. 1886. *Phantasms of the Living*. London: Rooms of the Society for Psychical Research, Trebner.

Moyer, A. E. 1998. "Simon Newcomb: Astronomer with an Attitude." *Scientific American* 279, no. 4: 88–93.

Oppenheim, J. 1985. *The Other World: Spiritualism and Psychical Research in England, 1850–1914*. New York: Cambridge University Press.

Spiritualism

BRAD CLARK

Spiritualism is the belief that communication with the dead can occur through a gifted intermediary called a medium. Mediumship is expressed in two basic forms, the mental and the physical. Mental mediums claim to receive psychic vibrations that can include mental images and messages from the dead. Related to this concept is a belief that some mediums can heal other individuals through spirit assistance. Physical mediums claim the ability to have spirits perform a range of strange phenomena. In darkened rooms, physical mediums have been said to levitate tables, materialize objects, present writing on sealed slate boards, produce eerie voices through floating trumpets, and present messages that many clients feel only the dead could know. Since the beginning of the history of Spiritualism, skeptics and former mediums have reported fraud and deceit behind the techniques used to convince believers.

Spiritualism was born in the United States in 1848 in the village of Hydesville, New York. Although talking with spirits was not a new idea among the world's cultures, it was unique in the Christian world. Christian theology had suppressed the idea of spirit communication by considering it a heresy. In the late 1700s, the writings of Emanuel Swedenborg and his description of the spirit world, along with Franz Anton Mesmer's experiments with hypnotism, primed an intellectual fascination in the subject. In particular, it was the Victorian interest in science that led to an explosion of Spiritualist demonstrations offered as proof of spirit contact.

Spiritualism blossomed as a religious response to the crisis of faith that grew in the mid-nineteenth century. It provided a new reason to believe for people troubled by the evidences offered for Christianity. Mediums claimed that they could produce scientific proof of life after death. Initially, participants were required to believe in nothing. They were simply asked to become investigators charged with observing demonstrations produced under so-called test conditions (Braude 1989, 4).

The movement began when two young sisters, Margaret and Katie Fox, claimed that the rapping sound heard in their house was produced by the spirit of a murdered peddler. Soon, neighbors came from all around to see the children and hear the sounds. At first, simple yes-or-no questions were asked of the spirit, but a regime was established before long whereby the entire alphabet was laboriously sounded out until a knock was heard, thus spelling out words one letter at a time. An older sister, Leah Fox Fish, arrived on the scene with a keen sense for marketing this phenomenon. She began renting out halls and charging admission to the demonstrations (Brandon 1983).

Newspaper accounts of this spectacle spread the story across the country. Soon, people were experimenting in their own homes, hoping to hear the same rapping sounds. Numerous men, women, and children from around the nation and across social lines were identified as having a facility for communicating with the dead. Within a few months, visiting Americans spread Spiritualism to Europe. Spiritualist churches sprang up in both large and small cities. Years later, Margaret Fox (1835?–1893) confessed that her mediumship was a fraud that began as a childhood prank and spiraled out of her control under the manipulation of her older sister, Leah; the sisters produced their rapping sounds by popping their toe joints (Fox Kane 1985). Later in life, as a penniless alcoholic, she recanted this confession and returned to supporting herself as a medium.

The French word *séance*, meaning "sitting," became the accepted name for meetings with mediums. The tedious letter-calling approach was replaced with a number of creative communication techniques. Some mediums found they could communicate with the spirit world if they were placed in a hypnotic trance. Originally, a hypnotist was used in this endeavor, but later, it became common for mediums to quickly place themselves into what they claimed was a trance. A number of these individuals became known as platform speakers. They often spoke at length of the beauty of the afterworld, commonly referred to as Summerland, after the world described by the Spiritualist Emanuel Swedenborg. Many of these speakers (and women in particular) often voiced the new, progressive ideals that were produced during the Victorian age (Braude 1989). A version of this type of performance still exists, but it is now called channeling. A channeler claims to allow the spirit of an entity to speak through his or her body.

Channeling was certainly inspired by the idea of spirit guides. Spiritualists claimed that spirit guides were teachers and the gatekeepers of spirit contact. Mediums often gave their clients detailed descriptions of messages that their spirit guides had for them. The guides were often Native American or from some exotic land. While claiming to be in a trance state, mediums would often adopt an accent when speaking for these spirit guides.

The dark room was the preferred ambiance for spirit contact, and the design of a séance was perfect for creating a setting where the bizarre was experienced firsthand. Participants arrived with the expectation that they would witness the inexplicable. For those who were mourning the loss of a loved one, the anticipation of receiving some message from the deceased made them especially gullible. The dark room, along with the medium's theatrical moaning and swaying, created a heightened sense of awareness, which the medium could manipulate to great psychological effect.

More dramatic physical manifestations soon became the rage, as people sought tangible proof of spirit contact. It became popular for mediums to tie two schoolhouse slate boards together and await written messages inside. Companies sprang up that supplied this underground industry with a range of trick apparatuses. Some sold slates with false fronts or with tools that enabled the medium to insert a small pencil between the slates.

Many mediums never needed to bother with mechanical aids. Their sleight-of-hand skills made it possible for them to perform the miracles that the paying audience wanted to see. Tricks of this sort included the materialization of "apports"—objects presented from the spirit world, typically flowers or small trinkets. Some mediums specialized in producing spirit paintings, in which a blank canvas was switched for one that was fully painted. Some photographers found they could charge higher rates when they included spirits in their portraits. These images were often created by making a double exposure or by having a confederate briefly step behind the person posing during the long exposure time. More than one spirit photographer was run out of town when it was discovered that the spirit in the portrait was

In 1935, the magician John Dunninger exposed the methods used by mediums. The illustration shows the multiple uses for a telescopic, or "reaching" rod. (Dunninger, Joseph, 1935, *Inside the Medium's Cabinet.* New York, David Kemp)

actually someone still living. In fact, many mediums faced criminal prosecution for fraud and, in some communities, were charged with witchcraft.

Some of the more dramatic occurrences of physical mediumship occurred within cabinets. Originally, these were literally full-size wooden boxes in which the medium was placed. From a space at the top of the box, strange things were seen to dart about. It became standard practice to have the medium tied securely within the box to create the "test conditions" that would prevent any nefarious control of the objects in the box. Often, musical instruments and bells were placed in the cabinet, and when the doors to the box were closed, the sounds of the instruments were heard. Following the Civil War, the Davenport brothers turned their cabinet performance into a popular touring stage show. Long after he retired, Ira Davenport admitted to the rope escapes and other techniques that were used in his famed performances (Houdini 1972, 21).

Later, the wooden cabinet was replaced with a more portable curtain cabinet. This setup was often created by simply hanging a couple of blankets or sheets across a rope in the corner of a room. The medium would sit within this enclosure, and the "sitters" were often instructed to hold hands and sing hymns while they waited for something otherworldly to occur. When the medium claimed to enter a trance state, ectoplasm would make its appearance from behind the curtain. It was believed that ectoplasm was a result of the spiritual realm briefly converting itself into a physical manifestation. This mysterious spiritual energy was said to need a conduit—the medium—and something to focus itself—the curtained cabinet. The ectoplasm might take the shape of a full-size figure, often draped in a transparent cover. Sometimes, it was found hanging out of the mouth or nose of the medium. It could also come from other orifices, including the female genitals. Pictures of faces from the spirit world were sometimes observed on the ectoplasm.

The "direct voice" of spirits was heard during some séances, communicated through a levitating tall cone of tin called a spirit trumpet. It was said that a voice box of ectoplasm allowed the spirit to speak. Often, luminous bands were attached to the trumpet so that its movement was visible in the dark. A number of clever methods were employed to create the illusion of the floating trumpet with its disembodied voice. The medium could connect a rubber tube to the trumpet and speak through it while moving the trumpet from side to side. A telescopic rod was sometimes used to move around a detached luminous band. In this case, the whispered voices heard in the dark room did not come from the trumpet. The illusion of the source of the sound worked on the same principle as that used by ventriloquists. Mediums took advantage of the fact that the source of sound is difficult to determine in the dark.

Sitters were warned that touching the ectoplasm could seriously injure or even kill the medium. In spite of the warning, many mediums were exposed when sitters grabbed at the object floating in front of them. Sometimes, a custom-designed apparatus covered with glowing phosphorus was found as the culprit. At other times, a confederate was revealed as the ghostly presence. The gauzelike ectoplasm that extruded from the medium's bodily orifices was often discovered to be exactly that—cotton gauze that was swallowed prior to the séance and then regurgitated and shaped as needed.

Magicians delighted in reproducing the effects of mediums and would often dedicate a portion of their shows to duplicating their mediums' tricks. The famed magician and escape artist Harry Houdini (1874–1926) devoted the latter part of his professional life to exposing the fraud and deception in Spiritualism. Early in his career, he had worked as a medium and had learned a number of the tricks employed in the field. It was only after the death of his mother that he became incensed at the inherent cruelty of those claiming to speak for the dead. Often wearing disguises, he attended the séances of every medium he could find. And every medium he examined used deceptive means to create the impression of spirit contact.

Houdini became part of a distinguished committee that investigated mediums for *Scientific American* magazine. This was one of the first times it was noted that a magician's presence was important to any investigative body researching paranormal claims. It was recognized that although academics had extensive training in their chosen disciplines, being an expert in one field did not make one an expert in methods of deception. As Houdini repeatedly demonstrated, even the brightest individuals could be easily deceived by a magician's tricks. He was occasionally surprised when audience members insisted that his performances were the result of his psychic ability, even after he assured them that they had seen a magic trick.

Spiritualism sparked scientific interest from the 1850s through the 1920s. A number of distinguished scientists examined various mediums, and many concluded that the effects that they witnessed were otherworldly. Interestingly, when these scientists became believers but later discovered fraud on the part of the medium, they accepted the excuse that the medium, when tired, would perform fake demonstrations. However, their inability to explain previous manifestations made them unwilling to change their position of acceptance.

The memory of strange occurrences in the séance room made believers of many. And it was precisely the little-recognized fallibility of memory that led many sincere individuals to recount events that were far from the truth. One researcher in 1886 compared descriptive reports written shortly after a sitting with those written by the same individual some weeks later. He found that the writers inflated selected elements and omitted key information in their later retelling of the story (Podmore 1963, 212). Further, they ignored seemingly insignificant happenings that were, in fact, keys to the deceptive techniques used by the medium. What was thought of as trivial or irrelevant was eliminated from the memory of the sitters. In one case, a second account of a séance neglected to mention that the medium was suffering from a cold, that he left the room to answer the door, and that the slate he was using slipped twice from his hands. Of course, the regular cough and sniffle of the medium would have served as an effective cover for the snap of a spring lock. Dropping the slate on the floor would have provided an opportunity for substitution or some other trickery. And leaving the room would have given the medium a chance to write out a message or glance at an encyclopedia.

More recent experiments have shown that a belief in the paranormal leads to a heightened level of inaccurate observation. In one study, researchers held fake séances at a convention for paranormal enthusiasts. Prior to the start of the experiment, the sitters were asked if they believed that genuine paranormal activity might sometimes occur during a séance. The participants were made aware of several objects in the room, some of which moved during the séance. At the end of the session, they were asked to complete a short questionnaire. The experiment found that believers in mediumship were more likely to report the movement of objects that, in fact, had remained stationary (Wiseman 1997, 265). It is likely that the many striking claims of Spiritualists are the result of this type of unconscious exaggeration.

Not all mediums were fraudulent. In his book *The Psychic Mafia*, M. Lamar Keene wrote of his experiences as a highly successful fraudulent medium from 1958 to 1971. Early in his career, Keene discovered that fraudulent mediums classified their peers in three categories—"open eyes," "shut eyes," and "open-to-yourself." The open-eyes mediums knew they were frauds and admitted it to others within the secret circle of other fraudulent mediums. The shut-eyes mediums genuinely thought they could pick up psychic vibrations. They were never let in on the tricks of the trade but were kept around because their obvious sincerity was good for public relations. The open-to-yourself medium recognized that some mediums were fraudulent but kept quiet about it because it strengthened the faith that others had in Spiritualism (Keene 1997).

It is likely that many of the shut-eyes mediums fit the pattern of the fantasy-prone personality. Such individuals, thought to make up about 4 percent of the general population, fantasize a large part of the time. They typically "see," "hear," "smell," "touch," and fully experience what they fantasize. For these individuals, their experiences seem fully real (Wilson and Barber 1983).

Keene reported that a sophisticated network of information sharing allowed mediums to reveal startling information to their clients. Keene worked at Camp Chesterfield in Indiana, a popular Spiritualist summer camp noted for its physical mediums. There, he learned that the camp maintained a large, secret library in the basement that held documentation on the camp's clients.

Most mediums can operate effectively without using any secret knowledge of their clients by employing a technique called "cold reading." This is a method of fortune-telling that is also used by astrologers, numerologists, and readers of tea leaves. Essentially, the cold reader gleans information from deductive reasoning and from his or her own unobtrusive observations. The medium will often rattle off a combination of vague and specific statements and look for a reaction. Eye movement, pupil dilation, the rhythm of breathing, the use of hands, body posture, and even unguarded comments will allow the medium to become ever more specific during the reading (Hyman 1977, 22). It seems clear that nonfraudulent shut-eyes mediums use cold reading, but they do so without any consciousness of the techniques they employ.

Today, the demonstrations of physical phenomena have nearly disappeared. Mediums who perform physical phenomena usually only do so before closed groups of steadfast believers, in an obvious attempt to eliminate the embarrassing exposures of the past. One group in England, the Noah's Ark Society, holds physical phenomena séances but only allows mediums to perform before its members. Skeptics are not welcome at these events, as it is claimed that their negative energy will adversely impact a medium's ability. The infamous Camp Chesterfield still advertises that mediums perform trumpet and apport séances (http://www.campchesterfield.net/medium.htm).

After its last popular peak in the 1920s, interest in Spiritualism is on the rise. Although only a small number of Spiritualist churches still exist, mediumship has captured the imagination of the popular culture. A number of mediums have written best-selling books and are active on the television talk show and lecture circuit. These mediums rely on cold reading as the basis for their acts, although many will use whatever information they can surreptitiously gather prior to a reading (see Shermer 1999, 54).

References:

Brandon, Ruth. 1983. *The Spiritualists: The Passion for the Occult in the Nineteenth and Twentieth Centuries.* New York: Alfred A. Knopf.

Braude, Ann. 1989. *Radical Spirits: Spiritualism and Women's Rights in Nineteenth Century America.* Boston: Beacon Press.

Fox Kane, Margaret. 1985. "Spiritualism Exposed: Margaret Fox Kane Confesses to Fraud." In *A Skeptic's Handbook of Parapsychology,* edited by Paul Kurtz, 225–233. Buffalo, NY: Prometheus Books.

Houdini, Harry. 1972. *A Magician among the Spirits.* New York: Arno Press.

Hyman, Ray. 1977. "Cold Reading: How to Convince Strangers That You Know All about Them." *Zetetic* 1, no. 2: 18–37.

Keene, M. Lamar. 1997. *The Psychic Mafia.* Buffalo, NY: Prometheus Books.

Podmore, Frank. 1963. *Mediums of the 19th Century.* Vol. 2. New Hyde Park, NY: University Books.

Shermer, Michael. 1999. *How We Believe: The Search for God in an Age of Science.* New York: W. H. Freeman.

Wilson, Sheryl C., and Theodore X. Barber. 1983. "The Fantasy-Prone Personality: Implications for Understanding Imagery, Hypnosis, and Parapsychological Phenomena." In *Imagery: Current Theory, Research and Applications,* edited by Anees A. Sheikh, 340–390. New York: Wiley.

Wiseman, Richard. 1997. *Deception and Self Deception: Investigating Psychics.* Buffalo, NY: Prometheus Books.

Stock Market Pseudoscience

JON BLUMENFELD

Technical, or "chart," analysis is any of a constellation of market trading strategies that rely completely on the historical price information of a security or group of securities to make predictions, with no reference to underlying (so-called fundamental) information, such as supply and demand, economic indicators, or current events. A chart is simply a graph of historical data, which may include any number of different pieces of information, such as opening prices, highs and lows of the day, and closing prices. More sophisticated charts may include secondary information calculated from the basic price information. Examples include moving averages, momentum indicators, and relative strength indexes (among many, many others). Some charts include information about commonly traded prices and volume of trading in order to capture psychologically important price levels.

Once a chart is made, a system is designed to make predictions based on past behavior—for example, by identifying certain shapes and features in the chart that have often been followed by major changes in "trend." A statistical analysis is commonly used to find a correlation between one or more "signals" and important changes in market behavior. Some systems attempt to make an analogy between market behavior and the behavior of some physical system, such as the variations of ocean waves or the phases of the Moon. The simplest systems consist of lines drawn directly onto the price chart of a security in an attempt to discover the timing of major market moves; more complicated systems incorporate mathematical tools such as Fourier analysis, Fibonacci numbers, chaos theory, and fractal geometry.

A hallmark of technical trading is that there is no room for subjectivity—the system makes all the decisions, with no need for human judgment. The field is marked by unsubstantiated claims of success; a track record of failure and associated rationalizations; a lack of peer-reviewed studies showing any positive results; and a collection of logical, mathematical, and statistical fallacies. Although there are plenty of instances of technical traders having success, there is no evidence that they are anything more than statistical artifact and the hidden use of standard, subjective, fundamental analysis. Technical traders fail to realize that the market would quickly overwhelm any truly successful system, rendering it useless.

According to the Managed Funds Association (MFA), the first technician in the field was Richard Donchian, who developed the initial "trend-following" systems in 1957. These systems are intended to indicate that a market is moving into a prolonged period of directional price movement—either up or down. Donchian's first system used moving averages; in the simplest of systems, trend lines are drawn directly onto the chart, and important price levels are read off in a fairly straightforward manner. When the first of

Stock market traders. (Yellow Dog Productions/The Image Bank)

these levels is breached, a move to the next level is indicated, and when the projected trend line is broken, a major change in direction is indicated.

Over the years, a plethora of derived indicators have been created, composite systems have been developed to compare groups of securities, and hand drawing has given way to computer-generated graphs, but the basic idea remains unchanged. The main theoretical justification for these systems is that the indicators give insight into market psychology, with the derived signal points showing price levels of major significance to market players.

Another type of technical analysis (which actually originated earlier than Donchian's trend-following systems) is the attempt to model markets after periodic physical systems, analogous to ones found in nature. Gann waves, Elliot waves, Fibonacci sequence–based systems [1, 1, 2, 3, 5, 8, etc. . . . where $f(n) = f(n-1) + f(n-2)$], and the more recent application of chaos theory and fractal mathematics generally fall into this category. Astrologically based systems have been growing in popularity in recent years, and the Market Technicians Association (MTA), a trade group, has recently recognized these systems as valid forms of technical analysis.

Although it is illegal for investment professionals to make exaggerated claims or guarantees of profitability, purveyors of technical systems are largely exempt from regulation unless they are actually offering to manage funds themselves. Most often, they are selling computerized systems for identifying trading signals, not actual trading advice. Claims (and guarantees) that winning trades are produced more than 90 percent of the time or annual returns of 1,000 percent or more are often advertised, but according to market regulators, the small investors or day traders who are most likely to use these systems lose money between 70 and 90 percent of the time.

It is common practice to attribute success to well-designed and strictly followed systems, whereas failure is written off through rationalizations. The most frequent reason cited for the failure of a system is the failure of the human trader to follow the system with absolute faith. Traders are constantly warned not to let their emotions lead them astray. Other common excuses are that the system itself is poorly designed or that the conclusions reached by the trader about signals are incorrect. In these days of high-tech automation, it is getting harder and harder to blame an individual system for failure while still maintaining the idea that technical analysis itself is valid.

Although some attempts have been made to draw a link between certain kinds of mathematical analysis and market behavior, economists and mathematicians generally agree that it is not possible to predict future market behavior from past data. The literature of the technical-analysis community reads like a combination of self-help manuals and get-rich-quick schemes, with just enough statistical analysis to provide a veneer of scientific-sounding words. Typical titles include *Self-Reliant Investing; Mind over Markets: Power Trading; Trading without Fear, How to Triple Your Money Every Year with Stock Index Futures;* and *Your Personal Computer Can Make You Rich in Stocks.*

Attempts to model or make theories about a system based only on data collected about past behavior are not, in and of themselves, necessarily pseudoscientific. Johannes Kepler derived his laws of planetary motion entirely from data recorded at the laboratory of Tycho Brahe, and it was not until some years later that Isaac Newton, with the inverse square law of gravity, confirmed Kepler's empirical laws theoretically. Unlike markets, however, the laws of planetary motion are relatively simple periodic functions with few variables.

It is always possible to take a given finite set of data points and derive equations that describe them. There are several mathematical tools for this, such as cubic splines and Fourier analysis. Although these tools can produce equations to match sets of data points, they are useless for predicting future points unless the data describe a function that is periodic, like the orbits of planets around the Sun. This periodicity has never been established for a financial market. Instead, most economists believe that markets are chaotic systems made up of many thousands of variables, extremely and unpredictably sensitive to any of them at any given time. There have been attempts to apply chaos theory to markets, but this is yet another misapplication of a much misunderstood technique. Chaos theory can help to illuminate the types of behavior possible in a system, but it is useless for making actual predictions.

Technical analysis often consists of an attempt to fit an equation to a set of points by a fine-tuning process known as "back-testing" and "optimization." This effort allows the technician to create an equation that fits the historical chart of a given market to any arbitrary degree of accuracy, but it establishes no link to any future performance of the market. In fact, it can be shown that there are many curves (actually, an infinite number) that can be fit to a given finite set of points, but none of them can be shown necessarily to make any kind of predictions about future movements.

Descriptions of the physical systems of the market, though seeming to come from a firmer theoretical footing, are subject to the same pitfalls, since the systems often have parameters that can be fine-tuned. It is possible to tweak a physical-description system so much that the original model is hardly recognizable. No physical system has ever been established as directly analogous to any market.

Many technicians are aware of these problems and try to design systems without back-testing and optimization. They make many attempts to find the best system for a market, repeatedly testing new systems against historical data. For instance, a simple system can be tested for buying and selling in response to certain external events, such as the phase of the Moon. The system can be tested for buying on the day of the full Moon and then selling on the day of the new Moon. This system can be compared to one that involves buying on the day *after* the full Moon and selling on the day *after* the new Moon and then two days after and three days after and so on. Some of these combinations will produce profits when applied to historical data, some will produce losses, and some winning combination will produce more profits than any other will. Through this analysis, technicians can claim that there is a weak correlation between any given market and the phase of the Moon, but this is a misunderstanding of statistics. In fact, the same analysis can be made with any periodic function (e.g., the phases of Venus, daily temperatures, the migration of birds), but the correlation is merely a statistical artifact and is useless for making predictions.

Despite these facts, the use of technical information is extremely widespread in the financial markets, and more important, some technicians have documented records of success. The most important reason for apparent success is that although subjective decisions are supposed to be eliminated from technical systems, "hidden" subjectivity does creep into the process. First, technicians believe that no system fits all market conditions and that the trader must choose a system based on the type of market currently observed. Markets may be trending up or down; they may also be "choppy," slow, or in any number of other states. Decisions about which system to use are highly subjective, as are decisions about the size of trades and when to cut losses or take profits. All of these subjective decisions are influenced by the trader's experience and knowledge about market conditions and current events, also known as "fundamental" information. The best "technical" traders may be "fundamental" traders in disguise.

Another reason for apparent success is simple luck. With many thousands of traders using technical systems, some are bound to have a degree of success merely by chance. Traders need to be right only a little more than 50 percent of the time to be successful, and many can succeed in this way for years.

Statistics on the success and failure of technical trading may themselves be misleading. For instance, it is widely claimed that nearly all technical traders with three-year or longer track records are winners, but this claim is biased by the fact that traders with losing records rarely have the opportunity to develop long-term track records.

Finally, some of the predictions of technical analysis, particularly those in the shortest time frame (minutes or hours rather than days, weeks, months, or years), become self-fulfilling prophecies. Many traders have access to similar short-term technicals, and in the absence of new fundamental information, these technicals may have some power. If 500 traders in the bond futures pit in Chicago all believe that a certain price is significant, that price takes on some temporary importance. In addition, some markets have traded in patterns at certain times. For example, it has long been known that certain food and energy commodities move in seasonal patterns, and traders who first noticed this were able to take advantage of that fact, but once information like this becomes widely known, the market tends to account for it in such a way as to make it impossible to make money from possessing the information. A popular maxim is that one of the functions of a marketplace is to destroy information—seasonals are an example of information that has long since been destroyed. Any other successful technical system would also quickly be eliminated.

References:

Kovachs, L. D. 1982. *Advanced Engineering Mathematics.* New York: Addison Wesley.

Northcote, Thomas. 1996. *Major Events in the History of the Managed Futures Industry.* Managed Futures Association. Chicago: Chicago Mercantile Exchange.

Rowntree, Derek. 1981. *Statistics without Tears: A Primer for Non-Mathematicians.* New York: Charles Scribner's Sons.

Subliminal Perception and Advertising

REBECCA RUSH

Subliminal perception is visual or auditory information that is allegedly discerned below the threshold of awareness and has the power to influence human behavior. Promoted in the late 1950s by James Vicary and popularized decades later through the books of Wilson Bryan Key, subliminal advertising relies on persuasive techniques that are neither seen nor heard and are perceived only by the subconscious. Subliminal messages can take the form of television or film images that appear on the screen for too short a duration to be consciously recognized, subauditory sounds that are buried beneath layers of music, and words or images embedded in pictures. In order for subliminal perception to take place, three factors must be present. First, the message must be deliberately and subtly inserted into some form of media. Then the information must be subconsciously perceived. Finally, the perception must influence the viewer or listener to act upon the message.

The concept of using subliminal perception to induce action gained attention in 1957 when Vicary, a market researcher, declared that movie patrons in Fort Lee, New Jersey, increased their consumption of Coca-Cola by 18 percent and popcorn by 58 percent after being exposed to the messages "Eat Popcorn" and "Drink Coca-Cola." These phrases were flashed on the screen during the film *Picnic* for a duration of 1/3,000 of a second every 5 seconds, and Vicary claimed the presence of these messages caused an increase in sales at the concession stand. The results of his study were never published, and in a 1962 interview with *Advertising Age,* Vicary recanted his findings as fraudulent. By that time, however, the power of subliminal messages was already being touted in Vance Packard's book *The Hidden Persuaders,* and governmental attention and further research were being focused on the possible uses of subliminal perception.

Banned by the National Association of Broadcasters in 1958, subliminal messages were believed to be a danger to the public, and concern over the matter provoked William Dawson, a Republican representative from Utah, to call for an investigation by the Federal Communications Commission (FCC). Meanwhile, early tests in the realm of subliminal perception yielded definitive results showing that the practice had little impact on television viewers. WTWO, a television station in Maine, inserted the message "If you have seen this message, write WTWO" in September 1957 for 1/80 of a second every 11 seconds. The station received no additional mail during the two-week trial. In February 1958, another television station, the Canadian Broadcasting Company, attempted to get viewers to use their phones during a broadcast by flashing the words "Telephone Now" 352 times in 30 minutes. Afterward, they contacted 500 viewers, only 1 of whom experienced an urge to use the phone; others, unaware of what message they had received, claimed to have felt hungry or thirsty during the program.

Subliminal advertising gained further recognition in the 1970s due to Wilson Bryan Key's book *Subliminal Seduction.* Key published the work in 1973 and followed it up with additional texts on the subject, all of which advanced the claim that advertisers were engaged in attempts to persuade consumers to purchase goods through the use of hidden images, or embeds. His work specifically detailed the use of subliminally perceived erotic imagery and suggestive words. Among his most notorious finds were the word *SEX* hidden in a number of locations (including Ritz crackers) and sexual images embedded in photographs of ice cubes, chicken nuggets, and a plate of clams. Key's ability to find licentious pictures in a multitude of seemingly innocent objects exemplifies the human tendency to find patterns and make meaning out of a series of random shapes and images. To train individuals to detect subliminal images such as clam-plate orgies and phallic ice cubes, Key would have them prepare by identifying the images that materialized as they stared at a sky filled with clouds.

The resurgence of a widespread belief in subliminal manipulation during the 1970s can be attributed, in part, to a growing distrust in authority. It was during this time that the Watergate scandal was made public, and Americans experienced increased suspicion in government, business, and the media. Similarly, public acceptance of the power of subliminal perceptions in the 1950s was influenced by a number of factors that played into apprehensions surrounding the Cold War. The movie *The Manchurian Candidate*, released in 1962, also fueled fears of mind-control and brainwashing techniques perpetrated on the American people in order to influence their behavior. By the 1990s, subliminal perception entered the New Age as consumers sought subliminal means of achieving self-help goals through specially produced cassette tapes.

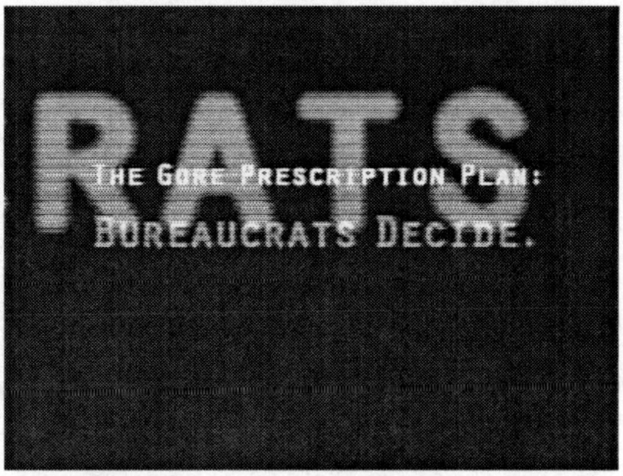

The word *RATS* appears in a frame in a Republican television ad that attacks Democratic nominee Al Gore's Medicare plan. (Reuters NewMedia Inc./CORBIS)

These self-help tapes marked a shift from public wariness concerning the use of subconscious messages to the acceptance of employing subliminal techniques for personal gain. These tapes claim to contain subliminal messages that will help listeners overcome smoking addictions, lose weight, and improve memory; retail outlets have also attempted to discourage shoplifters through broadcasting subliminal messages over public-address systems. The subauditory messages in subliminal recordings are so faint that they cannot be perceived and are often accompanied by music or nature sounds that drown out any audible message. In 1990, Anthony R. Pratkanis and his colleagues performed a study, supported by subsequent research, that concluded that listeners given cassettes assumed to contain hidden messages perceived the tapes to have been effective regardless of whether or not they contained actual subliminal messages. In this case, the tapes produced a placebo effect in which the desired behavior was influenced by expectations rather than subliminal methods.

Fears and misconceptions surrounding the potential of subliminal messages continue to infiltrate our culture, and claims have been made that sexual embeds are present in many

Disney films. These alleged subliminal images include the semblance of male genitalia in a tower on the cover of *The Little Mermaid* and a cloud of dust that appears to spell out *SEX* in the film *The Lion King*. Although these images, among others, may be attributed to playful graphic artists and therefore can be said to have been deliberately included, they are also readily apparent to the viewer and do not require perception below the threshold of awareness. In addition, the images fail to produce a message provoking action and instead appear to be either the work of animators seeking to leave their mark or a result of unintentional patterns that resemble a coherent image. Even more recently, during the 2000 presidential election campaign, the Republican Party was charged with using subliminal techniques in a television commercial that featured the word *RATS* following the phrase *BUREAUCRATS DECIDE*. Whether or not the word was included intentionally to encourage the voters to associate Democrats with rodents, the image remained on the screen long enough to be perceived by the conscious mind and therefore is not considered a subliminal message.

References:

Dixon, N. F. 1971. *Subliminal Perception: The Nature of a Controversy*. London: McGraw-Hill.

"FCC Information Bulletin on Subliminals." *Parascope*. URL: http://www.parascope.com/articles/0497/sublimdb.htm. (Accessed on January 29, 2001).

Haberstroh, Jack. 1994. *Ice Cube Sex*. Notre Dame, IN: Cross Cultural Publications.

Key, W. B. 1973. *Subliminal Seduction*. Englewood Cliffs, NJ: Signet.

——. 1980. *The Clam-Plate Orgy*. Englewood Cliffs, NJ: Prentice-Hall.

Onion, Amanda. "Mind Games: Subliminal Ads Mostly Ineffective, but Americans Think Otherwise." *ABC News*. URL: http://www.abcnews.com. (Accessed on January 29, 2001).

Packard, Vance. 1957. *The Hidden Persuaders*. New York: Pocket Books.

Pratkanis, A. R. 1992. "The Cargo-Cult Science of Subliminal Persuasion." *Skeptical Inquirer* 16: 260–272.

Pratkanis, A. R., and E. Aronson. 1992. *Age of Propaganda: The Everyday Use and Abuse of Persuasion*. New York: W. H. Freeman.

"The Subliminal Scares." *Parascope*. URL: http://www.parascope.com/articles/0397/sublim1.htm. (Accessed on January 29, 2001).

Sun Sign Astrology

GEOFFREY DEAN, IVAN W. KELLY,
ARTHUR MATHER, AND RUDOLF SMIT

These days, superstition is a boom industry and comes in many guises. Some are strange and unfamiliar, but everyone seems to know about sun signs, also called birth signs or star signs. Your sun sign is the sun's position in the tropical zodiac when you were born, and it is supposed to describe your character, your abilities, and your relationships. Sun sign astrology is an oversimplified astrology (see the "Astrology" entry in this encyclopedia) that became popular in the 1930s. It generally requires only a birth date regardless of year and is easy to commercialize, so it has become by far the most common kind of astrology in the Western world. It has no validity whatever—in sun sign astrology, the only thing that matters is whether it sells.

Sun Sign Columns

In Western countries, most newspapers and almost all women's magazines carry sun sign columns, which pretend to tell those born under each sun sign what their character or future is. There are two types of columns—*forecasts* ("Aquarius, romance improves after the 16th") and *delineations* or attributes of each sun sign ("Taureans are stubborn"). Delineations include compatibility ("Geminis and Librans make beautiful music together"). Sometimes, the two types of columns overlap ("Ariens are born to win").

Forecasts and their associated dial-a-horoscope phone lines are common in newspapers (daily and weekly forecasts), women's magazines (monthly forecasts), and sun sign annuals (yearly forecasts). Delineations are almost as common, appearing in weekend supplements, women's magazines ("secrets of your man's star sign"), and books. Typically, half of the astrology titles on display in New Age bookshops are on sun signs; for example, recent titles include *Sun Signs, Star Signs, Baby Signs, Cat Signs, Diet Signs, Fun Signs, Life Signs, Love Signs, Money Signs, Sex Signs,* and *Success Signs.*

Popularity

According to opinion polls, typically 50 percent of the population read sun sign columns at least sometimes, but only 5 percent take them seriously, so they are mostly seen as entertainment. Nevertheless, 1 percent read them often and take them very seriously, like horoscope junkies unable to exist without their daily fix. For such people, horoscopes are anything but entertainment.

Sun signs are a modern invention. If the history of astrology is represented by a loaf of bread, sun sign columns do not appear until halfway through the last slice, forecasts being generally unknown before the 1930s and delineations before the 1960s. Until then, the only sign that could be legitimately considered on its own was the rising sign, which was the original source of the word *horoscope*, from the Greek *horoskopos*, or watcher of the hour. But sun signs rapidly became part of Western culture. Today, they are frequently used to promote the sale of goods such as clothing, jewelry, pillows, curtains, tableware, TV dinners, soft drinks, posters, calendars, stationery, and especially women's magazines (but not men's magazines) and sun sign books. A sun sign supplement in a newspaper can boost sales by more than 10 percent.

The mass marketing of sun signs has tied them to fixed dates regardless of year; for example, Cancer's dates are June 22 to July 22. But such dates are only approximations. If the calendar year exactly matched the solar year, the dates on which the sun changed sign would be exactly the same from one year to the next. But because of the slight mismatch that leads to leap years, the dates can be a day off, which is why dates in sun sign books sometimes disagree. Those born near a cusp can look up their exact sun sign in an astrological ephemeris (a calendar of planetary positions), in the tables given in some do-it-yourself astrology books, or in some sun sign astrology books, such as that by Sasha Fenton (1992) (see any astrology bookstore). Some astrologers say the attributes of each sun sign change abruptly at a cusp; others say the attributes change gradually so that people born near a cusp are a mixture. But because sun signs have no validity, the difference is of no consequence.

Delineations

Sun sign delineations set out basic astrological tradition, and they tend to be our first contact with astrology. We hear or read what our sun sign is supposed to mean, compare it with what we see in ourselves, and proceed from there. But look at the meaning of each sun sign from Aries through Pisces—*assertive, possessive, versatile, sensitive, creative, critical, harmonious, secretive, adventurous, prudent, detached, impressionable.* (These meanings are sometimes expressed as *I am, I have, I think, I feel, I command, I analyze, I balance, I desire, I see, I use, I know, I believe.*) Because we are interested only in our own sign, we fail to notice that these traits are universal—everyone behaves in each of these ways at various times. Similarly, no matter what our sign is, we can always find matching behaviors, so we will conclude (wrongly) that sun sign astrology works. Furthermore, we tend to use only confirming strategies: if astrology says a person is extraverted, we tend to ask that person extraverted questions ("Do you go to parties?") rather than introverted questions ("Do you read books?"). Because even introverts occasionally do extraverted things, the answers cannot fail to confirm astrology. So we will again conclude (wrongly) that sun sign astrology works.

Forecasts

In contrast to delineations, sun sign forecasts bear no relation to any astrological tradition. They can be derived in various ways, from simple sign symbolism (so Leos can expect Leonian events) to planetary emphasis (so Mars currently in your sun sign might indicate a busy period). Or they can be pure invention, which explains why many forecasts have no discernible link with astrology ("The letter E is important this week" or "Peace and tranquility are worth a thousand gold pieces"). Regardless of how they are derived, these forecasts attract readers more by their style than by their astrology, that is, by their capacity for conveying maximum generality with maximum sincerity.

The Precession Argument

The most common argument against sun signs is that, due to precession, they are moving further and further away from the constellations that gave them their names. So today's Virgos are actually Leos, and in due course, they will be Cancerians, Geminis, Taureans, and so on, becoming Virgos again by roughly A.D. 26,000. But the argument is invalid. In Western astrology, the signs are measured in the tropical zodiac, not the sidereal zodiac of the constellations. Tropical signs begin at the vernal point (0 Aries), the first moment of spring, so it makes no difference where the constellations are. Nevertheless, if signs begin in springtime, they should reverse in the Southern Hemisphere. But astrologers ignore this complication; for them, the signs do not reverse. So in Australia and Brazil and South Africa, supposedly wintry Capricorns are born in the heat of summer. Perhaps astrologers hope that nobody will notice.

Validity of Sun Signs

Does using sun sign astrology add validity to sun sign forecasts and delineations, as compared to simply making them up? Or does it merely mislead readers into believing that their "thought for the day" in a forecast or delineation is more meaningful than one in, say, a desk calendar? The verdict of half a century of research is clear and consistent: sun sign astrology has no validity whatever (Fichten and Sunerton 1983; Culver and Ianna 1988; Dean and Mather 1996, 2000). Indeed, formal studies can be superfluous—columns have appeared on the wrong day due to a filing error or because old columns were being recycled to save money, but readers noticed no difference. And we need only look around us to see that people absolutely do not come in just twelve varieties.

Among astrologers, the verdict is less clear. In fact, since the 1960s, violent arguments over sun signs have periodically erupted in astrological journals. The arguments invariably repeat the same issues, ignore research findings, and therefore achieve nothing (Dean and Mather 1996, 2000). Some astrologers see sun signs as valid and good publicity; others see them as nonsense and exploitation. Critics point out that astrology can hardly be taken seriously when astrologers themselves show such a major division of opinion over such a basic issue.

Newspaper Disclaimers

In 1984, the U.S.-based Committee for the Scientific Investigation of Claims of the Paranormal (CSICOP) urged newspapers and magazines to label their sun sign columns with a disclaimer saying they were for entertainment only and had no basis in fact. The 1,200 U.S. newspapers with horoscope columns were slow to respond—by 1986, 0.5 percent had adopted a disclaimer, rising to 5 percent by 1994 but no further by 2000. In 1987, no New Zealand newspapers adopted the disclaimer when urged to do so by scientists, but one did add the caveat "for entertainment," and two major dailies did change the title of their columns to "Stars for Fun." This suggests that disclaimers will not be adopted unless brief and to the point.

Sun Signs and Self-Image

Interestingly, a weak but statistically significant link between sun sign and extraversion was reported in 1978, advance notice of which was hailed by astrologers as "possibly the most important development for astrology in this century" (Phenomena 1977). But the effect disappeared when people unfamilar with sun signs were tested, so the finding had a simple explanation—prior knowledge of astrology. Ask Sagittarians (who are supposedly sociable and outgoing) whether they like going to parties, and their answer might be tipped by astrology in favor of yes rather than no. The bias may be unconscious and very slight, but it is there nonetheless. When combined with the findings of national opinion polls, the results suggest that roughly one person in four believes sufficiently in astrology to measurably shift their self-image in the corresponding direction.

Popularity Revisited

Why are sun signs, a mere fragment of astrology, so hugely popular, and why are they remembered when so much other information about ourselves is forgotten? The reason may lie in our search for personal identity, the way in which we see ourselves in the world. Modern living is characterized by change, speed, and a loss of spiritual values. In the old days, our clues to finding a personal identity were taken from stable family and social settings. Today, this stability is greatly reduced, and traditional clues may well be less important than clues provided by films, TV, celebrities, and the occult. Whatever we may think of sun signs, they provide millions of people with a rich source of clues for constructing their identities—names (nothing impersonal here), personality, lifestyle, romance, occupation, everything. Even if the clues are false, the belief in their truth can make them true in their consequences, so they become a self-fulfilling prophecy in the same way that a sound bank can collapse if people believe it is unsound.

As well as providing clues for our personal identity, sun signs have other attractive features. They address ourselves and our relationships in a positive and nonjudgmental way; they help us talk about ourselves, thus creating closeness; they require only a birth date and are easy to learn; they are perceived to be mostly true; and they are highly available—only weather forecasts are more pervasive. Sun signs are popular because they fill a need, are dead simple, and appear to work. They are also big business. No other system comes close.

References:

Culver, R. G., and P. A. Ianna. 1988. *Astrology: True or False? A Scientific Evaluation.* Amherst, NY: Prometheus Books. A clear and very readable critique by astronomers, with much useful information and the results of their own extensive sun sign tests. Concludes that astrology (including sun sign astrology) is neither scientifically sound nor scientifically useful.

Dean, G., and A. Mather. 1996. "Sun Sign Columns: An Armchair Invitation." *Astrological Journal* 38: 143–155. An expanded version is at URL: http://www.astrology-and-science.com/. The authors survey the history of sun sign columns, the results of tests, and the disagreeing views of astrologers as expressed in astrological journals. To try to advance the debate, they invite astrologers and interested scientists to submit new ideas for testing sun sign columns.

——. 2000. "Sun Sign Columns: Response to an Invitation." *Skeptical Inquirer* 24, no. 5: 36–40. An expanded version is at URL: http://www.astrology-and-science.com/. The responses suggest that the negative results achieved to date are unlikely to change. Perhaps the most telling response is from Australian philosopher William Grey, one of the few philosophers to have interacted with astrologers and to have initiated a national survey of belief in astrology: "Astrologers have had plenty of opportunity to establish the validity of sun sign astrology via double-blind tests. That they have not done so is most easily explained by the hypothesis that they cannot do so. Sun sign astrology is not knowledge but epistemological hallucination" (p. 38).

Fenton, S. 1992. *Sun Signs: Discover Yourself and*

Others through Astrology. London: Aquarian. One of many similar books that describe at length how people born under each sun sign are supposed to look, love, work, and play. The author is a British consulting astrologer who at the time was the sun sign columnist for *Woman's Own*. She describes her book as "probably the simplest book I have ever written" (p. 11). The time and date of sun sign changes during 1930–1993 are listed on pp. 226–228.

Fichten, C. S., and B. Sunerton. 1983. "Popular Horoscopes and the Barnum Effect." *Journal of Psychology* 114: 123–134. Tests using 366 Canadian college students showed that daily and monthly sun sign columns were neither valid nor in agreement.

Phenomena. 1977. Editorial comment. *Phenomena: The Bulletin of Astrological News & Information* 1, no. 1: 1.

Occasionally, books appear that claim to validate sun signs. But in every case, a critical examination has revealed mistakes and procedural blunders. A best-selling example is:

Sachs, G. 1998. *The Astrology File: Scientific Proof of the Link between Star Signs and Human Behavior*. London: Orion Books. The author uses samples of up to several millions of cases—samples so huge that even the most trivial of errors (for example, in matching the sample to controls) becomes enormously inflated in statistical significance. So his conclusions are misleading. Otherwise, the book is readable and well set out. Two critiques that expertly reveal the book's mistakes and procedural blunders appear in *Correlation* 17, no. 1 (1998): 41–49, and also at URL: http://www.astrology-and-science.com/.

Synchronicity

CHRISTOPHER BONDS

Synchronicity is the name given by the Swiss psychiatrist Carl G. Jung (1875–1961) to the phenomenon of two or more events that seem to be connected but are not causally related; it is also called an acausal connecting principle. Jung (1951) grouped synchronistic phenomena into three categories: (1) the coincidence of a mental state (idea, feeling, image) with a simultaneous external event that corresponds in some meaningful way to the mental state; (2) the same correspondence of a mental state and a simultaneous external, meaningfully connected event, with the latter being outside of the observer's field of perception; and (3) a meaningful coincidence of a mental state with some future event.

To illustrate synchronicity, Jung (1951) told a story about a woman patient whose treatment had come to an impasse. During a therapy session, the woman told him about a dream she had in which someone had presented her with a golden scarab (the scarab beetle is an Egyptian symbol of rebirth). At that precise moment, Jung heard a tapping at the window behind him, and when he went to open it, a large beetle (a rose chafer) flew in. He caught it and handed it to her. The incident broke the patient's impasse so that treatment could proceed.

Jung believed that synchronicity had both a philosophical and a scientific basis. He cited many historical antecedents of synchronicity in alchemy, astrology, and Chinese philosophy. He also drew upon the philosophies of Gottfried Leibniz (1646–1716) and Johannes Kepler (1571–1630).

Paranormal phenomena interested Jung from the beginning of his career. The work of parapsychologist J. B. Rhine in the 1930s appeared to prove the existence of extrasensory perception (ESP), which for Jung suggested an empirical basis for synchronicity. Discussions with physicist Wolfgang Pauli added to his conceptualization of synchronicity as a necessary organizing principle of the universe that was equal but in opposition to causality. Jung's theory was highly controversial from the first, and he often complained about being misunderstood. Although he tried very hard to present a strong case for synchronicity, his ideas were and are far from convincing to the skeptic. The main reason, from the scientific standpoint, is the difficulty or impossibility of any experimental confirmation that synchronistic events are qualitatively different from ordinary coincidences. Another reason is that synchronistic experiences are too open to a variety of interpretations. Almost any coincidence can be a synchronistic event if the observer thinks it is. Nevertheless, synchronicity continues to have great appeal in the New Age movement today (a search of the Internet will easily show this to be true) because of its paranormal quality and because it postulates a hidden meaning to existence—a bond or connection between the psyche and the material universe—that can only be known by intu-

C. G. Jung. (Fortean Picture Library)

ition. But from the scientific point of view, synchronicity is neither testable nor falsifiable and must be considered pseudoscience.

Synchronicity is closely related to Jung's theory of the structure of the psyche. According to Jung, the deepest layer of the psyche, shared by all humans, is the collective unconscious. This layer is structured by patterns of instinctual behavior called archetypes. The archetypes can never be observed directly or made conscious; they function as a kind of source code for outer manifestations such as symbols, myth, and religion, among others. Synchronistic phenomena occur when an archetype is "activated" or when it exercises a particularly strong effect on conscious behavior or thought. This is most likely to happen when the person is at a crisis state of some kind (as was the woman in the scarab story) or is emotionally preoccupied with something. When this crisis or preoccupation is accompanied by a corresponding lowering of the consciousness threshold (as in a so-called trance state), the stage is set for a synchronistic event.

To empirically test for synchronicity, Jung analyzed the horoscopes of 400 married pairs, in three sets collected over time, to determine if an "acausal connection" existed between the horoscopes and actual marriages. Although he found no significant statistical evidence of a connection, he noted that in each of the three batches, the most frequent arrangement of planets was one of three that are important marriage indicators according to astrology. With the help of a statistician, Jung calculated that the probability of such a thing happening was so small as to be inconceivable (1 in 62,500,000), and therefore synchronistic. In Jung's words, "It is nothing but a chance result from the statistical point of view, yet it is *meaningful* on account of the fact that it looks as if it validated [the case for astrology]" (Jung 1955, 1958).

Unfortunately, the scientist is not permitted to place any weight on experimental results that *appear* to support something they in fact do not. From the scientific standpoint, there is no meaning to the coincidence. Furthermore, the excitement about the particular chart configurations that came out on top has a suspiciously post hoc quality (not uncommon in astrology!). Jung was aware of all this, yet he claimed that statistics are simply unable to account for such exceptional cases; therefore, he said, a new principle had to be devised.

One may ask why Jung, who considered himself an empiricist, was not content with more objective studies of supposedly synchronistic phenomena, which would address the issue of "meaningfulness" from the standpoint of the subject's observable mental state. But given Jung's personal and philosophical background, the nineteenth-century interest in Spiritualism and the occult, and the state of neuroscience at the time, it is easy to see how he could find the idea of synchronicity not only attractive but also necessary to his psychology.

References:

Bishop, Paul. 2000. *Synchronicity and Intellectual Intuition in Kant, Swedenborg, and Jung.* Lewiston, NY: Edwin Mellen Press.

Faber, M. D. 1998. *Synchronicity: C. G. Jung, Psychoanalysis, and Religion.* Westport, CT: Praeger.

Falk, Ruma. 1986. "On Coincidences." In *Science Confronts the Paranormal,* edited by Kendrick Frazier, 43–56. Buffalo, NY: Prometheus Books.

Gallo, Ernest. 1994. "Synchronicity and the Archetypes." *Skeptical Inquirer* 18, no. 4: 396–403.

Jung, Carl G. 1950–1955. "Letters on Synchronicity." In *The Collected Works of Carl G. Jung,* vol. 18, *The Symbolic Life,* edited by Sir Herbert Read, Michael Fordham, and Gerhard Adler, 502–509. New York: Pantheon.

——. 1951. "On Synchronicity." In *The Collected Works of Carl G. Jung,* vol. 8, *The Structure and Dynamics of the Psyche,* edited by Sir Herbert Read, Michael Fordham, and Gerhard Adler, 520–531. New York: Pantheon.

——. 1955. "Synchronicity: An Acausal Connecting Principle." In *The Collected Works of Carl G. Jung,* vol. 8, *The Structure and Dynamics of the Psyche,* edited by Sir Herbert Read, Michael Fordham, and Gerhard Adler, 417–519. New York: Pantheon.

——. 1958. "An Astrological Experiment." In *The Collected Works of Carl G. Jung,* vol. 18, *The Symbolic Life,* edited by Sir Herbert Read, Michael Fordham, and Gerhard Adler, 494–501. New York: Pantheon.

Jung, Carl G., and Wolfgang Pauli. 1955. *The Interpretation of Nature and the Psyche.* New York: Pantheon.

Koestler, Arthur. 1972. *The Roots of Coincidence.* New York: Random House.

Mansfield, Victor, Sally Rhine-Feather, and James Hall. "The Rhine-Jung Letters: Distinguishing Parapsychological from Synchronistic Events." *Art Matrix-Lightlink.* URL: http://www.lightlink.com/vic/rhine.html. (Accessed on June 2, 2001).

Therapeutic Touch

LARRY SARNER

Therapeutic Touch is a form of vitalism especially popular among nurses in the United States during the last quarter of the twentieth century. Weaving together mystical and pseudoscientific traditions from the mid-twentieth, nineteenth, and even late eighteenth centuries, nursing professor Dolores Krieger and Theosophist Dora Kunz developed the technique, commonly known by its initials, TT. Utilizing "intentionality" for healing, practitioners claimed to treat illness and relieve symptoms by passing their hands over and around the body of a sick person, rarely if ever touching it. From its initial introduction in the mid-1970s, the practice of TT grew in acceptance and influence within U.S. nursing schools (and, to a lesser extent, in the United Kingdom) until the mid-1990s, when it was affected by some schismatic offshoots and also was vigorously challenged by skeptics and academics. TT went into drastic decline immediately after the appearance of a 1998 paper about it in the *Journal of the American Medical Association* (*JAMA*) and is not much heard of today, just a few years later.

Though called Therapeutic *Touch,* the practice rarely involved any physical contact apparent to the outside observer. The very earliest practice of TT did involve actual contact between practitioner and patient as a sort of *laying on of hands.* However, within a couple of years after being introduced in nursing, the underlying vitalistic beliefs became dominant, and those beliefs suggested that the practice did not require physical contact. Rapidly thereafter, the "touch" became metaphysical, if not metaphorical.

Practicum

Therapeutic Touch practitioners were trained to perform the technique in three basic steps: centering, assessment, and intervention. (Some teachers subdivided one or more of these steps; others added some posttreatment behavior.)

Centering was a meditative act wherein the practitioner would focus on and form an "intention" to help and heal. This act was entirely inwardly directed and had no apparent interaction with the person who was to be treated, though the expectation was that as a result, the practitioner would be more "open" to receiving and transmitting the healing energies that are supposedly all around us. The meditation was expected to benefit the practitioner directly as well.

Assessment was an attempt to manually determine the locale of a disturbance in a client's vital force (variously termed *prana, qi, ki,* or *human energy field*). By moving the hands through the vital field, changes in "sensory cues" would reveal abnormalities in the vitalistic nature of the subject. Such abnormalities were usually described as a localized area of imbalance, congestion, or depletion of the vital force.

Intervention, or treatment, would necessarily follow successful assessment and adequate centering. In treatment, the hands were moved through the vital field again, this time with the intention of smoothing imperfections, removing blockages, or reenergizing depleted areas. Some practitioners would even flick "excess energy" from their fingertips. At this stage, the practitioner purportedly became a conduit for healing energies.

For two decades following the introduction of TT, practitioners viewed each phase of the process as critically important. Trainers and practitioners alike felt that they could sense "something" while moving their hands around another's body. The sensation was variously described in terms like *tingling, pulling, throbbing, hot, cold, spongy,* and *tactile as taffy*. Using such subjective sensations, the practitioner would try to diagnose medical problems in the assessment phase. During the intervention that followed, the practitioner was intentionally "repatterning" the vital field to bring its energy into "balance" throughout, which purportedly would stem disease and allow the body to heal itself. Belief in these sensations was at the core of TT practice before 1998. But in that year, a widely publicized paper appeared in *JAMA* with research that revealed the incapacity of trained practitioners to sense even the presence or absence of a person's vital field. That research suggested that assessment was illusory and that intervention probably was as well.

According to Krieger, Kunz, and others, successful treatment through Therapeutic Touch always required a genuine intentionality to heal, and they thought that a strong enough intention to help or heal could overcome other deficiencies in technique (such as failure to accurately assess). Every person was viewed as possessing an innate ability to form such intentionality, and the ability was believed to be harnessed with training and translated into an actual therapeutic outcome. But after the *JAMA* article appeared, Krieger and many others began publicly to downplay the importance of assessment and intervention, and they refocused the practice of TT onto centering and intentionality. They paid a price for doing so, however, by sacrificing their claims that TT was a scientifically valid therapy. Since *intention to heal* is fundamentally nonfalsifiable (that is, an untestable concept), the epistemological focus on it meant that TT became inscrutable and irredeemably pseudoscientific.

Claims for Healing

As with so many other alternative therapies, Therapeutic Touch had a long list of symptomatic conditions, illnesses, and disorders for which it was reputed to be effective. Indeed, the list was so long and diverse it was fairly called a panacea. By 1998, it was claimed to treat pain, nausea, diarrhea, headaches (both tension and migraine), arthritis, inflammation, fever, thyroid imbalances, decubitus ulcers, edematous legs, psychosomatic illnesses, Alzheimer's, AIDS, menstruation, premenstrual syndrome, asthma, autism, stroke, coma, multiple sclerosis, Raynaud's, measles, sundry forms of cancer and infection, and the effects of trauma. A couple of organs were said to be "insensitive" to treatment with TT (the pituitary and pancreas—hence, the notable lack of effectiveness with diabetes), but no one in the TT practitioner community publicly speculated on the reasons behind these exceptions.

Because the principal mechanism of action was believed to be the body healing itself, it was predictable that claims would be made that Therapeutic Touch would aid, stimulate, or enhance the body's natural functioning. Reports of this entailed reducing anxiety, promoting infant-parent bonding, increasing milk letdown during breast-feeding, seeding skin grafts, accelerating healing for wounds and broken bones, comforting end-of-life care, and aiding "social communication." Most dramatic was Krieger's unsupported claim that TT had restored vital signs to clinically dead infants by activating the "pendulum swing" of cellular proteins.

Clinicians often cite the placebo effect to account for reports of healing (though not for the restoration of life) from practices such as Therapeutic Touch. TT, however, was unique in that its proponents claimed that the placebo effect was irrelevant. The only belief system supposedly at work was that of the practitioner, not the patient. This was consistent with the dominant role of the healer's intentionality propounded by Krieger and her colleagues.

Although it was often stated that Therapeutic Touch acted only beneficially (thanks to intentionality), a few proponents raised certain cautions. For example, imparting too much energy to an infant could cause hyperactivity; in adults, overtreatment could result in lightheadedness, headaches, or accelerated growth of tumors. Overall, though, both the effects and the risks of overtreatment were considered quite small.

Etiology and Epistemology

Therapeutic Touch has its antecedents and, in fact, owes much to vitalistic traditions worldwide. For political and cultural reasons, Krieger introduced the practice to nursing as a latter-day form of Christianity's laying on of hands, and she attempted to impute intentionality as the mechanism for the miraculous cures attributed to that religious practice. She succeeded in getting a number of nuns and Catholic nursing orders to endorse and practice TT, but in time, it became clear that TT was not at all in that religious tradition. One must look elsewhere.

Because of the inseparability of Dora Kunz, onetime president of the American Theosophical Society, and the development and promotion of Therapeutic Touch (which is sometimes referred to as the Krieger-Kunz method), an obvious place to look for origins is in Theosophist belief and history. Two vitalistic elements of Theosophy are particularly relevant to TT praxis. First, under founder Madame Helena Blavatsky and her immediate successor, Annie Besant, Theosophy in the late nineteenth century attempted to resurrect the animal magnetism of Franz Anton Mesmer, after a century of disrepute. Second, Blavatsky imported Hindu Spiritualism into Europe (and Besant did the same for the United States); its traditions included a life-force concept known as prana. TT equated Mesmer's magnetic fluid with prana (or what Krieger called "pranic current"), and the assessment and intervention phases of TT were remarkably similar to Mesmer's practices for the treatment of hysteria. Centering was unquestionably a meditative practice patterned on an oriental model, sometimes described as yogic. This fusion of East and West (yin and yang, as it were) was very Theosophical.

Another tradition adapted by Krieger and others was the Chinese vitalism known as *qi* (in Japan, *ki*). In *traditional Chinese medicine* (TCM), healers attempt to direct qi through a force of will. The interruption or, alternatively, the enhancement of the flow of qi along defined *meridians* is at the heart of much TCM practice. (Acupuncture, for example, is predicated upon this.) In spite of her Buddhist preferences and prejudice toward things Indian, even Krieger found the similarity of TCM to TT practice too close to ignore, and she often talked in terms of meridians. However, never too far from her roots (together with Kunz), she fit meridians into a Theosophical context, conceptualizing TT's meridians as passing in and out of energetic centers in the body like lines of force emanating from the ends of a magnet; they identified these centers as *chakras,* the Vedic word for the foci of the life force.

During the early years, Krieger actively tried to disguise the mystical origins of Therapeutic Touch in order to get more willing acceptance of the practice in nursing and elsewhere. She and others proposed many scientific-sounding explanations for TT's mechanism of action. The first proffered was oxygen uptake by hemoglobin (the word *prana,* translated as "breath of life," was equated with oxygen), then the therapeutic value of skin-to-skin contact (which was lost when the noncontact version came along). Other concepts advanced were "electron transfer resonance," stereochemical similarities between hemoglobin and chlorophyll, electrostatic potentials influenced by healer brain activity, and unspecifiable concepts from quantum theory.

Though Krieger would later eschew all sciencelike explanations for Therapeutic Touch, overtly presenting the mysticism that underlies the practice, scientific-sounding explanations such as those mentioned earlier seemed a necessity at the beginning. Krieger found none better than the beliefs and teachings of nursing theorist Martha Rogers. Rogers was dean of the New York University (NYU) School of Nursing at the time Krieger got her Ph.D. there and while Krieger was on the teaching faculty. From that post, Rogers held sway over the largest and arguably most prestigious and influential of U.S. nursing schools; she used her position to further her own philosophy of nursing—the *science of unitary man* (later dubbed a more politically correct science of unitary human beings and more recently Rogerian science). In this "grand nursing theory," she speculated that humans were actually human energy fields (HEFs), and as such, a person's body extended beyond the skin. HEFs were constantly interacting with the surrounding environmental field (which included the HEFs of others). Adapting such ideas offered TT many entrées into nursing, but it was a purely political path. In truth, Rogerianism had no more factual or empirical basis than did TT, and it had no more real basis for being considered a scientific theory, much less an entire system (or science). That fact deterred neither Rogers nor her disciples (called Rogerians), including Krieger at the time, from using the label to promote their beliefs in various forms of "energetic" medicine. Right through the 1980s, Krieger often cited Rogers as providing a rationale for TT, and she made a special point of using the term *HEF* with the general public. For her part, Rogers oddly was lukewarm to the notion that TT was an example of applied Rogerian science, but Rogerians in general embraced TT as their own, unreservedly so after Rogers's death in 1994.

History

The Theosophical version of the laying on of hands was introduced to nursing by Dolores Krieger at a workshop for credulous nurse researchers hosted by the American Nurses' Association (ANA) at their 1973 annual convention. Before then, perhaps as early as 1965, it was a little-noted fringe activity engaged in by Theosophists at their retreat in Pumpkin Hollow, just north of New York City. Dora Kunz, who claimed to be a fifth-generation *sensitive* (that is, spirit medium or channeler), had been conducting "experiments" in the laying on of hands with a Hungarian expatriate by the name of Oskar Estabany. Estabany was the popular subject for many paranormal investigators in the 1960s, including Kunz and a new Ph.D. in nursing, Dolores Krieger. In 1971, Krieger and Kunz concluded they had found an objective measure of Estabany's abilities—

serum hemoglobin counts—and Krieger was able to get the ANA to listen in 1973. It was the beginning of a long association between the ANA and Krieger's followers. By early 1975, Krieger had a cover story on the newly renamed "Therapeutic Touch" in the ANA's flagship journal, the *American Journal of Nursing* (*AJN*).

Krieger immediately followed her 1973 paper with courses in the technique at NYU, playing off Martha Rogers's theory of nursing. She renamed the practice Therapeutic Touch because the term *laying on of hands* was considered an obstacle to acceptance by "curriculum committees and other institutional bulwarks of today's society" (Krieger 1975b). With the 1975 *AJN* appearance, TT course offerings were firmly established, and something of a cultlike following arose at NYU, where a detractor labeled them "Krieger's Krazies" (a label immediately co-opted by Krieger herself and turned to her advantage).

Another round of favorable articles appeared in *AJN* during 1979, just as the first of a series of TT-related dissertations were being approved at NYU. These articles and dissertations, along with a book by Krieger, revealed that since at least 1976, a variant of Therapeutic Touch, called *noncontact* TT, had been actively pursued by Krieger and her "Krazies." In 1982, one of these followers, Janet Quinn, got her Ph.D. with a seminal dissertation purportedly establishing with research that there was no difference between the contact and noncontact forms of TT. After that time, the contact version of TT faded from view and practice.

Throughout the 1980s, growth in the movement was the order of the day. Krieger and NYU landed research grants for investigations into Therapeutic Touch, all reporting its successful application to a wide range of illnesses and conditions. More books were to appear, first by Krieger and then by her disciples. Graduate students and new Ph.D.s fanned out into other nursing schools around the United States and Canada, with new TT-related doctoral dissertations and master's theses appearing wherever they went. For example, like a latter-day Johnny Appleseed, Janet Quinn went to the nursing faculties at the Medical University of South Carolina and then to the University of Colorado Health Sciences Center; each school immediately became a center for TT activism, with grants, papers, dissertations, and theses flowing liberally.

By the early 1990s, it could be fairly said that Therapeutic Touch was entrenched in nursing. One of the nursing profession's largest periodicals, *RN*, was publishing favorable articles about TT practice on a regular basis. Energy-field disturbance was recognized by the North American Nursing Diagnosis Association, with TT as the only treatment available. The Order of Nurses of Quebec declared TT was a bona fide nursing skill. One regulatory board, the Colorado State Board of Nursing, even declared it to be part of "mainstream" nursing. Continuing-education credits (CEUs) for nursing-license renewals were routinely granted for courses in and about TT. The National League for Nursing, the erstwhile credentialing agency for nursing schools in the United States, actively promoted TT with books, audiotapes, and videotapes. And the American Nurses' Association continued to hold workshops in TT, complete with CEUs. By middecade, numerous hospitals and private clinics nationwide and in the United Kingdom offered TT as treatment, and some 43,000 health professionals (mostly nurses) reportedly had been trained in this area, with at least half said to be actively practicing.

At its high-water mark, Therapeutic Touch was being taught in at least 100 nursing schools at colleges and universities in 75 countries, sometimes at the graduate level. Master's and doctoral degrees were being granted regularly on the basis of theses on TT. Some community colleges had frequent and extensive offerings for introducing new laypeople and professionals alike to the technique; the latter were able to get continuing-education credits toward license renewals. It was practiced by nurses in at least eighty hospitals in North America, some of which even went so far as to establish a "Department of Energy." The policies and procedures books of some institutions recognized TT as a nursing intervention, though often it was practiced without the permission or even knowledge of attending physicians. Krieger and Kunz established a trade group, Nurse Healers and Professional Associates Cooperative (NHPA), which at one time had more than 1,200 members.

Therapeutic Touch was as much a vehicle for the promotion of Theosophy as anything else, and Krieger acted as an evangelist, repeatedly stating that TT could be learned and practiced by anyone. This led her to resist the imposition of standards that might act as impediments to TT's ability to spread Theosophist philosophy. But not everyone who took up TT had Theosophist leanings, and Krieger's slackness in control inevitably led to copycats, heresies, and, as Krieger fought back, even schisms. An early Canadian follower, M. J. Bulbrook, joined with Krieger expatriate Janet Mentgen to start Healing Touch, an eclectic form of TT. Barbara Brennan coupled TT practice with her beliefs in auras (equated with human energy fields), spirit guides (channeling), and *hara* (intentionality) to create Brennan Healing Science (also known as Hands of Light). Chiropractic, always on the prowl to exploit fads for practice building, even constructed their own version of TT (Touch for Health).

Krieger at first tolerated all the apostasies, but she later regretted not dealing firmly with them as competitors. She had unkind words to say about Healing Touch in particular, after the latter beat out TT for exclusive certification offerings by the American Holistic Nursing Association. Krieger eventually attempted to use her established trade group, the NHPA, as a vehicle to impose guildlike control over the practice of TT (even establishing testing and certification of her own, which she had previously eschewed). But it was to no avail; the genie was already out of the bottle.

Research

Starting with the first public disclosure of Therapeutic Touch to the ANA and right through to the late 1990s, Krieger and Kunz claimed there was scientifically valid evidence for the practice. Indeed, through 1997, there were seventy-six reports of quantitative research into TT over twenty-five years, little of it in peer-reviewed journals (none in journals of the first rank, even within nursing). Of these, twenty-four were self-identified as unsupportive (statistically insignificant, small samples, inconclusive, or results negative to TT); included in that number were four failed attempts at replicating earlier experiments. As for the remaining fifty-two reports, at most only one may have demonstrated independent confirmation of any prior study. All of the quantitative research was clinical studies. In spite of the panacea-like claims for Therapeutic Touch, clinical research actually fell into just six broad categories: hematology, metabolic change, analgesia, effects on practitioners, and patients' mental states. Just a handful of papers reported on any other areas.

Krieger's first papers on her experiments in the early 1970s purported to show statistically that Therapeutic Touch elevated serum hemoglobin. The effect was deemed important because hemoglobin provides oxygen uptake for animal organisms, and prana is "intrinsic to the oxygen molecule." (She explained this was similar to the role of chlorophyll for plants, which did not surprise her given their "biochemical similarities.") Critics roundly attacked the report and the experiment: the samples were too small and not randomized; the statistical measures were inadequate and used inappropriately; hemoglobin is a measure not of blood's oxygen *uptake* but capacity; and the comparison with chlorophyll was "inexplicable." Moreover, subsequent studies found no significant relationship between TT and increased hemoglobin values or transcutaneous oxygen blood gas pressure. Krieger defended her statistics against her contemporaneous critics by attacking the Table of Random Numbers because it had been developed for use in the study of fertilizers, not human beings. "As any nurse knows," she concluded, "human beings are of a very different order of things" (Krieger 1975c). In due course, Krieger eventually acknowledged the difficulties with the study, but she defended herself as doing "the best I could with the knowledge available" (Krieger 1993).

In 1979, Patricia Heidt became the first of Krieger's Krazies to receive her doctorate from NYU with a dissertation showing TT's calming effect on anxious hospitalized patients. Though it, too, had its failings according to critics, it established academic respectability for the practice; it is not coincidental that the area of relaxation was the most researched in TT. That respectability was enhanced with the publication of Janet Quinn's dissertation in 1982, which had at least the appearance of replicating and confirming Heidt's results on a similar population.

Quinn's dissertation was arguably the most important publication for Therapeutic Touch in its history. Using a similar condition and population as did Heidt, Quinn replaced the laying-on-of-hands technique with the noncontact form of TT. The statistical analysis yielded results strikingly similar to Heidt's. In one stroke, apparently both TT's efficacy was established and the hands-off form of therapy was validated. There was also a third accomplishment: a placebo protocol had been introduced for noncontact TT.

Research design was a problem for all clinical researchers of Therapeutic Touch. Adequate control for the placebo effect vexed every one of them. Heidt had "casual touch" as a placebo, using pulse taking for her control. The hands-off TT investigated by Quinn required something more imaginative. Her alternative was simplicity itself: mimic TT with the same hand motions but prevent the placebo practitioners from forming the requisite intentionality by having them count backward by sevens from 100. It immediately became the standard placebo used in TT research. Within a few years, however, a series of failed replications had Quinn questioning whether that approach actually worked. Still a few more years after that, and another TT-Rogerian Ph.D., Thérèse Meehan, was dismissing it altogether. Even so, nothing was ever proposed to replace it, and TT research protocols continued to use it.

Beginning in the 1980s, grants flowed into TT research from public and private sources—the National Institutes of Health, the Public Health Service, the paranormalist Institute of Noetic Sciences, the eclectic Kellogg Foundation, and the Department of Defense. The latter awarded a notorious $335,000 grant for a burn-pain study that was tinged with scientific misconduct and ultimately yielded inconclusive results.

The results of Therapeutic Touch research were often disappointing and always controversial. From the beginning, critics were constantly sniping at the small sample sizes, the marginal statistical significances, the frequent appearance of "pilot" studies, and the use of phenomenological "qualitative" research (essentially anecdotes). In addition, all of the research investigated clinical effectiveness; fundamental research into the underlying concepts was never done. Without clear-cut evidence to back up their claims, TT supporters began cherry-picking from available studies. One such episode involved a shadowy figure from California who, in 1989, claimed to have discovered a promising effect of TT on the healing of wounds, attributing the effect to "subtle energies." This claim was widely circulated as the long-sought "proof" for both efficacy and underlying mechanism; Krieger even joined the organization that published the paper, the International Society for the Study of Subtle Energies and Energy Medicine (ISSSEEM). It turns out the study was just the first in a series of related investigations, and it ended in 1994 with the author declaring it "inconclusive." The first paper nonetheless continued to be enthusiastically cited by TT proponents.

Eventually, even some prominent advocates began to question what they were doing. In 1995, Meehan, who had widely published about TT, dismissed the entire corpus of TT research to that date: there was no convincing evidence beyond placebo for relaxation or anxiety relief, the effects on pain were unclear and needing replication, and the remaining claims were mere "speculation." She followed her critique with a prophetic prescription: "The [academic] charge of pseudoscience strikes at the heart of nurses' intellectual and scientific integrity and, sadly, is not unfounded. . . . The quest for truth about TT requires that proponents . . . engage in exacting scholarship and force themselves to search for evidence to refute their beliefs and experiences. This requires courage and self-discipline. But it must be done" (Meehan 1995).

Eschatology

Therapeutic Touch's web of influence began to unravel in Colorado, where Janet Quinn had been making substantial headway in having it accepted by the nursing establishment. In reaction to the Colorado Board of Nursing's endorsement of TT as mainstream nursing, a state senate committee held up reappointments of board members for a month until it secured a promise that the board would be more discreet in the future. Within a year, the board had repealed the continuing-education requirements that had been sustaining TT's and Healing Touch's professional visibility (the latter being headquartered in Colorado).

This setback was quickly followed by a challenge to the teaching of Therapeutic Touch at the School of Nursing at the University of Colorado. It survived, but a blue-ribbon investigative committee delivered a devastating critique that stripped the academic veneer from the practice: "There is not a sufficient body of data, both in quality and quantity, to establish TT as a unique and efficacious healing modality" (Claman et al. 1994).

There was a lingering hope that these setbacks could be reversed nationally when the dean of Colorado's School of Nursing (a vigorous TT supporter) became president of the National League of Nursing; it was hoped she could insinuate TT into nursing-school curricula through the league's accreditation function. But she had to resign during a crisis in which federal authorities were moving to strip the league of that valued function.

The coup de grace for Therapeutic Touch came on April 1, 1998. On that date, the *Journal of the American Medical Association* published a report on the experiments of an eleven-year-old Colorado girl, Emily Rosa, on TT. In a simple experiment first performed as a fourth-grade science-fair project, Rosa placed twenty-one TT practitioners behind a cardboard screen, with their hands extended through it; each subject was asked ten times to identify which of their hands the experimenter was hovering over with her own. Statistically, the subjects' answers were no better than guessing. TT practitioners were thus unable to show they could reliably detect the presence of a human being in a controlled setting. An exhaustive literature analysis was part of the published paper and revealed the thinness of the prior experimentation on TT, just as Meehan had feared. The massive publicity attending the publication in such a prestigious journal was devastating. Hundreds of print articles, TV appearances, and radio interviews exposed both TT and nursing to exceptionally unfavorable coverage regarding their lack of scientific rigor.

The *JAMA* paper was followed in July by another lengthy review of Therapeutic Touch research and practice by Meehan in the British *Journal of Advanced Nursing.* While toning down her 1995 rhetoric, she effectively confirmed the literature findings in the *JAMA* paper and concluded, as she had before, that TT was "intrinsically interrelated" with placebo. Coming when it did, Meehan's critique helped mute the outraged reaction of TT advocates to the *JAMA* affair.

TT never fully recovered from the series of blows coming from Colorado. After the *JAMA* appearance, a small number of pending clinical studies of TT were rushed into print, but no experimental refutation of the Rosa study was ever offered. (The Colorado-based ISSSEEM squelched an attempt to do so from within its

Reiki healing at 1997 Festival of Mind Body Spirit, London. (Guy Lyon Playfair/Fortean Picture Library)

ranks.) The failings of prior research were generally acknowledged through silence. Continuing-education offerings in TT dropped precipitously, as did courses in nursing schools. Nurses throughout the country challenged the use of TT in their institutions, and it was quietly dropped from policy and procedure books. Departments of Energy disappeared from hospitals that had them.

But while TT was apparently dying, vitalism in nursing was not. Reiki, a philosophically identical practice originating in Japan, tries to fill the vacuum left by TT. Ironically, the only thing that distinguishes Reiki from Therapeutic Touch is that it involves actual touch.

References:

Brennan, B. A. 1987. *Hands of Light: A Guide to Healing through the Human Energy Field.* New York: Bantam.

Claman, H. N., et al. 1994. *Report of the Chancellor's Committee on Therapeutic Touch.* Denver: University of Colorado Health Sciences Center.

Heidt, P. R. 1979. "An Investigation of the Effects of Therapeutic Touch on Anxiety of Hospitalized Patients." Ph.D. diss., New York University. Later published in part as "Effect of Therapeutic Touch on Anxiety Level of Hospitalized Patients." *Nursing Research* 30, no. 1 (January–February 1981): 32–37.

Krieger, D. 1973. "The Relationship of Touch, with Intent to Help or to Heal, to Subjects In-Vivo Hemoglobin Values: A Study in Personalized Interaction." In *Proceedings of the Ninth ANA Nursing Research Conference,* edited by the American Nurses' Association, 39–59. New York: American Nurses' Association.

——. 1975a. "Therapeutic Touch: The Imprimatur of Nursing." *American Journal of Nursing* 75, no. 5 (May): 784–787.

——. 1975b. "Therapeutic Touch: An Ancient but Unorthodox Nursing Intervention." *Journal of the New York State Nursing Association* 6, no. 2 (August): 6–10.

——. 1975c. "Response to Letter." *American Journal of Nursing* 75, no. 8 (August): 1282.

——. 1993. "Research Backs Therapeutic Touch" (letter). *Bristol Press,* April 21.

Meehan, M. T. C. 1995. "Quackery and Pseudo-Science" (letter). *American Journal of Nursing* 95, no. 7 (July): 17.

——. 1998. "Therapeutic Touch as a Nursing Intervention." *Journal of Advanced Nursing* 28, no. 1 (July): 117–125.

North American Nursing Diagnosis Association. 1995. "Energy Field Disturbance." In *Nursing Diagnosis: Application to Clinical Practice,* 6th ed., edited by L. J. Carpenito, 355–358. Philadelphia: Lippincott.

Quinn, J. F. 1982. "An Investigation of the Effects of Therapeutic Touch Done without Physical Contact on State of Anxiety of Hospitalized Cardiovascular Patients." Ph.D. diss., New York University.

Rosa, L. A. 1994. "Therapeutic Touch: Skeptics in Hand to Hand Combat over the Latest New Age Health Fad." *Skeptic* 3, no. 1 (Fall): 40–49.

Rosa, L. A., et al. 1998. "A Close Look at Therapeutic Touch." *Journal of the American Medical Association* 279, no. 13 (April 1): 1005–1010.

Sarner, L. W. 1998. "The 'Emily Event': Emily Rosa and the Therapeutic Touch Wars." *Skeptic* 6, no. 2 (Fall): 32–34.

Turner, J. G., et al. 1998. "The Effect of Therapeutic Touch on Pain and Anxiety in Burn Patients." *Journal of Advanced Nursing* 28, no. 1 (July): 10–20.

Wirth, D. P. 1989. "The Effect of Non-contact Therapeutic Touch on the Healing Rate of Full Thickness Dermal Wounds." *Subtle Energies* 1, no. 1: 1–20.

Wirth, D. P., et al. 1994. "Non-contact Therapeutic Touch and Wound Re-epithelialization: An Extension of Previous Research." *Complementary Therapies in Medicine* 2, no. 4 (October): 187–192.

Tunguska

ALAN HARRIS

Early in the morning of June 30, 1908, in a remote region of Siberia about 1,000 kilometers (600 miles) north of Irkutsk, an immense explosion occurred, centered about 8 kilometers (5 miles) over the forest below. Trees were flattened across an area approximately 50 kilometers in diameter, about the same size as the greater urban area of New York City. Trees near the epicenter of the event were ignited, and a forest fire covered the inner third or so of the area of devastation. Sonic booms were heard as far away as 1,000 kilometers; magnetic disturbances were recorded in the area; the barometric effect from the explosion was felt worldwide; and a seismic disturbance equal to a magnitude 4.5–5.0 earthquake was recorded. So sparsely populated was this region that apparently no one was killed. There were enough witnesses within a few hundred kilometers of the event to establish that the cause of the explosion came from the sky—apparently a giant meteoroid slamming into the atmosphere from the southeast.

At the time, there was little controversy over the nature of the event. The fact that "stones fall from the skies" was established a century earlier, and by 1908, it was well accepted. It was logical to assume that this was just an enormously larger-than-average meteoritic event. Unfortunately, these were troubled times in Europe generally and in Russia in particular, with World War I on the horizon and the Russian Revolution soon to follow. So, it was not until 1927 that the first scientific expedition, led by L. A. Kulik, visited the region. Kulik expected to map out the ground damage and collect meteorite specimens. In this and successive expeditions, he and his successor, E. L. Krinov, did the former in great detail. They discovered a large area where most trees had been blown down radially outward from an epicenter, with evidence for a forest fire covering a central area about equal to that inside the Washington, D.C., Beltway. But they failed to find any traces of meteoritic material. And thus began the mystery.

Briefly, the controversy can be summed up by the legal term *habeas corpus* ([you] have the body). All of the observed phenomena corresponded to expectations for a giant meteor, except that no "body" could be found. After years of searching for meteoritic debris, Krinov and other Russian researchers came to the conclusion that the Tunguska Cosmic Body (TCB) must have been a small comet rather than an asteroid. They reasoned that a comet might be composed mostly of ice and therefore would not leave physical traces in the form of meteorites. It was generally expected that if the offending body were made of stone, it would have left at least some fragments behind, and none were found. Following World War II, the "mystique" of Tunguska heightened. The atomic bomb demonstrated that explosions with the energy of the event at Tunguska (now reckoned to be between 10 and 20 megatons equivalent of TNT) could be of artificial origin, and the flurry of flying saucer reports beginning in 1947 led to speculation that Tunguska could have been the site of an explosion of a nuclear-powered alien spacecraft.

On February 12, 1947, another giant meteor struck, again in Siberia. This one, known as Sikhote-Alin after the nearby mountain range, was much smaller than the Tunguska event. Sikhote-Alin was first investigated only a few days after the event, over 300 eyewitnesses were interviewed, and a picture of the event was drawn by a landscape artist who happened to be doing a painting of the scene at the time. The documented characteristics of this event were very similar, although on a smaller scale, to Tunguska, with one exception—habeas corpus: this time, they had the "body." Sikhote-Alin was an iron meteorite fall, and it left numerous iron meteorites scattered over an area about 2 kilometers long and a few hundred meters wide. There can be no doubt of the nature of this event. One might expect the similarity between this and the prior event would have resolved the controversy over Tunguska, but instead, it only served to heighten the mystery because of the lack of a "body" at Tunguska.

In more recent times, a couple of oddities of modern physics have been added to the brew: suggestions that Tunguska was the impact of a "mini–black hole" or the impact of an antimatter "meteor." To be sure, none of these more far-out suggestions have been seriously embraced by the scientific community. The serious scientific controversy, such as it is, involves only the question of whether the cosmic body that entered the atmosphere in 1908 was

Felled trees as seen by Kulik, 1928. (N. A. Strukov/Tunguska Page of Bologna University: http://www.th.bo.infn.it/tunguska/)

Painting of the Sikhote-Alin fireball by observer, artist P. I. Medvedev. (NASA JSC photo S79-29470)

a stony body (an asteroid) or an icy one (a comet). The debate over the nature of the Tunguska Cosmic Body has been carried out at great length, both in time and words. Most of the early literature is in Russian. Krinov published a lengthy article in English in 1963 and a full book in 1966 describing the Tunguska and Sikhote-Alin events. Both are out of print but may be found in university libraries. Roy Gallant (1995) wrote a descriptive book for young readers, which, unfortunately, is also out of print. A more recent summary of the current status of Tunguska research was published by N. V. Vasilyev (1998 and Web site). This article was the introductory piece to the proceedings of a scientific meeting devoted to the study of the Tunguska event, held in Bologna, Italy, in 1996. These references provide a thorough description of the physical evidence surrounding the Tunguska event, including the estimates of the path of the incoming body, the visual and sonic phenomena, tree-fall and burn patterns, and the results of careful examination and sample collection from the sites.

A few additional facts have emerged from recent expeditions, perhaps most significantly the identification of traces of meteoritic material, including small dust grains embedded in tree sap dating from the time of the event and anomalous concentrations of the element iridium in the peat bog that lies directly under the explosion site, in a layer corresponding to the time of the event. Excess iridium is associated with cosmic impacts, most notably the one 65 million years ago that led to the extinction of the dinosaurs. This evidence quite conclusively argues that Tunguska was not an alien spaceship crash or the impact of a black hole or antimatter but instead a more ordinary arrival of a small asteroid or comet. Unfortunately, a comet, although containing some icy material, would also have plenty of cosmic dust, so the distinction between comet and asteroid is not resolved by this discovery.

In 1993, two scientific papers were published that would seem to lay to rest the comet-or-asteroid controversy. Christopher Chyba and colleagues (1993) and Jack Hills and M. Patrick Goda (1993) reported similar results based on computer calculations of the detailed physics on the atmospheric entry of high-speed projectiles. One can easily imagine the original purpose of such sophisticated computer codes and how they came to be openly available about that time. Both groups of researchers concluded that a cosmic body, entering Earth's atmosphere with about the expected speed and bringing with it the amount of energy (10 to 20 megatons) known to have been associated with the Tunguska event, would have to have the strength of a fairly hard stone. An object with the strength of ice would blow up far too high in the atmosphere, and an object of iron, like the Sikhote-

Alin meteoroid, would punch on through to the ground, leaving a crater. Only hard stone fits the observed explosion height. Furthermore, both groups found that the breakup of the incoming body, which would occur when the pressure of the atmosphere on the front face of the object exceeded its strength, would have been catastrophic and instantaneous; the meteoroid would have "exploded" into small particles only a few centimeters or smaller in size. Such tiny pieces—mostly dust or the size of sand and none larger than gravel—would be easily lost in the swampy bog beneath the explosion site in the nearly twenty years from the time of the event until the first expedition arrived to search for traces of the impacting object. As is often the case in scientific debates, these results have not fully settled the argument. With some exceptions, most Russian investigators still cling to a preference for a comet-like TCB, whereas those in the West mostly favor an asteroid-like TCB.

Further progress on this issue may have to wait until we have better knowledge of the physical properties of comets. But the scientific debate over whether the TCB was a comet or an asteroid is a mere quibble for specialists rather than a serious call for alternative, more bizarre explanations. Since the mid-1980s, we have become more acutely aware that we are living in a cosmic shooting gallery. There are about a million asteroids the size of the TCB in orbits crossing Earth's orbit around the Sun that are potentially capable of striking Earth. A collision of that size is expected every several hundred years; thus, the Tunguska event, about 100 years ago, is not at all unusual. That it occurred over land for at least a few people to see is a bit more unusual. For such an event to hit a heavily populated area would be even more rare, so it is not inconsistent with the historical record that no such massive catastrophe has ever been recorded. It is certain, though, that an explosion of 20 megatons of energy over a populated area could cause millions of deaths. Still larger impacts occur but less frequently. A giant impact of an asteroid or comet perhaps 15 kilometers (10 miles) in diameter is now thought to have ended the reign of the dinosaurs 65 million years ago. The crater scar has been found, about 200 kilometers in diameter and lying beneath later limestone sediments, on the tip of the Yucatán Peninsula. Based on the number of such large asteroids known (we can see and track things this large with telescopes from Earth), such a huge collision should occur only about once in 100 million years, which again is consistent with the last big one 65 million years ago. There is not much worry about such an event lying in our immediate future—we believe we have discovered and cataloged all the really big near-Earth asteroids (NEAs). There is concern, though, over NEAs just smaller than that, in the size range of one to a few kilometers in diameter. Some of these are not yet cataloged, and global consequences could occur if one were to hit Earth. It is estimated that impacts of this size occur about once in a million years and could cause climatic effects similar to a nuclear winter, including a global failure of agriculture for a year or two that would lead to massive famine worldwide. But such an event is a most extreme class of natural disaster, not only in terms of the numbers of fatalities it could cause but also in terms of the low frequency of occurrence. For this reason, cosmic impacts are all but ignored by the public as compared to earthquakes, volcanic eruptions, floods, and other natural disasters, despite the fact that the death rate (the number of deaths per event divided by the average time between occurrences) is comparable.

Finally, at the bottom end of the scale, meteoroids delivering kilotons of explosive energy arrive all the time; events with the energy of the Hiroshima atomic bomb occur about once a year. Why do we see no fatalities from these events? We can thank our atmosphere for that. Even hard stony meteoroids of this size explode high up in the atmosphere, so usually not even a muted shock wave reaches the ground. The smallest stony body that can cause any ground damage at all is only slightly smaller than the TCB. Below about 5 megatons of energy, the explosion occurs so high in the atmosphere that no damage occurs at the surface of Earth.

References:

Chyba, Christopher F., Peter J. Thomas, and Kevin J. Zahnle. 1993. "The 1908 Tunguska Explosion: Atmospheric Disruption of a Stony Asteroid." *Nature* 361: 40–44.

Gallant, Roy A. 1995. *The Day the Sky Split Apart.* New York: Atheneum Books. (See also http://www.galisteo.com/tunguska/docs.)

Hills, Jack G., and M. Patrick Goda. 1993. "The Fragmentation of Small Asteroids in the Atmosphere." *Astronomical Journal* 105: 1114–1144.

Krinov, E. L. 1963. "The Tunguska and Sikhote-Alin Meteorites." In *The Moon, Meteorites and Comets,* edited by B. M. Middlehurst and G. P. Kuiper, 208–234. Chicago: University of Chicago Press.

——. 1966. *Giant Meteorites.* Oxford: Pergamon Press.

Vasilyev, N. V. 1998. "The Tunguska Meteorite Problem Today." *Planetary & Space Science* 46: 129–150. (Condensed version is available at URL: http://www.galisteo.com/tunguska/docs/tmpt.html.)

University of Bologna Physics Department Tunguska Homepage. URL: http://www-th.bo.infn.it/tunguska/.

Tutankhamun's Curse

REBECCA BRADLEY

In 1922, Howard Carter and his patron, George Herbert, fifth earl of Carnarvon, made one of the richest archaeological finds of the twentieth century—the treasure-filled tomb of the pharaoh Tutankhamun in Egypt's Valley of the Kings. But did their discovery bring the curse of the dead king down upon their heads?

According to legend, an inscription over the tomb's outer door and a clay tablet found inside the antechamber were chillingly explicit: *Death shall come on swift wings to whoever toucheth the tomb of Pharaoh.* Psychic warnings were received from society fortune-tellers; dire omens were arrogantly ignored. And then, even as the tomb was opened, the curse claimed its first victim—Carter's canary was devoured by a cobra, the very creature worn as a protective deity on the pharaoh's brow. This warning, too, was ignored.

Scant weeks later, the curse claimed a second victim when Lord Carnarvon died abruptly in Cairo. At the very moment of his death, his beloved dog howled and fell dead in England, and the electricity failed across Cairo. Nor was this the end—the sad fate of Carnarvon (plus the dog and the canary) initiated a tragic series of deaths among those implicated in the tomb's discovery, from heavily involved archaeologists to peripherally involved relatives and visitors.

What was the cause of these misfortunes? To the media and a sensation-seeking public, it was clear the archaeologists had awakened supernatural forces that should have been left sleeping. But in the decades since, numerous writers seeking rational explanations have wondered whether more mundane forces may have been awakened—bacteria or fungus, radon or radioactivity, or even poisons left as booby traps by ancient priests. Some of these scenarios might sensibly explain a proportion of the observed deaths, but others are as problematic as the curse itself. All presuppose there is, in fact, a phenomenon to be explained. A careful examination of the legend and the evidence suggests this is not so.

Most of the canonical details display the symptoms of urban legend, existing in a suspicious number of variants or showing signs of having been "improved" in the telling. For example, by trawling through curse-related Web sites and books, we find that Carter's famous canary died at various locales and at different times—in the tomb, inside Carter's study, outside Carter's dighouse, on the day the staircase was uncovered, on the day the tomb was opened, on the day Carnarvon died, and so forth. We also learn that the electricity failure—not an uncommon occurrence in Cairo anyway—was brief, was possibly limited to the hotel or (sometimes) hospital in which Carnarvon died, and evidently took place some minutes *after* the earl's death. Stories about Carnarvon's dog generally neglect to take account of the time difference between Cairo and England; some add instead of subtract two hours, which only compounds the error.

The most compelling evidence—the series of mysterious deaths—evaporates on examination. Except for Carnarvon, who was chronically ill to begin with, few of the generally cited victims had any direct business with Tutankhamun; the media had a tendency to drag in any deaths with even the remotest connection to the tomb, no matter how absurd or tenuous. At the same time, most of the researchers intimately connected with the tomb clearance lived decades past their involvement, some reaching ripe old ages. Of the two who first entered the tomb with Lord Carnarvon, for example, Carter himself lived until 1939 and Lady Evelyn Herbert until 1980. Douglas Derry, who literally dismembered Tutankhamun's mummy during autopsy—surely a curse-worthy act if ever there was one—died in 1969 at the age of eighty-seven. A very few dignitaries or nonarchaeologist visitors died of various causes soon before or soon after a junket to the tomb; but the deaths of only a tiny handful of visitors out of the many thousands who drove Carter to distraction with their interruptions could imply that visiting the tomb was actually *good* for one's health.

Then there is the matter of the curse itself. Contrary to legend, no curse was inscribed over the door of the tomb; the much-quoted tablet did not exist and was most likely a journalistic invention. The "mummy's curse" motif, long a part of Western popular culture, appears to have been played up happily by the media to keep public interest on the boil when the slow pace of Carter's work did not provide enough dramatic copy to sell newspapers. There may also have been an element of malice—Carter's lack of "people skills," in particular, created resentment among the press and among certain archaeologists excluded from the find (primarily Arthur Weigall, who was slighted by Carter both as an Egyptologist and as a journalist). It has even been suggested that Carter and his British security guard concocted the story to discourage tomb robbers, but Carter's obvious disgust with the curse and its believers makes this hypothesis less than likely.

Finally, there is the very un-Egyptian nature of the curse to consider. Very few funerary maledictions have ever been found in Egypt, and those that have bear no resemblance to the popular modern idea of a mummy's curse. Strictly speaking, the canary story also fails in terms of Egyptian cultural logic: of the *two* tutelary goddesses on Tutankhamun's brow, vulture and cobra, it was the vulture who should have acted as the pharaoh's protector in that region of Egypt. Nor, in the light of what is known about ancient Egyptian attitudes, would Tutankhamun have any reason to curse Carter. On the contrary, the young pharaoh was fortunate in his discoverer. Carter preserved Tutankhamun's treasures in a careful, competent, and painstaking ten-year clearance, whereas many of his contemporaries might well have ripped through the tomb in weeks. Furthermore, to the ancient Egyptians, crucial conditions for immortality were that the name of the deceased be remembered by the living and that the soul be able to recognize its own image—and Carter did Tutankhamun a great favor in both respects.

In Carter's own words (1963, xxv), "The sentiment of the Egyptologist . . . is not one of fear, but of respect and awe. It is entirely opposed to the foolish superstitions which are far too prevalent among emotional people in search of 'psychic' excitement."

References:

Carter, Howard. 1963. *The Tomb of Tutankhamun.* Vol. 2. New York: Cooper Square Publishers.

Frayling, Christopher. 1992. *The Face of Tutankhamun.* London: Faber and Faber.

Reeves, Nicholas. 1997. *The Complete Tutankhamun.* London: Thames and Hudson.

UFOs

BARRY MARKOVSKY

UFO, the popular abbreviation for unidentified flying object, refers to any object that, from a given observer's perspective, is presumed to have floated or flown through Earth's atmosphere or outer space and is of uncertain nature and origin.

This definition implies several important points. First, what appears to be "unidentified" to an observer depends on what the observer already knows. A UFO for one observer may be an IFO—an *identified* flying object—for other observers, and the vast majority of UFOs that have been scrutinized by qualified investigators turn out to have rather mundane explanations. Often, observers simply lack one or more key pieces of information that otherwise would permit identification of the UFO.

Second, the UFO label applies equally well to sightings of objects on the ground, as long as the observer presumes that the object in question is or was capable of flight. In other words, it would be appropriate to refer to what appears to be a crashed flying saucer as a UFO even though it is not flying at the time.

Finally, the UFO label produces a residual category: it explains or gives meaning to an object only in the trivial sense of declaring that the object is *not* a member of any other category of objects previously known to the observer. Without further information, merely sighting a UFO cannot provide any validation of the extraterrestrial hypothesis (ETH) that UFOs provide evidence that intelligent alien beings have visited Earth.

Gallup public opinion polls in 1996 and 2001 found that between one-third and one-half of U.S. adults—as many as 100 million people in this country alone—believe UFOs or extraterrestrials (ETs) have visited Earth in some form. Perhaps it should not be very surprising that so many of us are willing to jump to the conclusion that alien intelligence lurks behind mysterious lights in the sky. After all, it is usually impossible to disprove outright such a belief, and few people have the time, expertise, and resources that would be required to solve any but the most implausible UFO mysteries. Additionally, the public generally is unaware of the great number of UFO cases that have been investigated scientifically by experts and subsequently shown to have prosaic explanations. Nevertheless, this does not explain why so many individuals are willing to make unwarranted inferences that invoke unknown forces and alien beings.

The published literature on UFOs and related subjects is enormous, and this relatively short examination must therefore be highly selective. Detailed accounts and case studies of UFO sightings are readily available elsewhere (see References), and so, instead of making a futile attempt to review them in any systematic way, this space is devoted to issues of particular relevance to a skeptical perspective on UFO claims.

Historical Overview

Undoubtedly, humankind has been noticing strange flying objects ever since we developed beliefs about what belongs "up there" and what does not. Recorded history is replete with stories of anomalous objects sighted in the heavens. By today's scientific standards, however, these tales do not hold up as accurate accounts of real phenomena. It is impossible to know the nature of the underlying phenomena when many descriptions so obviously have been colored by propaganda, popular fiction, religious fervor, and folklore.

Despite the absence of any recorded UFO sightings from antiquity, some ETH proponents treat certain ancient human artifacts as evidence. For example, they claim it is impossible that the pyramids of Egypt and the Americas could have been built by humans thousands of years ago with only primitive knowledge of engineering principles. ET proponents neglect to point out that research teams of archaeologists and engineers have devised construction techniques that were available to the ancients and can account for the pyramids and other artifacts. The same may be said of the "mystery" of how the large stone statues at Easter Island were raised from their original prone positions. Mystery-mongers also point to the large animal shapes and "runways" scratched into the Nazca Plains in Peru, asserting that such structures could only be used and appreciated by a culture with an advanced technology supporting flight. Although we cannot be certain of the reasons for the Nazca drawings, we do know that even the largest figures could still be appreciated from the ground and that they and others like them can be reproduced with very simple techniques for scaling up small drawings.

More recently, newspaper accounts from the latter nineteenth and early twentieth centuries described sporadic waves of sightings of strange objects in the sky. However, observers rarely presumed that the objects came from anywhere other than Earth. With the benefit of twenty-twenty hindsight, it appears that most of these sightings occurred under conditions that were ripe for mass delusion. An era of rapid technological development was under way, and the public was primed for wondrous breakthroughs. Heavier-than-air flight was the next frontier. Periodically, the mass media led the public to expect to see *something* in the skies, often by exaggerating prior sightings or by fabricating them completely. Moreover, nighttime viewing was the norm, and ambiguities produced by poor observational conditions likely added even more fuel to the collective imagination.

UFO sightings since the late 1940s have a distinct character when contrasted with these earlier reports. The modern UFO era can be said to have begun on June 24, 1947. On that date, pilot Kevin Arnold of Boise, Idaho, was flying a small plane near Mount Rainier, Washington, when he noticed a flash of light and nine disks appearing to be flying in a linear formation. He later spoke to a reporter from a local newspaper in Oregon, describing the objects as flying "like a saucer would if skipped over water." The Associated Press picked up the story, reporting the "saucer-like objects" as traveling at "incredible speed." The "flying saucer" label stuck, and public interest has remained high ever since, fed by a continuous flow of stories in books, magazines, newspapers, television, and movies and on the Internet.

The most notorious UFO case in history began to unfold soon after Arnold's sighting. Officers at Roswell Army Air Field issued a very unusual press announcement to local radio stations and newspapers, which was quickly picked up by the national media. The press release stated that a flying disk had landed on a ranch near Roswell, New Mexico, during the first week of July 1947. It indicated that the disk was picked up by the Intelligence Office of the 509th Bomb Group of the Eighth Air Force, Roswell Army Air Field, and subsequently loaned to higher headquarters. The events created only a short-lived public stir at the time, but they achieved wide notoriety when resurrected by Charles Berlitz and William Moore's 1980 book *The Roswell Incident.*

Many books, articles, and television documentaries have been written about the supposed Roswell UFO crash, most capitalizing on some combination of four interwoven claims. The first claim is that UFOs were sighted in the vicinity of Roswell in July 1947. The second is that prior to a restricted cleanup operation, several people witnessed a "debris field" about 75 miles from Roswell on a ranch managed by Mac Brazel. Some witnesses claim to have handled and even absconded with certain tiny pieces of physical evidence from that field, including a type of metallic foil, light structural members, and monofilament strands resembling fishing line. The third claim relates to first- and secondhand reports stating that the bodies of several alien beings were recovered from a crash site several miles from the debris field. Army officials ostensibly shipped the bodies to a secret location for examination. The fourth and final claim is that the U.S. government has persisted in covering up these events for many years, repeatedly claiming that the debris came from a downed weather balloon.

It is unlikely that the public ever will know the whole truth about the Roswell incident. For most people, however, *belief* in the Roswell crashed-saucer-and-dead-aliens claims depends less on what really happened than on the particular kinds of arguments and evidence to which one is exposed, as well as one's willingness to evaluate them critically. Consider the conclusions of *Roswell in Perspective* (Pflock 1994), probably the most thorough investigation of the Roswell incident to date. The author of that work noted that "we have only a highly speculative, highly tenuous link between anything seen in the sky and what was found by Mac Brazel and, possibly, elsewhere by the military (Pflock 1994, 61). Further, he stated, "It is beyond reasonable doubt that at least the great majority of what was recovered from the debris field was the remains of a Project Mogul flight" (Pflock 1994, 113). Although it was top secret at the time, we now know that Project Mogul involved launching arrays of specially constructed balloons that carried aloft electronic equipment for detecting enemy weapons. In addition, he wrote, "there is no proof . . . that there were bodies and that they were either alien entities or humans or other earthly creatures who had undergone something horrible. Proof must await more evidence" (Pflock 1994, 95). Although the only evidence for alien bodies comes from disputable testimonies, the book's author did admit to being "personally convinced" that bodies and wreckage of some kind were removed. The author concluded, "It also seems very likely that [military authorities] instituted a cover story in what turned out to be a highly successful attempt to keep Mogul under wraps" (Pflock 1994, 113). In sum, evidence for the story of a crashed saucer and dead aliens at Roswell is far poorer than one might expect based upon its contemporary portrayal in the mass media.

Media coverage of Roswell and other cases in the late 1940s paved the way for an upsurge in UFO claims with extraterrestrial overtones. By the early 1950s, many UFO reports had connotations of alien involvement. This trend surely was encouraged by several popular books published in 1950 that claimed alien life-forms were piloting flying saucers to Earth. By 1951, numerous newspapers and national magazines were publishing articles connecting UFOs with alien visitors. The phenomenon has continued unabated ever since.

Air force officers identify metallic fragments found by a farmer near Roswell, New Mexico, as pieces of a weather balloon. This is the basis of the Roswell incident, the supposed crash of an alien spacecraft. (Bettmann/CORBIS)

Kinds of UFO Claims

Writers and researchers have offered a variety of UFO typologies. The best known is the simple set of "close encounter" categories devised by astronomer J. Allen Hynek (see also Vallee 1990). A close encounter of the first kind (CE-I) is a basic UFO sighting with no physical evidence left behind. Most UFO sightings fall into this category. Close encounters of the second kind (CE-II) involve physical evidence or some form of interaction with the UFO. The debris field in the Roswell case is believed by many to be evidence of such an encounter. Finally, a close encounter of the third kind (CE-III) entails sighting or interacting with occupants of the UFO. This category could include seeing aliens through portholes in their mother ship, witnessing their bodies at a saucer crash site, or being subjected to a physical examination by them following abduction into their craft by a paralyzing beam of light.

Within each of these categories, there is a vast array of claims of highly variable detail and quality. Most CE-I sightings involve a moving, featureless bright object against a dark sky. Countless such cases have been captured in photographs, films, and videos. The problem with all of them, however, is that they carry insufficient information to determine the objects' true size, distance, and velocity. A dramatic example of this size-distance error is a video shot from a National Aeronautics and Space Administration (NASA) space shuttle in 1991, which has been broadcast nationally many times on a number of different television programs. The tape clearly shows a small bright object moving slowly through the field of view. It then suddenly changes direction and zips out of the picture. UFO proponents have declared the object to be an alien craft of considerable size, moving at high speed, and piloted by intelligent beings. NASA officials have a much simpler explanation: the object is an ice crystal drifting near the video camera, its video image "blooming" brightly in the direct sunlight. The change of direction was caused by the brief firing of a small rocket thruster whose flash is readily visible in the corner of the video image.

Many CE-I claims are far more elaborate and detailed than mere blobs of light. Philip J. Klass devoted a chapter of his book *UFOs: The Public Deceived* to a particular type of object described by observers with a dizzying array of adjectives, ranging from "like a giant jellyfish" to a seamless, silvery object with several tiers of windows and flashing, multicolored lights. Credible witnesses have been plentiful, including pilots and other highly educated observers. Here is an excellent example of how a key piece of information, such as a view from another vantage point, could have transformed UFOs into IFOs: Klass's chapter was about advertising airplanes and helicopters seen at night from oblique angles that prevent observers from reading the messages suspended in lights below them. Hundreds of these aircraft are in use in the United States, and they account for a remarkable number of UFO reports.

Most proponents of the ETH recognize that the great majority of UFO reports are attributable to mundane objects such as unusual cloud formations, the planet Venus, weather balloons, conventional aircraft, orbiting satellites, and meteors and space junk burning up on reentry. Venus in particular has produced a number of fascinating UFO stories. Especially on a Moon-less night, our sister planet can glow with a surprising luminosity. Seen through the trees from a moving automobile, for example, it may appear as a beacon from a spaceship tracking one's vehicle on a parallel course.

CE-II claims would go a long way toward establishing the extraterrestrial origins of UFOs—*if* the physical evidence were truly compelling. Jacques Vallee is a central figure among UFO investigators, known for his detailed and meticulous investigations. In his book *Confrontations*, he described a number of his investigations in various parts of the world, many of which involved the analysis of some kind of physical evidence left in the wake of a UFO sighting. On the surface, Vallee appeared to approach his subject skeptically and meticulously, time and again recognizing that his evidence failed to offer proof of extraterrestrial origins either for UFOs or for the ostensive artifacts some claim they have left behind. However, his lack of proof did not stop Vallee from making wild speculations that apparently he had come to believe. He suggested that "we are dealing with a yet unrecognized level of consciousness independent of man but closely linked to earth" (Vallee 1990, 99). Moreover, in the same paragraph in which he warned against jumping to the conclusion that UFOs represent advanced spacecraft from another planet, he assured us that "they promise to be much more: a challenge to many of our concepts in physics, perhaps a clue to the existence of unknown dimensions beyond spacetime." In other words, even though the physical evidence failed to demonstrate extraterrestrial origins, Vallee preserved the ETH via a conclusion that requires an even greater leap of faith.

Hundreds of CE-III claims were cataloged by Robert E. Bartholomew and George S. Howard in their *UFOs and Alien Contact*. These encounters fall into two major categories: alien contacts and alien abductions. Contactees believe that aliens have communicated with them—sometimes telepathically, sometimes through personal visits. Messages that contactees claim to have received from aliens tend to sound like lines from bad science-fiction movies. Sometimes, they are threatening ("Appear here tomorrow, or we will take your family!"), and at other times, they are kindly ("I come in peace"). Never have the aliens left behind artifacts, forwarding addresses, technological insights, or other information that would provide evidence of their extraterrestrial origins.

Alien abduction stories have become the most prevalent type of CE-III claim (and are sometimes given their own CE-IV classification), fueled by public fascination and the accommodating mass media. (See the "Alien Abductions" entry in this encyclopedia for a more extensive treatment.) Typically, the abductee recalls having been taken aboard an alien spacecraft and subjected to a sinister and highly invasive examination. Mysteriously, the aliens always manage to eliminate all evidence of incisions or other intrusions. The best known of these cases also assert that the aliens wiped out the abductee's conscious memories of the event, and so, the entire experience is suppressed until much later, when it emerges under hypnosis. If proponents are to be believed, then abduction by aliens must be declared a worldwide public health problem because they claim that literally millions have been abducted and mistreated, had their memories suppressed, and so remain inexplicably traumatized by their experience.

Factors that make alien abduction stories so convincing to believers fall well short of the sort of evidence that would be scientifically compelling. First, with no other physical evidence upon which to rely, the burden of proof rests on the individual testimonies of those who claim to have been abducted. This is immediately problematic because extensive research has shown that eyewitness testimony—especially under emotionally charged conditions—is highly unreliable. Second, the hypnotic state has proven to be especially conducive to elaborate fantasizing, combining elements from prior experiences (such as watching a science-fiction movie) with cues from the hypnotist. The misconception is that hypnosis reveals suppressed memories, but, in fact, it helps to create false ones. Third, proponents point out that there is an eerie sameness in the description of aliens across cases. This is far from the truth, especially in light of the wide variety of alien drawings that abductees have produced over the years. Even if it were true, it should not be too surprising that an image such as a large-eyed, small-mouthed, four-footed humanoid alien should appear in many drawings when such an image already has previously entered the public's consciousness through the mass media. Finally, there are alternative explanations for many abduction experiences, especially those that occurred at night, that do not invoke mysterious forces and entities. For instance, it turns out that a common condition known as sleep paralysis—a kind of dreamy, semiconscious state—can account for virtually all of the features of the alien abduction experience (Blackmore 1998).

From the standpoint of mainstream science, the lack of adequate evidence is a devastating problem for close-encounter claims. However, several additional issues pertaining to the ETH have not been addressed to the satisfaction of skeptics. These include problems caused by (1) logical flaws of some UFO arguments, (2) limitations imposed by the physics of space travel, and (3) human physiological and perceptual limitations. Some of the social factors that can contribute to the creation and maintenance of false beliefs will be considered here.

Logic of UFO Arguments

Earlier, it was noted that "UFO" is a residual category. People sometimes forget that just because you call something a UFO does not mean you possess any information about the object beyond the mere fact of its nonidentification. You still do not know what the object actually *is*, tempting though it may be to take that logical leap and infer extraterrestrial intelligence.

UFO supporters sometimes point to the fact that even the most comprehensive investigations fail to provide satisfactory explanations for all UFO sightings. This unexplained residual is then treated as adding support to the ETH, despite this also being an unwarranted logical leap. History shows that, rather than a given object being the first true UFO of extraterrestrial origin, it is far more likely that some crucial piece of information is missing or that existing information was misinterpreted.

One of the hallmarks of the scientific approach is the doctrine of falsifiability. Simply put, if it is not possible, in principle, to test and disprove a claim, then the claim is not scientific. However, there are a number of respected UFO investigators, known for their detailed and thorough analyses of the available evidence, who express their support for the ETH in the absence of clear findings that validate their position. Vallee, for instance, failing to find any physical evidence to support the ETH, argued, as noted earlier, that "we are dealing with a yet unrecognized level of consciousness." Perhaps this is true, except that there is no more evidence for this than for the alternative explanation that UFOs are the psychic projections of playful farm animals.

Another way to keep the ETH alive in spite of scant supportive evidence is by presuming that UFOs are "shy." Robert Sheaffer (1998) noted facetiously "their ability to select, on those rare occasions when they 'permit' a clear and detailed photograph to be taken, areas where there is *one and only one photographer* ready to snap their picture." He pointed out that in 1972, an unexpected, short-lived, genuine object from space appeared over areas of the western United States and Canada. Despite the sparse population in the viewing area, there was an extensive body of photographic and motion picture evidence from different vantage points, allowing the object to be identified—as a meteor. One also has to wonder why aliens would ever care to expend so much effort on staying just at the fringes of human and technological discernibility.

Several other logical issues also bear mention. First, when it comes to evidence, it is never true that quantity can substitute for quality. This is why it matters little how many eyewitness testimonies are gathered by proponents. There are simply too many ways that eyewitnesses are known to err, which throws into question all such accounts. Second, proponents of the ETH have been known to cite particular cases as supporting evidence long after they have been soundly debunked, which is an obvious misuse of evidence. Third, many UFO cases invoke a variety of mysterious correlated phenomena presumed to be caused by the UFO. These have included the deaths of animals or people, power failures, stalled vehicles, and other UFO sightings. However, with no prior constraint on what would constitute a mysterious correlate, there almost always will be *something* that one could dig up. But due to the lack of evidence to tie such events to the flying object, the correlation provides absolutely no added weight to the ETH. Finally, by the logic of the theory of evolution by natural selection, there is no chance that the most-sighted species of aliens would have evolved independently into forms that so closely resemble humans. The primate form evolved only once on Earth, out of hundreds of millions of species over the course of hundreds of millions of years. It is virtually impossible for that form to evolve again here, let alone on some alien planet with an entirely different evolutionary trajectory.

Physical Barriers

Some UFO claims, if they were true, would violate the laws of physics. Such laws forbid the kind of electromagnetic propulsion system that some have suggested explains the stunning feats of which UFOs are said to be capable. Scientists also point out that to reach Earth from anywhere outside our solar system would require either many thousands of years or else vehicles capable of traveling near the speed of light. Such spacecraft would need prohibitive quantities of fuel to reach such speeds, even if they had the technology to convert fuel into energy with perfect efficiency. Many other problems would have to be solved as well, such as developing a method to protect the spacecraft from the otherwise catastrophic effect of high-speed collisions with dust particles.

UFO proponents argue that aliens would have devised ways to circumscribe laws that only appear to be immutable to earthbound scientists. Some scientists have speculated that objects entering "wormholes" in space could travel immense distances instantaneously. This assumes that one could first find a conveniently located wormhole, that one's vehicle could withstand its tremendous gravitational and tidal forces, and that one could know in advance where in the universe one would emerge. For now, wormholes exist only in the realm of theory and so, in the absence of any actual evidence, cannot bolster the extraterrestrial hypothesis for the origins of UFOs.

Perception and Psychology

Perception is a complex, multistage process, most of which transpires unconsciously. Vision is the sensory mode that is most relevant to UFO-related beliefs, and a large body of scientific research attests to the feats and the foibles of the human sense of sight. The bottom line in this research is that, despite the fact that we have remarkable visual capabilities, there still are a great many ways that visual perceptions can mislead us. The problem with misperceptions is that rarely do we know when we are having them. Therefore, especially when viewing unfamiliar objects under less-than-ideal conditions, our confidence in what we have observed is not a reliable indicator of the underlying facts. Seeing may be believing, but that does not make it true.

The size-distance error mentioned previously is but one of many potential sources of misperception. Among the others is the autokinetic effect, whereby a stationary point of light against a dark background is seen to drift or dart about. The apparent motion looks absolutely real, but, in fact, it is due to unconscious eye movements. We see the movement parallax effect when looking at objects at different distances while we ourselves are in motion. Driving down the road, telephone poles seem to move rapidly in the direction opposite our car's motion, while the Moon seems to match our speed exactly. A surprising number of UFO cases involve the claim of being "chased" by Venus and other objects that appear to track the observer's motion because of movement parallax. President Jimmy Carter's famous UFO sighting turned out to be a glimpse of Venus. The full Moon illusion accounts for the apparent but illusory size differences of the Moon at the horizon versus the Moon high in the sky. The same effect can make the bright disk of a planet appear unexpectedly large when "floating" low in the sky, shining its "beam" through the trees.

Perception and psychology are closely related in the sense that all of our conscious perceptions are based upon interpretations of the sensations that we experience. As one prominent skeptic pointed out:

> The great failure of the pro-UFO movement has been its unwillingness to accept the fact that human perception and memory are not only unreliable under a variety of conditions (and these conditions are exactly those under which most UFOs are reported) but that perception and memory are also *constructive*. That is, perception is a function not only of the actual sensory stimulation that is picked up by the eye or ear but also a function of what we know and believe about the world, even if that knowledge and belief are wrong. (Hines 1988, 167–168)

The fact that we are so adept at inferring patterns serves us well in most situations. However, an expectancy effect occurs when the anticipation of a certain pattern leads us to perceive it whether or not it actually exists. That is why some early astronomers believed they saw canals on Mars, why observers may feel certain that they saw windows on UFOs that were later determined to be weather balloons or clouds, or why UFOs and ETs may seem inextricably linked.

Terence Hines noted that memory also is constructive, and this fact is borne out by a large body of research on the unreliability of eyewitness testimonies. The longer the lag between the perception and the recollection, the greater the opportunity for the memory to become embellished or otherwise altered. Consider the implications of this truth for the Roswell alien corpse witnesses who waited some three decades before going public.

Social Psychological Factors

People have a profound effect on one another's behaviors and beliefs, making social influence another potential source of belief in the ETH. Social psychologists have studied various kinds of influence, such as obedience, persuasion, and conformity—all of which have the potential to induce a belief in the absence of any direct experience or evidence. In short, one may believe that ETs have visited Earth because we perceive that others believe this. Research on paranormal beliefs has shown how readily this can happen (Markovsky and Thye 2001). In a situation in which the judgment is not clear-cut—as is often the case with UFO sightings—one stranger expressing the view that a paranormal event occurred can be sufficient to influence others. The effect is even stronger when the influencer is believed to be some type of expert or high-status person, even if he or she has no special skill relevant to *this* particular type of situation.

We cannot know for certain, but variations on the kind of passive social influence just described probably are a significant source of popular belief in the ETH. Seeing that others believe without reservation can be sufficient to influence those who otherwise may be indifferent. However, more active forms of social influence undoubtedly have an even greater impact. Television documentaries, tabloid news stories, magazine articles, popular books, and even personal acquaintances seldom merely report unadorned facts about UFOs. More often, they aim to persuade the viewer/reader/listener that something extraordinary has occurred.

Whether the attempt to persuade is done on a person-to-person basis or on a mass scale, among the most common techniques used are sharpening and leveling. Sharpening means emphasizing the gist of the message; leveling means leaving out information that seems inessential. The effect often is to radically alter the impression of the event that others receive. Facts that could serve as the key to unlocking the mystery are leveled because the person retelling the story found them uninteresting, whereas sharpening may enhance the mysteriousness of the claim. Both phenomena are evident in television programs on UFOs in which the evidence from classic cases is carefully sharpened through editing and investigations that would have provided mundane explanations are leveled.

Finally, it is also worth noting how emotional factors can play a role in the spread of UFO-related beliefs. For many, the prospect of being visited by alien beings carries with it a sense of wonder and exhilaration—expressions of which were captured with great effect in films such as *ET* and *Close Encounters of the Third Kind*. This emotional component is important for at least two reasons. First, rightly or wrongly (often wrongly), we use emotions to supplement or even to supplant rational judgment. That is, rather than suspending judgment when evidence is lacking, people will frequently use their emotions as a guide. If it feels good to believe, then believe we shall. Second, emotions are contagious and compelling. When a witness expresses emotions—apprehension, excitement, awe—while relating a close-encounter experience, members of the "audience" not only will perceive the story as more truthful but also, to some extent, will share the same emotional experiences. This contagion effect also underlies the UFO panics reported in the nineteenth and early twentieth centuries.

Organizational Involvement

As the reference to panics implies, the UFO phenomenon entails much more than individual observers pondering lights in the sky. However, whereas a social panic is relatively unorganized and short-lived, certain aspects of popular interest in UFOs are far more structured and enduring. In this regard, we find all the makings of a social movement, complete with organizations ranging from informal clubs to government-sponsored investigatory panels to national associations. With some important exceptions, these organizations presume that UFOs are guided by intelligent ETs or at least that it is highly likely this is so. The broader societal effects of having organized interest groups include increasing the legitimacy of the ETH, disseminating UFO claims more widely throughout the culture, and establishing mechanisms to make it easier for anyone to feel more personally connected with the search for evidence.

In the United States alone, local, state, and national groups and affiliates interested in UFO investigations number in the hundreds. The http://UFOINFO.com Web site includes listings in forty additional countries. Hundreds more UFO-related sites can be found by perusing the links available on these organizations' Web pages. Only a small number of organizations have achieved prominence, however, usually based upon longevity, size, and the involvement of researchers with scientific credentials.

Formed in 1952, the Aerial Phenomena Research Organization (APRO) was the first significant UFO interest group in the United States. Its members included a stable of academic consultants in a variety of disciplines. Founders Coral and James Lorenzen were quite convinced that UFOs were conducting mapping projects, and APRO pioneered the dissemination of stories of alien sightings. The organization closed down in 1988. In 1956, the National Investigations Committee on Aerial Phenomena (NICAP) was established. This UFO research organization was, for a time, the largest in the country, with numerous chapters. The group dissolved in the 1970s.

Ground Saucer Watch (GSW) was founded in 1957 by brothers William H. and J. A. Spaulding. Although now apparently inactive, this small group made a splash in 1977 when it filed a suit under the Freedom of Information Act against the Central Intelligence Agency (CIA). The group was convinced that the CIA was withholding secret information on government UFO investigations. The CIA complied with the suit as fully as possible under constraints imposed by national security concerns, and nothing at all extraordinary was revealed. Philip Klass made a strong case that those concerns had nothing to do with the ETH but rather related to the fear that "the USSR, with its growing fleet of long-range bombers and its newly acquired atomic bombs, could conceivably exploit UFO-mania within the U.S. to stage a surprise attack. The first eyewitness reports of approaching enemy bombers could too easily be dismissed as prosaic UFO reports, until the first atomic weapons begin to explode" (Klass 1983, 21). Because a small amount of information remained classified, however, conspiracy theorists have remained unconvinced.

Two groups that now dominate the UFO cultural scene are the Mutual UFO Network (MUFON), founded in 1969, and the Center for UFO Studies (CUFOS), begun in 1973 by astronomer J. Allen Hynek. MUFON rose as NICAP fell apart and key disaffected members switched allegiance. CUFOS has sought to elevate UFO investigation by limiting membership to established researchers; however, Robert Sheaffer (1996, 769) asserted that "since Hynek's death [in 1985], the scientific community has shown virtually no interest in the study of UFO reports."

Not surprisingly, the arm of the federal government most involved in UFO research has been the U.S. Air Force (USAF). Beginning in 1947 and for the next two decades, the USAF gathered data on thousands of UFO reports, its activities conducted under the designations Project Sign, Project Grudge, and Project Blue Book. With strong political pressure from influential UFO proponents, the USAF awarded a grant to the University of Colorado for a "scientific study of unidentified flying objects," to be conducted by a panel of experts headed by physicist Edward U. Condon. Neither the air force nor the independent "Condon Report" published in 1969 found any positive evidence in support of the ETH. (Project Blue Book files were released to the public in 1976 under the Freedom of Information Act.) Nevertheless, several hundred cases remained unexplained for lack of adequate information. For those individuals and groups willing to take the logical leap, these unexplained cases are sufficient to fuel continued belief in the ETH and in government cover-ups.

UFO Culture

A case could be made that, beyond particular organizations formed around interest in UFOs, the extraterrestrial hypothesis has found a stable niche in the ecology of public awareness. It is "locked in" in the sense that there is a critical mass of believers and promoters, sufficient to recruit new adherents and to sustain interest over time. Sociologist Erich Goode sees paranormalism and science as alternative cultures with fundamental differences in the way members view reality. Paranormalists take a commonsense approach to understanding phenomena such as UFOs. That is, the evidence of one's own impressions and inferences is taken as sufficient to form a belief: if it *seems* to be true, then it is true (for me, at least). Such thinking functions well in day-to-day situations, but science recognizes that common sense fails under certain conditions, and UFO sightings, stories about UFO sightings, and the UFO-ETH connection tend to fall neatly within those conditions. In a culture in which commonsense thinking is the norm and scientific thinking the exception, it should come as no surprise that such a large proportion of U.S. adults profess belief in alien visitations to Earth.

Both the scientific and the paranormal orientations have widespread bases of cultural support in this country. Scientific literacy and paranormal beliefs fluctuate from decade to decade, but neither is going to go away in our lifetimes. Because science views commonsense thinking as flawed and because those who employ common sense do not feel a need for any higher standards of evidence, UFOs are likely to persist as a cultural phenomenon even if proponents can amass no better evidence than that existing today.

References:

Bartholomew, Robert E., and George S. Howard. 1998. *UFOs and Alien Contact: Two Centuries of Mystery*. Amherst, NY: Prometheus Books.

Blackmore, Susan. 1998. "Abduction by Aliens or Sleep Paralysis?" *Skeptical Inquirer* 22, no. 3.

Center for UFO Studies. URL: http://www.cufos.org.

Condon Report. URL: http://www.ncas.org/condon.

Fund for UFO Research. URL: http://www.fufor.com.

Hines, Terence. 1988. *Pseudoscience and the Paranormal: A Critical Examination of the Evidence*. Amherst, NY: Prometheus Books.

Klass, Philip J. 1983. *UFOs: The Public Deceived*. Amherst, NY: Prometheus Books.

Markovsky, Barry, and Shane Thye. 2001. "Social Influence on Paranormal Beliefs." *Sociological Perspectives*.

Mutual UFO Network. URL: http://www.mufon.com.

Pflock, Karl T. 1994. *Roswell in Perspective*. Washington, DC: Fund for UFO Research.

Phillip Klass. URL: http://www.csicop.org/klassfiles.

Robert Sheaffer. URL: http://www.debunker.com/ufo.html.

Sheaffer, Robert. 1996. "Unidentified Flying Objects (UFOs)." In *The Encyclopedia of the Paranormal*, edited by Gordon Stein. Amherst, NY: Prometheus Books.

——. 1998. *UFO Sightings: The Evidence*. Amherst, NY: Prometheus Books.

Vallee, Jacques. 1990. *Confrontations: A Scientist's Search for Alien Contact*. New York: Ballantine Books.

Undeceiving Ourselves

GEOFFREY DEAN, IVAN W. KELLY, AND ARTHUR MATHER

"What a piece of work is man! How noble in reason! How infinite in faculties!" said Shakespeare's Hamlet. And for simple everyday living, Hamlet was right, otherwise we would not be here. But in more complex situations such as deciding the validity of pseudoscience X, Hamlet was wrong. In fact, "How poor in reason! How limited in faculties!" would be nearer the truth.

Blame it on evolution. We need more time to adapt to modern living. If human existence is represented by the height of a table, our entire recorded history is no thicker than a thin table mat; the two centuries of the Industrial Revolution are almost invisible, a mere postage stamp. We humans are designed for a world that no longer exists, one where our survival depended less on reason and more on blind reaction. A movement in the tall grass might be a tiger or the wind, but running was safer than reasoning. A man seeking the truth by reason did not live long.

There were other consequences. Children became programmed to learn quickly whatever they were told. They had to learn—and learn fast—that fire burns and dogs bite, or they would not survive. Adults became programmed to act on what seemed like a connection even if none actually existed. If eating a certain plant was followed by illness, the plant was avoided. The plant may not have caused the illness, but in times of plenty, nothing was lost by avoiding it. Progress became dependent on pattern recognition, the ability to make sense out of objects and people and situations. The ability was so decisive that we became driven to seek patterns even when none existed, as in seeing shapes among the stars. Then came language, which allowed the beliefs (true or false) arising from experience to survive and thus influence future society. So here we are today, trying to cope with science and pseudoscience by using a brain designed for survival in a quite different world. Consider what this means for beliefs in general.

Deceiving Ourselves

The credulity that helps children survive also helps them accept fantasies such as Santa Claus. As time goes on, they avoid conflicting beliefs by becoming more selective and by asking questions. As a result, most of us end up with much the same beliefs as our parents and the community. What determines most of our central beliefs is not our gender or intelligence or personality but our upbringing. Nevertheless, when we get emotional, as in anger or fear, information can easily bypass the reasoning parts of the brain. On a bad day, this can leave us with beliefs that we are compelled to follow even though they make no sense, such as compulsive hand washing or a fear of open spaces.

Is truth relevant? We like to think so, but society often sees truth or falsity as less important than believing. Faith is respected, skepticism is not. Disbelievers were once burned at the stake, and religion can still lead to war. It is faith, not reason, that kills, as happened at Jonestown in 1978 when 900 people died due to faith in their leader. Is logic relevant? Again, we like to think so, but logic in everyday life is often unrealistic. Nobody reasons logically to decide between strawberry and vanilla ice cream. And logic is often not justified anyway, simply because most errors are of little consequence.

In short, we are programmed to believe almost anything. What matters most is not truth or logic but content. Or, as Bertrand Russell said, what men want is not knowledge but certainty. For most of us, life becomes very difficult without the certainty provided by a belief system (any belief system). Thus, one of the few valid generalizations in social psychology is the "principle of certainty," which says when there is evidence both for and against a belief, most people show not low levels of certainty (which would be appropriate) but high levels of certainty either for or against (which is indefensible). For them, it is better to be wrong than uncertain.

Effects of Complexity

As things get more complex, as in pseudoscience X, they generally get more uncertain. So we reduce uncertainty by slotting cases into simplified pigeonholes. That is, in conformity with the principle of certainty, we opt for simple black or white rather than shades of gray. When information is lacking, we still use (invented) pigeonholes to fill in the gaps. We even *remember* via pigeonholes, thus distorting the original. As a result, we tend to make judgments by assumptions and similarities, we can find connections where none actually exist, and we are much less bothered by worthless data than we ought to be. All of these things are a legacy from our evolution. In fact, these things come so naturally that the liabilities they entailed went largely unnoticed until people tried to make a computer model. The liabilities arise when we want to find real connections and avoid mistakes, as when we first meet pseudoscience X.

To find connections where none actually exist, the only requirement is that our belief be established in advance (say, by reading about it), regardless of whether the belief is true or false. Suppose we believe that redheads are hot-tempered. If our subsequent experience of redheads is not clear-cut but rather vague, as is most likely, our belief cannot fail to be confirmed—we will see vague behavior as hot-tempered and vaguely red hair as genuinely red. Truth or falsity will not come into it. If it seems preposterous that your judgment could be affected by knowing the answer in advance, try making sense of this statement: "The trip was not delayed because the bottle shattered." The statement will seem vague and meaningless. But try again, this time thinking about christening a ship. The statement now seems crystal clear, and your belief that it is about a ship will seem amply confirmed. But the statement is actually about dropping a bottle of Coke on a hiking trip. So your judgment of a vague and unclear behavior was determined not by truth or falsity but by knowing the supposed answer in advance.

What if we have no prior beliefs? Here we can be led astray by another legacy from evolution, a potent learning process that occurs whenever something we do (believing in X, placing a bet) is followed by something else (feeling good, winning something). When the time interval is short, learning is automatic, and we can end up believing the two events are related when they are not. Even worse, our belief becomes very resistant to change if the event pair happens intermittently rather than all the time. Thus, intermittent winning at roulette encourages further bets because we see that losing does not deny winning, whereas fifty losses in a row persuades us to stop. Because the event pair can happen intermittently just by chance, we can end up believing all kinds of things that are actually false.

Our ability to make sense out of no sense is purely cognitive. It has nothing to do with having a fantasy-prone personality, which is characteristic of up to 4 percent of the population. Such people hallucinate and fantasize during a large part of their waking lives, but outwardly, they appear no different from anyone else. Indeed, as a group, they were not properly identified until 1983. Because they frequently confuse fantasy and reality, they tend to have paranormal beliefs and tend to be well represented among mystics, mediums, channelers, aura readers, and those who believe they have access to other realities. Comparisons can be revealing; for example, the incidence of such people in the population is many times higher than the incidence of astrologers or ufologists.

Judging Numbers

As it happens, we are quite good at things that require only counting. As marbles are drawn at random, we can estimate their average size or the proportion of red quite well. But once we start looking for links, such as between size and color, our ability disappears. For example, nurses were given the following data for a symptom and a disease:

Disease		Yes	No
Symptom	Yes	37	17
	No	33	13

Thus, 37 diseased patients showed the symptom, and 33 did not.

Are symptom and disease related in these data? The correct answer is no (in technical terms, the correlation is −0.02). But 80 percent of the nurses said yes, 7 percent said no, and the rest gave up. When asked to explain how they got their wrong answers, the majority of the nurses said the most common combination was yes/yes, therefore disease and symptom were related. They had ignored the other combinations, which show the opposite—the symptom is slightly more prevalent among those with no disease (17/30 = 0.57 versus 37/70 = 0.53). Similarly, if asked whether redheads are hot-tempered or if prayers are answered, hardly anyone considers even-tempered brunettes or non-prayed-for answers. Yet no link can exist unless redheads differ from brunettes in the incidence of temper and praying differs from not praying in the incidence of answers. In short, no conclusions are possible without data for all four combinations. So be suspicious when believers in pseudoscience X consider only yes/yes combinations, for example, only predictions that come true.

It gets worse. Once we move from data on paper to data drawn from memory, we become subject to further judgment errors such as the following, largely because memory is a process of reconstruction rather than retrieval:

- Vividness—we focus on vivid things, ignore dull things
- Representativeness—we focus on similarity, ignore actual occurrence
- Stereotypes—we use simplistic ideas, ignore actual observations
- Sample size—we ignore the huge sampling uncertainty of tiny samples
- Overconfidence—we tend to be overconfident in our judgments
- Overload—we cannot juggle more than about seven chunks of data at once

There are many other causes of judgment error. Furthermore, our judgment is not helped by the tedious effort required to avoid error as compared, say, to the notable lack of effort required to recognize faces, even though the latter calculations are far more complex—a result of our perceptual system having developed first. We can easily become oblivious to what really matters, so our judgment can be wrong in ways we never suspect. The demands of modern ideas have outrun our brains and minds. Unaided human judgment is simply incapable of dealing effectively with large amounts of complex information. We need help.

Now for the good news. We already have in place countermeasures against deceiving ourselves. They did not come quickly or easily, but they have been enormously successful. They are known as science. Or, as Nobel laureate Richard Feynman said, science is what we have learned about how not to fool ourselves. Of course, not everyone can be a scientist, but everyone can benefit from the following insights of science.

Undeceiving Ourselves

Barriers to Change. The problem is simple: we are generally unaware of our errors, and we are generally overconfident about our judgments, so it will seem implausible that our reasoning could be faulty. Especially if we are believers in pseudoscience X—no pseudoscience can afford to tolerate genuine science and error-free reasoning. We therefore have little incentive to change. But change we must.

Incentives to Change. Remember that judgment errors are pervasive even though most people are unaware of them. Unless a claim is supported by a tally of confirming and disconfirming cases, you can assume that judgment errors are alive and well. Consider emotional involvement. Hell hath no fury like a cherished belief under attack. Which is more desirable: feeling secure or being right? How much would it matter if your belief was wrong?

Beliefs versus Facts. Beliefs are just statements of opinion. You are free to agree or disagree. But when something is observed again and again, it is a fact. Facts are not beliefs. You cannot simply dismiss them. To do that, you have to fault the way the observations were made. To be a good observer requires training. No single person can be a final authority.

Being Critical. Ask believers in pseudoscience X the following questions; the aim is not to win but to learn. Why do you believe in X? This puts the burden of proof where it belongs—on the claimant. What evidence would you accept as creating problems for your belief? This is a potent question because it opposes the tendency to consider only confirming cases. Are there other explanations that could produce the same belief? This too is a potent question. Where did your idea come from? A credible source means the idea may be plausible even if the previous answers are unsatisfactory. Why should we believe in X? This restates the previous questions from our viewpoint.

No Information? Try to provide a plausible rival hypothesis. Thus, the "Draw-a-Person" personality test has been largely abandoned because the hypothesis "unusual person = inner conflicts" was displaced by the more plausible hypothesis "unusual person = lack of artistic ability." If you cannot think of a rival hypothesis, consider what might be the more plausible: X is true, or X is due to human judgment errors.

Open Minds. If X is possible but you have no evidence for or against, should you keep an open mind? The question is deceptive because the word *possible* is ambiguous. It can mean barely possible (if you jump off a cliff, it is possible that a freak wind will save you), or it can mean seriously possible (if you jump off a cliff, it is possible that you will die). Bare possibilities are vastly more numerous than could ever be studied, so only serious possibilities deserve an open mind. But an open mind requires us to tolerate uncertainty, which most of us find extremely difficult to do. Is the believer in X really open-minded? Be aware that believers use the open-mind idea to frustrate criticism. It works like this: for them, no possibility is off the rails, which (to them) confirms their open minds, whereas requests for evidence are for closed minds only. But their call for open minds is no more than a call to abandon all criticism; in effect, it provides the glue without which the pseudoscience might fall apart.

Data Snooping. If you snoop around in data looking for something interesting, then your judgment errors are bad news. Try the following remedies. Graph the results so you can see what is happening. Test the findings from half of the sample on the other half. Compare your results with those of similar or nearly similar studies. Replicate on fresh data or, if fresh data are unavailable, on random data.

Hostile Skeptics. Unfortunately, some skeptics are as intolerant of contrary views as any committed believer. Their ploys tend to be as follows. They keep raising the standards of evidence, or they find trivial flaws and claim they are fatal. Remedy: have them set the standards. They deny the case simply because it is impossible or unlikely. Remedy: have them give reasons. They make false claims such as "there are no cases of X," when in fact there are many cases. Remedy: be informed. They make accusations of incompetence or even fraud. Remedy: demand evidence. Point out that their argument is problematic if it leaves no room for people making honest mistakes.

Different Agendas. In the paranormal area, skeptics tend to focus on whether X is true, like gravity, so its truth is everything. But believers tend to focus on whether X is meaningful or beneficial in some way, like Santa Claus, so its truth may be of little consequence. Beware the difference.

Crooked Arguments. There are dozens of crooked arguments, some dignified by Latin names, which can be reduced to just two, each with a remedy. The first is making an irrelevant point. Remedy: so what? The second is ignoring a relevant point. Remedy: specify. These remedies make a good first defense against crooked arguments and one-sided opinions. Only X will save you (specify how). X is everyone's favorite (specify why). X is mentioned on TV (so what?).

Being Informed. Human judgment processes are an important area of psychological research. By 1970, there were more than 400 published studies; today, there are thousands, including dozens of books of which the more readable ones are listed under the "Further Reading" section at the end of this entry. They provide a rich resource for readers wishing to undeceive themselves.

Now for the Bad News

It seems self-evident that reading books and articles about undeceiving ourselves should improve our judgment skills. But the available studies suggest that the improvement is small. It is easy to see why—human judgment is such a vast topic that what we learn may not fit anything specific such as pseudoscience X. And even if it does fit, we may still have trouble with it; it is not easy to set aside our believe-anything-if-it-feels-good legacy from evolution. For example, it has been found that most people have trouble even with basic reasoning, such as providing sound evidence for their readily held opinions. So, despite our best intentions, merely reading the previous hints may leave us little better off. Fortunately, this is not the end of it, for what matters is not so much hints as practice, motivation, feedback, and being cautious.

We learn motor skills such as swimming and driving by practice and by learning from our mistakes. Swallowing water or hitting the curb gives us instant feedback on what to avoid. Undeceiving ourselves is basically the same process. Instead of moving our arms, we now have to move ideas, but the crucial component is no different—we learn by practice and by making mistakes, as when a doctor discovers that a supposed stomach cancer is actually an ulcer. So to succeed in undeceiving ourselves, we need constant practice, clear concepts, and clear feedback. A fuzzy concept means fuzzy feedback, so any unclarity is bad news. Imagine trying to learn if swallowing water or hitting the curb occurred at random.

But pseudosciences are typically fuzzy and full of unclarity. Worse, people can be highly intelligent in some areas and highly prone to judgment errors in others. So how can we achieve the required practice, motivation, feedback, and caution? The answer is to follow our reading of general works with the reading of works that specifically target pseudoscience X. Such works rarely sell well and may therefore be hard to find, but they are becoming increasingly available both in print and on the Internet. Undeceiving ourselves was never meant to be easy, but it has never been easier than now.

Further Reading:

The following works are readable and not technically difficult, but they may be less helpful than works that target a specific pseudoscience:

Dawes, R. M. 2001. *Everyday Irrationality: How Pseudoscientists, Lunatics, and the Rest of Us Systematically Fail to Think Rationally.* New York: Westview Press. A survey of everyday judgment errors that lead us astray. More technical than Gilovich (1991) but readable and with good examples.

Flesch, R. 1962. *The Art of Clear Thinking.* New York: Collier. Very readable, with many tests and puzzles for improving thinking habits. Shows how dozens of crooked arguments can be reduced to just two, each with a remedy (pp. 93–102). "Yes, clear thinking is rare. To approach it . . . we must be ready to sacrifice some of our personality and habits of thought as we face each new problem" (p. 225).

Gambrill, E. 1990. *Critical Thinking in Clinical Practice: Improving the Accuracy of Judgments and Decisions about Clients.* San Francisco: Jossey-Bass. How to reduce clinical errors in psychology, medicine, and the helping professions. Well organized and packed with information and many examples. Each chapter has a summary. A clinical practice is one that involves the treatment of clients.

Gilovich, T. 1991. *How We Know What Isn't So: The Fallibility of Human Reason in Everyday Life.* New York: Free Press. A nontechnical survey of the cognitive, social, and motivational processes by which even very bright people become convinced of the validity of false beliefs. Readable but rambling and poorly referenced.

Hogarth, R. M. 1987. *Judgement and Choice: The Psychology of Decision.* 2d ed. New York: Wiley. Emphasis is on business and management settings.

Piatelli-Palmarini, M. 1994. *Inevitable Illusions: How Mistakes of Reason Rule Our Minds.* New York: Wiley. A good, basic, nontechnical review of the research on cognitive illusions and why we tend to be impervious to the corrections offered by logic and evidence. Covers much the same ground as Gilovich (1991) but is better written.

Plous, S. 1993. *The Psychology of Judgment and Decision Making.* New York: McGraw-Hill. Clear, readable, fully referenced, the best of the recent books on the topic. Includes twelve pages listing all the judgment exercises used in the literature, which allows you to compare your judgments with those reported. If your interest is clinical judgments, then Gambrill (1990) is better.

Ruscio, J. 2002. "The Emptiness of Holism." *Skeptical Inquirer* 26, no. 2: 46–50. A critique of holistic claims. Conclusion: "Holism is an empty retreat from reality, a method by which pseudoscientists muddy rational thought, avoid clear and concise communication, and follow their own idiosyncratic beliefs to justify doing whatever they please in the name of all that sounds nice and feels good" (p. 50).

Sutherland, S. 1992. *Irrationality: Why We Don't Think Straight!* London: Constable. Surveys roughly 100 causes of irrational thinking. Nontechnical, very readable, abundant examples.

Witchcraft and Magic

JULIANNA YAU

Witchcraft and magic are collectives of practices such as rituals and spells, which are employed to cause or influence a desired outcome. Witchcraft is not always separable from religious rituals and is often heavily based on religious concepts. Forces such as spirits, gods, and demons are commonly attributed to acts of witchcraft, although modern-day practitioners of witchcraft prefer using terms such as *energy*, *will*, and *desire*. Witchcraft has been subject to disbelief because of the fallibility of evidence in support of its claimed effects. Critics find certain explanations of witchcraft more credible than others; the psychological and sociological effects of witchcraft are the only two factors that both believers and nonbelievers agree upon. The belief and practice of witchcraft have affected history in numerous parts of the world, and it is still being practiced in various forms around the globe.

Witchcraft is a part of virtually all human cultures, existing either within or alongside the culture's native religion. Although the exact practices differ among cultures and traditions, they all share similar structures. Acts of witchcraft can be categorized as either ritual or folk magic. Folk magic includes image magic, amulet/charm magic, healing magic, and ancestral veneration. Ritual magic, or ceremonial magic, is more formal and often involves one or more forms of folk magic. Folk-magic practices can sometimes take the form of rituals or ceremonies if one form is the dominant focus of the ritual.

Image magic is based on the belief that an image is linked to what it represents and that anything done to the image will result in a similar effect upon the original. Unmarked stuffed dolls in the shape of a human, known as poppets or voodoo dolls, are most commonly used in image magic. The doll is considered to represent a certain person and can be used to bring benefit or harm to that person. Recognized practices include wrapping or binding the doll with cloth or rope in hopes that the person will be physically, mentally, or otherwise restrained from being able to do harm. Marks can be made upon the doll, indicating injuries or illnesses to be cured or created.

Amulets and charms are objects kept near a person or place to attract the desired or repel the unwanted. Their power is believed to be mainly symbolic, either through drawing on the influence of what they represent or reminding the owner of their power. The composition and shape of the item are often specific to the tradition. Familiar amulets in North American society include clovers, rabbit feet, horseshoes, and various religious symbols.

Healing magic is used either as a replacement for or a supplement to oral or topical medicines when an illness has a strong psychological or spiritual cause. Similar to faith healing, it is meant to work without directly affecting the biological or chemical attributes of the subject. The systems of healing magic are based on concepts such as energy flow and spiritual unrest. Magical or religious rituals are often employed for exorcism, psychological therapy, or spiritual cleansing.

Ancestral veneration involves praying to the spirits of deceased family members and, in its magical form, asking for their assistance in completing a deed or in divination. Divination is the prediction of unknown and/or future events with the aid of tools such as tarot cards or casting sticks, and it is commonly employed by practitioners of witchcraft.

Sacrifices in magical rituals have been documented in several cultures throughout history. Other offerings such as food and valuables are also commonly found in magical rituals, especially when asking for the aid of spirits or gods. Sacrifices and offerings have been used to encourage spiritual favor. The use of offerings in rituals has continued to this day, although the practice of sacrifice has diminished considerably.

Witchcraft has often been dismissed as groundless because the link between the alleged influence on a problem and the desired effect cannot be proven to exist by current scientific methods or standards. Because witchcraft recognizes that multiple causes can affect a given outcome, practitioners often refute claims that witchcraft has no effect by stating that other factors may have prevented the desired outcome. These other factors include a range of scientific and superstitious causes. As the superstitious causes are highly fallible, they are often dismissed as unsound, resulting in the dismissal of witchcraft as a groundless enterprise.

Practitioners recognize the psychological and sociological impacts of witchcraft, but they deny any claims that these are the only effects of their practice. Critics, by contrast, will attribute *only* psychological and sociological impacts to witchcraft because they are the only effects that can be tested in controllable environments. But even in such environments, not enough factors are controllable to produce evidence with absolute certainty.

European societies in the Middle Ages were highly superstitious and merged the new Christian theology of that era with pre-Christian practices. People believed in phenomena such as poltergeists, spirits, and the Devil. Such superstitions were the backbone of the witchcraft practiced and understood by European societies in that period. The majority of Europeans believed in the power and existence of witchcraft, and historians have noted that this belief was enough to render witchcraft effective psychosomatically. Others attribute the effects of seventeenth-century witchcraft to hysteria, in both the medical and colloquial sense.

The most common practices of witchcraft in this period included amulet and image magic. The most powerful witches were considered to have been those who entered into a pact with the Devil and agreed to work for him in exchange for magical powers. All witchcraft was considered immoral, and those who were suspected of practicing witchcraft were persecuted. If someone was accused of witchcraft with supporting testimony, death was almost certain, for most attempts at proving innocence were futile, often interpreted as a lie to protect the witch. As a result of the persecutions, any actual practitioners either altered their rituals in order to go unobserved or took their practices underground to avoid suspicion.

Contemporary historians are often critical of Middle Age European testimonies about witchcraft because of the methods used during the witch persecution trials. It was common practice to extract confessions through various forms of torture, including the removal of fingernails, the insertion of needles, and other forms of forced mutilation. Historians have noted that under such circumstances, few would have withheld the confession the court desired. Others were deprived of sleep and/or food, which inevitably led to delusions and weakened physical and psychological resistance. Although these methods are currently viewed by most as inhumane and unscientific, they were common and widely acceptable practices at the time. Today, the term *witch trial* is often used as a colloquial reference to trials that use methods of torture and psychological manipulation in order to extract the desired information or confession.

When Europeans first encountered witchcraft outside of Europe, the practice was assumed to be a savage custom of unenlightened persons. Although European witches were persecuted, non-European witches were treated differently: efforts were made to dissuade them from their practices and convert them to Christianity. Like their European counterparts, the majority of these practitioners either altered their practices or infused the new knowledge with their existing practices. Because of the adaptability of the practitioners, the majority of magical practices have survived to this day.

Currently, witchcraft is one of the central aspects of the neopagan religion Wicca. Whereas the majority of modern practitioners in Europeanized countries are Wiccan, some choose to identify themselves only as witches or magicians. Methods of witchcraft and magic vary and still feature both ritual and folk magic. As explained by Wiccan practitioners, all acts of magic include the basic requirements of will, need, desire, and energy. Many will insist that the lack of any of these aspects will weaken the power of the act. Most consider disbelief a lack of will because the practitioner does not will the desired results to occur. Critics say that the desire for magic to be real will often cause the practitioner to ensure the occurrence of the expected outcome through nonmagical methods. Some practitioners agree but add that this is actually part of the act of magic—or the entire act itself.

Practitioners of witchcraft and magic are still persecuted, even though the majority of them maintain that they do not believe in the Devil and work only for positive change. This persecution varies in degree of physical and psychological violence. Even in the late 1990s, there were cases in which schools in North America prohibited students from wearing symbols connected to witchcraft. Practitioners of witchcraft in other cultures are still being dissuaded from their practices by nonbelievers.

References:

Hansen, Chadwick. 1969. *Witchcraft at Salem.* New York: G. Braziller.

Pinch, Geraldine. 1994. *Magic in Egypt.* Austin: University of Texas Press.

Russell, Jeffrey Burton. 1972. *Witchcraft in the Middle Ages.* London: Cornell University Press.

Starhawk. 1999. *The Spiral Dance: A Rebirth of the Ancient Religion of the Great Goddess.* New York: HarperCollins.

2

INVESTIGATIONS FROM SKEPTIC MAGAZINE

Acupuncture

DR. GEORGE A. ULETT, M.D.

"Acupuncture works!" Thus spake the consensus panel of the National Institute of Health and the Office of Alternative Medicine in November, 1997, dutifully recorded by the nation's press. But wait—the Chinese have known this for 3,000 years! The panel commented that, "The general theory of acupuncture is based on the premise that there are patterns of energy called Qi . . . throughout the body that are essential for optimal health." Blockage of this energy is thought to cause disease. Further, the panel described acupuncture as "a family of procedures . . . which involves . . . penetration of the skin . . . by thin, solid, generally metallic needles."

In his opening remarks, Wayne Jonas, head of the Office of Alternative Medicine, emphasized the need for scientific rigor in investigating this practice that is commonly described in unscientific terms. Given a scientific explanation, meaningful experiments could be conducted in the U.S. It is therefore most unfortunate that the panel paid little heed to the research findings reported by Dr. Pomeranz of Canada and Professor JiSheng Han of China, who clearly explained the endorphin mechanisms that have now replaced the ancient theories of an imaginary, mystical Qi. Thus, the panel missed a great opportunity to clarify for the public the confusion that now exists about use of the term "acupuncture," for there are dozens of kinds of acupuncture, including: laser acupuncture, ear acupuncture, acupressure, and Korean hand acupuncture, to name a few. Without exception all are based on magical theories that have never been proven.

It is a sad commentary on science education in this country when our citizens, aided by a sensation-seeking media, fall for such nebulous absurdities. China, where the metaphysical theories of acupuncture originated, is now becoming scientifically sophisticated and is moving away from these ancient pre-scientific folkways. Recently the Chinese government's Central Committee, together with the Chinese Academy of Science, proclaimed itself an advocate of the scientific attitude and in opposition to ignorance, superstition, and pseudoscience. Naive beliefs in the unproven explanations of Qi are weakening and so-called "masters of Qigong" are being unfrocked and subject to legal sanctions (Shen Zhenyu, 1997). In the U.S. however, Qigong lecturers are increasingly becoming a mainstay at seminars on alternative medicine and acupuncture.

At the November 3–5, 1997, Bethesda Consensus Meeting, findings from China supporting an evidence-based explanation of the mechanism of acupuncture were presented by the internationally recognized scientist Professor JiSheng Han from Beijing Medical University. He made it very clear that useful clinical acupuncture does not depend on belief in theories of Traditional Chinese Medicine and that it has nothing to do with meridian theory, treating blockages of Qi, yin/yang balancing, five element theory, or pulse diagnosis. Thirty years of careful research from his laboratory, with recent support from NIDA, has yielded evidence that replaces concepts of Qi with sound neurophysiological findings. To call this new practice "acupuncture" may appear somewhat misleading because "acu" means sharp and "puncture" implies piercing the skin. This new treatment does neither but, like the fabled Chinese Phoenix bird, this method has arisen from the ashes of the now obsolete Traditional Chinese Acupuncture (TCA). Although appropriately described as neuro-electric stimulation, this method falls under the umbrella term "acupuncture" because, increasingly in China and elsewhere, it is being accepted as a more effective way of treating conditions for which patients previously received TCA.

This new evidence-based acupuncture uses specific parameters of electrical stimulation and is done without needles. Its purpose is to enhance the gene expression of endorphins and other healing neuropeptides. This simple technique can be taught to physicians in a single brief session. Yet in some states today, physicians whose medical license permits surgery and other invasive treatments cannot legally administer simple scientific acupuncture treatments until they have received state certification following a mandatory several hundred hours of training in the occult metaphysical rituals of Chinese medicine. This problem is one that should have been addressed at the Bethesda meeting. Instead, the consensus panel suggested that certification examinations should also be offered in languages other than English.

PAIRS OF OPPOSITES (YIN/YANG)

YIN	YANG
earth	heaven
moon	sun
winter	summer
autumn	spring
female	male
cold	hot
inside	outside
dark	light
small	large
weak	strong
lower	upper
water	fire
night	day
right	left

Table 1

From Folklore to Fantasy

It is of interest to look at the history of how Traditional Chinese Acupuncture arose. Three thousand years ago ancient Chinese physicians serendipitously stimulated the skin sensors for pain, touch, and temperature by using, respectively, acupuncture needles, acupressure, and moxibustion. Together with herbs and a holistic approach to health care, this developed into the complex system of Chinese medicine. The theoretical explanations developed were meaningful to an agrarian people steeped in a folklore replete with concepts of cosmology and numerology.

The *Nei Ching, Yellow Emperor's Classic of Internal Medicine* (2697–2597 BCE), described the essential metaphysics of Traditional Chinese Medicine (TCM). In this, an hypothesized energy called Qi was thought to travel throughout the body in conduits called meridians. The main ones were 12 in number, like the months of the Chinese zodiac, each representing a major organ system of the body. In the *Nei Ching*, his minister explained to the Yellow Emperor that there are on these channels 365 hsueh or acupuncture points, "one

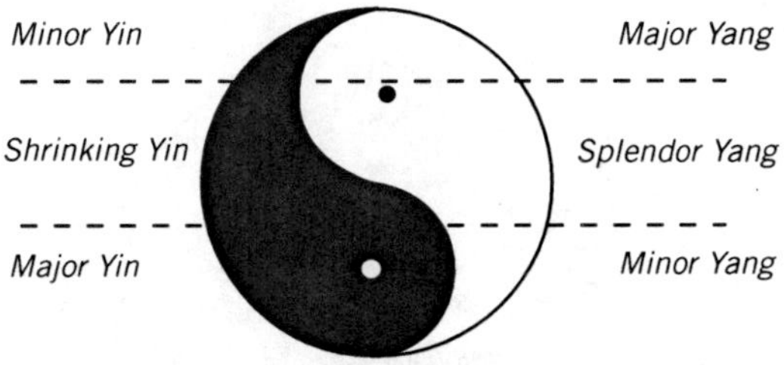

Figure 1

for each day of the year." At such points Qi was to be treated in order to supposedly relieve blockages believed to be the cause of illness.

The medical systems of all great civilizations have had knowledge of such hyperalgesic points but nowhere have they approached the systemic perfection and complexity of Chinese sinarterology ("meridian theory"). While modern medicine deals with material structures and tissues, classical Chinese medicine deals with theoretical functions to which physically demonstrated organs are only incidentally attached.

Treatment in TCM depends upon a knowledge of the complex philosophical laws governing the relationships of Yin and Yang (Table 1). In health there is a wholeness or balance of these forces, for as one increases the other decreases (Figure 1). Strength also varies with time of day and season of the year. These concepts influence the nature of the treatment and the prescription of foods and herbal medicines. Second only in importance to Yin and Yang are the five evolutive phases or elemental substances: wood, fire, earth, metal and water. As shown (Figure 2), the energy of these elements moves among the associated organs in circular fashion. Observing the direction of energy flow termed the sheng and ko cycles, traditional acupuncturists seek to tonify or sedate excesses or deficiencies of Qi.

The original ten organ systems relating to the five elements later became 12 by a splitting of the fire (heart) element into the "heart minister" (pericardium) and a vaguely described

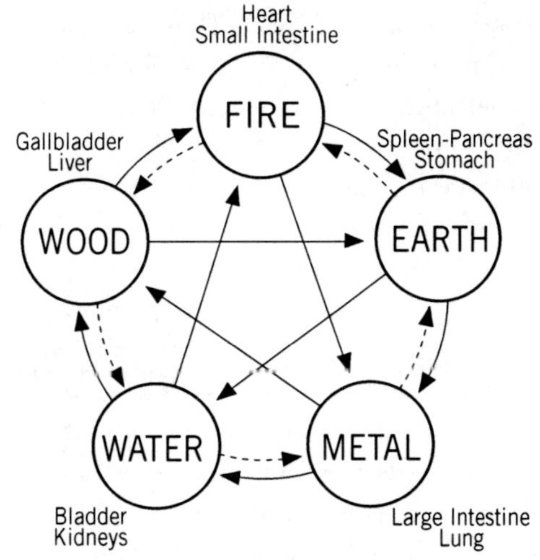

Figure 2

area, the "triple heater." This allowed for a balance between the left and right wrists necessary for pulse diagnosis. This is a method that claims the ability to discern six pulses on each wrist (Figure 3). These pulses represent six organs on the right wrist and six on the left wrist, detected by feeling 27 varieties of sensation by deep or shallow palpation. Pulse diagnosis is thought to indicate the location of blockages of Qi thus pointing to possible needling locations as points for treatment.

Through the centuries numerous changes and additions further complicated the meta-

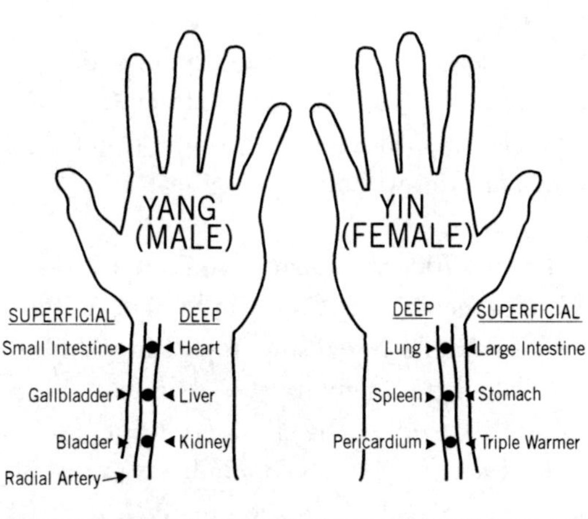

Figure 3

physical theories of TCM. There are two dozen forms of Qi flowing through an increased number of meridians with acupoints now believed to number near 1,000. Among the several dozen types of acupuncture treatment are included a number of microsystems such as auriculotherapy and Korean hand acupuncture. These posit the unproven belief that miniature maps of the body such as on the ear (Figure 4) or hand contain specific points representing individual organs of the body. Needle stimulation of these points is believed to directly affect the individual organs so represented. In addition to various types of needle stimulation, there is the use of colored lights, small magnets, and the burning of pellets of *Artemis vulgaris*, a procedure known as moxibustion.

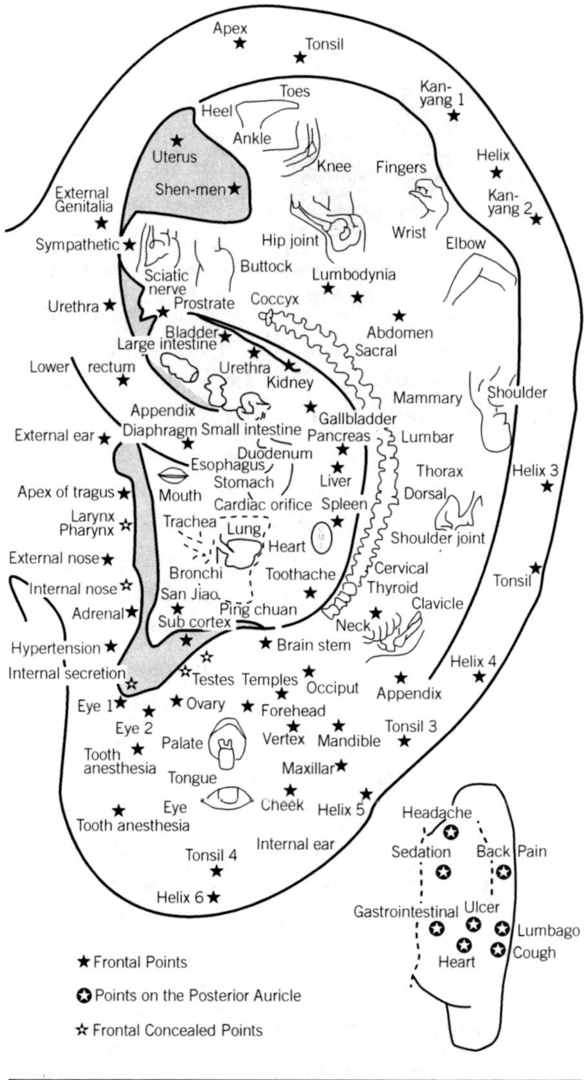

Figure 4

Despite its obscure metaphysical base, Chinese Medicine advanced throughout the ages on a par with and even exceeding some of the advances of medicine in other countries. The basic ancient metaphysics, however, remained unchanged despite remarkable advances in surgery, the discovery of vaccination and circulation of the blood before such knowledge was available in the West. The pre-scientific mystical belief in meridians as channels for Qi is still widely accepted and pulse diagnosis is practiced by those who ignore the anatomical knowledge that blood from the heart is pumped equally to the left and right radial arteries without passing near any 12 organ systems.

In the 1700s missionaries imported ideas of Western medicine to China, but as late as 1896 soldiers still harbored such superstitions as firing salvos to frighten the demons of the plague. The Boxer Rebellion (1900) marked a time of decline in the xenophobia that had effectively isolated China from many discoveries of Western science. Impressed by Western medicine, the Emperor in 1888 had banned the teaching of acupuncture in the Royal Medical Academy. Acupuncture resurfaced again in the 1940s when, after the Communist Revolution, Chairman Mao found himself faced with a population of millions and only a handful of Western-trained physicians. He solved this problem by reinstituting a nationwide program of Chinese folk medicine. The *Barefoot Doctor's Manual* became the bible by which thousands of Chinese "healers" were rapidly trained to spread TCM throughout the countryside. TCM worked well in China because it was nationalistic and deeply embedded in the Chinese culture.

Electricity and Acupuncture

The use of electricity to strengthen the acupuncture response has a long history. At the inception of acupuncture, vigorous manipulation of heavy needles in fibrous tissue may well have demonstrated the known piezoelectric effect and "pried off" a few electrons. It was not until 1765, however, that Gennai Hiraga of Envo, Japan reported the electrical stimulation of acupuncture needles. In France in 1825 Chevalier Sarlandiere described the application of an electric current from Leyden jars applied to acupuncture needles inserted for the treatment of rheumatic conditions. In the 1950s the Chinese reported the use of electro-acupuncture in order to produce the strongest possible analgesic effect for patients undergoing surgical procedures.

In 1971 President Nixon visited China. A *New York Times* reporter in his party, James Reston, experienced pain relief from acupuncture following emergency abdominal surgery. On his return to the United States he described the procedure in glowing terms that aroused wide interest, especially in patients seeking miracle cures for chronic diseases. China's ancient system of medicine thus received favorable recognition and TCM became a profitable export.

I had previously learned meridian theory acupuncture on a visit to Japan. This type of treatment helped some of my patients but I was chary of its mystical metaphysical explanations. There was some belief that acupuncture was a form of hypnosis. Our lab had been studying hypnosis and in 1972 we received an NIH grant to compare the effect of acupuncture and hypnosis on experimental pain. We found that acupuncture was not a form of hypnosis and that while needle acupuncture without electrical stimulation was somewhat effective, the addition of electricity increased effectiveness 100% (Figure 5; Parwatikar, et al., 1979). From studies of the literature, I then became convinced that instead of the many traditional acupuncture points, useful treatment could be given using only anatomically demonstrated motor points. I began an earnest search for a full scientific explanation of acupuncture. Data from the laboratories of Dr. Pomeranz of Canada (Pomeranz and Stux, 1979), and Professor Han of Beijing (Han, 1987), gave support for the publication in 1982 of my book *The Principles and Practice of Physiologic Acupuncture* (Ulett, 1982). Unfortunately, the AMA had been aware only of mysterious explanations of TCA and in 1974 declared acupuncture to be "quackery." This alienated U.S. physicians who were not disposed to learn a questionable technique despite the growing demand for it by their patients. Thus, the door was opened for the training of nonmedical persons to begin the practice of Chinese medicine under the name of "acupuncture." By 1995 there were reportedly over 20,000 "acupuncturists" in the U.S. Many are chiropractors, but only a few are

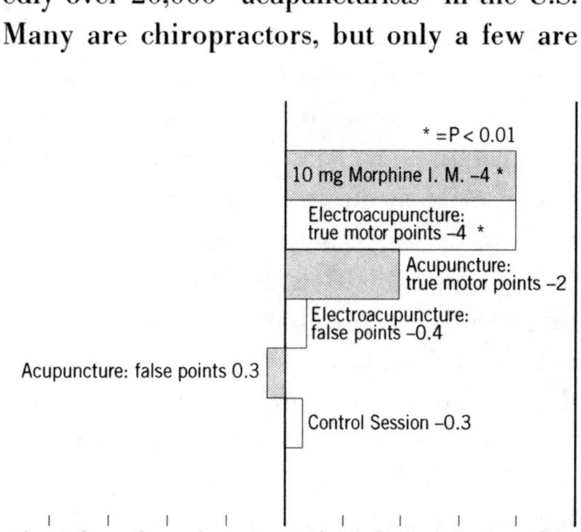

Figure 5

M.D.s or Doctors of Osteopathy. An increasing number of acupuncturists have no medical training. Many of them are combining other aspects of Oriental Medicine in their practice and hence a more accurate title would be "Oriental Medical Doctor" rather than "Licensed Acupuncturist."

The Evidence and the Effects

The schizophrenic nature of U.S. thinking about acupuncture is much in evidence. A popular book on acupuncture by Stux and Pomeranz (1987), entitled *Acupuncture: Textbook and Atlas,* is an example. Pomeranz is a recognized scientist who has promoted the idea that the acupuncture effect is mediated by endorphins. He announced at the Bethesda meeting that there are 17 well-documented lines of evidence supporting the fact that acupuncture works by an endorphin mechanism. In the first half of the book he gives research evidence for a scientific explanation of acupuncture. The remainder of the book seemingly ignores these facts and describes treatment methodology based on classical TCM meridian theory.

Such inconsistencies were prominent at the NIH consensus meeting. The panel's report that acupuncture is a useful treatment for a few specified medical conditions is tantamount to saying at a medical meeting, "drugs are useful" but without giving any indication of what drug, what dosage, or how it should be administered. Lacking at the conference was a clear description of the nature of the treatments given in terms of the "doses" of acupuncture and the guidelines followed for treating each patient. If the clinical studies presented were performed by "certified acupuncturists" who were consistent in their beliefs, they should have described their treatments with reference to curing the illnesses by unblocking Qi. Because Chinese medicine follows a holistic approach it stresses the individuality of each patient. The papers presented at the conference should have gone into detail about the manner of acupoints used for the treatment of each individual patient according to the ancient Chinese way of diagnosis. Also, it should have been reported if there were pulse changes after treatment indicating whether Qi blockages had been removed, or if there was a better balancing of Yin and Yang. Descriptions using such concepts of TCM would, of course, have rendered any meaningful scientific evaluation of treatment methodology impossible. The failure or inability of these presentations to describe how traditional Chinese acupuncture was used clearly indicates that the integration of traditional Chinese acupuncture into modern scientific medical practice is but wishful thinking. It is only by using factually supported scientific parameters of treatment, such as were presented by Dr. Han, that it would be possible to conduct meaningful research on acupuncture in the United States. Evidence-based acupuncture permits a description and control of variables that allow for replication and confirmation of research results by other investigators.

The Hidden Agenda

A hidden agenda item of the meeting was to support a continuation of the "cult of Qi." This became obvious when persons known to be teaching courses in meridian theory acupuncture discussed how 35 states have passed regulations requiring hundreds of hours of TCM training for acupuncture certification. Members of the American Association of Oriental Medicine (AAOM) have now compiled a TCM curriculum of 2,500 hours proposed as a national teaching standard. With the consensus panel's approval of acupuncture treatments,

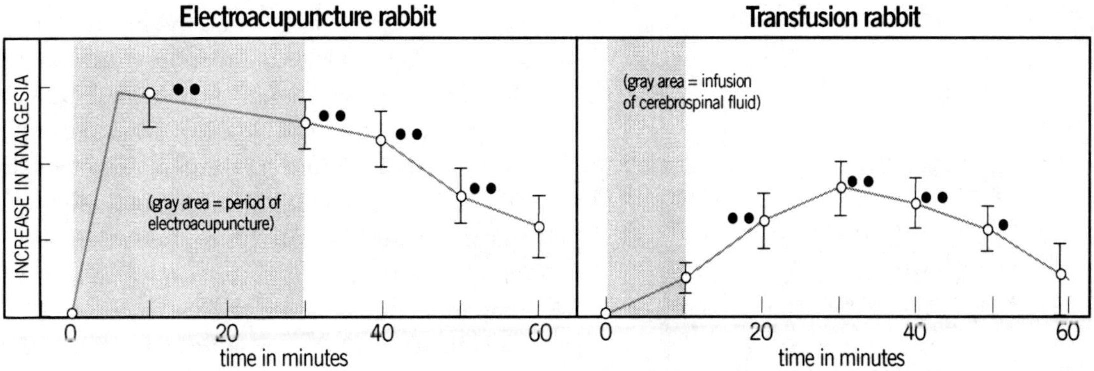

Figure 6

the AAOM's hand is strengthened to impose their occult beliefs on all 50 states. What a travesty this would be in view of the fact that scientific, needleless electro-acupuncture is now a simple evidence-based technique that can be taught to physicians in a single one-day seminar. The AAOM's mass training of "wanna-be doctors" would continue the teaching of the shamanistic cult of Qi and provide yet another costly and confusing deception of the medical public. Patients would be denied the benefits of receiving effective evidence-based scientific acupuncture treatments given by their own family physicians and adminstered only where medically appropriate.

To avoid such problems and to bring evidence based acupuncture within the domain of regular medical practice, the panel should have given greater credence to the outstanding scientific presentation by Professor JiSheng Han from Beijing College of Medicine. He described the results of 30 years of work from his laboratory. According to Han's now internationally acclaimed findings, acupuncture works primarily through neurochemical mechanisms that have been demonstrated in both animals and humans (Han and Terenius, 1982). Electrically induced acupuncture analgesia can be effected by a transfer of spinal fluid from a treated to an untreated animal (Figure 6). The neurochemicals involved include among others, Beta-endorphin, enkephalin and dynorphin. The gene expression of these neuropeptides has been demonstrated in studies of samples of human spinal fluid taken during acupuncture treatment. Specific frequencies of stimulation have been found to affect different neuropeptides. Two hertz stimulation increases the gene expression of endorphins while 100 hertz increases dynorphins (Figure 7). The endorphin mediated effect of electro-acupuncture can be blocked by naloxone. A cross tolerance between electro-acupuncture and morphine has been demonstrated. Parameters of electrical stimulation for maximum therapeutic effect were delineated, and it was shown that needles are no longer necessary; polymer conducting pads on the skin surface are sufficient (Figure 8). Dr. Han displayed the neuroelectric stimulator or HANS that he developed in his laboratory. He presented data derived from using this stimulator for 30 minutes prior to surgery, that showed the amount of gas anaesthetic to be reduced by as much as 50%. Using Han's method, stimulation of effective motor points on the hand can greatly relieve the unpleasant symptoms of withdrawal from drug addiction. My coworker has effectively used electrical stimulation with the HANS stimulator in his methadone clinic in Australia (Ulett and Nichols, 1996). Han's method is unlike the

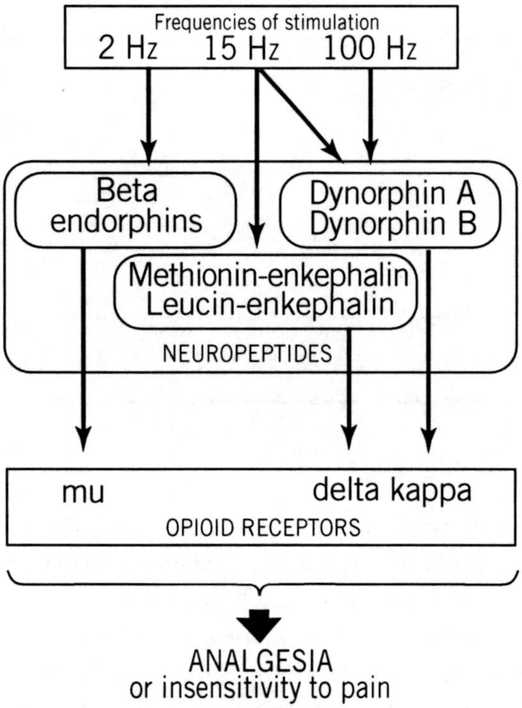

Figure 7

placebo method of ear acupuncture for addiction treatment widely used in the U.S. (Wells, et al., 1995). It is significant that the method of using unstimulated needles in the ear is routinely done within a treatment program resembling group therapy. Addictionologists seem to have somehow forgotten that Dr. Wen of Hong Kong, who introduced acupuncture treatment for addiction, specified the necessity for electrical stimulation (Wen and Cheung, 1973). Dr. Han's method emphasizes that it is the manner of electrical stimulation rather than any specific placement of needles that is the essence of acupuncture treatment.

Han's data in support of a scientific method of acupuncture were clearly presented at the consensus conference. The conclusion of the panel should not have been a weak endorsement of the ancient Chinese acupuncture method for some selected conditions. Rather it should have been a condemnation of the presenters for ignoring the scientific advances discussed by Drs. Pomeranz and Han that have been available in the literature for more than a decade. It is the method of evidence-based acupuncture that should have been used in the research studies presented. The panel should have pointed out to the public that in the U.S. many so-called "acupuncturists" are simply practicing what is in essence a form of pseudo-medicine. They neglect or are apparently unaware of the scientific facts now available that would allow them to move beyond the occult theories and rituals of Traditional Chinese Medicine. Should such persons wish to continue these irrational practices their licensure should indicate that they are practicing mystical Oriental medicine. Only in this way can the public identify practitioners of evidence-based, effective, neuroelectrical acupuncture from those whose treatments are more likely to be of a placebo nature. With the scientific data now available neuroelectric acupuncture can be seen as an example of an alternative practice that has become evidence-based medicine. It is a procedure for eliciting the gene expression of healing neurohormones by the electrical stimulation of motor points. It should be placed in the curriculum of all medical schools as a simple and readily taught useful procedure that does not require any special certification for physicians. The method is simple and painless without the dangers that can occur with invasive techniques. I have for years used Professor Han's technique in my clinical practice and have found it more effective than the ritualistic mystical procedures of TCM that I learned and practiced 25 years ago. It has been effective for many types of musculoskeletal pain both acute and chronic, for the relief of anxiety, addiction, and various psychosomatic conditions. Using my text, *The Biology of Acupunture* (Ulett and Han, 2002), with its illustrated atlas of motor points, I teach this method to physicians and medical students. Media hype to the contrary, there is no alternative to good medicine (Ulett, 1996).

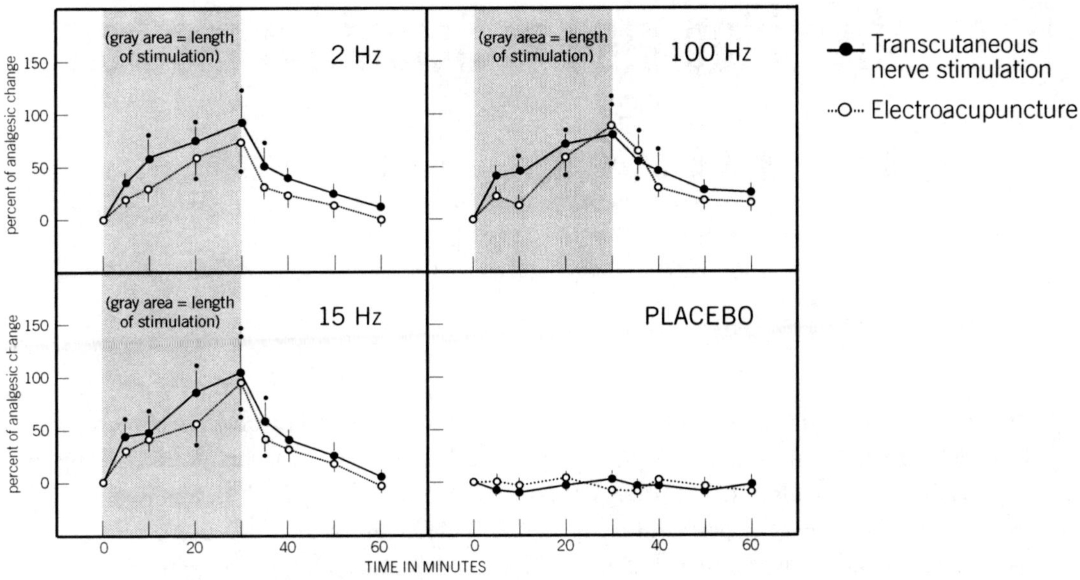

Figure 8

Modern evidence-based neuroelectric acupuncture should no longer be thought of as a part of alternative medicine but rather as a useful scientific technique for inclusion as part of the regular armamentarium of all practicing physicians.

References:

Han, J. S. 1987. *The Neurochemical Basis of Pain Relief by Acupuncture.* Beijing: Chinese Medical and Technology Press.

Han, J. S., and L. Terenius. 1982. "Neurochemical Basis of Acupuncture Analgesia." *Annual Review of Pharmacology and Toxicology* 22:193–220.

Parwatikar, S., M. Brown, J. Stern, G. Ulett, and I. Sletten. 1979. "Acupuncture, Hypnosis and Experimental Pain. I. Study with Volunteers. Acupuncture and Electrotherapeutics." *Res. Int. J.* 3:161–190.

Pomeranz, B., and Stux, G. (Eds.) 1979. *Scientific Basis of Acupuncture.* New York: Springer Verlag.

Shen Zhenyu. 1997. "China's Scientific Community Denounces Superstition and Pseudoscience." *Skeptical Briefs* 7:11: 18.

Stux, G., and B. Pomeranz. 1987. *Acupuncture: Textbook and Atlas.* Springer Verlag, Heidelberg, Berlin.

Ulett, G. 1982. *Principles and Practice of Physiologic Acupuncture.* St. Louis: Warren H. Green.

Ulett, G. 1996. *Alternative Medicine or Natural Healing.* St. Louis: Warren H. Green.

Ulett, G., and S. P. Han. 2002. *The Biology of Acupuncture.* St. Louis: Warren H. Green.

Ulett, G., and J. Nichols. 1996. *The Endorphin Connection.* Glebe, Australia: Fast Books, Wild and Woolley Pty. Ltd.

Wells, E., R. Jackson, R. Diaz, V. Stanton. 1995. "Acupuncture as an Adjunct to Methadone Treatment Services." *American Journal of Addiction* 4:198–214.

Wen, H. S. Cheung. 1973. "Treatment of Drug Addiction by Acupuncture and Electrical Stimulation." *Asian Journal of Medicine* 9:138–141.

Alternative Medicine v. Scientific Medicine

DR. HARRY K. ZIEL, M.D.

For the past four decades I have been practicing what is known as allopathic medicine, also called traditional or conventional medicine. For the past century it is this version of medical practice that has been in vogue in North America and is seen today as the standard of medical care. The medical methodology centers around a physician who takes the patient's chief complaint and health history, performs a physical examination, makes a provisional diagnosis, orders laboratory testing, requests referral if needed, outlines therapy, and prescribes medication and/or surgery. This relationship is based on scientific evidence that the proposed therapy is the optimal treatment. Medications prescribed are sanctioned by the Food and Drug Administration. Outcome data are screened and subjected to peer review through quality of care committees. Feedback to the physician of these outcome data with suggested recommendations, if any, for improvement in care closes the information loop.

Over the past decade, however, there has been a sharp increase in what is called complementary or alternative medicine (CAM) that is viewed as an accessory to or replacement of standard medical care. CAM care involves additional, nonconventional modes of therapy. Examples fall into four distinct categories:

1. Diet, Nutritional, and Lifestyle Changes: Nutritional counseling, herbal medicine, exercise routines, etc.
2. Mind-Body Control: Biofeedback, Psychotherapy, Hypnotherapy, Support Groups, Yoga, Tai-Chi, Meditation, Stress Reduction, etc.
3. Manual Healing: Osteopathic manipulation, Chiropractic manipulation, Physical therapy, Acupressure, etc.
4. Alternative Systems: Preventive Medicine, Acupuncture, Acupuncture with Electric Stimulation, Homeopathy, Naturopathy, Chinese Medicine, etc.

It's a war out there. Both sides hold strong opinions and clash over every aspect of healing. Physicians practicing traditional allopathic medicine (TAM) feel they've done their job when they make the diagnosis and prescribe the therapy. They might comment, "If my patient throws away my prescription once he's left my presence, the patient has only himself to blame for his failure to get well." On the other hand, the patient who discards the prescription might retort, "The doctor didn't listen. I know that he had 10 minutes set aside to see me and he's under the gun to complete a tight appointment schedule, but how can I know that he really knows my problem? How can I know that he is really doing his best for me?"

This noncompliant patient is today's fastest growing group of unsatisfied people. Who are these patients? The majority are white college-educated women, age 25 and up. Their annual family income exceeds $50,000. In 1997, the number of visits to CAM providers exceeded the number of visits to TAM providers. From 1990 to 1997, the number of visits to CAM providers increased by 47 percent, from 427 to 629 million visits.

Alarmingly, only 40 percent of patients who utilize CAM inform their regular physician that they are doing so. Dangerous drug interactions can result. For instance, when patients who fail to tell their physicians that they are taking herbs or megavitamins are then prescribed traditional coumadin anticoagulant medication, anticoagulant therapy regulation is pushed in and out of therapeutic range. Myocardial infarction, stroke, pulmonary embolus, hemorrhage from the kidneys, and other complications may occur.

In 1997 total payments to CAM providers reached $27 billion. This amount is similar to the 1997 out-of-pocket payment for all physician services in the United States. The increase in CAM payments is mostly attributable to more people seeking CAM care, not increased visits per individual. So CAM care is catching on and high cost doesn't appear to be an obstacle. To understand the reason we must take a brief look at the history of health care in the United States.

A Brief History of Medicine

Medical care in United States has evolved through several stages. At first health care was basically the art of giving tender loving care (TLC). Little in the way of technical skills, in contrast to today's practice, was available. Abscesses were incised and drained. Poultices were applied to festering wounds. Opium was given for pain. Decayed teeth were extracted.

Snake oil, cupping, bleeding, leeches, purges, nostrums, and potions (some poisonous) constituted the early physician's formulary. Since 85 percent of our maladies are self-limiting and are cured by nothing more than tincture of time, the less the patient saw the doctor, the better was his chance of recovery. Simply put, early physicians were capable of doing more harm than good. Only the very fittest survived the ministrations of these early physicians.

Into this grim medical culture came Samuel Hahnemann (1755–1843) who founded the school of homeopathic medicine. Because it was safer not to go to the doctor, Hahnemann's concept that medication given in the most dilute amounts had the greatest therapeutic value really did save lives. When a poison was diluted to less than trace amounts and was administered to an ill individual, the poison did no harm and did not interfere with the body's natural healing process. This is homeopathy's greatest claim to success. Homeopathic physicians did not interfere with mother nature's relentless attempt to restore good health.

It is said that there is no greater fool than he who fools himself. Homeopathic physicians were prime examples of this truth. Yet they did satisfy the primary rule of all medicine, Primum Non Nocere, first do no harm. As such, homeopathy represented an advance in medical care over the impediments to healing administered by early physicians.

The age of skepticism brought medicine's next major advance—allopathy. Anecdotes gave way to experiments. Drugs given to study groups had to show statistically significant better results than placebos given to a control group in order for the drug to be accepted for widespread public use. Ideally, which person gets a real drug and which person gets a sugar pill should not be known to either the patient or the project manager (a double blind study). Researchers test and retest results. Medical knowledge is cumulative. As newer and better medications become available, they replace earlier versions already in use.

Modern pharmaceutical companies market a whole host of drugs upon which the TAM or traditional physician draws to treat his patients. Diseases that carried death sentences at the turn of the century—pneumonia, tuberculosis, diabetes, hypertension, congestive heart failure, and myocardial infarction—now can be checked and cured, or controlled with effective medications. Anesthetics—local, regional, and general—are available to facilitate ever more venturesome surgeries. When Cesareans were first done without anesthesia, transfusion, or antibiotics (before 1900), it was a rare mother who survived. Today the mortality from Cesarean is less than 5 per 10,000 Cesareans (depending on the series reported).

Never before in history has medicine had the benefit of such extensive scientific testing and watchdog institutions as it has today. The Food and Drug Administration oversees the value of drugs and devices. Quality of care committees supervise ongoing care at the local level. Investigative review boards give approval for studies, and professional journals provide a means for peer review of study results.

Then why are increasing numbers of patients abandoning traditional medicine and turning up on the doorsteps of CAM? Why are they seeking alternative treatment?

References:

Special theme issue of the *Journal of the American Medical Association*, November 11, 1998, Vol. 280, No. 18:

Barnett, L. M. Book Review: *Kindly Medicine: Physio-Medicalism in America, 1836–1911*, by John S. Haller, Jr. *JAMA*, 1634–1635.

Eisenberg, D. M., Davis, R. B., Ettner, S. L., Appel, S., Wilkey, S., Van Rompay, M., Kessler, R. C. "Trends in Alternative Medicine Use in the United States, 1990–1997. Results of a Follow-up National Survey." *JAMA*, 280: 1569–1575.

Ernst, Edzard. 1998. "Harmless Herbs? A Review of the Recent Literature." *The American Journal of Medicine*, 104: 170.

Eskinazi, D. P. "Policy Perspectives Factors that Shape Alternative Medicine." *JAMA*, 280: 1621–1623.

Ezzo, J., Berman, B. M., Vickers, A. J., Linde, K. "Policy Perspectives Complementary Medicine and the Cochrane Collaboration." *JAMA*, 280: 1628–1630.

Fontanarosa, P. B., Lundberg, G. D. "Editorial Alternative Medicine Meets Science." *JAMA*, 280: 1618.

Iserson, K. V. Book Review: *Alternative Medicine and Ethics*, edited by James M. Humber and Robert F. Almeder (Biomedical Ethics Review). *JAMA*, 280: 1633.

Jonas, W. B. Editorial Alternative Medicine: "Learning from the Past, Examining the Present, Advancing to the Future." *JAMA*, 280: 1616.

Margolin, A., Avants, S. K., Kleber, H. D. "Policy Perspectives Investigating Alternative Medicine Therapies in Randomized Controlled Trials." *JAMA*, 280: 1626–1628.

Marwick, C. "Medical News and Perspectives Alterations are Ahead at the OAM." *JAMA*, 280: 1553–1554.

McPartland, J. M. Book Review: *Alternative Medicine: What Works*, by Adriane Fugh-Berman. *JAMA*, 280: 1635.

——. Book Review: *The Alternative Medicine Hand Book: The Complete Reference Guide of Alternative and Complementary Therapies*, by Barrie R. Casasileth. *JAMA*, 280: 1635–1636.

Mitka, M. "Medical News and Perspectives: FDA Never Promised an Herb Garden—But Sellers and Buyers Eager to See One Grow." *JAMA*, 280: 1554–1556.

Studdert, D. M., Eisenberg, D. M., Miller, F. H., Curto, D. A., Kaptchuk, T. J., Brennan, T. A. "Medical Malpractice Implications of Alternative Medicine." *JAMA*, 280: 1610–1615.

Sugarman, J. "Policy Perspectives Physicians' Ethical Obligations Regarding Alternative Medicine." *JAMA*, 280: 1623–1625.

Udani, J. "Resident Forum Integrating Alternative Medicine Into Practice." *JAMA*, 280: 1620.

Uzych, L. Book Review: *Complementary and Alternative Medicine: Legal Boundaries and Regulatory Perspectives*, by Michael H. Cohen. *JAMA*, 280: 1633–1634.

Special theme issue of *Skeptic*, Vol. 3, No. 1, 1994:

Jarvis, W. "Homeopathy: A Position Statement by the National Council Against Health Fraud." 50–57.

Raso, J. "Chasing Shadows on a Pitch-Black Wall: Spiritual Belief Systems Compete as Alternatives to Scientific Health Care." 58–69.

Rosa, L. "Therapeutic Touch: Skeptics in Hand to Hand Combat Over the Latest New Age Health Fad." 40–49.

Additional Journal Articles:

Astin, John. 1998. "Why Patients Use Alternative Medicine." *JAMA*, 279: 1548–1553.

Berman, B. M., Larson, D. B. 1992. "Alternative Medicine: Expanding Medical Horizons." US-GPO: x.

Eisenberg, D. M. 1997. "Advising Patients Who Seek Alternative Medical Therapies." *Ann. Inter. Med.* 127: 61–69.

Eisenberg D. M., et al. 1993. "Unconventional Medicine in the United States. Prevalence, Costs, and Patterns of Use." *NEJM*, 328: 246–252.

Fugh-Berman, A. 1996. "Alternative Medicine: What Works. A Comprehensive, Easy to Read Review of the Scientific Evidence, Pro and Con." Odonian Press: 5.

Gordon, N., et al. 1998. "Use of and Interest in Alternative Therapies Among Adult Primary Care Providers and Adult Members in a Large HMO Setting." *WJM*, Sept.

Koes, B., et al. 1996. "Spinal Manipulation for Low Back Pain: An Updated Systematic Review of Randomized Clinical Trials." *Spine*, 21; 24: 2860–2873.

NIH Consensus Statement on Acupuncture. 1997. Available by calling 888-NIH-CONSENSUS.

NIH Office of Alternative Medicine. 1997. "Clinical Practice Guidelines in Complementary and Alternative Medicine: An Analysis of Opportunities and Obstacles. Practice and Policy Guidelines Panel." *Arch. Fam. Med.*, 6.149–154.

Rosa, L., E. Rosa, L. Sarner, S. Barrett. 1998. "A Close Look at Therapeutic Touch." *JAMA*, April 1, 279: 1005–1010.

Shekelle, P., et al. 1992. "Spinal Manipulation for Low Back Pain." *Ann. Inter. Med.*, 192; 117: 590–598.

Ulett, G. 1996. *Alternative Medicine or Magical Healing*. St. Louis: Warren H. Green.

——. 1997. "Acupuncture's Secrets Revealed: From Cult of Qi to Evidence-Based Acupuncture." *Skeptic*, Vol. 5, No. 4: 46–51.

Vincent C., Furnham, A. 1996. "Why Do Patients Turn to Complementary Medicine?" *British Journal of Psychology*, 35: 37–48.

Wetzel, M. S., et al. 1998. "Courses Involving Complementary and Alternative Medicine Schools." *JAMA*, 280: 784–787.

Books:

Barrett, S., and R. E. Gots. 1998. *Chemical Sensitivity: The Truth About Environmental Illness*. Buffalo: Prometheus Books.

Barrett, S., and W. T. Jarvis (Eds.). 1993. *The Health Robbers*. Buffalo: Prometheus Books.

Cohen, Michael H. 1998. *Complementary and Alternative Medicine. Legal Boundaries and Regulatory Perspectives*. Johns Hopkins University Press.

Ernst, Edzard. 1996. *Complementary Medicine: An Objective Appraisal*. Butterworth-Heinemann.

Featherstone, C. 1997. *Medical Marriage*. Findhorn Press.

Helman, Cecil. G. 1994. *Culture, Health, and Illness*. Butterworth.

Helms, J. 1995. *Acupuncture Energetics*. Medical Acupuncture Publishers.

Leslie, C. 1976. *Asian Medical Systems*. Univ. Cal. Press.

Micozzi, M. 1996. *Fundamentals of Complementary and Alternative Medicine*. Churchill Livingstone.

Mowry, D. B. 1986. *The Scientific Validation of Herbal Medicine*. Keats Publishing.

Randi, J. 1989. *The Faith Healers*. Buffalo: Prometheus Books.

Raso, J. 1994. *Mystical Healing*. Buffalo: Prometheus Books.

——. 1996. *"Alternative" Healthcare: A Comprehensive Guide*. Buffalo: Prometheus Books.

Spencer, J. W., Jacobs, J. J. 1999. *Complementary/Alternative Medicine: An Evidence Based Approach*. Mosby.

Tyler, V. 1993. *The Honest Herbal*. Pharmaceutical Products Press/Hawthorn.

——. 1994. *Herbs of Choice: The Therapeutic Use of Phytomedicinals.* Pharmaceutical Products Press/Hawthorn.

Journals:

Alternative and Complementary Therapies. Mary Ann Liebert, Publ.

Alternative and Complementary Medicine. Mary Ann Liebert, Publ.

Alternative Therapies in Health and Medicine, Innovision Communications.

Consumer Reports, P.O. Box 53029, Boulder, CO 80322.

National Council Against Health Fraud Newsletter, P.O. Box 1276, Loma Linda, CA 92354.

Scientific Review of Alternative Medicine. Subscriptions: Prometheus Books, 59 John Glenn Dr., Amherst, NY 14228. 800/421-0351.

University of California at Berkeley Wellness Letter, P.O. Box 420148, Palm Coast, FL 32142.

Web Sites:

http://altmed.od.nih.gov (NIH, Office of Alternative Medicine)

http://consensus.nih.gov (NIH, Consensus Development Program)

Atlantis

The Search for the Lost Continent

PAT LINSE

Have you heard the story of the lost continent of Atlantis? Its people were beautiful, tall, athletic, and rich beyond compare. The buildings of its cities had walls covered with gleaming gold and roofs set with sparkling many-colored gems. Its streets were flower-lined canals crowded with ships flying silken banners that shimmered in the sun. But all was lost in a single horrible day of earthquakes and gigantic waves when the entire continent sank beneath the dark waves of the Atlantic ocean.

Thousands of books and over a half million internet web pages have described every aspect of Atlantean life. We can read about how they dressed, what they ate, what their art looked like, and what their music sounded like. We can read about their amazing technology now lost to modern humankind.

The Atlantis story as it is told today offers a compelling explanation for the rise of civilization. This wildly imaginative version is pieced together from bits of a dozen different Websites:

> The citizens of Atlantis had blue blood which gave them skin of a beautiful violet color and they often stood over 8 feet tall. They possessed advanced crystal technology that generated unlimited free energy. Crystals of different colors allowed them to completely control the weather and cure every disease. Their life span was over 800 years long. Some say that they came from planets beyond our own galaxy.
>
> They bred shorter inferior humans—us—as slaves to attend to their every need. But because their life was so easy they were bored. They set apart large areas of their continent as national parks where the humans were allowed to live. In these areas they staged violent storms, earthquakes and volcanos for their amusement, caring little about the unfortunate effects these disasters had on their slaves. Then one day their technology spun wildly out of control and the entire continent was destroyed and sank beneath the sea without a trace.
>
> But slaves who had formerly escaped from the continent and settled elsewhere in the world remembered the culture of their masters and tried to recreate it where ever they lived. They became the founders of the ancient civilizations that we today know as the Aztecs, the Mayans, the Incas, the Sumerians, the ancient Egyptians, and the ancient Chinese.

But no evidence of Atlantis has ever been found. Every mention of Atlantis can be traced back to a single source—the Greek philosopher Plato who first mentioned Atlantis over 2,500 years ago. Plato wrote a few pages describing the layout of the capital city and countryside and little else. The detailed information that fills the books and Websites is based on speculation, imagination, guesses, and even psychic channeling. No one paid much attention to Atlantis for a couple hundred years after Plato. Plato's own student, the philosopher Aristotle, was quoted as saying that Plato made up the story to make a point. Christian writers in the Middle Ages ignored Atlantis, considering it a pagan folk tale.

But about 2,000 years later, when Europe was shocked by the discovery of the American continents, a great excitement about Atlantis was kindled, and the possibility that it was more than a story was taken seriously. Both professional and amateur scientists alike began to search for traces of the lost continent.

A second surge of interest in Atlantis came in the 1800s when stories about Atlantis began to be combined with quests for spiritual knowledge.

Today the lessons of the story of the destruction of the corrupt pleasure-loving Atlanteans is more compelling than ever, because for the first time in history we actually have the technology to destroy civilization either through ecological disaster or nuclear war.

Who Was Plato and What Did He Write in His Dialogue about Atlantis?

Plato was a philosopher who lived 2,500 years ago in ancient Greece. Since he is the single source for the story, anyone searching for Atlantis has to start by looking carefully at what Plato wrote, and understand the context in which he wrote it.

But Plato was not writing history or geography when he mentioned Atlantis. He was writing about what the citizens in an ideal society should be like. He used Atlantis as an example of what can happen when a society becomes morally corrupt. Some people suppose that Plato wanted his readers to believe that Atlantis was real, and others argue that the story is just a way to make the moral lesson he was teaching more exciting and compelling.

Plato began his career writing plays. When he became a philosopher he invented a new form of writing that was like a short play, called a "dialogue." In a dialogue, several people who had different viewpoints argued about philosophical questions like "What is the best way to live an honorable life?" At the time, the study of "philosophy" (which means love of wisdom) also included questions about how the universe worked, so the characters in dialogues might also ask questions like "How were the stars created?" or "What happens to an object when it burns?" Plato felt that presenting several viewpoints was a way to promote a healthy mind and soul. He thought the process of thinking was as important as the conclusion that you might reach, and that exercising your mind was like going to the gym to exercise your body.

But the dialogue served as more than a way to make philosophical discussion dramatic. Plato lived in a conquered city where writing about how society should be organized was politically dangerous. Putting his opinions in the mouths of different characters gave him a measure of safety since no one could prove which character was expressing his view.

The Heroic Trinity

Plato was part of a group of three ancient Greek philosophers known as the Heroic Trinity. The first of the hero philosophers was Plato's teacher Socrates, Plato himself was the second, and Plato's student Aristotle was the third. Together they laid much of the foundation of modern Western thought.

Socrates was a remarkable character. Because he was disheveled and barefoot, a Greek playwright once joked that he was a disgrace to shoemakers. Yet his powerful personality, quick wit, and keen insights into human nature continue to inspire respect even today. Socrates tried to teach people to improve their thinking by challenging their views. This eventually created enemies and he was tried and sentenced to death by a vote of a committee of 30 citizens. Plato recorded the brilliant (but unsuccessful) speeches Socrates made in his own defense at his trial.

Aristotle was the first to analyze and classify things in a scientific manner. His influence dominated scientific thought so completely for 2,000 years that it eventually hindered scientific progress because people were reluctant to accept new discoveries when they contradicted Aristotle.

What Plato Wrote about Atlantis

Plato says the Atlantis story is about "the greatest action a people ever did." From all the excitement Atlantis has generated it would be reasonable to expect that these marvelous people would be the Atlanteans. But they are not. The story is actually about the "great and wonderful deeds" done by the people of *Athens* long ago! Plato wanted to inspire his fellow Athenians to value moral strength and become more like their noble ancestors because superior morality allowed the soldiers of a single city—Athens—to defeat an entire continent. From a dramatic standpoint, the more powerful and dangerous the Atlanteans were said to be, the more important Plato's moral instruction would seem. The two dialogues that mention Atlantis are named after their main speakers.

The Timaeus Dialogue

Socrates begins the dialogue by reminding Timaeus and Critias that he had asked them the day before to come up with examples of how an ideal society might act fighting in a great war. Critias says he knows a story that "by some mysterious coincidence" fits Socrates' idea exactly. He heard it when he was 10 from an aged poet. He has spent the night searching his memory, so that he could tell the whole story in detail.

Critias begins: "Listen Socrates to a tale which, though strange is certainly true, having been attested to by Solon." (Solon was a famous Greek leader and historian.) Solon visited Egypt where priests told him that the Greek Athenians knew nothing of their ancient history. They didn't even know about the greatest deed the Athenians ever performed—how the soldiers of Athens singlehandedly defeated a mighty empire located just outside the "Pillars of Hercules." The priests declared, "Solon, your country shone forth, in the excellence of her virtue and strength, among all human kind . . . when the rest fell off from her, being compelled to stand alone, after having undergone the very extremity of danger she defeated and triumphed over the invaders, and preserved from slavery those who were not yet subjugated, and generously liberated all the rest of us who dwell within the Pillars. But afterward there were violent earthquakes and in a single day and night all sank into the earth and the Island of Atlantis in like manner disappeared into the depths of the sea."

Then Timaeus speaks for most of the rest of the dialogue about how the universe was formed, and other matters not related to Atlantis.

The Critias Dialogue

The next day Socrates encourages Critias to tell his story in greater detail because, he says, details are required if the story is to seem believable. Critias promises to do just that.

Critias explains that although the story is from Egypt, the names in it are Greek because Solon carefully translated them. He says he has

Solon's original manuscript, which he has studied since childhood. (Which seems to contradict his earlier account about hearing the story from a poet and trying to recall it from memory. Could this be a hint from Plato that we should take what Critias says with a grain of salt?)

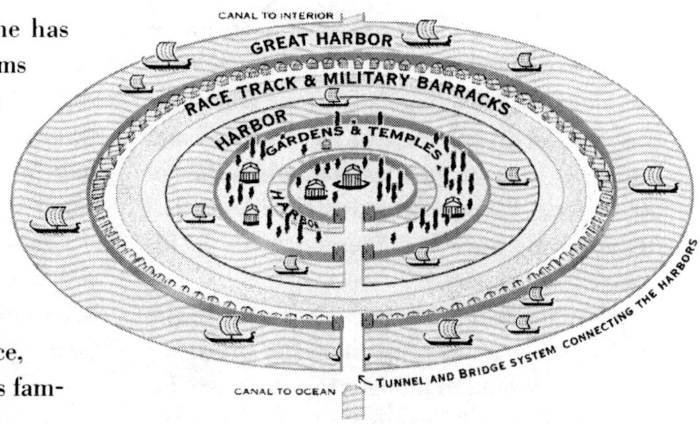

Diagram of Atlantean royal city

Half of Critias's speech discusses how the gods founded ancient Greece, and how the sea god Poseidon and his family founded Atlantis.

At last Critias gets around to describing Atlantis. It was a sunny island, marvelously beautiful, with rich forests for timber and a large level central plain which was overflowing with food crops, fruits, and flowers. The royal city was designed as a series of circular canals, lined with splendid palaces of white, red, and black stone set in fabulous patterns, shimmering with gold and the fiery glow of a valuable red metal called orichalch. In the very center of the circular canals stood a forbidden sanctuary surrounded by a golden fence. Poseidon himself had a huge temple completely covered with silver, with a roof of ivory, decorated with golden sculpture. A giant golden statue of Poseidon driving six winged horses, surrounded by 100 sea nymphs riding on dolphins filled the inner sanctuary up to the roof. Baths and pools fed by hot and cold springs were surrounded by gardens with every kind of beautiful and fruitful tree imaginable. A racecourse was built on the ground between two of the circular canals. And all of that was just the royal palace . . . !

The rich climate allowed the Atlanteans to raise a gigantic armed force. The area around the royal city provided 60,000 military officers, 10,000 chariots, 240,000 cavalry, 120,000 hoplites, 600,000 archers, slingers, stone and javelin throwers, and 24,000 ships. And this was from only 1 of the 10 divisions of the country!

Critias then goes back to discussing the political organization of the god-kings. He describes a bloody religious ritual where they catch and sacrifice a bull. He tells how Zeus, the king of all gods, becomes annoyed because the formerly noble Atlanteans have been corrupted by their great wealth. Zeus decided to punish them so he called all the other gods together in his house, "and when he had gathered them there he said . . ."

The dialogue ends there, right in the middle of a sentence!

Is It Possible That Plato Made Up Atlantis?

Many people who search for a real Atlantis argue that Plato couldn't have made up the Atlantis story because it contains too many realistic details.

But Plato lived both an exciting and dangerous life. He grew up during "The Golden Age of Athens"—one of the most amazing times in history when the arts flourished in a way that has barely been equaled since. Plato was highly educated and likely read manuscripts like "On Marvelous Things Heard" which told a story about a lush island that had been discovered, like Atlantis, "outside the Pillars of Hercules." As Plato grew out of his teens the

Golden Age came to an end when his city suffered a bitter military defeat. Plato sailed several times to the island of Sicily and lived in Syracuse, a city of fabulous architecture with splendid temples, a multi-level fort, and a covered boat canal. He knew about the circular harbor at Carthage that was controlled from a central island. He advised rulers and spent time in prison when one relationship turned bad. He watched a strange plague kill one out of three people in Athens.

Fighting Philosophers

Some modern authors misrepresent Plato's Atlantis as a peaceful paradise of spiritual people who had discovered the secret of eliminating war. But Plato said Atlantis was a gigantic military power that attacked without reason.

It is not surprising that soldiers were the heroes of Plato's story. Both Plato and his teacher Socrates were elite *hoplites*—upper class soldiers. The short squat Socrates was famous for his stamina and skill in battle. Plato probably grew up hearing about the recent Greek victory over the Persians. During his lifetime Athens won a war against the Carthaginians, lost an army in Sicily, and lost a war against a rival city Sparta.

Plato wrote about Atlantis when he was over 70, at the end of a rich life that would have given him plenty of material to draw upon. The Atlantis story was probably a combination of legends and bits of history woven together—whatever it took to create a memorable lesson.

Did Plato Expect People to Believe in Atlantis?

Plato often used myths and legends to illustrate a point. He expected his audience to recognize a parable (a story made up as a moral or religious lesson) when they heard it. Almost every parable he told starts with a statement that it is true. He himself explained that "We may liken the false to the true for the purpose of moral instruction."

In Plato's most famous dialogue, "The Republic," he suggested that on rare occasions it might be okay to tell what he called a "noble lie" to the lower classes for the purpose of creating a stable social order. Lower classes would be told that the gods created the present social order, making the rulers of gold, the military classes of silver, and the common working class from bronze. Plato was not a man who believed in total democracy for everyone. He himself was upper class and he also might have been suspicious of democracy because his dear friend and teacher Socrates was condemned to death by a democratically elected committee of citizens. (To his credit he felt that people of merit could rise above the class they were born into.)

What Would Plato Think about Today's Atlantis Stories?

Plato would be shocked to find his villains transformed into heroes. And he would be surprised to hear that Atlanteans are now said to possess superior wisdom, after he took pains to point out that they were destroyed because they made unwise choices.

The idea of a Golden Age is so appealing that it has overshadowed Plato's original lesson. A Golden Age and hidden wisdom are attractive because they are always located in the past or future, a very convenient arrangement for people who are dissatisfied with the present, but don't know what to do to improve it. Plato's life's passion was to discover the best way to organize society with the means available in the here and now, so he would likely disapprove of belief systems which place solutions out of reach in some mystical time.

Evidence for and against Atlantis Being an Accurate Historical Account

The Best Bet for a Real Atlantis

Around 3,500 years ago a massive volcanic explosion blew away the center of Santorini Island in the Aegean Sea near Greece, leaving a water-filled crater about 6 miles wide. In the 1960s on the islands that make up the crater's rim, archaeologists dug up the ruins of a luxurious city out of a layer of volcanic ash.

Could the disaster at Santorini have been the source of the Atlantis myth? The Santorini blast was one of the largest volcanic explosions known. The earthquakes, destructive waves, extensive ash fall, darkened skies, crop failure and resulting starvation must have been terrifying. It is possible that some memory of this event survived a thousand years of retelling until Plato's time.

Similarities of Santorini to Atlantis

- Land disappeared in an earthquake and was replaced by sea.
- The disaster destroyed an enemy of Athens. Atlanteans had attacked Athens; and Greek myths told of war between Minoans, the people who had settled Santorini, and the Greeks.
- Both Minoans and Atlanteans were seagoing traders.
- The Minoan island of Crete in particular had a sophisticated culture with large palaces, just like Atlantis.
- Both civilizations used rituals involving bulls in their religion.

At first the list of matches between Santorini and Atlantis is impressive. But some of these matches happen simply because Plato is describing a civilization like the one he lived in—dockyards, harbors, canals, chariots, hoplite warriors, shepherds with flocks, and temples for familiar gods.

Differences between Santorini and Atlantis

- The size is wrong. Atlantis was a continent, not a tiny island.
- The location is wrong. Atlantis was in the Atlantic Ocean, not the Aegean Sea.
- The date is wrong. Plato dated the Atlantis disaster thousands of years earlier than the Santorini explosion.
- The Minoans did not disappear after the Santorini disaster. Egyptian records show normal trade continued with the Minoans long after the Santorini explosion.

Plato did not have to depend on 1,000-year-old memories for the idea that a city could dis-

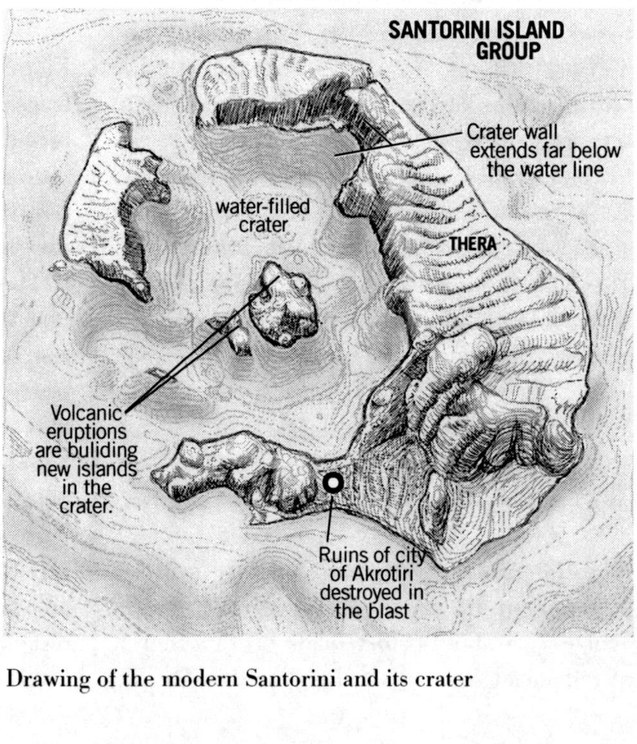

Drawing of the modern Santorini and its crater

Map of Greece

appear beneath the sea. In his own time two cities near Athens were destroyed by giant waves, sinking coastal land, and earthquakes.

As a soldier he surely had heard that a year before he was born, an earthquake and gigantic waves destroyed ships and a military outpost that the city of Athens built on the small island of Atalantë. This area continues to experience a sinking coastline, the last sudden sinking happening during a 1894 earthquake.

When Plato was 55, an earthquake destroyed the city of Helice, only 40 miles from Athens. Parts of the coastline sunk enough to submerge the trees that grew there. It was said that the waves that smashed into Helice swept the city so clean that people who arrived to bury the dead could find no one left to bury.

The key to how well a claim that someone has found Atlantis holds up is what they do with the details that *don't match* the theory. The writers who have investigated the possibility that Santorini could be Atlantis have often taken such negative evidence carefully into account.

For example, the impossible numbers relating to the date, size of the army, and land measurements in Plato's story all seem to be about ten times too large. Solon might have created this error if he confused the Egyptian or ancient Greek symbol for a hundred with the symbol for a thousand. If the numbers are all reduced by $^1/_{10}$ they make more sense.

Some have gone too far trying to make everything Plato said match something at Santorini. For example it has been claimed that Atlantis's ringed harbor was located in the now submerged center of the island, but there is no archaeological evidence of harbor ruins in Santorini's crater.

If Atlantis was inspired by Santorini, it is reasonable to assume that some legends might be mixed in with the facts, or that several events have been combined into one. Not every detail will have an explanation.

Arguments against Atlantis Being an Accurate Historical Account

- Plato's description was so exaggerated and spectacular that it was likely imaginary.
- There are no archeological ruins of cities anywhere on earth from the time that Plato claims Atlantis existed. Even if Plato was off by a zero, say 900 years rather than 9,000, there still are no corresponding ruins anywhere, especially in Greece and particularly in the Athens area itself. The earliest sophisticated civilizations were established 5000 to 6000 years ago.
- Egypt was supposedly conquered by Atlantis, but the Egyptians never wrote anything about it. The historian Herodotus actually met the same priests

that Solon supposedly spoke with, and he never mentioned hearing about Atlantis.

- Plato says the Atlanteans built a ditch that was 100 feet deep, 600 feet wide and an unbelievable 11,000 miles long. Nothing even a fraction as large has been found.
- There are no known forces that could destroy so large a land mass. True, parts of continents rise and fall, but the process takes millions of years. A comet or asteroid could destroy a large area, but no evidence has ever been found of all the other civilizations that Plato said traded with Atlantis.

Arguments for Atlantis Being an Accurate Historical Account

- Some legendary cities have turned out to be real, so Atlantis could be real.

The ancient Greek city of Troy is mentioned in almost every book about Atlantis because it is a mythical city that turned out to be real. This raises the author's hopes that Atlantis too might be found. Ubar, a wealthy incense trading post, was said to be lost beneath the desert sands of Saudi Arabia. The Koran, the holy book of the Muslims, said the people of Ubar were destroyed because they became corrupted by power and wealth. The city was said to have been swallowed up by the ground. With the help of Space Shuttle radar, ruins matching the story of Ubar were found—an incense trading city that had collapsed into a giant sinkhole. While no one can say for sure if the ruins actually were Ubar (no inscription with the actual name of Ubar was found), there is a good chance the site inspired at least some parts of the Ubar stories.

But just because some legendary cities have a basis in fact doesn't mean all legendary cities are real. Troy and Ubar both were mentioned in many different ancient writings. These stories did not seem to be copied from one another, which suggests that they preserved bits of real memories. All mentions of Atlantis, by contrast, can be traced back to just one source—Plato.

Where Was Atlantis?

- Many people have found places on earth that seem to match descriptions of Atlantis, so Atlantis must have been real.

Plato was very clear about where Atlantis was located. He said it was a large island as big as Libya and Asia combined (ancient names for north Africa and the middle east), in the Atlantic Ocean opposite the "Pillars of Hercules" (the Strait of Gibraltar). But no trace of a sunken city has ever been found under the Atlantic outside of the Strait of Gibraltar.

Strangely enough Plato goes on to state—as if his story explains a well-known fact—that ships can no longer sail in the Atlantic Ocean because the sunken continent is too near the surface. Historians explain Plato's lack of knowledge about the real nature of the Atlantic Ocean by pointing out that the Greeks of Plato's time knew almost nothing about the Atlantic because they were kept away from the area by a powerful seagoing enemy, the Carthaginians. Plato's mistake can be seen as another piece of evidence that his description of Atlantis was merely a fable.

But those who search for a real Atlantis point out that the term "Pillars of Hercules" might not mean a specific place. It was often used in a broad sense to stand for the limits of the known world. This allows them to locate Atlantis almost anywhere on the planet—and some have even placed it on Mars, the Moon and, of course, in "outer space"!

Next to Atlantis, the second most famous

Places Where People Think Atlantis Might Have Been

By far the most popular location for Atlantis is in the North Atlantic Ocean. But after that almost anything goes.

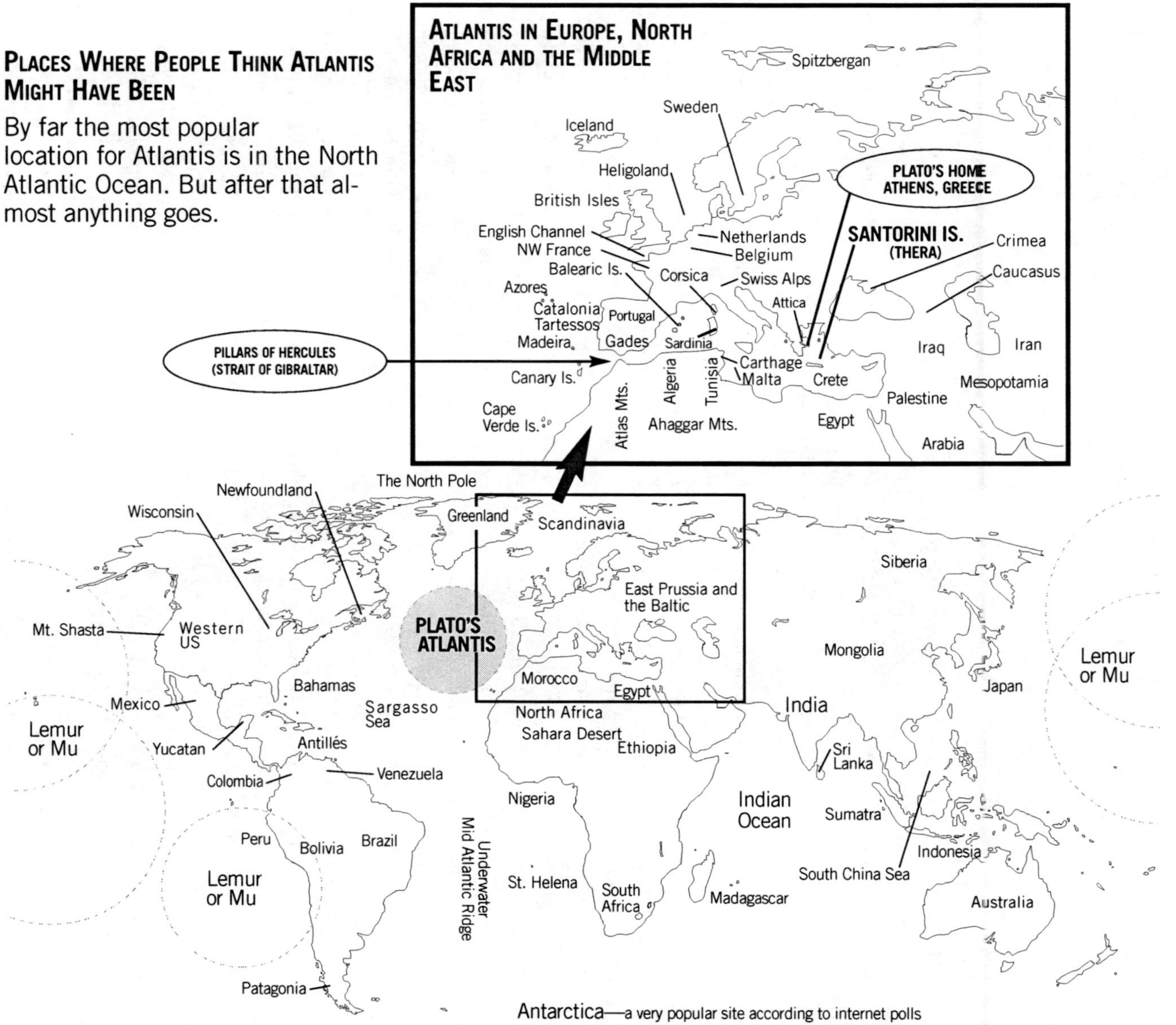

lost continent is Lemur, or Mu for short, said to have been located in the Pacific Ocean.

The large number of locations for the lost continents is the result of how easy it is for authors to find "evidence" that matches a detail or two mentioned by Plato. A pile of rocks, a word in a local language that starts with "A-T" or a local legend that suggests a golden age, plus a little imagination, and a lost continent is declared found.

By far the most popular location for Atlantis is in the North Atlantic Ocean. But after that almost anything goes.

The Legacy of the Atlantis Story

The antiquity of Plato's Atlantis story has given legitimacy to the stories of a golden age and perfect societies that it has spawned. Atlantis has been the inspiration for social movements and religions.

The name of the Greek god Atlas, after which the lost continent of Atlantis was named, lives on. The Atlas mountains of Morocco in northern Africa and the Atlantic Ocean are named after Atlas. In Plato's story he was one of 5 sets of twin sons born to Poseidon, the sea god who founded Atlantis. Poseidon divided Atlantis into 10 kingdoms, one for each son, and made Atlas chief king. (In other Greek stories Atlas was instead the son of the Titans [giants], and Poseidon was his son-in-law.) Different stories explain how Atlas got his famous task of holding the heavens and the earth apart. In one he is tricked into it, in another it is a punishment.

A book of maps came to be called an "atlas" because of the custom of illustrating an opening page with a picture of Atlas holding up the earth, or holding up a sphere representing the heavens.

In the 1500s many maps labeled at least one of the newly discovered American continents "Atlantis." As late as 1769 maps were made that showed the Americas divided into 10 sections—one section for each of Poseidon's 10 sons as described in Plato's dialogue. If things had gone a little differently and the name had stuck, Americans might be called Atlanteans.

The Mystical Search for Atlantis

The myth of Atlantis has become a template for people searching for answers to life's mysteries.

Ignatius Donnelly wrote a popular book in 1882 that inspired an entire movement: the search for a mystical Atlantis. He was a dreamer and an idealist. In 1857 he tried to found a utopian society in Minnesota, but it failed. He was elected to Congress and spent much of his time in Washington D.C. at the Library of Congress researching his theory that Atlantis was the original Garden of Eden. He thought the gods of the Greeks and many other peoples were confused memories of the real kings and queens of Atlantis. He said those who escaped the Atlantis disaster preserved memories of it which later became the story of Noah's flood.

Donnelly was responsible for popularizing the "golden age" myth—the idea that Atlantis was a superior civilization that was the source of all other civilizations.

Donnelly boasted that he presented his case like a lawyer—he collected evidence that supported his arguments and ignored evidence that didn't. This was a serious mistake because you can prove almost anything that way. Hundreds of Atlantis authors who followed Donnelly's example did just that and "proved" hundreds of contradictory theories, and "found" Atlantis almost everywhere.

Donnelly's highly influential book *Atlantis: The Antediluvian World* ("antediluvian" means "before the flood" [of the Bible]) has been reprinted over fifty times, and can still be purchased today.

In 1888 a colorful character, Madam Helena Petrovna Blavatsky, topped all other Atlantis authors by claiming her six volume work titled *The Secret Doctrine* was originally dictated in ancient Atlantis itself.

Madam Blavatsky and her followers didn't believe that human beings evolved from lower primates. She claimed creatures became increasingly more human by evolving through five stages that she called the five "root races." Her ideas reflected the strong racial prejudice of her day which considered non-Europeans inferior.

The first root race was a purely spiritual creature. Next came a slightly more solid jellyfish-like race. Stage three was an ape-like egg-laying "Lemurian" from a lost continent in the Pacific. They had animal-like snouts, and two eyes on the sides of their heads, and one eye in back. They lived during the time of the dinosaurs. Being stupid they mated with animals, producing the great apes. (Oops! Blavatsky didn't know there were no apes during the time of the dinosaurs!) However, enlightened beings from Venus civilized the Lemurians, and by 70 million years ago they had become quite human.

Root race four included Atlanteans, and the dark-skinned human racial groups. Jewish people were halfway between root race four and the last stage—root race number five. This most advanced group was of course Blavatsky's own group—European whites or Aryans.

Blavatsky and her followers are often considered the founders of the modern occult, spiritualist and New Age movements which combined western romanticism with eastern religious ideas. The blatantly racist root race theories resurface in many modern New Age beliefs that suggest dark skinned people such as the Egyptians or Aztecs were too stupid to build their own civilizations and needed the help of space aliens or lighter skinned people.

Edgar Cayce was a famous psychic who predicted that Atlantis would rise from the ocean, and most of the western United States would slide into the sea in 1968. His failed predictions did not inhibit other authors from making new predictions in the 1970s that the wicked state of California would snap off and plunge into the Pacific Ocean.

"Colonel" James Churchward thought Atlantis was only a colony of the Great Continent of Mu which floated over most of the area that is now the Pacific Ocean. He said Mu sank when the "gas pockets" that held up the continent exploded, leaving only the tops of its highest mountains above the water. These islands are known today as Hawaii, Samoa, Tahiti, and Easter Island. When Mu sank, massive worldwide earthquakes caused all civilization to collapse into savagery.

Chiropractic

Conventional or Alternative Healing?

DR. SAMUEL HOMOLA, D.C.

What should Chiropractic's role be in the future of medicine? Should chiropractors position themselves as primary care physicians and take advantage of the current popularity of alternative medicine, or should they limit their practice and focus to the area where scientific evidence has proven them most effective—relieving back pain?

When back pain occurs, most people first think of a chiropractor. Chiropractic manipulation and massage are the most commonly used forms of care for back pain. Each year, 20 million Americans (about one out of 13) use chiropractic services, most often for back pain. Back pain is the second most common reason expressed by patients for office visits to primary care physicians and the most common reason for office visits to orthopedic surgeons, neurosurgeons, and occupational medicine physicians. Estimates of total direct and indirect costs attributable to back pain are as high as $60 billion annually in the United States. In 1988, the cost of medical care for back pain was estimated to be $8 billion annually, with chiropractic care costing approximately $2.4 billion (paid by 5% of the population). In 1993, it was estimated that about 7% of adults in this country received chiropractic treatment in the past year. Chiropractic is the second most widely used form of alternative health care.

While most people go to a chiropractor for treatment of back pain, some go for the wrong reasons. According to statistics offered by the American Chiropractic Association, 94% of chiropractic patients are treated for neuromusculoskeletal conditions, 38% of which is for back pain, 28% for neck pain, 14% for headache, 8% for problems with the extremities, and 6% for other neuromusculoskeletal conditions. The remaining 6% are treated for viscerosomatic conditions, such as asthma, gastrointestinal disorders, and hypertension.

Although many chiropractors claim that spinal manipulation can improve health, most people view chiropractors as back specialists. RAND, an independent research organization, reported in 1991 that two-thirds of all patient visits for back pain are made to chiropractors, with chiropractors performing about 94% of all manipulation in the United States. A review of the literature supported the use of spinal manipulation for acute back pain not accompanied by neurological involvement or sciatic nerve irritation. RAND concluded that spinal manipulation conferred a short-term but significant benefit in pain relief.

In 1994, the Agency for Health Care Policy and Research (AHCPR) released a study called Acute Low Back Problems in Adults. Like RAND, this agency reported that spinal manipulation is helpful for patients experiencing acute lower back problems without radiculopathy or leg pain (sciatica) when used within the first month of symptoms. Both the RAND and the AHCPR reports confirmed what many Americans already knew—spinal manipulation helps relieve some types of back pain.

A second report released by RAND in 1996, The Appropriateness of Manipulation and Mobilization of the Cervical Spine, concluded that there is sufficient evidence to indicate that cervical spine manipulation or mobilization may improve range of motion and provide short-term relief for subacute or chronic neck pain and muscle tension headache. The report added, however, that 57.6% of reported indications for cervical manipulation were considered inappropriate, with 31.3% uncertain. Only 11.1% could be labeled appropriate. The rate of injury or stroke from cervical manipulation was estimated to be 1.46 per one million manipulations.

Seeking a Place in the Health-Care System

Paradoxically, with all indications pointing to acceptance and use of chiropractic as a method of treating back and neck pain, few chiropractors claim to be back specialists. Most claim to be primary care physicians, offering spinal adjustments to restore and maintain health by correcting subluxated (misaligned) vertebrae. Using the chiropractic subluxation theory to treat a broad scope of ailments, the chiropractic profession is seeking validation as an alternative healing method.

In 1996, the executive director of the Foundation for Chiropractic Education and Research (FCER) stated that "It has long been the position of FCER that to position chiropractors as 'back doctors' would be disastrous for the future of the profession and would only serve to limit the choice of treatments available to patients. Based on substantial anecdotal evidence supporting chiropractic intervention, FCER is funding research that investigates chiropractic treatment for the very ailments that Dr. Homola recommends that we back away from: colic, dysmenorrhea, and ear infection."

Definition by Consensus

Clearly positioning chiropractic as an unlimited form of primary care, the Association of Chiropractic Colleges, representing the 16 North American chiropractic colleges accredited by the Council on Chiropractic Education, met in July of 1996 and reached a consensus defining chiropractic:

> Chiropractic is a health care discipline which emphasizes the inherent recuperative power of the body to heal itself without the use of drugs or surgery. . . . The purpose of chiropractic is to optimize health. The body's innate recuperative power is affected by and integrated through the nervous system. . . . Chiropractic is concerned with the preservation and restoration of health, and focuses particular attention on the subluxation.

In 1997, the executive director of the FCER suggested that "If all the indicators are pointing to greater interest in and demand for natural health care on the part of the public, shouldn't chiropractic position itself to take advantage of this trend?"

A trend forecast for the chiropractic profession, commissioned by the FCER and funded by the National Chiropractic Mutual Insurance Company, was published in 1996 by Trends Research Institute. The forecast offered suggestions that would "Enable chiropractors to overcome their over-specialized public image and become recognized by the general public as primary care providers—providing a spectrum of valuable services capable of effectively treating a variety of common health problems." The study recommended that chiropractors focus upon weight management, vitamin counseling, and chronic ailments that tend to surface with age. "Weight management," the study concluded, "is a field of tremendous potential."

With neuromusculoskeletal conditions comprising 94% of the average chiropractor's practice, it is puzzling to see the chiropractic profession focusing its research on such conditions as infantile colic and ear infection. It is even more puzzling to see the chiropractic profession considering proposals to study weight management, vitamin counseling, and chronic ailments in order to gain recognition as primary care providers rather than build upon established recognition for what they do best—treating neck and back pain and other neuromusculoskeletal problems of mechanical origin.

Theory versus Reality

Part of the contradictory behavior of chiropractors and the misplaced priorities of the chiropractic profession can be explained by the theory defining chiropractic. As indicated by the 1996 consensus formulated by the Association of Chiropractic Colleges, all chiropractic colleges still teach that correction of vertebral subluxations plays an important role in the preservation and restoration of health. The public's perception of chiropractic, however, is such that few people (less than 7% of the population annually) go to a chiropractor for treatment of anything other than neck and back pain and tension headache.

Yet, the chiropractor is far from being a fully qualified back specialist. Since chiropractors do not prescribe pain medication and cannot perform surgical and invasive diagnostic procedures, their care of back pain is limited to uncomplicated problems that must be carefully selected if they are to treat the patient without the help of a medical practitioner.

According to the back-care guidelines formulated by the Agency for Health Care Policy and Research of the U.S. Department of Health and Human Services, spinal manipulation is effective for patients during the first month of acute lower back symptoms that are not accompanied by leg pain caused by nerve root involvement. And according to an article published in the October 8, 1998, issue of the *New England Journal of Medicine*, chiropractic manipulation was no more effective than the McKenzie method of physical therapy in treating patients with low back pain. Both of these methods—manipulation and physical therapy—had only marginally better outcomes than simply providing the patient with an educational booklet for self help. A subsequent study in the November 11, 1998, issue of the *Journal of the American Medical Association* reported that spinal manipulation is no more effective than massage in relieving episodic or recurring tension headache.

Since there is a risk of stroke or injury from cervical manipulation, such treatment should be used only when specifically indicated. According to RAND, only 11.1% of reported indications for cervical manipulation are considered to be appropriate. This means that most cervical manipulation done by chiropractors subjects the patient to unnecessary risk. In some cases, simple massage would be safer than manipulation.

While the treatment of back pain and other neuromusculoskeletal problems of mechanical origin forms the bulk of the average chiropractor's practice, and can be treated safely and effectively by a scientifically grounded and properly limited chiropractor, a chiropractor's diagnostic and treatment armamentarium is so limited that they can function only as specialists who must work with a narrow range of uncomplicated musculoskeletal problems. They must, or should, treat only those conditions that do not require use of pain medication, or those that require hospitalization for diagnosis and treatment such as patients who have severe pain, disease or infection, or patients who have been incapacitated by severe injury. As "back specialists," chiropractors can effectively treat only uncomplicated musculoskeletal problems for which manipulation and physical therapy are appropriate treatment.

An editorial in the October 8, 1998, *New England Journal of Medicine* asks "What Role for Chiropractic in Health Care?"

> That spinal manipulation is somewhat effective symptomatic therapy for some patients with acute low back pain is, I believe, no longer in dispute. . . . What about the use of spinal manipulation for other musculoskeletal problems? There is evidence from randomized clinical trials that spinal manipulation may be efficacious for some patients with neck pain. However, neither the efficacy of manipulation relative to that of other therapies nor the cost effectiveness of such therapy has been established. Moreover, the use of cervical manipulation arouses far greater concern about safety than the use of lumbar manipulation. . . .
>
> What is the role of chiropractic in health care? In 1979 Dr. Arnold Relman wrote an editorial for the *Journal* entitled "Chiropractic: Recognized But Unproved." Nearly 20 years later there appears to be little evidence to support the value of spinal manipulation . . . for nonmusculoskeletal conditions. For this reason, I think it is currently inappropriate to consider chiropractic as a broad-based alternative to traditional medical care. However, for some musculoskeletal conditions, chiropractic does provide some benefit to patients. The challenge for chiropractors is to demonstrate that they can achieve this benefit at a cost that patients or health insurers are willing to bear.

Chiropractic as an Alternative Healing Method

Although it appears that chiropractors have much to offer in the use of spinal manipulation and other physical treatment methods in the care of back pain and other musculoskeletal conditions, there is no evidence to indicate that the chiropractic profession is making any effort to define itself as a limited medical specialty—quite the contrary. Clinging to the fundamental definition of chiropractic (that adjusting vertebral subluxations to remove nerve interference will restore and maintain health), chiropractors continue to claim to be primary care providers, entitled by law and by definition to treat a broad scope of human ailments. This stand persists despite the fact that there have been no appropriately controlled studies that establish that spinal manipulation or any other form of somatic therapy represents a valid curative strategy for the treatment of any organic disease.

Many chiropractors claim that chiropractic manipulation is effective in the treatment of asthma—a "shining example" of what chiropractic can do in the treatment of disease. But when researchers conducted a randomized, controlled trial of chiropractic manipulation in the treatment of asthma, they concluded that "In children with mild or moderate asthma, the addition of chiropractic spinal manipulation to usual medical care provided no benefit."

After 43 years of practice as a chiropractor, I have concluded that it seems unlikely that research will support the use of chiropractic manipulation in the treatment of organic disease.

Presenting chiropractic as an alternative to medical care in the treatment of nonmusculoskeletal conditions may be a big mistake for the chiropractic profession. Chiropractors may be shooting themselves in the foot by claiming to be alternative primary care physicians. According to researchers at Yale University, an estimated 6.5% of the U.S. population in 1996 sought both conventional and unconventional health care; 1.8% used only unconventional services; 59% used only conventional care; and 32.2% used neither. This study reports less use of alternative medicine than estimated in previous studies. And it indicates that Americans who try alternative medicine generally use it to supplement or complement—not replace—traditional medical care.

Alternative medicine, while offering some treatment methods that may have value, is generally thought of as encompassing a variety of popular but unproven and nonsensical healing methods, such as homeopathy, foot reflexology, aromatherapy, applied kinesiology, therapeutic touch, Ayurvedic medicine, acupuncture, and 40 or 50 other questionable metaphysical healing methods sheltered under the umbrella of alternative care. *The Journal of the American Medical Association*, which devoted the entire November 11, 1998, issue to studies of specific applications of unproven healing methods, offered these editorial comments about alternative medicine:

> There is no alternative medicine. There is only scientifically proven, evidence-based medicine supported by solid data or unproven medicine, for which scientific evidence is lacking. . . .
>
> While acknowledging that many therapies used in conventional medical practice also have not been as rigorously evaluated as they should be, we agree that most alternative medicine has not been scientifically tested. However, for alternative therapies that are used by millions of patients every day and that generate billions of dollars in health care expenditures each year, the lack of convincing and compelling evidence on efficacy, safety, and outcomes is unacceptable and deeply troubling. We believe that physicians should become more knowledgeable about alternative medicine and increase their understanding of the possible benefits and limitations of alternative therapies. . . .
>
> However, until solid evidence is available that demonstrates the safety, efficacy, and effectiveness of specific alternative medicine interventions, uncritical acceptance of untested and unproven alternative medicine therapies must stop. Alternative therapies that have been shown to be of no benefit (aside from possible placebo effect) or that cause harm should be abandoned immediately. . . .
>
> For patients, for physicians and other health care professionals, and for alternative medicine practitioners—indeed, for all who share the goal of improving the health of individuals and of the public—there can be no alternative.

While alternative medicine is presently popular among the public, it is more of a fad than a science. It seems likely that as science investigates alternative healing methods, many will be abandoned and public support for various healing fads will diminish. Few chiropractors are able to view alternative health care as astutely and objectively as Craig Nelson, D.C., an associate professor at Northwestern Chiropractic College, who offered this commentary in a recent issue of the *Journal of the Neuromusculoskeletal System:*

> The recent enthusiasm for AHC [alternative health care] is largely a cultural rather than a scientific phenomenon. While in the short term this cultural shift may provide some new opportunities for the chiropractic profession and others, the public's present fascination with AHC will have little effect on the long-term vitality and growth of our profession. The pendulum will likely swing (and this may already have begun) away from the current fashionability of AHC.
>
> A far more important and durable trend than the current upswing in AHC is the recognition that health care providers must be held accountable for the safety and effectiveness of their practices. Orthodox methods of investigation and research will be the basis by which

> those practices are evaluated. The label, "alternative health care," will not excuse any profession from the requirement to demonstrate safety and effectiveness.

Time may be running out for support of irrational alternative healing methods. With patience wearing thin, some medical observers are speaking out. The editors of the *New England Journal of Medicine* echoed their colleagues at *JAMA:* "It is time for the scientific community to stop giving alternative medicine a free ride. There cannot be two kinds of medicine—conventional and alternative. There is only medicine that works and medicine that may or may not work. Once a treatment has been tested rigorously, it no longer matters whether it was considered alternative at the outset. If it is found to be reasonably safe and effective, it will be accepted. But assertions, speculation, and testimonials do not substitute for evidence. Alternative treatments should be subjected to scientific testing no less rigorous than that required for conventional treatments."

Until the chiropractic profession either proves or abandons the vertebral subluxation theory, it may continue to seek a free ride under the banner of alternative medicine. Since it does not appear likely that any scientific evidence can be produced to support the chiropractic theory, failure of the chiropractic profession to define and limit itself to a method of treating musculoskeletal conditions can only move it closer to the tenuous position of fringe healing methods that are unscientific and faddish, supported only by the whims of gullible and fickle supporters of metaphysical theories.

Ethical Responsibilities

Appropriate use of scientific spinal manipulation in the treatment of neck and back pain and related problems does not belong in the company of alternative methods that are unreceptive to scientific inquiry. Since 94% of all spinal manipulation in the United States is done by chiropractors, persons seeking spinal manipulation as a treatment for back pain must wade through a maze of nonsense to find a properly limited chiropractor. Physicians are often faced with the dilemma of deciding when and if they should refer a patient to a chiropractor.

Legally and ethically, it is a physician's responsibility to be honest and objective when offering advice about treatment of any kind. It is not enough to say that because an alternative treatment method is harmless it would be okay to use it. If a patient is led to believe that an unproven healing method might be useful in the treatment of illness, the patient might rely upon that treatment and delay use of a proven treatment method.

An alternative treatment method should be used only if there is reason to believe that the treatment might convey a specific benefit for the illness being treated or can complement a proven treatment method. No physician can ethically endorse an unproven treatment method that misleads the patient. A placebo effect provided by misinformation based on a false theory can do more harm than good, contributing to indiscriminate use of inappropriate treatment. A safe and useful placebo effect is best obtained from complementary care that is known to be based on a plausible theory that does not contradict the laws and principles of science.

While spinal manipulation used in the treatment of some types of neck and back pain and related neuromusculoskeletal problems is plausible and can be supported scientifically, there are many scientifically unsupportable chiropractic treatment methods that chiropractors claim will restore and maintain health. Physicians cannot and should not recommend such treatment since the theory underlying such treatment is not plausible and there is no controlled research or evidence to support the treatment.

Patients should understand that it is not in their best interest for physicians to remain silent about treatment methods that are false or misleading. A Seton Hall University Law professor offered this advice for physicians concerned about prescribing alternative healing methods:

> The physician-patient relationship requires the physician to be truthful with her patient. This duty of truthfulness circumscribes the provision of care that the physician has no scientific basis for believing the therapy will help the patient. The very fact that it is a physician who prescribes the therapy will endow the treatment with a false sense of legitimacy. Physician assent to the pursuit of unproven alternative treatments will, like managed care coverage, affirm patients' beliefs that innovative treatment offers them some real benefit, and perpetuate a system in which patients' ignorance results in their pursuit of care that they likely would not have elected had they been better informed.

Patients should learn to appreciate the skepticism and the advice of a physician who does not pull punches in analyzing treatment methods that are contrary to the principles of physics, chemistry, or biology. Some of these treatment methods are so nonsensical that they are an insult to human intelligence.

Most consumers are fairly well informed these days, but few of us have the means, the ability, or the time to search the scientific literature. Most of us rely upon our doctor to be honest and objective when seeking advice in matters that involve our health. But you cannot and should not rely totally upon your doctor in taking care of your body. You must make an effort to help yourself by maintaining a healthy skepticism while seeking reliable information on your own.

References

Angell M., J. Kassirer. 1998. "Alternative Medicine: The Risks of Untested and Unregulated Remedies." *N. Engl. J. Med.*, 339(12). Reprinted in *Skeptical Inquirer.* 1999; 23(1): 58–60.

Association of Chiropractic Colleges. 1996. "A Position Paper on Chiropractic." *J. Manipulative Physiol. Ther.*, 19(9): 633–637.

Baron J., P. Aker, E. Crowther, et al. 1998. "A Comparison of Active and Simulated Chiropractic Manipulation as Adjunctive Treatment for Childhood Asthma." *N. Engl. J. Med.*, 339(15): 1013–1020.

Bigos B., O. Bowyer, G. Braen, et al. 1994. "Acute Low Back Problems in Adults." Rockville, MD: Agency for Health Care Policy and Research. AHCPR publication 95-0642.

Boozang K. 1998. "Western Medicine Opens the Door to Alternative Medicine." *Am. J. Law and Medicine*, 24(2, 3): 185–204.

Bove G., N. Nilsson. 1998. "Spinal Manipulation in the Treatment of Episodic Tension-Type Headache." *J. Am. Med. Assoc.*, 280: 1576–1579.

Cherkin C., R. Deyo, M. Battie, J. Street, W. A. Barlow. 1998. "A Comparison of Physical Therapy, Chiropractic Manipulation, and Provision of an Educational Booklet for the Treatment of Patients with Low Back Pain." *N. Engl. J. Med.*, 339(15): 1021–1029.

Coulter I., E. Hurwitz, A. Adams, et al. 1996. *The Appropriateness of Manipulation and Mobilization of the Cervical Spine.* Santa Monica, CA: RAND.

Druss B., R. Rosenheck. 1999. "Association Between Use of Unconventional Therapies and Conventional Medical Services." *JAMA*, 282: 651–656.

Eisenberg D., R. Davis, S. Ettner, et al. 1998. "Trends in Alternative Medicine Use of the United States, 1990–1997: Results of a Follow-up National Survey." *J. Am. Med. Assoc.*, 280: 1569–1575.

Eisenberg D. M., R. C. Kessler, C. Foster, et al. 1993.

"Unconventional Medicine in the U.S." *N. Engl. J. Med.*, 328(4): 246–252.

Fontanarosa P., G. Lundberg. 1998. Editorial: "Alternative Medicine Meets Science." *J. Am. Med. Assoc.*, 280: 1618–1619.

Goertz C. 1998. ACA "Statistical Survey on Chiropractic Practice." *J. Am. Chiropractic Assoc.*, 35(11): 30–34.

Hawk C., M. Dusio. 1995. "A Survey of 492 U.S. Chiropractors on Primary Care and Prevention-Related Issues." *J. Manipulative Physiol. Ther.*, 18(2): 57–64.

Homola S. 1998. "Finding a Good Chiropractor." *Archives of Family Medicine*, 7(1): 20–23.

Homola S. 1999. *Inside Chiropractic: A Patient's Guide.* Amherst, NY: Prometheus Books.

Landmark Report on Public Perceptions of Alternative Care. 1998. Landmark Healthcare. Sacramento, CA; January, 1.

Nansel D., M. Szlazak. 1995. "Somatic Dysfunction and the Phenomenon of Visceral Disease Simulation: A Probable Explanation for the Apparent Effectiveness of Somatic Therapy in Patients Presumed to be Suffering From True Visceral Disease." *J. Manipulative Physiol. Ther.*, 18(6): 379–397.

Nelson C. 1998. "Understanding Alternative Healthcare." *J. Neuromusculoskeletal System*, 6(3): 95–99.

Seater S. 1996. "Letters to the Editor." *Chiropractic Technique*, 8(2): 93–94.

Seater S. 1997. "Chiropractic, Primary Care, and the Emerging Natural Healthcare Market." *J. Chiropractic Humanities*, 7(1): 42–45.

Shekelle P. 1998. Editorial: "What Role for Chiropractic in Healthcare?" *N. Engl. J. Med.*, 339(15): 1074–1075.

Shekelle P. G., A. H. Adams, M. R. Chassin, E. L. Hurwitz, R. B. Phillips, R. H. Brook. 1991. "The Appropriateness of Spinal Manipulation for Low-Back Pain: Project Overview and Literature Review." Santa Monica, CA: RAND, 34, 6–9.

Trends Research Institute. 1996. *The New Millennium Chiropractor: A Trend Forecast for the Chiropractic Profession.* Des Moines, IA: FCER, 1.

Christian Science as Pseudoscience

ROBERT MILLER

"Anyone who is not shocked by Quantum Physics doesn't understand it."
—Niels Bohr

"What did you do to the cat? It looks half dead."
—Schrödinger's wife

The Christian Science religion has long suffered membership losses and criticism for its stance on refusal of medical treatment. In recent years, however, the Christian Science Church has gained new momentum from its interpretation of quantum physics, a point of view supported by some physicists sympathetic to the church. Since most of its members hardly understand even classical physical science, let alone the "weirdness" posed by quantum theory (I should know since I was raised as a Christian Scientist), this coupling of religion with the world of the subatomic is somewhat shocking. In an unofficial Christian Science publication circulated in the Los Angeles area, a recent article entitled "A Thought From Quantum Physics" was published by David Carico, a Christian Scientist and physics professor at the University of Santa Clara. Dr. Carico's article contained the following statement:

> . . . according to physics these days . . . that which we call "matter" and perceive as a solid substance, actually has the same properties as a rainbow [they don't exist until they are observed]. Anything that suggests otherwise is an illusion. Here's how it works: The sheet of paper you're holding is made up of atoms, and these atoms are in turn made up of what we call "subatomic particles" (electrons, protons, neutrons, and others). But to describe the behavior of these subatomic particles, we have had to accept the fact that they are not things that are sitting there, in the paper, to be observed by you or not, as you choose. Rather, they can only be described as potential things—they're not actually in the paper, any more than a rainbow is sitting out there in the sky. They provide the potential for you to see a sheet of paper when you choose to look at the paper—just as the light and raindrops provide the potential to see that which you will call a rainbow when *you* choose to look in the sky. And since this applies to all subatomic particles, it also applies by inference to all atoms, and by further inference to all matter. That's pretty much it! Neat, huh?!? By the way—everything in the previous paragraphs is *physics* just as you

might find it in a physics book. I have not filtered any of it through Christian Science. (Carico, 1998, emphasis in original).

Of course, this particular angle on the attempt to merge quantum physics and religion is not exactly new. The best known attempt in recent years is probably Fritjof Capra's 1980 book, *The Tao of Physics*. Capra's book, in turn, has spawned many imitators, even influencing Gary Zukav's *The Dancing Wu Li Masters*, and some of Deepak Chopra's writings (see *Skeptic* Vol. 6, No. 2). But despite sharing its roots in Eastern thinking (Gardner, 1993), Christian Science is actually an unlikely religion to repackage quantum theory. For much of its history, the Christian Science church has rejected physics and other sciences as illusory, and thus relatively unimportant.

Christian Science and Its Doctrines

The First Church of Christ, Scientist (Christian Science's formal name) was founded in 1866 by Mary Baker Eddy (1821–1910), a semi-invalid, semi-illiterate who was a student of Phineas Quimby, a homeopathic faith healer and a major player in the spiritualist movement of the time. The *Oxford Dictionary of World Religions* describes the doctrine of the church as sharing "with Eastern religions a belief that ignorance is at the root of human unease—and thus of dis-ease: 'All reality is in God and his creation, harmonious and eternal. Therefore, the only reality of sin, sickness, or death is the awful fact that unrealities seem real to human, erring belief, until God strips off their disguise.'" Christian Science is best summed up in Ms. Eddy's statement: "Both sin and sickness are in error, and Truth is their remedy" (Gardner, 1993; Brenneman, 1990).

Ms. Eddy is promoted by the Church in a series of current lectures given at bookstores across the country to promote the "textbook" of the religion, *Science and Health with Key to the Scriptures*, as "a pioneer in the science of the mind body connection."

Enter Dr. Doyle, Physicist/Evangelist

The Christian Science Church has found a powerful ally for its interpretation of "reality" in Dr. Lawrence Doyle. Dr. Doyle is self-described as "principal investigator for NASA's SETI (Search for Extraterrestrial Intelligence) program," and the "senior astrophysicist at NASA's Ames Research Institute." I verified his credentials. Doyle also holds a position as visiting professor at Principia College in Illinois, a college run by the church, where he teaches courses on the history of science (Doyle, 1992). In a lecture given on August 28, 1997, in Long Beach, California, Dr. Doyle gave his brief and biased history of science, where Mary Baker Eddy's discoveries were highlighted as "the culmination of all science." In Doyle's words, "it is no accident that so many scientific discoveries, including quantum mechanics, were discovered in rapid succession after Mary Baker Eddy published *Science and Health* and founded the Church."

Doyle has, in the past, stated his unusual view of the history of science in other works and on interviews broadcast on National Public Radio. Eddy often compared herself to Columbus and Copernicus, and Doyle follows in this vein. After discussing the discoveries of Copernicus, Kepler, Newton, Maxwell, and Einstein, Doyle (1992) states: "In the latter half of the 19th century, Mary Baker Eddy discovered something, probably the most important discovery in the whole history of science. She made the discovery that reality is perfect, and spiritual, and that what you see as the material evidence is only a limited way of looking at what was really there."

In a 1992 videotape produced by the Christian Science Church, entitled "Infinity and Individuality," Doyle expanded on the place of the Christian Science Church, and its founder, in science. After describing how he places Mary Baker Eddy's books in his "scientific library," Doyle concludes:

> Among the amazing discoveries that I made through my study of the Bible and Science and Health is an understanding of infinity, and the significance of true individuality. It turns out, you cannot add all of the finite things and get to the infinite. Suddenly, I picked up the Bible, and read where Jesus said "I and my father are one." That jumped out at me, and I suddenly comprehended, "Infinity, of course, is one! There can only be one all."

Further insight into Doyle's ideas can be gleaned from his own description of his scientific training:

> In addition to being trained in astrophysics, I had early training in metaphysics, in Christian Science Sunday school. I was taught that God was Mind, Life, Love, Truth itself, Spirit itself, Soul, and that all these were ways of viewing the one source of the universe. So, it's only natural that as I do astrophysics, I would view the universe from a spiritual perspective. Mary Baker Eddy, the discoverer and founder of Christian Science, was kind of unusual in combining Christianity and Science. Christianity is the religion of unconditional love, and science is the study of unconditional truth. She said that love and truth were the same thing, and were synonyms for God.

This unique view of science leads Doyle to some unique conclusions regarding other matters traditionally at the forefront of religion's confrontation with modern science. Regarding his stance on evolution, for example, Doyle states that he "gets asked that a lot," but sidesteps a resolution of the issue by explaining "Christian Science states that you are created each and every day!" When pressed, he asserts that "this is not my field" (Doyle, 1997). Mary Baker Eddy never wrote on this subject, which may be why he is reluctant to offer an opinion.

Quantum Religion

In his 1997 lecture, Doyle began by reading Mary Baker Eddy's definition of matter from *Science and Health:*

> MATTER. Mythology; mortality; another name for mortal mind; illusion, intelligence, substance, and life in non-intelligence and mortality; life resulting in death, and death in life, sensation in the sensationless; mind originating in matter; the opposite of Truth; the opposite of Spirit, the opposite of God; that of which immortal Mind takes no cognizance; that which the mortal mind sees, feels, hears, tastes and smells only in belief. (Eddy, 1906)

The Christian Science church each week closes Sunday services with the church's Statement of Being, written by Eddy: "Matter is the unreal and temporal. . . . therefore, man is not material, he is spiritual" (Eddy, 1906). One would hope, perhaps, that "the most important discovery in the history of science" would be a little more clear, and slightly less redundant. But incredibly, for a physicist, Doyle goes further, adopting these statements to conclude "all matter is an illusion." As he boldly stated in his lecture, "I am saying that Quantum Physics proves that the moon only exists when you look at it!" This is a sly reference, which was likely lost on Doyle's audience, to Einstein's plaintive question to physicist Abraham Pais, asking if he "really believed that the moon only exists when you look at it" (Lindley, 1996).

Doyle then goes on to extrapolate that since matter only exists when you look at it, it is in fact "an illusion," ergo, Christian Science is proven. He also implied that, since matter is an illusion, material sickness, disease, etc., can be changed by substituting a "new reality" in its place (Doyle, 1997). (Actually, to be most consistent with Christian Science, just seeing the "truth" of God's creation would do, since God never makes a mistake, and does not allow "illusions" of imperfection to even exist.)

The Reality of Reality

This, then, is the puzzle posed by the "weirdness" of quantum physics upon which so many in religious and new age circles seize. According to the Heisenberg Uncertainty Principle, for example, you can't measure the position of a particle precisely without great uncertainty about either its velocity, or vice versa (Feynman, 1995). But, taking this to the extreme, if the existence of "reality" depends, for me, on my consciousness, and for you, on yours, how is it that we can agree on so much? As David Lindley, a former theoretical astrophysicist at Cambridge University and the Fermi National Accelerator Laboratory, and currently Associate Editor of *Science News*, stated in his book, *Where Does the Weirdness Go?*: "It has been seriously suggested that human consciousness or perception is somehow the key to the measurement problem . . . even a number of reputable physicists (including Wigner as well as his hardheaded mathematician friend John von Neumann) have leaned in this direction—more, perhaps, out of desperation than anything else."

So does this support the point of view of Christian Science that matter is an "illusion"? Well, as Lindley explains, it is somewhat misleading to say that "measurement affects the thing measured," because that implies that a quantum object was in some definite but unknown state, but was then disturbed by an act of measurement and is now in some other state. But the Copenhagen interpretation states otherwise, showing how measurement gives definition to quantities that were previously indefinite; there is no meaning that can be given to a quantity until it is measured (Lindley, 1996). The familiar examples of superposition—the two-slit experiment, electrons prior to spin measurement, or photons prior to a polarization measurement—apply to single objects, individual quantum pieces. But, in considering Christian Science's assertions regarding the "matter" of larger systems—ones composed of lots of photons, electrons, or atoms—do they behave in the same way? Not easily, it turns out, and only in superconductors (Lindley, 1996).

So is the moon not there when you look at it? Does it take a "measurement" of some sort to force the moon to take on a real location in space? Well, as Lindley and others have reasoned, the moon is there when no one's looking. Under the decoherence argument, any activity, including the rain of solar photons upon the moon, or the constant jiggling of atoms that make up the moon, forces enough of a physical process to constitute a "measurement," enough to get rid of superposed states. No actual observation is required, and the whole process carries on without any intervention of human action (Lindley, 1996).

There are other solutions, of course. One of the best known is that advanced by John Gribbin, author of *In Search of Schrödinger's Cat* and *Schrödinger's Kittens* and the *Search for Reality*. Gribbin describes the "many worlds" interpretation, embracing this possibility through the "transactional interpretation." The solution to the well-known puzzle posed by Schrödinger's famous cat and its fate under the Copenhagen interpretation of quantum mechanics would be different under each alternative system. Under decoherence, the cat would only be in an inconsistent half alive/half dead state for a split second, when the interaction of the atoms would force one state or another. Under "many worlds," new, branching universes would open for each possibility, finally forced into one "reality" when the measurement is made (Lindley, 1996; Gribbin, 1995; *Skeptic*, V. 3, #4, 1995).

However, neither of these possible solutions even comes close to "confirming" the tenets of Christian "Science." In Christian Science, Schrödinger's cat can never die, since nothing is "real" except for what God created, which doesn't include the illusion known as death. Poor cat—can't even die with dignity in this religion.

And what are we to make of Christian Science's published stance that "matter is unreal"? Even physics professors who are Christian Scientists, such as Dr. Carico, have to admit, contrary to Eddy's writings, that, "Yes, matter is real," in an attempt to shoe-horn Christian Science in through the back door of quantum theory (Carico, 1998). One member of the Mother Church told me that the new explanation is that "Matter is real. But standing behind every rock, every piece of matter, is the idea of that rock or that matter. Both are created by God" (Miller, 1998).

Richard Feynman concluded that "No one understands quantum theory" (Feynman, 1995). But one last point from what we do know is important. Subatomic particles, following quantum physics, act within rigidly prescribed limits. Scientific observers cannot just "make up" any inference they like about an electron. It either has a "spin" to the left or to the right, up or down. Once the measurement is made, each person can agree on the spin measurements, as fixed by the observation. That's it. Matter does not suddenly change, as in a dead person coming to life due to observation alone. The conclusions by the Christian Science Church regarding quantum physics, applying subatomic theory "by inference to all atoms, and by further inference to all matter" (in Carico's words), is not supported by the relationship between classical physics and quantum theory.

Mary Baker Eddy deserves credit, perhaps, for breaking ground in advancing the role of women in religion, and in her early observations regarding what science now knows as the placebo effect. But twisting quantum physics to support her theories not only misstates the patient observations of physics, but gives adherents of her religion a false sense of security in their view of the findings of science.

References:

Brenneman, R. 1990. *Deadly Blessings: Faith Healing on Trial.* Buffalo: Prometheus Books.

Capra, F. 1980. *The Tao of Physics.* New York: Bantam.

Carico, D. 1998. *A Thought from Quantum Physics.* Los Angeles: LA Christian Science Student Society Newsletter.

Doyle, L. 1992. *Infinity and Individuality.* Boston: Christian Science Publishing Society.

——. 1997. *Christian Science Symposium.* Long Beach, CA.

Eddy, M. B. 1906. *Science and Health with Key to the Scriptures.* Boston: Christian Science Publishing Society.

Feynman, R. 1995. *Six Easy Pieces.* Boston: Addison Wesley.

Gardner, M. 1993. *The Healing Revelations of Mary Baker Eddy: The Rise and Fall of Christian Science.* Buffalo: Prometheus Books.

Gribbin, J. 1984. *In Search of Schrödinger's Cat.* New York: Bantam Books.

——. 1995. *Schrödinger's Kittens and the Search for Reality.* Boston: Little, Brown.

Lindley, D. 1996. *Where Does the Weirdness Go?: Why Quantum Mechanics Is Strange, But Not As Strange As You Think.* New York: Basic Books.

Miller, R. 1998. Conversation with Margarita Miller, member, Christian Science Mother Church. March 31.

EMDR

Eye Movement Desensitization and Reprocessing

GERALD M. ROSEN,
RICHARD J. MCNALLY, AND
SCOTT O. LILIENFELD

While strolling through a park one day, Francine Shapiro noticed that certain of her troubling thoughts suddenly lost their distressing qualities. Curious about what had happened, Shapiro regenerated the mental images and again found them no longer upsetting. Attending closely to her behavior, she realized that her eyes had been spontaneously and rapidly shifting back and forth. Suspecting that rapid eye movements might possess hitherto untapped therapeutic powers, Shapiro began informal tests on her friends. She asked them to concentrate on a traumatic or disturbing memory and to track her finger visually as she moved it back and forth in front of their eyes. Her friends reported feeling better and their memories were no longer disturbing.

Shapiro's serendipitous observations inspired her to conduct a formal study in fulfillment of her doctoral dissertation in clinical psychology at the California-based, never-accredited, and now-defunct Professional School of Psychological Studies. This was an opportune time to conduct such a study. Posttraumatic Stress Disorder (PTSD) had been recognized in 1980 by the American Psychiatric Association as a distinct mental disorder resulting from exposure to extreme, terrifying events, such as combat and rape. People with PTSD often feel "numb," finding it difficult to experience positive emotions for loved ones, they are often hypervigilant and startle easily, and they typically "re-live" their terrible experiences in the form of nightmares and recurrent disturbing recollections of the trauma. This last PTSD symptom constituted the very phenomenon that Shapiro believed eye movements might help. What better way to test her discovery, therefore, than to apply the induction of eye movements to the treatment of traumatic memories associated with PTSD?

Shapiro treated 22 PTSD patients in her dissertation study by asking them to recall a traumatic memory and to follow her moving fingers with their eyes. She published her findings in 1989 in the *Journal of Traumatic Stress,* reporting that a single session of "Eye Movement Desensitization" abolished the distress associated with a traumatic memory in 100% of the patients. These results caught the attention of psychiatrists and clinical psychologists who were accustomed to spending much more time desensitizing the traumatic memories of their patients. Shapiro concluded that eye movements somehow accelerated the desensitization process associated with traditional imaginal exposure techniques. Further, she claimed that clinicians could "achieve complete desensitization of 75–80% of any individually treated trauma-related memory in a single 50-minute session," simply by reading her 1989 article. Shortly thereafter, Shapiro renamed her method "Eye Movement Desensitization and Reprocessing" or EMDR, a method that one enthusiast has claimed to be "the most revolutionary, important method to emerge in psychotherapy in decades." Others have declared EMDR to be nothing short of "amazing," "profound," and a "miracle" (see Shapiro and Forrest, 1997).

Although several popular expositions of EMDR have appeared (Gastright, 1995; Lilienfeld, 1996), the need to revisit this topic is pressing when one considers that the EMDR phenomenon thrives with unabated vigor and intensity, a full decade since its inception. EMDR workshops are offered throughout the world on an almost weekly basis. EMDR has been used to treat survivors of the Oklahoma City bombing, refugees in Bosnia and other areas ravaged by war, and communities where major natural disasters have occurred. EMDR enthusiasts have a professional organization to promote their method—the Eye Movement Desensitization and Reprocessing International Association (EMDRIA), and new applications of the method now can be discussed at the annual international EMDRIA conference. The National Institute of Mental Health has recently funded two major clinical outcome trials comparing EMDR with other treatments (e.g., cognitive-behavioral therapy and Prozac). Reflecting the major impact EMDR appears to be having in clinical psychology, the *Journal of Anxiety Disorders,* a leading peer-reviewed academic journal, recently featured a special issue (Spring, 1999) with 13 articles devoted entirely to EMDR. Clearly, this is a treatment method that warrants continued scrutiny.

Moving beyond popular expositions on EMDR, we examine in detail the most recent scientific evidence that bears on the method's efficacy, consider strategies employed by EMDR's proponents to deal with negative findings, and note historical parallels between EMDR and other controversial treatments. This scientific and historical analysis of EMDR may help shed light on a variety of other potentially pseudoscientific practices in the field of clinical psychology. In this respect, EMDR serves as a useful object lesson in the study of pseudoscience.

The Spread of EMDR

Shortly after completion of her doctorate, Shapiro founded the EMDR Institute, Inc. and promoted her admittedly experimental method in weekend training workshops held throughout the world. By 1990, the claim was made that EMDR had evolved into a complex methodology, so that Shapiro's original article could no longer suffice as an adequate description of the method. Instead, a more costly two-day workshop was required to be "certified." By 1991, even more training was required and a "Level II" workshop was mandatory for certification by the EMDR Institute, Inc. Thus, in the short span of two years, Shapiro had gone from her 1989 statement to an elaborate proprietary training model. These changes were reported to be motivated by clinical "experience," without any data to show that Level II training yielded better results than Level I, or the written instructions from 1989. Despite the absence of data, clinicians were told that using EMDR without appropriate training is dangerous, and training in authorized workshops became the minimum standard. By

1997, over 25,000 mental health professionals had accepted Shapiro's dictum and completed at least one of the workshops.

One of us completed both Level I and Level II training and experienced firsthand the revival-tent fervor of EMDR enthusiasts (Rosen, 1996). Shapiro spoke about an increasing range of applications for EMDR, the possibility of reversing cycles of violence in our society, and the use of EMDR in humanitarian efforts throughout the world. EMDR trainers went to Oklahoma City in the wake of the bombing, and volunteered their services in war ravaged Bosnia. Shapiro raised the possibility that EMDR might help people combat cancer or other terminal illnesses. An announcement for specialty workshops suggested the use of EMDR for everyday life issues, with such titles as, "Using EMDR to Help People Reach Their Peak at Work." Dr. Shapiro's vision of EMDR's contribution to our mental health and the world's future seemed boundless.

In addition to highly successful workshops, Shapiro began making headway in the academic and professional world of clinical psychology. She received the 1994 Distinguished Scientific Achievement Award from the California State Psychological Association, and she gave invited lectures for the American Psychological Association, the European Association for Behavioural and Cognitive Psychotherapies, and other international associations. A Committee within the American Psychological Association officially listed EMDR as "probably efficacious for civilian PTSD" after two randomized clinical trials with traumatized civilians found the method better than no treatment at all. These developments lent further credence to the novel notion that eye movements could assist in the cure of mental ailments. EMDR clearly represents a major event in the landscape of contemporary clinical psychology. Unfortunately, a careful analysis of published studies demonstrates that something is terribly amiss.

Playing Fair with Science

Studies on EMDR have found scant empirical support for the dramatic claims made by its enthusiasts. Most damaging are a number of recent well controlled studies that have consistently failed to find evidence that eye movements possess any therapeutic powers (see DeBell and Jones, 1997; Foa and Meadows, 1997; Lohr, Tolin and Lilienfeld, 1998; McNally, 1999a). Individuals who simply imagine a traumatic memory while staring straight ahead do as well as those who visually track the therapist's moving finger (Hazlett-Stevens, Lytle, and Borkevec, 1996). Other recent studies have compared EMDR with more traditional methods that expose clients to traumatic memories and desensitize anxiety. These studies find EMDR no more, and sometimes less, effective than the traditional methods of exposure (Devilly and Spence, 1999; Muris, Merckelbach, Van Haaften and Mayer, 1997). Findings like these suggest that Shapiro borrowed elements from extant methods, added the unnecessary ingredient of finger waving, and then took her technique on the road before science could catch up.

Shapiro's general response to negative findings has been to cite numerous publications that demonstrate EMDR's effectiveness and to point out that a Committee within the American Psychological Association (APA) has recognized the method as empirically supported and effective in treating civilian PTSD. Unfortunately, the Committee's standards only require that two studies demonstrate a method as more effective than no treatment. Anyone familiar with the history of psychotherapy will understand that this is not a stringent requirement, since almost all psychological interventions instill a sense of hope, demonstrate a placebo effect, and achieve effects greater than no treatment (Frank, 1973). Indeed, according to the APA Committee's criteria, both personal prayer and sugar pills should be considered empirically supported treatments, as these interventions have been found in multiple controlled studies to be more effective than the absence of treatment. More to the point, any treatment that adds an inert ingredient (such as eye movements) to an already developed treatment (such as exposure based therapies) can meet the criteria. In this context, one can view APA's listing of Shapiro's eye movement method as a reflection on the Committee's weak criteria, rather than a ringing endorsement of EMDR.

Shapiro also contends that studies failing to support EMDR have not employed the method faithfully, whereas studies demonstrating positive effects have had good "treatment fidelity." This position assumes that EMDR is effective; there are critical procedural components to which one must faithfully adhere; and treatment effects, though powerful, are not so robust that violations in procedural integrity can be ignored. The argument sounds reasonable at first; but, in fact, Shapiro has misused the concept of treatment fidelity by continually changing the procedures and levels of training which define faithful adherence to the method (Rosen, 1999). Thus, Shapiro originally claimed that simple written descriptions were sufficient to learn the method. By the time psychologists had implemented the 1989 written instructions and found no difference between EMDR and a no–eye movement control, it had become necessary to take Level I training. By the time Level I trained psychologists were finding in controlled studies that EMDR was not as effective as Shapiro had claimed, they were accused of being only half-trained because they had not taken the Level II workshop. Then Roger Pitman and his colleagues at Harvard Medical School received Level II training and compared the eye movements of EMDR with a control condition that employed finger tapping (Pitman, Orr, Altman, Longpre, Poire, and Macklin, 1996). By the time their finding of no difference between groups was published, Shapiro was claiming that "alternate forms of bilateral stimulation" could work just as well as eye movements. In other words, Pitman and his colleagues, some of the most prominent researchers in the field of PTSD, had just wasted their time comparing EMDR with itself!

The shifting procedures and training requirements for EMDR have created a seemingly endless catchup game for scientists. How can scientists test a method whose proponents insist on treatment fidelity for the induction of eye movements, then state that alternate tapping strategies are possible, next argue that various protocols must be followed, and then switch the decision rules for those protocols? How can scientists know they have been properly trained in a method when simple written descriptions first sufficed, then a Level I workshop was required, and then Level II training was the minimum standard? One can easily comprehend how the strategy adopted by Shapiro and other EMDR enthusiasts has created a slippery slope where refuted hypotheses constantly change, and the data never catch up. Like the Red Queen in Lewis Carroll's *Through the Looking Glass,* scientists who investigate the efficacy of EMDR are forced to keep running just to stay in the same place.

Lessons from History: Mesmer and Shapiro

If Shapiro had merely described EMDR as a minor variant of existing desensitization techniques using imagery, or if she had stopped to test the role of eye movements before marketing her method, we would not be writing this article today. However, instead of following this road in a manner consistent with the tenets of science, Shapiro demonstrated striking entrepreneurial zeal and parlayed her "discovery" into a spectacularly successful commercial enterprise. This is not the first time that a controversial method has been so promoted. In fact, the story of EMDR is eerily reminiscent of Franz Mesmer's discoveries in the late 18th century, with McNally (1999b) documenting no fewer than 17 striking parallels between the histories of the Mesmerism and EMDR movements.

Consider, for example, that some scholars hold that Mesmer creatively applied the scientific concepts of his day in a sincere effort to alleviate human suffering, thereby establishing the foundations of modern hypnosis. Others regard Mesmer as a cynical wheeler-dealer whose contributions to hypnosis were nothing more than the happy side effect of his entrepreneurial zeal. Shapiro has similarly provoked widely discrepant professional appraisals. Both Mesmer and Shapiro had their therapeutic epiphanies while walking outdoors; Mesmer was on a retreat in the Alps, Shapiro was strolling through a park. Like Shapiro, Mesmer established a lucrative institute for providing training, and insisted that trainees sign a document promising they would not impart their newly formed and powerful skills to others. Both were charismatic leaders who inspired the founding of professional societies to promote their therapies, and both offered pro bono treatment in the face of criticism that they were merely engaged in profit-making. Both Mesmerism and EMDR have been proclaimed useful for an astonishing range of ailments. Animal magnetism therapists touted the method for gout, blindness, deafness, scurvy, and paralysis in addition to psychosomatic problems. EMDR therapists claim their method is useful for paranoid schizophrenia, learning disabilities, eating disorders, grief, substance abuse, rage, guilt, multiple personality disorder, cancer, and even AIDS. Both Mesmer and Shapiro have challenged scientists to test their methods, and both have complained that disappointing results were attributable to poorly trained researchers. Although the striking parallels between Mesmer and Shapiro in no way impugn the scientific validity of EMDR, comparative analysis suggests that ingeniously promoted "miracle cures" are likely to share many properties.

Animal Magnetism Therapy and EMDR are two historically salient treatments that have stirred great excitement, only to fail the tests of time and science. There have been many others. A physician, James Walsh, wrote in 1923 on "The Story of the Cures that Fail," and noted the importance of a patient's faith in treatment. Dr. Walsh also observed that faddish techniques often parallel the latest developments in science. Thus, Mesmerism occurred at a time when early experiments on electricity had captured the imagination of the public. Likewise, references by Shapiro to "accelerated information processing" and "neuro networks" echo computer metaphors common today.

Historical lessons from cures that fail illustrate how novel treatments initially induce high levels of expectancies, but often lose their effectiveness over time. The history of failed therapies argues for caution and skepticism when bold new claims are made. Unfortunately, many of EMDR's proponents have made the very errors that Dr. Walsh so presciently warned us about. The continued acceptance and proliferation of EMDR among psychologists also represent a fundamental shift of attitude toward the most basic assumptions concerning burden of proof. Thus, before it was demonstrated convincingly that eye movements mattered, thousands of professionals started taking expensive workshops and waving their fingers. These psychologists, and the public, might consider a useful caution provided by James Oberg and reiterated by Carl Sagan (1995): it is a virtue to keep an open mind when evaluating new ideas, "just not so open that your brains fall out."

References:

DeBell, C. and R. D. Jones. 1997. "As Good as it Seems? A Review of EMDR Experimental Research." *Professional Psychology: Research and Practice*, 28, 153–163.

Devilly, G. J. and S. H. Spence. 1999. "The Relative Efficacy and Treatment Distress of EMDR and a Cognitive-Behavior Trauma Treatment Protocol in the Amelioration of Posttraumatic Stress Disorder." *Journal of Anxiety Disorders*, 13, 131–158.

Foa, E. B. and E. A. Meadows. 1997. "Psychosocial Treatments for Posttraumatic Stress Disorder: A Critical Review." *Annual Review of Psychology*, 48, 449–480.

Frank, J. D. 1973. *Persuasion and Healing: A Comparative Analysis of Psychotherapy*. Baltimore: Johns Hopkins University Press.

Gastright, J. 1995. "EMDR Works! Is That Enough?" *Cincinnati Skeptic*, 4, 1–3.

Hazlett-Stevens, H., R. A. Lytle, and T. D. Borkevec. 1996. "Efficacy of Eye Movement Desensitization in the Treatment of Cognitive Intrusions Related to Memories of a Past Stressful Event." Paper presented at the 30th Annual Convention of the Association for Advancement of Behavior Therapy, New York, N.Y. November.

Lilienfeld, S. O. 1996. "EMDR Treatment: Less Than Meets the Eye?" *Skeptical Inquirer*, 20, 25–31.

Lohr, M. M., D. F. Tolin, and S. O. Lilienfeld. 1998. "Efficacy of Eye Movement Desensitization and Reprocessing: Implications for Behavior Therapy." *Behavior Therapy*, 29, 123–156.

McNally, R. J. 1999a. "Research on Eye Movement Desensitization and Reprocessing (EMDR) as a Treatment for PTSD." *PTSD Research Quarterly*, 10 (1), 1–7.

McNally, R. J. 1999b. "EMDR and Mesmerism: A Comparative Historical Analysis." *Journal of Anxiety Disorders*, 13, 225–236.

Muris, P. H., H. Merckelbach, B. Van Haaften, and B. Mayer. 1997. "Eye Movement Desensitization and Reprocessing Versus Exposure in Vivo: A Single Session Crossover Study of Spider-Phobic Children." *British Journal of Psychiatry*, 171, 82–86.

Pitman, R. K., S. P. Orr, B. Altman, R. E. Longpre, R. E. Poire, and M. L. Macklin. 1996. "Emotional Processing During Eye Movement Desensitization and Reprocessing Therapy of Vietnam Veterans with Chronic Posttraumatic Stress Disorder." *Comprehensive Psychiatry*, 37, 419–429.

Rosen, G. M. 1996. "Level II Training for EMDR: One Commentator's View." *The Behavior Therapist*, 19, 76–77.

———. 1999. "Treatment Fidelity and Research on Eye Movement Desensitization and Reprocessing (EMDR)." *Journal of Anxiety Disorders*, 13, 173–184.

Sagan, C. 1995. *The Demon-Haunted World: Science as a Candle in the Dark*. New York: Random House.

Shapiro, F. 1989. "Efficacy of the Eye Movement Desensitization Procedure in the Treatment of Traumatic Memories." *Journal of Traumatic Stress*, 2, 199–223.

Shapiro, F., and M. S. Forrest. 1997. *EMDR: The Breakthrough Therapy for Overcoming Anxiety, Stress, and Trauma*. New York: Basic Books.

Walsh, J. J. 1923. *Cures: The Story of the Cures that Fail*. New York: D. Appleton.

Do Extraordinary Claims Require Extraordinary Evidence?

A Reappraisal of a Classic Skeptics' Axiom

THEODORE SCHICK JR.

It has been said by skeptics on virtually every talk show, published in nearly all skeptics' books, found on Skeptics Society stationery, printed on T-shirts and buttons, and has even appeared on a giant banner for a James Randi television special. "EXTRAORDINARY CLAIMS REQUIRE EXTRAORDINARY EVIDENCE." Paul Kurtz (1991, 50), Carl Sagan (1979, 73), and Martin Gardner (1983, 62), just to name a few, have turned this catchy phrase into a skeptical axiom. It stems from empiricist philosopher David Hume's essay *Of Miracles* (1777). There Hume claims that we can never be justified in believing that a miracle occurred because, by definition, the evidence for a miracle can never be stronger than the evidence against it. He writes (114–115):

> A miracle is a violation of the laws of nature. . . . Nothing is esteemed a miracle, if it ever happen in the common course of nature. . . . There must, therefore, be a uniform experience against every miraculous event, otherwise the event would not merit that appellation. And as a uniform experience amounts to a proof, there is here a direct and full *proof*; from the nature of the fact, against the existence of any miracles.
>
> Since miracles violate natural law, and since natural laws express regularities that are consistent with all past experience, the evidence in favor of miracles will never outweigh the evidence in favor of natural law.

The conception of evidence upon which this argument is based is what is known as an "atomistic" one. According to Hume, evidence comes in discrete packages (like an "atom"), each of which carries the same weight. In his *A Treatise of Human Nature,* for example, he tells us that "every past experiment has the same weight, and that 'tis only a superior number of them which can throw the balance on any side" (1739, 136). As a result, Hume believes that deciding between competing hypotheses is merely a matter of toting up the number of experiments in favor of each and determining which is the greater. "In all cases," he says, "we must balance the opposite experiments, where they are opposite, and deduct the smaller number from the greater, in order to know the exact force of the superior evidence" (1748, 111). For Hume, choosing among competing claims is simply a matter of weighing the evidence. The more extraordinary the claim—that is, the more of our past experience it conflicts with—the more evidence we need in order to accept it.

We now know however, that considerations of evidence alone are never enough to decide among competing claims because *all empirical claims are under-determined by their evidence.* In other words, the truth of an empirical claim cannot be established solely on the basis of its evidence because no one empirical claim follows from a body of empirical evidence, no matter how large. For any set of empirical data, any number of claims can be constructed to account for it, just as for any set of points on a Cartesian coordinate system, an infinite number of curves can be drawn through them. Because competing claims may be equally well supported by the same body of empirical evidence, any choice among them must appeal to factors other than evidential support.

Perhaps the best illustration of the evidential under-determination of empirical claims is provided by the problem of the external world described by Martin Gardner (1983, 12) this way:

> Everything we know about the world is based on information received through our senses. This world of our experience—the totality of all we see, hear, taste, touch, feel, and smell—is sometimes called our "phenomenal world. . . ." Charles S. Peirce invented a useful word for this phenomenal world. He called it the "phaneron." Let us admit at once that there is no way to prove to a solipsist [one who believes that they are the only thing that exists in the universe] that anything exists outside his or her phaneron, if by "prove" you mean the way you prove a theorem in logic or mathematics.

From the fact that there are certain sensations, it does not follow that there is an external world, for such sensations could exist even if there were no external world. Moreover, even if the existence of such sensations did entail the existence of an external world, they would not entail the existence of a physical world, for such sensations could be caused by any number of things, such as an evil demon, God, or an advanced neurophysiologist. Thus, the evidence of our senses, by itself, cannot justify any claim about the external world. If evidential support *were* the only criterion of rational choice, then the claim that our sensations are caused by physical objects would be no more reasonable than the claim that they are caused by an evil demon, and some accept this argument. But they are mistaken, for, as we shall see, the justification of empirical claims does not depend on sensory experience alone. There are additional criteria of rational choice.

One criterion we often appeal to in deciding among competing claims is *conservatism*—how well they fit with, or conserve, the established findings. Since the understanding yielded by an explanation is a function of the degree to which it systematizes and unifies our knowledge, any claim that contradicts previous findings is suspect. But the fact that a claim clashes with the received view cannot rule it out, for if it did we would never make any intellectual progress. To give up a well-established claim we need good reasons. Contrary to what the slogan in question would have us believe, however, we can have good reasons for accepting an extraordinary claim even if there is very little evidence in support of it, as witness the claim of Einstein.

According to Newton's theories of gravity and motion, space and time are absolute, and energy and mass are conserved. According to Einstein's theory of relativity, however, space and time are relative, and neither energy nor mass is conserved. So when it was first proposed, Einstein's theory was extraordinary because it lacked the virtue of conservatism.

Moreover, there was very little evidence to support it. The only fact that Einstein's theory could account for that Newton's could not was the precession of the perihelion of Mercury's orbit. Because Einstein's theory could explain more than Newton's, it had greater *scope* than Newton's. But the evidence in its favor was by no means extraordinary, especially early in its history.

Scope, however, was not the only thing that Einstein's theory had going for it. It also came to possess the virtue of fruitfulness, for it successfully predicted an unexpected phenomenon. The physicist Sir Arthur Eddington, an early convert to Einstein's theory, realized that if Einstein was right, space would be curved around massive objects (like the sun), and that the amount of the curvature could be measured by observing light rays that passed close to such objects. In 1919 he mounted an expedition to Africa to observe the sky during a total eclipse of the sun on May 29. He found that the position of stars whose light passed near the sun during the eclipse appeared to be shifted from their normal position by just the amount predicted by Einstein's theory.

Some might consider this evidence to be extraordinary, for it was unexpected. But to do so is to reject Hume's atomistic theory of evidence. In Hume's view, one experiment confirming Einstein's theory could not outweigh all of the experiments confirming Newton's, and hence could not be construed as extraordinary evidence. Moreover, most scientists did not consider the evidence provided by Eddington's eclipse experiment to be extraordinary. Stephen G. Brush (1989), for example, reports:

> [I]n the case of gravitational light bending most scientists ascribed essentially no weight to the mere circumstance that the phenomenon was predicted before it was observed. . . . The eclipse results put relativity much higher on the scientific agenda and provoked other scientists to try to give plausible alternative explanations. But light bending could not become reliable evidence for Einstein's theory until those alternatives failed, and then its weight was independent of the history of its discovery.

Thus, although Einstein's theory was fruitful, its successful and surprising predictions did not constitute extraordinary evidence in its favor.

What was it about Einstein's theory, then, that scientists found so compelling? For many, it was its simplicity, that is, its relative lack of independent assumptions. Einstein himself saw simplicity as the chief virtue of his theory: "I do not by any means find the chief significance of the general theory of relativity in the fact that it has predicted a few minute observable facts, but rather in the simplicity of its foundation and in its logical consistency" (1930). In fact, when Einstein received Eddington's cable noting the eclipse results he told one of his students, Ilse Rosenthal-Schneider (who had just asked him what if there had been no confirmation of the theory), "Then I would have been sorry for the dear Lord—the theory is correct." Other physicists shared this view, some going so far as to say that even if the first tests turned out negative, Einstein's theory should not be rejected (Lindsay and Margenau, 1936, 377). For these scientists, the depth of understanding afforded by the theory more than made up for its lack of supporting evidence. Newton himself explains the importance of simplicity this way (Beck and Holmes, 1968, 188):

> We are to admit no more causes of natural things than such as are both true and sufficient to explain their appearances. To this purpose the philosophers say that Nature does nothing in vain, and more is in vain when less will suffice. For Nature is pleased with simplicity, and affects not the pomp of superfluous causes.

To explain the importance of simplicity, however, it is not necessary to appeal to Nature's purposes or feelings. We need only recognize that the simpler a theory is, the more it systematizes and unifies our knowledge; and as a consequence the more it increases our understanding. Our belief that there is a physical world is justified because it meets the foregoing criteria of adequacy better than any of its competitors; that is, it provides the best explanation of our sense experience. Martin Gardner (1983, 24–25) explains:

> It is obvious that all we know about the world outside of us is what we infer from what is inside our skin, or rather inside our skull where the sensory inputs are interpreted. But the regularities of those inputs, such as the patterns of flying birds on our retinas, suggest the hypothesis that outside our eyes is a world independent of our inner experience. This hypothesis has enormous explanatory and predictive power. Moreover, it is a theory of extreme simplicity and therefore, by the principle of Occam's razor, preferable to more complex explanations.

Epistemologist Alan Goldman (1988, 204) concurs:

> It is clear that other grounds on which we generally prefer certain explanations to others rule out skeptical alternatives as preferable. Appeal to a demon who causes us to form false beliefs about appearances by confusing our applications of phenomenal terms is highly counterintuitive and leaves many questions that naturally arise regarding more precise causal antecedents, the nature, origin, and motives of this being, and so on, unanswered and seemingly unanswerable. Accepting such beliefs as true, on the other hand, leads to explanatory chains that encompass our deepest empirical theories. Thus, grounds of explanatory depth and fruitfulness lead us to prefer explanations that appeal to the truth of beliefs about appearances over skeptical alternatives.

Most extraordinary claims are put forward as explanations of something. Parapsychologists, for example, claim that the existence of psi energy best explains the fact that people sometimes score high on ESP tests; UFOlogists claim that the existence of alien spacecraft best explains the fact that people sometimes see unidentified objects in the sky; and astrologers claim that the existence of mysterious forces emanating from the stars and planets best explains why people sometimes believe that their horoscopes are accurate. Moreover, some argue that there is an extraordinary amount of confirming evidence for each of these hypotheses. An observation confirms a hypothesis in relation to a set of background beliefs, however, if and only if what is observed is more likely given the hypothesis and the background beliefs than it is given the background beliefs alone (Mackie, 1969, 27).

Thus, the hundreds of cases where people have scored above chance on ESP tests, the thousands of cases where people have seen unidentified objects in the sky, and the millions of cases where people believe their horoscopes are accurate, can be viewed as confirming instances of their respective claims. But even with all this confirming evidence, these claims are not rationally acceptable. Why? Because they do not provide the best explanation of that evidence. There are other explanations that are more conservative, more fruitful, more simple, and have greater scope. Since the adequacy of a claim is not determined by the evidence alone, extraordinary claims do not necessarily require extraordinary evidence.

Many believe that the principle that extraordinary claims require extraordinary evidence follows from Hume's maxim: "That no testimony is sufficient to establish a miracle, unless the testimony be of such a kind that its falsehood would be more miraculous than the fact which it endeavors to establish" (1777, 115–116; see also, *Skeptic*, Vol. 1, #2, 13, for a further discussion of this maxim). This principle, however, is problematic, for it contradicts Hume's earlier claim that a belief in miracles can never be justified. If we can never be justified in believing that a miracle has occurred, we can never be justified in believing that one thing is more miraculous than another. The ability to assign degrees of miraculousness presupposes the ability to identify miracles. Without the latter we cannot have the former. Given Hume's skepticism regarding miracles, then, his maxim is not very helpful.

Hume is right in claiming that a belief in miracles can never be justified. But he is wrong in claiming that the reason for this is that the amount of evidence for the lawfulness of nature will always outweigh the amount of evidence for a miracle. In his maxim, even Hume himself seems to recognize that this need not always be the case. The crux of the matter is not the quantity of evidence available, but the quality of the explanations offered. Supernatural explanations are simply not as good as natural ones. Not only are supernatural explanations less conservative, fruitful, and simple than natural ones, they also have less scope. In so far as nobody has any idea of what a supernatural force is or how it works, any appeal to the supernatural is an appeal to the incomprehensible. But we cannot explain the unknown by means of the incomprehensible. Supernatural explanations do not improve our understanding of a phenomenon. They simply mask the fact that we do not yet understand it.

When faced with a phenomenon we cannot explain, it is always more reasonable to assume that we do not know the operative natural laws than to assume that its cause is supernatural. Even St. Augustine recognized this: "A miracle happens not contrary to nature but contrary to our knowledge of nature" (415, XXI, 8). Our only hope of coming to a rational understanding of our experiences is to see them as the result of natural processes. The principle that should guide our thinking in this regard, then, is this: Just because we cannot explain something does not mean that it is supernatural.

If we replace the "miracle" in Hume's maxim with "extraordinary claim" it becomes: "No testimony is sufficient to establish an extraordinary claim unless the falsehood of the testimony is more improbable than the claim itself." What this maxim gives us is a necessary condition for sufficient evidence. It tells us that a body of evidence is sufficient for establishing a claim only when the truth of the evidence is more probable than the truth of the claim. While this is certainly true, it does not tell us when a body of evidence is sufficient. That is, it does not give us a sufficient condition for sufficient evidence. The analysis of theory choice developed here, however, gives us both necessary and sufficient conditions for sufficient evidence. In this view, there is sufficient evidence to establish a claim (extraordinary or otherwise) when and only when either the claim provides the best explanation of the evidence or the evidence provides the best explanation of the claim. Since an explanation is best when it meets the criteria of adequacy better than any other, we can agree with Thomas Kuhn that, "It is vitally important that scientists be taught to value these characteristics [the criteria of adequacy] and that they be provided with examples that illustrate them in practice" (1991, 261).

It is not possible to quantify how well a claim does with respect to any particular criterion of adequacy. There is no simplicity scale, for example, that can be used to assign a numerical value of simplicity to a hypothesis. Nor is it possible to rank the criteria in order of importance. At times we may rate conservatism more highly than scope, especially if the hypothesis in question is lacking in fruitfulness. At other times, we may rate simplicity higher than conservatism, especially if the hypothesis has at least as much scope as our current hypothesis. Doing well with respect to any particular criterion of adequacy is neither a necessary nor a sufficient condition for being a good hypothesis. So it is doubtful that there can be an algorithm for theory choice. Choosing among theories, like making judicial decisions, is a process that does not appear to be formalizable.

It might be thought that having judgments of rational acceptability rest on qualitative factors like conservatism, scope, simplicity, and fruitfulness rather than on quantitative factors like amount of evidence somehow undermines the objectivity of theory choice. This is not the case, however. Many distinctions that are not quantifiable are nevertheless perfectly objective. The point at which day turns into night or a hirsute person becomes bald cannot be precisely specified. But the distinctions between night and day or baldness and hirsuteness are as objective as they come. There are certainly borderline cases that reasonable people can disagree about, but there are also clear-cut cases where disagreement would be irrational. It would simply be wrong to believe that a person with a full head of (living) hair is bald. It would be equally wrong to believe that a theory that did not meet the criteria of adequacy as well as its competitors was the better theory.

The skeptical axiom under question is compelling both in its catchiness and its seemingly sound philosophical basis. Philosopher Paul Kurtz, for example, provides the following justification of the principle that extraordinary claims require extraordinary evidence: "Where a claim promises to overturn a whole body of data and hypotheses which we now accept on the basis of strong grounds, then before we accept it, we must have even stronger grounds to do so. . . . Thus before we can invoke miraculous or occult explanations that overturn well-established laws and regularities of experience and nature, we would need very strong evidence" (1991: 49–50). While the first statement is unexceptionable, however, the second is unacceptable. As we have seen, we can have strong grounds for accepting an extraordinary claim even if we do not have strong evidence for it.

The foregoing considerations indicate that extraordinary claims do not require extraordinary evidence. It can be reasonable to accept an extraordinary claim in the absence of extraordinary evidence as long as it provides the best explanation of the evidence available; that is, as long as it meets the criteria of adequacy better than any other explanation. For example, a jury can legitimately convict someone on the basis of a few pieces of circumstantial evidence as long as the defendant's guilt figures into the best explanation of that evidence. A better slogan for skeptics, then, would be this: Extraordinary claims require exemplary explanations.

Most extraordinary claims are not only not conservative (for they conflict with established findings), they are also not simple (for they postulate strange entities or complex conspiracies). Their scope is usually very limited (they usually explain only one particular phenomenon) and they often are not very fruitful (they often do not successfully predict any new phenomena). That is why it is normally unreasonable to accept them. But on those occasions when they do provide the best explanation of something, their acceptance is eminently rational.

References:

Augustine. 1950 [415]. *The City of God.* Modern Library. New York: Random House.

Beck, L. W. and Holmes, R. L. 1968. *Philosophic Inquiry.* Englewood Cliffs: Prentice Hall.

Brush, S. G. 1989. "Prediction and Theory Evaluation: The Case of Light Bending." *Science*, 246: 1124–1129.

Einstein, A. 1930. *Forum Philosophicum I*, 173. Quoted in Brush, *op. cit.*

Gardner, M. 1983. *The Whys of a Philosophical Scrivener.* New York: Quill.

Goldman, A. 1988. *Empirical Knowledge*. Berkeley: University of California Press.

Hume, D. 1902 [1748]. "Of Miracles." In *Enquiries Concerning the Human Understanding and Concerning the Principles of Morals*, 2nd ed., §10, ed. by L. A. Selby-Bigge. Oxford: Clarendon Press.

Hume, D. 1888 [1739]. *A Treatise of Human Nature*, ed. by L. A. Selby-Bigge. Oxford: Clarendon Press.

Kuhn, T. 1970. "Reflections on My Critics." *Criticism and the Growth of Knowledge*. (Imre Lakatos and Alan Musgrave, eds.) Cambridge: Cambridge University Press.

Kurtz, P. 1991. *The Transcendental Temptation*. Buffalo: Prometheus Books.

Lindsay, R. B. and Margenau, H. 1936. *Foundations of Physics*. New York: Wiley.

Mackie, J. L. 1969. "The Relevance Criterion in Confirmation Theory." *British Journal for the Philosophy of Science*. May 20.

Sagan, C. 1979. *Broca's Brain*. New York. Ballantine Books.

Facilitated Communication

GINA GREEN

The cliché that "there is nothing new under the sun" applies more than ever to the mental health profession today. We seem to be experiencing a myriad of new techniques to treat the developmentally disabled, Facilitated Communication being one of the most popular, yet in reality their underlying characteristics have been seen before. These components make up the structure of what might be considered a social movement:

- Assertions that a new technique produces remarkable effects are made in the absence of solid objective evidence, or what little evidence there is becomes highly overblown.
- Excitement about a possible breakthrough sweeps rapidly through the communities of parents, teachers, service providers, and others concerned with the welfare of individuals with disabilities.
- Eager, even desperate for something that might help, many invest considerable financial and emotional resources in the new technique.
- In the process, effective or potentially effective techniques are ignored.
- Few question the basis for the claims about the new treatment or the qualifications of the individuals making them.
- Anecdotal reports that seem to confirm the initial claims proliferate rapidly.
- Careful scientific evaluation to determine the real effects of the technique are not completed for some time, and can be made more difficult than usual by the well-known and powerful effects of expectancies.
- Some of these techniques have small specific positive effects, or at least do minimal harm.
- Eventually they fall out of favor, sometimes because they are discredited by sound research, sometimes simply because experience reveals their lack of efficacy, but probably most often because another fad treatment has come on the scene. Each retains some adherents, however, and some go relatively dormant for a while only to emerge again.

Parallel phenomena occur in other areas, such as treatments for AIDS, cancer, and various psychological problems. At present the field of developmental disabilities (especially autism) seems to be experiencing an epidemic of novel techniques, or "interventions," as they are called. Despite its parallels with other techniques, Facilitated Communication (FC) has probably had a greater impact than any other novel intervention in the history of treatment for persons with disabilities.

How FC Works

How does FC work? If you have never seen it in action it is quite a phenomenon to observe. Individuals with "severe communication impairments" (e.g., severe mental retardation, autism) are assisted in spelling words by "facilitators" (teachers or parents) who provide physical support, most often (at least initially) by holding their hand, wrist, or forearm while they point to letters on a keyboard or printed letter display. Right before your eyes, a mentally disabled person that just previously had virtually no communication skills, suddenly begins to spell out words, sentences, and whole paragraphs. Stories are told. Answers to questions are given. A child that did not appear to know the difference between a dog and an elephant can now be shown a series of pictures, correctly identifying them one by one, as his or her hand glides deftly over the keyboard, pecking out the correct letters. The assumption, of course, is that most of the words spelled in this fashion actually originate with the disabled partner and not the facilitator.

On its face, FC can seem simple and benign, and sometimes looks quite convincing. Its main proponents sometimes characterize FC simply as a strategy for teaching individuals to point in order to access systems like synthetic speech devices and keyboards to augment their communication. At the same time, however, they claim that it is a revolutionary means of unlocking highly developed literacy, numeracy, and communication repertoires in large numbers of individuals previously thought to have severe learning difficulties. For all the world it looks like a mental miracle, the kind of stuff they make movies about, as in "Awakenings."

The theory is that many such individuals do not have cognitive deficits at all, but instead have a presumed neuromotor impairment that prevents them from initiating and controlling vocal expression. Their average or even above average intelligence is locked away, awaiting release. The neuromotor disorder is also presumed to manifest itself in "hand function impairments" that make it necessary for someone else to stabilize the individual's hand and arm for pointing, and to pull the pointing hand back between selections to minimize impulsive or poorly planned responding. Candidates for FC are also presumed to lack confidence in their abilities, and so require the special touch and emotional support of a facilitator to communicate (i.e., a strap or device to hold the person's arm steady will not work).

FC thus has an almost irresistible appeal for parents, teachers, and other caring persons who struggle mightily to understand and communicate with individuals who often do not respond or communicated in return. But the very features that make FC so seductive, in combination with some other potent factors, have made it a topic of heated debate between believers and skeptics since its "discovery" in Australia nearly two decades ago.

Beginnings Down Under

It all began in the 1970s with Rosemary Crossley, a teacher in an institution in Melbourne in the Australian state of Victoria. She suspected that some of her young charges with severe cerebral palsy had far more ability than their physical impairments allowed them to demonstrate. When she gave them hand or arm support to help them point to pictures, letters, and other stimuli, Crossley became convinced that several of the children revealed literacy and math skills that they had somehow developed with little or no instruction, despite having lived most of their lives in an impoverished institutional environment.

Right away there was controversy about the technique that Crossley called Facilitated Communication Training. Two people were involved in creating the messages, and simple observation could not reveal how much each was contributing. Plus, many of the messages Crossley attributed to these institutionalized individuals defied plausibility. "Facilitated" accusations of abuse and expressions of wishes for major life changes (like leaving the institution) made it imperative to determine whether communications actually originated with the disabled individual or the facilitator. Matters were complicated by Crossley's emerging status as a heroine to many in the deinstitutionalization movement. Eventually, after a series of legal proceedings, a young woman with cerebral palsy with whom Crossley had developed a special relationship through FC was released from the institution to reside with Crossley. The institution was closed, and in 1986 Crossley started (with government financial support) the DEAL Centre (Dignity through Education and Language) to promote alternative communication approaches—principally FC—for individuals with severe communication impairments. Use of the method spread to programs in Victoria serving persons with various disabilities, accompanied by controversies about communications attributed to FC users on the basis of subjective reports.

Sufficiently serious issues arose to provoke formal statements of concern from professionals and parents in 1988, and a government-sponsored investigation in 1989. Despite Crossley's resistance to objective testing (on the basis that FC users refused to cooperate when their competence was questioned), some small-scale controlled evaluations were conducted in the course of that investigation. When the facilitator's knowledge about expected messages was well-controlled (more on this later), and the accuracy of messages was evaluated objectively, the effect disappeared. The disabled individuals were unable to communicate beyond their normal expectation. Instead, it appeared that the facilitators were authoring most FC messages, apparently without their awareness. These early studies suggested that FC was susceptible to a somewhat unusual kind of abuse: Allowing others to impose their own wishes, fears, hopes, and agendas on nonspeaking individuals.

A Social Movement Is Born

At about that time Douglas Biklen, a special education professor from Syracuse University, conducted a four week observational study of 21 DEAL clients said to be autistic, who were reported to engage in high-level discourse with the help of facilitators. Professor Biklen was already established as a leader in the "total inclusion movement," which seeks the full-time placement of all students with disabilities, regardless of their competencies and needs, in regular classrooms. The report describing his first qualitative study of FC, which Biklen said was begun "in an attempt not to test hypotheses but rather to generate them," appeared in the *Harvard Educational Review* in 1990. He reported that the communication of the individuals he observed (some of whom were being "facilitated" for the first time) was sophisticated in content, conceptualization, and vocabulary, and contained frequent references to feelings, wishes to be treated normally and to attend regular schools, and society's treatment of individuals with disabilities.

This was in sharp contrast with the well-documented difficulties in social, play, cognitive, and communication skills that constitute current diagnostic criteria for autism (not to mention that the diagnosis is difficult to make and is applied to individuals with a wide range of competencies and deficits in all those domains). In his seminal article, Biklen mentioned the controversy over the Australian findings, but asserted that informal "indicators that communication was the person's own were strong enough, in my view, to justify the continuing assumption of its validity."

Some of the indicators he reported observing were disabled individuals typing independently or with minimal physical contact with the facilitator; content (spelling errors, unexpected word usage, etc.) that appeared to be unique to each individual; and facial expressions or other signs that the individual understood the communication. He also noted that facilitators often could not tell who was doing the spelling and that they could be influencing the FC in subtle ways without their awareness, and that this could be a problem. Finally, on the basis of his uncontrolled observations and the reports of Crossley and other facilitators, Biklen decided that autism had to be redefined as a problem not of cognition or affect, but of voluntary motor control. He returned from Australia to establish the Facilitated Communication Institute (FCI) at Syracuse University, and the North American FC movement was underway.

The Movement Takes Off

Word of FC spread quickly with the help of several media reports of FC "miracles." The rate of information exchange increased geometrically, feeding the system and driving it forward. FC newsletters, conferences, and support networks contributed to the spread of astonishing success stories, along with examples of prose and poetry attributed to FC authors. The Syracuse FCI began training new facilitators in earnest, in workshops that lasted from a few hours to two or three days. At least two New England universities became satellite programs of the Syracuse FC Institute, as did numerous other private and public agencies that provided training and support for facilitators. Initiates (parents, paraprofessionals, and professionals in several disciplines) were often told that the technique was simple and required no special training. They were urged to train others, and to go out and try FC with disabled individuals. Thousands did. Soon FC was being heralded as a means of "empowering" individuals with severe disabilities to make their own decisions and participate fully in society. FC was rapidly becoming the Politically Correct treatment of choice.

Soon after publication of Biklen's article, special education personnel and parents around Syracuse, then throughout the U.S. and Canada, adopted FC enthusiastically. Scores of children were placed in regular classrooms doing grade-level academic work with "facilitation." Decisions about the lives of adults with severe disabilities—living arrangements, medical and other treatments, use of hearing aids, and so on—were based on "facilitated" messages without any attempt to verify authorship objectively. In many cases FC supplanted other communication modes, including vocal speech and augmentative communication systems, that do not require another person for message creation. Some psychologists, speech pathologists, and others began giving I.Q. and other standardized tests with "facilitation," changing diagnoses and program recommendations in accordance with the "facilitated" results. Suddenly "retarded" individuals were proclaimed to have average or above-average intelligence. "Facilitated" counseling and psychotherapy were promoted to help FC users deal with personal problems. Colleges and universities offered courses on FC. Millions of tax dollars were invested in promoting its widespread adoption, with little objective evaluation of its validity or efficacy.

Enter Psi, Exit Science

Not surprisingly, the experience of accomplishing a breakthrough and being part of a movement was a heady experience for many facilitators. Some, however, reported wondering all along whether the words being produced through FC were really coming from their disabled partners. Others who had serious doubts about the method from the outset found themselves under considerable pressure from parents, peers, and employers to adopt the method wholesale and without question. Reports that facilitators' private thoughts were being expressed through FC led some to conclude that individuals with autism must have telepathy—a view espoused by a professor of special education at the University of Wisconsin, among others.

Facilitators were also imbued, explicitly and implicitly, with a strong ideology that presents dilemmas for many who want to know who is really communicating in FC. Some components of the ideology include:

- Assume competence.
- Don't test.
- Prevent errors.
- Expect remarkable revelations in the form of hidden skills as well as sensitive personal information.
- Use circumstantial, subjective data to validate authorship.
- Avoid objective scrutiny.
- Emphasize "facilitated" over spoken or other communications.

Contradictory evidence from the controlled evaluations that had been conducted in Australia and those that emerged later in the U.S. were mentioned rarely, if at all, in FC training materials and newsletters. When that evidence was mentioned it was to criticize the evaluation methods and the people who employed them, and to explain away the results by saying essentially that FC could not be tested. In short, FC's validity was to be accepted largely on faith. With this, science was abandoned.

Concurrently Biklen, Crossley, and their colleagues published further reports of qualitative studies suggesting that FC was highly effective in eliciting unexpected literacy skills from large proportions of individuals with severe autism, mental retardation, and other disorders. Many of these individuals had received little instruction in reading and spelling, or if instruction had been attempted many had not appeared to learn very much. How, then, had they developed age-level or even precocious literacy skills? According to Biklen they acquired these skills from watching television, seeing their siblings do homework, and simply being exposed to words pervading the environment. Or perhaps some had actually been learning from instruction all along, but because their speech was limited they could not demonstrate what they learned.

How did they verify their claims? Biklen and his colleagues used participant observation and other methods employed by anthropologists, sociologists, and educators in field studies of cultures and social systems. The research was strictly descriptive, not experimental, and employed no objective measurement or procedures to minimize observer bias. Despite their acknowledgement of the real possibility of facilitator influence in FC, these studies did not control that critical variable.

Late in 1991 a few parents of students at the New England Center for Autism, where I serve as Director of Research, began pressing our program to adopt FC. They asked us to make rather drastic changes in their children's lives on the assumption that messages produced with FC represented the children's true wishes and competencies. Some were angry when we decided instead to use it only under conditions of a small-scale experimental study employing the kind of objective evaluation methods that we try to apply to all techniques. At that time we could find nothing about FC in the research literature, so we consulted respected colleagues around the country. Some (in California, surprisingly enough) had not heard of it yet. Others invoked a Ouija board analogy or

Clever Hans effect, and suggested that FC would be a short-lived fad. None knew of any objective evidence about FC. To our chagrin, we also encountered individuals with scientific training who were promoting the use of FC without considering the fundamental question about authorship.

The Sexual Abuse Component

The real possibility that "facilitated" words were those of the facilitators was not a cause for much concern as long as the process seemed benevolent. Few wished to throw a wet blanket on the euphoria created by reports of a breakthrough. But almost from the beginning, strange things began to happen: Some FC messages said—or were interpreted by facilitators to say—that disabled FC users had been abused by family members or caregivers. Often the abuse alleged was sexual, and many allegations contained extensive, explicit, pornographic details.

So many social movements have a sexual component in them, and FC is not different. Production of sex abuse allegations usually set in motion an inexorable chain of events. Beliefs about FC, the complexities inherent in the method, and the fact that the alleged victim may be seen as particularly vulnerable because he or she is disabled, now began to interact with the zealous pursuit that seems to typify investigations of sex abuse allegations. School or program administrators were notified, who in turn called in representatives of social services and law enforcement agencies. If the accused was a family member with whom the FC user resided, that person was either required to leave the home or the FC user was placed in foster care. If a parent was accused, both parents often faced criminal charges, one for perpetrating the alleged abuse, the other for knowing about it and failing to act. Often actions were initiated by social service workers to terminate parental custody or guardianship. If the accused was a school or program employee, they may have been suspended from their job or even fired. A long and trying ordeal was virtually guaranteed for all involved. An investigation began. Police interrogated the accused, and questioned the alleged victim through their facilitator. Other evidence was sought in the results of medical and psychological examinations of the alleged victim, and interviews with others who may have had information about the alleged events. A presumably independent facilitator was sometimes called in to try to corroborate the allegation, introducing another complexity: There appear to be no established safeguards or objective criteria for ensuring that independent facilitators in fact have no access to information about cases, nor for deciding what constitutes corroborating "facilitated" content.

False allegations have devastating emotional and financial effects on the accused and their families, but leaving individuals in situations in which they may be abused jeopardizes their physical and emotional welfare. It would seem that extreme caution and stringent rules of evidence should apply. A number of cases have arisen in which the only evidence was a "facilitated" allegation, although there have also been reports of cases in which corroborating evidence or confessions were obtained. When an allegation is made through FC, two separate but related questions must be addressed: Who made the allegation, and did the alleged events actually occur? Some courts and investigative bodies in Australia, the U.S., and Canada have decided that the first question must be answered by controlled testing of FC under conditions where independent observers can verify when the facilitator does and does not have information necessary to produce communications. If the FC user does not convey information accurately and reliably under those conditions, and there is no other solid evidence, the legal action is usually terminated. That has been the outcome of testing in every case of which I am aware, but by the time that determination has been made the accused have been traumatized for the better part of a year and have spent tens of thousands of dollars defending themselves. Solid corroborating evidence would certainly answer the second question—whether abuse occurred—but it does not follow logically that it answers the question about who authored the "facilitated" allegation.

Unfortunately, it wasn't until a number of false "facilitated" allegations of sexual abuse came to light that FC began to be scrutinized closely. As issues about the validity and reliability of FC were addressed in courtrooms all over the U.S., critical and questioning stories appeared in the print and electronic media. Concurrently (though somewhat slowly), results from a rapidly growing number of controlled evaluations began to be disseminated, and a few more skeptical voices were raised.

How to Test FC

The rationale for conducting controlled observations to determine authorship in FC is straightforward: If the disabled FC user is actually the source of the messages, then accurate and appropriate messages should be produced on virtually every opportunity when the facilitator has no knowledge of the expected message. Some controlled evaluations of FC have been mandated by legal questions like those just described, but a number were carried out by clinicians, researchers, and program administrators who simply wanted an objective empirical basis for making decisions about FC. Even James "The Amazing" Randi was consulted in the early stages of testing, some calling him in to make sure fraud and trickery were not involved, others because they genuinely wondered if psychic power was the cause. Randi's skepticism of the phenomenon was not welcomed by FC supporters. The first major American study was conducted by psychologist Douglas Wheeler and colleagues at the O.D. Heck Developmental Center in Schenectady, NY, who wanted objective evidence to convince skeptics that FC was valid.

How do you do a controlled study of FC? Recently I analyzed reports of 17 evaluations of FC that have appeared or have been accepted for publication in peer-reviewed professional journals, and eight presented at scientific conferences. The common and critical ingredients were:

1. Consent for participation.
2. Objective measures, i.e., use of independent, nonparticipating observers or judges, "blind" to the conditions in effect, who recorded data and/or evaluated the accuracy of FC output.
3. Maintenance of physical and emotional support by the facilitator.
4. With only a few exceptions, facilitator/FC user dyads who had been working together with apparent success for a considerable period before formal evaluations were conducted.
5. Familiar, common communication contexts (e.g., typical academic and language-development activities, discussing everyday events, naming or describing familiar pictures or objects).
6. Establishment of apparently successful FC in the evaluation context.
7. Control of information available to the facilitator.

The necessary control was established in a number of ways. In some studies, facilitators were simply asked to look at their partner and not the letter display, or were actually screened from the letter display. These kinds of tests were suggested by the observation that many facilitators focus intently on the letters while their partners look at the letters infrequently, if at all. Others presented visual stimuli like pictures, objects, or printed materials only to the FC user while the facilitator was screened from seeing them. Alternatively, spoken questions were presented only to the FC user while their facilitator wore earplugs or headphones playing masking noise. Several evaluations used a procedure described as "message passing": FC users were engaged in some familiar activities in the absence of facilitators, who then used FC to solicit descriptions of the activities. A couple of evaluations involved independent facilitators, unfamiliar with the FC user, who solicited information that was presumably unknown to the facilitator (e.g., the FC user's favorite food, a recent event in their life, names of family members, etc.).

The Results

The most telling evaluations used double-blind procedures, in which facilitators and their partners saw or heard different items on some trials, and the same item on other trials. Neither could tell what information their partner was receiving. Responses that corresponded to information presented to the facilitator and not to their partner provided direct evidence that facilitators were controlling those FC productions. Multiple tasks and control procedures were used by several investigators. Facilitators in all evaluations had been trained by leading proponents of FC, or by others who had had such training. They seemed representative of the general population of facilitators, including parents, paraprofessionals, teachers, speech pathologists, and other human service workers. The sample of FC users in these evaluations also appeared representative, comprising a total of 194 children and adults with autism, mental retardation, cerebral palsy, and related disorders.

None of these controlled evaluations produced compelling evidence that FC enabled individuals with disabilities to demonstrate unexpected literacy and communication skills, free of the facilitator's influence. Many messages were produced over numerous trials and sessions, but the vast majority were accurate and appropriate to context only when the facilitator knew what was to be produced. The strong inference is that facilitators authored most messages, although most reported that they were unaware of doing so. Sixteen evaluations found no evidence whatsoever of valid productions. A total of 23 individuals with various disabilities in nine different evaluations made accurate responses on some occasions when their facilitators did not know the answers, but most of those productions were commensurate with or less advanced than the individuals' documented skills without FC. That is, they were primarily single words and an occasional short phrase, produced on some trials by individuals whose vocal or signed communication exceeded that level, some of whom had documented reading skills before they were introduced to FC. For most of these individuals, there was clear evidence that on many other trials their facilitators controlled the productions. The controlled evaluations also demonstrated that most facilitators simply could not tell when and how much they were cueing their partners, emphasizing the importance of systematic, controlled observations for identifying the source of "facilitated" messages. The legal, ethical, and practical implications of these findings are obvious and serious. Together with the legal cases and critical media reports, they have made it a little more acceptable to voice skepticism about FC.

The Proponents Respond: Parallels with Psychics

Proponents of FC have criticized the controlled evaluations on several counts. The parallels of their responses to those received by James Randi when he tests psychics are startling. FC supporters, for example, argue that incorrect answers were due to lack of confidence, anxiety, or resistance on the part of FC users, who "freeze up" or become offended when challenged to prove their competence. Likewise, psychics claim they cannot perform in front of video cameras or in the presence of skeptics who make them anxious. In the case of FC, if this were true—if testing *per se* destroyed the FC process—participants in the controlled evaluations would not have responded at all, or would have produced inaccurate responses throughout, not just when their facilitators did not know the answers. Instead, many accurate words, descriptions, and other responses were produced, but for the most part only when facilitators knew what they were supposed to be.

Additionally, many evaluations took place in familiar surroundings in which individuals had engaged in FC for numerous sessions, with their regular facilitators and letter displays. Sessions typically were not conducted or were terminated if there were any signs of distress or unwillingness to continue. Few refusals were reported. Participants in most evaluations completed numerous trials and sessions over extended periods of time. Most appeared cooperative, even enthusiastic, throughout. Several evaluations were conducted in the context of typical FC sessions, using the same types of materials and questions to which participants had appeared to respond successfully. Questions were no more confrontational or intrusive (perhaps less so) than those often asked in regular FC sessions; in fact, many tasks were identical to those recommended for FC training, except that conditions were arranged so that facilitators could not know all the expected responses. Finally, if FC users simply become too anxious to communicate when challenged, one has to wonder how they are managing to perform in regular academic classrooms, on I.Q. and other tests, in front of TV cameras, and before large audiences at FC meetings. And how can they give "facilitated" testimony, under questioning by judges and attorneys (which is anxiety producing for anyone), as prosecutors in some sexual abuse allegation cases are now arguing is their right?

Another criticism of the controlled evaluations is that the facilitators were not familiar with their partners, were inadequately trained, or did not provide appropriate "facilitation." That is simply not true. As indicated in the summary above, the FC users' preferred facilitators participated with them in most evaluations. The only exceptions were two studies that assessed initial responsiveness to FC with facilitators and FC users who were "beginners" when the evaluation started, and a couple of legal cases in which unfamiliar facilitators were involved (who nonetheless "facilitated" successfully with the FC users before controlled testing began). Many facilitators were trained by leading proponents of FC. Most were encouraged to provide whatever physical and emotional support they wished during the evaluation. If they were not "facilitating" properly, few understandable communications would have been produced. Quite the opposite was true. There is a peculiar irony in this criticism, however, since proponents offer no specific guidelines or standards as to what constitutes sufficient training and experience for facilitators. Some facilitators have started using the method after reading an article, watching a videotape, or attending a brief workshop. When we began to take a look at FC at the New England Center for Autism, for example, our three speech-language pathologists were trained by Biklen in a two-day workshop. That appeared to be the norm at that time (late 1991). A further contradiction is that there are reports throughout the descriptive literature on FC that facilitators who were complete strangers had some individuals with severe disabilities "facilitating" sentences (more, in some cases) in their very first session.

Implausibilities and Inconsistencies

An oft-cited criticism of the controlled evaluations is that they required FC users to perform confrontational naming tasks, which proponents consider inappropriate because individuals with autism have global "word-finding" problems. This argument is implausible for several reasons. First, many evaluations did not require FC users to spell specific names; descriptions, copying, multiple-choice options, yes/no responses, and answers to open-ended questions were just some of the other kinds of responses solicited. Second, there is no solid evidence that such problems are exhibited by individuals with autism. It can be difficult to distinguish words that an individual presumably knows but cannot produce from words that they simply do not know, even with individuals who at one time had well-developed language (e.g., neurologic patients). This would seem to be even more difficult with individuals with autism. Even if this rationalization applied to individuals with autism, what accounts for the results with the many FC users who did not have autism? Additionally, at least three studies documented spontaneous oral naming responses by FC users with autism that were more accurate than their "facilitated" responses. That certainly goes against the "word-finding" hypothesis for those individuals.

Some FC proponents attribute negative findings to the supposition that most FC users are not experienced with the kinds of tasks presented to them in the controlled tests. This criticism is especially puzzling. By law, the skills of individuals with special needs must be evaluated on a regular basis, so most FC users have probably had a great deal of test experience. The tasks used in most controlled evaluations were like those used to teach and test academic and language skills in classrooms and training programs everywhere. In fact, many were precisely the kinds of activities that are recommended for FC training, on which the FC users in the controlled evaluations had been reported to perform very well. Again, if inexperience with the tasks were a plausible explanation, FC users should perform equally poorly when their facilitators did and did not know the expected answers. That was not the case in the controlled evaluations.

Finally, FC proponents are inconsistent in claiming that controlled testing undermines the FC user's confidence, while in the next breath they are quick to tout reports that some attempts at controlled evaluations have produced evidence of FC's validity. In other words, when the data contradict their claims, experiments are not valid; when the data support their claims, experiments are useful. A report from Australia (referred to as the IDRP report) said that three individuals with disabilities succeeded in "facilitating" the name of a gift they were given in the absence of their facilitators, but one was said to type his responses independently, without FC. The report provided no background information about the individuals, no details about the procedures, and described only one controlled trial completed by each individual. Another exercise described in a letter to the editor of a speech disorders journal claimed that four of five students thought to have severe language delays performed remarkably better with FC than without on a test of matching pictures to spoken words. The facilitator wore headphones but was not screened visually from the nearby examiner who was speaking the words, and no expressive communication was required of the FC users. At best, these exercises must be considered inconclusive, but they have been cited widely by proponents as scientific validations of FC. The contradiction inherent in arguing that controlled testing interferes with FC while endorsing exercises like these seems lost on them. The clear implication is that tests that appear to produce evidence supporting beliefs about FC are good, and tests that fail to do so are bad.

Silent Skeptics

If FC is so obviously not the mental miracle supporters claim it is, why does the movement continue to grow? Why hasn't the scientific community made a significant public statement against FC? A number of variables probably account for the initial and continuing reluctance of many skeptics to speak up. First, scientists in general are cautious about drawing conclusions without data. When FC first hit the disability community in North America, there were no objective data to be had. A rejoinder to Biklen's first report by Australian psychologists Robert Cummins and Margot Prior was submitted to the *Harvard Educational Review* early in 1991. Their paper summarized the results of controlled tests of the validity of FC and the legal and ethical problems it had engendered in Australia. It was not published until late summer 1992, and by that time the FC movement already had considerable momentum. Even then, many skeptics withheld judgment on the basis that the Australian data were limited. This was essentially our reasoning at the New England Center for Autism—that some individuals with autism might write or type better than they could speak (we knew a few), and that if there were some merit to the claims about FC, it would be revealed through careful research using objective methodology.

At the same time, however, we sensed something ominous in the rapidity and zeal with which FC was being applied, the resistance to critical scrutiny, and the antiscience stance of many adherents. Even as the dark side of the FC story began to unfold, relatively few in developmental disabilities who knew how to test the claims about FC experimentally wanted to get involved, perhaps thinking that the best response was to continue to do sound research in their own areas. Others did not to want to be seen as naysayers or debunkers.

Cummins and Prior, both with long histories of involvement in treatment and research in developmental disabilities, were among the first in Victoria to go public with their concerns about FC. Their expressions of skepticism and calls for caution were met with hostility and personal attacks from FC proponents in Australia, a scenario that has repeated itself in the U.S. That suggests another variable, in my opinion one of the most potent: It was (is) not Politically Correct in many circles to suggest that FC might not be all it appears, or even to call for objective evaluation to determine if it is. Those who do are likely to be labelled heretics, oppressors of the disabled, inhumane, negative, jealous of others' discoveries, "dinosaurs" who cannot accept new ideas, and out for financial gain.

The FC Future

Needless to say, considerable attention and acclaim have accrued to the leaders of the FC movement, but as the data and the harms have mounted, so has the criticism. Recent months have seen a marked shift in media coverage from the glowing reports of miracles that made almost no mention of objective evidence (e.g., *PrimeTime Live*) to stories about families for whom FC has been anything but a miracle. A documentary on the PBS investigative news program, *Frontline,* honed in on the implausibility and lack of empirical support for Biklen's initial claims, along with the emerging evidence from experimental evaluations showing overwhelmingly that most FC is facilitator communication.

The public position of Syracuse University officials appears to be that Professor Biklen's notions are simply provoking the furor and resistance that all radical new ideas encounter. Perhaps that is the case; time and objective data will tell. Time will most certainly be required for the legal system to do its part in determining the future of the FC movement. A number of cases involving "facilitated" sexual abuse allegations are in process at this writing. To my knowledge, there has been one conviction so far. Several individuals and families who have been cleared of false allegations have filed damage countersuits against the facilitators, school and program administrators, and social service agencies involved. On January 10, 1994, a civil suit was filed in Federal District Court for the northern district of New York seeking $10 million in damages on behalf of a family who were among the first victims of FC allegations in the U.S. Among the ten defendants are Douglas Biklen and Syracuse University.

Finally, if FC is not a mental miracle, is it sleight of hand? By this I do not mean there is intentional deceit on the part of the facilitators. Far from it. Most are genuine, honest, caring individuals who wish the best for their charges. Herein lies an explanation. The power of a belief system to direct thought and action is overwhelming. A full and complete explanation for the FC phenomenon is still forthcoming, but clearly there are parallels with the ideomotor responses that direct dowsing sticks and the Ouija board. As the facilitator gently directs the hand to begin typing, letters are formed into words and words into sentences. Just as with the Ouija board where elaborate thoughts seem to be generated out of thin air while both parties consciously try not to move the piece across the board, the facilitators do not appear to be conscious that it is them generating the communication. Even with the autistic child looking elsewhere, or not looking at all (eyes closed), the hand is still rapidly pecking out letters as if it were a miracle. Unfortunately there are no miracles in mental health. All of us wish FC were true, but the facts simply do not allow scientists and critical thinkers to replace knowledge with wish.

References:

Biklen, D. (1990) Communication unbound: Autism and praxis. *Harvard Educational Review,* 60, 291–315.

Biklen, D. (1992). Autism orthodoxy versus free speech: A reply to Cummins and Prior. *Harvard Educational Review,* 62, 242–256.

Biklen, D. (1993). *Communication unbound: How facilitated communication is challenging traditional views of autism and ability/disability.* New York: Teachers College Press.

Crossley, R. (1992a). Lending a hand: A personal account of the development of Facilitated Communication Training. *American Journal of Speech and Language Pathology,* May, 15, 18.

Crossley, R. (1992b). Getting the words out: Case studies in facilitated communication training. *Topics in Language Disorders,* 12, 46–59.

Cummins, R. A., & Prior, M. P. (1992). Autism and assisted communication: A response to Biklen. *Harvard Educational Review,* 62, 228–241.

Dillon, K. M. (1993). Facilitated Communication, autism, and ouija. *Skeptical Inquirer,* 17, 281–287.

Green, G. (forthcoming). The quality of the evidence. In H. C. Shane (Ed.), *The Clinical and Social Phenomenon of Facilitated Communication.* San Diego: Singular Press.

Green, G. (1993). Response to "What is the balance of proof for or against Facilitated Communication?" *AAMR News & Notes,* 6 (3), 5.

Green, G., & Shane, H. C. (1993). Facilitated Communication: The claims vs. the evidence. *Harvard Mental Health Letter,* 10, 4–5.

Hudson, A. (in press). Disability and facilitated communication: A critique. In T. H. Ollendick & R. J. Prinz (Eds.), *Advances in Clinical Child Psychology* (Vol. 17). New York: Plenum Press.

Jacobson, J. W., Eberlin, M., Mulick, J. A., Schwartz, A. A., Szempruch, J., & Wheeler, D. L. (in press). Autism and Facilitated Communication: Future directions. In J. L. Matson (Ed.), *Autism: Etiology, Diagnosis, and Treatment*. DeKalb, IL: Sycamore Press.

Jacobson, J. W., & Mulick, J. A. (1992). Speak for yourself, or . . . I can't quite put my finger on it! *Psychology in Mental Retardation and Developmental Disabilities*, 17, 3–7.

Mulick, J. A., Jacobson, J. W., & Kobe, R. H. (1993). Anguished silence and helping hands: Miracles in autism with Facilitated Communication. *Skeptical Inquirer*, 17, 270–280.

Homeopathy

WILLIAM JARVIS AND THE NATIONAL COUNCIL AGAINST HEALTH FRAUD

Homeopathy was devised by the German physician Samuel Hahnemann (1755–1843) as a reaction to practices based upon the ancient humoral theory which he labeled "allopathy." The term has been misapplied to regular medicine ever since. The cardinal principles of homeopathy include that (1) most diseases are caused by an infectious disorder called the psora; (2) life is a spiritual force (vitalism) which directs the body's healing; (3) remedies can be discerned by noting the symptoms that substances produce in overdose (proving), and applying them to conditions with similar symptoms in highly diluted doses (Law of Similia); (4) remedies become more effective with greater dilution (Law of Infinitesimals), and become more dilute when containers are tapped on the heel of the hand or a leather pad (potentizing). Homeopathy's principles have been refuted by the basic sciences of chemistry, physics, pharmacology, and pathology. Homeopathy meets the dictionary definitions of a sect and a cult—the characteristics of which prevent advances that would change Hahnemann's original principles. Most homeopathic studies are of poor methodological quality, and are subject to bias. Homeopathic product labels do not provide sufficient information to judge their dosages. Although homeopathic remedies are generally thought to be nontoxic due to their high dilutions, some preparations have proved harmful. The ostensible value of homeopathic products can be more than a placebo effect because some products have contained effective amounts of standard medications or have been adulterated. Only about half of the 300 homeopaths listed in the Directory of the National Center for Homeopathy are physicians. Others include naturopaths, chiropractors, acupuncturists, dentists, veterinarians, nurses or physician assistants. Homeopathy's appeal lies in its personal attention to patients. Homeopathy is a magnet for untrustworthy practitioners who pose a threat to public safety. A perverse belief in the "healing crisis" causes practitioners to ignore adverse reactions, or to value them as "toxins being expelled." The marketing of homeopathic products and services fits the definition of quackery established by a United States House of Representatives committee which investigated the problem (i.e., the promotion of "medical schemes or remedies known to be false, or which are unproven, for a profit"). The United States Food, Drug, and Cosmetic Act lists the Homeopathic Pharmacopeia of the United States as a recognized compendium, but this status was due to political influence, not scientific merit. The FDA has not required homeopathic products to meet the efficacy requirements applied to all other drugs, creating an unacceptable double standard for drug marketing. The Federal Trade

Commission has not taken action against homeopathic product advertising although it clearly does not meet the standards of truthful advertising generally applied to drugs. Postal authorities have not prosecuted mail-order product promoters that make unproven claims for mail fraud. Three states have established homeopathic licensing boards. Some of these have been administered by medical mavericks with a history of difficulties with former medical licensing boards.

Recommendations

The NCAHF advises consumers not to buy homeopathic products or to patronize homeopathic practitioners. Basic scientists are urged to be proactive in opposing the marketing of homeopathic remedies because of conflicts with known physical laws. Those who study homeopathic remedies are warned to beware of deceptive practices in addition to applying sound research methodologies. State and federal regulatory agencies are urged to require homeopathic products to meet the same standards as regular drugs, and to take strong enforcement actions against violators, including the discipline of health professionals who practice homeopathy. States are urged to abolish homeopathic licensing boards.

Origin

Homeopathy (derived from the Greek words homoios "similar" and pathos "suffering") *is* a sectarian healing system devised by Samuel Hahnemann (1755–1843), a German physician who rejected the harsh medical practices of his era which included bleeding, purging, vomiting and the administration of highly toxic drugs. Practices of the era were based on the ancient Greek humoral theory which attributed disease to an imbalance of four humors (blood, phlegm, and black and yellow bile) and four bodily conditions (hot, cold, wet, and dry) that corresponded to four elements (earth, air, fire, and water). Physicians attempted to balance the humors by treating symptoms with "opposites." For instance, fever (hot) was believed to be due to excess blood because patients were flushed; therefore, balance was sought by blood-letting in order to "cool" the patient. Hahnemann dubbed such practices "allopathy" (allos "opposite," pathos "suffering"), and sought to replace it with his "Law of Similia" that treated "like with like." Although medicine never accepted the label of allopathy, homeopaths continue to misrepresent physicians as allopaths to make their differences appear based upon conflicting ideologies rather than scientific pragmatism. Medical writers often refer to medical doctors as "allopaths" but their use of the term reflects an alternate definition of allopathy as "a system of medical practice making use of all measures proved of value in treatment of disease" (*Webster's New Collegiate Dictionary*). This definition is inconsistent with its root words "allos" and "pathos." Its duplicity aids those who wish to misrepresent medicine as ideologically allopathic (i.e., symptom suppression).

The Cardinal Principles of Homeopathy

The Psora and Vitalism

Hahnemann believed that 7/8ths of all diseases are due to an infectious disorder called the Psora (itch). In the words of Hahnemann's "Organon":

> This Psora is the sole true and fundamental cause that produces all the other countless

> forms of disease, which, under the names of nervous debility, hysteria, hypochondriasis, insanity, melancholy, idiocy, madness, epilepsy, and spasms of all kinds, softening of the bones, or rickets, scoliosis and chophouses, caries, cancer, fungus haematodes, gout-asthma and suppuration of the lungs, megrim, deafness, cataract and amaurosis, paralysis, loss of sense, pains of every kind, etc., appear in our pathology as so many peculiar, distinct, and independent diseases (Stalker, 1985).

Hahnemann believed that diseases represent a disturbance in the body's ability to heal itself and that only a small stimulus is needed to begin the healing process. He owed this to his faith in vitalism, which holds that life is a spiritual, nonmaterial process and that the body contains an innate wisdom that is its own healing force. A British homeopath explained its vitalism (Twentyman, 1982):

> Hahnemann . . . is . . . a child of the modern age of natural science, an adept in the chemistry of his day . . . But he can still hold a conviction that an immaterial vital entity animates our organism until death when the purely chemical forces prevail and decompose it. . . . This vital entity which he characterizes as immaterial, spirit-like, and which maintains in health the harmonious wholeness of the organism, is in fact the wholeness of it, can be influenced by dynamic causes. How does Hahnemann attempt to clarify the idea? He draws attention to phenomena like magnetic influences, the moon and the tides, infective illnesses and perhaps most importantly the influence of emotions and impulses of will on the organism (221–225).

Vitalism appeals to so-called "Holistic" or "New Age" medicine devotees, who prefer a metaphysical view of life processes, and readily accept homeopathy despite its scientific deficiencies.

Provings and the Law of Similia

Hahnemann's invention of homeopathy is reported to have originated with an experience in which he ingested a substantial dose of cinchona bark (the source of quinine) used to treat malaria. He noted that the symptoms he experienced were similar to those of malaria. He reasoned that since the remedy produced symptoms in overdose similar to the condition it was used to treat, this principle, his Law of Similia, could be used to discern the value of various medicines. He called this process proving a medicine. Promoters often misrepresent homeopathy as treating the "causes" rather than merely the "symptoms" of disease, but its reliance on the "proving" of remedies shows that homeopathy itself relies solely upon a symptom treatment.

Hahnemann's Law of Similia utilized the primitive view of monism that "nature is a unitary, organic whole with no independent parts" (*Webster's*) with inherent principles that like is like, like makes like, and like cures like. Monism is the basis of many ancient practices (e.g., eating the heart of a lion for courage), and holds that if one object resembles another they are alike in essence (like is like); idolatry in which carving a likeness of a god actually produces the god (like makes like); and folk medicine practices such as snakeroot being good for snakebite, because of their resemblance (like cures like). Hahnemann revived Paracelsus' Doctrine of Signatures, which declared that herbs would cure conditions or anatomical parts they resembled (Garrison, 1929, 206). The homeopathic Law of Similia, however, is unsupported by the basic sciences of physiology, pharmacology and pathology.

Law of Infinitesimal "Potentizing"

Hahnemann's Law of Infinitesimals holds that the smaller the dose of a medication, the more powerful will be its healing effects. He taught that substances could be potentized (i.e., their "immaterial and spiritual powers" released to make active substances more active, and inactive substances active). The process of potentizing involved the sequential dilution of remedial agents by succussion in which initial mixtures would be shaken at least 40 times, nine parts dumped, and nine parts of solvent added and shaken again. This process was repeated as many times as desired. Tapping on a leather pad or the heel of the hand was alleged to double the dilution—a notion that contradicts the laws of physics. Remedies are diluted to powers of ten and labeled with combinations of Arabic and Roman numerals (e.g., 3X = 1/1000, 4X = 1/10,000, 3C or 6X = 1/1,000,000, etc.). The fact that 19th-Century homeopathic remedies were dilute placebos made them preferable to the harsh concoctions being applied by the humoral practitioners.

According to the laws of chemistry, there is a limit to the dilution that can be made without losing the original substance altogether. This limit, called Avogadro's number (6.023×10^{23}), corresponds to homeopathic potencies of 12C or 24X (1 part in 1024). At this dilution there is less than a 50% chance that even one molecule of active material remains. Hahnemann himself realized that there was virtually no chance that any of the original substance remained at such high dilution, but explained it away in metaphysical terms. In addition to being contradicted by common sense, homeopathy's Law of Infinitesimals is invalidated by pharmaceutical dose-response studies.

Promoters claim that immunization and allergy desensitization verify homeopathy because they treat like with like, but neither meets the additional requirements of homeopathic theory and practice. Immunizations do not alleviate symptoms or cure. Neither immunization nor allergy desensitization grows stronger with dilution, nor can they be "potentized." Classical homeopaths proclaim that eating for relief of indigestion proved that like cures like, i.e., the Law of Similia. However, one does not obtain relief from indigestion by eating "potentized microdilutions" of the same food that was originally ingested. Other attempts to validate homeopathy such as the folksy value of "some of the hair of the dog that bit you" to relieve a hangover also fail to withstand close scrutiny.

Homeopathy and Science

Scientific medicine encompasses a collection of procedures, each of which must stand on its own as safe and effective for a specific purpose. History recounts examples of ancient healers doing the right thing for the wrong reason. Some bored holes in skulls (trephining) in order to liberate angry demons thought to be causing head pain, and in the process relieved intracranial pressure. This, however, does not validate the Demonic Theory. Also, foul-smelling swamps were drained on the basis of the miasmic theory, which taught that foul-smelling emanations from the Earth caused "bad air fever" (mal-air-ia). Further, Asclepian priests scraped spear shavings into the spear-wounds of warriors believing that the weapon that caused a wound would help in its healing (like-cures-like). Copper sulfate from the bronze spearheads may have inhibited infection. Just as doing these right practices for the wrong reasons did not validate the faulty theories upon which they were based, neither will the success of a "homeopathic" remedy comprehensively validate homeopathy's theory, pharmacology, and metaphysics.

Homeopathy clearly fits *Webster's* dictionary definitions of a cult: "A system for the cure of disease based on dogma set forth by its promulgator," and a sect: "a group adhering to a distinctive doctrine or a leader." Healing cults or sects cannot progress and retain their identity. Homeopathy is what Hahnemann said it was. To progress scientifically homeopathy would have to accept principles of pharmacology and pathology, which run counter to its "laws" of similia and infinitesimals, its potency theory, and notions of the psora and vitalism. By doing so, it would no longer be homeopathy but biomedicine.

Studies of Homeopathy

Controlled studies involving homeopathic remedies appear to divide along political lines. While the results of most studies do not support the use of homeopathic remedies, some ostensibly well-designed trials have yielded positive findings. Some of these, however, have been done by homeopaths, and their reports contain rhetoric that reflects bias strong enough to undermine confidence in the researchers' veracity. The best of these studies should be repeated by objective investigators with independent analyses of the homeopathic formulations employed to assure that they have not been adulterated with active medications.

A comprehensive review of experimental research in homeopathy was done by Scofield (1984). He concluded: "It is obvious from this review that, despite much experimental and clinical work, there is only little evidence to suggest that homeopathy is effective. This is because of bad design, execution, reporting, analysis and, particularly, failure to repeat promising experimental work and not necessarily because of the inefficacy of the system which has yet to be properly tested on a large enough scale. There is sufficient evidence to warrant the execution of well-designed, carefully controlled experiments." Scofield's most encouraging statement for homeopaths was that "homeopathy has most certainly not been disproved." However, Scofield ignored the scientific process. It is the absence of proof, not the absence of disproof, that is important. This is consistent with scientific dicta (based upon the statistical null hypothesis) that (1) no practice can be deemed safe or effective until proved to be so; and (2) the burden of proof is upon proponents.

A more recent meta-analysis of 107 controlled homeopathy trials appearing in 96 published reports also found "the evidence of clinical trials is positive but not sufficient to draw definitive conclusions because most trials are of low methodological quality and because of the unknown role of publication bias." They also concluded that there is a legitimate case for further evaluation of homeopathy, "but only by means of well-performed trials" (Kleijnen, 1991).

In 1988, a French scientist working at that country's prestigious INSERM institute claimed to have found that high dilutions of substances in water left a "memory," providing a rationale for homeopathy's Law of Infinitesimals. His findings were published in a highly regarded science journal, but with the caveat that the findings were unbelievable, and that the work was financed by a large homeopathic drug manufacturer (*Nature*, 1988). Subsequent investigations, including those by James Randi, disclosed that the research had been inappropriately carried out. The scandal resulted in the suspension of the scientist. Careful analysis of the study revealed that had the results been authentic, homeopathy would be more likely to worsen a patient's condition than to heal, and that it would be impossible to predict the effect of the same dose from one time to another (Sampson, 1989).

The sectarian nature of homeopathy raises serious questions about the trustworthiness of homeopathic researchers. Scofield appropriately stated: "It is hardly surprising in view of the quality of much of the experimental work as well as its philosophical framework, that this system of medicine is not accepted by the medical and scientific community at large." Two guiding rules required by skeptics of pseudoscience should be applied to homeopathic research, to wit: (1) extraordinary claims demand extraordinary evidence; and (2) it is not necessary to prove fraud, rather, the research must be done in such a manner that fraud is not possible.

Dubious Labeling

Recent years have seen an explosion of products labeled as "homeopathic." Among them are raw animal glands, herbal concoctions, and mineral remedies. Although some are reruns of old-time homeopathic preparations, others appear to be merely pretenders with high dilution their only homeopathic feature. For instance, homeopathic raw bovine testicles may be highly diluted, but in order to be truly homeopathic they should have been "proved" and potentized. To have been proved, healthy people should have been fed raw bovine testicles in moderate doses and the side-effects analyzed. Gland products are not representative of the kinds of therapeutic substances homeopaths have traditionally attempted to "prove," and it is unlikely that ingesting significant amounts of raw bovine testicles would produce any side effects. Such products appear to be intended to ward off regulatory enforcement action by merely labeling them "homeopathic," but such products do not meet the basic consumer protection principle of accurate labeling. Standard drug labeling informs consumers about the quantity of active ingredients per dose; homeopathic labeling only informs consumers about the number of serial dilutions of the remedy.

Questionable Safety

Although homeopathic remedies are generally thought to be nontoxic due to their high dilutions, some preparations have proved to be harmful. Perverse belief in the "healing crisis" can cause pseudomedical practitioners to misjudge adverse reactions as beneficial. Healing crisis is the theory that the body innately knows what is best for it. There is a corollary belief that adverse reactions to "natural remedies" are due to "toxins" being expelled, and that the worse these are, the worse would have been future diseases if not detoxified. Thus, believers are not alarmed by adverse reactions, and are encouraged to continue treating. At the same time, "allopathic" medicine is denigrated as the "suppressing of symptoms that represent the body's natural healing processes." Kerr and Yarborough (1986) reported a case of pancreatitis that developed in a patient ingesting a homeopathic remedy prescribed by a chiropractor. According to the authors, the manufacturer stated that 40–45% of persons taking the remedy experienced a healing crisis that included abdominal pain. Although classical homeopathy employed numerous extremely toxic substances in infinitesimal amounts, Kerr found that two of six homeopathic remedies ordered by mail contained "notable quantities" of arsenic. NCAHF doubts that homeopathic devotees would systematically report adverse effects.

Suspicious Effectiveness

Much has been made of the fact that a 24X dilution would no longer contain a single molecule of the original substance, and reported benefits are generally attributed to the placebo effect. However, many homeopathic dosages, although dilute, may contain enough of a substance to be effective. Homeopathic products also may work because of adulteration. Morice (1986, 862–863) reported that a homeopathic remedy called "Dumcap" appeared to be effective in treating asthma. Although labeled as containing "nux vomica" (strychnine), arsenic album (arsenic trioxide), Blatta onentalis (cockroach extract), and stramoni folic (stramonium), analysis revealed that the product was adulterated with therapeutic levels of the antiasthma, steroidal drugs prednisolone and betamethasone. Studies of homeopathic deemed unacceptable unless they have been monitored to assure that they were prepared according to homeopathic principles, their contents verified and dosage quantified, and secured to prevent tampering. As was stated above, simply labeling a product "homeopathic" does not guarantee that it does not contain a pharmacologically active dosage of an active substance (not all dilutions exceed Avogadro's number). To validate a specific homeopathic remedy, replication by others who have no vested interest in the results is required. To validate homeopathic theory, higher dilutions would also have to be shown to work better than higher concentrations. Thomas Paine, a signer of the United States' Declaration of Independence, is credited with establishing a principle for judging supernatural phenomena. He asked, "Is it easier to believe that nature has gone out of her course or that a man would tell a lie?"

Homeopathic Services

Census

The 1993 directory of the National Center for Homeopathy (Alexandria, VA) lists about 300 licensed practitioners. About half of these are physicians. The rest are mostly naturopaths, chiropractors, acupuncturists, veterinarians, dentists, nurses, or physician's assistants. A homeopathic marketing firm spokesperson believes that several hundred more consider themselves to be homeopaths, and that many conventional physicians utilize one or more homeopathic remedies (National Board of Chiropractic Examiners, 1993). However, no data have been published supporting these estimates. In 1991–2, 36.9% of chiropractors reported using homeopathic remedies in their practices.

A Haven for Untrustworthy Practitioners

Part of homeopathy's appeal is the personal attention paid to patients (Avina and Schneiderman, 1978). In practice, classical homeopaths emphasize taking 30 to 45 minutes with each patient, paying careful attention to the emotional state and administering only one remedy at a time. Classical homeopathy's close personal attention to patients, benign remedies, and special appeal to a select clientele make it seem innocuous if practitioners have the competence and good sense to recognize serious disorders and readily refer to other physicians. This, however, is not always the case.

Pseudosciences such as homeopathy, even if relatively benign, are magnets for cranks and charlatans. This poses a serious problem because untrustworthy or incompetent practitioners should not be granted the privilege of administering health care. True believing cranks may pose a more serious threat than con men because of their devotion to homeopathy's ideology. Their sincerity may make them more socially tolerable, but it can add to their potential danger. Irrational health care is never harmless, and it is irresponsible to create patient confidence in pseudomedicine. Although homeopathy may not pose a significant risk for a basically healthy patient, at some future time that same patient could face a situation where a life-or-death decision may swing on just such unwarranted confidence.

Some practitioners do not practice in homeopathy's classical manner, but use its "benign" reputation as a cover. A well-documented example occurred in Nevada. According to an expose by the *Las Vegas Review-Journal,* several maverick MDs who had been in serious legal difficulty in other states descended on Nevada and managed to get the State Legislature to set up a homeopathic licensing board with themselves in charge. However, none was actually practicing homeopathy. Rather, using an unapproved electronic device they practiced "energy medicine." When faced with the fact that they had deceived the State Legislature, proponents stated that they had used the more familiar term "homeopathy" because they feared that the legislators would not be able to grasp the new concept of "bioenergetics." The Nevada legislature rewrote the homeopathic practice act in 1987, specifically stating that Nevada homeopaths were limited to using substances prepared according to "the methods of Hahnemannian dilution and succussion, magnetically energized geometric pattern as defined in the official homeopathic pharmacopeia of the United States" (Hayslett, 1987).

It is difficult to believe that a physician could simultaneously sustain confidence in both homeopathy and scientific health care. It is common for homeopaths to misrepresent regular medicine as misguided to justify their unusual practices. Of special concern to NCAHF is the substitution of homeopathic preparations for standard immunizations. In 1989, an Idaho naturopath was prosecuted for selling homeopathic "immunization kits," which contained alcohol-and-water solutions and sugar pills. Defenders claimed that the homeopathic immunization products would "stimulate the immune system"; and that the FDA laboratory could not detect the active ingredients because they were so highly diluted with sugar.

Quackery

NCAHF is primarily concerned with homeopathy in the marketplace. It believes that marketing unproven homeopathic products and services precisely fits the definition of quackery: "A quack is anyone who promotes medical schemes or remedies known to be false, or which are unproven, for a profit" (Quackery, 1984). Dr. Kenneth Milstead, then Deputy Director of the FDA Bureau of Enforcement, stated (Young, 1968):

> It matters not whether the article is harmless or whether it gives some psychosomatic relief; whether it is cheap or whether it has value for other purposes; whether it is produced by an obscure firm or whether it is produced by a "reputable" firm—the promotion of it is still quackery.

Regulators Fiddle while Consumers Are Burned

Federal Regulation

For many years homeopathic product marketing was quiescent, but with the health fad boom of the 1970s and 1980s, promoters began touting homeopathic remedies. In 1985 the FDA estimated that between 50 and 60 companies were marketing such products in the United States (*FDA*, 1985). The 1938 Food, Drug, and Cosmetic Act contains a section that recognizes as "drugs" items listed in the Homeopathic Pharmacopeia of the United States. This was mainly due to the efforts of New York Senator Royal Copeland who was the foremost homeopathic physician of his day. In 1938, safety was the main issue, and the highly diluted homeopathic products seemed to pose no inherent danger. However, in 1962, the Kefauver-Harris Amendment was passed requiring that drugs be proved effective before distribution. A legal fight loomed as to whether or not homeopathic drugs were grandfathered by the law, but FDA did not press the issue. Instead, it permitted products aimed at common ailments to be marketed over-the-counter (OTC), and restricted those aimed at serious ailments to prescription only. This "passed the buck" to the states that regulate the practitioners who write the prescriptions, putting consumers at the mercy of maverick homeopathic physicians. It also sent a signal to marketers that it was open season on consumers with regard to OTC homeopathic products. The resulting marketplace growth increased the ability of trade groups to gain political support and made future regulatory action more difficult. Homeopathic claims of efficacy are unsubstantiated and violate the Federal Trade Commission (FTC) advertising standards, but the FTC has not acted against homeopathic advertising claims. Homeopathic remedies sold or transported by mail are subject to action by the U.S. Postal Inspectors, but few such actions have been taken.

State Regulation

Only Arizona, Connecticut, and Nevada have separate homeopathic licensing boards. At least two of these have included in prominent roles maverick medical doctors who have been in legal difficulties as regular physicians. Some state licensing boards permit licensed medical doctors to practice almost any kind of medicine they wish. Others, rightly in NCAHF's opinion, require that health care be held to rational and responsible standards. To its credit, the North Carolina Board of Medical Examiners revoked the license of the state's only practicing homeopath, concluding that he was "failing to conform to the standards of acceptable and prevailing medical practice." This resulted in a prolonged legal battle over the ability of a licensing board to impose standards of practice on its constituency. The state legislature eventually passed a law that limited the board's disciplinary power undermining the consumer protection aspects of responsible medicine.

Adopted February, 1994 by the National Council against Health Fraud.

RECOMMENDATIONS

To Consumers
Be aware that homeopathic products and services are marketed in a "buyer beware" situation at present. Homeopathic products are not required to meet the standards of effectiveness of drugs. Homeopathic services are poorly regulated. Physicians who practice homeopathy operate below the standards of responsible medicine. Some have backgrounds that raise serious questions about their honesty. Be aware that in some states that have homeopathic licensing boards the "foxes are guarding the chicken coops." Consumers should not entrust their health to physicians or nonphysicians who practice homeopathy.

To Basic Scientists
Homeopathy conflicts more with basic laws of physics, chemistry and pharmacology than with clinical medicine. Pharmacologists should be more proactive in opposing the marketing of homeopathic remedies. Because homeopathic theories contradict known physical laws, tests of homeopathic remedies require controls beyond those normally required of double-blind clinical trials including additional measures to show that fraud was not possible.

To the U.S. Food & Drug Administration
(1) Require that labels of homeopathic products indicate the precise amounts of ingredients in milligrams, micrograms, etc. (2) Require homeopathic products to meet the efficacy standards of all other drugs.

To the U.S. Federal Trade Commission
(1) Review advertising of homeopathic products in publications aimed at the public for

false and misleading claims. (2) Monitor and take action against advertisements in trade publications used to indoctrinate salespeople, who will in turn deceive consumers about the value of homeopathic products.

To U.S. Postal Inspectors
Prosecute distributors of homeopathic mail-order products that make unproven medical claims for mail fraud.

To State Legislators
Because homeopathy is scientifically indefensible: (1) Enact laws requiring that medical products sold within your state meet the standards of accurate labeling, truthful advertising, and premarketing proof of safety and effectiveness. (2) Abolish state licensing boards for homeopathy. (3) Do not allow homeopathy in the scope of practice of any health care provider.

To State Food & Drug Regulators
Take prompt regulatory action against manufacturers, wholesalers, and retailers of homeopathic products who violate the law.

To Medical Licensing Boards
(1) Discipline homeopathic practitioners for unprofessional conduct. (2) Prosecute nonphysicians engaging in homeopathy for practicing medicine without a license.

The National Council against Health Fraud is a private nonprofit, voluntary health agency that focuses upon health misinformation, fraud and quackery as public health problems. Its funding is derived primarily from membership dues and newsletter subscriptions. NCAHF unites consumers with health professionals, educators, researchers, attorneys, and others who believe that everyone has a stake in the quality of the health marketplace. NCAHF's positions on consumer health issues are based upon principles of science that form the basis of consumer protection law. These require: (1) full disclosure in labeling and other warranties (no secret formulas); (2) premarketing proof of safety and efficacy for products and services that claim to prevent, alleviate, or cure any disease or disorder; and (3) accountability for those who violate consumer laws. Its officers and board members serve without compensation. For more information, write: NCAHF, P.O. Box 1276, Loma Linda, CA 92354-1276; fax: 909-824-4838.

References:

Avina, Schneiderman. 1978. *West J. Med.* 128: 366–9.
Barrett, Jarvis. 1993. *The Health Robbers*. Buffalo: Prometheus Books.
Board of Science and Education. 1986. *Alternative Therapy*. British Med. Assoc.
FDA Consumer. 1985. March.
Garrison. 1929. *History of Medicine*. W. B. Saunders.
Hayslett. 1987. *Las Vegas Review Journal*. July 5.
Kerr, Yarborough. 1986. *New Engl. J. Med.* 314: (25): 1642–3.
Kerr, J. 1986. *Taxicol Clin. Toxicol*, 24: 451–459.
Kleijnen, Knipschild. 1991. *Brit. Med. J.* 302: 316–23.
Morice. 1986. *The Lancet*, April 12.
National Board of Chiropractic Examiners. 1993. Job Analysis of Chiropractic.
Nature. 1988. 333: 816.
Quackery: A $10 Billion Scandal. 1984. U.S. House of Representatives, 98th Congress, 2nd Session, Comm. Publ. #98-435, May 31.
Sampson. 1989. *Skeptical Inquirer*, Fall.
Scofield. 1984. *The Brit. Homeo. J.* 73: 161–226.
Stalker, Glymour. 1985. *Examining Holistic Medicine*. Buffalo: Prometheus Books.
Twentyman. 1982. *Royal Soc. of Health J.* 102 (5): 221–225.
Young. 1968. *The Medical Messiahs*. Princeton.

Immortality

The Search for Everlasting Life

STEVEN B. HARRIS

Since the 1960s when the late Joseph Campbell's *The Hero with a Thousand Faces* began to be read on college campuses, and most especially since the Bill Moyers PBS interviews with Campbell (1988) which made his work still more popular, many people have begun to look at mythology in a new light. We all know, or thought we knew, what a myth was—one of those weird stories that people in other cultures tell. Our stories, by contrast, were called "religion," or "scripture," and were not weird at all. In fact, Joseph Campbell (tongue firmly in cheek) once defined myth as "someone else's religion."

Myth is not only religion, of course, but something more inclusive. Myth might broadly encompass such things as rituals and beliefs, but most especially myth is the collection of primitive stories that we tell ourselves in order to have a narrative psychological framework with which to deal with the world. In the largest sense, myth includes (but is not limited to) any story which answers the difficult questions of life, such as: Who am I? Where did I come from? Where am I going? What is the far future going to be like? What is expected of me? Who are the heroes? What's going to happen to me when I die?

In life it is important to answer these questions (even if the answer is insupportable fantasy), since excessive worry about them may detract from basic survival efficiency. We know from recent psychological experiments, for instance, that compared with objective assessment, people with normal "healthy" mental outlooks consistently overestimate their own abilities and strengths; whereas people who are depressed are far more realistic in such judgments. Why would human nature saddle us with a normal mental state which gives us an unrealistic view of the world? The answer may lie in the fact that anxiety saps strength and ruins performance (as many an Olympic athlete has discovered). Anxiety is so bad that sometimes it is worth a small cost in objectivity to be rid of it.

A major function of myth (and of a large part of human culture) is to relieve anxiety by answering unanswerable questions. Karl Marx once said that religion is the opiate of the masses, but perhaps what he would have said today (given modern pharmacology) is that religion is the Valium of the masses. The same can be said of superstition. Superstition, in fact, is also just another name for other people's religion.

Of course, there is also much art in myth. Myths may not be factual, but that does not mean that in some sense they are not true. As Professor Campbell reminds us, all metaphors are (in the narrow sense) lies. After all, the moon is not really a ghostly galleon, tossed on cloudy seas. Myths are metaphors—metaphors for something that cannot be said any other way; they are stories that speak to a basic and very old part of the human consciousness—the part of the consciousness that holds basic cultural programming.

Mythic stories (to adopt a technical metaphor) are a little like the programming in the "read-only memory" chips of a computer; they represent programming that is more or less permanent. Once you are culturally programmed the first time, you are stuck with it for good, and after that (i.e., after a certain age), any new cultural myths will sound foreign and alien to you. As any missionary can attest, mythic re-programming is often not completely successful because of this effect. The same effect appears when people lose faith later in life—we remember Bertrand Russell's famous thesis that Catholic atheists are quite different sorts of people than Protestant atheists.

The Mortal Hero

Much of cultural programming is in stories, and since the time of James Joyce's introduction of the idea of the "monomyth," it has been argued that there are only a few basic stories, and all good tales are variations on these. The basic love story, for example, in all its permutations, never seems to tire if told well. There are also basic creation myths, including a cycle of myths involving feminine forces and goddesses (as Robert Graves reminds us) which seem to be important in artistic inspiration. Finally, from the masculine side, there are stories of the hero, an often semidivine and usually male adventurer who is on a quest or a journey, and who must win a victory of some kind before returning home with the power that he has won. (The traditional hero, being at once both masculine and admirable, is presently out of fashion in many university English departments, but Campbell's paradigms seem to work best for the science fiction themes we will cover.)

Although the hero is often semi-divine, it is a feature of many hero tales that he be at least partly human, and thus mortal. It is important to note that the rules of conduct are manifestly different for Gods; Gods are beyond morality in myth, and many of the Greek myths about divine behavior (especially as retold later in Latin) are as amoral as modern TV soap opera. Morality and the question of "The Good," however, are important for mortal humans (who have only a limited time to learn from mistakes), and thus the tale of the mortal hero is often a morality play. Hero tales are often stories of the mortal human who manages, as a hero, to make of himself something more. Given this fact, one of the most popular and one of the oldest of the hero myths is that of the hero who seeks the boon of immortality. We will now examine how this myth is played out in religion, science, and science fiction.

Resurrection and the Hero

We suspect that tales of resurrection have been around for as long as there have been people. Neanderthal graves have been found with food and tools in them, and we are led inexorably to the idea that these things were included in the grave because it was thought that the deceased might one day need them. From this we infer that Neanderthals had some form of language, since it would seem impossible to communicate something as abstract as "life-after-death" with a few grunts and barks. By this loose chain of reasoning we can guess that even Neanderthals had a culture, and that culture told immortality stories.

The oldest written story known is a more than 5,000-year-old Sumerian tale of a hero in search of immortality—the story of Gilgamesh the King. Heros are often semi-divine as well as royal, and King Gilgamesh is 2/3 god and 1/3 man. Gilgamesh's human part presumably confers mortality on him, and in one of the Gilgamesh tales he realizes that he is one day going to die, and so starts out looking for the secret of life. After he finds immortality he foolishly loses it, and thus Gilgamesh becomes one of the first tragic heros.

Almost every culture has its tale of the divine but mortal hero in search of the gift of immortality (although the hero is usually more successful than Gilgamesh), for example, Adonis, Tammuz, Dionysus, etc. One of the most important myths, however, is that of the Egyptian Osiris, a god who comes to Earth to be a teacher, and here gets assassinated and dismembered (if heroes are fully divine, they are often still vulnerable). Later, after being reassembled by his divine brother Horus, Osiris goes on to become God of the Dead. His sacred name is thereafter used in the ritual in which the dead of Egypt make the journey through the underworld to be immortally reunited with the breath of life. Egypt is the first society we know of to link the ideas of immortality and resurrection with human technology—in this case the technology of mummification—but the application of the technology was ritualistic and thoroughly religious.

The biblical Pharisees believed in the resurrection of the dead, and the myth of the resurrected hero was, according to the Gospel of Matthew, present in Palestine in the time of Jesus. According to Matthew (16:14), Jesus asks the disciples who people are saying that he (Jesus) is, and they reply in part that some people think that he is really John the Baptist. John the Baptist had already been beheaded by this time (Matt. 14), so the disciples are obliged to repeat the popular myth of a popular hero getting killed and coming back to life to work miracles. And all this is before the crucifixion of Jesus. Thus, anyone who takes the testimony of the New Testament literally must also admit that mythic folk-stories of the return of a popular dead figure were then widespread, just as they are now.

Resurrected hero stories seem to occur in all cultures. When the Roman Catholic church made it to the New World in the 16th century, some of the resurrection myths the natives were telling were so close to the Christian one that some of the Jesuits listening to them were convinced they were the work of the devil. A more Jungian view is that these archetypal stories are reflections of the way the human collective unconscious is constructed; or, if you prefer modern neurophysiology, the networking of the neural architecture. In any case, if we do not have a God-shaped place in our souls, we at least may have a resurrected-hero myth in our psychological make-up.

Mal-resurrection and the Anti-hero

It is interesting to examine what happens mythologically when the resurrected individual is not a hero, and no official religious process is involved. There has always been a darker side to resurrection stories. It may be expected that Kings and demigods return from death; but people do not always want the same for their more mundane elderly relatives, particularly in areas of scarce land or resources. Here we have a source of anxiety, with which it is the social function of myth to deal. In mythology, the newly dead (unless royal) are always dangerous unless properly dealt with, and are apt to give trouble to the living in various ways until they have completely decayed to safe bone. It has been popular in many cultures worldwide, in fact, to ritually treat a new corpse in various ways to insure that it stays in the grave and does not become a revenant.

Originally, many mal-resurrection stories and myths probably had their origin in misunderstanding of what happens to an unembalmed human body after burial. Today we know that the natural decay process sometimes results in bloated corpses which look fatter, which may exhibit a discharge of blood from the mouth, and have skin changes which appear more life-like, rather than less. Unsophisticated people, on seeing these changes, apparently were apt to infer that the corpse had been out and about, and feasting on blood. A collection of such stories later loosely inspired an enduring personification of evil immortality and resurrection—Stoker's *Dracula* (1897).

The walking mummy of the Karloff movie is of course closely related to the vampire. In mythic terms, resurrection from the dead is possible, but without a standard religious mechanism, or at least a royal or divine hero-patron (such as Osiris or Jesus), such resurrections in myth are evil, and can be expected to produce monsters. In the case of the vampire and the mummy, the result is a living dead man who is not the original person, but rather a transformed and murderous demon. In fiction, as in myth, the general message to the common public about coming back from the dead is: "Do not try it without the religious seal of approval."

Immortality through Resurrection and Resuscitation

Before we return to mal-resurrection, we must consider a second theme—that of technology and medical progress. The critical element in science fiction is the speculative impact of technology on individuals and culture, and it is technical progress and its implications which have, more than anything else, made the mythic vampire and his cousins more immediate in our time. Dracula and The Mummy are rather late figures in the history of horror, and as immortal personifications of mal-resurrection, both are recognizably literary grandchildren of Mary Shelley. Long before Shelley and the birth of science fiction, however, came certain developments in the science of resuscitation which made people think differently about resurrection.

Historically, there is some suggestion of mouth-to-mouth resuscitation in the Bible (II Kings 4). Although the story appears a bit garbled, like the story of the resuscitation of the child before it in I Kings 17, both stories contain descriptive elements of chest compression, and there is clearly something more than mysticism going on in the account. For centuries, however, the Western world made little progress in the matter. In the middle ages, when much of the advancement of medical science was in Moslem hands, Arabic medical books told of a little-known secret which had been passed down from midwife to midwife: if one blew into the nostrils of a stillborn infant, sometimes it began breathing. We know that Arab physicians also did some experimenting with attempting to resuscitate corpses with bellows, but word of this work was not widespread either.

Then, in the middle of the 15th century, everything was changed by the invention of the cast-metal movable-type printing press. Suddenly, written knowledge became relatively cheap to own because the work to manufacture it was now drastically less. Science, whose treasure-trove was a wealth of experimental detail which did not lend itself well to oral tradition, was particularly benefitted by printing. In fact, partially linked to this important device was not only the Renaissance and Reformation, but the Scientific Revolution.

One of the earliest influential books of the Scientific Revolution was Andreas Vesalius' atlas of the human body, where (among many other things) Vesalius describes techniques for resuscitating asphyxiated dogs with bellows. Similarly, Paracelsus, an alchemist and one of the great physicians of his age, was also said to have attempted the resuscitation of a corpse using bellows, a trick he perhaps picked up from Arabic medical writings. Physicians eventually learned that simple mouth-to-mouth resuscitation sometimes worked on recently asphyxiated adults as well as it did on newborns.

By the 1740s, several cases of successful mouth-to-mouth resuscitation had been reported, the most famous of which was Tossach's 1744 report of the resuscitation of a clinically dead coal miner (no breath or heartbeat) who had been suddenly overcome after descending into a burned-out mine. By the 1760s, in the wake of such reports, a number of groups advocating the resuscitation of drowned persons had sprung up in Europe. In 1774 a society was founded in London to promulgate the idea that "dead" people in some cases were not dead. Called, after a bit of experimentation, the "Society for the Recovery of Persons Apparently Drowned," it quickly evolved into the Humane Society (still later, with official patronage and funding, the Royal Humane Society, which it remains to this day). The Humane Society advocated techniques which were highly advanced. Three months after the society's founding, as an example, a society member had the opportunity to minister to a 3-year-old child named Catherine Sophie Greenhill, who had fallen from an upper story window onto flagstones, and been pronounced dead at the scene. The society member, an apothecary named Squires, was on the scene within twenty minutes, and history records that he proceeded to give the clinically dead child several shocks through the chest with a portable electrostatic generator. This treatment caused her to regain pulse and respiration, and she eventually (after a time in coma) recovered fully.

The resuscitation of little Catherine Greenhill was probably the first successful cardiac defibrillation, and it followed earlier suggestions by American scientist Benjamin Franklin and others that electricity might possibly be used to "revivify" the human body. And so it proved able to do in selected circumstances. By 1788, a royal silver medal was awarded to Humane Society member Charles Kite, who was by this time not only advocating the resuscitation of victims in cardiac arrest with bellows and nasolaryngeal intubation, but had also developed his own electrostatic revivifying machine which used Leyden jar capacitors in a way exactly analogous to the DC capacitative countershock of the modern cardiac defibrillator.

The enlightened state of the late 18th century as regards resuscitation was not to last. From the very first, dark images from the human psyche began to gather in resistance to the new ideas. Technology never intervenes in a major way into human life without creating new anxieties and a certain amount of social backlash. Resuscitation had its problems. To begin with, the discovery that "death" was not a sure and objective state did not exactly sit well in the public mind. Charles Kite was of the opinion that not even putrefaction was a sure sign of permanent death, since it might also be due to advanced scurvy! The public was wondering: if one could be mistaken for dead, like Shakespeare's Juliet, when one was in fact resuscitatable, did that imply you could be buried alive? It did.

The result of this realization was a psychological terror familiar from Edgar Allan Poe's "The Premature Burial." Poe, however, popularizing the problem for early 19th-century America, was actually late to the controversy. In 18th-century Europe the fear of premature burial or dissection was not just the preoccupation of macabre writers—whole classes of people were affected, albeit in different ways. Upper class persons took to fitting coffins and crypts with special signalling devices which could be used to alert the outside world in case the occupant should inexplicably revive. The lower classes had their own special problems, too, since anatomical dissection (long a part of the punishment for heinous crimes because it denied the malefactor an intact bodily identity or a grave) had now taken on a special meaning. To wit: it killed.

Resurrection in Science and Fiction

With scientific resuscitation, technology had finally intruded into the macabre. The horrific potential of the new electromechanical resuscitative technology had its first fruitful literary influence on Mary Shelley, a teenager who in late 1816 had first set out to write a ghost story, but had instead ended up producing Frankenstein (1818), a cautionary tale of the technological resuscitation of a monster composed of pieces of corpses by a medical experimenter. "Frightful it must be," writes Shelley of her vision of the monster in an 1831 introduction to the book, "for supremely frightful would be the effect of any human endeavour to mock the stupendous mechanism of the Creator of the world." Given the spirit of the times Shelley's story touched a public nerve as though with one of the new electrical machines, and Frankenstein's monster was an instant sensation. In keeping with its archetypal nature, the tale, completed while Shelley was still only nineteen, remains her most famous and enduring work.

After the Frankenstein sensation, something strange happened. Shortly after the publication of Shelley's famous story, the new medicine began to go out of favor, and the science of resuscitation began to suffer on both the technical and mythological fronts. It happened for several reasons. Mouth-to-mouth resuscitation was discarded for bellows, which, in turn, were discarded for technical reasons. Electrical resuscitation fared no better than mechanical "respiration" (ventilation). The new phenomenon of electricity early-on was transformed into a quack cure by the practice of "galvanism" (passing mild shocks through the body in an attempt to cure disease) and its reputation accordingly tarnished. Then, and perhaps even more devastatingly, the charming new electricity was in turn transmuted into a powerful and dangerous force by the giant alternating current transformers of George Westinghouse (maligned from the first for their deadliness in a rival Edison PR campaign) and also by the newfangled American electric chair (1890). Technologies may suffer from social stigmas as well as people. Mary Shelley had originally not specified the method of the revivification of her monster, but Shelley's group of literary friends (as she tells us) had been discussing galvanism a few hours before the vision of the artificial monster came to her in a nightmare. By 1931, in the new electrified America, Dr. Frankenstein's monster came into the movies electrically charged, and soon the electric chair was producing its own monsters in the cinema (e.g., Boris Karloff's *The Walking Dead,* 1936).

For more than a century after Shelley (and indeed to this day) Frankenstein colored resuscitation as it appears in science fiction. An exception is Edgar Allan Poe's 1845 story "Some Words with a Mummy," which is social commentary rather than horror. The mummy of the title is resurrected by galvanism, and is one of a race of ancient Egyptians who have perfected suspended animation, and have used it to travel rapidly through time at pleasure, as tourists and revisionist historians. As such the tale is one of the first positive fictional treatments of suspended animation.

Poe had an antecedent for the idea, for "Some Words with a Mummy" echoes some much earlier optimistic thoughts on the subject by Dr. John Hunter (1728–1793) who had, in the year 1766, experimentally frozen live fish in an attempt to prove the idea that human beings might be able to see the far future by being intermittently frozen for long periods (the fish died and Hunter soon abandoned the idea). Another scientist to take an interest in suspended animation was Hunter's transatlantic contemporary Benjamin Franklin. Franklin not only foresaw advanced treatments for aging as a result of science, but in a 1773 letter to his friend Jacques DuBorg, the inventor wished that he might be preservatively embalmed "with a few friends," in order to see eventually what might become of his beloved America in the far future. Franklin thus is not only one of the first men to speculate about seeing the future in such a scientific way, but he is also the first to see that such thoughts inevitably move one to want to take some of your social network with you for company. Poe's story and the private 18th century views of Hunter and Franklin stand in contrast to the much more common and much more alienating views of long delayed revival of individuals, a time-travel-to-the-future genre which perhaps can be said to begin with Washington Irving's dark and poignant "Rip Van Winkle" (1820), and which continues with H.G. Wells' time traveler and sleeper.

Poe's other exploration of attempts to bypass the immediate effects of death, written at about the same time as "Some Words with a Mummy," is more typically macabre. In "The Facts in the Case of M. Valdemar" (1845), the Frenchman Valdemar dies while under a deep hypnotic trance. So deep is the trance that, although heartbeat and breathing have stopped, Valdemar's tongue still obeys commands—"I was sleeping, but now I am dead," he states in one of the most famous lines of the genre. For seven months this state of suspended animation continues in Poe's tale, with the dead body (save for the horribly moving tongue) locked in rigor mortis, but basically unchanged. Finally, at the end of the story, the experimenters decide to end the trance, and the hypnotized man turns, in less than a minute, into a "nearly liquid mass" of decay.

In the long-delayed and unnaturally rapid decay of Poe's released hypnotic subject, we recognize the traditional fate of staked vampires, those other escapees of traditional mortality. As in Rider Haggard's *She*, Wilde's *Portrait of Dorian Grey*, and Hilton's *Lost Horizon*, slowing or arrest of nature's aging or dying process in fiction often runs up a kind of cosmic credit card bill which may later become due all at once, with dire consequences. Such themes suggest a cultural psychological heritage which views death and decay as inevitable forces which, like some bottled-up natural flow or pressure, are apt to produce explosive and terrible results if held in abeyance even temporarily.

To be sure, this kind of universal debt does not accrue to the original monster in Shelley, which does not age. In Frankenstein, rather, the price which the monster pays for its artificial life is alienation and social ostracism (it is horribly ugly). The monster also suffers neglect and abandonment by its only "parent"—its creator. With few exceptions, however, secularly resurrected figures in fiction since the beginning of the genre have usually paid a more direct kind of price for their existences. The same is true of those who direct the reanimation, as well, although the ignorant sometimes escape the ultimate price (as in W. W. Jacobs' 1902 story "The Monkey's Paw").

The next major comment on scientific reanimation of the dead is from that gentle but slightly unhinged dropout from life, H. P. Lovecraft. Lovecraft's first professional sale, "Herbert West, Reanimator" (1922), is his tribute to Shelley, though it would be some time in Lovecraft's own writing before he would be able to explore the psychology of horror as deftly as Shelley did. "Herbert West, Reanimator" is a straightforward story of a young medical student of a materialist bent who seeks to reanimate corpses by chemical means. He is only partially successful—his reanimated beings are murderous, even if they were good people in life (one of the demonic monsters is a late kindly and philanthropic Dean of Medicine). Like Shelley, Lovecraft carefully never gives any of his reanimated corpses what it takes to be human: those bodies that are whole behave as animals, and those which have human intelligence and understanding are horribly mutilated. And like Shelley's, Dr. West's resurrections, are mal-resurrections; West, as creator of the beings, is inevitably destroyed by them.

The Sociology of Resuscitation and Resurrection

Possibly for some escapist reason, in Lovecraft's own heyday the Great Depression had triggered a spate of American films about horror, and in many cases their content was quite scientific and the lead scientist usually a biologist. (It was not until 1945 that the smock of the mad scientist passed from biologist to physicist. Recall that it is said the First World War was fought by the chemists, the Second by the physicists.) *Frankenstein* starred Boris Karloff (1931), who also played the title role in *The Mummy* (1932). A few years later (after the success of Universal's *Son of Frankenstein*), Columbia Pictures made a quintet of Karloff horror movies (1939–42) with even more explicit themes of scientific life-prolongation or resuscitation. In *The Man They Could Not Hang* (1939) Karloff plays a doctor who has discovered a way to place humans into suspended animation with an artificial heart machine. In the script, the authorities mistake a suspended man for dead (the "Juliet problem" again) and Karloff is sentenced to death for murder. After he is hanged, a student uses the same machine to resuscitate him. The resuscitated Karloff is evil and vengeful, however, and soon sets about killing the people who convicted him—another scientific resurrection that failed to do anyone any good.

A positive view of scientific resuscitation and life prolongation does not occur in the movies until the great Robert Wise film *The Day the Earth Stood Still* (Twentieth Century-Fox, 1951). This movie is the tale of a human-like alien named "Klaatu" who visits Earth in a flying saucer (that looks remarkably like the "real" UFOs that began appearing soon after), accompanied by a giant robot named Gort. While trying to deliver a warning to humanity, Klaatu is killed by the army. In the film's climax Klaatu's body is recovered by Gort, and then resuscitated with the aid of machinery inside the saucer. Klaatu, now risen from the dead, is free to deliver his message and ascend to the heavens.

The Day the Earth Stood Still not only delivers a political message about the threat of nuclear war, it presents deliberate and shameless biblical allegory—the resurrected hero myth recast in science fiction terms. Klaatu is to be understood as a Christ figure who is sent from the heavens to warn mankind of its sins. (As a particularly poignant touch his Earthly pseudonym is "Mr. Carpenter!") Although Klaatu's coming is attended by wondrous events, his wish for a meeting with the political leaders of the world is rejected. Like Christ among the common folk, Klaatu now finds himself in the home of an ordinary citizen. His uncommonness is all too apparent, however; Klaatu's teaching of the famous Einstein-figure Professor Barnhardt (Sam Jaffe) is as much a personal self-revelation as that of the boy Jesus in the temple confounding the Rabbis. Eventually Klaatu does go public, but being high priest of technology, he eventually demonstrates his power not by calming the water, but by calming and silencing the world's machines by neutralizing all electricity—the day the Earth stood still.

In keeping with the allegory, Klaatu is finally betrayed and murdered for his trouble by the very people that he came to warn. His body is taken to a jail cell (in lieu of a tomb), and there guarded by soldiers. The cell is opened by a mechanical servant in place of an angel, and there is finally the resurrection by Gort. (Patricia Neal is the Mary Magdalene figure, asking the questions for us.) Eventually, message of warning delivered, Klaatu ascends into the heavens.

In many ways *The Day the Earth Stood Still* is not a typical science fiction movie of its time. Alien beings from space are not seen in this film as marauding monsters. Even more intriguing is the idea that high technology, as manifested in space transportation, would naturally be expected to go hand in hand with youth-prolongation (Klaatu is 78 but looks 35; his people live twice as long as Earthlings). High technology is linked with advanced resuscitation capability, but not with horror. This is archetypally a bit odd, and possibly in consequence historically it did not go without controversy. Screenwriter Edmund H. North's script for the film (itself an adaptation of a 1940 Harry Bates short story titled "Farewell to the Master") originally called for the alien Klaatu to simply be resuscitated by Gort and thereafter to go about his functionally immortal business. Unfortunately, the Breen Censorship Board (an autocratic self-censorship mechanism of the movie industry especially active during the cold war years) was scandalized at the idea of Gort the Robot bringing Klaatu to life, saying "Only God can do that!" North's protestation that the movie was science fiction and that the action in question involved genuinely unearthly alien technologies got nowhere. Eventually, a compromise was worked out: Klaatu was to invoke deity (in the final script Klaatu asserts rather piously that the power of life and death belongs only to the "Almighty Spirit"); and he was also to issue a statement admitting mortality (in the final script we find that the life conferred by the saucer machine is good only "for a limited period," which "no one can tell"—an obvious compromise with the censors. With these changes, the Breen Board, apparently satisfied that it had protected the public from the un-American idea of scientific immortality, withdrew its ban. The scene in which Klaatu explains that scientific resurrection is (in effect) not all it is cracked up to be remains as a monument to popular resistance to the idea of casting scientific progress in any form resembling God.

The Day the Earth Stood Still is considered one of a handful of contenders for best science fiction movie ever made. This honor is at least partly a result of the film's reworking of the old resurrection myth. The power of this particular theme may be gauged by the fact that the record box-office opening movie of all time, *E.T.: The Extraterrestrial* (Universal, 1982), pulls exactly the same psychological strings as *The Day the Earth Stood Still* (as does the even later "*E.T.* rip-off" *Starman*). In *E.T.*, we see the heavenly being visiting Earth with magic life-restoring powers (the glowing finger). Again, there is an unenlightened government sending squads of soldiers chasing after the visitor, who all the while is more content to spend his time with common folk and children. Again we see the visitor's death and technological resurrection (the difference being that in 1982 they had cardiac defibrillation, which was included). And again there is the ascension to the heavens, this time to the heavenly parents, since E.T. was only a child.

Cryonics: A Modern Prometheus

Horror writers seem to have a love of the cold, and both Shelley and Poe (*The Narrative of Arthur Gordon Pym of Nantucket*) employ a frozen backdrop to good effect. Later authors follow in the same tradition, and the first writer to go so far as to employ cryogenic preservation for monsters is H. P. Lovecraft. In Lovecraft's novella *In the Mountains of Madness* (1931) an antarctic expedition unearths frozen half-animal/half-vegetable creatures dating from an earlier age. In a scene which has since become hackneyed (but Lovecraft did it first!) a scientist dissects one creature while the others are allowed to thaw, unattended. The result is carnage. Later it transpires that the monsters are an extinct intelligent species who long ago created all life on Earth. This created life includes not only the familiar forms that led to man, but also a race of servant monsters which (as the story progresses) end up turning upon their creators, Frankenstein-style. In Lovecraft, even the monsters are troubled with monsters!

Lovecraft may have been not only the first writer to consider the cold as a method of preserving horrific creatures, but also "dead" humans who refuse to be done with life. In "Cool Air" (1928), which obviously owes a great debt to Poe's "Valdemar," Lovecraft tells us of a physician-scientist who, because of a very curious illness, must keep his rooms at all times at low temperature. The narrator befriends the doctor, but eventually finds that his new acquaintance has not only begun to exhibit a strange odor, but (moreover) is requiring lower and lower temperatures as time goes on. Eventually the air-conditioning fails, and while the narrator is off trying to get a replacement part, the good doctor dissolves in the manner of monsieur Valdemar. It turns out that he has been clinically dead for 18 years, but has kept himself preserved by means of the cold.

Does Lovecraft now generally get credit for the cryonics idea? One of Lovecraft's stories ("The Whisperer in Darkness," 1930) uses the device of having creatures from another planet remove human brains and place them into mechanical supports for shipment across outer space. This treatment (according to the story) makes them functionally immortal, and is also used to excellent effect as a device for horror as these Earthlings find themselves kidnapped, removed from their bodies as naked brains kept alive by machinery, and taken away into space by fungoid creatures from Pluto.

The Blurred Line between Science and Science Fiction

Would the far future be worse than death? We know that, in the real world, by 1935 *Time Magazine* was featuring the predictions of a Hollywood clinical chemist named Ralph S. Willard, who was claiming to be able to freeze monkeys and resuscitate them. Willard proposed to use the process on convicts in order to store them more cheaply, and even on jobless people (until times got better), would-be suicides (until a cure had been found for depression), and on those curious about the future. Today we are certain that Willard was a humbug, but before he disappeared into the mists of science fiction history we saw him one more time, acting as technical consultant to a Boris Karloff film entitled *The Man with Nine Lives* (1940), co-written by the same man who wrote *The Man They Could Not Hang*. Again we see the scientist who is conducting experiments in human suspended animation. Again there are the authorities who visit the lab of the mad scientist, see a frozen man, and decide that a murder has occurred. This time, however, the scientist is able to take revenge before he can be sent to jail; his solution is to lock himself and the visiting authorities (the coroner, the D.A., and the Sheriff) into a freezer in the basement of his island laboratory, where all undergo cryonic suspension. Ten years later the lab is re-discovered, and the suspendees all revived by another researcher. Again, the experience of resuscitation from sleep/death has turned scientist into mad scientist (the mal-resurrection) and he begins to kill his fellow suspendees in a series of cryonics experiments. In the end the police arrive and put an end to him.

The history of the real practice of cryonics is less dramatic, at least at the beginning. Heedless of Boris Karloff's fate, a young soldier took up the idea of cryonics in the 1940s. While recovering from war wounds, Robert C. W. Ettinger read "The Jameson Satellite," and in 1948 wrote a cryonics science fiction story (*The Penultimate Trump!*) in which he first suggested the idea of a man dying of old age deliberately being frozen to wait for advances in human rejuvenation technology. Ettinger eventually went on to become a college physics teacher. Finally, in 1962, in a full length book titled *The Prospect of Immortality* (eventually re-published by Doubleday in 1964), Ettinger argued formally for a cryonics program to begin in the non-fiction world.

By the early 1970s it was known that some small crustaceans and worms, and even mammalian embryos, could be cooled in liquid nitrogen or helium to the point where all metabolism stopped, and there stored as long as anybody liked. Here was structure, but no function. Ettinger argued that because frozen organisms could be revived, "life" was not something that had necessarily disappeared simply because things did not run. Ettinger's view of death was that organisms are like automobiles; thus an organism which is not functioning may not be "dead" (in the sense of permanence) if whatever caused the failure to function is repairable. The only criteria that mattered in revival were the same criteria which one would employ in order to know whether one could repair a damaged automobile: What was the original structure? Did enough structure remain that one could infer what was, from what is? Did one have the tools to effect such repairs?

Ettinger argued that we do not have such tools today, but that we may have such tools tomorrow. Today's "dead" people might be resuscitatable by the standards of the future. Thus, we now probably conduct many autopsies on people who are, by the standards of the future, only very sick. If such people could be delivered to the future reasonably intact and undecayed (as by cryogenic preservation), and if future physicians were also able to repair the damage which was caused by freezing, then it would make sense to freeze people now who had been pronounced "dead," just in case something could be done for them later. In 1965, an early devotee of Ettinger suggested that the process be called "cryonics," and so it came to be. The word is now in common use.

The line between science and science fiction became further blurred on December 15, 1966, when Walter Elias Disney died of lung cancer. Reporters who covered the death had earlier in the day also happened to cover another press conference, coincidentally announcing the formation of the Cryonics Society of California (the first cryonics society on the West coast). Somewhere in all of the melee, the story surfaced that Disney himself had been frozen. Though it is almost certain that there was nothing to the rumor, Disney apparently once expressed interest in the concept of cryonics. What makes the story interesting is not so much the rumor's truth or falsehood, but rather its astonishing power. It was a rumor of amazing vitality that went so far as to insinuate itself as fact into at least one biography of Disney, even though there was not a shred of physical evidence to support it. To this very day, the idea that the great animator awaits "reanimation" somewhere in cold storage may still come up in casual conversation anywhere. In fact, this factoid is the only thing that most people in this country "know" about cryonics.

In the Disney story we see that some of the essential elements are present for a particular archetypal pattern. There is the element of (possible) resurrection and attempt to beat death. Plus there is the fact that Disney was a hero to most Americans—a man who symbolized magic, wonder, imagination, kindness, daring, love of children, and (not incidentally) great wealth. He had ruled over his own Magic Kingdom, Castle, and Land. That a man with such personal power should make a try for the elixir of life was a story that fit well into the collective unconscious. There was simply something about the tale that made it "go," even as there also seems to be about modern myths that such public heros as John F. Kennedy (King Arthur of his own Camelot) or Elvis Presley (The King of Rock and Roll) have somehow managed to beat death and are off in the wings somewhere, waiting to return.

The result of all this was that cryonics received its maximum press from the Disney death in December of 1966. When later a non-famous man actually did made arrangements to be frozen at "death," and followed through with the process (January, 1967), the news and the *LIFE Magazine* story were overridden in most of the country by the fatal Apollo spacecraft fire. The first man ever frozen to cryogenic temperatures was Professor James Bedford of Glendale College, who remains unchanged today, 25 years later, submerged in liquid nitrogen at 320 degrees below zero at the laboratories of the Alcor Life Extension Foundation in Riverside, California. Since 1967, 62 people have followed Bedford's example.

In film, the fate of cryonically preserved people is generally bad. Individuals who are involuntarily cryonically suspended may be allowed to get away with only a severe case of alienation (*Caveman*, 1984; *Late for Dinner*, 1991), but it is clear that anyone who deliberately attempts to cheat death is in for the full Frankenstein treatment. In 1985, a made-for-TV movie called *Chiller* (directed by Wes Craven) featured a cryogenically suspended man who is revived, after which it is discovered that (very much in the style of Lovecraft) the revived one has returned without a soul, and is now utterly evil. When Richard Kobritz, the executive producer of *Chiller*, was asked how the writers had finally come up with the plot for CBS (which wanted to do a horror movie with a cryonics slant), Kobritz stated, "Why, we just asked everybody we knew what bothered them most about the cryonics idea." Mythically, cryonics seems in some ways to have been the recipient of a great deal of the backlash against life-extension and resuscitation caused by half a century of mal-resurrection horror films and stories.

Because of the unique world view of cryonicists, some actual encounters between real-world authorities and cryonicists have played out as though scripted in a horror film. In late 1987, for instance, when an elderly woman in poor health died and was frozen at the Alcor laboratory in Riverside, there was an investigation into the death. In 1930s B-movie fashion, the Alcor laboratory was visited by police and coroners looking for a body which they considered dead, but which cryonicists considered in suspension and possibly still revivable. Early in 1988, several cryonicists went to jail briefly for failure to produce the elderly woman's cryogenically preserved remains, which had been hidden by her son against the possibility of autopsy. The action throughout was generally in keeping with the fine old "mad-scientist" genre in which the crazy researcher sees something more in the clinically dead body than do the "proper" authorities. In the Riverside case, the authorities never did get the remains and finally had to close the case.

Some of the "Juliet problem" of the modern Riverside cryonicists, of course, was inevitable, as we have seen from our fictional and historical discussion. To the cryonicist, someone whose heart has just stopped, but who has not yet suffered brain decay, is not necessarily permanently dead, but rather simply metabolically disadvantaged (or if you will, "flexionally disabled," or "thermally different"—choose your own politically correct term!). In any case, cryonicists do not consider fresh corpses as "things," but rather as sick people (indeed, "patients"). At present writing cryonics remains legal in California, following a series of court battles between cryonics organizations and the State, culminating with a final appellate court decision (June, 1992). The California Board of Public Health had originally taken the odd public position that cryonics was illegal because there was no "cryonics" box to check on the standard paper form which the State of California used to keep track of the disposition of human remains. It soon became clear, however, that more philosophical and perhaps visceral problems worried the State. In one appeal before the court, for example, the State attorney acting for the California Department of Health Services asked: "Should cryonically suspended people be considered dead, or should a separate category of suspended people be created? How should such people be registered in official records? What happens to the estate and the assets of the 'decedent' after the decedent is put in cryonic suspension? What would happen to such estate and assets if and when cryonic suspension is successful and the decedent is restored to life? Whose identity is the person to assume or be assigned and what of the record of the person's death?"

Science, Religion, and Immortality

From almost the beginning of the Scientific Revolution, the emerging technology of resuscitation began to suggest that the process by which human beings go out of existence is as much of a gradual and hard-to-define thing as the process by which they enter it. From the beginning of human culture a set of stories or myths has allowed mankind to deal with threatening changes such as death, and such stories have come to be modified in the scientific age to allow humans to deal philosophically with a limited amount of resuscitation. Along the way, however, there have been plenty of nightmares.

In matters religious, moral, and philosophical, a fundamentalist can be thought of as a person who has little tolerance for ambiguity. Fundamentalists in many spheres are often Aristotelians—binary thinkers who can see only black and white in a world of continuous analog changes and shades of gray. In matters of death, the role of the fundamentalist is played by the vitalist, and by the legal views of the modern State (legal thinking is usually binary/Aristotelian in positing that all actions are intrinsically either legal or illegal). Such people reject the ambiguity which is suggested by resuscitation or cryonics.

It is my thesis that historically, many malresurrection stories have arisen as fundamentalist or vitalist reactions to the ambiguity in death which has been gradually introduced by science since the middle of the 18th century. From riots over dissections, to public worries over being buried alive, to the difficult-to-explain failure of resuscitative techniques to catch on in medicine for more than a century after they were invented, to modern attempts to suppress cryonics by the State of California, the anxieties and the stresses of vitalists have shaped the way in which resuscitation from a long period of clinical death might be viewed by society.

In the literature of science fiction, from Frankenstein to Poe to Lovecraft to Stephen King, scientific or secular resurrection and resuscitation are rarely seen in a positive light. Occasionally, non-horror scientific resurrection stories have had to fight censorship simply because they failed to add enough of the Frankenstein voice (e.g., *The Day the Earth Stood Still*). So strong has the literary tradition of horror in scientific life-extension become since Frankenstein, in fact, that even traditionally positive stories of resurrection have since been re-cast by modern authors in darker terms: the walking mummy, for instance, is a re-working of ancient Egyptian religious belief regarding a technological resurrection, and even in Nikos Kazantzakis' *Last Temptation of Christ* the traditional Lazarus tale has mutated into a mal-resurrection.

As a society, we have tales of "out of body" experiences that let us cope mythically with short term resuscitations—most of these "just-so" stories involve having the soul jerked back and forth between the body and some kind of anteroom to Heaven (e.g., the popular film *Flatliners*). Such stories work well enough to allow even vitalists to deal with the realities of everyday medicine. It is probable, however, that the mythic structure which lets us deal with such true-life situations is due shortly to come under more strain. Consider the following:

On June 10, 1988, a two-and-a-half-year-old girl fell into a mountain stream of melting-snow runoff near her home in Utah, was quickly swept beneath the surface, and drowned. Her mother called rescue operations, who arrived and could not locate the body, but managed to dam off the flow to the side stream which contained it. Over time the water level gradually fell, until eventually (an hour later) one of the girl's arms was uncovered 60 feet downstream, where the body had wedged underwater near a rock. The little girl had been under water for 66 minutes; she was retrieved cold and with eyes open—no pulse, no heartbeat. Given CPR, she was transported to a nearby medical center in Salt Lake City and resuscitated with the aid of a heart-lung machine. Although she had been clinically dead for over an hour, she recovered completely save for a slight residual tremor.

There is no reason to believe that an hour represents the limit for resuscitation from hypothermic clinical death. One authoritative text believes that the ultimate limit even "in the warm" may be as long as an hour, long enough to put us in the realm of *The Day the Earth Stood Still*. Experimental dogs have already been revived in good health from longer than four hours at the temperature of ice. Even these figures are to be regarded as applying only in the context of how far into the future our present knowledge of physiology will let us reasonably peer. What the ultimate limit is, only the future can tell. It is in the hope that the limits are wide that a few cryonicists are frozen every month in the United States.

Whatever the limit turns out to be, our speculative fiction and our myths must find some way to explain it to us at the emotional level; that is the reason we create them. Science fiction, in its ceaseless speculation about the boundaries of technology and human experience, will surely play a pivotal role in how we accept radical new resuscitation and life extension technologies, and how we live with them. Science fiction, hopefully, will escape entirely from the fundamentalists in this, and will remain free to explore all possible answers and all possible questions. That may be difficult to do, given mankind's long history of telling stories in one particular way, but we owe it to ourselves to try.

The Liquefying "Blood" of St. Januarius

JAMES RANDI

"The prejudice of credulity may, in some measure, be cured by learning to set a high value on truth."
—Isaac Watts, English theologian and author, 1674–1748

In the year C.E. 305, an Italian chap now known to the Catholic church as Saint Januarius is said to have been martyred by decapitation. We are told that an enterprising bystander witnessing this festive event had the foresight to bottle some of the resulting blood and also to save the head of the unfortunate fellow.

Much time elapsed after the martyr passed. Then, in 1337, just about the time when relics-of-the-saints were becoming very popular among competing archbishops (the infamous Shroud of Turin popped up at that time as well), the Cathedral of Naples announced that the head of Januarius and the vial of his blood, recently rediscovered, were going on display. (Mind you, the head was not actually shown. A silver urn said to contain it was shown, as it is even to this day. It seems no one has ever troubled—or dared—to look inside the urn. But then, faith is a wonderful thing.)

It was 52 years later that the archbishop of Naples disclosed another wonderful fact: under certain limited circumstances, he said, the red-brown congealed blood in its lavishly mounted reliquary would miraculously liquefy if the congregationalist's prayers were earnest enough. And, for the past six centuries, this popular wonder has been regularly exhibited at the cathedral to the never-failing astonishment of the public. So established is the event that a group of local women are specifically charged with leading the enthusiastic praying. In the ceremony, the reigning archbishop reverently inverts the bottle, the congregation prays fervently, the process is repeated many times, and eventually the "blood" becomes a bright red, freely flowing liquid.

What are those "limited circumstances?" Well, only the archbishop, they say, can cause the transformation; others are evidently not worthy. The miracle only occurs on special feast days. The liquefication, which never fails, can take anywhere from a few minutes to a few days to occur. Really? The unfortunate fact is that the substance in that reliquary has often liquefied during the process of cleaning and polishing the device, while it is being handled—by quite ordinary folks—on any day of the year.

Three Italian chemists recently became curious about this wonder. Doctors Luigi Garlaschelli, Franco Ramaccini, and Sergio Della Sala looked into the remote possibility that perhaps a hoax was afoot. Unable to directly examine the substance due to its sanctified nature, the team had to content themselves with examining the very unsatisfactory evidence presented in various pamphlets on sale at the cathedral. They saw that the church had permitted selected scientists to view the relic by indirect means, resulting in an infuriatingly incomplete data set. Iron, an element present in hemoglobin, had been detected during the previous inadequate examination; the resulting conclusion jumped to was that the substance in the reliquary was indeed real blood, and any doubt that might have existed among the faithful was banished.

The team of chemists was unsatisfied with the conclusion. They reasoned that if they could replicate the observed effect, there might be some cause to doubt the validity of the miracle. Using materials that were available locally from the slopes of nearby Mount Vesuvius, and utilizing procedures that were well-known to medieval workers, the team eventually produced a liquid that in every way matched the liquid in the reliquary. It is the correct red-brown color, it coats the interior surface of the container in the same way, it gels solidly, liquefies, and becomes a translucent, bright blood red when jarred, shaken or repeatedly inverted. It re-solidifies after a few hours if undisturbed.

Dr. Garlaschelli, who is presently at the department of organic chemistry at the University of Pavia, has generously sent me samples of his product to examine. I now use it as part of my lecture demonstration, and I often get angry reactions from audience members who challenge me about whether the replication of a miracle by ordinary means disproves the miraculous nature of the original phenomenon. The answer is, of course, no. But all the doubt could be nicely resolved if the present archbishop of Naples would allow close examination of the "blood" of Saint Januarius. I'm not holding my breath waiting for such permission.

Remember that in 1978, when the Shroud of Turin was finally properly tested, the claims of the skeptics were firmly established to be true. Of course, the results were denied by the Shroud fans, who had suddenly discovered that carbon-dating does not apply to religious relics. Were the "blood" of Saint Januarius to be properly examined, no doubt similar new scientific discoveries would be announced.

The remarkable fact about all such matters is that the faithful persist in accepting them as miracles despite the absence of any supporting evidence, the questionable manner in which they are produced, the fact that far more parsimonious explanations are available, and/or the strong evidence that a hoax exists. One defender of the Shroud of Turin, the Reverend David Sox, for example, commented after the definitive investigation was completed that for him "'Forgery' and 'authenticity' are essentially meaningless terms." Perhaps for you, sir; not for me.

A final observation on the Saint Januarius "miracle": little advertised is the fact that in the Naples area there have also been, in the past, similar liquefying-blood miracles claimed for seven other saints! These have now been dropped by the church, and Saint Januarius is the only one being promoted. Umm. . . .

Psychoanalysis as Pseudoscience

KEVIN MACDONALD

In *The Ghost in the Machine*, Arthur Koestler describes the closed cognitive matrix that defines paranoid Conspiracy Theories: (1) they claim to represent a universal truth, capable of explaining all phenomena; (2) they cannot be logically or empirically refuted because all potentially damaging information must be interpreted in terms of the theory; and (3) any criticism is met by a counter-offensive that shifts the argument to the subjective motivations of the critic. Koestler concludes that such theories employ "sophisticated methods of casuistry, centered on axioms of great emotive power, . . . indifferent to the rules of common logic" and become "a kind of Wonderland croquet, played with mobile hoops" (263).

One usually thinks of conspiracy theories as being held by people who are poorly educated, downwardly socially mobile, and/or mentally maladjusted. A good example describing these processes is Richard Hofstadter's *The Paranoid Style in American Politics*, which details the history of Heartland America's off-again on-again love affair with xenophobia and Conspiracy Theory. However, my purpose here is to examine the moral and intellectual legacy of psychoanalysis and show how well it fits Koestler's definition and therefore constitutes a very long-lived and influential Conspiracy Theory—but one tailored to the prejudices of elites and cosmopolitans rather than those of the "plain folks down home."

Cult Characteristics

In 1911 Freud disciple and psychoanalyst Eugen Bleuler left the movement, concluding "this 'who is not for us is against us,' this 'all or nothing,' is necessary for religious communities and useful for political parties. I can therefore understand the principle as such, but for science I consider it harmful" (in Gay 1987, 144–145). This observation by an insider says a lot about what psychoanalysis had become—a cult-like religion.

The apex of the authoritarian, antiscientific institutional structure of psychoanalysis was the secret committee of handpicked loyalists sworn to uphold psychoanalytic orthodoxy described by Phyllis Grosskurth in *The Secret Ring: Freud's Inner Circle and the Politics of Psychoanalysis* (1991, 15):

> By insisting the Committee must be absolutely secret, Freud enshrined the principle of confidentiality. The various psychoanalytic societies that emerged from the Committee were like Communist cells, in which the members vowed eternal obedience to their leader. Psychoanalysis became institutionalized by the founding of journals and the training of candidates; in short an extraordinarily effective political entity.

There were repeated admonitions for the Committee to present a "united front" against all opposition, for "maintaining control over the whole organization," for "keeping the troops in line," and "reporting to the commander" (Grosskurth, 97). Consider Otto Rank's astonishing letter of 1924 in which he attributes his heretical behavior in questioning the Oedipal complex to his own neurotic unconscious conflicts, he promises to see things "more objectively after the removal of my affective resistance," and is thankful that Freud "found my explanations satisfactory and has forgiven me personally." Grosskurth notes how "Freud seems to have acted as the Grand Inquisitor, and Rank's groveling 'confession' could have served as a model for the Russian show trials of the 1930s." Freud viewed the entire episode as a success; Rank had been cured of his neurosis "just as if he had gone through a proper analysis" (Grosskurth, 1991, 167–168).

The staunch Freud disciple Fritz Wittels (1924) decried the "suppression of free criticism within the Society . . . Freud is treated as a demigod, or even as a god. No criticism of his utterances is permitted." He tells us that Freud's Drei Abhandlungen zur Sexualtheorie is "the psychoanalyst's Bible." This is no mere figure of speech. The faithful disciples regard one another's books as of no account. They recognize no authority but Freud's; they rarely read or quote one another. When they quote it is from the Master, that they may give the pure milk of the word" (142–143). Freud "had little desire that [his] associates should be persons of strong individuality, and that they should be critical and ambitious collaborators. The realm of psychoanalysis was his idea and his will, and he welcomed anyone who accepted his views" (134). The others were simply expelled. All of the major figures around Freud appear to have been extremely submissive personalities who absolutely revered Freud as father figure. Indeed, the members appear to have self-consciously viewed themselves as loyal sons to Freud the father-figure (complete with sibling rivalry as the "brothers" jockeyed for position as the "father's" favorite), while Freud viewed his close followers as his children, with power to interfere in their personal lives (Hale, 1995, 29).

Ernest Jones, Freud's worshipful biographer and the official head of the movement after Jung's defection, "grasped the fact that to be a friend of Freud's meant being a sycophant. It meant opening oneself completely to him, to be willing to pour out all one's confidences to him" (Grosskurth, 48). Masson (1990, 152) suggests that "Jones believed that to disagree with Freud (the father) was tantamount to patricide (father murder)." When Sandor Ferenczi, a central figure in the inner circle of psychoanalysis during the 1920s, disagreed with Freud on the reality of childhood sexual abuse, Jones called him a "homicidal maniac" (152).

Regarding Ferenczi, Grosskurth notes that "(t)he thought of a disagreement with Freud was unbearable . . ."; "There were occasions when he rebelled against his dependency, but always he returned repentant and submissive" (54–55). Similarly, Masson (1990) describes Kurt Eissler, the closest confidant of Anna Freud's inner circle in the 1960s, by saying that "What he felt for Freud seemed to border on worship." He held one thing sacred, and hence beyond criticism: Freud" (121–122). It was common among the disciples to imitate Freud's personal mannerisms, and even among analysts who did not know Freud personally, there were "intense feelings, fantasies, transferences, identifications" (Hale, 1995, 30).

Evidence for the essentially cult-like character of psychoanalysis is the unique role of disciples who are able to trace themselves back to Freud in a direct line of descent. "The idea of being a chosen disciple, privileged to have direct contact with the master, has survived and is continued in the procedures of many of the training programs of the institutes" (Arlow &

Brenner, 1988, 5). "The intensely filial relationships to Freud of the first generation were gradually replaced by a highly emotional relationship to a fantasied Freud, still the primal founder, but also to organizations, to peers, to superiors in the institute hierarchy—above all—to the training analyst, the training analyst's analyst, and, if possible, back to Freud and his circle became a determinant of psychoanalytic prestige" (Hale, 1995, 32).

Unlike most sciences, there is a reverence for what one might term the sacred texts of the movement—Freud's writings—both in teaching and in the current psychoanalytic literature. Arlow and Brenner (1988) note that *Studies of Hysteria* and *The Interpretation of Dreams* are almost 100 years old, but continue to be standard texts in psychoanalytic training programs. They also describe "the recurrent appearance in the analytic literature of articles redoing, extending, deepening, and modifying Freud's early case histories" (5). Indeed, it is remarkable to scan psychoanalytic journal articles and find how many of those references are to Freud's work written well over 60 years ago. In examining six issues of *Psychoanalytic Quarterly* from 1988–1989, I found 92 references to Freud in 33 articles. Only four had no references to Freud, and of these, one had no references at all and one had only one reference.

The continued use of Freud's texts in instruction and the continuing references to Freud's work would not be conceivable in real science. While Darwin is venerated for his scientific work as the founder of the modern science of evolutionary biology, studies in evolutionary biology only infrequently refer to his writings because the field has moved so far beyond his work. *The Origin of Species* and *The Descent of Man* are important texts in the history of science, but are not used for current instruction. Moreover, central features of Darwin's account, such as his views on inheritance, have been completely rejected by modern science. With Freud, however, there is continuing fealty to the master, at least within an important subset of the movement.

Besides Rank, other deviators—Fleiss, Adler, Jung, Ferenczi—were also diagnosed as suffering from a variety of psychiatric disorders and as needing further psychoanalysis in order to see the light. Freud "never tired of repeating the now notorious contention that the opposition to psychoanalysis stemmed from 'resistances'" arising from emotional sources (Esterson, 1993, 216). He attributed Jung's defection to "strong neurotic and egotistic motives" (in Gay, 1988, 481). Even Peter Gay, the psychoanalytic loyalist and historian of the movement, writes that "These ventures into character assassination are instances of the kind of aggressive analysis that psychoanalysts, Freud in the vanguard, at once deplored and practiced. This . . . was the way that analysts thought about others, and about themselves." The practice was "endemic among analysts, a common professional deformation" (1988, 481).

This practice continues to this day. A common thread of the letters sent by the many aggrieved psychoanalysts in response to Frederick Crews's critical articles in the *New York Review of Books* was that they were "composed in a state of bitter anger by a malcontent with a vicious disposition" (293). Crews' Freud bashing was typically explained in terms of botched transferences and Oedipal complexes gone awry. Another recent case is that of Jeffrey Masson (1990) who suffered similar questionings of his sanity for challenging the central Freudian dogma of the Oedipal complex.

Psychoanalysis, unlike scientific theory, but very much like certain religious or political movements, has essentially been immune from attacks leveled at it either from inside or outside the movement. Insiders who dissented from central doctrines were simply expelled and often went on to found their own psychoanalytically oriented sects, typically with the same disregard for canons of scientific method as the parent religion. There is a long line of such expelled dissenters in the history of psychoanalysis, and the list continues to lengthen with the recent expulsion of Jeffrey Masson. Moreover, the central core of loyalists that has always existed in psychoanalysis functions to preserve the image of Freud as a heroic scientist to the point that many of Freud's papers have been locked away from the prying eyes of scholars for periods extending as far ahead as the 22nd century.

Thought Control

The entire Freudian enterprise appears more and more like an authoritarian religious cult than a scientific movement. Indeed, several authors have pointed out that psychoanalysis has many features in common with brainwashing (Bailey, 1960, 1965; Salter, 1996). Frank Sulloway (1979b) describes the indoctrination characteristic of training analyses in which any objection by the analyst is viewed as a resistance to be overcome. And even Shelly Orgel (1990), who remains a defender of the psychoanalytic faith, writes of the feelings of many contemporary analysts that their analysts had behaved aggressively toward them, turning them into devoted and passive followers of their highly idealized analyst.

Jeffrey Masson (1990) provides fascinating insight into psychoanalysis as thought control and aggression. Masson's training analysis involved a completely one-sided relationship in which the analyst had all of the power and in which the trainee was expected to put up with any and all indignities. Leaving the training analyst would have meant giving up psychoanalysis because the training analyst would claim that the trainee was unfit for a career as a psychoanalyst. The result of the analysis was an idealization of the training analyst and loyal support of the training analyst's writings. Masson was more or less blackmailed into agreeing to include his own training analyst's name on a paper he was writing or be forced to reenter analysis. Masson comments that "Being in such an analysis is like growing up with a despotic parent" (86), since the qualities it requires in the prospective analysts are meekness and abject obedience.

I suggest that the inculcation of passive and devoted followers via the aggression and thought control represented by psychoanalysis has always been an important aspect of the entire belief system. At a deep level, the fundamentally pseudoscientific structure of psychoanalysis implies that disputes cannot be resolved in a scientific manner, with the result that, as John Kerr (1992) notes, the only means of resolving disputes involves the exercise of personal power. The result was that the movement was doomed to develop into a mainstream orthodoxy punctuated by numerous sectarian deviations originated by heretics who were expelled from the movement. These offshoots then replicated the fundamentally irrational pseudoscientific structure of all psychoanalysis inspired movements: "(E)ach major disagreement over theory or therapy seemed to require a new validating social group, a psychoanalytic tradition that recent splits within Freudian institutes seem only to confirm" (Hale, 1995, 26). Perhaps the most bizarre such offshoot was the movement initiated by Wilhelm Reich, well-covered in Joel Carlinsky's *Skeptic* (2:3) article "Epigones of Orgonomy."

The problem continues. Crews (1995) describes recent scholarship on psychoanalysis that shows not only that psychoanalysis was never more than a pseudoscience but that Freud engaged in scientific fraud when developing his theories. Allen Esterson's (1993) *Seductive Mirage: An Exploration of the Work of Sigmund Freud* demonstrates convincingly that

Freud's patients did not volunteer any information on seduction or primal scenes at all. The seduction stories which provide the empirical basis of the Oedipal complex were in fact a construction by Freud who then interpreted his patients' distress on hearing his constructions as proof of the theory. Freud then deceptively obscured the fact that his patients' stories were reconstructions and interpretations based on his a priori theory. He also retroactively changed the identity of the fancied seducers from nonfamily members (servants, etc.) to the fathers that his Oedipal story required.

Now 100 years after its inception, the theories of the Oedipal complex, childhood sexuality, and the sexual etiology of the neuroses remain without any independent empirical validation and play no role whatever in mainstream developmental psychology. From an evolutionary point of view the idea that children would have a specifically sexual attraction to their opposite sex parent is highly implausible, since such an incestuous relationship would result in inbreeding depression (MacDonald, 1986). The proposal that boys desire to kill their fathers conflicts with the general importance of paternal provisioning of resources in understanding the evolution of the family (MacDonald, 1988; 1992a, b): Boys who had succeeded in killing their fathers and having sex with their mothers would not only be left with genetically inferior offspring, they would also be deprived of paternal support and protection. Modern developmental studies indicate that many fathers and sons have very close, reciprocated affectional relationships beginning in infancy, and the normative pattern in Western societies is for mothers and sons to have very intimate and affectionate, but decidedly nonsexual relationships. Most domestic violence takes place betwen genetically related individuals (Daly and Wilson).

The continued life of these concepts in psychoanalytic circles is testimony to the continuing unscientific, religious nature of the entire enterprise. Indeed, Kurzweil (1989, 89) notes that "In the beginning, the Freudians tried to 'prove' the universality of the Oedipus complex; later on, they took it for granted. Ultimately, they no longer spelled out the reasons for the pervasiveness of childhood sexuality and its consequences in the cultural monographs: they all accepted it."

There is also increasing attention paid to the ethical dimensions of psychoanalysis as Freud himself practiced it. Freud seems to have been remarkably indifferent to his patients' suffering, but his ethical lapses extend far beyond a lack of empathy. Crews recounts the case of Horace Frink, an American psychoanalyst who was having an affair with a bank heiress. Freud diagnosed Frink as a latent homosexual(!) and advised him to divorce his wife and marry the heiress, with the stated aim of tapping into the heiress' funds for a financial contribution to psychoanalysis. To make the plan work, the heiress had to divorce her husband as well. All of this came about, but the two abandoned spouses were devastated and soon died, Frink's new wife sued for divorce, and Frink himself sank into depression and repeated attempts at suicide.

Then there is the case of Dora Bauer. Freud diagnosed the teenaged Dora as suffering from hysteria for refusing to have a sexual relationship with a married man, Herr K., as a sort of quid pro quo so that her father would continue to have an affair with Herr K.'s wife. Crews comments that "In short, a sexually and morally uninhibited [Dora] rounded into psychic trim by Freud, would have been of service to both her father and Herr K., the two predatory males who, unlike any of the women in the story, basked in the glow of Freud's unwavering respect" (52). The Dora case is typical also in that the patient's diagnosis was based entirely on preconceived ideas and circular reasoning in which the patient's negative emotional response to the psychoanalytic hypothesis was construed as evidence for the hypothesis.

Recovered Memory Therapy

Another new wrinkle is that psychoanalysis has had a very pernicious effect on psychotherapeutic practice, in particular the phenomenon of the Recovered Memory Therapy (RMT). At the time when Crews's articles originally appeared in the *NYRB,* Crews was content to claim only a genealogical relationship between psychoanalysis and RMT. He now documents a much closer relationship between the two movements. A significant number of psychoanalysts are now rejecting the orthodox psychoanalytic theory that claims of infantile sexual abuse are illusory manifestations of Oedipal desires. These renegade psychoanalysts are in fact now adopting Freud's earlier seduction theory of 1896 in which neurosis was conceptualized as the result of actual sexual abuse—a theory which Freud developed in the same manner as he developed the Oedipal story that replaced it: by making suggestions to patients and doggedly persisting in his explanation until the patient acknowledged the truth of the psychoanalytic explanation. Crews emphasizes that there is no end to the possible harmful social and moral influences of such a theory in the hands of its pseudoscientific practitioners, including bankruptcy, breaking up of families, and imprisonment of family members.

Because of its belief in the reality of memories of childhood sexual abuse, the RMT movement must be viewed as a psychoanalytic heresy. As with all of the previous psychoanalytic heresies, however, RMT shares a commitment to a methodology that results in self-validation of theoretical claims. Unverifiable phenomena have been at the very center of psychoanalysis and its intellectual offspring from the beginning. The following quote from Freud is an exemplar of the type of attitude that carries over into the RMT movement (in Crews, 209):

> The work keeps on coming to a stop and they keep on maintaining that this time nothing has occurred to them. We must not believe what they say, we must always assume, and tell them, too, that they have kept something back. . . . We must insist on this, we must repeat the pressure and represent ourselves as infallible, till at last we are really told something. . . .

The therapist may suppose that the patient had experienced sexual trauma even without any external evidence or memory of the event. Recovered Memory Therapists, in the words of one such practitioner, "must validate the patient's belief that abuse occurred, or risk reenacting the role of denying parent, which may have enabled the abuse in the first place" (Crews, 25). The technique ensures validation and indeed finds a moral rationale for insisting on validation. But it cannot provide even the beginnings of a search for truth.

The case of Eileen Franklin Lipsker, in the news recently when her accused father was released from jail, is particularly fascinating because Lipsker has recently "remembered" several other crimes that could not have possibly occurred. Even before this turn of events, however, Lipsker had developed increasingly bizarre "memories" about her father, including a murder that no one else, including the police, had heard about, and a supposed rape by Eileen's godfather that was aided by the father. The "memories" were gradually elaborated as a result of the suggestions of a psychotherapist and their veracity attested to by Lenore Terr, a professor of psychiatry at the University of California–San Francisco. Terr used the aura of science surrounding her academic affiliation to convince the jury that an expert like herself could distinguish authentic from nonauthentic repressed memories.

Then there is the fantastic case of the Ingram family of Olympia, Washington, in which Paul Ingram confessed to a myriad of crimes whose memory he thought he had completely repressed, including repeatedly raping both his daughters and one son, getting his daughters to perform sexual favors for his friends, torturing the girls, getting his wife to have sex with animals, and murdering and cannibalizing babies at Satanic rituals. The truly remarkable thing about this example is the willingness of people to be convinced of the bizarre and impossible.

A condition that greatly facilitates people's credulity is the belief among a significant number of professionals in psychology that such repressed memories are commonplace. No fewer than five psychologists and counselors encouraged Ingram in his hallucinations. However, a skeptical psychologist finally asked Ingram about a completely fictitious accusation that Ingram had encouraged his children to have sex while he watched. Sure enough, the next day Ingram came up with a highly detailed repressed memory of watching his children have sex. Ingram, who pleaded guilty to the crimes, after belatedly coming to believe in his innocence, is now serving 20 years in prison for six counts of child molestation.

Like psychoanalysis itself, RMT has become a political movement bent on enforcing an official orthodoxy. Indeed, given the history of psychoanalysis it is not in the least surprising that RMT would likewise be an authoritarian political movement. Judith Lewis Herman, a leading proponent of RMT, claims that "Advances in the field occur only when [women] are supported by a political movement powerful enough to legitimate an alliance between investigators and patients and to counteract the ordinary social processes of silencing and denial" (Crews, 160). RMT has been behind lengthening the statutes of limitations in some states to periods of 30 years or more to provide enough time for repressed memories of crimes to surface. And, as with any such political movement, it seems superfluous to note that big money is involved, in the case of RMT ranging from fees for therapy, the publication industry, and the litigation industry spawned by this movement.

The Connection to the Left

Much of Crews's recent work on psychoanalysis and RMT was originally published in the prestigious *New York Review of Books (NYRB)*. The *NYRB* has long been a bastion of the intellectual left and, as Crews notes, publication of such material in such a publication is "almost like pet owners who had negligently or maliciously consigned their parakeet to the mercies of an ever-lurking cat" (Crews, 1995, 288). Publications like the *NYRB* have been instrumental in propagating psychoanalytic and similar doctrines as scientifically and intellectually reputable for decades, and there is the suggestion that had Crews published his articles in a less visible and less politicized medium they could have been safely ignored as has commonly been the practice over the long history of psychoanalysis.

There is a long and interesting association between psychoanalysis and the political and cultural left. Support of radical and Marxist ideals was common among Freud's early followers, and leftist attitudes have been common in later years among psychoanalysts (Hale, 1995, 31; Kurzweil, 1989, 36, 284), as, e.g., among the groups in Berlin and Vienna during the post–World War I era (Kurzweil, 1989; 46–47); in the post revolutionary Soviet Union where all of the top psychoanalysts were Bolsheviks and Trotsky supporters and were among the most powerful political figures in the country (Chamberlain, 1995); and in America from the 1920s to the present (Torrey, 1992, 33, 93ff; 122–123). If Crews is correct in his analysis of the institutional structure of psychoanalysis as an authoritarian political movement—and he certainly is—one is left with the conclusion that one of the century's major intellectual and cultural forces was nothing more than a highly disciplined political movement masquerading as science.

Psychoanalysis has proved to be a veritable treasure trove of ideas for those intent on developing radical critiques of Western culture, beginning with Freud's own *Totem and Taboo* and *Civilization and Its Discontents*. Crews provides an excellent account of how Freud tended to make dogmatic claims about the source of his patients' unhappiness based on nothing more than his own suggestions. His failure to follow even the minimum standards of scientific or rational intellectual inquiry extended to his cultural writings as well. Freud's wider speculations on human culture rest on a number of extremely naive, prescientific conceptualizations of human sexual behavior and its relation to culture. Particularly outrageous was Freud's "primal horde" story of how over many generations sons had killed their fathers in order to mate with their mothers until Oedipal guilt had forced them to repress this activity. The theory is not only completely speculative as it attempts to explain a nonexistent phenomenon—the Oedipal complex—it also requires Lamarckian inheritance, a theory that, at least by the time of *Civilization and Its Discontents* (where the doctrine was reaffirmed), had been completely rejected by the scientific community.

Freud's Armageddon

While Freud's was a self-consciously speculative theory, his speculations clearly had an agenda. Rather than provide speculations which reaffirmed the moral and intellectual basis of the culture of his day, his speculations were an integral part of his war on culture—so much so that he viewed *Totem and Taboo* as a victory over Rome and the Catholic Church (Rothman and Isenberg, 1974). In Freud's eyes he was the Carthaginian general Hannibal fighting the evil Romans that to him represented Western civilization. Peter Gay notes that Freud was proud of his enemies—the persecuting Roman Catholic Church, the hypocritical bourgeoisie, the obtuse psychiatric establishment, the materialistic Americans—so proud, indeed, that they grew in his mind into potent specters far more malevolent and far less divided than they were in reality. He likened himself to Hannibal, to Ahasuerus, to Joseph, to Moses, all men with historic missions, potent adversaries, and difficult fates (Gay, 1988, 604). Freud described this "Hannibal fantasy" as "one of the driving forces of [my] mental life" (Sulloway 1979a).

In this regard, it is interesting to note that *Totem and Taboo* and *Civilization and Its Discontents* present the view that the restrictions on sexual behavior, so apparent as an aspect of Western culture during Freud's life, are the source of art, love, and even civilization itself. Freud's conceptions of the origins and function of sexual repression in Western societies contain, as Peter Gay (329) notes, some of Freud's "most subversive conjectures." Neurosis and unhappiness are the price to be paid for civilization because neurosis and unhappiness are the inevitable result of repressing sexual urges.

Freud appears to have been well aware that his conjectures were entirely speculative. Freud was "amused" when *Totem and Taboo* was termed a "just so" story by a British anthropologist in 1920, and stated only that his critic "was deficient in phantasy," apparently a concession that the work was indeed fanciful. Freud stated that "It would be nonsensical to strive for exactitude with this material, as it would be unreasonable to demand certainty." Similarly, Freud described *Civilization and Its*

Discontents as "an essentially dilettantish foundation" on which "rises a thinly tapered analytic investigation." And Freud was well aware that his attack on religion in *The Future of an Illusion* was scientifically weak, describing it by noting that "the analytic content of the work is very thin" (Gay, 330, 543, 524).

Freud's countercultural writings scarcely exhaust the mischief wreaked by psychoanalysis. The works of Herbert Marcuse, Norman Brown, Wilhelm Reich, Jacques Lacan, Erich Fromm, and a host of neoFreudians come to mind immediately, but this barely scratches the surface. Psychoanalysis influenced thought in a wide range of areas, including sociology, child rearing, criminology, anthropology, literary criticism, art, literature, and the popular media.

In fact Freud's ideas have often been labeled as subversive. Indeed, "[Freud himself] was convinced that it was in the very nature of psychoanalytic doctrine to appear shocking and subversive. On board ship to America he did not feel that he was bringing that country a new panacea. With his typically dry wit he told his traveling companions, 'We are bringing them the plague'" (Mannoni, 1971, 168).

Peter Gay terms Freud's work generally "subversive" (1987, 140), his sexual ideology in particular "deeply subversive for his time" (148); and his *Totem and Taboo* as containing "subversive conjectures" (327) in its analysis of culture. Rothman and Isenberg (1974) convincingly argue that Freud actually viewed the *Interpretation of Dreams* as a victory against the Catholic Church and that he viewed *Totem and Taboo* as a successful attempt to analyze the Christian religion in terms of defense mechanisms, primitive drives, and neurotic symptomatology. Gay notes that "while the implications of Darwin's views were threatening and unsettling, they were not quite so directly abrasive, not quite so unrespectable, as Freud's views on infantile sexuality, the ubiquity of perversions, and the dynamic power of unconscious urges" (144).

And the contrast between Freud and Darwin as scientists could scarcely be more clear. Darwin spent years patiently collecting his data and was hesitant to publish his work, agreeing to do so only after another scientist, Alfred Russel Wallace, came up with similar ideas. Freud, on the other hand, conducted his career more like a military general bent on conquering an enemy, as Gay concluded (1987, 145):

> While Darwin was satisfied with revising his work after further reflection and absorbing palpable hits by rational critics, while he trusted the passage of time and the weight of his argumentation, Freud orchestrated his wooing of the public mind through a loyal cadre of adherents, founded periodicals and wrote popularizations that would spread the authorized word, dominated international congresses of analysis until he felt too frail to attend them and after that through surrogates like his daughter Anna.

Psychoanalysis has a lot to atone for. The contemporary upsurge of victims of RMT and the long line of individual victims like Horace Frink and Dora Bauer are only a small part of its moral wreckage. The fact that the *NYRB* published Crews's attacks on psychoanalysis may be a vital sign that the life of psychoanalysis as an underpinning of the intellectual left is weakening. The *NYRB* is only one of many elements of the vast media and intellectual network that has supported psychoanalysis throughout the century, but all signs are that psychoanalysis has become an intellectual and scientific embarrassment to all save the truest of true believers. The fact that its scientific stature has been utterly discredited in such a prestigious forum and by someone who is sympathetic to the cultural influences it has generated suggests that psychoanalysis may well have lost its political punch.

But don't expect either psychoanalysis or RMT to die soon. Because they are fundamentally religious and political rather than scientific, such movements have a life of their own, and will expire only when they are perceived as no longer serving the personal or political interests of their advocates.

References:

Arlow, J. A., & Brenner, C. 1988. "The Future of Psychoanalysis." *Psychoanalytic Quarterly*, LVII, 1–14.

Bailey, P. 1960. "Rigged Radio Interview with Illustrations of Various 'Egoideals.'" *Perspectives in Biology and Medicine*, 4, 199–265.

——. 1965. *Unserene: A Tragedy in Three Acts*. Springfield, IL: Charles Thomas.

Caton, Hiram. (Ed.) 1990. *The Samoa Reader: Anthropologists Take Stock*. Lanham, MD: University Press of America.

Chamberlain, L. 1995. "Freud and the Eros of the Impossible." *Times Literary Supplement*, August 25, pp. 9–10.

Cioffi, Frank. 1969. "Wittgenstein's Freud." In *Studies in the Philosophy of Wittgenstein*. P. Winch (ed.). London: Routledge & Kegan Paul.

——. 1970. "Freud and the Idea of a Pseudoscience." In *Explanation in the Behavioural Sciences*. R. Borger & F. Cioffi (eds.). Cambridge University Press.

——. 1972. "Wollheim on Freud." *Inquiry*, 15, 171–186.

Crews, F., et al. 1995. *The Memory Wars: Freud's Legacy in Dispute*. New York: New York Review.

Daly, Martin, and Wilson, Margo. 1988. *Homicide*. New York: Aldine.

Ellenberger, Henri. 1970. *The Discovery of the Unconscious*. New York: Basic Books.

Esterson, Allen. 1993. *Seductive Mirage: An Exploration of the Work of Sigmund Freud*. Chicago: Open Court.

Gay, Peter. 1987. *A Godless Jew: Freud, Atheism, and the Making of Psychoanalysis*. New Haven: Yale University Press.

——. 1988. *Freud: A Life for Our Time*. New York: W. W. Norton.

Grosskurth, Phyllis. 1991. *The Secret Ring: Freud's Inner Circle and the Politics of Psychoanalysis*. Boston: Addison-Wesley.

Grünbaum, Adolf. 1984. *The Foundations of Psychoanalysis*. Berkeley: University of California Press.

Hale, N. G. 1995. *The Rise and Crisis of Psychoanalysis in the United States: Freud and the Americans, 1917–1985*. New York: Oxford University Press.

Hofstadter, R. 1965. *The Paranoid Style in American Politics and Other Essays*. New York: Knopf.

Kerr, John. 1992. *A Most Dangerous Method: The Story of Jung, Freud, and Sabina Spielrein*. New York: Knopf.

Koestler, A. 1967. *The Ghost in the Machine*. New York: Macmillan.

Kurzweil, E. 1989. *The Freudians: A Comparative Perspective*. New Haven: Yale University Press.

MacDonald, Kevin. 1986. "Civilization and Its Discontents Revisited: Freud as an Evolutionary Biologist." *Journal of Social and Biological Structures*, 9, 213–220.

——. 1988. *Social and Personality Development: An Evolutionary Synthesis*. New York: Plenum.

——. 1992a. "Warmth as a Developmental Construct: An Evolutionary Analysis." *Child Development*, 63, 753–773.

——. 1992b. "A Time and a Place for Everything: A Discrete Systems Perspective on the Role of Children's Rough and Tumble Play in Educational Settings." *Early Education and Development*, 3, 334–355.

MacMillan, Malcolm. 1991. *Freud Evaluated: The Completed Arc*. The Hague: Elsevier North Holland.

Mannoni, Octave. 1971. *Freud*. R. Belice (trans.). New York: Pantheon Books.

Masson, Jeffrey M. 1984. *The Assault on Truth: Freud's Suppression of the Seduction Theory*. New York: Farrar, Straus, and Giroux.

——. 1990. *Final Analysis: The Making and Unmaking of a Psychoanalyst*. Reading, MA: Addison-Wesley Publishing Co.

McGrath, W. J. 1974. "Freud as Hannibal: The Politics of the Brother Band." *Central European History*, 7, 31–57.

Orgel, Shelly. 1990. "The Future of Psychoanalysis." *Psychoanalytic Quarterly*, LIX, 1–20.

Rothman, Stanley, and Isenberg, P. 1974. "Sigmund Freud and the Politics of Marginality." *Central European History*, 7, 58–78.

Salter, Frank. 1996. "A Comparative Analysis of Brainwashing Techniques." In Irenaus Eibl-Eibesfeldt and Frank Salter (Eds.), *Ideology, Warfare, and Indoctrinability*. Proceedings of the Symposium on Ideology, Warfare, and Indoctrinability sponsored by the Max Planck Society at Ringberg Castle, Germany, January 12, 1995; in press.

Sulloway, Frank. 1979a. *Freud: Biologist of the Mind*. New York: Basic Books.

——. 1979b. "Freud as Conquistador." *The New Republic* (August), 25–31.

Torrey, E. F. 1992. *Freudian Fraud: The Malignant Effect of Freud's Theory on American Thought and Culture*. New York: HarperCollins.

Wittels. F. 1924. *Sigmund Freud: His Personality, His Teaching, & His School*, trans. E. and C. Paul. London: George Allen & Unwin.

Psychotherapy as Pseudoscience

TANA DINEEN

A housewife, unhappy with her life, decides to seek therapy to deal with her loneliness and frustration. The therapist arranges to see her weekly, discussing topics ranging from her childhood memories and her parents, to her isolated adult life and her passionless marriage. A year later, she is still unhappy with her life but she is "happy with her therapy," claiming that finally someone listens to her and understands her problems.

Has therapy worked? By her standard, it has; she likes her therapist and she believes that her therapist understands her. Her therapist is happy with a reliable and talkative client, who pays her bills and supports his practice. But has anything really changed? The housewife remains isolated, her marriage is still passionless; her life is essentially no different from what it was when she began treatment.

She and her therapist would likely argue that her life is different because she thinks about it differently and is on the road to recovery. Some would say that, because she is now more in touch with her feelings and is working through her unconscious material, the therapy fees are dollars well spent.

This example is just one of millions of psychotherapy cases that begin each year with people seeking help for mild and diffuse forms of dissatisfaction with life, unfulfilled goals, unrealized expectations, and unmet dreams. Some cases are spiced up with the recovered memories of abuse, or the diagnosis of mental disorders such as Depression, Post Traumatic Stress Disorder, Shopoholism, or Internet Addiction. Often an air of scientific professionalism is achieved through the use of techniques with titles such as Hypnosis, Eye Movement Desensitization and Reprocessing (EMDR), Cognitive-Behavioral Therapy, or Neuro Linguistic Programming (NLP).

But in the end are any of these treatments any more effective than talking with a friend or just getting on with life? Do they do anything more than give the client a sense of self-importance and worth, and an inner glow that comes from being the center of attention? Does therapy really make any difference? Is psychotherapy worth the money that individuals, insurance companies and governments pay? Before we decide, let's examine some information that the "Psychology Industry," a term which I will define shortly, would prefer to keep hidden from the public.

By way of background, I am a licensed psychologist who has, undeniably, broken ranks. Five years ago, I forced myself to step back and take a cold hard look at my profession. I am still a psychologist by license but I am not practicing. What I see being done under the name of psychology is so seriously contaminated by errors in logic, popular notions, and personal beliefs, and it is doing so much harm to people, that I find myself in this strange role of working to curb the pervasive influence of my own chosen profession. Long ago I lost any expectation that the necessary corrective actions would come from within the profession; so I find myself speaking most often now to people outside my profession—to philosophers, ethicists, the clergy, educators, criminologists, and lawyers, hoping to find among them skeptics who are willing to think critically about America's love affair with psychology. As Noam Chomsky observed: "One waits in vain for psychologists to state the limit of their knowledge."

Recently, addressing a conference on professional ethics, I discussed the relationship between the consumer, or the client/patient, and the service provider, or the psychologist. I suggested that the Psychology Industry is selling consumers a bill of goods, that psychological services are in many ways a scam, and that psychological treatment is a modern psychic version of snake oil. From the witches' brews of ancient times to the traveling medicine shows, from copper bracelets to Kickapoo Indian Oil, society has always had an abundance of secret concoctions and panaceas to cure all of its ailments. For instance, the discovery of radium by the Curies began the Mild Radium Therapy movement, particularly popular among American socialites, and precipitated a lucrative trade in radium-based belts, hearing aids, toothpaste, face cream, and hair tonic. Most popular of all was Radiothor, a glow-in-the-dark mineral water which carried promises of a cure for more than 150 maladies.

Psychotherapy may well be nothing more than one of these concoctions. While snake oil had no effective agent, it did have sufficient common alcohol to make people feel better until their ailments naturally went away. Similarly, psychotherapy has no effective agent, but people, like the woman described above, buy it, believe in it, and insist that it works because it makes them feel better about themselves for a while. This change, if it can be called that, may well be derived from nothing more than the expression of concern and caring, and not from specialized treatment worthy of payment.

I do not mean to suggest that psychotherapy is a premeditated scam. Most psychologists, I think, genuinely believe that they are helping people, a view supported by the professional organizations and licensing bodies. Information does exist that would cause them to question their assumption; however, most remain focused on selling their services, marketing their products, making a living and feeling good about themselves. They ignore the data and, thus, manage to maintain a belief, tantamount to a faith, in what they are selling.

The Psychology Industry

In the Fall of 1993 after spending an afternoon discussing what was happening in psychology with a colleague of mine, Sam Keen—the former editor of *Psychology Today*—I half jokingly asked whether he thought that psychologists might one day start questioning those beliefs and leaving the profession in the way that dissenting priests had, some time ago, begun to leave the Church. He paused, thought for a moment, and then replied: "Not a chance. There's too much money in it."

On one level this summarizes what I mean by the Psychology Industry. Over 30 years ago, I walked into my first psychology class at McGill University in Montreal. My professor, Donald Hebb, was one of the most respected neuropsychologists of the century. He was fond of saying something which I have only recently come to appreciate. He kept insisting that psychology must be "more than common sense." Psychologists are obliged to go beyond what people commonly believe, to test out notions and see if they stand up under scrutiny. He insisted on science—on investigation, on the continuous questioning of beliefs. For almost three decades, I worked as a clinician, trying to apply the knowledge from my discipline. But psychology has changed. Humble curiosity has given way to an arrogant certainty. It seems that psychologists have discovered that questions don't pay, but answers do. What seemed once a responsible profession is now a big business whose success is directly related to how many people buy what it sells.

This is why I now speak of The Psychology Industry. When people think of industries, they tend to think of automobiles, computers, cosmetics, or entertainment; of easily identifiable products with price-tags, warranties, and trademarks. Such industries are visibly defined by their products and by their boundaries. The Psychology Industry, being much broader, less defined (or definable), is much harder to pin down. At its core, along with the traditional mental health professions of psychology, psychiatry, psychoanalysis, and clinical social work, is a fifth psychological profession: psychotherapy. No longer can clear distinctions be made between them; so, what I call the Psychology Industry comprises all five of these and it encompasses, as well, the ever-expanding array of psychotherapists and the counselors and advisors of all persuasions, whether licensed, credentialed, proclaimed, or self-proclaimed. This view is consistent with that of the American Psychological Association (APA): "The general public often has difficulty in understanding the differences between professional psychologists and other types of psychologists, between professional psychologists and psychiatrists, between psychologists and counselors, or between psychologists and a variety of other professionals who deal with emotional, health, and behavioral problems" (Fox, 1994, 49). As well, this term acknowledges that around the edges of the industry are others whose work, whether it involves writing, consulting, lecturing, or even movie-making, relies on the Psychology Industry which, in turn, benefits from their promotion of all things psychological.

Evidence of this current success and growth of the Psychology Industry can be seen in the number of Americans who have become users. In the early 1960s, only 14% of the U.S. population (25 million of a total 180 million) had ever received psychological services. By 1976, the estimate had risen to 26%, by 1990 33% (65 million of 250 million). In 1995 the APA stated that 46% of the U.S. population (128 million) had seen a mental health professional. Some predict that by the year 2000 users will be the majority, constituting perhaps as much as 80% of the population.

While some might consider this to be evidence of a profound national need of epidemic proportions, it can equally, and more accurately, be seen as an indication of the subtle but highly effective marketing techniques used by the Psychology Industry to generate the demand required to meet the ever-increasing supply of psychologists. As Jerome Frank, in his classic book on psychotherapy, *Persuasion and Healing*, observed decades ago (8):

> Ironically, mental health education, which aims to teach people how to cope more effectively with life, has instead increased the demand for psychotherapeutic help. By calling attention to symptoms they might otherwise ignore and by labeling those symptoms as signs of neurosis, mental health education can create unwarranted anxieties, leading those to seek psychotherapy who do not need it. The demand for psychotherapy keeps pace with the supply, and at times one has the uneasy feeling that the supply may be creating the demand.

While it is difficult to get an accurate reading of the total number of psychologists because of their diversity and the lack of any accountability or control over who represent themselves as psychologists or therapists, estimates are that the number has risen by 2000% since 1970. The following figures give some indication of the growth in one sector of the Industry licensed psychologists. There has been a steady increase in licensed doctoral psychologists and an even more rapid growth in APA membership. When these numbers are related to U.S. Census population data to show the number of licensed psychologists per 10,000 population, the increase in supply is dramatically evident. These licensed psychologists, however, constitute only one quarter of those who refer to themselves as "psychologists" and less than five percent of the estimated total number of the people who are actually selling psychological services. Using this broader definition, there is at least one psychologist for every 250 people in America.

What becomes immediately apparent is that not only has supply kept up with demand, it has, in fact, exceeded it, creating the need for greater marketing of psychological services and for the development of new "products" and the expansion of the markets. Figures showing the gross income of the Psychology Industry are impossible to come by, again because of its diverse sales force. However, when the data of the 1987 National Medical Expenditures Survey, the most recent of its kind, are extrapolated to 1995, 88.2 million outpatient psychotherapy visits were made by Americans for a total cost of 4.7 billion dollars. Considering that approximately half of the Psychology Industry consists of psychologists who cannot receive third-party insurance payments, it is not unreasonable to assume that both of these figures are much larger, probably in the region of 175 million visits at a total cost of $9 billion. This figure accounts only for direct patient services and does not include the cost for other services such as expert testimony in courts, which is a major growth area for the Industry.

Psychotherapy

Psychotherapy is the most visible and popular aspect of the Psychology Industry. When people think of psychology, they generally think of psychotherapy: the couch, telling secrets, reporting dreams, the psychotherapist saying "Uh-huh." Examining issues raised in the struggle between managed care and psychological practitioners, it is safe to say that there are two central issues:

1. Money—funding is being limited both with regard to the length of treatment and also to how much will be paid per session.
2. Control—psychologists are resisting the imposition of case managers or assessors who control and approve their services.

The Psychology Industry is arguing for unlimited funding and control, basing its position on two simple statements:

1. Psychotherapy works!
2. Long-term therapy works better!

To evaluate these claims, I direct your attention to two major studies addressing these points, starting with their conclusions. One study I will call the CR Study concluded:

1. Psychotherapy works: "our groundbreaking survey shows that psychotherapy usually works."
2. Long-term therapy makes a difference: "Longer psychotherapy was associated with better outcomes."

The other study I will call the FB Study concluded:

1. Psychological services may not work: "Clinical services . . . very effectively delivered . . . in a higher quality system of care that were nonetheless ineffective. A very impressive structure was built on a very weak foundation."
2. Longer term treatment isn't better: "more is not always better."

Based on these summaries, which study do you think the Psychology Industry chooses to publicize and promote? And why? Obviously, if I were in the business (as I used to be) I would want to tell everyone about the first, the CR Study, and hide the second, the FB Study. Let us examine both with an eye to what is revealed and what is concealed; or, as with snake oil, what is on the label and what is in the bottle.

The CR Study

CR stands for *Consumer Reports*, the magazine that talks about how satisfied consumers are with their vacuum cleaners and toasters. In November 1994, it reported on a "candid, in-depth survey" of its readers regarding their satisfaction with psychotherapy in an article entitled: "Mental Health: Does Therapy Help?" Martin Seligman, the psychologist who was the consultant to the project and is now the President of the APA, described the results in a companion article in the flagship journal of the APA, as sending "a message of hope for other people dealing with emotional problems," and as establishing a "new gold standard" for the evaluation of psychotherapy effectiveness. Before accepting his endorsement, let me just draw your attention to how the survey was done and how the results were interpreted.

The CR report and Seligman's article were based on the results of a supplement to the 1994 annual automobile survey sent to all 180,000 subscribers. Readers were asked to respond "if at any time over the past three years [they had] experienced stress or other emotional problems for which [they] sought help from any of the following: friends, relatives, or a member of the clergy; a mental-health professional like a psychologist, counselor, or psychiatrist; your family doctor; or a support group." It was, in the usual style of CR, a consumer satisfaction survey. It did not ask respondents objective, factual questions such as how much alcohol they drank before going for help as compared to after, or how many fights they had then and are having now with their spouses, or how often they thought of suicide then as compared to the past month. Nor did it seek independent verification of the self-reports. Instead it asked readers how much better they felt and how much they thought therapy had helped them. It was these responses that became distorted and translated into "convincing evidence that therapy can make an important difference."

Despite the broad invitation, only approximately 7,000 (3.9%) responded to the mental health survey; of these, 4,000 (2.2%) reported seeing a mental health professional, family doctor, or attending a support group; the remaining 3,000 (1.6%) had talked to a friend, relative, or clergy. For reasons that they will not make public, CR chose to ignore the experiences of this latter group of 3,000, and to attend only to the 4,000, with particular emphasis on the 2,900 (1.6%) who saw mental health professionals.

Seligman admits that this response rate, which, for some reason, he elevates to 13%, is "rather low absolutely." In fact, the response rate is only 2.2%, far lower than his figure and a rate which CR even admitted to being "very low." As well, this small sample consisted of individuals who were mostly middle class, well educated, predominantly female and with a median age of 46; thus, it was not representative of the United States as a whole or even of the general CR readership. Seligman dismisses this sampling problem by "guessing" that it is representative of those "who make up the bulk of psychotherapy patients," never giving further thought as to what this may mean both for the data and about the upper middle class nature of psychotherapy. In most other cases, such a low return rate and skewed population would have rendered a study invalid, not acceptable for publication and, therefore, not warranting any further analysis or comment.

But these inherent problems did not stop *Consumer Reports,* Seligman, or the Psychology Industry from proceeding to draw sweeping conclusions about the worth of psychotherapy. In reference to our two questions, this is how they analyzed their meager data.

Does Therapy Work?

Seligman's authoritative answer to this question is yes: "The overall improvement rates were strikingly high across the entire spectrum of treatments and disorders in the CR study."

Both the CR article and the subsequent marketing material from APA claim that nine out of 10 people were helped at least "somewhat" by psychotherapy. But, for psychotherapy to work, one needs people with problems. Such is not the case here. Over half of the respondents (58.2%) said that they felt "so-so," "quite good," or even "very good" before treatment. Seligman apparently doesn't scratch his head at this point and wonder whether these people are therapy junkies. Rather he views them as "being sick" and not knowing it, referring to them as "'subclinical' in their problems" and falling "one symptom short of a full-blown 'disorder.'" From a common sense, non-psychologized perspective, wouldn't these people be considered normal, "okay," or even in "great shape"? And wouldn't one wonder whether, for them, psychotherapy was more recreational than therapeutic? And, if so, how does one really know whether treatment is even appropriate, let alone whether it works?

To further add to the confusion, Seligman states, in support of his claim that "therapy works," that 64% of those receiving six months or less of therapy reported that their problems were resolved. However, his own chart would seem to indicate that, when an average is calculated across disciplines, only 30% of the people reported that treatment "made things a lot better" with respect to their specific problems. One is left wondering how is it possible that 64% reported that their problems were resolved when only 30% said that their problems were improved? Remember that the APA and CR both say that psychotherapy helped 9 out of 10 people. Given that Seligman failed to identify these inconsistencies when he declared the results to be "clear-cut" proof of effectiveness, he leaves one wondering how many other instances of misinformation exist in his article, and in that of CR.

Whether the figure is 30, 64, or 90 percent, CR and Seligman assume that the reported improvement in people's feelings while they were seeing a mental health professional was attributable to the psychotherapy. But can we accept this assumption? If people are given an antibiotic and their colds go away in a few weeks, can we conclude that the antibiotic cured the cold? We can't because we know that most people naturally get over a cold in a week or two. So too, we know that, like the common cold, the stresses and emotional upsets in life usually abate over time. Decades ago, the late Hans Eysenck demonstrated that, over time, people show comparable improvement with or without treatment. As well, the conclusion that therapy made the people better disregards the well-known phenomenon of "regression to the mean" which takes into account the high probability that people seek treatment at a time when they feel particularly bad and that, at a later point in time, they are likely to feel better. As Dawes points out, if "people enter therapy when they are extremely unhappy, they are less likely to be as unhappy later, independent of the effects of therapy itself. Hence, this 'regression effect' can create the illusion that the therapy has helped to alleviate their unhappiness, whether it has or not. In fact, even if the therapy has been downright harmful, people are less likely to be as unhappy later as when they entered therapy" (Dawes, 1994, 44).

To determine whether therapy was really effective for those in the CR survey, a comparison group is needed of people with similar problems who did not receive treatment. Such a group did exist, but for unexplained reasons CR chose to ignore those 3,000 respondents who spoke to friends, relatives, and the clergy. Although both groups did describe their emotional state at the time they filled out the survey, which would have given some indication of the effect of time, neither CR nor Seligman was willing, when repeatedly asked, to provide any further information or clarification or even to reveal whether these groups were similar. On all occasions, they refused, claiming that this data was proprietary and would not be analyzed or released. Seligman, in private communication, has made conflicting comments, on one occasion saying that he too would like to see the data, and on another attempting to assure this author that there was nothing of substance to be found there.

We are left wondering about Seligman's role and why CR will not report on this crucial data. If, in fact, professional treatment was superior to lay help, would not both parties want the public to know this, and if it is not more effective, does not CR, and the APA, have the responsibility to consumers to inform them that people are no more satisfied by paid services than by ones that are free?

Is Long-Term Psychotherapy Better?

The handling of the data with regard to this question can best be addressed by comparing two graphs. The first one, from Seligman's own article, visually suggests that the answer is obvious; the longer the therapy, the better the outcome. Seligman, in fact, stated: "long-term therapy produced more improvement than short-term therapy. This result was very robust . . ." (Seligman, 1995, 968).

But wait a minute! Notice the Y axis, the vertical one that measures the improvement. It is truncated so that it begins at 190, not 0 where you would expect it to start. The visual effect is to lead us to think that the change is small at the beginning and significantly greater over time. However, if this chart is accurately drawn this dramatic effect disappears, showing us that most of the "improvement" (80%), whatever that might mean, actually takes place in the first months; and further treatment of up to two years and more contributes only a further 20%.

The FB Study

FB stands for the Fort Bragg Demonstration Project, funded at a cost of $80,000,000 of public funds (Bickman, 1996; Bickman, et al. 1995). Cast in such glowing terms as: "a national showcase," "a truly unique opportunity," and "state of the art," this study was intended to show that "a continuum of mental health and substance abuse services is more cost-effective than services delivered in the more typical fragmented system" (DeLeon and Williams, 1997, 551). Where the CR survey suffered from the multitude of methodological problems and continues to be criticized for its numerous flaws, the only criticism lodged against the FB Project was that it had not been replicated, a weakness which was overcome by the results of a similar study in Stark County, Ohio, with similar findings at the six-month and two-year follow-ups.

The FB Project offered in-patient and outpatient services to the more than 42,000 child and adolescent dependents in the Fort Bragg catchment area for more than five years, from June 1990 to September 1995. This group of children was from middle and majority (an estimated 68%) of the children who are covered by private health insurance. And, most important, unlike the CR survey, the FB survey evaluated treatment effectiveness and outcome, not relying merely on reports of consumer satisfaction. Instead of questionable retrospective self-reports, this project relied on independent psychometric measures systematically taken both during and after treatment.

As such, the Project provided what psychologist Leonard Bickman, its senior researcher, described as "a rare opportunity to examine both costs and clinical outcomes in a careful and comprehensive evaluation of the implementation of an innovative system of care" which psychologists predicted would increase accessibility to treatment, improve results through individualized case management, and reduce overall costs.

However, what it found was that, despite better access, greater continuity of care, fewer restrictions on treatment, and more client satisfaction, the cost was higher and the clinical results no better than those at the comparison site: not at all what the Psychology Industry had either expected or wanted! Even though users expressed satisfaction about their treatment, there was no concurrent evidence of effectiveness, supporting the opinion that "satisfaction" is not a measure of effectiveness. In summarizing the significance of these results, Bickman and others drew the following conclusions:

1. The assumption that clinical services are in any way effective might very well be erroneous. Citing the lack of clinical outcomes as "the most unanticipated finding," Bickman stated that "these results should raise serious doubts about some current clinical beliefs" about the effectiveness of psychological services. He continued that "although substantial evidence for the efficacy of psychotherapy under laboratory-like conditions exists, there is scant evidence of its effectiveness in real-life community settings. For children and adolescents, the picture is even more disappointing. We have no evidence for the effectiveness of innovative community-based treatments such as home-based care or day treatment" (Bickman, et al. 1997, 1543–1548).

   This conclusion gains strength in light of the fact that there have been very few studies which have evaluated the effectiveness of treatment in real-world settings, and when these are analyzed, they show an average effect size very close to zero (Weisz, et al., 1995, 688–701). In another major study designed to seek out such evidence, Bickman's colleague, Bhar Weiss, carefully examined the effect of two years of traditional child psychotherapy as it is typically delivered in out-patient settings. What he found was not the expected benefits but rather no effect at all (Weiss, 1997).
2. Longer treatment results in higher costs without corresponding significant results. The Psychology Industry argues strenuously against the model that allows others, such as Managed Care Systems, to tell them what treatment they should provide or how long they should do it. Yet, the Fort Bragg data shows that what psychologists call their "experienced clinical judgment" was not cost-effective and led to a higher proportion of children being in treatment longer. "Six months after starting treatment, 41% at the Demonstration site were still receiving services compared to 13% at the Comparison site," even though most of the limited change that did occur was evidenced in the first six months with greatly diminishing returns after that time.

   Stating that "more is not always better," Bickman attributes these excessive costs to the unlimited access of psychologists to funds. "The Demonstration costs were much higher

> ($7,777/treated child) than the Comparison ($4,904/treated child) . . . The costs of treating the average child were higher because of longer time spent in treatment, greater volume of traditional services, heavy use of intermediate services, and higher per-unit costs" (Bickman, 1996, 694). Feldman agrees, stating that "the study demonstrates that in an unmanaged system of care when services and benefits become rich so do providers" (Feldman, 1997, 560).

The Seligman and *Consumer Reports* position that "longer is better" and that the public is suffering when limits are imposed on the length of therapy, is weakened by the data from the Fort Bragg Project. As Hoagwood, from the National Institute of Mental Health, said when referring to this Project, "the belief that simply providing more services will lead to improved outcomes has been shown to be delusional" (Hoagwood, 1997, 548).

The Psychology Industry's Reactions

What is the reaction of the Psychology Industry? If the Psychology Industry is scientifically and ethically motivated, then it would have to address at least the flaws and numerous confusions in the CR survey and the doubts raised by the Fort Bragg Project. On the other hand, if it is motivated by profit, the strategy would be to ignore Fort Bragg and enthusiastically endorse *Consumer Reports*.

While a senior executive in APA candidly identified the CR survey as a marketing and political tool, a phone call to the APA evoked a different reply from its Practice Directorate, responsible for promoting the practice of psychology and providing the public education program, a multi-million-dollar initiative designed to sell psychology to the American public. Ignoring that the APA's own President, Seligman, was a consultant to the survey, the individual at the Directorate stated that they "refer to that study whenever (they) can because it is particularly credible because it was done by an independent party." What about the Fort Bragg Project? When asked, they claimed that they knew nothing about it and expressed no interest, but added "in helping to educate the public, the CR study is best because it is written a lot more in consumer language because it is written for a magazine read by the general public. By disseminating to the public the information that is published in a journal, it needs to be transformed in a way that would be easily readable by the average person out there. That's the beauty of the CR piece." (Personal communication with the office of the APA Practice Directorate.)

As for the Fort Bragg Project, it will not be touted, as the *Consumer Reports* survey is, in their public education campaign. It is unlikely that clinicians will reduce or limit their treatment to conserve the scarce resources and limited insurance benefits. It is unlikely that it will be referred to by practicing psychologists when they speak of their worth or importance. It is unlikely that it will affect the way psychological services are developed or funded. It is unlikely that it will change the beliefs of those within the Industry for, although the Fort Bragg study is well designed, well implemented, well analyzed, and produces results that are about as clear-cut as can be imagined, it doesn't support the current claims of the Psychology Industry. "In the end," as Sechrest and Walsh put it, "what it comes down to is whether professional psychology is going to be guided by its dogma or its data" (Sechrest and Walsh, 1997, 536) or, put somewhat differently, whether it will use science to guide its action or misuse science to sell its products.

I do not want you to infer that these are the only two studies ever conducted or that the conclusions of the Fort Bragg project are being expressed now for the first time. One of the first of the studies to address the issue of psychotherapy effectiveness was conducted in 1952 by the British psychologist Hans Eysenck. He compared the outcome for patients who had received eclectic psychotherapy with those of people who had received no treatment. The results for the first group indicated that 64% showed improvement, a finding which initially seemed supportive of psychotherapy, for it was presumed that these patients would have remained the same or become worse if not treated. However, to everyone's dismay, Eysenck then took a look at the untreated group and discovered that 72% of them had improved by the second year. (Curiously, Eysenck's figure of 64% improvement in the treated group is the same figure that Seligman reported 45 years later as showing improvement from therapy. This leaves one wondering if the data protected by CR's claim of proprietary rights also shows similar outcome with no treatment.) Despite the lack of any specific treatment Eysenck's second group showed an overall 90% recovery in five years. In a subsequent, more extensive, study in 1965, he concluded that psychotherapy was unessential to a patient's recovery: "We have found that neurotic disorders tend to be self-limiting, that psychoanalysis is no more successful than any other method, and that in fact all methods of psychotherapy fail to improve on the recovery rate obtained through ordinary life experiences and nonspecific treatment" (Eysenck, 1965).

Some have challenged these findings, claiming that they were unfair, or not sufficiently controlled to be considered scientific. Whether or not these criticisms have weight, Eysenck's studies served as a gauntlet challenging others to more closely examine the claims of psychotherapy. Thus began what Ellen Herman (1995) has described as a sub-industry within psychology—the psychotherapy evaluation business which still flourishes despite no firm proof that psychotherapy works. And psychotherapy continues to be promoted despite the many well-conducted studies which find little evidence for its specific effectiveness, and even data that suggest that it could be harmful.

Let me draw your attention to one of these, a classic study which examined the results of a number of other studies. In a review of therapy factors that account for significant client progress, Lambert calculated the percent of improvement that could be attributed to each of several variables (Lambert, 1986). He found that "spontaneous remission" (improvement of the problem by itself without any treatment) accounted for 40%, an additional 15% of the change resulted from placebo effects (which he referred to as "expectancy controls," that is, that the patient expected to get better no matter what was done), while a further 30% improved as the result of common factors in the relationship, such as trust, empathy, insight, and warmth. Only 15% of the overall improvement could be attributed to any specific psychological intervention or technique. Based on these findings one could conclude that 85% of clients would improve with the help of a good friend, and 40% without even that.

The Noble Lie

If these findings are true, then why is psychotherapy given so much credit? To answer this I think that we need to return to our image of psychotherapy as snake oil. Like its forerunner, psychotherapy can make people feel satisfied, if sometimes only briefly, because they have been listened to and made to feel important. Like snake oil salesmen, psychologists have a good sales spiel. Consider the following statement by Kottler, the author of *On Being a Therapist* and numerous other books on psychotherapy (108):

> Telling clients that we can help them is assuredly fuel; even if it is not strictly true. . . . By communicating confidence, however false it might feel, we establish hope and motivation in the client. We would lose clients very quickly if after every bungled interpretation . . . we muttered "Oops, I blew that one." We would never get a client to come back if we were completely honest with them . . . the client may need to believe in this lie.

Some forms of deception and lying have always been a part of psychological practice, sometimes in the form of suggestive therapies, sometimes in the declarative but unfounded statements of psychologists, and sometimes in misleading advertising.

When confronted by moral objections to the deception of patients, Pierre Janet, a contemporary of Freud's, responded:

> I am sorry that I cannot share these exalted and beautiful scruples. . . . My belief is that the patient wants a doctor who will cure; that the doctor's professional duty is to give any remedy that will be useful, and to prescribe it in the way in which it will do most good. Now I think that bread pills are medically indicated in certain cases and that they will act far more powerfully if I deck them out with impressive names. When I prescribe such a formidable placebo, I believe that I am fulfilling my professional duty (Janet, 1925, 338).

Janet's (and Kottler's) assumption was that patients want and need to be treated as children by paternal and protective, if not always honest, therapists, and that it is in the best interest of these patients to lie, for "there are some to whom, as a matter of strict moral obligation, we must lie."

Thus deception, justified in terms of benefit to the user, has become an acceptable practice and a cornerstone of the Psychology Industry. For as Kottler wrote: "Certain lies may therefore be necessary, if not therapeutic. If lying to a client, deliberately or unintentionally, is unethical since it promotes deceit and deception, perhaps it is just as unethical to be completely truthful."

Whether expressed in terms of creating positive expectations which are believed to be essential for a good therapy outcome, or fostering unconditional acceptance and positive regard, or giving unquestioning support to a claim of abuse, the Noble Lie has become acceptable in the Psychology Industry. It has simply become an aspect of doing business. When Dan Sexton, Director of the National Child Abuse Hot Line, was questioned in this regard, he responded (Sexton, 1989):

> I'm not a law enforcement person, thank God! I'm a psychology person, so I don't need the evidence. I come from a very different place, I don't need to see evidence to believe . . . I don't care what law enforcement's perspective is, that's not my perspective. I'm a mental health professional. I need to find a way to help survivors heal to the trauma that they had as children and to help support other clinicians who are trying to help survivors and victims of this kind of crime.

For these, and many other psychologists, it doesn't matter whether facts are true or whether what they say is honest, what matters is that the consumers believe them. Alan Scheflin, a lawyer and law professor, in addressing a conference on hypnosis and psychotherapy, went even further when he encouraged psychologists to consider it their ethical responsibility to intentionally deceive their clients (Scheflin, 1995):

> The point I want to make is the assumption that implanting false memories is wrong I would like to raise the issue of whether we are right to say it is wrong. . . . When we get through the false memory issue perhaps we

> can start to debate the serious question . . . that therapists are in fact social influence purveyors and it is your job to use those techniques. And hypnosis will lead the way into the social influence literature. And then we can start to talk about the ethics of using false memories therapeutically.

Thus, to Scheflin and to the many psychologists who gave him a standing ovation, the end justifies the means even if the means is to mislead, deceive, and lie to the user, and to create a false history of her or his life. Perhaps another reason that Scheflin got such a rousing round of applause was that he was promising psychologists that soon "there will be a point—though there has not been one yet" when they would find the power that "would make therapists more effective in treating the problems of the patient" (Scheflin, 1994, 202). His message was that the power to change people, to create not only good memories but good (albeit false) identities, was soon to be discovered; that although psychologists may feel insecure about their abilities, they need not worry because the techniques to influence, persuade, and change people were being developed. His message was encouraging to the many psychologists who carry on their daily practice of professional deception, projecting an image of themselves as confident and self assured so that their clients will be satisfied customers.

The results of psychotherapy research are so equivocal and fraught with doubt that no one can honestly say even that it has a positive effect, let alone say that more is better. As well, there are sufficient data to warrant the caution that in some, if not many, instances it may even be harmful, actually increasing, prolonging and even creating the problems it is thought to alleviate. The recent exposés of recovered memory therapy may be showing merely the tip of the iceberg of damaging effects.

At best, psychotherapy may be the simple provision of human caring, empathy, sense of worth and source of optimism; the "purchase of friendship." And, like other nostrums, selling it may require enthusiastic exaggeration, and unscrupulous deception.

While I leave you now to make your own decision as to whether psychotherapy is the "snake oil of the 90s," I want to draw your attention to Seligman's own words of warning to consumers, written some years before the CR study and his election to the presidency of the APA. While he may now prefer to ignore his statement, it speaks loudly and clearly on this matter (Seligman, 1994, 8):

> Making up your mind about self-improvement courses, psychotherapy, and medication . . . is difficult because the industries that champion them are enormous and profitable and try to sell themselves with highly persuasive means: testimonials, case histories, word of mouth, endorsements . . . all slick forms of advertising.

References:

Bickman, Leonard. 1996. "A Continuum of Care: More Is Not Always Better." *American Psychologist*, 51(7).

——, Guthrie, P. R., Foster, E. M., Lambert, E. W., Summerfelt, W. T., Breda, C. S., and Heflinger, C. A. 1995. *Evaluating Managed Mental Health Services: The Fort Bragg Experiment.* New York: Plenum.

——, Summerfelt, William Thomas, and Noser, Kelly. 1997. "Comparative Outcomes of Emotionally Disturbed Children and Adolescents in a System of Services and Usual Care." *Psychiatric Services*, Dec. 48(12).

Consumer Reports. 1995. "How Is Your Doctor Treating You?" February: 81–88.

——. 1995. "Mental Health: Does Therapy Help?" November: 734–739.

Dawes, Robyn M. 1994. *House of Cards: Psychology and Psychotherapy Built on Myth.* New York: Macmillan.

DeLeon, Patrick H., and Williams, Janice G. 1997.

"Evaluation Research and Public Policy Formation: Are Psychologists Collectively Willing to Accept Unpopular Findings?" *American Psychologist*, 52(5).

Dineen, Tana. 1998. *Manufacturing Victims: What the Psychology Industry is Doing to People*. Montreal: Robert Davies Multimedia Publishing.

Eysenck, Hans J. 1952. "The Effects of Psychotherapy: An Evaluation." *Journal of Consulting Psychology*, 16, 319–324.

——. 1965. "The Effects of Psychotherapy." *International Journal of Psychiatry*, 97–168.

Feldman, Saul. 1997. "The Fort Bragg Demonstration and Evaluation." *American Psychologist*, 52(5).

Fox, Ronald E. 1994. "Training Professional Psychologists for the Twenty-First Century." *American Psychologist*, 49(3).

Frank, Jerome D. 1961 (1973). *Persuasion and Healing: A Comparative Study of Psychotherapy*. Baltimore, MD: The Johns Hopkins University Press.

Hagen, Margaret A. 1997. *Whores of the Court: The Fraud of Psychiatric Testimony and the Rape of American Justice*. New York: Regan Books, HarperCollins.

Herman, Ellen. 1995. *The Romance of American Psychology: Political Culture in the Age of Experts*. Berkeley: University of California Press.

Hillman, James, and Ventura, Michael. 1992. *We've Had a Hundred Years of Psychotherapy and the World Is Getting Worse*. San Francisco: HarperCollins.

Hoagwood, Kimberly. 1997. "Interpreting Nullity: The Fort Bragg Experiment–A Comparative Success or Failure?" *American Psychologist*, 52(5), 548.

Huber, Peter W. 1991. *Galileo's Revenge: Junk Science in the Courtroom*. New York: Basic Books.

Hughes, Robert. 1993. *Culture of Complaint: The Fraying of America*. New York: Oxford University Press.

Janet, Pierre. 1925. *Psychological Healing: A Historical and Clinical Study*. Translation by Eden and Cedar Paul. Vol. II. New York: Macmillan.

Kottler, J. 1987. *On Being a Therapist*. San Francisco: Jossey Bass Publ.

Lambert, M. J. 1986. "Some Implications of Psychotherapy Outcome Research for Eclectic Psychotherapy." *International Journal of Eclectic Psychotherapy*, 5(1), 16–44.

Revel, Jean-Francois. 1991. *The Flight from Truth*. New York: Random House.

Scheflin, Alan. 1994. "Hypnosis 1994 and Beyond." *Hypnos*, 21(4).

——. 1995. "Plenary Address at the Frontiers of Hypnosis Conference." The Fourth National Assembly of the Federation of Canadian Societies of Clinical Hypnosis, Banff, Canada, May.

Sechrest, L., and Walsh, M. 1997. "Dogma or Data: Bragging Rights." *American Psychologist*, 52(5), 536.

Seligman, M. 1994. *What You Can Change and What You Can't*. New York: Knopf.

——. 1995. "The Effectiveness of Psychotherapy: The *Consumers Reports* Study." *American Psychologist*, 50(12): 965–974.

Sexton, D. 1989. "Gaining Insights into the Complexity of Ritualistic Abuse. The Eighth National Conference on Child Abuse and Neglect." Tape #28.

Showalter, Elaine. 1997. *Hystories: Hysterical Epidemics and Modern Media*. New York: Columbia University Press.

Smail, David. 1987. *Taking Care: An Alternative to Therapy*. London: J. M. Dent and Sons.

Weiss, Bhar. 1997. "The Effectiveness of Traditional Child Psychotherapy." Unpublished report of the Center for Mental Health Policy, Vanderbilt University.

Weisz, J. R., Donenberg, G. R., Hans, S. S., and Weisz, B. 1995. "Bridging the Gap between Lab and Clinic in Child and Adolescent Psychotherapy." *Journal of Consulting and Clinical Psychology*, 63, 688–701.

Zilbergeld, Bernie. 1983. *The Shrinking of America: Myths of Psychological Change*. Boston: Little, Brown and Co.

Pyramids

The Mystery of Their Origins

PAT LINSE

Every year 2 million tourists travel to northern Africa to a limestone plateau a few miles west of Egypt's mighty Nile river. There they gaze in wonder at ancient ruins that have amazed people for almost 5000 years—the Pyramids of Giza.

The ruins at Giza have been a tourist destination since the time of the Roman empire. Interest in the pyramids and the culture of ancient Egypt reached new highs in Europe when Napoleon Bonaparte invaded Egypt with his army in 1789. Although his military objective failed, he had also brought with him a small army of scholars, surveyors, and artists to study the ruins of Egypt. The excitement their work produced in Europe inspired still more research and exploration. One of the discoveries of the Napoleonic expedition was the famed Rosetta stone that allowed the ancient picture writing of the Egyptians to be read for the first time in 1,500 years.

The ability to read what the ancient Egyptians had to say about themselves provided the most accurate view of ancient Egyptian society that anyone had seen for many centuries.

Today ancient Egyptian culture continues to fascinate us. Scholars studying the ruins still make worldwide headlines with every major discovery. Countless books, television documentaries, and even a number of blockbuster films have featured ancient Egypt. There is even a pyramid (symbolizing durability) printed on the back of every U.S. one dollar bill.

Why Do People Seem So Fascinated by Pyramids?

The sheer size of the pyramids alone was enough to attract attention and inspire wonder. For much of its history the largest of them, the Great Pyramid of Giza, was the tallest structure in the world. In the ancient world only the legendary lighthouse at Alexandria was said to have been taller. It was not until 4,500 years later in 1889, with the invention of steel beam construction, that the Eiffel tower of Paris, France, rose higher. The Great Pyramid is still one of the most massive structures ever built. Just how big was it? The Great Pyramid was 480 feet high. It is said to have been built from 2½ million blocks of limestone, each averaging 2½ tons (5,000 pounds).

The great age of the pyramids fascinates people, and lends a veneer of ancient wisdom to any belief system that claims to share the knowledge of the civilization that built them. The pyramids were already 1,000 years old during the time that the biblical story of Moses and the exodus from Egypt was thought to have taken place. They were already 2,500 years old during the time of Christ. It is not surprising they have always been considered one of the wonders of the world.

The fact that the pyramids are pure geometric shapes also seems to fascinate people.

The Great Pyramid from the top

If they had been built in the form of an enormous perfect cube, or a giant sphere, people would probably find them equally mysterious. A pure shape seems to have a hidden relationship to the rest of the universe because various aspects of it can be described by mathematical formulas. Modern architects often use pure shapes such as these when they want to express profound or futuristic themes.

Many people around the world have built

(Above) Futuristic architecture at the 1939 New York World's Fair

(Above center) A tower in the sacred district of the capital of Medieval Sri Lanka. (Above right) Mayan temple–Guatemala. (Center left) Silbury Hill, a 4500 year old artificial mound–England. (Bottom left) Aztec "Pyramid of the Moon" –Mexico

pyramids for religious purposes, often with the idea of building a platform to get closer to the heavens. Some people assume that all pyramid builders must have gotten their idea from the same place. But it is a universal human intuition that gods are to be found in the sky. The pyramid shape is simply the most practical way to build a high structure if your building method consists of piling up rocks or earth. Even nature prefers a pyramid shape. When erosion wears landforms down into mountains, a stable triangle shape is the result.

We know from what they wrote that the Egyptians attached many symbolic meanings to their pyramids.

The Evolution of the Pyramid

Did the Egyptians need help from space aliens to build the pyramids? Some authors say they did. They claim that the largest of the pyramids appeared suddenly in a primitive culture that had never built anything like them.

This is simply not true. The concept of the pyramid developed slowly along with religious ideas about the afterlife. Once the idea was perfected, the Egyptians did not continue to build gigantic pyramids. The largest of the pyramids were built during weather cycles that generated large food surpluses. Cycles of drought made Egypt too poor to build large monuments. Also, the Egyptian pharaohs eventually lost faith in the ability of the frequently robbed pyramids to protect their mummies. They began to be buried instead in hidden underground chambers.

Before Pyramids

A mound symbolized rebirth to the Egyptians who saw fresh new plants sprout every year from the muddy high ground that emerged from flood waters of the Nile river. The burial mound eventually developed into a squarish structure called a "mastaba." The mastaba was a house for the spirit of the dead person to live in.

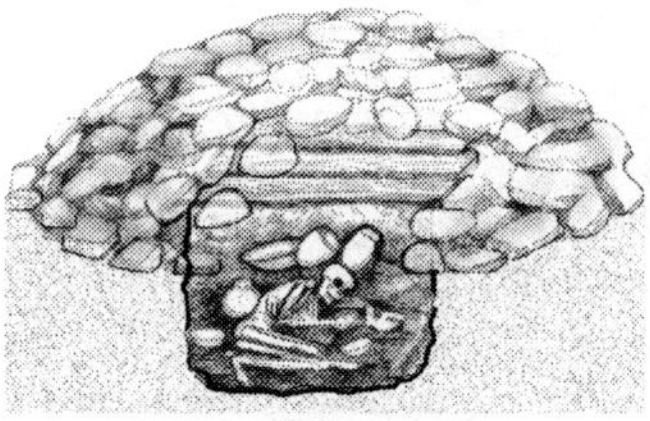

1. X-ray view of an early burial mound.

2. X-ray view of a mastaba. Mastabas could be quite elaborate with rooms and courtyards for the spirit.

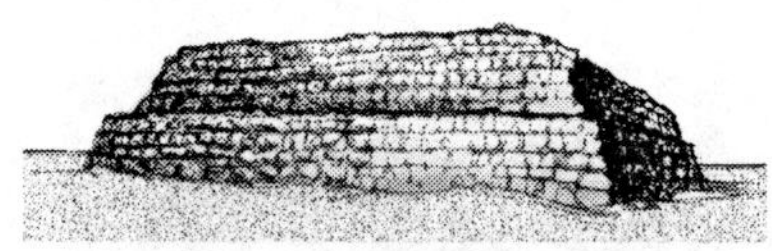

3. A two-layered mastaba.

4. The ruins of a royal burial site—a gigantic many-roomed mastaba surrounds a stepped mound that covers the tomb.

5. The Step Pyramid—The First Pyramid

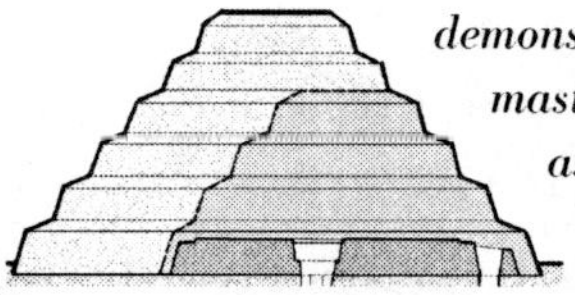

The interior structure of the Step Pyramid demonstrates the leap from a stepped mastaba to a pyramid shape. It started as a mastaba enclosed in a courtyard. This was later enlarged to a four-level mastaba, and then it was rebuilt as a 6-level mastaba. The rock under the Step Pyramid's courtyard is honeycombed with 400 rooms connected by 3 1/2 miles of tunnels!

6. Sneferu's Three Pyramids

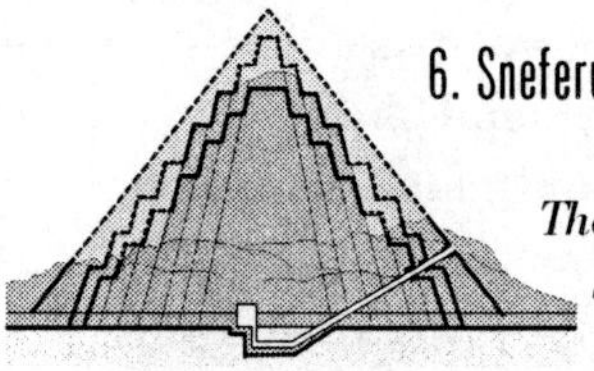

The pyramid at Meidum—Sneferu's first pyramid, now in ruins, was originally a step pyramid. He later rebuilt it with straight sides after he finished the Bent and Red pyramids.

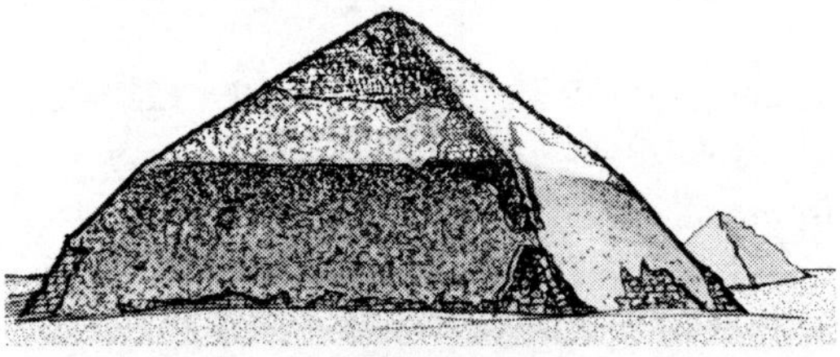

7. The Bent Pyramid—Sneferu's builders changed the steep angle of the sides when the pyramid structure began to fail.

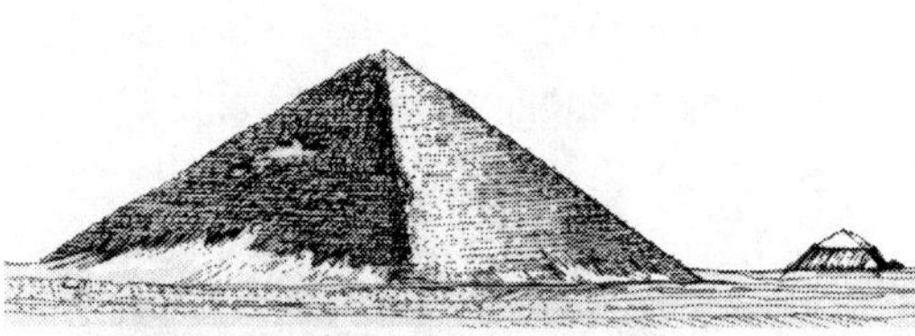

8. The Red Pyramid—Sneferu learned from his mistakes and this is a carefully planned and well-built pyramid.

9. *Left to right: (Left) Menkaures' small pyramid at Giza placed more emphasis on its temple rather than the pyramid's size. (Center) Khafre, the son of Khufu, built the middle pyramid and the Sphinx, a lion statue with the head of a man. (Right) The Great Pyramid, built by Sneferu's son Khufu, is the largest ever built at 480 feet tall.*

After Giza

Pyramids were built for another 1000 years, but none achieved the fame of the ones at Giza. Later pyramids often had cores of mudbrick or poor quality stone. When their outer shell of protective limestone was stolen, they gradually crumbled into unrecognizable blobs.

Small pyramids became popular as part of private tombs two hundred years after the last royal pyramids were built. For 800 years no royal pyramids were built in Egypt. The practice was then revived by the Nubian King Piye, who ruled all of Egypt in the 25th dynasty. The Nubians built almost 200 pyramids.

The Giza Pyramids—They've Been Robbed!

Almost every pyramid was soon robbed of the treasure put in its chambers for use in the afterlife. Pharaohs eventually lost faith in the pyramid's ability to protect their mummies, and they began to use hidden grave sites. The pyramids were also eventually robbed of their outer shell of high quality building stone. The Giza pyramids had an inner core of stone, so they kept their basic shape.

10. *The ruined mudbrick core of Amenemhet III's first pyramid hides an elaborate tunnel system.*

11. *Private tombs*

12. *Nubian pyramids*

Why the Egyptians Built the Pyramids

The Pyramid Complex

When people in our culture hear that the pyramids were "the tombs of the pharaohs" they often suppose that they were like our grave stones—a marker in a quiet place with an occasional visitor.

But the pyramid was only one part of a large group or "complex" of structures that served as much more than a tomb. These structures were a religious site where rituals were performed to transform the dead Pharaoh's body into a mummy, and to prepare him for a god-like existence in the afterlife. The Egyptian concept of a ghost or soul was more elaborate than ours and separate ceremonies were necessary for each of the soul's three distinct parts. To support a pharaoh's activities in the afterlife, hundreds of priests took turns continuing the daily rituals long after he died. An entire town was necessary to keep up this activity.

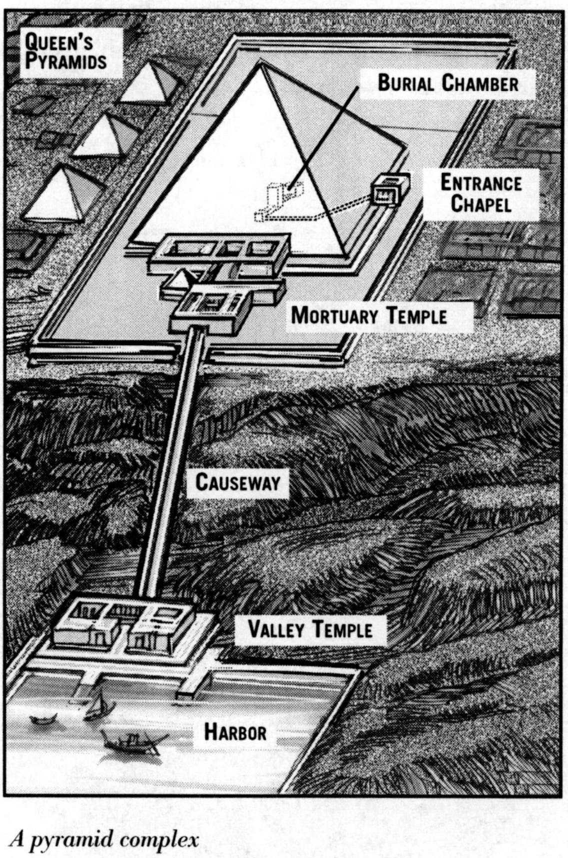

A pyramid complex

The pyramids were called "Houses of Eternal Life" and they were built to ensure that the pharaoh would become immortal. Egyptians believed that if the pharaoh became immortal, all of the people that lived during his reign would be immortal as well. This is a critical point to understand. The promise of immortality for the common people was probably what led them to organize and channel nearly all the surplus wealth of their society into building pyramid complexes. During its construction, a pyramid was the focal point of Egyptian society. From the evidence so far uncovered of complete towns surrounding some of the pyramid sites, it appears that these areas also served as an administrative center or capital of the country.

Egyptians named their pyramids in a way that showed that the god-like characteristics of the pharaoh became associated with the structure itself. Here are some examples of names given to pyramids (the words in italics are the names of pharaohs): *Menkaure* is Divine; Pure are the Places of *Userkaf;* The Perfection of *Pepi* is Established; *Pepi* is Established and Living; *Amenemhet* Lives.

Cracking a Real Code: Hieroglyphics and the Rosetta Stone

In 1822 a Frenchman named Jean-François Champollion played a major role in rediscovering how to read the mysterious Egyptian picture-letters known as hieroglyphics.

Hieroglyphics were difficult to decode because they were not a simple phonetic alphabet like ours, nor were they written in a single direction. A good working knowledge of hieroglyphics required knowing 200 to 400 symbols out of the approximately 3000 symbols available. Some pictures stood for individual let-

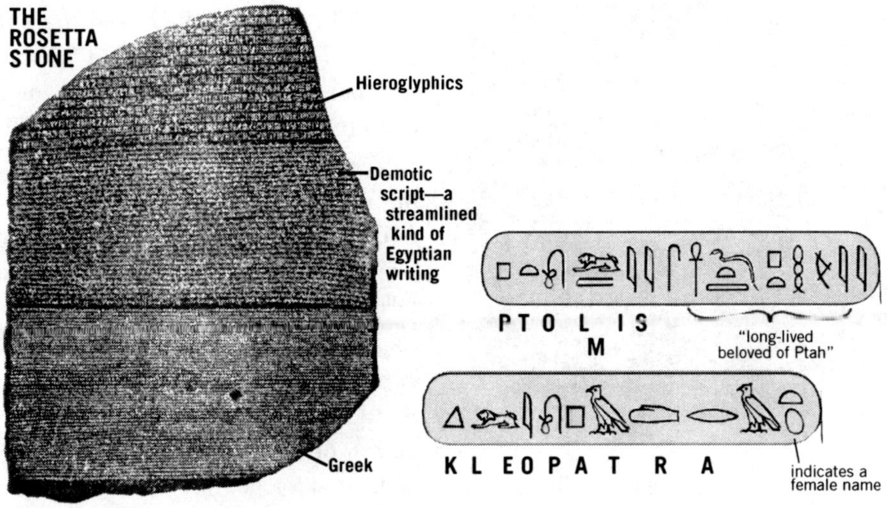

The Rosetta Stone and the Ptolemy and Cleopatra cartouches

ters. Others stood for whole words, phrases, or syllables. Other symbols were placed at the end of a word to further define it.

Hieroglyphics could also be written in any direction—up, down, left or right. The symbols that show people or animals indicate in which direction the writing is to be read. You start from the direction the figures are facing.

Champollion cracked the code using the Rosetta stone because it had the same inscriptions carved on it in three languages. One section was hieroglyphics and another was Greek—a language which Champollion could read. He guessed that rounded boxes known as cartouches within the hieroglyphic section contained names. He suspected the cartouches contained the name Ptolemy (spelled in Greek as "Ptolemaios") since that Pharaoh's name appeared many times in the Greek section.

Another inscription on another stone contained a cartouche that he suspected spelled out the name "Cleopatra." Selecting the letters that both names had in common—"L," "O," "P," and "T," he looked to see if the hieroglyphic symbols matched. They did. That also gave him clues as to what the other symbols might be, and he was on his way to unraveling a 1500 year old mystery.

What Did the Pyramid Symbolize?

The pyramid shape was rich in religious meanings for the ancient Egyptians. One of their earliest stories about the creation of the world involved a primeval mound that emerged from bottomless waters. This is not a surprising creation image for a people who were used to months of flooding, who waited for the land to reemerge, and for plants to be regenerated. An ancient sun god was said to have set this creation in motion, so the pyramidal mound was associated from earliest times with both the earth and sky.

Some Egyptologists think the pyramid shape itself also represented the sun's rays as they

Sun's rays over the pyramid

Pyramid interior with text on walls

are seen breaking through clouds. The sun's rays were described as a ramp that the pharaoh used to ascend to heaven. This image perfectly matches the concept of a pyramid as a means by which a pharaoh is transformed from an earthly to a heavenly being.

The Pyramid Texts are writing on the walls and ceilings of the chambers inside some pyramids. They explain the Egyptian concept of death and the afterlife, and the purpose of the pyramid. They are some of the earliest religious writings known and include myths, spells, hymns, and poems, as well as burial ceremonies.

The Nile and Its Pyramids

There are about 300 pyramids strung along the length of the Nile river. They start at the point where the Nile fans out before emptying into the Mediterranean Sea, and continue up the Nile to the ancient black kingdom of Nubia, where about 2/3 of them are found.

To understand ancient Egypt you have to understand what the Nile meant to the culture.

The Nile was a green ribbon of life cutting through a barren desert landscape. The dense population of ancient Egypt depended on the yearly cycle of floods rather than on rainfall to water their crops. The Nile started to rise in mid-July and slowly covered the entire river plain with its smooth glass-like waters until October. The fields were enriched as silt settled out of the standing water.

The Nile provided a bountiful living that allowed the Egyptians to take time out for spiritual considerations such as concern for the afterlife. They were able to organize the workforce to build their monuments because their society was already organized around the irrigation systems that were necessary to bring flood water into their fields, hold the water, and to drain it out again after it had dropped its load of fertile silt.

The flood months also allowed the farmers spare time to work on the pyramids without disrupting the harvest. The periods of Egyptian history that are famous for magnificent cultural achievement—like the Fourth Dynasty when the Giza pyramids were built—had long periods of stable and consistent flooding. For the most part the Nile performed reliably and gave the ancient Egyptians a comfortable life. However, too many years in a row of low flooding and dry fields brought starvation. Too much water left too long in the fields delayed planting, which meant that crops did not form seed before they were dried up by the hot summer sun.

Historically, periods of floods that were too high or too low are associated with periods of social decline and collapse in ancient Egypt.

Ancient Theories

The antiquity and great size of the Giza monuments lent authority to anyone who successfully claimed an association with their builders.

- One thousand years after they were built Pharaoh Tuthmosis IV claimed that as a prince he had fallen asleep in the shade of the Sphinx and it had spoken to him

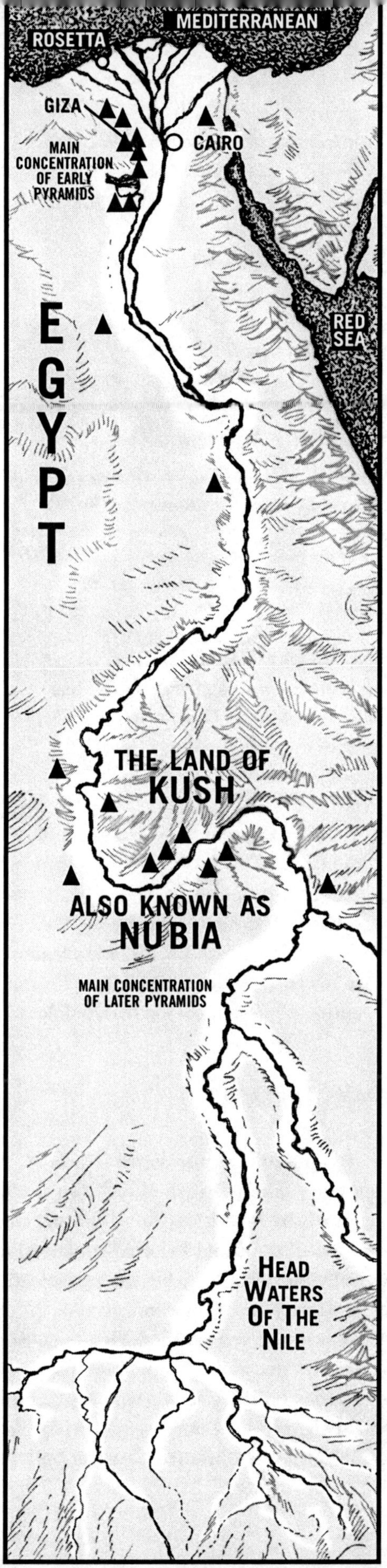

to tell him that he would someday be King. Since his older brother was actually first in line to be Pharaoh, he needed this miracle story to reinforce his claim to the throne.

- Two thousand years after the Giza pyramids were built, a cult of priests still worshiped the Pharaohs that built them as gods.
- Two thousand six hundred years after they were built, the Jewish historian Josephus mistakenly claimed that Hebrew slaves had provided the labor.
- In a blend of Greek, Egyptian and Middle Eastern lore, it was claimed that an ancient king or, in some versions, the Greek god Hermes built the pyramids to hide secret knowledge from the unworthy and to protect that knowledge from a great flood.
- In later centuries, the Arabs said fabulous objects were hidden inside—weapons that never rusted, glass that could bend without breaking, and a vase that always poured water and never ran empty (a marvel indeed to desert dwellers).
- Of course treasure was also said to be hidden there—objects made of rubies, gold and precious stones—a rumor based perhaps on the knowledge of real treasure buried with mummies.
- Christian medieval Europe incorrectly thought the Pyramids were the grain storage bins of Joseph of biblical fame.

Modern Alternative Theories

Amazing stories about the pyramids are still told today. While these alternate theories disagree about who planned the pyramids, they all agree that it was not the ancient Egyptians.

Map of the Nile and its pyramids

They argue that the Egyptians were too "primitive" to build the pyramids without help. The theories usually center around the Great Pyramid, which is said to contain advanced knowledge hidden in its measurements—knowledge that the Egyptians could not possibly have known about.

Oddly enough these alternative theories came from the same historic event that inspired increased scientific knowledge of Egypt's past—Napoleon's 1789 invasion. Two camps of opposing viewpoints arose—those in the academic/scientific world, and those who sought alternative or spiritual explanations.

Though many of the promoters of alternative theories rejected the discipline of science, they still liked to look scientific by using measurements and mathematical calculations to "prove" their ideas.

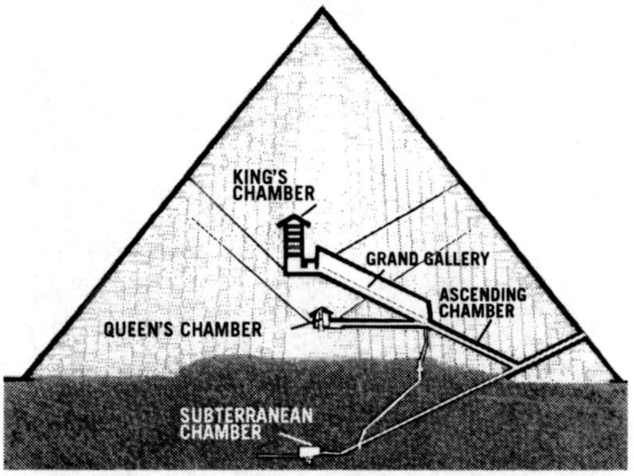

Pyramidology analyzes the interior chambers of the Great Pyramid. The floor of the Queen's chamber in the Great Pyramid meets the angled floor of the ascending chamber at a point 33 1/2 inches before the end of the passage—this represents the date of the Crucifixion, exactly 33 1/2 years after Christ's birth. If this measurement is taken as the hypotenuse of a right triangle, the base of that triangle is 30 inches—the exact number of years from Christ's birth to his baptism.

The 1800s—Pyramidology

Pyramidologists used elaborate measurements to discover hidden codes built into the halls, chambers and dimensions of the Great Pyramid. They tried to link every measurement to events in Christianity in an attempt to prove that their unique interpretation of what the Bible said was literally true. The two examples of this kind of thinking below reached dramatically different conclusions.

- Noah and his sons built the Great pyramid of Giza under the guidance of God. It is really "the Bible written in stone." Measurements of its chambers and passages are a record of the past and can also be used to predict the future, Christ's Second Coming, and the end of the world. (The Pyramidology movement was inspired by John Taylor's book *The Great Pyramid: Why Was It Built and Who Built It,* written in 1859.)
- The Great Pyramid of Giza was built by the devil. (This was the idea of Joseph T. "Judge" Rutherford, a leader of the early Jehovah Witness movement, in *The Watch Tower,* Nov. 28, 1928 in response to his predecessor Charles Taze Russell's enthusiastic support for Pyramidology.)

Pyramidology and Racism

The pyramidology movement gained respectability when British Royal Astronomer Charles Piazzi Smyth declared that the measurements of the Great Pyramid contained not only religious prophecy, but all mathematical and geological knowledge conceivable.

Of course the self-satisfied Victorians could not imagine the idol-worshipping brown people of Africa building such a wonder. They decided their own ancestors were the Israelis of the Bible who had invaded Egypt to build the

Great Pyramid under God's direction. The idea that the ancient Anglo-Saxons rather than the Jews were the Biblical God's real chosen people is still used today by some white supremacist groups to justify racism.

The 1900's Superior Civilization Theories

Theories of the 1900s also found hidden codes. The gods and kings of ancient stories were replaced with beings of superior intelligence who possessed advanced technology. Many of these theories sound like science fiction because they are often a reaction to the same rapid technological changes and social anxieties that inspired Sci Fi itself.

These contemporary stories express modern hopes and fears. They suggest the secret codes are warnings about the need to change our ways because of moral decay, the coming end of the world, nuclear war, or environmental pollution.

- A spaceship of defeated survivors of a galactic war escaped to our solar system and interbred with our primitive ape-like ancestors to create a new species—us! They provided us with pyramid building technology—laser beams to cut huge stones, anti-gravity machines to move them, and radioactive paste to fuse them together. (From *Chariots of the Gods?* by Erich von Däniken in 1968.)
- Structures on Mars and Earth are a great cosmic blueprint created by alien civilizations. Mathematical relationships connect Giza to an area on Mars called Cydonia where there are pyramids and the Sphinx-like face on Mars. (From *The Mars Mystery* written by Graham Hancock in 1998.)
- People from the lost continent of Atlantis used advanced technology (lost when Atlantis sank into the ocean) to build the pyramids.
- Egyptian priests used psychic power—now lost to our modern scientific mindset—to move the great stones.

These current theories still depend upon some of the old discoveries of pyramidology that relate pyramid measures to astronomy, geology, and mathematics. The famous best-selling book *Chariots of the Gods?* claimed that the height of the pyramid multiplied by a billion equals the distance from the earth to the sun, or 98,000,000 miles. (Actually the earth makes an oval path around the sun so the distance between them varies. That distance is usually averaged as 93—not 98—million miles.)

Another "proof" states that if you start at the north pole and draw a line straight down through the Great Pyramid to the south pole, your line will travel over more land than anywhere else on the globe. (Why placing a pyramid on a line that can travel over the most land mass of anywhere on earth is so significant has never been explained.)

But a favorite way to prove an advanced intelligence built the Great Pyramid is to point

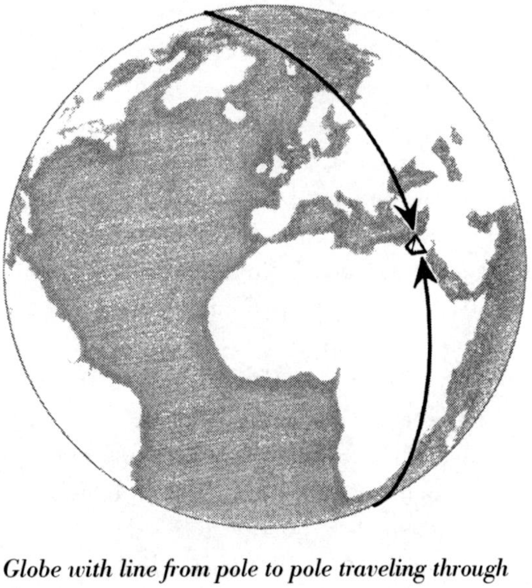

Globe with line from pole to pole traveling through Giza.

out that the distance around the base of the Great Pyramid divided by twice its height gives a special number called "pi"—3.14. First of all just because a number is discovered in pyramid measurements doesn't mean it was deliberately hidden there. (This is explained below in the section called "The Number Game.") Also, pi was well known in ancient cultures, although they had a slightly less accurate number for it than we do. Babylonians used 3.125 for pi, and the ancient Egyptians used 3.1605.

Technical stuff about pi (also written as the Greek letter "π"):

- The distance around the outside of any circle is about 3.14 times greater than the distance straight through the center of that circle. This number was important for ancient builders and surveyors who needed to be able to calculate the area of circles to compare them to square spaces. Pi fascinates people because it never comes out exactly even. Mathematicians love the game of finding ever more accurate values for it. The last we heard, a computer calculated pi to over 51 billion decimal points. Many ancient cultures used an even 3 for pi. Modern engineers usually don't need more than 7 decimal points, and scientists rarely need more than 20.
- Some claim to get a 6-decimal figure for pi—3.141592—from Great Pyramid measurements, but this is not as amazingly accurate as it seems because it was calculated from estimates of the thickness of the pyramid's missing outer layer and top.
- The 5th century Greek historian Herodotus claimed a different mathematical ratio was built into the pyramid—that the area of each lateral face equals a square whose sides are as long as the pyramid is tall—this coincidently produces a ratio close to pi.

The Number Game

The custom of manipulating numbers to discover hidden meanings is called Numerology. It is so easy to come up with startling coincidences that "hidden" numerical relationships should not be used to prove the existence of helpful space aliens or unknown advanced civilizations. Finding these relationships is really a game of "Pick and Choose."

Mathematician Martin Gardner demonstrated how easy it is to find a pattern within a bunch of unrelated numbers. He analyzed the Washington Monument to see if he could "discover" the property of fiveness to it:

> Its height is 555 feet and 5 inches. The base is 55 feet square, and the windows are set at 500 feet from the base. If the base is multiplied by sixty (or five times the number of months in a year) it gives 3,300, which is the exact weight of the capstone in pounds. Also, the word "Washington" has exactly ten letters (two times five). And if the weight of the capstone is multiplied by the base, the result is 181,500—a fairly close approximation of the speed of light in miles per second.

He then joked "it should take an average mathematician about 55 minutes to discover the above 'truths.'"

You can find amazing "coincidences" by measuring your own home. On my first try I discovered that the length of my house times 10,000 is the same as the distance to the sun divided by the number of days in a year! I started out by dividing the distance to the sun by the days in a year just because it sounded like an impressive (but actually meaningless) astronomical fact. I instantly saw that it more or less matched the length of one side of my house, give or take a few zeros. So I added the "times 10,000" to get a match! The secret of numerology is to just keep manipulating numbers until you get amazing-sounding matches.

How the Egyptians Built the Pyramids

Alternate theory people say that archeologists tell us the pyramids appeared suddenly in the historical record with nothing to indicate that anything came before them. They say no one knows how the pyramids were built, and that modern engineers failed when they tried to build one.

But archeologists have tested different pyramid building theories. Egyptologist Mark Lehner successfully cut stone blocks and built a small experimental pyramid in just 6 weeks. He used ancient techniques, with only one exception. To speed work, stonecutters were allowed to use hand tools made of the more durable iron instead of softer copper tools like the Egyptians used.

Remnants of ancient ramps taught archeologists how they should be contructed. The modern experimental reconstruction of a pyramid ramp shown below allowed a team of 20 men to easily drag a 2-ton stone on a wooden sledge. To make the trackway, stone walls were built along each side of the ramp and the middle was filled in with stones, clay, and gypsum (the same material used to make plaster of paris). Wooden planks were spaced along the roadbed and water was poured in front of the sledge runners to help them slide. The road

Reconstruction of a pyramid ramp

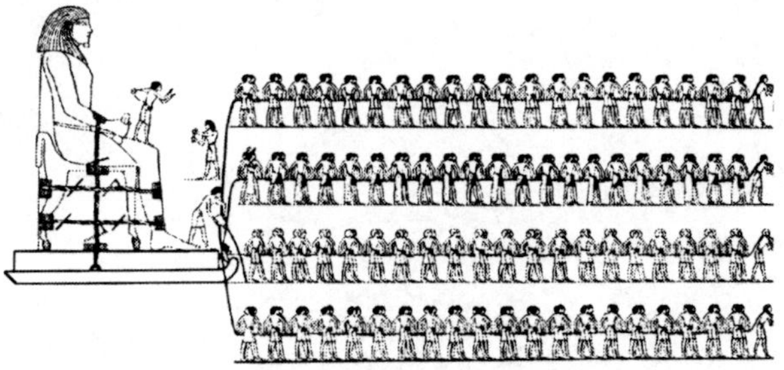

Egyptian drawing of a moving statue

surface could be covered with clay to make it even more slippery.

Alternate theory people also claim the Great Pyramid is placed with a perfection primitive people could not have achieved on their own—its base nearly perfectly level and square, and its sides are precisely oriented toward North, South, East, and West. The Egyptians, however, were a sophisticated, highly organized society with specialists who had generations to perfect the tools and mathematics required to survey and level. Experience with the annual Nile flood led them to develop these skills because fields had to be re-surveyed every year when field boundaries were wiped out by mud. There are many simple ways to find north and achieve level surfaces that do not require advanced technology.

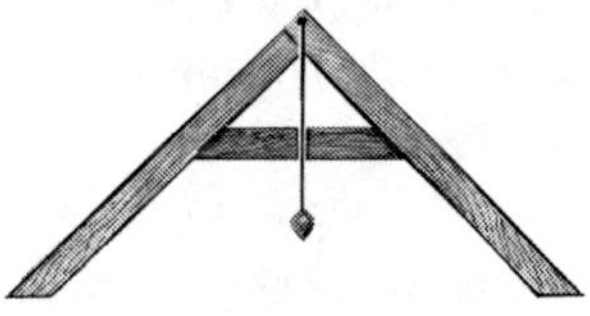

The Square Level: a tool for creating a level surface. When the feet of the "A" are exactly level, the centerweight hangs through the exact middle of the "A".

Finding True North—Three Methods

If you have ever spent much time watching the sky you know that as the hours pass, the sun and stars appear to move in an arc across the

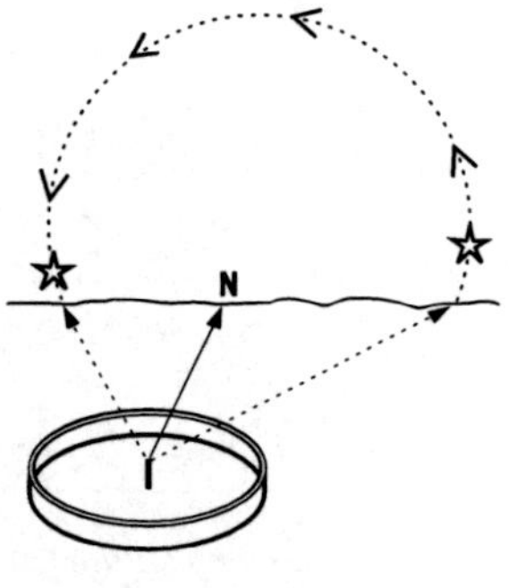

Finding north

sky from east to west—and the center point around which they seem to rotate is north.

- To find north with a star: mark the point where a star rises in the east and the point where it sets in the west. North will be exactly in the middle of these two marks. This measurement is very accurate if it is taken from inside a high circular wall that creates a level artificial horizon. Observations are made through a notch in a tall stick placed in the center of the circle. The wall can be made perfectly level by building a dam of clay around the top of the wall to hold water, and trimming the top of the wall to the water line.
- To find north from the sun's movement: place a tall stick straight into the ground. Mark its shadow three hours before noon and three hours after noon. North will be in the middle of these marks.
- There was no single star that could be used to find true north during the time the Great Pyramids were built, but there were two stars that circled the pole point. The Egyptians could have dropped a weighted line to find out when one star lined up exactly above the other—this would have given them true north.

But this last method was accurate only at the time the Great Pyramids were built, because the earth wobbles slightly on its axis, causing the position of the stars to gradually shift over long periods of time. Later pyramids vary slightly from true north to the same degree that these two stars have drifted away from exactly circling true north, suggesting that this was a method the Egyptians actually used.

How the Stones Were Cut and Moved without Modern Machinery

To lay a foundation for the claim that unknown advanced civilizations or visitors from space built the pyramids, alternate theory people claim that the Egyptians couldn't have cut and moved the gigantic stones that make up the pyramids.

The Egyptians left us many clues about how they worked. The pyramids were built on a limestone plateau to the west of the Nile partly for spiritual reasons—the western horizon was associated with passage to the afterlife. But the plateau was also chosen because it provided both a firm foundation to build on, and quarry sites for the millions of limestone blocks needed for the core of the structures. Canals from the nearby Nile allowed barges of higher quality stone to be floated in—granite for the interior chambers and fine grained limestone for the outer casing.

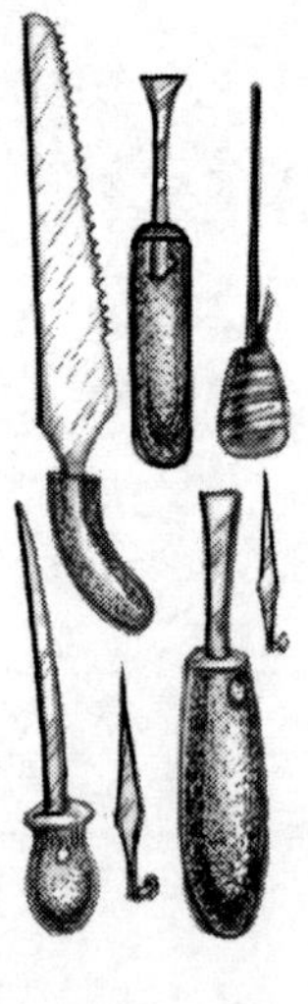

Copper drills, chisels, saws, and awls were similar to today's hand tools. Worn blades like these were found in the rubble between the stones of the Great pyramid.

Stoneworkers with rounded diorite tools

Notched granite in ancient quarries

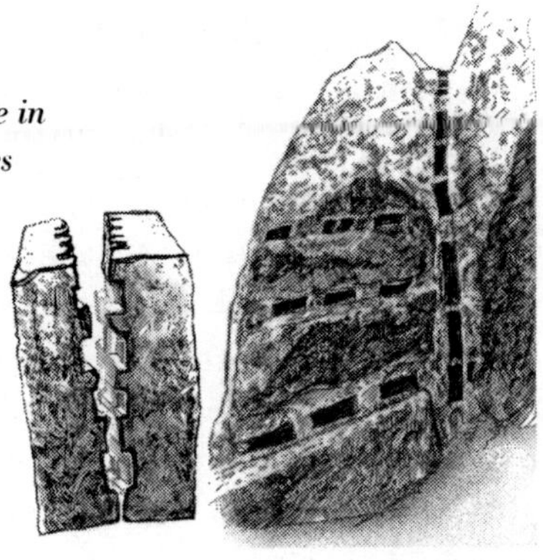

Two possible ramp styles

Remains of an ancient ramp

The ancient pyramid quarries are filled with half-finished blocks, and abandoned and broken tools that show how the Egyptians worked. Copper saws and chisels were used on the softer limestone. Granite could be sawed with the help of quartz abrasives or shattered by applying heat and then cold water. Granite was also shaped by pounding with rounded tools made of an extremely hard stone called diorite.

While it's hard for us to imagine cutting huge blocks of granite from quarries by slowly bashing out channels around them, the Egyptians had both the time and manpower to work that way. Notches were cut in granite and used to split the stone loose with wooden levers.

Roads, Ramps and Levers

Archeologists have discovered the remains of roads used to drag stones from quarries to pyramids.

Ramps that were used to raise stones up onto pyramids are found still in place against abandoned projects. The beginning of the Great Pyramid's ramp was found at a nearby limestone quarry. No one is sure how the blocks were placed at the very top of the pyramids. Some think wooden levers were used. Others think that stones were moved into place with a system of spiral ramps.

Testing Pyramid Power Theories

The pyramid shape is believed by some to attract or generate special kinds of energy that can:

• sharpen razor blades • purify water • preserve food • polish jewelry or coins • improve wine • heal patients • mummify pets • stop aging • improve memory • keep milk fresh • extend shelf-life of medicines • improve sexual desire • aid concentration • reduce stress & tension • prevent rust • heighten charges of psychic energy • enhance meditation • mummify meat & eggs • preserve flowers • stimulate plant growth • increase relaxation • improve coffee • calm children • heal cuts, bruises & burns • reduce toothache & headache pain • improve certain fruit juices.

Our hypothesis is that pyramid power is a folktale inspired by the fame of the well-preserved ancient mummies, and the thrill of supposing that you have re-discovered ancient hidden knowledge unknown to science.

If pyramid power really existed it would be wonderful indeed. It would open up a whole new branch of science and inspire new technology. But no scientific tests to date have managed to detect it.

It's easy enough to test pyramid power yourself:

Construct a small four sided cardboard pyramid, and get a cardboard box of about the same size. Place them side by side with an identical plate of perishable food under each shape. Or perhaps you might want to try tarnished pennies to see if they become polished under the pyramid. (Pyramid power advocates recommend that one side of the pyramid be placed facing due north, and that the pyramid be kept away from devices like TV's and radios.) Leave the box and pyramid alone for a few days to give the power time to work. Run the experiment at least 10 times. This is very important—running an experiment too few times could give you chance results.

Each time you run the experiment have someone judge which plate of food, if any, is fresher, or which pennies are shinier. Record the results of each test. By guessing, your judge should pick the pyramid plate about half the time just by chance, even if there is no such thing as pyramid power. So to detect pyramid power the judge must consistently pick the pyramid plate more than half the time.

When you present the plates to your judges, secretly mark them so you don't lose track of which is which. It is *very important* that the judges can't tell which plate was covered by which shape so they won't be influenced by pro– or anti–pyramid power feelings. The judges should also be alone when they make their decision so that no one influences them one way or the other. Having these two safeguards against unintended cheating is an important invention of science called the "double blind" experiment. Both the person making the judgement about which plate of food is fresher, or which pennies are shinier, and anyone else in the room should be blind as to which plate was under the pyramid. The problem with many experiments that pyramid power believers have done is that they were not double blind, and they didn't run enough trials to avoid chance results.

Satanic Ritual Abuse

JEFFREY VICTOR

Since the publication of my book *Satanic Panic: The Creation of a Contemporary Legend* in June, 1993, I have had some surprises. On the positive side, the book was awarded the 1994 H. L. Mencken Award by the Free Press Association, which is given yearly for books about individual rights and abuses of power. On the negative side, my publisher and I were slapped with a million dollar lawsuit by two therapists whose activities in promoting the satanic cult scare I reported in the book.

My book investigated four social conditions through which the satanic cult scare is manifested: 1) false accusations of satanic cult ritual child abuse made by adult psychotherapy patients and by very young children; 2) community over-reaction to the teenage pseudo-satanism fad; 3) community rumor-panics in response to stories about secret, criminal satanic cults; and 4) censorship campaigns aimed at supposed satanic influences in children's books and elementary school books. The most harmful of these conditions has been false accusations of ritual child abuse. They have put innocent people in prison, separated children from their parents, destroyed many good people's reputations, and caused bitter controversy.

This article offers a sketch of major developments over the last two years in the evolution of the satanic cult scare, with a focus upon the controversy surrounding claims about ritual child abuse. Going beyond description to analysis, it offers some insights learned from the current satanic cult scare about the origins of similar moral panics.

Claims about satanic cult ritual child abuse (SRA) arise from the convergence of two different moral panics: the child sexual abuse scare and the satanic cult scare. Sociologists use the term "moral panic" to refer to a social condition in which a great many people in a society over-react to a newly perceived threat to their well-being from social deviants, even though the actual threat is either non-existent or greatly exaggerated (Goode and Ben-Yehuda, 1994). Unlike an episodic panic, such as the "War of the Worlds" panic of 1938, a moral panic is long-lasting and gives rise to organizations, laws, and procedures to combat the perceived threat. Moral panics are usually accompanied by moral crusades against the social deviants and their perceived "evil" influences in society. Examples of past moral panics include the European witch-hunt, outbreaks of anti-Semitic persecutions, the white slavery scare, and the 1950's Red Scare in the U.S.

Public Opinion and Satanic Cults

Some scholars have suggested that public opinion during moral panics goes through certain phases of development (Penrod, 1952). It begins with an emerging perception of the threat among scattered opinion leaders, who try to spread an awareness of the threat to the general public. Then, in the second phase, there is an explosion of concern about the threat, expressed through the mass media, which enables the concern to be shared by a wide audience. Eventually, the threat is defined as "real" by the largest possible number of receptive people. At that point, skepticism about the extent of the threat begins to surface, resulting in resistance to fearful appeals and widening social controversy about the over-reaction. Afterwards, the exaggerated concern withers away, except among a few marginalized zealots.

The alarmist propaganda promoted by anti-satanist moral crusaders since the early 1980s has been effective in constructing a phantom threat from imaginary evil-doers. Widespread grass-roots belief in the danger of satanic cult crime continues to persist. A 1994 national survey reported in *Redbook* magazine (Ross, 1994) found that 70% of Americans "believe that at least some people who claim that they were abused by satanic cults as children but repressed the memories for years are telling the truth." Moreover, "32% say that the FBI and the police ignore evidence because they don't want to admit the cults exist."

However, American public opinion about the threat from secret, criminal satanic cults is currently in the process of growing increasingly more receptive to skeptical analysis. My evidence for this perception is that there have been an increasing number of mass media pieces offering a skeptical examination of claims about satanic cult crime, including ritual child abuse. Over the last year or two, there have been several segments of television documentaries and many popular magazine articles critical of accusations of satanic ritual abuse. More importantly, skeptical pieces of investigative journalism have appeared in several mainstream and Christian religious publications. The impact has been to stir concern about wild allegations of satanic cult crime. The gatekeepers of mass media ideas had previously regarded SRA allegations as being the nonsense of TV talk-shows and country "rednecks." It would appear that critical analysis sells only after tabloid sensationalism has saturated the market for dramatic horror stories.

Probably the most widely influential piece of investigative journalism was a two part article by Lawrence Wright published in *The New Yorker*, in 1993, under the title "Remembering Satan." (Wright later published his work in a book by the same title.) The piece focused on the 1988–89 case of Paul Ingram in Washington State. Ingram was a very active fundamentalist Christian, deputy sheriff, and Republican county chairman, who was accused by his two daughters of sexually abusing and torturing them during satanic cult meetings (and sacrificing babies, as well). During five months of questioning by police, by his pastor, and by two psychotherapists, Ingram gradually described vague "memories" of participating with his wife and poker-playing buddies in a satanic cult, which for 17 years sexually abused his daughters in ritual group orgies. Ingram's vague memories came forth after his pastor told him that Satan was the deceiver of the mind and that he might not be able to remember sexually abusing his daughters due to Satan's power over his mind.

The psychotherapists employed questionable quasi-hypnotic procedures with Ingram to aid recall of his "faulty" memory. These procedures included progressive relaxation, meditative Christian prayer, and visualization exercises. Ingram later recanted his elaborate confessions of bizarre criminal activity, but too late to avoid being convicted and sentenced to 20 years in prison. His wife and poker-playing buddies were not prosecuted, but their lives were ruined.

The Politics of Public Opinion

In the arena of competing claims about a potential threat to society, political authority is often more important in providing credibility for claims than is scientific authority, at least in the short run of events. As I noted in my book, four states had passed laws against a variety of "ritualistic" (satanic cult) crimes by 1991, thereby lending credibility to the purported threat. Since that time, serious political struggles have taken place in California and Utah, as believer groups have lobbied state legislatures to recognize a threat from secret criminal satanic cults and pass special laws against ritual child abuse. The Utah state legislature spent $250,000 for a task force investigation of satanic cult crimes against children and could find no persuasive legal evidence to support the claims. A similar government supported task force in California came to similar conclusions.

The case for skepticism about conspiracy theories concerning satanic cult ritual abuse was supported by two significant government reports, one from the United Kingdom and the other from the Netherlands, both countries where the plague of satanic cult panic was brought by American sources. In 1994, an official report from the U.K. Department of Health came to the conclusion that there was no evidence to justify allegations of the ritual abuse of children by secret satanic cults, in any of the 84 cases it investigated in which children had been taken away from their parents (La Fontaine, 1994). The report noted that "The alleged disclosures of satanic abuse by younger children were influenced by adults." In addition, the report concluded that fear of satanic ritual abuse had been spread by the evangelical Christians and by American and British professional ritual abuse "specialists," whose qualifications were never verified. As recently as January, 1995, eight English men and women were cleared of charges of ritual child abuse and released from prison, after the court determined that several children had simply made up accusations of satanic cult crime against them (*Daily Mail Reporter*, 1995).

A government report from the Ministry of Justice of the Netherlands came to very similar conclusions, after a thorough investigation of SRA accounts in that country (Netherlands Ministry of Justice, 1994). The report further suggested three complementary explanations for the epidemic of false accusations. In some cases, SRA accusations could be a replacement for other, genuine traumas (a "screen memory"). In addition, therapists and child protection workers who strongly believe in the existence of satanic ritual abuse may inadvertently prompt these stories from their patients ("suggestion effects"). Finally, SRA accounts may also be a manifestation of the satanic cult urban (contemporary) legend; threat-filled rumor stories repeated so frequently and so widely that they are regarded, without question, as being true.

The Conflict between Scientists and Therapists

Law enforcement agencies and courts commonly rely upon the presumptive expertise of psychotherapists and child-protection social workers in attempts to distinguish between true and false accusations of sexual child abuse. What these professionals interpret as being indicators of true and false accusations is crucial in making arrests and obtaining convictions. Psychotherapists and child-protection social workers are increasingly thrust into the legal arena and attributed authority to influence the determination of people's guilt or innocence of crime, based upon their interpretations of psychological indicators. Whether or not they want that authority (and most do not seek it), these professionals have become agents of social control in the arena of sexual child abuse.

Believing therapists continue to promote their claims in seminar training programs about SRA and in popular culture books. (This is so in the U.S., as well as in other mainly English-speaking countries.) The newly revised edition of the most widely sold popular self-help book about recovering from child sexual abuse, *The Courage to Heal* (1994), included a special section on "Facing Sadistic Ritual Abuse." The influence of this book is shown in a national study of self-help books, which found that *The Courage to Heal* was the most widely used and recommended book by mental health workers for the topic of child sexual abuse (Santrock, Minnett and Campbell, 1994). Diverse kinds of "therapists" and "counselors" of varying amounts of education and training continue to treat "survivors" of SRA. (However, skepticism appears to be increasing most rapidly among those with the most advanced education.) Several investigative journalists have told me that a few mental hospitals and clinics are making a thriving business out of treating "survivors" of SRA.

At the same time, there are significant signs of changing opinion among therapists who previously "went public" about their belief in the existence of secret satanic cults which sexually abuse children, causing multiple personality disorder and loss of memories of the abuse. Several very prominent therapists have been back-pedaling about their previous claims. (I will discreetly refrain from mentioning any names.) Now, they caution about "overreacting" to the problem and suggest that the threat may have been "over-estimated." Moreover, a semantic shift can be found in the current claims of previous believers. They now use the term "ritual abuse" or "sadistic abuse" and speak about "cults," avoiding the satanic adjective.

Skeptical psychotherapists have finally found receptive audiences. They have been aided by the incredibly rapid growth of the False Memory Syndrome Foundation. Its membership consists mainly of middle aged and older parents whose adult children have accused them of child sexual abuse, based upon memories "recovered" during psychotherapy. The FMS Foundation has played a crucial role in facilitating scientific communication between therapists and researchers who are skeptical about the sudden, sharp increase in sexual abuse accusations supported only by long-lost memories. It has been very effective in gathering scientific and legal information about the phenomenon of false memories and disseminating that information to the mass media and to concerned professionals. Since its establishment in March, 1992, by a few families in Philadelphia, the FMS Foundation has been contacted by about 16,000 affected families and has grown to a paying membership of about 2,500 families plus 500 professionals. The FMS Foundation has a professional advisory board of 44 psychotherapists and behavioral scientists (on which I was invited to participate). An early survey of FMS parents found that about 18% were accused of SRA. A reasonable estimate of the number of families in the United States affected by SRA accusations is that they are in the thousands.

Another important development is the growth in the number of people who have retracted their memories of childhood sexual abuse. About 300 retractors have contacted the FMS Foundation for help. Some of them have organized a support group with a newsletter.

The struggle between SRA-believing therapists and skeptics is expressed in professional journals and newsletter articles, in newspaper interviews and, most importantly, in conflicting expert witness testimony in court. While conflicting opinion among therapists is great, there is an even greater cleavage between psychotherapists and behavioral science researchers. Few, if any, research psychologists and sociologists accept claims about the existence of organized, secret satanic cult criminals.

The Psychology of Belief

It is axiomatic that science can't prove the non-existence of something. Science cannot prove the non-existence of demons which take over people's souls, or aliens from UFOs who kidnap people and erase their memories, or past lives which influence our memories. Science can't prove the non-existence of secretive yeti, or big-foot creatures, or Loch Ness monsters. Therefore we should not anticipate that any research finding will determine that secret, criminal satanic cults do not exist. However, the logic of scientific evidence can be used to develop alternative interpretations to claims of truth based upon misperceptions of events.

The most important recent research study of SRA accusations is a large scale, national research project carried out by Gail Goodman, under the auspices of the National Center on Child Abuse and Neglect (Goodman, et. al. 1994). Goodman's study investigated a large national sample of clinical psychologists, psychiatrists and clinical social workers, asking about experiences with child and adult patients who claimed to be victims of satanic ritual abuse and other religion-related child abuse. Unfortunately, there is insufficient space here to present its many useful findings. In brief, the research could not find a single case of alleged child sexual abuse where there was clear corroborating evidence for the existence of a well-organized inter-generational satanic cult which tortured children and committed murders. On the other hand, the research did find evidence that in isolated cases, individual perpetrators did employ references to Satan to intimidate child victims (Goleman 1994). The report concludes: "Our research leads us to believe that there are many more children being abused in the name of God than in the name of Satan" (1994, 14). One important specific finding was that the therapists reported a total of 43 "repressed memory cases," cases in which adult patients had "recovered" lost memories of satanic ritual abuse in psychotherapy. Significantly, the therapists all reported that they believed their patients' memory accounts, even though there was no corroborating evidence for the crime.

Rogers and Brodie (1993) carried out a smaller survey of the beliefs of 53 child protection social workers from a county in southern California, who held master's degrees and had from three to 15 years of work experience. The survey found that 45% of the social workers agreed with claims that: "satanic ritualistic abuse involves a national conspiracy or network of multi-generational perpetrators where babies, children and adults are sexually assaulted, physically mutilated, or killed." Moreover, they also found that: "Nearly half believe there are thousands of missing children in the U.S. every year who simply disappear and their bodies are never recovered, who may have been victims of SRA." Concerning adult memories of SRA, Rogers and Brodie found that: "23% agreed that half or more of those who repress memories of childhood sexual abuse may also be victims of SRA."

A scientific understanding of false memories of SRA may be a key to understanding the social dynamics which create false memories in general. As a sociologist who is studying psychotherapists, it seems to me that the social psychology of belief is more central to an understanding of false memories than is the psychology of individual memory.

The Legal Backlash against Psychotherapists

A number of malpractice lawsuits have recently been brought against therapists by accused parents and/or retractors, many of which involve accusations of SRA. These lawsuits may help to determine the legal status of the presumptive "expertise" of therapists in identifying child sexual abuse from "recovered" memories. About 50 retractors have filed or are planning to file malpractice lawsuits against their former therapists, according to FMS Foundation sources. False memories of satanic ritual abuse are surprisingly common in these cases and occur in about 50% of them. The usual psychiatric diagnosis is multiple personality disorder, supposedly caused by satanic cult torture and "mental programming."

One important case is that of a Houston, Texas, family which was torn apart by recovered memories of SRA. The case involves Dennis Schwiderski, an executive in an oil company (Waterhouse, 1994). His wife, Kathryn, became convinced that she was a member and victim of a satanic cult since her childhood and that she sexually and physically abused her own children. After seeking therapy for depression, she was committed to long-term treatment in a Houston hospital for seven years, where she was diagnosed as having multiple personality disorder, caused by repressed memories of satanic ritual abuse. Her two teenage daughters and a son were similarly diagnosed and also became convinced that they were victims of SRA. The cult activity supposedly included rape, torture, human sacrifice and cannibalism. Mr. Schwiderski was investigated by a grand jury and the case was dropped for lack of evidence. He and his wife divorced after he had spent $328,000 of his own funds on treatment for his family, in addition to three million dollars of insurance money. Now, she has recanted her memories and both she and Mr. Schwiderski, along with one daughter, are suing the hospital and a collection of therapists for negligence and insurance fraud. The defendants, some of whom are prominent therapists, stand by their diagnosis and claim that their treatment was justified, because the Schwiderski family were members of a satanic cult.

The legal battles over recovered memories are being fought on several other fronts. Some accused parents have brought third-party lawsuits against their adult children's psychotherapists, without being joined by children who have retracted their memories. At this time, about 25 such lawsuits have been filed or are being planned by parents, according to sources at the FMS Foundation. So far, one landmark case has been won by a parent: the Ramona case in Napa, California. (The Ramona case, however, did not involve SRA allegations.)

On the other side of the legal struggle, some adult children are bringing civil lawsuits or criminal charges against their parents, based upon "recovered memories" they continue to believe. An FMS Foundation legal survey found 432 families where a lawsuit against parents was either threatened, or pending in court. Surprisingly, about one-third (N=149) of these situations involve SRA accusations against parents, and in 76 of these SRA situations, accusations were also made against other people in addition to the parents. At the time of this writing, not a single SRA case has been lost in court by the parents.

To Believe or Not Believe the Children

A second source of ritual abuse accusations is those voiced by children. Beginning with the McMartin Preschool case and other similar cases in the early 1980s, there have been a great number of people accused by children of bizarre acts of torture and sexual molestation. Most of the accused are child care workers, often young women. Many of these people have been imprisoned on the sole basis of children's accusations. The most well-known example is that of Margaret Kelly Michaels, who was accused of repeatedly sexually abusing 19 children while she was working at a New Jersey child care center. She spent five years in jail before her conviction was finally overturned by a New Jersey Court of Appeals in 1993.

An important recent SRA case was the trial of Dale Akiki, a deformed and retarded adult who was accused of sexually abusing and ritually torturing children while he worked as a teacher's aide at a San Diego church's child care center (Fine, 1994). The accusations came from more than a dozen children, aged three and four. The accusations arose during a local satanic panic in the San Diego area, during a period when a government-appointed ritual abuse task force was active in publicizing concerns about secret satanic cults. Satanic cult rumors spread wildly during the time that the police were publicly investigating allegations against at least six families. The children who accused Akiki were subjected to intense questioning by parents and child protection workers. After spending two and a half years in jail, Akiki was finally acquitted in November, 1993.

These kinds of cases are not rare. Other recent cases where local satanic panics gave rise to children's accusations of satanic ritual abuse and the arrest of innocent people include those in Martensville, Saskatchewan, and Gilmer, Texas. The Edenton, North Carolina case, featured on PBS-TV's "Frontline" (1993) program, began under similar circumstances. The conviction of the child care workers in that case is now being appealed and a decision by a three-judge panel is expected soon.

In some cases, accusing children, now grown into young adults, have belatedly confessed that they made up the stories in response to the intense questioning of child protection workers and therapists. One such example is that of Andrew Myers, who in 1994 admitted to a reporter that he and a group of five other children had lied about ritual sex abuse in accusations made against his parents and 22 other people 10 years before in the famous Jordan, Minnesota case (Marcotty, 1994).

The deeper significance of all these legal cases is that they reveal the increasing intrusion of para-medical, psychiatric expertise into criminal investigations and into efforts to distinguish between true and false accusations of crime. Efforts to verify evidence of crime in recovered "repressed" memories, in personality symptoms of emotional trauma and in the uncorroborated accusations of children are all manifestations of a much broader social change in the nature of social control.

What Can Be Learned about Moral Panics?

The outline of a rational alternative to the satanic cult conspiracy theory is now clear. Isolated cases were misperceived and distorted through the lens of the contemporary legend about satanic cults, creating a moral panic. The widely circulating satanic cult rumors distorted the judgment of believing child protection workers, psychotherapists and police. They influenced the false memories of many patients in psychotherapy, and were used in false accusations by some children. Paradoxically, the rumors are also used by a few child molesters to manipulate children.

However, a key sociological question remains. What underlying social conditions enabled false accusations of satanic ritual abuse to spread so rapidly throughout society and to become regarded as credible by some agents of the legal system? Any adequate explanation cannot be founded upon personality traits, such as superstition, scientific ignorance, suggestibility, or maliciousness. Collective behavior is a product of social dynamics and not a product of personality traits. The popular explanation of attributing moral panics to "contagious hysteria"—meaning a form of temporary, collective psychopathology—simply trivializes these conditions as anomalies. The hysteria explanation ignores the fact that moral panics are quite "normal" (recurrent) events in social systems.

Sociologists are interested in the social conditions which lead to the false labeling of people as social deviants and false accusations of crime. One extensive analysis of incidents of false accusations found that three social conditions increase the prevalence of false accusations: (1) perception of a widespread threat; (2) competition for authority; and (3) faulty techniques of investigation (Klemke and Tiedeman, 1990). I have added a fourth condition: (4) ideological fervor. These conditions may help to explain why false accusations of ritual abuse have rather suddenly surfaced and spread. They are conditions which cause moral panics in general:

1. Perception of a Threat—A widespread perception of a threat to society from some kind of newly detected social deviants can easily give rise to indiscriminate false accusations. The newly perceived threat which has given rise to false accusations of SRA is the widespread belief that child sexual abuse is much more common than previously thought, and that there exists, hidden in society, a great many secret satanic cult criminals who sexually abuse children.

Although there is no evidence that well-organized, secret satanic cults exist, there is abundant evidence that sexual child abuse is much more common than had previously been believed. However, this fact doesn't negate the possibility that the potential threat from child molesters has been grossly over-estimated, causing a rush to judgement in cases of many falsely accused parents and child care workers.

2. Competition for Authority—Sociologists recognize that authority figures play a key role in defining forms of deviance and in providing legitimacy to claims about new threats to society. Institutional authorities do not automatically regard all claims about threats to society as being equally credible. However, when new forms of authority begin to emerge and compete for power over a jurisdiction, those new authorities may be tempted to use the perceived threat to expand their power. They may over-reach their expertise. The stage is set for the spread of false accusations.

In the 20th century, the social authority to define and interpret deviant behavior has gradually shifted from religious and political authorities to medical and "mental illness" authorities (Conrad and Schneider, 1992; Conrad, 1992). Past examples include the medicalized definitions of alcohol abuse, illegal drug use, homosexual behavior and child abuse. One consequence is that increasingly, the presumptive "expertise" of medical and paramedical authorities (meaning therapists and child protection social workers) is relied upon by other authorities: police, judges and juries.

In this context, the social legitimation of false accusations of SRA can be seen as a result of the expansion of paramedical-psychiatric authority into the domain of detecting criminals who sexually abuse children. Previously, this was primarily the domain of legal authorities and their investigative methods.

3. Faulty Techniques—When authorities employ inadequate instruments for distinguishing between true and false accusations, false accusations are easily regarded as being credible. When ambiguous indicators are used to identify criminals, the net is thrown widely, and an accusation itself becomes enough evidence for guilt. Then, when authorities replicate their errors again and again with their faulty techniques, they are likely to believe that the replication confirms their diagnosis or conclusions. (The frequent replication of SRA accounts from patients is the main "evidence" which believing therapists say convinces them.)

Faulty techniques in the investigation of child sexual abuse include the use of highly ambiguous lists of symptoms to identify child sexual abuse; similarly ambiguous lists used to identify the long-range effects of child sexual abuse in adult personality; faulty techniques to investigate accounts drawn from lost memories ("memory recovery" techniques designed to uncover "repressed" memories); and inadequate techniques to corroborate accusations of crime based upon memory accounts. Simply put, scientific techniques for distinguishing between true and false accusations of child sexual abuse have not yet been developed, even though many therapists may claim that expertise. A host of excellent books has recently been published which document these faulty psychotherapy techniques and their harmful consequences, including those by Ofshe and Watters (1994), Pendergrast (1995), Wakefield and Underwager (1994), and Yapko (1994). Accusations of SRA are subject to these same faulty techniques. Therapists who are confident in these techniques are caught in a dilemma. Either they must validate bizarre SRA accusations as credible, or regard them as being mere anomalies and downplay them. (Recognizing the existence of faulty techniques of investigation does not imply that police investigations of child sexual abuse are futile. Child sexual abuse is a horrendous crime. Unfortunately, the means for detecting it are highly fallible.)

4. Ideological Fervor—When authorities are motivated by ideological fervor, critical analysis commonly falls by the wayside. Fervent ideologists usually have a vested interest in uncovering evil-doers in society. They are likely to feel that false accusations in pursuit of their ideological goals are unfortunate, isolated incidents. Some ideologists may even tolerate false accusations, out of worry that investigating these incidents might discredit the whole fabric of their ideology.

Many of the agents of social control, clerics, police, therapists and child protection workers, who have been most receptive to SRA stories, have been either fundamentalists or feminists. Atrocity stories about satanic cult crimes are most likely to be believed and circulated by people having either a fundamentalist Christian or a radical feminist ideology, as my research and other reports have made clear. This curious mixture has also fueled past moral crusades against prostitution, alcohol and pornography. So it should not be surprising that some fervent fundamentalists and feminists can accept the same illusions. (Skeptics need to be careful not to paint all feminists and Christian religionists with a broad brush. Some feminists and some Christian traditionalists have been particularly skeptical about claims of satanic cult crime and have published important pieces of critical analysis. See: Perrin and Parrott, 1993.)

Past moral panics have been facilitated by social conditions very similar to the ones which now account for the ritual child abuse witch hunt. During the anti-Communist "Red" scare of the 1950s, some national political authorities, such as Joseph McCarthy and J. Edgar Hoover, encouraged a variety of paralegal Communist hunter groups to search out and identify thousands of secret, "un-American subversives," using ambiguous indicators of subversive activity. Nationalistic anti-communism fueled the search for inner enemies, and even the professionalism of police agencies was corrupted by it. The Stalinist-Communist purges of the 1930s grew out of similar conditions, even though it can be said that top-level authorities more directly orchestrated the moral panic. Similar conditions facilitated the classic European witch hunts. There was widespread fear of secret, conspiratorial witches, who supposedly practiced black magic. Secular witchfinders arose as specialized experts in detecting witches, in competition with church authorities. Ambiguous tests were used by the witchfinders to detect witches, so that the accused were almost automatically found guilty (Currie, 1986). The ideology of the orthodox Christian religion fueled the Inquisition's search for any kind of potential heretic (or skeptic).

Skeptics are common targets during moral panics. The reasons are clear. They are critical of appeals to fear from unsubstantiated threats. They question conventionally accepted authority. They demand careful, rational techniques of investigation. And they are wary of ideological extremism of any kind, religious or political, right or left. So skeptics, watch your step.

References:

Bass, E. and L. Davis. 1994. *The Courage to Heal* (Third Edition). New York: HarperCollins.

Conrad, F. and J. W. Schneider. 1992. *Deviance and Medicalization: From Badness to Sickness.* Philadelphia: Temple University Press.

Conrad, P. 1992. "Medicalization and Social Control." *Annual Review of Sociology*, 18: 209–232.

Currie, E. 1986. "Crimes Without Criminals: Witchcraft and its Control in Renaissance Europe." *Law and Society Review*, 3, 1:7–32.

Daily Mail Reporter. 1995. "Our Nightmare Year, Accused of Satanism and Child Sexual Abuse." *The Daily Mail.* (U.K.) Jan. 13: 26.

Fine, J. 1994. "Seeking Evil." *California Lawyer.* July, 50–55 & 90–92.

Frontline. 1993. "Prisoners of Silence." PBS-TV, Oct. 19.

Goleman, D. 1994. "Proof Lacking for Ritual Abuse by Satanists." *New York Times*, Oct. 31: A13.

Goode, E. and N. Ben-Yehuda. 1994. *Moral Panics: The Social Construction of Deviance.* Cambridge, MA: Blackwell.

Goodman, G., with Q. Jianjian, B. Bottoms, and P. Shaver. 1994. "Characteristics and Sources of Allegations of Ritualistic Child Abuse," Final Report to the National Center on Child Abuse and Neglect. (Available from: Dr. Gail Goodman, Department of Psychology, University of California, Davis, Davis, CA 95616.)

Klemke, L. and G. Tiedeman. 1990. "Toward an Understanding of False Accusation: The Pure Case of Deviant Labeling." In: *Deviant Behavior: Readings in the Sociology of Norm Violations.* Ed. by C. Bryant. New York: Hemisphere Pub., 266–285.

La Fontaine, J. S. 1994. *The Extent and Nature of Organized and Ritual Abuse: Research Findings.* United Kingdom Department of Health Report. London (U.K.): HMSO Publications.

Marcotty, J. 1994. "Ten Years After/The Legacy of Jordan." *Minneapolis Star-Tribune.* Oct. 16: 14A–15A.

Netherlands Ministry of Justice. 1994. Report of the Working Group on Ritual Abuse. Den Haag, The Netherlands: Direction of Constitutional and Criminal Law. (English translation by J. W. Nienhuys.)

Ofshe, R. and E. Watters. 1994. *Making Monsters: False Memories, Psychotherapy, and Sexual Hysteria.* New York: Scribners.

Pendergrast, M. 1995. *Victims of Memory: Incest Accusations and Shattered Lives.* Hinesburg, VT: Upper Access Books.

Penrod, L. 1952. *On the Objective Study of Crowd Behavior.* London (U.K.): H. K. Lewis & Co.

Perrin, R. and L. Parrott. 1993. "Memories of Satanic Ritual Abuse." *Christianity Today.* June 21: 18–23.

Rogers, M. and L. Brodie. 1993. "Front Line Providers in Investigation of Child Abuse: Personal background, experience, knowledge base and attitudes of social workers regarding child sexual abuse survivors and recovered memories." Paper presented at the conference of the False Memory Syndrome Foundation. (Available from 17662 Irvine Blvd., Tustin, CA 92680.)

Ross, A. 1994. "Blame It on the Devil." *Redbook*, June: 86–89 & 110, 114, 116.

Santrock, J., A. Minnett and B. Campbell. 1994. *The Authoritative Guide to Self-Help Books.* New York: Guilford.

Victor, J. S. *1993. Satanic Panic: The Creation of A Contemporary Legend.* Chicago: Open Court.

Wakefield, H. and R. Underwager. 1994. *The Return of the Furies: An Investigation into Recovered Memory Therapy.* Chicago: Open Court.

Waterhouse, R. 1994. "There'll Be the Devil to Pay." *The Independent.* (U.K.) Oct. 17.

Wright, Lawrence 1994. *Remembering Satan.* New York: Knopf.

Yapko, M. 1994. *Suggestions of Abuse: True and False Memories of Childhood Sexual Abuse.* New York: Simon and Schuster.

Science and God

BERNARD LEIKIND

Modern science offers scant comfort for those who believe that God plays an active role in the physical operations of the universe. Beginning some three centuries ago, researchers have gradually extended the realm of natural law and simultaneously reduced the kingdom of supernatural powers. Today even the beginnings of life and the beginnings of the universe, God's traditional domains, are subjects of scientific study.

One result of this extension of natural law is that some believers no longer claim that God rules nature. God's manifestations, they might say, are not the forces and materials upon which we may all stub our toes. God exists as spiritual ideas and feelings within our minds and hearts and appears to us through our thoughts and actions. This makes God responsible for the good and the bad in human behavior. I have no quarrel with these believers, although I see no way to tell if their views are true or false. To me these ideas seem an unneeded metaphor for the source of human character and behavior.

How many still adhere to more traditional views? There must be many. In various forms, these traditional believers hold that God created the universe, as in the Genesis stories; or that God runs the universe, minding every sparrow fluttering in a tree; or that God intervenes in the natural world to adjust the otherwise smooth operation of natural law, saving this baby here and striking that plane from the sky there. They believe that the world expresses God's inscrutable purposes.

These believers find that scientific knowledge, cosmology, fundamental physics, chemistry, biology, anthropology, and psychology, undermine their views on every front. Religious knowledge, which professes absolute truth, fails while science, which professes fallibility, succeeds. Any open minded seeker must agree that tradition and revelation cannot provide us with reliable guides to the natural world.

Physics and Biology Take over God's Job

Until the rise of science in the 18th century, Westerners believed the biblical accounts of God's creation and operation of the world. They believed that God was in the details. Newton, an eccentrically religious man, taught us that the heavens and Earth were one, governed by a single, marvelous, all-embracing law. God no longer managed the flight of every butterfly. Instead, he established a law of beautiful simplicity and set the world free to run its course. To many believers and to most scientists of Newton's time, God had created and energized the world's marvelous mechanism but did not involve himself in its daily operations. This clockwork universe was and still is profoundly unsatisfactory to those who prefer an interventionist god who can be induced to watch over and protect them.

In any case, most thought that life fundamentally differed from the world's gears, axles, pulleys, and grease. God, they believed, had instilled a spiritual essence in humans that distinguished their substances and souls from lower lifeforms and from inanimate matter. The collapse of this élan vital doctrine began in the 19th century. We can date this event to the 1820s with Wohler's synthesis of urea. Life's chemicals were just chemicals. Believers retreated again. If God used the laws of physics and chemistry when he created life, at least he must have designed its many forms. Ignoring our appendixes, bad backs, and baldness, believers now asserted that the perfection of natural design reflected God's all-embracing purposes and the perfection of his methods. He chose to make the okapi and the platypus, the mudfish and the bumblebee just so. How else could it have happened?

Darwin taught us how. His theory of natural selection explained the mysterious fact of evolution and the riotous variety of life. Subsequent developments in biology have confirmed and extended the truths that Darwin proposed. Unknown and perhaps even unimaginable to Darwin, discoveries such as classical genetics and the genetic code have corroborated his proclamation of the unity of life. We now know that humans barely differ from chimpanzees and that our most prized accomplishments, such as language and culture, are merely one end of a continuum that extends from animals to us.

Natural Law Rules Everywhere for All Time

Physics and astronomy, in the meantime, were expanding their territory. Einstein's theories extended Newton's laws to universal scales while quantum mechanics brought the reach of scientific law to the tiniest objects. Let us consider one of the amazing recent developments in astrophysics, cosmology, and fundamental physics that confirms the universal reach of scientific knowledge.

Using quantum theory, relativity, fluid dynamics, and other sciences, astrophysicists study the structure of stars. Unfortunately, we have only one relatively nearby example, if you consider 96 million miles nearby, and we can only see its surface. Astronomy is an observational science, and the time scales over which stars change are millions of times longer than the lives of astronomers and even of astronomy. Astronomers are like naturalists studying a forest and imagining from a week's observations of trees, young and old, the life cycles of all trees. Unlike the woodsmen, who can chop down the trees to look inside, astronomers can only look at stars from a distance. Connecting their ideas with experiments and observations whenever they could, modern researchers now understand the composition and life history of stars.

Recently a supernova appeared in the Magellanic Clouds, smallish galaxies near our Milky Way. Although astronomers see a few supernovas each year in distant galaxies, this was the first close one since 1572. A supernova is the death of a large star. Its nuclear fuel exhausted, the star no longer resists the inward pull of gravity with thermal energy and radiation pressure. According to theory, in amazing and rapid sequence the core collapses and explodes. An outward-bound shock wave blasts away the now unsupported outer layers. Rebounding inward the shock wave crushes the interior, which may collapse to form a neutron star or a black hole. The entire event may last only minutes. This star-stuff maelstrom radiates immense quantities of energy. For a few days or weeks, a supernova may give off as much energy as an entire galaxy of 10 billion normal stars. No astronomer was lucky enough to have been looking at this star at the moment its light began to brighten. An observant astronomical observatory technician in Chile noticed the new star while stepping outside, perhaps for a smoke.

Astrophysicists predict that a supernova's tremendous burst of light accompanies an even larger burst of neutrinos, the ghost particles of physics. These particles, which have hardly any properties at all and which barely interact with anything, must have zoomed off in all directions with nearly the speed of light early in the immense collapse. The astrophysicists' calculations about collapsing stars showed that the supernova must have emitted its neutrino blast before its light reached its brightest. They called their colleagues at neutrino observatories deep below the Earth's surface in the United States, Europe, and Japan. Neutrinos are so elusive that these massive detectors, huge tanks of cleaning fluid or water, must collect neutrino evidence for months. "Look in your tanks," the astrophysicists said. "You have already made a great discovery." They were right. Each of the observatories had detected a few 10s of neutrinos at about the same time.

Consider this achievement. Using theories from nearly every part of physics, special and general relativity, quantum mechanics, fluid mechanics, thermodynamics, nuclear physics, atomic physics, and elementary particles, scientists had predicted the events in a star's death throes. The stuff of the star transformed itself under extreme conditions and complexity never duplicated on Earth. If any of these theories had been in error by much, this prediction would have failed. The supernova explosion provided us with a test of virtually all of physics. This property of our theories—that evidence from many sources combines and confirms itself—is a major reason why working scientists believe they are approaching the truth about nature. This also shows us that the laws of nature known to us on Earth must be the same hundreds of thousands of light years away and must have been the same when that star exploded hundreds of thousands of years ago.

Science Pushes God's Tasks into the Distant Past

Beginning with the discovery of deep time by early 19th century geologists, scientists have pushed to successively earlier epochs the moment when, as in the Sidney Harris cartoon about a complicated mathematics proof, ". . . a miracle occurs."

In biology, this miraculous moment is the time, billions of years ago, when chemicals somehow crossed the line from just chemicals to living chemicals. In this matter lack of evidence hampers us, and it may be that our evidence of this imperceptible and distant event will never be sufficient to eliminate all but one theory. Even if we had been there, we might not have been able to notice the slight difference between definitely dead chemicals and definitely alive chemicals. We would not have seen anything spectacular enough to class as a miracle.

Our problem is not that we have no ideas and so need a miracle. Our problem is that we have too many good ideas and the right one may still not be among them. Even if, through hard thinking and good experiments, we succeed in creating life from inanimate chemicals, how could we confirm that we had found what had happened on the early Earth?

In a sense, the creation of life may be easy. We now have evidence of ancient algae from more than 3.5 billion years ago. The Earth had only become cool enough for liquid water to exist a few hundred million years before. So, to reach the stage of algae in the allotted time, life's most primitive forms must have begun as soon as liquid water became possible. Doing better than pond scum, on the other hand, must be difficult. It took more than two billion years for more complicated life to appear.

In physics, the miraculous moment is the Big Bang. The entire universe, all of everything, even space and time themselves, appeared from nothing. How could this be? No one knows. Will we ever know? Until recently, most physicists thought not. The very conditions at the beginning—the so-called singularity—seemed to destroy the validity of the known laws of physics. Not only were physicists resigned to failure, they were distressed by the idea of a creation. It smelled too much like the Garden of Eden.

In his book *God and the Astronomers*, Dr. Robert Jastrow cites distinguished physicists expressing their discomfort at the thought that the universe had somehow sprung into existence. Although the Big Bang differed from the biblical story in every detail but the critical one of creation itself, the religious took solace and some scientists despaired. Dr. Jastrow pictures the scientists climbing the mountain of nature's truths. Exhausted, they barely crawl to the top. They are surprised to find a convention of theologians. "What took you so long? We have been here all along."

What is the Big Bang and what is the evidence for it? In the 1920s astronomers discovered that the color of the light sent to us from distant galaxies was redder than the light from nearer ones. The more distant the galaxy the more the shift of light from short wavelengths to longer wavelengths. The astronomers tried many ways to account for this reddening. For example, intergalactic dust can redden star light in the same way that the eruptions of Mt. St. Helen's and Mt. Pinatubo reddened sunsets. After the astronomers accounted for all known reddening causes one remained: the distant galaxies were moving away from us. This recession velocity lowers the light's frequency in the same way that the pitch of a car horn sounds lower when the car moves away from us than it does when it stands next to us.

Like raisins in bread baking in an oven, the galaxies are sailing apart from each other, and the farther apart they are the faster they are separating. Running the film backwards, so to speak, astronomers calculate that 10 or 20 billion years ago everything was in the same place. Distance measurements are among the most difficult and controversial in astronomy. That is why age estimates for the universe have only one significant figure and the range covers a factor of two. To be brief, astronomers have had to adjust their time scales every decade or so. At first, the adjustments made the universe older. You may have read the recent newspaper accounts to the effect that new measurements of the universe's age show that the universe is younger than its oldest stars. Whatever the actual age turns out to be, this controversy does not contradict the idea that everything was once in one place, or, put another way, that every place was one place. The dispute has to do with how fast to run the film backwards.

By the 1960s, when Dr. Jastrow wrote his book, two other powerful lines of evidence had persuaded astronomers that the Big Bang was real and that various proposals for an eternal universe were unworthy. One of these lines of evidence was the discovery of the so-called three degree blackbody radiation. This faint microwave radiation, which comes to us from intergalactic space, finds its only natural explanation as the remnant radiation from the exceedingly hot, dense early universe cooled by the expansion. A third line of evidence that supports the Big Bang idea is that astronomers can calculate from supposed conditions of the earliest state the amount of primordial helium, lithium, and a few other light elements.

These three pillars—cosmic expansion, remnant background radiation, and primordial elements—form the foundation of what is now a massively supported theory called the Big Bang. During the past 30 years, the Big Bang theory has passed many scientific tests. Like carpenters laying in additional crossbracing, astronomers have solved problems posed by the theory, and they have made predictions subsequently confirmed. Over the years the Big Bang theory has withstood many storms. And yet, from clouds looming over this triumph of natural law a mocking voice still calls out, "Where did it all come from? Explain that!"

Is Supernatural Intervention Necessary?

The remnant three degree microwave radiation comes to us equally from all directions. This tells us that when the radiation last hit something, a few hundred thousand years after the beginning, the things that it hit were uniformly distributed throughout the young universe. If this last opaque material were uniform then astronomers could not explain how gravitation could have produced the clumps that later formed galaxies. They have been looking for nonuniformities in the radiation.

Recently, they found them. Berkeley professor George Smoot announced that, after many years of searching, his detectors had measured tiny variations, less than one hundredth of one per cent, in the microwave background. He proclaimed "We have seen the mind of God!" With this overblown metaphor Professor Smoot, who probably does not believe that God created the universe, sent physicists' eyes rolling. For at almost the same time other physicists have begun to find the tools and the language to ponder the uncaused formation of the universe from nothing.

This extraordinary advance arises from a startling confluence of our theories of the microscopic world, quantum mechanics and elementary particle theory, and our theory of the universe as a whole, general relativity. The ideas involved are speculative. This area of investigation is still an exciting melee where imagination counts as much as careful calculation and observation.

I will try to explain some of these ideas to you, but do not quote me about this. Everyone's ideas might be completely different next year. The significance of these ideas is not whether they are right or wrong, but that the realm of the last miracle now seems within the reach of science.

Quantum mechanics is our most fundamental theory about the microscopic world. This powerful, deep, accurate, and beautiful theory teaches us that the world of the tiny is radically different from the world of our everyday experience. One remarkable difference is that tiny things have vague properties. Usually scientists explain the Heisenberg Uncertainty Principle, which embodies this vagueness, by saying that you cannot measure a particle's position and velocity at the same time as accurately as you wish. Measuring one very carefully will disturb the measurement of the other. This leaves readers with the view that the particle has a particular position and velocity, but that the scientists cannot measure it. A more truthful statement of the Uncertainty Principle would be that a particle does not have a position or velocity. The scientist and the particle together create the fuzzy position or velocity that the measuring instrument reports.

Forgive me for not going into the details. The upshot of this microworldly vagueness is that "nothing" itself has properties. For if "nothing" were exactly zero, it would violate the Uncertainty Principle that every tiny thing is vague. Physicists call this "nothing" the vacuum, and the vacuum has ghostly properties. Particles and their antiparticle brethren spring into existence and vanish again. They have to do this so quickly that we cannot directly see them. If we could see them, it would not be a vacuum, but if they were not there the vacuum would be exactly "nothing" and it would violate the vagueness rule.

Some of you may think that this is worse than theologians considering whether an omnipotent God can make an object too big for God to move. The difference is that this fuzzy nothing has effects that we can compute and measure. A few decades ago, in a theoretical and experimental tour de force, physicists calculated and observed the Lamb shift, as it is called. The properties of atoms are different by a tiny amount because of the fuzzy vacuum, so we know that quantum things pop in and out of existence.

General relativity is our most fundamental theory about the entire universe, about gravitation, and about space and time. Einstein taught us that these things inextricably entwine. General relativity is not a quantum theory, and physicists believe that every theory at its base has to be a quantum theory. What happens when you apply quantum mechanics to general relativity? No one today knows, but we can make some guesses.

Space and time themselves must come in tiny indivisible chunks. Professor John Wheeler, a famous relativist, illustrated one of his papers with a close-up of a sponge. "This is a picture of space-time at the smallest scales," he wrote.

In popular speech the shortest possible time is a New York minute. It is the time that elapses between a stoplight turning green and the cabby behind you honking his horn. The time is: 0.00000000000000000000000000000000000000000005 seconds. Physicists call this the Planck time. What is a New York inch? It is the distance light travels in a New York minute, which is a number with 10 fewer zeros. If space and time come in chunks the chunks are tiny.

What happens to matter when it squeezes into such tiny spaces and when things happen at such short times? No one knows very much about this, but we think that it must happen to the stuff inside a black hole and we guess that the universe must have been that small once. Students of quantum gravity think that, just as quantum particles flicker in and out of existence from nothingness, so must quantum time and space. What does it mean for a tiny bubble of space and time to come into existence? The nothing of quantum relativity produces universes. It's a little hard to talk about how big these bubbles are, since everything, including space and time, are inside them, but most are tiny and short lived. Some, however, in the fuzzy way of quantum theories, are bigger and last longer. The laws of physics themselves appear within the bubbles and may differ from one bubble to the next. Do not forget that we are engaging in informed speculation.

Even now, within black holes, irresistible gravitation is crushing matter to the quantum nothingness from which universes can appear. Perhaps, some physicists have speculated, universes themselves evolve. Those that have the right laws and properties to produce new universes remain. We are here because this particular universe has the right properties to produce black holes and so must its ancestor universes. Evidently, to produce many black holes a universe must be big enough and powerful enough to last long enough for it to be possible that life can evolve somewhere within it.

These wonderful speculations are different from metaphysics. They stand upon strong theories and solid knowledge, but they are at the frontiers of human knowledge. Are they true even there? No one knows, but unlike revelation, these ideas are subject to critical tests—experimental and observational confirmation.

One brief matter remains. Would a quantum relativistic bubble universe have a cause? It would not. Quantum events, such as the decay of a radioactive nucleus, the spontaneous creation of elementary particles, or the measurement of a fuzzy quantum property, have no causes. The doctrine that quantum events have causes yet unknown to us is called the theory of hidden variables. Albert Einstein, acknowledged the accuracy of quantum mechanics, but hoped someone would clear up the quantum world's fuzziness. Amazingly, in the last decade, experiments have shown us in a powerful and general way that to wish for hidden variables is a forlorn hope.

The Uncaused Universe

Dear reader, you have followed me a long way. I have suggested to you that the last miracle, the creation of the universe, may not remain a supernatural event for long. When science finally solves the origin of the universe, the last reason for belief in the supernatural will vanish but the mystery will remain. Let us face the facts with courage. The universe is without cause and without purpose. This assertion throws many believers into a black funk. "What is life for?" they ask. "How can I live without knowing that my life has a purpose?" Cause and purpose are not properties of the universe like mass and momentum. They are creations of the human mind. That fact is the source of our glory and of our despair. We are responsible to ourselves, to our peers, and to future generations for the consequences of our actions, insofar as we can foresee them.

Science and Its Myths

WILLIAM F. MCCOMAS

For almost 25 years the National Science Board has surveyed the American public as part of its Science and Engineering Indicators study to determine the state of interest in and awareness of fundamental issues in the sciences and technology. The results of the most recent study were recently released with the conclusion that the "level of interest in science and technology in the U.S. has remained relatively stable over the past 16 years with approximately 40 percent of Americans expressing a high level of interest in science . . ." (National Science Board, 1996). This finding is encouraging, but when probing more deeply it becomes clear that even though Americans may be interested in science, they have no clear idea how science functions at a procedural level.

One element of the survey examined what individuals think about how science is conducted. The study designers formulated a series of questions aimed at classifying respondents' positions on a four-level hierarchy of understanding the nature of science. Those at Level I understand that science is concerned with the development and testing of theory. Those lacking this degree of sophistication regarding science, but still having an awareness that experiments require a control group, would be classified as Level II. Individuals at Level III do not have the comprehension of those in the higher two groups but still see scientific findings based on a foundation of careful and rigorous comparison with precise measurements. Those lacking any appreciation of the nature of science at any of these levels are classified as Level IV.

When the surveys were analyzed, those with higher levels of education and with more science and mathematics education were likely to be in the groups with the most sophisticated understanding about how science functions. Considering all of the 2000 adult respondents, two percent were at Level I, 22 percent were at Level II, 13 percent were at Level III and 64 percent were at Level IV. This finding is sobering. Even as measured by this assessment of the basic nature of science elements contained in this study, over 60 percent of the American public effectively has no knowledge of how science operates.

As a science educator with almost 20 years of service both in high schools and universities, I do not find this conclusion surprising. Even though the goal that students understand the nature of science is typically cited as equal in importance to acquiring science content (as outlined by the AAAS in 1993 and the National Research Council in 1996), it is clear that students and adults simply do not appreciate even the most basic aspects of the operation of science. The reasons are astoundingly simple.

Decades ago, educator and philosopher Joseph Schwab (1962) criticized science instruction by calling it the "rhetoric of conclusions." Then and now, science content is emphasized to the exclusion of process both in texts and in science classes. Science teachers have no opportunity to learn how science functions and, not surprisingly, do not emphasize that aspect of science to their students. Furthermore, those educators who would like to share something of the pageant of science with students can only turn to textbooks that frequently underrepresent the process element of science or, in many instances, actually misrepresent the way in which knowledge is generated in science.

This article discusses many of the "myths of science" that are common in science textbooks, in classroom discourse and in the minds of adult Americans. At one level, myths can be entertaining, but when fact and fiction blur, myths lose their entertainment value and serve to block understanding. As a professor of science education at the University of Southern California, I have discovered that such is the case with the myths of science.

Joseph Campbell (1968) proposed that the similarity among many folk myths worldwide is due to a subconscious link between all peoples. As engaging as this notion is, no such link can explain the myths of science. Misconceptions about science are most likely due to the lack of philosophy of science content in teacher education programs, the failure of such programs to provide and require real science research experiences for pre-service teachers, and the generally shallow treatment of the nature of science in the pre-college textbooks to which teachers might turn for guidance.

As Stephen Jay Gould points out in "The Case of the Creeping Fox Terrier Clone" (1988), science textbook writers are among the most egregious purveyors of myth and inaccuracy. The "fox terrier" refers to the classic comparison used to express the size of the dawn horse, tiny precursor to the modern horse. This comparison is unfortunate for two reasons. Not only was this horse ancestor much bigger than a fox terrier, but the fox terrier breed of dog is virtually unknown to today's American students. The major criticism leveled by Gould is that once this comparison took hold, no one bothered checking its validity or utility. Through time, one author after another simply repeated the inept comparison and continued a tradition making many science texts virtual clones of each other on this and countless other points.

In an attempt to provide a more realistic view of science and point out issues on which science teachers should focus, this article presents and discusses 15 widely held, yet incorrect, ideas about the nature of science. There is no implication that all students, or most teachers for that matter, hold all of these views to be true, nor is the list meant to be the definitive catalog. Cole (1986) and Rothman (1992) point out additional misconceptions worthy of consideration. However, years of science teaching and the review of countless texts have substantiated the validity of the inventory presented here.

Myth 1: Hypotheses Become Theories Which in Turn Become Laws

This myth deals with the general belief that with increased evidence there is a developmental sequence through which scientific ideas pass on their way to final acceptance. Many believe that scientific ideas pass through the hypothesis and theory stages and finally mature as laws. A former president expressed his misunderstanding of science by saying that he was not troubled by the idea of evolution because it was, in his words, "just a theory." The president's misstatement is the essence of this myth; an idea is not worthy of consideration until "lawness" has been bestowed upon it.

The problem created by the false hierarchical nature inherent in this myth is that theories and laws are very different kinds of knowledge. Of course there is a relationship between laws and theories, but it is not the case that one simply becomes the other, no matter how much empirical evidence is amassed. Laws are generalizations, principles or patterns in nature, while theories are the explanations of those generalizations (Rhodes and Schaible, 1989; Homer and Rubba, 1979; Campbell, 1953).

For instance, Newton described the relationship of mass and distance to gravitational attraction between objects with such precision that we can use the law of gravity to plan space flights. During the Apollo 8 mission, astronaut Bill Anders responded to the question of who was flying the spacecraft by saying, "I think Isaac Newton is doing most of the driving right now" (Chaikin, 1994, 127). His response was understood by all to mean that the capsule was simply following the basic laws of physics described by Isaac Newton centuries earlier.

The more thorny, and many would say more interesting, issue with respect to gravity is the explanation for why the law operates as it does. At this point, there is no well accepted theory of gravity. Some physicists suggest that gravity waves are the correct explanation, but with clear confirmation and consensus lacking, most feel that the nature of gravity still eludes science. Interestingly, Newton addressed the distinction between law and theory with respect to gravity. Although he thought he had discovered the law of gravity, he refrained from speculating publicly about its cause: "I have not been able to discover the cause of those properties of gravity from phenomena, and I frame no hypothesis . . . it is enough that gravity does really exist, and act according to the laws which we have explained . . ." (1720, 547).

Myth 2: Scientific Laws Are Absolute

This myth involves two elements. First, even if individuals understand that scientific laws are equal in importance to theories, they rarely appreciate that all knowledge in science is tentative, occasionally seeing "proof" in science as identical to proof in mathematics. The issue of tentativeness is part of the self-correcting aspect of science but one that those who would fault science frequently ignore. Creationists, for instance, are quick to criticize science by pointing to the fossil tooth discovered in Nebraska early in this century. Initially, the tooth was considered to have come from a primitive human, but it was later found to be that of an extinct pig. Scientists made both the initial mistake and the later correction, but those who would like to fault science only discuss the error and never the resolution.

Another aspect of this myth stems from the realization that there are different basic kinds of laws—deterministic and probabilistic. Although both types of laws are as tentative as any scientific knowledge, the laws of the physical sciences are typically deterministic in that cause and effect are more securely linked, while the laws in biology usually have a probability factor associated with them. In the life sciences it is typical to see limitations placed on the application of laws. For example, Mendel's laws of inheritance work only with single gene pairs and not even with all such pairs. This issue has called some to question if there are any laws in biology. My response would be that there are laws in the life sciences, but the rules for their application are somewhat different from those applied in the physical sciences.

Myth 3: A Hypothesis Is an Educated Guess

The definition of the term hypothesis has taken on an almost mantra-like life of its own in science classes. If a hypothesis is simply an educated guess, as students typically assert, the question remains, "an educated guess about what?" The best answer for this question must be, that without a clear view of the context in which the term is used, it is impossible to tell.

The term hypothesis has at least three definitions, and for that reason should be abandoned and replaced, or at least used with caution. For instance, when Newton said that he framed no hypothesis as to the cause of gravity, he was saying that he had no speculation about an explanation of why the law of gravity operates as it does. In this case, Newton used the term hypothesis to represent an immature theory.

As a solution to the hypothesis problem, Sonleitner (1989) suggested that tentative or trial laws be called generalizing hypotheses with provisional theories referred to as explanatory hypotheses. Another approach would be to abandon the word hypothesis altogether in favor of terms such as speculative law or speculative theory. With evidence, generalizing hypotheses may become laws and speculative theories become theories, but under no circumstances do theories become laws. Finally, when students are asked to propose a hypothesis during a laboratory experience, the term now means a prediction. As for those hypotheses that are really forecasts, perhaps they should simply be called what they are, predictions.

Myth 4: A General and Universal Scientific Method Exists

The notion that a common series of steps is followed by all research scientists must be among the most pervasive myths of science given the appearance of such a list in the introductory chapters of many precollege science texts. The steps listed for the scientific method vary somewhat from text to text but usually include: a) define the problem, b) gather background information, c) form a hypothesis, d) make observations, e) test the hypothesis and f) draw conclusions. Some texts conclude their list of the steps of the scientific method with communication of results as the final element.

The universal scientific method is one of science education's most pervasive "creeping fox terriers." The multi-step list seems to have started innocently enough when Keeslar (1945a, b) prepared a list of a number of characteristics associated with scientific research such as establishing controls, keeping accurate records, making careful observations and measurements. This list was refined into a questionnaire and submitted to research scientists for validation. Items that were highly ranked were put in a logical order and made part of the final list of elements associated with the investigation of scientific problems. This list was quickly adopted by textbook writers as the description of how science is done. In time the list was reduced from 10 items to the six mentioned above, but in the hands of generations of textbook writers, a simple list of characteristics associated with scientific research became a description of how all scientists work.

Another reason for the widespread belief in a general scientific method may be the way in which results are presented for publication in research journals. The standardized style makes it appear that scientists follow a standard research plan. Medawar (1990) reacted to the common style exhibited by research papers by calling the scientific paper a fraud since the final journal report rarely outlines the actual way in which the problem was investigated.

Those who study scientists at work have shown that no research method is applied universally (Carey, 1994; Gibbs and Lawson, 1992; Chalmers, 1990 and Gjertsen, 1989). The notion of a single scientific method is so pervasive that many students must be disappointed when they discover that scientists do not have a framed copy of the steps of the scientific method posted above each laboratory workbench.

Close inspection will reveal that scientists approach and solve problems by using imagination, creativity, prior knowledge, and perseverance. These, of course, are the same methods used by all effective problem-solvers. The lesson to be learned is that science is no different from other human endeavors where puzzles are investigated. Fortunately, this is one myth that may eventually be displaced since many newer texts are abandoning or augmenting the list in favor of discussing the various methods of science.

Myth 5: Evidence Accumulated Carefully Will Result in Sure Knowledge

All investigators, including scientists, collect and interpret empirical evidence through the process called induction. This is a technique by which individual pieces of evidence are collected and examined until a law is discovered or a theory is invented. Useful as this technique is, even a preponderance of evidence does not guarantee the production of valid knowledge because of what is called the "problem of induction."

Induction was first formalized by Francis Bacon in the 17th century. In his 1620 book, *Novum Organum,* Bacon advised that facts be accumulated without bias to reach a conclusion. The method of induction he suggested is in part the principal way by which humans traditionally have produced generalizations that permit predictions.

The problem is that it is both impossible to make all observations pertaining to a given situation and illogical to secure all relevant facts for all time, past, present and future. However, only by making all relevant observations throughout all time could one say that a final valid conclusion had been made. On a personal level, this problem is of little consequence, but in science the problem is significant. Scientists formulate laws and theories that are supposed to hold true in all places and for all time, but the problem of induction makes such a guarantee impossible.

The proposal of a new law often begins through induction as facts are heaped upon other relevant facts. Deduction is useful in checking the validity of a law. For example, if we postulate that all swans are white, we can evaluate the law by predicting that the next swan found will also be white. If it is, the law is supported (but not proved as will be seen). Locating a black swan will cause the law to be rejected.

The nature of induction itself is another interesting aspect associated with this myth. If we set aside the problem of induction momentarily, there is still the issue of how scientists make the final leap from the mass of evidence to the conclusion. In an idealized view of induction, the accumulated evidence will simply result in the production of a new law or theory in a procedural or mechanical fashion. In reality, such is not the case. The issue is far more complex and interesting. The final creative leap from evidence to scientific knowledge is the focus of the next myth of science.

Myth 6: Science and Its Methods Provide Absolute Proof

The general success of the scientific endeavor suggests that its products must be valid. However, a hallmark of science is that it is subject to revision when new information is presented or new analyses conducted. Tentativeness and the concomitant lack of dogmatism are one of the points that differentiate science from other forms of knowledge. Accumulated evidence can provide support, validation and substantiation for a law or theory, but never prove those laws and theories to be true (see Homer and Rubba, 1978; and Lopushinsky, 1993).

The problem of induction argues against absolute proof in science, but there is another element of this myth worth exploring. In actuality, the only truly conclusive knowledge produced by science results when a notion is falsified. What this means is that no matter what scientific idea is considered, once disconfirming evidence begins to accumulate, at least we know that the notion is untrue. Consider the example of the white swans again. One could search the world and see only white swans, and arrive at the generalization that "all swans are white." However, the discovery of one black swan has the potential to overturn, or at least result in modifications of, this proposed law of nature. Finding yet another white swan does not prove anything; its discovery simply provides some comfort that the idea has merit. Whether scientists routinely try to falsify their notions as has been recommended by philosopher of science Karl Popper, and how much contrary evidence it takes for a scientist's mind to change are fascinating issues.

Myth 7: Science Is Procedural More Than Creative

We accept that no single guaranteed method of science can account for the success of science, but realize that induction (the collection and interpretation of individual facts providing the raw materials for laws and theories) is the foundation of most scientific endeavors. This awareness brings with it a paradox. If induction itself is not a guaranteed method for arriving at conclusions, how do scientists develop useful laws and theories?

Induction makes use of individual facts that are collected, analyzed and examined. Some observers may perceive a pattern in these data and propose a law in response, but there is no logical or procedural method by which the pattern is suggested. With a theory, the issue is much the same. Only the creativity of the individual scientist permits the discovery of laws and the invention of theories. If there truly was a single scientific method, two individuals with the same expertise could review the same facts and likely reach identical conclusions. There is no guarantee of this because the range and application of creativity are a personal attribute.

Unfortunately, many of the methods used to teach science actually work against the creative element. The majority of laboratory exercises, for instance, are simple verification activities. The teacher discusses what is going to happen in the laboratory, the manual provides step-by-step directions and the student is expected to arrive at a particular answer. Since this approach is the antithesis of the way in which science actually operates, such a portrayal must seem dry, clinical and uninteresting to many students. In her 1990 book, *They're Not Dumb, They're Different,* Shiela Tobias argues that many capable and clever students reject science as a career because they are not given opportunities to see it as an exciting and creative pursuit. Science itself may be impoverished when students who feel a need for creative outlet eliminate it as a potential career because of the way it is taught.

Myth 8: Science & Its Methods Can Answer All Questions

Philosophers of science have found it useful to refer to the work of Karl Popper (1968) and his principle of falsifiability to provide an operational definition of what counts as science. Popper suggested that only those ideas that are potentially falsifiable are scientific ideas.

For instance, the law of gravity states that

more massive objects exert a stronger gravitation attraction than do objects with less mass when distance is held constant. This is a scientific statement because it could be falsified if newly discovered objects operate differently with respect to gravitational attraction. In contrast, the core idea among creationists is that species were placed on Earth fully formed by some supernatural force. Obviously, there is no scientific method by which such a belief could be shown to be false. Since this special creation view is impossible to falsify, it is not scientific and the term "creation science" is an oxymoron. Creation science is a religious belief and as such does not require that it be falsifiable. Hundreds of years ago thoughtful theologians and scientists carved out their spheres of influence and expertise and have since coexisted with little acrimony. Today, only those who fail to understand the distinction between science and religion confuse the rules, roles, and limitations of these two important world views.

It should now be clear that some questions simply must not be asked of scientists. During one of the recent creation science trials, for instance, science Nobel laureates were asked to sign a statement about the nature of science to provide some guidance to the court. Seventy-two of these famous scientists responded resoundingly to support such a statement; after all they were experts in the realm of science (Klayman, Slocombe, Lehman & Kaufman, 1986). Later, those interested in citing expert opinion in the abortion debate asked scientists to issue a statement regarding their feelings on this issue. Wisely, few participated. Science cannot answer the moral and ethical questions engendered by the matter of abortion. Of course, scientists as individuals have personal opinions about many issues, but as a group, they should remain silent if those issues are outside the realm of scientific inquiry. Science as a discipline simply cannot conclusively answer moral, ethical, aesthetic, social, and metaphysical questions.

Myth 9: Scientists Are Particularly Objective

Scientists are no different in their level of objectivity than are other professionals. They are careful in the analysis of evidence and in the procedures applied to arrive at conclusions. With this admission, it may seem that this myth is valid, but contributions from both the philosophy of science and psychology reveal that there are at least three major reasons that make complete objectivity impossible.

Many philosophers of science support Popper's (1963) view that science can advance only through a string of what he called conjectures and refutations. In other words, Popper recommends that scientists should propose laws and theories as conjectures and then actively work to disprove or refute their ideas. Popper suggests that the absence of contrary evidence, demonstrated through an active program of refutation, will provide the best support available. It may seem like a strange way of thinking about verification, but the absence of disproof is considered support. There is one major problem with the idea of conjecture and refutation. Popper seems to have proposed it as a recommendation for scientists, not as a description of what scientists do. From a philosophical perspective the idea is sound, but there are no indications that scientists actively practice programs to search for disconfirming evidence.

Another aspect of the inability of scientists to be objective is found in theory-laden observations, a psychological notion (Hodson, 1986). Scientists, like all observers, hold myriad preconceptions and biases about the way the world operates. These notions, held in the subconscious, affect the ability of everyone to make observations. It is impossible to collect and interpret facts without any bias. There have been countless cases in the history of science in which scientists have failed to include particular observations in their final reports. This occurs, not because of fraud or deceit, but because of the prior knowledge possessed by the individual. Certain facts either were not seen at all or were deemed unimportant based on the scientists' prior expectations. In earlier discussions of induction, we postulated that two individuals reviewing the same data might not reach the same conclusions. Not only does individual creativity play a role, but the issue of personal theory-laden observation further complicates the situation.

This lesson has clear implications for science teaching. Teachers typically provide learning experiences for students without considering their prior knowledge. In the laboratory, for instance, students are asked to perform activities, make observations and then form conclusions. There is an expectation that the conclusions formed will be both self-evident and uniform. In other words, teachers anticipate that the data will lead all pupils to the same conclusion. This could only happen if each student had exactly the same prior conceptions and made and evaluated observations using identical schemes. This does not happen in science nor does it occur in the science classroom.

Related to the issue of theory-based observations is the allegiance to the paradigm. Thomas Kuhn (1970), in his ground-breaking analysis of the history of science, shows that scientists work within a research tradition called a paradigm. This research tradition, shared by those working in a given discipline, provides clues to the questions worth investigating, dictates what evidence is admissible and prescribes the tests and techniques that are reasonable. Although the paradigm provides direction to the research, it may also stifle or limit investigation. Anything that confines the research endeavor necessarily limits objectivity. While there is no conscious desire on the part of scientists to limit discussion, it is likely that some new ideas in science are rejected because of the paradigm issue. When research reports are submitted for publication, they are reviewed by other members of the discipline. Ideas from outside the paradigm are liable to be eliminated from consideration as crackpot or poor science and thus will not appear in print.

Examples of scientific ideas that were originally rejected because they fell outside the accepted paradigm include the sun-centered solar system, warm-bloodedness in dinosaurs, the germ theory of disease, and continental drift. When the idea of moving continents was first proposed early in this century by Alfred Wegener, it was vigorously rejected. Scientists could not accept an idea for which there was no explanatory mechanism and was so contrary to the traditional teachings of their discipline. Continental drift was finally accepted in the 1960s with the proposal of plate tectonics as a mechanism to explain how continental plates move (Hallam, 1975 and Menard, 1986). This fundamental change in the Earth sciences, called a revolution by Kuhn, might have occurred earlier had it not been for the strength of the prior paradigm.

It would be misleading to conclude a discussion of scientific paradigms on a negative note. Although the examples provided do show the contrary aspects associated with paradigm-fixity, Kuhn would argue that the blinders created by allegiance to the paradigm help keep scientists on track. His review of the history of science demonstrates that paradigms are responsible for far more successes in science than delays.

Myth 10: Experiments Are the Principal Route to Scientific Knowledge

Throughout their school science careers, students are encouraged to associate science with experimentation. Virtually all hands-on experiences that students have in science class are called an experiment even if it would be more accurate to refer to these exercises as technical procedures, explorations, or activities. True experiments involve carefully orchestrated procedures along with control and test groups usually with the goal of establishing a cause and effect relationship. Of course, true experimentation is a useful tool in science, but it is not the sole route to knowledge.

Many noteworthy scientists have used non-experimental techniques to advance knowledge. In fact, in a number of science disciplines, true experimentation is not possible because of the inability to control variables. Many fundamental discoveries in astronomy are based on extensive observations rather than experiments. Copernicus and Kepler changed our view of the solar system using observational evidence derived from lengthy and detailed observations frequently contributed by other scientists, but neither performed experiments.

Charles Darwin's investigatory regime was frequently more similar to qualitative techniques used in the social sciences than to the experimental techniques associated with the natural sciences. Darwin recorded his extensive observations in notebooks annotated by speculations and thoughts about those observations. Although Darwin supported the inductive method proposed by Bacon, he was aware that observation without speculation or prior understanding was both ineffective and impossible. Techniques similar to Darwin's have been widely used by scientists Goodall and Fossey in their primate studies. Scientific knowledge is gained in a variety of ways including observation, analysis, speculation, and library investigation, in addition to experimentation.

Myth 11: Conclusions in Science Are Reviewed for Accuracy

Frequently, the final step in the traditional scientific method is that researchers communicate their results so that others may learn from and evaluate their work. When completing laboratory reports, students are frequently told to present their methods section so clearly that others could repeat the investigation. The conclusion that students will likely draw from this requirement is that professional scientists are also constantly reviewing each other's experiments to check up on each other. Unfortunately, while such a check-and-balance system would be useful, the number of findings from one researcher which are checked by others is vanishingly small. In reality, most scientists are simply too busy and research funds too limited for this type of review.

It is interesting to note that when scientific experiments are repeated it is usually because a scientific conclusion attacks the prevailing paradigm. In the recent case of cold fusion, scientists worldwide dropped what they were doing to try to repeat the findings provided by Fleishman and Pons. In fairness, these two scientists not only assailed the conventional wisdom but did so in a press conference rather than in a peer-reviewed journal.

The result of lack of oversight has recently put science itself under suspicion. The pressures of achieving tenure, accruing honors, and gaining funds do result in instances of outright scientific fraud, but fortunately such cases are quite rare. However, even without fraud, the enormous amount of original scientific research published, and the pressure to produce new information rather than reproduce others' work, dramatically increase the possibility that errors will go unnoticed.

An interesting corollary to this myth is that scientists rarely report valid but negative results. While this is understandable given the space limitations in scientific journals, the failure to report what did not work is a problem. Only when those working in a particular scientific discipline have access to all of the information regarding a phenomenon, both positive and negative, can the discipline progress.

Myth 12: Acceptance of New Scientific Knowledge Is Straightforward

This misconception addresses the belief that when a more accurate interpretation for the evidence is produced, it will immediately be accepted by the scientific community. Nothing could be farther from the truth, as we have seen in at least one previous myth. A new idea that is not too far from the expectations of scientists working in a particular field would probably gain entry into scientific journals without much trouble—particularly if it comes from someone working in that field. However, if the idea is a significant breakthrough or revolution in Kuhn's use of the term, particularly if it is counter-intuitive or comes from outside the discipline, its acceptance is by no means quick and easy.

The lesson to be learned from this myth is that science is at its heart a human activity. Humans are the producers of new knowledge and also the arbiters of what counts as new knowledge. While nothing like a vote takes place when a new idea is proposed, the peer-review system acts as a gatekeeper to new ideas. Those notions that cannot find a place in the journals will never have a chance to be accepted or denied. Even those new visions of reality that do make it into the journals still have to pass what might best be called the "conference test" if they are to be accepted. Discrepant notions are the talk of professional conferences where they are debated not only in the meeting halls but during dinner and over drinks. As an example, consider the current debate about the origin of modern humans. One view suggests that modern humans arose in various places around the world from ancestral stock, while a competing story places the origin of modern humans squarely in Africa from which they migrated to displace the more primitive human forms living elsewhere. The story is best told in James Shreeve's wonderful book, *The Neanderthal Enigma* (1995), in which he discusses the evidence, personalities, and politics that have directed the conversation about which view should prevail. In many cases, the acceptance of a new scientific idea might be as much a matter of the dynamics of personalities as the strength of the arguments.

Myth 13: Models Represent Reality

This may be one myth that is shared by both scientists and lay persons alike and is related to the distinction between the philosophical views of realism and instrumentalism. Realism is a position that what science produces not only works and permits the production of accurate predictions but actually represents and/or describes nature as known by some omniscient entity. Of course, one of the central limitations of science is that the true nature of reality can never be known because there is no omniscient entity to ask. Science developed, at least in part, to answer questions about the natural world and get as close to "the truth" as possible, but no bell rings or light blinks to tell scientists that they have found the truth. Another philosophical precept is that as long as scientific ideas function properly and are consonant with all of the evidence it does not matter whether they correspond with reality or not. The ideas are useful and descriptive and that should be the end of it.

With this distinction between realism and instrumentalism in mind, we can now turn to the idea of a scientific model. Although no survey has ever been taken on this issue, it seems logical that scientists do believe that they are not just producing useful ideas but that their ideas and descriptions correspond to a reality external to the scientists themselves. Certainly the average person believes this to be true. It is doubtful that anyone seriously questions the model suggested by the kinetic molecular theory of matter which views atoms and molecules as tiny discrete balls that have elastic collisions thus explaining whole ranges of phenomena. Never mind that no one has ever seen these tiny balls or witnessed their impacts. The model works, it permits both predictions and explanations, and therefore must be true. A realist would say that it is true, while an instrumentalist would say it does not matter as long as there is something to be gained from keeping the idea in mind.

The story may be apocryphal, but it is commonly repeated among science educators that when students were once asked what color atoms were, the answer was closely linked to the textbook in use by those students. If the book illustrated atoms as blue, then that was the color students would assign to atoms when asked. It would probably serve us well to think of models as "useful fictions," but it is doubtful that more than a few keep this warning in mind. After all, what got Galileo in trouble was not that he adopted and supported a sun-centered model, but that he taught that the model was the truth.

Myth 14: Science and Technology Are the Same

A common misconception is the idea that science and technology are much the same. In fact, many believe that television, rockets, computers, and even refrigerators are science. However, one of the hallmarks of science is that it is not practical, while refrigerators certainly are. The pursuit of knowledge for the sake of knowledge alone is called pure science, while its exploitation in the production of a commercial product is applied science.

Today, most investigators are working on problems that are at least in part directed from outside their laboratories. Scientists typically blend the quest of pure science in order to solve a technological challenge. In many ways the distinction between pure and applied science is not crucial, but it is interesting to explore what motivates scientists to work on their problems. Few scientists have the luxury to pursue any goal they choose since most scientific work is funded by organizations with an agenda. This funding relationship is not necessarily damaging, but the freedom experienced by the pure scientists of the Elizabethan and Victorian Ages is long gone.

Myth 15: Science Is a Solitary Pursuit

Most would likely accept the premise that science builds on prior work, but that essentially great scientific discoveries are made by great scientists. Even the Nobel prizes recognize the achievements of individual scientists rather than research teams, therefore science must be a somewhat solitary and individual pursuit. Sociologists of science who study scientists at work have shown that only rarely does a scientific idea arise in the mind of a lone individual which is then validated by that individual alone and accepted by the scientific community. The process is much more like a negotiation than the revelation of truth. Scientists work in research teams within a community of like-minded investigators. Many problems in science are simply too complex for a single individual to pursue alone due to constraints of time, intellectual capital, and funding.

Conclusions

The message from the Science and Engineering Indicators Study and from an evaluation of these myths is simple. We must rethink science education. Both students and those who teach science must focus on the nature of science itself rather than just its facts and principles. School science must give students an opportunity to experience science and its processes, free of the legends, misconceptions and idealizations inherent in the myths about the nature of the scientific enterprise. There must be increased opportunity for both beginning and experienced teachers to learn about and apply the real rules of the game of science, accompanied by careful review of textbooks to remove the "creeping fox terriers" that have helped provide an inaccurate view of science and its nature. Only by clearing away the mist of half-truths and revealing science in its full light, with knowledge of both its strengths and limitations, will all students appreciate the true pageant of science and be able fairly to judge its processes and products.

References:

American Association for the Advancement of Science. 1993. *Benchmarks for Science Literacy.* New York: Oxford University Press.

Bacon, F. 1952 (1620). *The New Organon.* In R. M. Hutchins (Ed.), *Great Books of the Western World: Vol. 30. The Works of Francis Bacon* (107–195). Chicago: Encyclopedia Britannica, Inc.

Campbell, J. 1968. *The Hero with a Thousand Faces.* Princeton, NJ: Princeton University Press.

Campbell, N. 1953. *What Is Science?* New York: Dover Publications.

Carey, S. S. 1994. *A Beginner's Guide to Scientific Method.* Belmont, CA: Wadsworth Publishing Company.

Chaikin, A. 1994. *A Man on the Moon: The Voyages of the Apollo Astronauts.* New York: Viking Press.

Chalmers, A. 1990. *Science and Its Fabrication.* Minneapolis, MN: University of Minnesota Press.

Cole, K. C. 1986. "Things Your Teacher Never Told You about Science: Nine Shocking Revelations!" *Newsday Magazine,* March 23: 21–27.

Gibbs, A. and Lawson, A. E. 1992. "The Nature of Scientific Thinking as Reflected by the Work of Biologists and by Biology Textbooks." *American Biology Teacher,* 543: 137–152.

Gjertsen, D. 1989. *Science and Philosophy Past and Present.* New York: Penguin Books.

Gould, S. J. 1988. "The Case of the Creeping Fox Terrier Clone." *Natural History,* 96(1): 16–24.

Hallam, A. 1975. "Alfred Wegener and the Hypothesis of Continental Drift." *Scientific American,* 2322: 88–97.

Hodson, D. 1986. "The Nature of Scientific Observation." *School Science Review,* 68242: 17–28.

Horner, J. K. & Rubba, P. A. 1979. "The Laws Are Mature Theories Fable." *The Science Teacher,* 46(2): 31.

——. 1978. "The Myth of Absolute Truth." *The Science Teacher,* 45(1): 29–30.

Keeslar, O. 1945a. "A Survey of Research Studies Dealing with the Elements of Scientific Method." *Science Education,* 29(4): 212–216.

——. 1945b. "The Elements of Scientific Method." *Science Education,* 29(5): 273–278.

Klayman, R. A., Slocombe, W. B., Lehman, J. S. and Kaufman, B. S. 1986. Amicus Curiae Brief of 72 Nobel Laureates, 17 State Academies of Science, and 7 Other Scientific Organizations, in Support of Appellees. In support of Appellees. *Edwards v. Aguillard,* 85–1513.

Kuhn, T. S. 1970. *The Structure of Scientific Revolutions.* 2nd edition. Chicago: University of Chicago Press.

Lopushinsky, T. 1993. "Does Science Deal in Truth?" *The Journal of College Science Teaching,* 23(3): 208.

Medawar. P. B. 1990 (1963). "Is the Scientific Paper a Fraud?" In P. B. Medawar, *The Threat and the Glory.* New York: HarperCollins.

Menard, H. W. 1986. *The Ocean of Truth: A Personal History of Global Tectonics.* Princeton, NJ: Princeton University Press.

National Research Council 1996. *The National Science Education Standards.* Washington, DC: National Academy Press.

National Science Board 1996. *Science and Engineering Indicators: 1996.* Washington, DC: U.S. Government Printing Office (NSB 96-21).

Popper, K. R. 1968. *The Logic of Scientific Discovery* 2nd ed. revised. New York: Harper Torchbooks.

——. 1963. *Conjectures and Refutations: The Growth of Scientific Knowledge.* New York: Harper and Row.

Rhodes, G. and Schaible 1989. "Fact, Law, and Theory: Ways of Thinking in Science and Literature." *Journal of College Science Teaching,* 18(4): 228–232 & 288.

Rothman, M. A. 1992. *The Science Gap.* Buffalo: Prometheus Books.

Schwab, J. J. 1962. "The Teaching of Science as Inquiry." In J. J. Schwab & P. F. Brandwein (Eds.), *The Teaching of Science.* Cambridge, MA: Harvard University Press.

Shreeve, J. 1995. *The Neanderthal Enigma: Solving the Mystery of Modern Human Origins.* New York: William Morrow and Company.

Sonleitner, F. J. 1989. "Theories, Laws and All That." *National Center for Science Education, Newsletter,* 9(6): 34.

Tobias, S. 1990. *They're Not Dumb, They're Different: Stalking the Second Tier.* Tucson, AZ: The Research Corporation.

Science and Religion

MASSIMO PIGLIUCCI

"The most common of all follies is
to believe passionately in the palpably not true.
It is the chief occupation of mankind."
–H. L. Mencken

The relationship between science and religion (S&R), and even the one between skepticism and religion, is warming up. At least, that is the feeling one gets from a cursory look at recent happenings, from the publication of books and articles in popular magazines about science "finding" God, to the frantic activities of the Templeton Foundation "for the furthering of religion." Two scientists–Paul Davies, and most recently Freeman Dyson–received the one-million dollar Templeton Prize for "progress in religion," the single largest cash prize in history. S&R is not just warm, it's hot!

Thus, the time is ripe for a skeptical analysis of the subject, which, to me, seems muddled by two basic sources of confusion: (1) we need to separate logical/philosophical arguments from those that are either pragmatic or concern freedom of speech; (2) we have to acknowledge that there are many more possible positions on the S&R question than are usually considered, and that a thorough understanding of the whole gamut is necessary to make any progress. This article presents an analysis of both these sources of confusion and an attempt at a classification scheme of the available positions. Since there is no such thing as completely objective reporting, I will advocate my own position as well.

What the Discussion Is and Is Not About

Lest I be accused of being a "rabid atheist" let me make my position clear: I am an atheist in the sense of someone who does not think there is any good reason to believe in a supernatural entity that created and somehow supervises the universe. I do not know that such an entity does not exist, but until extraordinary evidence is provided to substantiate such an extraordinary claim, I relegate God to the same realm as Santa Claus. Rabid I am not, if by that one means an attitude of unreasonable adherence to a doctrine more accepted than carefully considered. My interest in religion comes out of my personal journey into finding out how things really are. Since I am an educator who believes that helping people think critically will result in a better society, I must also react against other people's attempt to curtail my freedom of thought and speech.

Let me briefly examine three components

of the science and religion debate and attempt to separate them as clearly as possible.

1. The relationship between science and religion is a legitimate area of philosophical inquiry which must be informed by both religion (theology) and science.
2. S&R discussions, especially in the United States, carry practical consequences that do not affect science and religion in an equal manner.
3. Discussing S&R has repercussions on the cherished value of freedom of speech for scientists, skeptics, and religionists.

Point 1 is the only point that really should be up for discussion, because it is the only one in which one can seriously engage in free inquiry and reach general conclusions (regardless of whether such conclusions will be shared by a majority). Unfortunately it is often confused with Points 2 and 3 by both believers and nonbelievers.

Point 2 boils down to the fact that attacks on religion are considered politically incorrect—the remarks by Minnesota Governor Jesse Ventura resulted in his popularity dropping 28 points overnight in a poll. Scientists are especially aware of the fact that their research funding depends almost entirely on public financing through various federal agencies such as the National Science Foundation and the National Institutes of Health. Since federal funding is controlled by politicians, who in turn have a tendency to respond to every nuance of their constituency as gauged by the latest poll (Jesse Ventura being an exception), it follows that no matter what your opinion as a scientist on matters of the spirit, it is wiser to stick to your job and avoid upsetting your prince and benefactor.

This is all the more so because of two other things we know about scientists: the overwhelming majority of them do not believe in a personal God (about 60% of general scientists and a staggering 93% of top scientists), and the reason they become scientists is to pursue questions for which science is a particularly good tool. Most of these questions are rather more mundane than the existence of God.

The result of this odd mix is that while most prominent scientists do not believe in a personal God because of their understanding of science and of its implications, they must come out in public with conciliatory statements to the effect that there is no possible contradiction between the two.

The resolution to Point 2 is that there is a philosophical, if not scientific, contradiction between science and religion (see below), but it is not in scientists' interest to start an unholy war that they would lose (given the religious and political climate of the United States). Therefore, if asked, one could answer with the universally convenient "no comment" and live at peace with one's conscience.

Point 3 is rarely raised directly within the S&R debate, but it clearly lurks behind some of the responses one gets when talking or writing about it. Let me make it as clear as possible: no self-respecting scientist or educator—believer or nonbeliever—would want to limit the freedom of speech or expression of any party, including religionists or creationists. There is a fundamental, if rarely fully appreciated, distinction between openly criticizing a position, which is part of the very idea of free speech, and attempting to coerce people into believing what you think is true, or limiting their ability to believe and practice what you think is not true. While religious fundamentalists often do not respect this distinction, most religious progressivists, agnostics and atheists do. It should therefore be clear that discussions about science and religion, or evolution and creationism, deal with free inquiry and education, and in no sense are meant to limit anybody's free speech. Asking to limit what is taught in a science classroom to what is pertinent to that science is sound educational policy, not censorship.

The Many Facets of Science and Religion

In order to continue our discussion of the legitimate philosophical, scientific, and religious aspects of the science and religion quagmire we need a frame of reference to guide us. What I present here is an elaboration on a classification scheme proposed by Michael Shermer. Shermer suggests that there are three worldviews, or "models," that people can adopt when thinking about science and religion. According to the same worlds model there is only one reality and science and religion are two different ways of looking at it. Eventually both will converge on the same final answers, within the limited capabilities of human beings to actually pursue such fundamental questions. The conflicting worlds model asserts that there is only one reality (as the same world scenario also acknowledges) but that science and religion collide head on when it comes to the shape that reality takes. Either one or the other is correct, but not both (or possibly neither, as Immanuel Kant might have argued). In the separate worlds model science and religion are not only different kinds of human activities, but they pursue entirely separate goals. Asking about the similarities and differences between science and religion is the philosophical equivalent of comparing apples and oranges. "These are two such different things," Shermer told Sharon Begley in *Newsweek*'s cover story "Science Finds God," "it would be like using baseball stats to prove a point in football."

Using Shermer's model as a starting point for thinking about S&R, I realized that something is missing. One cannot reasonably talk about the conflict between science and religion unless one also specifies what is meant by religion or God (usually there is less controversy on what is meant by science, though some philosophers and social scientists would surely disagree). So what makes Shermer's picture incomplete is the very important fact that different people have different Gods. I am not referring to the relatively minor variations of the idea of God among the major monotheistic religions, but to the fact that God can be one of many radically different things, and that unless we specify which God we are talking about, we will not make any further progress.

My tentative solution to the problem is therefore presented in Figure 1. Here the panoply of positions concerning the S&R debate is arranged along two axes: on the abscissa we have the level of contrast between science and religion, which goes from none (same worlds model) to moderate (separate worlds) to high (conflicting worlds). On the ordinate is the "fuzziness" of the concept of God, which ranges from a personal God who intervenes in everyday human affairs to the concept of a Naturalistic God who acts only through the laws of physics, to the most esoteric position of deism characterized by a God who created the universe but did not interfere with it since, or even no God (nontheism).

These conceptions of God may take many forms. However, the common denominator to the belief in a personal God is the idea that (S)He intervenes in individual lives, performs miracles, or otherwise shows direct concern for us mortals. A naturalistic God, on the other hand, is a bit more detached: if (S)He intervenes at all it is through the tortuous ways of the natural laws that (S)He himself designed for this universe. Finally, the God of deism does not interfere, even indirectly, in human affairs, but simply answers the fundamental

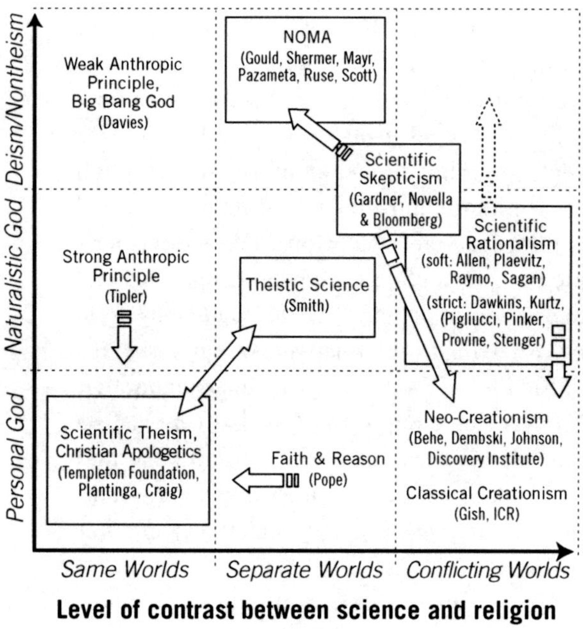

Figure 1

question of why there is something instead of nothing.

Big Bangs, Anthropic Principles, and Christian Apologetics

Figure 1 shows what personalities as diverse as physicists Paul Davies and Frank Tipler, conservative Christian apologist Alvin Plantinga, and science-religion crusader John Templeton have in common, as well as where they differ. Sir John Templeton is a British citizen native of Tennessee, and he has invested $800 million of his personal fortune into furthering a better understanding of religion through science. The Templeton Foundation has sponsored a panoply of activities resulting in articles, books, and conferences whose goal is to "discover spiritual information."

According to Sir John, science has made incredible progress in discovering truths about the natural world. Ergo, its powerful methods should be useful to religion in order to augment our knowledge of God and matters spiritual. And Templeton is putting his money where his mouth is by funding several scientific projects (at the rate of hundreds of thousands of dollars each) as well as by awarding the Templeton Prize, which is financially heftier than the Nobel.

Examples of the science-to-religion connection that Templeton envisions are illuminating. His Foundation has given hard cash to Pietro Pietrini of the National Institute of Neurological Disorders and Stroke to study "Imaging brain activity in forgiving people" ($125,000); Lee Dugatkin of the University of Louisville was awarded $62,757 for research on "Evolutionary and Judaic approaches to forgiving behavior." Herbert Benson of Harvard was aided in answering the question "Does intercessory prayer help sick people?," while Frans de Waal of Emory University was given funds for studying "forgiveness" among primates.

Templeton's efforts (but not necessarily those of all the researchers who are receiving his money) fall into what can be termed scientific theism, that is, the idea that one can scientifically investigate the mind of God. This particular position within the science and religion universe is actually a very old and revered one, having its roots in classical Christian Apologetics à la St. Thomas Aquinas and continuing today through the efforts of individuals like Plantinga and William Craig.

If, however, one believes in a more remote kind of God but wishes to retain the concept of science and religion uncovering the same truth, the choice is not limited to scientific theism. Two other positions are possible, depending on whether one subscribes to a naturalistic or to a deistic God, the Strong Anthropic Principle and Weak Anthropic Principle, the latter also known as the "God of the Big Bang." Of course, throughout this discussion the actual position of individuals within my framework may be different from what I suggest here, either because the boundaries between categories are fuzzy rather than well delineated, or because I may have misunderstood particular individuals' positions based on their writings.

The Weak Anthropic Principle says that there is very little variation in the known constants and laws of physics that could be tolerated if the universe were to be a place friendly to life as we know it. As is, this is a rather trivial observation, but if one wants to read philosophical implications into it, then it is a small leap of faith to claim that the universe was created because life had to exist. From here, there is another small logical gap to the Strong Anthropic Principle, which infers an intelligent designer with a purpose behind the whole shebang. Several physicists and cosmologists have played with different versions of the Anthropic Principle, including Frank Tipler (one of the original proponents of the principle) and Paul Davies, whose exact position on the matter is a bit more difficult to ascertain, but whose awkward combination of a connection with the Templeton Foundation and very careful speculative writings on cosmology puts him squarely in the upper left corner of my diagram.

The anthropic principle is difficult to counter on purely philosophical grounds, other than it seems to be begging the question and somehow reverses the direction of causality (a general cause is inferred from the observation of a particular result of that cause). Furthermore, it is not useful as a scientific hypothesis, since all it says is that we are here because we are here. The Principle has, however, been effectively attacked on positive scientific grounds by showing that many more possible universes could support some sort of life, an attack that has weakened the "improbable" argument on which the Principle is based. A more fatal blow might come in the near future from superstring theory, the current working hypothesis for the reconciliation of the theories of relativity and of quantum mechanics.

While all these positions are compatible with Shermer's "same worlds" scenario, it is clear that a scientist feels more and more comfortable the more one moves toward the upper end of the ordinate in my diagram, that is, the more fuzzy and distant the concept of God becomes (notice that one can adopt a Strong Anthropic Principle scenario and slip toward a personal God at the same time, as indicated by the arrow in the figure). This observation in and of itself, I think, points toward a fundamental degree of discomfort between science and religion.

Gould, the Pope, and Huston Smith

When we examine the portion of the graph in Figure 1 that falls in the area identified by Shermer as the domain of the "separate worlds" model, we deal with a range of characters that go from agnostic evolutionary biologist Stephen Jay Gould (Harvard) and nontheist Eugenie Scott (National Center for Science Education) to the Pope himself, passing through the ambiguous position of the charismatic Huston Smith, the acclaimed author of *The World's Religions*. Let's see how this variation is again accounted for by the different concepts of God these positions reflect.

Several scientists, philosophers, and skeptics, including Shermer, Scott, Mayr, Pazameta, and Michael Ruse, loosely fall into the position outlined by Gould as NOMA, or Non-Overlapping Magisteria (although Ruse is mildly critical of some aspects of this position). NOMA says that science deals with facts, religion with morality; the first focuses on what is, the latter on what ought to be. Citing what in philosophy is known as the "naturalistic fallacy"—one cannot derive what ought to be from what is—Gould concludes that science and religion are forever separate. Another way to look at NOMA has been articulated by Eugenie Scott when she pointed to the distinction between methodological and philosophical naturalism. According to Scott, science adopts naturalism as a convenient tool for conducting research, in a methodological sense. In order to deny the existence of God, however, one has to be a naturalist in the philosophical sense of the term, that is, one has to conclude that the physical world is all there is. Ergo, science cannot inform us as to the existence of God, because naturalism is not a scientific conclusion, but an assumption of the scientific method. If science does not have anything to say about God (and obviously, says Scott, religion is incapable of informing science about the natural world), then NOMA logically follows.

Scott's reasoning is more sophisticated than Gould's, though they share several points. The main commonality is the fact that NOMA defenders are really using the concept of a rather distant God detached from the everyday functioning of nature, since even Gould (and certainly Scott, who makes a living out of valiantly battling creationism) admits that a personal God is in direct contradiction with the scientific evidence. A naturalistic God is marginally compatible with NOMA, but both Gould and Scott seem to be rather uncomfortable with that notion.

I have criticized Gould's position in detail elsewhere and I will therefore only summarize my objections to NOMA here and then briefly turn to Scott's argument. As far as I can see there are at least three points where NOMA fails: (1.) NOMA applies to the very special concept of God that a deist would feel comfortable with, not to what most people think of as "God." Hence, NOMA cannot heal the current schism in our society between religionists and secularists, contrary to what Gould claims. (2.) The naturalistic fallacy can be challenged. For one thing, why shouldn't we use "what is" as at least a rough guide to "what ought to be"? At the very least we should treat this as an open question. Also, science can certainly inform us about the consequences of "what is" so that we can better determine what ought to be to further our own happiness, and science does a much better job at it than religion, whose conclusions are derived from ancient authorities with little knowledge of nature and of human psychology and sociology. (3) It is certainly not true that morality (or, more properly, ethics) is the sole domain of religion, since ethical philosophy has also been providing us with a rational way of discussing our behaviors and their social impact.

Scott's distinction between methodological and philosophical naturalism is certainly more valid than Gould's Solomonic separation between science and religion. A full critique of her position is available online, but the gist of the counterargument has been clearly articulated by Will Provine. Essentially, you can't have your cake and eat it too. Methodological naturalism is not independent, but derived from philosophical naturalism. Therefore, naturalism is an essential component of science not just as a practical device, but because it is part of the very fabric of the scientific method. For example, when scientists apply either Occam's razor (a preference for explanations that make use of a minimum number of necessary theoretical constructs) or Hume's dictum (a preference for less "miraculous" explanations), they are practicing a particular philosophy. Science cannot be divested of such philosophy without losing its nature. This point is seized upon by creationists such as Phillip Johnson, who accuse science of being a religion. Provine's very reasonable rejoinder is that science does indeed make a leap of faith, but that such leap is infinitesimal compared to the leap made by religionists. Furthermore, science's leap—unlike religion's—has produced tangible miracles, such as the laptop computer and a doubled lifespan in the last century.

Moving down the God axis in Figure 1 we come to what I have termed "theistic science" (as opposed to scientific theism). It is not exactly clear how well Smith fits into this category, but his position is the closest I could find to represent the land between NOMA and the Pope (notice the diagonal arrow bridging theistic science and scientific theism, which could represent two sides of the same coin). Smith argues against scientism, an idea that can be defined in different ways. I would argue that scientism is the concept that science can and will resolve every question or problem in any realm if given enough time and resources. I don't think that even the most grant-hungry professional researchers readily subscribe to it, but I know of individuals who seem to.

Smith, however, thinks of scientism (for example in a lecture delivered at Oak Ridge, TN in 1998) as the idea that the scientific method is the best way to investigate reality. According to Smith there are other ways, including intuition and religious revelation. The important point is that these alternatives are not available within science, thereby excluding certain aspects of "reality" from scientific investigation. Smith is joined by Alvin Plantinga in the scientific theism corner, particularly evident in his request that the National Association of Biology Teachers modify their definition of evolution by dropping such philosophically (and politically) loaded words as "impersonal" and "unguided" when referring to the process of natural selection.

While the area occupied by theistic science is borderline and intermixed with different degrees of scientific theism and NOMA (and I do not know which specific mix Smith would prefer), the general idea is that according to theistic science it is perfectly sensible to say that there is a God as well as a physical universe. The distinctive point of theistic science is that the God behind the universe works in very subtle ways and entirely through natural laws, so that it is impossible (or at least very difficult) to infer his presence (unlike the case of the Anthropic Principle, where an intelligent designer is the only possible conclusion).

As the reader can see, then, the center of the diagram in Figure 1 is a rather gray area from which one can easily move to almost any other position by introducing one or more qualifiers. If applied to evolution in particular, theistic science translates into theistic evolution, where evolutionary theory is by and large correct (therefore science is on solid ground), but it includes the added twist that evolution is the (rather inefficient and clumsy) way God works. This is what Barry Lynn (Americans United for the Separation of Church and State) may have meant when he concluded the 1997 PBS *Firing Line* debate for the evolution side by suggesting that the Word (God) in the beginning may simply have been "Evolve!"

The Pope's position assumes the personal God of Catholics but it includes an element of fuzziness as well, and it is accompanied by an arrow pointing left in Figure 1 because one could think of it as a variation of the same worlds model that does not go quite as far as scientific theism à la Templeton. Pope John Paul II has expressed himself twice in the last few years on the relationship between science and religion. In a letter written to the Pontifical Academy of Sciences, he first declared that Christians should not reject the findings of modern science, including evolutionary theory. This is because, in his words, "Truth cannot contradict Truth" (which is why this position could be construed as leaning toward the left side of the diagram).

However, the Pope drew a line at the origin of the human soul, which of course had to be injected directly by God. This creates a rather abrupt discontinuity because it introduces an arbitrary dualism within the process of human evolution, a stratagem with which science does not sit very well, as Richard Dawkins pointed out. John Paul II's more recently published "Fides et Ratio" argues for the fact that science and faith can be used to uncover parallel realities for which each is best equipped, similar to what Gould states as the foundations of NOMA. It is because of this position and the implied dualism that I situated the Pope toward the center of the diagram.

Within the separate (or almost separate) worlds, therefore, one can go from essentially no conflict between science and religion if no god or a deistic God is considered, to a position that is logically possible but increasingly inconsistent with both Occam's razor and Hume's dictum. Depending on how much importance one accords to the philosophical foundations of science, this area of the Science-Religion space can be more or less comfortably inhabited by moderate scientists or moderate religionists.

The Many Faces of Creationism

The lower right corner of Figure 1 is characterized by two positions whose exponents have a lot in common but who despise each other almost as much as they are opposed to everything else that populates the R&S conceptual space. I am referring to "classical" creationism as embodied, for example, by Duane Gish and his colleagues at the Institute for Creation Research, and to the "neo-creationism" movement well represented by Michael Behe (1996), William Dembski (1998), Phillip Johnson (1997) and other associates of the "Discovery Institute."

No matter what kind of creationist you are, you are very likely to believe in a personal God and in a fundamental conflict between science and religion (or at least, so it seems from the array of publications within both the classical and neo-creationist camps). The main difference between Gish's group and Johnson's ensemble is that the latter is more sophisticated philosophically and makes a more slick use of scientific terminology and pseudoscientific concepts. They are also much more politically savvy, though they do not enjoy the grassroots support of classical creationists because they ironically tend to be seen by most people as "too intellectual."

Essentially, most neo-creationists (among whom there is quite a bit of variation) do not believe in a young Earth, accept micro-evolution (though recently so do some classical creationists), don't believe in the literal truth of the Bible, and don't even call themselves creationists—the preferred term for their version of things is "intelligent design" (some even go so far as to avoid stating just who this intelligent designer might be).

While debunking classical creationism is nowadays not too trying an intellectual exercise, neo-creationists are quite something else. Behe's book on "irreducible complexity" makes the point that the molecular machinery of living organisms is so complex and necessitates all of its parts working in synchrony that it must have been designed. A good rebuttal has to span from David Hume's devastating critique of the generalized version of the argument from design to modern findings on the evolution of specific biochemical pathways. Dembski's reasoning that intelligent design can be inferred by excluding all other alternative hypotheses on probabilistic reasoning entirely misses the more parsimonious explanation of unintelligent design (i.e., natural selection) to account for biological history and diversity. Finally, Johnson's main thrust that science is really a philosophical enterprise with no better claim to reality than religion can be dealt with by using Provine's argument about philosophical naturalism discussed above.

The Twin Souls of Skepticism

Last, but not least, let's consider the two main versions of modern skepticism, which have produced a lively debate within the skeptic community and which represent the forefront of rational thinking about science and religion. I am referring to what in Figure 1 are labeled "scientific skepticism" and "scientific rationalism," positions associated with people such as Carl Sagan, Will Provine, and Richard Dawkins (the fact that my name falls in one of those fields merely reflects the influence that these people have had on my thinking).

First, notice that both skeptical positions are rather unusual, in that they span more than one quadrant, diagonally in the case of scientific skepticism, vertically for scientific rationalism. Scientific skepticism is the position that skepticism is possible only in regard to questions and claims that can be investigated empirically (i.e., scientifically). For example, Novella and Bloomberg state that "Claims that are not testable are simply outside the realm of science." However, scientific skepticism immediately embarks on a slippery slope that the same authors acknowledge in their article. They admit that "Testable religious claims, such as those of creationists, faith healers, and miracle men are amenable to scientific skepticism," so that religion is not entirely out of the scope of skeptical inquiry.

Furthermore, they acknowledge that there is no distinction in principle between religion and any other kind of nonsense believed by all sorts of people: "There is no distinction between believing in leprechauns, alien abductions, ESP, reincarnation, or the existence of God—each equally lacks objective evidence. From this perspective, separating out the latter two beliefs and labeling them as religion—thereby exempting them from critical analysis—is intellectually dishonest." That is, scientific skepticism converges toward scientific rationalism (see below) when one considers personal gods that intervene in everyday life, but moves toward a NOMA-like position if God is defined in a distant and incomprehensible fashion.

One of the most convincing arguments adduced by scientific skeptics to keep religion out of skeptical inquiry is that a believer can always come up with unfalsifiable ad hoc explanations of any inconsistency in a religious belief. While this is certainly true, is this not an equally valid critique of, say, skeptical inquiry into paranormal phenomena? After all, how many times have we heard the "true believers" saying that the reason a medium failed a controlled test is because of the negative vibrations produced by the skeptic? Nicholas Humphrey, in his excellent *Leaps of Faith*, even reports that paranormalists have come up with a negative theory of ESP that "predicts" that the frequency of genuine paranormal phenomena is inversely proportional to attempts at empirically investigating them! This sounds like religious believers' attempts to save their cherished mythology.

As much as one might question scientific skepticism on the basis of more or less subtle philosophical points, there is of course another, more practical side to this position, which also makes for a convergence toward NOMA. As Novella and Bloomberg honestly admit, it is a matter of resources: "This single issue, which is not central to our purpose, could potentially drain our resources, monopolize our public image, and alienate many potential skeptics." This is, unfortunately, very true. It is also true that the skeptic community cannot and should not require any article of faith (such as unbelief in God) from any of its members. However, we do require that there are no sacred cows. Anything and everything must be the subject of free inquiry and skeptical investigation. To allow otherwise, for practical or any other kind of reasons, is an intellectual travesty. On the other hand, what can and should be admitted is that God and religion truly do represent only one facet of the universe of interest to skeptics, and that skeptical analyses of the God question may or may not be fruitful. Therefore, let us proceed with caution, but proceed nevertheless.

Within the framework of scientific rationalism one arrives at the belief in the nonexistence of God, not because of certain knowledge, but because of a sliding scale of methods. At one extreme, we can confidently rebut the personal Gods of creationists on firm empirical grounds: science is sufficient to conclude beyond reasonable doubt that there never was a worldwide flood and that the evolutionary sequence of the Cosmos does not follow either of the two versions of Genesis. The more we move toward a deistic and fuzzily defined God, however, the more scientific rationalism reaches into its toolbox and shifts from empirical science to logical philosophy informed by science. Ultimately, the most convincing arguments against a deistic God are Hume's dictum and Occam's razor. These are philosophical arguments, but they also constitute the bedrock of all of science, and cannot therefore be dismissed as non-scientific. The reason we put our trust in these two principles is because their application in the empirical sciences has led to such spectacular successes throughout the last three centuries.

Admittedly, the scientific rationalist is on less firm ground the more she moves vertically up in Figure 1. But this is not a fatal blow because no reasonable skeptic asserts her positions as definitive truths. All we are saying is "show us." The main reason I prefer scientific rationalism to scientific skepticism (which is more akin to the philosophical position known as empiricism and espoused by many English philosophers between the 17th and 19th centuries) comes down to a matter of which tradeoffs one is more willing to accept. Scientific skepticism trades off the breadth of its inquiry (which is limited) for the power of its methods (which, being based on empirical science, are the most powerful devised thus far). Scientific rationalism, on the other hand, retains as much of the power of science as possible, but uses other instruments—such as philosophy and logic—to expand the scope of its inquiry. As a scientist I have been trained within scientific skepticism; as a somewhat rational human being, I yearn for the wide horizons of scientific rationalism.

Different Beliefs for Different Folks

It should be obvious from this survey that there are many ways to slice the science-religion question, certainly more than I have discussed or can even think of. As mentioned at the onset, the point is not to censure any particular position, but rather to explore their differences from a logical as well as a psychological perspective. In fact, it is almost as interesting to debate the question as it is to wonder why some people subscribe to one point of view or to another (which is the main point of Shermer's 1999 book).

I have already stated my personal preference for scientific rationalism on the grounds that it is highly compatible with the empirical evidence and makes very reasonable assumptions where the evidence is lacking. However, scientific skepticism, NOMA, and even some very weak forms of the anthropic principle are certainly difficult to definitely exclude, and enough intelligent people adopt them to provide some pause for reflection. The more we move from the upper right to the lower left corner of Figure 1, however, the more difficult one's position becomes to defend empirically or rationally, all the way down to the innumerable absurdities embedded in Christian apologetics.

The two axes of Figure 1 define the degree of personality of the god one believes in and the conflict one feels between that concept of god and the world as science uncovers it. As such, this diagram defines a series of fuzzy, slowly intergrading areas of thinking that can help us both understand the relationship between science and religion and the human protagonists of this debate. Where you see yourself and others in Figure 1 is bound to shape your life trajectory in this world and your interactions with other people. What happens beyond this world is anybody's guess, but mine is: nothing.

References:

Allen, S. 1999. "Two Mind-Sets." *Skeptical Inquirer* 23(4), 47–49.

Begley, S. 1998. "Science Finds God." *Newsweek*, July 20, 46–52.

Behe, M. J. 1996. *Darwin's Black Box*. New York: Free Press.

Casti, J. L. 1989. *Paradigms Lost*. New York: Avon Books.

Davies, P. 1993. *The Mind of God*. New York: Simon and Schuster.

Dawkins, R. 1999. "You Can't Have it Both Ways: Irreconcilable Differences?" *Skeptical Inquirer* 23(4), 62–64.

Dembski, W. A. 1998. *The Design Inference*. Cambridge University Press.

Gish, D. T. 1995. *Evolution: The Fossils Still Say No!* North Santee, CA: Institute for Creation Research; http://www.icr.org/.

Gould, S. J. 1997. "Nonoverlapping Magisteria." *Natural History*, March, 16–22.

Gould, S. J. 1999. *Rocks of Ages*. New York: Ballantine.

Greene, B. 1999. *The Elegant Universe*. New York: W. W. Norton.

Holden, C. 1999. "Subjecting Belief to the Scientific Method." *Science* 284, 1257–1259.

Hume, D. 1779. *Dialogues Concerning Natural Religion*. Edinburgh: Gilbert Elliot.

Humphrey, N. 1996. *Leaps of Faith*. New York: Basic Books.

John-Paul-II. 1997. "Message to the Pontifical Academy of Sciences." Reprinted in *Quarterly Review of Biology* 72: 381–383.

Johnson, P. 1997. *Defeating Darwinism by Opening Minds*. InterVarsity Press, Downers Grove, IL.

Kitty, A. 1998. "Objectivity in Journalism: Should We be Skeptical?" *Skeptic* 6(1), 54–61.

Kurtz, P. 1999. "Should Skeptical Inquiry be Applied to Religion?" *Skeptical Inquirer* 23(4), 24–28.

Larson, E. J. and L. Witham. 1997. "Scientists are Still Keeping the Faith." *Nature* 386: 435–436.

Larson, E. J. and L. Witham. 1998. "Leading Scientists Still Reject God." *Nature* 394: 313.

Ledo, M. 1997. *Bible Bloopers: Evidence that Demands a Verdict, Too! A Skeptic Answers Josh McDowell*. Atlanta: AFS.

Leikind, B. J. 1997. "Do Recent Discoveries in Science Offer Evidence for the Existence of God?" *Skeptic* 5(2): 66–69.

Letter to *Skeptic* 6(4), 28.

Mayr, E. 1999. "The Concerns of Science." *Skeptical Inquirer* 23(4), 65.

McIver, T. 1996. "A Walk Through Earth History: All Eight Thousand Years." *Skeptic* 4(1), 32–41.

Moore, G. E. 1903. *Principia Ethical*. Cambridge University Press.

Novella, S. and Bloomberg, D. 1999. "Scientific Skepticism, CSICOP, and the Local Groups." *Skeptical Inquirer* 23(4), 44–46.

Palevitz, B. A. 1999. "Science and the Versus of Religion." *Skeptical Inquirer* 23(4), 32–36.

Palevitz, B. A. and R. Lewis. 1999. "Short Shrift to Evolution?" *The Scientist* 2/1, 11.

Pazameta, Z. 1999. "Science vs. Religion." *Skeptical Inquirer* 23(4), 37–39.

Pigliucci, M. 1998. "A Case against God: Science and the Falsifiability Question in Theology." *Skeptic* 6(2), 66–73.

Pigliucci, M. (in press). "Chance, Necessity, and the New Holy War against Science." A review of W. A. Dembski's *The Design Inference*. *BioScience*.

Pigliucci, M. 1999. "Gould's Separate 'Magisteria': Two Views." A review of *Rocks of Ages* by S. J. Gould. *Skeptical Inquirer* 23(6), 53–56.

Pinker, S. 1999. "Whence Religious Belief?" *Skeptical Inquirer* 23(4), 53–54.

Provine, W. 1988. "Scientists, Face It! Science and Religion are Incompatible." *The Scientist* 9(5), 10.

Raymo, C. 1998. *Skeptics and True Believers*. New York: Walker and Co.

Scott, E. C. 1999. "The 'Science and Religion Movement.'" *Skeptical Inquirer* 23(4), 29–31.

Shanks, N. and K. H. Joplin. 1999. "Redundant Complexity: A Critical Analysis of Intelligent Design in Biochemistry." *Philosophy of Science* 66: 268–282.

Shermer, M. 1997. *Why People Believe Weird Things*. New York: W. H. Freeman.

Shermer, M. 1998. "Is God Dead?" *Skeptic* 6(3), 80–87.

Shermer, M. 1999. *How We Believe: The Search for God in an Age of Science*. New York: W. H. Freeman.

Smith, H. 1991. *The World's Religions*. San Francisco: HarperCollins.

Smith, H. 1993. *The Common Vision of the World's Religions Forgotten Truth*. Harper, San Francisco.

Stenger, V. J. 1996. "Cosmythology: Was the Universe Designed to Produce Us?" *Skeptic* 4(2): 36–41.

Tipler, F. J. 1995. *The Physics of Immortality*. New York: Doubleday.

Trott, R. 1994. "Debating the ICR's Duane Gish." talk.origins (accessed: 1/16/98), web page, www.talkorigins.org/faqs/debating/gish.html.

Wilson, E. O. 1998. *Consilience*. New York: Knopf.

Skepticism and Credulity

Finding the Balance between Type I and Type II Errors

BILL WISDOM

René Descartes in his *Meditations on First Philosophy*, William James in *The Will to Believe*, and others have noted that with respect to our intellectual lives we have the distinct though related goals of acquiring true beliefs and avoiding false ones. Accordingly, we can err in two different ways. We commit a Type I Error when we fail to believe a truth; we commit a Type II Error when we believe a falsehood. Since we want to avoid errors of both kinds, the ideal strategy would prevent the commission of either error. But there is no such perfect strategy. Nor can we guarantee freedom from error by "playing it safe"—by withholding our judgment on some proposition. For if that proposition happens to be true, we thereby commit a Type I error.

There is a simple strategy we could adopt to avoid the commission of all Type I errors—believe everything! Believing everything, we would be sure to believe every truth. But of course we would also believe every falsehood—and hence would commit every Type II error possible.

Similarly, there is a strategy we could adopt to avoid the commission of all Type II errors—believe nothing! Believing nothing, we would never believe a falsehood. But of course we would also believe no truths—and hence would commit every Type I error possible.

Unable to eliminate the risk of error entirely, we would at least like to minimize it. But how? A single strategy to minimize error seems no easier to come by than a single strategy to eliminate it. A suggestion comes from what, in fact, we do when confronted with some specific candidate for belief—call it proposition P. We are guided by a preliminary decision—often but not always a quick decision—about the cost of being wrong in either of the two ways available. Not knowing whether P is true or false, I ask myself which would be worse on this occasion: failing to believe P if it were true (i.e., committing a Type I error) or believing P if it were false (i.e., committing a Type II error). Sometimes one and sometimes the other seems considerably "worse" or in some sense more costly, and we adjust our policy on that occasion accordingly.

If we feel that it would be worse to commit a Type I than a Type II error—worse to miss the truth—then we allow ourselves to believe P on relatively slight evidence lest we fail to believe true P (willingly running the risk of believing false P). For example, Pascal says: "I would have far more fear of being mistaken, and of finding that the Christian religion was true, than of being mistaken in believing it true" (Pascal, 85). That is, with respect to the proposition "the Christian religion is true," committing a Type I error is far worse, he says (considering the prospect of eternal torment), than committing a Type II error (presumably living a finitely long life of wasted virtue). Accordingly, Pascal recommends that we accept Christianity without any supporting evidence at all. (Earlier in the text he gives the details of what has become known as "Pascal's Wager.")

If, on the other hand, we feel that it would be worse on some particular occasion to commit a Type II than a Type I error—worse to hold a false belief—then we require a lot of evidence before we assent, lest we believe false P (willingly running the risk of not believing true P). For example, we hope that the responsible authorities approach in this spirit the proposition "this new drug [pesticide, infant formula, . . .] is safe." If they commit a Type I error then profits may be lost. But if they commit a Type II error—they believe that it is safe while it is not—lives are lost. Consumers would prefer they make a Type I error until there is very strong evidence of its safety.

The issue on any particular occasion, then, is this: how much evidence must I have on behalf of P before I am willing to believe it; and how ready or reluctant am I to believe P? The greater I take the relative cost of a Type II over a Type I error to be (i.e., the more reluctant I am to believe), the more or stronger evidence I require to overcome that reluctance. And conversely, the more costly I think a Type I error to be, the readier I am to believe P on relatively weak grounds.

This analysis of what happens on a specific occasion provides the basis for a distinction between what might be called intellectual "personality-types." There are some folks who in general have relatively low standards of evidence who believe "at the drop of a hat." They behave as if they regard Type I errors as worse than Type II . . . as if in their desire to believe truths they are willing to accumulate a lot of false beliefs as well. These people are called credulous or gullible. And we might generously think of credulity as a policy for minimizing error generally—at least errors of the more serious sort.

On the other hand, there are folks who in general have relatively high standards of evidence—who display a reluctance to believe until overwhelmed by evidence. They behave as if they regard Type II errors as worse than Type I—as if they have, in the colorful phrase of William James, a "preponderant private horror of becoming a dupe" (18). These people are called skeptical. And we might think of skepticism as a policy for minimizing error generally—at least errors of the more serious sort. I shall address two types of thinking that help us distinguish between extreme skepticism and extreme credulity, through the writings of René Descartes, William Clifford, and William James. Skepticism, while not universally held, might seem to be the more respectable stance of the two. People are not ridiculed for being skeptical as they are for being gullible. Indeed skepticism has its advocates, while it would seem nobody recommends gullibility. But this is a feature of our particular age. In other times, it seems, the credulous—at least people credulous with respect to the most important matters—were honored, while the skeptical were burned. So it might be interesting to survey some of the things that have been said for and against skepticism and gullibility respectively.

Because so much has been said about the importance of keeping your standards of evidence high, I shall address only two skepticisms that are specially important for this study. Descartes is important as an early modern exponent of our distinction between two ways of erring. Clifford is important for his influence on James, whose views we shall go on to consider.

Descartes and Clifford

In an effort to establish a firm foundation for his ideas René Descartes, in his *Meditations on First Philosophy* (1640), sets out to find which if any propositions are indubitable. This he does by raising his standards of evidence as high as possible—he will not believe anything that can be doubted, or anything that could be false. In the present terminology, he begins with the determination to commit no Type II errors, even if that leaves him believing nothing.

Certain propositions soon turn out to be indubitable and hence utterly believable on his account: that he exists, that he is a thinking being, that God exists, and that God is not a deceiver. The non-deceptive character of God then provides Descartes with a guarantee that "whatever I perceive very clearly and distinctly is true" (87). Because God is not a deceiver, says Descartes, He has given us no mental faculty the exercise of which could lead us into incorrigible error. But there could be no stronger assurance of the truth of something than our clear and distinct perception of it. So we could not correct our false assent to something clearly and distinctly perceived. So what is clearly and distinctly perceived cannot be false. We should confine our belief to what is clearly and distinctly perceived, and thereby avoid all Type II errors.

Of course by keeping our standards so high, we will commit a lot of Type I errors—we will fail to believe a lot of truths. But Descartes provides a recipe for reducing the number of Type I errors. By critical refinement, we should bring to clarity and distinctness of apprehension more and more of our concepts, and thus bring to indubitability more and more candidates for our belief—thus continually reducing the number of Type I errors, while still preventing the occurrence of Type II errors entirely. (While this material is spread through much of Meditations, the hard core of it is in the last few paragraphs of Meditation Four and the first paragraph of Meditation Five. Descartes says: "Even if I have no power to avoid [Type I] error in the first way . . . , which requires a clear perception of everything I have to deliberate on, I can avoid [Type II] error in the second way, which depends merely on my remembering to withhold judgment on any occasion when the truth of the matter is not clear.")

Descartes's position neatly fits our general picture of a skeptic. He proceeds from the very beginning of his Meditations precisely as if he regards Type II error—believing a falsehood—as far worse than Type I error—missing a truth. He is willing to forego indefinitely many truths rather than allow the slightest falsehood into the body of his beliefs. As time goes on, he will assent to new propositions, but only after they too have been brought to indubitability—i.e., each has met the very highest standards of evidence. (In practice Descartes is not quite this severe: he recognizes probabilities—propositions that are good bets though not indubitable. In these cases he acknowledges that it is prudent to behave as if they were true. But with respect to his philosophy, "the task now in hand does not involve action but merely the acquisition of knowledge" [79].)

In *The Ethics of Belief* (1877) William K. Clifford argues that "it is wrong always, everywhere, and for anyone, to believe anything upon insufficient evidence" (77). Though clearly hostile to religious belief, he nonetheless uses terms like "sinful" and "evil" for "belief . . . given to unproved and unquestioned statements" (74), even if those statements are true. We have a "duty to mankind . . . to guard ourselves from such beliefs as from a pestilence" (75). As an extreme skeptic Clifford demands that we keep our standards of evidence very high, recognizing that in doing so we will cut ourselves off from a lot of true beliefs.

As bad as unjustified belief is, "a greater and wider evil arises when the credulous character is maintained and supported, when a habit of believing for unworthy reasons is fostered and made permanent" (76). That character, that habit, is particularly to be found in "those simple souls . . . who have been brought up from the cradle with a horror of doubt" (77). For Clifford, credulity is the antithesis of his own position.

Believing Because It Is Absurd

It seems that the dominant Christian position today considers it legitimate or appropriate (though not necessary) for the believer to exercise reason—i.e., to apply logic to the material of experience. This view holds that it is not irrational to be a Christian, and perhaps even that the exercise of reason can lead one to, or support belief in, the main tenets of Christianity. Such is one familiar reading of Psalm 19:1: "The heavens declare the glory of God; and the firmament sheweth his handiwork"—i.e., the observable world provides ample evidence of the existence and nature of God. But however strong or old this attitude toward reason, there is a contrary attitude that goes back to the very earliest Christian documents, where God's wisdom reigns supreme: "It is written, I will destroy the wisdom of the wise, and will bring to nothing the understanding of the prudent. . . . Hath not God made foolish the wisdom of this world? For after that in the wisdom of God the world by wisdom knew not God, it pleased God by the foolishness of preaching to save them that believe. . . . The foolishness of God is wiser than men. . . . Not many wise men after the flesh . . . are called: but God hath chosen the foolish things of the world to confound the wise . . ." (1 Corinthians 1: 19–27).

The lesson is unavoidable: to the natural mind, the Christian message is foolish, irrational. This attitude is noteworthy in that the Psalmist says precisely that the atheistic position is the foolish one (14:1, 53:1). "Your faith," says Paul, "should not stand in the wisdom of men, but in the power of God" (2:5).

Some 40 years later, a similar attitude toward reason was put in the mouth of the risen Jesus himself. Many people have not read or have forgotten the punchline to the story of Doubting Thomas. Told that Jesus has risen from the dead, Thomas says that he will not believe it until he can see and feel Jesus' wounded body. When Jesus presents himself to Thomas and invites empirical investigation Thomas says, "My Lord and my God." Jesus' rejoinder is interesting: "Because thou hast seen me, thou hast believed: blessed are they that have not seen, and yet have believed" (John 20: 28–29). Thomas is not blessed for his belief, which however belated is at least justified; the Lord's blessing is reserved for those with unjustified, groundless belief.

It is with this kind of scriptural backing that Tertullian says, perhaps a hundred years after John's gospel: all you need to do is believe what Jesus taught, however implausible. Once you have accepted Jesus' teaching, close your mind to everything else and stop thinking. "With our faith, we desire no further belief. For this is our palmary faith, that there is nothing which we ought to believe besides" (346). "[I]t is really better for us not to know a thing, because He has not revealed it to us, than to know it according to man's wisdom . . ." (354). In his *History of Philosophy*, Wilhelm Windelband concludes that "with Tertullian, the content of revelation is not only above reason, but also in a certain sense contrary to reason. . . . The gospel is not only incomprehensible, but is also in necessary contradiction with worldly discernment: credible est quia ineptum est; certum est, quia impossibile est—credo quia absurdum" (225). It is believable because it is foolish, it is certain because it is impossible—I believe it because it is absurd.

William James and Pragmatic Belief

The most elaborate and I think the strongest defense of credulity comes from William James (1842–1910), one of America's most distinguished and influential philosophers and psychologists. His position is laid out most fully in *The Will to Believe* (1896), but parts of it appear in *Reflex Action and Theism* (1881), *The Sentiment of Rationality* (1882), and *Is Life Worth Living?* (1895). (All citations below will be from the 1956 edition that includes all of these writings.)

James argues that under certain broad circumstances we are entitled to hold beliefs and (most importantly) religious beliefs that are absolutely groundless. He calls The Will to Believe a "justification of faith . . . in religious matters," understanding faith as belief held without rational support. Before examining his argument we must deal with two red herrings.

At a number of points James offers an odd definition of faith. Faith, he says, is the adoption of "a believing attitude . . . in spite of the fact that our merely logical intellect may not have been coerced" (1956, 1–2). "Faith means belief in something concerning which doubt is still theoretically possible; . . . faith is the readiness to act in a cause the prosperous issue of which is not certified to us in advance" (90).

I find this odd because on this definition nearly all beliefs about matters of fact, including the best-confirmed laws of nature, would be held on faith. Surely that is the wrong way to use the word faith. It is clear that we all must have "faith" in this sense. But James does not run his argument thus; his defense of credulity does not depend on redefining the word faith in this bizarre way. So we can disregard this definition.

James also alleges that the scientist must hold on faith "the proposition . . . that the course of nature is uniform. That nature will follow tomorrow the same laws that she follows today is . . . a truth which no man can know; but in the interests of cognition as well as of action we must postulate or assume it" (91). On this he is simply wrong: we need not postulate or assume the uniformity of nature. We certainly hope that nature is orderly. Proceeding on that hope, we would soon discover that we were wrong if the laws of nature changed or ceased. So this "justification of faith" also fails, since the uniformity of nature is not a necessary article of faith.

James' key argument is in *The Will to Believe*, which I would urge everyone to read. It is only 30 pages long (in my edition) and well worth the effort. But for those who have not read the essay (and are willing to take my word for it), I'll summarize his argument. It is based on a number of key definitions.

An hypothesis is "anything that may be proposed to our belief." A live hypothesis is one "which appeals as a real possibility to him to whom it is proposed." An option is a "decision between two hypotheses." A living option is one "in which both hypotheses are live ones." "If I say, . . . 'Either call my theory true or call it false,' your option is avoidable. . . . You may decline to offer any judgment as to my theory. But if I say, ['Either accept my theory or don't'], I put on you a forced option, for there is no standing place outside of the alternative. Every dilemma based on a complete logical disjunction, with no possibility of not choosing, is an option of this forced kind." An option is momentous when (a) the opportunity it represents is unique, (b) the stake is significant, and

(c) the decision is irreversible. "For our purposes we may call an option a genuine option when it is of the forced, living, and momentous kind" (2–4). "When I say 'willing [or passional] nature', . . . I mean all such factors of belief as fear and hope, prejudice and passion, imitation and partisanship, the circumpressure of our caste and set" (9). James concludes: "The thesis I defend is, briefly stated, this: Our passional nature not only lawfully may, but must, decide an option between propositions, whenever it is a genuine option that cannot by its nature be decided on intellectual grounds . . ." (11).

The rest of his argument depends crucially on our earlier distinction between Type I and Type II errors: "Believe truth! Shun error!—these . . . are two materially different laws; and by choosing between them we may end by coloring differently our whole intellectual life. We may regard the chase for truth as paramount, and the avoidance of error as secondary; or we may, on the other hand, treat the avoidance of error as more imperative, and let truth take its chance. . . . These feelings of our duty about either truth or error are in any case only expressions of our passional life" (18), since rational grounds cannot be given for preferring Type I to Type II errors or vice versa.

In other words, there can be no rational grounds for preferring either skepticism or gullibility. When evidence one way or the other is unavailable (or evenly balanced), I am obliged to decide a forced option on passional grounds—i.e., on other than intellectual or rational grounds—and hence I am fully entitled to use "the subjective method, the method of belief based on desire" (97). That is, I am entitled to believe something simply because I would like it to be true—or for any other reason. The skeptic would—and he notes that Clifford does—recommend the suspension of judgment in such a case. But to follow that advice is to risk a Type I error: losing the truth and its attendant benefits—just as the believer risks a Type II error: believing falsely. "Dupery for dupery, what proof is there [or could there be] that dupery through hope [of being right: a Type II error] is . . . worse than dupery through fear [of being wrong: a Type I error]?" (27).

James applies these principles to religious belief and the ultimate question of God's existence. He first explains "the religious hypothesis": "Religion says essentially two things. First, she says that the best things are the more eternal things, the overlapping things, the things in the universe that throw the last stone, so to speak, and say the final word. 'Perfection is eternal,'—this phrase . . . seems a good way of putting this first affirmation of religion, an affirmation which obviously cannot yet be verified scientifically at all. The second affirmation of religion is that we are better off even now if we believe her first affirmation to be true" (25–26). In case you do not yet understand exactly what religion says, James adds: "[The] feeling . . . that by obstinately believing that there are gods . . . we are doing the universe the deepest service we can seems part of the living essence of the religious hypothesis" (28).

First, James takes himself to be addressing only those who regard the option as living: if the religious hypothesis makes no appeal whatsoever to your belief, there is no point in proceeding. Second, he says that the religious option is momentous: "We are supposed to gain, even now, by our belief, and to lose by our non-belief, a certain vital good" (26). So, third, if the option is forced, it is genuine. Further, it surely "cannot by its nature be decided on intellectual ground." So, if the religious option is forced, it legitimately may, because it must, be decided on passional grounds—it would be fully responsible to adopt (or reject) the religious hypothesis on whatever whim might move us.

James guarantees that the option is forced by posing it as he does: "either accept the religious hypothesis or don't." Further he regularly alleges that the failure to accept the religious hypothesis is tantamount to outright rejection. To simplify the point, suppose that the religious hypothesis were "There is a God." He wants us to regard the suspension of judgment as tantamount to rejection—to regard the agnostic position as indistinguishable from atheism. If he can do that, he may be able to lure into the theist fold those who find outright atheism distasteful. "It is often practically impossible to distinguish doubt from dogmatic negation. . . . He who commands himself not to be credulous of God, of duty, of freedom, of immortality, may again and again be indistinguishable from him who dogmatically denies them. . . . Who is not for is against. The universe will have no neutrals in these questions. In theory as in practice, dodge or hedge, or talk as we like about a wise skepticism, we are really doing volunteer military service for one side or the other" (109). James's argument is clever, charming, enticing. What is there to say? I believe that there are at least three things that together should remove whatever allure the argument might have.

1. Do not be tricked by his effort to exhibit the suspension of judgment as virtual rejection of the religious hypothesis. This sleight-of-hand is accomplished by his particular way of forcing the religious option. But his is not the only way. I could also force the religious option by demanding of you: "Reject God or don't!" Sensing the inadequacy of the evidence, you are unwilling to reject the God hypothesis, so you don't. But that's tantamount to theism! Who is not against is for. The universe will have no neutrals in these questions. This of course is preposterous. If by this simple trick of logic I can simultaneously exhibit the suspension of judgment in the absence of evidence as both theism and atheism, then we have to reject James's maneuver as a rhetorical ploy with no probative force at all.
2. James insists that his authorization of groundless belief applies only when an option "cannot by its nature be decided on intellectual grounds." On the other hand, "in our dealings with objective nature we obviously are recorders, not makers, of the truth. . . . Throughout the breadth of physical nature facts are what they are quite independently of us . . ." (20). In such cases, James insists that the only responsible thing to do is to seek out the relevant evidence and suspend judgment until its dictates are clear. But on a view of justification, with which I am largely sympathetic, an allegedly substantive claim that cannot even in principle be confirmed or disconfirmed is empty nonsense, and hence not a candidate for my belief at all, since it does not say anything. So for me at least—and, I would think, for any rational person—an option "that cannot by its nature be decided on intellectual grounds" cannot be a living option because the hypothesis involved can make no appeal to my belief, there being nothing to believe. So for people like me at least, there can be no "genuine option that cannot by its nature be decided on intellectual grounds"; so there can be nothing for James's permission to work on.
3. Regarding God and religion, if a claim "that cannot by its nature be decided on intellectual grounds" is literal nonsense James's religious hypothesis is worse—it is sheer gibberish. I defy anyone to make sense of it:

   First [religion] says that the best things are the more eternal things, the overlapping things, the things in the

> universe that throw the last stone, so to speak, and say the final word. "Perfection is eternal,"—this phrase . . . seems a good way of putting this first affirmation of religion. . . . The second affirmation of religion is that we are better off even now if we believe her first affirmation to be true. [The] feeling . . . that by obstinately believing that there are gods . . . we are doing the universe the deepest service we can, seems part of the living essence of the religious hypothesis.

I think I know why he puts "the religious hypothesis" in this untenable way. He must make sure that it "obviously cannot . . . be verified scientifically at all." The more clearly sensible the religious hypothesis is, the more obviously susceptible it will be to rational investigation, and hence the less eligible for groundless adoption.

The following clearly recognizable version of a religious hypothesis is not gibberish: "The natural world was brought into existence about 10,000 years ago by the will of the one eternal, omnipotent, omniscient, benevolent Being, whose son Jesus died and arose from the dead about 2000 years ago." That makes perfectly good sense: I've got a good idea of what the whole sentence and each part of it means. Why does James not use this or some similarly intelligible version of the religious hypothesis? Because this and the other clearly intelligible formulations are clearly falsifiable and clearly false. In the choice between what is false and what is unintelligible, the proponent of religious belief will (almost) always opt for the unintelligible.

So my final criticism is that James's religious option cannot be genuine because his religious hypothesis cannot be live: it can make no appeal to my belief—or anyone's.

References:

Clifford, William K. 1947 (1877). "The Ethics of Belief." *The Ethics of Belief and Other Essays*. London: Watts & Co.

Descartes, R. 1988 (1640). "Meditations on First Philosophy." *Descartes: Selected Philosophical Writings*. J. Cottingham, R. Stoothoff, D. Murdoch (Trans.). Cambridge University Press.

James, W. 1956. *The Will to Believe and Other Essays in Popular Philosophy, and Human Immortality*. New York: Dover Publications.

Pascal, B. 1941 (1670). "Pensées." W. F. Trotter (Trans). *Pensées and The Provincial Letters*. New York: The Modern Library.

Tertullian (A.D. 160–230). 1966. *Prescription against Heretics: Greek and Roman Philosophy after Aristotle*. Jason L. Saunders (Ed.). New York: The Free Press.

Windelband, W. 1958 (1892). *A History of Philosophy, Vol. 1: Greek, Roman, and Medieval*. James H. Tufts (Trans.). New York: Harper.

Thought Field Therapy

DAVID X. SWENSON

"It is beyond amazing. I have never seen any treatment so powerful. . . . It is extraordinarily powerful, in that clients receive nearly immediate relief from their suffering and the treatment appears to be permanent." So read testimonials on promotional brochures and websites for one of a new cluster of unusual psychotherapies, called "power therapies" due to their alleged rapid and strong results.

These treatments are becoming mainstream, as measured by their attraction of thousands of licensed social workers, psychiatrists, psychologists, and other health professionals to training workshops. These practitioners return home to apply these techniques and their accompanying beliefs on vulnerable, traumatized people, and often charge large fees for it. In September, 1994, a federal grant of $355,225 was given to the University of Alabama, Birmington Burn Center to test the use of Therapeutic Touch (TT) to manipulate the "human energy field" and heal injuries (Turner, 1994). The technique, which, like other "energy" therapies, has no clear empirical support, claims to be taught at 75 schools and universities, practiced at 95 health facilities, and has been taught to more than 48,000 health care professionals in 75 countries (O'Mathna, 1998). The influence has been strong enough to lead the North American Nursing Diagnostic Association to list "energy-field disturbance" among its other diagnostic categories. Many managed care organizations provide coverage for certain alternative therapies, and 58% of national HMOs planned to do so by the end of 1998 (Blecher, 1997).

Although Barnum had a name for them, what is leading fairly conventional therapists to unquestioningly adopt unconventional methods and abandon much of the scientific inquiry on which their professions are based? This paper will consider some of the forces that lead therapists to use these approaches, examine the assumptions that underlie the practice, and identify some of the flaws in the practice that should be thoroughly researched before they are applied to clients. One of the fastest growing appears to be Thought Field Therapy (TFT), and it will be the focus of discussion and representative of the other, similar approaches.

Drivers of the Quick Fix

From the pay for service platform of the 1970s, the health care system has shifted to a highly regulated managed care platform in the 1980s and 1990s. The result is a dramatic change in the management, cost, and delivery of services. Capitation (per head maximum cost per year) and Diagnosis Related Groups (DRGs) became the basis for reimbursement from the government. The payment was the same whether the patient stayed in the hospital for two days or two weeks. Health care had to find ways to recover the lost dollars due to managed care, and health care became a business. One of the ways to cut costs was to limit the number of sessions for which psychotherapy patients could be reimbursed. This meant that mental health centers and independent providers needed to promote themselves based on short term and effective treatments, and provide high turnover in patients to maintain volume.

Enter the brief and short-term therapies. In contrast to traditional psychotherapies of the past that allowed clients the luxury of having their problems discovered, uncovered, or allowed to emerge, these new therapies took clients at their word: whatever they presented as the problem was all that would be treated. If there were other problems, or the problem changed, the case would require reauthorization for further treatment. Such approaches as Ericksonian hypnosis, solution focused therapy, and crisis intervention became popular. It was crisis intervention and the need for rapid treatment of trauma that became the launching pad for the power therapies.

Crisis intervention involves having brief but intense contact with victims of crisis or trauma for the purpose of calming them, reducing risk, and returning them to their precrisis level of functioning. Without effective intervention, people make adjustments around the trauma but often carry their symptoms with them in the form of posttraumatic stress disorder, phobias, sleep disturbances, relationship problems, and the like. As many as 25% of the survivors may develop more serious post traumatic stress disorder as a consequence (Wylie, 1996). Many of these people seek counseling or support, but some have not found it effective. They become vulnerable to the power therapy claims of "immediate resolution of the problem" and "permanent cure." Therapists, eager to find powerful and effective short term means to work with people—both to end their suffering and to increase turnover—are also vulnerable to these claims. Although most professionals have had course work covering research design, statistics, and the scientific method, a large part of their preferred activity is involved with treatment rather than research. Many therapists are very feeling oriented, they relate extremely effectively with clients, and they are often interested in experimenting loosely with new techniques. A poll of participants on the Internet Trauma Forum reflected the attitude: they were asked about the Tapas Acupuncture Technique (see below), and their opinion was reported as, "Well, it's no wackier than TFT, so why not try it?" (Wylie, 1996, 36).

The Power Therapies

The power therapies are so called owing to their claimed remarkable facility in resolving severe or persistent emotional disorders and trauma. One could equally argue that most of them could have derived the name from their connection with bodily or spiritual energy systems on which they rely for explanation (Pearl and Tayar, 1996; Stenger, 1998). The 1990s has seen a plethora of unconventional and "alternative" approaches become popular—promoted widely without clear evidence of their effectiveness other than by testimonials and poorly designed studies.

Some of the more unusual treatments are based on the belief that the body's energy system has become blocked, toxified, weakened, or otherwise disrupted, thereby causing symptoms of physical and/or psychological nature. Acupuncture and acupressure are best known, and though many aspects have been clearly debunked, evidence continues to emerge that some healing mechanism may be involved. It is on tradition and the possibility of such a healing mechanism that the power therapies proliferate.

- Thought Field Therapy (TFT) uses sequences of finger tapping on acupressure points, and combinations of sensory activities (repeating statements, counting, rolling the eyes, humming a tune) while thinking of the distressing situation.
- Touch for Health (TH) or Applied Kinesiology (AK—which actually has no relationship to the discipline of kinesiology) uses muscle testing (usually pulling down on an extended arm) to test where in the body energy is blocked. It then uses acupressure to release and balance the body's energy.
- Therapeutic Touch (TT) is not touch at all, but a "manipulation of the body's aura" by passing hands over the person's body. This method has been widely used in the nursing profession.
- Tapas Acupressure Technique (TAT) appears to be a recent and briefer derivative of TFT that involves using three fingers of one hand to apply gentle pressure to facial acupressure points, while the other hand supports the back of the head. Positive statements may also be used as well as asking where the problem is "stored" in one's body or life.
- Ear Tapping Desensitization and Remobilization (ETDR) is one of the newest treatments on the alternative scene, and involves using one hand to tap acupressure points on the ear.
- Emotional Freedom Technique (EFT) was developed as a branch of TFT, and involves tapping near the end points of energy meridians. It has one tapping sequence for all physical and emotional problems, in contrast to 10–15 sequences in TFT.

These approaches claim to heal a host of diverse physical and psychological conditions. Thought Field Therapy claims to heal the following disorders: trauma (effects of rape, abuse, crime, war), phobias, anxiety, addictions, grief, physical pain, panic, obsessive-compulsive disorders, eating disorders, depression, chronic anger, guilt, self sabotage, food addictions, rejection, sexual problems, fibromyalgia, migraine headaches, and love pain (Callahan, 1998; Edwards, 1997). There is even a suggestion that it might cure cancer (Callahan and Callahan, 1996).

Some of these alternative approaches are also receiving acceptance from respected figures in psychotherapy and emergency services. For example, Charles Figley, editor of *Traumatology* journal and a leading figure in international trauma services, has stated, "It is extraordinarily powerful, in that clients receive nearly immediate relief from their suffering and the treatment appears to be permanent" (Danzig, 1998). Several articles in the journal have been devoted to case studies and discussion of the procedure. To the journal's credit, they also featured a thorough critique of TFT (Hooke, 1998). A comparison of TFT with other power therapies was also published in the *Family Therapy Networker*—a major publication read by family therapists—in which TFT was reviewed favorably.

Thought Field Therapy

TFT development. A self-described "pioneer" in cognitive and behavior therapies, Callahan (1997) reports having developed TFT over the past two decades. He notes that the sources that led to his formulation of the treatment included the Chinese discovery of energy meridians, and the later refinement of acupressure by George Goodheart, DC, a chiropractor who promoted Applied Kinesiology or Touch for Health. Callahan claims that the success that propelled him into promoting TFT was the "Case of Mary" (Callahan, 1997).

This client had been suffering from severe phobia of water that had restricted her life and caused much discomfort. After several months of traditional cognitive and desensitization treatments, and noting that Mary felt sick to her stomach when near the water, Callahan felt inspired to try tapping the acupressure points under her eyes associated with the stomach meridian. Immediately and miraculously, Mary commented, "It's gone . . . I don't think I'm afraid of the water anymore." Bolstered by this success, Callahan set about to find points related to other disorders and use these in his treatment of clients. This case has since become a classic in TFT.

Energetic assumptions. Foundational to Callahan's claims is the belief in and use of an intangible bodily energy that perfuses the person, travels along channels or meridians, and may become blocked by physical or emotional illness. The task of the therapist is to unblock these energy buildups, or "perturbations" as Callahan refers to them. Some healers go to extravagant means to explain and account for these energy fields that they claim to manipulate (Bunnell, ND). Callahan borrows heavily on three sources for his argument. He cites the longevity of acupuncture in Chinese medicine, as well as the use of energy concepts in many ancient civilizations. He and other power therapists also cite the seminal work of Harold Saxton Burr, author of *The Fields of Life* (1972). Burr, a researcher at Yale University School of Medicine, claimed to have discovered "electrodynamic energy fields" that surround all living things. Finally, and with some strain, research and opinion are cited from quantum physics which postulate that "thought field is similar to the concepts of a gravitational or electromagnetic field in physics" (Schwartz, ND). They play loosely with these ideas, claiming that since "matter and energy are interchangeable, thoughts and feelings are influenced by the energy in our bodies."

The TFT literature is replete with references to popular scientists like David Bohm, Rupert Sheldrake, and Roger Penrose. In addition, several kinds of instruments claim to measure the energy field, such as the "biometer"—a "sophisticated form of galvanic skin response meter" that "registers resistance to negative energy," for only $795 (Long, 1998). Some supporters even say that "the energy system is electro-mechanical, electro-optical, electro-acoustic, electromagnetic and can be directly engineered just like VCRs and computers" (Craig, ND). Applying concepts from particle physics to behavior does not deter TFT theorists, although the explanations become somewhat convoluted and abstract (Callahan, NDb):

> One can understand the relevance of the usage of "active information" in that the microstate of the perturbations creates the macro state that the person feels when depressed, angry, anxious, etc. Psychotherapy is the transformation (or subsumption) of this micro state which results in the elimination of the negative emotion.

Although this may be a nice metaphor, there is no evidence that events in the micro world of particle physics are represented in the macro world of humans. Finally, Callahan denies much of what is currently known in established neuropsychology by claiming, "the role of the amygdala . . . [deep brain structure known to regulate emotion] is not only not fundamental in generating disturbing emotions, it is not even in the right ball park" (Callahan, 1998b). He prefers, instead, to attribute emotion to the energy field.

In contrast to the claims of TFT, current research has not confirmed the existence of human energy fields related to illness or in response to treatment factors. For example, the use of Applied Kinesiology (the foundation of Callahan's work) in assessing nutrient status found that it was no more useful than random guessing (Kenney, Clemens, and Forsythe, 1988). Other studies and reviews clearly state that there is insufficient evidence to justify reliance on energy concepts as a modality for treatment (Basser, 1998; Richardson, 1986; Prance, 1988; Skrabanek, 1984). Kirlian photography, so often cited as "proof" of life energy fields, has been found to be an artifact of the photographic method and other factors such as moisture, barometric pressure, and temperature (Carroll, 1998).

Algorithms: the formulary for treatment. Callahan claims that perturbations in the thought field can be corrected by tapping specific sequences of acupressure points, or algorithms. He claims to have discovered several tapping sequences which reduce the symptoms of a variety of conditions. Prior to tapping, clients are asked to rate their level of discomfort on a 10-point SUDS, or "Subjective Units of Discomfort Scale." The success of treatment is determined by a reduction in this self report. For example, a simple phobia can be treated by having the client thinking of or becoming "tuned" to the fear, then tapping bilaterally under the eye, under the arm, on the collarbone, on the "gamut" spot on the back of each hand, and repeating the tapping sequence. A reduction in SUDS from 10 to 2 would be considered a success. In other treatments designed to "activate the left and right brain," the tapping may be supplemented by asking the client to roll the eyes, repeat phrases, hum brief tunes, and count. Callahan (NDa) cautions that using the recipe-like algorithms will not result in as high a success rate as using the advanced technique of "causal diagnosis," however, which:

> . . . reveals the fundamental and deepest causes of a problem, causes which are demonstrably more fundamental than the brain, the amygdala, the nervous system, chemistry, early childhood experiences, beliefs, or cognitive factors, and, thus makes it possible to effectively address the fundamental causes. These causal elements . . . are universal and applicable to all cultural groups and even very young children and animals. There is no treatment in psychology that comes close to the power of treatment informed by CT-TFT Causal Diagnosis.

Although Callahan is not specific in his popular literature about the details of this technique (one must order his tapes or attend his training), it appears that the advanced diagnostic uses the muscle testing procedure used in Applied Kinesiology—Callahan's foundational discipline.

Psychological Reversal (PR). When treatments do not work as rapidly as expected, TFT has an explanation: the polarity of energy in the body has become reversed, thereby preventing treatments from working. As evidence of this "180 degree [reversal] . . . a person will say South when they mean North," "up" when they mean "down," or experience "temporary dyslexia" (Callahan and Callahan, 1996). Purportedly the correction of PR approximately doubles the success rate (Callahan, NDb). Clients who have addictive urges are instructed to perform the algorithm on themselves about 20 times a day to avoid another polarity reversal.

The Apex Problem: When they don't appreciate the cure. Callahan apparently found TFT in a sticky position when some clients would be "successfully treated" but "unable to see that therapy did the job." Callahan concluded that it was just too good to be true, and they failed to see the obvious results, or "forgot" that they had the problem. As a component of training, TFT therapists are prepared to accept that "we predict that the client will report improvement and further predict that he is not likely to credit the therapy for improvement" (Callahan and Callahan, 1996). Gary Craig, developer of EFT, suggests that good business marketing dictates that practitioners present the unusual approach in a way that "fits in with client beliefs," thereby bypassing their objections (Craig, NDb). Callahan apparently concurs: when responding to a criticism that as many as half of people refuse the unusual technique, he suggests that "It's all in the presentation" (Callahan, NDb).

TFT Training. That TFT has not been accepted as a conventionally substantiated technique has not stopped people from eagerly seeking training. Callahan's office estimates that over 3,000 professionals have been trained in the technique worldwide, 90% of whom have advanced degrees (Callahan, personal communication). Three levels of training are offered: Level One Algorithm Training involves two one-day training periods covering theory, phobias, trauma, addictions, anger, stress, psychological reversal, depression, pain, peak performance, obsession, rage, panic attacks, and deep psychological reversal. Level Two Diagnostic Training involves muscle testing procedures derived from Applied Kinesiology. Level Three is reserved for advanced training in "Voice Technology" that involves taking snippets of voice recordings from phone conversations, and analyzing them for energy blockages.

Callahan is reluctant to provide details of the equipment used, but denies that it is similar to voice stress analysis, and states that the particulars are proprietary (personal communication). The levels of training have been packaged in several formats, including day-long workshops as well as home training by audio and videotapes. One promotion for initial training listed cost at $150 for a day, but advanced training costs as much as $10,000, though it has recently been repriced lower (McWorter, 1998). Training by Callahan costs about $500 for a home study course, $3000 for six month phone support, and an astounding $100,000 for full Voice Technology training (personal communication).

TFT Evidence: Clinical vs. Empirical

The claims of TFT are indeed miraculous compared to other forms of psychotherapy. Callahan reports that the TFT algorithm for anxiety "eliminates the addictive urge, regardless of the addictive substance, in about 90% of addicts" (Callahan, 1998, 22). One handout of testimonials distributed at a training workshop states that its success rate is anywhere from 80–97%! The same handout claims it is quick, painless, drug free, non-invasive, simple to use, gives immediate relief, provides permanent symptom resolution, can be taught to anyone, does no harm, does not require talking about problems, and is faster than the more "traditional methods of long term therapy [that] will no longer be accepted by either the patient or insurance companies." Quite a set of claims!

Most of the claims for the efficacy of TFT are clinical—single case reports—or even client testimonials about the effects of treatment. While these are interesting, they are not considered strong scientific evidence because they do not follow sound protocol for testing a claim. This type of reporting is also the basis of a current controversy between clinicians and researchers (Baldwin, 1997). Callahan takes pride in his open demonstrations of TFT effectiveness, and he suggests other practitioners are reluctant to subject their treatments to this level of public scrutiny. The stage for some of his tests is literally that—a videotape is distributed in Level One training that shows his appearance on several popular television talk shows (e.g., *Leeza Gibbons*, *CNN*, *Evening Magazine*, *Tom Snyder*) working with volunteers from the audience who presumably have a phobia. Unfortunately, the tapes do not show him working with them in detail—they only express their phobic complaint and then there is a brief scene in which they confront their fears which he interprets as indicating some degree of success.

In a detailed examination of TFT, an article in the journal *Traumatology* (Hooke, 1998) comments on the suggestive results, but strongly criticizes methodological weaknesses, lack of control groups, and insufficient reporting of data and methodology. After reviewing the few studies that have been conducted and interviewing TFT therapists and Callahan himself, Hooke concluded that the studies have significant flaws that prevent conclusions from being drawn about the efficacy of TFT. In addition, he challenges the assumption of the energy that is supposed to underly TFT and the connections between that energy and the tapping methods. He suggests several alternative hypotheses that do not require these extraordinary assumptions about TFT, and cautions clinicians to be judicious in using and charging for an experimental technique. In a rebuttal to Hooke, Callahan gives some unusual examples of sharks and platypuses using electro-sensing apparatus in their noses, as well as the finding of microgranules of magnetite in the human brain, as evidence that bioenergy exists. Callahan also attempts to refute some of the confounding information provided by TFT therapists who provided information to Hooke—at the very least, this suggests that the TFT training is not monitored very well.

In spite of the challenges to research methodology in TFT, Callahan goes one step further and questions whether conventional scientific methods and standards are appropriate for the study of TFT. In a Level One training handout titled "A new paradigm for determining causality," it is proposed that new criteria for establishing scientific credibility and value are needed. Citing articles from the *Journal of Consciousness* (Bilodeau, 1996) and *Alternative Therapies* (Keine and von Schon-Angerer, 1998), TFT practitioners argue that the results of TFT are "so dramatic and obvious," that traditional designs that involve control and experimental groups, experimental placebo, randomization, and double-blind are unnecessary. They state that the entire pattern or gestalt of the procedure as well as the intention of the practitioner produces an obvious outcome, and proposing to study TFT using traditional research and statistical methods is "ridiculous." Callahan firmly states that his technique is not a function of placebo, distraction, demand characteristics, repetition, hypnosis, affirmations, conditioning, additives, or symptom substitution (Callahan, NDc).

In support of these claims, Callahan and others have initiated a few studies of the effectiveness of TFT—not on the underlying mechanisms, however. Leonoff (1996) reports a nearly unheard of 97% success rate using the Voice Technology technique and that it was replicated with an equal 97% success rate, in a study by Callahan a decade earlier. In these two studies, people called in to a popular radio program and were treated by the therapists. When asked about the characteristics of those who might not be helped by TFT, Callahan stated that those numbers comprise "only one of one-hundred, and we could even reduce that to less than one percent" (Callahan, personal communication). Such studies obviously have a nonrandom sample (talk show callers), who only present self report, who operate under social pressure to perform or receive attention, and who have not had the nature of their conditions determined.

Perhaps one of the better studies conducted, a "controlled placebo double-blind study" on the effectiveness of TFT in treating acrophobia by Carbonell and Figley (ND), was not reported in detail, published in a peer reviewed journal, or given a complete citation in a TFT workshop handout. Reportedly, 49 college students who had a fear of heights were identified by self report and a screening measure. They were randomly assigned to control and experimental groups. All subjects were treated for psychological reversal, and then given a specific algorithm or a placebo algorithm. Interestingly, both groups showed some improvement, with the TFT group showing "statistically significant" improvement (Carbonell, ND). Nonetheless, the control group's improvement appears to challenge Callahan's claim that suggestion or placebo does not operate in TFT. In addition, the application of PR to the control group contaminates the design—they still received a treatment.

The use of self report in establishing a SUDS level is a key component in TFT. All of the few studies reported use a change in SUDS as major evidence of the efficacy of TFT. Self report, however, is notoriously amenable to influences of context, suggestion, charisma of the researcher, etc. Although "screening measures" (no statistics given on validity or reliability of the instrument) were used to confirm self report of the presence of discomfort, this was not established by more thorough interview, history taking, and diagnosis—conventional means of establishing the presence, type, and severity of a disorder. In some of the radio and television shows, volunteers merely had to claim that they had a phobia in order to participate in the demonstration. There is clearly a need for better measurements of pre- and post-treatment behaviors before such treatments can be legitimately claimed to work.

Many of the clinical examples, especially those conducted on radio and television programs, are probably influenced by the "demand characteristics" of the demonstration. Demand characteristics refers to the implicit "demand" on the responder to meet the expectations of, or to please, the audience or therapist. Performing for a popular audience, or even as a demonstration "client" in a training workshop or conference, places certain pressures on people to show results. Control of this intervening influence does not appear to be seriously considered in the studies of TFT researchers.

Conclusion

TFT and related power therapies are proliferating at a remarkable rate, driven by changes in the health care system, client needs, and therapist demands. While it is important to innovate in treatment and discover more effective modalities, it is critical to thoroughly research the efficacy and mechanisms of these new techniques.

An important aspect of a theory is that it should be falsifiable. The structure of TFT suggests that it is designed to be unfalsifiable. When disconfirming evidence is discovered, it is explained away by Psychological Reversal, the Apex Problem, the client simply needing more treatments, or saying the practitioner is not applying the technique properly or is confounding it with additional techniques, or by the client's corrupting effective treatment by ingesting toxic substances that further disrupt the energy field. In some studies where clients apparently did not respond, there were no attempts noted to discover the reason. Allergies to ingested foods or inhaled odors are also purported to be "toxins" that can unknowingly reverse polarity, thereby reducing treatment effects. In addition, TFT argues that standard research criteria do not apply to their methodology.

A more serious concern involves the ethics of proliferating training and promotion for a treatment before sufficient research has been conducted to establish it as legitimate and safe. Callahan noted that his claims have been investigated by the California Board of Psychology, but the concern was dropped when he provided a five minute cure to a driving phobia that the investigator had (personal communication). Practitioners of TFT claim that the procedure does no harm. However, distracting a client from other forms of treatment that may be more useful does not qualify as a beneficial effect. Placebos can also cause "nocebo" or untoward effects by exacerbating preexisting conditions (Thorpe, 1994). In addition, a meta learning (implicit learning occurring by way of the method used) appears to expose the client to a new world view—one operating by subtle, invisible, yet powerful energies inside and outside the body that influence thinking, emotional well being, and behavior. Such a world view, instilled in a conventional person, may unexpectedly lead to concerns about the nature of reality. This is a secondary risk that has not been discussed in the literature or presented as a concern in training.

Although some of the results of TFT are interesting and suggestive, the serious methodological limitations, use of unconventional explanations (e.g., energy) that have not been supported by research, flawed arguments, and minimizing or demeaning established research findings undermine proper evaluation. Given the large numbers of clients on whom the technique reportedly works, it is unfortunate that well designed and controlled studies do not take advantage of these numbers. With the current interest level, it is likely that more research will be conducted, hopefully with better attention given to design and peer review.

References:

Baldwin, D. V. 1997. "Innovation, Controversy, and Consensus in Traumatology." *Traumatology*, 3(1). URL: http://psy.uq.edu.au/PTSD/trauma/T037.html

Basser, S. 1998. "Acupuncture: The Facts." *The Skeptic*, 13(2), 27. URL: http://www.skeptics.com.au/journal/acufacts.htm

Bilodeau, D. J. 1996. "Physics, Machines and the Hard Problem." *Journal of Consciousness*, 3(5/6).

Blecher, M. B. 1997. "Gold in Goldenseal." *Hospital Health Networks*, 71, October, 50–52.

Bunnell, T. (ND). "A Tentative Mechanism for Healing." URL: http://www.positivehealth.com/Bakissue/bunnell.htm

Burr, H. S. 1972. *The Fields of Life*. New York: Ballantine.

Callahan, R. J. (NDa). "Approved Curriculum Training." URL: http://www.mrx.com/Tftsch. html

Callahan, R. J. (NDb). "What is TFT?" URL: http://www.mrx.com/TFTDISC.html

Callahan, R. J. (NDc). "What TFT is Not: Fallacious Explanations for TFT."

Callahan, R. J. 1997. "Thought Field Therapy: The case of Mary." *Traumatology*, 3(1), URL: http://psy.uq.edu.au/PTSD/trauma/T039.html

Callahan, R. J. 1998a. *Thought Field Therapy: Algorithms Level 1 Training Manual.*

Callahan, R. J. 1998b. "Response to Hooke's Review of TFT." *Traumatology*, 3(2). URL: http://rdz.acor.org/lists/trauma/v3i2art4.html

Callahan, R. J., and J. Callahan. 1996. *Thought Field Therapy (TFT) and Trauma: Treatment and Theory*. Indian Wells, CA: The Callahan Techniques.

Carbonell, J. L. (ND). "An Experimental Study of TFT and Acrophobia." URL: http://tftrx.com/expstudy.htm

Carroll, R. T. 1998. "The Skeptics Dictionary." URL: http://wheel.ucdavis.edu/~btcarrol/skeptic/kirlian.html

Craig, G. H. (NDa). "A Physicist Speaks." http://www.emofree.com/physicist.htm

Craig, G. H. (NDb). "Re-evaluating the Apex Problem." URL: http://www.emofree.com/re-evalu.htm

Danzig, V. 1998. "Thought Field Therapy: Training Center of La Jolla." URL: http://www.thoughtfield.com/quotes.htm

Edwards, D. 1997. "Thought Field Therapy." URL: http://www.discoverarchway.com/html/thought_field_therapy.html

Hooke, W. 1998. "A Review of Thought Field Therapy." *Traumatology*, 3(2). URL: http://rdz.acor.org/lists/trauma/v3i2art3.html

Keine, H., and T. von Schon-Angerer. 1998. "Single-case Causality Assessment as a Basis for Clinical Judgment." *Alternative Therapies*, 4(1), 41–47.

Kenney, J. J., R. Clemens, and K. D. Forsythe. 1988. "Applied Kinesiology Unreliable for Assessing Nutrient Status." *Journal of the American Dietetic Association*, 88, 698–704.

Leonoff, G. 1996. "Phobia and Anxiety Treatment by Telephone and Radio: Replication of Callahan's 1986 Study." URL: http://mrx.com/tft4std.html

Long, A. 1998. "Brief Power Therapies and the Biomonitor." URL: http://www.clearingtech.net/fsj/sample3.html

McWorter, J. 1998. "Thought Field Therapy: A Novel Treatment for Trauma." *San Diego Therapist Newsletter.* URL: http://www.camft-sandiego.org/thotfield.html

O'Mathna, D. P. Spring, 1998. "Therapeutic Touch: What Could be the Harm?" *The Scientific Review of Alternative Medicine.* URL: http://hcrc.org/sram/harmtt.html

Pearl, S., and S. Tayar. 1996. "Alternative Medicine: Science or Pseudoscience?" April. URL: http://aries. phys.yorku.ca/~mmd r/1800/alternative.html

Prance, S. E. 1988. "Research on Traditional Acupuncture—Science or Myth?: A Review." *Journal of the Royal Society of Medicine*, 81(10), 588–590.

Richardson, P. H. 1986. "Acupuncture for the Treatment of Pain: A Review of Evaluative Research." *Pain*, 24(1), 15–40.

Schwartz, R. A. (ND). "Thought Field Therapy (TFT)?" URL: http://www.solutionmind.com/schwarz.m.html

Skrabanek, P. 1984. "Acupuncture and the Age of Unreason." *The Lancet*, 1, 1169–1171.

Stenger, V. J. 1998. "Reality Check: The Energy Fields of Life." CSICOP On-Line, June. URL: http://www.csicop.org/sb/9806/reality-check.html

Thorpe, B. 1994. "Touch: A Modality Missing From Your Practice?" *Advanced Nursing Practice*, 2, 10.

Turner, J. G. 1994. "The Effect of Therapeutic Touch on Pain and Infection in Burn Patients." Tri-Service Nursing Research Grant #MDA 905-94-Z-0080 and Application #N94-020.

Wylie, M. S. 1996. "Going for the Cure." *Family Therapy Networker*, July/August, 21–37.

Velikovsky

Cultures in Collision on the Fringes of Science

DAVID MORRISON

In 1950, Dr. Immanuel Velikovsky published *Worlds in Collision (WIC)*, a remarkable book that challenged the fundamental tenets of a number of academic fields. Velikovsky, a psychoanalyst living in New York, sought to replace the current paradigms of physics, astronomy, geology, and geophysics. He also challenged prevailing notions of ancient history, anthropology, and archeology. Generally, persons with such ideas are dismissed as cranks and their work is ignored, but Velikovsky's book was published by the prestigious Macmillan house and accompanied by laudatory endorsements. He was described as "an international scholar" comparable to Newton, Darwin and Einstein, and his book was "a magnificent piece of historical scholarly research." While scientists denounced it, *WIC* rose to the bestseller lists. Thus began the "Velikovsky affair," one of the most prominent 20th century examples of the collision between the cultures of science and letters. Even after it became abundantly clear that Velikovsky was wrong, many persons condemned the scientific community for their hostile reaction toward him and his work.

The Velikovsky affair raises many issues that are pertinent today—it illuminates profound differences in assumptions and methodology that set scientists and their work apart from the rest of society, and it illustrates barriers to communication. Fortunately Velikovsky dealt with ancient history, and the debates that surrounded him had little practical significance. But next time we may not be so lucky.

The Fundamentals of *Worlds in Collision*

Velikovsky's primary thesis was that ancient myths and legends described real events of a cataclysmic nature that the Earth experienced in the millennia that preceded the rise of classical Greece. He concluded that cross-cultural concordances among these legends demonstrated that these traumatic events were global in scale and the result of astronomical agents. He then identified the specific astronomical causes, which involved the ejection of Venus from Jupiter and repeated near-collisions among Venus, Mars and Earth—events that changed the Earth's rotation and orbit and precipitated global floods, volcanic eruptions, meteoric bombardment, and other catastrophes. Finally, in an effort to explain this aberrant planetary behavior, Velikovsky proposed that electromagnetic forces are (or were) of critical importance in determining planetary motions.

All this is presented in a book of 401 pages, amply stocked with scholarly-looking footnotes and a long index. However, the style is popular and has a curious old-world ring to it. Many of the references, for example in quoting translations from ancient middle-eastern languages, are from 19th-century sources that have long since been superceded. *WIC* appears to have sprung fully developed from the brow of its creator. It is not an extension or a summary of previous work, because Velikovsky did not publish articles in professional journals. He had no collaborators, and he gives little credit to such predecessors as Ignatius Donnelly, whose 1883 book *Ragnarok* has many similarities to *WIC*. Further obscuring his purpose, Velikovsky asks the reader to "consider for himself whether he is reading a book of fiction or non-fiction, whether what he is reading is invention or historical fact."

Many critics described the book as exceedingly well written, and certainly it had enthusiastic readers. Yet to a scientist it seems vague, illogical, and obtuse. Velikovsky is inconsistent, for example, in describing the chemistry of the atmosphere of Venus and Mars in various places as "hydrocarbon," "carbohydrate," "petroleum gases," and "of the nature of carbon" (whatever that means). His many references to electromagnetic phenomena, always presented qualitatively (without numbers or equations), are incomprehensible, as in the following: "The cessation of the diurnal rotation could also be caused—and most efficiently—by the earth's passing through a strong magnetic field; eddy currents would be generated in the surface of the earth, which . . . would slow down the earth or bring it to a rotational stasis . . . if the interaction with the magnetic field caused the earth to renew its spinning . . ." etc.

One of the most mystifying passages is found in the final pages of *WIC*, where Velikovsky suddenly introduces the idea that the solar system is like an atom, and hypothesizes that if the forces between sun and planets were primarily electrical the planets might be able to jump from one orbit to another like the quantum changes in the electrons of an atom. This stunning *non-sequitur* seems to be a serious suggestion, not just a reversal of the common pedagogical analogy that compares electron orbits to miniature planetary systems.

Henry Bauer, whose book *Beyond Velikovsky* is the definitive history and analysis of the subject, summarizes *WIC* thus: "Velikovsky displays a lack of understanding of chemistry, physics, and astronomy . . . [He] is not only ignorant of the facts . . . his whole approach is not that of the scientist . . . he does not weigh his evidence . . . he does not adduce independent tests of validity . . . [his] ideas about natural science are not worth taking seriously." Yet "he discusses these subjects in a manner that would convey, to a layman, an apparent familiarity with these fields." It is this difference in response—the easy dismissal by scientists, while non-scientists found him to be plausible and even compelling—that makes Velikovsky relevant today to our understanding of the relationship between scientists and the public.

Initial Responses

Worlds in Collision was extravagantly praised at the time of its publication by authors Eric Larrabee in *Harpers*, John Lear in *Colliers*, and Fulton Oursler in *Readers Digest*, among others. Reviewers who questioned some parts of the book were able to find other sections praiseworthy. No reviewer was expert in all the areas spanned by Velikovsky's "interdisciplinary synthesis." Carl Sagan recalled that he spoke to one "distinguished professor of semitics" who told him that "'the Near-Eastern and Biblical scholarship is of course nonsense, but I was impressed by the astronomy'. I had rather the opposite view."

Although Velikovsky sought the overthrow of half a dozen different disciplines, it was only the astronomers who fought back, led by Harlow Shapley, the distinguished Director of Harvard College Observatory. Shapley, who had met Velikovsky and had seen an early draft, privately pressured Macmillan to drop the book on the grounds that such a publication would undercut the company's credibility as an academic publisher. This threatened boycott was successful, and Macmillan transferred the account to Doubleday, which had no textbook division. In defense of Macmillan, it should be noted that the publisher did obtain several reviews from scientists in advance of the publication of *WIC*. I have read three of these reviews, and all of them said, in effect, that the science was very probably wrong but that the book was fun to read and would probably sell well. None of these reviewers advised that *WIC* should not be published.

Cecelia Payne-Gaposhkin, another senior astronomer at Harvard, wrote an early review without having read the book. She asked, rhetorically, "Is this scientific age so uncritical, so ignorant of the nature of evidence, that any considerable number of people will be fooled by a sloppy parade of jargon . . . The road to fame and fortune is clear. Never mind logic; never mind the precise meaning of words or the results of exact research. Employ the vocabulary of a dozen fields of learning. Use a liberal sprinkling of Biblical phrases. . . ." Other reviews were equally scathing, with several of the reviewers admitting to not having read the entire book, or even a substantial part of it. "The [scientific] statements are all so completely at variance with known principles . . . It is unnecessary to waste . . . space . . . pointing out the numerous errors." "Sincere musing of a man . . . unfamiliar with the details and general principles of the physical sciences." "The kindest judgement is to class him in the select army of hoaxters." "The screwball fringe."

There is no doubt that the scientific community felt threatened by Velikovsky's popularity, and they fought back within their own professional sphere. Doubleday's ads for *WIC* were rejected by *Science* and *Scientific American*, and both refused to print rebuttals from Velikovsky. (Both journals were following their usual policies against accepting ads for pseudoscience or printing rebuttals to book reviews.) These actions, and the astronomers' outspoken attacks on *WIC*, have been widely criticized—especially Shapley's threat to Macmillan. Velikovsky himself assumed the mantle of an academic martyr as well as heretic, hounded by organized science. Bauer later concluded that the astronomers "had acted unethically, and all [Velikovsky's] critics would henceforth stand in danger of being tarred with the same brush . . . for suppressing ideas, assaulting academic freedom, acting dogmatically and in authoritarian fashion." Bauer believed that "literally inexcusable steps had been taken to prevent an expression of opinion." Carl Sagan, one of Velikovsky's most outspoken critics, worried that "there is no excuse for any attempt to suppress new ideas, least of all by scientists committed to the free exchange of ideas."

Unethical or Realistic?

I believe these often-repeated accusations of unethical behavior deserve reanalysis. There is ample evidence that Velikovsky was little more than a crank, something that was evident to the astronomers from even a cursory look at his book. Except for its publication by a mainstream publishing house and wide praise by non-academic reviewers, his work was not so different from the unsolicited manuscripts on cosmology written by well-meaning laypersons without a whiff of understanding of relativistic theory that I and most astronomers receive routinely. Even in the most freedom-loving society, no one has a right to be published, let alone to be read, as witnessed by the many book manuscripts or papers submitted and rejected in peer review every day.

Let us also consider the charge that Velikovsky's academic freedom was threatened. Academic freedom is the right of university faculty and researchers to teach and carry out their research free from restrictions imposed by their institutions or government bodies. But Velikovsky was not on any faculty—he lived in the town of Princeton, New Jersey, but had no association with Princeton University or any other academic institution. He was always free to submit papers to technical journals, but he chose not to do so, because he rejected on principle the practice of peer review. Years later, he submitted a note to *Science*. As he himself described, "My answer . . . was returned for rewriting after one or two reviewers took issue with my statement that the lower atmosphere of Venus is oxidizing. I had an easy answer to make . . . but I grew tired of the prospect of negotiating and rewriting."

The reception Velikovsky received from astronomers and other critics was what any nonscientist might expect who strayed into areas he did not understand. It seems to me that one's reaction on these issues depends on one's assessment of the true value of Velikovsky's scholarship. If you believe that his thesis could be correct or might stimulate other work, then it is reasonable to upbraid his critics and call for dispassionate scholarly discourse. But if you conclude that his ideas are nonsense, then no such dialogue is required, and issues of "fairness" and "academic freedom" are irrelevant.

Irrelevant or not, accusations of unethical behavior by the scientific community colored all future discussion of Velikovsky. Some rallied to his support not because they thought he was correct, but because they felt he had been denied a hearing by organized science. The focus of debate shifted from *WIC* to "the Velikovsky affair," as noted by Alfred de Grazia in *American Behavioral Scientist:* "The Velikovsky case has little to do with the correctness or otherwise of his theories. What is in question is the entire reception system that science uses in dealing with innovation . . . Who determines scientific truth? What is their warrant? How do they do so? . . . In the end, some judgement must be passed upon the behavior of science and, if adverse, some remedies must be proposed."

Velikovsky Confirmed by NASA

Velikovsky might have quietly faded away in the 1960s, were it not for discoveries that seemed to confirm some of his predictions. Velikovsky supporter Lynn E. Rose wrote the following glowing testimony in *Pensée:* "Velikovsky's theory is by now a near-classic case of a successful empirical analysis. The theory was eminently open to testing, since it entailed a number of important consequences not yet verified, and many of these were incompatible with rival theories. Succeeding years witnessed the verification of a great many of those consequences and the disconfirmation of none. By all the usual canons of sound methodology the theory should now be accepted as a successful one; that is, one that may be regarded as very probably true." Velikovsky himself wrote in 1974 that "My work today is no longer heretical. Most of it is incorporated in textbooks." What a terrific story of a "comeback kid," and what a satisfying deflation of the pompous defenders of scientific orthodoxy. This was how the story was played. But was it correct?

Velikovsky's claimed successes nearly all involved planetary exploration. Surprisingly, terrestrial geology, to which he had devoted an entire book (*Earth in Upheaval*), apparently did not provide an appropriate testing ground.

The predictions most often cited were radio emissions from Jupiter, the high temperature and cloud composition of Venus, and the nature of the lunar surface. He received a big boost when physicist V. Bargmann and astronomer L. Motz agreed to his request that they write a letter to *Science* noting his predictions concerning Jupiter, Venus and the terrestrial magnetosphere. They stated that "Although we disagree with Velikovsky's theories, we feel impelled . . . to establish Velikovsky's priority of prediction of these . . . points and to urge, in view of these prognostications, that his other conclusions be objectively re-examined." Let's look briefly at each of these predictions.

Within a few years of the publication of *WIC*, researchers in the young field of radio astronomy identified Jupiter as a strong source of "non-thermal" radio emissions, later attributed to energetic plasma in the magnetosphere of Jupiter. Velikovsky had predicted that Jupiter might be a source of radio emission, analogous to radio emission from the Sun and stars. He thought Jovian radio emissions could originate in turbulent motions within a hot electrically charged atmosphere, still seething from the expulsion of Venus. He thus claimed this discovery as a confirming test of his theory. In fact the emission is unrelated to the atmosphere of Jupiter, but rather is due to Jupiter's strong magnetic field and the ions trapped within it. The presence of a hot interior for Jupiter, discovered later (1969), does not seem to have been much discussed in the Velikovsky literature.

Even more widely advertised was Velikovsky's successful prediction that Venus is "hot," which he attributed to the recent violent birth of the planet together with its subsequent close passages by the Sun, its intense tidal interaction with Mars and Earth, and various electrostatic discharges. This hot Venus is not strictly speaking a prediction. In *WIC* Velikovsky cited as evidence the observed absence of cooling of Venus' clouds at night, from which he inferred that the clouds were heated primarily from below. He also predicted an excess of radiated heat and an observable cooling of the planet over time. Within a few years, radio astronomers had measured the surface temperature of Venus and found a blistering 400+ degrees C, a quite unexpected discovery. The surface source of the emission was confirmed by Mariner 2 in 1962, and temperature was measured directly by Soviet landers beginning in 1970. So far this seems like a vindication of Velikovsky. But there was no net excess of energy radiated from the planet (since the thick atmosphere effectively contained the high surface temperature). The total radiated energy was in balance with absorbed sunlight, and it did not decline over time. Further, the high surface temperature had an alternative explanation from an atmospheric greenhouse effect.

The third topic dealt with the chemistry of Venus. Velikovsky predicted a hydrogen-rich atmosphere (since Venus had been born out of Jovian material), hydrocarbon clouds, and even "petroleum fires burning on the surface." This claim received inadvertent support in 1963 when a NASA publicist erroneously stated that Mariner 2 had found evidence of hydrocarbon clouds, a quite amusing story of human fallibility later recounted by Sagan. The Velikovsky camp trumpeted success when the hydrocarbon clouds were announced and then steadfastly refused to believe the subsequent retraction. Some of his more paranoid supporters even accused NASA of fabricating the retraction solely to undercut Velikovsky. The true composition of the clouds was not determined until 1973, more than a decade later. They are made primarily of sulfuric acid particles, consistent with the oxidizing chemistry of the atmosphere of Venus.

Finally, there were a number of predictions about the Moon, all deriving from its role in close encounters between Earth, Venus and Mars. Velikovsky summarized these predictions in an op-ed article in the *New York Times* on the eve of the Apollo lunar landings. He asserted that the lunar surface and been molten 3500 years ago, that the lunar craters were primarily formed by bubbles in the molten crust, that the ray craters were the result of immense interplanetary discharges, and that the lunar surface today would show strong evidence of remanent magnetism and localized radioactivity resulting from these discharges. He even warned of danger to the astronauts from intense lunar radioactivity. In fact, the surface of the Moon consists of ancient rocks that were last molten more than three billion years ago. The small remanent magnetism is an artifact of the lunar magnetic field of three billion years ago when the lava solidified. There is no excess radioactivity, and the craters are conclusively shown to be the result of impacts. Yet Velikovsky claimed success with his predictions, since Apollo had found remanent magnetism, and he explained away the dating of lunar samples as some undefined consequence of the electric discharges. To his death, Velikovsky believed that the lunar craters were congealed lava bubbles. Thus, strangely, he could not accept the role of impacts in shaping planetary surfaces, the one true form of planetary-scale catastrophism that has emerged from our exploration of the planetary system by spacecraft.

So what is the score? Velikovsky was correct about the high surface temperature of Venus (although he never quantified his prediction, saying only that "Venus is hot"), but apparently for the wrong reasons, since there was no net energy excess. In the case of Jupiter's radio emissions, he correctly predicted a more important role for electromagnetic effects than was thought at the time, but the radio noise from Jupiter is unrelated to the hot, electrically charged atmosphere hypothesized by Velikovsky. On the chemistry of Venus, he was completely wrong, as he was also in his predictions for the Moon, whose geological activity virtually ceased more than 3 billion years ago, providing strong disconfirmatory evidence of his theory of recent planetary encounters.

Does this analysis of Velikovsky's discoveries match the paean by Sidney Willhelm, who wrote that "Velikovsky's correct diagnosis . . . can only mean that Velikovsky himself is the foremost scientist of the twentieth century [and] among the foremost thinkers of all time"? I think not, despite Velikovsky's own self-evaluation in the peroration of his AAAS lecture in 1974 where he said "my *Worlds in Collision*, as well as *Earth in Upheaval*, do not require any revisions, whereas all books on terrestrial and celestial science of 1950 need complete rewriting. . . . None of my critics can erase the magnetosphere, nobody can stop the noises of Jupiter, nobody can cool off Venus, and nobody can change a single sentence in my books."

The AAAS Debate

Thanks to his claims of successful predictions, Velikovsky's star was rising in the early 1970s. He received many invitations to lecture at universities, even at his old nemesis, Harvard. He also spoke at NASA's Ames and Langley Research Centers. Quite a number of faculty and students (mostly from the humanities and social sciences) took up his cause, and a journal (*Pensée*) devoted to his ideas began publication.

Against this backdrop, Carl Sagan of Cornell University and other astronomers decided to devote a symposium at the 1974 AAAS meeting to Velikovsky. Sagan wrote that "I and some other colleagues in the AAAS have advocated a regular set of discussions . . . of hypotheses which are on the borderlines of science and which have attracted substantial public interest. The idea is not to attempt definitively to settle such issues, but rather to illustrate the process of reasoned disputation, and perhaps to show how scientists approach a problem. . . ."

Donald Goldsmith of SUNY Stony Brook, one of the symposium organizers, wrote that "the stated commitment of the AAAS to the sharing of scientific ideas with the public, together with the public interest in his theories, provided sufficient reason to hold a symposium." Owen Gingerich of Harvard, another organizer, later recalled: "I remember two reasons for organizing it. First the Velikovsky supporters were arguing that scientists were close-minded and unwilling to listen to their good arguments, and we felt something should be done to defuse this claim by giving them a public platform. Secondly, and for me more important, my students were hearing a lot of pro-Velikovsky news, and no respectable astronomers were willing to take the time of day to explain to the general public why they didn't take his scenario seriously . . . I don't think there was any effort to convert the hard-core Velikovskyites, but simply to make arguments available to a broad general public." Velikovsky himself, however, took the invitation as a victory. He wrote that "the astronomers are on the defensive. . . . They asked me to participate in the AAAS meeting. I did not ask."

The Velikovsky symposium was the most popular event of the 1974 AAAS meeting, drawing a crowd of nearly 1500. Since the principal speakers, Velikovsky and Sagan, both exceeded their time allocations, the symposium was continued in a special evening session. Although there were seven speakers, attention focused on Velikovsky and Sagan. Velikovsky, then in his late 70s, projected a vigorous image, tall, imperious, and confident. Sagan, little more than half as old, was equally articulate, confident, and charismatic. Their papers and subsequent comments provide an excellent overview of most of the astronomical areas of dispute between Velikovsky and "establishment science," but little on ancient history or archeology. Velikovsky presented a succinct summary of his theory, with emphasis on his successful predictions. He received a standing ovation for his proud assertion that not one word of his writings needed revision. Sagan focused on 10 tests of Velikovsky taken from *WIC*. Their presentations are in proceedings volumes.

Velikovsky's supporters took little satisfaction in the outcome. They had come expecting to hear a reasoned scientific discourse conducted among equals. Instead they were hit with Sagan's debunking, aimed not at them but at the general public. As Leroy Ellenberger later wrote, "Sagan's analysis of *WIC* was not designed to appeal to the interested, informed layman who was interested in Velikovsky, yet also amenable to a reasoned, valid critique. Sagan's analysis contained errors in physics that were never corrected." Velikovsky himself accused the symposium organizers of bias, with "no pursuit of scientific debate in mind. . . . The scientific and semi-scientific press showed by its reports that it was orchestrated—the very sentences, and the very same errors of fact and number, appeared simultaneously in many reviews."

Sagan intended his "10 problems" to provide a definitive answer to Velikovsky as well as an example of how scientists analyze new hypotheses. However, Velikovsky and his followers considered Sagan's paper to be an unforgivable catalog of errors. It may be useful, therefore, to assess Sagan's 10 problems from the perspective of 25 years later. In doing so, I will use two terms common in the space sciences. One is "back-of-the-envelope" or "rough order of magnitude" estimates, abbreviated ROM. These are simplified calculations to obtain a very approximate numerical solution. Often a ROM estimate is sufficient to reject an implausible hypothesis. Second is the concept of the "strawman"—a simplified version of an idea that is used as a first rough estimate. Both ROMs and strawman arguments appear extensively in Sagan's critique.

Problem 1: The ejection of Venus by Jupiter. Velikovsky had not explained how or when the Venus-comet got loose on a planet-crossing orbit, but he did say it was ejected from the Jupiter system. (Later he would suggest that Jupiter split apart as a result of interactions with Saturn.) Sagan analyzed a strawman in which Venus is ejected from Jupiter like a bullet shot from a cannon. He used a ROM calculation to show that the energy of such an expulsion is more than sufficient to melt the proto-Venus and probably to splatter it all over the solar system. Unfortunately, he used a slightly wrong value for the escape velocity from Jupiter. This did not invalidate his ROM argument, but it undercut the credibility of the entire exercise for the non-science audience, who usually expect calculations by scientists to be precise.

Problem 2: Repeated collisions among the Earth, Venus, and Mars. Velikovsky had asserted that multiple collisions occurred between these three planets during roughly one millennium ending about 700 BCE. Sagan performed a ROM calculation of the probabilities of repeated planetary near-encounters. Since Velikovsky provided no information on the orbital dynamics that would make these events happen, Sagan tested a strawman in which the events are stochastic (unrelated), showing that the odds against such a series of near-collisions are absurdly high (one in 10^{23}). But Sagan did not consider coupled or resonant orbits, which would invalidate his strawman. His is also a *post hoc* probability calculation—after the fact, almost any specific sequence of events seems improbable, as Velikovsky correctly stated in his rebuttal.

Problem 3: The Earth's rotation. Velikovsky asserted that the Earth's rotation changed dramatically about 3000 years ago; in his preferred scenario it actually stopped, then began rotating again in the opposite direction. Sagan raised many valid objections to the idea that tidal or electromagnetic forces could have stopped the Earth's rotation, let alone start it up again. These are among the principal flaws in Velikovsky's scenario.

Problem 4: Terrestrial geology and lunar craters. In Velikovsky's theory, the Earth suffered extreme geological disruption from the close passes of Venus and Mars. Sagan noted many contradictions between Velikovsky's scenario and the geological record. There was not a general eruption of terrestrial volcanoes a few thousand years ago, mountains were not thrown up, and the lunar surface was not melted.

Problem 5: Chemistry and biology of the terrestrial planets. Sagan pointed out that Venus's oxidizing chemistry is inconsistent with its supposed Jovian origin and noted many other problems in Velikovsky's chemistry, such as the composition of the Martian polar caps. Velikovsky responded by quoting old astronomical authorities in support, but that is beside the point, since these references had since been proved wrong.

Problem 6: Manna. Velikovsky concluded that manna (edible carbohydrates) fell on the Earth from Venus, perhaps manufactured by microorganisms out of the hydrocarbons of the comet's tail. Sagan set up a strawman in which the Venus-comet shed manna over the entire inner solar system, and he used a ROM calculation to show that the quantity of manna exceeded the entire mass of the Earth—a *reductio ad absurdum.* The exercise doesn't prove much, since Velikovsky never postulated a model to explain the production of manna, but it went over well with audiences and Sagan, like Velikovsky, was a showman.

Problem 7: The clouds of Venus. Sagan, who was one of the world's experts on the atmosphere of Venus, effectively demonstrated that Velikovsky's ideas on this subject were completely at odds with the facts, concluding "Velikovsky's idea that the clouds of Venus are composed of hydrocarbons or carbohydrates is neither original or correct." Velikovsky's reply stressed that hydrocarbon clouds had been suggested by others, but again this is beside the point—by 1974 we knew the clouds were sulfuric acid, although Velikovsky could not accept that fact.

Problem 8: The temperature of Venus. Again Sagan was on solid ground, speaking as one of the originators of the greenhouse model for the atmosphere of Venus. Velikovsky categorically rejected the greenhouse model as "contradicting the second law of thermodynamics" and apparently believed it was a fabrication designed solely to repudiate his theory. He also continued to assert, in contradiction to the astronomical data, that Venus emitted more energy than it absorbed from the Sun. There was no contest here, with all the facts on Sagan's side. Unfortunately, Sagan added a quantitative appendix on the heating of Venus during a close passage by the Sun that makes no sense to me and has been widely criticized, undercutting his temperature argument.

Problem 9: The craters of Venus. Sagan noted that the presence of craters on Venus (recently discovered by cloud-penetrating radar) contradicted the claimed youth of Venus. This is at best a weak uniformitarian sort of argument, based on an assumption of roughly constant impact rates to form the craters. However, Velikovsky thought the craters resulted from recent interplanetary electrical discharges and did not accept the idea of widespread impact cratering in the planetary system. Neither perspective is very edifying.

Problem 10: The circularization of the orbit of Venus and nongravitational forces in the solar system. Sagan pointed out that there is no evidence that electromagnetic forces play any role in planetary dynamics, and that even if such other forces were at work it would be extremely difficult to change an elongated orbit into a circle (and Venus has the most circular orbit of any planet). These are sound arguments, and neither Velikovsky nor his supporters provided a coherent theory to rationalize the planetary motions that were central to his theory.

My own judgment is that Sagan's critique would have been stronger without Problems 1, 2, 6, 9, and Appendix 3. But I can understand his use of strawman models and ROM calculations. One of the frustrations of dealing with Velikovsky is his vagueness and lack of quantitative reasoning. In the absence of any specific scenarios or models from Velikovsky, Sagan substituted his own strawman versions and showed how absurd they are. In their rebuttals, Velikovsky and his supporters repeatedly said that Sagan had misrepresented their positions, but they did not offer any real alternatives. Sagan wanted to illustrate scientific thinking and show how hypotheses could be tested quantitatively. But this meant nothing to Velikovsky. His supporters delighted in finding minor errors in Sagan's paper (and he made quite a few), but they missed the big picture.

The AAAS debate and subsequent publication were successful from the perspective of the scientist-organizers, but they infuriated Velikovsky's supporters. Instead of serious scientific discussion, Sagan aimed his presentation at journalists and the public, seemingly delighting in making Velikovsky look ridiculous. As a consequence, the AAAS debate actually strengthened the stature of Velikovsky among his supporters. From the AAAS meeting until his death in 1979, Velikovsky presided over a number of "scientific symposia" devoted to his work and saw the publications of thousands of pages of "scientific papers" defending his theory.

During this period, Jim Warwick, the pioneering radio astronomer from the University of Colorado, and I independently tried to connect with the Velikovskians. We both spoke at a Velikovsky Symposium held at McMaster University in Canada in 1974. Warwick discussed the source of Jovian non-thermal radio emission, giving Velikovsky some credit for his "prediction" but placing the issue in context. As he later wrote, "I tried to emphasize that someone in the early fifties or late forties who surmised that a hot object (by hypothesis, Jupiter) might be a source of intense non-thermal radio waves was far out on a limb, but not completely out of his gourd. For these remarks I was almost scalped alive by the participants of the symposium, though I felt I was being generous to Velikovsky, almost to a fault."

My McMaster paper focused on the predictions from *WIC* that had been contradicted by the planetary exploration of the past quarter century, especially those dealing with Venus and the Moon. I was booed twice, once when I mentioned the greenhouse effect, another time when I showed the most recent Mariner 10 photographs of impact craters on Mercury. Velikovsky himself did not attend my talk and was unwilling to speak to me. As a former student of Sagan's, I was considered his stooge, undercutting any credibility I might otherwise have had.

Meanwhile, the definitive answer on Velikovsky was emerging from terrestrial science. Rather than astronomers' arguments about what might have happened to Venus or the Moon, direct information became available on our own planet's history. First were data from dendrochronology, the use of tree rings to determine past growing conditions (and hence climate), which were extended back more than 4000 years, into the period in which the Velikovskian global catastrophes were supposed to have happened. Then the early 1980s saw analyses of Greenland ice cores that provided annual values for global average temperature and volcanic dust and sulfuric acid going back tens of thousands of years. There was no evidence of any volcanic or climatic catastrophes at or near the times proposed by Velikovsky from his interpretation of ancient myths. In 1984 Leroy Ellenberger, until then a disciple of Velikovsky, published a devastating summary of this evidence in *Kronos*, the Velikovskian journal of the moment. From that time forward only the most fanatical and closed-minded of the Velikovsky circle could continue to defend him as "the greatest scientist of our time."

The Real Catastrophist Revolution in Science

Ironically, the year of Velikovsky's death, 1979, saw the keystone work that heralded a new perspective on Earth history, one much more open to catastrophist ideas. Already Gene Shoemaker and other planetary scientists had established the important role of impact cratering in planetary history, while Stephen Jay Gould and other biologists had published evidence of punctuated equilibrium—a stepwise history of evolutionary change. In 1979 Luis and Walter Alvarez and their colleagues made the critical identification of extraterrestrial material at the KT boundary—evidence that the impact of a comet or asteroid about 15 km in diameter had triggered the mass extinction that ended the age of the dinosaurs. Within a few years the idea of short-term, catastrophic changes in geological and biological history had become acceptable, ending a century in which strict uniformitarianism held sway almost unchallenged. In this new perspective, the course of biological evolution on Earth was critically linked with the planet's astronomical environment.

Was this new acceptance of catastrophist ideas related to the Velikovsky debates of the previous 30 years? Presumably the scientists who were now leading this revolution were aware of Velikovsky and his theory. Had he influenced them? Some of Velikovsky's supporters argued that he should be credited with success on the broad issues of a catastrophic Earth history even if he was wrong in his specifics. The suggestion was now made that Velikovsky had been the prophet of this new attitude toward planetary history.

The opposite hypothesis is also possible—that Velikovsky with his crazy ideas tainted catastrophism and discouraged young scientists from pursuing anything that might be associated even vaguely with him. Velikovsky himself hinted at this interpretation in his AAAS talk when he said "I may have even caused retardation in the development of science by making some opponents cling to their unacceptable views only because such views may contradict Velikovsky."

Rather than debate this issue on philosophical grounds, I decided to ask a group of scientists who have been leaders in establishing the new paradigms in which occasional violent events, such as asteroid impacts, play a significant role in planetary and biological history. I sent my questions to 25 of these scientists, and received 23 answers. As noted in the table, very few of them claimed any influence on their own scientific careers, but nearly half thought that Velikovsky had an overall negative effect by tainting catastrophist thinking and holding it up to ridicule.

Results of poll on possible influences of Velikovsky on main-stream science (23 of 25 replying):

1. At the time you began your research in these areas, were you familiar with Velikovsky and his "theory"?
   Yes: 18 No: 5

2. At that time, had you read *Worlds in Collision?*
   Yes: 7 Partially: 8 No: 8

3. Did Velikovsky and his ideas influence your interest in research on more catastrophist concepts in Earth and planetary science, either positively or negatively?
   Positive: 1 None: 16 Negative: 5

4. Do you think that Velikovsky and his ideas had any significant influence on the acceptance of catastrophist ideas in Earth and planetary science over the past half-century, either positive or negative?
   Positive: 0 None: 14 Negative: 9

George Wetherill (Carnegie Institution of Washington geophysicist, authority on planetary formation, dynamics, and evolution):

> I was a graduate student at the University of Chicago at the time *Worlds in Collision* was published, and I was asked my opinion of it by nonscientific students whom I knew socially. I browsed through a copy they showed me, and learned enough about his ideas to explain why I felt them to be of no scientific value . . . Velikovsky and his ideas had no influence at all on my thinking about scientific phenomena.

Walter Alvarez (UC Berkeley geologist, originator of the impact theory of the KT extinction, author of *T Rex and the Crater of Doom*):

> [Velikovsky did not influence science] in any positive ways. I considered him part of the problem we faced in getting a hearing for the KT impact hypothesis, because his ideas, which were incompatible with the laws of physics, had confirmed many geologists in their view that people working on extraterrestrial causes for events in Earth history were not doing good science.

David Raup (U Chicago paleontologist, authority on mass extinctions and evolution, author of *Extinction: Bad Luck or Bad Genes*):

> [Velikovsky's] reputation added to my general feeling of unease with catastrophism . . . [For the field generally] I suspect his influence was substantial—but almost entirely negative.

Richard Muller (UC Berkeley physicist, originator of the Nemesis hypothesis, author of *Nemesis*):

> As someone who was deeply involved in the controversies at the time, I feel very strongly that having Velikovsky pave the path definitely did not help! . . . It was an annoyance to answer some colleagues who would bring up Velikovsky, and ask what I thought about him. I tried reading some of his books at that point, but found them so annoying (because of their apparent disinterest in truth) that I never finished more than about 5 pages.

Jay Melosh (U Arizona geologist, authority on the physics of impacts, author of *Impact Cratering*):

> I was fully aware (and embarrassed) by his "theories" . . . Any influence was purely negative. I had to continually explain to audiences that, although some of the recent work I was doing sounded a little like his ideas, there was no connection and the time scales for the proposed catastrophes are totally different.

Peter Ward (U Washington paleontologist, authority on impacts and craters, author of *Rare Earth*):

> I read parts of Velikovsky as a sophomore in college. I remember not finishing it because I had a very good astronomy background and knew bunk when I read it. I was busy with more important stuff, so I read parts, laughed, and moved on. . . . I think it is so fringe that it had no effect on the positive "neocatastrophism" that is so useful to our science today.

Norm Sleep (Stanford U geophysicist, authority on the impact frustration of life on early Earth):

> His effect was, if anything, negative. . . . One of his followers was friends with an MIT student when I was there. The follower seemed to be a true believer and kept citing things from the Velikovsky book and demanding a conventional explanation of poorly cited data that had once puzzled some geologist. The reasoning went: if you can't find a conventional perfect explanation in two minutes that satisfies me, then my crazy theory and only it must be right.

Jack Hills (Los Alamos National Lab physicist, authority on planetary impacts and dynamics):

> I [first encountered] *Worlds in Collision* in the astronomy section of my public library in 1958, when I was in the 8th grade. I opened it up at random and read a section where he had Venus passing near the Earth to produce the parting of the Red Sea and other nonsense. I spent no more than 15 minutes reading the book. I put the book back on the shelf. I recall being very indignant that it should be in the astronomy section.

Don Yeomans (Jet Propulsion Lab planetary scientist, authority on solar system dynamics, author of *Comets*):

> Within the scientific community, I don't think his ideas were taken seriously enough to directly influence any research directions. However, his ideas were well known and endlessly discussed within the popular press. . . . For me, the most memorable aspect of the Velikovsky affair was the zeal with which those outside the scientific community attacked scientists who pointed out the absurdities in Velikovsky's ideas. I remember being struck by how strident amateur scientists were in railing

against what they perceived as the narrow-minded, elitist, scientific establishment.

Michael Rampino (New York U geochemist, authority on terrestrial impact cratering and extinctions):

> My general feeling is that Velikovsky added nothing positive to the debates on catastrophism. No reputable astronomers or geologists took him seriously. His geology and planetary science were completely wrong, and I found that he used mostly out-of-date references for geological "mysteries" that had long since been cleared up. I have found that some scientists were impressed by the historical references, but agreed that the science was bunk, while historians criticized the history and chronology, but thought that the "science" was exciting. He was primarily a negative factor, often used to make the new catastrophism debates seem silly. [As a put-down], some scientists would say to me, "That sounds very Velikovskian."

Communication and Miscommunication

The most relevant aspects of Velikovsky from a 21st century perspective are lessons concerning communication and miscommunication. Much of the fuel that maintained the Velikovsky affair was rooted in fundamental differences between natural science and social science, or more generally between science and letters. These groups did not speak the same language or adopt the same standards of evidence. In general, scientists saw Velikovsky as a crank and could not understand why others honored his scholarship, while those from the other culture could not understand how scientists could reject his entire hypothesis on the basis of a cursory look.

The flaw in this analysis arises from the ordering of these four theses, from the general to the particular. That may be the way some philosophers think, but it is not science. Science beings with the particulars and gradually works toward more general statements, perhaps ultimately arriving at widely accepted laws or theories. The generalizations, however, depend on the validity of the particulars. In the case of *WIC*, the particulars are the legends that describe synchronized global catastrophes and link these events to struggles between planets/gods. If the global catastrophes did not happen (as subsequently shown by tree-ring and ice-core data), or if the connection between myths and actual planetary encounters is false, then the entire structure collapses. The generalizations (theses 1 and 2) are indefensible without the data to support them. As we have seen, Velikovsky had no role in the development of revolutionary, more catastrophe-tolerant paradigms for Earth science (such as the connection between asteroid/comet impacts and mass extinctions), even though these scientific revolutions were taking place at the same time as the Velikovsky debates.

Science faces a major challenge in public communication, especially when we try to go beyond facts and explain how scientists think and make choices. The wrangling between scientists and other scholarly communities over Velikovsky understandably confused the issues. Velikovsky took his case directly to his readers, bypassing the normal processes of scientific review and debate. The initial reaction of the astronomers was to condemn *WIC* as not only wrong, but as something that should not even have been published. Yet the book sold well and had many supporters. A good case can be made that the derisive arguments from authority presented by the astronomers were counterproductive. There may be lessons here for our current confrontations with such popular pseudoscience topics as alien abductions and creationism.

Let me mention three anecdotes to illustrate the ineffectiveness of the scientific critiques in influencing public attitudes. I will start with a personal story. I first read *WIC* in the early 1950s, when I was in Junior High School in Illinois. The book was shelved in our local public library under "science." Neither the librarian nor my science teachers had any idea that this was not legitimate scholarship. Nearly two decades later, when Leroy Ellenberger encountered the book in 1969, he "got intrigued by the cover puffery and started to read it . . . nobody I knew was familiar with the book, and when I checked out reviews and critiques in the library I was not impressed by what I found. Thus, I was ripe for picking." Finally, recall Harvard's Owen Gingerich's comment in 1973 that "my students were hearing a lot of pro-Velikovsky news, and no respectable astronomers were willing to take the time of day to explain to the general public why they didn't take his scenario seriously." From these and other examples it seems that Velikovsky won the publicity wars. I wonder if it would be different today, where Amazon.com and other Internet sources make book reviews and other commentary more readily available.

Velikovsky was an effective speaker and advocate for his ideas. Much like some proponents of creationism today, he carefully controlled the "turf" on which he fought. Velikovsky ensured that the focus of any meetings or presentations was on him and his theory. He made no contributions to regular scientific conferences, and with few exceptions he did not appear on the same platform with his critics. Even at meetings organized to promote his theories, he did not engage in dialog with the other presenters, but waited to make his entrance as the acclaimed leader and keynote speaker. Viewing the universe through the singular filter of his own theory, he devoted the last 30 years of his life to defending the inherent accuracy of what he had written in the 1940s.

Part of his style can be attributed to the European scholarly tradition in which he grew up. A Byelorussian Jew, he was educated in Moscow, worked for a time in Berlin, and was prominent in the Zionist movement. As Bauer writes, "he was of his time and of his place." Bauer compares Velikovsky to some of his (Bauer's) teachers, also from middle Europe, who were men "of both the scientific and literary cultures" who represented "the hierarchical, authoritarian circumstances of continental scholarship, whose normal everyday self-possession and dignity of manners would seem to us, in our present place and at this distance in time, unwarrantably and unbearably arrogant." I have had a similar experience in dealing with physicist Edward Teller, the "father of the H Bomb" and former Director of Livermore National Laboratory. Teller is equally imperious and authoritarian, and he is treated with abject deference by his former students and protégés. These are all examples of a culture far removed from current American science, with its informality, open exchange of opinion, and encouragement of collaboration.

There are still those who praise Velikovsky as a scholar even though they recognize that his theory of cosmic catastrophes was wrong. Bauer writes that "The substance of his claims includes much interesting and instructive material about history, legend, human belief, religion; a great deal that can reasonably be believed by any rational person who chooses to do so." But I wonder: can someone who was so oblivious to his own ignorance of even the most basic fundamentals of the physical sciences be trusted as a reliable source on history, legend, belief, and religion?

I am not competent to judge these matters, but I will quote mid-eastern archeologist William Stiebing of the University of New Orleans, one of the few non-astronomers to seriously analyze Velikovsky's scholarship. He wrote that "a historian should find out as much as possible about the background and transmission of the sources he uses. Velikovsky generally ignores this fundamental principle of responsible historical scholarship. . . . Problems with Velikovsky's methodology and with his interpretations of various texts are enough to cause almost all mythologists, anthropologists, and ancient historians to reject his thesis. . . . Whatever rewriting of ancient history may be done as a result of future discoveries, it is a safe bet that it will not resemble [Velikovsky's] scenario." If Velikovsky was wrong in so many particulars, how can his grand synthesis be of much value either?

It seems clear that physical scientists have a great deal to learn about communicating with our colleagues from other disciplines as well as with the public. In criticizing Velikovsky, the astronomers in particular often came across as intolerant and closed-minded. It was so obvious to us that Velikovsky's ideas were crackpot that we did not explain our reasoning to others in a convincing way. By contrast, our colleagues in the geological sciences simply ignored him. I do not know which was the better strategy.

Finally, however, I can agree with Bauer that the process of confronting Velikovsky has been educational and sometimes constructive. We can all learn from mistakes as well as successes. I will close with four additional quotes from some of the pioneers in planetary catastrophism:

Peter Schultz (Brown University planetary geologist, authority on impact processes, editor of *Geological Implications of Impacts of Large Asteroids and Comets with the Earth*):

> One effect of Velikovsky's work was to force me to take greater care to identify and clearly distinguish between observations, interpretations, extrapolations, reasoned speculations, and just plain pseudoscience. When you see such ideas expressed to the extremes, you are better able to see the signs earlier on. And it forced me to be aware of how scientific studies could be manipulated into pseudoscience scenarios.

Bill Hartmann (Planetary Science Institute planetary scientist and artist, originator of impact theory for origin of the Moon, author of *Moons and Planets*):

> I was very interested in the controversy, which was still hot while I was in grad school, and in what Velikovsky said and did—and in the meta-question of how should he have known he was on the wrong track.

Brian Toon (U Colorado atmospheric physicist, authority on the environmental effects of impacts):

> Velikovsky influenced me by showing me how the public is so easily fooled by pseudoscience.

Clark Chapman (South-West Research Institute planetary scientist, authority on asteroids and impact hazards, author of *Cosmic Catastrophes*):

> I think that Velikovsky and his followers, and their debates with people like [Morrison] and Sagan, have been influential in sharpening the understanding of distinctions between science, pseudoscience, and other non-scientific intellectual disciplines as paths toward truth. This has been interesting, even valuable, and wouldn't have happened hadn't Velikovsky and his followers been so determined to argue their case.

Acknowledgments:

I gratefully acknowledge the detailed analysis of these issues published in 1984 by Henry H. Bauer. His documentation provides an essential foundation to any analysis of Velikovsky. I also thank Leroy Ellenberger for many years of stimulating discussion and debate. Once a member of Velikovsky's inner circle, he later became a leading opponent. He has helped me to try to look at these issues from the other side and to appreciate how poorly the scientific critics communicated with the public.

References:

Bargmann, V., and L. Motz. 1962. "Letter to the editor." *Science* 138: 1350.

Bauer, Henry H. 1984. *Beyond Velikovsky: The History of a Controversy.* University of Illinois Press.

Bauer, H. H. 1995. "Velikovsky's Place in the History of Science." *Skeptic,* Vol. 3, No. 4: 52.

de Grazia, Alfred. 1966. *The Velikovsky Affair: The Warfare of Science and Scientism.* University Books.

Ellenberger, Leroy. "Still Facing Many Problems." *Kronos* X:1, 1984, and X:3, 1985.

Rose, Lynn E. 1976. "The Censorship of Velikovsky's Interdisciplinary Synthesis," in *Velikovsky Reconsidered.* New York: Doubleday.

Sagan, Carl. 1977. *Scientists Confront Velikovsky,* edited by Donald Goldsmith, Cornell University Press.

Stiebing, William. 1984. *Ancient Astronauts, Cosmic Collisions, and Other Popular Theories about Man's Past.* Buffalo: Prometheus Books.

Velikovsky, Immanuel. 1950. *Worlds in Collision.* New York: Doubleday.

Velikovsky, Immanuel. 1977. "My Challenge to Conventional Views in Science," in *Velikovsky and Establishment Science.* Kronos Press.

Willhelm, Sidney. "The Velikovskian Upheaval: A Temporocentric Challenge," in *Velikovsky and Establishment Science,* p. 60.

Witchcraft and the Origins of Science

DR. RICHARD OLSON

When Michael Shermer reviewed the second volume of my *Science Deified* (*Skeptic* 1992), he began with an interesting passage from Robert Pirsig's *Zen and the Art of Motorcycle Maintenance* (1974) that I would like to use as an introduction to this essay on the status of beliefs in spirit phenomena and witchcraft during the second half of the 17th century. In this passage, Pirsig's protagonist explains to his son why he does not believe in ghosts (1974, 38–39):

> They are unscientific. They contain no matter and have no energy and therefore according to the laws of science, do not exist except in people's minds. Of course, the laws of science contain no matter and have no energy either and therefore do not exist except in people's minds . . . It's best to refuse to believe in either ghosts or the laws of science.

The reason this passage jars us into thought is that it applies currently accepted criteria for what it means to be an object in the world, and uses those to reject the existence of ghosts; then it plays a mind game on us by somehow applying the same criteria to statements which everyone is presumed to assent to and arguing that if we shouldn't believe in ghosts, we shouldn't believe in science either.

The usual expectation among American intellectuals—certainly among those who view themselves as in the least bit skeptical—is that anyone who believes in "science" will not believe in such creatures of superstition as ghosts, spirit phenomena, or "witches." Indeed, the first paragraph of the first chapter of the first edition of Garvin McCain and Erwin Segal's immensely popular *The Game of Science* begins with the claim that we no longer believe in witches precisely because we believe in science:

> Why don't you believe in witches? That question may seem ridiculous but our ancestors, who were probably as bright as we are, did believe in them, and acted accordingly. Why are we so different and superior? The evidence for or against witches is no better than it was 400 years ago. For us, it is almost impossible to believe in witches; for our ancestors, it was equally difficult to deny their existence. Our new beliefs exist, in part, due to the development of "scientific attitudes" (McCain and Segal, 1969, 3).

Though this statement certainly reflects what most American intellectuals believe, there is a strange historical irony contained in it and in Pirsig's intentionally perverse argument that if one doesn't believe in ghosts, one shouldn't believe in scientific laws either.

What I want to argue is that beliefs in witches, ghosts, and demons were heavily under attack and on the wane in England at the very beginning of the 17th century before the rise of what we would usually identify as modern scientific attitudes. But witchcraft beliefs and beliefs in other spirit phenomena underwent a remarkable revival among British intellectuals during the period after the Restoration of James II to the throne in 1660; and this revival of demonological beliefs was directly and self-consciously attached to the rise of modern scientific attitudes among the men who were members of the Royal Society of London. So at least for a time it may be true to say that men actually came to believe in witches as a result of the development of scientific attitudes. In this case, the reverse of Pirsig's argument was taken with deadly seriousness by Joseph Glanvill, who argued that if one believed in the methods of modern science, one should also believe in ghosts and witches. It is probably also true (though here the issue is more complicated) that certain arguments in favor of witchcraft made mid-17th-century intellectuals more favorably disposed to the new science than they would otherwise have been and that a general belief in spirit phenomena, for which witchcraft stood as a symbol (Schafer, 1969, 55–85). In order to explain how and why the rise of modern science became tied to beliefs in spirit phenomena in mid-17th-century England, I think we need to discuss briefly a continental phenomenon at the end of the 16th century, and look at the impact it had on early 17th-century English religious developments.

Early Criticism of Belief in Demonic Possession

A serious and concerted attack on beliefs in witchcraft and demonic possession had been launched at the end of the 16th century in connection with a series of spectacular exorcisms that were quite literally staged before thousands of witnesses in France between 1566 and 1599. The goal of the Catholic priests who carried out these exorcisms was to promote the reconversion of French Protestant Huguenots to Catholicism by demonstrating the power of the true Catholic religion; and they seem to have had substantial success.

Understandably, these Catholics' claims were widely challenged by Protestant propagandists; but ironically, they were also strongly challenged by the French Catholic Crown as well; for during the 1580's and 1590's, public exorcisms were stirring up religious passions just at a time when the French Crown, through the Edict of Nantes, was trying to calm religious hostilities and establish official tolerance for Protestantism. As a consequence, in 1598, Henry IV ordered the physician Michael Marescot and a group of medical colleagues to investigate the popular claims to demonic possession of one Marthe Brosier in the expectation that they could establish that her "possession" was either a mis-diagnosis of a natural disease such as epilepsy or hysteria, or that they could prove it to be a deliberate fraud. Marescot's *Discourse véritable sur le faict de Marthe Brosier de Romorantin prétendue démoniaque* appeared in 1599, to be translated immediately into English. The overall verdict of Marescot's investigation was stated in a memorable line: "Nothing from the devil, much counterfeit, a little from disease" (Walker, 1981, 35).

Without totally denying the possibility of demonic possession, Marescot and his colleagues were able to establish to their own satisfaction, that of the king, and that of many readers, that in one of the most celebrated cases of "possession," an initially deluded and psychologically unbalanced woman had been exploited by her family and by a group of Catholic clergy, for both financial gain and for the seditious purpose of stirring up anti-Huguenot sentiment. In the process, Marescot reviewed a series of experimental tests for legitimate possession which had become widely accepted by the late 16th century:

1. possessed persons were supposed to be able to understand and speak languages of which they had no prior knowledge;
2. possessed persons were supposed to be able to discern secrets and predict events of which they could have no natural knowledge—i.e., they had clairvoyance;
3. possessed persons had abnormal bodily strength and insensitivity to pain; and
4. possessed persons expressed revulsion at holy things, especially the reading of scripture or contact with holy water or other blessed objects (Walker, 12).

Under investigation by Marescot and four other physicians, it was shown that Marthe Brosier could understand neither Latin nor Greek, as her advocates had claimed; that she had no reaction to holy water that was passed off as ordinary water, but that she convulsed when she was given plain water that she was told was blessed; that she showed no special clairvoyant powers; and that when she was read passages from the *Aeneid*, expecting them to be biblical passages, she showed dramatic signs of disturbance. Finally, though during her fits Marthe could withstand the pain of the "deep pricking of long pins" in her hands and neck without showing discomfort, Marescot did find her responsive to normal sources of pain when not in convulsions; and he identified her reactions in this matter as typical of "melancholic" persons (Walker, 34–35, 38).

Just a few years later, the English physician Edward Jordan, who was consulted in two cases of supposed demonic possession, published a treatise whose title discloses its major conclusions: *A briefe Discourse of a Disease called the Suffocation of the Mother. Written upon occasions which hath been of late taken therby, to suspect possession of an evil spirit, or some such like supernatural power. Wherein it is declared that diverse strange actions and passions of the body of man, which in the common opinion are imputed to the Divell, have their true natural causes, and do accompany this disease* (1603). In this work Jordan identified almost all of those symptoms that had been traditionally identified with demonic possession and witchcraft—especially insensibility, convulsions, and fits brought on by the presence of particular persons or artifacts with symptoms of hysteria. Thus, by the early years of the 17th-century there was a substantial medical literature which simultaneously denied the existence of possession and attacked virtually all of the traditional tests for its existence.

Anglicans Attack Demonology to Defend Their Religious Interests

Early 17th-century Anglican attitudes toward demonic possession and witchcraft were shaped primarily by the existence of this medical literature, in response to the Continental Catholic propaganda, and in response to a series of cases in which both an English Jesuit priest, William Weston, and a Puritan preacher, John Darrell, claimed to have cast demons out of a number of possessed children between 1585 and 1598 (Walker, 43–73). Weston's activities were commenced in 1585, but it was not until 1602 that a formal inquiry was held regarding his exorcisms. Darrell's castings out of devils began in 1596; but in 1598, he was tried in London, condemned for fraudulent practices and both deposed from the ministry and sent to prison for at least a brief stay (Walker, 64).

In fact, Darrell's case seems to have been part of a major anti-Puritan campaign by Archbishop John Whitgift, his Bishop in London, Richard Bancroft, and Bancroft's chaplain, Samuel Harsnett. Like the French Catholic exorcisms of the late 16th-century, Darrell's spectacular success casting out devils was drawing much favorable attention for his religion; but Darrell's demonics did most of the French examples one better by using their clairvoyance to name witches whom Darrell subsequently had arrested (Walker, 63). As a popular and visible Puritan, Darrell drew Whitgift and Bancroft's serious attention; and they apparently decided to discredit him by trying him for fraud. According to evidence given by William Sommers, the last of those he had dispossessed, Darrell taught several of his demonics how to simulate their symptoms, and at least in one case, i.e., that of Sommers, he even suggested the fraud to the victim (62–64). Sommers later recanted his evidence and there were apparently any number of irregularities in the trial, including a refusal to allow Darrell to speak; so the "trial" was continued in a series of publications for the next five years.

The major Anglican arguments were presented in Harsnett's *A Discovery of the Fraudulent Practices of John Darrell* (1599) and in John Deacon and John Walker's *Dialogical Discourses of Spirits and Devils* (1601–1602). In *The Trial of Maist Darrell* (1599), the Puritans responded by offering a largely scriptural defense of their claim that possession was possible and that it could be eliminated by appropriate prayers to God (Walker, 67–68). But they also complained about the procedures used in Darrell's trial and they argued (quite rightly at the time) that the Anglican prosecutors of Darrell were more interested in destroying Puritanism than in eradicating Catholicism, otherwise they would have tried Weston the Jesuit. To this claim, Whitgift and Bancroft responded by ordering an investigation of Weston's claims and Harsnett responded by publishing *A Declaration of Egregious Popish Impostors, to withdraw the Harts of Her Majesty's Subjects from the Truth of Christian Religion Professed in England, Under the Pretense of Casting out Devils Practiced by Edmunds, Alias Weston, a Jesuit* (1603). Circumstances had conspired to give middle-of-the-road Anglican apologists an opportunity to simultaneously discredit both the Catholic and Puritan opposition by attacking their claims of dispossession. But in order to do so, the Anglicans had to act incidentally to undermine belief in both demonic possession and in witchcraft by almost completely accepting the medical views of Marescot, Jordan, and their colleagues. One of their most important converts was James I, who had defended beliefs in possession and witchcraft in his famous *Daemonology* of 1597, but who had turned into a strong opponent of witch persecution by 1616 (Shapiro, 1983, 199). Technically, neither Harsnett nor Deacon and Walker denied the possibility of witchcraft or dispossession, although Harsnett probably doubted the existence of either. What they did do was offer an explanation of how melancholia and hysteria might cause persons to believe in both as well as a demonstration that in many cases, men like Weston and Darrell exploited those beliefs and used fraudulent techniques to delude people into believing in their power to exorcise or to dispossess persons who were possessed. The major concern which had held Harsnett and others back from taking an even stronger stance against belief in witchcraft and possession at the beginning of the 17th century was laid out in the dedication of *The Trial of Mr. Darrell:*

> Atheists abound in these days and witchcraft is called into question. Which error is confirmed by denying dispossession and both these errors confirm atheists mightily. . . . If neither possession nor witchcraft (contrary to what has been so long generally and confidently affirmed), why should we think that there are devils? If no devils, no God (Walker, 71, 72).

Puritans thus warned the readers of Anglican tracts that demonology and witchcraft were proof against atheistic materialism.

Demonic Power Becomes a Natural Phenomenon

In order to protect themselves from the claim that their attacks on possession and witchcraft were simultaneously denials of the fallen angel and of God, early 17th-century Anglican apologists insisted that the devil might indeed involve himself in human affairs, but that if he did, it must be through the use of natural rather than supernatural powers (Shapiro, 200–204). In John Cotta's words:

> Though the divel indeed, as a spirit, may do, and doth many things above and beyond the course of some particular natures: yet doth he not, nor is able to rule or command over general Nature, or infringe or alter her inviolable decrees in the perpetual and never interrupted order of all generations; neither is he generally master of universal Nature, but Nature [is] master and commander of him. For Nature is nothing else but the ordinary power of God in all things created, among which, the divel being a creature, is contained, and therefore subject to that universal power (Clark, 1984, 360).

One critical consequence of the "naturalization" of presumed demonic powers was that it brought the study of demonic activities clearly within the realm of natural knowledge. Thus, Francis Bacon argued in *De Argumentis Scientarium* that well established "narratives of sorceries, witchcrafts, charms, dreams, divinations, and the like" should be included as legitimate data in natural histories in order to establish "in which cases and how far effects attributed to superstition participate in natural causes" (cited in Clark, 355).

Even though he remained formally open-minded regarding the existence of witches and demons, when Bacon chose to discuss particular issues, he, like other Anglicans, explained beliefs in witchcraft as arising out of the misinterpretation of natural phenomena. Thus, for example, in the Sylva sylvarum, he argued that the hallucinogenic effects of some ointments produced a mistaken belief in real transvection (human flight) and metamorphoses; so that when women charged as witches confessed to being transformed into animals and transported to witches' sabbaths, they were mistakenly reporting their hallucinations as reality.

For most of the first half of the 17th century, while the twin threats of Puritanism and Catholicism seemed more immediate and critical to the Anglican cause than philosophically based atheism, Anglican intellectuals continued to express strong skepticism regarding specific claims of spirit phenomena and to insist that what had traditionally been attributed to supernatural influences was actually accomplished through natural ones. This was particularly true because as Puritanism and dissent became ever stronger, popular attempts at witch persecution intensified, and established authorities became ever more fearful of the religious enthusiasm which underlay them.

Atheism Reverses Attitudes about Spirit Phenomena

The problem faced by Anglican religious figures changed dramatically with the publication of a series of frightening works by Thomas Hobbes. Philosophical materialism and atheism had been a minor, though growing, problem in early 17th-century England. But the publication of Hobbes's *Leviathan* in 1651, *De Corpore* in 1655, and *A Physical Dialog, or a Conjecture about the Nature of the Air* in 1661, deflected attention from Catholicism and Sectarianism alike, and made Hobbesian Atheism the new chief target of moderate Anglican apologetic literature.

Whether Hobbes was really an atheist is a topic on which scholars might differ—though just for the record, I believe he was—but no one can doubt that he was a bitter enemy of what he called priestcraft—or the authority of religious persons. Hobbes believed that priests had usurped power that rightly belonged to the secular sovereign. In order to justify his attacks on priestcraft he turned to a set of arguments that had been used by materialist philosophers, such as the atomist, Epicuros, in antiquity. According to the ancient materialists and Hobbes, priests exploit a natural human fear of the unknown to convince people that invisible powers and agents are at work in the world and that they (the priests) alone have the power to intercede on people's behalf to control these "spirits." "Who," wrote Hobbes, "that is in fear of ghosts, will not bear great respect to those who can make the holy water that drives them from him" (cited in Shapin and Shaffer, 1985, 96). Similarly, he wrote: "By their demonology, and the use of exorcism, and other things appertaining thereto, the priests keep, or think they keep, the people in awe of their power and lessen the dependence of subjects on the sovereign power of their country." Since it was the false belief in spirits, made possible by ignorance about the causes of events, that gave the clergy its power, according to Hobbes, the most effective way to fight the power of the clergy was first, to demonstrate that spirits, or "incorporeal substances," do not exist; and second, to demonstrate that all phenomena can, and indeed, must be explicable solely in terms of matter in motion.

To undermine belief in immaterial spirits, Hobbes developed a logical argument that depended very heavily on ideas which owe their existence to Aristotelian philosophy. The meaning of the term "substance," he argues, is derived from our experiences of physical bodies or "corps." The term "incorporeal substance" or "immaterial substance" is thus self-contradictory. To accomplish the second part of his goal, Hobbes purported to be able to give a completely materialist account of all natural philosophy. But in doing so he departed from ancient atomism in a way that turns out to play a major role in linking witchcraft and the experimental philosophy of the royal society.

The ancient atomists had posited the existence of atoms and void space, claiming that atoms move freely through the void. Descartes, however, defined Matter as that which has dimensions; and from this definition—which Hobbes accepted—it followed that there can be no void; because any space, no matter how small, has dimensions and therefore must contain matter.

Note here for future reference, Hobbes uses precisely the same kind of argument to deny the possibility of spirits and to deny the possibility of empty space. The question of whether empty space exists, like the question of whether immaterial spirits exist, is not to be answered empirically. Both questions are to be answered by a purely rational analysis of definitions.

Hobbes's claim regarding spirits was, quite rightly, seen as an attack on almost all fundamental Christian beliefs, for it denied not only the existence of demons and witches, but also the immateriality and hence the immortality of the human soul. And if this weren't enough, Hobbesian Materialism took on an additional troubling aspect during the later civil war period when it was adopted by Richard Overton, a notorious political radical and one of the founders of the Levellers sect.

Joseph Glanvill and the Scientific Defense of Witches

This now brings me to the central events of my story—events connected with a moderate Anglican apologist who became both a defender and a member of the Royal Society of

London in 1662—a man by the name of Joseph Glanvill.

Although Glanvill had a longstanding interest in spirit phenomena stemming from his commitment to the Cambridge Platonist doctrine of pre-existent souls, and though he had begun his investigations into the appearances of apparitions as early as 1662, it was not until 1666 that he published the first version of his often improved and expanded treatise on witchcraft, *Some Philosophical Considerations Touching on Witches and Witchcraft.* A friend, Justice of the Peace Robert Hunt, had tried to prosecute a coven of witches during 1664 in Somersetshire; but the local gentry were so skeptical that they mocked his efforts. In response, Hunt, who knew of Glanvill's interests, sent the depositions from the accused witches along with a description of the gentry's repose to Glanvill, and Glanvill responded with a refutation of the most common reasons for disbelief (Jobe, 1981, 346–347).

Glanvill begins by explaining what is at stake if the belief in witches should be abandoned. Borrowing his theme from the earlier anti-Anglican defenders of Robert Darrell, he writes:

> He that thinks there is no witch, believes a devil gratis, or at least upon inducements which he is likely to find himself disposed to deny when he pleases. And when men are arrived at this degree of dissidence and infidelity, we are beholden to them if they believe either Angel or Spirit, Resurrection of the Body or Immortality of Souls. These things hang together in a chain of connection, at least in these men's hypotheses; and it is but a happy chance if he that has lost one link, holds another (Glanvill, 1676, 2).

The central doctrines of religion are thus being endangered by those who do not believe in witches.

Secondly, Glanvill immediately seeks to identify the disbelief in witches with the Hobbesian attack on experimental philosophy. The question of whether witches exist or not, he argues, is a question of fact; and as such it can only be settled by appeal to authority or sensory evidence. There are thousands of eye- and ear-witnesses who have attested to "things done by persons of despicable power and knowledge, beyond the reach of art and ordinary nature," and these include not only "vulgar" persons, but "wise and grave discerners . . . when no interest could oblige them to agree together in a common lie." Unfortunately, he argues, no amount of empirical evidence could convince those who do not believe in witches, "since those that deny the being of witches, do it not out of ignorance of these heads of argument . . . but from an apprehension that such a belief is absurd, and the things, impossible. . . . Upon these presumptions they condemn all demonstrations of this nature, and are hardened against conviction" (Glanvill, 1676, 3).

For Glanvill, then, the key issue was whether one placed greater confidence in well attested experiences or in metaphysical claims regarding the possibility of the existence of certain kinds of entities. It is not reasonable, he insists, "first to presume the thing impossible, and thence to conclude that the fact cannot be proved: On the contrary, we should judge of the action by the evidence, and not the evidence by our fancies about the action. This is proudly to exalt our own opinions above the clearest testimonies and most sensible demonstrations of fact: and so to give the lie to all mankind, rather than distrust the conceits of our bold imaginations" (Glanvill, 5–6). Given his belief in the limitations of human reason and the inability of humans to possess more than probable knowledge of any causal account of any phenomenon, Glanvill says that humans have no right to insist upon the impossibility of anything. The most they can legitimately claim is that they cannot conceive or imagine how the actions in question take place, and this inability to conceive "only argues the weakness and imperfection of our knowledge and apprehensions; not the impossibility of those performances" (Glanvill, 7).

Precisely the same kind of argument was being carried on simultaneously between Robert Boyle and Thomas Hobbes regarding the evacuated space created by Boyle in his air pumps. Hobbes denied that the space could be empty because he was committed to a conception of space derived from Descartes. According to this conception, matter, or body, is defined by extension, so that any extended region must contain matter, and a vacuum is literally impossible. Boyle, whose notions of matter and space were derived from atomist notions, was unwilling to fight on Hobbesian ground. Whether or not extension could exist without a material substance underlying, it was technically an undecidable question and therefore beyond the bounds of natural philosophy for Boyle. The key question was whether well attested experiments justified the claim that the evacuated region was empty of ordinary corpuscles of air; and he and his allies were convinced that they did.

The resolutions of the problem of how to decide whether witches exist and how to decide whether the receiver of an air pump could be evacuated were understood by all parties to the 17th-century debates to be clearly linked with one another. No one, least of all Glanvill and Boyle, doubted that every reader convinced by Glanvill's arguments about witches would also be driven toward acceptance of Boyle's arguments about the phenomenon of the air pump, and vice versa.

Neither Glanvill and his allies, nor Boyle and his allies, wanted to encourage credulity and a lack of critical analysis of experience or experiments. To have argued that any individual's factual claims should be blindly accepted would have been, in their common view, to play into the hand of religious enthusiasts and philosophical charlatans. Instead, both sought to encourage "diffidence and backwardness of assent" to any such claims and to encourage the careful empirical investigation of all (Boyle, 1772, vol. 1, ccxx–ccxxii).

As early as January 19, 1663, Glanvill had begun to investigate claims of spirit phenomena when he and a gentleman friend traveled to Tedworth in Wilshire, where a "drumming" spirit was said to haunt the house of a Mr. Mompesson. The two men first interviewed the servants and several neighbors, including two local ministers of impeccable reputation, all of whom had been present when the spirit made noises or threw objects about the house. Then they themselves experienced the noises that the spirit produced and tried to discover "if there were any trick, contrivance, or common sense of it," but they could find nothing; so Glanvill was persuaded that "the noise was made by some Daemon or Spirit" (Glanvill, 1689, 329). Glanvill delayed publication of his account of the "Drummer of Tedworth" at Mr. Mompesson's request until the strange phenomena ceased. (He was concerned that the spirit would become angered by the books!) In 1668, however, it became the first of 28 different detailed accounts of spirits and witches which Glanvill published as appendices to his philosophical treatments of witchcraft and demons in order to reliably establish the evidence for their existence. Summarizing his account of the drummer, Glanvill lays out a litany of criteria which such an account ought to have in order to be credible support for the belief in spirits:

> [The phenomena] are strange enough to prove themselves effects of some invisible extraordinary Agent, and so demonstrate that there are spirits, who sometimes sensibly intermeddle in our affairs. And I think they do it with clearness of evidence. For these things

> were not done long ago or at a far distance, in an ignorant age, or among a barbarous people. They were not seen by two or three only of the Melancholic or superstitious, and reported by those that made them [to] serve the advantage and interest of a party. They were not the passages of a day or night, nor the vanishing glances of an apparition; but these transactions were near and are public, frequent, and of diverse years continuance, witnessed by multitudes of competent and unbiased attestors, and acted in a searching and incredulous age: Arguments enough, one would think, to convince any modest and capable reason (Glanvill, 1689, 338).

From the comments of Samuel Pepys, who had found the earlier versions of Glanvill's Witchcraft essay "unconvincing," it is fairly clear that the accounts of actual spirit events increased the impact of his arguments, making them, in Pepys' view, "worth reading indeed" (Cope, 1956, 14). Whatever other impact they had, these "ghost stories" certainly made best sellers out of numerous editions of Glanvill's *Saducismus Triumphantus* and stimulated a whole tradition of dramatic and fictional treatments of spirit phenomena.

In the 1668 *A Blow at Modern Saducism*, which saw the first appearance of Glanvill's account of the drummer of Tedworth, Glanvill also attempted to recruit the Royal Society to help in investigations of spirits and thus in support for the true religion:

> Did the Society direct some of its wary and luciferous enquiries towards the world of spirits, believe we should have another kind of Metaphysics, than those [that] are taught by men that love to write great volumes and to be subtle about nothing? For we know not anything of the world we live in, but by experiment and the phenomena; and there is the same way of speculating immaterial nature, by extraordinary events and apparitions, which possibly might be improved to notices not contemptible, were there a cautious and faithful history made of those certain and uncommon appearances. At least it would be standing evidence against SADDUCISM, to which the present age is so unhappily disposed, and a sensible argument of our Immortality (cited in Prior, 1932, 182).

While the Royal Society offered no official response to Glanvill's request, many members contributed directly to Glanvill's collection of Spirit relations. Boyle sent a report of an Irish Witch, whom he had investigated, and confirmed his first-hand support of an earlier account of a demonic possession at Mascon in France, for example. And John Beale sent him letters on the possible effects of witchcraft on butter production. Perhaps more importantly, many Royal Society members began to incorporate spirits into their laboratory world (Schaffer, 1987, 55–85).

It is not clear to me which group benefitted more from the mutually supportive arguments of Anglican demonologists and experimental natural philosophers after 1666. On the one hand, Glanvill and his Anglican colleagues, such as Henry More, reached a far wider audience; and many persons who welcomed Glanvill's "defence" of traditional Christian beliefs in the immortality of the soul were probably swayed toward a sympathy for experimental philosophy. On the other hand, experimental philosophers, as a group, probably had a more profound impact in legitimizing Glanvill's views among intellectuals. In any event, for at least a couple of decades after the Restoration, the belief in ghosts and witches—which had begun to decline in the late 16th and early 17th century—returned as a serious and popular topic for polemical discussions; and those who argued in favor of beliefs in spirit phenomena simultaneously drew arguments from and promoted experimental science (Jobe, 1981, 343–356).

References:

Boyle, Robert, 1772. *The Works of the Honourable Robert Boyle*, ed., Thomas Birch. Six Volumes, 2nd ed. London, J. & F. Rivington.

Clark, Stuart, 1984. "The Scientific Status of Demonology," in Brian Vickers, ed., *Occult and Scientific Mentalities in the Renaissance*. Cambridge, Cambridge University Press.

Cope, Jackson I., 1956. *Joseph Glanvill: Anglican Apologist*. St. Louis, Washington University Studies.

Glanvill, Joseph, 1676. *Essays on Several Important Subjects*. London, S. Lownds.

Glanvill, Joseph, 1689. *Saducismus Triumphatus: or, Full and Plain Evidence Concerning Witches and Apparitions*. London, S. Lownds.

Jobe, Thomas Harmon, 1981. "The Devil in Restoration Science: The Glanvill-Webster Witchcraft Debate," *Isis*, 72, 343–356.

McCain, Garvin, and Segal, Erwin M., 1969. *The Game of Science*. Belmont, Brooks/Cole.

Pirsig, Robert M., 1974. *Zen and the Art of Motorcycle Maintenance: An Inquiry into Values*. New York, William Morrow and Company.

Prior, Moody E., 1932. "Joseph Glanvill, Witchcraft, and Seventeenth Century Science," *Modern Philology*, 30, 167–193.

Shapin, Steven, and Shaffer, Simon, 1985. *Leviathan and the Air Pump: Hobbes, Boyle, and the Experimental Life*. Princeton, Princeton University Press.

Shapiro, Barbara, 1983. *Probability and Certainty in Seventeenth-Century England*. Princeton, Princeton University Press.

Walker, D. P., 1981. *Unclean Spirits: Possession and Exorcism in France and England in the Late Sixteenth and Early Seventeenth Centuries*. Philadelphia, University of Pennsylvania Press.

Witches and Witchcraft

CLAYTON DREES

Most Americans are comfortable with the religious pluralism our Constitution guarantees and our society seems to encourage. We routinely—and for the most part disinterestedly—live and work daily with people of many faiths. Such acceptance of religious diversity has not always been the case, however, in the evolving European culture that in many ways gave first expression to much of our own. Ever since the adoption of Christianity as the official religion of the late Roman Empire, leaders of both church and state have sought to define and enforce their own versions of religious "truth" for all believers. Those who have disagreed with such "truths," or who have opposed ecclesiastical and political rulers on other, often non-religious grounds, have frequently suffered denunciation and persecution as heretics. Even in our own religiously diverse America of today, some who believe differently from most have also suffered various indignities, perhaps as a result of the legacy that our nation has inherited from Christian Europe.

It is the intent of this article to demonstrate that religious "truth" is a relative concept, that one person's "truth" is another person's "error." The idea that religious truth and error are relative to one's viewpoint can be clearly seen through an examination of the culture of medieval Europe. From the 11th-century onward, charges of "heresy" or theological error were used not only against religious dissenters, but against social, political and economic non-conformists as well. It will be seen that charges of heresy, although rooted in the devout faith that characterized the medieval mind, were often employed as weapons to further the interests of Europe's ecclesiastical and secular leaders. In the end, it is hoped that naturally skeptical readers will question claims to *exclusive* religious "truth," a form of bigotry which has, it is sad to note, been a frequent cause of mankind's inhumanity to man in history. At the same time, it is also hoped that concepts like "heresy" and "exclusive truth" will be exchanged for those of "diversity" and "tolerance" in attitudes toward religion in our modern world.

A classic example of the relativity of religious truth in medieval Europe, the well-known case of Jeanne D'Arc (Joan of Arc), serves to illustrate the main premise of this article. Jeanne was a young French peasant girl from Domremy in Lorraine whose claim to have heard the "voices" of SS. Michael, Catherine and Margaret won her an interview with the Dauphin of France in 1428. After convincing skeptical French leaders that her voices had commanded her to lead armies against the invading English, she inspired several French victories before her capture by the enemy in 1430. Jeanne's captors naturally interpreted her voices as demonic, especially since they had been responsible for successful French attacks against English positions. They declared that she had "fallen into diverse errors and diverse crimes of schism, idolatry, invocations of devils and numerous others," and so burned her at the stake in 1431 as a "relapse and heretic" (Pernoud, 230). A Papal court, however, posthumously found Jeanne D'Arc innocent of all heresy in 1456, and she was subsequently canonized a patron saint of France by the Roman Catholic Church in 1920.

This case graphically illustrates the problem that confronts modern historians of religion and heresy. Two hostile branches of the same 15th-century Roman Church interpreted Jeanne's voices differently, though each claimed at the time to have done so correctly. Both sides, in fact, believed that their own interpretation was so "true" that the French were willing to entrust their armies to her leadership while the English were quite willing to burn her alive. The point is that the heretic of the English was the saint of the French, and that religious truth and error are relative to one's point of view. Given the ambiguities apparent in the case of Jeanne D'Arc, let us now examine the roots of Christian understandings of heresy as they evolved in the European Middle Ages.

Throughout the history of Christianity, heresy seemingly represented a far more significant threat to the religious and political order of Europe than did Judaism, Islam or even witchcraft. Heretics were considered dangerous because, unlike Jews or Muslims who had never been Christians in the first place, or witches who were supposed victims of supernatural influences, heretics were thought to have consciously renounced Christian precepts for contrary religious beliefs. English Bishop Robert Grosseteste of Lincoln attempted to define heresy in the early 13th-century when he wrote that "heresy is an opinion chosen by human faculties, contrary to sacred scripture, openly held, and pertinaciously defended. Hairesis in Greek, choice in Latin" (Peters, 4, 167). This strict definition highlighted the negative choice element of heresy, but limited it to theological deviations from scripture alone. Grosseteste thus ignored repudiations of the creeds, canon laws and the teachings of the early Church Fathers, all of which also formed a large part of official Christian dogma. French theologian John of Brevicoxa's description of heresy in 1375 was more inclusive:

> [Heresy] consists of those assertions which are shown to be incompatible with holy scripture, or with apostolic doctrine not found in holy scripture, or with truths inspired by or revealed to the Church, together with other truths which one cannot rationally deny (Peters, 301).

This definition, with its "other truths," now identified heresy as any challenge to Church teachings regardless of their basis in scripture. Since medieval churchmen like Robert Grosseteste and John of Brevicoxa equated Christianity with all that was good and orderly in European culture, they also viewed heresy as a single, organized, monolithic attack on Christendom as a whole. This attitude dominated clerical thinking on heresy in the Middle Ages, and had its roots deep in the scriptural and traditional foundations of the Church.

The New Testament was the theological cornerstone upon which Christianity rested, especially after an orthodox canon of Biblical books was agreed upon at the Council of Carthage in A.D. 397. New Testament gospels and epistles contained many pronouncements on religious dissent that survived to shape the attitudes and define the coercive policies of later medieval officials. The parable of the banquet in Luke, for example, to which the host invited the poor and crippled in place of his reluctant guests, advised readers to "force people to come in to make sure my house is full" (Luke 14: 15–24). This injunction was later interpreted to sanction coercive measures against non-conformists who remained outside the universal house of Christendom. The greatest number of New Testament warnings against heresy came from Pauline Epistles, addressed to fledgling Christian congregations in the midst of theological dispute. The Pauline author cautioned that anyone following "a different version of the Good News . . . is to be condemned" (Galatians 1: 6–10). He went on to sanction severe treatment of religious and social deviants, for "the authorities are there to serve God; they carry out God's revenge by punishing wrongdoers" (Romans 13: 4–5). The Pauline writer prescribed "the raging fire that is to burn rebels" (Hebrews 10: 26–8). He was certain that good Christians would "do all you can to preserve the unity of the Spirit" because "there is one Body, one Spirit" (Ephesians 4: 3–4). These definitions of corporate Christian society, with their accompanying directives against those who rejected it, did much to justify later persecutions in a Church that owed a great deal of its early formation to Pauline influence.

Patristic pronouncements and legal precedents against heresy followed their Biblical antecedents in quick succession. Tertullian of Carthage, the gifted second century Christian apologist (and a Montanist "heretic" himself), was the first to identify the Greek term hairesein as "choice which a man exercises either to establish [false doctrines] or to adopt them" (Peters, 29–31). The first recorded execution for heresy took place in A.D. 385. In that year clerics in southern France delivered Priscillian, an ascetic priest who believed in both good and evil Gods as a Manichaean dualist, over to Emperor Maximius for beheading (Sulpicius Severus, 252–4). This event also represented the first cooperative act of the Church and government alliance that would soon, in the medieval period that followed, use accusations of heresy to justify conviction and then execution of religious, political and social non-conformists.

Perhaps the most influential theologian of the Middle Ages was Bishop Augustine of Hippo (A.D. 354–430), a former Manichaean himself who underwent an emotional conversion to become one of the early Church's most ardent apologists. His statement on heresy and universal Christian society, based upon the banquet parable in Luke 14 and on corporate political theory, set a philosophical standard:

> If the Church . . . exercises, as she ought, the power which she has received by a divine gift, together with religion and faith, and if those who are found in the highways and hedges, that is, in heresies and schisms, are compelled to come in, she is not to be blamed for compelling them, but they for waiting to be compelled. The banquet of the Lord is the unity of the Body of Christ, not only in the sacrament of the altar, but also in the bond of peace (1955, 166).

Augustine invited governmental assistance against heretics, for "a devout emperor preferred . . . by stringent religious laws to force those who carried the standard of Christ against Christ to return to Catholic unity, under stress of fear." His main point was "not whether anyone is being forced to do something, but what sort of thing he is being forced to do, whether it is good or bad" (1955, 168). The Bishop of Hippo claimed that all coercive means were justified if heretics could be induced to embrace his spiritual "City of God," a "perfectly ordered and perfectly harmonious fellowship" of universal Christian faith. If dissenters remained contumacious, then Augustine's fellowship was just in its punishment, without guilt, of "those who do not belong to this City of God," because "the unrighteous man's grief in his punishment is more appropriate than his rejoicing in sin" (1863, 353, 330).

These recommendations were preserved and inscribed in late Roman legal practice when Emperors Theodosius II and Justinian commissioned their respective law codes in the

5th- and 6th-centuries A.D. With the inclusion of rather ferocious laws against heretics in these legal collections, the coercive support of the secular state now backed the religious condemnations of the Church. It remained, however, for the Church itself to assemble its own supporting body of canon laws on heresy, for it fell to Europe's spiritual leaders to determine what constituted heresy in the first place. This process began in earnest once the Papacy and the Holy Roman Emperors restored amicable relations following the Investiture Controversy of the 11th-century. Spurred in part by a sudden increase in religious dissent across Europe, the Church moved quickly to define doctrine and develop methods by which it might detect the presence of heresy. Gratian, a monk at Bologna, compiled his Decretum in 1140 to collect the wisdom of the Fathers, the Church councils and the popes in one compendium of canon law. Gratian defined heresy as freely chosen, obstinately defended opinion held contrary to the dogma of Church teaching (Tierney, 13). With this very specific description in place, the Third Lateran Council met in 1179 under the aegis of Pope Alexander III to back Gratian's canon with coercive power. Realizing that the proper role of ecclesiastical authority was not physical punishment, the council urged that secular rulers "vigorously oppose such pests and defend with arms the Christian people" (Peters, 169). This dangerous decree sanctioned the use of organized force against religious non-conformists, a weapon soon utilized by Pope Innocent III in 1208 when he called the Albigensian "Crusade" against Cathar dissidents in southern France. The final bulwark against heresy was added to canon law in 1231 by Pope Gregory IX, whose decretal Ille Humani Generis established the first centralized agency—the Inquisition—to ferret out and try suspected heretics. Gregory authorized the order of Dominican friars to be sent as judges into different districts . . . to seek out diligently those who are heretics or are infamed of heresy . . . [and] to proceed against them according to our statutes recently promulgated" (Peters, 197).

With the creation of the Papal Inquisition, the Church possessed all the tools it needed to suppress disaffected elements in Christendom and secure its position at the top of European society.

What, indeed, frightened the medieval Church enough to generate such repressive measures against individuals and groups which were, after all, mostly fellow believers in Christ? Many modern scholars agree that ecclesiastical leaders saw in heresy a unified, monolithic threat to a Christian society that they dominated morally, politically and, in many ways, financially (Christie-Murray, 10–11). Churchmen in the Middle Ages enjoyed an elite position in society. Their moral injunctions rang out from cathedral and parish pulpits every day. In the feudal system that dominated the European political and social scene for many centuries, high-ranking prelates served kings as vassals and advisors. Since the Church owned between one-quarter and one-half of the landed wealth in medieval Europe, bishops and abbots were important feudal magnates who wielded much political and economic clout in their communities and kingdoms. It was therefore natural that such important members of society should want to defend their elite position at all costs, and it was also natural that they should turn to their allies, the secular rulers of society, to help them do it.

Several modern historians have suggested other, supporting explanations for this defensive Church posture. R. I. Moore believes that medieval clerics, monopolizing power and wealth in their service to both Church and king, deliberately inaugurated harsh persecution of all social outcasts, including not only heretics, but Jews and lepers as well. This allowed them to focus social criticism away from their own privileged positions and thus to ensure continued popular obedience to their dictates (67, 113–6). Norman Cohn and Lester K. Little identified a medieval hierarchy that was anxious to retain its control over a society in the midst of an economic and political transformation. As towns and their new merchant citizens appropriated more wealth and political power from the landed nobility after the 10th-century, dispossessed or "rootless" poor lost their traditional economic ties and became willing converts to radical ideas that criticized the changing social order (Cohn, 14–5). Those who benefitted from this new order—clerical administrators, urban entrepreneurs, royal bureaucrats—generally remained loyal to official Church teachings. Those left behind by social and economic changes, however—day laborers, small-scale artisans and younger sons of the nobility—tended to wander into social disobedience and so became targets of repression as heretics (Little, 144–5).

Thus, those accused of heresy were seen to challenge not only the doctrines and authority of the Church, they also assailed the very soul of medieval Europe in the context of a universally "orthodox" Christian culture. The term orthodox comes to us from the Greek words orthos, meaning "true," and doxos, meaning opinion. It is generally used to describe officially approved or, in some cases, the mainstream religious beliefs of a society or culture. Karl F. Morrison of Rutgers University has described orthodoxy in terms of cultural tradition, or the unique experience that gives any society its cohesion. Such cultural tradition required that all members of the social group participate in its rituals and beliefs through consensus recording of, and adherence to, that tradition's principles. As societies changed, however, tension arose between conservative and innovative elements that pulled at the cultural tradition and excluded some members of the social group from the shifting consensus. Societies in such transition typically enacted laws and established rigid dogmas to retain their unifying cohesion, a process which often led to rejection and then persecution of nonconformists (350–4). This process was at work in 11th-century Europe, when Christendom's unifying cultural tradition was being described as a corpus Christi mysticum, or a universal and orthodox "body" of believers under the leadership of Christ. Two centuries later, this corporate association had evolved into a description of the Church hierarchy and its governing role in Christian society. The same tradition was soon put to use by Europe's emerging monarchies to assert their own sovereignty over cohesive "corporations" of obedient subjects (Kantorowicz, 195–201). Religious and social misfits therefore stood outside the tradition-based orthodoxy of the Church *and* royal governments, challenging the rule of the established hierarchy with their own versions of "true opinion." Such persons or groups were often those who had also lost their social and economic identity in a changing medieval world. Thus, they seemed to pose a dangerous threat to all that Christian society represented, and so drew accusations of heresy from its rulers.

There are many examples of groups in the Middle Ages that seemed—at least to anxious ecclesiastical officials—to pose such a cohesive threat to universal Christendom. The Cathar heretics, mentioned briefly above, are one such group. The Cathari, so-called because of the spiritual catharsis that seemed to accompany their austere simplicity, may have originated in eastern Europe as descendants of the fourth century Manichaean dualists. They appeared in western Europe, possibly via missionary activity from Bulgaria, in the 11th- and early 12th-centuries to terrify churchmen who first encountered them. Cathars believed in two deities, a wicked God who had created the inferior physical world and its concomitant evil, and a good God, represented by a spectral Jesus Christ, who taught rejection of earthly cares for spiritual purity. They maintained their own priests, perfecti, who had received the consolamentum, or laying-on-of-hands purification ritual. Cathar clerics lived ascetic lifestyles as they wandered and preached throughout Europe. Because of their spiritual example, they apparently attracted a fairly wide following among dissatisfied Christians in the Rhine valley and especially in southern France (Christie-Murray, 104–8).

Eleventh century clerics like Gerard of Cambrai and Wazo of Liege, who first confronted the strange dualists in their dioceses, called them "manichees" because their beliefs so resembled those of the ancient heretics. Both Gerard and Wazo employed caritas, or caring dissuasion, to convince the suspects of their "errors" (Wakefield and Evans, 82–93). As we have seen, however, potestas or coercive measures were soon enacted and employed against Cathar dualists who seemed increasingly organized and militant by the turn of the 13th-century. With Cathars winning support even among the nobility of southern France, and erecting defensive forts in the mountains, the Albigensian Crusade (named for the Cathar stronghold of Albi) was launched against them by Pope Innocent III between 1208 and 1229 (Oldenbourg, 1–6).

On the surface, this largely Church version of the Cathars and their movement seems to justify the harsh methods and awful butchery that Pope Innocent's crusade unleashed against them. The Cathars did indeed seem organized on a European-wide scale; coupled with their military preparations, they must certainly have appeared to be the unified, monolithic threat to Christian society that ecclesiastical and secular leaders so feared. And yet, closer examination reveals some flaws in this picture. The prospect of Cathar insurgency became so real to medieval churchmen that any religious or social deviants, be they dualists or not, were often identified in the prosecuting records as Cathars. For example, Guibert of Nogent was present during an examination of heretics he called "Manicheans" at Soissons in 1114. Although he claimed in his memoirs that these "heretics" denied the sacraments and practiced free sex, Guibert made no mention of dualist beliefs or perfecti clerics (Peters, 72–3). On another occasion, King Henry II of England himself had a group of 30 foreign-born publicani (another name for Cathars) interrogated at Oxford in 1166. Although the leader of the suspected heretics, Gerard, claimed that his followers were "themselves Christians, and reverent toward apostolic doctrine," King Henry's examining officials still found them guilty of renouncing sacraments, marriage and Church unity (William of Newburgh, 131–3). The English priests did not, however, charge them with dualist beliefs. Such casual mis-identifications as these tend to give modern researchers a distorted view of the size and threat of genuine dualist dissent in Europe. Another interpretive problem involves the real motives for the Albigensian Crusade. Pope Innocent probably intended that the expedition concentrate on suppressing the Cathars, but Robert E. Lerner and other scholars have persuasively argued that those who led the campaign had other ideas. The real aims of King Philip IV of France, who supported the Crusade, and Simon de Montfort, who was its principal commander, were to annex the semi-independent county of Toulouse to the French Crown and to carve out feudal fiefdoms for themselves (Lerner, 192–3). It would seem, then, despite the very real presence of Cathar dissent in Europe, that the threat it represented may well have been exaggerated and that the Albigensian Crusade was pursued for more than purely religious reasons. Such revelations as these cast doubts upon any treatment of heresy that depended upon religious "truth" for its justification and employed legal force as its means of execution.

Another example of medieval heresy that seemingly posed a larger and more cohesive threat to Christian society than it perhaps really did involved the Lollards of England. The Lollards were originally followers of Oxford University theologian John Wyclif (1330–84). Wyclif was condemned by English prelates in the late 1370s for his radical views on the sacraments, his denials of relics and pilgrimages, and his rejection of Church property and papal authority. He had powerful friends at the court of King Edward III, and so escaped heresy prosecution to retire to a country parish and a peaceful death. Had his teachings remained within the academic confines of Oxford University, where a certain amount of theological debate was permitted, there would probably have been little popular religious dissent to worry Church officials in 15th-century England. His ideas did not remain at Oxford, however, but leaked beyond its protective walls to thousands of semi-literate and illiterate artisans and peasants. They in turn modified Wyclif's complex theological precepts to suit their own understanding and serve their own economic and social agenda (Lander, 116–9).

The Lollards, called "lollers" because they reputedly mumbled continuous prayers to themselves, lived in pockets all over the midlands and southern regions of England. They aroused fear in churchmen because they had created a Wycliffite and therefore erroneous version of the Bible, in vernacular English. Lollards were also thought to have taught heresy in schools, and even to have conspired to overthrow the English Church and Crown. When the Peasant Revolt erupted in East Anglia and Kent in 1381, both ecclesiastical and secular authorities suspected that Wyclif's radical ideas had somehow influenced the rebels (Aston, 273–6). Another abortive rising in 1414 under suspected Lollard Sir John Oldcastle was all that Church and government leaders needed to prove the existence of kingdom-wide "rumors, congregations and insurrections . . . by them which were of the sect of heresy commonly called Lollardy." The Archbishop of Canterbury formed an alliance with king and Parliament to prosecute Lollard dissidents throughout the realm and to "burn before the people in a prominent place" any who might prove unrepentant (*Statutes*, 181–3, 125–8).

Once again, recalling the Cathars in southern France two centuries before, ecclesiastical authorities seemed justified in their suppression of potential Lollard conspiracy and revolt in 15th-century England. Many modern scholars, however, question whether Lollards were organized or even in communication with one another at all, and so criticize Church and Crown persecution of them as organized rebels. K. B. McFarlane has argued that, since little record of Lollard contact survives the 15th-century, their religious non-conformity was in fact a vulgarized, individualized, almost unrelated stepchild of the complicated scholastic theology of John Wyclif. As such, Lollards never represented a unified sect, nor did they pose the dangerous social and political threat that English authorities imagined and worked so hard to subdue (179–86). J. A. F. Thomson has also downplayed the Lollard menace to society, especially after the failed Oldcastle revolt of 1414, which he claims was too small to have had any chance of real success (1–3, 249–53). Why, then, did the English Church and Crown feel it necessary to prosecute thousands and burn hundreds of Lollards? It is true that leaders in medieval Europe pictured the Christian society over which they ruled as a cohesive, homogeneous polity. Any threatened division of this polity therefore meant a schism in God's perfect design for mankind and, having been charged with carrying out the divine will, churchmen and kings felt it their duty to extirpate heresy. It is also true, however, that Lollards represented a challenge to the authority that those churchmen and kings commanded in society. If enough believers adopted Lollard teachings as their own, Church and Crown stood to lose the power—and the landed wealth —that they largely monopolized throughout medieval Europe. Although such motives were probably subconscious in the minds of arguably devout Christian officials, they nonetheless added urgency to the need to portray and persecute scattered religious dissenters as organized Lollard subversives.

The Cathar and Lollard stories are two important examples of religious reaction made more desperate by the perception of a political, social and economic menace. Other examples of "heretical" persecution, for reasons other than threatened insurrection, abound in surviving medieval records. Sometimes, charges of heresy were used by popes—the highest ecclesiastical authority in western Christendom—as convenient legal weapons against purely political opponents whose religious beliefs were perfectly orthodox. Such was the case when Pope Gregory VII (1073–85) struggled for ten years against his nemesis, Holy Roman Emperor Henry IV (1056–1106), over which of them would command greater authority in European affairs. Their so-called Investiture Contest saw Pope Gregory accuse the Emperor, who had appointed his own bishops according to centuries-old custom, of being "a promoter and partner of heresies." Henry IV had never denied any of the formal teachings of the Church, and so was not technically a true heretic. On the other hand, Pope Gregory actually forgave another, more clearly heretical individual, Berengar of Tours, who had challenged the doctrine of eucharistic transubstantiation (Gregory VII, 140, 166 and 170–1). In this case, heresy charges were employed by an anxious pope primarily to undermine the political opposition of a powerful secular ruler. Gregory VII's interpretation of heresy was thus arbitrary and relative to his own political purpose.

In other instances, people were accused of heresy when their honest attempts to reform what they saw as Church worldliness did not meet with the approval of the ecclesiastical authorities. Such people were the 12th- and 13th-century Waldensians. In 1173 a Lyons merchant named Valdes sought to imitate Christ by abandoning his wealth for a life of wandering, preaching and begging. His followers spread out across southern Europe but, in their zeal to bring more apostolic spirituality to Christendom, they began to question Church land ownership, participation in warfare, and even the institution of the priesthood itself. This, of course, was too much for the leaders of the Church. Pope Lucius III condemned the Waldensians as heretics in his bull Ad Abolendam in 1184, and the sect was ruthlessly hunted down until only a few survived the Middle Ages by taking refuge in mountain retreats in northern Italy (Peters, 139–41). In contrast to the Waldensians, the Franciscan order fared much better. Francis of Assisi, a near contemporary of Valdes, also grew disenchanted with his father's wealth and so chose to imitate Christ as an itinerant preacher. Perhaps possessed of better political instincts than Valdes, Francis wrote out a "rule" or code of conduct for his followers, and then submitted it to Pope Innocent III in 1210 for approval (Little, 146–52). Thus the Franciscans, whose program was not radically different from that of the early Waldensians, became a recognized order of friars that enjoyed the blessings, not the enmity, of the universal Church. The principal difference between the two movements was papal sanction, a clear indication of the fine line that divided today's heretic from tomorrow's saint in the medieval world.

Sometimes, the Church introduced its own innovations in doctrine and, when religious conservatives clung to old teachings, these once-orthodox believers were "left behind" as heretics. This happened to a group of simple peasants or rustici at Chalons-sur-Marne in 1024. They were examined for heresy because they had refused to accept the official new "truth" of infant baptism and real presence in the Eucharist (Moore, 17). Other individuals, more social outcasts than doctrinal non-conformists, were also attacked as heretics. One such was Tanchelm of Utrecht, who reportedly preached free love before a statue of the Virgin Mary while soliciting donations in buckets at Flemish roadsides. Peter of Bruys denounced infant baptism and destroyed crosses, but his real offense seems to have been his suit of animal skins and his fragrant need (even by medieval standards) to bathe. He was lynched by an angry mob and burned to death in 1130 (Christie-Murray, 101–2). Another group of "heretics," the 11th-century Patarines of Milan, are more difficult to categorize. These reformers actually allied themselves with Papal interests at a time when Gregory VII, Alexander II and Urban II were attempting to clean up certain clerical abuses in the Church. They were declared heretics, however, by the archbishops of Milan and their aristocratic allies, the traditional rulers of the town. The largely bourgeois Patarines were apparently trying to establish a merchant-dominated commune in Milan that would exclude archepiscopal and landed elements from power (Martines, 16–19). Their "heresy" was thus more political and economic in nature than it was theological, while their condemnation as "heretics," given Papal approval of their activities, was clearly subjective. It would seem once again, even within the Church itself, that the heretic of one ecclesiastical faction sometimes proved to be the saint of another.

A final example will serve to illustrate the relative nature of heresy, and indeed, of orthodoxy, in medieval Europe. Many readers are probably familiar with the fascinating story of Peter Abelard. This great 12th-century theologian and scholastic philosopher himself fell afoul of Church leaders and was condemned as a heretic. Abelard's most inflammatory treatise was *Sic et Non* (Yes and No) which, through its skillful use of syllogisms and dialectical proofs, examined the veracity of the teachings of the Church Fathers. His self-professed intent was not to deny Church doctrine, but to "banish fearful anxiety and all uncertainties" by reconciling it with reason (Abelard, 270). Church authorities disagreed with his approach, however, and censured Abelard at a council at Soissons in 1121 for his aberrant understandings of the Trinity. No less a figure than Bernard of Clairvaux, a giant of 12th-century piety and orthodoxy, followed this condemnation with one of his own at a debate with Abelard at Sens in 1140. But were Peter Abelard's writings really heretical, as his critics claimed? As we have seen, he believed that they were not, a conclusion that was later substantiated in the thought of the great 13th-century theologian Thomas Aquinas. How, then, did Abelard attract such hostility from a Church he professed to love and, indeed, served as a Cluniac monk in his final years?

The answer to this question may lie in Abelard's arrogance, which certainly infuriated his professors and colleagues in the academic community of Paris. He was, by most accounts, the brightest mind to have appeared in some time in the city. He may well have generated ill-will among scholars who, then as now, were embarrassed when confused or made to look bad by their students. The key to Abelard's "heresy" may also have been his illicit love affair with Heloise, a young woman whom he seduced and secretly married before he was found out by her outraged family. As a cleric in the early stages of holy orders, Abelard's behavior on this occasion was not strictly forbidden, but it was certainly frowned upon by a Church that considered him incontinent and immoral (Gilson, 1–19). Thus, Peter Abelard's intellectual ego and his uncontrollable passion were probably as responsible for the accusations of heresy he attracted as were his innovative, if not technically deviant, pronouncements on the Trinity and the Church Fathers. Later thought by many to have been the father of medieval scholastic philosophy,

Peter Abelard was a heretic because his originality, his arrogance and his passion precluded sainthood.

It should be apparent from this discussion that heresy was used as a weapon by both Church and secular rulers to suppress a wide variety of religious, social, economic and especially political dissent in medieval Europe. Despite attempts to limit heresy in canon law to conscious rejections of orthodox doctrine, the charge became a convenient means by which to condemn potentially disruptive enemies and, because it was viewed as an offense against God, to rally public support for such condemnations. We have also seen that many of those who were prosecuted and even executed as heretics were in fact not religious non-conformists at all. In many cases, accused heretics did profess some variant form of Christianity, but these dissidents were usually isolated as individuals or divided into small cells with little or no unifying organization. The cohesive "threat" that they seemingly posed to Christian society—the device used to justify their persecution—was thus often exaggerated. The harsh suppression of suspected heretics may, in fairness, have been due to the wish of Europe's leaders to perform their God-given duty as defenders of Christian society. It must be remembered that the medieval mind was almost obsessively preoccupied with faith and piety as the means toward everlasting salvation. Thus, any divergence from the "truth" of the teachings that guaranteed that salvation was naturally considered dangerous "error." Still, it has been argued by historians that other, possibly subconscious motives inspired the ruthlessness with which heretics were hunted down and destroyed. Those who flouted religious authority simultaneously undermined the credibility of that authority among the faithful throughout an otherwise universally obedient Christendom. The doubts that dissenters sowed, if allowed to circulate among unlettered believers, threatened to foment popular revolt and topple the Christian clergy from the privileged positions they enjoyed. It therefore became critical that ecclesiastical and secular rulers silence any form of theological—or social, political and economic—disaffection so that they could preserve their power and wealth in a united Christian world.

Although medieval Europe is in many ways far removed from our daily experience in modern America, the tribulations of those times force us to ask some important questions about religion in our own culture. First, how universally "true" is any form of belief when its precepts serve more of humankind's economic and political agenda than God's divine plan? Second, how "wrong" can other believers be if their faith brings them spiritual satisfaction and gives meaning to their lives? Are there not many scriptures and many creeds among us, all of which claim and indeed speak truth to those who hold them sacred? Since surely one person's heresy is another's truth, it seems that tolerance through open-minded understanding is our best chance to avoid the religious bigotry that has so often been the reason for mankind's worst inhumanity to man in the past.

References:

Abelard, Peter. 1974. *The Letters of Abelard and Heloise.* Translated and edited by Betty Radice. Harmondsworth: Penguin Books.

Aston, Margaret. 1976. "Lollardy and Sedition, 1381–1431." Hilton, R. H., Editor. *Peasants, Knights and Heretics: Studies in Medieval English Social History.* Cambridge: Cambridge University Press.

Augustine of Hippo. 1863. *De Civitate Dei.* Leipzig: B. G. Teubner.

Augustine of Hippo. 1955. "Letter 185: Augustine to Boniface." *Fathers of the Church: A New Translation XXX.* New York: Fathers of the Church.

Christie-Murray, David. 1989. *A History of Heresy*. Oxford: Oxford University Press.

Cohn, Norman. 1970. *Pursuit of the Millennium*. Oxford: Oxford University Press.

Gilson, Etienne. 1960. *Heloise and Abelard*. Ann Arbor: University of Michigan Press.

Gregory VII. 1932. *The Correspondence of Pope Gregory VII*. Edited and translated by Ephraim Emerton. New York: Columbia University Press.

Kantorowicz, Ernst H. 1957. *The King's Two Bodies: A Study in Mediaeval Political Theology*. Princeton: Princeton University Press.

Lander, J. R. 1969. *Conflict and Stability in Fifteenth Century England*. London: Hutchinson University Library.

Lerner, Robert E. 1965. "The Uses of Heterodoxy: The French Monarchy and Unbelief in the Thirteenth Century." *French Historical Studies* IV, No. 2.

Little, Lester K. 1978. *Religious Poverty and the Profit Economy in Medieval Europe*. Ithaca: Cornell University Press.

Martines, Lauro. 1979. *Power and Imagination: City-States in Renaissance Italy*. Baltimore: Johns Hopkins University Press.

Moore, R. I. 1987. *The Formation of a Persecuting Society: Power and Deviance in Western Europe, 950–1250*. Oxford: Basil Blackwell.

Oldenbourg, Zoe. 1990. *Massacre at Montsegur: A History of the Albigensian Crusade*, New York: Dorset Press.

Pernoud, Regine. 1988. *Joan of Arc: By Herself and Her Witnesses*. New York: Dorset Press.

Peters, Edward, editor. 1980. *Heresy and Authority in Medieval Europe*. Philadelphia: University of Pennsylvania Press.

Statutes of the Realm II. 1815. London: His Majesty's Stationery Office.

Sulpicius Severus. 1949. "St. Martin and the Condemnation of Priscillian." *Fathers of the Church: A New Translation VII*. New York: Fathers of the Church.

Tierney, Brian. 1982. *Religion, Law and the Growth of Constitutional Thought, 1150–1650*. Cambridge: Cambridge University Press.

Wakefield, Walter L., and Evans, Austin P., editors and translators. 1965. *Heresies of the High Middle Ages*. New York: Columbia University Press.

William of Newburgh. 1884. *Historia Rerum Anglicarum*. Edited by Richard Howlett for Rolls Series. London: Longman and Company.

CPSIA information can be obtained at www.ICGtesting.com
Printed in the USA
BVOW02*2016020714

357936BV00007B/66/P